Springer Series on Atomic, Optical, and Plasma Physics

Volume 130

The Springer Series on Atomic, Optical, and Plasma Physics covers in a comprehensive manner theory and experiment in the entire field of atoms and molecules and their interaction with electromagnetic radiation. Books in the series provide a rich source of new ideas and techniques with wide applications in fields such as chemistry, materials science, astrophysics, surface science, plasma technology, advanced optics, aeronomy, and engineering. Laser physics is a particular connecting theme that has provided much of the continuing impetus for new developments in the field, such as quantum computation and Bose-Einstein condensation. The purpose of the series is to cover the gap between standard undergraduate textbooks and the research literature with emphasis on the fundamental ideas, methods, techniques, and results in the field.

Yoshiaki Kato · Kunioki Mima · Sergei Bulanov

High Power Laser and Plasma Science

Springer

Yoshiaki Kato
Institute of Laser Engineering
Osaka University
Osaka, Japan

Kunioki Mima
Institute of Laser Engineering
Osaka University
Osaka, Japan

Sergei Bulanov
ELI-ERIC, ELI Beamlines
Dolni Brezany, Czech Republic

ISSN 1615-5653 ISSN 2197-6791 (electronic)
Springer Series on Atomic, Optical, and Plasma Physics
ISBN 978-3-032-08929-8 ISBN 978-3-032-08930-4 (eBook)
https://doi.org/10.1007/978-3-032-08930-4

This Springer imprint is published by the registered company Springer Nature Switzerland AG
The registered company address is: Gewerbestrasse 11, 6330 Cham, Switzerland

To
Reiko Kato
Kazuko Mima
Nadezhda Bobrova

Preface

Since invention of the laser in 1960, high power laser and plasma science have evolved with close cooperation between these fields. The laser fusion research conducted on large scale in many countries has spurred development of high-power lasers, from kJ to MJ. This development led to significant progress in the understanding of high-temperature, high-density plasmas, from linear and nonlinear laser-plasma interactions to plasma processes, which are closely related to space plasmas. The laser fusion research has achieved the scientific breakeven in 2022.

Strong impact to the plasma science was brought by the chirped pulse amplification (CPA), that made it possible to generate ultrashort, high-power laser pulses with compact facilities. With the CPA, femto- and atto-second science and high-field laser science have emerged. In the extremely high field created by the CPA lasers, the electrons acquire relativistic energies under intense laser field, opening many new fields, such as generation of high-energy photons, electrons, ions and neutrons, and investigation of nonlinear processes in vacuum.

The authors of this book have been engaged in research and education on high-power laser, plasma physics, laser fusion, and high-field science. In these experiences, we think that there is a large gap between the education programs in standard universities and the knowledge required for research and development in this field. Although there are many excellent books and review articles, it will be useful to provide an introductory book on high-power laser and plasma science to fill this gap, so that students and young scientists will become familiar with this field smoothly. This book will be also useful to overview the rapidly evolving field of high-power lasers and plasma science. Since each part is written as a comprehensive description, each part can be used also as a textbook in the graduate courses on lasers and plasma physics.

This book is comprised of four parts. Part I *Introduction* presents recent topics in high power laser and plasma science and summarises the contents of Part II, Part III, and Part IV. Part II *High Power Lasers* describes fundamental concepts, technologies, and recent advances in high power lasers. It may be used by students, scientists, and engineers who are working with high-power lasers. Part III *Laser Plasmas and Inertial Fusion Physics* is expected to be useful for the graduate students and the scientists who are starting the research on laser plasmas physics and laser fusion. This part contains introduction to plasma kinetics theory, electro-magnetic phenomena, heat transport, laser fusion, hydrodynamic instabilities, radiation transport, and laser neutron source. Part IV *High Field Science with Lasers* may be used by the students and young physicists working in the field of theoretical laser plasma physics, because it contains an introduction to theory of charged particle acceleration in laser plasma interaction and presents the methods of describing nonlinear processes in relativistic laser plasma. Parts II, III, and IV have been written mainly by Yoshiaki Kato, Kunioki Mima and Sergei Bulanov, respectively, with mutual support to complete the contents.

Hamamatsu, Japan — Yoshiaki Kato
Osaka, Japan — Kunioki Mima
Dolni Brezany, Czech Republic — Sergei Bulanov

Acknowledgements In writing this book, we are indebted to many scientists working in various fields; high power lasers, plasma physics, high-field science, high energy density physics, and laser fusion. In appreciation of these pioneering works, we hope that the dynamic advances in these fields are properly conveyed, at least partly, to the readers of this book.

We would like to express our special thanks to our colleagues we have worked with and to our friends for helping us in writing this book: Thomas M. Antonsen, Yasunobu Arikawa, Hiroshi Azechi, Gennadyi Bagdasarov, Pavel Bakule, Christopher P. J. Barty, Tom Blackburn, Marco Borghesi, Stepan Bulanov, Valery Yu. Bychenkov, Hongbo Cai, Francesco Califano, E. Mike Campbell, Uddhab Chaulagain, Liming Chen, Ming Chen, Pisin Chen, Il Woo Choi, Hiroyuki Daido, Galina Dudnikova, Eugenia Echkina, John Emmett, Eric Esarey, Timur Zh. Esirkepov, Daniela Farina, Shunsuke Fujioka, Kazuhisa Fujita, Vladimir A. Gasilov, Leonida Antonio Gizzi, Zhang Gong, Gabriele Maria Grittani, Yan Jun Gu, Hideaki Habara, Prokopis Hadjisolomou, Ryohei Hanayama, Karen Hatsagortsyan, X. T. He, Akira Hiruma, John Holtzrichter, Tomonao Hosokai, Javier Honrubia, Roman Hvezda, Igor N. Inovenkov, Jiro Itatani, Vyacheslav Ivanov, Natsumi Iwata, Yasukazu Izawa, Tae Moon Jeong, Martin Jirka, Tomoyuki Johzaki, Monma Jovanovic, Masaki Kando, Satyabrata Kar, Susumu Kato, Tetsuya Kawachi, Shigeo Kawata, Mike Key, Christopher Keitel, Vladimir S. Khoroshkov, Chul Min Kim, Hyung Taek Kim, Toyoaki Kimura, Yasuaki Kishimoto, Hiromitsu Kiriyama, Onrej Klimo, Ryosuke Kodama, James Koga, Kiminori Kondo, David Kolenaty, Georg Korn, Marcel Lamac, Carlo Maria Lazzarini, Wim P. Leemans, Yuxin Leng, Kirill V. Lezhnin, Ruxin Li, Jian-Xing Li, Zhaoyang Li, Tatiana V. Liseykina, C. S. Liu, Maurizio Lontano, Sebastian Lorenz, Tammy Ma, Brian MacGowan, Alexander Macleod, Anatoly Maksimchuk, Daniele Margarone, Mattias Marklund, Dennis Matthews, Martin Matys, Howard Milchberg, Noriaki Miyanaga, Alexander Molodozhentsev, Alessio Morace, Yoshitaka Mori, Gerard Mourou, Masakatsu Murakami, Hideo Nagatomo, Sadao Nakai, Tatsufumi Nakamura, Masahiro Nakatsuka, Chang Hee Nam, Natalia Naumova, Jaroslav Nejdl, Michal Nevrkla, Peter Nickles, Takeshi Nishikawa, Katsunobu Nishihara, Hiroaki Nishimura, Mamiko Nishiuchi, Takayoshi Norimatsu, Shinichiro Okihara, Olga G. Olkhovskaya, Satoshi Orimo, Francesco Pegoraro, Manuel J. Perlado, Goeff Pert, Tatiana A. Pikuz, Alexander S. Pirozhkov, Reinhart Poprawe, Stuart Ramsden, Christopher Ridgers, Jan Ridky, George Rocca, Lorenzo Romagnani, Nikolai N. Rosanov, Bedrich Rus, Brandon Russell, Shuji Sakabe, Hiroshi Sakagami, Tokayoshi Sano, Javier Sanz, Pavel Sasorov, Gianluca Sarri, Carl Schroeder, Stéphane Sebban, Yasuhiko Sentoku, Rashid Shaisultanov, Zheng-Ming Sheng, Isao Shimamura, Hiroyuki Shiraga, Luis Silva, Milos Skoric, Alzbeta Spadova, Toshihiko Taguchi, Toshiki Tajima, Hideaki Takabe, Greg Tallents, Kazuo Tanaka, Alex Thomas, John Trenholm, Akihiko Tsuboi, Petr Valenta, Roberto Versaci, Maria Vranic, Pavel Vrba, Miroslava Vrbova, Vitaly Vshivkov, Shiji Wang, Tianyun Wei, Scott Wilks, Takeshi Yabe, Wenchao Yan, Akifumi Yogo, Kunio Yoshida, Jie Zhang, Alexei Zhidkov, and Iliia Zymak.

We are also indebted to our former mentors and colleagues whom we have lost: Ximing Deng, Anatoly Ya. Faenov, Akira Hasegawa, Teruo Hiruma, Prediman Kaw, Ciaran Lewes, Zunqi Lin, Takayasu Mochizuki, Nikolay B. Narozhnyi, David Neely, Kyoji Nishikawa, Alexander S. Sakharov, Wolfgang Sandner, Boris B. P. Stoicheff, Hiroshi Takuma, Chiyoe Yamanaka, and Tatsuhiko Yamanaka.

Hamamatsu, Japan Yoshiaki Kato
Osaka, Japan Kunioki Mima
Dolni Brezany, Czech Republic Sergei Bulanov

Competing Interests The authors have no competing interests to declare that are relevant to the content of this manuscript.

About Kunioki Mima

Our co-author, esteemed colleague and good friend Kunioki Mima passed away on January 5, 2025, at the age of 79. Professor Mima (三間 圀興, Mima Kunioki, August 17, 1945–January 5, 2025) was an outstanding expert in the fields of plasma physics, controlled nuclear fusion and nonlinear wave theory. He has graduated from Kyoto University. For many years, he was affiliated with Osaka University, where he was the director of the Institute of Laser Engineering.

His name was familiar to everyone working in the field of plasma physics. Professor Mima was Fellow of the American Physical Society and the recipient of such renowned awards as the John Dawson Award for outstanding achievements in plasma physics research, the Edward Teller Award and the Hannes Alfvén Award of the European Physical Society in recognition of the global significance of his work. One of his remarkable results is the Hasegawa-Mima equation, named after Akira Hasegawa and himself, which describes turbulence in magnetized plasmas and the dynamics of vortices in a dispersive medium.

It was a special experience and pleasure to work with Kunioki Mima on this book. His main impact on the book is Part III, which deals with plasma physics and controlled thermonuclear fusion. Mima continued to work on the book while he was seriously ill.

We will keep warm memories of Kunioki Mima in our thoughts.

Sergei Bulanov
Yoshiaki Kato

Contents

About the Authors

Yoshiaki Kato has received Ph.D. in Applied Physics at the University of Tokyo. He was at R&D division of JEOL, Ltd. and the Department of Physics, University of Toronto. Then he has worked at the Institute of Laser Engineering (ILE), Osaka University (1975–1998), Kansai Photon Science Institute, Japan Atomic Energy Research Agency (JAERI) (1998–2005), Quantum Beam Application Division, Japan Atomic Energy Agency (JAEA) (2005–2007), and Graduate School for the Creation of New Photonics Industries (GPI) at Hamamatsu City (2007–2020). Since 2020, he has been a Guest Professor at ILE, Osaka University.

He was engaged in research on laser spectroscopy, high power laser, x-ray laser, ultrashort pulse laser, laser-driven particle beam, and industrial applications of high-power lasers. At ILE, he has developed the first multi-beam Nd-doped phosphate glass high-power laser for laser fusion research and initiated random phasing of high-power laser for uniform target irradiation and plasma instability suppression. He has received the APS Award for Excellence in Plasma Physics and the Daiwa Anglo-Japanese Foundation DAIWA AWARD for UK-Japan joint research on x-ray lasers. He has received The Order of the Sacred Treasure, Gold Rays with Neck Ribbon in 2018. He is Professor Emeritus of Osaka University, Fellow of APS, and Fellow of The Laser Society of Japan. In 2025, he was elected President of MATSUO Foundation for promotion of basic science.

He has chaired OECD Global Science Forum Committee on Compact High-Intensity Short-Pulse Lasers, which led to establishment of The International Committee on Ultra-intense Lasers (ICUIL) as a Working Group of The International Union of Pure and Applied Physics (IUPAP) in 2003, and the Asian Intense Laser Network which organizes Asian Symposium on Intense Laser Science (ASILS). He has been sponsoring The Optics and Photonics International Congress (OPIC), held every year in Yokohama Japan.

He used to enjoy playing soccer. He is now enjoying farming in Hamamatsu.

Kunioki Mima has received Ph.D. in Physics at Kyoto University. Then he moved to the Faculty of Science of Hiroshima University. After two years, he moved to the Faculty of Engineering Osaka University to start research on laser plasma physics.

During this period, he was invited to Bell Telephone Laboratory, Murray Hill, New Jersey, and UCLA, California, in USA for two years. He has worked at the Institute of Laser Engineering Osaka University (1975–2009), the Graduate School for the Creation of New Photonics Industries (2009–2020) and Institute of Fusion Nuclear, University Politecnica de Madrid in Spain (2009–2014).

In these organizations, he has worked on research and education of plasma nonlinear waves and turbulence in plasmas, laser plasma interaction and radiation hydrodynamics, free electron laser, fusion target design, fusion reactor design, and relativistic laser plasmas. At Bell Telephone Laboratory, he invented the model equation so called "Hasegawa-Mima equation" to describe the anomalous transport in MCF plasmas and planetary-atmospheric dynamics.

He has received APS Award for Excellence in Plasma Physics, Hannes Alfven Medal and Award from European Physical Society for the theory of fusion plasma turbulence, Osaka Science Prize for nonlinear plasma theory, ANS Edward Teller Medal for laser fusion research, Science and Technology Prize of MEXT, Japan, for fast ignition research, and The Order of the Sacred Treasure, Gold Rays with Neck Ribbon in 2024. He is Professor Emeritus of Osaka University, Guest Professor of ILE, Osaka, Fellow of APS and Fellow of The Laser Society of Japan.

He was enjoying mountain climbing with his colleagues, students, and his wife, with a dream of climbing the Japanese top 100 mountains.

Professor Kunioki Mima has deceased on fifth of January 2025 at the age of 79.

Sergei Bulanov has received Ph.D. in Theoretical Physics and Astrophysics at the Moscow Institute of Physics and Technology and the Dr.Sci. degree in Plasma Physics at A. M. Prokhorov Institute of General Physics of Russian Academy of Sciences. He worked at the Department of Theoretical Physics in P. N. Lebedev Institute of Physics and at the Plasma Physics Laboratory in A. M. Prokhorov Institute of General Physics in Moscow. Then he worked at the Kansai Photon Science Institute (JAERI, JAEA and QST). Currently he works as a head of department in the ELI-Beamlines in Czech Republic being associated with the KPSI-QST.

In these organizations, he has been engaged in the research on cosmic ray astrophysics and solar flare physics, plasma discharge physics, high power microwave and laser radiation interaction with matter, charged particle acceleration and fundamental physics. He has received the State Prize Award of the USSR, the Hannes Alfven Medal and Award from European Physical Society, and The Order of Rising Sun, Gold Rays with Rosette.

He used to enjoy hunting, doing wrestling, cross-country and down-hill skiing and practicing kyudo (Japanese archery) at Nara Budo Association.

Part I
Introduction

Chapter 1
Recent Progress in High Power Laser and Plasma Science

As an introduction to this book, we introduce several recent topics in high power laser and plasma science.

1.1 Ignition of Laser Fusion

Achieving the fusion ignition has been a major milestone in the laser-driven inertial confinement fusion since its invention by J. Nuckolls in 1961 and opening to the public in 1972 (Nuckolls et al., 1972). In August 2021, self-burning of the fusion reaction has been achieved (Abu-Shawareb et al., 2022; Zylstra et al., 2022) at the US National Ignition Facility (NIF) (Spaeth et al., 2016), exceeding the Lawson criterion for ignition (Fig. 1.1, marked by N210808) where fusion self-heating exceeds all energy losses (Lawson, 1957). In this achievement, fusion energy of 1.37 MJ was generated with the laser energy delivered to the target of 1.92 MJ, corresponding to a target gain G of 0.72. In December 2022, with further refinement of the target and slight increase in the laser energy, the fusion ignition has been achieved with $G = 1.5$ where 3.15 MJ fusion energy was generated with 2.05 MJ laser energy (Fig. 1.1, marked by N221204) (Abu-Shawareb et al., 2024).

Based on the successful achievement of the self-burning of fusion reaction at LLNL, "Basic Research Needs Workshop" on the Inertial-confinement Fusion Energy (IFE) has been held in June 2022 for harnessing fusion for our energy needs as a coordinated, public–private partnership program (Ma & Betti, 2022). With venture capital investments and supports from public programs, close to ten startup companies on IFE have been established in USA, Europe, Australia, and Japan based on various fusion schemes and various drivers (Kramer, 2023), joining to the 20+ startups engaged in magnetic-confinement fusion energy (MFE) (Holland, 2022). This US program is followed in Germany by holding Expert Panel of the Inertial

Y. Kato et al., *High Power Laser and Plasma Science*, Springer Series on Atomic, Optical, and Plasma Physics 130, https://doi.org/10.1007/978-3-032-08930-4_1

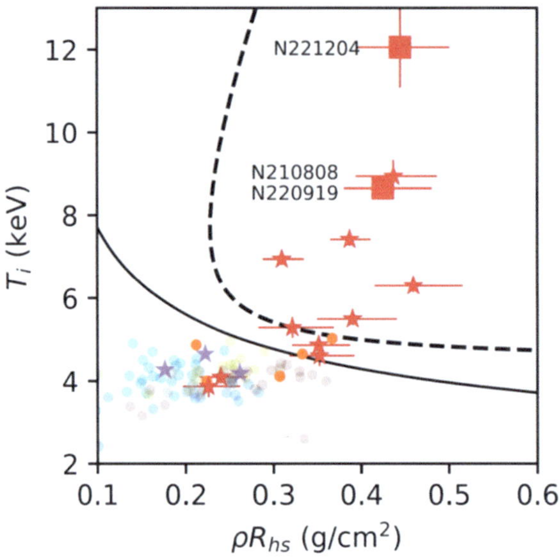

Fig. 1.1 Achievement of the fusion burning in NIF. This diagram shows the ion temperature T_i and the areal density ρR_{hs} of the compressed core, where each mark corresponds to each NIF shot displayed by different shapes and colors for various target designs. The boundary of the burning plasma is characterized by two curves: the solid curve shows the boundary where self-heating surpasses the initial mechanical work to form the plasma, and the dashed curve shows the boundary where alpha heating exceeds the radiation and thermal conduction losses. Target gain of 1.5 was achieved at N221204, where the ion temperature has reached 12 keV. Reprinted from Pak et al. (2024), with permission from The American Physical Society and the author(s). © 2024 The American Physical Society

Fusion (Haefner et al., 2022), leading to starting a new large-scale program toward developing the fusion power plant by 2040 (Stark-Watzinger, 2024).

1.2 Ultra-High Intensity Lasers and High Field Science

The Nobel Prize in Physics 2018 was awarded to Gerard Mourou and Donna Strickland for the invention of chirped pulse amplification (CPA), which enabled generating extremely high optical field with compact, ultrashort-duration, ultrahigh-power lasers (Strickland & Mourou, 1985). With the CPA laser, theoretical and experimental investigations on high field science have rapidly developed (Mourou et al., 2006), such as laser wake-field electron acceleration (Tajima & Dawson, 1979), laser-ion acceleration (Ziegler et al., 2024), nonlinear Compton scattering (Mirzaie et al., 2024), and gamma-flash generation (Pirozhkov et al., 2024).

The CPA has also opened atto-second science, where electron dynamics in matter are revealed with the femto-second and atto-second pulses generated with the CPA lasers and its high-order harmonics in the extreme ultraviolet region. The Nobel Prize in Physics 2023 was awarded to the pioneers in this field: Pierre Agostini, Ferenc Krausz and Anne L'Huillier "for experimental methods that generate attosecond pulses of light for the study of electron dynamics in matter".

Entering the twenty-first century, over one-hundred ultrahigh intensity CPA lasers have been developed worldwide (Barty, 2020), including 10-petawatt (PW)-peak-power lasers. Further research and development are in progress toward realization of exawatt (EW)-power lasers (Danson et al., 2019). As shown in Fig. 1.2, the peak power of the CPA laser has increased rapidly in these ~ 20 years, reaching 10 PW with a Ti:sapphire laser at ELI Nuclear Photonics (ELI-NP) in 2020 (Lureau et al., 2020). Construction of 100 PW laser SEL based on optical parametric chirped pulse amplification (OPCPA), together with development of an X-ray free electron laser (XFEL), is in progress at Shanghai Institute of Optics and Fine Mechanics and Shanghai Synchrotron Radiation Facility (Peng et al., 2021), where laser induced birefringence in vacuum will be probed with XFEL (Ahmadiniaz et al., 2025). To open zepto-second exawatt physics, a new scheme called post compression is developed in Russia and other countries (Mourou et al., 2014).

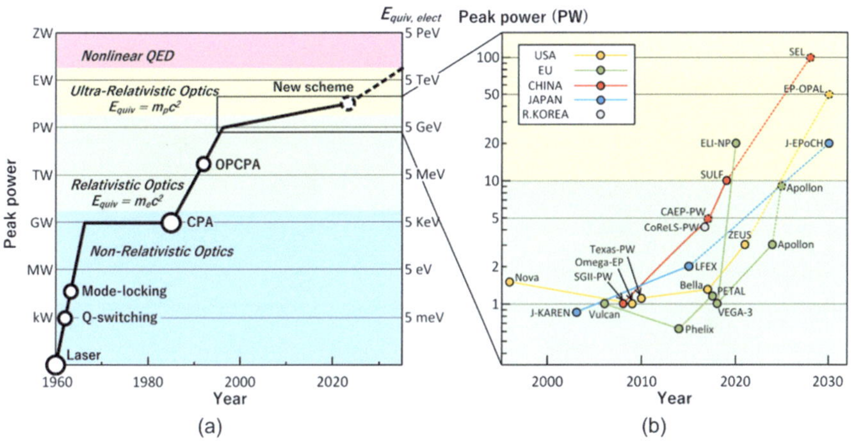

Fig. 1.2 Evolution of short-pulse, high-power lasers. **a** Peak power versus year. The electron quiver energy (ponderomotive energy of an electron) when the laser is focused close to the diffraction limit is shown on the right-hand side. **b** Peak power of high-power lasers in several countries and regions, with CPA lasers in solid circles and OPCPA lasers in dotted circles. The lasers under construction or planned are shown by dotted lines. Detailed accounts of these lasers are described in Part II (Chap. 7). Figure prepared by Zhaoyang Li and Yoshiaki Kato

1.3 High-Efficiency, Repetitive High-Power Solid-State Lasers

Repetitively operated high power lasers are required for IFE, high field science, and industrial applications. The important factors for repetitive operation are high pumping efficiency, efficient heat removal, and laser damage to optical components.

High pumping efficiency is achieved in solid-state lasers with high-power laser diode (LD) pumping, with an overall wall plug efficiency exceeding 20%. Innovative amplifier schemes have been developed for efficient heat removal (Fig. 1.3), with demonstrated performance of 100 J, 10 Hz operation with the face-pumped gas-cooled multi-slab amplifiers (Mason et al., 2018) and 10 J, 100 Hz operation with the active mirrors (Ogino et al., 2022). Although significant progress has been achieved on understanding and improvement of the laser damage in the NIF development (Manes et al., 2016), improvement of the laser damage of optical components is still required in ultra-high peak power lasers, especially diffraction gratings for pulse compression in CPA lasers, and in repetitively operated lasers. The damage-resistant gaseous optics under development may solve this problem, enabling high-power and high-intensity lasers very compact (Michine & Yoneda, 2020).

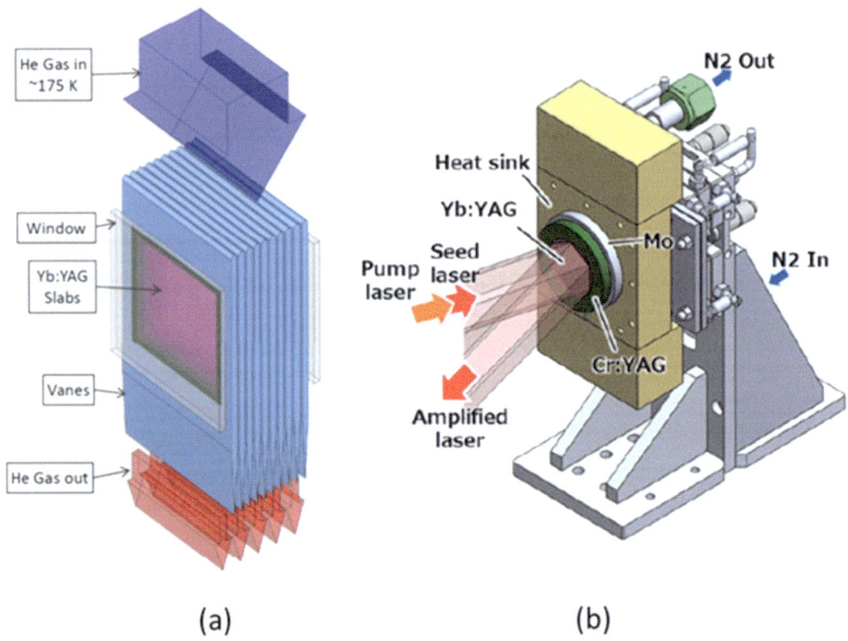

Fig. 1.3 Configurations of LD-pumped, high-repetition, high-energy solid-state lasers: **a** face-pumped gas-cooled multi-slab amplifier. Reprinted from Ertel et al. (2011). © 2011 Optical Society of America. **b** Active mirror. Reprinted from Ogino et al. (2022). © 2022 Optica Publishing Group

1.4 Intense Coherent Radiation Sources

The wavelength of coherent radiation has been extended to both shorter and longer wavelengths with the high-power lasers. The soft-X-ray laser in laser-produced multi-charged ions was initially realized with a multi-kJ laser (Matthews et al., 1985), but is now generated at 5 Hz by pumping with a compact 1-J laser (Larotonda et al., 2004). High-order harmonics in gases have been extended to 1.6 keV by pumping with mid-infrared femtosecond lasers (Popmintchev et al., 2021).

In another novel scheme of "relativistic flying mirror" shown in Fig. 1.4, a laser pulse of optical wavelength is converted to soft X-ray radiation where the frequency of a source pulse is upshifted by the Doppler effect upon reflection from a plasma mirror moving close the velocity of light in vacuum (Kando et al., 2007; Pirozhkov et al., 2007).

At a shorter wavelength, a 1.54-angstrom (8.0 keV) atomic X-ray laser has been realized by inner-shell ionization of cupper by intense pumping with an X-ray free electron laser (Yoneda et al., 2015). In place of using a high energy accelerator for lasing in hard X-ray region, a compact X-ray free electron laser is being developed, where lasing at 27 nm has been achieved with mono-energetic 490 MeV electrons generated by laser wake-field acceleration (Wang et al., 2021). On the long wavelength side, high-intensity single-cycle THz pulses have been generated by optical rectification in a nonlinear crystal, opening an unexplored field of nonlinear transport phenomena in solids (Hirori et al., 2011).

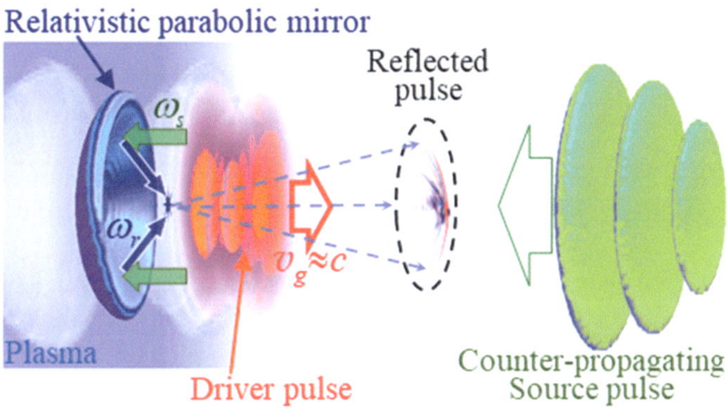

Fig. 1.4 Frequency up-conversion of an optical laser ω_s (source pulse, moving from right to left) to a soft X-ray pulse ω_r by reflection from a plasma mirror which moves from left to right at a velocity close to light. This relativistic flying mirror is produced in a plasma with a high-intensity laser pulse (driver pulse) which moves from left to right. The reflected X-ray pulse is focused to a small size of the X-ray wavelength when the relativistic flying mirror is parabolic-shaped. Courtesy of A. S. Pirozhkov and T. Esirkepov

1.5 Introduction to Part II: High Power Lasers

In high-power lasers, important parameters for applications in science and industry are; wavelength, energy, peak power, duration, pulse shape, focusing, efficiency, repetition rate, reliability, and cost. Laser scientists and engineers strive to control these parameters to fit specific application needs, under constraints of physics, materials, fabrication, electronics, and environment. In the first four chapters in Part II, we introduce basic frameworks which are required in the design and use of the high-power lasers. Then, in Chaps. 6 and 7, we focus on the architectures, performances and prospects of high energy pulsed lasers and high-power, ultrashort-pulse lasers, respectively. Emphases are placed on recent progress and new ideas in high-energy and high-power lasers.

Part II is organized as follows.

In Chap. 2, Propagation of Laser Light, we first describe propagation of an optical wave in a free space and in materials including material dispersion. This is followed by introduction to several optical components which are important in high power lasers. Then propagation of ultrashort pulses is described, with introduction to the chirped pulse amplification and methods for dispersion control to generate transform-limited ultrashort pulses. In the end of this chapter, we will briefly introduce the relativity which is required in applications of high intensity lasers.

In Chap. 3, Light Amplification by Stimulated Emission of Radiation, introduction to quantum mechanics and the particle nature of light are described. Then we derive probabilities for the spontaneous and stimulated emission of light which are fundamental to the laser action.

In Chap. 4, Generation, Amplification and Propagation of a High-Power Laser Beam, concepts of laser oscillation and amplification are first introduced. This is followed by a description of a basic equation on propagation of a high-power laser beam, including nonlinear refractive index and material dispersion. This equation is used to design and analyze the high-power lasers. Then we present more specific and important aspects of the high-power lasers: solid-state laser materials and laser damage.

In Chap. 5, Nonlinear Optical Processes in High Power Lasers, nonlinear optical processes such as frequency conversion and optical parametric amplification are first described. Then self-focusing and self-phase modulation arising from the third-order nonlinear susceptibility are described. In the final section, we consider how a few-cycle duration, intense light pulses propagate in dispersive media.

In Chap. 6, High-Energy Pulsed Lasers, high-energy pulsed lasers and progress in laser fusion and high-energy density science research are briefly reviewed. Then architecture and control in the National Ignition Facility at Lawrence Livermore National Laboratory are described in some detail as a good example of high-precision, high-energy laser system. Furthermore, we review progress in the development of high-efficiency, repetitive-operated, high-energy lasers in two approaches, laser-diode pumped solid-state lasers and excimer laser, as candidates for the inertial fusion energy driver.

In Chap. 7, High-Power, Ultrashort-Pulse Lasers, development of ultrashort pulse lasers with peak powers from TW to PW and toward EW is briefly reviewed. Then detailed account of the high-power CPA Ti:sapphire laser is presented, describing how various parameters such as peak power, pulse shape, pulse contrast, and focusing are controlled. This is followed by the description of high-energy CPA Nd:glass lasers which are used for fast ignition of laser fusion and intense X-ray generation for high-density plasma diagnostics. Then several approaches to ultra-high power pulse generation are reviewed: chirped-beam CPA, optical parametric chirped pulse amplification (OPCPA), post compression, and coherent beam combining. In the final section, progress in the repetitive, high-power short-pulse lasers is reviewed as an approach to realize highly efficient, ultrahigh-power, ultrashort pulse lasers.

Basic Books for Part II

Electromagnetic Theory and Optics

Landau, L. D. and Lifshitz, E. M. (1980), "The Classical Theory of Fields," 4-th Edition, Elsevier.
Jackson, John D. (1999), "Classical Electrodynamics," 3-rd ed. John Wiley, New York.
Born, M. and Wolf, E. (1999), "Principles of Optics: Electromagnetic Theory of Propagation, Interference and Diffraction of Light," 7-th ed. Cambridge, London.

Quantum Mechanics and Quantum Optics

Dirac, P. A. M. (1988), "The Principles of Quantum Mechanics," 4-th ed. Oxford Science Publications, Oxford.
Heitler, W. (2010), "The Quantum Theory of Radiation," Dover Publications.
Loudon, R. (2010), "The Quantum Theory of Light," 3rd. ed., Oxford.

Lasers

Siegman, A. E. (1986), "Lasers," University Science Books, Palo Alto.
Yariv, Amnon (1989), "Quantum Electronics," 3-rd ed. John Wiley.
Träger, F., ed. (2007), "Springer Handbook of Lasers and Optics," Springer, New York.
Koechner, Walter (2006), "Solid-State Laser Engineering," 6-th ed. Springer, New York.

Nonlinear Optics

Shen, Y. R. (1984) "The Principles of Nonlinear Optics," Wiley, New York.
Boyd, R. W. (2020), "Nonlinear Optics," 4-th ed. Academic Press, London.
Agrawal, G. P. (2019), "Nonlinear Fiber Optics," 6-th ed. Academic Press, Oxford.
Dmitriev, V. G., Gurzadyan, G. G., Nikogosyan, D. N. (1999), "Handbook of nonlinear optical crystals," 2-nd ed. Springer, New York.
Sutherland, R. L. (2003), "Handbook of nonlinear optics," 2nd Ed., Marcel Dekker, New York.

1.6 Introduction to Part III: Laser Plasma Physics and Laser Fusion

With the discovery of electrons in 1895 and of the atomic theory of matter, plasma science took off in the beginning of the twentieth century. Plasma science consists of electromagnetism, statistical mechanics, fluid mechanics, and atomic physics. The plasma is often called the fourth state of matter, which is composed of the ionized gases and perhaps the most abundant form of matter, making up nearly 99.9% of the observable universe. In the 1920s, Irving Langmuir discovered collective plasma oscillations in a laboratory gas discharge. Until the late 1950s, plasma science developed along the two distinct lines. One line was a byproduct of the ionospheric, solar terrestrial and astrophysical research. The pioneering works of plasma physics are described in the monographs by Spitzer (1962) and Chen (1984). A new approach to space science was started by the successful launching of Sputnik, a satellite to orbit the earth, in 1957. Together with the development of space science, the other line is the nuclear fusion research which has stimulated the progress of plasma science since 1955.

The research on physics of hot dense plasmas produced by high power lasers started in the beginning of 1960 soon after the invention of the laser. As shown in Fig. 8.1 of Part III, there are various plasmas in nature and laboratories in the temperature and density space. In nature, the stars like the Sun and the Sirius consist of hot dense plasmas, the aurora borealis is low density magnetized plasma, the thunder lightening is partially ionized plasmas, and so on. In laboratories, neon lamps, arc discharge, liquid metal, magnetic-confinement fusion plasmas, laser produced plasmas and so on are familiar to us. The laser-produced plasmas extend over a wide range which includes ideal plasma, strongly coupled plasma, degenerate plasma, laser fusion plasma and relativistic plasma as shown in Fig. 8.1 of Part III and in Fig. 1.5. In Fig. 8.1 of Part III, the broken line $\Gamma = 1$ is the boundary between the strongly coupled plasma and the ideal plasma and the solid line $k_B T = \varepsilon_F$ is the boundary between the classical plasma and the degenerate plasma (Ichimaru, 2004). The physical meanings of these boundaries are described in Chap. 8 of Part III.

When an intense laser irradiates a solid target, the target is ablated, and a hot dense plasma is created on the surface as shown in Fig. 1.5. The laser deposits the energy near the laser cut-off (C) and the absorbed energy is carried into the solid to produce the hot dense plasma. The heat conduction front is called the ablation front where the solid material evaporates and expands into the vacuum. The temperature and the density at the ablation front are several 100 eV and almost solid density, respectively. So, the pressure at the ablation front reaches a few 10's Mbar, which is higher by one order of magnitude than the pressure of the core of the Earth. The ablation pressure drives a shock wave inside the solid target to produce a warm dense plasma or a hot dense plasma, where free electrons strongly coupled with atoms to generate X-ray radiation. So, the shock wave propagation is the subject of radiation hydrodynamics. On the other hand, the temperature of the plasma around the laser cut-off density is heated up to 10 keV. This relatively low density and high temperature plasma is

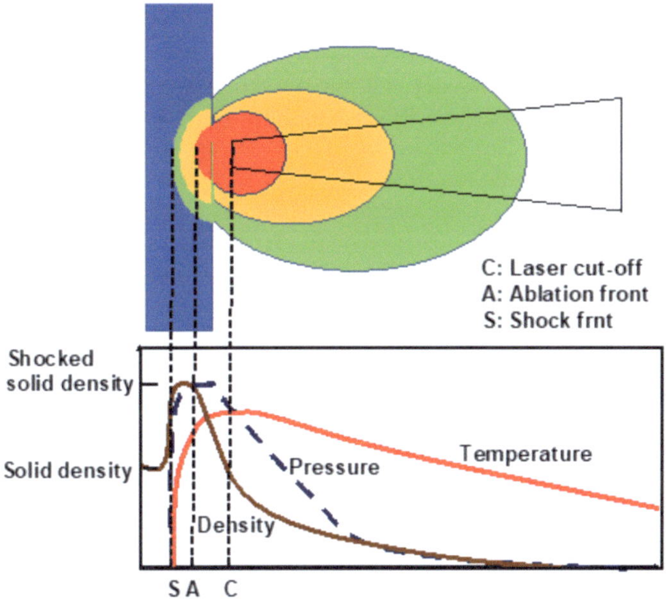

Fig. 1.5 Laser-produced plasma when a laser irradiates a solid target from the right. The upper diagram shows the temperature contour of the laser plasma. The temperature, density and pressure profiles along the laser axis are shown by the red line, brown line, and black broken line respectively, in the lower diagram

called corona plasma according to the similarity to the solar corona. As described above, the laser produced plasma extends over quite wide range in the temperature, density, and pressure space, where various nonlinear wave phenomena take place.

In Part III, we describe non-relativistic laser plasma physics with a temperature below a few 10's keV, and their applications like laser fusion and neutron sources, while Part IV describes the relativistic plasmas related to high field science, which is shown by the red area in Fig. 8.1 of Part III. High density and high-pressure plasmas indicated by the yellow area in this figure show the density and temperature of the laser plasmas. The plasma in this area is called hot dense plasma or warm dense plasma and is closely related to the laser fusion plasma which extends across the classical–degenerate ($k_B T = \varepsilon_F$) and the ideal–nonideal ($\Gamma = 1$) boundaries. From the laser plasma research, which was started in 1960s right after the invention of the laser, the laser fusion (Atzeni & Meyer-ter-Vehn, 2004; Nuckolls et al., 1972), the nonlinear wave phenomena (Galeev & Lipatov, 1984; Kruer, 1988), the plasma X-ray radiation, the high-pressure material physics and so on have been opened.

In 1972, Nuckolls et al. (1972) proposed nuclear fusion with implosion. The implosion means the instantaneous compression of a spherical ball. The implosion by the ablation pressure has been applied for compression and heating of fusion capsule. In 1986, ILE, Osaka University demonstrated high fusion neutron yield up to 10^{13} with the ablative implosion (Yamanaka & Nakai, 1986). After 25 years, the neutron yield has reached a few times 10^{16} with NIF at LLNL, USA (Hurricane et al., 2014; Le Pape et al., 2018; Zylstra et al., 2022).

Laser plasma physics includes both the microscopic processes of femtosecond and sub-micrometer scales, and the macroscopic processes of multi-nanosecond and millimeter scales. The microscopic processes are dominated by the kinetic effects; the electromagnetic field interacting with electrons and ions, and the atomic processes. On the other hand, the macroscopic processes are dominated by the radiation hydrodynamics where the electron thermal transport, ablation processes, shock waves, hydrodynamic instabilities, the radiation heat transport, self-generation of electro-magnetic fields and so on, take place.

In simulation studies, microscopic processes are investigated by the particle in cell (PIC) simulation and the molecular dynamic simulation. The macroscopic processes are investigated by the radiation hydrodynamic simulation together with the data basis on transport coefficients and the equation of state (EOS) which are made by the plasma kinetic theory, the atomic physics and the statistical mechanics.

Part III is organized as follows.

In Chap. 8, we introduce briefly the history of development of plasma science. Examples of plasma phenomena in the universe and laboratory are shown. In Chap. 9, the plasma kinetic equation, in particular, the Fokker–Planck equation and the Vlasov–Maxwell equations are derived. In Chap. 10, we describe laser plasma initiation process, electromagnetic wave (laser) propagation, and inverse-bremsstrahlung (collisional laser absorption) in nonuniform plasma.

In Chap. 11, we describe theory, simulation and experiments on laser absorption processes which include the linear mode conversion (resonance absorption), surface electron acceleration and hot electron generation. In Chap. 12, we describe the nonlinear wave phenomena related to strong laser plasma interactions, such as stimulated scatterings, parametric instabilities, and hot electron generation.

Chapter 13 is devoted to the radiation hydrodynamics of laser ablation, namely, the ablation pressure scaling laws and the electron and radiation heat transport. In Chap. 14, we describe the introduction to laser fusion research with a brief history, gain scaling laws, implosion physics, alternative laser fusion concepts such as fast ignition and shock ignition, and the present status of the laser fusion research. In Chap. 15, we briefly describe approaches to uniform laser irradiation which is important for uniform target compression. Chapter 16 describes hydrodynamic instabilities, another important subject for uniform compression.

Chapter 17 is devoted to the radiation transport and the plasma atomic processes. Finally, in Chap. 18, we describe the laser neutron source as an example of laser-plasma application.

Basic Books for Part III

Physics of Plasmas

Chen, F. F. (1984) "Introduction to Plasma Physics and Controlled Fusion", Plenum Press, New York.

Eliezer, S., Ghatak, A., Hora, H. (2002), "Fundamentals of Equations of State", World Scientific, New Jersey.
Ginzburg, V. L. (1964), "Propagation of Electromagnetic Waves in Plasmas", Pergamon, New York.
Landau, L. D. and Lifshitz, E. M. (1965) "Quantum Mechanics", Pergamon, New York, 1965) 2nd ed.
Lifshitz, E. M. and Pitaevskij, L. P. (1981), "Course of Theoretical Physics: Physical Kinetics", Oxford, Pergamon.
Mima, K. and Nishikawa, K. (1984), "Basic Plasma Physics II," ed. Galeev A.A. and Sudan R.N., Chapter 6.5, p. 452.
Spitzer, L., Jr. (1962) "Physics of Fully Ionized Gases," second revised edition, Dover Publications, New York.
Tajima, T., Mima, K., Baldis, H. (2000), "High Field Science", Kluwer Academic, Plenum Publishers, New York.
Zel'dovich, Ya, and Raizer, Yu. (1966), "Physics of Shock Waves and High Temperature Hydrodynamic Phenomena", Academic Press, New York & London.

Laser Plasma Interaction

Bolton, P. R., Pardoni, K., Schreiber, J. (2018), "Part I: Laser-Driven Particle Acceleration and Associated Energetic Photon and Neutron Generation: Current Understanding and Basic Capability, Part II: Applications of Laser-Driven Particle Acceleration", CRC Press, Taylor & Francis Group, Boca Raton.
Eliezer, S., Mima, K., ed. (2009) "Applications of Laser-Plasma Interactions", CRC press, Taylor & Francis Group, Boca Raton.
Kruer, W. L. (1988), "The physics of laser plasma interactions", Addison Wesley, Frontiers of Physics Series Vol. 73.
Liu, C. S., Tripati, V. K., and Eliasson, B. (2019) "High-Power Laser-Plasma Interaction", Cambridge University Press.
Takabe, H. (2020), "The Physics of Laser Plasmas and Applications—Volume 1: Physics of Laser Matter Interaction," Springer Series in Plasma Science and Technology, Springer Nature, Cham, Switzerland.

Laser Fusion

Atzeni, S. and Meyer-ter-Vehn, J. (2004), "The Physics of Inertial Fusion", Clarendon Press, Oxford.
Basov, N. G. and Krokhin, O. N. (1964), "The conditions of plasma heating by the optical generator radiation", Proc. 3rd Internat. Conf. Quant. Electron., Paris, 1963, ed. P. Grivet and N. Bloembergen Columbia University Press, New York.
Lindl, J. (1998), "Inertial Confinement Fusion—The Quest for Ignition and Energy Gain using Indirect Drive", Springer, New York.
Miharas, D. and Miharas, B.W. (1984), "Foundation of Radiation Hydrodynamics", Oxford University Press, New York.
Mima, K., Nishiguchi, A., Baldis, H., Takabe, A. and Yamanaka, C. (1994), "Laser Plasma Theory and Simulation", ed. V. S. Letokhov, C. V. Shank, R. Shen, and H. Walther, Harwood Academic Publisher.

Nakai, S. and Takabe, H. (1996), "Principles of inertial confinement fusion—physics of implosion and concept of inertial fusion energy", Rep. Prog. Phys. **59**, 1071.
Takabe, H. (2024), "The Physics of Laser Plasmas and Applications—Volume 2: Fluid Models and Atomic Physics of Plasmas" Springer Series in Plasma Science and Technology, Springer Nature, Cham, Switzerland.

1.7 Introduction to Part IV: High Field Science by Lasers

In the limit of extremely strong electromagnetic wave intensity its interaction with a single charged particles leads to the particle acceleration and in the electromagnetic radiation emission via nonlinear Thomson (in the classical electrodynamics regime) and Compton (when the quantum mechanics effects come into play) scattering resulting in the emission of high energy photon emission.

Since, when the extremely intense laser pulse with the corresponding electric field above the interatomic field irradiates the target, its material becomes almost instantaneously ionized during the time interval in the order of the electromagnetic wave period, the laser in fact interacts with the plasma. Laser-plasma interaction is highly nonlinear, which depends on a few such parameters as the target density, the scale of the plasma inhomogeneity, the laser pulse amplitude, duration, polarization, etc.

In the case of the ultra-short pulse laser propagation in the low-density plasma transparent for the laser radiation, it excites nonlinear plasma waves known as the wake waves having a regular structure in what concerned the electron density and electric field distribution.

The wake wave propagates with the velocity close to speed of light in vacuum thus providing the conditions for resonant electrons to be accelerated up to high energy.

During the laser pulse interaction with the overdense plasma target, when the plasma density is above critical value, i.e. the plasma is opaque for electromagnetic wave, the laser energy first is transferred to the plasma electrons in the region in vicinity of the target–vacuum interface. The high energy electrons then propagate through the target with a part of them leaving the target, thus producing strong charge separation electric field. This electric field can accelerate the ions. Such compact laser ion accelerators can have a broad range of various applications ranging from hadron therapy in oncology and the controlled nuclear fusion within the framework of the Fast Ignition concept to the hadron-hadron colliders required in fundamental physics.

Coming back to the case of the laser interaction with underdense plasma targets, we can see that in nonlinear plasma waves excited by the ultra-short pulse laser the electron density modulations have a form of thin layers moving also with the velocity close to a speed of light in vacuum. For a counter-propagating electromagnetic wave, the high-density relativistic electron layers act as the relativistic flying mirrors. The reflected at the relativistic flying mirror light has the frequency upshifted according to the double Doppler effect, corresponding to the X-ray range of a photon energy.

The laser light interacting with the high electron density walls of the cavity formed at the front and in the wake behind the laser pulse generates the high order harmonics via the relativistic mechanisms. High-efficiency relativistic high order harmonics are emitted also during the laser irradiation of solid targets.

The X-ray sources of coherent, powerful, high brightness radiation are required for a wide range of applications in material sciences, medicine and biology, e.g. for macro molecular imaging. The intensity of focused high power X-ray bunches is expected to reach an unprecedented level allowing for testing the physics of nonlinear quantum electrodynamics processes.

Enhancing the electromagnetic radiation intensity either in an extensive way by increasing the laser energy and power or in an intensive manner by using the relativistic flying mirror concept, will bring the laser-matter interaction into the extreme-field-science regimes. In the first novel regime, the radiation friction will change completely the laser-matter interaction paradigm, when the laser energy transformation of high efficiency into hard electromagnetic radiation energy will result in the high-power gamma-ray flash emission, providing, in particular, an instrument for nuclear physics, material sciences, and medicine. The quantum electrodynamics effects will also soon come into play. The gamma-photons interacting with the high amplitude laser field can produce the electron–positron pairs. Subsequent acceleration by the laser light of electrons and positrons with gamma photon emission will lead to the avalanche-type processes with the abundant electron–positron-gamma plasma creation.

When the electromagnetic radiation intensity will further increase approaching the limit of critical field of quantum electrodynamics, the vacuum loses the property of been a void substance, starting to behave as a nonlinear optical medium with the refraction index depending on the electromagnetic wave amplitude. The nonlinear QED vacuum properties will be seen in the light birefringence, in the high order harmonic generation, and in the photon-photon scattering. In this limit the electron–positron pair creation by strong field from the vacuum will allow for testing the quantum electrodynamics in the nonperturbative regime (Fig. 1.6).

Part IV is organized as follows.

In Chap. 19, we give an overview of the key dimensionless parameters determining the regimes of the plasma formation from the gas irradiated by strong electromagnetic radiation. Then we present the parameters characterizing relativistic charged particle dynamics in the high intensity electromagnetic wave, radiation friction effects and the effects of quantum mechanics. At the end of this chapter, we introduce the parameters for plasma and electromagnetic waves propagating in the collisionless plasma and principle of qualitative scaling in modelling astrophysical phenomena under terrestrial laboratory conditions.

Chapter 20 is devoted to theoretical description of the electron dynamics in the field of strong electromagnetic and electrostatic waves within the framework of classical and quantum mechanics.

In Chap. 21, we present the description of the finite-amplitude Langmuir and electromagnetic waves in collisionless plasmas. The nonlinear Langmuir wave breaking process is addressed in detail due to its fundamental importance in the application

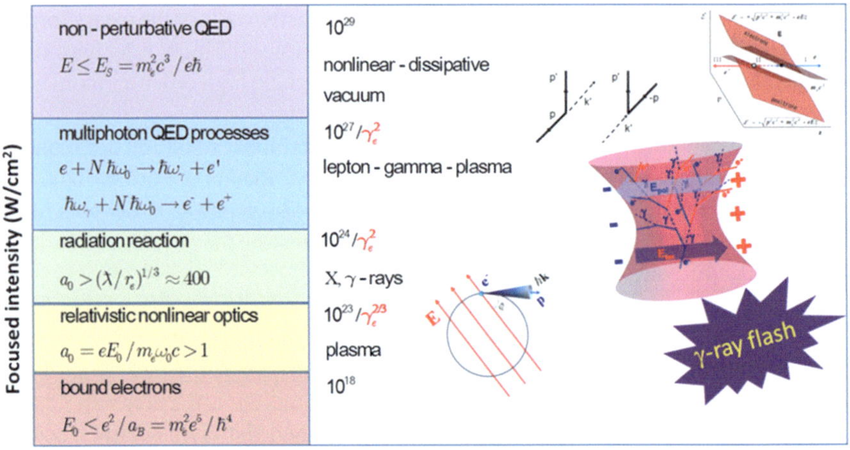

Fig. 1.6 Extreme field limits in high-intensity laser interaction with matter and vacuum

to the laser electron acceleration, which is discussed in Chap. 24. Then the theoretical consideration is presented of relativistic electromagnetic solitons and electron vortices.

Chapter 22 presents consideration of the laser radiation interaction with overdense plasma including the theoretical results concerning the electromagnetic wave propagation in inhomogeneous plasma. Detailed description of nonlinear Langmuir oscillations, electron and ion acceleration and the laser energy absorption in the plasma resonance region and at the plasma–vacuum interface including description of the surface waves are presented.

Chapter 23 contains theoretical analysis of electromagnetic radiation emission by the electron interacting with an electromagnetic wave via the Thomson, in classical limit, and via the Compton scattering in the quantum regime. An attention is paid to the radiation friction effects when the radiating ultrarelativistic electron interacts with single electromagnetic wave and with multiple strong electromagnetic waves.

Chapter 24 is devoted to the laser electron accelerators whose operation is based on the Laser Wake Field Acceleration paradigm (Tajima & Dawson, 1979). Here we discuss the properties of the wake wave generation in the wake left behind the ultrashort laser pulse propagating in the collisionless plasma. The nonlinear wake wave breaking is considered in regards with the electron injection into the acceleration phase. Then we present the scaling of the Laser Wake Field Acceleration and discuss multi-stage laser electron configurations. The energy scaling of the electrons generated in the direct laser acceleration regime is presented. At the chapter end we describe the capillary discharge physics in connection with the laser pulse guiding and the active lensing of high energy electron beams.

Chapter 25 contains discussions of laser ion acceleration mechanisms. Here we describe the Coulomb explosion and Target Normal Sheath Acceleration and the ion acceleration by the laser radiation pressure, including transverse instability of thin foil target and the ion acceleration in an extended plasma. An approach towards formation of high-quality ion beams using double layer thin foil target it presented there.

In Chap. 26, the concept of Relativistic Flying Mirror is presented. This concept is known as Relativistic Catoptrics. Relativistic flying mirrors in laser plasmas are thin, dense electron or electron–ion layers accelerated by high intensity electromagnetic waves to velocities close to the speed of light in vacuum. The reflection of the electromagnetic wave from the relativistic mirror results in energy and frequency changing. In the counter-propagation configuration, the frequency of the reflected wave is upshifted by the factor proportional to the squared gamma-factor of the mirror. This scientific area promises the development of sources of ultrashort X-ray pulses in the attosecond range. In this section we discuss basic properties of the electromagnetic wave reflection from the relativistic mirrors moving with sub-luminal and super-luminal velocity, from the accelerating mirrors, and from oscillating mirrors. We present an approach to-wards calculating the reflection coefficient of various plasma mirrors. Then we consider the case of the "photon acceleration" during the light interaction with nonlinear plasma waves and the concept of Burst Intensification by Singularity Emitting.

Chapter 27 is devoted to consideration of hard photon sources formed in the relativistic laser plasmas. Here we consider two types of the X-ray and gamma-ray sources. In the first case, we describe the high-power gamma-ray flash generation in the laser pulse interaction with inhomogeneous plasma targets when the gamma-ray emission occurs due to the nonlinear Thomson scattering or/end multiphoton Compton scattering. A gamma-ray burst in a laser plasma is also the result of a sequence of processes in which high-energy electrons in the plasma are accelerated by a strong electromagnetic field and then interact with the strong electromagnetic field, emitting high-energy photons. It can be considered as an analogue of synchrotron-self-Compton radiation in an ultra-intense electromagnetic field. The second type of hard photon source is based on the Relativistic Flying Mirror concept discussed in Chap. 26. We present the results of theoretical analysis and computer simulations of the relativistic flying mirror formed in the nonlinear wake waves and its experimental demonstration.

In the final Chap. 28, we present an overview of extreme field limits when the effects of quantum electrodynamics become not negligible in the interaction of high intense electromagnetic radiation with matter and vacuum. We address here such the properties of the nonlinear quantum vacuum as the vacuum polarization and the electron–positron pair creation from vacuum.

Basic Books for Part IV

High-Field Science and Nonlinear Wave Theory

Arnold, V. I., (1978), Mathematical Methods of Classical Mechanics, (Springer, Berlin).

Berestetskii, V. B., Lifshitz, E. M., and Pitaevskii, L. P., (1982), Course of Theoretical Physics, Vol. 4: Quantum Electrodynamics (Pergamon, Oxford).

Berezinsky, V. S., Bulanov, S. V., Ginzburg, V. L., Dogiel, V. A., and Ptuskin, V. S., (1990), Astrophysics of Cosmic Rays (North Holland/Elsevier, Amsterdam, 1990).

Bulanov, S. V., (2000), Introduction to Nonlinear Physics. (Scuola Normale Superiore, Pisa).

Dittrich, W. and Gies, H., (2000), Probing the Quantum Vacuum. Perturbative Effective Action Approach in Quantum Electrodynamics and its Application, Springer Tracts in Modern Physics Vol. 166 (Springer-Verlag, Berlin, Heidelberg).

Dodd, R. K., Eilbeck, J. C., Gibbon, J. D., and Morris, H. C., (1982), Solitons and Nonlinear Wave Equations (Academic Press, London).

Ginzburg V. L., (1962), Propagation of Electromagnetic Waves in Plasma, (Gordon and Breach Science Publishers, Inc.).

Ginzburg V. L., (1979), Theoretical Physics and Astrophysics, (Pergamon).

Kadomstev, B. B., Cooperative effects in plasmas, in Reviews of Plasma Physics, edited by V. D. Shafranov (Kluwer Academic/Plenum Publishers, New York, 2001), Vol. 22, p. 1.

Konopinski, E. J., (1981), Electromagnetic Fields and Relativistic Particles (McGraw Hill Book Company, N. Y.).

Kruer W. (1988), The Physics of Laser Plasma Interactions, Addison Wesley Publishing Company.

Lamb, H., (1932), Hydrodynamics (Cambridge: The Univ. Press).

Landau, L. D., and Lifshitz, E. M., (1976), Course of Theoretical Physics, Vol. 1. Mechanics (Pergamon, Oxford).

Landau, L. D., and Lifshitz, E. M., (1980), Course of Theoretical Physics, Vol. 2: The Classical Theory of Fields, (Pergamon, Oxford).

Landau, L. D., and Lifshitz, E. M., (1984), Course of Theoretical Physics, Vol. 8: Electrodynamics of Continuous Media (Oxford: Pergamon Press).

Landau, L. D., and Lifshitz, E. M., (1987), Course of Theoretical Physics, Vol. 6: Fluid Mechanics (Oxford: Pergamon Press).

Liu, C. S., Tripathi, V. K., Eliasson, B., (2019), High-Power Laser-Plasma Interaction, (Cambridge University Press).

Poston, T. and Stewart, I., (1978), Catastrophe Theory and its Applications (London: Pitman).

Sedov, L. I., (1959), Similarity and dimensional methods in mechanics (Academic Press, New York).

Sedov, L. I., (1971), A course in continuum mechanics. Volumes. I–IV. (Wolters-Noordhoff Publishing, Netherlands).

Vereshchagin, G. V. and Aksenov, A. G., (2017), Relativistic Kinetic Theory with Applications in Astrophysics and Cosmology (Cambridge University Press, New York).
Whitham, G. B., (1974), Linear and Nonlinear Waves (Wiley, New York).
Zakharov, V. E., Manakov, S. V., Novikov, S. P., and Pitaevskii, L. P., (1984), Theory of Solitons (Consultants Bureau, New York).

References

Abu-Shawareb, H., et al. (2022). Lawson criterion for ignition exceeded in an inertial fusion experiment. *Physical Review Letters, 129*, 075001.
Abu-Shawareb, H., et al. (2024). Achievement of target gain larger than unity in an inertial fusion experiment. *Physical Review Letters, 132*, 065102.
Ahmadiniaz, N., Bähtz, C., Benediktovitch, A., et al. (2025). Towards a vacuum birefringence experiment at the Helmholtz international beamline for extreme fields. *High Power Laser Science and Engineering, 13*, e7.
Atzeni, S., & Meyer-ter-Vehn, J. (2004). *The physics of inertial fusion.* Clarendon Press.
Barty, C. P. J. (2020). *Ultrahigh intensity lasers worldwide.* The International Committee on Ultra-High Intensity Lasers. https://www.icuil.org/about-icuil/committee.html
Chen, F. F. (1984). *Introduction to plasma physics and controlled fusion.* Plenum Press.
Danson, C. N., Haefner, C., Bromage, J., et al. (2019). Petawatt and exawatt class lasers worldwide. *High Power Laser Science and Engineering, 7*, Article e54.
Ertel, K., Banerjee, S., Mason, P. D., et al. (2011). Optimising the efficiency of pulsed diode pumped Yb:YAG laser amplifiers for ns pulse generation. *Optics Express, 19*, 26610.
Galeev, A. A., & Lipatov, A. S. (1984). Plasma processes in cometary atmospheres. *Advances in Space Research, 4*, 229.
Haefner, C. L., et al. (2022). *Memorandum laser inertial fusion energy.* BMBF-Inertial Fusion Expert Commission.
Hirori, H., Doi, A., Blanchard, F., et al. (2011). Single-cycle terahertz pulses with amplitudes exceeding 1 MV/cm generated by optical rectification in $LiNbO_3$. *Applied Physics Letters, 98*, 091106.
Holland, A. (2022). *The global fusion industry in 2022—Fusion companies survey.* Fusion Industry Association.
Hurricane, O., Callahan, D. A., Casey, D. T., et al. (2014). Fuel gain exceeding unity in an inertially confined fusion implosion. *Nature, 506*, 343.
Ichimaru, S. (2004). *Statistical plasma physics: Volume II, condensed plasmas.* CRC Press.
Kando, M., Fukuda, Y., Pirozhkov, A. S., et al. (2007). Demonstration of laser-frequency upshift by electron-density modulations in a plasma wakefield. *Physical Review Letters, 99*, 135001.
Kramer, D. (2023). NIF success gives laser fusion energy a shot in the arm. *Physics Today.*
Kruer, W. L. (1988). *The physics of laser plasma interactions.* In *Frontiers of Physics Series* (Vol. 73). Addison Wesley.
Larotonda, M. A., Luther, B. M., Wang, Y., et al. (2004). Characteristics of a saturated 18.9 nm tabletop laser operating at 5 Hz repetition rate. *IEEE Journal of Selected Topics in Quantum Electronics, 10*, 1363.
Lawson, J. D. (1957). Some criteria for a power producing thermonuclear reactor. *Proceedings of the Physical Society. Section B, 70*, 6.
Le Pape, S., Berzak Hopkins, L. F., Divol, L., et al. (2018). Fusion energy output greater than the kinetic energy of an imploding shell at the National Ignition Facility. *Physical Review Letters, 120*, 245003.

Lureau, F., Matras, G., Chalus, O., et al. (2020). High-energy hybrid femtosecond laser system demonstrating 2 × 10 PW capability. *High Power Laser Science and Engineering, 8*, Article e43.

Ma, T., & Betti, R. (2022). *Report of the 2022 fusion energy sciences basic research needs workshop.* U.S. Department of Energy, Office of Science. https://events.bizzabo.com/IFEBRN2022/home

Manes, K. R., Spaeth, M. L., Adams, J. J., Bowers, M. W., et al. (2016). Damage mechanisms avoided or managed for NIF large optics. *Fusion Science and Technology, 69*, 146.

Mason, P., Baneerjee, S., Smith, J., et al. (2018). Development of a 100 J, 10 Hz laser for compression experiments at the High Energy Density instrument at the European XFEL. *High Power Laser Science and Engineering, 6*, Article e65.

Matthews, D. L., Hagelstein, P. L., Rosen, M. D., et al. (1985). Demonstration of a soft X-ray amplifier. *Physical Review Letters, 54*, 110.

Michine, Y., & Yoneda, H. (2020). Ultrahigh damage threshold optics for high power lasers. *Communications on Physics, 3*, 24.

Mirzaie, M., Hojbota, C.I., Kim, D.Y. et al., (2024). All-optical nonlinear Compton scattering performed with a multi-petawatt laser. *Nat Photon, 18*, 1212.

Mourou, G., Mironov, S., Khazanov, E., et al. (2014). Single cycle thin film compressor opening the door to zeptosecond-exawatt physics. *The European Physical Journal Special Topics, 223*, 1181.

Mourou, G., Tajima, T., & Bulanov, S. V. (2006). Optics in the relativistic regime. *Reviews of Modern Physics, 78*, 309.

Nuckolls, J., Wood, L., Thiessen, A., & Zimmerman, G. (1972). Laser compression of matter to super-high densities: Thermonuclear (CTR) applications. *Nature, 239*, 139.

Ogino, J., Tokita, S., Kitajima, S., et al. (2022). 10-J, 100-Hz conduction-cooled active mirror laser. *Optics Continuum, 1*, 1270.

Pak, A., Zystra, A. B., Baker, K. L., et al. (2024). Observations and properties of the first laboratory fusion experiment to exceed a target gain of unity. *Physical Review E, 109*, 025203.

Peng, Y., Yu, L., Wang, X., et al. (2021). Overview and status of station of extreme light toward 100 PW. *The Review of Laser Engineering, 49*, 93.

Pirozhkov, A. S., Ma, J., Kando, M., et al. (2007). Frequency multiplication of light back-reflected from a relativistic wake wave. *Physics of Plasmas, 14*, 123106.

Pirozhkov, A. S., Sagisaka, A., Ogura, K., et al. (2024). Demonstration of the brightest nano-size gamma source. arXiv:2410.06537 (physics).

Popmintchev, T., Chen, M.-C., Popmintchev, D., et al. (2021). Bright coherent ultrahigh harmonics in the keV X-ray regime from mid-infrared femtosecond lasers. *Science, 336*, 1287.

Spaeth, M. L., Manes, K. R., Kalanter, D. H., et al. (2016). Description of the NIF laser. *Fusion Science and Technology, 69*, 25.

Spitzer, L. Jr. (1962). *Physics of fully ionized gases* (2nd revised ed.). Dover Publications.

Stark-Watzinger, B. (2024). *Förderprogramm Fusion 2040—Forschung auf dem Weg zum Fusionskraftwerk (Funding programme fusion 2040—Research on the way to a fusion power plant).* BMBF (Federal Ministry of Education and Research).

Strickland, D., & Mourou, G. (1985). Compression of amplified chirped optical pulses. *Optics Communications, 56*, 219.

Tajima, T., & Dawson, J. M. (1979). Laser electron accelerator. *Physical Review Letters, 43*, 267.

Wang, W., Feng, K., Ke, L., et al. (2021). Free electron lasing at 27 nanometres based on a laser wakefield accelerator. *Nature, 595*, 516.

Yamanaka, C., & Nakai, S. (1986). Thermonuclear neutron yield of 10^{12} achieved with Gekko XII green laser. *Nature, 319*, 757.

Yoneda, H., Inubushi, Y., Nagamine, K., et al. (2015). Atomic inner-shell laser at 1.5-ångström wavelength pumped by an X-ray free-electron laser. *Nature, 524*, 446.

Ziegler, T., Göthel, I., Assenbaum, S., et al. (2024). Laser-driven high-energy proton beams from cascaded acceleration regimes. *Nature Physics, 20*, 1211.

Zylstra, A. B., Hurricane, O. A., Callahan, D. A., et al. (2022). Burning plasma achieved in inertial fusion. *Nature, 601*, 542.

Part II
High Power Lasers

Chapter 2
Propagation of Laser Light

Abstract In this chapter, we consider propagation of laser light in a linear region based on the wave optics, and the material dispersion: frequency dependence of the refractive index in complex expression. In Sect. 2.1, basic properties of a Gaussian beam are derived in paraxial wave approximation of an electromagnetic wave, with addition of the angular momentum of a laser beam including application to high field science and material processing. In Sect. 2.2, basic properties of the optical components used in high power lasers are described. In Sect. 2.3, we consider propagation of an ultrashort-duration laser pulse in a dispersive media, i.e., how the amplitude and phase change during propagation due to material dispersion. Also described are the dispersion properties of several optical components required to control propagation of ultrashort laser pulses. In Sect. 2.4, we briefly introduce relativity which is an extension of the electromagnetic theory and essential in high field science. The equations and constants are expressed in the International System of Units (SI Units). Vectors and tensors are indicated by bold letters.

Keywords Maxwell equations · Laser · Gaussian beam · Angular momentum of light · Dispersion · Ultrashort pulse · Chirped pulse amplification · Pulse compression · Pulse expansion · Chirped mirror

2.1 Propagation of an Optical Wave

2.1.1 Maxwell Equations

The macroscopic Maxwell equations describe relations between the electric and magnetic fields: electric field $\boldsymbol{E}(\boldsymbol{r}, t)$ (V/m), magnetic flux $\boldsymbol{B}(\boldsymbol{r}, t)$ (V s/m^2 = T), electric displacement $\boldsymbol{D}(\boldsymbol{r}, t)$ (A s/m^2 = Q/m^2), magnetic field strength $\boldsymbol{H}(\boldsymbol{r}, t)$ (A/m), charge density $\rho(\boldsymbol{r}, t)$ (A s/m^3 = Q/m^3), and current density $\boldsymbol{j}(\boldsymbol{r}, t)$ (A/m^2). These quantities, which are space-averaged over an atomic scale, are functions of the position $\boldsymbol{r} = (x, y, z)$ and time t. The SI units of these quantities are given in parentheses. Denoting the divergence and the curl by $\nabla\cdot$ and $\nabla\times$ respectively,

Y. Kato et al., *High Power Laser and Plasma Science*, Springer Series on Atomic, Optical, and Plasma Physics 130, https://doi.org/10.1007/978-3-032-08930-4_2

$$\nabla \cdot \boldsymbol{D}(\boldsymbol{r}, t) = \rho(\boldsymbol{r}, t), \tag{2.1}$$

$$\nabla \cdot \boldsymbol{B}(\boldsymbol{r}, t) = 0, \tag{2.2}$$

$$\nabla \times \boldsymbol{E}(\boldsymbol{r}, t) = -\frac{\partial \boldsymbol{B}(\boldsymbol{r}, t)}{\partial t}, \tag{2.3}$$

$$\nabla \times \boldsymbol{H}(\boldsymbol{r}, t) = \boldsymbol{j}(\boldsymbol{r}, t) + \frac{\partial \boldsymbol{D}(\boldsymbol{r}, t)}{\partial t}. \tag{2.4}$$

Equations (2.1), (2.2), (2.3) and (2.4) express, respectively, the Coulomb's law, absence of a "magnetic charge" (magnetic monopole), the Faraday's induction law, and the Ampere's law with addition of the displacement current $\partial \boldsymbol{D}/\partial t$, with which this set of equations become consistent with the continuity equation of the charge:

$$\frac{\partial \rho(\boldsymbol{r}, t)}{\partial t} + \nabla \cdot \boldsymbol{j}(\boldsymbol{r}, t) = 0. \tag{2.5}$$

Response of a material to the electric field $\boldsymbol{E}$ is described by the electric displacement $\boldsymbol{D}$ as:

$$\boldsymbol{D} = \varepsilon_0 \boldsymbol{E} + \boldsymbol{P}, \tag{2.6}$$

where $\varepsilon_0 (= 8.8542 \times 10^{-12}\,\mathrm{A\,s\,V^{-1}\,m^{-1}})$ is the permittivity of free space. The electric polarization $\boldsymbol{P}$ [A s/m^2] is the dipole moment per unit volume, which is expressed in vector and scholar notations respectively as:

$$\begin{aligned} \boldsymbol{P} &= N\langle \boldsymbol{\mu} \rangle \equiv \varepsilon_0 \boldsymbol{\chi}^{(1)} \cdot \boldsymbol{E}, \\ P_i &= \varepsilon_0 \sum_{j=x,y,z} \chi_{ij}^{(1)} E_j, \end{aligned} \tag{2.7}$$

where the suffix (i and j) in the second equation refers to the Cartesian coordinates x, y, or z. Here $\langle \boldsymbol{\mu} \rangle$ is an ensemble-averaged dipole moment of atoms induced by the time-varying electric field $\boldsymbol{E}(\boldsymbol{r}, t)$ and N is the number density of the atoms. $\boldsymbol{\chi}^{(1)}$ is the linear electric susceptibility, which is a tensor since $\boldsymbol{P}$ is not always parallel to $\boldsymbol{E}$ in anisotropic media. The electric displacement $\boldsymbol{D}$ is thus written as

$$\boldsymbol{D} = \varepsilon_0 \left(1 + \boldsymbol{\chi}^{(1)}\right) \boldsymbol{E} \equiv \boldsymbol{\varepsilon} \cdot \boldsymbol{E}, \tag{2.8}$$

where $\boldsymbol{\varepsilon}$ is the permittivity tensor. Here we consider light propagation in an optically isotropic media and describe $\boldsymbol{\chi}^{(1)}$ and $\boldsymbol{\varepsilon}$ in scalars as $\chi^{(1)}$ and ε, respectively. (Light propagation in optically anisotropic media is considered in Sect. 2.2.4.)

Similar to the relation between $\boldsymbol{P}$ and $\boldsymbol{D}$, the magnetization $\boldsymbol{M}$ [A/m] is expressed by the magnetic flux $\boldsymbol{B}$ as

$$\boldsymbol{B} = \mu_0 \boldsymbol{H} + \boldsymbol{M} \equiv \mu \boldsymbol{H}, \tag{2.9}$$

where $\mu_0 (= 4\pi \times 10^{-7}\ \mathrm{V\,s\,A^{-1}\,m^{-1}})$ and μ are the permeabilities in free space and in material, respectively. The magnetic permeability of most of the dielectric media is equal to μ_0 since magnetization is not induced by the optical field. Equations (2.8) and (2.9) are called constitutive relations of the Maxwell equations.

2.1.2 *Light Propagation in a Free Space*

In vacuum and a dielectric media where $\boldsymbol{j} = 0$ and $\rho = 0$, Eqs. (2.1)–(2.4) yield a simplified set of equations:

$$\begin{aligned} &\nabla \cdot \boldsymbol{E} = 0, \quad \nabla \cdot \boldsymbol{B} = 0, \\ &\nabla \times \boldsymbol{E} = -\frac{\partial \boldsymbol{B}}{\partial t}, \quad \nabla \times \boldsymbol{B} = \varepsilon\mu \frac{\partial \boldsymbol{E}}{\partial t}. \end{aligned} \tag{2.10}$$

Operating curl to these equations and noting $\nabla \times (\nabla \times \boldsymbol{A}) = \nabla(\nabla \cdot \boldsymbol{A}) - \nabla^2 \boldsymbol{A}$ for an arbitrary vector $\boldsymbol{A}$, we obtain,

$$\nabla^2 \boldsymbol{E} - \varepsilon\mu \frac{\partial^2 \boldsymbol{E}}{\partial t^2} = 0, \quad \nabla^2 \boldsymbol{B} - \varepsilon\mu \frac{\partial^2 \boldsymbol{B}}{\partial t^2} = 0. \tag{2.11}$$

A general solution of these partial differential equations for each Cartesian component of $\boldsymbol{E}$ and $\boldsymbol{B}$ is written as

$$u(z, t) = u(z - ct) + u(z + ct), \tag{2.12}$$

where z denotes a representative spatial coordinate, and c is given by

$$c = \frac{1}{\sqrt{\varepsilon\mu}} = \frac{1}{\sqrt{\varepsilon_0\mu_0}} \sqrt{\frac{\varepsilon_0\mu_0}{\varepsilon\mu}} = \frac{c_0}{n}, \tag{2.13}$$

with

$$\begin{aligned} c_0 &= \frac{1}{\sqrt{\varepsilon_0\mu_0}}, \\ n &= \sqrt{\frac{\varepsilon\mu}{\varepsilon_0\mu_0}} \cong \sqrt{\frac{\varepsilon}{\varepsilon_0}} \quad (\text{when } \mu \cong \mu_0). \end{aligned} \tag{2.14}$$

Equation (2.12) shows that $\boldsymbol{E}$ and $\boldsymbol{B}$ propagate in space toward $+z$ and $-z$ directions with a velocity c, which is described as $c = c_0/n$ where $c_0 (= 2.9979 \times 10^8\ \mathrm{ms^{-1}})$ is the light velocity in vacuum and n is called a refractive index. Introducing the wavevector $\boldsymbol{k}$

$$\boldsymbol{k} = \frac{n\omega}{c_0}\hat{\boldsymbol{s}}, \tag{2.15}$$

where $\hat{\boldsymbol{s}}$ is a unit vector toward the propagation direction, the general solution of (2.11) is described as a superposition of sinusoidal waves:

$$\begin{aligned}\boldsymbol{E}(\mathrm{r}, t) &= \boldsymbol{E}_0 \cos(\boldsymbol{k}\cdot\boldsymbol{r} - \omega t),\\ \boldsymbol{B}(\mathrm{r}, t) &= \boldsymbol{B}_0 \cos(\boldsymbol{k}\cdot\boldsymbol{r} - \omega t).\end{aligned} \tag{2.16}$$

We find from (2.10),

$$\hat{\boldsymbol{s}}\cdot\boldsymbol{E}(\boldsymbol{r}, t) = 0 \text{ and } \hat{\boldsymbol{s}}\cdot\boldsymbol{B}(\boldsymbol{r}, t) = 0. \tag{2.17}$$

These relations show that both $\boldsymbol{E}$ and $\boldsymbol{B}$ are perpendicular to the propagation direction $\hat{\boldsymbol{s}}$, i. e., the optical waves are transverse waves.

From (2.10) we obtain two independent sets of solutions of (E_x, B_y) and (E_y, B_x):

$$\frac{1}{c^2}\frac{\partial E_x}{\partial t} = -\frac{\partial B_y}{\partial z}, \quad \frac{\partial B_y}{\partial t} = -\frac{\partial E_x}{\partial z}. \tag{2.18}$$

$$\frac{1}{c^2}\frac{\partial E_y}{\partial t} = -\frac{\partial B_x}{\partial z}, \quad \frac{\partial B_x}{\partial t} = \frac{\partial E_y}{\partial z}. \tag{2.19}$$

The first set (E_x, B_y) is called x-polarized wave, and the second set (E_y, B_x) is called y-polarized wave. Noting $\omega = ck$, (2.18) leads to

$$E_x = cB_y = \sqrt{\frac{\mu}{\varepsilon}}H_y. \tag{2.20}$$

The E/H ratio given in (2.20) has a dimension of resistivity with $E/H = 376\ \Omega$ in vacuum. The propagation of an x-polarized plane wave is shown schematically in Fig. 2.1.

In general, the polarization state of light is described as a superposition of x-polarized and y-polarized waves, which is expressed in vector representation as

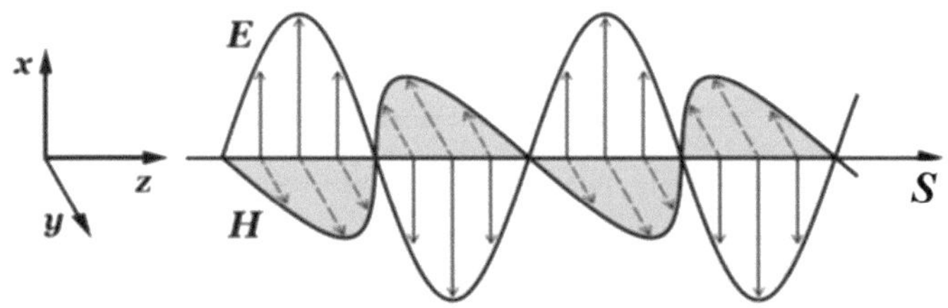

Fig. 2.1 An x-polarized plane optical wave propagating in the z-direction. The energy flow shown by the Poynting vector $\boldsymbol{S}$ propagates in the z-direction

$$\begin{pmatrix} E_x \\ E_y \end{pmatrix} = \begin{pmatrix} E_{x0}\cos(kz - \omega t) \\ E_{y0}\cos(kz - \omega t + \delta) \end{pmatrix}. \tag{2.21}$$

Here E_{x0} and E_{y0} are the amplitudes of the x- and y-components of the electric field, respectively, and δ is the phase difference between the x and y components. The trajectory of the (E_x, E_y) vector in the x–y plane, which depends on the amplitudes and δ, is elliptic in general. When $E_{x0} = E_{y0}$ and $\delta = \pi/2$ (or $\delta = -\pi/2$) the trajectory becomes circular, rotating right-handed (or left-handed) when observed from the $+z$ direction.

2.1.3 Energy Density and Flux

The energy conservation equation of the electromagnetic field is derived by multiplying $\boldsymbol{H}$ to (2.3) and $\boldsymbol{E}$ to (2.4), adding these equations, and noting $\nabla \cdot (\boldsymbol{A} \times \boldsymbol{B}) = \boldsymbol{B} \cdot (\nabla \times \boldsymbol{A}) - \boldsymbol{A} \cdot (\nabla \times \boldsymbol{B})$ for arbitrary vectors $\boldsymbol{A}$ and $\boldsymbol{B}$;

$$\boldsymbol{j} \cdot \boldsymbol{E} + \frac{\partial}{\partial t}\left[\frac{1}{2}(\varepsilon \boldsymbol{E} \cdot \boldsymbol{E} + \mu \boldsymbol{H} \cdot \boldsymbol{H})\right] = -\nabla \cdot (\boldsymbol{E} \times \boldsymbol{H}). \tag{2.22}$$

The first term on the left-hand side corresponds to the ohmic heating. We can rewrite (2.22) as

$$\frac{\partial u}{\partial t} + \nabla \cdot \boldsymbol{S} = -\boldsymbol{j} \cdot \boldsymbol{E}. \tag{2.23}$$

In this equation,

$$u = \frac{1}{2}(\varepsilon \boldsymbol{E} \cdot \boldsymbol{E} + \mu \boldsymbol{H} \cdot \boldsymbol{H}) \tag{2.24}$$

is the energy density (J/m^3) of the optical wave. It is composed of the electric energy density $u_e = \varepsilon \boldsymbol{E} \cdot \boldsymbol{E}/2$ and the magnetic energy density $u_m = \mu \boldsymbol{H} \cdot \boldsymbol{H}/2$. Due to (2.20), $u_e = u_m$ for a plane wave in a homogenous medium, resulting in

$$u = u_e + u_m = 2u_e = 2u_m. \tag{2.25}$$

The vector

$$\boldsymbol{S} = \boldsymbol{E} \times \boldsymbol{H} \tag{2.26}$$

is the energy flux (W/m^2) (called *Poynting vector*) of the optical wave, which has a magnitude $|\boldsymbol{S}| = cu$ and is directed toward $\boldsymbol{E} \times \boldsymbol{H}$ as shown in Fig. 2.1. When an optical wave is completely reflected by a material at normal incidence, it applies a pressure to the material with a magnitude $p = 2\varepsilon E_0^2 = 2u$ per unit area. Since the reflection is accompanied with a change in the momentum of the optical wave, the momentum density of the optical wave, which is equal for the incident and the reflected waves, is given by

$$\boldsymbol{p} = \frac{u}{c}\hat{\boldsymbol{k}} = \frac{1}{c^2}\boldsymbol{E} \times \boldsymbol{H} = \frac{1}{c^2}\boldsymbol{S}. \tag{2.27}$$

The energy flux, averaged in time longer than the period of the light wave, is called the intensity of the optical wave. Using the amplitude of the electric field $E_0(\mathrm{r}, t)$ defined in (2.16), the intensity of a plane optical wave propagating in the z-direction is written as,

$$I(\boldsymbol{r}, t) = \frac{1}{2}\varepsilon_0 c|E_0(\boldsymbol{r}, t)|^2. \tag{2.28}$$

Here, $\boldsymbol{r}$ is the coordinate normal to z. The intensity is often expressed in unit of W/cm^2. The power of an optical wave, expressed in W, is the intensity integrated over the area normal to the propagation direction:

$$\mathcal{P}(t) = \int_A I(\boldsymbol{r}, t)dA. \tag{2.29}$$

For pulsed lasers where the power changes in time, often "peak power" is used to characterize the laser beam. The energy of a pulsed laser beam, expressed in J, is the power integrated over the pulse duration:

$$\mathcal{E} = \int \mathcal{P}(t)dt. \tag{2.30}$$

Note that energy, intensity, and momentum of an optical wave take continuous values, since they depend only on the amplitude of the electric field. This contrasts with the quantum description of radiation, where an optical wave is a flow of "photons", taking discrete values in unit of $\hbar\omega$. In this description, the radiation intensity is expressed as $I = cn\hbar\omega$, where c is the light velocity, n is the photon number density, $\hbar = h/2\pi$ is the Planck constant, and $\hbar\omega$ is the energy of each photon, as we will describe in Sect. 3.1.

A charged particle is accelerated by an optical field by the Lorentz force as $mdv/dt = qE_0$, where m and q are the mass and the charge of the particle, respectively, and v is the velocity. From this equation, the peak velocity of the particle accelerated in one optical cycle is given approximately as $v_0 \sim qE_0/m\omega$. The ratio $a_0 = v_0/c$, called *normalized vector potential*, is used to characterize the field amplitude:

$$a_0 \equiv \frac{q}{mc\omega}E_0 = \frac{q}{mc}A_0. \tag{2.31}$$

A charged particle is accelerated close to the relativistic velocity in one laser cycle when $a_0 \sim 1$. From (2.28), a_0 for an electron is expressed with the laser wavelength λ and the intensity I, in units often used in experiment, as

$$a_0 = \sqrt{\frac{q^2}{2\pi^2 \varepsilon_0 m^2 c^5} \lambda^2 I_0} \simeq 0.86\,\lambda\ (\mu\text{m}) \sqrt{I_0\ (10^{18}\,\text{W/cm}^2)}. \tag{2.32}$$

The relativistic intensity (where $a_0 = 1$) for an electron is $I = 2.11 \times 10^{18}$ W/cm^2, with the corresponding electric field amplitude of $E_0 = 4.0 \times 10^{12}$ V/m at $\lambda = 0.8\,\mu$m (wavelength of the Ti:sapphire laser). The relativistic intensity, which is inversely proportional to the particle mass, is 3.87×10^{21} W/cm^2 for a proton.

In the laser field, an electron has time-averaged quivering energy, called *poderomotive potential energy*:

$$U_p = \left\langle \frac{1}{2} m v^2 \right\rangle = \frac{q^2 E_0^2}{4 m \omega^2}. \tag{2.33}$$

In a practical unit, the electron quiver energy is expressed as U_p (eV) $= 9.337 \times 10^{-14} I$ (W/cm^2) $\cdot \lambda^2$ (μm). When the field intensity is spatially nonuniform, *pondermotive force* acts to a charged particle from the high intensity region to the lower intensity region as

$$\boldsymbol{F}_p = -\nabla U_p(\mathbf{R}, t). \tag{2.34}$$

2.1.4 Vector and Scalar Potentials

Here we introduce the vector and scalar potentials of the electromagnetic wave, which will be used in considering the interaction of light with an atom in Sect. 3.2.1. In the electrostatics, the electric field is expressed with a scalar potential ϕ as $\boldsymbol{E} = -\nabla\phi$. Generalizing this relation, we express the magnetic field with a vector potential. Considering $\nabla \cdot \boldsymbol{B} = 0$, the magnetic field can be expressed with a vector potential $\boldsymbol{A}$ as

$$\boldsymbol{B} = \nabla \times \boldsymbol{A}. \tag{2.35}$$

With (2.35), the third equation of (2.10) becomes $\nabla \times (\boldsymbol{E} + \partial \boldsymbol{A}/\partial t) = 0$. Noting $\nabla \times \nabla\phi = 0$, we can write $\boldsymbol{E}$ as

$$\boldsymbol{E} = -\nabla\phi - \frac{\partial \boldsymbol{A}}{\partial t}. \tag{2.36}$$

Since $\nabla \times (\nabla \eta) = 0$ holds for an arbitrary function $\eta(x, y, z)$, the relations (2.35) and (2.36) are maintained even when the vector and scalar potentials are replaced by $\boldsymbol{A}' = \boldsymbol{A} + \nabla \eta$ and $\phi' = \phi - \partial \eta / \partial t$, respectively. Using this arbitrariness, we can transform the potentials $\boldsymbol{A}$ and ϕ to satisfy the following relation, called *Lorentz condition*,

$$\frac{1}{c^2} \frac{\partial \phi}{\partial t} + \nabla \cdot \boldsymbol{A} = 0. \tag{2.37}$$

Using (2.31) and $\nabla \times \nabla \times \boldsymbol{A} = \nabla \cdot \nabla \boldsymbol{A} - \Delta \boldsymbol{A}$ in the Maxwell equations, $\boldsymbol{A}$ and ϕ satisfy the following equations,

$$\nabla^2 \boldsymbol{A} - \frac{1}{c^2} \frac{\partial^2 \boldsymbol{A}}{\partial t^2} = -\mu_0 \boldsymbol{J}, \tag{2.38}$$

$$\nabla^2 \phi - \frac{1}{c^2} \frac{\partial^2 \phi}{\partial t^2} = -\frac{\rho}{\varepsilon_0}. \tag{2.39}$$

The Maxwell equations are now expressed by (2.38) and (2.39).

In vacuum where $\rho = 0$ and $\boldsymbol{J} = 0$, the scalar potential can be set as $\phi = 0$ by making the gauge transformation to ϕ and $\boldsymbol{A}$. Then the following set of equations is obtained:

$$\nabla \cdot \boldsymbol{A} = 0, \tag{2.40}$$

$$\nabla^2 \boldsymbol{A} - \frac{1}{c^2} \frac{\partial^2 \boldsymbol{A}}{\partial t^2} = 0. \tag{2.41}$$

In vacuum, $\boldsymbol{E}$ and $\boldsymbol{B}$ are derived from the vector potential $\boldsymbol{A}$ simply as

$$\boldsymbol{E} = -\frac{\partial \boldsymbol{A}}{\partial t}, \tag{2.42}$$

$$\boldsymbol{B} = \nabla \times \boldsymbol{A}. \tag{2.43}$$

The vector potential $\boldsymbol{A}$ is not directly measurable, whereas $\boldsymbol{E}$ and $\boldsymbol{B}$ are measurable from the force to a charged particle. Aharonov and Bohm (1959) have predicted that the phase of a charged particle will be affected by the vector potential, called the Aharonov–Bohm effect. Tonomura (1993) has experimentally shown the Aharonov–Bohm effect by measuring an interference pattern of two electron waves which have propagated through different paths in the vector potential, clearly demonstrating the existence of the vector potential.

2.1.5 Gaussian Beam

Let us consider a paraxial wave propagating in the z direction (Fox & Li, 1961). The electric field (2.16) is described by taking the real part of the complex expression as

$$\boldsymbol{E}(\mathrm{r}, t) = \boldsymbol{E}_0 \cos(\boldsymbol{k} \cdot \boldsymbol{r} - \omega t) = \mathrm{Re}\left[\boldsymbol{E}_0 \mathrm{e}^{i(\boldsymbol{k}\cdot\boldsymbol{r} - \omega t)}\right] = \mathrm{Re}\left[\hat{\boldsymbol{e}} A(\boldsymbol{r}, t) \mathrm{e}^{i(\boldsymbol{k}\cdot\boldsymbol{r} - \omega t)}\right], \quad (2.44)$$

where $\hat{\boldsymbol{e}}$ is a unit vector in the polarization direction, and $A(\boldsymbol{r}, t)$ is a slowly varying amplitude of the electric field. (Note $A(\boldsymbol{r}, t)$ in this equation is not a vector potential.) Assuming that the temporal change of the pulse envelope is slow compared to the optical cycle, the wave propagation is described from (2.11) and as

$$\nabla^2 A(\boldsymbol{r}) + k^2 A(\boldsymbol{r}) = 0. \quad (2.45)$$

In the paraxial wave approximation $\left|\partial^2 A/\partial z^2\right| \ll |k \partial A/\partial z|$, which holds when the change of $A(\mathrm{r}, t)$ in the z-direction is small relative to the wavelength, the following equation is obtained:

$$\frac{\partial^2 A}{\partial x^2} + \frac{\partial^2 A}{\partial y^2} + 2ik\frac{\partial A}{\partial z} = 0. \quad (2.46)$$

The solution of $A(x, y, z)$ in the rectangular coordinate is described by the Hermite–Gaussian function,

$$A_{m,n}^{\mathrm{HG}}(x, y, z) = \frac{w_0}{w(z)} H_m\left(\frac{\sqrt{2}x}{w(z)}\right) H_n\left(\frac{\sqrt{2}y}{w(z)}\right) \exp\left(-\frac{r^2}{w(z)^2}\right) \times \exp\left\{-i\left(k\frac{r^2}{2R(z)} - \Phi(m, n, z)\right)\right\}, \quad (2.47)$$

where $r^2 = x^2 + y^2$.

In (2.47), the Gaussian distribution $A(x, y, z) \sim \exp\left(-r^2/w(z)^2\right)$ is superposed with $H_m(x)$, the mth order Hermite polynomials, whose low orders are given by:

$$H_0(x) = 1, \ H_1(x) = 2x, \ H_2(x) = 4x^2 - 2, \ \cdots . \quad (2.48)$$

The field $A_{m,n}^{HG}(x, y, z)$ is called the m, n-mode, or $\mathrm{TEM_{mn}}$ (transverse electric and magnetic mode). The 0,0-mode is the fundamental mode which is called a Gaussian beam. Higher order modes, for example the 2,1-mode $u_{2,1}(x, y, z) \propto (2x^2 - 1)y\exp\left(-r^2/w(z)^2\right)$ has 2 nodes in the x-direction and 1 node in the y-direction, resulting in 6 bright regions in the intensity distribution. The high order modes of $m, n \geq 1$ have larger spot sizes and larger divergences compared with the fundamental mode.

The amplitude of the TEM_{00} mode expressed by (2.47) has a Gaussian distribution $\exp(-r^2/w(z)^2)$ with a *spot size* $w(z)$:

$$w^2(z) = w_0^2\left[1 + \left(\frac{\lambda z}{\pi w_0^2}\right)^2\right] = w_0^2\left[1 + \left(\frac{z}{z_0}\right)^2\right], \tag{2.49}$$

where $z_0 \equiv \pi w_0^2/\lambda$ is called the *Rayleigh range*. The spot size $w(z)$ takes the minimum value w_0 at $z = 0$, becomes $w = \sqrt{2}w_0$ at $z = z_0$, and increases asymptotically as $w(z) \sim w_0(z/z_0)$ at $z \gg z_0$. Since the intensity is proportional to the square of the amplitude,

$$I(r) = I_0(z)\exp\left[-2r^2/w^2(z)\right], \tag{2.50}$$

The spot size $w(z)$ corresponds to the radius where the intensity is $1/e^2$ of the peak intensity $I_0(z)$ at $r = 0$. The power $\mathcal{P}(r)$ within the circular area of radius r is

$$\mathcal{P}(r) = \int_0^r I(r)2\pi r dr = \frac{\pi w^2(z)I_0(z)}{2}\left[1 - \exp(-2r^2/w^2(z))\right]. \tag{2.51}$$

Since the total power is given by $\mathcal{P}(r = \infty) = \pi w^2(z)(I_0(z)/2)$, 86.5% of total power is contained within a circular area of radius $w(z)$.

Noting that a spherical wave of radius $R = \sqrt{x^2 + y^2 + z^2}$ is described as

$$A(x, y, z) = \frac{e^{ikR}}{R} \simeq \frac{1}{R}\exp\left[ik\left(z + \frac{x^2 + y^2}{2R}\right)\right], \tag{2.52}$$

where $x^2 + y^2 \ll z^2$ is assumed, we find that the field given in (2.47) has a spherical wave-front whose radius of curvature is given by

$$R(z) = z\left[1 + \left(\frac{\pi w_0^2}{\lambda z}\right)^2\right] = z\left[1 + \left(\frac{z_0}{z}\right)^2\right]. \tag{2.53}$$

The wavefront is planer at z = 0 since $R(z = 0) = \infty$ and becomes $R = 2z_0$ at $z = z_0$. The distance $2z_0$ between $z = \pm z_0$ is called a *confocal parameter*, or depth of focus which is important in various optical systems.

The full-angle beam divergence of the TEM_{00} mode is inversely proportional to w_0 as

$$\Theta_0 = 2\lim_{z\to\infty}\frac{w(z)}{R(z)} = \frac{2\lambda}{\pi w_0} \sim \frac{\lambda}{w_0}. \tag{2.54}$$

The major parameters of the Gaussian beam are schematically shown in Fig. 2.2.

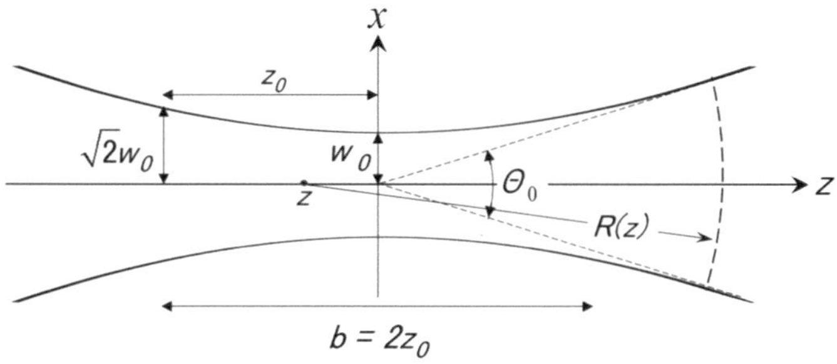

Fig. 2.2 Schematic diagram of a Gaussian beam, showing the beam radius $w(z)$ by solid curves, the minimum beam radius w_0, the Rayleigh range z_0, the confocal parameter b, the wave-front curvature $R(z)$, and the full-angel beam divergence Θ_0

The term Φ in (2.47), called Gouy phase, is a shift from the plane wave due to the spherical wavefront:

$$\Phi(m, n; z) = (m + n + 1)tan^{-1}\left(\frac{z}{z_0}\right). \tag{2.55}$$

The Gouy phase changes from $-\pi/2$ to $+\pi/2$ when z changes from $-\infty$ to $+\infty$ (and from $-\pi/4$ to $+\pi/4$ from $z = -z_0$ to $z = +z_0$) for the fundamental mode ($m = n = 0$). It increases by (m + n + 1) times for higher order modes. This phase term is important, for example, in considering the phase matching condition in high-order harmonic generation.

The beam diameter D and the full angle divergence Θ of higher-order modes are larger than those of the TEM_{00} mode by approximately M times as

$$D = MD_0, \tag{2.56}$$

$$\Theta = M\Theta_0, \tag{2.57}$$

where $D_0 = 2w_0$. Although these parameters change when the beam is transported, the beam-product parameter $D\Theta$ (often expressed in mm-mrad) is conserved:

$$D\Theta = M^2 D_0 \Theta_0 = M^2(4\lambda/\pi). \tag{2.58}$$

The parameter M^2 is used to evaluate the quality of a laser beam, by comparing it with a diffraction-limited Gaussian beam where $M^2 = 1$.

2.1.6 Light with Angular Momentum

An optical wave has an angular momentum in addition to the "kinetic" momentum. For a plane optical wave, the momentum density $\boldsymbol{p} = (\boldsymbol{E} \times \boldsymbol{H})/c^2 = \boldsymbol{S}/c^2$ has a component only along the propagation direction z (Eq. 2.27). Therefore, the density of the angular momentum $\boldsymbol{m} = \boldsymbol{r} \times \boldsymbol{p}$ does not have a component in the propagation direction. This situation changes for the paraxial wave due to its curved wavefront, where the momentum density $\boldsymbol{p}$ has small components perpendicular to the propagation direction and the angular momentum density $\boldsymbol{m}$ has components in the propagation direction.

Beth (1936) has measured the mechanical torque due to the angular momentum of circularly polarized light and shown that this torque is ascribed to the spin of the photons, called spin angular momentum, where the spin is represented by $\sigma = +1$ or $\sigma = -1$ corresponding to the right- or left-hand circular polarization, respectively. In addition to the spin angular momentum, Allen et al. (1992) have shown that a Laguerre–Gaussian (LG) beam has an orbital angular momentum (OAM) l.

Since these pioneering works, the light with OAM has been applied to broad fields (Padgett, 2017; Yao et al., 2011), such as optical trapping (Kuga et al., 1997), optical manipulation of microscopic objects (Paterson et al., 2001), optical communication, and quantum information processing. The OAM is also applied to high intensity lasers, as described later in this section.

The solution of Eq. (2.46) for the paraxial wave is also described in the cylindrical coordinate (r, ϕ, z) by the Laguerre–Gaussian function

$$A_{p,l}^{\mathrm{LG}}(r,\phi,z) = C_{p,l}^{\mathrm{LG}} \frac{w_0}{w(z)} \left(\frac{\sqrt{2}r}{w(z)}\right)^{|l|} L_p^{|l|}\left(\frac{2r^2}{w^2(z)}\right) \exp\left[-\frac{r^2}{w^2(z)}\right] \times \exp(-il\phi) \exp\left\{-i\left[k\frac{r^2}{2R(z)} - (2p+|l|+1)\tan^{-1}\left(\frac{z}{z_R}\right)\right]\right\}, \tag{2.59}$$

where $C_{p,l}^{\mathrm{LG}} = \sqrt{2p!/(\pi(p+|l|)!}$ is the normalization constant, $L_p^{|l|}$ is the associated Laguerre polynomial, r is the radial coordinate, and ϕ is the azimuthal angle. The radius of the Gaussian beam $w(z)$ and the radius of curvature of the wave front $R(z)$ are same as the Hermite -Gaussian description given by (2.49) and (2.53), respectively. The azimuthal mode index l is an integer, and p is the radial mode index with $p = 0, 1, 2, \ldots, |l|$. Due to the $\exp(-il\phi)$ term, the phase rotates l times when ϕ changes by 2π, resulting in the spiral wavefront.

Figure 2.3 shows the intensity, phase, and wavefront of the 0,0-, 1,1-, and 1,2-modes (the first and the second letter corresponds to p and l, respectively) of the LG $A_{p,l}^{\mathrm{LG}}$ beam (Morita, 2016). The light with an orbital angular momentum is also called an optical vortex since the phase rotates helically having a singular point at $r = 0$ where the intensity becomes zero for $l \geq 1$.

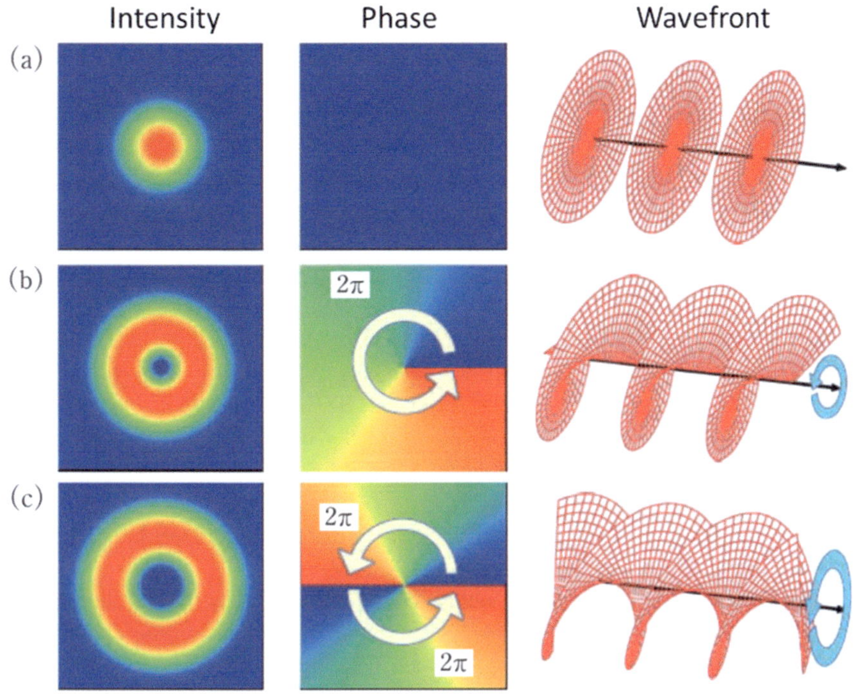

Fig. 2.3 The intensity, phase, and wavefront of **a** 0,0-, **b** 1,1- and **c** 1,2-modes of the LG $u_{p,l}^{\mathrm{LG}}$ beam. The 0,0 mode is a Gaussian beam. In the higher order modes, phase is rotated with ϕ by l times, and the intensity changes along the axial direction resulting in p rings in the intensity distribution. Reprinted from Morita (2016), with permission from The Japan Society of Applied Physics. © 2016 The Japan Society of Applied Physics

Since the Poynting vector $\boldsymbol{S}$ is perpendicular to the wavefront, the momentum density $\boldsymbol{p}$, which is proportional to $\boldsymbol{S}$, has components in the $\boldsymbol{r}$ and $\boldsymbol{\phi}$ directions for the spiral beams. Deriving $\boldsymbol{B}$ from $\boldsymbol{E}$ by $\nabla \times \boldsymbol{E} = -\partial \boldsymbol{B}/\partial t$, the time averaged momentum density is described as (Allen & Padgett, 2000; Padgett et al., 1995)

$$\langle \overline{\boldsymbol{p}}(\boldsymbol{r}, t)\rangle_T = \frac{\varepsilon_0}{2}\omega\left\{\frac{kr}{R(z)}|A(\boldsymbol{r})|^2\hat{\boldsymbol{r}} + \left[\frac{l}{r}|A(\boldsymbol{r})|^2 - \frac{1}{2}\sigma\frac{\partial|A(\boldsymbol{r})|^2}{\partial r}\right]\hat{\boldsymbol{\phi}} + k|A(\boldsymbol{r})|^2\hat{\boldsymbol{z}}\right\}. \tag{2.60}$$

Here $\hat{\boldsymbol{r}}$, $\hat{\boldsymbol{\phi}}$, and $\hat{\boldsymbol{z}}$ are unit vectors in the radial, azimuthal, and propagation directions, respectively. In the right-hand side of this equation, the first term is due to the wavefront curvature of a Gaussian beam, the last term due to the Poynting vector of a plane wave, and the third term with σ due to the circular polarization (spin) of the optical wave. The second term with l is characteristic to the optical vortex, by which the energy flow has a component in the azimuthal direction.

The time-averaged z-component of the angular momentum density $\boldsymbol{m} = \boldsymbol{r} \times \boldsymbol{p}$ is derived from this momentum density as

$$\langle \overline{m}_z(\boldsymbol{r}, t) \rangle_\mathrm{T} = \frac{\varepsilon_0}{2}\omega \left[l|A(\boldsymbol{r})|^2 - \frac{1}{2}\sigma r \frac{\partial |A(\boldsymbol{r})|^2}{\partial r} \right]. \tag{2.61}$$

The first and the second terms in the right-hand side of this equation are due to the orbital and the spin angular momentum, respectively. The ratio of the total angular momentum $\langle \overline{M}_z \rangle_\mathrm{T}$ obtained by integrating (2.61) over the beam area to the total energy density $\langle W \rangle_\mathrm{T}$ is expressed as

$$\frac{\langle \overline{J}_z \rangle_\mathrm{T}}{\langle W \rangle_\mathrm{T}} = \frac{\hbar(l + \sigma)}{\hbar\omega}. \tag{2.62}$$

This equation shows that each photon has a spin angular momentum (SAM) $\hbar\sigma$ and an orbital angular momentum (OAM) $\hbar l$. Since OAM is determined independently from SAM, the optical vortex is a parameter added to the laser beam, leading to new possibilities in the application of the optical beam.

There are several methods to generate laser beams carrying the orbital angular momentum. A general method is to create a computer-generated hologram which is formed by adding a helical phase profile and a linear phase ramp. The forked diffraction grating thus created produces a helically phased beam in the first diffraction order when illuminated with a Gaussian beam (Arlt et al., 1998; Yao et al., 2011). One of the advantages of this approach is that it is possible to generate LG beams of complex amplitude and phase distributions with a programmable, commercially available spatial light modulator (SLM) (Ando et al., 2009). Due to the intensity limit of SLMs caused by the laser-induced damage to the liquid crystals, it is necessary to amplify the generated OAM beam for high power laser applications.

Alternatively, a LG beam can be generated directly by conversion of the Hermite-Gaussian (HG) beam (including the fundamental mode) with several methods; a cylindrical lens with its axis tilted to the x–y axis of the HG beam (Beijersbergen et al., 1993), and a spiral phase plate where the thickness is varied gradually in the azimuthal direction (Beijersbergen et al., 1994; Sueda et al., 2004). Radially or azimuthally polarized beam without wavefront rotation, called cylindrical vector (CV) beam, is generated from a linearly polarized Gaussian beam with a spatially varying retarder (SVR), shown in Fig. 2.4a (Machavariani et al., 2007). A linearly polarized LG beam is generated by a combination of half wave plates and an SVR in a configuration shown in Fig. 2.4b (Marrucci et al., 2006; Tokizane et al., 2009). Utilizing the achromatic property of the latter configuration, high-power, ultrashort optical vortex pulses have been generated (Chen et al., 2022; Yamane et al., 2012).

The high-power OAM and CV beams are expected to bring new possibilities to high field science: generation of relativistic vortex electron beam (Jiang et al., 2021; Vieira et al., 2018), enhanced emission of betatron radiation (Martins et al., 2019), and generation of mega-gauss magnetic field in a plasma solenoid (Nuter et al.,

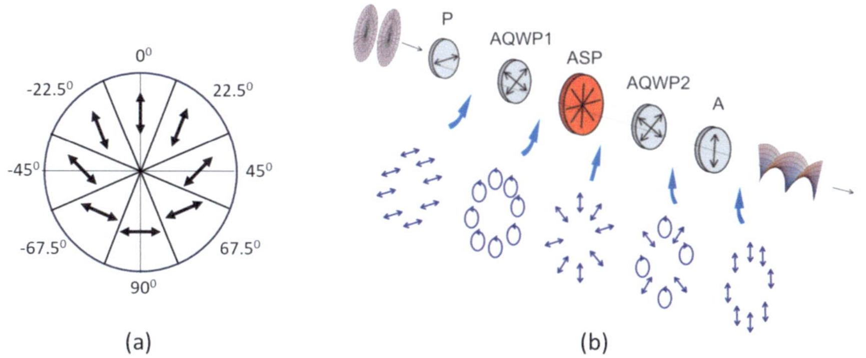

Fig. 2.4 Examples to generate an LG beam. **a** A spatially varying retarder (SVR) to generate a radially or azimuthally polarized beam, composed of eight sectors of λ/2 retardation plates with their crystal axes properly oriented. **b** An axially polarized beam, generated with a polarizer (P), a λ/4 plate (AQWP1) and an axially symmetric polarizer (ASP), is transformed into an LB beam with the second λ/4 plate (AQWP2) and an analyzer (A). Reprinted from Tokizane et al. (2009). © 2009 Optical Society of America

2018) with the OAM beams, and intense gamma-ray emission with tightly focused radially polarized CV beam due to intense longitudinal field component at the focus (Hadjisolomou et al., 2021).

In material processing, the angular momentum and the polarization of the laser beam are transferred to the surface structures of the processed material (Fig. 2.5). An LG beam of 30-ns duration produces chiral nanoneedles on the ablated surface where the chiral direction depends on the helical direction of the laser beam (Toyoda et al., 2012). High aspect-ratio micro-diameter holes have been drilled with the vector beams of ps duration (Kraus et al., 2010). Also, highly ordered complex nanostructures have been formed with fs-duration vector beams, enabling fabrication of complex surface structures that mimic the morphological features of various species found in nature (Skoulas et al., 2017). For industrial implementation of these approaches, development of high-power vector beams is in progress, realizing 1.7 kW average power and 5.8 mJ pulse energy with 7.8 ps duration by amplification with two-stage thin-disk multi-pass amplifiers (Loescher et al., 2021).

2.1.7 Dispersion

Material dispersion affects light propagation in linear and nonlinear regions. The electric field $\boldsymbol{E}(\boldsymbol{r}, t)$ expressed in (2.44) is decomposed into its frequency spectrum $\tilde{\boldsymbol{E}}(\boldsymbol{r}, \omega)$ by the Fourier transformation:

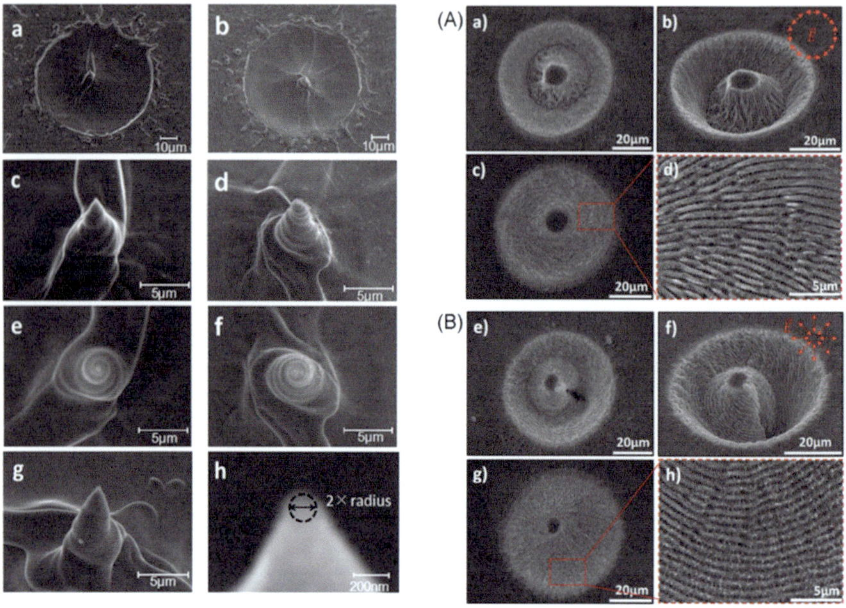

Fig. 2.5 Scanning electron microscope images of the metal surfaces irradiated with the optical vortex (8 photos on the left), and the vector beam (8 photos on the right), respectively. (Left photos) A tantalum surface irradiated with four shots of optical vortices of 20 ns-duration, 1064 nm, 0.5 mJ pulses with the total angular momentum $J = L + S$ of -2 (clockwise) in (**a**, **c**, **e**), and $+2$ (counterclockwise) in (**b**, **d**, **f**), respectively. A sharp nanoneedle is formed at the center of a ~ 65 μm diameter crater in both cases of (**a**, **b**), but with opposite chirality in (**a**, **c**, **e**) and (**b**, **d**, **f**), respectively. The nano needles are viewed at 25° in (**c**, **d**) and from the top in (**e**, **f**), respectively. **g**, **h** Nanoneedles with chirality disappearance formed with 2 mJ laser irradiation. Reprinted from Toyoda et al. (2012). © 2012 American Chemical Society. (Right photos) A Ni surface structure created by 1000-pulse irradiation of 170 fs, 1 kHz, 1026 nm laser of **a**–**d** azimuthal and **e**–**h** radial polarizations, respectively. These are top views except for **b**, **f** of 45° views. **d**, **h** Magnified images showing the periodic textures whose directions are perpendicular to the laser polarization. Reprinted from Skoulas et al. (2017). © 2017 Springer Nature

$$\boldsymbol{E}(\boldsymbol{r}, t) = \frac{1}{2\pi} \int_{-\infty}^{\infty} \tilde{\boldsymbol{E}}(\boldsymbol{r}, \omega) \mathrm{e}^{-i\omega t} d\omega. \tag{2.63}$$

Alternatively, $\tilde{\boldsymbol{E}}(\boldsymbol{r}, \omega)$ is derived from $\boldsymbol{E}(\boldsymbol{r}, t)$ by the inverse Fourier transformation:

$$\tilde{\boldsymbol{E}}(\boldsymbol{r}, \omega) = \int_{-\infty}^{\infty} \boldsymbol{E}(\boldsymbol{r}, t) \mathrm{e}^{i\omega t} dt. \tag{2.64}$$

Since $\boldsymbol{E}(\boldsymbol{r}, t)$ is a real function, its complex Fourier components $\tilde{\boldsymbol{E}}(\boldsymbol{r}, \omega)$ satisfy the following relation: $\tilde{\boldsymbol{E}}(\boldsymbol{r}, -\omega) = \tilde{\boldsymbol{E}}^*(\boldsymbol{r}, \omega)$. Other electromagnetic quantities are expressed in a similar manner.

The macroscopic polarization $\boldsymbol{P}$ originates from the ensemble-averaged microscopic polarization as $\boldsymbol{P}(\boldsymbol{r}, t) = N\langle\boldsymbol{\mu}(t)\rangle$, where $\boldsymbol{\mu}(t)=q\boldsymbol{r}(t)$ is the atomic dipole moment, q is the charge and $\boldsymbol{r}(t)$ is the spatial coordinate of an electron. The causality condition exists between the electric field and the polarization: the polarization is induced only after the electric field is applied due to finite response time of the electrons.

The polarization is thus described as:

$$\boldsymbol{P}(\boldsymbol{r}, t) = \varepsilon_0 \int_{-\infty}^{t} \boldsymbol{\chi}^{(1)}(t - t') \cdot \boldsymbol{E}(\boldsymbol{r}, t - t')dt'. \tag{2.65}$$

Writing this relation in terms of the frequency as $\tilde{\boldsymbol{P}}(\boldsymbol{r}, \omega) = \varepsilon_0 \tilde{\boldsymbol{\chi}}^{(1)}(\omega) \cdot \tilde{\boldsymbol{E}}(\boldsymbol{r}, \omega)$, the Fourier transform of the linear electric susceptibility $\tilde{\boldsymbol{\chi}}^{(1)}(\omega)$ is expressed as

$$\tilde{\chi}^{(1)}(\omega) = \int_{-\infty}^{\infty} \chi^{(1)}(t) \exp(i\omega t)dt = \chi'(\omega) + i\chi''(\omega). \tag{2.66}$$

The second term $i\chi''(\omega)$ arises from the phase delay of polarization.

The response of an atom to the electric field is derived by considering the interaction of an optical wave with an atom which can be modelled as an oscillator with a resonance frequency ω_0. Alternatively in quantum mechanics, the interaction is analyzed with the density matrix for the two atomic levels (Sargent et al., 1978; Yariv, 1989). Considering an x-polarized optical wave, the x-component of the polarization is described as

$$P_x(\omega) = \frac{(N_l - N_u)\left|\mu_{ul}^x\right|^2 E_x(\omega)}{\hbar} \cdot \frac{\omega_0 - \omega + i(\Delta\omega/2)}{(\omega_0 - \omega)^2 + (\Delta\omega/2)^2}. \tag{2.67}$$

Here the subscripts l and u refer to the lower- and the upper-levels of an atom, respectively, N_l and N_u are the number densities of the atoms in l and u, μ_{ul}^x is the matrix element of the x-component of the electric dipole moment between l and u, $\hbar\omega_0$ is the energy difference between l and u, and $\Delta\omega$ is the spectral broadening (full width at half maximum) of these levels. Using a normalized Lorentzian line shape function,

$$g(\omega) = \frac{(\Delta\omega/2\pi)}{(\omega - \omega_0)^2 + (\Delta\omega/2)^2}, \quad \int_{-\infty}^{\infty} g(\omega)d\omega = 1, \tag{2.68}$$

the real and imaginary parts of the electric susceptibility (2.66) is written as

$$\chi'(\omega) = -\frac{(N_l - N_u)\pi \left|\mu_{ul}^x\right|^2}{\varepsilon_0 \hbar} \cdot \frac{\omega - \omega_0}{\Delta\omega/2} g(\omega) = -\frac{\omega - \omega_0}{\Delta\omega/2} \chi''(\omega), \tag{2.69}$$

$$\chi''(\omega) = \frac{(N_l - N_u)\pi \left|\mu_{ui}^x\right|^2}{\varepsilon_0 \hbar} \cdot g(\omega). \tag{2.70}$$

Figure 2.6 shows the frequency dependences of $\chi'(\omega)$ and $\chi''(\omega)$ by the solid curves in red and blue, respectively, for $N_l > N_u$.

Referring to Eq. (2.6), we separate the total polarization into polarization of the base material $\boldsymbol{P}$ and polarization due to resonance with a specific atom in the medium $\boldsymbol{P}_{\text{res}}$:

$$\boldsymbol{D} = \varepsilon_0 \boldsymbol{E} + \boldsymbol{P} + \boldsymbol{P}_{\text{res}} \equiv \varepsilon \boldsymbol{E} + \varepsilon_0 \chi(\omega) \boldsymbol{E}, \tag{2.71}$$

where ε is the permittivity of the base material. The permittivity $\hat{\varepsilon}(\omega)$ including the resonant component is,

$$\hat{\varepsilon}(\omega) = \varepsilon + \varepsilon_0 \chi(\omega) = \varepsilon + \varepsilon_0 \chi'(\omega) + i\varepsilon_0 \chi''(\omega). \tag{2.72}$$

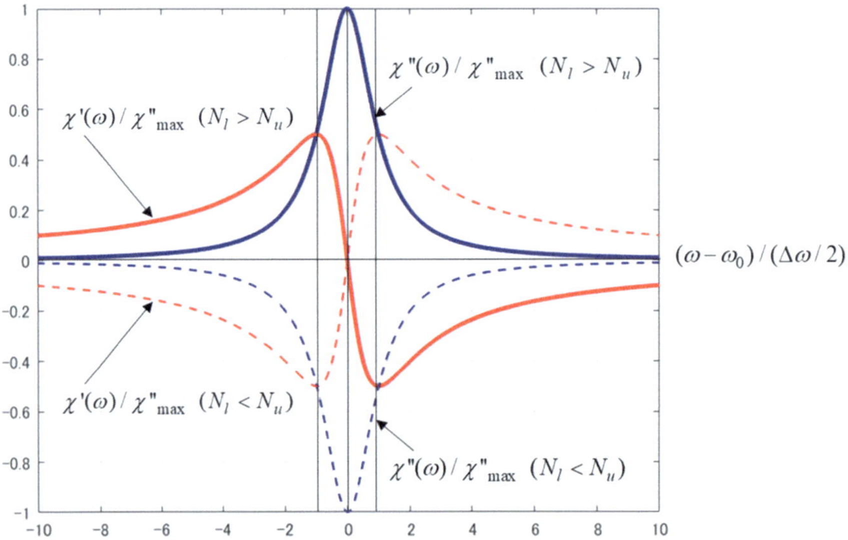

Fig. 2.6 Frequency dependences of $\chi'(\omega)$ (shown in red) and $\chi''(\omega)$(shown in blue) for **a** $N_l > N_u$ (solid curves) and **b** $N_u > N_l$ (dotted curves). Horizontal axis is in unit of $(\omega - \omega_0)/(\Delta\omega/2)$. The vertical axes of $\chi'(\omega)$ and $\chi''(\omega)$ are normalized by $\chi''_{\text{max}}(\omega)$

Describing a plane wave propagating in the z-direction as (2.44) and assuming $|\chi| \ll 1$, the wavenumber is expressed as a complex quantity $\hat{k}(\omega)$

$$\hat{k}(\omega) = \omega\sqrt{\mu\varepsilon_t} \cong k\left[1 + \frac{\chi'(\omega)}{2n_0^2}\right] + i\frac{k\chi''(\omega)}{2n_0^2}, \tag{2.73}$$

where $n_0 = \sqrt{\varepsilon/\varepsilon_0}$ is the refractive index of the base material. Denoting

$$\Delta k \equiv \frac{k\chi'(\omega)}{2n_0^2}, \quad \gamma \equiv \frac{k\chi''(\omega)}{n_0^2}, \tag{2.74}$$

the electric field is described as

$$E_x(z,t) = \mathrm{Re}\left[E_{0x}e^{i(\hat{k}z-\omega t)}\right] = \mathrm{Re}\left[E_{0x}\exp\{i[(k+\Delta k)z - \omega t] - (\gamma/2)z\}\right]. \tag{2.75}$$

We find from this equation that the resonant part of the polarization results in the phase delay $(\Delta k)z$ and the attenuation of the optical wave with a power attenuation coefficient γ in the frequency region close to the resonance.

The refractive index is also described by a complex quantity as

$$\hat{n} = \sqrt{\mu\hat{\varepsilon}} = \frac{c}{\omega}\hat{k} = n(1 + i\kappa), \tag{2.76}$$

where κ is called the attenuation index (or extinction coefficient). The real part of the refractive index changes from n_0 to $n = n_0 + \Delta n$ as

$$n = \sqrt{\frac{\varepsilon'}{\varepsilon_0}} = \sqrt{\frac{\varepsilon}{\varepsilon_0}}\left(1 + \frac{\varepsilon_0}{\varepsilon}\chi'\right)^{1/2} \simeq n_0 + \frac{\chi'}{2n_0} \equiv n_0 + \Delta n, \quad \Delta n = \frac{\chi'}{2n_0}. \tag{2.77}$$

From the frequency dependence of $\chi'(\omega)$ shown in Fig. 2.6, the refractive index increases with ω, i.e. $dn'/d\omega > 0$, in the frequency region of $\omega < \omega_0 - \Delta\omega/2$. This property, called "normal dispersion", is applicable to optical propagation in transparent dielectric media. The refractive index of a transparent material in normal dispersion region is often expressed by a Sellmeier's dispersion formula,

$$n^2 \cong 1 + \sum_i \frac{A_i\lambda^2}{\lambda^2 - \lambda_i^2} \cong a + \frac{b}{\lambda^2 - \lambda_0^2} - c\lambda^2, \tag{2.78}$$

where λ is the wavelength in μm. The constants in the latter formula for many nonlinear optical crystals are given in Dmitriev et al. (1999).

The frequency region of $\omega_0 - \Delta\omega/2 < \omega < \omega_0 + \Delta\omega/2$ is called "anomalous dispersion" where $dn'/d\omega < 0$. At higher frequencies of $\omega > \omega_0 + \Delta\omega/2$, the refractive index becomes smaller than the refractive index of the base material, i.e., $n' < n$. For short wavelength optical waves such as in EUV and X-rays, the refractive index becomes smaller than 1 since $\omega \gg \omega_0$ holds, where ω_0 represents binding energies of the outer-shell electrons of atoms constituting the medium. Anomalous dispersion is revealed in the EUV and X-ray regions at the frequencies close to the excitation frequencies of the inner shell electrons (Attwood, 1999). In the infrared and far infrared regions on the other hand, the vibrational and rotational excitations in solids and molecules in gases contribute to the dispersion, resulting in deviation from the dispersion curve owing to electronic excitation.

From Eqs. (2.70) and (2.74), the attenuation coefficient is expressed as

$$\gamma(\omega) = \frac{(N_l - N_u)\pi \left|\mu_{ul}^x\right|^2 \omega}{n^2 \varepsilon_0 c\hbar} g(\omega). \tag{2.79}$$

When $N_l > N_u$, $\gamma(\omega)$ is positive, i.e., the optical wave is attenuated. When $N_u > N_l$ on the other hand, the signs of $\chi'(\omega)$ and $\chi''(\omega)$ are inverted as shown by the dotted curves in Fig. 2.6. In this case, $\gamma(\omega)$ becomes negative, i.e., the optical wave is amplified with the power gain coefficient given by (2.79). In laser devices, this "population inversion" is created to amplify an optical wave, as described in Sect. 3.2.2.

2.1.8 Kramers–Kronig Relations and Analytic Signal

The relations between the real and imaginary parts of the electric susceptibility given in Eqs. (2.69) and (2.70) are expressed in a more general form known as the Kramers–Kronig relations (Jackson, 1999, Sect. 7.10; Kramers, 1927; Kronig, 1926), which is derived from the causality (Toll, 1956):

$$\chi'(\omega) = \frac{1}{\pi} P.V. \int_{-\infty}^{+\infty} \frac{\chi''(\omega')}{\omega' - \omega} d\omega',$$

$$\chi''(\omega) = -\frac{1}{\pi} P.V. \int_{-\infty}^{+\infty} \frac{\chi'(\omega')}{\omega' - \omega} d\omega'. \tag{2.80}$$

Here "P.V." denotes the Cauchy principal value. Since $\chi''(\omega)$ is an odd and $\chi'(\omega)$ is an even function of ω due to $\chi(-\omega) = \chi^*(\omega)$, Eq. (2.80) is rewritten, by multiplying $(\omega'+\omega)/(\omega'+\omega)$ to the integrant, as an integral only in the non-negative frequencies:

$$\chi'(\omega) = \frac{2}{\pi} P.V. \int_{0}^{+\infty} \frac{\chi''(\omega')\omega'}{\omega'^2 - \omega^2} d\omega',$$

$$\chi''(\omega) = -\frac{2\omega}{\pi} P.V. \int_{0}^{+\infty} \frac{\chi'(\omega')}{\omega'^2 - \omega^2} d\omega'. \tag{2.81}$$

These equations reflect the situation that all information is contained in the non-negative frequency components arising from the causality. The Kramers–Kronig relations show that $\chi'(\omega)$ is derived from $\chi''(\omega)$ and vice versa. For example, the frequency dependence of the refractive index is calculated from the measured data of the spectral absorption.

The Kramers–Kronig relations is known as Hilbert transform in mathematics and as analytic signal in information processing. A signal with no negative-frequency component is called an analytic signal (Smith, 2007). A real-valued function $u(t)$ is Hilbert-transformed to another real-value function $\hat{u}(t)$ as

$$\hat{u}(t) = u(t) * \frac{1}{\pi t} = \frac{1}{\pi} P.V. \int_{-\infty}^{\infty} \frac{u(t)}{t - \tau} d\tau, \tag{2.82}$$

which is a convolution of $u(t)$ with $1/\pi t$. Writing the Fourier transform of $u(t)$ as $U(\omega)$, the Hilbert transform has the effect to shift the phase of $U(\omega)$; the negative frequency components by $+\pi/2$ (+ 90°) and the positive frequency components by $-\pi/2$. Noting $i = \mathrm{e}^{+i\pi/2}$, this operation is written as multiplying $-i \cdot \mathrm{sgn}(\omega)$ to $U(\omega)$ where $\mathrm{sgn}(\omega)$ is the sign function.

The analytic signal of $u(t)$ is a complex-valued function composed of $u(t)$ and $\hat{u}(t)$ as

$$u_a(t) = u(t) + i\hat{u}(t). \tag{2.83}$$

The Fourier transformation of $u_a(t)$ is

$$U_a(\omega) = \mathcal{F}[u_a(t)] = \begin{cases} 2U(\omega) & \omega > 0, \\ U(\omega) & \omega = 0, \\ 0 & \omega < 0. \end{cases} \tag{2.84}$$

This equation shows that $U_a(\omega)$ is composed only of the positive frequency part of $U(\omega)$. That is, analytic signal $u_a(t)$ is the inverse transform of the positive frequency part of the Fourier spectrum of the signal $u(t)$. These relations are shown schematically in Fig. 2.7.

The analytic signal is expressed in an exponential form as

$$u_a(t) = u(t) + i\hat{u}(t) \equiv A(t) \exp[i\theta(t)]. \tag{2.85}$$

In this equation, $A(t)$ and $\theta(t)$ correspond to the amplitude and phase at time t, respectively. These quantities and the instantaneous frequency $\omega(t)$ are derived from $u_a(t)$ as

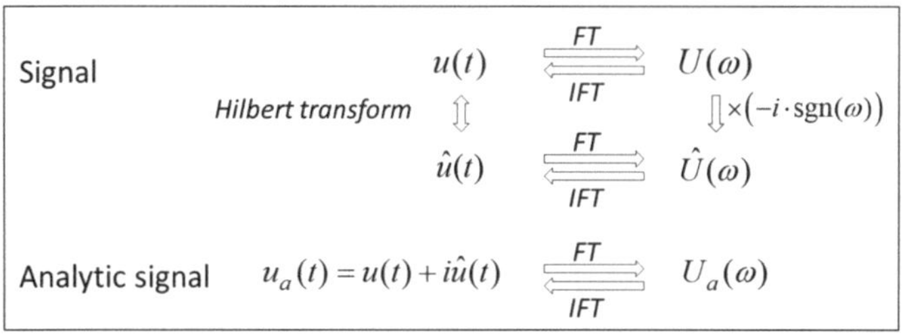

Fig. 2.7 Formation of an analytic signal with Hilbert transform. Here *FT* and *IFT* refer to Fourier transformation and inverse Fourier transformation, respectively. Hilbert transform of $U(\omega)$ is obtained by multiplying $U(\omega)$ with $-i\mathrm{sgn}(\omega)$ where $\mathrm{sgn}(\omega) = 1, 0, -1$ for $\omega => 0, 0, < 0$, respectively

$$\begin{aligned} A(t) &= \sqrt{u(t)^2 + \hat{u}(t)^2}, \\ \theta(t) &= \tan^{-1}[\hat{u}(t)/u(t)], \\ \omega(t) &= \frac{d}{dt}\theta(t). \end{aligned} \tag{2.86}$$

For a given signal of $u(t) = A(t)\cos[\phi(t)]$, its Hilbert transform is $\hat{u}(t) = A(t)\sin[\phi(t)]$. Then the analytic signal is derived as

$$u_a(t) = A(t)\cos[\phi(t)] + iA(t)\sin[\phi(t)] = A(t)\exp[i\phi(t)]. \tag{2.87}$$

When the signal $u(t)$ exhibits complicated temporal variations due, for example, to superposition of many waves of different frequencies and phases, the instantaneous values of the amplitude, phase and frequency of the signal are derived from the analytic signal. We will use this concept in Sect. 2.3.

2.2 Optical Components for High Power Lasers

2.2.1 Refraction and Reflection

Consider a plane wave in a (x, z) plane which is incident at boundary I from medium 0 to medium 1 with the refractive indices of n_0 and n_1, respectively, as shown in Fig. 2.8. We denote the incident, reflected and transmitted fields with the suffices i, r and t, respectively. The following boundary conditions are derived from Maxwell's equations when $\boldsymbol{j} = 0$; x and y components of $\boldsymbol{E}$ and $\boldsymbol{H}$ are continuous and z components of $\boldsymbol{D}$ and $\boldsymbol{B}$ are continuous above and below the boundary plane at $z = 0$ (Born & Wolf, 1999):

$$E_{x,0} = E_{x,1}, \quad H_{x,0} = H_{x,1}, \quad E_{y,0} = E_{y,1}, \quad H_{y,0} = H_{y,1}$$

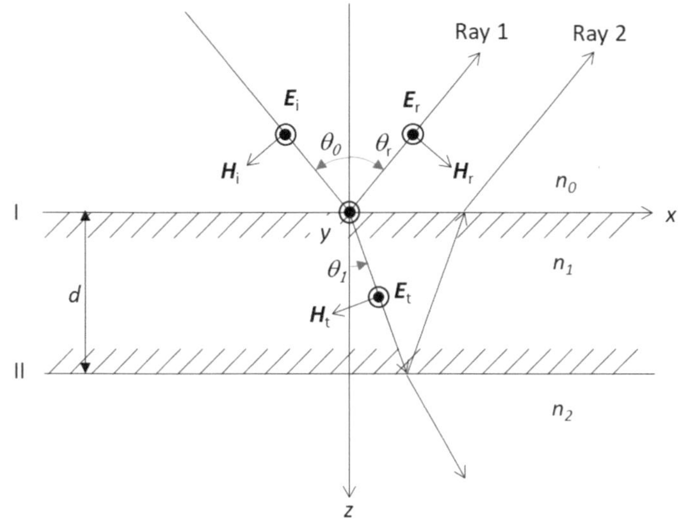

Fig. 2.8 Reflection and refraction of light at boundaries I and II. This figure shows the case when the incident field is *s*-polarized, i.e., the electric field is in the *y*-direction which is outward from the *x*–*z* plane

$$D_{z,0} = D_{z,1}, \quad B_{z,0} = B_{z,1}. \tag{2.88}$$

Referring to Fig. 2.8, we denote the incidence, reflection, and refraction angles, measured from the surface normal as θ_0, θ_r and θ_1, respectively. Noting that the plane wave propagates in the medium 0 and 1 with the phase velocities of $v_0 = c/n_0$ and $v_1 = c/n_1$ respectively, the *Snell' Law of refraction* is obtained

$$\begin{aligned} \theta_0 &= \theta_r \\ n_0 \sin\theta_0 &= n_1 \sin\theta_1. \end{aligned} \tag{2.89}$$

From the boundary condition and the amplitude ratio of $\boldsymbol{E}$ and $\boldsymbol{H}$ in (2.20), the amplitude reflection coefficient and the amplitude transmission coefficient, known as *Fresnel equations*, are derived:

$$\left.\begin{aligned} r_\parallel &\equiv \left(\frac{E_r}{E_i}\right)_\parallel = \frac{n_1\cos\theta_0 - n_0\cos\theta_1}{n_1\cos\theta_0 + n_0\cos\theta_1} = \frac{\tan(\theta_0-\theta_1)}{\tan(\theta_0+\theta_1)}, \\ r_\perp &\equiv \left(\frac{E_r}{E_i}\right)_\perp = \frac{n_0\cos\theta_0 - n_1\cos\theta_1}{n_0\cos\theta_0 + n_1\cos\theta_1} = \frac{\sin(\theta_0-\theta_1)}{\sin(\theta_0+\theta_1)}, \\ t_\parallel &\equiv \left(\frac{E_t}{E_i}\right)_\parallel = \frac{2n_0\cos\theta_0}{n_0\cos\theta_1 + n_1\cos\theta_0} = \frac{2\sin\theta_1\cos\theta_0}{\sin(\theta_0+\theta_1)\cos(\theta_0-\theta_1)}, \\ t_\perp &\equiv \left(\frac{E_t}{E_i}\right)_\perp = \frac{2n_0\cos\theta_0}{n_0\cos\theta_0 + n_1\cos\theta_1} = \frac{2\sin\theta_1\cos\theta_0}{\sin(\theta_0+\theta_1)}. \end{aligned}\right\} \tag{2.90}$$

Here $\|$ and $\perp$ denote the electric field components in and perpendicular to the incidence plane, respectively. The $\|$ and $\perp$ polarization components are called *p*-polarized (or TM for transverse magnetic) and *s*-polarized (or TE for transverse electric). The amplitudes E_i, E_r and E_t in the above equations refer to the values at the boundary with E_i, E_r in medium 0 and E_t in medium 1.

Since the intensity is given by $I = (1/2)\varepsilon_0 nc|E|^2$, the incident and transmitted fluxes at the boundary are

$$J^{(i)} = (1/2)\varepsilon_0 n_0 c|E_i|^2 \cos\theta_0, \quad J^{(r)} = (1/2)\varepsilon_0 n_0 c|E_r|^2 \cos\theta_r,$$
$$J^{(t)} = (1/2)\varepsilon_0 n_1 c|E_t|^2 \cos\theta_1. \tag{2.91}$$

Therefore, the reflectance and the transmittance are given by

$$R = \frac{J^{(r)}}{J^{(i)}} = |r|^2, \quad T = \frac{J^{(t)}}{J^{(i)}} = \frac{n_1 \cos\theta_1}{n_0 \cos\theta_0}|t|^2. \tag{2.92}$$

At normal incidence,

$$R = \left(\frac{n-1}{n+1}\right)^2, \quad T = \frac{4n}{(n+1)^2}, \tag{2.93}$$

where $n \equiv n_1/n_0$.

The reflectance of the *p*-polarized light $r_\|$ becomes zero when $\theta_0 + \theta_1 = \pi/2$, i.e., $\tan\theta_0 = n$. At this incidence angle of $\theta_B = \tan^{-1} n$, called *Brewster angle*, a transparent material transmits the *p*-polarized component of light without reflection loss. A plane-parallel optical glass placed at the Brewster angle in a laser cavity is used to control the laser oscillation to *p*-polarization.

When $n_0 > n_1$, the refraction angle of the transmitted beam becomes $\theta_1 = \pi/2$ at the critical incidence angle of $\sin\theta_c = n_1/n_0$. When the incidence angle exceeds the critical angle, the light is totally reflected. Although there is no energy flow in the z-direction in this case, the wave penetrates beyond the boundary into material 1. Writing $n_0/n_1 \equiv n_{01}$, the penetrated wave is given by

$$\boldsymbol{E}_t = \boldsymbol{E}_{0t} \exp\left[-i\omega\left(t - \frac{x\sin\theta_0}{n_{01}v_1}\right)\right] \exp\left(-\frac{\omega z}{v_1}\sqrt{\frac{\sin^2\theta_0}{n_{01}^2} - 1}\right), \tag{2.94}$$

where $\boldsymbol{E}_{0t}$ is the field at the boundary and v_1 is the phase velocity in medium 1. Noting that $\omega/v_1 = k_1$ where k_1 is the wavenumber in medium 1, (2.94) shows the existence of an *evanescent wave* which propagates in the *x*-direction and its amplitude decreases exponentially with *z*. The penetration depth is approximately given by $1/k_1 = \lambda_1/2\pi$. The evanescent wave is important, for example, in optical fibers, photonics crystals, and near-field microscopes.

2.2.2 *Diffraction*

2.2.2.1 Fresnel and Fraunhofer Diffraction

The propagation of a laser beam is always associated with diffraction. A Gaussian beam preserves its field distribution during propagation in a free space. However, the optical field is often truncated by the apertures of various optical elements, affecting the intensity distribution due to diffraction of the optical wave. When an optical wave passes through an aperture, the field propagates to the geometrical shadow of the aperture. In general, the (x, y, z) components of the electric and magnetic fields are coupled with each other due to interaction with the aperture material. Since this coupling effect extends only over a few wavelengths at the aperture, we can describe propagation of the field as a scalar value without considering the coupling between the field components when the aperture area is large compared with the wavelength.

From the wave Eq. (2.45), the field distribution of the diffracted wave is derived in a simplified form known as the *Huygens–Fresnel principle*. Referring to Fig. 2.9, we describe the field at P_0 on the aperture plane as $u_0(\xi, \eta)$. The field $A_1(x, y, z_1)$ at P_1 on the observation plane located at z_1 is given by

$$A_1(x, y, z_1) = \frac{1}{i\lambda} \iint_{\Sigma} A_0(\xi, \eta) \frac{\exp(ikr_{01})}{r_{01}} \cos\theta ds. \tag{2.95}$$

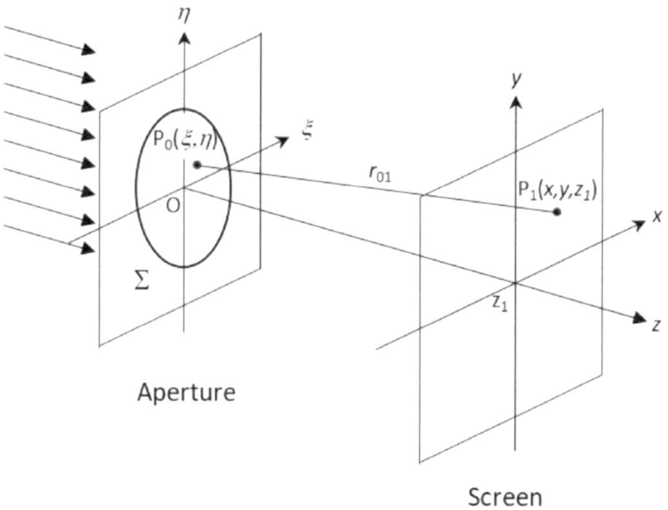

Fig. 2.9 Coordinates at the aperture (ξ, η) and the screen (x, y, z_1) in calculating the field on a screen at $z = z_1$ after passing through an aperture Σ

Here r_{01} is the distance between P_0 and P_1, θ is the angle between r_{01} and the z-axis with $\cos\theta = z_1/r_{01}$, and the integral is taken over the aperture Σ. Equation (2.95) shows that $A_1(x, y, z_1)$ is expressed as a superposition of the spherical waves originating from P_0 on the aperture.

Approximating r_{01} as

$$r_{01} = \sqrt{z_1^2 + (x-\xi)^2 + (y-\eta)^2} \approx z_1\left[1 + \frac{1}{2}\left(\frac{x-\xi}{z}\right)^2 + \frac{1}{2}\left(\frac{y-\eta}{z}\right)^2\right], \tag{2.96}$$

and retaining only the lateral dependences in the phase term, (2.95) becomes

$$A_1(x, y, z_1) = \frac{e^{ikz}}{i\lambda z_1}\iint_{-\infty}^{\infty} A_0(\xi,\eta)\exp\left\{\frac{ik}{2z_1}\left[(x-\xi)^2 + (y-\eta)^2\right]\right\}d\xi d\eta. \tag{2.97}$$

This expression of the Fresnel diffraction is used to obtain the field distribution in the near field of $z_1 \le 2D^2/\lambda$ where D is the linear dimension of the aperture. The right-hand side of (2.97) is a convolution of two functions $A_0(\xi, \eta)$ and $h(x, y)$,

$$A_1(x, y, z_1) = \iint_{-\infty}^{\infty} A_0(\xi,\eta)h(x-\xi, y-\eta)d\xi d\eta, \tag{2.98}$$

with the convolution kernel of

$$h(x, y) = \frac{e^{ikz_1}}{i\lambda z_1}\exp\left\{\frac{ik}{2z_1}\left(x^2 + y^2\right)\right\}. \tag{2.99}$$

The spatial frequency distribution of a laser beam at the screen is calculated by the Fourier transform of (2.98). Applying the convolution theorem to (2.98), which states that the Fourier transform of the convolution of two functions is the product of the Fourier transforms of the individual functions, we obtain

$$\mathcal{F}\left[A_1(x, y, z_1)\right] = \mathcal{F}[A_0(\xi,\eta)]\cdot\mathcal{F}\left[h(x, y)\right] = \mathcal{F}[A_0(\xi,\eta)]\cdot e^{ikz_1}e^{-i\pi\lambda z_1(f_x^2+f_y^2)}, \tag{2.100}$$

where $f_x = x/\lambda z_1$ and $f_y = y/\lambda z_1$ are the spatial frequencies of the field distribution at the observation plane. The field distribution at the screen is derived by taking the inverse Fourier transform of (2.100) as

$$A_1(x, y, z_1) = \mathcal{F}^{-1}\left\{\mathcal{F}[A_0(\xi,\eta)]\cdot e^{ikz_1}e^{-i\pi\lambda z_1(f_x^2+f_y^2)}\right\}. \tag{2.101}$$

Thus the field distribution after propagation is calculated in the following three steps: (a) Fourier transform the field distribution before propagation $F[A_0(\xi, \eta)]$, (b) derive $F\left[A_1(x, y, z_1)\right]$ by multiplying the phase factor to $F[A_0(\xi, \eta)]$, and (c) inverse Fourier transform $F\left[A_1(x, y, z_1)\right]$.

The field distribution at the far field of $z > 2D^2/\lambda$, called the *Fraunhofer diffraction* pattern, is obtained from (2.97), by approximating the phase term $(x^2 + y^2)/2z$ as constant:

$$A_1(x, y, z) = \frac{e^{ik\left[z+(x^2+y^2)/2z\right]}}{i\lambda z} \int\int_{-\infty}^{\infty} A_0(\xi, \eta) \exp\left[-\frac{ik}{z}(x\xi + y\eta)\right] d\xi d\eta. \quad (2.102)$$

The integral in this expression is a Fourier transform of the field distribution at the aperture, evaluated at the spatial frequencies $f_x = x/\lambda z, f_y = y/\lambda z$. Although the condition of $z > 2D^2/\lambda$ is not easily met, the far field distribution is obtained at the focus of a lens since a positive lens of a focal length f applies phase delay proportional to the square of the radius as $e^{-ik(x^2+y^2)/2f}$ to the input field and cancels the phase term in (2.102) at $z = f$.

2.2.2.2 Diffraction by a Grating

In ultrashort-duration, high-power lasers, diffraction gratings are used to expand and compress the pulses as described in Sect. 2.3.1. A diffraction grating diffracts light into specific directions where the diffracted beamlets from the periodic structure constructively interfere (Born & Wolf, 1999). Considering a reflection grating shown in Fig. 2.10, we denote the groove spacing by d, the total number of the grooves irradiated by light by N, and the incidence and the diffraction angles by θ_i and θ_d, respectively. Here θ_i and θ_d are measured from the surface normal, and θ_d is taken positive when the diffractive ray is in the opposite side of the incident ray relative to the surface normal.

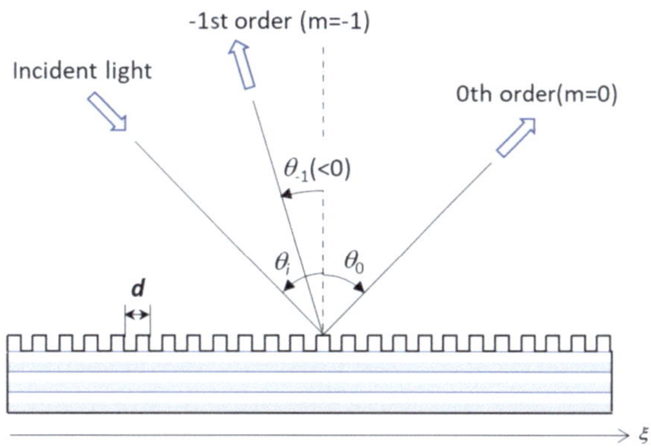

Fig. 2.10 Diffraction grating with groove spacing d. The diffracted beams of the 0th (m = 0) and the − 1st (m = − 1) order are shown. The sign of the diffraction angle θ_0 is positive, whereas θ_{-1} is negative since the − 1st order is diffracted to the same side of the incident beam (Littrow configuration)

Integrating the waves diffracted from all grooves over the grating area, the intensity distribution of the diffracted wave is

$$I(p) = |U(p)|^2 = I^{(0)}(p)\frac{\sin^2(Nkdp/2)}{\sin^2(kdp/2)}, \quad (2.103)$$

where $k = 2\pi/\lambda$ is the wavenumber and $p = \sin\theta_d - \sin\theta_i$. The intensity distribution $I(p)$ becomes the maximum with $I(p) = I^{(0)}(p)N^2$ when $kdp/2 = m\pi$, i.e.

$$\sin\theta_d - \sin\theta_i = m\frac{\lambda}{d} \quad (m = 0, \pm 1, \pm 2, \ldots). \quad (2.104)$$

Here m is called the diffraction order. This condition corresponds to $(2\pi)m$ phase difference between the incident and diffracted waves, since $d \cdot p/\lambda$ is the path difference between the incident and diffracted wave from each groove in unit of the wavelength.

The function $I^{(0)}(p)$ expresses the diffraction by each groove. For example, in the laminar-type grating shown in Fig. 2.10, it is possible to concentrate the diffracted wave only on the – 1st order by optimum design of the groove depth relative to the groove period for a particular wavelength and polarization of the incident light. We note that a coupled-wave theory considering the coupling of $\boldsymbol{E}$ and $\boldsymbol{H}$ based on Maxwell's equations is necessary in the analysis for $d/\lambda \leq 10$ (Moharam et al., 1995).

The 0th order corresponds to the specular reflection where $\theta_d(m = 0) = \theta_i$. As N increases, the peak intensity of the diffracted wave increases as N^2, and the directional broadening of the diffracted wave decreases as $1/N$. When the grating period d is comparable to the wavelength λ, the incident beam can be diffracted only to the $m = -1$ order on the same side of the incident beam relative to the grating normal. The condition of $\theta_d \approx -\theta_i$ is called Littrow configuration. Instead of using θ_d which is a negative value in Littrow, the notation of $\theta_D \equiv -\theta_d > 0$ is also used. With this notation, the diffraction condition for $m = -1$ is written as

$$\sin\theta_D + \sin\theta_i = \frac{\lambda}{d} = \frac{2\pi c}{\omega d}. \quad (2.105)$$

2.2.3 *Interference*

2.2.3.1 Optical Interferometers

Various types of interferometers are used in characterization and application of the lasers. Figure 2.11 shows three types of interferometers: Michelson interferometer,

Fizeau interferometer, and two-beam interferometer. With a Michelson interferometer, extremely small changes in the optical path are measured, for example, due to the gravitational wave (Abbott et al., 2016). The surface-profiles and the optical homogeneities of various optical components are accurately measured with the Fizeau interferometer (b). Holographic gratings are fabricated with a two-beam interferometer (c).

In these interferometers, an incident laser beam is split into two beams and combined after passing through different optical paths. The change in the optical path length in one of the paths is observed as a change in the interference pattern of the superposed beam. Let us consider a superposition of two electric fields $\boldsymbol{E}_1$ and $\boldsymbol{E}_2$:

$$\begin{aligned} \boldsymbol{E}_1(\boldsymbol{r}, t) &= (1/2)\big[\boldsymbol{E}_{10}(\boldsymbol{r}, t)\exp[i(\boldsymbol{k}_1 \cdot \boldsymbol{r} + \phi_1 - \omega t)] + c.c\big], \\ \boldsymbol{E}_2(\boldsymbol{r}, t) &= (1/2)\big[\boldsymbol{E}_{20}(\boldsymbol{r}, t)\exp[i(\boldsymbol{k}_2 \cdot \boldsymbol{r} + \phi_2 - \omega t)] + c.c\big], \end{aligned} \tag{2.106}$$

where $\boldsymbol{k}_i$ and ϕ_i $(i = 1, 2)$ are the wavevectors and the initial phases of these fields, respectively. The intensity of the superposed field $\boldsymbol{E} = \boldsymbol{E}_1 + \boldsymbol{E}_2$ is

$$\begin{aligned} I &= (1/2)\varepsilon_0 c|\boldsymbol{E}_0|^2 = (1/2)\varepsilon_0 c\big[|\boldsymbol{E}_{10}|^2 + |\boldsymbol{E}_{20}|^2 + 2\langle \boldsymbol{E}_1 \boldsymbol{E}_2\rangle\big] \\ &\equiv I_1 + I_2 + J_{12}. \end{aligned} \tag{2.107}$$

Here $I_i = (1/2)\varepsilon_0 c|\boldsymbol{E}_{i0}|^2$ $(i = 1, 2)$ is the intensity of each field. The interference term J_{12} is

$$\begin{aligned} J_{12} &= \varepsilon_0 c\langle \boldsymbol{E}_1 \boldsymbol{E}_2\rangle = \varepsilon_0 c|\boldsymbol{E}_{10}| \cdot |\boldsymbol{E}_{20}|\cos\delta, \\ \delta &= (\boldsymbol{k}_1 - \boldsymbol{k}_2)\boldsymbol{r} + (\phi_1 - \phi_2). \end{aligned} \tag{2.108}$$

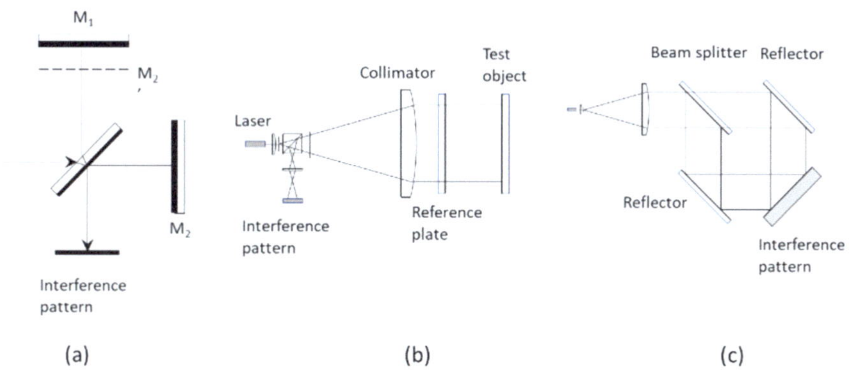

Fig. 2.11 Configurations of three typical interferometers: **a** Michelson interferometer, **b** Fizeau interferometer, and **c** two-beam interferometer for fabrication of holographic gratings

With the change in the relative phase δ, the intensity of the superposed field changes sinusoidally as

$$I = I_1 + I_2 + 2\sqrt{I_1 I_2}\cos\delta \tag{2.109}$$

The two fields constructively interfere at $\delta = 0, \pm 2\pi, \pm 4\pi, \ldots$ resulting in the maximum intensity $I_{\max} = I_1 + I_2 + 2\sqrt{I_1 I_2}$, whereas destructively interfere at $\delta = \pm\pi, \pm 3\pi, \pm 5\pi, \ldots$ resulting in the minimum intensity $I_{\min} = I_1 + I_2 - 2\sqrt{I_1 I_2}$. The contrast of the interference pattern is

$$\gamma = \frac{I_{\max}}{I_{\text{avr}}} = 1 + \frac{2\sqrt{I_1 I_2}}{I_1 + I_2} \tag{2.110}$$

For a special case of $I_1 = I_2 = I_0$, the interference pattern shows the maximum contrast of $\gamma = 1$ with $I_{\max} = 4I_0$ and $I_{\min} = 0$. A coherent, intense laser beam with the maximum intensity of $I_{\max} = \mathrm{N}^2 I_0$ is generated by coherently combining N laser beams.

2.2.3.2 Multilayer Dielectric Coatings

Multilayer dielectric coatings are used in lasers as high-reflectivity mirrors, anti-reflection surfaces, polarizing beam splitters, and chirped mirrors for dispersion compensation in ultrafast lasers. In addition to these optical properties, high damage threshold is required in high-power lasers. In this subsection, basic design of the multilayers is presented and several issues in the multilayers for high-power and high-intensity lasers are briefly described.

The optical property of the multilayer is calculated as a superposition of the reflections at the interfaces of the multilayer. Consider a linearly polarized wave is incident from the air to a dielectric film of thickness d coated on a substrate as shown in Fig. 2.8. We denote the refractive indices of the air, thin film, and substrate as n_0, n_1 and n_2, respectively, and denote the air-film boundary as I and the film-substrate boundary as II. As shown in Eq. (2.89), $n_0 \sin\theta_0$ is preserved in all layers: i.e., $n_0 \sin\theta_0 = n_1 \sin\theta_1 = n_2 \sin\theta_2$. We denote the incident, reflected and transmitted waves by the suffix of i, r, and t, respectively.

From the boundary condition (2.88), the components of $\boldsymbol{E}$ and $\boldsymbol{H}$ tangential to the boundary are continuous across the boundary. When an incident wave is s-polarized (i.e., $\boldsymbol{E}$ is vertical to the x–z plane), the E-fields at above and below the boundary are related as

$$E_I = E_{iI} + E_{rI} = E_{tI} + E'_{rII}, \tag{2.111}$$

where E'_{rII} is the reflected wave from the boundary II. Since $i\boldsymbol{k} \times \boldsymbol{E} = i\omega\mu_0\boldsymbol{H}$ from the third equation of (2.10), $\boldsymbol{H}$ is polarized in the incidence plane with its amplitude given by $H = \sqrt{\varepsilon_0/\mu_0} nE$. The continuity of H at the boundary I is

$$H_I = \sqrt{\frac{\varepsilon_0}{\mu_0}}(E_{iI} - E_{rI})n_0 \cos\theta_0 = \sqrt{\frac{\varepsilon_0}{\mu_0}}\left(E_{tI} - E'_{rII}\right)n_1 \cos\theta_1. \tag{2.112}$$

At the boundary II where a reflected wave does not exist, these conditions become

$$E_{\mathrm{II}} = E_{i\mathrm{II}} + E_{r\mathrm{II}} = E_{t\mathrm{II}},$$
$$H_{\mathrm{II}} = \sqrt{\frac{\varepsilon_0}{\mu_0}}(E_{i\mathrm{II}} - E_{r\mathrm{II}})n_1 \cos\theta_1 = \sqrt{\frac{\varepsilon_0}{\mu_0}}E_{t\mathrm{II}}n_2 \cos\theta_2. \tag{2.113}$$

The difference in the optical paths of Ray-1 and Ray-2 in Fig. 2.8 is

$$\Lambda = 2\frac{n_1 d}{\cos\theta_1} - (2d\tan\theta_1)n_0 \sin\theta_0 = 2n_1 d\cos\theta_1. \tag{2.114}$$

Therefore, a wave in traversing the film in a single path undergoes a phase shift by $\phi_1 \equiv k_0(2n_1 d\cos\theta_1)/2 = k_0 n_1 d\cos\theta_1$. Thus, the fields at the boundaries I and II are related as

$$E_{i\mathrm{II}} = E_{t\mathrm{I}}e^{+i\phi_1}, \quad E_{r\mathrm{II}} = E'_{r\mathrm{II}}e^{-i\phi_1}. \tag{2.115}$$

Combining these equations, the fields at the boundaries I and II are expressed by the *characteristic matrix* $\mathrm{M_I}$ as

$$\begin{bmatrix} E_1 \\ H_1 \end{bmatrix} = \begin{bmatrix} \cos\phi_1 & -i\sin\phi_1/\left(\sqrt{\varepsilon_0/\mu_0}n_1\cos\theta_1\right) \\ -i\sin\phi_1\cdot\left(\sqrt{\varepsilon_0/\mu_0}n_1\cos\theta_1\right) & \cos\phi_1 \end{bmatrix}\begin{bmatrix} E_{1I} \\ H_{1I} \end{bmatrix}$$
$$\equiv \mathcal{M}_{1,s}\begin{bmatrix} E_{\mathrm{II}} \\ H_{\mathrm{II}} \end{bmatrix}. \tag{2.116}$$

Following the similar steps, the characteristic matrix for the *p*-polarized incident wave is obtained:

$$\mathcal{M}_{1,\mathrm{p}} = \begin{bmatrix} \cos\phi_1 & -i\sin\phi_1/\left(\sqrt{\varepsilon_0/\mu_0}n_1/\cos\theta_1\right) \\ -i\sin\phi_1\cdot\left(\sqrt{\varepsilon_0/\mu_0}n_1/\cos\theta_1\right) & \cos\phi_1 \end{bmatrix}. \tag{2.117}$$

The characteristic matrix $\mathcal{M}$ of a N-multilayer dielectric film coated on a substrate is obtained by replacing the substrate with another thin layer and repeating the procedures:

$$\begin{bmatrix} E_1 \\ H_1 \end{bmatrix} = \mathcal{M}_1\mathcal{M}_2 \ldots \mathcal{M}_{\mathrm{N}} \begin{bmatrix} E_{\mathrm{N}+1} \\ H_{\mathrm{N}+1} \end{bmatrix} \equiv \mathcal{M} \begin{bmatrix} E_{\mathrm{N}+1} \\ H_{\mathrm{N}+1} \end{bmatrix} \equiv \begin{bmatrix} m_{11} & m_{12} \\ m_{21} & m_{22} \end{bmatrix} \begin{bmatrix} E_{\mathrm{N}+1} \\ H_{\mathrm{N}+1} \end{bmatrix}. \tag{2.118}$$

Since E_1 and H_1 at the boundary I are given in (2.111) and (2.112), (2.118) reads

$$\begin{bmatrix} E_{i\mathrm{I}} + E_{r\mathrm{I}} \\ (E_{i\mathrm{I}} - E_{r\mathrm{I}})Y_0 \end{bmatrix} = \mathcal{M} \begin{bmatrix} E_{t\mathrm{N}} \\ E_{t\mathrm{N}} Y_s \end{bmatrix}, \tag{2.119}$$

where $\mathrm{Y}_0 = \sqrt{\varepsilon_0/\mu_0}n_0 \cos\theta_0$ and $\mathrm{Y}_s = \sqrt{\varepsilon_0/\mu_0}n_s \cos\theta_s$.

The amplitude reflection and transmission coefficients of the N-stack multilayers are derived from these equations as

$$r = \frac{E_{r\mathrm{I}}}{E_{i\mathrm{I}}} = \frac{(m_{11} + m_{12}\mathrm{Y}_\mathrm{s})\mathrm{Y}_0 - (m_{21} + m_{22}\mathrm{Y}_\mathrm{s})}{(m_{11} + m_{12}\mathrm{Y}_\mathrm{s})\mathrm{Y}_0 + (m_{21} + m_{22}\mathrm{Y}_\mathrm{s})}, \tag{2.120}$$

$$t = \frac{E_{t\mathrm{N}}}{E_{i\mathrm{I}}} = \frac{2\mathrm{Y}_0}{(m_{11} + m_{12}\mathrm{Y}_\mathrm{s})\mathrm{Y}_0 + (m_{21} + m_{22}\mathrm{Y}_\mathrm{s})}, \tag{2.121}$$

where m_{ij} is the *ij*-element of the matrix $\mathcal{M}$. Thus, the optical property of a multilayer film is calculated using the characteristic matrix of each film.

Let us consider the reflection coefficient r of a monolayer coated on a substrate. Using the characteristic matrix for a single layer (2.116), it is given as

$$r = \frac{r_{01} + r_{12}e^{2ik_0h}}{1 + r_{01}r_{12}e^{2ik_0h}}, \tag{2.122}$$

where r_{01} and r_{12} are the reflection coefficients of the first boundary and the second boundary, respectively:

$$r_{01} = \frac{n_0 \cos\theta_0 - n_1 \cos\theta_1}{n_0 \cos\theta_0 + n_1 \cos\theta_1}, \quad r_{12} = \frac{n_1 \cos\theta_1 - n_s \cos\theta_s}{n_1 \cos\theta_1 + n_s \cos\theta_s}. \tag{2.123}$$

The reflectance R is

$$R = |r|^2 = \frac{r_{01}^2 + r_{12}^2 + 2r_{01}r_{12}\cos(2k_0d)}{1 + r_{01}^2r_{12}^2 + 2r_{01}r_{12}\cos(2k_0d)} \tag{2.124}$$

The reflectance takes the extreme values at $2k_0d = m\pi$ where $m = 0, 1, 2 \ldots$. At normal incidence, this relation takes a simple form:

$$nd = m\frac{\lambda}{4} \tag{2.125}$$

When m takes odd integers, the reflectivity is given by

$$R = \left(\frac{n_0 n_2 - n_1^2}{n_0 n_2 + n_1^2} \right)^2 \tag{2.126}$$

The reflectivity becomes 0 when $n_1 = \sqrt{n_0 n_2}$. In general, the reflectivity becomes minimum when $n_1 < n_2$, whereas the reflectivity becomes maximum when $n_1 > n_2$. This relation holds also for the multilayer by replacing n_1 with the refractive index of the layer adjacent to the substrate.

The multilayer films are typically composed of a stack of a pair of low refractive index (n_L) and high refractive index (n_H) films of quarter-wavelength optical thickness (QWOT). For example, a multilayer film composed of m-stacks of *HL* layers coated on a glass substrate is designated as $g(HL)^m a$, where g and a stand for glass and air, respectively.

In laser applications, the laser damage of the multilayer is a severe problem. Often the damage starts from the high-index layer due to higher absorption coefficient of the high-index materials. Calculation of the electric field distribution in the quarter-wave stack shows that the field has peaks at the interfaces between the low- and high-index layers. Often non-quarter wave stacks are fabricated to reduce the field in high-index layers.

As an example, Fig. 2.12 shows two designs of a double-wavelength laser mirror which has high reflectance at both 694.3 and 1064 nm (Kimura et al., 2007). Distributions of the refractive index and the electric fields of the two wavelengths as a function of distance from the surface are shown, where laser beam is incident from left at normal incidence. The mirror is composed of (a) double QW-stack and (b) non-QW stack of high- ($n_H = 2.1$) and low- ($n_L = 1.5$) refractive index layers coated on a glass substrate. In (a), the 694.3 nm light is reflected by the 694.3 nm QW-stack which is close to the surface, whereas the 1064 nm light is reflected by the 1064 nm stack far from the surface. The field distribution shows that in (a) the 1064 nm field has high amplitude in the 694.3 nm QW-stack, whereas in (b) the fields of both wavelengths are evenly distributed with lower maximum amplitudes in the multilayers. From this field distribution, it is expected that (b) has higher damage threshold, which has been proven by the damage test.

2.2.4 *Birefringence of Anisotropic Crystals*

Anisotropic crystals and electro-optic materials are used for nonlinear frequency conversion and controlling the laser light, respectively. As shown in Eq. (2.8), the electric displacement $\boldsymbol{D}$ is written as $\boldsymbol{D} = \boldsymbol{\varepsilon} \cdot \boldsymbol{E}$, where $\boldsymbol{\varepsilon}$ is the permittivity tensor of the material. This relation is expressed as

$$D_k = \sum_{l=x,y,z} \varepsilon_{kl} E_l \equiv \varepsilon_{kl} E_l, \tag{2.127}$$

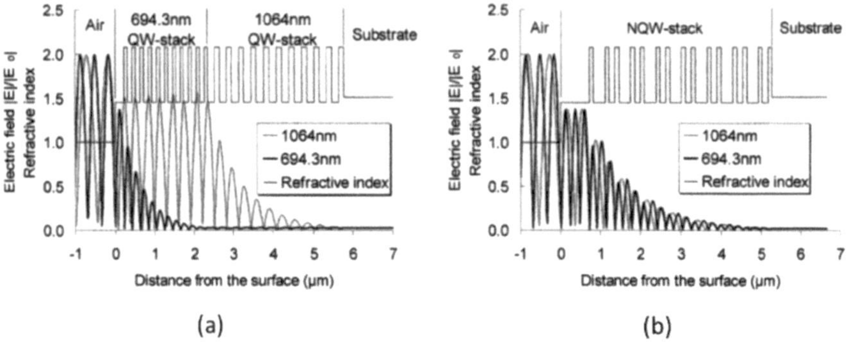

Fig. 2.12 Distribution of the electric fields of 694.3 and 1064 nm light in the dielectric multilayers composed of **a** double QW-stacks and **b** non-QW stack, both designed for high reflectance at 694.3 and 1064 nm. The refractive indices of the multilayers, substrate and air are shown in the insets. Reprinted from Kimura et al. (2007), with permission from The Laser Society of Japan. © 2007 The Laser Society of Japan

where the subscript refers to the Cartesian coordinates (x, y, z) which are fixed with respect to the crystal axes. The energy conservation of the electromagnetic wave for isotropic materials is also held in anisotropic systems where the energy density u and the energy flux (Poynting vector) $\boldsymbol{S}$ are written as

$$u = u_e + u_m = \frac{1}{2}\sum_{k,l} E_k \varepsilon_{kl} E_l + \frac{1}{2}\mu H^2, \tag{2.128}$$

$$\boldsymbol{S} = \boldsymbol{E} \times \boldsymbol{H} = v_t u \hat{\boldsymbol{t}}. \tag{2.129}$$

Here u_e and u_m are the energy densities of the electric field and the magnetic field, respectively. The energy flux $\boldsymbol{S}$ propagates with the ray velocity v_t to the direction $\hat{\boldsymbol{t}}$, a unit vector perpendicular to $\boldsymbol{E}$ and $\boldsymbol{H}$.

These relations are valid when the dielectric tensor is real and symmetric with respect to the suffices k and l, i.e., $\varepsilon_{kl} = \varepsilon_{lk}$. We can diagonalize the symmetric tensor ε_{kl} with respect to the new coordinate x, y, z called principal dielectric axes as

$$\begin{bmatrix} D_x \\ D_y \\ D_z \end{bmatrix} = \begin{pmatrix} \varepsilon_x & 0 & 0 \\ 0 & \varepsilon_y & 0 \\ 0 & 0 & \varepsilon_z \end{pmatrix} \begin{bmatrix} E_x \\ E_y \\ E_z \end{bmatrix}. \tag{2.130}$$

In this coordinate, $\boldsymbol{D}$ and $\boldsymbol{E}$ are parallel and the components of $\boldsymbol{D}$ satisfy the following relation,

$$\frac{D_x^2}{\varepsilon_x} + \frac{D_y^2}{\varepsilon_y} + \frac{D_z^2}{\varepsilon_z} = 2w_e \equiv C. \tag{2.131}$$

The optical properties of the crystals are categorized into 3 groups, where each group is composed of the symmetry classes as shown below.

I	Isotropic	$\varepsilon_x = \varepsilon_y = \varepsilon_z$	Cubic
II	Uniaxial	$\varepsilon_x = \varepsilon_y \neq \varepsilon_z$	Trigonal, tetragonal, hexagonal
III	Biaxial	$\varepsilon_x \neq \varepsilon_y \neq \varepsilon_z$	Orthorhombic, monoclinic, triclinic

In a uniaxial crystal, $n_x = n_y \neq n_z$, where the refractive index n_i is defined as $\varepsilon_i \equiv n_i^2 \varepsilon_0\,(i = x, y, z)$, and z is called an optical axis (c-axis). A uniaxial crystal is called "positive" when $n_z > n_x = n_y$ and "negative" when $n_z < n_x = n_y$, respectively.

Let us consider a monochromatic wave of frequency ω whose wavefront propagates in the direction $\hat{\boldsymbol{s}}$ ($\hat{\boldsymbol{s}}$ is a unit vector normal to the wavefront) with phase velocity $v_p = c/n$ where n is the refractive index. The electric and magnetic fields and their flux densities $\boldsymbol{E}$, $\boldsymbol{D}$, $\boldsymbol{H}$ and $\boldsymbol{B}$ propagate also as $\exp[i(\boldsymbol{k}\cdot\boldsymbol{r}) - i\omega t]$ where $\boldsymbol{k} = \omega(n/c)\hat{\boldsymbol{s}}$. Since the derivatives of these vectors with respect to time and space result in $\partial/\partial t = -i\omega$ and $\nabla\times = i\omega(n/c)\hat{\boldsymbol{s}}\times$, respectively, we obtain the following relations from the Maxwell equations,

$$n\hat{\boldsymbol{s}} \times \boldsymbol{H} = -c\boldsymbol{D}, \quad n\hat{\boldsymbol{s}} \times \boldsymbol{E} = \mu_0 c\boldsymbol{H}. \tag{2.132}$$

Eliminating $\boldsymbol{H}$ from these equations,

$$\boldsymbol{D} = -\left(\frac{n^2}{\mu_0 c^2}\right)\hat{\boldsymbol{s}} \times (\hat{\boldsymbol{s}} \times \boldsymbol{E}) = n^2\varepsilon_0[\boldsymbol{E} - \hat{\boldsymbol{s}}(\boldsymbol{E}\cdot\hat{\boldsymbol{s}})]. \tag{2.133}$$

Equation (2.132) shows that, $\boldsymbol{E}$, $\boldsymbol{D}$ and $\hat{\boldsymbol{s}}$ are in a plane orthogonal to $\boldsymbol{H}$, and $\boldsymbol{D}$ is perpendicular to $\hat{\boldsymbol{s}}$. When the wave propagates at an oblique angle to the optical axis, $\boldsymbol{E}$ is perpendicular to the propagation direction $\hat{\boldsymbol{t}}$ which is different from $\hat{\boldsymbol{s}}$. With (2.128) and (2.133), the energy density is written as

$$u = u_e + u_m = \frac{1}{2}\boldsymbol{E}\cdot\boldsymbol{D} + \frac{1}{2}\boldsymbol{B}\cdot\boldsymbol{H} = -\frac{n}{2}\boldsymbol{E}\cdot(\hat{\boldsymbol{s}} \times \boldsymbol{H}) + \frac{n}{2}(\hat{\boldsymbol{s}} \times \boldsymbol{E})\cdot\boldsymbol{H} = \frac{n}{c}\hat{\boldsymbol{s}}\cdot\boldsymbol{S}. \tag{2.134}$$

Here the energy flux $\boldsymbol{S}$ propagates as $\boldsymbol{S} = v_t u\hat{\boldsymbol{t}}$. The group velocity v_t is faster than the phase velocity since $v_p = c/n = v_t\hat{\boldsymbol{t}}\cdot\hat{\boldsymbol{s}} = v_t\cos\alpha$, where α is the angle between $\hat{\boldsymbol{t}}$ and $\hat{\boldsymbol{s}}$.

Rewriting (2.131) by normalizing D_i as $D_i/\sqrt{\varepsilon_0 C} = i\ (i = x, y, z)$, we obtain

$$\frac{x^2}{n_x^2} + \frac{y^2}{n_y^2} + \frac{z^2}{n_z^2} = 1. \tag{2.135}$$

Figure 2.13 shows an ellipsoid defined by this equation, known as an *index ellipsoid* (or *ellipsoid of wave normal*, *indicatrix*), where x, y, and z are crystal axes. This

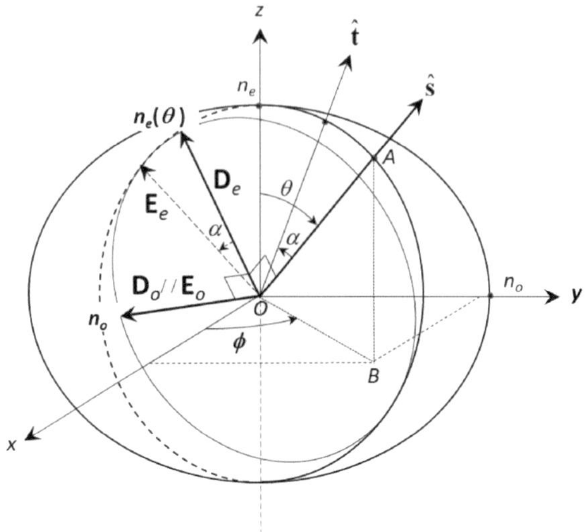

Fig. 2.13 An index of ellipsoid for a negative uniaxial crystal. Shown is the propagation of a plane wave with its wavefront propagating in the direction $\hat{s}$, at an angle θ from the z-axis and ϕ from the x-axis. An ellipse formed by cutting the ellipsoid at O with a plane normal to $\hat{s}$ is shown by the shaded area. There are two independent waves, orthogonally polarized shown by $\boldsymbol{D}_o$ and $\boldsymbol{D}_e$, with different phase velocities determined by the refractive indices n_0 and $n_e(\theta)$, respectively. The Poynting vector of the extraordinary ray is directed toward $\hat{\boldsymbol{t}}$ at an angle α from $\hat{s}$

figure shows an ellipsoid of a negative uniaxial crystal, where the refractive indices are often denoted as ordinary $n_o \equiv n_x = n_y$ and extraordinary $n_e \equiv n_z$, respectively.

Let us consider a plane wave propagating in a uniaxial crystal in the direction $\hat{s}$, which makes an angle θ from the z-axis and an azimuthal angle ϕ from the x-axis, respectively. (A vector in the direction $\hat{s}$ is shown by OA, and OB is its projection to the x–y plane.) When the ellipsoid of wave normal is cut by a plane crossing the origin O and normal to $\hat{s}$, an ellipse is formed as shown by the shaded area in Fig. 2.12. The displacement $\boldsymbol{D}$ is in this shaded plane since it is normal to $\hat{s}$.

Equations (2.133) and (2.135) show that a plane wave propagating in the direction $\hat{s}$ has two independent solutions which are orthogonally polarized and propagate with different phase velocities v_p. One wave, named "ordinary ray", is polarized along the minor axis of the ellipse, as shown by $\boldsymbol{D}_o$, and propagates with $v_p = c/n_o$ i.e., with the refractive index n_o. Since the electric field of the ordinary ray $\boldsymbol{E}_o$ is parallel to the displacement $\boldsymbol{D}_o$, the energy flux of the ordinary ray $\boldsymbol{S}_o$ propagates in the same direction as $\boldsymbol{k}$ with the velocity $v_t = v_p\ v_t$ which does not depend on θ.

Another wave, called "extraordinary ray", is polarized along the major axis, shown by $\boldsymbol{D}_e$. It propagates with a phase velocity $v_p = c/n_e(\theta)$, where the refractive index $n_e(\theta)$ changes from n_0 at $\theta = 0$ to n_e at $\theta = 90$:

$$\frac{1}{n_e^2(\theta)} = \frac{\cos^2\theta}{n_o^2} + \frac{\sin^2\theta}{n_e^2}. \tag{2.136}$$

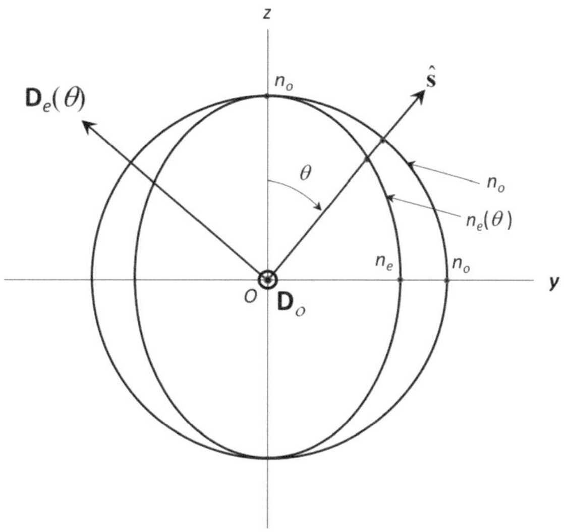

Fig. 2.14 Index surface of a negative uniaxial crystal ($n_e < n_o$), which shows the refractive indices of the ordinary ray (o) and the extraordinary ray (e) propagating in the direction $\hat{s}$

The electric field of the extraordinary ray $\boldsymbol{E}_e$ lies in the ellipse containing OA, at an angle $\cos \alpha = \hat{s} \cdot \hat{t}$ to $\boldsymbol{D}_e$, as shown in Fig. 2.12. Therefore, the flux of extraordinary ray propagates in the direction $\hat{t}$, with a velocity $v_t = v_p / \cos \alpha$. The oscillation direction of the electric field $\boldsymbol{E}_e$ is normal to $\hat{t}$, which is tilted from the wavefront by α.

Another useful presentation of the birefringence is the *index surface* (or *normal surface*) shown in Fig. 2.14, which is an intersection of the index ellipsoid with the z–$\hat{s}$ plane. The index surface shows the refractive indices of the rays propagating to $\hat{s}$, for the ordinary ray o and the extraordinary ray e, respectively. The ellipse marked by $n_e(\theta)$ corresponds to Eq. (2.136). In this diagram, the polarizations of the extraordinary ray $\boldsymbol{D}_e(\theta)$ and the ordinary ray $\boldsymbol{D}_o(\theta)$ are in- and out- of the y–z plane, respectively. The index surface is used for determining the phase matching condition for frequency conversion in nonlinear optics, as described in Sect. 5.2.

2.3 Ultrashort Laser Pulse

In this section, the basic concept of the chirped pulse amplification (CPA) is presented as an introduction to Chap. 7 on high-power ultrashort-pulse laser systems. After brief introduction to CPA, temporal and spectral changes of ultrashort laser pulses due to propagation in dispersive materials are given. Then basic optical systems to control the ultrashort pulses used in CPA lasers are described. Detailed account of generation, propagation and applications of ultrashort-duration laser pulses are given, for example, in Diels and Rudolph (2006), Wollenhaupt et al. (2007), and Agrawal (2019).

2.3.1 Chirped Pulse Amplification

An ultrashort pulse of femto-second duration is generated by coherently superposing optical waves of evenly spaced frequencies over a broad bandwidth. When an ultrashort pulse is amplified with a broad-band laser amplifier, the attainable laser power is limited by the optical damage to the laser material caused laser intensification due to nonlinear self-focusing of the laser beam. To avoid this limitation, Mourou has proposed a new scheme named "chirped pulse amplification (CPA)" (Pessot et al., 1987; Strickland & Mourou, 1985).

In this scheme, an ultrashort pulse is stretched in time by ~ 1000, thus decreasing the peak power before amplification. This stretched pulse is amplified without being limited by the laser damage. After amplification, the pulse is re-compressed to the original pulse width, resulting in an extremely high-power ultrashort pulse. Pulse stretching is achieved by up-chirping the frequency of an ultrashort pulse, where a low-frequency component of the laser pulse precedes the high-frequency components. Pulse compression is achieved by applying frequency down-chirping, i.e., a reverse process of pulse stretching. The CPA scheme is shown schematically in Fig. 2.15, where a pair of diffraction gratings is used for pulse stretching and pulse compression, as described in the following sections.

Although this scheme appears straightforward in principle, extreme care is required to control the phase of an ultrafast pulse with high precision (within ~

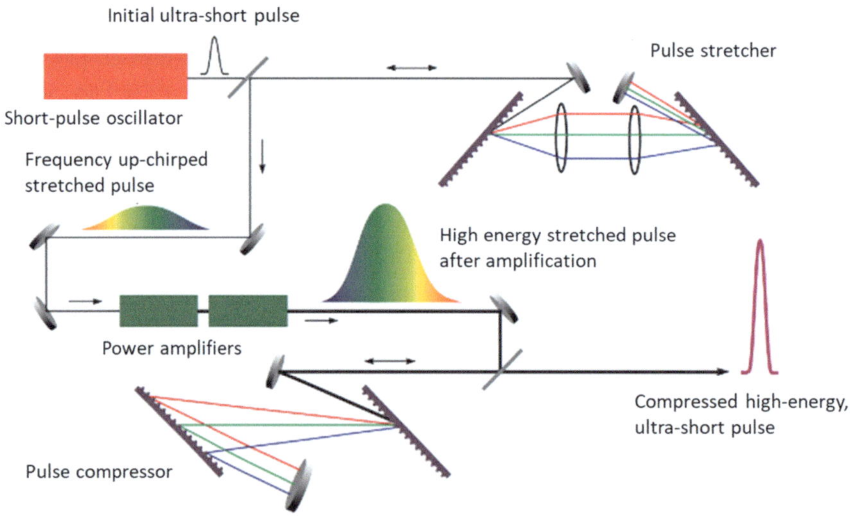

Fig. 2.15 A schematic diagram of the chirped-pulse amplification (CPA). In this diagram, a pair of gratings is used for stretching and compression of the pulse, where the laser beam is double-passed by shifting its vertical position using corner-reflector mirrors. Modified from Chirped pulse amplification in Wikipedia (September 2023). https://en.wikipedia.org/wiki/Chirped_pulse_amplification

1/10th of 2π radians) over broad spectral range during propagation and amplification by considering the dispersions of optical materials and components. In the following, we describe how an ultrafast pulse is affected by the dispersion and how to compensate for the dispersion.

2.3.2 *Pulse Shape Changes Due to Phase Dispersion*

Let us consider how the intensity profile of an ultrafast pulse is affected by the dispersion of the spectral phase. We describe an optical pulse with an electric field at a fixed position in space as

$$E(t) = A(t)\cos[\phi(t)]. \tag{2.137}$$

Here $A(t)$ is an instantaneous amplitude (pulse envelope) which we assume to be maximum at $t = 0$. $\phi(t)$ is an instantaneous phase comprised of three terms,

$$\phi(t) = \omega_0 t + \phi_0 + \phi_1(t), \tag{2.138}$$

where ω_0 is a carrier frequency, ϕ_0 is a time-independent phase, and $\phi_1(t)$ is an additional time-varying phase. The phase ϕ_0, called a carrier envelope phase, determines the relative position between the envelope and the oscillating field, with $E(t)$ taking the maximum value of $A(t = 0)$ when $\phi_0 = 0$ and $\phi_1(t) = 0$.

Referring to Eq. (2.85), we express the electric field $E(t)$ by its analytic signal $E_\mathrm{a}(t)$:

$$E_a(t) = E(t) + \hat{E}(t) = A(t)\cos[\phi(t)] + iA(t)\sin[\phi(t)] = A(t)\mathrm{e}^{i\phi(t)}. \tag{2.139}$$

The field $E(t)$ is the real part of $E_a(t)$. The instantaneous frequency at t is derived from the phase as

$$\omega(t) = \frac{\mathrm{d}\phi(t)}{\mathrm{d}t} = \omega_0 + \frac{\mathrm{d}\phi_1(t)}{\mathrm{d}t}. \tag{2.140}$$

Due to various processes contributing to $\phi_1(t)$, an ultrashort laser pulse is comprised of many frequency components around the carrier frequency ω_0.

Let us consider a typical pulse of Gaussian temporal profile without time-varying phase:

$$E_a(t) = A(t)e^{-2\ln 2(t^2/\Delta t^2)}e^{i\omega_0 t}, \tag{2.141}$$

where Δt is a pulse width (full width at half maximum, FWHM) of the intensity $I(t) \propto |E(t)|^2$. The spectral distribution of this pulse is

$$E_a(\omega) = A(t)\Delta t\sqrt{\frac{\pi}{2\ln 2}}\mathrm{e}^{-2\ln 2\left[(\omega-\omega_0)^2/\Delta\omega^2\right]}, \tag{2.142}$$

where $\Delta\omega = 4\ln 2/\Delta t$ is the FWHM width of the spectral intensity $I(\omega) \propto |E_a(\omega)|^2$. Expressing this relation with frequency ν, we obtain the time-bandwidth product,

$$\Delta t \cdot \Delta\nu = \frac{2\ln 2}{\pi} = 0.441. \tag{2.143}$$

A laser pulse which has the minimum time-bandwidth product is called a Fourier-transform-limited (FTL) pulse or a bandwidth-limited pulse. The minimum value of the time-bandwidth product $\Delta\nu \cdot \Delta t$ depends on the pulse shape in time: 0.315 for a hyperbolic secant pulse, 0.441 for a Gaussian pulse, and 0.886 for a square pulse. As a numerical example, a FTL Gaussian pulse of 800 nm wavelength and $\Delta t = 10$ fs (FWHM), corresponding to 3.8 optical cycles, has a frequency bandwidth of $\Delta\nu = 4.4 \times 10^{13}\ \mathrm{s}^{-1} = 44$ THz (i.e., $\Delta\nu/\nu = 0.12$). When the phase changes in time, the time-bandwidth product becomes larger than these minimum values.

The temporal pulse shape $E_a(t)$ is derived from the inverse Fourier transform of the spectral distribution $E_a(\omega)$. Let us consider how the pulse shape of an FTL pulse changes by the spectral dispersion. We expand the spectral phase $\phi(\omega)$ around ω_0 into a Taylor series,

$$\begin{aligned}\phi(\omega) = \phi_0 + \phi_1(\omega-\omega_0) + \frac{1}{2!}\phi_2(\omega-\omega_0)^2 + \frac{1}{3!}\phi_3(\omega-\omega_0)^3 \\ + \frac{1}{4!}\phi_4(\omega-\omega_0)^4 + \cdots,\end{aligned} \tag{2.144}$$

where $\phi_0 = \phi(\omega_0)$ and

$$\phi_m = \left(\frac{\mathrm{d}^m\phi}{\mathrm{d}\omega^m}\right)_{\omega=\omega_0} \quad (m = 1, 2, \ldots) \tag{2.145}$$

The coefficients $\phi_1, \phi_2, \phi_3, \phi_4$, expressed in units of [fs, fs^2, fs^3, fs^4] respectively, are called the group delay, the second-order dispersion (SOD), the third-order dispersion (TOD), and the fourth-order dispersion (FOD), respectively.

The zeroth-order term ϕ_0 corresponds to the absolute phase in the time domain. The first-order term ϕ_1 corresponds to the time shift of the envelope. By taking up to the second-order term in (2.142), the spectral distribution is

$$E_a(\omega)\big|_{\phi_2} = A(t)\frac{\Delta t}{2\sqrt{\ln 2}}\mathrm{e}^{-2\ln 2\left[(\omega-\omega_0)^2/\Delta\omega^2\right]}\mathrm{e}^{-i\frac{1}{2}\phi_2(\omega-\omega_0)^2}. \tag{2.146}$$

The temporal profile of the field amplitude, derived from the inverse Fourier transform of (2.146), is expressed in terms of ϕ_2 and the pulse width Δt_0 (FWHM) for $\phi_2 = 0$ as,

$$E_a(t)\big|_{\phi_2} = A(t)\frac{1}{\sqrt{2\pi}\gamma^{1/4}}e^{-2\ln 2\left(t^2/\Delta t_{\phi_2}^2\right)}e^{i(\omega_0+at)t-i\varepsilon}, \tag{2.147}$$

where

$$\Delta t_{\phi_2} = \sqrt{\gamma}\Delta t_0,\ \ \gamma = 1 + \left(4\ln 2\cdot\frac{\phi_2}{\Delta t_0^2}\right)^2,\ \ a = 8(\ln 2)^2\frac{\phi_2}{\gamma\Delta t_0^4},$$
$$\varepsilon = \frac{1}{2}\arctan\left(4\ln 2\frac{\phi_2}{\Delta t_0^2}\right). \tag{2.148}$$

Equation (2.147) shows that the instantaneous frequency changes in time as $\omega = \omega_0 + at$ with a rate a [rad/s^2] and the pulse width broadens from Δt_0 to $\Delta t_{\phi_2} = \sqrt{\gamma}\Delta t_0$. When a is positive, the frequency is up-chirped, i.e., ω increases with time. The pulse width broadening due to ϕ_2 is estimated as

$$\Delta t_{\phi_2}^2 = \Delta t_0^2 + \frac{(4\ln 2)^2}{\Delta t_0^2}\phi_2^2 \tag{2.149}$$

Note that a shorter-duration pulse broadens faster since the second term of the right-hand side is inversely proportional to Δt_0^2.

With the third-order dispersion ϕ''' (TOD), the spectral distribution $\tilde{E}(\omega)\big|_{\phi'''}$ is expressed as

$$E_a(\omega)\big|_{\phi'''} = A(t)\frac{\Delta t}{2\sqrt{\ln 2}}e^{-2\ln 2\left[(\omega-\omega_0)^2/\Delta\omega^2\right]}e^{-i\frac{1}{6}\phi'''(\omega_0)\cdot(\omega-\omega_0)^3}. \tag{2.150}$$

The temporal profile of the electric field with the third order dispersion $E(t)\big|_{\phi_3}$ is analytically solved for a Gaussian distribution and expressed by an Airy function (McMullen, 1977; Miyagi et al., 1979). The field amplitude normalized by $\sqrt{P_0}$ (P_0 is the peak power) with only the third-order dispersion, i.e., without other effects such as second order dispersion, linear attenuation, and nonlinear effects, is given (Agrawal, 2019) by

$$U(t) = 2^{5/6}\pi^{1/2}\frac{T_0}{|\phi_3|^{1/3}}\exp\left[\frac{T_0^4/(3\phi_3) - t}{\phi_3/T_0^2}\right]Ai\left[\frac{T_0^4/(2\phi_3) - t}{2^{-1/3}|\phi_3|^{1/3}}\right]. \tag{2.151}$$

Here $T_0 = \Delta t_0/(2\sqrt{\ln 2}) = 0.601\Delta t_0$ is the half width at $1/e$ intensity. Equation (2.151) is a product of an exponential term and an Airy function $Ai(t)$ which oscillates in time with varied period and amplitude.

We show in Fig. 2.16 the dependences of the pulse shape ($U(t)^2$) on ϕ_3 and Δt_0. In Fig. 2.16a ϕ_3 is varied from 30 fs^3 to 3000 fs^3 for a 10-fs pulse, whereas in Fig. 2.16b Δt_0 is varied from 5 to 20 fs for $\phi_3 = 1000$ fs^3. We find that the pulse shape becomes asymmetric accompanied by a shift of the pulse peak and an oscillatory structure when $\phi_3 \geq (\Delta t_0)^3$. Since ϕ_3 is positive in this example, post pulses are generated

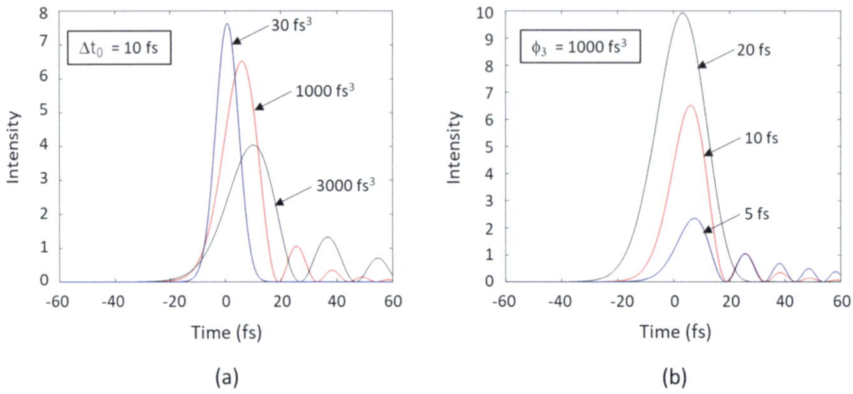

Fig. 2.16 Dependences of the pulse shape **a** on TOD of 30 fs^3, 1000 fs^3 and 3000 fs^3 respectively for a 10-fs pulse, and **b** on the pulse width of 5 fs, 10 fs and 20 fs for TOD of 1000 fs^3. The vertical scale for the intensity is in arbitrary unit

because instantaneous frequency increases at the rising and trailing parts of the pulse, resulting in the time delay of these parts relative to the central part of the pulse. The oscillatory structure arises from the interference of these delayed waves. When the sign of TOD is negative, pre-pulses are generated instead of post-pulses due to the third-order dispersion.

2.3.3 Propagation of an Ultrashort Pulse in Transparent Material

When an ultrafast pulse propagates in a transparent material, the pulse moves at a slower velocity, accompanied with pulse broadening and pulse shape change, due to the material dispersion. The phase of an optical pulse propagating in a material over a distance L is given by

$$\phi(\omega, L) = \omega t - kL. \tag{2.152}$$

When there is no dispersion, the phase is written as $\phi(\omega, L) = \omega(t - L/v_p)$ where $v_p = \omega/k$ is the phase velocity. To consider the dispersion, we expand $k(\omega)$ in a Taylor series around ω_0 as

$$k(\omega) = k_0 + k_1 \cdot \Delta\omega + \frac{1}{2!}k_2 \cdot \Delta\omega^2 + \frac{1}{3!}k_3 \cdot \Delta\omega^3 + \frac{1}{4!}k_4 \cdot \Delta\omega^4 + \cdots, \tag{2.153}$$

where $\Delta\omega = \omega - \omega_0$, $k_0 = k(\omega_0)$ and

$$k_m = \left(\frac{\mathrm{d}^m k}{\mathrm{d}\omega^m}\right)_{\omega=\omega_0} \quad (m = 1, 2, \ldots). \tag{2.154}$$

Note that $k_m L$ $(m = 1, 2, \ldots)$ correspond to ϕ_m $(m = 1, 2, \ldots)$. Considering up to the second-order term of this expansion, and writing

$$k_1 = \frac{dk}{d\omega} \equiv \frac{1}{v_g}, \tag{2.155}$$

Equation (2.152) reads

$$\begin{aligned}\phi(\omega, L) &\cong (\omega_0 + \Delta\omega)t - \left[k_0 + k_1 \cdot \Delta\omega + \frac{1}{2}k_2 \cdot \Delta\omega^2\right]L \\ &= (\omega_0 + \Delta\omega)t - \left[\frac{\omega_0}{v_p} + \frac{1}{v_g}\Delta\omega + \frac{1}{2}k_2 \cdot \Delta\omega^2\right]L \\ &= \omega_0\left(t - \frac{L}{v_p}\right) + \Delta\omega\left(t - \frac{L}{v_g}\right) - \frac{1}{2}k_2 \cdot \Delta\omega^2 L. \end{aligned} \tag{2.156}$$

In the last equation, the first term shows that the optical wave at the carrier frequency ω_0 propagates with the phase velocity v_p. The second term shows that the pulse peak moves with the *group velocity* $v_g = d\omega/dk$, since the optical waves of all frequency components take the same phase at $t = L/v_g$, forming a pulse peak at $L = v_g t$. The time that the pulse has taken in propagating L with the group velocity v_g is called a *group delay* $T_g = L/v_g$.

Noting $k(\omega) = (\omega/c_0)n(\omega)$, the group velocity is written in terms of the refractive index $n(\omega)$ as

$$\frac{1}{v_g} = \frac{\mathrm{d}k}{\mathrm{d}\omega} = \frac{1}{c_0}\frac{\mathrm{d}(n\omega)}{\mathrm{d}\omega} = \frac{1}{c_0}\left(n + \omega\frac{\mathrm{d}n}{\mathrm{d}\omega}\right) = \frac{1}{c_0}\left(n - \lambda\frac{\mathrm{d}n}{\mathrm{d}\lambda}\right) \equiv \frac{n_g}{c_0}. \tag{2.157}$$

Here $n_g = c_0/v_g$ is called a *group velocity refractive index* (or *group index*). In the normal dispersion region, the group velocity is less than the phase velocity since $dn/d\omega > 0$.

The third term in (2.152), called *group velocity dispersion* (*GVD*) or *group delay dispersion* (*GDD*), describes the pulse broadening due to difference in the group velocities among the waves of different frequencies:

$$GVD = \frac{\mathrm{d}T_g}{\mathrm{d}\omega} = \frac{L}{c_0}\left(2\frac{\mathrm{d}n}{\mathrm{d}\omega} + \omega\frac{\mathrm{d}^2 n}{\mathrm{d}\omega^2}\right) = \frac{\lambda^3 L}{2\pi c_0^2}\frac{\mathrm{d}^2 n}{\mathrm{d}\lambda^2}. \tag{2.158}$$

The GVD is also expressed by the *group-velocity dispersion parameter D* defined by

$$D = \frac{d}{d\lambda}\left(\frac{1}{v_g}\right) = -\frac{\omega}{\lambda L}\frac{d^2\phi}{d\omega^2} = -\frac{\omega}{\lambda L}GVD. \tag{2.159}$$

The propagation distance over which the pulse broadens twice the original pulse width Δt_0 is called a *dispersion length* L_D. Referring to Eq. (2.149), this condition is given by $\Delta t_{\phi''} = 2\Delta t_0$, leading to

$$L_D = \frac{\sqrt{3}}{4\ln 2}\frac{(\Delta t)^2}{|k''|} = 0.62\frac{(\Delta t)^2}{|k''|}. \tag{2.160}$$

The fourth and the fifth terms in (2.153) correspond to the third-order dispersion (*TOD*) and the fourth-order dispersion (FOD), respectively, during propagation in the material:

$$TOD = \phi_3 = \frac{L}{c_0}\left(3\frac{d^2n}{d\omega^2} + \omega\frac{d^3n}{d\omega^3}\right) = \frac{-\lambda^4 L}{4\pi^2 c_0^3}\left(3\frac{d^2n}{d\lambda^2} + \lambda\frac{d^3n}{d\lambda^3}\right). \tag{2.161}$$

$$FOD = \phi_4 = \frac{\lambda^5 L}{8\pi^3 c_0^4}\left(12\frac{d^2n}{d\lambda^2} + 8\lambda\frac{d^3n}{d\lambda^3} + \lambda^2\frac{d^4n}{d\lambda^4}\right). \tag{2.162}$$

Due to the TOD, propagation of an ultrafast pulse is accompanied with the pulse asymmetry and additional broadening. We can introduce a dispersion length associated with the TOD as

$$L_D' = \frac{(\Delta t)^3}{|k'''|}. \tag{2.163}$$

As a numerical example, let us consider an ultrafast pulse of 800 nm wavelength propagating in a fused quartz, which has dispersion parameters of $T_g = 4900$ [fs/mm], $GVD = 40\left[\mathrm{fs}^2/\mathrm{mm}\right]$, and $TOD = 28\left[\mathrm{fs}^3/\mathrm{mm}\right]$. Noting that L_D and L'_D scale with $(\Delta t)^2$ and $(\Delta t)^3$ respectively, the characteristic lengths become $L_D =$ 1. 55 mm and $L'_D = 35.7$ mm for a 10-fs pulse, whereas $L_D = 14$ mm and $L'_D =$ 960 mm for a 30-fs pulse. This example suggests that the dispersion has to be well controlled, especially in generating an FTL pulse of shorter than ~ 30 fs.

2.3.4 Dispersion Properties of Basic Optical Elements

Dispersion compensation is required for controlling high-intensity ultrashort laser pulses. In this section, we describe dispersion properties of four optical elements often used in CPA lasers. A pair of prisms (Sect. 2.3.4.1) is used to generate a transform-limited pulse in the front-end of the CPA lasers, since it is capable of compensating TOD in addition to GVD. A pair of parallel diffraction gratings (Sect. 2.3.4.2) is used for pulse compression of a positively chirped pulse, whereas a pair of anti-parallel diffraction gratings (Sect. 2.3.4.3) is used for pulse stretching by generating a positive frequency chirp to an input ultrashort pulse. Instead of using a pair of diffraction gratings, pulse compression is also implemented with a chirped mirror (Sect. 2.3.4.4), which is more compact and has a higher damage threshold compared with the diffraction gratings. Detailed accounts of the dispersion compensation in high-power ultrafast laser systems are described in Chap. 7.

2.3.4.1 Dispersion Control with Prisms

With prisms and diffraction gratings, different frequency components are separated by angular dispersion. The angular dispersion is always associated with the two effects: pulse front tilt and negative group velocity dispersion. We first consider the tilt of the pulse front accompanied with the angular dispersion. The phase front (surface of constant phase) propagates with the phase velocity $v_p = \omega_0/k$ whereas the pulse front (surface of constant intensity) propagates with the group velocity $v_g = d\omega/dk$, which is slower than the phase velocity in normal dispersion materials.

When an optical pulse is refracted with a prism, the pulse front becomes tilted by an angle γ relative to the phase front due to difference in the phase velocity and the group velocity, as shown in Fig. 2.17a. The tilt angle is derived by calculating the time delay between the phase front and the pulse front as

$$\tan\gamma = \left(\frac{a}{b}\right)\lambda\frac{dn}{d\lambda}, \tag{2.164}$$

where a is the base length of the prism, b is the beam width after refraction, and n and $dn/d\lambda$ are the refractive index and its dispersion, respectively (Bor, 1985). Since dispersion of the refraction angle ε of a prism is (Born & Wolf, 1999, Sect. 5.7)

$$\frac{d\varepsilon}{d\lambda} = \left(\frac{a}{b}\right)\frac{dn}{d\lambda}, \tag{2.165}$$

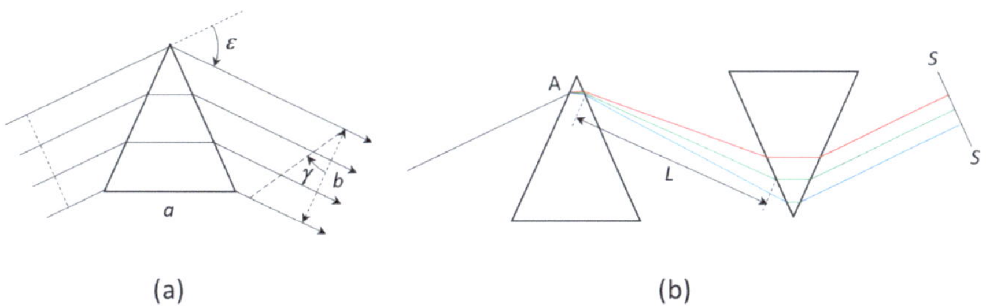

Fig. 2.17 Dispersion with prisms. **a** Pulse front tilt γ with respect to the phase front. **b** Angular dispersion and collimation with a pair of prisms, where the wavelength is indicated in color with red for the long wavelength

the tilt angle is expressed by the dispersion of the refraction angle as

$$\tan\gamma = \lambda\frac{d\varepsilon}{d\lambda}. \tag{2.166}$$

As described in Sect. 2.3.3, the frequency is up-chirped when GVD > 0 and down-chirped when GVD < 0. Since the angular dispersion is always accompanied with negative GVD (Martinez et al., 1984), optical elements with angular dispersion can be used to compensate for the dispersion of up-chirped optical waves.

Let us consider two identical isosceles prisms arranged in parallel with opposite directions as shown in Fig. 2.17b (Fork et al., 1984). The first prism refracts the rays of different frequency components at different angles, larger refraction at higher frequency, whereas the second prism collimates the rays to the same direction. The total spectral phase $\phi_m(\omega) = k(\omega)L$ is derived by integrating the phase over the entire path starting from the entrance point A to the final plane B normal to the phase front for each frequency component.

The group velocity dispersion GDV is obtained by the second derivative of the phase as $GVD = \phi_2$, which is expressed in terms of the wavelength as (Diels & Rudolph, 2006)

$$\phi_2 = \frac{\lambda_0^3}{2\pi c^2}\left[L_g\frac{d^2n}{d\lambda^2}\bigg|_{\lambda_0} - \left(4L + \frac{L_g}{n^3}\right)\left(\frac{dn}{d\lambda}\bigg|_{\lambda_0}\right)^2\right]. \tag{2.167}$$

In this equation, L_g is the physical length of the ray in the two prisms, L is the optical path length between the two prisms, and derivatives are taken at the central wavelength (or frequency) of the spectrum indicated by λ_0 (or ω_0). The first term in this equation is due to the material dispersion of the glass which is positive in normal dispersion region. The second term, which arises from the angular dispersion during propagation in vacuum (or air) and the glass, is negative irrespective of the sign of $dn/d\lambda$. Since $L \gg L_g$ in most cases, the second term is larger than the first term, resulting in the negative GVD.

The third order dispersion $TOD = \phi_3$, which affects the temporal pulse shape, is calculated by deriving GVD with ω. When the angular dispersion in the glass is neglected, the TOD for a Brewster angle prism is:

$$\phi_3 \approx \frac{\lambda_0^4}{4\pi^2c^3}\left\{12L\left[n'^2\left(1 - \lambda_0 n'(n^{-3} - 2n)\right) + \lambda_0 n' n''\right] - L_g(3n'' + \lambda_0 n''')\right\}, \tag{2.168}$$

where n', n'' and n''' are derivatives of n with respect to λ taken at λ_0.

It is possible to adjust GVD by translating one of the prisms perpendicular to its base, which changes the optical length in the glass. By changing the geometrical configuration and the glass material, the prism can be also used to adjust TOD which is crucial for generating a transform limited pulse of less than ~ 20 fs width. For the mode-locked Ti:sapphire and $CrLiSrAlF_6$ oscillators, it was shown that the round-trip GVD and TOD can be completely compensated with a pair of prisms in a cavity (Lemoff et al., 1992, 1993).

A transverse displacement of spectral components at the output plane can be avoided by returning the beam using a retroreflector at the *S–S* plane or by placing another set of two prisms in a configuration symmetric to the *S–S* plane.

2.3.4.2 Pulse Compression with Parallel Diffraction Gratings

In the chirped pulse amplification laser systems, a pair of parallel diffraction gratings shown by G_1 and G_2 in Fig. 2.18 is used for compression of a positively chirped time-expanded pulse to a short pulse (Treacy, 1969). A plane wave diffracted by G_1 propagates to G_2 with the phase front normal to the propagation direction. The G_1 diffracts different frequency components with larger diffraction angles (larger θ_d) for longer wavelengths (lower frequencies), whereas G_2 collimates the waves of different frequency components to the same direction. This device applies a negative frequency chirp as described below, thus compensating for a positive frequency chirp of the incident wave.

Let us consider a plane wave with frequency $\omega = \omega_0 + \Omega$, where ω_0 is a center frequency, which is diffracted at A and propagates to $P(x, z)$ where z-axis is in the propagation direction of the ω_0 wave and x-axis is normal to z. Writing the wave vector at ω as $\boldsymbol{k}(\omega)$, the phase shift from A to P is given by

$$\phi = \boldsymbol{k}(\omega) \cdot \overline{AP} = (\omega/c) \cdot [z \cos\alpha + x \sin\alpha]. \tag{2.169}$$

In this equation, α is the difference of the diffraction angles at ω and ω_0 which can be expanded in terms of Ω as

$$\begin{aligned}\alpha = \theta_d(\omega) - \theta_d(\omega_0) &= \left(\frac{\partial\theta_d}{\partial\omega}\right)_0 \cdot \Omega + \frac{1}{2}\left(\frac{\partial^2\theta_d}{\partial\omega^2}\right)_0 \cdot \Omega^2 + \frac{1}{6}\left(\frac{\partial^3\theta_d}{\partial\omega^3}\right)_0 \cdot \Omega^3 + \cdots \\ &\equiv \theta_{d1}\Omega + \frac{1}{2}\theta_{d2}\Omega^2 + \frac{1}{6}\theta_{d3}\Omega^3 + \cdots .\end{aligned} \tag{2.170}$$

Here derivatives are taken at $\omega = \omega_0$ and expressed in abbreviated notations in the last line. By expanding $\cos\alpha$ and $\sin\alpha$ in terms of α, the phase shift is written in a power series of Ω as

$$\begin{aligned}\phi(\omega) &= \frac{1}{c}(\omega_0 + \Omega) \cdot [z\cos\alpha + x\sin\alpha] \\ &\cong \frac{\omega_0}{c}z + \frac{1}{c}(z + \theta_{d1}\omega_0 x)\Omega + \frac{1}{2c}\left\{-(\theta_{d2})^2\omega_0 z + (2\theta_{d1} + \theta_{d2}\omega_0)x\right\}\Omega^2 \\ &\quad + \frac{1}{6c}\left\{-3\theta_{d1}[\theta_{d1} + \theta_{d2}\omega_0]z + \left[-(\theta_{d1})^3\omega_0 + 3\theta_{d2} + \theta_{d3}\omega_0\right]x\right\}\Omega^3 + \cdots .\end{aligned} \tag{2.171}$$

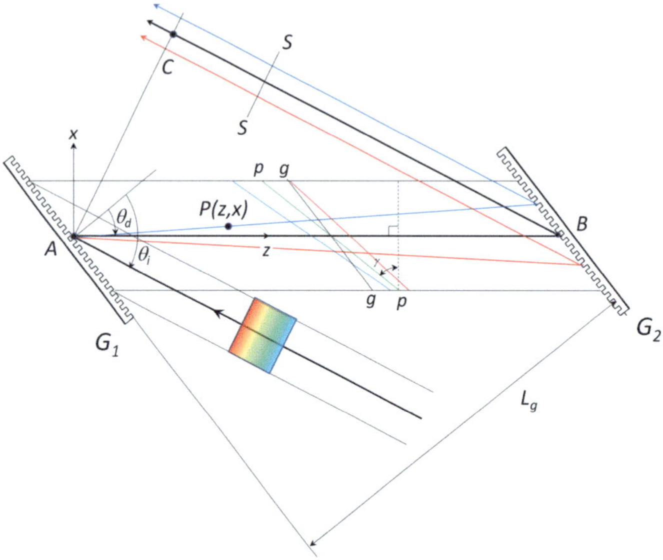

Fig. 2.18 A schematic of a pulse compressor with two parallel diffraction gratings G_1 and G_2, with a positive-chirped pulse incident to G_1 at an incidence angle θ_i. The diffraction angle at the center frequency ω_0 is denoted by θ_d, which is defined positive for the − 1 order. The lines *AB* and *BC* show the ray at ω_0, whereas the diffracted beams of $\Omega = \omega - \omega_0$ are shown in blue ($\Omega > 0$) and red ($\Omega < 0$), respectively. The *z*-coordinate is along *AB* and *x* is normal to *z*. The line *gg* is parallel to the grating surface along which the pulse width is constant. The pulse width decreases as it propagates due to compensation of the positive chirp by the negative GVD of the grating. The pulse front is shown by *pp* which is tilted by γ from the phase front. The pulse width in the pulse front is not uniform as indicated by the width of the pulse front

Noting the relation $\phi(\omega) = \phi_0 + \phi_1 \cdot \Omega + (1/2)\phi_2 \cdot \Omega^2 + (1/6)\phi_3 \cdot \Omega^3 + \cdots$, the coefficients of Ω, Ω^2 and Ω^3 correspond to ϕ_1, ϕ_2 and ϕ_3, respectively. Since the angular dispersion is derived as $\theta' = -2\pi c(\omega_0^2 d \cos\theta_D)$, ϕ_1 becomes

$$\phi_1 = \frac{1}{c}(z + \theta_{d1}\omega_0 x) = \frac{1}{c}\left(z - \frac{\lambda}{d\cos\theta_d}x\right). \tag{2.172}$$

Recalling that ϕ_1 corresponds to the pulse front delay as $\phi_1 = T_g = \overline{AP}/v_g$, where v_g is the group velocity, the above relation shows that the pulse front is delayed as $T_g = (1/c) \cdot [z - (\lambda/d\cos\theta_D)x]$. Thus, the pulse front is tilted with respect to the phase front by an angle γ, as shown in Fig. 2.17;

$$\tan\gamma = \frac{\lambda}{d\cos\theta_d}. \tag{2.173}$$

Since $d\theta_d/d\lambda = 1/(d\cos\theta_d)$, the tilt angle is expressed as $\tan\gamma = \lambda(d\theta_d/d\lambda)$, in the same form for a prism.

Noting that θ_{d2} is given by

$$\theta_{d2} = \frac{2\pi c}{d\omega^2 \cos\phi_d}\left(\frac{2}{\omega} + \tan\theta_d \cdot \theta_{d1}\right), \tag{2.174}$$

ϕ_2 is derived as

$$\begin{aligned}\phi_2 = \left.\frac{d^2\phi}{d\omega^2}\right|_{\omega_0} &= \frac{1}{c}\left[-(\theta_{d1})^2\omega_0 z + (2\theta_{d1} + \theta_{d2}\omega_0)x\right] \\ &= -\frac{\lambda_0^3}{2\pi c^2 d^2 \cos^2\theta_d}(z + x\tan\theta_d).\end{aligned} \tag{2.175}$$

Since $z + x\tan\theta_d$ corresponds to the grating surface, this equation shows that ϕ_2 is constant in the surface parallel to the grating and increases linearly with z.

As described in Sect. 2.3.2, the chirp rate a defined by $\omega = \omega_0 + at$ is related to ϕ_2 as $a = \phi_2/2$. Since the sign of ϕ_2 is negative, the grating dispersion applies a negative chirp to an incident pulse. When an incident pulse is positively chirped, the chirp is gradually compensated as the pulse propagates from A to B, eventually reaching to a transform limited short pulse (assuming that TOD and higher order terms are small) when the propagation distance L satisfies the following condition:

$$L = -\frac{4\pi c^2 d^2 \cos^2\theta_d}{\lambda^3}a. \tag{2.176}$$

We have shown that the pulse width of a transform limited pulse increases from Δt_0 to $\Delta t_{\phi''}$ as $\Delta t_{\phi_2}^2 = \Delta t_0^2 + [16(\ln 2)^2/\Delta t_0^2]\cdot\phi_2^2$ due to ϕ_2 (see Eq. 2.146). Alternatively, the pulse width of a long-duration positive-chirped laser pulse will decrease gradually as it propagates from A to B, resulting in a transform-limited short pulse at G_2 when the phase is perfectly compensated. The change of the pulse width is shown schematically in Fig. 2.18.

With reference to Fig. 2.18, we consider a beam starting at A on G_1, diffracted at B on G_2, and reaches to C. The total phase delay along the path is

$$\phi(\omega) = \frac{\omega}{c}\overline{ABC}(\omega) - 2\pi\frac{L_g\tan(\theta_d)}{d} = \frac{\omega L_p}{c}[1 + \cos(\theta_i + \theta_d)] - 2\pi\frac{L_g\tan(\theta_d)}{d}. \tag{2.177}$$

Here L_g is the distance between G_1 and G_2 normal to the grating surface and $L_p = L_g/\cos\theta_d$ is the optical path length between the gratings. (Note that θ_d is negative in Littrow configuration.) The second term on the right-hand-side of this equation, which is 2π times the number of grooves, arises from the 2π phase difference per groove at diffraction, compensating for the phase shift by the first grating.

From Eq. (2.169) and the diffraction formula, the group delay $T_g = d\phi/d\omega$ is calculated as

$$T_g = \phi_1 = \frac{L_p}{c}[1 + \cos(\theta_i + \theta_d)]. \tag{2.178}$$

Since this expression is equal to the first term of Eq. (2.174), this equation shows that the pulse front tilt is compensated by G_2, resulting in the pulse front parallel to the phase front.

The GVD is derived by differentiation of this equation:

$$GVD = \phi_2 = -\frac{4\pi^2 c}{\omega_0^3 d^2} \cdot \frac{L_p}{\cos^2\theta_d} = -\frac{\lambda_0^3}{2\pi c^2 d^2} \cdot \frac{L_p}{\cos^2\theta_d} \tag{2.179}$$

As expected, this GVD is equal to ϕ_2 with $z = L_p$. The third-order dispersion TOD and the fourth-order dispersion FOD are derived from GVD as

$$TOD = \phi_3 = -\phi_2 \frac{3\lambda_0}{2\pi c}\left(1 + \frac{\lambda_0}{d}\frac{\sin\theta_d}{\cos^2\theta_d}\right), \tag{2.180}$$

$$FOD = \phi_4 = \phi_2 \frac{3\lambda_0^2}{2\pi^2 c^2}\left(4 + 8\frac{\lambda_0}{d}\frac{\sin\theta_d}{\cos^2\theta_d}\right) + \frac{\lambda_0^2}{d^2}\left[1 + \tan^2\theta_d\left(6 + 5\tan^2\theta_d\right)\right]. \tag{2.181}$$

The signs of GVD, TOD and FOD of the compressor are negative, positive, and negative, respectively. From Eq. (2.178), the TOD can be adjusted by changing the grating constant and the incidence angle.

Since the spectral components are displaced on G_2, the optical wave near the edge of the grating does not contain whole spectral components, leading to pulse broadening. This spectral displacement can be compensated for by using another pair of diffraction gratings placed in a symmetric configuration after the first grating pair or using a retro reflector for returning the beam back through the same optical path. The dispersion is doubled in these configurations.

The misalignment tolerances in the grating pulse compressor at various bandwidths and beam sizes have been analyzed in detail by Webb et al. (2019). These works have treated the optical field as a plane wave. Since a Gaussian beam has a spherical wavefront, wave-optical description is necessary to accurately analyze the angular dispersion at the grating (Diels et al., 2006; Martinez, 1986).

2.3.4.3 Pulse Stretching with Anti-parallel Diffraction Gratings

A flexible layout using a 4f optics for pulse stretching and compression was developed by Martinez (1987). Referring to Fig. 2.19a, a 4f optics is composed of two lenses L_1 and L_2 of focal length f separated by $2f$. With this system, an image of an object placed at O with a distance f upstream of L_1 is formed with unit magnification $m = -1$ at I at a distance f downstream of L_2, where the negative sign is due to image reversion. With this configuration, the field distribution including amplitude and phase at the object plane O is reproduced at the image plane I. Therefore, the field distribution downstream of the image plane is calculated as if the 4f optics does not exist, except for the change of sign in the transverse coordinates (x and y). (This property is used in high power lasers to relay the laser field over a long distance from the front end to the final optics by using multi-stage image relay optics for minimizing the diffraction, as described in Sect. 4.2.)

For the z coordinate upstream of L_1, the sign is defined as positive upstream of O and negative between O and L_1. For L_2, the sign is defined as positive downstream of I and negative between I and L_2. With the 4f optics, an object at d_1 (>0) upstream of O forms its image with unit magnification at $d_1{}'$ (<0), where $d_1{}' = -\, d_1$. Alternatively, an object at d_1 (<0) forms its image with unit magnification at $d_1{}'$ (>0), where $d_1{}' = -\, d_1$.

Referring to Fig. 2.19b, when a diffraction grating G_1 is placed at $d_{G1}{}'$ (<0), its image $G_1{}'$ is formed at $d_{G1}{}'$ (>0), where $d_{G1}{}' = -\, d_{G1}$, with the inverted distribution of G_1. Let us consider that the second grating G_2 is placed at $d_{G2}{}'$ (<0) with an angle anti-parallel to G_1. Then various frequency components of the diffracted beam propagate parallel to each other. The spectral displacement of the diffracted beam can be compensated for by returning the beam back the same path using a retroreflector, for example with a roof mirror (RM), with which the dispersion is doubled.

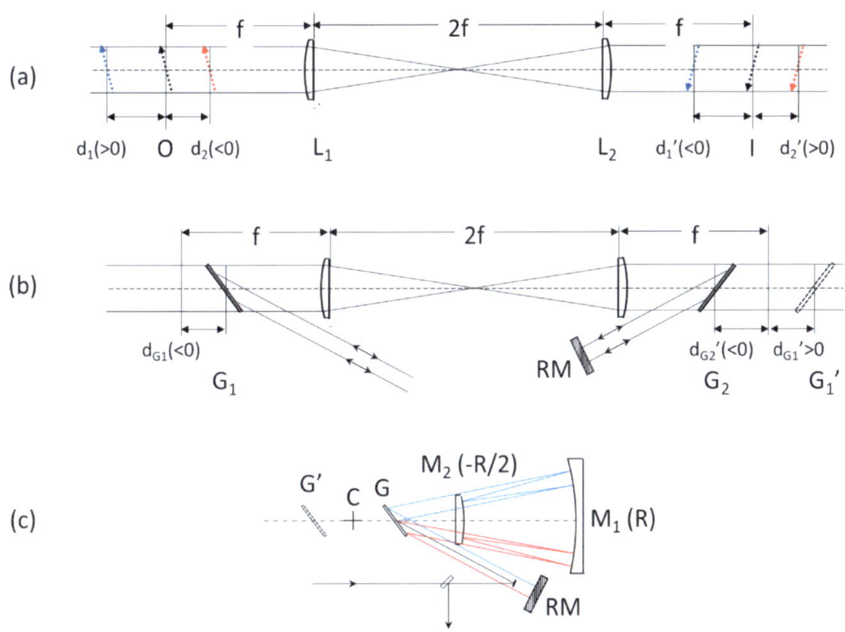

Fig. 2.19 Configurations of the pulse stretcher with 4f optics. **a** Imaging properties of a 4f optics. **b** Pulse stretcher using anti-parallel gratings with 4f optics. **c** Pulse stretcher based on an Öffner triplet telescope configuration. (The rays of long and short wavelengths are shown in red and blue, respectively)

The field at G_2 is obtained by reversing the field distribution upstream from $G_1{}'$ to G_2 by an effective distance $z_{eff} = d_{G1} + d'_{G2} = -(|d_{G1}| + |d'_{G2}|)$. Since the sign of the phase is reversed, the dispersion of this optical system for double path is given by

$$GVD = \phi_2 = -\frac{\lambda_0^3}{\pi c^2 d^2} \cdot \frac{z_{eff}}{\cos^2 \phi_d} = \frac{\lambda_0^3}{\pi c^2 d^2} \cdot \frac{|d_{G1}| + |d'_{G2}|}{\cos^2 \phi_d}. \tag{2.182}$$

Since GVD is positive in this case, due to the negative sign of z_{eff}, this layout can be used for pulse stretching, as demonstrated by Pessot et al. (1987). Alternatively, the GVD of this system becomes negative when G_2 is located downstream of F_2 and $z_{eff} = d_{G1} + d_{G2} > 0$ is satisfied.

To avoid the chromatic aberration of the 4f lenses, reflective pulse stretchers composed of cylindrical mirrors (Barty et al., 1993, 1994) and an axially symmetric Öffner triplet telescope (Cheriaux et al., 1996; Öffner, 1971) were developed. Referring to Fig. 2.19c, a combination of a concentric concave spherical mirror M_1 and a convex spherical mirror M_2 form an imaging optics of unit magnification. The distance between the grating G and its image G' corresponds to the effective distance z_{eff}. In this configuration, an incident pulse is dispersed by the grating G, reflected three times by M_1 and M_2, and diffracted again by G. The beam is retro-reflected by a roof mirror *RM* and returns in an elevated path for double-passing the stretcher to double the pulse expansion ratio. When the radius of curvature R_2 of M_2 is half that of M_1, i.e., $R_2 = -R_1/2$, spherical and astigmatism as well as coma and chromatic aberrations are canceled when the grating is located at C, the center of curvature of both mirrors. However spherical aberration remains when G is located away from C as shown in Fig. 2.19c. A general method to calculate the dispersion properties of complex and aberrated stretcher-compressors is given in Druon et al. (2008).

2.3.4.4 Dispersion Control with a Chirped Mirror

The multilayer design in Fig. 2.11 shows the wavelength dependence of the penetration depth in the multilayers, deeper for the longer wavelength. Using this property, a *chirped multilayer coating* was invented for dispersion control of ultra-short pulses (Szipöcs et al., 1994). With the chirped mirror, negative group-delay dispersion (GDD), i.e., smaller group delay at higher frequencies, is produced over broad frequencies, by gradually changing the thickness of the QW stacks along the depth from the surface.

An improved design of the chirped mirror, named *double-chirped mirror* (DCM), was invented to reduce oscillations in the group delay (by ~ 40 fs), which arises from interference of the reflections from the front and back sections of the mirror (Kärtner et al., 1997). This reflection was reduced by impedance-matching of the multilayer structures, by gradually changing the thickness of the high-index layers together with the period of the Bragg structures. With detailed analyses (Matuschek et al., 1999), DCMs with custom-tailored dispersion characteristics have been designed, with

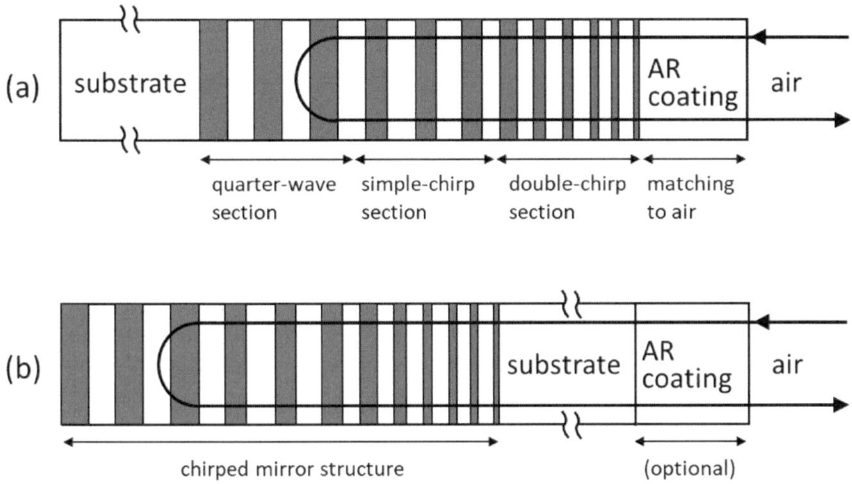

Fig. 2.20 Schematic layouts of **a** standard double-chirped mirror (DCM) and **b** back-side-coated (BASIC) chirped mirror, where light is incident from the right-hand side. In both configurations, the chirped mirror is composed of double-chirp section, simple-chirp section, and quarter-wave section. Reprinted with modification from Matuschek et al. (2000), with permission from Springer Nature; permission conveyed through Copyright Clearance Center, Inc. © 2000 Springer Nature

which higher-order dispersion components are compensated for over broad frequencies required for generating a few-cycle optical pulses. To reduce the remaining ~ 20 fs^2 GDD oscillations arising from interference with the anti-reflection coating on the front side, further improvement in DCM was achieved by coating the DCM on the backside surface of a substrate thus separating DCM from the anti-reflection coating, as shown in Fig. 2.20. With this design, the GDD oscillation as small as ~ 2 fs^2 has been realized over 220 THz bandwidth (Matuschek et al., 2000).

2.4 Relativity

The concept and technology based on relativity is indispensable for our daily life; for example, the Global Positioning System (GPS) maintains its high accuracy by correcting the difference in the elapsed time between the ground surface and the satellite. In high field science with high-intensity lasers, electrons (and ions) are accelerated close to the speed of light, leading to relativistic effects in the laser-matter and laser-vacuum interaction. The laser is also an important tool for investigating the gravitational effects: the "gravitational wave astronomy" has been opened with the laser interferometer, and a new field of "relativistic geodesy" has been opened with the atomic clock for measuring the gravitational frequency shift. In this section, basic equations of the relativity are introduced and a few examples of laser application to measure and use gravity are described. (For detailed accounts of relativity, refer to textbooks, for example, Landau and Lifshitz (1980).)

2.4.1 Special Theory of Relativity and Lorentz Transformation

The special theory of relativity is based on two postulates (Einstein, 1905a, 1905b):

(1) The laws of physics are identical in all inertial frames.
(2) The speed of light in free space is invariant for all observers in inertial frames.

The inertial frame refers to a system where an object without force upon it does not accelerate, i.e., an object remains at rest or moves at a constant velocity. Many important consequences are derived from these postulates. This theory is named "special" since this theory is applicable only to the inertial frame.

We consider two reference frames S and S′, with S′ moving with velocity V relative to S in the x-direction. Then from the assumption that the speed of light takes the same value c in S and S′, the following relations are derived between the spatial and time coordinates (x, y, z, t) in S and (x', y', z', t') in S′:

$$x' = \frac{x - Vt}{\left(1 - V^2/c^2\right)^{1/2}} = \gamma(x - Vt), \quad y' = y, \quad z' = z, \tag{2.183}$$

$$t' = \frac{t - \left(V/c^2\right)x}{\left(1 - V^2/c^2\right)^{1/2}} = \gamma[t - (\beta/c)x]. \tag{2.184}$$

Here β and γ are given by

$$\beta \equiv \frac{V}{c}, \tag{2.185}$$

$$\gamma \equiv \frac{1}{\left(1 - \beta^2\right)^{1/2}} \equiv \frac{1}{\left(1 - V^2/c^2\right)^{1/2}}. \tag{2.186}$$

These transformations between the inertial frames are called *Lorentz transformation*. We note that $0 \leq \beta \leq 1$ and $\gamma \geq 1$ since $0 \leq V \leq c$.

Many important consequences are derived from the Lorentz transformation. For example, the length L in frame S becomes shorter in the moving frame S′ as,

$$L' = \frac{L}{\gamma} = L\left(1 - \beta^2\right)^{1/2}. \tag{2.187}$$

Writing the time lapses in the frames S and S′ as $\tau = t_2 - t_1$ and $\tau' = t_2' - t_1'$, respectively, the time lapse becomes longer in S′ as,

$$\tau' = \gamma\tau = \frac{\tau}{\left(1-\beta^2\right)^{1/2}} \tag{2.188}$$

That is, the clock moves slower in the moving frame. The velocity is transformed as

$$v_x' = \frac{v_x - V}{1 - \beta v_x/c}, \tag{2.189}$$

$$v_{y,z}' = \frac{v_{y,z}}{\gamma(1 - \beta v_x/c)}. \tag{2.190}$$

In the limit of $V = 0$, $v_x' = v_x - V$, $v_y' = v_y$, and $v_z' = v_z$ as expected.

2.4.2 Relativistic Momentum and Energy

As postulated in (1), the physical laws take identical forms in all reference frames. For example, the momentum $\boldsymbol{p} = M_0\boldsymbol{v}$ in the rest frame is expressed as $\boldsymbol{p} = M\boldsymbol{v}$ in the moving frame when we describe $\boldsymbol{p}$ and $\boldsymbol{v}$ as the momentum and the velocity in the moving frame, respectively.

The momentum of a particle is written as

$$\boldsymbol{p} \equiv \frac{M_0\boldsymbol{v}}{\left(1 - v^2/c^2\right)^{1/2}}. \tag{2.191}$$

We may interpret this relation that the rest mass M_0 increases in the moving frame as

$$M(v) = \frac{M_0}{\left(1 - v^2/c^2\right)^{1/2}} = M_0\gamma. \tag{2.192}$$

As a numerical example, $\gamma = 2.3$ when $\beta = 0.9$, i.e., the mass increases by $2.3\times$ of the rest mass when it is accelerated to 90% of the light velocity.

When we define the total energy of a particle by $E \equiv M_0c^2\gamma$, and noting the relations $p^2 = M_0^2c^2\beta^2\gamma^2$ and $\gamma^2 - \beta^2\gamma^2 = 1$, E is described as:

$$E^2 - p^2c^2 = M_0^2c^4. \tag{2.193}$$

The total energy E thus reads:

$$E = \sqrt{M_0^2c^4 + p^2c^2}. \tag{2.194}$$

If the particle is at rest, the energy is given by $E = M_0c^2$; that is, the energy and the mass are equivalent. The stars in the universe keep their activities by converting the mass to energy via nuclear fusion reactions.

When $pc \ll Mc^2$, E is written as

$$E = M_0c^2 + \frac{1}{2}\frac{p^2}{M_0}. \tag{2.195}$$

The second term corresponds to the kinetic energy of the particle. When $pc \gg Mc^2$, which is valid for high energy particles, the energy is simply expressed as

$$E = cp \tag{2.196}$$

Since the momentum of a photon is given by $p = \hbar\omega/c = E/c$, (2.194) implies that the mass of the photon is zero.

2.4.3 General Theory of Relativity

Albert Einstein has developed the general theory of relativity which is applicable to any (general) frames of references, including the objects under acceleration (Einstein, 1916, 1920).

The general theory of relativity is based on the following two postulates:

(1) The laws of physics take identical forms in all coordinate systems.
(2) It is possible to counteract the gravitational force by an apparent force due to acceleration motion (Equivalence principle).

When a particle is placed in a gravitational field, its movement is described by the equation of motion:

$$m_I\boldsymbol{\alpha} = m_G\boldsymbol{g}. \tag{2.197}$$

where $\boldsymbol{\alpha}$ is the acceleration, m_I is the inertial mass, $\boldsymbol{g}$ is the gravitational acceleration, and m_G is the gravitational mass. It is not obvious that this equation is valid since the origins of m_I and m_G are different. The equivalence principle that the inertial mass is equal to the gravitational mass $m_I = m_G$ has been validated in the experiment by Eötvös et al. (1922) and more recently on a satellite with 10^{-15} precision (Touboult et al., 2022),

Einstein has developed the general theory to include the gravitational field generated by the distributed mass. This theory is described by the Einstein equation:

$$R_{\mu\nu} - \frac{1}{2}g_{\mu\nu}R + \Lambda g_{\mu\nu} = \frac{8\pi G}{c^4}T_{\mu\nu}. \tag{2.198}$$

The right-hand side of this equation is the energy–momentum tensor due to the mass distribution $T_{\mu\nu}$. The left-hand side of this equation describes the geometrical property of the space–time formed by the mass distribution given on the right-hand side of this equation. The property of the space–time is characterized by the metric tensor $g_{\mu\nu}$ which determines the distance between two adjacent locations in the 4-dimentional space–time coordinate. The curvature of spacetime is described by R. The term Λ, called cosmological constant, was added later by Einstein to keep the universe in a static state. The relation of the cosmological constant to the expanding universe found by Edwin Hubble in the astronomical observation (Hubble, 1929) is still an important subject in cosmology.

The Einstein equation shows that the spacetime is curved by the mass distribution and light is deflected from the straight path. This prediction was soon verified by Arthur Eddington in 1919 (Dyson et al., 1920), confirming the validity of the Einstien's general theory of relativity.

One of the consequences of the general theory of relativity is the gravitational time dilation. Time moves slowly when the clock is close to a massive body, compared with the clock placed far away. The time dilation due to the gravitational effect between the clock on earth and the clock in the GPS satellite is 38 μs in one day. High accuracy of the GPS is maintained by taking this time dilation into operation. The general theory of relativity has been accurately tested in the laboratory using two optical lattice atomic clocks of 18-digits precision (Takano et al., 2016) and in space using hydrogen masers (Herrmann et al., 2018). More recently, the gravitational shift was tested outside the laboratory at a broadcasting tower (634-m high Tokyo Skytree) with two transportable atomic clocks, one at the ground level and another at 450 m above the ground level (Takamoto et al., 2020). As shown in Fig. 2.21, the gravitational red shift of 21.18 Hz due to height difference has been accurately measured, demonstrating the feasibility of field applications such as monitoring spatiotemporal changes of geopotentials which might lead to volcano eruptions and earthquakes, by forming "Internet of Clocks".

2.4.4 *Gravitational Wave Telescope*

The gravitational wave generated by the change of mass distribution due to merger of two massive blackholes was measured for the first time in 2016 at the advanced LIGO with the laser interferometer shown in Fig. 2.22 (Abbot et al., 2016). A Michelson interferometer of 4 km arm length is adjusted for destructive interference, with which a signal is detected when the relative arm length changes at the arrival of the gravitational wave. The change in the interferometer distance due to slight displacement of the mirrors by the gravitational wave is given by

$$\Delta\xi_{GW} \approx 4 \times 10^{-18}\left(\frac{h}{10^{-21}}\right)\left(\frac{\xi_0}{4\,\mathrm{km}}\right)\ [\mathrm{m}], \tag{2.199}$$

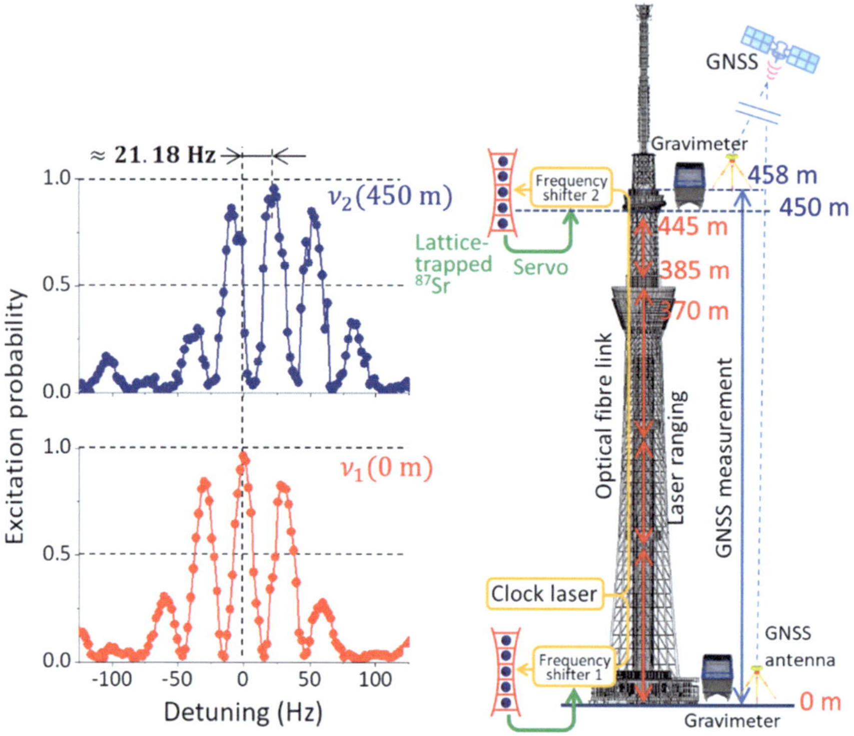

Fig. 2.21 Height measurement using two optical lattice clocks at 450 m height difference placed in the observatory floor of the Tokyo Skytree (ν_2) and the ground floor (ν_1), connected with an optical fiber link. The measured frequency difference of 21.18 Hz agrees well with the measurements by Global Navigation Satellite System (two GNSS antennas) and laser ranging (connecting four separate segments) complemented by spirit levelling and gravimeters, demonstrating the feasibility of the optical lattice clock for field applications. Courtesy of Prof. Hidetoshi Katori

where $\xi_0 \approx 4$ km is the baseline length of the interferometer and h is the amplitude of the gravitational wave. This displacement is typically smaller than the nuclear radius. A small change of $h \leq 0.1 \times 10^{-21}$ has been detected by reducing the optical losses of the optical components of the interferometer, increasing the laser power stored in the interferometer cavity, and reducing the mechanical and thermal noises. The measured amplitude and the time dependence of the gravitational wave were in very good agreement with the theoretical calculation for the two blackhole merger where a mass equivalent to 3 times the mass of the sun was released in 0.1 s as the gravitational wave.

Various projects on the gravitational wave telescope are in progress, in USA, Europe, and Japan, forming a network to locate the source of the gravitational wave. Furthermore, the space interferometer LISA (Laser Interferometer Space Antenna) with a triangular geometry of 2.5×10^6 km base length is planned to be launched in 2034 by The European Space Agency (ESA).

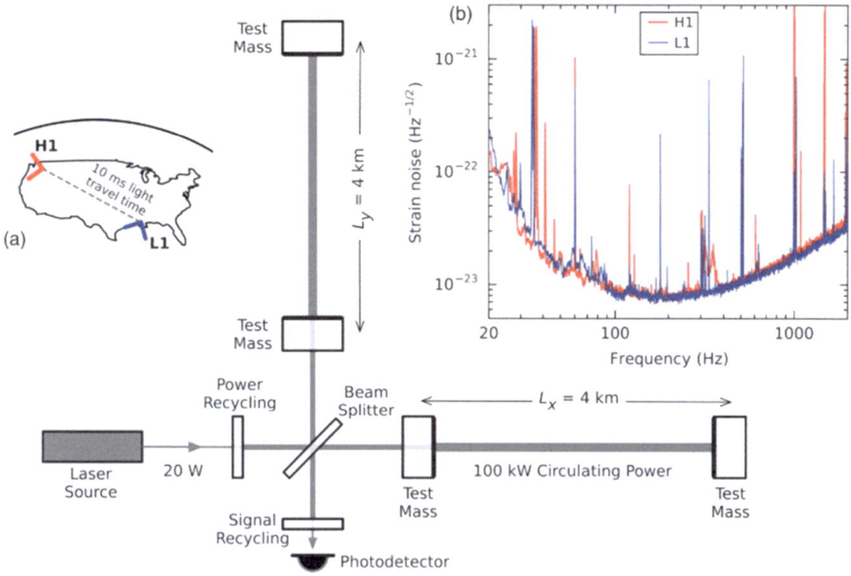

Fig. 2.22 Advanced LIGO gravitational wave detector, which is a Michelson interferometer consisting of two orthogonal arms of 4 km length, each with a high finesse interferometer to increase the laser power in each arm by 300 times. Together with the partially transmitting power-recycling-mirror which increases the power by 35 times, the circulating power in each arm reaches to 100 kW with 20 W incident power. The gravitational wave signals were detected by taking the coincidence between the outputs of the two detectors located in the west (H1) and east (L1) of the USA, as shown in (**a**). Recently, the strain signal noise shown in the inset (**b**) has been further reduced by 1.6–1.9 times by incorporating frequency-dependent squeezing in the detectors (Ganapathy et al., 2023). Reprinted from Abbott et al. (2016). © 2016 The American Physical Society

References

Abbott, B. P., et al. (2016). Observation of gravitational waves from a binary black hole merger. *Physical Review Letters, 116*, 061102. https://doi.org/10.1103/PhysRevLett.116.061102

Agrawal, G. P. (2019). *Nonlinear fiber optics* (6th ed.). Academic Press.

Aharonov, Y., & Bohm, D. (1959). Significance of electromagnetic potentials in the quantum theory. *Physical Review, 115*, 485.

Allen, L., Beijersbergen, M. W., Spreeuw, R. J. C., et al. (1992). Orbital angular momentum of light and the transformation of Laguerre–Gaussian laser modes. *Physical Review A, 45*, 8185.

Allen, L., & Padgett, M. J. (2000). The Poynting vector in Laguerre–Gaussian beams and the interpretation of their angular momentum density. *Optics Communications, 184*, 67.

Ando, T., Ohtake, Y., Matsumoto, N., et al. (2009). Mode purities of Laguerre–Gaussian beams generated via complex-amplitude modulation using phase-only spatial light modulators. *Optics Letters, 34*, 34.

Arlt, J., Dholakia, K., Allen, L., et al. (1998). The production of multi-ringed Laguerre–Gaussian modes by computer-generated holograms. *Journal of Modern Optics, 45*, 1231.
Attwood, D. (1999). *Soft X-rays and extreme ultraviolet radiation, principles and applications.* Cambridge University Press.
Barty, C. P. A., Gordon, C. L., III., & Lemoff, B. E. (1994). Multiterawatt 30-fs Ti:sapphire laser system. *Optics Letters, 19*, 1442.
Barty, C. P. J., Lemoff, B. E., & Gordon, C. L., III. (1993). Generation, measurement, and amplification of 20-fs high-peak-power pulses from a regeneratively initiated self-mode-locked Ti:sapphire laser. *Proceedings of SPIE, 1861*, 6.
Beijersbergen, M. W., Alen, L., van der Veen, H. E. L. O., et al. (1993). Astigmatic laser mode converters and transfer of orbital angular momentum. *Optics Communications, 96*, 123.
Beijersbergen, M. W., Coerwinkel, R. P. C., Kristensen, M., et al. (1994). Helical-wavefront laser beams produced with a spiral phaseplate. *Optics Communications, 112*, 321.
Beth, R. A. (1936). Mechanical detection and measurement of the angular momentum of light. *Physical Review, 50*, 115.
Bor, Z., & Rácz, B. (1985). Group velocity dispersion in prisms and its application to pulse compression and travelling-wave excitation. *Optics Communications, 54*, 165.
Born, M., & Wolf, E. (1999). *Principles of optics: Electromagnetic theory of propagation, interference and diffraction of light* (7th ed.). Cambridge University Press.
Chambaret, J. P., Le Blanc, C., Cheriaux, G., et al. (1996). Generation of 25-TW, 32-fs pulses at 10 Hz. *Optics Letters, 21*, 1921.
Diels, J.-C., & Rudolph, W. (2006). *Ultrashort laser pulse phenomena. Fundamentals, techniques, and applications on a femtosecond time scale* (2nd ed.). Elsevier.
Dmitriev, V. G., Gurzadyan, G. G., & Nikogosyan, D. N. (1999). *Handbook of nonlinear optical crystals* (2nd ed.). Springer.
Druon, F., Hanna, M., Lucas-Leclin, G., et al. (2008). Simple and general method to calculate the dispersion properties of complex and aberrated stretchers-compressors. *Journal of the Optical Society of America B, 25*, 754.
Dyson, F. W., Eddington, A. S., & Davidson, C. R. (1920). A determination of the deflection of light by the sun's gravitational field, from observations made at the solar eclipse of May 29, 1919. *Philosophical Transactions of the Royal Society A, 220*, 291.
Einstein, A. (1905a). Zur Elektrodynamik bewegter Körper. *Annalen der Physik, 4*(322), 891.
Einstein, A. (1905b). Ist die Trägheit eines Körpers von seinem Energieinhart abhängig? *Annalen der Physik, 4*(323), 639.
Einstein, A. (1916). Die Grundlage der allgemeinen Relativitätstheorie. *Annalen der Physik, 354*, 769.
Einstein, A. (1920). Relativity—The special and general theory. In R. W. Lawson (Trans.). Mathen & Co. (2020 edition by Quaternion).
Eötvös, L., Pekár, D., & Fekete, E. (1922). Beiträge zum Gesetz der Proportionalität von Trägheit and Gravität (Contributions to the law of proportionality of inertia and gravity). *Annalen der Physik, 373*, 11.
Fork, R. L., Martinez, O. E., & Gordon, J. P. (1984). Negative dispersion using pairs of prisms. *Optics Letters, 9*, 150.
Fox, A. G., & Li, T. (1961). Resonant modes in a maser interferometer. *The Bell System Technical Journal, 40*, 453.
Ganapathy, D., Jia, W., Nakano, M., et al. (2023). Broadband quantum enhancement of the LIGO detectors with frequency-dependent squeezing. *Physical Review X, 13*, 041021.

Hadjisolomou, P., Jeong, T. M., Valenta, P., et al. (2021). Gamma-ray flash generation in irradiating a thin foil target by a single-cycle tightly focused extreme power laser pulse. *Physical Review E, 104*, 015203.
Herrmann, S., Finke, F., Lülf, M., et al. (2018). Gravitational redshift test using eccentric Galileo satellites. *Physical Review Letters, 121*, 231101.
Hubble, E. (1929). A relation between distance and radial velocity among extra-galactic nebulae. *Proceedings of the National Academy of Sciences, 15*, 168.
Jackson, J. D. (1999). *Classical electrodynamics* (3rd ed.). Wiley.
Jiang, C., Wang, W. P., Weber, S., et al. (2021). Direct acceleration of an annular attosecond electron slice driven by near-infrared Laguerre–Gaussian laser. *High Power Laser Science and Engineering, 9*, e44.
Kärtner, F. X., Matuschek, N., Schibli, T., Keller, U., et al. (1997). Design and fabrication of double-chirped mirrors. *Optics Letters, 22*, 831.
Kimura, S., & Izawa, T. (2007). Dielectric multilayer mirrors for high power lasers. *Review of Laser Engineering, 35*, 142. (in Japanese).
Kramers, H. A. (1927). La diffusion de la lumiere par les atomes. *Atti del Congresso Internazionale di Fisica (Transactions of Volta Centenary Congress) Como, 2*, 545.
Kraus, M., Ahmed, M. A., Michalowski, A., et al. (2010). Microdrilling in steel using ultrashort pulsed laser beams with radial and azimuthal polarization. *Optics Express, 18*, 22305.
Kronig, R. de L. (1926). On the theory of the dispersion of X-rays. *Journal of the Optical Society of America, 12*, 547.
Kuga, T., Torii, Y., Shiokawa, N., et al. (1997). Novel optical trap of atoms with a doughnut beam. *Physical Review Letters, 78*, 4713.
Landau, L. D., & Lifshitz, E. M. (1980). *The classical theory of fields* (4th ed.). Elsevier.
Lemoff, B. E., & Barty, C. P. A. (1992). Generation of high-peak-power 20-fs pulses from a regeneratively initiated, self-mode-locked Ti:sapphire laser. *Optics Letters, 17*, 1367.
Lemoff, B. E., & Barty, C. P. A. (1993). Cubic-phase-free dispersion compensation in solid-state ultrashort-pulse lasers. *Optics Letters, 18*, 57.
Loescher, A., Rocker, C., Graf, T., et al. (2021). Azimuthally polarized picosecond vector beam with 1.7 kW of average output power. *Optics Letters, 46*, 3492.
Machavariani, G., Lumer, Y., Moshe, I., et al. (2007). Efficient extracavity generation of radially and azimuthally polarized beams. *Optics Letters, 32*, 1468.
Marrucci, L., Manzo, C., & Paparo, D. (2006). Optical spin-to-orbital angular momentum conversion in inhomogeneous anisotropic media. *Physical Review Letters, 96*, 163905.
Martinez, O. E. (1986). Grating and prism compressors in the case of finite bean size. *Journal of the Optical Society of America B, 3*, 929.
Martinez, O. E. (1987). 3000 times grating compressor with positive group velocity dispersion: application to fiber compensation in 1.3–1.6 μm region. *IEEE Journal of Quantum Electronics, 23*, 59.
Martinez, O. E., Gordon, J. P., & Fork, R. L. (1984). Negative group-velocity dispersion using refraction. *Journal of the Optical Society of America A, 1*, 1003.
Martins, J. L., Vieira, J., Ferri, J., et al. (2019). Radiation emission in laser-wakefields driven by structured laser pulses with orbital angular momentum. *Scientific Reports, 9*, 9840.
Matuschek, N., Gallmann, L., Sutter, D. H., Steinmeyer, G., & Keller, U. (2000). Back-side-coated chirped mirrors with ultra-smooth broadband dispersion characteristics. *Applied Physics B, 71*, 509.
Matuschek, N., Kärtner, F. X., & Keller, U. (1999). Analytical design of double-chirped mirrors with custom-tailored dispersion characteristics. *IEEE Journal of Quantum Electronics, 35*, 129.
McMullen, J. D. (1977). Chirped-pulse compression in strongly dispersive media. *Journal of the Optical Society of America, 67*, 1575.
Miyagi, M., & Nishida, S. (1979). Pulse spreading in a single-mode fiber due to third-order dispersion. *Applied Optics, 18*, 678.

Moharam, M. G., Grann, E. B., Pommet, D. A., et al. (1995). Formulation for stable and efficient implementation of the rigorous coupled-wave analysis of binary gratings. *Journal of the Optical Society of America A, 12*, 1068.
Morita, R. (2016). Generation of optical vortices and their application. *Oyo-Butsuri, 85*, 396. (in Japanese).
Nuter, R., Korneev, P., Thiele, I., et al. (2018). Plasma solenoid driven by a laser beam carrying an orbital angular momentum. *Physical Review E, 98*, 033211.
Öffner, A. (1971). Unit power imaging catoptric anastigmat. U.S. patent 3,748,015.
Padgett, M. J. (2017). Orbital angular momentum 25 years on (invited). *Optics Express, 25*, 11265.
Padgett, M. J., & Allen, L. (1995). The Poynting vector in Laguerre–Gaussian laser modes. *Optics Communications, 121*, 36.
Paterson, L., MacDonald, M. P., Arlt, J., et al. (2001). Controlled rotation of optically trapped microscopic particles. *Science, 292*, 912.
Pessot, M., Maine, P., & Mourou, G. (1987). 1000 times expansion/compression of optical pulses for chirped pulse amplification. *Optics Communications, 62*, 419.
Sargent, M., III., Scully, M. O., & Lamb, W. E., Jr. (1978). *Laser physics*. Westview Press.
Skoulas, E., Manousaki, A., Fotakis, C., et al. (2017). Biomimetic surface structuring using cylindrical vector femtosecond laser beams. *Scientific Reports, 7*, 45144. https://doi.org/10.1038/srep45114
Smith, J. O. (2007). *Mathematics of the discrete Fourier transform (DFT) with audio applications* (2nd ed., online book, 2007 edition). http://ccrma.stanford.edu/~jos/mdft/
Strickland, D., & Mourou, G. (1985). Compression of amplified chirped optical pulses. *Optics Communications, 56*, 219.
Sueda, K., Miyaji, G., Miyanaga, N., et al. (2004). Laguerre–Gaussian beam generated with a multilevel spiral phase plate for high intensity laser pulses. *Optics Express, 12*, 3548.
Szipöcs, R., Ferencz, K., Spielmann, C., & Krausz, F. (1994). Chirped multilayer coatings for broadband dispersion control in femtosecond lasers. *Optics Letters, 19*, 201.
Takamoto, M., et al. (2020). Test of general relativity by a pair of transportable optical lattice clocks. *Nature Photonics, 14*, 411.
Takano, T., et al. (2016). Geopotential measurements with synchronously linked optical lattice clocks. *Nature Photonics, 10*, 662.
Tokizane, Y., Oka, K., & Morita, R. (2009). Supercontinuum optical vortex pulse generation without spatial or topological-charge dispersion. *Optics Express, 17*, 14517. https://doi.org/10.1364/OE.17.014517
Toll, J. S. (1956). Causality and the dispersion relation: Logical foundations. *Physical Review, 104*, 1760.
Tonomura, A. (1993). *Electron holography*. Springer.
Touboult, P., Metris, G., Rodrigues, M, et al. (2022). MICROSCOPE mission: Final results of the test of the equivalence principle. *Physical Review Letters, 129*, 121102.
Toyoda, K., Miyamoto, K., Aoki, N., et al. (2012). Using optical vortex to control the chirality of twisted metal nanostructures. *Nano Letters, 12*, 3645. https://doi.org/10.1021/nl301347j
Treacy, E. B. (1969). Optical pulse compression with diffraction gratings. *IEEE Journal of Quantum Electronics, 5*, 454.
Vieira, J., Mendonca, J. T., & Quere, F. (2018). Optical control of the topology of laser-plasma accelerators. *Physical Review Letters, 121*, 054801.
Webb, B., Guardalben, M. J., Dorrer, C., et al. (2019). Simulation of grating compressor misalignment tolerances and mitigation strategies for chirped-pulse-amplification systems of varying bandwidths and beam sizes. *Applied Optics, 58*, 234.

Wollenhaupt, M., Assion, A., & Baumert, T. (2007). Femtosecond laser pulses: Liner properties, manipulation, generation and measurement (Chap. 12). In F. Träger (Ed.), *Springer handbook of lasers and optics* (pp. 937–986). Springer.
Yamane, K., Toda, Y., & Morita, R. (2012). Ultrashort optical-vortex pulse generation in few-cycle regime. *Optics Express, 20*, 18986.
Yao, A. M., & Padgett, M. J. (2011). Orbital angular momentum: Origins, behavior and applications. *Advances in Optics and Photonics, 3*, 161.
Yariv, A. (1989). *Quantum electronics* (3rd ed.). Wiley.

Chapter 3
Light Amplification by Stimulated Emission of Radiation

Abstract LASER is an abbreviation of Light Amplification by Stimulated Emission of Radiation. The research on maser and laser was named "quantum electronics", expressing the concept that these are new electronic amplifiers in the microwave and optical frequency regions which are enabled by the quantum–mechanical concept of stimulated emission of radiation. Although we can describe laser light as a superposition of coherent electromagnetic waves which are formalized by the electromagnetic theory, the quantized photons play essential roles, for example, in the spontaneous emission which starts the laser action, creating entangled photon pairs for quantum teleportation, and reducing the quantum noise of the detectors. In this chapter, the concepts of photons and quantum mechanics are briefly introduced, followed by quantization of the optical field. Then absorption, spontaneous emission, and stimulated emission of photons is described. Finally, the gain coefficient for light amplification is derived as the basic parameter in laser operation.

Keywords Laser · Photon · Photoelectric effect · Compton scattering · Blackbody radiation · Schrödinger equation · Stimulated emission · Spontaneous emission · Gain coefficient · Transition probability

3.1 From Waves to Particles

3.1.1 Photon

A closed cavity which confines radiation at all frequencies is called a *blackbody*. Radiation emitted from a small hole of a blackbody is termed as *blackbody radiation*. Due to thermal equilibrium between the cavity wall and the radiation inside the cavity, the spectrum of the blackbody radiation is determined only by the temperature T. In the nineteenth century (and even today), measuring very high temperature inside a blast furnace was very important in the steel industry for producing high quality steels. The Rayleigh–Jeans formula describing the spectrum of the blackbody radiation, derived from the electromagnetic theory, agreed with the experimental observation

Y. Kato et al., *High Power Laser and Plasma Science*, Springer Series on Atomic, Optical, and Plasma Physics 130, https://doi.org/10.1007/978-3-032-08930-4_3

only at long wavelength regions. By making thorough consideration, Max Planck has derived the Planck's law which was consistent with the observed spectrum of blackbody radiation over the whole spectral region.

The electromagnetic wave confined in a cavity is characterized by its *mode*, which specifies polarization, propagation direction, frequency, and phase. The energy density of the electromagnetic field in a cavity is derived by multiplying the average energy of the mode with the number of modes in a unit volume (mode density) and integrating over possible modes. Consider an electromagnetic wave in a linear cavity of length L. With a periodic boundary condition of $\boldsymbol{E}(x = 0) = \boldsymbol{E}(x = L)$ where $n\lambda = L$ (n is an integer), the wavenumber $k = 2\pi/\lambda$ takes discrete values as $k = (2\pi/L)n$. More generally, when an electromagnetic wave is confined in a cubic cavity of length L, the x, y, z components of the wave vector $\boldsymbol{k}$ (k_x, k_y, k_z) are written as

$$k_x = \frac{2\pi}{L}n_x, \quad k_y = \frac{2\pi}{L}n_y, \quad k_z = \frac{2\pi}{L}n_z. \quad (n_{x,y,z} = 0, \pm 1, \pm 2, \ldots) \tag{3.1}$$

Any field in the cavity is described as a superposition of the cavity modes which are specified by the mode number (n_x, n_y, n_z) and the polarization. Since $k = (2\pi/L)\sqrt{n_x^2 + n_y^2 + n_z^2}$, the number of modes between k and $k + dk$ is obtained by calculating the numbers of (n_x, n_y, n_z) contained in a 1/8-th of the spherical shell of radius k and the thickness dk as

$$N_{k\sim k+dk}dk = 2 \times 4\pi \left(\frac{kL}{2\pi}\right)^2 \left(\frac{L}{2\pi}\right) dk = \frac{L^3}{\pi^2}k^2 dk, \tag{3.2}$$

where the factor 2 arises from two polarizations of the electromagnetic field. This expression can be rewritten in terms of the angular frequency ω as

$$N_{\omega\sim\omega+d\omega}d\omega = \frac{L^3}{\pi^2 c^3}\omega^2 d\omega. \tag{3.3}$$

By treating the electromagnetic wave as a harmonic oscillator, Lord Rayleigh has applied the law of equipartition of energy, which states that the average energy of each harmonic oscillator is $k_\mathrm{B}T$ at thermal equilibrium, where k_B is the Boltzmann constant. The energy density of the electromagnetic field is then given by

$$E_{\omega\sim\omega+d\omega}d\omega = \frac{\omega^2}{\pi^2 c^3}k_\mathrm{B}T d\omega. \tag{3.4}$$

This equation, called *Rayleigh–Jeans formula*, agreed with the observed blackbody radiation at low frequencies, but was not valid to all frequencies since it diverges at high frequencies.

Instead of using $k_\mathrm{B}T$ for the average energy of each mode of the electromagnetic field, Max Planck has assumed that the energy of the electromagnetic field takes discrete values with a unit of $\hbar\omega$:

$$\varepsilon_i = n_i \hbar\omega, \tag{3.5}$$

where n_i is a positive integer and $\hbar = h/2\pi$ is the reduced Planck constant. This assumption is different from the electromagnetic theory where the energy density depends only on the field amplitude as shown in Eq. (2.24).

The probability that this mode takes the energy ε_i is given by the Boltzmann distribution,

$$N_i = C\exp\left(-\frac{\varepsilon_i}{k_B T}\right), \tag{3.6}$$

where k_B is the Boltzmann constant and C is a normalization factor.

Then the average energy of the electromagnetic field with frequency ω is

$$\bar{\varepsilon} = \frac{\sum_{i=0}^{\infty} N_i \varepsilon_i}{\sum_{i=0}^{\infty} N_i} = \frac{\hbar\omega \sum_{n_i=0}^{\infty} n_i \exp\left(-\frac{n_i\hbar\omega}{kT}\right)}{\sum_{n_i=0}^{\infty} \exp\left(-\frac{n_i\hbar\omega}{kT}\right)} = \frac{\hbar\omega}{\exp\left(\frac{\hbar\omega}{kT}\right) - 1}. \tag{3.7}$$

Multiplying this average energy with the number density of the modes of frequency ω, the energy density of radiation reads

$$U(\omega, T) = \frac{\hbar\omega^3}{\pi c^3} \frac{1}{e^{\hbar\omega/k_\mathrm{B}T} - 1}. \tag{3.8}$$

This equation is called *Planck formula*.

Since the blackbody radiation is emitted to all directions with the speed of light c, the spectral radiance [W/(sr m^2)] is

$$I(\omega, T) = \frac{c}{4\pi} U(\omega, T) = \frac{\hbar\omega^3}{4\pi^2 c^2} \frac{1}{e^{\hbar\omega/k_\mathrm{B}T} - 1}. \tag{3.9}$$

Integrating Eq. (3.9) with ω and using the integration formula $\int_0^\infty \left\{x^3/(e^x - 1)\right\}dx = \pi^4/15$, the radiance emittance, i.e., the radiation power emitted per unit area [W/m^2] is derived as

$$E(T) = \int_0^\infty I(\omega, T)d\omega = \frac{\hbar}{4\pi^2 c^2} \int_0^\infty \frac{\omega^3}{e^{\hbar\omega/k_\mathrm{B}T} - 1} d\omega = \sigma T^4, \tag{3.10}$$

where $\sigma = 5.670 \times 10^{-8}$ W/m^2 K^4 is the *Stefan–Boltzmann constant*:

$$\sigma = \frac{\pi^2 k_\mathrm{B}^4}{60 c^2 \hbar^3} = \frac{2\pi^5 k_\mathrm{B}^4}{15 c^2 h^3}. \tag{3.11}$$

As seen in (3.11), the radiance emittance $E(T)$ increases with T^4.

The radiant spectral brightness is written in terms of the wavelength λ as

$$U(\lambda, T)d\lambda = \frac{16\pi^2\hbar c}{\lambda^5}\frac{1}{e^{2\pi\hbar c/\lambda k_B T} - 1}d\lambda. \tag{3.12}$$

The wavelength λ_{max} where the spectral distribution $U(\lambda, T)$ takes the maximum value is obtained by taking the derivative of Eq. (3.12) as

$$\lambda_{max} = \frac{b}{T}, \tag{3.13}$$

where $b = 2.897729 \times 10^{-3}$ mK. This relation is known as *Wien's displacement law*. The radiant spectral brightness $U(\lambda, T)$ at several temperatures is shown in Fig. 3.1. With this formula, we can estimate the temperature of the material if the emitting surface is close to the blackbody. For example, λ_{max} is 9.66 μm, 2.90 μm, 1.61 μm, and 483 nm at 300 K (room temperature), 1000 K (melting temperature of aluminum), 1800 K (melting temperature of ion), and 6000 K (the surface temperature of the sun), respectively.

Albert Einstein has extended Planck's assumption to the experimental observation of the photo-electric effect, proposing that the electromagnetic wave is composed of photons, each with the energy given by $\varepsilon = \hbar\omega$. In analogy with the electromagnetic theory where the electromagnetic wave has a momentum $p = u/c$ as shown in Eq. (2.27), each photon is expected to have a momentum given by

$$p = \varepsilon/c = \hbar\omega/c = \hbar k. \tag{3.14}$$

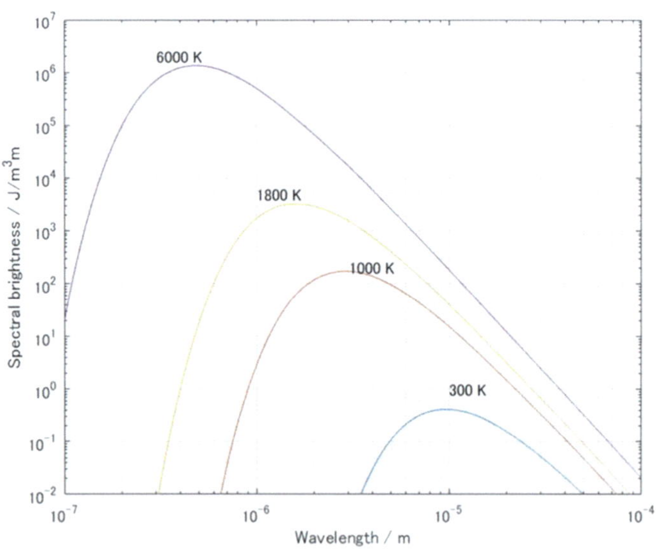

Fig. 3.1 Radiant spectral brightness of the blackbody radiation at temperatures of 300, 1000, 1800 and 6000 K

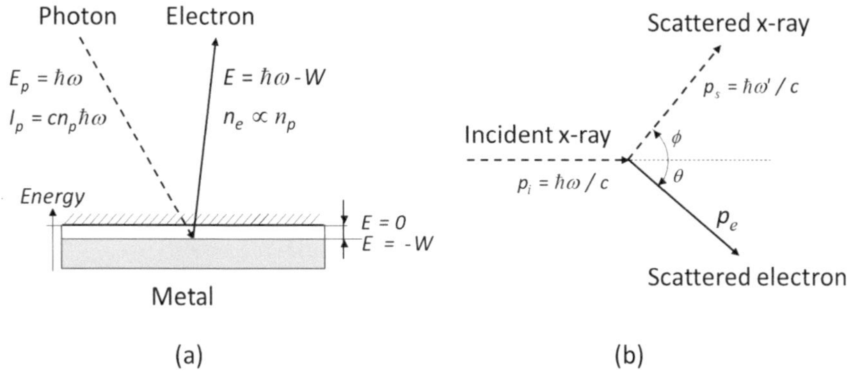

Fig. 3.2 Experiments demonstrating the quantum property of light: **a** photoelectric effect on photon energy, and **b** Compton effect on photon momentum

When the radiation is composed of many modes which we denote by α, the energy density u, the momentum $\boldsymbol{P}$ and the intensity I of the radiation are given respectively as,

$$u = \sum_{\alpha} n_{\alpha} \hbar\omega_{\alpha}, \tag{3.15}$$

$$\boldsymbol{P} = \sum_{\alpha} n_{\alpha} \hbar\boldsymbol{k}_{\alpha}, \tag{3.16}$$

$$I = \sum_{\alpha} n_{\alpha} c\hbar\omega_{\alpha}, \tag{3.17}$$

where summation is taken over the modes α. The momentum of photon given by (3.16) was verified by A. Compton in 1922 in the experiment on inelastic scattering of X-rays by the electrons in crystals. In this experiment called *Compton effect*, the angular dependence of the electron energy and momentum was clearly explained by the momentum conservation of the photon and the electron (Fig. 3.2).

3.1.2 *Quantum Mechanics*

3.1.2.1 Wave and Particle

The wave-particle duality of radiation has inspired Louis de Broglie to propose a hypothesis that the particles such as an electron may have the wave property. In analogy to the relation between the momentum and the wavelength of a photon shown in Eq. (3.14), it was postulated that an electron moving with the velocity v will show the wave property with its wavelength given by

$$\lambda = 2\pi\hbar/p = 2\pi\hbar/m_e v, \tag{3.18}$$

where m_e is the electron mass and $p = m_e v = m_e(dx/dt)$ is the momentum of an electron. This relation was demonstrated by de Broglie in the electron diffraction experiment in 1924, proving the particle-wave duality of the electron.

In parallel with the quantum theory of the electromagnetic wave by A. Einstein, Niels Bohr has established a new model of an atom for explaining the spectrum of the hydrogen atom observed by J. Balmer, known as Balmer series:

$$\frac{1}{\lambda} = R_\infty\left(\frac{1}{2^2} - \frac{1}{n^2}\right) \quad \text{for } n = 3, 4, 5, \ldots, \tag{3.19}$$

where λ is the wavelength of the absorption or emission lines and R_∞ is the Rydberg constant:

$$R_\infty = \frac{m_e e^4}{64\pi^3\varepsilon_0^2\hbar^3 c}. \tag{3.20}$$

Bohr has postulated that an electron orbiting around the nucleus takes only discrete angular momentum:

$$m_e v r = n\pi\hbar, \quad (n = 1, 2, 3, \ldots), \tag{3.21}$$

where r is the orbit radius and v is the velocity. He has also postulated that radiation is emitted when an electron makes a transition from a high energy state E_2 to a lower energy state E_1 with the frequency given by

$$\hbar\omega = \Delta E = E_2 - E_1, \tag{3.22}$$

where ΔE is the energy difference between the higher state and the lower state. The Bohr's model worked well for analyzing the spectra of the hydrogen atom, but it was not satisfactory for analyzing complex spectra of various atoms.

3.1.2.2 Schrödinger Wave Equation

The quantum theory was developed during 1925–1930 by extensive works of W. K. Heisenberg, E. Schrödinger, P. Dirac, and many others. The concept of quantum mechanics is expressed in the following equation:

$$H|\psi\rangle = E|\psi\rangle. \tag{3.23}$$

Here $|\psi\rangle$ describes the state of the system (such as atoms and photons), where $| >$ is called "ket" and its complex conjugate $< |$ is called "bra" (Dirac, 1988). The *wavefunction* ψ describes the wave nature of the system. $\hat{H}$ is the *operator* which corresponds to the action to measure a physical quantity of this state, and E is the quantity that is obtained in this measurement. In this example, $\hat{H}$ is the operator of the energy (Hamiltonian) and E is the energy of this system. The merit of the quantum mechanics is that the expectation value of any physical quantity is calculated by using the operator of this quantity once the wave function of the state is determined.

When ψ is composed of several states ψ_i which takes an energy E_i, ψ is expressed as a linear combination of these states:

$$|\psi\rangle = c_1|\psi_1\rangle + c_2|\psi_2\rangle + \cdots + c_N|\psi_N\rangle. \tag{3.24}$$

Here it is assumed that all wavefunctions such as ψ and ψ_i are normalized:

$$\left|\int d\boldsymbol{r}\psi^*(\boldsymbol{r}) \cdot \psi(\boldsymbol{r})\right|^2 \equiv |\langle\psi \mid \psi\rangle|^2 = |\langle\psi_i \mid \psi_i\rangle|^2 = 1. \tag{3.25}$$

The quantum mechanics is based on the *probabilistic interpretatio*n, where the probability P_ϕ that the state $|\phi\rangle$ is found in $|\Psi\rangle$ is given by

$$P_\phi = |\langle\phi \mid \psi\rangle|^2 = \left|\int d\boldsymbol{r}\phi^*(\boldsymbol{r}) \cdot \psi(\boldsymbol{r})\right|^2 \tag{3.26}$$

For example, the probability that energy E_i is measured in the system ψ given by Eq. (3.23) is obtained as

$$P_i = |\langle\psi_i \mid \psi\rangle|^2 = \left|c_j\right|^2. \tag{3.27}$$

When the wavefunction ψ_i is written as a function of the spatial coordinate of the electron $\boldsymbol{r}$, the spatial distribution of the electron is

$$P(\boldsymbol{r})_i = |\psi_i(\boldsymbol{r})|^2 \tag{3.28}$$

When two physical quantities are not independent of each other, i.e., they cannot be measured simultaneously without affecting each other, they are called non-exchangeable. For example, the position x and the momentum p of a particle are non-exchangeable. The operators of these quantities, x and p for example, satisfy the following *commutation relation*:

$$\left[x, p\right] \equiv xp - px = i\hbar. \tag{3.29}$$

Determination of x and p are accompanied with the uncertainties Δx and Δp (standard deviations), called *uncertainty relation*, given by

$$\Delta x \Delta p \geq \frac{\hbar}{2}. \tag{3.30}$$

When a particle propagates in the x-direction, the wavefunction of the particle takes the following form:

$$\Psi(x,t) = C\exp\left[i\left(\frac{2\pi x}{\lambda} - \omega t\right)\right], \tag{3.31}$$

where λ and ω are the wavelength and the angular frequency of the wave, respectively. Using the relations $p = 2\pi\hbar/\lambda$ and $E = \hbar\omega$, derived from analogy with the photons, Eq. (3.31)is written as:

$$\Psi(x,t) = C\exp\left[\frac{i}{\hbar}(px - Et)\right]. \tag{3.32}$$

From this equation, the following relations are obtained:

$$i\hbar\frac{\partial\Psi}{\partial t} = E\Psi, \tag{3.33}$$

$$-\hbar^2\frac{\partial^2\Psi}{\partial x^2} = p^2\Psi = 2m[E - V(x)]\Psi. \tag{3.34}$$

In deriving (3.34), the relation $E = p^2/2m + V(x)$ is used, where $V(x)$ is the potential energy of the particle.

From these relations, the *time dependent Schrödinger equation*, written in the three-dimensional spatial coordinates, is derived:

$$i\hbar\frac{\partial\Psi}{dt} = -\frac{\hbar^2}{2m}\nabla^2\Psi + V\Psi. \tag{3.35}$$

In the stationary state, the spatial part of the wave function $\psi(\boldsymbol{r})$ follows the following equation:

$$\left[-\frac{\hbar^2}{2m}\nabla^2 + V(\boldsymbol{r})\right]\psi(\boldsymbol{r}) = \hat{H}\psi(\boldsymbol{r}) = E\psi(\boldsymbol{r}). \tag{3.36}$$

Here $\hat{H}$ is the Hamiltonian and E is the energy of the state $\psi(\boldsymbol{r})$.

Noting that the first term on the left-hand side of (3.36) corresponds to the kinetic energy $p^2/2m$, the operator of the momentum is given by:

$$p_x = -i\hbar\frac{\partial}{dx}, \quad p_y = -i\hbar\frac{\partial}{dy}, \quad p_z = -i\hbar\frac{\partial}{dz} \tag{3.37}$$

3.1.2.3 Hydrogen Atom

As an example of application of the Schrödinger Eq. (3.36), we consider a hydrogen atom which is composed of one electron and a proton. The potential V depends only on the electron-nucleus distance r as

$$V(\boldsymbol{r}) = -\frac{e^2}{4\pi\varepsilon_0 r}. \tag{3.38}$$

In this case, the wavefunction is written in the spherical coordinate by the product of the functions of ρ, θ, and ϕ:

$$\psi_{n,l,m} = C\rho^l e^{-\rho/2} L_{n+l}^{2l+1}(\rho) P_l^{|m|}(\cos\theta) e^{im\phi}, \tag{3.39}$$

where $\rho = 2r/na_H$ with $a_H = \varepsilon_0 h^2/\pi\mu e^2$(the first Bohr radius), $\mu = m_e M/(m_e + M) \approx m_e$ is the reduced mass of electron m_e and proton M, $L_{n+l}^{2l+1}(\rho)$ is the associated Laguerre polynomial, and $P_l^{|m|}(\cos\theta)$ is the Legendre polynomial.

The wavefunction of each eigenstate (n, l, m) is specified by the principal quantum number n ($n = 1, 2, 3, \ldots$), the angular moment quantum number l ($l = 0, 1, 2, \ldots, n - 1$) and the magnetic quantum number m ($m = 0, \pm 1, \pm 2, \ldots, \pm l$). The $n = 1$ state is the ground state, $n = 2$ the first excited state, $n = 3$ the second excited state, etc. Since the electrons are fermions, each state specified by (n, l, m) is occupied only up to two electrons, one with up-spin and another with down-spin. The energy of each eigenstate depends only on n as

$$E_n = -\frac{\mu e^4}{8\varepsilon_0^2 h^2}\left(\frac{1}{n^2}\right) \approx -hcR_\infty\left(\frac{1}{n^2}\right), \tag{3.40}$$

where $R_\infty = \left(\mu e^4\right)/\left(8\varepsilon_0^2 h^3 c\right)$ is the Rydberg constant.

The energy level diagram of a hydrogen atom is shown in Fig. 3.3. Also shown in this diagram are intense lines observed in the emission and absorption spectra of a hydrogen atom: Lyman series denoted by Ly-α, β, γ due to np-$1s$ transitions (n = 2, 3, 4, …), and Balmer series denoted by H-α, β, γ due to ns-$2p$ and nd-$2p$ transitions (n = 3, 4, 5, …). Defining *parity* as the plus or minus sign of a physical quantity when the spatial coordinate is inverted (for example, x is inverted to $-x$), the s and d states have even parity ($P = 1$), whereas the p state has an odd parity ($P = -1$). These intense lines arise from *allowed transitions* where the parities of the upper level and the lower level have different signs. Noting that the parity of the electric field is $P = -1$, the change in the parity of the atomic states in the electric dipole transition is $\Delta P = -1$ since the whole process should be independent of the spatial inversion.

The degenerate levels of $l = 1, 2, 3, \ldots$ are split into fine structures due to spin–orbit interaction (L-S coupling), which arises from the interaction between the magnetic moments of the orbiting electron and the electron spin. In the atoms of $Z \geq 2$, where

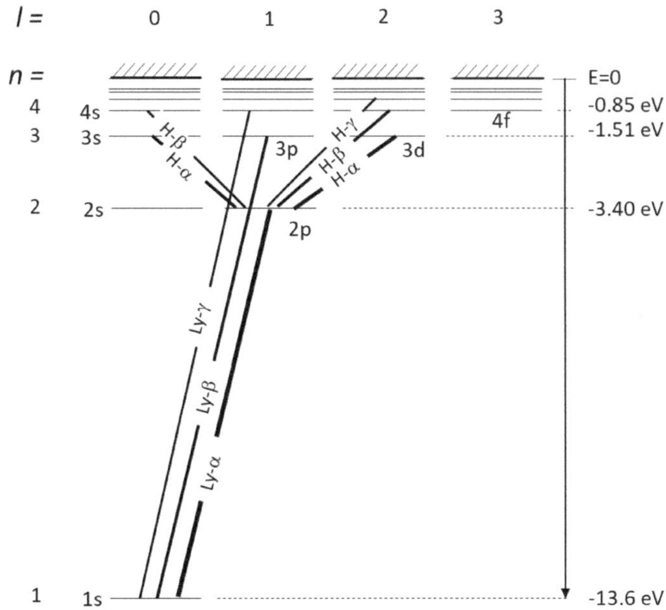

Fig. 3.3 Energy levels of a hydrogen atom

Z is a nuclear charge, many electrons are bound by the nucleus and interact with each other. The energy states of these atoms, molecules and condensed states result in various properties of the materials.

3.1.3 Quantization of Radiation

In the quantum mechanics described in the previous section, the wave properties of particles are described by the wave equation and radiation is treated as an electromagnetic wave. In this subsection, we introduce a framework for "quantization of radiation" which is required to describe the quantum nature of radiation.

An optical wave is written as a superposition of sinusoidal waves, each characterized by the mode of the wave. We show in the following that the energy of each optical mode is written in the same form as the energy of a harmonic oscillator.

Referring to Sect. 2.1.4, the electric field $\boldsymbol{E}$ and the magnetic field $\boldsymbol{B}$ are expressed by the vector potential $\boldsymbol{A}$ as $\boldsymbol{E} = -\partial \boldsymbol{A}/\partial t$ and $\boldsymbol{B} = \nabla \times \boldsymbol{A}$, respectively. Since $\boldsymbol{A}$, which is a real quantity, propagates as a transverse wave, it is expanded into Fourier components by assuming a boundary condition that the field is periodic in a cube of length L:

$$\boldsymbol{A}(\boldsymbol{r}, t) = \frac{1}{L^{3/2}} \sum_{\boldsymbol{k}} \sum_{\lambda=1,2} \hat{\boldsymbol{e}}_{\boldsymbol{k}}^{(\lambda)} \left[A_{\boldsymbol{k},\lambda}(t) e^{i\boldsymbol{k}\cdot\boldsymbol{r}} + A_{\boldsymbol{k},\lambda}^{*}(t) e^{-i\boldsymbol{k}\cdot\boldsymbol{r}} \right], \tag{3.41}$$

where λ denotes the two polarizations of the transverse waves, $\hat{\boldsymbol{e}}_{\boldsymbol{k}}^{(\lambda)}$ is a unit vector in the polarization direction, and $\boldsymbol{k}(k_x, k_y, k_z) = (\pi/L)(n_x, n_y, n_z)$ is a wave vector of each mode specified by positive integers (n_x, n_y, n_z).

Introducing real variables $q_{\boldsymbol{k},\lambda}(t)$ and $p_{\boldsymbol{k},\lambda}(t)$ in place of the complex variable $A_{\boldsymbol{k},\lambda}(t)$ as

$$\begin{aligned} q_{\boldsymbol{k},\lambda}(t) &= \sqrt{2\varepsilon_0} \left[A_{\boldsymbol{k},\lambda}(t) + A_{\boldsymbol{k},\lambda}^{*}(t) \right], \\ p_{\boldsymbol{k},\lambda}(t) &= -i\sqrt{2\varepsilon_0}\omega \left[A_{\boldsymbol{k},\lambda}(t) - A_{\boldsymbol{k},\lambda}^{*}(t) \right], \end{aligned} \tag{3.42}$$

the total energy of the electromagnetic field is expressed in terms of $\boldsymbol{A}$ as

$$\begin{aligned} W &= \int \frac{1}{2}(\varepsilon_0 \boldsymbol{E}^2 + \mu_0 \boldsymbol{H}^2) d\boldsymbol{r} = 4\varepsilon_0 \sum_{\boldsymbol{k}} \sum_{\lambda=1,2} \omega_{\boldsymbol{k}}^2 A_{\boldsymbol{k},\lambda}^{*}(t) A_{\boldsymbol{k},\lambda}(t) \\ &= \frac{1}{2} \sum_{\boldsymbol{k}} \sum_{\lambda=1,2} \left[p_{\boldsymbol{k},\lambda}^2(t) + \omega_{\boldsymbol{k}}^2 q_{\boldsymbol{k},\lambda}^2(t) \right]. \end{aligned} \tag{3.43}$$

This equation shows that the energy of the optical wave is described in a similar form as the harmonic oscillator which is given by $E = p^2/(2m) + \omega^2 q^2$, where q is the position, p is the momentum, m is the mass, and ω is the resonance frequency of the harmonic oscillator, respectively.

Applying quantum mechanical formulation to the optical field, W, $q_{\boldsymbol{k},\lambda}(t)$ and $p_{\boldsymbol{k},\lambda}(t)$ are transferred to the operators as H, $q_\alpha(t)$, and $p_\alpha(t)$, respectively, where the mode specified by $(\boldsymbol{k}, \lambda)$ is denoted by α:

$$H = \frac{1}{2} \sum_{\alpha} \left[p_\alpha^2(t) + \omega_\alpha^2 q_\alpha^2(t) \right]. \tag{3.44}$$

Let us introduce operators $a_\alpha^\dagger$ and a_α defined by

$$\begin{aligned} a_\alpha^\dagger &= \frac{1}{\sqrt{2\hbar\omega_\alpha}} (\omega_\alpha q_\alpha - i p_\alpha), \\ a_\alpha &= \frac{1}{\sqrt{2\hbar\omega_\alpha}} (\omega_\alpha q_\alpha + i p_\alpha). \end{aligned} \tag{3.45}$$

Noting that $[q_\alpha, p_\alpha] = i\hbar$, these operators satisfy the following relation,

$$[a_\alpha, a_\alpha^\dagger] = a_\alpha a_\alpha^\dagger - a_\alpha^\dagger a_\alpha = 1. \tag{3.46}$$

We define the number operator N_α, its eigen function $|n_\alpha\rangle$ and its eigenvalue n_α as

$$N_\alpha = a_\alpha^\dagger a_\alpha,$$
$$N_\alpha|n_\alpha\rangle = n_\alpha|n_\alpha\rangle. \tag{3.47}$$

Since N_α is an Hermite operator, n_α is a real quantity. N_α satisfies the following commutation relations with a_α and $a_\alpha^\dagger$:

$$[a_\alpha, N_{\alpha'}] = \delta_{\alpha,\alpha'} a_\alpha,$$
$$\left[a_\alpha^\dagger, N_{\alpha'}\right] = -\delta_{\alpha,\alpha'} a_\alpha^\dagger, \tag{3.48}$$

where $\delta_{\alpha,\alpha'} = 1\ (\alpha = \alpha')$ and $\delta_{\alpha,\alpha'} = 0\ (\alpha \neq \alpha')$. There relations lead to

$$N_\alpha a_\alpha|n_\alpha\rangle = (a_\alpha N_\alpha - a_\alpha)|n_\alpha\rangle = (n_\alpha - 1)a_\alpha|n_\alpha\rangle,$$
$$N_\alpha a_\alpha^\dagger|n_\alpha\rangle = (n_\alpha + 1)a_\alpha^\dagger|n_\alpha\rangle. \tag{3.49}$$

Recalling that N_α is the operator to extract the eigenvalue of $|n_\alpha\rangle$, the first equation of (3.49) shows that $a_\alpha|n_\alpha\rangle$ is proportional to $|n_\alpha - 1\rangle$, i.e., $a_\alpha|n_\alpha\rangle = C_-|n_\alpha - 1\rangle$, where C_- is a normalization constant. That is, the operator a_α reduces the eigenvalue of the $|n_\alpha\rangle$ state by 1, creating the $|n_\alpha - 1\rangle$ state. Likewise, the second equation of (3.49) shows that $a_\alpha^\dagger$ increases the eigenvalue by 1, creating the $|n_\alpha + 1\rangle$ state, i.e., $a_\alpha^\dagger|n_\alpha\rangle = C_+|n_\alpha + 1\rangle$. These relations are expressed as

$$a_\alpha|n_\alpha\rangle = C_-|n_\alpha - 1\rangle = \sqrt{n_\alpha}|n_\alpha - 1\rangle,$$
$$a_\alpha^\dagger|n_\alpha\rangle = C_+|n_\alpha + 1\rangle = \sqrt{n_\alpha + 1}|n_\alpha + 1\rangle. \tag{3.50}$$

Here the coefficients $\sqrt{n_\alpha}$ and $\sqrt{n_\alpha + 1}$ are determined from the absolute squares of the right-hand sides of these equations:

$$\langle n_\alpha - 1\,|n_\alpha - 1\rangle = \langle n_\alpha\left|a_\alpha^\dagger a_\alpha\right|n_\alpha\rangle = n_\alpha,$$
$$\langle n_\alpha + 1\,|n_\alpha + 1\rangle = \langle n_\alpha\left|a_\alpha a_\alpha^\dagger\right|n_\alpha\rangle = \langle n_\alpha\left|a_\alpha^\dagger a_\alpha + 1\right|n_\alpha\rangle = n_\alpha + 1. \tag{3.51}$$

Since these equations are positive, n_α is positive satisfying $n_\alpha \geq 0$ and takes only integers since otherwise negative states are created by repetitively applying the a_α operator to the non-integer state.

The operators $a_\alpha^\dagger$ and a_α are called a *creation operator* and an *annihilation operator*, respectively; $\hat{a}_\alpha^\dagger$ creates one photon in n_a-state and generates $(n_a + 1)$-state, whereas $\hat{a}_\alpha$ annihilates one photon in n_a-state and generates $(n_a - 1)$-state.

The Hamiltonian of the optical field is thus expressed as

$$H = \sum_\alpha \hbar\omega_\alpha \left(a_\alpha^\dagger a_\alpha + \frac{1}{2} \right) = \sum_\alpha \hbar\omega_\alpha \left(N_\alpha + \frac{1}{2} \right). \tag{3.52}$$

The term $(1/2)\hbar\omega_\alpha$, called *zero-point energy*, is the energy of the zero-point vibration due to quantum mechanical uncertainty of the position and the momentum at the ground state. By subtracting the zero-point energy which is an infinite constant, the energy is expressed as

$$H = \sum_\alpha \hbar\omega_\alpha a_\alpha^\dagger a_\alpha = \sum_\alpha \hbar\omega_\alpha N_\alpha. \tag{3.53}$$

Applying this Hamiltonian to the field $|\{n_\alpha\}\rangle$,

$$H|\{n_\alpha\}\rangle = \sum_\alpha \hbar\omega_\alpha N_\alpha |\{n_\alpha\}\rangle = \sum_\alpha n_\alpha \hbar\omega_\alpha |\{n_\alpha\}\rangle. \tag{3.54}$$

This equation shows that the field specified by α is composed of n_α photons each with energy $\hbar\omega_\alpha$.

As described in Sect. 2.1.3, the momentum of the optical wave is given by $\boldsymbol{g} = (1/c^2)\boldsymbol{E} \times \boldsymbol{H}$. Following the steps like quantization of the energy, the operator of the momentum $\boldsymbol{P}$ and its eigen values are described as,

$$\begin{aligned} \boldsymbol{P} &= \sum_\alpha \hbar\boldsymbol{k}_\alpha \left(a_\alpha^\dagger a_\alpha + \frac{1}{2} \right) = \sum_\alpha \hbar\boldsymbol{k}_\alpha \left(N_\alpha + \frac{1}{2} \right), \\ \boldsymbol{P}|\{n_\alpha\}\rangle &= \sum_\alpha n_\alpha \hbar\boldsymbol{k}_\alpha |\{n_\alpha\}\rangle. \end{aligned} \tag{3.55}$$

These relations show that a photon has an energy of $\hbar\omega$ and a momentum $\hbar k$.

The optical field is specified by the number of photons in each mode denoted by a suffix j:

$$\Psi_R = \big| n_1, n_2, \ldots, n_j, \ldots \big\rangle. \tag{3.56}$$

The energy of this field is given by $W_R = \sum_j W_j = \sum_j n_j \hbar\omega_j$. In ordinal light such as the thermal radiation, the photons are distributed over many modes, the number of photons in each mode is mostly zero, and the distribution changes rapidly in time. In contrast, the photon distribution of the laser light is concentrated in small numbers of modes. The extreme case where the photons are concentrated in only one mode is described as $\Psi_R = |0, 0, \ldots, n_\lambda \gg 1, 0, 0, \ldots\rangle$. (This state is possible since the photon is a boson particle.)

In the photon *number state* (called *Fock state*) where the number of the photons in a mode is fixed, the wave character of the field is lost since the spatial distribution of the photons is not determined. The particle-wave duality is expressed by the uncertainty relationship between the photon number Δn and the phase $\Delta\phi$:

$$\Delta n \cdot \Delta\phi \geq \frac{1}{2}. \tag{3.57}$$

This relation shows that we can measure the phase of the field with an accuracy $\Delta\phi$ when the photon number has some uncertainty Δn. The minimum value of the uncertainty of Eq. (3.57) is obtained for a *coherent state* $|\alpha\rangle$:

$$|\alpha\rangle = e^{-|\alpha|^2/2} \sum_{n=0}^{\infty} \frac{\left(\alpha \hat{a}^{\dagger}\right)^n}{n!} |0\rangle. \tag{3.58}$$

The photon number distribution in $|\alpha\rangle$ has a Poisson distribution:

$$P_n = |\langle n \mid \alpha\rangle|^2 = \frac{e^{-\alpha} |\alpha|^{2n}}{n!}. \tag{3.59}$$

The average $\langle\hat{n}\rangle$ and the variance $\left\langle \Delta n^2 \right\rangle$ of n in $|\alpha\rangle$ have the same value:

$$\langle\hat{n}\rangle = \langle\alpha \left| \hat{a}^{\dagger}\hat{a} \right| |\alpha\rangle = |\alpha|^2, \tag{3.60}$$

$$\left\langle \Delta n^2 \right\rangle = \langle\alpha \left| (\hat{n} - \langle\hat{n}\rangle)^2 \right| |\alpha\rangle = |\alpha|^2. \tag{3.61}$$

The coherent state corresponds to a wave packet of sinusoidal electric field which moves in space.

3.2 Absorption, Emission, and Amplification of Light

In this section we consider interaction of radiation with an atom, where the radiation is described by the photon number state. The probabilities for absorption and emission are derived from which the gain coefficient of laser amplification is derived.

3.2.1 Interaction of Light with an Atom

3.2.1.1 Interaction Hamiltonian

The equation of motion of a particle of mass m and charge q under the *Lorentz force* of the electromagnetic field is expressed in terms of the vector potential $\boldsymbol{A}$ and the scalar potential ϕ, introduced in Sect. 2.1.4;

$$m\frac{d^2\boldsymbol{r}}{dt^2} = q(\boldsymbol{E} + \boldsymbol{v} \times \boldsymbol{B}) = q\left(-\nabla\phi - \frac{\partial \boldsymbol{A}}{\partial t} + \boldsymbol{v} \times (\nabla \times \boldsymbol{A}) \right). \tag{3.62}$$

This equation of motion is derived from the Lagrangian

$$L = T - U = \frac{1}{2}m|\boldsymbol{v}|^2 - q \cdot \phi(\boldsymbol{r}) + q\boldsymbol{v} \cdot \boldsymbol{A}, \tag{3.63}$$

where $T = (1/2)m|\boldsymbol{v}|^2$ is the kinetic energy and $U = q \cdot \phi(\boldsymbol{r}) - q\boldsymbol{v} \cdot \boldsymbol{A}$ is the generalized potential energy.

The Hamiltonian corresponding to this Lagrangian is

$$H = \frac{1}{2m}(\boldsymbol{p} - q\boldsymbol{A})^2 + q\phi. \tag{3.64}$$

The first term on the right-hand side is the kinetic energy of the particle under the electromagnetic field and the second term is the electrostatic energy due to Coulomb interaction between the charged particles. This equation shows that the momentum of the particle is given by $\boldsymbol{p} - e\boldsymbol{A}$ in the electromagnetic field.

The Hamiltonian operator of the total system including the radiation field is:

$$H = \frac{1}{2m}(\boldsymbol{p} - q\boldsymbol{A})^2 + e\phi + \sum_k \sum_{\lambda=1}^{2} \hbar\omega_k a_\alpha^\dagger a_\alpha \tag{3.65}$$

where λ specifies the polarization of the field. Since we can show that $\boldsymbol{p}$ and $\boldsymbol{A}$ are interchangeable, i.e., $[\boldsymbol{p}, \boldsymbol{A}] = 0$, the interaction of radiation with the particle is described as,

$$H_I = -\frac{q}{m}\boldsymbol{A}(\boldsymbol{r}) \cdot \boldsymbol{p} + \frac{e^2}{2m}\boldsymbol{A}(\boldsymbol{r}) \cdot \boldsymbol{A}(\boldsymbol{r}) \equiv H_1 + H_2 \tag{3.66}$$

The term H_1 linear to $\boldsymbol{A}$ corresponds to either absorption or emission of one photon, whereas H_2 corresponds to two photon processes such as light scattering and two-photon absorption or emission. In this section, we consider only the first term H_1.

The total Hamiltonian is the sum of the noninteracting Hamiltonian H_0 which is composed of the Hamiltonians of the electron H_e and the radiation H_R, and the interaction Hamiltonian H_I:

$$\begin{aligned} H &= H_0 + H_I, \\ H_0 &= H_e + H_R, \\ H_e &= \frac{\mathbf{p}^2}{2m} + \phi(\boldsymbol{r}), \quad H_R = \sum_k \sum_{\lambda=1}^{2} \hbar\omega_k a_\alpha^\dagger a_\alpha. \end{aligned} \tag{3.67}$$

Denoting the eigenstates of the electron and the radiation by $|\psi_j\rangle$ and $|n_\alpha\rangle$, respectively, these states have the eigen energies of $E_j + \sum_\alpha n_\alpha \hbar\omega_\alpha$:

$$H_o|\psi_j\rangle|n_\alpha\rangle = \left(E_j + \sum_\alpha n_\alpha \hbar\omega_\alpha\right)|\psi_j\rangle|n_\alpha\rangle. \tag{3.68}$$

Noting that the operator $\boldsymbol{A}$ is written using a_α and $a_\alpha^\dagger$ as

$$\boldsymbol{A}(\boldsymbol{r}, t) = \sum_\alpha \sqrt{\frac{\hbar}{2\varepsilon_0 V \omega_k}} \boldsymbol{\varepsilon}_\alpha^{(\lambda)} \left[a_\alpha^+ e^{-i\boldsymbol{k}\cdot\boldsymbol{r}} + a_\alpha e^{i\boldsymbol{k}\cdot\boldsymbol{r}}\right], \tag{3.69}$$

the interaction Hamiltonian reads

$$H_I = -\frac{q}{m}\sum_\alpha \sqrt{\frac{\hbar}{2\varepsilon_0 V \omega_k}} (\boldsymbol{\varepsilon}_\alpha \cdot \boldsymbol{p})\left[a_\alpha^+ e^{-i\boldsymbol{k}\cdot\boldsymbol{r}} + a_\alpha e^{i\boldsymbol{k}\cdot\boldsymbol{r}}\right]. \tag{3.70}$$

In this interaction Hamiltonian, we approximate the exponential terms as 1 by keeping only the first term of the expansion $e^{-i\mathbf{k}\cdot\mathbf{r}} = 1 - i\mathbf{k}\cdot\mathbf{r} - (\mathbf{k}\cdot\mathbf{r})^2/2 + \cdots$, since the spatial variation of the electric field, characterized by its wavelength $\lambda = 2\pi/k$, is small compared with the spatial distribution of the electrons in the atom which is of the order of 0.1 nm. (This approximation, called *electric dipole approximation*, is not valid in the X-ray region where the wavelength becomes comparable to the size of the atom.)

3.2.1.2 Transition Probability

Let us consider interaction of an optical field with an atom. The total Hamiltonian is the sum of a noninteracting Hamiltonian H_0 and an interaction Hamiltonian H_I:

$$H = H_0 + H_I. \tag{3.71}$$

The states of the atom and the optical field change in time due to the interaction. This process is described by the time-dependent Schrödinger equation Eq. (3.35):

$$i\hbar\frac{\partial}{\partial t}|\psi(t)\rangle = H|\psi(t)\rangle. \tag{3.72}$$

Let us assume that the interaction starts at $t = 0$, where $|\psi(t \le 0)\rangle = |\psi_i\rangle$. After $t = 0$, the final state is mixed gradually with the initial state due to the interaction. The transition probability is obtained by calculating the amount of mixing of the final state. By treating the interaction H_I as a small perturbation, that is, $H_0 \gg H_I$, the transition probability from the initial state $|\psi_i\rangle$ to the final state $|\psi_f\rangle$ is calculated by the following formula (often called as *Fermi's golden rule*),

$$w_{fi} = \frac{2\pi}{\hbar}|M_{fi}|^2\delta(E_f - E_i). \tag{3.73}$$

Here

$$M_{fi} = \langle f|H_{\mathrm{I}}|i\rangle \tag{3.74}$$

is the matrix element of $\hat{H}_I$ between the initial and the final state, and $\delta(E_f - E_i)$ is the delta function which conserves the energy in the transition from the initial state to the final state.

3.2.2 *Absorption, Emission, and Amplification*

In this subsection, we consider absorption and emission of photons by atoms and determine the gain coefficient of the laser.

3.2.2.1 Absorption

We first consider an absorption process shown in Fig. 3.4a. An atom in a lower electronic state $|l\rangle$ interacts with an optical field of $|n_\alpha\rangle$ state where there are n_α photons in the α mode. We denote this initial state as $|l; n_\alpha\rangle$. When the photon energy is close to the transition energy, the atom is excited to an upper state $|u\rangle$ by absorbing one photon in the α mode, resulting in the optical field $|n_\alpha - 1\rangle$. This final state is denoted as $|u; n_\alpha - 1\rangle$. The matrix element for the transition from the initial state to the final state is derived from Eqs. (3.70) and (3.74) as

$$\begin{aligned} M_{fi} &= \langle u; n_\alpha - 1|H_I|l; n_\alpha\rangle \\ &= -\frac{q}{m}\sum_\alpha \sqrt{\frac{\hbar}{2\varepsilon_0 V\omega_k}}\langle u; n_\alpha - 1\big|(\boldsymbol{\varepsilon}_\alpha \cdot \boldsymbol{p})\big[a_\alpha^+ + a_\alpha\big]\big|l; n_\alpha\rangle \\ &= -\frac{q}{m}\sum_\alpha \sqrt{\frac{\hbar n_\alpha}{2\varepsilon_0 V\omega_k}}\langle u|(\boldsymbol{\varepsilon}_\alpha \cdot \boldsymbol{p})|l\rangle \end{aligned} \tag{3.75}$$

The factor $\sqrt{n_\alpha}$ arises from $\langle n_\alpha - 1|a_\alpha|n_\alpha\rangle = \sqrt{n_\alpha}$.

The momentum $\boldsymbol{p}$ is operated to the electrons. Noting $H_e = \boldsymbol{p}^2/2m + \phi(\boldsymbol{r})$ and the relation $\left[x_m, \boldsymbol{p}^2\right] = \sum_n \left[x_m, p_n\right]p_n + \sum_n p_n\left[x_m, p_n\right] = 2i\hbar p_m$, $\boldsymbol{p}$ is written as

$$\boldsymbol{p} = -\frac{im}{\hbar}\left[\boldsymbol{r}, \boldsymbol{p}^2/2m + \phi(\boldsymbol{r})\right] = -\frac{im}{\hbar}[\boldsymbol{r}, H_e]. \tag{3.76}$$

The matrix element in Eq. (3.75) becomes,

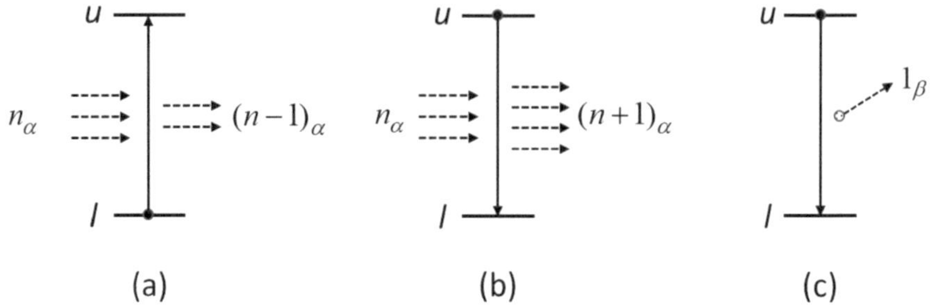

Fig. 3.4 Schematic diagrams of the absorption and emission processes **a** absorption, **b** stimulated emission, and **c** spontaneous emission. In these diagrams, n_α refers the number of the photons in the α-mode and u and l refer the upper and lower atomic levels, respectively

$$\begin{aligned}\langle u|(\boldsymbol{\varepsilon}_\alpha \cdot \boldsymbol{p})|l\rangle &= -\frac{im}{\hbar}\langle u|(\boldsymbol{\varepsilon}_\alpha \cdot [\boldsymbol{r}, H_e])|l\rangle \\ &= -\frac{im}{\hbar}\langle u|((\boldsymbol{\varepsilon}_\alpha \cdot \boldsymbol{r})H_e - H_e(\boldsymbol{\varepsilon}_\alpha \cdot \boldsymbol{r}))|l\rangle \\ &= -\frac{im}{\hbar}(\boldsymbol{\varepsilon}_\alpha \cdot \boldsymbol{r}_{ul})(E_l - E_u) = im\omega_0(\boldsymbol{\varepsilon}_\alpha \cdot \boldsymbol{r}_{ul}),\end{aligned} \tag{3.77}$$

where $\boldsymbol{r}_{ul} = \langle u|\boldsymbol{r}|l\rangle$, $\boldsymbol{\varepsilon}_\alpha$ is a unit vector in the polarization direction and $\hbar\omega_0 = E_u - E_l$.

Using the matrix element, Eq. (3.77), the transition probability (3.73) is derived:

$$\begin{aligned}w_{fi} &= \frac{2\pi}{\hbar}|M_{fi}|^2\delta(E_f - E_i) \\ &= \frac{\pi q^2}{\varepsilon_0 V}\sum_\alpha \frac{n_\alpha \omega_0^2}{\omega_\alpha}|\boldsymbol{\varepsilon}_\alpha \cdot \boldsymbol{r}_{ul}|^2\delta(\hbar\omega_0 - \hbar\omega_\alpha) \\ &= \frac{\pi q^2 \omega_0 n_\alpha}{\varepsilon_0 \hbar V}|\boldsymbol{\varepsilon}_\alpha \cdot \boldsymbol{r}_{ul}|^2\delta(\omega_0 - \omega_\alpha)\end{aligned} \tag{3.78}$$

Here the relation $\delta(E_f - E_i) = \delta(E_u - E_l - \hbar\omega_\alpha) = \delta(\hbar\omega_0 - \hbar\omega_\alpha) = \delta(\omega_0 - \omega_\alpha)/\hbar$ is used.

The attenuation coefficient of the incident light for a medium with an atomic number density in the lower state of N_l is derived from this transition probability. Since the intensity I_α is written in terms of the photon number density n_α/V as $I_\alpha = c\hbar\omega n_\alpha/V$, the absorbed power of the optical wave is

$$\frac{dI_\alpha}{dz} = \frac{P_{abs}}{V} = N_l\hbar\omega \cdot w_{fi} = \frac{N_l\pi\omega_0}{\varepsilon_0 c\hbar}|\boldsymbol{\varepsilon}_\alpha \cdot \boldsymbol{\mu}_{ul}|^2 g(\omega)I_\alpha. \tag{3.79}$$

Here the electric dipole moment $\boldsymbol{\mu}_{ul} = q\boldsymbol{r}_{ul}$ is introduced and $\delta(\omega_0 - \omega_\alpha)$ is replaced by a normalized spectral distribution function $g(\omega)$ to include broadening of the atomic levels. Assuming that the atoms are oriented in the same direction relative to the polarization direction $\boldsymbol{\varepsilon}_\alpha$, the attenuation coefficient $\gamma(\omega)$ defined by $dI_\alpha/dz = \gamma(\omega)I_\alpha$ is derived as

$$\gamma(\omega) = \frac{N_l \pi \omega_0}{\varepsilon_0 c \hbar} |\boldsymbol{\varepsilon}_\alpha \cdot \boldsymbol{\mu}_{ul}|^2 g(\omega). \tag{3.80}$$

This expression agrees with Eq. (2.79).

3.2.2.2 Spontaneous Emission, Stimulated Emission, and Light Amplification

The emission processes are shown in Fig. 3.4b, c. An atom in an upper electronic state $|u\rangle$ interacts with an optical field of $|n_\alpha\rangle$ state. We denote this initial state as $|u; n_\alpha\rangle$. Due to interaction, the atom makes a transition to a lower state $|l\rangle$ by emitting one photon in the α mode, resulting in the optical field $|n_\alpha + 1\rangle$. This final state is denoted as $|l; n_\alpha + 1\rangle$. The matrix element for the transition from the initial to the final state is

$$\begin{aligned} M_{fi} &= \langle l; n_\alpha + 1|H_1|u; n_\alpha\rangle \\ &= -\frac{q}{m}\sum_\alpha \sqrt{\frac{\hbar}{2\varepsilon_0 V \omega_k}} \big\langle l; n_\alpha + 1\big|(\varepsilon_\alpha \cdot \mathbf{p})\big[a_\alpha^+ + a_\alpha\big]\big|u; n_\alpha\big\rangle \\ &= -\frac{q}{m}\sum_\alpha \sqrt{\frac{\hbar(n_\alpha + 1)}{2\varepsilon_0 V \omega_k}} \langle l|(\varepsilon_\alpha \cdot \mathbf{p})|u\rangle. \end{aligned} \tag{3.81}$$

Note that the factor $\sqrt{n_\alpha + 1}$ arises from $\big\langle n_\alpha + 1\big|a_\alpha^\dagger\big|n_\alpha\big\rangle = \sqrt{n_\alpha + 1}$.

The transition probability is derived in the same procedures as the absorption of photon:

$$w_{fi} = \frac{\pi q^2 \omega_0 (n_\alpha + 1)}{\varepsilon_0 \hbar V} \sum_\alpha |\boldsymbol{\varepsilon}_\alpha \cdot \mathbf{r}_{ul}|^2 \delta(\omega_0 - \omega_\alpha). \tag{3.82}$$

The transition probability for photon emission is proportional to $n_\alpha + 1$, leading to two major consequences.

The first consequence, important in terms of the laser, is that the transition probability increases almost linearly with n_α when $n_\alpha \gg 1$. That is, a photon is emitted in the α mode in proportional to n_α, before the atom in $|u\rangle$ makes transition in other processes such as spontaneous emission and relaxation to other states. In this stimulated emission (or induced emission) process, shown schematically in Fig. 3.4b, the optical wave is amplified since the photon is emitted into the same mode as the incident photons, thus increasing the photon number in the α mode. The laser takes advantage of this property by confining the optical wave in an optical cavity for exponential amplification of the optical wave. Denoting the atomic number density in the upper state as N_u, the gain coefficient is described in the same form as in Eq. (3.80), except N_l is replaced by N_u. Since the actual gain is the difference between the gain and the attenuation, the gain coefficient is given by

$$\gamma(\omega) = \frac{(N_u - N_l)\pi\omega_0}{\varepsilon_0 c\hbar} |\boldsymbol{\varepsilon}_\alpha \cdot \boldsymbol{\mu}_{ul}|^2 g(\omega) = \Delta N \cdot \sigma(\omega). \tag{3.83}$$

Here $\Delta N = N_u - N_l > 0$ is the population inversion density and $\sigma(\omega)$ is the *stimulated emission cross section,*

$$\sigma(\omega) = \frac{\gamma(\omega)}{N_u - N_l} = \frac{\pi\omega_0}{\varepsilon_0 c\hbar} |\boldsymbol{\varepsilon} \cdot \boldsymbol{\mu}_{ul}|^2 g(\omega). \tag{3.84}$$

When the degeneracies of the energy levels are taken into account, ΔN reads as $\Delta N = (N_u/g_u) - (N_l/g_l)$, where g_u and g_l are the degeneracies of $|u\rangle$ and $|l\rangle$ states, respectively. A schematic diagram of amplification of an optical wave by stimulated emission in a population-inverted medium is shown in Fig. 3.5.

The second major consequence of (3.82) is that a photon is emitted without an incident photon, i.e. $n_\alpha = 0$, as shown schematically in Fig. 3.4c. Incoherent light such as the sunlight and LED light are generated by this spontaneous emission process. The spontaneously emitted photons become the noise source to the laser oscillation since these photons are emitted with random phases different from the phase of the laser mode.

The total spontaneous emission probability is obtained by integrating over all modes $\alpha = (\mathbf{k}, \lambda)$ that the photons are spontaneously emitted. Noting that the number of modes contained in a spherical shell of radius k and thickness dk is $k^2 dk d\Omega V/(2\pi)^3$ where $d\Omega = \sin\theta d\theta d\phi$ is a solid angle in the spherical coordinate, the spontaneous emission rate A_{ul} is

$$\begin{aligned} A_{ul} &= \int_k \int_{\theta=0}^{\pi} \int_{\phi=0}^{2\pi} \frac{\pi\omega_0}{\varepsilon_0 \hbar V} \frac{Vk^2 dk}{(2\pi)^3} \sin\theta d\theta d\phi \sum_{\lambda=1,2} |\boldsymbol{\varepsilon}_\lambda \cdot \boldsymbol{\mu}_{ul}|^2 \delta(\omega_0 - \omega) \\ &= \frac{\omega_0^3}{8\pi^2 \varepsilon_0 c^3 \hbar} \int_{\theta=0}^{\pi} \int_{\phi=0}^{2\pi} \sum_{\lambda=1,2} |\boldsymbol{\varepsilon}_\lambda \cdot \boldsymbol{\mu}_{ul}|^2 \sin\theta d\theta d\phi. \end{aligned} \tag{3.85}$$

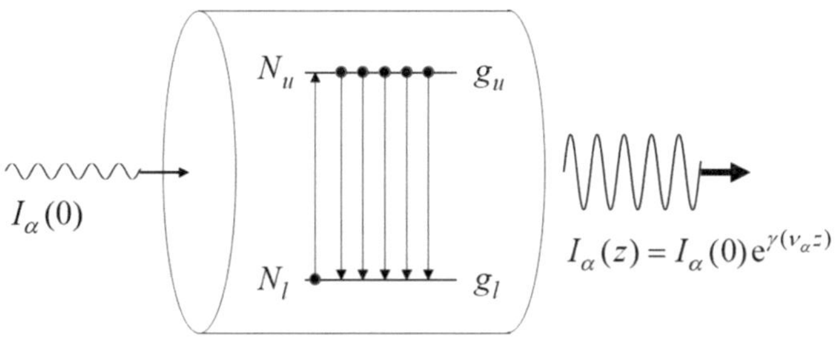

Fig. 3.5 Amplification of an optical wave by stimulated emission in a population-inverted medium

Taking the propagation direction of the optical wave as z and the angle between z and $\boldsymbol{\mu}_{ul}$ as θ, the following relation holds:

$$\sum_{\lambda=x,y} |\boldsymbol{\varepsilon}_\lambda \cdot \boldsymbol{\mu}_{ul}|^2 + |\hat{z} \cdot \boldsymbol{\mu}_{ul}|^2 = |\boldsymbol{\mu}_{ul}|^2$$
$$\rightarrow \sum_{\lambda=x,y} |\boldsymbol{\varepsilon}_\lambda \cdot \boldsymbol{\mu}_{ul}|^2 = |\boldsymbol{\mu}_{ul}|^2\left(1 - \cos^2\theta\right) = \sin^2\theta|\boldsymbol{\mu}_{ul}|^2. \tag{3.86}$$

Using this relation in (3.85) and noting $\int_0^\pi \sin^3\theta d\theta = 4/3$, we obtain

$$A_{ul} = \frac{\omega_0^3|\boldsymbol{\mu}_{ul}|^2}{3\pi\varepsilon_0 c^3\hbar} = \frac{1}{\tau_{ul}}, \tag{3.87}$$

where τ_{ul} is the spontaneous emission lifetime due to the u-l transition. Since τ_{ul} is inversely proportional to ω^3, the excited states with high excitation energies have shorter lifetime, decaying quickly to the lower states. Multiplying A_{ul} with $\hbar\omega$, the spontaneous emission power

$$S = \frac{4}{3}\frac{\omega_k^4}{c^3}\left|\boldsymbol{\mu}_{fi}\right|^2 \tag{3.88}$$

increases also with ω^4. Using the spontaneous emission rate A_{ul} and noting $|\boldsymbol{\varepsilon}_\alpha \cdot \boldsymbol{\mu}_{ul}|^2 = (1/3)|\boldsymbol{\mu}_{ul}|^2$, the gain coefficient is often expressed as

$$\gamma(\omega) = (N_u - N_l)\frac{\pi^2 c^2}{\omega_0^2} g(\omega) A_{ul}. \tag{3.89}$$

Reference

Dirac, P. A. M. (1988). *The principles of quantum mechanics* (4th ed.). Oxford Science Publications.

Chapter 4
Generation, Amplification and Propagation of a High-Power Laser Beam

Abstract In this chapter, basic concepts of generation, amplification, and propagation of a high-power laser beam are presented in Sects. 4.1 and 4.2. We describe basic formulation of laser beam propagation including nonlinear refractive index and spectral dispersion, which is used to design high-energy lasers and to analyze high-intensity laser beam propagation. Then we describe two important subjects in high-power lasers: solid-state laser materials for high-energy and high-power lasers (Sect. 4.3), and laser-induced damage of optical components (Sect. 4.4).

Keywords Laser oscillation · Saturation · Resonator · Transverse mode · Longitudinal mode · Mode locking · Carrier envelope phase · Wavefront aberration · B-integral · Spatial filtering · Lanthanide ions · Transition-metal ions · Laser ceramics · Nd · Yb · Laser damage

4.1 Laser Oscillator

A laser oscillator consists of an amplifier and a resonator which provides positive feedback to selected modes for oscillation. By properly controlling the cavity modes, the laser oscillates with various properties: continuous-wave oscillation for stable frequency generation, Q-switched oscillation for high energy pulse generation, and mode-locked oscillation for ultrafast pulse generation.

4.1.1 Laser Oscillation

A simplified layout of a laser oscillator is shown in Fig. 4.1. An optical amplifier is placed in an optical resonator (called optical cavity) which provides feedback at the resonant modes of the cavity. When the gain of the amplifier exceeds the optical loss, the light starts to oscillate at the resonant modes (Fox & Li, 1961; Kogelnik & Li, 1966).

Y. Kato et al., *High Power Laser and Plasma Science*, Springer Series on Atomic, Optical, and Plasma Physics 130, https://doi.org/10.1007/978-3-032-08930-4_4

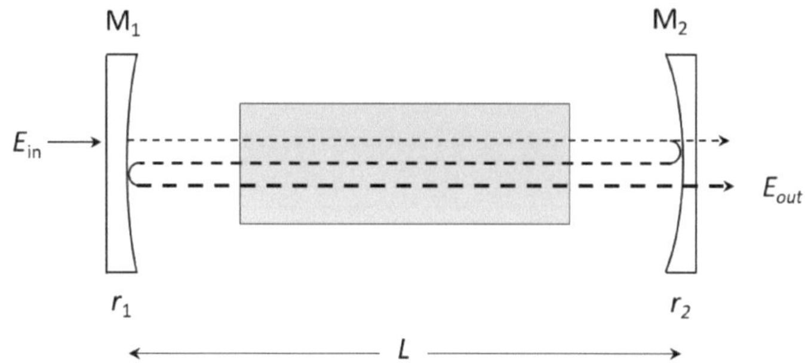

Fig. 4.1 A simplified layout of a laser oscillator, which consists of an optical cavity formed by the mirrors M_1 and M_2 and a gain material (shaded)

With reference to Fig. 4.1, we assume that an optical wave E_{in} enters the cavity through mirror M_1. To simplify the expressions, we assume that the cavity is filled uniformly with an amplifying medium. After single-pass propagation inside the cavity, the wave is amplified as $G = \exp[ikL + (\gamma/2)L]$, where $k = 2\pi n/c$ is the wavenumber, n is the refractive index of the material in the cavity, c is the light velocity in vacuum, L is the cavity length, and $\gamma/2$ is the amplitude gain. The optical wave emitted from this cavity through the mirror M_2 is expressed as

$$E_{out} = E_{in}t_1t_2G\left[1 + r_1r_2G^2 + \left(r_1r_2G^2\right)^2 + \cdots\right] = \frac{t_1t_2G}{1 - r_1r_2G^2}E_{in}. \tag{4.1}$$

Here $r_i = \sqrt{R_i}$ is the amplitude reflectance, R_i the power reflectance, and $t_i = 1 - r_i$ the amplitude transmittance of the mirrors M_1 and M_2, respectively.

When the following condition is satisfied, the optical field starts oscillation without the incident field:

$$G^2r_1r_2 = r_1r_2\exp\left[i4\pi(\nu/c)L + \gamma L\right] = 1, \tag{4.2}$$

where ν is the frequency. Taking the real and imaginary parts, following conditions are obtained:

$$\gamma_{th} = -\ln(r_1r_2)/L = -\ln\sqrt{R_1R_2}/L, \tag{4.3}$$

$$\nu_q = \frac{c}{2L}q \quad (q\text{: integer}). \tag{4.4}$$

Equation (4.3) shows that the laser oscillation starts when the gain coefficient per unit length becomes equal to the power loss per unit length in the cavity, which we express by α. In addition to the transmission loss through the cavity mirrors, there are various factors that contribute of the cavity loss. Expressing the absorption and scattering loss rate by κ and the diffraction loss by δ_d, α is given by

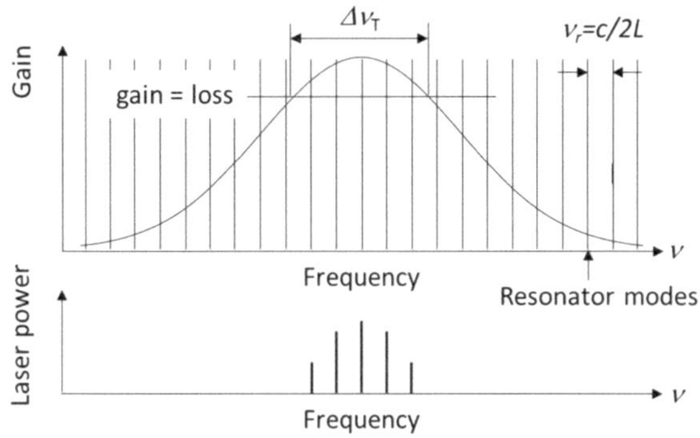

Fig. 4.2 Schematic diagram of the laser oscillation condition. The upper diagram shows the gain curve, which is proportional to the spectral distribution, the loss of the cavity (horizontal line), and the frequencies of the resonator modes. The lower diagram shows that the laser oscillates at the resonant frequencies where the gain exceeds the cavity loss

$$\alpha = \kappa - \ln\sqrt{R_1R_2}/L - \ln(1-\delta_d)/L. \tag{4.5}$$

Rewriting (4.4) as $(\lambda/2)q = L$, we find that the laser oscillates at frequencies where the standing waves with q nodes are formed in the cavity. The resonant frequencies are distributed with equal spacing of $\nu_r = \nu_q - \nu_{q-1} = c/(2L)$. (We will use ν_r for the repetition frequency of the mode-locked laser, as described later.)

The oscillation condition is schematically depicted in Fig. 4.2, where the resonant frequencies, the gain curve which is proportional to the spectral distribution function $\gamma(\nu)$, and the frequency distribution of the laser oscillation are shown. The laser oscillates at the resonant frequencies in a frequency region where the gain exceeds the loss of the cavity. In this example, the laser oscillates at 5 frequencies, although these modes often compete each other resulting in unstable oscillation. (The effect of anomalous dispersion due to atomic transition is described in Sect. 4.1.6.)

The quality of a resonator is expressed by the quality factor Q:

$$Q = \frac{\omega_0 U}{S} = \frac{\omega_0 U}{-dU/dt}, \tag{4.6}$$

where ω_0 is the resonant angular frequency, U is the energy stored in the resonator, and S is the energy loss rate. Applying this relation to the optical cavity, the energy stored in the cavity decreases exponentially during the round trip in the cavity as

$$\begin{aligned} U &= U_0\exp\left[-(\omega_0/Q)t\right] = U_0\exp\left[-(\omega_0/Q)(2L/c)\right] \\ &= U_0\exp(-2\alpha L). \end{aligned} \tag{4.7}$$

Therefore, Q is related to the power loss per unit length α as:

$$Q = \frac{\omega_0}{c\alpha} = \frac{2\pi}{\lambda\alpha}. \tag{4.8}$$

Taking the Fourier transformation of $E(t) = E_0 \exp[i\omega_0 t - (\omega_0 t/2Q)]$, we obtain the spectral distribution of the electric field in the cavity:

$$E(\omega) = \frac{1}{\sqrt{2\pi}} \int_{-\infty}^{+\infty} E(t)\mathrm{e}^{-i\omega t} dt = \frac{1}{\sqrt{2\pi}} \frac{i}{\omega - \omega_0 + i(\omega_0/2Q)}. \tag{4.9}$$

The power spectrum has a Lorentzian distribution:

$$|E(\omega)|^2 = \frac{1}{2\pi} \frac{1}{(\omega - \omega_0)^2 + (\omega_0/2Q)^2} \equiv \frac{1}{2\pi} \frac{1}{(\omega - \omega_0)^2 + (\delta\omega/2)^2}, \tag{4.10}$$

which has a full width at half maximum of $\delta\omega = \omega_0/Q$. Thus, Q expresses the quality of the resonator as

$$Q = \omega_0/\delta\omega = \nu_0/\delta\nu. \tag{4.11}$$

The ratio of the mode spacing ν_r to the line width $\delta\nu$ is called *finesse*, corresponding to the sharpness of the resonance:

$$F = \nu_r/\delta\nu. \tag{4.12}$$

As an example, we consider a resonator of $L = 0.1$ m operating at $\lambda = 1$ μm with the energy loss only by the mirrors of $R_1 = 1$ and $R_2 = 0.90$. This resonator has a Q-value of $Q = 1.19 \times 10^7$, $\nu_r = c/(2L) = 1.5$ GHz, $\delta\nu = 25$ MHz, and F = 60. As shown in this example, very sharp resonance is attained with the optical cavities.

4.1.2 Saturation Intensity and Output Power

Light of frequency $\nu = (E_u - E_l)/h$ is amplified in the transition from an upper-state u to a lower-state l in a medium with population inversion, where E_u and E_l are the energies of the upper and lower states, respectively. Let us consider a four-level system, shown in Fig. 4.3, where an atom in the ground state g is pumped to an excited state e with the rate W_p. Assuming that the decay from e to u is very fast, the rate equations for the number densities of the upper-state N_u and the lower-state N_l are described as,

$$\frac{dN_u}{dt} = W_p N_g - \left(N_u - \frac{g_u}{g_l} N_l\right)\sigma n c - \frac{N_u}{\tau_u}, \tag{4.13}$$

$$\frac{dN_l}{dt} = \left(N_u - \frac{g_u}{g_l}N_l\right)\sigma nc + \frac{N_u}{\tau_u} - \frac{N_l}{\tau_l}, \tag{4.14}$$

where n is the photon number density, σ is the stimulated emission cross section, g_u and g_l are the degeneracies of the levels u and l respectively, and $1/\tau_u$ and $1/\tau_l$ are the decay rates from u and l to lower states respectively.

When the lower-level decays to the ground level very rapidly, i.e., $\tau_l \ll \tau_u$, the population inversion density ΔN becomes $\Delta N \equiv N_u - (g_u/g_l)N_l \cong N_u$. Writing the total number density as $N \cong N_u + N_g$, the rate equation reads

$$\frac{d\Delta N}{dt} = W_p(N - \Delta N) - \sigma nc\Delta N - \frac{N_u}{\tau_u}. \tag{4.15}$$

At the steady state where $d\Delta N/dt = 0$, ΔN depends on the intensity $I = h\nu nc$ as,

$$\Delta N = \frac{\Delta N_0}{1 + I/I_s}. \tag{4.16}$$

Here the initial inversion density ΔN_0 and the saturation intensity I_s are given by:

$$\Delta N_0 = \frac{W_p N}{W_p + 1/\tau_u}, \tag{4.17}$$

$$I_s = \left(W_p + 1/\tau_u\right)\frac{h\nu}{\sigma}. \tag{4.18}$$

The gain coefficient γ given by Eq. (3.89) becomes:

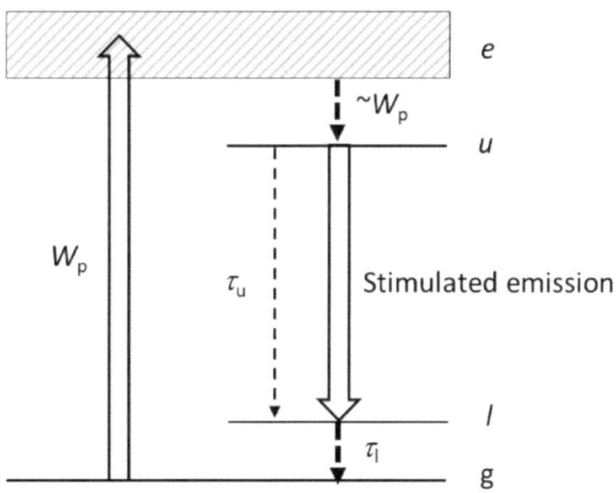

Fig. 4.3 A four-level system where light is amplified by the stimulated emission from u to l

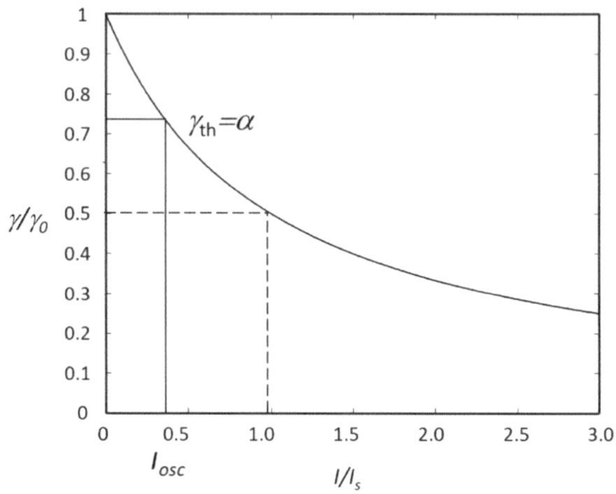

Fig. 4.4 Dependence of γ on the laser intensity I. The laser oscillates at I_{osc}, where the gain γ is equal to the cavity loss α. The gain γ becomes half of the initial value γ_0 at $I = I_s$

$$\gamma = \Delta N\sigma = \frac{\Delta N_0\sigma}{1 + I/I_s} = \frac{\gamma_0}{1 + I/I_s}, \tag{4.19}$$

where $\gamma_0 = \Delta N_0\sigma$ is the small signal gain coefficient. From Eqs. (4.16) and (4.19), ΔN and γ decrease as I increases since the transition due to the stimulated emission increases proportional to the laser intensity I.

At the steady state, stable laser oscillation is obtained at the threshold condition where the gain and the loss are balanced, i.e., $\gamma_{\mathrm{th}} = \alpha$. Referring to Fig. 4.4, the oscillation starts at $\gamma > \alpha$. However, γ decreases gradually as I increase until the balance is reached at I_{osc} where $\gamma_{\mathrm{th}} = \alpha$. The intensity I_{osc} becomes higher with the larger population inversion $(\Delta N)_0$.

Let us consider the optimum condition to obtain the highest output power available from the oscillator. The gain at the steady state is fixed to the threshold value γ_{th}, which is determined by the cavity loss α. Including the diffraction loss into κ, the intensity inside the cavity is given by

$$I = \left(\frac{\gamma_0}{\gamma_{th}} - 1\right)I_s = \left(\frac{\gamma_0}{\kappa - \ln\sqrt{R_1R_2}/L} - 1\right)I_s \tag{4.20}$$

Noting that $-\ln\sqrt{R_1R_2} = -\ln\sqrt{1-T} \simeq T/2$ for a cavity for $R_1 = 1$ and $R_2 = 1 - T$, and denoting the beam area by A, the power inside the cavity is:

$$P_{\mathrm{cavity}} = IA = I_sA\left(\frac{\gamma_0}{\kappa + T/2L} - 1\right) = I_sA\left(\frac{2\gamma_0L}{2\kappa L + T} - 1\right) \tag{4.21}$$

The output power transmitted through the mirror is:

$$P_{\mathrm{out}} = TP_{\mathrm{cavity}} = I_sA\left(\frac{2\gamma_0L}{2\kappa L + T} - 1\right)T. \tag{4.22}$$

In these equations, $2\gamma_0 L$ and $2\kappa L$ correspond to the round-trip gain and the round-trip loss, respectively.

The dependences of P_{cavity} and P_{out} on T are plotted in Fig. 4.5. As shown in Eq. (4.21), we can store very high power in the cavity if the cavity loss and the output coupling are reduced. As T increases from 0, P_{cavity} decreases rapidly and P_{out} increases up to the optimum transmittance, which is derived from $dP_{out}/dT = 0$ as:

$$T_{opt} = 2L\left(\sqrt{\gamma_0 \kappa} - \kappa\right) = 2\kappa L\left(\sqrt{\gamma_0/\kappa} - 1\right). \tag{4.23}$$

The output power at this optimum condition is:

$$P_{out,opt} = 2I_s AL\left(\sqrt{\gamma_0} - \sqrt{\kappa}\right)^2. \tag{4.24}$$

Referring to Eqs. (3.83), (3.84) and (3.87), the transition with large electric-dipole moment μ_{ul} has a short radiative lifetime τ, large stimulated-emission cross-section σ, and high gain coefficient γ. This allowed transition is suitable to start the oscillation, although it is accompanied with low output power since the saturation intensity decreases as $I_s \propto 1/\sigma$.

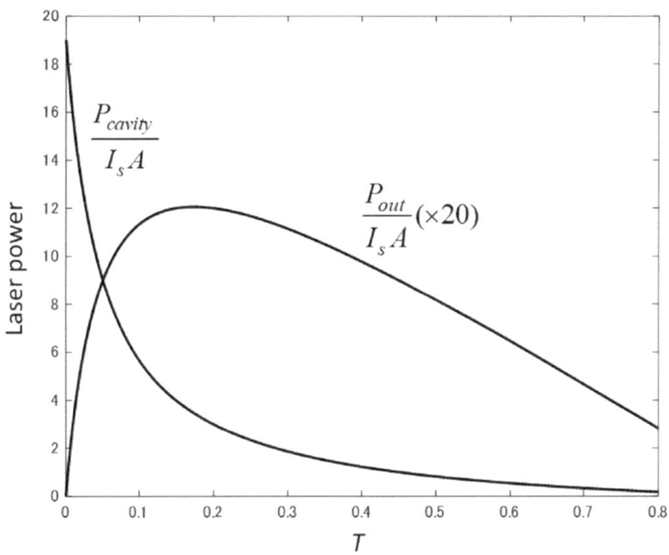

Fig. 4.5 Dependences of the power inside the cavity P_{cavity} and the output power P_{out} on the transmittance of the cavity mirror T. Both of P_{cavity} and P_{out} are normalized by $I_s A$, and P_{out} is multiplied by 20. Parameters in this example are $2\gamma_0 L = 1$ and $2\kappa L = 0.05$

4.1.3 Resonant Modes in an Optical Cavity

Let us consider a Gaussian beam described in Sect. 2.1.5. Referring to Fig. 2.2, when we place a pair of reflectors at z_1 and z_2 on the optical axis with the curvature radii of $R(z_1)$ and $R(z_2)$ which are equal to the wavefront radii of the Gaussian beam at z_1 and z_2, respectively, a standing optical field is formed due to multiple reflections in the optical resonator. In another word, the field distribution in a resonator is determined by specifying the curvatures of the mirrors and the distance between them.

As an example, we consider a symmetric resonator where two concave mirrors with same radius of curvature R are placed at $z_{1,2} = \pm(L/2)$. Introducing these conditions in Eq. (2.51), we obtain $z_0^2 = (L/2) \cdot (R - L/2)$. The spot size at z = 0 is described with R and L as

$$w_0 = \left(\frac{\lambda}{\pi}\right)^{1/2} \left(\frac{L}{2}\right)^{1/4} \left(R - \frac{L}{2}\right)^{1/4}. \tag{4.25}$$

In a special case of a *confocal resonator*, where two concave mirrors with a radius of curvature b is placed at $z = \pm(b/2)$, simple relations are obtained for the beam radius:

$$w_0 = \sqrt{\frac{\lambda b}{2\pi}}, \quad w\left(z = \pm\frac{b}{2}\right) = \sqrt{2} w_0. \tag{4.26}$$

The parameter b is called a *confocal parameter*. Often the *Rayleigh length* $z_\mathrm{R} = b/2$ is used as the approximate range where the spot size is kept near the minimum value ω_0:

$$z_\mathrm{R} = b/2 = \frac{\pi w_0^2}{\lambda}. \tag{4.27}$$

Using z_R, the spot size at z is written as

$$w(z) = w_0 \sqrt{1 + \left(\frac{z}{z_R}\right)^2}. \tag{4.28}$$

A standing wave is formed when the phase shift between the mirrors is an integral multiple of π.

$$kL - (m + n + 1)\left(tan^{-1}\frac{z_2}{z_0} - tan^{-1}\frac{z_1}{z_0}\right) = q\pi \quad (q = 1, 2, 3, \ldots). \tag{4.29}$$

Expressing this condition with the frequency ν,

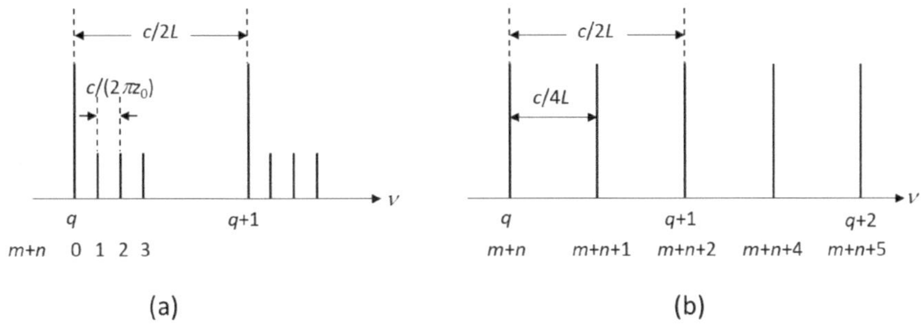

Fig. 4.6 The resonance frequencies of **a** a symmetric near-planer resonator and **b** a confocal resonator

$$\nu = \frac{c}{2L}\left[q + \frac{1}{\pi}(m+n+1)\left(tan^{-1}\frac{z_2}{z_0} - tan^{-1}\frac{z_1}{z_0}\right)\right] \equiv \nu_{q,mn}, \tag{4.30}$$

where $\nu_{q,mn}$ is the resonance frequency specified by the longitudinal mode number q and the transverse mode numbers m and n. For a symmetric resonator composed of two mirrors whose curvature radii R are larger than the cavity length L, (4.30) becomes

$$\nu_{q,mn} = \frac{c}{2L}\left\{q + \frac{L}{\pi z_0}(m+n+1)\right\}. \tag{4.31}$$

The resonant frequencies of the (q, m, n) modes are shown in Fig. 4.6a. The resonant frequencies of the TEM_{00} mode ($m = n = 0$) are distributed with equal spacing of $\nu_r = \nu_q - \nu_{q-1} = c/(2L)$, whereas the resonant frequencies of the transverse modes are offset from the longitudinal modes.

For a confocal resonator, where $z_0 = b/2$, the resonance condition (4.31) reads,

$$\nu_{q,mn} = \frac{c}{2L}\left(q + \frac{m+n+1}{2}\right). \tag{4.32}$$

The resonant frequencies of the confocal resonator are shown in Fig. 4.6b. In this case, the longitudinal and the transverse modes degenerate, resulting in the frequency separation half of the longitudinal mode separation. From this property, the confocal resonator is often used for high-resolution spectral analyses.

In the laser oscillator where light is amplified by amplifying material, the dispersion due to atomic transition is added to the optical propagation. In this case, the wavenumber k is expressed by Eq. (2.71). Considering the phase term, a phase of $\Delta k = k\chi'(\omega)/2$ is added to k, where $\chi'(\omega)$ is given by Eq. (2.67). Including Δk, the resonance condition (4.29) for a symmetric resonator reads:

$$(k + \Delta k)L - 2(m+n+1)tan^{-1}\left(\frac{z}{z_0}\right) = q\pi \quad (q = 1, 2, 3, \ldots). \tag{4.33}$$

Expressing this condition with the angular frequency ω,

$$\omega\left[1+\frac{\chi'(\omega)}{2}\right]=(2\pi)\frac{c}{2L}\left\{q+\frac{2}{\pi}(m+n+1)tan^{-1}\left(\frac{z}{z_0}\right)\right\}\equiv\omega_{q,mn}, \tag{4.34}$$

where $\omega_{q,mn}$ is the resonant angular frequency when $\Delta k = 0$. The real and imaginary parts of $\chi(\omega)$ are related as

$$\chi'(\omega)=-\chi''(\omega)\frac{\omega-\omega_0}{\Delta\omega/2}. \tag{4.35}$$

Noting $\chi''(\omega) = -n_{NR}^2\gamma/k = -c\gamma/\omega$ from Eq. (2.72) and the gain coefficient is balanced to the cavity loss coefficient as $\gamma = \alpha$ at the steady state, Eq. (4.34) reads

$$\omega\left[1-\chi''(\omega)\frac{\omega-\omega_0}{\Delta\omega}\right]=\omega\left[1+\frac{c\alpha}{\omega}\frac{\omega-\omega_0}{\Delta\omega}\right]=\omega_{q,mn}. \tag{4.36}$$

Noting $c\alpha = \omega_0/Q = \delta\omega$, we derive

$$\frac{\omega-\omega_{c,qmn}}{\delta\omega}+\frac{\omega-\omega_0}{\Delta\omega}=0, \tag{4.37}$$

where $\delta\omega$ is the full width of the cavity resonance and $\Delta\omega$ is the full spectral width of the laser transition. Since $\Delta\omega >> \delta\omega$, this relation is written as

$$\omega\simeq\omega_{q,mn}+\frac{\delta\omega}{\Delta\omega}\left(\omega_0-\omega_{q,mn}\right). \tag{4.38}$$

This equation shows that the laser oscillates near the cavity resonance frequency $\omega_{q,mn}$, but shifts slightly towards the laser transition center frequency ω_0. This effect is called *frequency pulling*.

The spectral width of the laser emission is limited by the phase fluctuation due to spontaneous emission which adds random phase to the coherent phase of the stimulated emission (Schawlow & Townes, 1958):

$$\delta\nu_L=\frac{4\pi h\nu(\delta\nu_c)^2}{P}, \tag{4.39}$$

where $\delta\nu_c$ is the frequency width of the cavity resonance and P is the laser power. For example, the linewidth of the laser emission is $\delta\nu_L = 8.0 \times 10^{-4}$ Hz for a laser operating at 1 W output power with a laser cavity of $\delta\nu_c = 25$ MHz, corresponding to a laser cavity of 10-cm length and 90% effective reflectance including various losses. Since $\delta\nu_L/\nu_L \sim 10^{-18}$ in this case, this example shows that a very narrow-band optical wave is generated with laser oscillation. However, the actual laser frequency is affected by many factors such as mechanical vibration and thermal drift of the laser cavity and laser medium. By controlling these effects, an optical wave with a frequency stability of 10^{-18} level has been achieved by locking the laser frequency to an atomic transition (for example, Katori et al., 2003).

The laser output stays constant, in principle, at the value where gain and loss are balanced. In actuality, the laser power fluctuates due to various factors: mode competition, relaxation oscillation, pump intensity fluctuation, mechanical vibration, thermal instability, and degradation of optical components. Thorough designs are required in mechanical, thermal and electronics engineering to realize the laser with excellent power and frequency stability over short and long terms.

4.1.4 Transverse Modes

When a laser medium is pumped over a large area, which is often the case for side-pumped lasers, the laser tends to oscillate in high-order transverse modes. With the high-transverse mode oscillation, high laser power is obtained due to large mode-volume, but the beam-product parameter $D\Theta$ of the laser beam increases due to large M^2 as described in Eq. (2.56).

When the laser oscillates in several longitudinal and/or transverse modes simultaneously, the oscillation tends to become unstable due to *mode competition*, where these modes interfere each other producing spatial and temporal modulations in the laser oscillation. Control of the transverse modes is necessary to produce a high-quality laser beam in terms of temporal, spectral and spatial properties.

Oscillation in high-order transverse modes is suppressed by placing an aperture in the laser cavity with the diameter slightly larger than the spot size of the fundamental mode. Another method which became available due to laser diode pumping of solid-state lasers is to axially pump the laser medium so that the pumping area matches with the spot size of the TEM_{00} mode. This method, which is implemented in solid-state lasers and fiber lasers, is useful also to increase the overall efficiency of the laser.

In any of these configurations, high optical uniformities are required to the optical components in the laser cavity and downstream, to minimize deleterious effects such as wavefront distortion, birefringence, absorption, and scattering under strong pumping conditions.

4.1.5 Longitudinal Modes

The laser oscillates at the cavity resonance frequencies where the gain exceeds the losses. When the laser oscillates at multiple longitudinal modes, these oscillations interfere each other, resulting in the "beat" in the temporal waveform. For example, with the cavity of $L = 0.1$ m, the mode spacing is $\Delta\nu = 1.5$ GHz, resulting in the temporal modulation of 0.67 ns. When many longitudinal and transverse modes oscillate simultaneously, the temporal modulation leads to fluctuations in the temporal waveform and the output power due to the mode competition.

There are several approaches to control the longitudinal modes. A simple but effective method is to place an etalon in the laser cavity, slightly tilted from normal incidence. The etalon, which is composed of parallel reflectors, acts as a frequency-dependent transmitter due to interferences of the waves reflected at both surfaces. As an example, a 1-cm thick plane parallel plate of fused silica (refractive index of $n = 1.46$) coated on both surfaces with the reflectivity of $R = 10\%$ has a mode spacing of $\nu_r = 9.6$ GHz and a finesse of $F = 0.49$. Although the modulation of the transmittance is not large, it can select only one longitudinal mode when its transmission peak is adjusted to the resonant frequency of the laser transition.

Another reliable approach is to inject a stable, single-frequency laser beam into the laser cavity. Then the laser oscillation builds up from this frequency since the stimulated emission rate of this mode is higher due to high photon density than the spontaneous emission rates of other modes.

Another method is to use a very short cavity so that the cavity spacing is larger than the spectral width of the laser transition. In the ring-cavity Nd:YAG laser named NPRO for non-planer ring-oscillator (Kane & Byer, 1985), the mode spacing exceeds the spectral width of the Nd:YAG, since the cavity length is only a few millimeters. In this design, unidirectional oscillation is achieved by suppressing the counterpropagating wave using the Faraday effect by applying the magnetic field on the crystal. Higher power is obtained by unidirectional oscillation since the hole burning caused by the standing wave is avoided. This laser is used as an oscillator of the high-stability laser for gravitational detection (see Sect. 2.4.4).

4.1.6 Mode Locking

The resonance frequencies of the laser cavity are distributed at equal separations given by $\omega_r = 2\pi(c/2L)$ where L is the cavity length, with additional offset due to dispersion at the laser transition frequency as described in Sect. 4.1.3. By locking the longitudinal modes, the laser oscillates simultaneously at equally spaced frequencies with the same phase. Simplifying that these modes have same amplitudes, the electric field is given by:

$$\begin{aligned} E(t) &= E_0 \sum_{n=1}^{N} e^{i(\omega_0 + n\omega_r)t} = E_0 e^{i\omega_0 t} \sum_{n=1}^{N} \cos(n\omega_r t) \\ &= E_0 e^{i\omega_0 t} \frac{\sin(N\omega_r t/2)}{\sin(\omega_r t/2)}. \end{aligned} \tag{4.40}$$

The formula $\sum_{j=0}^{n} \cos(\varphi + j\alpha) = \sin[(n+1)\alpha/2] \cdot \cos(\varphi + n\alpha/2)/\sin(\alpha/2)$ is used in deriving this equation. The temporal pulse shape of this field is,

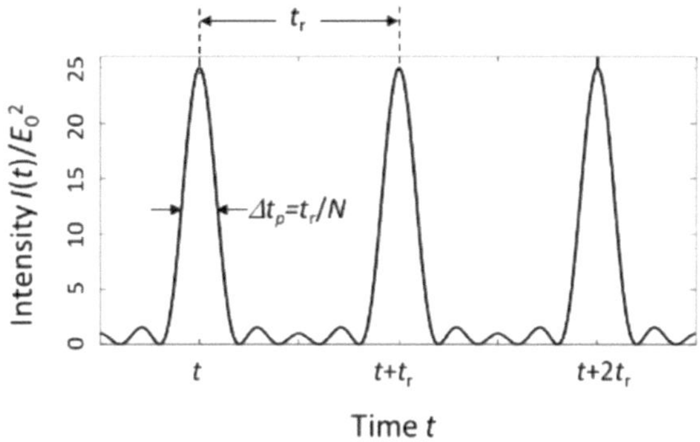

Fig. 4.7 Temporal pulse shape of a mode-locked pulse for N = 5

$$I(t) = |E(t)|^2 = E_0^2 \frac{\sin^2(N\omega_r t/2)}{\sin^2(\omega_r t/2)} = E_0^2 \frac{\sin^2(N\pi t/t_r)}{\sin^2(\pi t/t_r)} \tag{4.41}$$

An example of this pulse shape for $N = 5$ is shown in Fig. 4.7. The intensity takes the maximum value at $\sin^2(\pi t/t_r) = 0$, i.e., $t/t_r = n$ $(n = 0, 1, 2, \ldots)$. The pulse interval t_r, the pulse width of each pulse Δt_p, and the peak intensity of each pulse $I(t)_{peak}$ are given respectively as

$$t_r = \frac{2\pi}{\omega_r} = \frac{1}{\nu_r} = \frac{2L}{c}, \tag{4.42}$$

$$\Delta t_p = \frac{2L}{Nc} = \frac{1}{N\nu_r} = \frac{t_r}{N}, \tag{4.43}$$

$$I(t)_{\text{peak}} = E_0^2 N^2. \tag{4.44}$$

The pulse interval t_r is equal to the round-trip time of the laser pulse in the laser cavity since a single short laser pulse circulates in the cavity. The pulse width Δt_p is the inverse of the total bandwidth $\Delta\nu_{\text{total}} = N\nu_r$, corresponding to the uncertainty relation between time and energy. The peak intensity increases as the square of the number of the oscillating modes N^2 due to constructive interference of the in-phase oscillation of N modes.

Very short and intense laser pulses are generated by locking the longitudinal modes of broad-band laser medium. For example, the titanium-doped sapphire laser ($Ti{:}Al_2O_3$) has a gain over 660–1180 nm centered at ~ 800 nm, corresponding to the frequency bandwidth of $\Delta\nu_{total} \sim 2 \times 10^{14}\ s^{-1}$. Ultrashort laser pulses of less than 6 fs duration have been generated with a mode-locked Ti:sapphire laser, which is only 2.2 cycles of the 800 nm optical wave (Sutter et al., 1999).

The mode-locking is achieved either by passive or active mode-locking. In the passive mode-locking, a saturable absorber is placed in the cavity close to the total-reflectance mirror, with which only high intensity pulses are selected from the random pulses generated by spontaneous emission since the saturable absorber transmits preferentially the intense pulses. As the saturable absorber, organic dyes, semiconductors and more recently carbon nanotubes and graphene are used. The major requirements to the saturable absorbers are broad bandwidth, low saturation intensity, distinct saturation property, short recovery time, and durability to intense radiation and high temperature. Another method for the passive mode-locking is the Kerr lens mode-locking (KLM) of the Ti:sapphire and other lasers, where the laser crystal is used as an intensity dependent converging lens due to nonlinear self-focusing in the crystal (Spence et al., 1991). Selection of a high intensity pulse is achieved by optimizing the cavity configuration to preferentially select the small beam waist of the self-focused laser beam. The KLM has advantages of low optical loss, very fast response, and laser-wavelength independent, although it is sensitive to laser configuration and alignment for obtaining stable mode-locking.

In the active mode-locking, either a phase modulator (FM) or an amplitude modulator (AM) is placed in the laser cavity, both operated at the modulation frequency ν_{m} which is exactly equal to the axial mode separation of $\nu_{\mathrm{r}} = c/2L$. Since the side bands to each axial mode generated by the FM or AM modulation have the same frequencies of the adjacent modes, all modes are coupled together and oscillate in the same phase.

4.1.7 Optical Frequency Comb and Carrier Envelope Phase Control

A mode-locked laser oscillates at multiple frequencies which are equally spaced at the axial mode separation of $\nu_{\mathrm{r}} = c/2L$. The equal-spaced optical frequencies are called an *optical frequency comb* and depicted by thick solid lines in Fig. 4.8. When ν_{r} is determined accurately, the frequency comb can be used as a frequency ruler which can span over a broad spectral region. Noting this property, a mode-locked short pulse laser was applied to high resolution spectroscopy for the first time by Eckstein et al. (1978).

Considering the mode-locked laser in detail, the oscillation frequencies of the mode-locked pulses are given by,

$$\nu_n = n\nu_{\mathrm{r}} + \nu_{off}, \tag{4.45}$$

where $\nu_{\mathrm{r}} = c/2L$ is equal to the repetition frequency of the mode locked laser pulses, which is 1 GHz for a laser cavity of $L = 15$ cm for example, n is a large integer of the order of 10^5, and ν_{off} is the frequency offset between the laser oscillation frequency ν_n and the cavity resonance frequency $n\nu_{\mathrm{r}}$ as shown in Fig. 4.8. This relation implies that the laser frequency ν_{n} is accurately determined when the laser repetition rate is locked exactly equal to the cavity resonance ν_{r} and the frequency offset ν_{off} is controlled to zero or some fixed value.

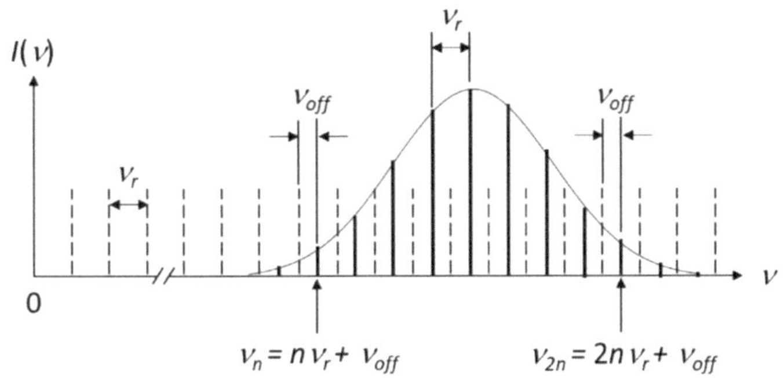

Fig. 4.8 The frequency comb of a mode-locked laser (solid lines) and the resonant frequencies of the laser cavity (dotted lines). The offset of these frequencies is shown by ν_{off}. Here ν_n and ν_{2n} are the laser frequencies of the nth order and the $2n$th order laser oscillations, respectively

A novel self-referencing technique to measure and control the frequency offset ν_0 was realized by Holzwarth et al. (2000) and Jones et al. (2000) by noting the relation,

$$2\nu_n - \nu_{2n} = 2(n\nu_{rep} + \nu_{off}) - (2n\nu_{rep} + \nu_{off}) = \nu_{off}, \tag{4.46}$$

where $\nu_{2n} = 2n\nu_r + \nu_{off}$ is the laser frequency of the $2n$th mode. This relation shows that ν_{off}, which is in the radio-frequency region, is measured as the beat frequency when the second harmonic of the nth mode oscillation $2\nu_n$ is mixed with the $2n$th mode oscillation ν_{2n}. The broad coherent source covering a full octave frequency region, which is required for this method, was generated by propagating a Ti:sapphire laser pulse centered at 820 nm in a photonic crystal silica fiber (Knight et al., 1996; Ranka et al., 2000). Noting that a broadband continuum pulse generated by self-phase modulation is phase-locked to the incident field (Bellini & Hänsch, 2000), a phase-coherent continuum stretching from 510 to 1125 nm was generated by the four-wave mixing in the near-zero dispersion photonic crystal fiber. The pulse repetition frequency ν_r and the frequency offset ν_{off} were controlled by forming a closed loop to feedback the laser pulses to the reflector of the laser cavity. The optical frequency comb thus realized direct link between the frequency standard Cs atomic clock of microwave frequency to accurate atomic clocks in the optical region in one step (Diddams et al., 2000).

The frequency offset ν_{off} of the laser oscillation from the cavity mode results in the shift of the carrier wave from the envelope of the mode-locked pulses with the relation of $\Delta\varphi \cdot \nu_{rep} = 2\pi\,\nu_{off}$ (Reichert et al., 1999). The electric field of the mode-locked laser including the frequency offset in Eq. (4.40) is expressed as

$$E(t) = E_0 e^{i(\omega_0 t - \Delta\varphi)} \frac{\sin(N\omega_r t/2)}{\sin(\omega_r t/2)} \tag{4.47}$$

The dependence of the electric field on the phase shift $\Delta\varphi$ is shown in Fig. 4.9. When the phase shift $\Delta\varphi$ is offset from 0, the position of the peak of the electric field

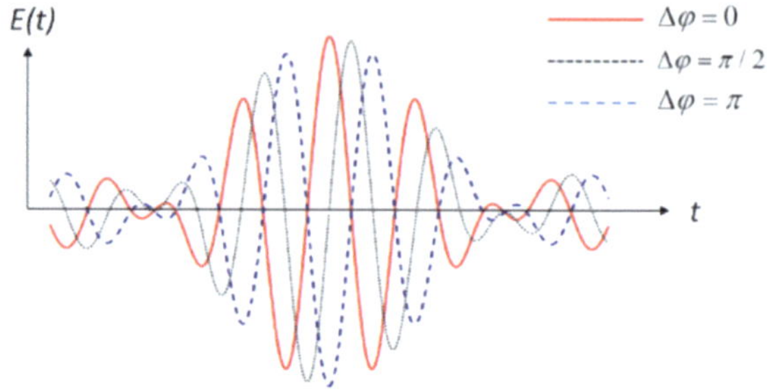

Fig. 4.9 Dependence of the electric field on the phase of the carrier wave relative to the pulse envelope, shown for the phase shifts of $\Delta\varphi = 0,\ \pi/2,$ and π, respectively

shifts from the peak of the pulse envelope, reversing sign at $\Delta\varphi = \pi$. The electric field of an ultrashort pulse becomes unstable when $\Delta\varphi$ or equivalently ν_{off} changes from pule to pulse. Stabilization of the frequency offset to $\nu_{\text{off}} = 0$, called *carrier envelope phase control*, is necessary to generate stable ultrashort pulses, as well as to use the frequency comb for various applications such as precision-spectroscopy and high-order harmonic generation at short wavelengths.

4.2 Propagation of a High-Power Laser Beam

High energy pulsed lasers are composed of many elements for generation, amplification, frequency conversion, focusing, and control of a laser beam. In this section, we consider amplification and propagation of a laser beam in a laser system composed of multiple-stage amplifiers. We restrict our discussion to laser pulses longer than ~ 1 ps duration corresponding to a spectral width of $\Delta\nu \leq 1$ THz, where pulse propagation is not affected by dispersion. The propagation of ultrashort laser pulses is considered in Chap. 7.

4.2.1 Layout of High-Power Laser System

A schematic layout of a multistage laser system for scientific and industrial applications is shown in Fig. 4.10. An oscillator to generate controlled laser pulses is isolated with an optical isolator to decouple the oscillator from the following amplifier stages. The laser pulse is then amplified by a pre-amplifier and main amplifiers. To keep the fluence below the saturation fluence of the amplifier materials and/or below the damage threshold of the optical components, the laser beam is spatially expanded with beam expanders which are also used for spatial filtering and image relaying as described later. The amplified laser beam is wavefront-corrected with an adaptive optics, frequency converted with nonlinear crystals (not shown in this figure) and focused onto a target with an aspheric lens or a reflector. The optical

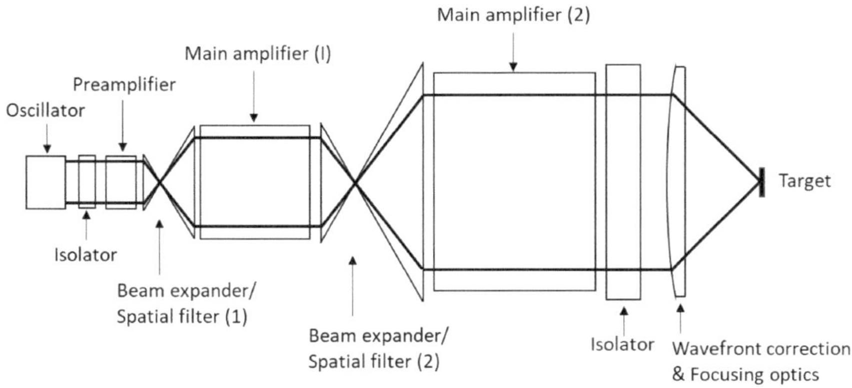

Fig. 4.10 A schematic layout of a multiple-stage laser system

coupling between the amplifiers and the target reflection is prevented with optical isolators.

In high energy lasers for laser fusion research, for example, a laser system is composed of many amplifier chains in parallel to generate high laser energy. In the complex laser systems, the laser beam often propagates over hundreds of meters from the oscillator to the target. In this section, we describe several basic concepts in amplification and propagation of the laser beam (Holzrichter et al., 1982).

4.2.2 Amplification of a Laser Beam

A laser beam is amplified by extracting the energy stored as population inversion E_{st}, which is given using Eq. (4.17) as

$$E_{st} = h\nu \Delta N_0 = h\nu \frac{W_p N}{W_p + 1/\tau_u}. \tag{4.48}$$

Here N is the number density of the lasing atoms, ΔN_0 is the initial population inversion density, W_p is the pumping rate to the excited state, and τ_u is the spontaneous emission lifetime of the upper laser level. Although the stored energy increases with τ_u, extraction of the stored energy becomes more difficult since the saturation fluence increases.

We shall consider propagation of a laser pulse in an amplifier with the initial inversion density ΔN_0. The energy stored in the amplifier is efficiently extracted when the laser fluence becomes close to the saturation fluence, where the temporal pulse shape changes due to the saturation effect. The laser pulse is initially amplified exponentially as $I(z) = I_0 \exp(g_0 z)$, where $g_0 = \sigma \Delta_0$ is the small signal gain coefficient. When the laser intensity increases, gradually approaching to the saturation intensity

I_s, the gain decreases as $g = g_0(1 + I/I_s)$. Since the pulse shape changes drastically in a multiple-stage amplifier system, it is necessary to design the pulse shape at the front-end to generate a pulse shape required for target irradiation by calculating the amplification property. The spectral distribution also changes due to spectral dependence of the gain saturation when the spectral broadening is inhomogeneous.

Describing the photon number density as n, the time dependences of ΔN and n are described, for a 4-level system, as

$$\frac{\partial \Delta N}{\partial t} = -nc\sigma \Delta N, \tag{4.49}$$

$$\frac{\partial n}{\partial t} = nc\sigma \Delta N - c\frac{\partial n}{\partial z}. \tag{4.50}$$

The spatial and temporal development of the photon density $n(x, t)$, called *Frantz-Nodvik equation*, is derived from these coupled nonlinear equations (Frantz & Nodvik, 1963) as

$$\frac{n(z, t)}{n_0(t - z/c)} = \left\{1 - \left(1 - e^{-\sigma \Delta N z}\right) \exp\left[-\sigma n_0 c\left(t - \frac{z}{c}\right)\right]\right\}^{-1}, \tag{4.51}$$

where n_0 is the initial photon density. Considering a square laser pulse of duration t_{p}, the fluence is given by $E = n(h\nu)ct_p$. Integrating Eq. (4.50) in time, the energy gain passing through an amplifier of length $z = l$ is given by (Koechner, 2006)

$$G = \frac{E_{\mathrm{s}}}{E_{\mathrm{in}}} \ln\left\{1 + \left[\exp\left(\frac{E_{\mathrm{in}}}{E_{\mathrm{s}}}\right) - 1\right]G_0\right\} \tag{4.52}$$

where E_{in} is the fluence before amplification and E_{s} is the saturation fluence,

$$E_{\mathrm{s}} = \frac{h\nu}{\gamma\sigma} = \frac{J_{\mathrm{st}}}{\gamma g_0}, \tag{4.53}$$

where $\gamma = 1 + g_u/g_l \approx 1$ for a four-level system.

There are several issues related to high gain amplifiers: pre-lasing and parasitic oscillation due to amplified spontaneous emission. Pre-lasing is the lasing in a single amplifier or coupling of multiple amplifiers during pumping, exhausting the stored energy before arrival of the main laser pulse. This occurs when $R_1 R_2 \exp(2\gamma l) \geq 1$ is satisfied, where R_1 and R_2 are reflectivity of the surfaces which are close to normal to the laser beam axis. For example, pre-lasing will take place when two amplifiers placed in tandem with a gain-length product of $\gamma l = 3$ are faced with two surfaces of only $R = 0.25\%$. Various methods are used to avoid the pre-lasing; tilting the surfaces from normal to the laser beam axis, eliminating scattering of the optical surfaces, applying antireflection coatings to the optical surfaces, and placing optical isolators between the high-gain amplifiers and the target. Even if the pre-lasing does not take place, the amplified spontaneous emission could affect the target before arrival of the main laser pulse since the total energy is high enough to modify the surface property. Optical shutters such as Pockels cells are used to remove the spontaneous emission outside the time window of the laser pulse.

Parasitic oscillation is the lasing due to amplified spontaneous emission within high-gain amplifiers, such as laser rods, active mirrors, thin disks, and large-aperture disks. The parasitic oscillation takes place along the diagonal direction of the active disks or around the side surface of the amplifier as peripheral modes. The parasitic oscillation is prevented by applying absorbing material (liquids, adhesives, and glasses), whose refractive index is very close the refractive index of the laser material, to reduce reflection at the side surfaces and absorb the spontaneous emission.

4.2.3 Spatial Beam Profile

Propagation of a laser beam is controlled to transport the laser beam through the laser system and to irradiate the target with the required irradiation distributions such as diffraction-limited focusing for high intensity irradiation or uniform irradiation over a large area for material processing.

A laser beam with a flat-top beam profile is preferable for efficient extraction of the energy stored in the amplifier and to reduce laser-induced damage to optical elements. We define a filling factor as a measure of the uniformity of the beam profile:

$$F = \frac{\Phi_{av}}{\Phi_0} = \frac{\int_S \Phi(\mathbf{r})d\mathbf{r}/S}{\Phi_0}, \tag{4.54}$$

where Φ_{av} is the fluence averaged over the beam area S, and Φ_0 is the peak value of the fluence. The beam area is defined by $S = \pi w^2$ where w is the beam radius at $\Phi(r) = (1/e^2)\Phi_0 = 0.135\ \Phi_0$ for a Gaussian beam. The filling factor is $F = 1$ for a flat-top beam and $F = 0.865$ for a Gaussian beam as described in Sect. 2.1.5.

The flat-top beam results in sharp ring structures due to Fresnel diffraction at the sharp edge of the beam. A better beam profile is a flat-top distribution in the center, decreasing gradually to 0 at the edge of the beam. This beam profile is modeled by a super-Gaussian:

$$\Phi_n(r) = \Phi_0 \exp\left[-2(r/w)^n\right], \tag{4.55}$$

where $n = 2$ corresponds to the Gaussian beam and$w$ defines the beam radius. The fluence distribution of the super-Gaussian beam Φ_n(r) with n = 2, 6, 10 and 20 are shown in Fig. 4.11. As n increases, the beam profile becomes closer to the flat-top with lager filling factor.

Laser beam truncation by apertures such as lenses and amplifiers produce diffraction ripples in the downstream (Campillo et al., 1973). Truncation with a hard aperture at 1% of the peak intensity results in ~ 20% intensity modulation, as estimated from Eq. (2.98) for the intensity distribution of two-beam interference. Various types of

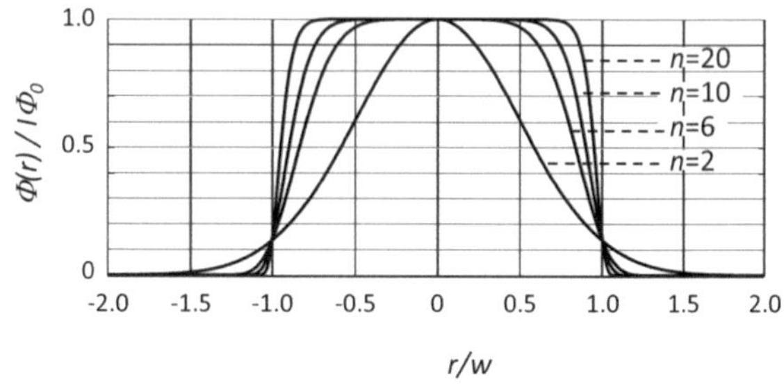

Fig. 4.11 The fluence distribution of the super-Gaussian beams with $n = 2$, 6, 10 and 20

"soft apertures" have been developed to produce a super Gaussian beam, with a serrated aperture (Auerbach et al., 1994) most widely used.

The intensity nonuniformity due to the Fresnel diffraction can be reduced by *image relaying* with the configuration shown in Fig. 4.12 (Hunt et al., 1977, 1978). In this configuration, a pair of lenses L_1 and L_2, with focal lengths of f_1 and f_2 respectively, are used to form an image of A located d_1 upstream of L_1 onto B located d_2 downstream of L_2. The distance d_2 is given by $d_2 = m(f_1 + f_2 - md_1)$ where $m = f_2/f_1$ is the magnification of the pair of lenses. (As a typical example, $d_2 = f_2$ when $d_1 = f_1$.) The image at B can be further relayed to C by the second optics formed by the lenses L_3 and L_4. Since the field distribution at A is reproduced at B and further at C, the laser beam can be propagated over a long distance without being affected by the Fresnel diffraction, if the beam is not truncated by the apertures.

The intensity distribution of the laser beam is often accompanied with the modulations of high spatial frequencies arising from scatterings by small particles, inhomogeneities of optical components, and nonlinear self-focusing (Kuroda & Kato, 1981). Describing the typical scale length of the amplitude or phase nonuniformity of the laser beam by l, the spatial frequency f is given by $f = 1/l$. The nonuniformities of high spatial frequencies, which lead to small scale self-focusing and optical damage to laser components, can be removed by *spatial filtering* by placing a pinhole at the focus of L_1 as shown in Fig. 4.12 (Hunt et al., 1978).

Referring to Eq. (2.100), the field distribution $A_f(x_f, y_f)$ at the focal plane of F_1 is

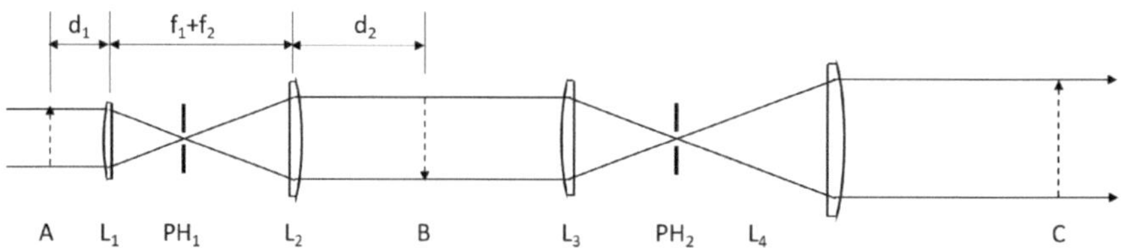

Fig. 4.12 Image relaying and spatial filtering. The field distribution at A is relayed to B and to C by the optics formed by a pair of lenses. During the beam transport, high spatial frequency components are filtered by the pinholes located at the focal points of the lenses

$$A_f(x_f, y_f) = \frac{\exp(-ikf)}{i\lambda f} \int\int_{-\infty}^{+\infty} A_{in}(x, y) \exp\left[i\frac{2\pi}{\lambda f}(xx_f + yy_f)\right] dxdy$$
$$= \frac{\exp(-ikf)}{i\lambda f} A_{in}(f_x, f_y), \tag{4.56}$$

where $A_{in}(x, y)$ is the incident field at L_1, f is the focal length, and $A_{in}(f_x, f_y)$ is the Fourier transform of $A_{in}(x, y)$. The spatial frequencies (f_x, f_y) are written by the coordinates (x_f, y_f) as

$$f_x = \frac{x_f}{\lambda f}, \quad f_y = \frac{y_f}{\lambda f}. \tag{4.57}$$

A pinhole of radius r_f placed at the focal position removes high spatial frequency components of the incident beam with the cut-off frequency $f_{cut} = r_f/\lambda f$. The growth of the small-scale intensity nonuniformities is thus suppressed by the spatial filter. The spatial filter is evacuated to prevent air breakdown at the focus.

4.2.4 Wavefront Aberration

A laser beam without any aberration can be focused to a small beam whose size is determined by the diffraction limit. For example, a Gaussian beam of diameter D is focused to a gaussian beam of radius r given by

$$r = \frac{2\lambda f}{\pi D} = 0.635\lambda F, \tag{4.58}$$

where f is a focal length of the lens and $F = f/D$ is the F-number. The super-Gaussian beam is focused to a slightly larger spot (Hunt et al., 1976).

When the incident beam has wavefront aberration W, the beam is focused to a larger spot with reduced peak intensity, where W is defined as the separation of the wavefront from the reference sphere. The intensity at the center of the beam decreases as

$$i \equiv \frac{I_W}{I_{W=0}} = 1 - \left(\frac{2\pi}{\lambda}\right)^2 \langle \Delta W^2 \rangle = 1 - \left(2\pi \frac{\sigma_W}{\lambda}\right)^2, \tag{4.59}$$

where $\langle \Delta W^2 \rangle = \langle W^2 \rangle - (\langle W \rangle)^2$ is the mean square and $\sigma_W = \sqrt{\langle \Delta W^2 \rangle}$ is the standard deviation of the aberration, and average is taken over the beam area. The tolerance of $i \geq 0.8$ is called Maréchel criterion, which corresponds to $\sigma_W \leq \lambda/14$ (Born &

Wolf, 1999, Sect. 9.3). The Maréchel criterion is also expressed in terms of the peak-to-valley (p–v) of the wavefront error as $W_{p-v} \leq \lambda/4$. W_{p-v} and σ_W are related as $W_{p-v} \approx 3\sigma_W$ if the aberration has a Gaussian distribution.

In general, a laser system is composed of many optical elements, which may be grouped into reflection optics and transmission optics. When we denote the aberrations of the reflection and transmission optics as $\sigma_{W,r}$ and $\sigma_{W,t}$, respectively, and considering that these aberrations are not correlated each other, the total aberration of the system is

$$\left(\sigma_{W,total}\right)^2 = N_r\left(\sigma_{W,r}\right)^2 + N_t\left(\sigma_{W,t}\right)^2 + \left(\sigma_{W,nl}\right)^2, \tag{4.60}$$

where N_r and N_t are the total numbers of the reflection and transmission optical elements, respectively. The aberration $\sigma_{W,nl}$ is the phase retardation due to nonlinear refractive index, which is described in the next subsection. If the laser beam is focused to twice the diameter of the diffraction-limited beam, the peak intensity is reduced to $i = 0.25$, corresponding to the total aberration of $\sigma_{\text{W,total}} = \lambda/7.3$. As a rule of thumb, the aberration of each component is kept below $\sigma_{W,comp} \approx \sigma_{W,total}/\sqrt{N}$ where N is the total number of optical components in a beam line.

The reflection optics with a peak-to-valley (p–v) surface error of $W_{p-v,r}$ results in the aberration of $\sigma_{W,r} \approx 2W_{p-v,r}/3 = W_{p-v,r}/1.5$ since the optical length is twice of W_{p-v} in reflection. The p–v aberration of a transmission optics is

$$\sigma_{W,t} \approx \left[(n-1)\left(W_{p-v,t1} + W_{p-v,t2}\right) + l\delta n_{p-v}\right]/3, \tag{4.61}$$

where $W_{p-v,t1}$ and $W_{p-v,t2}$ are the p–v aberrations of the two surfaces, n is the refractive index, δn_{p-v} is the p–v variation of the refractive index due to inhomogeneity of the optical materials, and l is the length of the material. This estimation shows that optical materials with high optical homogeneities and accurate surface finish are required for high power laser optics.

In actual laser systems, especially in repetitively operated lasers, large aberration arises due to thermal and mechanical loadings to the materials, such as nonuniform pumping, nonuniform cooling, thermal lensing, and stress birefringence. Various types of lasers have been designed to reduce these effects under high power operation (Koechner, 2006).

4.2.5 Nonlinear Refractive Index and B-Integral

Due to nonlinear polarization, the refractive index of a transparent material increases from n_0 to n at high laser intensities as (Boyd, 2020)

$$n = n_0 + \delta n = n_0 + \gamma_{\text{nl}} I = n_0 + \overline{n}_2 |E|^2 = n_0 + \frac{3\chi^{(3)}}{n_0}|E|^2. \tag{4.62}$$

Here the nonlinear refractive index is expressed in terms of the intensity I and the electric field amplitude E with $I = (\varepsilon_0 c/2)|E|^2$ (Eq. 2.28). The coefficient $\gamma_{\rm nl}$ is called the nonlinear refractive coefficient (also called nonlinear refractive index), $\overline{n}_2$ is the nonlinear refractive index, and $\chi^{(3)}$ is the third-order nonlinear susceptibility (see Sect. 4.1). These coefficients are related as follows:

$$\gamma_{\rm nl} = \frac{4}{n_0 \varepsilon_0 c}\overline{n}_2 = \frac{12}{n_0^2 \varepsilon_0 c}\chi^{(3)}. \tag{4.63}$$

The $\gamma_{\rm nl}$ and $\overline{n}_2$ are converted as $\gamma_{\rm nl}$ [cm^2/W] $= 4.19 \times 10^{-3}$ (n_2/n_0) [esu].

Since the optical path length increases due to increase in the refractive index, the laser-irradiated region acts as a positive lens, resulting in self-focusing of the laser beam. The total nonlinear phase shift due to increase in the refractive index is called *B-integral*:

$$B = \frac{2\pi}{\lambda}\int_0^L \gamma_{\rm nl} I(z)dz, \tag{4.64}$$

where λ is the laser wavelength in the medium.

The nonlinear refractive coefficients of fused silica, phosphate laser glass, Al_2O_3, YAG and $LiNbO_3$ are $\gamma = 2.1$, 2.8, 3.2, 6.9 and 9.1 (in unit of 10^{-16} cm^2/W), respectively (Koechner, 2006, Table 4.2). For $\lambda = 1$ μm and $I = 10$ GW/cm^2, the refractive index of a fused silica increases by $\delta n = \gamma_{\rm nl} I \approx 2 \times 10^{-6}$, resulting in the B-integral of $B \sim 0.12$ after propagation in a fused silica for $L = 1$-cm.

When a Gaussian beam propagates in a nonlinear medium over a length l, the phase at the beam center $r = 0$ is delayed by amount of $\Delta\varphi \simeq \gamma_{\rm nl} I_0 l$ relative to $r = w_0$, where I_0 is the intensity at $r = 0$ and w_0 is the beam radius. Approximating the delayed phase by a spherical wave, the laser beam will be focused at $f \simeq w_0^2/(2\gamma_{\rm nl} I_0 l)$. Once the self-focusing starts, the focal length decreases due to increase in the intensity at the beam center, resulting in faster focusing during propagation and catastrophic focusing. At high power, the laser beam breaks up into small filaments, called *small-scale self-focusing* and *beam breakup*. (Detailed descriptions are given in Sect. 5.4.2.) The high spatial frequency components due to the beam break-up are removed with spatial filters in multi-stage-amplifier lasers to prevent catastrophic damages to the optical components. A rule of thumb for the allowed B-integral is $B \sim 2$ between the spatial filters.

4.2.6 Propagation of a High-Power Laser Pulse

Let us consider propagation of a high-power laser pulse, propagating in the z-direction, including the nonlinear refractive index (Boyd, 2020; Shen, 1984). Starting from Eq. (2.11), we include nonlinear polarization $\boldsymbol{P}_{\rm NL}$ in the electric displacement $\boldsymbol{D}$

as $\boldsymbol{D} = \boldsymbol{\varepsilon} \cdot \boldsymbol{E} + \boldsymbol{P}_{\mathrm{NL}}$ (see Sect. 5.1.1). The nonlinear polarization which contributes to the optical Kerr effect (self-focusing) is $P_{\mathrm{NL}} = (3/4)\varepsilon_0 \chi^{(3)} |E|^2 E$. The wave equation is then written as,

$$\left(\frac{\partial^2}{\partial x^2} + \frac{\partial^2}{\partial y^2}\right) E(\boldsymbol{r}, t) + \frac{\partial^2}{\partial z^2} E(\boldsymbol{r}, t) - \frac{n_0(\omega)^2}{c_0^2} \frac{\partial^2 E(\boldsymbol{r}, t)}{\partial t^2} = \mu_0 \frac{\partial^2 P_{\mathrm{NL}}}{\partial t^2}. \tag{4.65}$$

Here $n_0^2(\omega)/c_0^2 = \varepsilon\mu$ includes the material dispersion arising from the linear polarization $\boldsymbol{P}_{\mathrm{L}}$. The lefthand side of this equation describes the linear propagation and the righthand side is the contribution from the nonlinear polarization.

First, we consider propagation in the linear region without including the right-hand side of this equation. Separating the dependence of $E(\boldsymbol{r}, t)$ on the transverse and longitudinal coordinates, we explicitly describe the dependence on z, although dependence on x and y are implicitly included. For this case, the electric field with the carrier frequency ω_0 and the wavenumber $k_0 = n_o(\omega_0)\omega_0/c_0$ is written as

$$E(z, t) = A(z, t) \exp(ik_0 z - i\omega_0 t), \tag{4.66}$$

where the envelope $A(z, t)$ varies slowly with respect to z and t (slowly varying envelope approximation: SVEA).

Equation (4.65) can be analyzed in the frequency domain by taking the Fourier transformation of $E(z, t)$;

$$\begin{aligned} \tilde{E}(z, \omega) &= \int_{-\infty}^{\infty} E(z, t) \exp(i\omega t) dt \\ &= \int_{-\infty}^{\infty} A(z, t) \exp[ik_0 z + i(\omega - \omega_0)t] dt + c.c \\ &= \tilde{A}(z, \omega - \omega_0) \exp(ik_0 z) + \tilde{A}^*(z, \omega + \omega_0) \exp(-ik_0 z) \\ &\cong \tilde{A}(z, \omega - \omega_0) \exp(ik_0 z). \end{aligned} \tag{4.67}$$

Here $\tilde{A}(z, \omega)$ is the Fourier transformation of $A(z, t)$. Considering only the laser pulse moving to $+z$ direction, the second term in the third line (moving toward $-z$ direction) is not included in the last line. Noting $k(\omega) = n(\omega)\omega/c_0$, Eq. (4.65) is written in the frequency domain as

$$\left(\frac{\partial^2}{\partial x^2} + \frac{\partial^2}{\partial y^2}\right) \tilde{E}(z, \omega) + \frac{\partial^2 \tilde{E}(z, \omega)}{\partial z^2} + k^2 \tilde{E}(z, \omega) = 0. \tag{4.68}$$

The second term of this equation is

$$\frac{\partial^2 \tilde{E}(z, \omega)}{\partial z^2} \simeq \left[2ik_0 \frac{\partial \tilde{A}(z, \omega - \omega_0)}{\partial z} - k_0^2 \tilde{A}(z, \omega - \omega_0)\right] \exp(ik_0 z), \tag{4.69}$$

where the second-order derivative has been dropped under SVEA. Equation (4.68) now reads,

$$\left(\frac{\partial^2}{\partial x^2}+\frac{\partial^2}{\partial y^2}\right)\tilde{A}(z,\omega-\omega_0)+2ik_0\frac{\partial\tilde{A}(z,\omega-\omega_0)}{\partial z} + \left(k^2-k_0^2\right)\tilde{A}(z,\omega-\omega_0)=0. \tag{4.70}$$

With approximation of $k^2-k_0^2\simeq 2k_0(k-k_0)$, we expand k with ω as,

$$k=k_0+\Delta k_{\rm nl}+k_1(\omega-\omega_0)+\frac{1}{2}k_2(\omega-\omega_0)^2+\cdots. \tag{4.71}$$

Here $k_1=1/v_{\rm g}$ where $v_{\rm g}$ is the group velocity, and $k_2=\left(d^2k/d\omega^2\right)_{\omega=\omega_0}$ is the group velocity dispersion.

$$\left(\frac{\partial^2}{\partial x^2}+\frac{\partial^2}{\partial y^2}\right)\tilde{A}(z,\omega-\omega_0) + 2ik_0\left[\frac{\partial}{\partial z}-ik_1(\omega-\omega_0)-\frac{1}{2}ik_2(\omega-\omega_0)^2\right]\times\tilde{A}(z,\omega-\omega_0)=0. \tag{4.72}$$

Now we return to the time domain by taking inverse Fourier transformation of each term of this equation $\tilde{A}(z,\omega-\omega_0)$:

$$\frac{1}{2\pi}\int_{-\infty}^{\infty}\tilde{A}(z,\omega-\omega_0)\exp[-i(\omega-\omega_0)t]d(\omega-\omega_0)=A(z,t),$$
$$\frac{1}{2\pi}\int_{-\infty}^{\infty}(\omega-\omega_0)\tilde{A}(z,\omega-\omega_0)\exp[-i(\omega-\omega_0)t]d(\omega-\omega_0)=i\frac{\partial}{\partial t}A(z,t),$$
$$\frac{1}{2\pi}\int_{-\infty}^{\infty}(\omega-\omega_0)^2\tilde{A}(z,\omega-\omega_0)\exp[-i(\omega-\omega_0)t]d(\omega-\omega_0)=-\frac{\partial^2}{\partial t^2}A(z,t). \tag{4.73}$$

Including the nonlinear polarization on the righthand side, and writing the group velocity $k_1=1/v_{\rm g}$ explicitly, Eq. (4.65) becomes,

$$\left(\frac{\partial^2}{\partial x^2}+\frac{\partial^2}{\partial y^2}\right)A(z,t)-2ik_0\frac{\partial A(z,t)}{\partial z}-i\frac{2k_0}{v_g}\frac{\partial A(z,t)}{\partial t}+k_0k_2\frac{\partial^2A(z,t)}{\partial t^2} =\frac{3}{4c^2}\chi^{(3)}\frac{\partial^2}{\partial t^2}\left[|A(z,t)|^2A(z,t)\right]. \tag{4.74}$$

Since a laser pulse moves with the group velocity v_g, we can simplify this equation in a moving frame with a retarded time τ and the z-coordinate ξ

$$\tau = t - \frac{z}{v_g}, \quad \xi = z, \tag{4.75}$$

and describe the amplitude in this frame as $A(x, y, \xi, \tau) = A(x, y, z, t)$. By noting

$$\begin{aligned}\frac{\partial A(z,t)}{\partial z} &= \frac{\partial A(\xi,\tau)}{\partial z} + \frac{\partial A(\xi,\tau)}{\partial \tau}\frac{\partial \tau}{\partial z} = \frac{\partial A(\xi,\tau)}{\partial z} - \frac{1}{v_g}\frac{\partial A(\xi,\tau)}{\partial \tau},\\ \frac{\partial A(z,t)}{\partial t} &= \frac{\partial A(\xi,\tau)}{\partial z}\frac{\partial z}{\partial t} + \frac{\partial A(\xi,\tau)}{\partial \tau}\frac{\partial \tau}{\partial t} = \frac{\partial A(\xi,\tau)}{\partial \tau},\end{aligned} \tag{4.76}$$

the pulse propagation in the moving frame is written as,

$$\begin{aligned}&\left(\frac{\partial^2}{\partial x^2} + \frac{\partial^2}{\partial y^2}\right)A(\xi,\tau) - 2ik_0\frac{\partial A(\xi,\tau)}{\partial z} + k_0k_2\frac{\partial^2 A(\xi,\tau)}{\partial \tau^2}\\ &= \frac{3}{4c^2}\chi^{(3)}\frac{\partial^2}{\partial t^2}\left[|A(\xi,\tau)|^2 A(\xi,\tau)\right].\end{aligned} \tag{4.77}$$

In analogy with the form of the time-dependent Schrödinger Eq. (3.35), Eq. (4.77) is called a nonlinear Schrödinger equation (NLSE), where the righthand side of this equation corresponds to the potential V in (3.35). Equation (4.77) is used in considering propagation of extremely short, intense laser pulses in Sect. 5.5.

4.2.7 Numerical Simulation of High-Power Laser Pulse Propagation

Quantitative analysis of complex laser systems is important in assessment and improvement of the performance of the laser system and designing a new laser system with improved performance and cost effectiveness. A comprehensive numerical simulation of the laser beam propagation in high-energy Nd:glass laser systems for laser fusion research was reported by Simmons et al. (1981), where essential processes such as amplification, beam propagation, image relaying, spatial filtering and focusing on target, are considered. Since this analysis is applicable to various types of laser systems, a brief description of this work is given in the following.

Referring to Eq. (4.77), we omit the third term on the left-hand side since the group velocity dispersion is small in the glass laser due to its small bandwidth. Considering that the nonlinear polarization oscillates with ω, the righthand side of this equation is written as

$$\mu_0 \frac{\partial^2 P_{NL}}{\partial t^2} = -\mu_0 \omega^2 P_{NL} = -\mu_0 \omega^2 \cdot \frac{3\varepsilon_0}{4} \chi^{(3)} |A|^2 A = -\frac{k^2}{4n_0} (\gamma_{nl} I) A, \tag{4.78}$$

where $k = n_0 \omega / c_0$. (Temporal change of the amplitude A is considered in Sect. 5.5 on propagation of ultrashort pulses.) By adding the amplitude gain coefficient α on the right-hand side, Eq. (4.77) reads:

$$\left(\frac{\partial^2}{\partial x^2} + \frac{\partial^2}{\partial y^2} \right) A - 2ik_0 \frac{\partial A}{\partial z} = -\frac{k_0^2}{4n_0} (\gamma_{nl} I) A + ik\alpha A. \tag{4.79}$$

The left-hand side of this equation describes the wave propagation in a free space with SVEA, and the right-hand side describes the self-focusing (the first term) and amplification (second term). The spatial and temporal evolution of the pulse is calculated by treating the quantities A, α and I as functions of the spatial coordinates (x, y, z) and the retarded time τ.

Instead of solving the differential Eq. (4.79) in temporal and spatial steps, Simmons et al. developed a code (MALAPROP) where the field A is calculated by sequentially calling each element in the laser chain and applying "operation" to the field to derive the field after each element. Diffraction in each element is neglected by treating each element as "thin".

4.2.7.1 Beam Propagation in a Free Space

The beam propagation in a free space is calculated by Fourier transformation of the field amplitude as described in Sect. 2.2.2.1. The field distribution at the observation plane $A_1(x, y)$ is calculated from the field distribution at the source plane $A_0(x, y)$ as:

$$A_1(x, y) = F^{-1}\left\{ F\left[A_0(x, y)\right] \cdot e^{ikz} e^{-i\pi\lambda z (f_x^2 + f_y^2)} \right\}. \tag{4.80}$$

4.2.7.2 Amplification and *B*-Integral

Transmittance at each surface is calculated using Fresnel reflection formula (2.90). Elliptical- or rectangular-shaped Nd-doped glasses, placed at Brewster's angle to the beam axis, are used in the amplifiers of high-energy laser systems. The nonlinear phase shift is calculated by operating the B-integral defined by (4.64) to the incident beam $A_0(x, y)$

$$A(x, y) = A_0(x, y) \exp\left[iB(x, y)\right]. \tag{4.81}$$

The beam propagation equation for the amplifier is,

$$\frac{\partial A}{\partial z} = ik\gamma' IA + \frac{1}{2}\left(\frac{\ln G}{L'} - \beta\right)A, \tag{4.82}$$

where γ' and L' are the effective nonlinear refractive coefficient and the optical path length, respectively, considering the beam propagation in the tilted amplifier, and β is the bulk loss coefficient. By lumping the bulk losses to equivalent Fresnel reflections, integration of (4.82) with respect to z yields

$$A(x, y; \tau) = A_0(x, y; \tau)G^{1/2}(x, y; \tau)\exp\left[iB(x, y; \tau)\right], \tag{4.83}$$

where $B(x, y; \tau)$ is

$$B(x, y; \tau) = k\gamma' L' I_0(x, y; \tau)\left[\frac{G(x, y; \tau) - 1}{\ln G(x, y; \tau)}\right]. \tag{4.84}$$

Writing the intensity and the saturation fluence as $I = nch\nu$ and $E_s = h\nu/\sigma$, respectively, the amplification factor $G(x, y; \tau) = n(x, y; \tau)/n_0(x, y; 0)$ given by Eq. (4.52) is described as

$$G(x, y; \tau) = \frac{G_0 f(x, y; \tau)}{1 + G_0\left[f(x, y; \tau) - 1\right]}, \tag{4.85}$$

where

$$f(x, y; \tau) = \exp\left[(n_0 E_s)^{-1}\int_{-\infty}^{\tau} I_0(x, y; t)dt\right]. \tag{4.86}$$

For long optical elements, such as rod amplifiers, the optical elements are divided into thin segments with a small signal gain of $G_0 \leq 2$.

4.2.7.3 Spatial Filtering

The field distribution at the focal plane of the spatial filter $A_f(x_f, y_f)$ is given by (4.56). Therefore, the intensity distribution is,

$$I_f(x_f, y_f) = (\lambda f)^{-2}\left|A_{in}(f_x, f_y)\right|^2, \tag{4.87}$$

where,

$$\left|x_f\right| = \lambda f\left|f_x\right|, \quad \left|y_f\right| = \lambda f\left|f_y\right|. \tag{4.88}$$

The field after passing through a pinhole of radius r_p is given by

$$A_{f'}(f_x, f_y) = \begin{cases} 0, & \text{for } |f_p| > r_p/(\lambda f) \\ A_f(x_f, y_f), & \text{for } |f_p| < r_p/(\lambda f) \end{cases} \quad (4.89)$$

where $f_p = \left(f_x^2 + f_y^2\right)^{1/2}$. The field after the output lens is obtained by multiplying the distance-dependent phase factor and taking the inverse transformation as,

$$A(x, y) = \iint \left\{A_S(f_x, f_y) \exp\left[-i\pi\lambda d\left(f_x^2 + f_y^2\right)\right]\right\} \exp[2\pi i(f_x x + f_y y)] df_x df_y, \quad (4.90)$$

where $d = -f_{in}(1 + M^{-1})$. Finally, the transverse scale is expanded by M and the intensity is reduced by M^{-2}.

Other factors are also considered in this code: propagation in air which contributes to nonlinear phase shift and stimulated Raman and Brillouin scatterings, and the spatial amplitude and phase modulations due to noise sources in the laser system such as dust particles and damages of the optical components. These factors become important in large-scale lasers as described in Sect. 6.2.

4.3 Laser Materials for High-Energy and High-Power Lasers

Since the demonstration of the ruby laser (Cr^{3+}:Al_2O_3) in 1960, various kinds of solid-state lasers have been developed for high-power and high-energy lasers. We focus our attention on lanthanide-ions and transition-metal-ions for high-energy and high peak-power lasers. These ions are doped in crystals, ceramics, glasses and fibers, and pumped by flashlamps, solid-state lasers, and high-power laser diodes (LDs). The overall performance of the lasers is determined by proper combination of these factors. In this section, we describe mainly spectroscopic properties of these ions doped in several host materials. Comprehensive reviews on solid state lasers can be found, for example, in Koechner (2006), Träger (2007, Chap. 11) and Sorokina (2003). High-energy/high-power laser systems with these ions are described in Chaps. 6 and 7.

4.3.1 Introduction

In general, laser material with high saturation fluence is suitable for high-energy pulsed lasers due to its high stored-energy density, leading to compact laser system if we can efficiently extract the stored energy. On the other hand, laser material with broad bandwidth and relatively high saturation fluence is suitable for high-peak power, ultrashort-pulse lasers. In Fig. 4.13, we show the wavelength and the saturation fluence of typical solid-state lasers: lanthanide-ion lasers (in green) and transition-metal-ion lasers (in red, deep red, brown, and violet as described in the figure caption). The narrow-band lasers are shown by filled circles. The wavelength coverages of broad-band lasers are shown by horizontal lines, with the main lasing wavelengths by filled circles. In addition to the solid-state lasers operating in the near-infrared (0.78–2.5 μm) region, some of the solid-state lasers operating in the mid-infrared (2.5–25 μm) regions are included due to their importance in many applications such as medicine, environment, and agriculture.

In this subsection, we describe how these laser parameters are related to the spectroscopic properties of these laser materials. In addition to the spectroscopic properties, other properties are also important for high-power lasers: high thermal conductance and high mechanical strength of the host material for repetitive high-power operation, and the growth capability of large-size laser materials for scaling to high-energy/high-power laser systems.

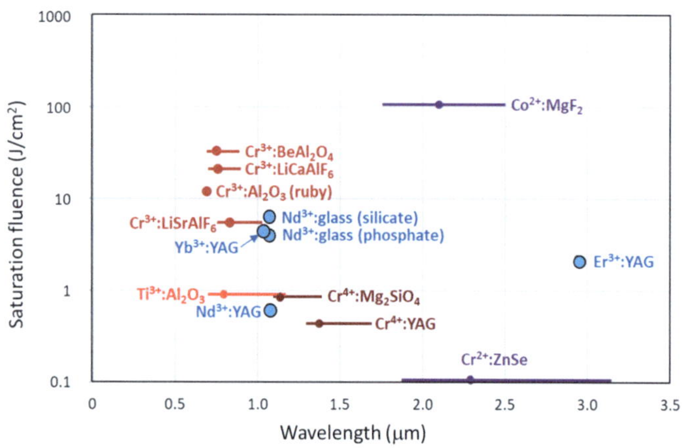

Fig. 4.13 Lasing wavelengths and saturation fluences of various solid-state lasers with lanthanide and transition-metal ions. For broadband lasers, the lasing regions are shown by horizontal lines, with their main lasing wavelengths in circles. (The lasing regions are not symmetric to the main wavelength, since this graph is plotted with the wavelength, not in frequency.) For visibility, different lasing ions are marked by different colors: lanthanides (Nd^{3+}, Yb^{3+} and Er^{3+}) in blue, $Ti^{3+}:Al_2O_3$ in red, Cr^{3+} in deep red, Cr^{4+} in brown, and Cr^{2+} and Co^{2+} in violet. Here the saturation fluence E_{sat} is derived, by assuming the laser to be a four-level system, from the stimulated emission cross section σ and the photon energy hv as $E_{sat} = hv/\sigma$, which is written in reduced units as $E_{sat}\left(\mathrm{J/cm^2}\right) = 20/\lambda(\mu\mathrm{m}) \cdot \sigma\left(10^{-20}\,\mathrm{cm^2}\right) \cdot n_2 n_2$

4.3.2 Lanthanide-Ion Lasers

4.3.2.1 Spectroscopic Properties of Lanthanide Ions

The lanthanides refers to 15 elements of the atomic number A = 57 ~ 71, i.e., La, Ce, Pr, Nd, Pm, Sm, Eu, Gd, Tb, Dy, Ho, Er, Tm, Yb and Lu, which have the electronic configurations of $4f^n5s^25p^65d^16s^2$ (n = 0 ~ 14). (The lanthanides are also called rare-earth elements, which include scandium (Sc), and yttrium (Y) in addition to the lanthanides.) The trivalent ions of the lanthanides M^{3+} form chemical bonds by their $5d^1$, $6s^2$ (and 4f in some cases) electrons with neighboring atoms. Since the 4f states are shielded by a closed shell formed by the $5s^25p^6$ electrons, the energy levels of the unpaired 4f electrons are not strongly affected by the crystal field, resulting in absorption and fluorescence spectra with narrow spectral widths in crystals and glasses. Although the electric dipole transition between the electronic states in the 4f-shell is parity forbidden in free M^{3+} ions, the transition becomes weakly allowed due to mixing with the opposite parity states (5d and 5f states) by perturbation of the odd parity terms in the crystal field, resulting in relatively long radiative lifetime of a few-hundred microseconds for the excited states. Due to these properties, lasing in the ultraviolet, visible, and near- and mid-infrared regions have been obtained with the lanthanides doped in various crystals and glasses.

Figure 4.14 shows a simplified energy diagram (Dieke diagram) of the trivalent lanthanide ions in $LaCl_3$ (Dieke, 1968), together with the number of 4f electrons n_{4f} and the number of unpaired electrons $n_{\text{u-p}}$. The energy level is designated as $^{2S+1}L_J$, where the orbital quantum numbers $L = 0, 1, 2, 3, 4, 5, 6$ are expressed by S, P, D, F, G, H, I, respectively, $2S + 1$ is the multiplicity of the level due to the spin quantum number S, and $J = L + S$ is the total angular momentum.

Let us examine the energy level of Nd^{3+} as an example, where $S = 3/2$ and $2S + 1 = 4$ since Nd^{3+} has 3 unpaired electrons. The Coulomb interaction splits different L terms into lower (^4I) and upper (^4F, ^2S, ^4S) levels separated by $\geq 10^4$ cm^{-1}. The spin–orbit interaction splits different J terms ($J = 9/2$, $11/2$, etc.) into $2S + 1$ levels separated by ~ 10^3 cm^{-1}. The crystal field further splits each $^{2S+1}L_J$ manifold into $2J + 1$ sublevels by ~ 10^2 cm^{-1}. Note that the energies and the radiative and non-radiative transition rates depend on the host crystals where the lanthanide ions are embedded in.

Transitions of the pumping (upward lines in green), lasing (downward lines in red), and nonradiative decay to lower levels (downward dashed line in orange) for Nd^{3+} and Yb^{3+} are also shown. When the upper levels (enclosed by a green square in Fig. 4.14) are pumped from the ground state $^4I_{9/2}$ by visible light, these excited sates quickly decay to the lowest level of the excited states $^4F_{3/2}$ due to interaction with lattice vibration (phonons). In another example, the Er^{3+}:YAG laser at 2.8 μm shown in Fig. 4.13 is due to the transition between the excited states, from $^4I_{11/2}$ to $^4I_{13/2}$, as shown by a downward red line.

Also shown in this diagram with slanting dashed lines in orange are energy transfer from one ion to another to increase the pumping efficiency: for example, from Yb

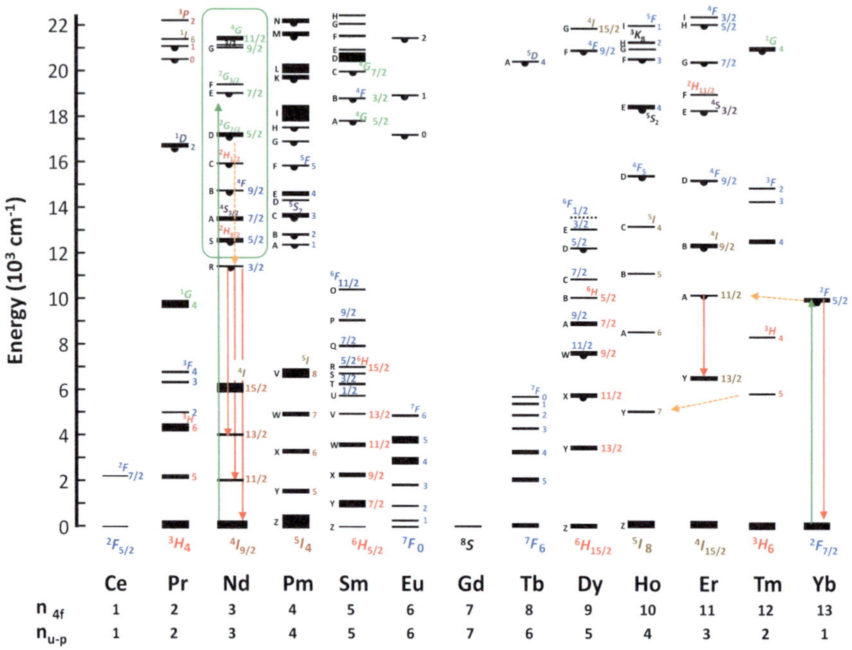

Fig. 4.14 Simplified energy diagram (Dieke diagram) of trivalent lanthanide ions. The quantum states with different L are shown with different colors. The numbers of the 4f electrons n_{4f} and the un-paired 4f electrons $n_{\text{u-p}}$ are shown below this diagram. The transitions of the pumping (upward in green), lasing (downward in red), and non-radiative decay (downward dashed line in orange) are shown for Nd^{3+} and Yb^{3+}. Also shown by the slanting dashed lines in orange are energy transfers from Yb to Er and from Tm to Ho

$^2F_{5/2}$ to Er $^4I_{11/2}$ and from Tm $^3H_4$ to Ho $^5I_7$. We note that various lanthanide ions are important for generating laser emission and intense fluorescence in the ultraviolet, visible, and near- and mid-infrared regions (Sorokina, 2003; Träger, 2007; Chap. 11).

4.3.2.2 Nd- and Yb-Doped Crystals

Figure 4.15 shows more detailed energy level diagrams of (a) Nd^{3+}:YAG and (b) Yb^{3+}:YAG, with the wavelengths (in nm) of the pumping and laser transitions. Here we show only the pumping to low excited states with laser diodes, where high quantum efficiency (photon-energy ratio of the lasing to the pumping) is achieved. In Nd^{3+}, the $^4I_{11/2}$ state (~ 2000 cm^{-1} higher than the ground state) is not thermally populated at room temperature (208.5 cm^{-1} at T = 300 K) and decays non-radiatively to the ground state in less than 1 ns, thus forming a four-level system. Lasing is obtained in the transition from $^4F_{3/2}$ to $^4I_{11/2}$ at 1064 nm, as well as in other transitions when the lasing at 1064 nm is suppressed.

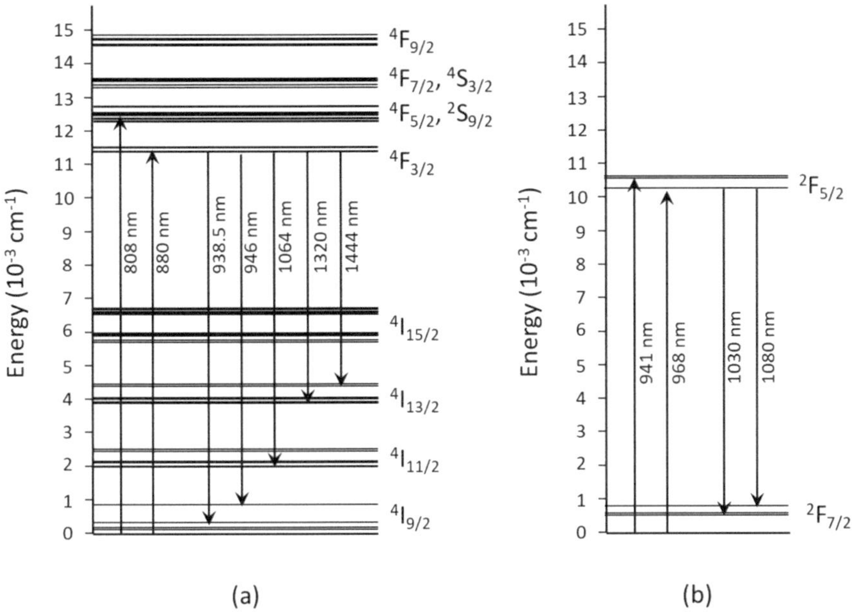

Fig. 4.15 Energy level diagrams of **a** Nd:YAG and **b** Yb:YAG, showing the pumping and lasing transitions

The energy level of Yb^{3+} is much simpler, composed of $^2F_{7/2}$ and $^2F_{5/2}$ states since Yb^{3+} has only one unpaired hole. The laser transition takes place from $^2F_{5/2}$ to $^2F_{7/2}$ when $^2F_{5/2}$ is pumped at 941 nm wavelength. The Yb laser works as a quasi-three-level system, since the lower laser level, which is 512 cm^{-1} above the ground state, is partially populated (5.5%) at room temperature. The efficiency of the Yb laser increases by intense pumping to reduce the ground-state population or by operating at low temperature to reduce the thermal population of the lower laser level.

A very small quantum defect of 6% is expected by pumping at 968 nm, called zero-phonon line, with a laser diode of very narrow linewidth (2.5 nm at 300 K and ~ 0.3 nm at 150 K) and high wavelength stability (De Vido et al., 2020). We will show in Sect. 6.3.3 that such narrow-band laser diodes are available.

The Yb^{3+} has several favorable properties for high-power lasers: (1) No excited state absorption (ESA) for pumping and lasing radiations, (2) Very high quantum efficiency (92 or 94%) which results in reduced heating of the host medium, (3) Long radiative lifetime of the upper state (951 μs in YAG at room temperature) leading to high stored-energy density, (4) Broad emission bandwidth (86 cm^{-1} in Yb:YAG and 120 cm^{-1} in Yb:Er:glass) enabling generation and amplification of sub-picosecond pulses, and (5) Yb^{3+} allows high doping concentration in Y-based host crystals such as YAG and YLF due to small ionic radius of Yb^{3+}, leading to reduction in the amplifier thickness and the size of the laser system.

Efficient lasing has been achieved at room temperature with a Yb:YAG crystal by intense LD pumping at 968 nm to decrease the population of the ground state and the lower laser level (Lacovara et al., 1991). It is possible to reduce the population of the lower laser level to almost zero by cooling the Yb-doped crystals. Efficient lasing has been achieved even with low intensity pumping by cooling the Yb-doped crystals to ~ 70 K: quantum-defect-limited operation with 90% slope efficiency with Yb:YAG (Shoji et al., 2004) and sub-ps (795 fs) pulse generation with a Yb:YLF CPA laser (Kawanaka et al., 2003). In addition to achieving efficient lasing, cryogenic cooling of Yb-doped crystals is also advantageous for high power laser systems: thermal conductivity is ~ 7× higher, thermal expansion coefficient is ~ 5× less, and thermo-optical coefficient dn/dT is ~ 6× less than those at room temperature, leading to ~ 3× higher thermal shock resistance with Yb:YAG (Kawanaka et al., 2005).

High optical-quality, large crystals are required for high-power lasers. Typical laser crystals such as YAG and YLF are grown by the Czochralski method where a host material with proper impurities is melted in a high temperature crucible. A small seed crystal is dipped in the molten material at the top surface, rotated, and pulled upwards slowly while the material temperature is gradually lowered. The major difficulty in growing large single crystals with this method is the formation of optical inhomogeneity (called "core") at the center of the cylindrical crystal due to growth of the solid–liquid interface from the center to the peripheral. Growth of large core-free garnet crystals became possible recently with an improved Czochralski method, where the liquid–solid interface is kept flat from the top to downwards during the crystal growth (Košelja et al., 2016). With this method, YAG crystals doped with Nd, Yb and other lanthanides have been grown up to 125 mm in diameter and 150 mm in length with small concentration gradient along the length (Fig. 4.16).

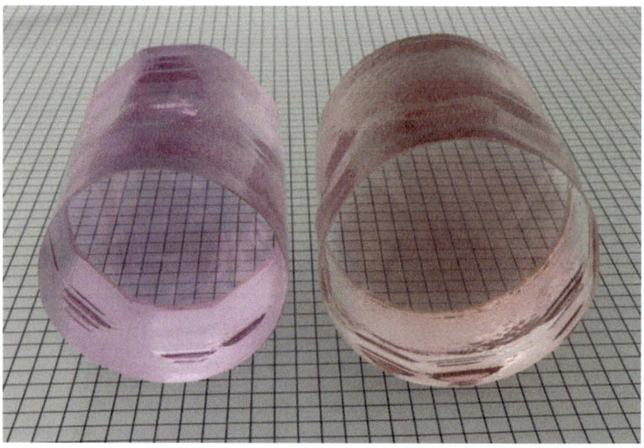

Fig. 4.16 Laser crystals of cylindrical shapes: (left) Nd:YAG (0.2 at. % Nd) and (right) Yb:YAG (1 at. % Yb), both with 110 mm in diameter and 140 mm in length. The scale of the grid is 1 cm × 1 cm. Courtesy of CRYTUR Inc.

4.3.2.3 Laser Ceramics

Laser ceramics doped with lanthanides have been developed as a complementary material to laser crystals. Translucent ceramics (polycrystalline alumina) were developed in the 1950s for high-pressure sodium lamp bulbs (Coble, 1961a, 1961b). The laser ceramics were created in the 1990s by reducing internal losses of the polycrystalline ceramics, arising from absorption due to impurities and scatterings by residual air bubbles, different crystal phases and segregation at the crystal boundaries, to the utmost limit (Ikesue et al., 1995).

Fabrication of laser ceramics is categorized, by the difference in the starting material, into a powder synthesis method (Ikesue et al., 1995) and a liquid synthesis method (Lu et al., 2002). In the powder synthesis method for fabricating Nd:YAG ceramics as an example, powders of Al_2O_3, Y_2O_3 and Nd_2O_3 are milled, dried, cold-pressed, and sintered, resulting in the ceramics composed of Nd:YAG crystal grains of typically 20 ~ 50 μm in diameter. In the liquid synthesis method, solutions of Al^{3+}-, Y^{3+}- and Nd^{3+}-chlorides are mixed and dried to form powders composed of nano-size Nd:YAG particles. These powders are then milled, slip-cast, and sintered in vacuum at ~ 1700 °C, which is lower than the melting temperature of a YAG crystal (1930 °C), resulting in the ceramics composed of a few-μm diameter grains. Note that the grain size which is comparable to or less than the laser wavelength is preferrable to reduce the Mie scattering by the crystal grains. Detailed account of fabrication processes of oxide laser ceramics is described, for example, in Kravchenko and Kopylov (2013).

The laser ceramics has several unique properties: (a) It is possible to fabricate large, high quality laser ceramics, (b) Higher doping is possible since the lattice deformation due to high doping is confined in each crystal grain, resulting in high optical uniformity in a macroscopic scale, and (c) Transparent ceramics can be fabricated with various materials for which large, high-quality single crystals are difficult to grow due, for example, to too high melting temperature for the crucibles used in the crystal growth. For example, transparent ceramics of sesquioxide (RE_2O_3, where RE refers to rare-earth elements) whose melting temperature is over 2400 °C has proven to be useful for high-power laser (Furuse et al., 2018; Kong et al., 2005) and short pulse laser (Tokurakawa et al., 2007a, 2007b), as well as for magneto-optical devices (Furuse and Yasuhara, 2017; Slezak et al., 2019). Transparent laser ceramics of *non-cubic* Nd-doped fluorapatite (Nd:FAP) crystal with an average grain size of 140 nm has been also fabricated by liquid phase synthesis with spark plasma sintering (Furuse et al., 2019).

In Fig. 4.17 we show (a) a photograph of 0.4 at. % Nd:YAG ceramic plate of 115 × 115 × 22 mm in size, and (b) a magnified view of this plate observed with a field emission scanning electron microscope (FE-SEM) (Yagi et al., 2007). The latter image shows that an average grain size is about 2 μm in diameter and the scattering centers such as defect, pore, and second phase are not observed. The scattering loss of this sample has been measured to be less than the scattering loss of a state-of-art single crystal counterpart. More recently, Fujioka et al. (2025) have succeed in fabricating $Nd:CaF_2$ transparent ceramics which is suitable for application in high-average power ultrashort-pulse lasers (see Sect. 7.5.1 for more details).

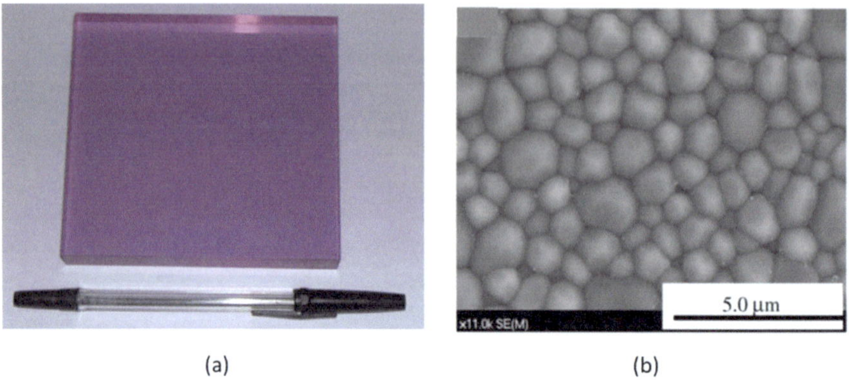

Fig. 4.17 **a** Photograph of a 0.4 at % Nd:glass ceramic plate (115 × 115 × 22 mm) and **b** magnified view (× 11,000) of this plate with an FE-SEM. Reprinted from Yagi et al. (2007), with permission from Elsevier: permission conveyed through Copyright Clearance Center, Inc. © 2007 Elsevier

Repetitively operated high-energy pulsed lasers have been realized by LD pumping of cryogenic Yb:YAG ceramic amplifiers with gas-cooled multi-slabs (David et al., 2018; Mason et al., 2018) and liquid-nitrogen-cooled active mirrors (Ogino et al., 2022) as described in Sect. 6.3. For high average-power laser applications, quasi-continuous laser operation with ~ 100 kW power over several seconds have been demonstrated by LD-pumping of thin Nd:YAG ceramic slabs in tiled (Marmo et al., 2009) and single-aperture (Klimek & Mandl, 2011) configurations.

4.3.2.4 Nd-Doped Glasses

In this subsection, we focus on Nd-doped glasses for high-energy pulsed lasers, as an example of the lanthanide-doped glasses. There are two important parameters to evaluate laser materials for application to high-energy and high-power lasers: stimulated emission cross section σ and nonlinear refractive index n_2. In general, a laser material with smaller stimulated emission cross section σ is suitable for high energy lasers since it can store larger energy E_{st} as population inversion ΔN_0. On the other hand, small σ results in lower gain coefficient γ, requiring longer amplifier length which leads to self-focusing and higher cost of the laser system. Small σ also leads to higher saturation intensity I_s, which should be kept below the damage thresholds of optical materials. The optimum value of σ is determined in terms of performance and cost of the total laser system.

The intensity distribution of a high-power laser beam becomes nonuniform due to diffraction and nonlinear phase shift of the field amplitude, resulting in damage of the optical components and irradiation nonuniformity on the target. This nonlinear effect is characterized by the B-integral defined in Eq. (4.64). Since the B-integral has to be less than a certain value, 2 rad for example, the maximum intensity and thus the maximum power of the laser output is inversely proportional to the magnitude of n_2, favoring a laser material with smaller n_2 for high power operation.

In the Nd-doped glasses, the Nd ions are contained in oxide glasses, which are composed of glass former oxides (borate B^{3+}, silicate Si^{4+}, phosphate P^{5+}, germanate Ge^{4+}), modifier oxides (Li^+, Na^+, K^+, Cs^+; Mg^{2+}, Ca^{2+}, Sr^{2+}, Pb^{2+}, Ba^{2+}), and intermediate oxides (Be^{4+}, Al^{3+}, Ti^{4+}). In the amorphous glasses, the crystal-field induced splitting of the ${}^{2S+1}L_J$ manifolds depend on the site of the ion, resulting in inhomogeneous spectral broadening and longer spontaneous emission lifetime compared to these parameters in crystals. The silicate glass is formed by a randomly connected three-dimensional network, whereas the phosphate glass is formed by two-dimensional chain structures. The larger radiative decay rate of the Nd-phosphate than the Nd-silicate is ascribed to larger crystal field asymmetry due to the chain structure of the phosphate glass (Izumitani et al., 1982) ${}^{2S+1}L_J$.

The laser glasses are useful as a laser material in the following respects: isotropic and high-quality laser glasses are fabricated in various shapes and sizes such as a few-μm diameter fibers, cm-diameter rods, and meter-size elliptical disks; it can be doped with various ions at high concentration without concentration quenching; it is applicable to sub-picosecond pulse generation due to its broad spectral width. The major disadvantage is the low thermal conductivity of the glasses, making it difficult for high-repetition and high-average-power laser operation.

The properties of the Nd-doped silicate, phosphate and fluorophosphate glasses are shown in Table 4.1 (Izumitani, 1984). Comparing the Nd-silicate and Nd-phosphate glasses, the Nd-phosphate has a larger stimulated emission cross section and smaller spectral broadening than the Nd-silicate. The saturation fluence estimated by $E_s = h\nu/\sigma$ is 7.0 J/cm^2 for Nd-silicate and 4.5 J/cm^2 for Nd-phosphate glasses, respectively. The nonlinear refractive index of Nd-phosphate is ~ 30% smaller than Nd-silicate. Since the phosphate glass LHG-8 is a thermal, i.e., the temperature change of the optical path length dS/dT is close to zero, the laser system is less affected by the temperature increase due to thermal deposition during operation. Noting that the temperature change of the optical path length is given by $dS/dT = (n-1)\alpha + dn/dT$, this property arises from the negative value of the thermal coefficient of the refractive index dn/dT which compensates for the positive thermal expansion coefficient α as shown in Table 4.1.

By modelling an atom with a classical nonlinear oscillator, Boling et al. (1978) derived an empirical relationship of the nonlinear refractive index n_2 (electronic contribution) with the refractive index n, for linearly polarized light, which fits well to various optical glasses:

$$n_2 = \frac{\left(n^2+2\right)^2\left(n^2-1\right)^2}{16\pi N \hbar\omega_0 n}. \tag{4.91}$$

The quantities N and $\hbar_0\omega$ in this equation are determined by fitting the refractive index of the relevant glass with the Lorentz-Lorenz equation. Equation (4.91) shows

Table 4.1 Properties of Nd-doped silicate, phosphate and fluorophosphate laser glasses

Properties	Glass type		
	Silicate (LSG-91H)	Phosphate (LHG-8)	Fluorophosphate (LHG-10)
Laser property			
Nd_2O_3 (wt. %)	3.1	3	2.4
Peak wavelength (λ_p, nm)	1062	1054	1051
Linewidth ($\Delta\lambda$, Å)	274	218	265
Stimulated emission cross section (σ_p, 10^{-20} cm^2)	2.7	4.2	2.7
Fluorescence lifetime (μs)	300	315	384
Optical property			
Refractive index (n)	1.5498 (1052 nm)	1.52005 (1054 nm)	1.46078 (1051 nm)
Nonlinear refractive index (n_2, 10^{-13} esu)	1.58	1.13	0.61
Thermal property			
Thermal expansion coeff. (α, 10^{-6}/°C)	9	11.2	14.5
Thermal coeff. of n (dn/dT, 10^{-6}/°C)	1.6	− 5.3	− 7.7
Thermal coeff. of optical length (dS/dT, 10^{-6}/°C)	6.6	0.6	− 1
Thermal conductivity (25 °C, kcal/m°C)	0.89	0.5	0.64
Specific heat (cal/g °C)	0.15	0.18	
Mechanical property			
Density (g/cm^3)	2.81	2.83	3.64
Knoop hardness	590	321	361
Young's modulus (kg/mm^2)	8890	5110	7250
Chemical property			
Water solubility (wt. loss %, 100 °C, 1 h)	0.036	0.13	0.034
Acid solubility (wt. loss %, HNO_2, 100 °C, 1 h)	0.039	0.47	0.337

The properties of LG-670 and LG-760 of Schott are close to LSG-91H and LHG-8 of Hoya, respectively that n_2 is smaller in glasses of smaller n, with dependence stronger than linear relationship. The nonlinear refractive index of the silicate and phosphate laser glasses are 1.58 and 1.13 in 10^{-13} esu, respectively.

The ratio of the stimulated cross section and the nonlinear refractive index σ/n_2 can be used as a figure of merit of a laser material for generating high-power laser pulses, since amplifier length is reduced with larger σ and a laser pulse of higher power is generated with the same aperture with smaller n_2. Since σ/n_2 of the phosphate glass is 2.2 times higher than the silicate glass, it is expected that approximately $2\times$ higher power can be generated with phosphate laser glass of the same aperture. This was proven by demonstration of the Nd-phosphate laser systems for high-power operation (Bunkenburg et al., 1981; Kato et al., 1981; Yamanaka et al., 1981). Although the fluorophosphate laser glass has an extremely small nonlinear refractive index as shown in Table 4.1, its development is not pursued due to its low laser damage threshold. The Nd:glass laser can be also used to generate short pulses of 150 fs duration with the "mixed glass" (i.e., amplifiers composed of phosphate and silicate laser glasses) for broadening the amplification bandwidth utilizing different peak wavelengths of these glasses (Chériaux et al., 2019; Hays et al., 2007).

The optical and laser properties of various laser glasses have been analyzed extensively by Krupke (1974), Jacobs and Weber (1976) and others, applying the Judd–Ofelt model for calculating the line strengths between the $4f^n$ states (Judd, 1962; Ofelt, 1962). In this method, the following properties are first measured using small samples of laser glasses: absorption spectrum (peak wavelength of the absorption band and integrated absorption coefficient), fluorescence spectrum, fluorescence lifetime, and linear refractive index. Using the measured data, various properties required to evaluate the laser glasses are derived: spontaneous emission probability and fluorescence branching ratio from an initial manifold $\left|(S', L')J'\right\rangle$ to a final manifold $\left|(\overline{S}, \overline{L})\overline{J}\right\rangle$, radiative lifetime and radiative quantum efficiency of the $\left|(S', L')J'\right\rangle$ manifold, and the peak induced-emission cross section for the ${}^4F_{3/2} \rightarrow {}^4I_{11/2}$ transition. Small-signal gain coefficients of various laser glasses in disk amplifiers were measured under identical conditions and compared with the calculation using the spectroscopic data and the spectra of the flashlamps for pumping, verifying the usefulness of the spectroscopic approach for optimum design of high-energy glass laser systems (Linford et al., 1979).

4.3.3 Transition-Metal-Ion Lasers

Transition metal refers to the elements in the 3rd, 4th, 5th, and 6th row of the periodic table which have partially filled d-sub-shells. Lasing has been observed with the transition metals in the third row, i.e., Ti, V, Cr, Mn, Fe, Co, Ni and Cu, which take the ground state electronic configurations $3d^n4s^2$, where n increases from 2 for Ti to 9 for Cu. The two 4s electrons and one ~ several d electrons are ionized to form ionic bonding with neighboring cations in a crystal, leaving the transition-metal atoms in ionized states. Unlike lanthanides where the 4f electrons are shielded by a closed shell formed by the $5s^2 5p^6$ electrons, the 3d-electrons of the transition metal ions are not shielded, resulting in the electronic states which are strongly affected by the crystalline field of the host materials (Tanabe & Sugano, 1954a, 1954b).

Among many possible combinations of the transition metal ions with the host crystals, lasing has been observed in a limited number of the transition-metal ions of Ti^{3+}, V^{2+}, Cr^{2+}, Cr^{3+}, Cr^{4+}, Mn^{5+}, Fe^{2+}, Co^{2+}, and Ni^{2+} with proper combination with host crystals, for which high population inversion density and weak excited state absorption are realized. First we focus on the Ti-doped sapphire (Ti:sapphire, Ti^{3+}:Al_2O_3) laser which is suitable for generating ultrashort-duration, ultrahigh-peak-power pulses (Moulton, 1986). In Ti^{3+}:Al_2O_3, Ti^{3+} substitutes for Al^{3+} located at the center of an octahedron formed by six O^{2-} ions. The energy level of the Ti^{3+}:Al_2O_3 is the simplest among the transition metal ions since there is only one 3d electron left in the d-shell. The crystal field to Ti^{3+} is decomposed into a cubic field and a trigonal field, where the cubic field is far stronger than the trigonal field. Under the cubic field, the fivefold degenerate 3d state is split into two states: a triply degenerate ${}^2T_2$ ground state and a doubly degenerate ${}^2E$ upper state with an energy separation of $\Delta = E({}^2E) - E({}^2T_2) \sim 19{,}000\ \text{cm}^{-1}$. The trigonal field and the spin–orbit interaction further split the ${}^2T_2$ and ${}^2E$ levels into three and two sublevels, respectively.

Figure 4.18 shows a schematic energy level diagram of Ti^{3+}:Al_2O_3 where the horizontal axis is the configurational coordinate Q. Due to coupling of the electron with the lattice, the octahedral crystal structure is deformed to stabilize the electronic states to lower energies (static Jahn–Teller effect). In Ti^{3+}:Al_2O_3, the Jahn–Teller stabilization energy E_{JT} is 119 cm^{-1} for ${}^2T_2$ and 3080 cm^{-1} for ${}^2E$. The vibronic potential surface of ${}^2T_2$ is composed of three independent paraboloids, one for each of the three degenerate states. The rapid transition among these paraboloids arises due to the dynamic Jahn–Teller effect (MacFarlane et al., 1968).

The ${}^2T_2$ and ${}^2E$ states have the lowest energy at the configuration coordinate of Q_{g} in ${}^2T_2$ and Q_{u} in ${}^2E$, respectively. Since the upper state is more strongly affected by the crystal field, $Q_{\text{u}} > Q_{\text{g}}$. This results in the radiative transition to the vibronic levels of ${}^2T_2$, leading to the fluorescence spectrum covering a broad spectrum of 660 ~ 1180 nm with a peak at 790 nm, which does not overlap with the absorption spectrum. In the Ti^{3+}:Al_2O_3 laser, the ${}^2E$ state, optically pumped from the ${}^2T_2$ ground state at $Q_{\text{g}} \sim 0$, relax quickly to $\sim Q_{\text{u}}$ creating population inversion to the ${}^2T_2$ ground state. Although the electric dipole transition within a 3d electron shell is parity forbidden, it becomes weakly allowed due to the trigonally-symmetric component of the crystal field. The ${}^2E$ state is weakly coupled with the ${}^2T_2$ state via phonons, reducing its radiative lifetime from 3.85 μs at $\leq$ 200 K to 3.2 μs at room temperature. Due to this short radiative lifetime, the Ti:sapphire laser is often pumped by pulsed lasers such as the second harmonic of Nd:YAG and Nd:glass lasers. The ${}^2E$–${}^2T_2$ transition has peak stimulated emission cross section of $4.1 \times 10^{-19}\ \text{cm}^2$ for $E \| c$ and $2.0 \times 10^{-19}\ \text{cm}^2$ for $E \perp c$ and saturation fluence of 0.9 J/cm^2. There is no excited-state absorption in the spectral region of the pump and fluorescence bands owing to the simple $3d^1$

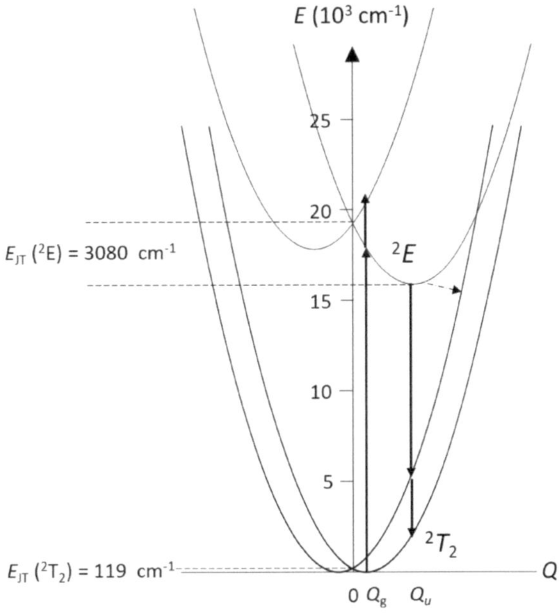

Fig. 4.18 Schematic energy diagram of Ti^{+3}:Al_2O_3 laser, where Q is the configurational coordinate. The $^2T_2$ and ^2E states take the lowest energies at Q_g and Q_u, due to the Jahn–Teller effect, with the stabilization energies of $E_{JT}(^2T_2)$ and $E_{JT}(^2E)$ for $^2T_2$ and ^2E, respectively. Radiative transition takes place from the ^2E vibronic level to the empty $^2T_2$ level, as shown by the vertical solid line. Nonradiative decay from ^2E to $^2T_2$ is shown by a slanting dashed line

configuration, resulting in high quantum efficiency of 80% at room temperature and lasing over the whole fluorescence spectral region.

Another example of the transition-metal-ion laser is the Cr^{3+} laser, where Cr^{3+} substitutes for Al^{+3} located in host crystals such as alexandrite ($BeAl_2O_4$), $LiCaAlF_6$, $LiSrAlF_6$, and sapphire (Al_2O_3). The Cr^{3+} has three 3d electrons in the d shell ($3d^3$). The energy level diagram of d^3 isoelectronic sequences is shown in Fig. 4.19 (Tanabe & Sugano, 1954b). In this diagram, the energy E of the electronic states, measured from the energy of the lowest-energy state, is plotted as a function of the crystal field strength D_q, both normalized by the Racah parameter B (a measure of the electrostatic repulsion between electrons). In cubic crystalline field, its lowest energy state ^4F is split into a triply degenerate $d\varepsilon$ of lower energy ($^4A_2$ $(d\varepsilon^3)$) and a doubly degenerate $d\gamma$ of higher energy ($^4F_2$ $(d\varepsilon^3 d\gamma^1)$), where the energy of the $^4F_2$ state increases with the field strength D_q. On the other hand, the upper state ^2E, which originates from ^2G of a free ion, is less affected by the crystal field. The energy of $^4F_2$ is lower than ^2E at weak crystal field ($Dq/B \leq 2.1$), whereas $^4F_2$ becomes higher than ^2E at higher crystal field ($Dq/B \geq 2.1$).

In Cr^{3+}:Al_2O_3 (ruby) where the field strength is high, the lowest excited state is ^2E which has a long radiative lifetime of 3 ms since the transition from ^2E to $^4A_2$ is parity forbidden, making the ruby laser suitable for single-frequency, high-energy pulsed operation. In the Cr^{3+}-$BeAl_2O_4$, $LiCaAlF_6$, and $LiSrAlF_6$ crystals where the field strength is $Dq/B \approx 2.1$, the $^4F_2$ state becomes slightly below the ^2E state. In these lasers, the stimulated emission takes place from $^4F_2$ to $^4A_2$ with ^2E acting as a reservoir of $^4F_2$. Although the true spontaneous emission lifetime of $^4F_2$ is ~ 6.6 μs, the effective lifetime becomes longer as 67, 170 and 260 μs for $LiSrAlF_6$, $LiCaAlF_6$

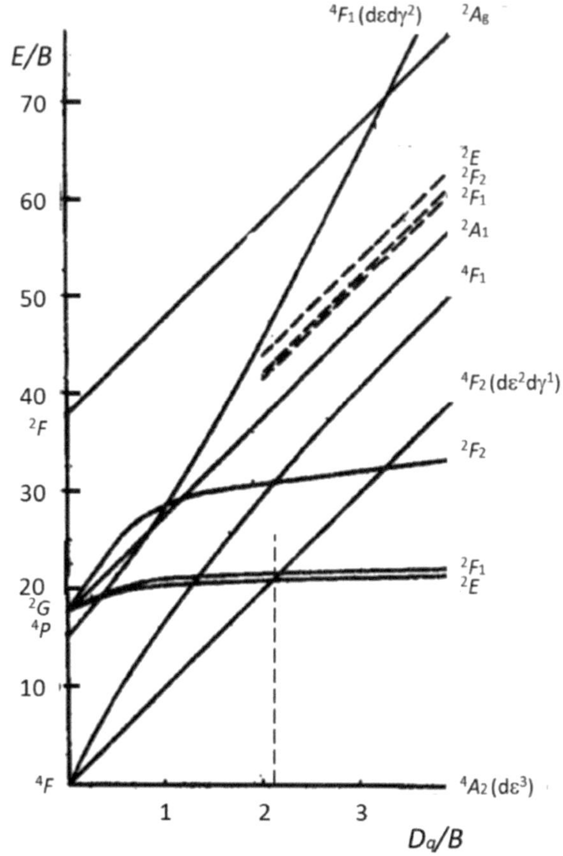

Fig. 4.19 Energy level diagram of d^3 isoelectronic sequences in a crystal field of cubic symmetry. This diagram shows dependence of the level energy E on the crystalline field strength D_q, both normalized by the Racah parameter B. The vertical dotted line at $D_q/B \sim 2.1$ shows where the lowest energy state changes from $^4F_2$ to ^2E with an increase in D_q. Reprinted from Tanabe and Sugano (1954a, 1954b), with permission from The Japan Physical Society. © 1954 The Japan Physical Society

and $BeAl_2O_4$, respectively. The stimulated emission cross section is also affected by the $^4F_2$–^2E coupling, varying as 4.8, 1.3 and 0.8 in unit of 10^{-20} cm^2 for $LiSrAlF_6$, $LiCaAlF_6$ and $BeAl_2O_4$, respectively. The tuning range of the Cr-doped lasers (780–1010, 720–840, and 700–820 nm for $LiSrAlF_6$, $LiCaAlF_6$ and $BeAl_2O_4$, respectively) is affected by the excited state absorption, resulting in smaller bandwidth compared with Ti:sapphire of 660–1180 nm, as shown in Fig. 4.13.

Various combinations of the transition-metal-ions and the host crystals have been investigated to increase efficiency and wavelength regions, including Cr^{4+}:Mg_2SiO_4, Cr^{4+}:YAG, Cr^{2+}:ZnSe, and Co^{+2}:MgF_2 whose tuning ranges cover near- and mid-infrared regions as shown in Fig. 4.13.

4.4 Laser Damage

4.4.1 Introduction

The laser damage has been studied extensively for scientific and commercial lasers (Brown, 1981; Koechner, 2006). Laser damage is also a serious issue in high power laser systems, since large optical components cannot be easily replaced to new ones due to time and cost constraints. In developing the National Ignition Facility (NIF) at the Lawrence Livermore National Laboratory (LLNL), systematic investigation on the laser damage was undertaken to enable operation of NIF over many shots keeping its initial performance of 1.8 MJ, 500 TW at 351 nm, and at higher performance over longer period (Baisden et al., 2016; Manes et al., 2016; Spaeth et al., 2016a).

As one of the outcome of this investigation, NIF is now safely operated at mean fluence of ~ 9 J/cm^2 and peak fluence of ~ 20 J/cm^2 at 351 nm at the final focusing assembly, far higher than the typical fluence for damage growth of ~ 4 J/cm^2 at 351 nm. In spite of this significant advance, the laser damage remains to be an important issue in other types of lasers: for example, diffraction gratings for pulse compression in high peak power CPA and OPCPA lasers, and multilayer-coated mirrors in high-repetition, high-average-power lasers. In this section we describe advances in improving the laser damage achieved at NIF, which have aspects common with other types of lasers. Recent approach of damage-resistant gas optics to manipulate high fluence laser beams is described in the final section.

4.4.2 Optical Components in High Power Laser Systems

Many optical components are installed in high-energy laser systems, whose typical layout is shown in Fig. 4.10. Examples of optical components which are susceptible to laser damage in these laser systems, including the components for high peak-power lasers, are listed in Table 4.2. The laser damage of these components depends on various parameters: transmission or reflection optics, laser parameters (wavelength, pulse width, pulse shape, polarization, repetition rate), environment (temperature, humidity, cleanliness), and system design (reflection from other components and target, holographic effect, and parasitic amplification), some of which are described in the following.

4.4.3 Estimation of Laser Damage During Laser Operation

Laser damage, which starts with small defects, grows to large sizes by repeated laser irradiation, leading often to damage of other optics and performance degradation of the laser system. Therefore, it is important to determine the threshold of laser

Table 4.2 Typical optical components in high power lasers, which are susceptible to damage due to laser irradiation at high fluence

Laser system	Device	Material	Wavelength (nm)
High energy	Amplifier	Nd:glass	1053
	Reflectors	Multilayer coating	1053
	Window	Fused silica (flat)	1053, 527, 351
	Spatial filter	Fused silica (curved)	1053
	Faraday isolator	RE-doped glass, crystal	1053
	Freq. conv.	KDP, DKDP	1053, 527, 351
High peak power	CPA amplifier	Ti:sapphire	800 (660–1180)
	OPCPA amplifier	DKDP, BBO	800 (660–1180)
	Compressor	Grating	800 (660–1180)
	Compressor	Chirped mirror	800 (660–1180)

fluence below which the laser can be operated without being affected by the damage formation. The methods to determine the laser-induced damage threshold are defined by ISO 11254-2:2011 (International Standards Organization, 2011). In this method, an optical component is irradiated by either a single (1) or multiple (S) shots on a same site, called 1-on-1 and S-on-1, respectively. Existence of the damage is examined after irradiation with a Nomarski microscope of μm spatial resolution. By irradiating the sample on different sites with the same fluence and changing the fluence level stepwise, the damage probability is plotted versus the laser fluence. The damage threshold fluence is determined by extrapolating the plot to the zero probability.

Since this method requires sample irradiation with large number of shots, a more time-saving method has been developed to enable determining the damage probability of many optical components within a constrained time for component inspection (Carr et al., 2006). In this method, the damage initiation density $\rho(\phi)$ as a function of the fluence ϕ is obtained in a single shot, where an optical component is irradiated with a large area, high energy (~ 10 J) laser pulse whose fluence distribution is digitally recorded. After irradiation, the optics under test is moved to an automatic scanning microscope that records each damage site with its coordinate assigned to the laser fluence at this coordinate. By plotting the number of the pinpoint damage sites versus the fluence, $\rho(\phi)$ is determined. To take representative data, measurements are made over a surface area of one ~ several cm^2 for surface damage or one ~ several mm^3 for bulk damage.

Since the laser damage is a probabilistic process due to random spatial distribution of the damage precursors and laser fluence is not uniform, the damage "threshold" cannot be determined uniquely. Instead, the number of expected damage sites formed by laser irradiation is calculated by convolving $\rho(\phi)$ with the fluence distribution profile $PDF(\phi)$ as

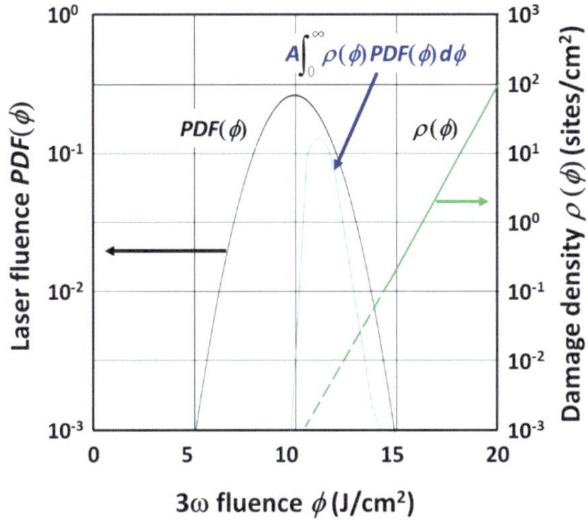

Fig. 4.20 Estimating the number of expected damage sites on an optical component by convolving measured values of the damage initiation density $\rho(\phi)$ (line in green) with the laser fluence distribution $PDF(\phi)$ (curve in black), resulting in $\rho(\phi)PDF(\phi)$ (curve in blue). This example is for a polished fused silica surface of a NIF final optic irradiated with 3ω laser beam of 9.6 J/cm^2 average fluence and 15% contrast. In this example, 2.4 damage sites are expected after one laser shot. Reprinted with modification from Manes et al. (2016), with permission from Taylor & Francis: permission conveyed through Copyright Clearance Center, Inc. © 2016 Taylor & Francis

$$A\int_0^\infty \rho(\phi)PDF(\phi)d\phi, \tag{4.92}$$

where A is the area of the laser beam and $PDF(\phi)$ is normalized as $\int_0^\infty PDF(\phi)d\phi = 1$. An example of this process is shown in Fig. 4.20. The operation condition of a laser system to keep the damages to within allowed number can be determined by measuring $\rho(\phi)$ of each optics and measuring and/or simulating the fluence distribution $PDF(\phi)$ along the laser beam path.

4.4.4 Initiation and Growth of Laser Damage

In this subsection, we describe the formation and growth of damage sites in optical materials (bulk damage) and at the surfaces (surface damage).

4.4.4.1 Bulk Damage

In dielectric materials, small damages are initially formed by electron avalanche (Yablonovitch & Bloembergen, 1972). In this process, free electrons are supplied by ionization of shallow traps arising from lattice defects, voids, color centers, etc. that are present in almost all optical materials. (Probability of multiphoton absorption is lower than this process, since the band gap of a fused silica, for example, is ~ 9 eV, whereas the photon energy at 351 nm is 3.5 eV.) The free electrons with an initial density of ~ 10^{10} cm^{-3} (Smith et al., 1976) are accelerated with high electric field to an energy sufficient to ionize bound electrons in a lattice. The secondary electrons are also accelerated by the laser field to ionize other bound electrons. The electron avalanches are generated in the material and a plasma with ~ 10^{18} cm^{-3} density is formed. The plasma is heated rapidly by inverse bremsstrahlung absorption to a high temperature above the melting point of the material, leading to formation of irreversible damage sites. The threshold intensity for damage formation measured on fused silica is 11×10^{10} W/cm^2, i.e., 4.7×10^2 J/cm^2 for a 1.064 μm-wavelength laser of 4.7 ns pulse width (Smith, 1978). Since this fluence is higher than the laser fluence of high-power lasers by more than 1-order of magnitude, the bulk damage has small probability in fused silica and in other transparent dielectric materials, unless the laser is focused, for example, by self-focusing or ghost focusing of a reflected laser beam.

The large Nd-doped-glass slabs are manufactured with a continuous melting system where the laser glasses are melted in platinum-lined crucibles at high temperatures (~ 1000 °C). The metallic platinum Pt^0, which may be included in the glass during this process, are oxidized and dissolved into the glass matrix in the form of highly stable $(Pt^{4+}Cl^{-}{}_6)^{2-}$ anion by addition of chlorine-containing gasses (Suratwala et al., 2004). Although not frequent, small (~ 10 μm) metallic platinum particles which remain in the laser glass may grow to large-size cracks by laser irradiation at high fluence.

Large crystals of KDP are used for second-harmonic generation and plasma-electrode Pockels cell (PEPC), and deuterated KDP (DKDP or KD*P) for third-harmonic generation. In NIF, these crystals are cut from a boule of $55 \times 55 \times 55$ cm^3 at different angles relative to the crystal axis. The large KDP and DKDP crystals are grown by the rapid growth technique, developed by Zaitseva et al. (1991, 1997) and improved for faster speed and better quality at LLNL (Baisden et al., 2016).

The KDP and DKDP crystals contain many (millions of) damage-precursor sites which are formed during the crystal growth process. Although these sites grow to large damages when directly irradiated at high laser fluence, they can be conditioned by pulsed-laser irradiation with gradually increasing fluence stepwise. This process, called laser conditioning, results in significant increase in damage resistance. By raster scanning the crystal for conditioning with optimal parameters (351 nm, 500 ps, up to 5 J/cm^2), a DKDP crystal can be used at high fluence of ~ 12 J/cm^2 at 351 nm (Adams et al., 2007). The bulk conditioning is also effective to reduce the damage density at the exit surface of the crystal.

4.4.4.2 Surface Damage

Many transmission optics are used in high power lasers at high fluence and various wavelengths, as shown in Table 4.2. Optical surfaces of these optics are more susceptible to laser damage than in the bulk, due to absorption and field intensification caused by small particles and defects associated with surface finishing processes. Extensive studies undertaken at LLNL have identified the origins of the damage-precursors at the surface and led to the methods to remove the precursors, resulting in the damage density sufficiently small (less than 10^{-3} sites/cm^2) for large optics (~ 1000 cm^2) to be used at a high fluence of ~ 18 J/cm^2 at 351 nm (Spaeth et al., 2016b).

Brief accounts of these studies are described in the following. Surfaces of optical components, made of fused silica for example, are prepared by finishing (grinding and polishing) and post processing (cleaning, etching, and coating). Many microcracks in the surface layer produced during grinding are removed during polishing by chemical and/or plastic flow mechanisms. Extensive care is required to eliminate large particles in the polishing slurry, since these particles produce deep scratches which become laser damage precursors. Even with careful grinding and polishing, it was found by photoluminescence technique (Laurence et al., 2009) that large number of disorders and defects exist in the surface layer. Furthermore, sub-bandgap absorption at the fractures in a thin defect layer at the surface was found to be the dominant source of laser damage initiation, whereas less evidence was found for damage initiation by displaced or densified material and by fluence intensification (Miller et al., 2010).

Although the defect layer at the surface has been removed up to ~ 200 nm by chemical etching with fluoride-based etching solutions, the damage threshold did not improve much as expected. It was found that new damage precursors ("invisible" precursors) are produced during etching by the etch-related silica-based precipitates $(NH_4)_2SiF_6$, for example, that are formed preferentially on the fracture surfaces and cusps (Bude et al., 2014). Etching removal of the surface layer up to ~ 30 μm with improved etching processes (etchant solution and agitation) has resulted in remarkable reduction of the damage initiation probability by ~ 5 orders of magnitudes (Suratwala et al., 2011). With these improvements in the surface preparation, the damage density at high laser fluence has decreased dramatically to allow the laser fluence close to ~ 20 J/cm^2 at 351 nm. With a typical laser fluence distribution of 1.8-MJ 3ω NIF shot, the expected number of damage sites is less than 1.6 over the whole beam area of 40×40 cm^2.

By irradiation of intense laser light, the temperature of the damage precursor increases rapidly, followed by rapid heating of the surrounding glass matrix to ~ 2200 K due to runaway absorption arising from strong absorption of the heated region, and expansion of the heated area during laser irradiation time. Once damage is formed on surface, the damage size grows to larger sizes by irradiation with multiple laser shots, increasing the possibility of damage formation in other optics. It was found that exit-side damage is more threatening than front-side damages to the laser system; the density of the front-surface damage sites is much lower (~ 1%) and the size is much smaller than the damages on the exit surface. Systematic experiments

(Norton et al., 2004) have shown an empirical relation that the lateral size of the exit surface damage grows exponentially from the initial diameter D_0 to the diameter D_N after N shots as

$$D_N = D_0 \exp[\alpha(\phi)N]. \tag{4.93}$$

Here $\alpha(\phi)$ is an exponential growth coefficient, which depends on the fluence ϕ (in unit of J/cm^2) as $\alpha(\phi) = 0.0392 \cdot (\phi - 5) \cdot \Phi(\phi - 5)$ where $\Phi(\phi - 5)$ is a step function. As an example, when the damage site is irradiated at 10 J/cm^2 for 50 shots, the size increases by almost 100 times. In more detail, the growth coefficient and its threshold fluence depend on various factors such as fluence, wavelength and duration of the laser pulse, and the size of the damage site.

High-reflectivity mirrors are also susceptible to surface damage. Typically, high reflectivity mirrors for near-infrared to visible region are composed of 30–40 pairs of approximately quarter-wave HfO_2/SiO_2 layers. These layers are deposited on a substrate by several methods, such as electron-beam vacuum evaporation (EB), ion-assisted deposition (IAD), and ion beam sputtering deposition (IBSD). Laser damage to multilayers results from small particles, called nodules, which are included in the multilayer before or during the coating. The nodules are produced by various processes, for example, inadequate surface cleaning, small particles floating in the coating chamber, ejecta from the e-beam sources, etc. We note also that the optical surfaces are often contaminated during storage even in a clean environment due to adsorption of organic molecules, leading to decrease in the damage threshold (Bien-Aime et al., 2009; Jitsuno et al., 2013; Ling et al., 2013).

High damage resistance of ~ 25 J/cm^2 for a 1ω, 3 ns pulse has been obtained by laser conditioning, where the surface contaminants and nodules are ejected from the multilayer by laser irradiation at the fluence just above the ejection threshold, creating very smooth pits (Fornier et al., 1994). Even higher damage resistance of ~ 100 J/cm^2 has been obtained by "planarization", where a flat surface is created before multilayer coating by repeating alternate process of silica coating over contaminant particles on the substrate and etching away half of the silica coating (Stolz et al., 2014).

In NIF, all optics in the laser system are monitored for damage formation. The damage sites found in this monitoring are repaired off-line using a CO_2 laser to form a conical shape, by which the damage does not grow to a larger size and the damage to other optics is prevented. By routinely recycling the repaired optics, the laser system is operated by keeping its initial high performance (Spaeth et al., 2016b).

4.4.5 Damages Due to Optical Coupling and Stimulated Scattering

In high power pulsed laser systems, laser damage can occur due to various effects which do not depend on the local intensity at each optics. These effects include optical coupling between the optical components, and nonlinear effects such as stimulated scattering and filamentation in the optics and propagating medium (Manes et al., 2016).

The laser damage is sometimes generated on unexpected components by the backward propagating beam arising from reflection from the target and/or optical components in the beamline. Even if the reflectance from these components is low, the fluence of the backward beam can exceed the damage threshold due to amplification in the amplifiers where the stored energy remains due to incomplete energy extraction, beam diameter reduction during backward propagation, and ghost focusing by curved surfaces of the focusing optics on the beamline. It is necessary to carefully analyze the laser system by tracing the backward propagation, and to protect the laser system from backward propagation with optical isolators such as a Faraday rotator and a Pockels cell switch.

Laser damage can be also produced due to "hot imaging" when there is a scattering source on a plane such as contamination and damage. With reference to Fig. 4.21, an incident plane wave and a diverging wave originating from a scatterer overlap at an optic located at z from the scatter, forming a Fresnel diffraction ring on the optic. The optic acts as a Fresnel lens due to nonlinear superposition of an incident plane wave and a diverging wave. The Fresnel lens generates a converging wave and forms a conjugate image of the scattering source at $f = z$. The on-axis intensity at the conjugate plane due to interference of the converging wave and an incident plane wave is $I(0) = 1 + B^2$ for an opaque scatterer and $I(0) = (1 + B)^2$ for an transmitting scatterer, where B is the B-integral. Since $B \sim 2$ in typical glass laser systems, the intensity becomes more than 5 times higher, leading to damage to the optic at the conjugate plane. With a nonlinear crystal, the hot image can be also produced at a different imaging position by nonlinear interference of the 2ω plane wave generated in the crystal and a 1ω diverging wave, for example.

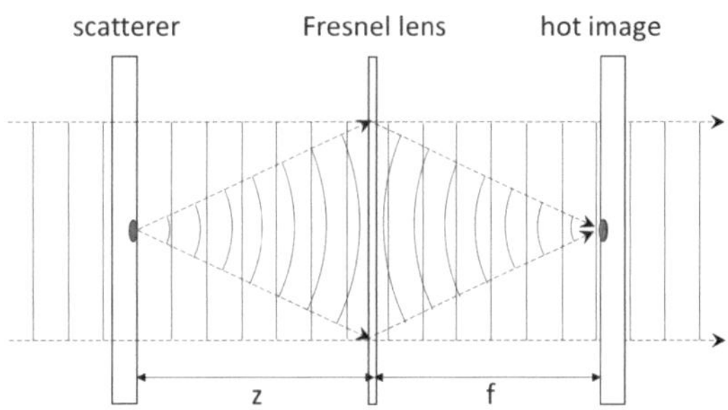

Fig. 4.21 Formation of a hot image due to a Fresnel lens which is formed by nonlinear superposition of an incident plane and a diverging wave from a scatterer plane. Modified from Manes et al. (2016)

Laser energy depletion and laser-induced damage can take place by stimulated Brillouin scattering (SBS), where the energy of the incident light is efficiently converted to an acoustic wave due to nonlinear three-wave coupling. In the 1960s and 70s when the ruby laser was used as a high-power pulsed light source, damages were produced occasionally in the ruby crystals due to intense acoustic wave when the laser was operated close to single longitudinal mode. In high-power pulsed laser systems such as NIF, transverse stimulated Brillouin scattering (TSBS) can take place in large optics, where the scattered light grows in a transverse direction (normal to the laser propagation direction) due to long gain length over the beam diameter. Laser damages were produced in various optics by several processes: filamentation of the scattered light in the optics, and the shock wave generated at the bezel due to ablation by the intense TSBS light (Manes et al., 2016). The decay time of the acoustic wave τ generated by TSBS in a fused silica is ~ 9.9 ns for 1053 nm laser wavelength and ~ 1.1 ns for 351 nm laser wavelength, respectively. The SBS can be suppressed by reducing the laser pulse width to shorter than τ, or broadening the laser bandwidth to over $\Delta\nu = 1/2\pi\tau$ (Murray et al., 1989).

Deterioration of the spatial distribution of a high-power laser beam, which results in reduction of low-spatial-frequency focusable energy, is induced by stimulated rotational Raman scattering (SRRS) of nitrogen and oxygen molecules in air due to long (~ 100 m) propagation path from the laser output to the target chamber. For example, ~ 1% energy is converted to S(8) and S(10) rotational Raman lines of nitrogen at an intensity-length product ($I_L L$) of 12 TW/cm with 2.5-ns-duration square pulses (Henesian et al., 1985). Measurements at various laser facilities have shown that the practical upper limit for propagation in air of a 1ω laser beam is $I_L L \leq 10\,\mathrm{TW/cm}$ (Manes et al., 2016). The relaxation time of the rotational Raman scattering due to dephasing collisions in air at standard pressure and temperature is 120–130 ps, requiring spectral broadening of ~ 300 GHz at 1ω. Since this large spectral broadening is not suitable for efficient frequency conversion with KDP and DKDP crystals, the long beam lines of high-power lasers are often filled with an inert gas (argon) in place of air.

4.4.6 Damage-Resistant Gaseous Optics

Solid state optics are damaged permanently once small damages are formed. The diffraction grating required for pulse compression is one of the weakest components in high-power ultrashort pulse lasers (Bonod & Neauport, 2016). High power laser beams can be manipulated with plasma optics since the plasma is already ionized, such as plasma mirrors (Thaury et al., 2007) and plasma gratings (Sheng et al., 2003; Shi et al., 2011). Another approach is a gaseous optics where a high-power laser beam is diffracted with a diffraction grating formed in a neutral gas (Michine & Yoneda, 2020).

Referring to Fig. 4.22, an atmospheric-pressure gas composed of O_2 and O_3 (ozone) molecules generated in an ozonizer section, with an ozone fraction varied in 1

~ 10% for optimization, flows to an interaction region as a laminar flow. At the interaction region, the O_3 molecules are decomposed by a 248 nm-wavelength KrF excimer laser pulse of 15 ns duration. Through photo-dissociation O_3 and recombination of the dissociation product O^* with O_2,

$$O_3 + h\nu \rightarrow O_2 + O^*, \quad O^* + O_2 \rightarrow O_3 \tag{4.94}$$

the absorbed laser energy is converted to the thermal energy of the O_2 and O_3 molecules. When two excimer beams are interfered to form an interference pattern, a periodic thermal structure is formed in the gas. A periodic density structure is created in the gas, ~ 50–60 ns after the uv pulse irradiation, through conversion of the thermal energy into the sound wave.

The transmission grating of 10-mm depth works as a volume grating, diffracting a laser beam incident normal to the flow direction with 96% efficiency to the first diffraction order when the conditions are optimized. The damage threshold of the ozone-mixed gas has been measured to be 1.6 kJ/cm^2 for a laser pulse of 532 nm, 6 ns duration, i.e., ~ 2-orders higher than the present solid-state optics.

This gaseous optics can be used not only to redirect a laser beam but also to focus a beam by forming a spherical volume grating in a gas, with which the laser optics

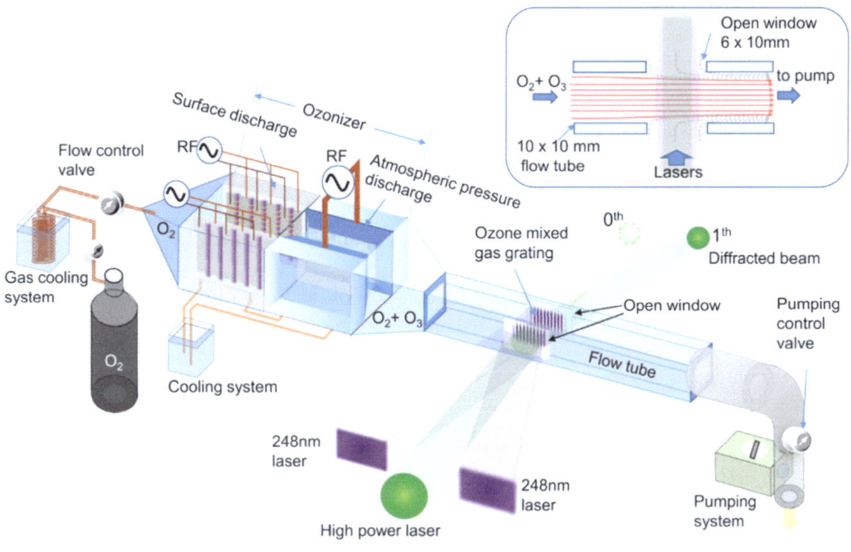

Fig. 4.22 Diffraction optics produced by dissociation of ozone molecules (O_3) by irradiation of two interfering 248 nm laser beams. The O_3 is generated from O_2 with EUV radiation from corona and atmospheric discharges. The O_2 and O_3 mixed gas flows in a 10 × 10 mm rectangular tube as a laminar flow by pumping with constant flow velocity. The KrF laser beams for grating formation and a high-power laser beam for diffraction are incident to the gas through an open window of 6 × 10 mm area in the middle of the flow tube. Reprinted from Michine and Yoneda (2020). © 2020 Springer Nature

can be protected from the target debris. It could be also used for pulse compression of ultrashort-duration, high intensity pulses if a grating of thinner depth is created with high density gases. Although further developments are required, it may become possible to manipulate high power laser beams with the compact damage-resistant gaseous optics, enabling significant down-sizing of high-energy, high-power laser systems.

References

Adams, J. J., Jarboe, J. A., Carr, C. W., et al. (2007). Results of sub-nanosecond laser-conditioning of KD_2PO_4 crystals. *Proceedings of SPIE, 6403*, 64031M.

Auerbach, J. M., & Karpenko, V. P. (1994). Serrated-aperture apodizers for high-energy laser systems. *Applied Optics, 33*, 3179.

Baisden, P. A., Atherton, L. J., Hawley, R. A., et al. (2016). Large optics for the National Ignition Facility. *Fusion Science and Technology, 69*, 295.

Bellini, M., & Hänsch, T. W. (2000). Phase-locked white-light continuum pulses: Toward a universal optical frequency-comb synthesizer. *Optics Letters, 25*, 1049.

Bien-Aime, K., Belin, C., Gallais, L., et al. (2009). Impact of storage induced outgassing organic contamination on laser induced damage of silica optics at 351 nm. *Optics Express, 17*, 18703.

Boling, N., Glass, A., & Owyoung, A. (1978). Empirical relationships for predicting nonlinear refractive index changes in optical solids. *IEEE Journal of Quantum Electronics, 14*, 601.

Bonod, N., & Neauport, J. (2016). Diffraction gratings: From principles to applications in high-intensity lasers. *Advances in Optics and Photonics, 8*, 156.

Born, M., & Wolf, E. (1999). *Principles of optics: electromagnetic theory of propagation, interference and diffraction of light* (7th ed.). Cambridge University Press.

Boyd, R. W. (2020). *Nonlinear optics* (4th ed.). Academic Press.

Brown, D. C. (1981). *High-peak-power Nd:glass laser systems*. Springer.

Bude, J., Miller, P., Baxamusa, S., et al. (2014). High fluence laser damage precursors and their mitigation in fused silica. *Optics Express, 22*, 5839.

Bunkenburg, J., Boles, J., Brown, D. C., Eastman, J., et al. (1981). The omega high-power phosphate glass system: Design and performance. *IEEE Journal of Quantum Electronics, 17*, 1620.

Campillo, A. J., Pearson, J. E., Shapiro, S. L., & Terrell, N. J., Jr. (1973). Fresnel diffraction effects in the design of high-power laser systems. *Applied Physics Letters, 23*, 85.

Carr, C. W., Feit, M. D., Nostrand, M. C., et al. (2006). Techniques for qualitative and quantitative measurement of aspects of laser-induced damage important for laser beam propagation. *Measurement Science & Technology, 17*, 1958.

Chériaux, G., Gaul, E., Anipenkov, R., et al. (2019). KJ-10 PW class laser system at 1 shot a minute. *SPIE Proceedings, 10898*, 1089806.

Coble, R. L. (1961a). Sintering crystalline solids. I. Intermediate and final state diffusion models. *Journal of Applied Physics, 32*, 787.

Coble, R. L. (1961b). Sintering crystalline solids. II. Experimental test of diffusion models in powder compacts. *Journal of Applied Physics, 32*, 793.

David, S. P., Jambunathan, V., Lucianetti, A., et al. (2018). Overview of ytterbium based transparent ceramics for diode pumped high energy solid-state lasers. *High Power Laser Science and Engineering, 6*, Article e62.

De Vido, M., Wojtusiak, A., & Ertel, K. (2020). High-resolution absorption measurement at the zero phonon line of Yb:YAG between 80 K and 300 K. *Optical Materials Express, 10*, 717.

Diddams, S. A., Jones, D. J., Ye, J., et al. (2000). Direct link between microwave and optical frequencies with a 300 THz femtosecond laser comb. *Physical Review Letters, 84*, 5102.

Dieke, G. H. (1968). *Spectra and energy levels of rare-earth ions in crystals*. Wiley-Interscience.
Eckstein, J. N., Ferguson, A. I., & Hänsch, T. W. (1978). High-resolution two-photon spectroscopy with picosecond light pulses. *Physical Review Letters, 40*, 847.
Fornier, A., Cordillot, C., Ausserre, D., et al. (1994). Laser conditioning of optical coatings: Some issues in the characterization by atomic force microscopy. *Proceedings of SPIE, 2114*, 355.
Fox, A. G., & Li, T. (1961). Resonant modes in a maser interferometer. *Bell System Technical Journal, 40*, 453.
Frantz, L. M., & Nodvik, J. S. (1963). Theory of pulse propagation in a laser amplifier. *Journal of Applied Physics, 34*, 2346.
Fujioka, K., Matsumoto, Y., Tamaru, Y., et al. (2025). Fabrication, spectroscopic characteristics, and lasing performance of Nd,La:CaF_2 transparent ceramics. *Optical Materials Express, 15*, 890.
Furuse, H., Morita, K., Yasuhara, R., et al. (2019). Development of laser optical elements by spark plasma sintering technique. *The Review of Laser Engineering, 47*, 448. (in Japanese).
Furuse, H., Nakasawa, S., Yoshida, H., et al. (2018). Transparent ultrafine Yb^{3+}:Y_2O_3 laser ceramics fabricated by spark plasma sintering. *Journal of the American Ceramic Society, 101*, 694.
Furuse, H., & Yasuhara, R. (2017). Magneto-optical characteristics of holmium oxide (Ho_2O_3) ceramics. *Optical Materials Express, 7*, 27.
Hays, G. R., Gaul, E. W., Martinez, M. D., et al. (2007). Broad-spectrum neodymium-doped laser glasses for high-energy chirped-pulse amplification. *Applied Optics, 46*, 4813.
Henesian, M. A., Swift, C. D., & Murray, J. R. (1985). Stimulated rotational Raman scattering in nitrogen in long air paths. *Optics Letters, 10*, 565.
Holzrichter, J. F., Eimerl, D., George, E. V., Trenholme, J. B., Simmons, W. W., & Hunt, J. T. (1982). High power pulsed lasers. *Journal of Fusion Energy, 2*, 5.
Holzwarth, R., Udem, T., Hänsch, T. W., et al. (2000). Optical frequency synthesizer for precision spectroscopy. *Physical Review Letters, 85*, 2264.
Hunt, J. T., Glaze, J. A., Simmons, W. W., & Renard, P. A. (1978). Suppression of self-focusing through low-pass spatial filtering and relay imaging. *Applied Optics, 17*, 2053.
Hunt, J. T., Renard, P. A., & Nelson, R. G. (1976). Focusing properties of an aberrated laser beam. *Applied Optics, 15*, 1458.
Hunt, J. T., Renard, P. A., & Simmons, W. W. (1977). Improved performance of fusion lasers using the imaging properties of multiple spatial filters. *Applied Optics, 16*, 779.
Ikesue, A., Kinoshita, T., Kamata, K., et al. (1995). Fabrication and optical properties of high-performance polycrystalline Nd:YAG ceramics for solid-state lasers. *Journal of the American Ceramic Society, 78*, 1033.
International Standards Organization. (2011). *Lasers and laser-related equipment—Test methods for laser-induced damage threshold—Part 2: Threshold determination*. The International Organization for Standardization, ISO 21254-2:2011.
Izumitani, T. (1984). *Optical glasses*. Kyoritsu-Publishing. (in Japanese).
Izumitani, T., Toratani, H., & Kuroda, H. (1982). Radiative and nonradiative properties of neodymium doped silicate and phosphate glasses. *Journal of Non-Crystalline Solids, 47*, 87.
Jacobs, R. R., & Weber, M. J. (1976). Dependence of the $^4F_{3/2} \rightarrow {}^4I_{11/2}$ induced-emission cross section for Nd^{3+} on glass composition. *IEEE Journal of Quantum Electronics, 12*, 102.
Jitsuno, T., Murakami, H., Kato, K., et al. (2013). Recent progresses on insights of laser damage mechanisms and influence of contamination in optics. *Proceedings of SPIE, 8786*, 87860B.
Jones, D. J., Diddams, S. A., Ranka, J. K., et al. (2000). Carrier-envelope phase control of femtosecond mode-locked lasers and direct optical frequency synthesis. *Science, 288*, 635.
Judd, B. R. (1962). Optical absorption intensities of rare-earth ions. *Physical Review, 127*, 750.
Kane, T. J., & Byer, R. L. (1985). Monolithic, unidirectional single-mode Nd:YAG ring laser. *Optics Letters, 10*, 65.
Kato, Y., Yoshida, K., Kuroda, J., & Yamanaka, C. (1981). 3.4-TW performance of a Nd:phosphate glass laser with output aperture of 20 cm. *Applied Physics Letters, 38*, 72.

Katori, H., Takamoto, M., Pal'chikov, V. G., et al. (2003). Ultrastable optical clock with neutral atoms in an engineered light shift trap. *Physical Review Letters, 91*, 173005.
Kawanaka, J., Tokita, S., Nishioka, H., et al. (2005). Dramatically improved laser characteristics of diode-pumped Yb-doped materials at low temperature. *Laser Physics, 15*, 1306.
Kawanaka, J., Yamakawa, K., Nishioka, H., et al. (2003). 30-mJ, diode-pumped, chirped-pulse Yb:YLF regenerative amplifier. *Optics Letters, 28*, 2121.
Klimek, D. E., & Mandl, A. (2011). Nd:YAG ceramic ThinZag® high power laser development (Chap. 9). In H. Injeyan & G. D. Goodno (Eds.), *High-power laser handbook*. McGraw Hill.
Knight, J. C., Birks, T. A., Russell, P. S. J., & Atkin, D. M. (1996). All-silica single-mode optical fiber with photonic crystal cladding. *Optics Letters, 21*, 1547; *Optics Letters, 22*, 484.
Koechner, W. (2006). *Solid-state laser engineering* (6th ed.). Springer, NY.
Kogelnik, H., & Li, T. (1966). Laser beams and resonators. *Proceedings of the IEEE, 54*, 1312.
Kong, J., Tang, D. Y., Zhao, B., et al. (2005). 9.2-W diode-end pumped Yb:Y_2O_3 ceramic laser. *Applied Physics Letters, 86*, 161116.
Košelja, M., Rus, B., Kubát, J., & Houžvička, J. (2016, October 31). *Laser material research in the Czech Republic stimulated by the ELI-beamlines project*, ASSL Boston, paper AM3A.6.
Kravchenko, V. B., & Kopylov, Y. L. (2013). Oxide laser ceramics. In B. Denker & E. Shklovsky (Eds.), *Handbook of solid-state lasers—Materials, systems and applications*. Woodhead Publishing.
Krupke, W. F. (1974). Induced-emission cross sections in neodymium laser glasses. *IEEE Journal of Quantum Electronics, 10*, 450.
Kuroda, J., & Kato, Y. (1981). Diffraction effect due to optical noise sources in a glass laser system. *Japanese Journal of Applied Physics, 20*, 1493.
Lacovara, P., Choi, H. K., Wang, R. L., et al. (1991). Room-temperature diode-pumped Yb:YAG laser. *Optics Letters, 16*, 1089.
Laurence, T. A., Bude, J. D., Shen, N., et al. (2009). Metallic-like photoluminescence and absorption in fused silica surface flaws. *Applied Physics Letters, 94*, Article 151114.
Linford, G. J., Saroyan, R. A., Trenholm, J. B., & Weber, M. J. (1979). Measurements and modelling of gain coefficients for neodymium laser glasses. *IEEE Journal of Quantum Electronics, 15*, 510.
Ling, X., Wang, G., Zhao, Y., et al. (2013). Laser-induced damage of the optical coatings due to organic contamination in vacuum. *Applied Surface Science, 270*, 346.
Lu, J., Ueda, K., Yagi, H., et al. (2002). Neodymium doped yttrium aluminum garnet ($Y_3Al_5O_{12}$) nanocrystalline ceramics—A new generation of solid state laser and optical materials. *Journal of Alloys and Compounds, 341*, 220.
MacFarlane, R. M., Wong, J. Y., & Sturge, M. D. (1968). Dynamic Jahn–Teller effect in octahedrally coordinated d^1 impurity systems. *Physical Review, 166*, 250.
Manes, K. R., Spaeth, M. L., Adams, J. J., Bowers, et al. (2016). Damage mechanisms avoided or managed for NIF large optics. *Fusion Science and Technology, 69*, 146.
Marmo, J., Injeyan, H., Komine, H., et al. (2009). Joint high power solid state laser program advancements at Northrop Grumman. *Proceedings of SPIE, 7195*, 719507.
Mason, P., Baneerjee, S., Smith, J., et al. (2018). Development of a 100 J, 10 Hz laser for compression experiments at the high energy density instrument at the European XFEL. *High Power Laser Science and Engineering, 6*, Article e65.
Michine, Y., & Yoneda, H. (2020). Ultrahigh damage threshold optics for high power lasers. *Communications on Physics, 3*, 24. https://doi.org/10.1038/s42005-020-0286-6
Miller, P. E., Bude, J. D., Suratwala, T. I., et al. (2010). Fracture-induced subbandgap absorption as a precursor to optical damage on fused silica surfaces. *Optics Letters, 35*, 2702.
Moulton, P. F. (1986). Spectroscopic and laser characteristics of Ti:Al_2O_3. *Journal of the Optical Society of America B, 3*, 125.
Murray, J. R., Smith, J. R., Ehrlich, R. B., et al. (1989). Experimental observation and suppression of transverse stimulated Brillouin scattering in large optical components. *Journal of the Optical Society of America B, 6*, 2402.

Norton, M. A., Donohue, E. E., Hollingsworth, W. G., et al. (2004). Growth of laser-initiated damage in fused silica at 527 nm. *Proceedings of SPIE, 5273*, 236.
Ofelt, G. S. (1962). Intensities of crystal spectra of rare-earth ions. *The Journal of Chemical Physics, 37*, 511.
Ogino, J., Tokita, S., Kitajima, S., et al. (2022). 10-J, 100-Hz conduction-cooled active mirror laser. *Optics Continuum, 1*, 1270. https://doi.org/10.1364/OPTCON.456554
Ranka, J. K., Windeler, R. S., & Stentz, A. J. (2000). Visible continuum generation in air–silica microstructure optical fibers with anomalous dispersion at 800 nm. *Optics Letters, 25*, 25.
Reichert, J., Holzwarth, R., Udem, T., & Hänsch, T. W. (1999). Measuring the frequency of light with mode-locked lasers. *Optics Communication, 172*, 59.
Schawlow, A. L., & Townes, C. H. (1958). Infrared and optical masers. *Physical Review, 112*, 1940.
Shen, Y. R. (1984). *The principles of nonlinear optics*. Wiley.
Sheng, Z.-M., Zhang, J., & Umstadter, D. (2003). Plasma density gratings induced by intersecting laser pulses in underdense plasmas. *Applied Physics B, 77*, 673.
Shi, L., Li, W., Wang, Y., et al. (2011). Generation of high-density electrons based on plasma grating induced Bragg diffraction in air. *Physical Review Letters, 107*, Article 095004.
Shoji, T., Tokita, S., Kawanaka, J., et al. (2004). Quantum-defect-limited operation of diode-pumped Yb:YAG laser at low temperature. *Japanese Journal of Applied Physics, 43*, L496.
Simmons, W. W., Hunt, J. T., & Warren, W. E. (1981). Light propagation through large laser systems. *IEEE Journal of Quantum Electronics, 17*, 1727.
Slezak, O., Yasuhara, R., Vojna, D., et al. (2019). Temperature-wavelength dependence of Verdet constant of Dy_2O_3 ceramics. *Optical Materials Express, 9*, 2971.
Smith, W. L. (1978). Laser-induced breakdown in optical materials. *Optical Engineering, 17*, 175489.
Smith, W. L., Bechtel, J. H., & Bloembergen, N. (1976). Picosecond laser-induced damage morphology: Spatially resolved microscopic plasma sites. *Optics Communication, 18*, 592.
Sorokina, I. T. (2003). Crystalline mid-infrared lasers. In I. T. Sorokina & K. L. Vodopyanov (Eds.), *Topics in applied physics solid-state mid-infrared laser sources* (Vol. 89, p. 255). Springer.
Spaeth, M. L., Manes, K. R., Kalanter, D. H., et al. (2016a). Description of the NIF laser. *Fusion Science and Technology, 69*, 25.
Spaeth, M. L., Wegner, P. J., Suratwala, T. I., et al. (2016b). Optics recycle loop strategy for NIF operations above UV laser-induced damage threshold. *Fusion Science and Technology, 69*, 265.
Spence, D. E., Kean, P. N., & Sibbet, W. (1991). 60-fsec pulse generation from a self-mode-locked Ti:sapphire laser. *Optics Letters, 16*, 42.
Stolz, C. J., Wolfe, J. E., Adams, J. J., et al. (2014). High laser-resistant multilayer mirrors by nodular defect planarization. *Applied Optics, 53*, A291.
Suratwala, T. I., Campbell, J. H., Miller, P. E., et al. (2004). Phosphate laser glass for NIF: Production status, slab selection, and recent technical advances. *Proceedings of SPIE, 5341*, 102.
Suratwala, T. I., Miller, P. E., Bude, J. D., et al. (2011). HF-based etching processes for improving laser damage resistance of fused silica optical surfaces. *Journal of the American Ceramic Society, 94*, 416.
Sutter, D. H., Steinmeyer, G., Gallmann, L., et al. (1999). Semiconductor saturable-absorber mirror–assisted Kerr-lens mode-locked Ti:sapphire laser producing pulses in the two-cycle regime. *Optics Letters, 24*, 631.
Tanabe, Y., & Sugano, S. (1954a). On the absorption spectra of complex ions. I. *Journal of the Physical Society of Japan, 9*, 753.
Tanabe, Y., & Sugano, S. (1954b). On the absorption spectra of complex ions. II. *Journal of the Physical Society of Japan, 9*, 766.
Thaury, C., Quere, F., Geindre, J.-P., et al. (2007). Plasma mirrors for ultrahigh intensity optics. *Nature Physics, 3*, 424.
Tokurakawa, M., Shirakawa, A., Ueda, K., et al. (2007b). Diode-pumped sub-100 fs Kerr-lens mode-locked $Yb:Sc_2O_3$ ceramic laser. *Optics Letters, 32*, 3382.

Tokurakawa, M., Takaichi, K., Shirakawa, A., et al. (2007a). Diode-pumped 188 fs mode-locked Yb:Y_2O_3 ceramic laser. *Applied Physics Letters, 90*, Article 071101.

Träger, F. (Ed.). (2007). *Springer handbook of lasers and optics*. Springer.

Yablonovitch, E., & Bloembergen, N. (1972). Avalanche ionization and the limiting diameter of filaments induced by light pulses in transparent media. *Physical Review Letters, 29*, 907.

Yagi, H., Yanagitani, T., Takaichi, K., et al. (2007). Characterizations and laser performances of highly transparent Nd^{3+}:$Y_3Al_5O_{12}$ laser ceramics. *Optical Materials, 29*, 1258.

Yamanaka, C., Kato, Y., Izawa, Y., Yoshida, K., et al. (1981). Nd-doped phosphate glass laser systems for laser-fusion research. *IEEE Journal of Quantum Electronics, 17*, 1639.

Zaitseva, N. P., De Yoreo, J. J., Dehaven, M. R., et al. (1997). Rapid growth of large scale (40–55 cm) KH_2PO_4 crystals. *Journal of Crystal Growth, 180*, 255.

Zaitseva, N. P., Smolskii, I. L., & Rashkovich, L. N. (1991). Study of rapid growth of KDP Crystals by temperature lowering. *Soviet Physics Crystallography, 36*, 113.

Chapter 5
Nonlinear Optical Processes in High Power Lasers

Abstract When an intense light interacts with a material, the material responds nonlinearly to the field amplitude. In this section, basic concepts of nonlinear optical processes are presented in Sect. 5.1. Then several nonlinear processes important in high power lasers are described: frequency conversion (Sect. 5.2), optical parametric amplification including optical parametric chirped pulse amplification (Sect. 5.3), self-focusing and self-phase modulation (Sect. 5.4), and propagation of extremely short, intense laser pulses (Sect. 5.5).

Keywords Nonlinear optical susceptibility · Nonlinear coupling of optical waves · Optical parametric amplification · Phase matching · Nonlinear refractive index · Self-focusing · Self-phase modulation

5.1 Nonlinear Optical Processes

5.1.1 Nonlinear Polarization

The electric displacement $\boldsymbol{D}$ is expressed as $\boldsymbol{D} = \varepsilon_0\boldsymbol{E}+\boldsymbol{P}=\varepsilon\boldsymbol{E}$ where $\boldsymbol{E}$ is the electric field and $\boldsymbol{P}=\varepsilon_0\boldsymbol{\chi} \cdot \boldsymbol{E}$ is the polarization. The optical susceptibility $\boldsymbol{\chi}$ describes the response of the material to the electric field. In the following we mainly focus our discussion on polarization due to electrons, since these are important in short-duration high power lasers. Polarization due to atomic motions which have slower response times, leading to the Raman and Brillouin scatterings, are not described, although same formalisms are applicable.

When the amplitude of the electric field increases, the electron motion deviates from linear response to $\boldsymbol{E}$. Expanding $\boldsymbol{\chi}$ with respect to $\boldsymbol{E}$ and separating $\boldsymbol{P}$ into linear and nonlinear terms, $\boldsymbol{P}_L$ and $\boldsymbol{P}_{NL}$,

$$\boldsymbol{P} = \varepsilon_0(\boldsymbol{\chi}^{(1)} \cdot \boldsymbol{E} + \boldsymbol{\chi}^{(2)}\!:\!\boldsymbol{E}\cdot\boldsymbol{E} + \boldsymbol{\chi}^{(3)}\,\vdots\,\boldsymbol{E}\cdot\boldsymbol{E}\cdot\boldsymbol{E} + \cdots) \equiv \boldsymbol{P}_{\mathrm{L}} + \boldsymbol{P}_{\mathrm{NL}};$$
$$\boldsymbol{P}_{\mathrm{L}} = \varepsilon_0\boldsymbol{\chi}^{(1)} \cdot \boldsymbol{E},$$
$$\boldsymbol{P}_{\mathrm{NL}} = \varepsilon_0(\boldsymbol{\chi}^{(2)}\!:\!\boldsymbol{E}\cdot\boldsymbol{E} + \boldsymbol{\chi}^{(3)}\,\vdots\,\boldsymbol{E}\cdot\boldsymbol{E}\cdot\boldsymbol{E} + \cdots). \tag{5.1}$$

Y. Kato et al., *High Power Laser and Plasma Science*, Springer Series on Atomic, Optical, and Plasma Physics 130, https://doi.org/10.1007/978-3-032-08930-4_5

Here $\boldsymbol{\chi}^{(1)}$, $\boldsymbol{\chi}^{(2)}$ and $\boldsymbol{\chi}^{(3)}$ are the first- (linear), second- and third-order susceptibilities expressed by 2nd, 3rd, and 4th-rank tensors, respectively. Writing the dielectric constant in the linear regime as $\boldsymbol{\varepsilon} = \varepsilon_0(1 + \boldsymbol{\chi}^{(1)})$, $\boldsymbol{D}$ is described as

$$\boldsymbol{D} = \varepsilon_0 \boldsymbol{E} + \boldsymbol{P}_{\mathrm{L}} + \boldsymbol{P}_{\mathrm{NL}} = \varepsilon_0(1 + \chi^{(1)})\boldsymbol{E} + \boldsymbol{P}_{\mathrm{NL}} = \boldsymbol{\varepsilon} \cdot \boldsymbol{E} + \boldsymbol{P}_{\mathrm{NL}}. \tag{5.2}$$

Similarly to the linear polarization expressed in Eq. (2.65) where finite response time of the electrons are considered, the nonlinear polarization is written as

$$\begin{aligned} \boldsymbol{P}_{\mathrm{NL}}(t) = \varepsilon_0 \Bigg[& \iint \boldsymbol{\chi}^{(2)}(t - t_1, t - t_2) : \boldsymbol{E}(\mathrm{t}_1)\boldsymbol{E}(\mathrm{t}_2)\mathrm{d}\mathrm{t}_1\mathrm{d}\mathrm{t}_2 \\ & + \iiint \boldsymbol{\chi}^{(3)}(t - t_1, t - t_2, t - t_3) \vdots \boldsymbol{E}(t_1)\boldsymbol{E}(t_2)\boldsymbol{E}(t_3)dt_1dt_2dt_3 \Bigg]. \end{aligned} \tag{5.3}$$

Specifying the polarization component (x, y, or z) of P, E and χ in the suffix, Fourier transformation of (5.3) leads to

$$\begin{aligned} P_i^{(1)}(\omega) &= \varepsilon_0 \sum_j \chi_{ij}^{(1)}(\omega)E_j(\omega), \\ P_i^{(2)}(\omega_3) &= \varepsilon_0 \sum_{jk} \chi_{ijk}^{(2)}(-\omega_3; \omega_1, \omega_2)E_j(\omega_1)E_k(\omega_2), \\ P_i^{(3)}(\omega_4) &= \varepsilon_0 \sum_{jkl} \chi_{ijkl}^{(3)}(-\omega_4; \omega_1, \omega_2, \omega_3)E_j(\omega_1)E_k(\omega_2)E_l(\omega_3), \end{aligned} \tag{5.4}$$

where the numbers in the superscripts refer to linear, second-order and third-order polarizations, respectively.

The negative and positive signs of the frequencies in $\chi^{(2)}$ and $\chi^{(3)}$ express the relations $\omega_3 = \omega_1 + \omega_2$ and $\omega_4 = \omega_1 + \omega_2 + \omega_3$, respectively. Since the driving fields have both positive and negative frequency components, the second order polarization $P_i^{(2)}(\omega_3)$, for example, comprises many combinations of frequencies such as $\omega_1 + \omega_2$, $\omega_1 - \omega_2$, $\omega_2 - \omega_1$, $2\omega_1$, $2\omega_2$ and $0\ (= \omega_j - \omega_j,\ j = 1,\ 2)$. Among these components, only one of them is generated with appreciable intensity when it satisfies the phase matching condition $\boldsymbol{k}_3 = \boldsymbol{k}_1 + \boldsymbol{k}_2$ where $\boldsymbol{k}_{\mathrm{i}}$ is the wave number of the ω_{i} wave, as described in the following.

The sum- and difference-frequency generation processes are used to generate coherent light from long to short wavelengths or from short to long wavelengths for various scientific and industrial applications. Examples of nonlinear optical processes are shown in Fig. 5.1 using the energy level diagrams: (a) sum-frequency generation $\omega_3 = \omega_1 + \omega_2$, (b) difference frequency generation $\omega_3 = \omega_1 - \omega_2$, (c) parametric frequency conversion $\omega_p = \omega_s + \omega_i$, (d) third harmonic generation $3\omega = \omega + \omega + \omega$, (e) stimulated Stokes Raman scattering $\omega_S = \omega_p - \omega_v$ and (f) stimulated anti-Stokes Raman scattering $\omega_{AS} = \omega_p + \omega_v$, where ω_v is the frequency of the optical vibration. Among various nonlinear optical processes, we focus on several processes closely

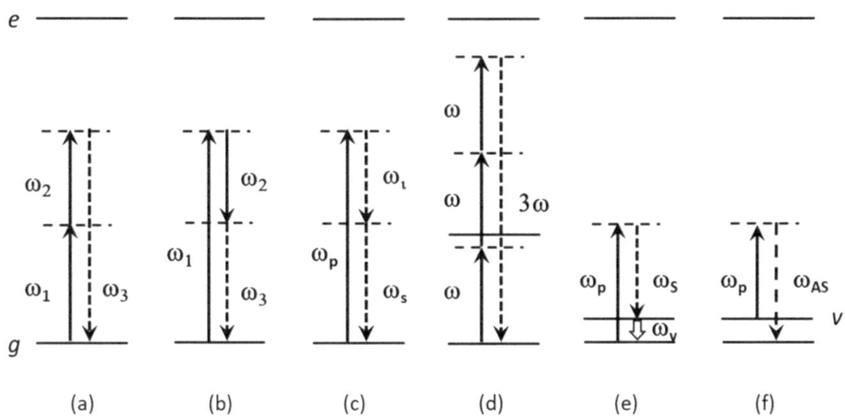

Fig. 5.1 Examples of nonlinear optical processes: **a** sum-frequency generation, **b** difference frequency generation, **c** optical parametric frequency conversion, **d** third harmonic generation, **e** stimulated Stokes Raman scattering, and **f** stimulated anti-Stokes Raman scattering. The solid lines show the energy levels of the ground state, the excited state, and the vibrational level, marked by *g*, *e* and *v*, respectively. The virtual states are shown by dotted lines

related to high power lasers: harmonic generation, optical parametric amplification, self-focusing, and self-phase modulation.

5.1.2 Nonlinear Optical Susceptibilities

For order-of magnitude estimation of the atomic response to an optical field, let us compare the strengths of the atomic field and the laser field. The electric field of a hydrogen atom E_{at} which acts on an electron is estimated as $E_{at} = e/(4\pi\varepsilon_0 a_0^2) = 5.14 \times 10^{11}$ V/m, where a_0 is the Bohr radius (orbital radius of an electron in the ground state), e is the electron charge, and ε_0 is the dielectric constant of vacuum. The laser field E is expressed by the intensity as $I = (1/2)\varepsilon_0 c E^2$. Taking a pulsed laser of 1-J energy, 10-ns pulse-width and 1-cm^2 beam area as an example, the intensity is $I = 10^8$ W/cm^2 and the corresponding field strength is $E = 2.74 \times 10^7$ V/m. Although the perturbation of the laser field to an atom is very small since $E \sim 5 \times 10^{-5} E_{at}$, the nonlinear optical field such as the second harmonics emitted from each atom become comparable to the incident field due to coherent superposition of these fields when the phase matching condition is satisfied.

The magnitude of the second-order nonlinear optical susceptibility is estimated by assuming that the nonlinear polarization becomes comparable to the linear polarization at $E \sim E_{at}$. From the condition $\varepsilon_0 \chi^{(2)} E_{at}^2 \cong \varepsilon_0 \chi^{(1)} E_{at}$, $\chi^{(2)}$ is estimated as $\chi^{(2)}/\chi^{(1)} \approx 1/E_{at} \cong 2 \times 10^{-12}\,(m/V) = 2\,(pm/V)$. Similarly, the third-order nonlinear optical susceptibility is estimated as $\chi^{(3)}/\chi^{(1)} \approx 1/E_{at}^2 \cong 3.8 \times 10^{-24}$ m^2/V^2 (Armstrong et al., 1962).

Miller (1964) has found an empirical rule that the ratio of the nonlinear susceptibility to the linear susceptibility is remarkably constant. The Miller' rule is expressed for the frequency up-conversion of generating $\omega_3 = \omega_1 + \omega_2$ from ω_1 and ω_2 as

$$\frac{\chi^{(2)}(\omega_1 + \omega_2, \omega_1, \omega_2)}{\chi^{(1)}(\omega_1 + \omega_2)\chi^{(1)}(\omega_1)\chi^{(1)}(\omega_2)} \equiv \delta \approx \text{constant} \tag{5.5}$$

The relation (5.5) shows that we can estimate the approximate value of the nonlinear susceptibility from the linear susceptibility. Garrett and Robinson (1966) have validated the Miller's rule by an analysis of the electron motion in an anharmonic potential, deriving the relation $\delta \sim (ma\varepsilon_0^2)/(N^2e^3)$ where N is the atomic number density and max^2 is the restoring force to the electron at the coordinate x in an anharmonic potential characterized by a.

Exact expressions of the nonlinear optical susceptibilities were derived by Armstrong et al. (1962) by quantum mechanical analysis of the nonlinear interaction processes. For example, referring to the energy level diagram Fig. 5.1a, the three-wave interaction is described by the transitions (i) from the ground state g to the first virtual state, (ii) from the first to the second virtual state, and finally (iii) from the second virtual state to g. When the photon energies are far from resonance to the excited states, the second-order nonlinear optical susceptibility for this process is expressed as

$$\begin{aligned}\chi^{(2)}(-\omega_3; \omega_1, \omega_2) &= \frac{N}{\varepsilon_0 \hbar^2} \sum_{u,u'} \mu_{lu}\mu_{uu'}\mu_{u'l} \\ &\times \left[\frac{1}{(\omega_{ul} - \omega_1 - \omega_2)(\omega_{u'l} - \omega_1)} + \frac{1}{(\omega_{ul} + \omega_2)(\omega_{u'l} - \omega_1)} \right. \\ &\left. + \frac{1}{(\omega_{ul} + \omega_2)(\omega_{u'l} + \omega_1 + \omega_2)} \right],\end{aligned} \tag{5.6}$$

with additional three terms by interchanging ω_1 and ω_2 in the right-hand side of (5.6). Here the energy levels denoted by l, u and u' refer to the initial, the first virtual and the second virtual states respectively, ω_{ul} and $\omega_{u'l}$ are the angular frequencies of the u-l and u'-l transitions, and μ_{ul}, $\mu_{u'u}$, $\mu_{lu'}$ are the matrix elements of the electric dipole moment between these states. This expression is useful in evaluating the contributions of the excited states in terms of symmetry and energy, especially when one of the optical fields is close to resonance with the excited state.

In general, the nonlinear susceptibility has many terms. In the case of three wave interaction described by the second equation of (5.4), there are six tensors corresponding to the sum- and difference-frequency generations of $\omega_3 = \omega_1 + \omega_2$, $\omega_1 = \omega_3 - \omega_2$ and $\omega_2 = \omega_3 - \omega_1$,

$$\begin{aligned}&\chi_{ijk}^{(2)}(-\omega_3; \omega_1, \omega_2), \quad \chi_{ijk}^{(2)}(-\omega_3; \omega_2, \omega_1), \quad \chi_{ijk}^{(2)}(-\omega_1; \omega_3, -\omega_2), \\ &\chi_{ijk}^{(2)}(-\omega_1; -\omega_2, \omega_3), \; \chi_{ijk}^{(2)}(-\omega_2; \omega_3, -\omega_1), \; \chi_{ijk}^{(2)}(-\omega_2; -\omega_1, \omega_3)\end{aligned} \tag{5.7}$$

and six additional tensors where each frequency is replaced by its opposite-sign frequency. Since the indices *i, j, k* can take *x, y,* and *z* independently, there are 324 (12 $\times 3^3$) tensors to describe the interaction. However, the number of independent tensors is reduced to 81, since polarization is a real quantity and there are redundancies in the summation:

$$\chi_{ijk}^{(2)}(-\omega_3;\omega_1,\omega_2)=\chi_{ikj}^{(2)}(\omega_3;-\omega_2,-\omega_1)^*,$$
$$\chi_{ijk}^{(2)}(-\omega_3;\omega_1,\omega_2)=\chi_{ikj}^{(2)}(-\omega_3;\omega_2,\omega_1). \tag{5.8}$$

When the optical frequencies are far from resonance to excited states, the linear and nonlinear susceptibilities can be treated as real quantities, since the time-delayed response is neglected. In this case, the frequency arguments in the nonlinear susceptibility can be interchanged freely when the corresponding cartesian indices are interchanged simultaneously:

$$\begin{aligned}\chi_{ijk}^{(2)}(-\omega_3;\omega_1,\omega_2)&=\chi_{ikj}^{(2)}(-\omega_3;\omega_2,\omega_1)=\chi_{jik}^{(2)}(-\omega_1;\omega_3,-\omega_2)\\&=\chi_{jki}^{(2)}(-\omega_1;-\omega_2,\omega_3)=\chi_{kij}^{(2)}(-\omega_2;\omega_3,-\omega_1)=\chi_{kji}^{(2)}(-\omega_2;-\omega_1,\omega_3).\end{aligned} \tag{5.9}$$

This condition, called full permutation symmetry, leads to 27 independent tensors.

Furthermore, when we approximate that the susceptibility does not depend on the frequency, the indices can be permuted without permuting the frequencies. In this case, known as Kleinman condition, we can write $\chi_{ijk}^{(2)}$ without specifying the frequency (Kleinman, 1962). Often *d*-coefficients defined by $d_{ijk}=(1/2)\chi_{ijk}^{(2)}$ are used in place of the second-order nonlinear susceptibilities. Since the last two indices of d_{ijk} are symmetric due to the Kleinman condition, *jk* is expressed by the contracted notation *l* as

$$\begin{array}{lcccccc} jk: & 11 & 22 & 33 & (23,32) & (31,13) & (12,21)\\ l: & 1 & 2 & 3 & 4 & 5 & 6\end{array} \tag{5.10}$$

Using the contracted matrix d_{il}, the nonlinear polarization for the sum frequency generation process $\omega_3=\omega_1+\omega_2$ is described as

$$\begin{bmatrix}P_x(\omega_3)\\P_y(\omega_3)\\P_z(\omega_3)\end{bmatrix}=\frac{\varepsilon_0}{2}\begin{bmatrix}d_{11}&d_{12}&d_{13}&d_{14}&d_{15}&d_{16}\\d_{21}&d_{22}&d_{23}&d_{24}&d_{25}&d_{26}\\d_{31}&d_{32}&d_{33}&d_{34}&d_{35}&d_{36}\end{bmatrix}\times\begin{bmatrix}E_x(\omega_1)E_x(\omega_2)\\E_y(\omega_1)E_y(\omega_2)\\E_z(\omega_1)E_z(\omega_2)\\E_y(\omega_1)E_z(\omega_2)+E_z(\omega_1)E_y(\omega_2)\\E_x(\omega_1)E_z(\omega_2)+E_z(\omega_1)E_x(\omega_2)\\E_x(\omega_1)E_y(\omega_2)+E_y(\omega_1)E_x(\omega_2)\end{bmatrix} \tag{5.11}$$

The permutation between the index I and the indices j and k in d_{il} leads to further reduction of independent components. Since $d_{12} = d_{xyy} = d_{yxy} = d_{26}$ holds, for example, the d-matrix is reduced to 10 independent components:

$$\begin{bmatrix} d_{11} & d_{12} & d_{13} & d_{14} & d_{15} & d_{16} \\ d_{16} & d_{22} & d_{23} & d_{24} & d_{14} & d_{12} \\ d_{15} & d_{24} & d_{33} & d_{23} & d_{13} & d_{14} \end{bmatrix} \tag{5.12}$$

The number of non-zero components in d_{il} is reduced when the structural symmetry is considered, leaving only a small number of elements in most of nonlinear materials.

The d-coefficients of several nonlinear crystals often used in frequency conversion are shown in Table 5.1 (Sutherland, 2003, Chap. 5). Also shown in this Table are the refractive indices for ordinary and extraordinary rays at the fundamental and the second-harmonic wavelengths. In this Table, the components of the tensor d_{il} are given in terms of the principal axes of the crystal. The coefficient d_{eff} refers to the effective nonlinear optical coefficients in the second harmonic generation for the type-I (t-I) and type-II (t-II) phase matching configurations, as described in Sect. 5.2.

Table 5.1 Second-order nonlinear optical coefficients d at 1064 nm, and the ordinary and extraordinary refractive indices, n_o and n_e, at the wavelength of 1064 nm (ω) and 532 nm (2ω), respectively, of several nonlinear crystals with their symmetry classes

Material	Symmetry class	$n(\omega)$	$n(2\omega)$	d (pm/V)	Refs.
$LiNbO_3$	3m	$n_o = 2.234$; $n_e = 2.155$	$n_o = 2.325$; $n_e = 2.233$	$d_{31} = -5.95$; $d_{33} = -33.4$	(a)
BBO	3m	$n_o = 1.6544$; $n_e = 1.5422$	$n_o = 1.6743$; $n_e = 1.5548$	$d_{22} = 2.2$; $d_{\mathrm{eff}} = 2.0$ (t-I), 1.4 (t-II)	(b)
LBO	mm2	$n_X = 1.566$; $n_Y = 1.591$; $n_Z = 1.606$	$n_X = 1.579$; $n_Y = 1.607$; $n_Z = 1.621$	$d_{24} = 0.74$; $d_{31} = 0.8$–1.3; $d_{33} \sim 0$	(c)
KDP	$\bar{4}2m$	$n_o = 1.494$; $n_e = 1.460$	$n_o = 1.512$; $n_e = 1.471$	$d_{36} = 0.39$	(d)
CLBO	$\bar{4}2m$	$n_o = 1.485$; $n_e = 1.435$	$n_o = 1.498$; $n_e = 1.445$	$d_{36} = 0.7$; $d_{\mathrm{eff}} = 0.38$ (t-I); $d_{\mathrm{eff}} = 0.67$ (t-II)	(e)

In these examples, mm2 (orthorhombic) is a biaxial crystal, and 3m (trigonal) and 4m (tetragonal) are uniaxial crystals. References: (a) Choy and Byer (1976); (b) Shoji et al. (1999); (c) Lin et al. (1990); (d) Dmitriev et al. (1999); (e) Shoji et al. (2001)

5.2 Frequency Conversion

5.2.1 Nonlinear Coupling of Optical Waves

The interaction of optical waves in nonlinear dielectrics has been analyzed extensively by Armstrong et al. (1962). Including the nonlinear polarization $\boldsymbol{P}_{NL}$ in the Maxwell equation (2.11), a nonlinear Maxwell equation is described as,

$$\nabla^2 \boldsymbol{E} - \varepsilon\mu \frac{\partial^2 \boldsymbol{E}}{\partial t^2} = \mu \frac{\partial^2 \boldsymbol{P}_{NL}}{\partial t^2} \tag{5.13}$$

Let us consider nonlinear interaction of three waves of frequencies of $\omega_1, \omega_2, \omega_3$ (where $\omega_1 < \omega_2 < \omega_3$) with $\omega_3 = \omega_1 + \omega_2$. We describe the three fields as transverse plane waves propagating in the z-direction:

$$\begin{aligned} E_j^{(\omega_1)}(z,t) &= {}^1\!/_2\left[E_{j0}^{(\omega_1)}(z)e^{i(k_1 z - \omega_1 t + \phi_1)} + c.c\right] \\ E_k^{(\omega_2)}(z,t) &= {}^1\!/_2\left[E_{k0}^{(\omega_2)}(z)e^{i(k_2 z - \omega_2 t + \phi_2)} + c.c\right] \\ E_i^{(\omega_3)}(z,t) &= {}^1\!/_2\left[E_{i0}^{(\omega_3)}(z)e^{i(k_3 z - \omega_3 t + \phi_3)} + c.c\right] \end{aligned} \tag{5.14}$$

where $E_0(z)$ is the phaser amplitude of $E(z, t)$, ϕ_i.is the initial phase of the ω_i wave ($i = 1, 2, 3$), and the subscripts i, j, k refer to x or y. With slowly varying amplitude approximation of $\left(dE_i^{\omega}/dz\right)k \gg \left(d^2E_i^{\omega}/dz^2\right)$ and $\partial/\partial t = -i\omega$, the wave equation (5.13) for the ω_3 field yields

$$\begin{aligned} &\left(\frac{k_3^2}{2}E_{i0}^{(\omega_3)} + ik_3\frac{dE_{i0}^{(\omega_3)}}{dz}\right)e^{i(k_3 z - \omega_3 t + \phi_3)} + c.c. \\ &= \omega_3^2\mu_0\varepsilon_3\left[\frac{E_{i0}^{(\omega_3)}}{2}e^{i(k_3 z - \omega_3 t + \phi_3)} + c.c.\right] - \mu_0\frac{\partial^2}{\partial t^2}\left[P_{NL}^{(\omega_3)}(z,t)\right]_i \end{aligned} \tag{5.15}$$

where ε_3 is the dielectric constant at ω_3.

The i component of the second order nonlinear polarization is described as

$$(P_{NL}^{(\omega_3)})_i = \frac{1}{2}\varepsilon_0 d_{eff} E_j^{(\omega_1)} E_k^{(\omega_2)} \tag{5.16}$$

Here d_{eff} denotes an effective nonlinear dielectric tensor, transformed from d_{IJK} defined in a crystal coordinate (X,Y,Z) to the laboratory coordinate (x,y,z) as

$$d_{eff} = \hat{\mathbf{e}}^{(\omega_3)} \cdot \mathbf{d} : \hat{\mathbf{e}}^{(\omega_1)} \cdot \hat{\mathbf{e}}^{(\omega_2)} \tag{5.17}$$

where $\hat{\mathbf{e}}^{(\omega_i)}$ is a unit vector in the direction of polarization of the ω_i wave. Noting $\omega_3^2\mu_0\varepsilon_3 = k_3^2$, Eq. (5.15) yields

$$ik_3 \frac{dE_i^{(\omega_3)}}{dz} e^{i(k_3 z+\phi_3)} = \varepsilon_0\mu_0\omega_3^2 d_{eff} E_j^{(\omega_1)} E_k^{(\omega_2)} e^{i[(k_1+k_2)z+(\phi_1+\phi_2)]} \tag{5.18}$$

Rewriting,

$$\frac{dE_i^{(\omega_3)}}{dz} = -i\frac{\omega_3}{c_0 n_3} d_{eff} E_j^{(\omega_1)} E_k^{(\omega_2)} e^{-i[(k_3-k_1-k_2)z-(\phi_3-\phi_1-\phi_2)]}, \tag{5.19}$$

where n_3 is the refractive index at ω_3 and c_0 is the light velocity in vacuum. Similar equations are derived for the ω_1 and ω_2 waves:

$$\begin{aligned} \frac{dE_j^{(\omega_1)}}{dz} &= -i\frac{\omega_1}{c_0 n_1} d_{eff} E_i^{(\omega_3)} E_k^{*(\omega_2)} e^{+i[(k_3-k_1-k_2)z-(\phi_3-\phi_1-\phi_2)]}, \\ \frac{dE_k^{(\omega_2)}}{dz} &= -i\frac{\omega_2}{c_0 n_2} d_{eff} E_i^{(\omega_3)} E_j^{*(\omega_1)} e^{+i[(k_3-k_1-k_2)z-(\phi_3-\phi_1-\phi_2)]}. \end{aligned} \tag{5.20}$$

The energy transfer between the three waves depends on the relative phase of these waves, $\Phi = \phi_3 - \phi_1 - \phi_2$. Analysis of the three-wave coupling (Armstrong et al., 1962) shows that, in sum frequency generation, ω_1 and ω_2 are efficiently converted to ω_3 when $\Phi = -\pi/2$. On the other hand, in parametric amplification, ω_3 is efficiently converted to ω_1 and ω_2 when the phase of the ω_3 advances the phases of ω_1 and ω_2 by 90-degrees, i.e., $\Phi = \pi/2$.

In sum frequency generation, where ω_3 is generated by ω_1 and ω_2 pump waves, the intensity of ω_3 is calculated by integrating (5.19) from $z = 0$ to L with the initial condition $E_i^{\omega_3}(0) = 0$. Noting $(e^{-ix} - 1)\cdot(e^{ix} - 1) = 4\sin^2(x/2)$, the intensity at L is

$$\begin{aligned} I_i^{(\omega_3)}(L) &= \frac{\varepsilon_0 c_0 n_3}{2} E_i^{(\omega_3)}(L) E_i^{(\omega_3)*}(L) \\ &= 2\left(\frac{\mu_0}{\varepsilon_0}\right)^{1/2} \frac{1}{n_3}\omega_3^2 (d_{eff})^2 \left|E_j^{(\omega_1)}\right|^2 \left|E_k^{(\omega_2)}\right|^2 L^2 \frac{\sin^2(\Delta kL/2)}{(\Delta kL/2)^2} \\ &= 8\left(\frac{\mu_0}{\varepsilon_0}\right)^{3/2} \frac{\omega_3^2 (d_{eff})^2 L^2}{n_1 n_2 n_3} I_j^{(\omega_1)} I_k^{(\omega_2)} \frac{\sin^2(\Delta kL/2)}{(\Delta kL/2)^2}, \end{aligned} \tag{5.21}$$

where n_i is the refractive index at ω_i. Due to unequal phase velocities of the interacting waves, the phase mismatch between the three waves Δk arises:

$$\Delta k = k_3 - k_1 - k_2. \tag{5.22}$$

In the following, we consider second harmonic generation as an example of the sum frequency generation, where $\omega_1 = \omega_2 \equiv \omega$, $\omega_3 \equiv 2\omega$ and the phase mismatch is expressed as $\Delta k = k_i^{(2\omega)} - k_j^{(\omega)} - k_k^{(\omega)}$. (Note that the fundamental waves can have either the same ($j = k$) or different ($j \neq k$) polarizations in satisfying the phase matching condition, as described later.) When the phase matching is satisfied, i.e., $\Delta k = 0$, the induced polarization at 2ω propagates with the same phase velocity as the driving ω waves, resulting in continued growth of the second harmonic wave. Equation (5.21) shows that the second harmonic conversion efficiency $\eta_{2\omega} = I_i^{(2\omega)} \Big/ \left(I_j^{(\omega)} I_k^{(\omega)} \right)$ increases with $d_{eff}^2 L^2 I^{(\omega)}$.

When $\Delta k \neq 0$, although the fundamental wave is converted to the second harmonic in propagation of the fundamental wave, the phase slip between the second harmonic and the fundamental wave increases gradually, becoming π radian at a distance $L_c = \pi \big/ |\Delta k|$, called a coherence length. In propagation over L_c, the second harmonic wave is converted back to the fundamental wave, resulting in full restoration of the fundamental wave at $L = 2L_c = 2\pi/|\Delta k|$.

Assuming that the fields are described as plane waves of infinite beam sizes and the energy losses due to scattering, absorption, etc. are small, the conversion efficiency including the pump wave depletion and the phase mismatch, is expressed in terms of the Jacobi elliptic sine function $\mathrm{sn}(u, k)$ (Sutherland, 2003):

$$\begin{aligned} \eta_{2\omega}(L) &= k\ \mathrm{sn}^2(u, k); \\ u &= \left[\sqrt{1 + \left(\Delta k L_{NL} \big/ 4\right)^2} + \left(\Delta k L_{NL} \big/ 4\right) \right] \left(L \big/ L_{NL}\right), \\ k &= \left[\sqrt{1 + \left(\Delta k L_{NL} \big/ 4\right)^2} - \left(\Delta k L_{NL} \big/ 4\right) \right]^2. \end{aligned} \tag{5.23}$$

Here L_{NL} is a nonlinear scale length over which approximately 67% conversion efficiency is obtained in the second harmonic generation when $\Delta k = 0$:

$$L_{NL} = \frac{1}{4\pi\, d_{eff}} \sqrt{\frac{2\varepsilon_0 n(\omega)^2 n(2\omega) c \lambda(\omega)^2}{I^{(\omega)}(0)}}, \tag{5.24}$$

where $n(\omega)$ and $n(2\omega)$ are the refractive indices at ω and 2ω, respectively, $\lambda(\omega)$ is the wavelength of the ω wave and $I^{(\omega)}(0)$ is the initial intensity of the pumping beam. For example, when a laser pulse 1-μm wavelength with an intensity of 10^8 W/cm^2 is converted to its second harmonic in a KDP crystal, the nonlinear scale length is $L_{NL} \sim 2.7$ cm.

The elliptic function $\mathrm{sn}(u, k)$ converges to $\tanh(u)$ in perfect phase matching where $k = 1$, since $\mathrm{sn}(u, 1) = \tanh(u)$. Whereas it converges to $\sin(u)$ for large phase mismatch where $k = 0$, since $\mathrm{sn}(u, 0) = \sin(u)$. The plot of the second harmonic conversion efficiency over the propagation length with several phase mismatch conditions is shown in Fig. 5.2. The conversion efficiency increases as $\tanh(u)$ for $\Delta k = 0$ up to ~ 100% at a crystal length of $L \geq 3L_{\mathrm{NL}}$. However, when there is a phase mismatch of $\Delta k \cdot L_{\mathrm{NL}} \approx 1$ for example, the conversion efficiency peaks at $L \sim 1.5\, L_{\mathrm{NL}}$ with ~ 60% and decreases at longer crystal lengths. With a

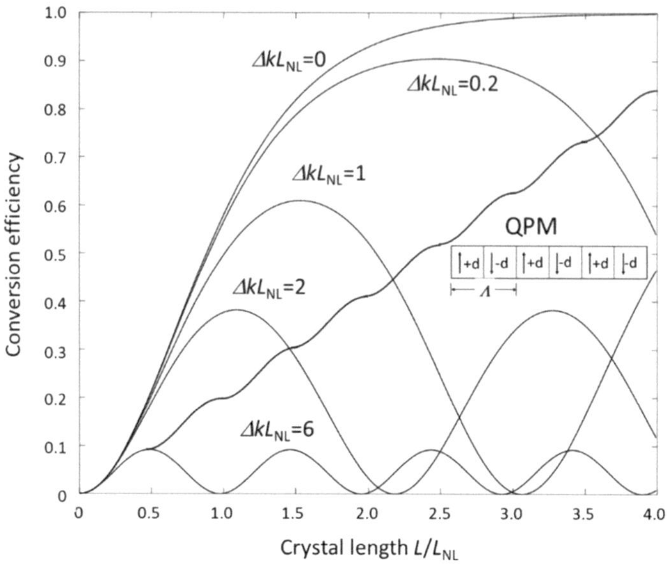

Fig. 5.2 Second harmonic conversion efficiency versus crystal length L, plotted for several phase mismatch conditions of $\Delta k\, L_{LN} = 0, 0.2, 1, 2$, and 6, respectively. Here Δk is phase mismatch and L_{LN} is a nonlinear length where approximately 67% conversion efficiency is obtained for perfect phase matching. Also shown is the conversion efficiency for the quasi-phase-matching (QPM) with periodically poled nonlinear crystals as depicted in the inset

phase mismatch, the conversion efficiency oscillates with the crystal length due to energy exchange between the fundamental and the second harmonic waves, as seen in Fig. 5.2. (The energy exchange between the pump wave and the signal wave also takes place in other coherent nonlinear processes such as sum-frequency generation and optical parametric amplification.)

The decrease in the conversion efficiency due to the phase mismatch in the birefringence phase matching (BPM) is alleviated by "quasi-phase-matching (QPM)" (Armstrong et al., 1962). Instead of using a single nonlinear crystal whose optical axis is oriented, for example, vertically, QPM uses a periodically poled material which is composed of multiple crystals whose optical axes are inverted periodically with a period Λ as shown in the inset at middle right of Fig. 5.2. Taking the conversion efficiency curve of $\Delta k \cdot L_{NL} = 6$ for a single-crystal BPM as an example, the conversion to the second harmonic starts to decrease at $L = 0.5\, L_{NL}$ where the phase slip between 2ω and ω waves become π. If the crystal orientation is reversed at this point, which is equivalent to the change of sign of d_{eff}, this phase slip is recovered along the propagation, resulting in the growth of the 2ω wave as shown by the stepwise increasing curve marked by QPM.

With the QPM, it becomes possible to use nonlinear crystals whose birefringence is not large enough to satisfy the phase matching condition, and to use the large nonlinear coefficient which cannot be utilized in BPM (Fejer et al., 1992). The QPM requires alternating the crystal axis with a micrometer-scale period. The breakthrough in fabrication of the QPM structure was achieved by periodically applying a static electric field to invert the orientation of ferroelectric domains in a thin lithium niobate ($LiNbO_3$) waveguide for frequency conversion of low-power laser diodes (Yamada et al., 1993). Various approaches were undertaken to apply this poling method to develop tunable light sources mostly in the infrared regions with optical parametric amplification (Pasiskevicius, 2013). For high power laser application, large-aperture (5 mm) periodically poled QPM devices were developed with MgO-doped $LiNbO_3$ (Ishizuki & Taira, 2005) and MgO-$LiTaO_3$ (Ishizuki & Taira, 2010). With the MgO-$LiTaO_3$ QPM of = 32.5 μm period and 39 mm length, high energy optical parametric oscillation of 118 mJ output with ~ 70% slope efficiency has been realized. To extend the wavelength coverage to visible and ultraviolet with QPM, a QPM device with crystal quartz is being developed to utilize its high transparency over broad spectral region (0.15–2.μm) and high damage threshold, although its nonlinear coefficient ($d_{\mathrm{eff}} = 0.35$ pm/V) is ~ 1% of $LiNbO_3$. The periodic laminar structure of quartz has been constructed by stacking thin crystal plates with alternating polarity (Ishizuki et al., 2018) and by stress-induced polarity inversion of quartz (Ishizuki & Taira, 2020).

The phase matching condition expressed as $k_i^{(\omega_3)} = k_j^{(\omega_1)} + k_k^{(\omega_2)}$ for the three-wave coupling of plane waves is expressed more generally as $\boldsymbol{k}^{(\omega_3)} = \boldsymbol{k}^{(\omega_1)} + \boldsymbol{k}^{(\omega_2)}$ for coherent coupling of non-parallel waves. This phase matching condition is equivalent to the conservation of photon momentum $\boldsymbol{p}^{(\omega_3)} = \boldsymbol{p}^{(\omega_1)} + \boldsymbol{p}^{(\omega_2)}$ since the photon momentum $\boldsymbol{p}$ is proportional to the wave vector $\boldsymbol{k}$ as $\boldsymbol{p} = \hbar\boldsymbol{k}$. The conversion from the ω_1 and ω_2 waves to the ω_3 wave is the process where one ω_3 photon is created due to annihilation of one ω_1 and one ω_2 photon, respectively. Since the photon number is conserved and the intensity is proportional to the photon number density n as $I = n\hbar\omega$, the following Manley-Rowe relation holds:

$$\frac{I_3}{\omega_3} = \frac{I_1}{\omega_1} + \frac{I_2}{\omega_2} \tag{5.25}$$

This relation expresses the maximum conversion efficiency in three-wave coupling processes.

5.2.2 *Second- and Third-Harmonic Generation*

The phase matching condition for the second-harmonic generation (SHG) is described as $\boldsymbol{k}^{(2\omega)} = 2\boldsymbol{k}^{(\omega)}$, where the wavenumber is $k^{(\omega)} = n^{(\omega)}\omega/c$. The phase matching with collinearly propagating waves is called *scalar phase matching*. The phase matching condition of SHG for the scalar phase matching is described in terms of the refractive index as $n^{(2\omega)} = n^{(\omega)}$. Although this condition is not realized in normal dispersion media where $n^{(2\omega)} > n^{(\omega)}$, it is possible to achieve this condition by using optically anisotropic crystals.

With optically anisotropic crystals, the refractive index of an extraordinary ray depends on the polarization and the propagation direction as given by Eq. (2.136) for a uniaxial crystal. With a uniaxial crystal, we can find a phase matching angle with both the negative ($n_e < n_o$) and positive ($n_e > n_o$) crystals. Taking SHG with a uniaxial negative crystal as an example, there are two ways to achieve the phase matching; type I where the fundamental waves have the same polarization, and type II where the fundamental waves have orthogonal polarizations. The phase matching conditions in type I and type II are described as:

$$\begin{aligned} n_e^{(2\omega)} &= n_o^{(\omega)}, && \text{(Type I)} \\ n_e^{(2\omega)} &= \tfrac{1}{2}\left[n_o^{(\omega)} + n_e^{(\omega)}\right]. && \text{(Type II)} \end{aligned} \tag{5.26}$$

Referring to Sect. 2.2.4, Fig. 5.3 shows the index surfaces of a negative uniaxial crystal for the optical waves of ω (shown in red) and 2ω (shown in blue). The matching conditions given by Eq. (5.26) are marked by the closed circles for type I on the right-hand side and for type II on the left-hand side of this figure, respectively.

As an example, let us consider a potassium dihydrogen phosphate (KH_2PO_4, KDP) crystal which is used for frequency up-conversion of high-energy lasers such as for laser fusion research, since it is possible to grow high-quality, very-large (almost half-meter aperture) KDP crystals. With KDP which is a uniaxial negative crystal, phase matching is possible with *ooe* for type I or *oee* for type II configurations, where the first two letters refer to the polarizations of the fundamental waves and the third letter the polarization of the second harmonic wave, respectively. Type I phase matching is possible by orienting the polarization of the ω wave parallel to the ordinary axis of the crystal. Type II phase matching is possible by orienting the polarization of the ω wave 45-degrees to the ordinary axis for equal amplitudes in the ordinary and extraordinary axes.

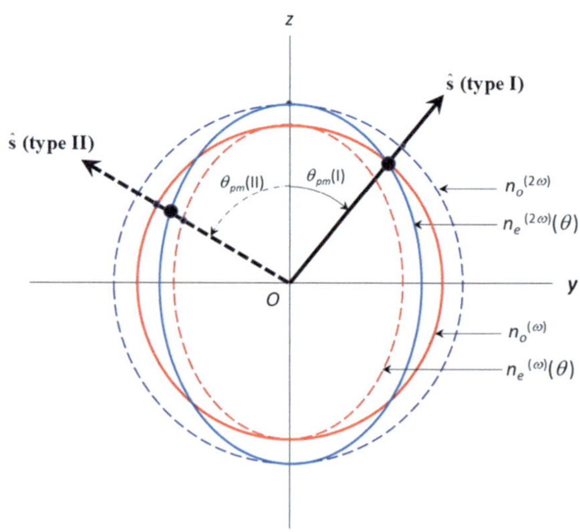

Fig. 5.3 Index surfaces of a uniaxial negative crystal for ω (shown in red) and 2ω (shown in blue). The phase matching angle θ_{pm} for SHG are shown; for type I phase matching on the right-hand side and for type II phase matching on the left-hand side

Using Eq. (2.136) and the ordinary and extraordinary refractive indices of KDP at ω and 2ω shown in Table 5.1, we find that type I phase matching is achieved at an angle of $\theta_{pm}(I) = 40.9°$ for a fundamental wave of 1053 nm. (The type II phase matching angle of KDP is $\theta_{pm}(II) = 59°$ for a fundamental wave of 1053 nm.)

Referring to Eq. (5.17), the effective nonlinear optical susceptibility for the second harmonic conversion is given by $d_{eff} = \hat{\mathbf{e}}^{(2\omega)} \cdot \mathbf{d} : \hat{\mathbf{e}}_1^{(\omega)} \cdot \hat{\mathbf{e}}_2^{(\omega)}$, where the two pumping waves are denoted by the suffix 1 or 2. Considering type-I SHG with a KDP crystal and referring to Fig. 2.13, when the second-harmonic wave propagates at an angle θ_{pm} from the optical-axis (z-axis) and in a plane at ϕ from the x-axis of the crystal, the (x,y,z)-components of the polarizations of the second harmonic and the fundamental waves are

$$\begin{aligned} \hat{\mathbf{e}}^{(2\omega)} &= (-\cos\theta\cos\phi,\ -\cos\theta\cos\phi,\ \sin\theta), \\ \hat{\mathbf{e}}_1^{(\omega)} = \hat{\mathbf{e}}_2^{(\omega)} &= (\sin\phi,\ -\cos\phi,\ 0). \end{aligned} \tag{5.27}$$

The KDP crystal, belonging to the $\overline{4}2m$ point group, has two nonvanishing d elements:

$$\begin{aligned} d_{14} &= d_{XYZ} = d_{XZY} = d_{YXZ} = d_{YZX}, \\ d_{36} &= d_{ZXY} = d_{ZYX}, \end{aligned} \tag{5.28}$$

where $d_{14} = d_{36}$ if the Kleinman condition is applicable. The effective nonlinear susceptibility is therefore,

$$\begin{aligned} d_{eff} &= \begin{bmatrix} -\cos\theta\cos\phi \\ -\cos\theta\sin\phi \\ \sin\theta \end{bmatrix}^T \begin{bmatrix} 0\,0\,0\ d_{14} & 0 & 0 \\ 0\,0\,0\ 0 & d_{14} & 0 \\ 0\,0\,0\ 0 & 0 & d_{36} \end{bmatrix} \begin{bmatrix} \sin^2\phi \\ \cos^2\phi \\ 0 \\ 0 \\ 0 \\ -2\cos\phi\sin\phi \end{bmatrix} \\ &= -d_{36}\sin\theta\sin 2\phi. \end{aligned} \tag{5.29}$$

For $\theta_{pm} = 40.9°$ and $\phi = 45°$, $d_{eff} = -0.65 d_{36} = -0.28\,(pV/m)$. For type II SHG with KDP, the effective nonlinear susceptibility is $d_{eff} = -d_{14}\sin\theta\sin 2\phi$.

Frequency conversion to shorter wavelengths is required in many applications; shorter wavelength lasers are absorbed at higher densities in laser produced plasmas or can be focused to smaller spots in lithographic applications. For the third harmonic generation (THG) of high-power lasers in laser fusion research, for example, the fundamental and the second harmonic waves are mixed using two KDP crystals in the type-I/type-II configuration, where SHG is in type I and THG is in type II configurations, respectively, as shown in Fig. 5.4. In this configuration, we can directly mix the orthogonally polarized ω and 2ω waves generated in SHG in type I

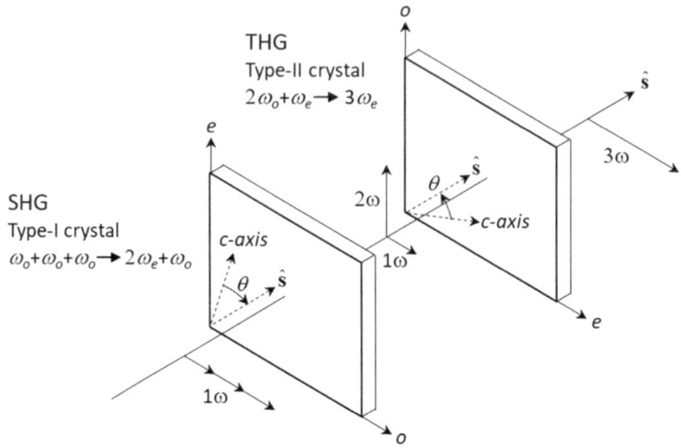

Fig. 5.4 Layout of the third harmonic generation by mixing the fundamental and the second harmonic waves. The second harmonic wave is generated in the first crystal in type-I configuration, where equal photon numbers of ω and 2ω waves exit after the crystal. The third harmonic wave is generated with the second crystal in the type-II configuration by mixing ω and 2ω photons. The c-axis of the first crystal is in the vertical plane composed of $\boldsymbol{e}$ and $\hat{s}$ vectors, whereas it is in the horizontal plane, also composed of $\boldsymbol{e}$ and $\hat{s}$ vectors, in the second crystal, respectively

configuration. In collinear propagation, the phase matching condition $\boldsymbol{k}^{(3\omega)} = \boldsymbol{k}^{(\omega)} + \boldsymbol{k}^{(2\omega)}$ is satisfied when,

$$n_e^{(3\omega)} = \frac{1}{3}\left[n_e^{(\omega)} + 2n_o^{(2\omega)}\right]. \tag{5.30}$$

Since one each of ω and 2ω photon is used to generate one 3ω photon, the maximum THG efficiency is obtained when the second harmonic wave is generated with 2/3 energy efficiency in the SHG, as shown in Fig. 5.4.

5.3 Optical Parametric Amplification

In optical parametric amplification (OPA), a light wave is amplified by nonlinear optical process, instead of amplifying light with population-inverted atoms. In OPA, a high-intensity coherent *pump* wave at frequency ω_p is converted to a lower frequency coherent *signal* wave at ω_s and a lower frequency *idler* wave at ω_i. In this process, the photon energy conservation $\hbar\omega_p = \hbar\omega_s + \hbar\omega_i$ and the photon momentum conservation $\hbar\boldsymbol{k}_p = \hbar\boldsymbol{k}_s + \hbar\boldsymbol{k}_i$ hold, leading to $\omega_p = \omega_s + \omega_i$ (where $\omega_p > \omega_s > \omega_i$) and the phase matching condition $\boldsymbol{k}_p = \boldsymbol{k}_i + \boldsymbol{k}_s$. Due to three-wave coupling, the signal wave and the spontaneously emitted idler wave are amplified, automatically satisfying the energy and momentum conservations.

The OPA has several unique properties compared to conventional amplifiers.

(a) Broad gain bandwidths are obtained for amplification of the signal and idler beams, when the phase matching condition is satisfied by proper choice of nonlinear crystals and matching angles.
(b) The broad-bandwidth capability can be used to generate coherent waves in near-infrared and mid-infrared regions where laser materials are not easily available with solid-state lasers. The coherent waves in these regions are generated as a frequency tunable oscillator called optical parametric oscillator (OPO), or as an optical parametric generator (OPG). In OPO, which is pumped by a pulsed or continuous-wave laser, a nonlinear crystal is placed in an optical cavity resonant to either or both signal and idler wave.
(c) Due to its broad bandwidth capability, OPA can be used to amplify a frequency chirped pulse for generating extremely short and high intensity optical pulses, not attainable with conventional amplifiers. This approach is called optical parametric chirped-pulse amplification (OPCPA). In OPA, multiple independent pump pulses can be used to amplify the signal wave while preserving the coherence of the signal wave.
(d) The OPA is suitable for high-repetition amplification due to small heat accumulation in the amplifier, since the nonlinear crystal acts only for frequency conversion.
(e) The OPA is often used as a preamplifier for reducing pre-pulses and amplified spontaneous emission, since amplification takes place only during the pump pulse.
(f) High gain is obtained in OPA in short propagation distance and with large aperture amplifiers without suffering from parasitic oscillation and amplified stimulated emission in the amplifiers since gain occurs only during the pulsed pumping.

In this subsection, basic concepts of OPA relevant to application to OPCPA are described. The high power OPCPA laser systems are described in Sect. 7.4.2.

5.3.1 Gain of Optical Parametric Amplification

Denoting the pump, signal and idler waves by p, s, and i respectively, coherent coupling of the three waves with $\omega_p = \omega_3$, $\omega_s = \omega_2$, and $\omega_i = \omega_1$ propagating in the z-direction is described by Eqs. (5.19) and (5.20). In these equations, $\Delta k = (\boldsymbol{k}_p - \boldsymbol{k}_s - \boldsymbol{k}_i) \cdot \hat{z}$, and the electric field of each wave is the field component satisfying the phase matching condition. These coupled wave equations can be solved under the slowly varying envelope approximation including the pump wave depletion and the phase mismatch (Byer, 1975). The intensity gain of the signal wave for intense pump wave, i.e.,$\lambda_p P_p \gg \lambda_s P_s$ is given by,

$$G = \frac{I_s(L)}{I_s(0)} = 1 + \frac{\Gamma^2}{\Gamma^2 - (\Delta k/2)^2} \sinh^2\left(L\sqrt{\Gamma^2 - (\Delta k/2)^2}\right),$$
$$\Gamma^2 = \frac{8\pi^2 d_{eff}^2 I_p}{n_p n_s n_i \lambda_s \lambda_i \varepsilon_0 c_0}. \tag{5.31}$$

Here I_p is the pump intensity, d_{eff} is the effective nonlinear optical coefficient, ε_0 is the vacuum permittivity, and $n_{p,s,i}$ and $\lambda_{s,i}$ are the refractive indices and the wavelengths of the pump, signal and idler waves, respectively. This equation shows that the gain coefficient of the signal wave is Γ when $\Delta k = 0$, whereas it decreases to $\Gamma\sqrt{1 - (\Delta k/2\Gamma)^2}$ when $\Delta k \neq 0$. When the gain is larger than the phase mismatch, i.e., $\Gamma \gg \Delta k/2$, the gain is given by

$$G = \frac{1}{4}\frac{\exp\left(2\Gamma L\sqrt{1 - (\Delta k/2\Gamma)^2}\right)}{1 - (\Delta k/2\Gamma)^2} \approx \frac{\exp(2\Gamma L)}{4}. \tag{5.32}$$

The plot of the intensity gain of the signal wave versus the propagation distance is shown in Fig. 5.5. The signal wave is amplified exponentially, accompanied by the growth of the idler wave (not shown in this figure). When the photon numbers of the pump and signal waves become equal, i.e., $\lambda_p P_p = \lambda_s P_s$, the conversion efficiency starts to oscillate due to reverse energy exchange from the signal and idler waves to the pump wave, like SHG shown in Fig. 5.2.

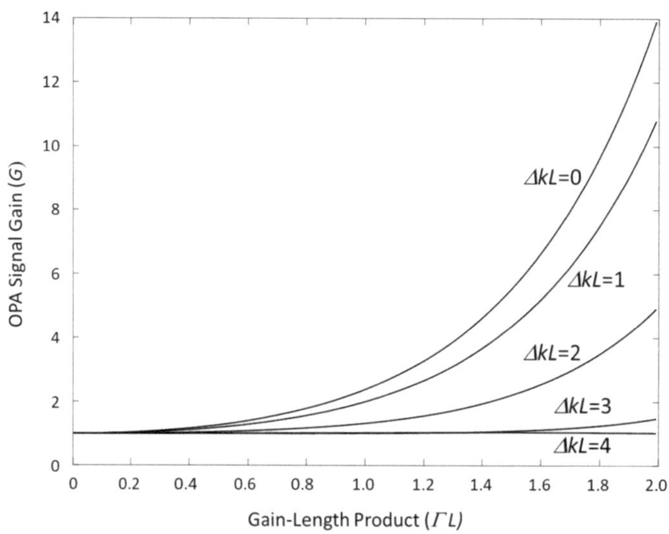

Fig. 5.5 Power gain G of the signal wave versus the gain-length product, plotted for several phase mismatch conditions. Here Γ is the small signal gain coefficient, Δk is the phase mismatch, and L is the propagation distance

5.3.2 *Phase Matching for OPA*

5.3.2.1 Collinear Phase Matching

Similarly to SHG, the phase matching is possible for OPA with uniaxial crystals in both type I or type II configurations. Noting that the wavenumber k is described by $k = n(\omega/c_0)$, where c_0 is the light velocity in vacuum, the type I phase matching with a uniaxial negative crystal in collinear propagation $k_p = k_s + k_i$ is expressed as

$$n_{\mathrm{e}}^{(\omega_{\mathrm{p}})}(\theta)\omega_{\mathrm{p}} = n_{\mathrm{o}}^{(\omega_{\mathrm{s}})}\omega_{\mathrm{s}} + n_{\mathrm{o}}^{(\omega_{\mathrm{i}})}\omega_{\mathrm{i}}, \tag{5.33}$$

where $n_e^{(\omega_p)}(\theta)$ is the extraordinary refractive index of the pumping wave which depends on the angle θ as Eq. (2.134), and $n_o^{(\omega_s)}$ and $n_o^{(\omega_i)}$ are ordinary refractive indices of the signal and idler waves, respectively.

We show in Fig. 5.6 the phase matching diagrams, for type I phase matching with a uniaxial negative crystal in the (ooe) configuration, where the locus of k, which corresponds to $n\omega/c_0$, are shown for the pump (e), signal (o), and idler (o) waves. Figure 5.6a is for collinear propagation and Fig. 5.6b is for noncollinear propagation, which is described later.

The phase matching angle θ_{pm} is determined by the crossing point of the locus of $\boldsymbol{k}_p(\theta)$ with $\boldsymbol{k}_s + \boldsymbol{k}_i$. In collinear propagation, Fig. 5.6a suggests that the phase matching angle increases as ω_s increases, since the radius of the $k_s + k_i$ increases. (Although increase in ω_s is accompanied with decrease in ω_i, the increase in $n_o^{(\omega_s)}$ is larger than the decrease in $n_o^{(\omega_i)}$ since we are assuming $\omega_s > \omega_i$.) In the type II phase matching, not shown in Fig. 5.6, either the idler or signal wave is extraordinary, i.e., (eoe) or (oee).

5.3.2.2 Noncollinear Phase Matching

In OPA (and other wave-coupling processes), it is possible to satisfy the phase matching condition $\boldsymbol{k}_p = \boldsymbol{k}_s + \boldsymbol{k}_i$ in noncollinear configuration where the pump, signal, and idler waves propagate at different directions, as shown in Fig. 5.6b. The phase matching angle for noncollinear propagation $\theta_{pm}(ncl)$ is satisfied over $\theta_{pm}(col) \leq \theta_{pm}(ncl) \leq [180° - \theta_{pm}(col)]$ where $\theta_{pm}(col)$ is the collinear phase matching angle, since the relation $n_e^{(\omega_p)}(\theta)\omega_p \leq n_o^{(\omega_s)}\omega_s + n_o^{(\omega_i)}\omega_i$ holds only in this region. A configuration of the wavevectors of the pump, signal, and idler waves for noncollinear phase matching is shown in Fig. 5.7. The signal wave $\boldsymbol{k}_s$ is amplified at an angle α from the propagation direction of the pump wave $\boldsymbol{k}_p$ and an idler wave $\boldsymbol{k}_i$ is emitted at an angle Ω from the signal wave. Projecting the wavevectors along and perpendicular to $\boldsymbol{k}_p$, the phase matching is described as

$$k_p \cos\alpha = k_s + k_i \cos\Omega,$$

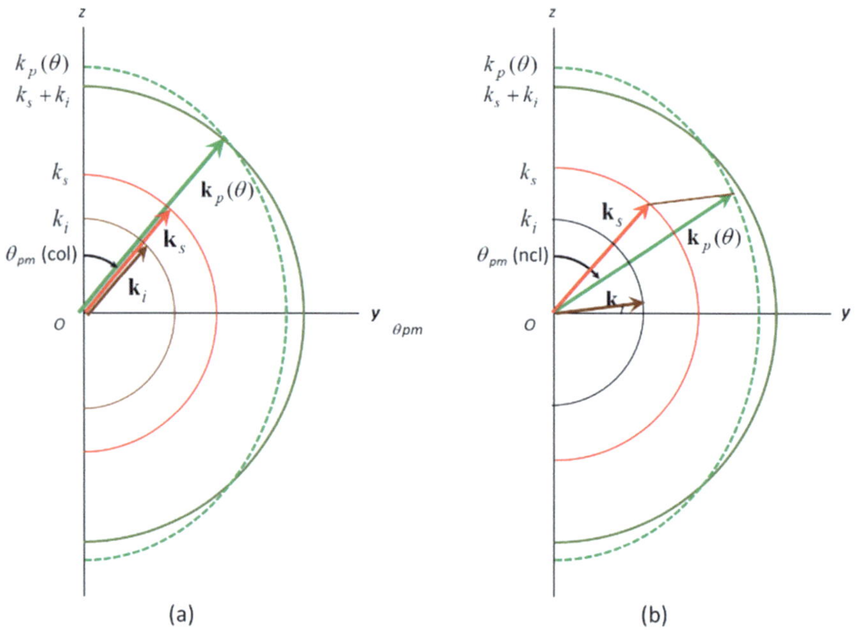

Fig. 5.6 Phase matching diagram for optical parametric amplification with a uniaxial negative crystal, expressed in terms of k_j, where j refers to either p, s or i. **a** is for collinear propagation and **b** is for noncollinear propagation. The solid and dotted curves correspond to the ordinary and extraordinary waves, respectively. The colors qualitatively indicate the wavelengths with green for pump, red for signal, and brown for idler waves. The arrowed solid lines show the wavevectors of the three waves satisfying the phase matching at the phase matching angles shown by θ_{pm}

$$k_p \sin\alpha = k_i \sin\Omega. \tag{5.34}$$

From these equations, the angles α and Ω are derived as

$$\cos\alpha = \frac{k_{\mathrm{p}}^2 + k_{\mathrm{s}}^2 - k_{\mathrm{i}}^2}{2k_{\mathrm{p}}k_{\mathrm{s}}}, \quad \cos\Omega = \frac{k_{\mathrm{p}}^2 - k_{\mathrm{s}}^2 - k_{\mathrm{i}}^2}{2k_{\mathrm{s}}k_{\mathrm{i}}}. \tag{5.35}$$

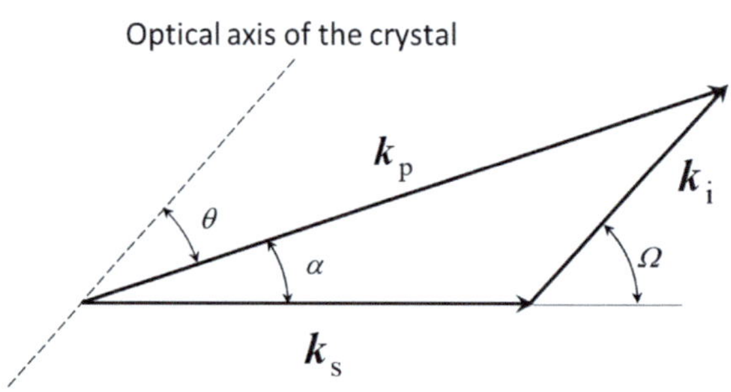

Fig. 5.7 Phase matching diagram for noncollinear OPA. The wavevectors of the pump, signal, and idler waves are shown by $\mathbf{k}_p$, $\mathbf{k}_s$, and $\mathbf{k}_i$, respectively

The type I phase matching angle of noncollinear OPA for a uniaxial negative crystal is given by

$$\tan^2\theta_{\mathrm{pm}}^{\mathrm{ooe}}(\mathrm{ncl}) = \frac{1-U}{W-1};$$
$$U = \frac{A^2+B^2+2AB\cos\Omega}{C^2},\ \mathrm{W} = \frac{A^2+B^2+2AB\cos\Omega}{F^2}, \tag{5.36}$$
$$A = n_{\mathrm{o}1}/\lambda_1,\ \ B = n_{\mathrm{o}2}/\lambda_2,\ \ C = n_{\mathrm{o}3}/\lambda_3,\ \ F = n_{\mathrm{e}3}/\lambda_3,$$

where subscripts 1, 2, 3 refer to idler, signal, pump waves, n_{o} and n_{e} are ordinary and extraordinary refractive indices, respectively, and Ω is the angle between $\boldsymbol{k}_i$ and $\boldsymbol{k}_s$ (Dmitriev et al., 1999, Sect. 2.5). Equation (5.36) provides the phase matching angle for collinear propagation $\theta_{pm}^{ooe}(col)$ when $\Omega = 0$.

5.3.3 *Gain Bandwidth of Optical Parametric Amplification*

Phase matching over broad frequencies is required in OPCPA for amplification of ultrashort-duration signal pulses. In the near infrared region, a phase matching over ~ 100 nm bandwidth is required to amplify a 10-fs pulse and ~ 200 nm to amplify a few-fs pulse. In OPCPA, a signal wave is amplified when the short-duration pump, signal and idler pulses propagate with the spatial overlap of these pulses. The spatial overlap is kept when these pulses propagate with equal group velocities in the signal-wave propagation direction and with small group velocity dispersions (GVDs) so that the short pulse-widths are maintained during propagation (Cerullo & De Silvesti, 2003; Cerullo & Manzoni, 2007).

Alternatively, the gain bandwidth is derived in the frequency domain by considering the dispersion of the phase matching condition (Dabu, 2010; Lozhkarev et al., 2005). Since a short-duration pulse is composed of broad frequency spectrum, the phase mismatch arises at the frequencies far from the phase matching frequency. Noting that the parametric gain G decreases with increase in the phase mismatch Δk as given in Eq. (5.31), the frequency bandwidth $\Delta\omega$ of the optical parametric amplification is determined from the relation between Δk and $\Delta\omega$.

The phase matching condition is satisfied at a particular set of frequencies of the pump, signal and idler waves as

$$\omega_p - \omega_{s0} - \omega_{i0} = 0,$$
$$\Delta k^{(0)} = k_p(\omega_p) - k_s(\omega_{s0}) - k_i(\omega_{i0}) = 0. \tag{5.37}$$

We consider a situation where the pump pulse is close to monochromatic. When the signal frequency changes by $\Delta\omega$ as $\omega_s = \omega_{s0} + \Delta\omega$, the idler frequency changes to $\omega_i = \omega_{i0} - \Delta\omega$. The dispersion of the wavevector mismatch is described by the Taylor series expansion of Δk with $\Delta\omega$:

$$\begin{aligned}
\Delta k &= \Delta k^{(0)} + \Delta k^{(1)} + \Delta k^{(2)} + \Delta k^{(3)} + \Delta k^{(4)} + \ldots; \\
\Delta k^{(0)} &= k_p(\omega_p) - k_s(\omega_{s0}) - k_i(\omega_{i0}) = 0, \\
\Delta k^{(1)} &= \left(\frac{\partial \Delta k}{\partial \omega_s}\right) d\omega_s \approx -\left(\frac{\partial k_s}{\partial \omega_s} - \frac{\partial k_i}{\partial \omega_i}\right)\Delta\omega = -\left(\frac{1}{v_{gs}} - \frac{1}{v_{gi}}\right)\Delta\omega, \\
\Delta k^{(2)} &= \frac{1}{2!}\left(\frac{\partial^2 \Delta k}{\partial \omega_s^2}\right)(d\omega_s)^2 \approx -\frac{1}{2!}\left(\frac{\partial^2 k_s}{\partial \omega_s^2} + \frac{\partial^2 k_i}{\partial \omega_i^2}\right)(\Delta\omega)^2, \\
\Delta k^{(3)} &= \frac{1}{3!}\left(\frac{\partial^3 \Delta k}{\partial \omega_s^3}\right)(d\omega_s)^3 \approx -\frac{1}{3!}\left(\frac{\partial^3 k_s}{\partial \omega_s^3} - \frac{\partial^3 k_i}{\partial \omega_i^3}\right)(\Delta\omega)^3, \\
\Delta k^{(4)} &= \frac{1}{4!}\left(\frac{\partial^4 \Delta k}{\partial \omega_s^4}\right)(d\omega_s)^4 \approx -\frac{1}{4!}\left(\frac{\partial^4 k_s}{\partial \omega_s^4} + \frac{\partial^4 k_i}{\partial \omega_i^4}\right)(\Delta\omega)^4.
\end{aligned} \tag{5.38}$$

The v_{gs} and v_{gi} in the third equation are the group velocities of the signal and idler waves, respectively.

With using the first-order mismatch term $\Delta k^{(1)}$ in Eq. (5.31), the parametric gain frequency bandwidth (full width at half width) is estimated, under the large gain approximation $\Gamma L \gg 1$, as:

$$\Delta\nu^{(1)} = \frac{2(\ln 2)^{1/2}}{\pi}\left(\frac{G}{L}\right)^{1/2} \frac{1}{\left|{}^{1}\!/_{v_{gs}} - {}^{1}\!/_{v_{gi}}\right|}. \tag{5.39}$$

This equation shows that the bandwidth increases when the group velocities of the signal and idler waves are close to each other. Referring to Fig. 5.7 for non-collinear phase matching, $\Delta k^{(1)} = 0$ when $v_{gs} = v_{gi}\cos\Omega$. The amplification bandwidth achieved when $\Delta k^{(1)} = 0$ is called "broad band".

When $\Delta k^{(1)} = 0$ is satisfied, the amplification bandwidth significantly broadens, and becomes limited by the second-order mismatch term $\Delta k^{(2)}$, resulting in

$$\Delta\nu^{(2)} = \frac{2(\ln 2)^{1/4}}{\pi}\left(\frac{G}{L}\right)^{1/4} \frac{1}{|(GVD)_s + (GVD)_i|}, \tag{5.40}$$

where $(GVD)_{s,i} = \partial^2 k_{s,i} / \partial\omega_{s,i}^2$ are the group velocity dispersions of the signal and idler waves, respectively. The amplification bandwidth for $\Delta k^{(1)} = \Delta k^{(2)} = 0$ is called "ultra broad band". The amplification bandwidths with higher order dispersion terms $\Delta k^{(3)}$, $\Delta k^{(4)}$ are given in Dabu (2010).

For each nonlinear crystal, the amplification bandwidth of OPCPA depends on many parameters: type of phase matching (type I, type II; collinear and non-collinear geometry), pump wavelength, ordinary and extra-ordinary refractive indices with their dispersions over broad spectral regions, and the pump intensity. In the following we consider a DKDP crystal which has very broad bandwidth capability.

The amplification bandwidths achievable with KDP (KH_2PO_4), DKDP (KD_2PO_4) and partially deuterated KDP (DKDP; $KH_{2(1-x)}D_{2x}PO_4$) crystals have been studied to develop an ultrashort-pulse OPCPA laser. Lozhkarev et al. (2005) have evaluated the first order and the second order dispersions ($\partial k / \partial\omega$ and $\partial k^2 / \partial\omega^2$) of the ordinary and extraordinary refractive indices of p-DKDP by creating Sellmeier equations which fit the published data of the refractive indices of KDP and DKDP. Using these parameters, the deuteron fraction for ultrabroadband amplification which satisfies $\Delta k^{(1)} = \Delta k^{(2)} = 0$ for non-collinear type I (*oo-e*) phase matching has been calculated, resulting in the maximum gain bandwidth at degenerate three-wave interaction where $\omega_s = \omega_i = \omega_p/2$ in the collinear phase matching.

Evaluating the optimum conditions for the ultrabroadband gain in the collinear degenerate phase matching, Dabu has derived that the gain bandwidth of 280 nm is obtained with a 58%-deuterated DKDP of 40 mm length by pumping at 526.5 nm wavelength (second harmonic of a Nd:glass laser) at the intensity of ~ 1 GW/cm^2. In a small angle non-collinear phase matching for separation of the signal and the idler waves, slightly smaller bandwidth of 255 nm is expected with a signal-idler wavevector angle Ω of 18.2 mrad (24 mrad outside the crystal), requiring ~ 20 m propagation distance for separation of 400 mm-diameter signal and idler beams (Dabu, 2010, 2021).

Fujioka et al. (2015) have grown DKDP single crystals with several deuteration fractions, shown in Fig. 5.8a, and determined the ordinary and extraordinary refractive indices over 0.4–1.6 μm wavelengths by the minimum deviation method using the DKDP prisms cut out from the grown crystals. With these refractive indices, dependence of the phase matching angle on the signal wavelength was determined as shown in Fig. 5.8b for the non-colinear angle α of 0.2° and 0.4°, respectively. Dependence of the gain, calculated from Eq. (5.31), on the signal wavelength shows that a gain bandwidth of 490 nm (FWHM) is achieved with a deuteration fraction of 68%, where the phase matching internal angle θ_{pm} is 38° and the signal-idler angle is 3.5 mrad. These results show that DKDP is capable of amplifying a-few-cycle duration pulse by pumping with the second harmonic of the Nd:glass laser.

5.3.4 Self-phase-Stabilization of an Idler Wave

As described in Sect. 5.2.1, the pump wave is efficiently converted to the signal and idler waves when the relative phase of these waves satisfies the following relation,

$$\Phi = \phi_p - \phi_s - \phi_i = \pi / 2. \tag{5.41}$$

From this relation, the phase of the idler wave ϕ_i is constant when the phase difference of the pump and the signal wave $\phi_p - \phi_s$ is constant. Since the white-light continuum generated by self-phase modulation is found to be completely phase-locked to the driving pulse (Bellini & Hänsch, 2000), $\phi_p - \phi_s$ will stay constant when a broadband pulse generated by self-phase modulation is used as a signal wave, even if the phase of the pump pulse fluctuates.

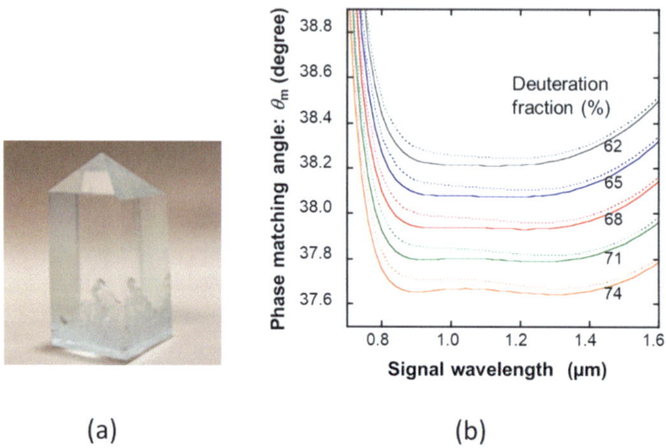

Fig. 5.8 Determination of the bandwidth of OPCPA with a DKDP crystal. **a** A DKDP crystal of 69.5% deuteration fraction with dimensions of 50 × 50 × 102 mm^3, where crystal c-axis is in the vertical direction. Dependence of the refractive index dispersion on the deuteration fraction was determined using the prisms cut out from the crystals with several deuteration fractions. **b** The dependence of phase matching angle on the signal wavelength for the deuteration fractions of 62–74% for the non-collinear angle α of 0.2° (solid lines) and 0.4° (dotted lines). Courtesy of Prof. Kana Fujioka

By utilizing this relation, Baltuska et al. (2002) have demonstrated generation of an idler pulse by optical parametric amplification (OPA) whose carrier-envelope phase is automatically stabilized in the presence of phase fluctuation of the input pulse without using phase-control devices. For extending the high-order harmonics to shorter wavelengths and to generate isolated attosecond-pulses, the phase-stabilized idler pulses of near infrared region have been amplified as a high-energy driver for HHG. The CEP-stable idler wave was amplified with two-stage OPAs to generate mJ-energy, few-optical-cycle pulses in the near infrared region (Mucke et al., 2009; Vozzi et al., 2008). In another application, the CEP-stable, high-temporal contrast, ultrafast laser pulses at 910 nm generated by second-harmonic conversion of 1820 nm idler wave have been used as the seed laser of a 100 PW OPCPA laser system (Shao et al., 2020).

5.3.5 *Multiple Beam Pumping*

The noncollinear configuration is suitable to pump an optical parametric amplifier with multiple pump beams since it is possible to find phase matching condition for different beams by properly adjusting the angle and frequency of the pump beams. An interesting and important property of the optical parametric amplification is that it is possible to use multiple pump beams to amplify one signal beam, without affecting the coherence of the signal beam arising from interferences of the pump beams.

When a signal beam is pumped by multiple pump beams denoted by j ($j = 1, 2, \ldots$), an idler beam is generated for each pump beam which satisfy the following relations:

$$\begin{aligned} \boldsymbol{k}_{pj} &= \boldsymbol{k}_s + \boldsymbol{k}_{ij}, \\ \Phi &= \phi_{pj} - \phi_s - \phi_{ij} = \pi/2. \end{aligned} \tag{5.42}$$

The requirement for the multiple beam pumping is that each pump beam must be phase matched to the signal beam as expressed in the first equation of (5.42). On the other hand, the phase of the signal beam ϕ_s is maintained even if the phase of the pump beam ϕ_{pj} varies among different beams, since an idler wave is generated for each pump wave with a phase ϕ_{ij} automatically satisfying the relation $\phi_{ij} = \phi_{pj} - \phi_s - \pi/2$.

Dubietis et al. (1998) have proven this property in theory and experiment in three-beam-pumping, demonstrating that the energy transfer from three pump pulses is equivalent to that observed with a single pump pulse with a signal beam intensity equal to the summation of the three pump pulses. In addition to increasing the pump energy to increase the signal gain, the multiple beam pumping has been applied to broaden the gain bandwidth with various configurations: phase matching at different spectral regions with different beams (Zeromskis et al., 2002), pumping with multiple frequency beams (Herrmann et al. 2010; Tamošauskas et al., 2008) and pumping with converging beams (Harimoto & Yamakawa, 2005; Johnson et al., 2011).

5.3.6 Optical Parametric Chirped-Pulse Amplification (OPCPA)

In the conventional optical parametric amplification (OPA), a nonlinear crystal is pumped with a pump beam with a pulse width similar to that of a signal beam, as shown in Fig. 5.9a. When the duration of the signal pulse becomes shorter, this approach has a limit in increasing the power due to damage to optical components by the high-power pump beam. To overcome this limit, Dubietis et al (1992) has proposed to apply the CPA concept to OPA to generate a high-power ultrashort laser pulse without causing damage to optical components during amplification.

This approach was extended by Ross et al. (1997) as a new method to generate an ultrashort-duration, ultrahigh-power laser pulse, named as optical parametric chirped pulse amplification (OPCPA). In a typical high peak-power OPCPA system, a frequency-chirped pulse of ~ 1-ns duration is amplified as a signal beam by OPA in a large-aperture nonlinear crystal, pumped by high-energy laser beams of ~ 1-ns duration, as shown in Fig. 5.9b. The amplified signal pulse is recompressed, producing an ultrashort-duration, high peak power laser pulse. Note that an idler pulse is also amplified and compressed to generate a high peak power pulse.

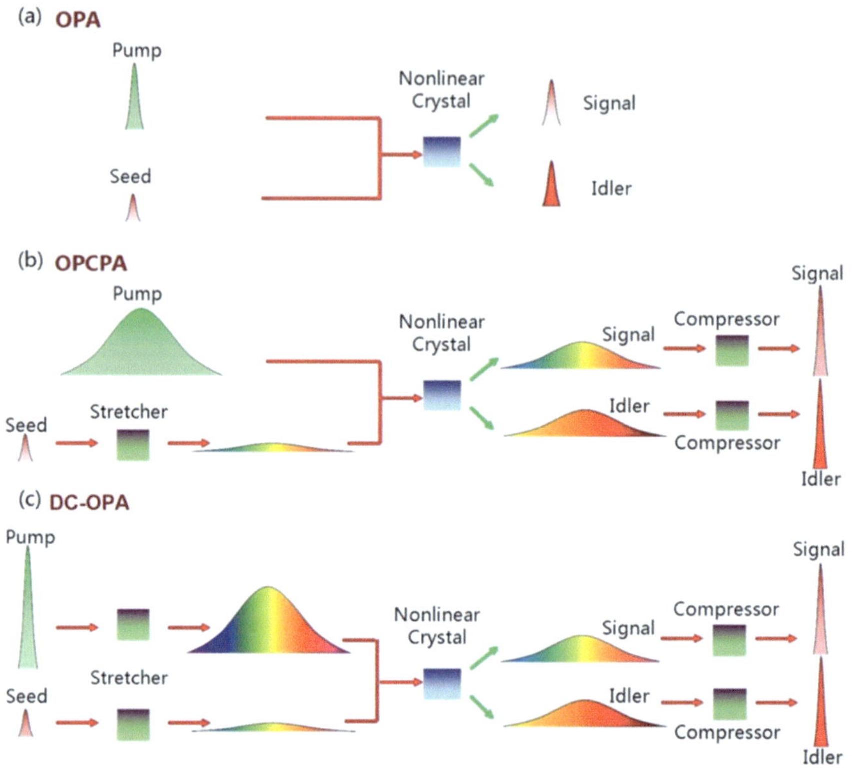

Fig. 5.9 Schematic layouts for generation of high-intensity, ultrashort pulses based on **a** optical parametric amplification (OPA): **b** optical parametric chirped-pulse amplification (OPCPA) and **c** dual-chirp optical parametric amplification (DC-OPA). Reprinted from Zhang et al. (2011). © 2011 Optical Society of America

The major advantage of OPCPA over the CPA lasers using broadband amplifiers such as Ti:sapphire is that the gain bandwidth of OPCPA can exceed the bandwidth of laser amplifiers, enabling generating a-few-cycle optical pulses, when phase matching methods described in Sect. 5.3.4 are properly implemented. Another important aspect of OPCPA is that there is no thermal loading in the amplifier crystal, resulting in high-quality amplified beam without thermal-induced optical aberration.

Detailed study on analysis and optimization of OPCPA for generating high power and ultrashort pulses is given in Ross et al. (2002). This study has investigated the efficiency and bandwidth of OPCPA in the presence of non-flat temporal and spatial profiles, the phase relations of the signal, idler and pump pulses, and the effect of group velocity mismatch and misalignment tolerance for both collinear BBO-OPA and noncollinear BBO- and KDP-OPCPA. Considering the practical limitations such as availability of large aperture nonlinear crystals and damage thresholds of diffraction gratings for pulse compression, this study led to a prototype design of a 163 PW, 22 fs OPCPA system with a 34 cm × 34 cm aperture KDP crystal in the last stage pumped with multiple beams of a high-power Nd:glass laser. Various approaches undertaken worldwide to develop OPCPA laser systems for generation of ultrahigh power, ultrashort-duration pulses are described in Sect. 7.4.2.

In the OPCPA, the pump pulse is narrow-band, and the seed pulse is broad-band with chirping. A "chirp compensation scheme" has been proposed to increase the gain over a broad band by applying a chirp on a long-duration pump pulse to keep phase matching with the chirped signal pulse over the whole pulse duration at a fixed crystal angle (Ross et al., 2002). As an extension of this scheme, "dual-chirped optical parametric amplification (DC-OPA)" has been developed to generate high-power pulses in the near-infrared region, with the configuration shown in Fig. 5.9c, mainly as the driving source for high-order harmonic generation in soft x-ray region (Fu et al., 2015; Zhang et al., 2011). This scheme is suitable to use a high-energy Ti:sapphire laser for both seed pulse generation and optical parametric amplification of the seed pulse.

With this scheme, CEP-stable 10.4 fs (sub-two-cycle) duration, > 100 mJ energy, 10-TW peak power pulses centered at 1.7 μm wavelength were generated at 10 Hz in three-stage OPA using type-I BiB_3O_6 (BiBO) crystals (Xu et al., 2022). In an upgraded scheme named advanced DC-OPA, the amplification bandwidth was increased to over one octave, resulting in generation of 8.58 fs duration (1.05 cycles) pulses with 50-mJ energy and 6-TW peak power (Xu & Takahashi, 2024). In this system, a CEP-stable, broadband seed laser pulses in the mid-infrared of 1.4–3.1 μm, generated by difference frequency generation (DFG) of spectrally broadened pulses from a gas cell, are amplified with a preamplifier and three-stage OPAs, each composed of type-I BiB_3O_6 (BiBO) and type-I $MgO:LiNO_3$ (MgO:LN) nonlinear crystals to amplify different spectral regions, which are pumped with 750 mJ Ti:sapphire laser pulses, and finally compressed with a bulk compressor (a 40-mm long sapphire). The spectral phase of the compressed pulses is fed back to the acousto-optic programmable dispersive filter (AOPDF) placed in the frontend, which adjusts the spectral phase of the seed pulses to minimize the spectral phase dispersion to generate transform-limited amplified pulses. It is envisioned that even shorter duration 6.2 fs (0.75 cycle) pulses could be generated with using three crystals for each OPA stage to further broaden the spectral region for amplification.

5.4 Self-focusing and Self-phase Modulation

The refractive index of materials depends on the laser intensity due to nonlinear susceptibility. The intensity-dependent refractive index leads to various processes, affecting propagation of the laser beam. In this subsection, we describe two nonlinear processes which are important in high power lasers, i.e., self-focusing and self-phase modulation.

5.4.1 Nonlinear Refractive Index

The intensity dependence of the refractive index is described by the nonlinear refractive coefficient γ_{nl} as,

$$n = n_0 + \delta n = n_0 + \gamma_{nl} I. \tag{5.43}$$

Here n_0 is the refractive index of the medium and I is the laser intensity. The increase in the refractive index $\delta n = \gamma_{nl} I$ is induced by various processes with different response times (shown in parentheses); electronic polarization (10^{-16} s), molecular orientation (10^{-12} s), electrostriction (10^{-9} s), saturated atomic absorption (10^{-8} s), thermal effects (10^{-3} s), and photorefractive effect (material dependent, 10^{-6} s for compound semiconductor). In the following we consider only electronic polarization, which is important in high-power, short-pulse solid-state lasers due to its fast response, although other processes have larger nonlinear refractive indices.

5.4.2 Self-focusing and Beam-Breakup

As described in Sect. 4.2.5, a laser beam is self-focused due to an increase in the refractive index at high laser intensity. Since self-focusing of the laser beam leads to laser-induced damage to optical materials, the self-focusing is detrimental in high-power solid-state lasers (Brown, 1981; Lukishova et al., 2009). On the other hand, self-focusing plays important roles in applications such as laser-ranging, where a femtosecond laser pulse propagates in an air over a long distance forming filamentary structures (Couairon & Mysyrowicz, 2009).

The nonlinear increase in the refractive index is expressed either in terms of intensity I or in terms of time-average of the squared field amplitude $\langle E^2 \rangle$:

$$n = n_0 + \gamma_{nl} I = n_0 + n_2 \langle E^2 \rangle. \tag{5.44}$$

Here n_2 is the nonlinear refractive index and γ_{nl} is the nonlinear refractive coefficient (or nonlinear refractive index). Often nonlinear property of γ_{nl} is given in SI unit and n_2 in esu. These are related as $\gamma_{nl}\,[\mathrm{cm^2/W}] = 4.19 \times 10^{-3} (n_2 / n_0)\,[\mathrm{esu}]$ (Koechner, 2006). In the following, we use γ_{nl} in SI unit.

Self-focusing of a laser beam leads to three different effects: (a) (whole beam) self-focusing where a laser beam is focused in a nonlinear material, (b) self-trapping where a laser beam is trapped and propagates as a small filament, and (c) beam-breakup where a large beam breaks up into many filaments.

First, we describe whole beam self-focusing, which takes place when the focusing due to increase in the refractive index exceeds the beam divergence by diffraction. Considering a simplified model (Chiao et al., 1964) where a circular beam of diameter d with uniform intensity distribution propagates in a medium whose refractive index is given by Eq. (5.43). Referring to Fig. 5.10a, the laser beam is confined in a cylinder formed by the laser beam due to total reflection, when the grazing angle of the ray to the cylinder wall ϕ is smaller than $\phi_c = (\pi/2) - \theta_c$ where θ_c is the critical angle given by $\sin\theta_c = n_0/(n_0 + \delta n)$. Since $\phi_c \ll 1$ and $\delta n \ll 1$, $\sin\theta_c = \cos\phi_c \simeq 1 - \phi_c^2/2$ and $n_0/(n_0 + \delta n) \simeq 1 - \delta n/n_0$. Equating these, we obtain

$$\phi_c \simeq \sqrt{\frac{2\delta n}{n_0}} = \sqrt{\frac{2\gamma_{nl} I}{n_0}} \tag{5.45}$$

Referring to Fig. 5.10b, the divergence angle (half angle) of a circular beam of diameter d due to Fraunhofer diffraction is given by $\phi_{diff} = 1.22\lambda_o/n_0 d$ where λ_o is the laser wavelength in vacuum. Here ϕ_{diff} corresponds to the radius of the first dark ring of the Airy pattern. Equating ϕ_c and ϕ_{diff}, and expressing I in terms of the laser power as $P = (\pi d^2/4)I$, we obtain the critical power for self-focusing;

$$P_{cr} = \frac{\pi(1.22)^2\lambda_0^2}{8 n_0 \gamma_{nl}}. \tag{5.46}$$

Taking a fused silica as an example where $\gamma_{nl} = 2.1 \times 10^{-16}\,\mathrm{cm}^2/\mathrm{W}$ and $n_0 = 1.45$, the critical power for $\lambda_0 = 1\,\mu m$ is estimated as $P_{cr} \approx 19\,MW$. We note that the critical power does not depend on the beam diameter.

At $P \geq P_{cr}$, the laser beam is self-focused and forms a filamentary structure along the beam path, leading to dielectric break-down of the medium. When the laser beam propagates in gaseous media such as air, the beam is defocused due to formation of low-density plasmas by ionization of molecules. In this case, the laser beam is "self-trapped" by repeating focusing and defocusing over a long distance (Chin et al., 2005).

At $P \gg P_{cr}$, a laser beam breaks up into many filaments, called "beam break-up", each filament comprising $P \geq P_{cr}$. Considering the growth of field perturbation in

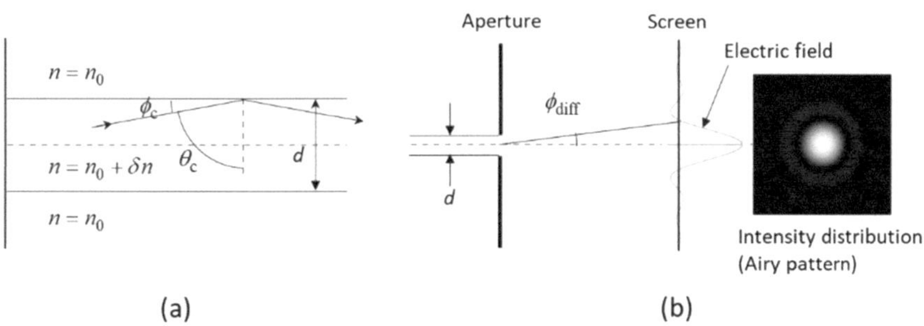

Fig. 5.10 **a** Self focusing of a laser beam due to total reflection in a cylinder formed by the laser beam itself due to nonlinear increase in the refractive index. **b** Fraunhofer diffraction of a laser beam of diameter d, where the diffraction angle ϕ_{diff} is the angle to the first dark ring of the diffraction pattern (Airy pattern) on a screen at a far field

the laser beam, high spatial frequency perturbations have a larger growth rate, since the equivalent focal length is shorter for small scale non-uniformities. However, the growth rate will decrease at very high spatial frequencies due to diffraction.

Bespalov and Talanov (1966) and Suydam (1994) analyzed the instability of the field amplitude by introducing a small perturbation a to the field as $A = A_0 + a$ in the paraxial wave equation, Eq. (4.64), for the case of $\alpha = 0$ (i.e., without amplification), where A_0 is the field amplitude without perturbation. Expressing the spatial frequency of the perturbation by K, the growth rate g of the perturbation is derived as

$$g = K\left(\gamma_{nl} I - \frac{K^2 \lambda_0^2}{16\pi^2}\right)^{1/2}. \tag{5.47}$$

The dependence of g on K is shown in Fig. 5.11a. The growth rate g initially increases linearly with K, becomes maximum at $K = K_m$, and decreases to $g = 0$ at a critical wavenumber $K = K_c$. From (5.47), K_m is given by

$$K_m = (2\pi/\lambda_0)(2\gamma_{nl} I)^{1/2}, \tag{5.48}$$

and $K_c = \sqrt{2} K_m$. The maximum growth rate g_m is

$$g_m = (2\pi/\lambda_0)\gamma_{nl} I = \frac{\lambda_0}{4\pi} K_m^2. \tag{5.49}$$

Note that $g_m L$ is equal to the B-integral given in (4.64) where L is the thickness of the medium. That is, the maximum growth rate of the lateral amplitude perturbation corresponds to the nonlinear phase shift B. It was demonstrated that a laser beam breaks up into many filaments with the spatial frequency expressed by Eq. (5.48) (Campillo et al., 1973). We note that the instability growth is also expressed in terms of the angle between the perturbation propagation and the z-axis $\theta = K/K_0$ where K_0 is the wavenumber of the pumping beam. The noise grows within the critical angle $\theta_c = K_c/K_0$.

As an example of the actual spatial intensity distribution of a high-power laser beam, Fig. 5.11b shows a near field pattern at the output of a phosphate glass laser,

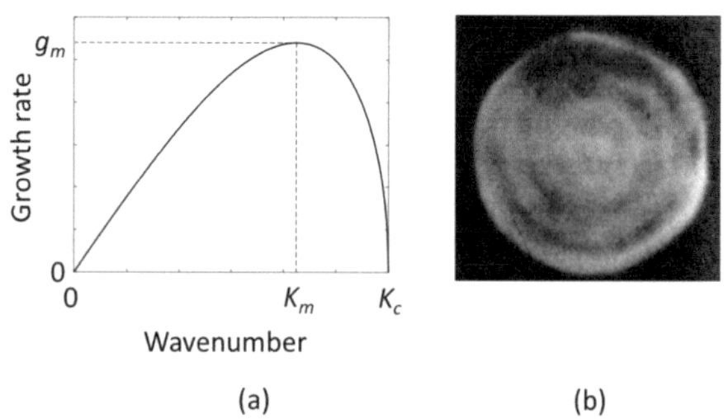

Fig. 5.11 **a** Dependence of the growth rate of the amplitude perturbation due to small-scale self-focusing on the spatial frequency of the perturbation. **b** A near field pattern of a high-power laser beam at the intensity of 19 GW/cm^2, where beam breakup is not yet taking place

where the beam diameter is 20 cm, output energy is 346 J and the pulse width is 100 ps. The intensity at the high fluence area is ~ 19 GW/cm^2, estimated from the measured peak-to-average fluence ratio of 1.60. The beam pattern is quite uniform without initiation of the small-scale beam-breakup (Kato et al., 1981).

The B-integral through propagation in the final amplifier, composed of three Brewster-angled elliptical disks of 20-cm aperture, is estimated to be 2.2, assuming the laser beam has propagated through the laser and optical glasses at the intensity of 19 GW/cm^2. However, this B-integral is over-estimated, since the laser intensity, and therefore the growth rate g_m, gradually increases during propagation in the amplifier (Brown, 1981; Lukishova et al., 2009). This effect leads to a decrease in the overall B-integral for high gain amplifiers. Accounting for the small signal gain of the final amplifier of G ~ 3.2, the actual B-integral is B ~ 1 for the data shown in Fig. 5.11b.

5.4.3 Self-phase Modulation

When an intense laser pulse propagates in a transparent optical material with a nonlinear refractive coefficient γ_{nl}, the instantaneous frequency changes in time, resulting in the frequency broadening (Shimizu, 1967). The phase of the optical wave is delayed due to nonlinear increase in the refractive index, which is called self-phase modulation:

$$\Phi_{NL}(t) = -\gamma_{nl} I(t)\omega_0 L/c. \tag{5.50}$$

The instantaneous frequency of the optical wave changes as

$$\omega(t) = \omega_0 + \delta\omega(t) = \omega_0 + \frac{d}{dt}\Phi_{NL}(t) = \omega_0 - \frac{\gamma_{nl}\omega_0 L}{c}\frac{d}{dt}I(t). \tag{5.51}$$

Considering a Gaussian laser pulse $I(t) = I_0 \exp\left(-t^2/\tau_0^2\right)$ where τ_0 is the half width at $1/e$ of the peak intensity, the nonlinear frequency shift is

$$\delta\omega(t) = -\frac{\gamma_{nl}\omega_0 L I_0}{c}\frac{d}{dt}\exp^{-(t^2/\tau_0^2)} = \frac{2\Phi_{NL}^{max}}{\tau_0^2} t \exp^{-(t^2/\tau_0^2)}, \tag{5.52}$$

where $\Phi_{NL}^{\max} = \gamma_{nl}\omega_0 L I_0/c$. When $\gamma_{nl} > 0$ and the response is instantaneous, the instantaneous frequency is initially down-shifted at the rising part of the pulse, then up-chirped at $-0.71 < t/\tau_0 < 0.71$, and down-chirped at the trailing part of the pulse as shown in Fig. 5.12b. The maximum frequency shift, derived from $d\delta\omega(t)/dt = 0$, is reached at $t = \pm\tau_0/\sqrt{2}$ with the value,

$$\delta\omega_{\max} = -\frac{2\Phi_{NL}^{\max}}{\tau_0^2} t \exp^{-(t^2/\tau_0^2)}\Big|_{t=\tau_0/\sqrt{2}} = 0.86\frac{\Phi_{NL}^{\max}}{\tau_0}. \tag{5.53}$$

The self-modulated light extends from $\omega_0 - \delta\omega_{max}$ to $\omega_0 + \delta\omega_{max}$. Intensity modulations comprised of regular periodic structures arise from constructive or destructive interference of the two waves generated at different times (t_1 and t_2 in Fig. 5.12b) (Stolen & Lin, 1978). The number of the modulation peaks is given approximately by $M \approx (\Phi_{NL}^{max}/\pi) + (1/2)$ (Cebeddu et al., 1970). The modulation patterns expected for various inputs of transform-limited, up-chirped and down-chirped pulses are given in Finot et al. (2018).

The frequency of the self-modulated light is up-chirped almost linearly near the center of the pulse $t \approx 0$. The chirp rate a defined by $\omega = \omega_0 + at$ is

$$a = \frac{2\Phi_{NL}^{max}}{\tau_0^2} = \frac{2n_2\omega_0 L I_0}{c\tau_0^2}. \tag{5.54}$$

The chirp rate a is proportional to the second-order dispersion coefficient ϕ_2. Due to the spectral dispersion, the frequency down-shifted component propagates faster

Fig. 5.12 **a** Time dependence of the intensity of a Gaussian pulse. **b** Time dependence of the frequency of a Gaussian pulse due to self-phase modulation

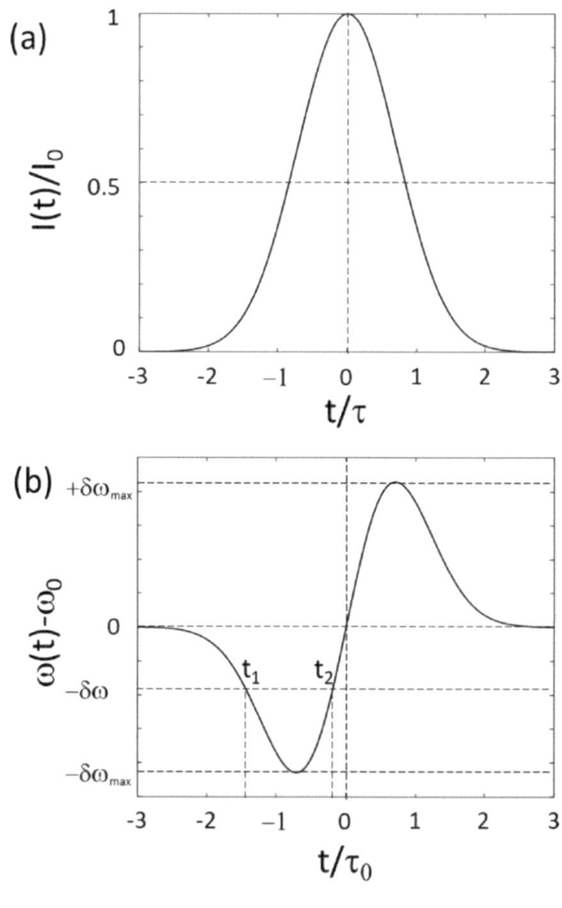

than the up-shifted component, resulting in the temporal broadening of the pulse, in addition to the pulse broadening due to linear spectral dispersion with the rate given in Eq. (2.148).

We note that various dispersion compensation methods such as the grating pair and the chirped mirrors compensate for the linear-chirped part of the pulse. However, the phase changes at the rising and trailing parts of the pulse are not compensated by these methods, resulting in sub-pulses to the compressed pulse. These sub-pulses can be eliminated, for example, by temporally selecting the central part of the pulse (Yamakawa et al., 1991).

5.5 Propagation of Extremely Short, Intense Laser Pulses

Extremely short duration (a few-cycle), extremely high-power lasers are becoming available as described in Sect. 7.4. For describing propagation of short-duration, intense laser pulses, we have derived Eq. (4.77) based on the slowly varying envelope approximation (SVEA) $|\partial_z A| \ll k_0|A|$ (where ∂_z is the partial derivation with respect to z and k_0 is the wavenumber). However, it is necessary to investigate the applicability of this equation, and what will be expected when the laser pulse becomes close to a few-cycle duration. In this section, we will introduce theoretical investigations on several issues: pulse shape, soliton formation, and self-focusing. It is shown that a few-cycle duration solitons will propagate stably, and the growth of self-focusing instability will be suppressed when an extremely short duration pulse propagates at the wavelength region of anomalous dispersion.

5.5.1 *Formulation of Ultrashort Pulse Propagation*

In the nonlinear Schrödinger equation (NLSE) [Eq. (4.77)], the material dispersion is included up to the second order, which is not sufficient for ultrashort pulses. Brabec and Krausz (1997) has solved the NLSE and expressed the solution in the moving frame as:

$$\begin{aligned} \boldsymbol{E}(\boldsymbol{r}_\perp, \xi, \tau) &= A(\boldsymbol{r}_\perp, \xi, \tau)\exp[-i\omega_0\tau + i\psi(\xi)] + c.c., \\ \psi(\xi) &= \psi_0 + (k_0 - \omega_0 k_1)\xi = \psi_0 + \omega_0\left(\frac{1}{v_{ph}} - \frac{1}{v_g}\right)\xi \end{aligned} \tag{5.55}$$

Here v_{ph} and v_g are the phase and group velocities, respectively, and $\psi(\xi)$ is the carrier envelope phase, where ξ is the moving coordinate (4.75). This equation is valid for $\left|\partial_\xi E\right| \ll k_0|E|$, termed *slowly evolving wave approximation* (SEWA), which is satisfied when the envelope A and the carrier phase $\psi(\xi)$ do not vary significantly over the laser wavelength. The latter condition is valid when the difference between the group velocity and the phase velocity is small compared to the phase velocity. The SEWA is applicable to pulses over broad range, from many cycles to a few cycles.

Figure 5.13 shows the change of the electric field and its envelope of a laser pulse during propagation, when a sech-shaped pulse of 2.67 fs duration (FWHM), corresponding to one optical cycle at 0.8 μm, is propagated in a fused silica at an intensity of 4×10^{13} W/cm^2. Figure 5.13a shows the pulse profile at a propagation distance of 20 μm, when the group velocity dispersion (GVD) is assumed to be zero. An optical shock is created at the trailing edge of the pulse envelope due to strong frequency change comparable to the carrier frequency by the optical Kerr effect. Figure 5.13b shows the pulse profile at the propagation distance of 80 μm, where the GDV of fused silica is included. The pulse duration is elongated, the frequency is up-chirped due to normal material dispersion, and the shock is moderated by GVD.

Instead of describing the field in terms of its envelope and phase, Bespalov et al. (2002) has calculated the electric field directly from the wave equation (4.65), first without diffraction and nonlinear polarization. To cover the broad spectral range required for ultrashort pulses, the following dispersion equation is introduced into the term $\partial_t^2\left[n_0(\omega)^2\boldsymbol{E}\right]/c_o^2$:

$$n_0(\omega)^2 = N_0^2 + 2c_0N_0\left(a\omega^2 + a_1\omega^4 + \cdots - b\omega^{-2} - b_1\omega^{-4}\cdots\right), \tag{5.56}$$

where N_0, a, a_1, ..., b, b_1... are the empirical constants of the optical material. This equation describes the material dispersion over a broad spectral range (Born & Wolf, 1999). In introducing Eq. (5.56) into Eq. (4.65), the frequency is converted to the time coordinate as $\omega \leftrightarrow i\partial_t$ and $1/\omega \leftrightarrow -i\int_{-\infty}^{\tau} d\tau'$ [see Eq. (4.73)]. Equation (4.65) is numerically solved, with modifications if required to incorporate

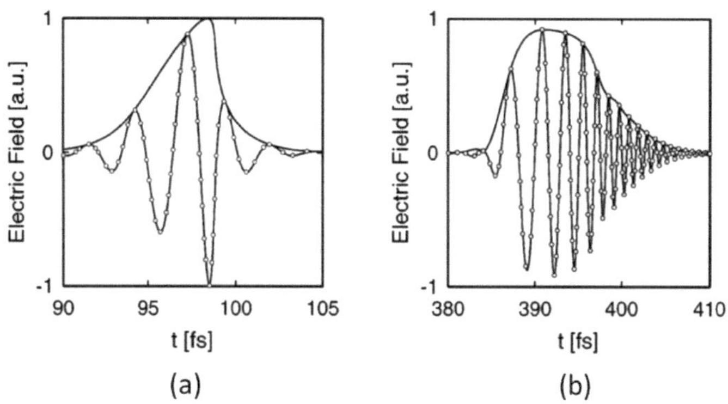

Fig. 5.13 Propagation of a single-cycle pulse of 0.8 μm center wavelength in a fused silica at an intensity of 4×10^{13} W/cm^2, showing the electric field and its envelope at the propagation distances of **a** 20 μm, and **b** 80 μm, respectively. The GVD is assumed to be zero in (**a**). The curves in full line are obtained by direct numerical calculation of the nonlinear Maxwell equation and the open circles by analytical equations. Reprinted from Brabec and Krausz (1997), with permission from The American Physical Society and the author(s). © 1977 The American Physical Society

various processes such as another dispersion formula, stimulated Raman scattering, two-photon absorption, diffraction, etc.

5.5.2 *Soliton Formation with Extremely Short Pulses*

In considering the nonlinear wave propagation, the solitary wave (soliton) is an important part in terms of theory and applications. In the framework of NLSE, a soliton is formed in the medium of anomalous dispersion since the positive frequency chirp due to the optical Kerr effect is counter-balanced by the negative frequency chirp due to the anomalous dispersion. The propagation of stable solitons in optical fibers was first suggested by Hasegawa and Tappert (1973) and experimentally demonstrated by Mollenauer et al. (1980). Due to successive theoretical and technical innovations, optical soliton communication is now available for high-speed, long distance information transmission systems (Nakazawa et al., 1995). Decreasing the pulse duration from the present 10 ps soliton to femto-second soliton is investigated for higher speed transmission (Kurokawa et al., 1994).

It is not obvious if solitons are formed when the duration of the input becomes only a few cycles. In analyzing the pulse propagation of extremely short pulses, Skobelev et al. (2007) have assumed a model material, generalizing the optical property of a fused silica, which has two absorption bands at ω_1 and ω_2 where the material is transparent in $\omega_1 < \omega < \omega_2$. For a fused silica without any impurity, the wavelengths of the nearest absorption lines corresponding to the absorption bands at ω_1 and ω_2 are $\lambda_1 = 9.1$ μm due to molecular vibration of SiO_2 and $\lambda_2 = 0.116$ μm due to electronic excitation, respectively. Due to these bands, the real part of the permittivity is described as

$$\varepsilon_r = \varepsilon_0 + a\omega^2 - \frac{b}{\omega^2}, \tag{5.57}$$

which is a truncated version of Eq. (5.56). The dispersion terms $a\omega^2$ and $-(b/\omega^2)$ arise from the high- and low-frequency bands, respectively. The group velocity dispersion $k''(\omega) = d^2(\omega\varepsilon_r^{1/2}/c)\big/d\omega^2 = (3a\omega - b/\omega^3)/(c\varepsilon_0^{1/2})$ becomes zero at $\omega_{cr} = (b/3a)^{1/4}$, corresponding to the critical wavelength of $\lambda_{cr} = 1.27\,\mu m$ for a fused silica. The material has anomalous dispersion in $\omega < \omega_{cr}$, zero dispersion at $\omega = \omega_{cr}$, and normal dispersion in $\omega > \omega_{cr}$.

Figure 5.14a shows the evolution of the electric field (*a1*) and its spectrum (*a2*) during propagation for an input pulse of a few-cycle duration with a center frequency $\omega_0/\omega_{cr} = 0.667$, i. e., in the anomalous dispersion region, whereas a small fraction of the spectrum (~ 16% of energy) is in the normal dispersion region of $\omega > 1.5\omega_0$. As the pulse propagates in z, the input pulse separates into two parts. One part propagates stably with its temporal and spectrum profiles converging to constant profiles, reaching to a few-cycle pulse duration with the center frequency slightly red-shifted to $\omega_{out(1)} = 0.92\omega_0$. With this pulse, the spectrum is entirely in the anomalous GVD region, containing approximately 70% of the input pulse energy. This pulse satisfies the conditions required for the soliton formation. Another part of the pulse is quickly separated from the first part and dissipated as the pulse propagates. This part of the pulse originates from the normal dispersion of the input spectrum.

In Fig. 5.14a, the input pulse energy is close to the minimum energy to form one soliton, which is denoted by $N = 1$. Figure 5.14b shows the evolution of a pulse for the input pulses of $N = 2.04$ (*b1*) and $N = 4.02$ (*b2*), respectively, for an idealized medium of no dispersion. In both cases, new pulses are formed by splitting from the original pulse and propagate stably as new solitons with slower velocities. The total number of the solitons is equal to N. During propagation, the original pulse becomes shorter, reaching to a single cycle at $z = 60$ as shown in the inset of (*b1*), and keeps a few-cycle duration over long distances.

We note that the fused silica used in optics have an absorption band at 2.73 μm due to SiOH stretching and its overtones at 2.24 and 1.37 μm arising from H_2O impurity (Moore & Smith, 2002). Although the transmittance is strongly affected by impurity, the dispersion property is not much affected since OH concentration is less than ~ 10 ppm wt.% in purified silica glasses. In the optical fibers for information transmission, the water content is minimized to reduce the absorption loss in very long propagation distance.

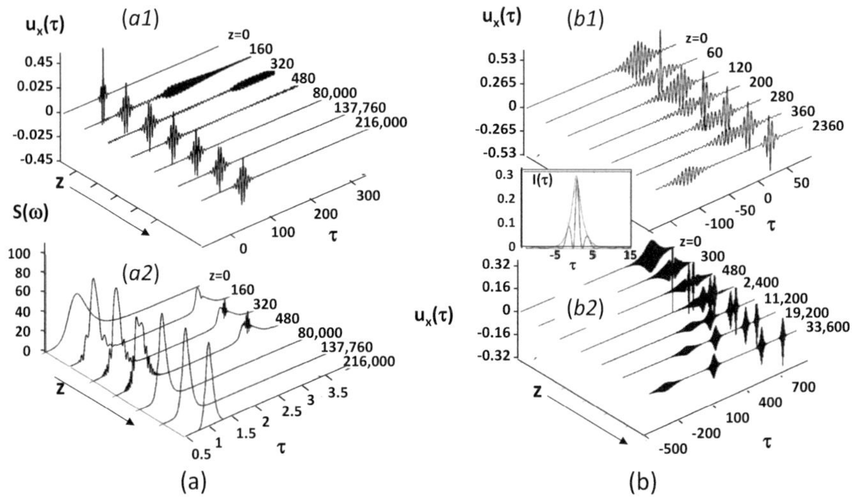

Fig. 5.14 Propagation of a few-cycle input pulse of **a** an input energy close to a single soliton formation ($N = 1$) and **b** higher energies with $N = 2$ (*b1*) and $N = 4$ (*b2*), respectively. In (**a**), the electric field and the spectrum are shown in (*a1*) and (*a2*), respectively. In (**b**), sub-pulses are split from the main pulse, resulting in multiple solitons with its number determined by N. Reprinted from Skobelev et al. (2007), with permission from The American Physical Society and the author(s). Note that the texts in the original figure are enlarged for visibility. © 2007 The American Physical Society

5.5.3 Self-focusing of Extremely Short Pulses

We have seen in the previous sections that an extremely short pulse propagates as a soliton in the anomalous dispersion media with its duration reaching close to the single cycle oscillation. When the spatial distribution is included in this propagation, multiple filaments are formed from uniform intensity distribution. Bergé et al. (2010) have calculated propagation of a high intensity pulse in silica (a) in the normal dispersion region of 355 nm and (b) in the anomalous dispersion region of 1550 nm, with the results shown in Fig. 5.15. In both cases the incident beam has a super-Gaussian profile with 1-mm beam radius and 50 fs pulse width, with the mean intensity of 0.46 TW/cm^2 in (a) and 0.73 TW/cm^2 in (b), respectively. In addition to the large-scale beam breakup, small-scale filaments are formed and grow to high intensities in both the normal and anomalous dispersion regions, although detailed filament structures are different. Due to self-focusing, the beam collapses and forms the plasma at the propagation distance of z_c ~ 2.45 mm in (a) and 8 mm in (b).

Balakin et al. have analyzed the pulse propagation in more detail and shown that the beam breakup instability does not grow when the initial pulse duration is less than ~ 10 optical cycles (Balakin et al., 2016). Implementing the time derivative in the right-hand side of Eq. (4.77), the nonlinear Schrödinger equation is expressed, in a moving frame of $\tau = (t - z/v_g)/\omega$ with dimensionless variables, as

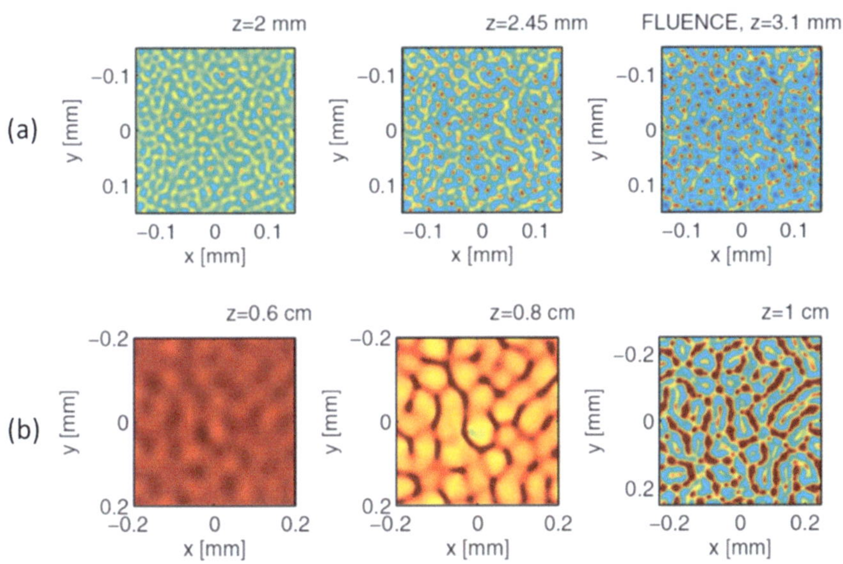

Fig. 5.15 Spatial fluence distributions at the central part of the laser beam at different propagation distances shown in each figure. The wavelength of the laser beam is 355 nm in (**a**) and 1550 nm in (**b**). Reprinted from Bergé et al. (2010), with permission from The American Physical Society and the author(s). © The American Physical Society

$$\left(\frac{\partial^2\Psi}{\partial x^2}+\frac{\partial^2\Psi}{\partial y^2}\right)+i\frac{\partial\Psi}{\partial z}+\frac{\partial^2\Psi}{\partial\tau^2}+|\Psi|^2\Psi-2i|\Psi|^2\frac{\partial\Psi}{\partial\tau}+=0, \tag{5.58}$$

which is called the derivative nonlinear Schrödinger equation (DNLS). Here the first and the second terms express the pulse propagation in SVEA, the third term the group velocity dispersion, and the fourth and fifth terms derived from the right-hand side of Eq. (4.77). The fourth term corresponds to the optical Kerr effect which leads to self-focusing and beam breakup. The fifth term corresponds to the intensity-dependent group velocity and instability suppression as described in the following.

To analyze the stability of the perturbations, the normalized field Ψ is expressed with the perturbation v as

$$\Psi(z,\tau,r_\perp)=[u_0+v(z,\tau,r_\perp)]\exp\left(iu_0^2z+2iu_0^2\tau\right). \tag{5.59}$$

Here u_0 is the field amplitude and $r_\perp$ is the radial coordinate. Writing the perturbation as $v(z,\tau,r_\perp)\propto\exp(\Gamma z+i\Omega\tau+ik_\perp r_\perp)$ where $k_\perp$ is the wavenumber of the radial perturbation and Ω is the angular frequency, the spatial growth rate of the field perturbation Γ is

$$\Gamma=-2iu_0^2\Omega\pm\sqrt{\left(\Omega^2+k_\perp^2\right)\left(2u_0^2-\Omega^2-k_\perp^2\right)}. \tag{5.60}$$

The second term of this equation shows that the perturbation grows exponentially during propagation when $\Gamma>0$ which holds in the range of $0\le\sqrt{\Omega^2+k_\perp^2}\le\sqrt{2}u_0$. In this expression, the first term Ω corresponds to the modulational instability where frequency sidebands are generated. The second term $k_\perp\le\sqrt{2}u_0$ corresponds to the beam breakup described in Sect. 5.4.2. Referring to Eq. (5.49), the maximum growth rate is $\Gamma_{\max}=u_0^2$ in this dimensionless unit.

From the imaginary term of Eq. (5.60), the velocity of the instability $v_{inst}(u_o)$ is derived as $1/v_{inst}(u_o)=\partial\mathrm{Im}(\Gamma)/\partial\Omega=-2u_0^2$. Since this is negative, the perturbation is delayed from the pulse, i.e., the breakup instability becomes convective. The critical pulse duration τ_{cr} below which the perturbation growth is suppressed is estimated approximately by the condition that the delay time of the perturbation is half of the pulse duration. This condition is expressed as $z_{cr}/v_{inst}(u_o)\approx z_{cr}2u_0^2\approx\tau_{cr}/2$, here z_{cr} is the critical distance of pulse propagation for instability suppression. Considering that the perturbation grows as $\exp(\Gamma z)\le\exp(\Gamma_{\max}z)$, we estimate the critical distance z_{cr} by $z_{cr}\Gamma_{\max}\approx 15$. From these relations, the pulse duration below which the perturbation growth is suppressed is $\tau_p\le\tau_{cr}=60/\omega\approx 10/\nu$ in dimensional unit, which corresponds to about ten optical cycles. This critical pulse duration does not depend on the intensity since both of $1/v_{inst}(u_o)$ and $\Gamma_{\max}$ are proportional to u_0^2. We note that this instability suppression is the nonlinear process and does not depend on material dispersion.

For more quantitative evaluation, we show in Fig. 5.16 the dynamics of the wave packet intensity calculated by the wave equation (5.58) in (z,x,τ) coordinates, without (a) and with (b) the last term of this equation, i.e., the instability suppression term.

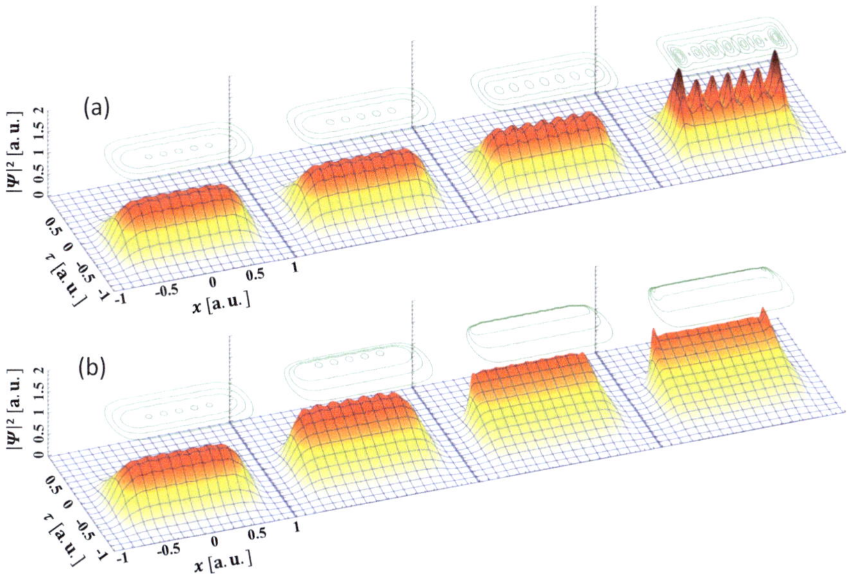

Fig. 5.16 Dynamics of the wave packet of a few-cycle duration moving to the z-direction **a** without and **b** with the last term of Eq. (5.58), which corresponds to the instability suppression. Vertical axis is the intensity $|\Psi(z, x, \tau)|^2$, x-axis is the lateral coordinate of the pulse, and the pulse front is negative in time τ. The green lines show intensity contours. Reprinted from Balakin et al. (2016), with permission from The American Physical Society and the author(s). Note that the texts in the original figure are enlarged for visibility. © 2016 The American Physical Society

Without the last term, the beam breakup in the x-direction increases as the pulse propagates in the z-direction. With the last term included, the growth of the beam break-up is suppressed since the high-intensity part moves to the rear side of the pulse (positive in time), forming a shock-front structure at the rear edge of the pulse.

With the beam breakup instability suppressed, propagation of an extremely short pulse including 2-D diffraction has been analyzed by Balakin et al. (2017). Figure 5.17 shows the dynamics of a circularly polarized field with an initial pulse duration of 10 optical cycles with single soliton energy ($N = 1$) propagating in medium of anomalous group velocity dispersion. In Fig. 5.17a–f, upper traces show contours of the field amplitude $|u(z, \tau, r)|$ in (z,r) plane with its maximum amplitude $u_{\max}$, and lower traces in blue show evolution of the field u on the beam axis, both at six propagation distances. In the lower traces, (g) and (h) show, respectively, the field u and its envelope $|u|$, and the spectral distribution $S(\omega)$, both at $z = 0$ (dotted lines in blue) and $z = 2700$ (lines in red), whereas (i) shows the pulse duration τ/τ_0 (line in red) and the pulse energy W/W_0 (dotted line in blue) during propagation.

These diagrams show that the pulse is self-compressed in time, accompanied with gradual decrease in the beam radius (i. e., the beam is self-focused) as the pulse propagates, where the pulse keeps its pancake shape, i.e., longitudinal pulse width is always shorter than the radial width. At $z = 2700$ where the pulse collapses, the

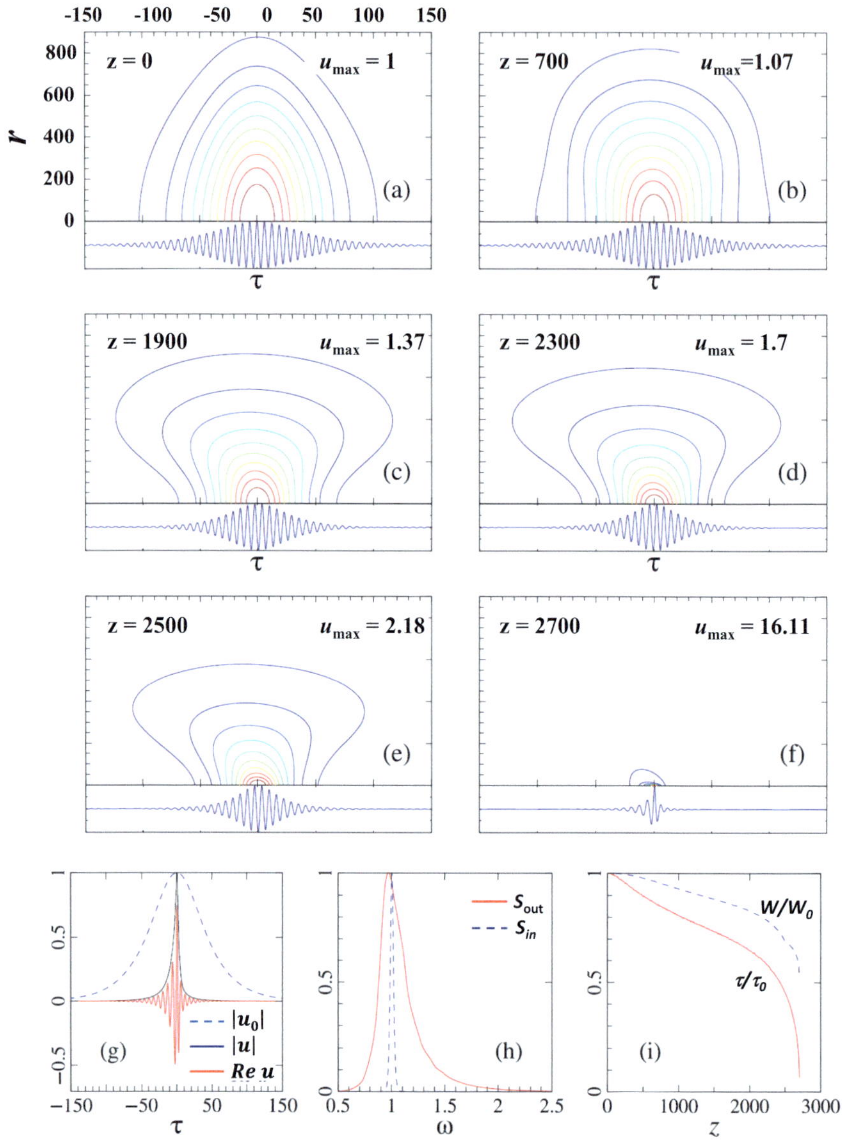

Fig. 5.17 Dynamics of a circularly polarized, single-soliton pulse of ten-cycle duration propagating in a medium of anomalous dispersion. **a–f** The field contours in the (z, r) plane (upper traces) and the field distribution on the axis (lower traces) at the propagation distances from $z = 0$ to 2700. (**g**) and (**h**) show, respectively, the pulse shapes and the spectra of the initial and final pulses. **i** Changes of the pulse duration and the pulse energy during propagation. Reprinted from Balakin et al. (2017). Note that the texts in the original figure are enlarged for visibility. © 2017 IOP Publishing

r.m.s. pulse duration becomes $\tau_{pulse} = 1.1\,T_0$ where T_0 is one optical cycle and the peak intensity $|u_{\max}|^2$ increases by ~ 230 times of the initial value. Therefore, a ten-cycle, single soliton pulse is compressed to a single-cycle pulse and the intensity is enhanced by 230 times, with 55% energy transmittance. In terms of practical parameters, an initial pulse of 100 fs duration, 15 mJ energy, and 1.5 mm beam radius (corresponding to 150 GW peak power) will be compressed to a pulse of 7 fs duration, 7 mJ energy and 1 TW peak power, after passing through a 1.5 cm thick YAG plate.

When the input pulse energy is increased corresponding to multiple solitons ($N \geq 2$), the pulse is self-compressed at shorter propagation distances, forming multiple solitons with complex temporal and spatial structures. The instability suppression and the pulse self-compression described in this section, which have not been experimentally demonstrated, will have important implications for the generation of very high-power, extremely short pulses.

References

Armstrong, J. A., Bloembergen, N., Ducuing, J., & Pershan, P. S. (1962). Interactions between light waves in a nonlinear dielectric. *Physical Review, 127*, 1918.

Balakin, A. A., Kim, A. V., Litvak, A. G., et al. (2016). Extreme self-compression of laser pulses in the self-focusing mode resistant to transverse instability. *Physical Review A, 94*, Article 043812.

Balakin, A. A., Litvak, A. G., Mironov, V. A., et al. (2017). Self-compression of soliton-like laser pulses in the process of self-focusing. *Journal of Optics, 19*, Article 095503.

Baltuska, A., Fuji, A., & Kobayashi, T. (2002). Controlling the carrier-envelope phase of ultrashort light pulses with optical parametric amplifiers. *Physical Review Letters, 88*, Article 133901.

Bellini, M., & Hänsch, T. W. (2000). Phase-locked white-light continuum pulses: Toward a universal optical frequency-comb synthesizer. *Optics Letters, 25*, 1049.

Bergé, L., Mauger, S., & Skupin, S. (2010). Multi-filamentation of powerful optical pulses in silica. *Physical Review A, 81*, Article 013817.

Bespalov, V. G., Kozlov, S. A., Shpolyanskiy, Yu. A., et al. (2002). Simplified field wave equations for the nonlinear propagation of extremely short light pulses. *Physical Review A, 66*, Article 013811.

Bespalov, V. I., & Talanov, V. I. (1966). Filamentary structure of light beams in nonlinear liquids. *JETP Letters, 3*, 307.

Born, M., & Wolf, E. (1999). *Principles of optics: Electromagnetic theory of propagation, interference and diffraction of light* (7th ed.). Cambridge University Press.

Brabec, T., & Krausz, F. (1997). Nonlinear optical pulse propagation in the single-cycle regime. *Physical Review Letters, 78*, 3282.

Brown, D. C. (1981). *High-peak-power Nd:glass laser systems*. Springer.

Byer, R. L. (1975). Optical parametric oscillators. In H. Rabin & C. L. Tang (Eds.), *Quantum electronics: A treatise, nonlinear optics, part B* (pp. 587–702). Academic Press.

Campillo, A. J., Pearson, J. E., Shapiro, S. L., & Terrell, N. J., Jr. (1973). Fresnel diffraction effects in the design of high-power laser systems. *Applied Physics Letters, 23*, 85.

Cebeddu, R., Polloni, R., Sacci, C. A., et al. (1970). Self-phase modulation and "rocking" of molecules in trapped filaments of light with picosecond pulses. *Physical Review A, 2*, 1955.

Cerullo, G., & De Silvestri, S. (2003). Ultrafast optical parametric amplifiers. *Review of Scientific Instruments, 74*, 1.

Cerullo, G., & Manzoni, C. (2007). Solid-state ultrafast optical parametric amplifiers. In A. Sennaroglu (Ed.), *Solid-state lasers and applications* (pp. 437–472). CRC Press.

Chiao, R. Y., Garmire, E., & Townes, C. H. (1964). Self-trapping of optical beams. *Physical Review Letters, 13*, 479.

Chin, S. L., Hosseini, S. A., Liu, W., et al. (2005). The propagation of powerful femtosecond laser pulses in optical media: Physics, applications, and new challenges. *Canadian Journal of Physics, 83*, 863.

Choy, M. M., & Byer, R. L. (1976). Accurate second order susceptibility measurements of visible and infrared nonlinear crystals. *Physical Review B, 14*, 1693.

Couairon, A., & Mysyrowicz, A. (2009). Self-focusing and filamentation of femtosecond pulses in air and condensed matter: simulations and experiments. In: R. W. Boyd, S. G. Lukishova, & Y. Shen (Eds.), *Self-focusing: Past and present*. Topics in Applied Physics (Vol. 114, p. 297). Springer.

Dabu, R. (2010). Very broad gain bandwidth parametric amplification in nonlinear crystals at critical wavelength degeneracy. *Optics Express, 18*, 11689.

Dabu, R. (2021). Optical parametric amplification at critical wavelength degeneracy—A proposed approach for 100-PW class femtosecond laser development. *OSA Continuum, 4*, 1658.

Dmitriev, V. G., Gurzadyan, G. G., & Nikogosyan, D. N. (1999). *Handbook of nonlinear optical crystals* (2nd ed.). Springer.

Dubietis, A., Jonusauskas, G., & Piskarskas, A. (1992). Powerful femtosecond pulse generation by chirped and stretched pulse parametric amplification in BBO crystal. *Optics Communications, 88*, 437.

Dubietis, A., Danielius, R., Tomasauskas, G., & Piskarskas, A. (1998). Combining effect in a multiple-beam-pumped optical parametric amplifier. *Journal of the Optical Society of America B: Optical Physics, 15*, 1135.

Fejer, M. M., Magel, G. A., Jundt, D. H., et al. (1992). Quasi-phase-matched second harmonic generation: Tuning and tolerances. *IEEE Journal of Quantum Electronics, 28*, 2631.

Finot, C., Chaussard, F., & Boscolo, S. (2018). Simple guidelines to predict self-phase modulation patterns. *Journal of the Optical Society of America B, 35*, 3143.

Fu, Y., Takahashi, E. J., & Midorikawa, K. (2015). High-energy infrared femtosecond pulses generated by dual-chirped optical parametric amplification. *Optics Letters, 40*, 5082.

Fujioka, K., Fujimoto, Y., Tsubakimoto, K., et al. (2015). Partially deuterated potassium dihydrogen phosphate optimized for ultra-broadband optical parametric amplification. *Journal of Applied Physics, 117*, Article 093103.

Garrett, C. G. B., & Robinson, F. N. H. (1966). Miller's phenomenological rule for computing nonlinear susceptibilities. *IEEE Journal of Quantum Electronics, QE-2*, 328.

Harimoto, T., & Yamakawa, K. (2005). Proposal for ultrabroadband phase-matching optical parametric chirped pulse amplification with a diverged pump beam. *Japanese Journal of Applied Physics, 44*, 3962.

Hasegawa, A., & Tappert, F. (1973). Transmission of stationary nonlinear optical pulses in dispersive dielectric fibers. I. Anomalous dispersion. *Applied Physics Letters, 23*, 142.

Herrmann, D., Homann, C., Tautz, R., Scharrer, M., et al. (2010). Approaching the full octave: Noncollinear optical parametric chirped pulse amplification with two-color pumping. *Optics Express, 18*, 18752.

Ishizuki, H., & Taira, T. (2005). High-energy quasi-phase-matched optical parametric oscillation in a periodically poled MgO:$LiNbO_3$ device with a 5 mm × 5 mm aperture. *Optics Letters, 30*, 2918.

Ishizuki, H., & Taira, T. (2010). High energy quasi-phase matched optical parametric oscillation using Mg-doped congruent $LiTaO_3$ crystal. *Optical Express, 18*, 253.

Ishizuki, H., & Taira, T. (2020). Polarity inversion of crystal quartz using a quasi-phase matching stamp. *Optics Express, 28*, 6505.

Ishizuki, H., Yahia, V., & Taira, T. (2018). Characteristics of crystal quartz for high-intensity, sub-nanosecond wavelength conversion. *Optical Materials Express, 8*, 1260.

Johnson, P. J. M., Prokhorenko, V. I., & Miller, R. J. D. (2011). Enhanced bandwidth noncollinear optical parametric amplification with a narrowband anamorphic pump. *Optics Letters, 36*, 2170.
Kato, Y., Yoshida, K., Kuroda, J., & Yamanaka, C. (1981). 3.4-TW performance of a Nd:phosphate glass laser with output aperture of 20 cm. *Applied Physics Letters, 38*, 72.
Kleinman, D. A. (1962). Nonlinear dielectric polarization in optical media. *Physical Review, 126*, 1977.
Koechner, W. (2006). *Solid-state laser engineering* (6-th ed.). Springer.
Kurokawa, K., Kubota, H., & Nakazawa, M. (1994). Femtosecond soliton interactions in a distributed erbium-doped fiber amplifier. *IEEE Journal of Quantum Electronics, 30*, 2220.
Lin, S., Sun, Z., Wu, B., et al. (1990). The nonlinear optical characteristics of a Li_3BO_5 crystal. *Journal of Applied Physics, 67*, 634.
Lozhkarev, V. V., Freidman, G. I., Ginzburg, V. N., et al. (2005). Study of broadband optical parametric chirped pulse amplification in a DKDP crystal pumped by the second harmonic of a Nd:YLF laser. *Laser Physics, 15*, 1319.
Lukishova, S. G., Senatsky, Y. V., Bykovsky, N. E., & Scheulin, A. S. (2009). Beam shaping and suppression of self-focusing in high-peak-power Nd:glass laser systems. In: R.W. Boyd, S. G. Lukishova, & Y. R. Shen (Eds.), *Self-focusing: Past and present*. Topics in Applied Physics (Vol. 114, p. 191). Springer.
Miller, R. C. (1964). Optical second harmonic generation in piezoelectric crystals. *Applied Physics Letters, 5*, 17.
Mollenauer, L. F., Stolen, R. H., & Gordon, J. P. (1980). Experimental observation of picosecond pulse narrowing and solitons in optical fibers. *Physical Review Letters, 45*, 1095.
Moore, L. A., & Smith, C. M. (2002). Fused silica as an optical material. *Optical Material Express, 12*, 3043.
Mucke, O. D., Sidorov, D., Dombi, P., Pugzlys, A., et al. (2009). Scalable Yb-MOPA-driven carrier-envelope phase-stable few-cycle parametric amplifier at 1.5 μm. *Optics Letters, 34*, 118.
Nakazawa, M., Kimura, Y., Suzuki, K., et al. (1995). Soliton transmission at 20 Gbit/s over 2000 km in Tokyo metropolitan optical network. *Electronics Letters, 31*, 1478.
Pasiskevicius, V. (2013). Nonlinear crystals for solid-state lasers. In B. Denker & E. Shklovsky (Ed.), *Handbook of solid-state lasers*. Materials, systems and applications (pp. 139–168). Woodhead Publication.
Ross, I. N., Matousek, P., Towrie, M., Langley, A. J., & Collier, J. L. (1997). The prospects for ultra-short pulse duration and ultrahigh intensity using optical parametric chirped pulse amplifiers. *Optics Communications, 144*, 125.
Ross, I. N., Matousek, P., New, G. H., et al. (2002). Analysis and optimization of optical parametric chirped pulse amplification. *Journal of the Optical Society of America B, 19*, 2945.
Shao, B., Li, Y., Peng, Y., Wang, P., et al. (2020). Broad-bandwidth high-temporal-contrast carrier-envelope-phase-stabilized laser seed for 100 PW lasers. *Optics Letters, 45*, 2215.
Shimizu, F. (1967). Frequency broadening in liquids by a short light pulse. *Physical Review Letters, 19*, 1097.
Shoji, I., Nakamura, H., Ito, R., et al. (2001). Absolute measurement of second-order nonlinear-optical coefficients of $CsLiB_6O_{10}$ for visible-to-ultraviolet second-harmonic wavelengths. *Journal of the Optical Society of America B: Optical Physics, 18*, 302.
Shoji, I., Nakamura, H., Ohdaira, K., et al. (1999). Absolute measurement of second-order nonlinear-optical coefficients of β-BaB_2O_4 for visible to ultraviolet second-harmonic wavelengths. *Journal of the Optical Society of America B: Optical Physics, 16*, 620.
Skobelev, S. A., Kartashov, D. V., & Kim, A. V. (2007). Few-optical-cycle solitons and pulse self-compression in a Kerr medium. *Physical Review Letters, 99*, Article 203902.
Stolen, R. H., & Lin, C. (1978). Self-phase modulation in silica optical fibers. *Physical Review A, 17*, 1448.
Sutherland, R. L. (2003). *Handbook of nonlinear optics* (2nd ed.). Marcel Dekker.
Suydam, B. R. (1994). Self-focusing of very powerful laser beams II. *IEEE J. Quant. Electr, QE-10*, 837.

Tamošauskas, G., Dubietis, A., Valiulis, G., & Piskarskas, A. (2008). Optical parametric amplifier pumped by two mutually incoherent laser beams. *Applied Physics B, 91*, 305.

Vozzi, C., Manzoni, C., Calegari, F., Benedetti, E., et al. (2008). Characterization of a high-energy self-phase stabilized near-infrared parametric source. *Journal of the Optical Society of America B: Optical Physics, 25*, B112.

Xu, L., & Takahashi, E. J. (2024). Dual-chirped optical parametric amplification of high-energy single-cycle laser pulses. *Nature Photonics, 18*, 99.

Xu, L., Ishii, N., Itatani, J., et al. (2022). 100-mJ class, sub-two-cycle, carrier-envelope phase-stable dual-chirped optical parametric amplification. *Optics Letters, 47*, 3371.

Yamada, M., Nada, N., Saitoh, M., et al. (1993). First-order quasi-phase matched $LiNbO_3$ waveguide periodically poled by applying an external field for efficient blue second-harmonic generation. *Applied Physics Letters, 62*, 435.

Yamakawa, K., Shiraga, H., Kato, Y., et al. (1991). Prepulse-free 30-TW, 1-ps Nd:glass laser. *Optics Letters, 16*, 1593.

Zeromskis, E., Dubietis, A., Tamosauskas, G., & Piskarskas, A. (2002). Gain bandwidth broadening of the continuum-seeded optical parametric amplifier by use of two pump beams. *Optics Communications, 203*, 435.

Zhang, Q., Takahashi, E., Mucke, O. D., Lu, P., & Midorikawa, K. (2011). Dual-chirped optical parametric amplification for generating few hundred mJ infrared pulses. *Optics Express, 19*, 7190. https://doi.org/10.1364/OE.19.007190

Chapter 6
High-Energy Pulsed Lasers

Abstract High-energy pulsed lasers have been developed since the birth of the ruby laser in 1960. Among various solid-state and gaseous lasers, the high energy Nd-doped glass laser has been developed most extensively due mainly to its suitability for laser fusion research in terms of high-energy generation, temporal and spatial pulse shape controllability, and short-wavelength conversion. These developments have led to the demonstration of the laser fusion ignition, with a target gain of $G = 1.5$ in 2022 at the National Ignition Facility (NIF), Lawrence Livermore National Laboratory (Pak et al. in Phys Rev E 109:025203, 2024). In parallel, laser development is moving toward repetitive, high-energy pulsed lasers required for realization of laser fusion power plants. In Sect. 6.1, a brief overview of the high-energy pulsed solid-state lasers for laser fusion and high-energy density science is presented. Then a detailed account of NIF, a high-precision and high-energy laser system, is described in Sect. 6.2. Section 6.3 describes laser-diode-pumped high-average-power pulsed lasers and high-power laser diodes which are applicable to laser fusion, high intensity science, and industrial applications. Section 6.4 describes development of high-energy, pulsed excimer lasers, which have several unique properties as laser fusion driver.

Keywords Laser fusion · Fusion ignition · National Ignition Facility · Nd-doped glass laser · High power laser diodes · Laser-diode pumped solid state laser · DPSSL · High-power excimer laser

6.1 High-Energy Pulsed Lasers for Laser Fusion and High Energy Density Science

6.1.1 Introduction

Since completion of the National Ignition Facility (NIF) at the Lawrence Livermore National Laboratory (LLNL) in 2009, the laser fusion experiments at NIF have advanced remarkably, with recent achievements of the alpha-particle self-heating of

Y. Kato et al., *High Power Laser and Plasma Science*, Springer Series on Atomic, Optical, and Plasma Physics 130, https://doi.org/10.1007/978-3-032-08930-4_6

inertially confined plasma (Hurricane et al., 2016) and the burning plasma exceeding the Lawson criterion for ignition in inertial confinement fusion (ICF) (Abu-Shawareb et al., 2022; Zylstra et al., 2022). Following these demonstrations, the target gain larger than one has been achieved at NIF in December 2022, where fusion energy of 3.15 MJ was produced with 2.05 MJ laser energy, corresponding to a target gain of 1.5 (Abu-Shawareb et al., 2024; Hurricane et al., 2024; Kritcher et al., 2024; Pak et al., 2024; Rubery et al., 2024), followed by a higher target gain of 1.7 in 2023. Based on the success of the fusion ignition, coordinated activities in the public and private sectors have been started toward realization of inertial fusion energy (IFE) in US (Ma & Betti, 2022). In this Section, research and development of high-energy pulsed lasers for laser fusion and high energy density science are reviewed with emphasis on the high-energy Nd:glass lasers.

Research on laser plasma interaction and nuclear fusion in high-temperature plasmas was started soon after demonstration of a ruby laser (Maiman et al., 1961). A large-scale program on the inertial confinement fusion (ICF) research was started in USA in 1972, when J. Nuckolls et al. have shown that hydrogen will be compressed to more than 1000 times liquid density by laser-driven implosion, leading to efficient thermonuclear burn of deuterium and tritium fuel and realization of fusion power reactors (Nuckolls et al., 1972). The ICF programs are undertaken also in other countries to pursue the laser fusion as a source of inexhaustible energy (Basov et al., 1972; Batani et al., 2023; Kirillov et al., 2000; Miquel et al., 2016; Yamanaka et al., 1981a). The laser fusion research has also stimulated evolution of high energy density (HED) science; a broad interdisciplinary field including fusion science, equation of states, atomic, nuclear and particle physics, earth, planetary and astrophysics, and high-energy particle and radiation generation (Drake, 2009).

The ICF research has been pursued with various types of pulsed-power drivers; high-energy pulsed lasers (solid-state laser, excimer laser, CO_2 laser, and iodine laser), magnetized liner inertial fusion (Gomez et al., 2020), and heavy ion beams (Seidl et al., 2013). Among these drivers, the solid-state laser, especially Nd-doped glass laser, has been developed to a high degree of integrity due to its high stored energy density and its capability for short wavelength conversion.

The shot rate of the high-energy Nd:glass lasers is limited by the cooling time of the laser glasses which are pumped by high-energy flashlamps. This situation is changing due to the development of high-power laser diodes (LDs). A 100 J, 10 Hz, 1kW average power solid state laser has been demonstrated by LD pumping of Yb:YAG multi-slabs cooled by cryogenic helium gas flow (Mason et al., 2015). Development of high-energy, high-rep lasers with various schemes are also undertaken, as described in Sect. 6.3.

6.1.2 Overview of High-Energy Pulsed Lasers

For laser fusion and HED science research, various types of high-energy pulsed lasers have been developed in many countries as listed in Table 6.1. These lasers include

Nd:glass, CO_2, KrF excimer, and iodine lasers. The laser energy has increased from the initial 1 Joule level to MJ level in these 50 years. Following completion of the 1.8 MJ, 192-beam NIF at LLNL in 2009, high-energy pulsed lasers with similar architectures have started operation; Laser Mega-Joule (LMJ) at CEA in France and Shengwang-III (SG-III) at LFRC in China. The operating laser wavelength has shifted from the initial 1 μm to shorter wavelengths of 526 and 351 nm by frequency up-conversion. Furthermore, high-peak power, short-pulse beamlines have been added to high-energy Nd:glass lasers for laser fusion and high energy density science by introducing chirped-pulse amplification (CPA). (In Table 6.1 only nanosecond-operation high-energy lasers are shown, whereas high-energy CPA Nd:glass lasers are described in Sect. 7.3.)

In Fig. 6.1 we show evolution of the laser energy of the Nd:glass lasers at several laboratories in these 50 years, with the photographs or bird's-eye view of several facilities in Fig. 6.2.

6.1.3 Progress in Laser Fusion and High-Energy Density Science

The advances in architecture and performance of high-energy lasers reflect the progress in the scientific understanding on the indirect-drive (Lindl, 1995; Lindl et al., 2004, 2014) and direct-drive (Craxton et al., 2015) laser fusion and HED science, as described briefly in the following.

(a) Target irradiation: The fusion target is irradiated either directly by multiple laser beams (direct-drive) or indirectly by x-rays converted from the laser in a closed cavity called "hohlraum" (indirect-drive). In direct drive, non-isentropic compression schemes such as fast ignition (Kodama et al., 2001; Tabak et al., 1994) and shock ignition (Betti et al., 2007) are also pursued. The laser beam configurations are very different in these schemes: two bundles in the indirect-drive, spherical irradiation in the direct-drive and shock ignition, and spherical irradiation plus a high-intensity heating beam in the fast ignition, as shown in Fig. 5.3. The HED experiments also require various irradiation conditions suitable for each experiment.

(b) Laser energy: The laser energy required to achieve fusion ignition with the indirect-drive has increased from the initial ~ 10 kJ to ~ MJ, achieving the fusion ignition with a fusion gain of 1.5 with 2.05 MJ laser energy (Abu-Shawareb et al., 2024). Although it is expected that ignition will be achieved with less driver energy with the direct-drive, shock ignition, and fast ignition, much experimental and theoretical investigations are necessary for quantitative evaluation of each scheme.

(c) Laser wavelength: The requirement for the wavelength of the laser driver has changed from the initial 1 μm to shorter wavelengths in ultraviolet region. With long-wavelength lasers such as Nd:glass (1.06 μm) and carbon dioxide (10.6 μm)

Table 6.1 High-energy pulsed laser facilities of ns operation for laser fusion (LF) and high-energy–density science (HED)

Country	Institute(1)	Laser	Objective(2)	Facility	Pulsewidth(3)	Energy (kJ)(4)	Beams	Wavelemgth (nm)(5)	Operation(6)	References
USA	LLNL	Nd:glass	LF-id	Janus	ns	0.01	2	1064	1974	
		Nd:glass	LF-id	Argus	ns	2	2	1064	1976	
		Nd:glass	LF-id	Shiva	ns	10	20	1064	1977	Speck et al. (1981)
		Nd:glass	LF-id	Nova	ns	100	10	1053	1984	Campbell et al. (1986)
		Nd:glass	LF-id	NIF	ns	1800	192	351	2009	Spaeth et al. (2016a, 2016b)
	LLE	Nd:glass	LF-d	OMEGA	ns	2	24	351	1980	Soures et al. (1983)
		Nd:glass	LF-d	OMEGA 60	ns	30	60	351	1995	Boehly et al. (1997)
	LANL	CO_2	LF	Helios	ns	10	8	10,600	1978	Ladish (1979)
	NRL	KrF	LF-d	Nike	ns	3	56	248	1996	Obenschain et al. (2015)
Japan	ILE	Nd:glass	LF-d	G-IV	ns	2	4	1053	1974	Kato et al. (1978)
		Nd:glass	LF-d	G-MII	ns	2	2	1053	1980	Kato et al. (1981)
		Nd:glass	LF-d	G-XII	ns	20	12	2	1983	Yamanaka et al. (1981a, 1981b)
		CO_2	LF-d	LEKKO-VIII	ns	10	8	10,600	1980	Yamanaka et al., (1981a, 1981c)

(continued)

Table 6.1 (continued)

Country	Institute[1]	Laser	Objective[2]	Facility	Pulsewidth[3]	Energy (kJ)[4]	Beams	Wavelemgth (nm)[5]	Operation[6]	References
France	CEA	Nd:glass	LF-id	Octal	ns	2	8	1064	1978	Andre et al. (1979)
		Nd:glass	LF-id	Phebus	ns	2	20	1053	1986	Buresi et al. (1985)
		Nd:glass	LF-id	LMJ	ns	450[7] (1800)	56[7] (240)	351	2019	Miquel et al., (2016) Denis et al. (2021)
	LULI	Nd:glass	HEDS	nano2000	ns	1/b	2	1053	2006	Zou et al. (2008)
UK	CLF	Nd:glass	HEDS	Vulcan	ns	2.6	8	1053	1984	Ross et al. (1981)
	AWE	Nd:glass	LF/HEDS	Orion	ns	5	10	351	2013	Hopps et al. (2015)
					ps	(1/b)	2	1054		
Russia	VNIIEF	Iodine	LF	ISKRA V	ns		12	1315	1990	Kirillov et al. (2000)
China	NLHPLP	Nd:glass	LF/HEDS	SG-II	ns	0.75/b	8	1053/351	2000	Zhu et al. (2018)
		Nd:glass	LF/HEDS	SG-II 9th	ns	5	1	1053	2006	Zhu et al. (2018)
		Nd:glass	LF/HEDS	SG-II UP	ns	3/b	8	351	2014	Zhu et al. (2018)

(continued)

Table 6.1 (continued)

Country	Institute[1]	Laser	Objective[2]	Facility	Pulsewidth[3]	Energy (kJ)[4]	Beams	Wavelemgth (nm)[5]	Operation[6]	References
	LFRC	Nd:glass	LF-id	SG-III	ns	180	48	351	2015	Zheng et al. (2016)

Acronyms and Notes: (1) LLNL—Lawrence Livermore National Laboratory, LLE—Laboratory for Laser Energetics, University of Rochester, LANL—Los Alamos National Laboratory, NRL—Naval Research Laboratory, ILE—Institute of Laser Engineering, Osaka University, CEA—Commissariat a 'l'Energie Atomique et aux énergies alternatives, LULI—Laboratory for the User of Intense Lasers, CLF—Central Laser Facility, Rutherford Appleton Laboratory, AWE—Atomic Weapons Establishment, VNIIEF—Russian Federal Nuclear Center, NLHPLP—National Laboratory on High Power Laser and the Physics, LFRC—Laser Fusion Research Center, Chinese Academy of Engineering Physics. (2) LF—LF-id: indirect-drive laser fusion, LF-d—direct-drive laser fusion. (3) Operation pulse width. (4) Laser energy in kJ for ns operation. (5) Laser wavelengths of 1064 nm and 1053 nm refer to Nd:silicate and Nd-phosphate glass lasers, respectively. 526 and 351 nm refer to the second- and third harmonics of 1053 nm. (6) Starting year of facility operation. (7) Performance in 2019 (from Denis et al., 2021)

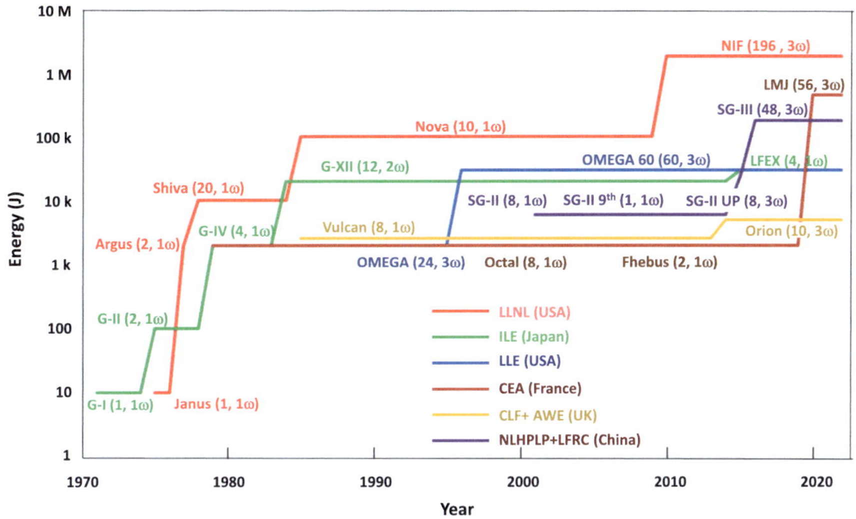

Fig. 6.1 Evolution of the laser energy of Nd:glass laser facilities since 1970: LLNL (red), CEA (deep red), ILE (green), LLE (blue), CLF and AWE (yellow), and NLHPLP and LFRC (purple). The facility names, the number of beams and the laser frequencies (in parentheses) are shown in the labels, where 1ω, 2ω and 3ω refer to the fundamental, the second harmonic, and the third harmonics of 1 micrometer wavelength, respectively

lasers, it was found that high-energy supra-thermal electrons (hot electrons) are generated in the interaction with low-density plasmas, leading to preheating of cryogenic fuels. The short wavelength laser light is absorbed at high plasma densities with high absorption efficiency, resulting in high ablation pressure (Garban-Labaune et al., 1981). Excimer lasers are inherently suitable as a laser fusion driver due its short wavelength and broad bandwidth, if high-power, high-energy excimer lasers are realized (Obenschain et al., 2015).

(d) Precise control of laser performance: With deeper understandings on various processes in laser fusion, it became clear that very precise control of laser parameters is necessary to achieve fusion ignition. Laser system is required to be flexible to realize the optimum conditions of various parameters such as the pulse shape, laser energy, laser wavelength, and focusing.

(e) Coherence reduction: In the direct-drive laser fusion, reduction of coherence of the laser beam is required for uniform irradiation of spherical targets and to reduce plasma instabilities (Kato et al., 1984; Lehmberg et al., 1987). It was found that coherence reduction is also important to control energy balance between the laser beams in indirect- and direct-drives, by reducing crossed beam energy transfer (CBET) arising from coupling of intersecting laser beams via laser-plasma interaction (Hüller et al., 2020).

(f) High-energy short-pulse lasers: Generation of high-energy, short-duration laser pulses became possible by implementing chirped-pulse amplification to high-energy Nd:glass lasers. The PW-class lasers with ~ 1-ps pulse duration have

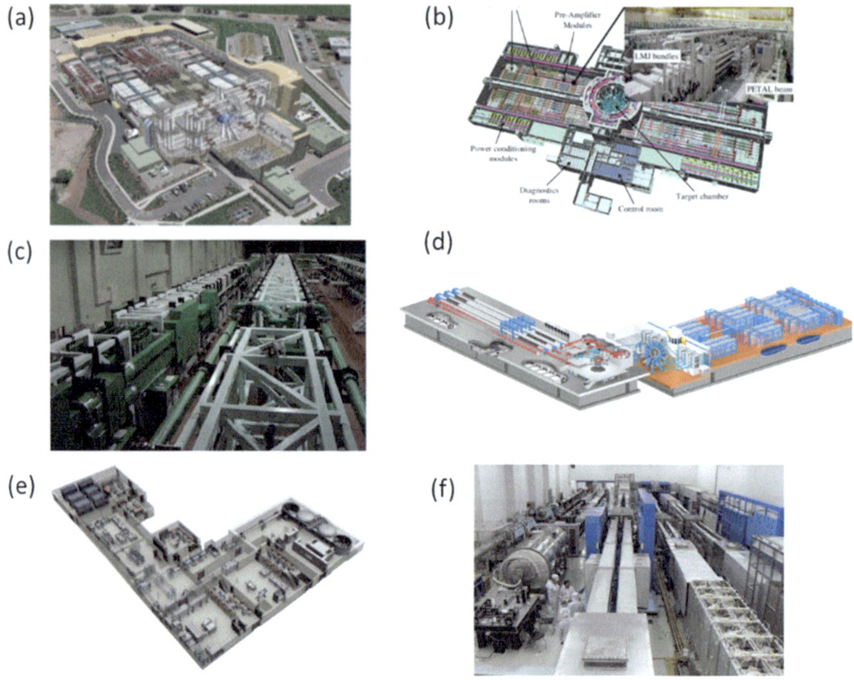

Fig. 6.2 High-energy pulsed lasers. **a** National Ignition Facility at LLNL. Reprinted from Moses et al. (2016), with permission from Taylor & Francis; permission conveyed through Copyright Clearance Center, Inc. © 2016 Taylor & Francis. **b** Laser Megajoule and PETAL at CEA. Reprinted from Miquel et al. (2016). © 2016 IOP Science, **c** Gekko-XII and FIREX at ILE (https://www.ile.osaka-u.ac.jp/eng/. © ILE Osaka University), **d** Omega and Omega-EP at LLE (https://www.lle.rochester.edu/omega-laser-facility-2/. © LLE University of Rochester), **e** Vulcan PW at Central Laser Facility (https://www.clf.stfc.ac.uk/Pages/Vulcan.aspx. © Central Laser Facility, Rutherford Appleton Laboratory), and **f** Shenguang-II at NLHPLP (http://english.siom.cas.cn/aboutsiom/Division_1/jointlaboratoryonhighpowerlaserandphysics/labintroduction_22236/. © National Laboratory for High Power Laser and Plasma, Chinese Academy of Sciences)

been developed for fast ignition, advanced radiography, and HED science as described in Sect. 7.4.

6.1.4 Architecture of High-Energy Nd:Glass Lasers

High-energy Nd:glass lasers are composed of multiple beam-line power amplifiers, where a laser beam generated at the front-end section is split into multiple beams. Each beam is expanded and amplified in several stages and further expanded to reduce the laser fluence at the target chamber where nonlinear crystals for frequency conversion and focusing optics are located, as shown in Fig. 4.10.

In Nd:glass lasers, the main power amplifiers are disk amplifiers, where elliptically shaped Nd-doped glasses placed at the Brewster angle to the beam propagation direction are pumped by high-energy xenon flashlamps (McMahon et al., 1973). Although laser glasses and flashlamps are cooled by dry nitrogen (or dry air), it takes more than one hour to recover them to a room temperature due to low thermal conduction of the laser glasses.

6.2 National Ignition Facility—High-Precision, High-Energy Laser System

The National Ignition Facility (NIF) was developed to achieve ignition of laser-driven inertial confinement fusion and to perform experiments on high energy density science (Moses et al., 2016). (Here "ignition" is defined as "gain greater than unity", i.e., the fusion energy exceeds the laser energy.) The NIF is the culmination of high-power laser developments at LLNL. It surpasses previous lasers in terms of laser performance and comprehensiveness for precision, flexibility, and operation. In this subsection we describe several unique features of NIF with emphasis on laser technology, since these features will be valuable as reference for developing high performance, complex laser systems for various applications. Detailed accounts of NIF are described in the *Special Issue on Development of the National Ignition Facility and Transition to a User Facility for the Ignition Campaign and High Energy Density Scientific Research* in Fusion Science and Technology **46** [1] 2016.

6.2.1 *Fusion Ignition with NIF*

A schematic diagram of the ignition target in indirect-drive laser fusion is shown in Fig. 6.3c (Lindl et al., 2014). The NIF laser is composed of 196 laser beams which are propagated in units of four beams (called "quad") from the laser to the target chamber, thus forming 48 quads in total. The 24 quads enter the hohlraum (empty cavity) through two laser entrance holes, i.e., 12 quads through each entrance hole, and irradiate the inner surface of the hohlraum. Since the hohlraum wall is made of high-z material, it acts as a blackbody cavity for the x-rays. The fuel capsule is irradiated with the x-rays with a radiation temperature reaching ~ 300 eV. The hohlraum is filled with low pressure helium gas to control the plasma instabilities and the irradiation symmetry by mitigating expansion of the wall plasma.

In entering the hohlraum, the 12 quads from one of the entrance holes form two inner cones of 23.5° and 30° (2 quads each) and two outer cones of 44.5° and 50° (4 quads each), where the angles are from the hohlraum axis. The inner cones and the outer cones irradiate the cavity wall close to and far from the fuel target (capsule). The laser wavelength of the inner cones is slightly detuned from that of the outer cones,

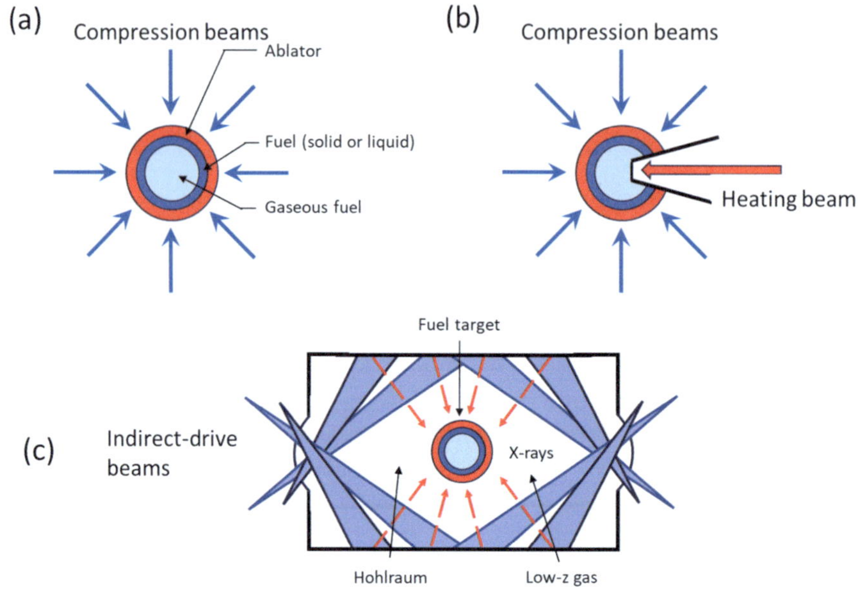

Fig. 6.3 Configurations of laser-driven inertial confinement fusion. **a** Direct-drive where a fuel capsule is irradiated directly by multiple laser beams. **b** Fast ignition where a high-density compressed fuel is heated with high-energy electrons or ions generated with a high-intensity heating laser beam. **c** Indirect drive where a fuel capsule placed in a cavity (hohlraum) is irradiated with x-rays generated at the cavity wall by irradiation with multiple indirect laser beams

to transfer the laser energy from the outer to inner cones via stimulated Brillouin scattering in the plasma for controlling the irradiation symmetry on target (Michel et al., 2010).

The fuel target is composed of an ablator, which is a spherical shell made of low-z material doped with higher-z material for preheat protection. A deuterium–tritium (DT) liquid or solid layer is formed inside the capsule at cryogenic temperature. The DT fuel is supplied through a fill tube of small diameter (< 10 μm) into the capsule, which is supported by thin plastic films in the cavity. The laser pulse, composed typically of four pulses, is precisely designed so that the shocks produced by the first three pulses converge to the capsule center simultaneously for adiabatic compression with the main drive pulse.

Based on this target design and the NIF conceptual designs, major parameters of the performance and configuration of the laser system were defined (Spaeth et al., 2016a, 2016b). For the indirect-drive laser fusion experiments, NIF is required to generate 1.8 MJ energy and 500 TW peak power at 0.35 μm on target, within 20 ns pulse-duration, focused on a cylindrical target in two cones per hemisphere with a $<$600 μm spot size and with beam smoothing.

Construction of the NIF building was started in 1997 and the NIF laser was completed in 2009. The National Ignition Campaign (NIC) was established in 2005 to develop wide range of capabilities during 2009–2012 such as target fabrication, cryogenic layering, plasma diagnostic systems, experimental platforms and other NIF infrastructures, with the primary goals of (a) performing integrated ignition experiments that would lead to ignition and burn, and (b) transition of NIF from project completion to a routinely-operating facility (Lindl et al., 2014).

By utilizing the NIF capabilities, the fusion yield has increased gradually: 26 kJ in 2013 due to alpha-particle self-heating achieved by increasing the foot pulse to reduce ablation-front instability (Hurricane et al., 2016); 53 kJ in 2017 exceeding the kinetic energy by using a high-density carbon capsule (Le Pape et al., 2018); 56 kJ in 2019 by increasing the capsule size and improving irradiation symmetry via cross-beam energy transfer (CBET) (Zylstra et al., 2021a); ~ 160 kJ in 2021 with finer CBET control and modification of the cylinder shape (Zylstra et al. 2021b); 1.35 MJ in August 2021 achieving the burning plasma with a fusion gain of ~ 0.7 (Zylstra et al., 2022); and 3.15 MJ in December 2022 achieving fusion ignition with a fusion gain of 1.5 (Abu-Shawareb et al., 2024).

6.2.2 *Architecture of NIF*

Figure 6.4 shows an overall layout of NIF, where 48 quads comprising 196 laser beams are grouped into four clusters (12 quads in each cluster) and placed in two laser bays (two clusters each in Laser Bay 1 and Laser Bay 2). The 48 quad beamlines are redirected at the switchyard and directed to a target chamber located in Target Area.

All large-aperture components of the laser and the final optics have a clear aperture of 40 cm × 40 cm, since each laser beam of NIF is square-shaped of ~ 37 cm × 37 cm which is compatible with the maximum size of the KDP crystals for 3ω conversion. To irradiate the target with 1.8 MJ energy at 3ω, each beam delivers ~ 9.4 kJ at the output of the final optics assembly. The final optics are exposed to a laser beam at an average fluence of > 8.5 J/cm^2 at 3ω, and the laser beamline optics are exposed at > 10 J/cm^2 at 1.05μm. The NIF is expected to keep this performance in routine operation over its operational lifetime of at least 30 years. The output energy of NIF has been increased gradually and is now operated at over 2 MJ at 3ω, enabling to achieve fusion ignition.

The NIF has been successful in operating the whole system of laser, target fabrication and plasma diagnostics since 2009, generating significant results on laser fusion and high energy density science. In the following several important features of NIF are described which support these outcomes.

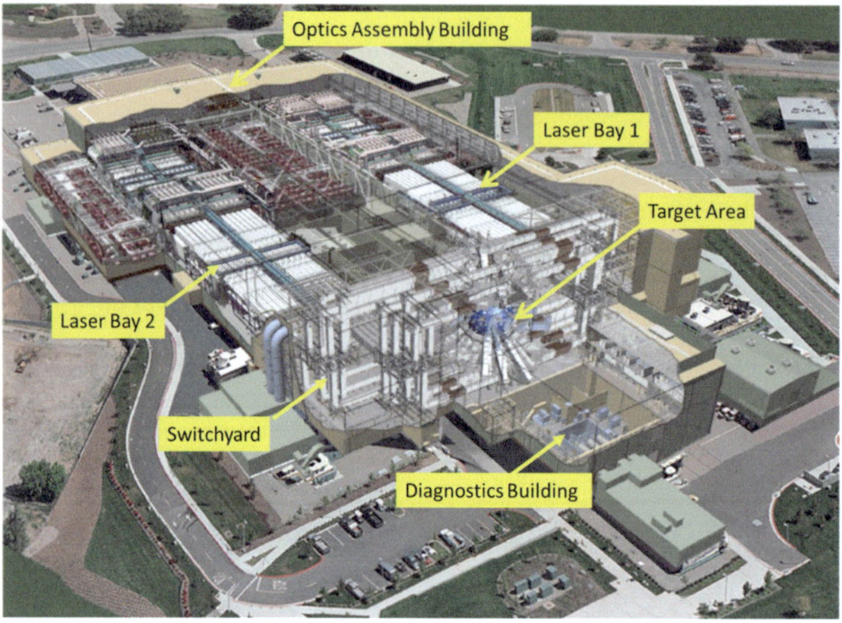

Fig. 6.4 A cutaway view of the NIF laser facility. The NIF laser is installed in Laser Bay 1 and Laser Bay 2. The front end of NIF is in an area between these Laser Bays. The output beams are directed to the Target Area via Switchyard. Experimental data are collected and analyzed in Diagnostic Building. Optical components are assembled and recycled at Optics Assembly Building. The footprint of the NIF building is approximately 150 m in length × 95 m in width. Reprinted with modification from Moses et al. (2016), with permission from Taylor & Francis; permission conveyed through Copyright Clearance Center, Inc. © 2016 Taylor & Francis

6.2.3 NIF Laser System

6.2.3.1 Beam Propagation in NIF

Figure 6.5 shows a layout of one of the 196 laser beams, comprising laser amplifiers, transport mirrors in the switchyard, final optics assembly, and the target. A well-controlled laser pulse prepared in the front end (not shown in this figure) is injected into the transport spatial filter close to its focal point and amplified by a power amplifier (comprising 5 laser disk glasses). The amplified pulse is injected into a laser cavity composed of two cavity mirrors by switching the laser polarization with a plasma electrode Pockels cell (PEPC). The laser pulse is amplified by the main amplifier (comprising 11 laser glasses) by 4 passes in this cavity. After ejection from the amplifier cavity, the laser pulse is amplified again by the power amplifier and sent to the final optics assembly through the transport spatial filter. The wavefront aberration is corrected by a deformable mirror which is one of the cavity mirrors. The large-aperture PEPC is switched with a glow-discharge plasma electrode which

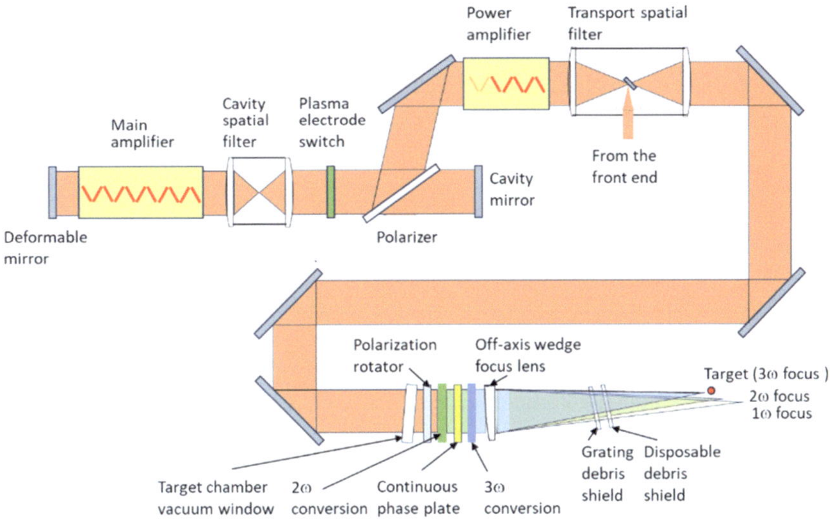

Fig. 6.5 Schematic layout of the laser system and the final optics assembly. A laser beam from the front end is injected into the amplifier chain at the focus of the transport spatial filter. The amplified pulse is frequency converted and focused on the target with the final optics assembly

uniformly applies ~ 15 kV in ~ 100 ns to a z-cut 1-cm thick KDP crystal (Rhodes et al., 1995).

This laser architecture enables the unification of optical component specifications leading to reductions in optics storage and fabrication time. It also enables simplification of the mechanical layout enabling parts replacement in a bundle as a unit, resulting in a reduction of the maintenance time.

At the final optics assembly, shown in the lower half in Fig. 6.5, the 1ω laser beam is converted to 3ω by first generating 2ω with a KDP (2ω conversion) and then mixing 1ω and 2ω with a deuterated KDP (DKDP) in type I/type II configuration (3ω conversion, see Sect. 5.2.2). The 3ω laser beam is focused with an off-axis wedged focus lens to a point which is separated from the 1ω and 2ω focus points due to color dispersion. A continuous phase plate is inserted to control a beam profile and to smooth the beam on target as described later. The polarization of the laser beam could be rotated with a polarization rotator for added smoothing by polarization randomization. A grating debris shield protects up-stream optics from the target debris, while splitting a small fraction of the 3ω beam for diagnostics. Each of these optics is housed in a lightweight unit for rapid exchange.

6.2.3.2 Control of Laser Damage

Unique strategies are implemented to keep operating the laser at its full specification, or even higher when necessary, over its operational lifetime (Spaeth et al., 2016a,

2016c). Mainly three approaches are taken in this strategy: (a) reducing the damage probability in fabricating the optics, (b) shadowing of damage sites, and (c) optics recycle loop to reuse the optics as much as possible by repairing the damage in short term without purchasing new optics.

(a) **Reducing the Damage Probability**

The final optics components are most susceptible to laser damage since they are exposed to high fluence at short wavelengths. The fused-silica optics and KDP/DKDP in the final optics assembly are fabricated to reduce the damage density to less than ~ 0.03 sites/cm^2, equivalent to 50 damage sites over 50 cm × 50 cm aperture at the NIF 3ω fluence (see Sect. 4.4). To achieve this level, inclusion-free fused-silica substrates are finished by griding, polishing and chemical treatment to minimize damage-inducing particles and scratches at and near the surface. All KDP/DKDP crystals are laser conditioned by raster-scanning with a 355 nm laser. These optics are coated with a silica sol–gel anti-reflection coating of high damage threshold (Baisden et al., 2016).

(b) **Shadowing of Damage Sites**

Although highly damage resistant optics are installed in NIF, damages are still generated since the laser is routinely operated at high IR and UV fluences. Once laser damage is generated even in small sizes, the damage size grows exponentially with the number of shots under repeated irradiation. The damage in an optic also leads to damages to other optics due to diffraction and focusing caused by the damage sites. It is important to regularly monitor the growth of the damage before it becomes unrepairable or dangerous to other optics.

The optics in the final optics assembly in NIF are inspected after each laser shot with a damage inspection system, with which location and size of every damage are measured with a spatial resolution of ~20 μm. When the size of the damage grows to an unrepairable size (typically ~300 μm), a part of the laser beam irradiating the damage site is blocked. A programmable spatial shaper is placed in the front end of the laser beam to shadow the damage site from laser irradiation to prevent further growth of the damage (Spaeth et al., 2016a). The programmable spatial shaper is an optically addressable light valve, illuminated by ultraviolet light through a digitally addressable liquid crystal modulator, to generate a shadow with soft edge not to induce Fresnel diffraction. When the beam area covered by multiple shadows becomes a significant fraction of the total beam area, the optics module containing the damaged optics is removed and exchanged by a repaired or new optics.

(c) **Optics Recycle Loop**

The damaged optics removed from the final optics assembly is decontaminated, if necessary, cleaned, and inspected. If the damage size is within ~ 1 mm, the damage site is repaired by CO_2 laser ablation, where fused quartz is ablated by irradiation of tightly focused, high irradiance, short duration (~10 μs), 10.6 μm-wavelength laser pulses to form a conical shape, with which a laser beam is refracted without leading to damage formation in this and other optics. With an automated station, this repair process takes only a few minutes.

The optics which cannot be repaired at NIF are sent to the vendor for refinish or replaced with new optics. The optics recycling loop has been operating at a throughput of 40 optics per week, sufficient to keep operating NIF at the fluence above the damage thresholds of various optics (Spaeth et al., 2016b).

6.2.3.3 Front End of NIF

The front end of NIF is called an injection laser system (ILS) (Wisoff et al., 2004). The NIF has 48 ILS units for the 48 quads so that each quad can be operated independently at different laser specifications to meet various requirements. Each ILS is composed of a master oscillator room (MOR), a regenerative preamplifier, a four-pass preamplifier, an input sensor package, and a preamplifier beam transport system (PABTS).

Each MOR installs three master oscillators operated at 3 different wavelengths, one for the outer cone and two for the inner cones, so that the laser beam is sent to a specified cone with proper wavelength. The laser beam from the oscillator is amplitude-modulated at 3 GHz to prevent the damage due to transverse stimulated Brillouin scattering in downstream large optics. The laser beam is also frequency-modulated at 17 GHz for beam smoothing as described later. A laser pulse with arbitrary temporal pulse shape is generated with a 2-stage electro-optic modulator, providing a total pulse width up to 35 ns for each quad. In addition, short (~ 88 ps) impulses are provided to the shaped pulse to generate fast rise-time pulses.

A ~ 1 nJ laser pulse from the master oscillator is amplified to ~ 10 mJ with a regen with high energy stability of better than 0.5% rms for all pulse shapes. The output from regen is spatially expanded by 30 × and shaped with a serrated soft aperture and a beam shaping mask to pre-compensate for spatial gain nonuniformities in the downstream amplifiers. The location of the beam shaping mask is the starting point for relaying the beam profile throughout the beam propagation. The programmable spatial shaper for shadowing the damage sites in the final optics assembly and other optics is also placed at this point.

The ~ 10 mJ output from regen is amplified to ~ 6 J with a four-pass flashlamp-pumped Nd:glass preamplifier. A diffraction grating may be inserted, replacing the end mirror of the 4-pass cavity, to provide angular dispersion for beam smoothing on target. The amplified pulse from the preamplifier is monitored with input sensors and split into 4 beams to be injected into the four main amplifiers of a quad after adjusting the timing between the beams.

6.2.3.4 Beam Smoothing

Uniform target irradiation is required in the direct-drive as well as indirect-drive laser fusion. To mitigate irradiation nonuniformities near the focal region of a coherent laser beam arising from various factors such as diffraction, interference and aberration, Kato et al (1984) has proposed using a random phase plate, which consists of a two-dimensional array of transmitting areas each of which applies a phase shift randomly chosen between 0 and π radian. With the random phase plate, irradiation profile at the far field is a superposition of envelopes determined by the array pattern and the speckles due to spatial coherence of the laser beam. Several approaches have been proposed to reduce the speckle by rapidly scanning the speckles through modulation of the phase in time and space: induced spatial incoherence (Lehmberg & Obenschain, 1983), beam smoothing by spectral dispersion (SSD, Skupsky et al., 1989), SSD with random phase plate (Rothenberg, 1997), and three-directional spectral dispersion (Miyaji et al., 2002). (Detailed accounts of these methods are described in Part II, Chap. 9.)

In NIF, 1-D SSD is implemented for beam smoothing, where the laser beam is frequency-modulated at 17 GHz and spectrally dispersed with a reflection grating in the four-pass preamplifier located in the front end. The phase plate used in NIF is a continuous phase plate (CPP), which is an aggregation of small weak positive and negative lenses (Dixit et al., 1996). By properly designing the surface profile of the CPP, it is possible to control the shape, size, and irradiation distribution at the far field, for example, for optimizing the irradiation on the laser entrance hole of the hohlraum. The topological surface of the CPP is fabricated with high precision by polishing a fused quartz substrate with magneto-rheological fluid (MRF) finishing (Baisden et al., 2016).

6.2.3.5 Laser Operation at Specified Performance

The operation of NIF to deliver laser outputs to the target at specified parameters for each quad, such as pulse shape, energy, power, and power balance, is supported by a software named Laser Performance and Operations Model (LPOM) (Shaw et al., 2004, 2008). When the laser output parameters are specified, the LPOM calculates the settings for the shots for the whole laser system from the master oscillator to the target, as well as the settings for various diagnostics along the beamlines. LPOM also analyzes the shot results and archives the laser performance data for every shot to improve the laser operation over long periods of time.

The laser beam propagation throughout the whole laser system to the target is calculated with a code with the algorithm described in Sect. 4.2.6. Detailed information for all elements such as the measured properties of the optics and the data from monitoring stations are used as input parameters for calculation. Since the output performance (pulse shape and energy) is specified for each quad, the beam propagation is calculated backward from the target to the laser frontend. The calculation of the beam propagation is described by a series of transfer functions and inverse transfer functions that relate to the pulse input and output at each section. For example, the inverse transfer function of the Frantz–Nodvik equation is written as

$$\varphi_{In} = \varphi_s \ln\left\{\frac{1}{G_0}\left[\exp\left(\frac{\varphi_{Out}}{\varphi_s}\right) - 1\right] + 1\right\} \tag{6.1}$$

where φ_{In} and φ_{Out} are the input and output fluences, φ_s is the saturation fluence, and G_0 is the small signal gain. With this relation, the input fluence is determined from the output fluence.

6.3 Diode-Pumped High-Average-Power Lasers

Operation of the high-energy lasers described above are restricted to ~ several shots in a day, limited mainly by wavefront distortion due to heat deposition in the laser glasses. The repetitively pulsed diode-pumped solid-state lasers (DPSSLs) became available due to recent progress in performance and cost reduction of high-power laser diodes. The high-power DPSSLs, initially developed toward realizing a driver for inertial fusion energy systems, are now developed for broader applications: laser material processing, high-brightness electromagnetic and particle generation, and ultrafast high-intensity laser pumping. In this section, we first describe general descriptions on pumping efficiency and thermal effects in solid state lasers. Then various architectures for realization of repetitively pulsed high-power DPSSLs are introduced. Recent progress in new-type high-power laser diodes are described in the final section.

6.3.1 *Pumping Efficiency and Thermal Effects*

Major limitations to the repetition rate of the high-energy pulsed lasers are related to the pumping efficiency and the thermo-optical effects due to heat deposition in the laser material (Koechner, 2006). We consider a case where energy stored in a diode-pumped amplifier is extracted by stimulated emission with a signal laser beam. Figure 6.6a shows a general configuration for amplification of an input laser beam in a laser material which is pumped by multiple LD beams. The power flow for this case from the electrical input to the laser output is shown in Fig. 6.6b. The pump beam from the laser diodes (electrical efficiency η_p) is transported with transport optics (transport efficiency η_t) and absorbed in the laser material (absorption efficiency η_a), where the absorbed energy is stored as the population inversion. The efficiency for creating the population inversion is determined by the Stokes factor, or quantum efficiency, $\eta_S = (h\nu_L)/(h\nu_p) = \lambda_p/\lambda_L$ where the suffix p and L refer to the pump and signal laser respectively, and the branching ratio η_b, ratio of the upper-laser-state and the pump-band. The energy stored as population inversion is extracted by the

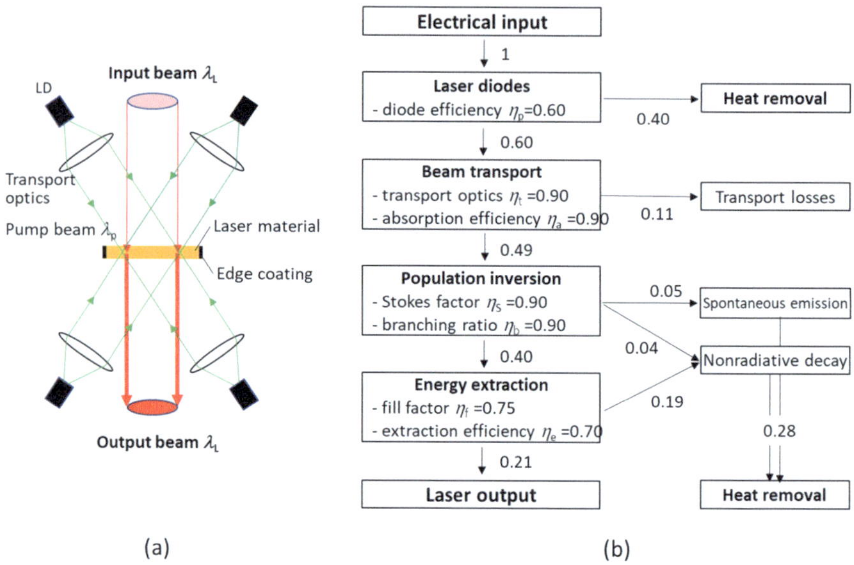

Fig. 6.6 **a** A general configuration for amplification of an input laser beam in a laser material pumped by multiple LD beams. **b** Power flow from the electrical input to the laser output corresponding to the layout shown in (**a**), where the efficiency parameters are given with feasible values for estimating the individual and overall efficiencies. Also shown are the energy flow which are dissipated as heat and radiation

signal laser beam of a fill factor $\eta_f \cong A_L/A_p$ (area ratio of the input laser beam and the pumped region) with the extraction efficiency η_e which is determined by the laser fluence and the saturation fluence (see Sect. 4.2).

With laser diode pumping, it is possible to achieve higher efficiencies compared with flashlamp pumping due to several factors: high intrinsic efficiency of the laser diodes (high η_p), good matching of the diode wavelength to the pump band which is close to the upper laser state (high η_a, η_s and η_b), efficient pump absorption by multi-passing the pump beam in the amplifier (high η_a), and optimum configurations for the pump-beam transport (high η_t) and the stored-energy extraction (high η_f and η_e). In Fig. 6.6b, the power flow is estimated, normalized by the electrical input, using feasible values of the efficiency parameters for a Yb:YAG amplifier which has large η_s and η_b. (Detailed investigation is necessary for each case to evaluate exact values of these parameters.) In this diagram, the efficiency of each unit written in the square is relative to the input to this unit, whereas the overall efficiencies relative to the electrical input are written adjacent to the arrows outside the squares. In this example, an optical-to-optical efficiency (laser output power / laser diode power) of 35% and an electrical efficiency (laser output power / electrical input) of 21% are expected for this amplifier.

The energies not extracted as the laser output are lost as heat and radiation, as shown schematically in the right-hand side of Fig. 6.6b with typical values for each

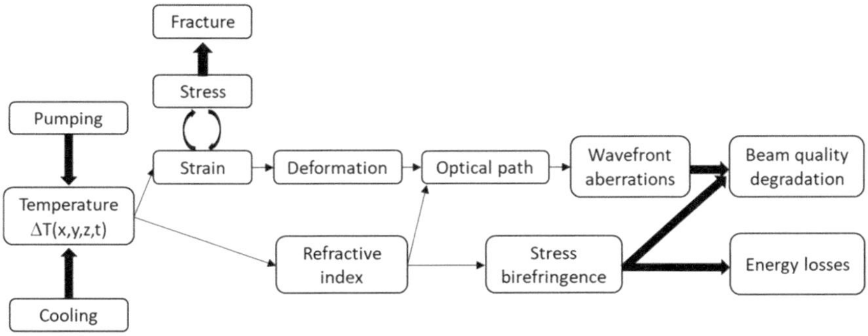

Fig. 6.7 Thermal effects in solid-state lasers showing the relations of the mechanical and optical properties caused by spatially-nonuniform temperature increase in the laser material

step. The figures described adjacent to the arrows are relative to the electrical input. Approximately 40% of the electrical energy is deposited as the thermal energy in the laser diodes. Part of the laser diode energy (~ 11% of the electrical input) is lost during beam transport due to reflection and scattering in the transport optics and the laser medium. Approximately 28% of the electrical input energy is lost in the laser medium, in which the energy loss by nonradiative decay (23%) is converted to the thermal energy in the laser material, whereas the spontaneous emission in the laser material (5%) is absorbed in the edge coating surrounding the laser material when we can neglect the spontaneous emission emitted out of the laser material.

The thermal energy deposited in the laser material and the edge coating result in various adverse effects to the laser beam: thermal lensing, stress-induced birefringence, aberration of the laser beam, and fracture of the laser material. Efficient heat removal is an important consideration in designing the repetitively pulsed high-power lasers. The opto-mechanical effects due to temperature increase in the laser material and edge coating are shown in Fig. 6.7 (Chénais et al., 2006).

The time-dependent temperature distribution $T(x, y, z, t)$ due to heat deposition $S(x, y, z, t)$, where z is the beam propagation direction and t is time, is determined by the energy-conservation law,

$$\rho C_p \frac{\partial T(x, y, z, t)}{\partial t} - K_{th}(T)\nabla^2 T(x, y, z, t) = S(x, y, z, t) = \eta_h \alpha I(x, y, z, t) \quad (6.2)$$

where ρ is the density, C_p is the specific heat capacity, K_{th} is the thermal conductivity, α is the absorption coefficient at the laser wavelength, and η_{h} is the fraction of the absorbed energy transferred to heat. The laser intensity at z, $I(x, y, z, t)$, is calculated by considering the saturation effect along propagation over z, as discussed in Sect. 3.2. For a LD face-pumped laser amplifier of Yb-doped crystals, Tamer et al. (2019) have shown that the temperature distribution in the amplifier numerically calculated with this model using proper material parameters agrees well with the experimental measurements to within 0.5 °C.

The heat deposited in the laser material is removed by close contact with a "radiator" whose temperature is kept low. The absolute temperature inside the laser crystal is determined by specifying the boundary condition where the thermal flux across the boundary is continuous:

$$\boldsymbol{j}_q \cdot \boldsymbol{n} = -K_c \frac{\partial T}{\partial n} = H(T - T_c) \tag{6.3}$$

where $\boldsymbol{j}_q$ is the thermal density flux, $\boldsymbol{n}$ is a unit vector normal to the boundary surface, H is the heat transfer coefficient, and T_c is the temperature of the radiator. Various approaches are taken for efficient cooling to realize high heat transfer coefficient H over a large area, as described in the following sections.

When the temperature distribution in the material is determined, we can calculate the thermal stresses and strains. The strain tensor ε_{ij} is related to the stress tensor σ_{ij} and the thermal expansion as,

$$\varepsilon_{ij} = S_{ijkl}\sigma_{kl} + \alpha_{ij}\Delta T \tag{6.4}$$

where S_{ijkl} is the compliance tensor, α_{ij} is the thermal expansion coefficient tensor, and ΔT is the temperature increase due to pumping. The laser material starts to fracture when the stress exceeds the tensile stress at a location of the highest stress and/or the lowest tensile stress.

Considering a laser slab of refractive index n and thickness d, the optical path increases from nd to $nd + \Delta(nd)$ where $\Delta(nd) = nd\left[\left(\partial n/\partial T\right)_\varepsilon/n + \alpha_l\right]\Delta T$, in which $\left(\partial n/\partial T\right)_\varepsilon$ is the thermal coefficient of the refractive index at constant strain and α_l is the linear thermal expansion coefficient. Due to non-uniform spatial distribution of the heat source and the cooling flux, aberrations are generated to the laser beam passing through the laser medium, such as second-order radial (thermal lensing), spherical, astigmatism, and higher-orders. In the case that the laser beam is reflected at the back surface, such as active mirrors and thin disks, the wavefront of the laser beam is affected directly by the deformation of the back surface.

The refractive index of a material is described by an index ellipsoid (or indicatrix) where principal dielectric axes are related to the crystal symmetry (see Sect. 2.2.4). When the material is under stress, the index ellipsoid is perturbed due to stress birefringence (photo-elastic effect). Considering an optically isotropic material (for simplicity) with the principal axes of the stress tensor chosen as x and y, the index ellipsoid becomes elliptical in this plane with the refractive indices n_x and n_y whose difference is given by

$$n_y - n_x = \frac{1}{2} n_0^3 (q_{11} - q_{12})(\sigma_{xx} - \sigma_{yy}) \tag{6.5}$$

where n_0 is the refractive index without stress, σ_{xx} and σ_{yy} are the stress tensors, and q_{11} and q_{12} are the stress-optical coefficients (Born & Wolf, 1999, Chapter XV). (Note that contracted notations are used for the subscripts of q_{ij} as described in

Sect. 5.1.2.) The refractive index anisotropy results in the phase difference between the x- and y-polarized waves as

$$\delta = \frac{2\pi}{\lambda} d\left(n_y - n_x\right) = \frac{\pi d}{\lambda} n_0^3 (q_{11} - q_{12})(\sigma_{xx} - \sigma_{yy}) \tag{6.6}$$

When this material is placed between a polarizer and an analyzer with the same polarization directions but angled to the stress-tensor axes, the intensity distribution of a laser beam becomes nonuniform after passing through the material. This stress birefringence is important in the design and experimental evaluation of the amplifiers for repetitive high-power lasers.

6.3.2 Various DPSSL Architectures

Various DPSSL architectures have been invented to maximize the laser power, repetition rate and beam quality, and to achieve high efficiency and low thermal effects. In this section, we describe typical DPSSL architectures: gas-cooled multi-slab amplifiers, active mirrors, and thin disks.

6.3.2.1 Face-Pumped, Gas-Cooled Multi-slab Amplifier

The face-pumped, gas-cooled multi-slab amplifier refers to the architecture where multiple slabs of a laser material are pumped at their broad surfaces by laser diodes and cooled by rapid gas flow through their slab faces. This architecture was developed at the Lawrence Livermore National Laboratory (LLNL) as a driver for inertial-confinement fusion energy reactors, with the initial performance goals of 10% electrical efficiency, 10 Hz repetition rate, 100 J energy/pulse at 2–10 ns pulse length, 1.047 μm wavelength, and 10^7 shot life (Krupke, 1989; Orth et al., 1996; Payne et al., 1998).

In a prototype system named Mercury, seven slabs of Yb-doped strontium fluoroapatite (Yb:S-FAP, $Yb^{3+}:Sr_5(PO_4)_3F$) crystals, with dimensions of 4 cm × 6 cm × 0.75 cm, are face-pumped by four of 100 kW laser diode arrays in a configuration similar to Fig. 6.6a. The Yb:S-FAP slabs, separated by 1-mm-wide gas-cooling channels, are cooled by turbulent subsonic helium gas flow at room temperature which removes heat with 1–3 W/cm^2 rate from the faces of each slab. Helium is used for cooling due to its high thermal conductance and small perturbation to the laser beam due to its low refractive index. Since LDs are power-limited, the Yb:S-FAP is chosen due to its long radiative lifetime (1.12 ms). The energy stored in the slabs are extracted by a laser beam in four passes at an average fluence of 8 J/cm^2 at the final pass. With a front-end and two amplifier units, Mercury has operated at 55 J output at 10 Hz for 7 h (Bayramian et al., 2007).

The Mercury output was frequency-doubled with an yttrium calcium oxyborate (YCOB) crystal, generating 31.7 J/pulse at 10 Hz (average power of 317 W). The HAPLS (High-repetition-rate Advanced Laser System), outgrowth of Mercury, was used as a pump laser for a 10 Hz, PW-class high-average-power Ti:sapphire laser system at Extreme Light Infrastructure Beamlines in Czech Republic (Bayramian et al., 2008). In the HAPLS, a gas-cooled multi-slab Nd:glass amplifier is used to pump a 1-PW CPA Ti:sapphire laser with the performance of 30 J, 30 fs and 10 Hz (Häfner et al., 2017).

The face-pumped, gas-cooled multi-slab amplifier architecture was applied to develop next generation high-energy PW-class lasers at the Central Laser Facility of Rutherford Appleton Laboratory in UK. In the 10 Hz, 100-J laser system named DiPOLE100, developed in cooperation with HiLASE in Czech Republic (Divoky et al., 2014), multiple Yb:YAG ceramic slabs are diode-pumped and face-cooled by a helium gas flow at a cryogenic temperature (Mason et al., 2015). In addition to the intrinsic favorable properties of Yb:YAG such as high quantum efficiency and relatively long fluorescence lifetime, the cryogenic operation of Yb:YAG provides added advantages over room temperature operation such as reduced population of the lower lasing level and excellent thermo-mechanical properties of the host crystal as described in Sect. 4.3.2.2.

The DiPOLE architecture is based on the analyses by Ertel et al. (2011) to determine the optimum parameters for the high optical-to-optical efficiency with a face-pumped, cryogenic gas-cooled multi-slab Yb:YAG amplifier with the configuration shown in Fig. 6.8. In this concept, multiple Yb:YAG slabs, each surrounded by Cr^{4+}-doped YAG absorptive cladding and higher doping toward the middle, are cooled by rapid cryogenic helium gas flow on their faces. These faces are used for LD pumping and energy extraction of the laser beam. The optimum parameters for the pump (spectrum, intensity, duration, number of passes) and the laser slab (doping distribution among the slabs to balance the pump absorption and heat loading, slab thickness) were derived by considering the atomic processes (absorption and emission spectra, quantum defect, fluorescence decay, amplified spontaneous emission, and reabsorption of the laser emission from the lower laser level) and the practical limitations (pump brightness and the damage threshold).

The main amplifier of DiPOLE100 consists of six square-shaped Yb:YAG ceramic slabs of 120 mm width and 8.5 mm thickness with the Yb doping of 0.4, 0.6 and 1.0 at. % from the outside to the center, respectively. The slabs are pumped on-axis from both sides, each with 940-nm LD of 280 kW peak power and up to 1.2 ms duration at 10 Hz repetition rate. Each LD beam is centered at the amplifier, with a 79 mm square flat-top-profile. A laser pulse of 1030 nm wavelength with a square-shape of 75 mm × 75 mm extracts the stored energy in four passes, resulting in the output energy of 107 J when pumped with LDs of 506 J energy and 0.9 ms duration, corresponding to an optical-to-optical efficiency of 21%, at a repetition rate of 1 Hz which was limited due to insufficient cryo-temperature of 175 K, higher than a design value of 150 K (Banerjee et al., 2016). With improved cooling to 150 K, the 100 J, 10 Hz (1 kW) operation with 22% efficiency over 4.3×10^4 shots was demonstrated (Mason et al., 2017), enabling development of a 1 kW laser system (DiPOLE100X) with

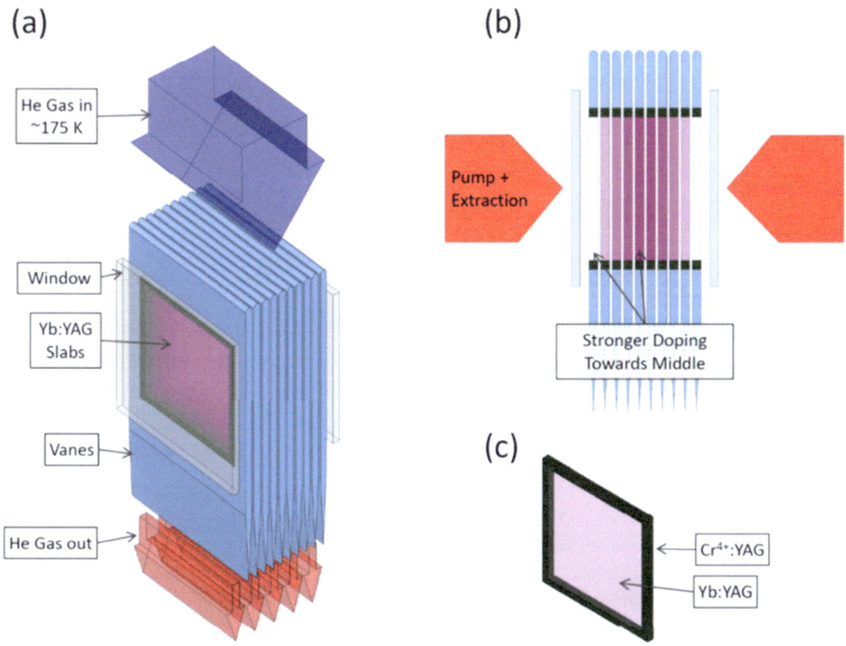

Fig. 6.8 Concept of a face-pumped, helium-gas-cooled multi-slab Yb:YAG amplifier, with overall view (**a**), side view (**b**), and individual amplifier slab (**c**). Reprinted from Ertel et al. 2011. © 2011 Optical Society of America

full capability for pulse shaping and wavefront control for high density compression experiments at the European X-ray Free Electron Laser facility (Mason et al., 2018). The scale-up of the DiPOLE architecture was demonstrated to higher energies of 150 J at 1 Hz (Banerjee et al., 2020) and 253 J at 0.2 Hz (Sekine et al., 2022).

Also demonstrated is the frequency up-conversion of the DiPOLE100 with large aperture LBO crystals to 525 nm and 343 nm, with average power of 597 W (59.7 J, 10 Hz) and 65 W (65.0 J, 1 Hz) and the harmonic conversion efficiencies of 66% and 68%, respectively (Phillips et al., 2021). Including the 940 nm pump-to-1030 nm laser efficiency of 22%, the overall pump-to-343 nm conversion efficiency is 15%.

6.3.2.2 Active Mirrors

Another approach to generating high energy laser pulses is the active mirror, which was developed originally at the Laboratory for Laser Energetics, University of Rochester for laser fusion research (Abate et al., 1981; Brown et al., 1981). It is composed of a large disk-shaped Nd:glass of several-10-cm in diameter and several-cm in thickness with its periphery surrounded by a fluorescence-absorbing cladding. A laser beam enters into the laser disk from an anti-reflection-coated front surface

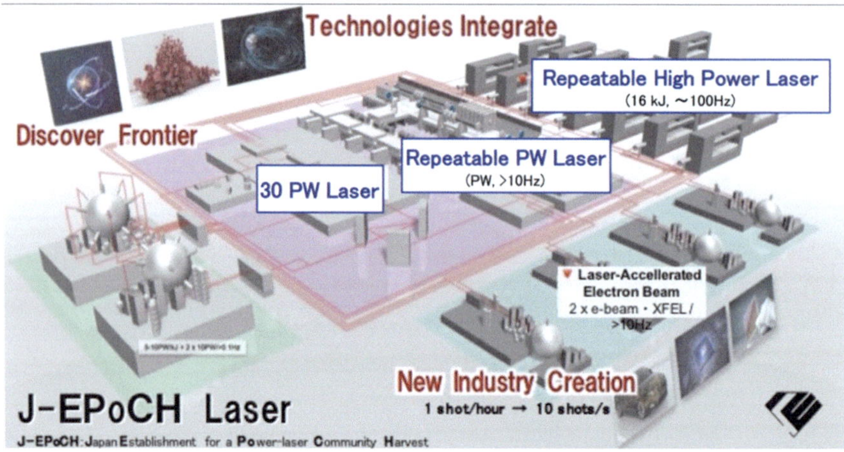

Fig. 6.9 A conceptual design of the multi-purpose high-repetition laser facility J-EPoCH, which is comprised of 160 beams of 100J/100Hz laser modules to provide high-energy, high-peak-power and high-repetition lower power laser beams for energy, science and industrial applications. Curtesy of the Institute of Laser Engineering (ILE), Osaka University https://www.ile.osaka-u.ac.jp/ja/research/project/high-powerlaser/index.html. © ILE Osaka University

and is reflected at a high-reflection-coated back surface for double-pass amplification. The laser disk is optically pumped by flashlamps and its heat is removed with a liquid coolant, both through the back-side surface.

A repetitive-pulsed high-energy laser system based on the LD-pumped active mirrors (Kawanaka et al., 2008) is developed as a part of the next generation multi-purpose laser facility J-EPoCH at the Institute of Laser Engineering, Osaka University (Kodama, 2021). Figure 6.9 shows a conceptual design of J-EPoCH, where 160 beams of 100 J/100Hz laser modules (named SENJU) provide twelve 8 kJ/ns beams for high density science, one 30 PW beam for high field science, and many fs and ns beams for discovery science, technology integration, and industry creation.

As the first step of the J-EPoCH development, a 10-J, 100-Hz, 1 kW average power laser based on active mirrors has been demonstrated (Ogino et al., 2022). In this laser system, whose layout is shown in Fig. 6.10a, 100 Hz, 530 mJ, 10 ns, 32-mm diameter (hexagonal shaped) laser pulses, generated in the front end and the preamplifiers, are amplified to 100 Hz, 10-J, 10-ns pulses with double-passed six-stage active-mirror amplifiers. In each active mirror, shown in Fig. 6.10b, a Yb:YAG ceramic slab of 45- (heads1 ~ 4) or 50-mm (heads 5 and 6) diameter and 7-mm thickness with Yb concentration of 1.0% or 0.6%, is optically bonded to a thick Cr^{4+}:YAG edge-cladding to absorb ASE, and is housed in a copper cooler-jacket. This unit is mounted via a molybdenum plate to a copper heat sink, which is cooled with flowing liquid nitrogen at a temperature of 77 K.

Each Yb:YAG ceramic is pumped with three 940 nm laser diodes with a total peak power of 83 kW in 0.5 ms at 100 Hz. The pump light is almost completely absorbed in a single pass through the 7-mm-thick Yb:YAG. It was found from simulation and

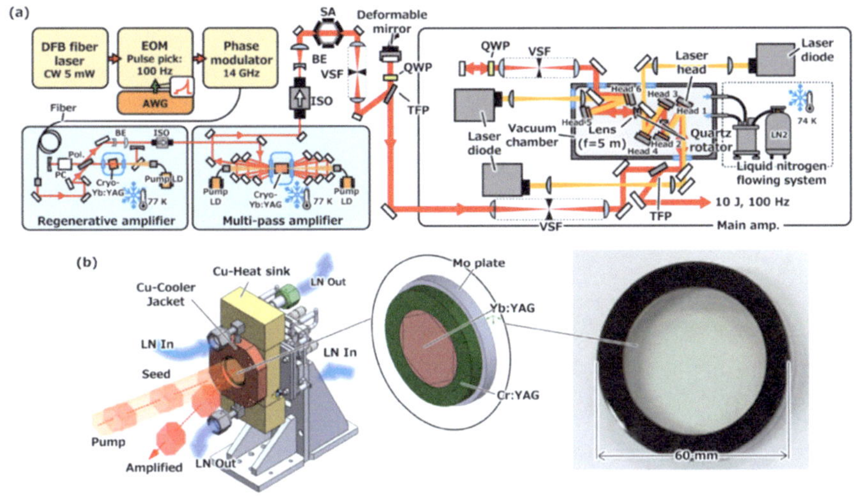

Fig. 6.10 **a** System layout of the 10J/100Hz laser where cryogenically cooled Yb:YAG ceramic active mirrors are used in the main amplifiers. Seed pulses generated in the front-end and amplified with regenerative and multi-pass amplifiers are amplified with the main amplifier composed of 6-stage active-mirror amplifiers. **b** Configuration of the active mirror amplifier, where a Yb:YAG ceramic with Cr:YAG edge coating is housed in a Cu-cooler jacket and mounted on a Cu heat sink for cooling with a flowing liquid nitrogen. Reprinted from Ogino et al. (2022). © 2022 Optica Publishing Group

experiment that the cooler jacket around the Yb:YAG/Cr:YAG slab is very effective to remove the heat deposited in the edge-cladding, enabling to keep the temperature of the gain medium to 98 K with uniform distribution. An optical-to-optical efficiency of 23.6% was realized with this laser. Based on this result, a 100 J, 100 Hz laser is being developed with the ceramic Yb:YAG active mirrors of larger apertures.

Active mirror architecture is also used to generate higher-repetition laser pulses (Calendron et al., 2018). At the Colorado State University, 1 kHz, 1.1 kW, 4.5 ps duration laser pulses have been generated with high beam quality (M^2 ~ 1.35) by compressing chirped pulses generated with the double-pass two-stage cryogenically cooled Yb:YAG active mirror amplifiers with an optical-to-optical efficiency of ~ 36% (Reagan et al., 2018; Wang et al., 2020). In this work, a square-shaped 30 mm × 30 mm × 2 mm thick, 3 at. % Yb:YAG crystal was optically bonded to an Cr^{4+}:YAG cladding on the four lateral faces and the front face of the assembly was bonded to an undoped YAG cap to reduce ASE and add mechanical stability to the assembly.

Thermal behavior of this laser system has been analyzed with comprehensive simulation and compared with detailed measurements of the temperature distribution, the wavefront distortion, and the focal shift (Chi et al., 2019). The thermal distribution was determined by calculating the absorption of the pump beam in the Yb:YAG crystal, which was then dissipated by radiative and nonradiative processes.

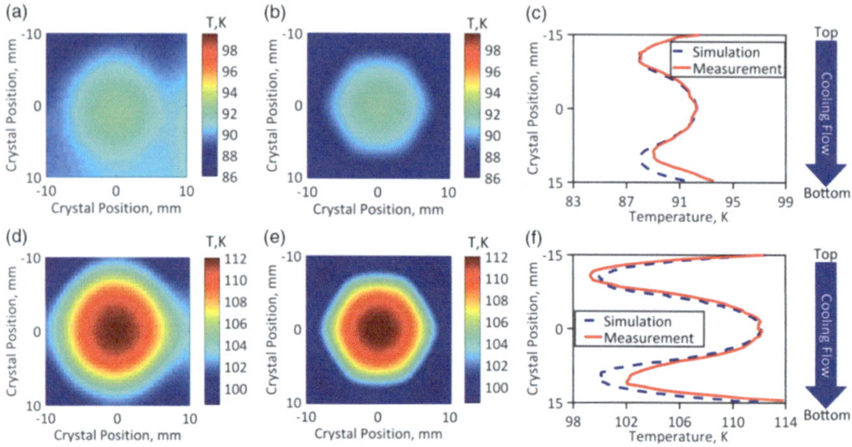

Fig. 6.11 Temperature distributions in the Yb:YAG slab pumped at 444 W (upper figures) and 1010 W (lower figures) pump powers, respectively. **a**, **d** The measured and **b, e** the simulated distributions. **c**, **f** The vertical cut of the measured (red) and simulated (blue) temperature distributions. Cooling flow direction is from top to bottom. Reprinted from Chi et al. (2019), with permission from Optical Society of America. © 2019 Optical Society of America

The pump-induced optical path difference was calculated from the thermal expansion, temperature dependence of the refractive index, and electronic contribution due to population inversion. Figure 6.11 shows the measured and the simulated temperature distributions at different pump powers, showing quantitative agreement between them. These figures reveal that the temperature near the disk edge is higher than at the center due to heating of the Cr^{4+}:YAG edge-cladding by strong ASE emitted from the laser-irradiated region of the Yb:YAG crystal.

6.3.2.3 Thin Disks

The thin-disk laser was invented by Giesen et al. (1994), where a very thin laser disk (typically 100–200 μm thickness and 10–20 mm diameter) is mounted directly onto a diamond heatsink, which is cooled by water flow, and pumped by laser diodes from the front-side. Using folding optics, the pump light is passed through the disk many times for sufficient absorption in the thin disk (Erhard et al., 2000). Owing to efficient heat removal from the disk back surface and close to one-dimensional temperature distribution in the disk, the thin disk laser is suitable for operation at multi-kW average power at ~ kHz repetition rate with excellent beam quality.

Figure 6.12 shows an example of the pumping scheme of the thin disk laser. A small-divergence LD pump beam emitted from a fiber output is reflected by a parabolic reflector (PR) at ① and imaged on the thin disk. The partly-absorbed pump beam reflected from the thin disk is reflected by PR at ②, forming a parallel beam, reflected by a set of the prism mirrors to a circularly shifted position and sent back

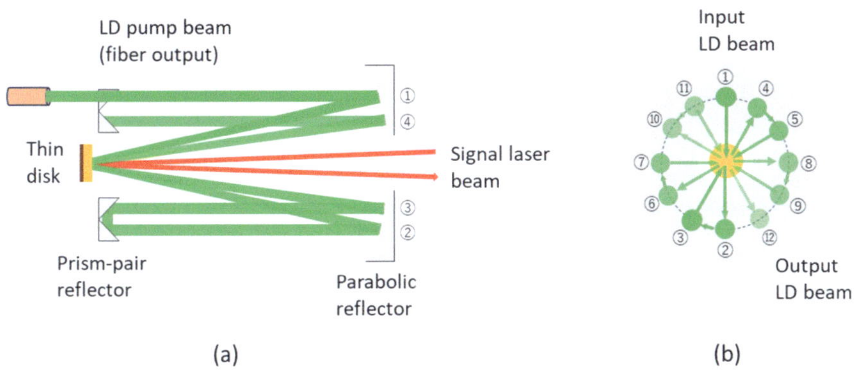

Fig. 6.12 An example of the optical configuration for pumping a thin disk amplifier with an LD beam: **a** is a side view and **b** is an end view

to PR at ③. Repeating this process, the LD beam is absorbed many times (six times in Fig. 6.21) on the thin disk. The pump beam could be double-passed for better absorption by placing a reflector at the output ⑫.

The energy stored in the disk is extracted efficiently with a signal beam by several schemes: continuous-wave (cw) oscillation, cavity damping, regenerative amplification, mode-locked oscillation, and multi-pass amplification. In cw operation of the Yb:YAG thin disk laser, up to 16 kW output power with a beam quality of 8 mm·mrad at 1030 nm has been realized. With the cavity damping, the energy stored in the cavity at saturation fluence is switched out with a pulse width of the cavity round-trip time (typically ~ 30 ns duration) and 100 mJ energy at 250 kHz maximum repetition rate with an average power of up to 2 kW (Saraceno et al., 2019; Schmidt & Schaefer, 2018).

Shorter duration pulses can be generated with thin-disk regenerated amplifiers, where seed pulses injected in the cavity are amplified in multiple passes through the gain medium and switched out with a Pockels cell switch. By injection of 1.3 ns, ~ 0.5 mJ chirped pulses into a stable ring cavity which contains two thin-disk Yb:YAG amplifiers, output pulses of 1 kW average power (200 mJ/5 kHz) with $M^2 = 1.10$ were generated. By compression of these chirped pulses with a pair of multilayer dielectric gratings, 1.0 kW-average power pulses of 1.1 ps duration at 5 kHz (pulse energy 206 mJ) has been generated (Nubbemeyer et al., 2017). With the multi-pass amplifier where a signal beam is amplified in multi-passes through the think disk heads (Negel et al., 2015), the 200 mJ pulses from the regenerative amplifier were further amplified to 720 mJ at 1 kHz with 920 fs duration for the laser light guiding project (Herkommer et al., 2020).

The thin-disk laser is also suitable for mode-locked oscillation to generate fs pulses of high-average power at MHz repetition rate. By mode-locking with a Semiconductor Saturable Absorber Mirror (SESAM) in the laser cavity (Keller et al., 1996), 350-W average power with 940-fs, 39-mJ pulses at 8.88-MHz repetition rate and 37-MW peak power has been delivered (Saltarelli et al., 2019). Alternatively with

Kerr-lens mode-locking where an aperture is placed near the cavity mirror, 270-W average power with femto-second (210–330 fs) pulses at > 10-MHz repetition rate and 38-MW peak power has been delivered (Brons et al., 2014).

6.3.3 High Power Laser Diodes

In this subsection, principles and structures of high-power laser diodes (LDs) used for pumping high-power solid-state lasers are described. Then several types of LDs are introduced with emphasis on the edge-emitting LD where ~ 10 kW peak power has been generated with a single LD module. Recent progress in the vertical-cavity surface emitting laser (VCSEL) and the photonic-crystal surface-emitting laser (PCSEL) for high-power operation are briefly presented. Finally major issues on high-power LDs for application to future IFE systems are given.

6.3.3.1 Edge-Emitting Laser Diode

A laser diode (semiconductor laser) incorporates many functions in small structures to obtain efficient lasing with low injection current. In Fig. 6.13, we show schematic diagrams of (a) energy level, (b) refractive index, and (c) device configuration of an edge-emitting double-heterostructure GaAs–$Al_xGa_{1-x}As$ injection laser, with which continuous lasing at room temperature was obtained for the first time (Hayashi et al., 1970). In this double heterostructure laser (DH laser), an active layer (GaAs) is sandwiched between an n-type $Al_xGa_{1-x}As$ and a p-type $Al_xGa_{1-x}As$ layer, where x is the atomic fraction of Ga replaced by Al in GaAs. With the increase in the Al fraction, the bandgap (energy difference) between the conduction- and valence-bands increases, which is accompanied with the decrease in the refractive index. Since the refractive index of the active layer is higher than the refractive indices of the $Al_xGa_{1-x}As$ layers, the light wave is confined in the active layer in the vertical direction (*y*-direction in Fig. 6.13). These layers are formed by molecular-beam-epitaxy (MBE) or metal–organic-chemical-vapor-deposition (MOCVD) in monolithic processes.

By applying forward voltage, population inversion is created in the active layer due to injection of the electrons from the n-layer and the holes from the p-layer, leading to photon emission due to electron–hole recombination and lasing when the gain exceeds the losses. Since the gain coefficient in the active layer reaches to ~ 100 cm^{-1} due to large stimulated-emission cross section (~ 10^{-15} cm^2) and high carrier density (> 10^{18} cm^{-3}), the lasing is obtained with a gain length of less than 1 mm (typically 300–500 μm) even with small (~ 30%) feedback by a Fabry–Perot cavity formed by cleaved crystal surfaces or a distributed Bragg reflector (DBR).

A schematic of the $Al_xGa_{1-x}As$ laser is shown in Fig. 6.13c. The thickness of the active layer d is typically ~0.1 μm to reduce the threshold current for laser oscillation, whereas the laser light propagates also outside the active layer as an evanescent wave. The threshold current of the DH laser has been further reduced

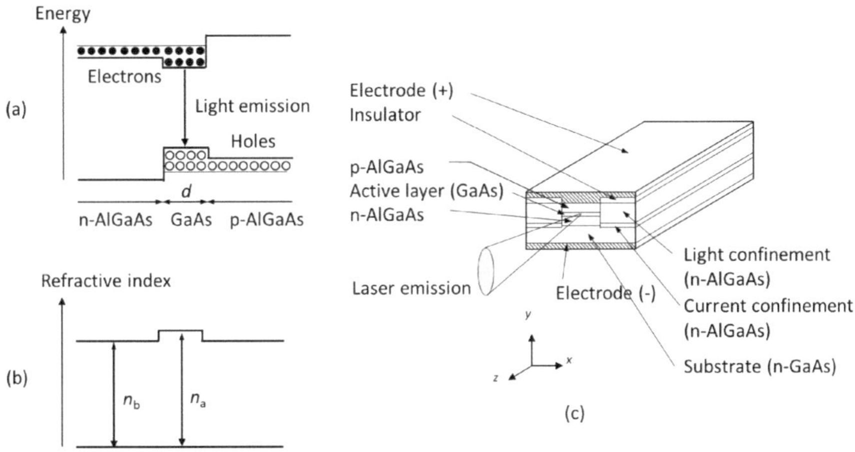

Fig. 6.13 Schematic diagrams of **a** the energy level, **b** the refractive index, and **c** the device configuration of a double-heterostructure GaAs–$Al_xGa_{1-x}As$ injection laser. In (**a**) and (**b**), the horizontal axis corresponds to the *y*-direction in (**c**). Laser emission is confined in the GaAs active layer which has a higher refractive index than the refractive indices of the $Al_xGa_{1-x}As$ layers sandwiching the active layer

(factor > 10) by introduction of a quantum well (QW) structure in the active layer having a thickness of less than the electron de Broglie wavelength ~ 30 nm, given by

$$\lambda_d = \frac{h}{p} = \frac{h}{\sqrt{2m_e E}} = \frac{h}{\sqrt{2m_e k_B T}} \tag{6.7}$$

where p is the momentum, $m_e = 0.067m_0$ is the effective mass in GaAs, and $E = k_B T$ is thermal energy of the electrons. When the thickness of the gain region is ~ 10 nm, the energy levels of the electrons (and holes) become discrete due to quantization of the wavefunctions and the electrons (and holes) are confined in the quantum well, resulting in higher gain and stable transition energy against the temperature and current changes. Since the gain of a single quantum well (SQW) laser is limited, multiple quantum-well (MQW) lasers are used to further reduce the threshold current and to increase the laser power (Thompson & Kirby, 1973; Tsang, 1981).

Propagation of the optical wave is confined also in the horizontal direction (x in Fig. 6.13c) by forming burying layers of lower refractive index at both sides of the active layer. Injection current is restricted in the active layer for higher efficiency by forming an insulating layer above the burying layer. Due to the rectangular shape of the active layer, the laser is emitted with larger divergence ($\sim 30°$) in the vertical direction and smaller divergence ($\sim 10°$) in the horizontal direction.

The wavelength of the $Al_xGa_{1-x}As$ laser can be tuned between 880 and 750 nm by changing the Al fraction in the active layer from x = 0 to x = 0.18. It is possible to match the wavelength, for example, to the Nd absorption band of 803 nm. (The Al fraction of the barrier layer $Al_yGa_{1-y}As$ in this case is chosen higher than that of the active layer, i.e., y > x.) Alternatively, $In_xGa_{1-x}As$ laser, where Ga is partly replaced by In, is suitable for pumping Yb at 940 nm since it can be tuned between 900 and 1100 nm by increasing the indium fraction.

A high-power LD is composed of a monolithic linear array of multiple DH lasers which are arrayed along the x-direction, containing 10–50 laser diodes in ~ 10 mm (Streifer et al., 1988). This LD bar module is stacked in the y-direction to form a two-dimensional array of the laser diodes of ~ 10 mm × 50 mm area, with which 10 kW peak power at 940 nm is emitted in pulsed operation of 400 μs duration and 25 Hz repetition rate when driven by 45 V and 380 A power, corresponding to a wall-plug efficiency of ~ 57% (datasheet of Pulsed Laser Diode Module, Hamamatsu Photonics K.K.).

The output power is limited by several factors: catastrophic optical damage (COD) of the output facet, laser power saturation arising from the heat deposition, and output kink in the current-output curve due to change in the transverse modes. The COD arising from the interface states of small energy separation at the output facet is avoided by decreasing the power density and/or forming a large bandgap region near the facet, enabling output power of ~ 1 W from each emitter of 100 μm-wide and 1 μm-thick area, corresponding to a power density of 1 MW/cm^2. Since electrical energy is dissipated in small areas of the LD bars, heat management is very important to efficiently remove the heat to reduce the wavelength shift and prevent the heat-induced LD damage (Beach et al., 1992).

Control of the laser wavelength of the LD stack is required to reduce the overall spectral broadening, since the wavelength emitted from each LD varies from one element to another and the wavelength shifts with current and temperature changes. Operation of an LD stack with small bandwidth and high spectral stability has been achieved, for example, with a configuration shown in Fig. 6.14 (Zheng et al., 2014). In this configuration, the laser beam diverging along the fast axis (*y*-direction) is collimated to 1.3° by microcylindrical lenses placed in front of the LD bars. An external cavity is formed with the back facet of the laser diodes and a volume Bragg grating (VBG) (Volodin et al., 2004) at the front side. Figure 6.14b shows that the laser wavelength is kept stable, and the laser bandwidth stays narrow (0.37 nm) even with changes in the cooling temperature and the current (not shown in this figure), whereas the laser wavelength changes significantly without the VBG. The LD with a VBG cavity is suitable for pumping the zero-phonon line of Yb to achieve high quantum efficiency (see Sect. 4.3.2.2).

6.3.3.2 Surface-Emitting Laser Diodes

With surface-emitting lasers, a laser beam is emitted perpendicular to the active layer of a laser diode, in contrast to the laser emission from the edge of the active layer. The vertical-cavity surface-emitting lasers (VCSELs), which are used in optical LAN and many applications, is extended to generate ~ 1 kW power with a small VCSEL 2D array. Another type of surface-emitting laser, the photonic-crystal

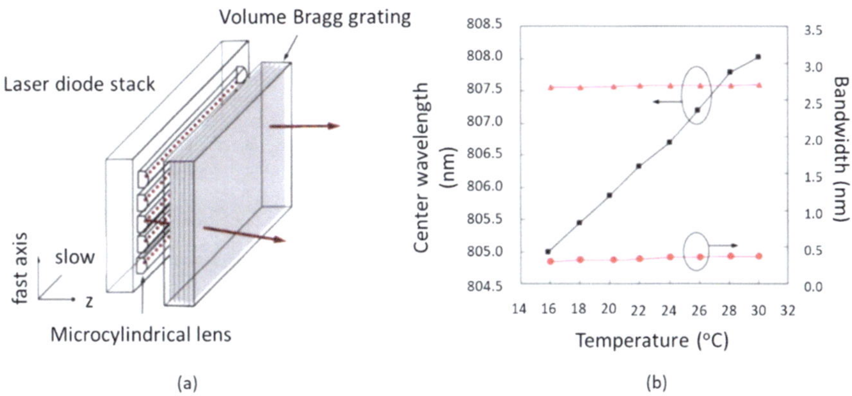

Fig. 6.14 **a** A schematic diagram of a LD stack with microcylindrical lenses for collimation in the fast axis and a volume Bragg grating (VBG) in front of the LD stack to form an external laser cavity. **b** Dependence of the center wavelength of the LD stack on the cooling temperature with the VGB cavity (red triangles) and free running without VBG (black squares). Also shown is the temperature dependence of the bandwidth of the LD stack with the VBG cavity (red circles). Reprinted from Zheng et al. (2014), with permission from The Japan Society of Applied Physics. © 2014 The Japan Society of Applied Physics

surface-emitting laser (PCSEL), is also scaled to 100-W-to-1 kW single-mode operation. In this subsection, we briefly describe recent progresses in VCSEL and PCSEL for high-power operation.

(a) **Vertical-Cavity Surface-Emitting Laser (VCSEL)**

The vertical-cavity surface-emitting laser (VCSEL) was first demonstrated at liquid nitrogen temperature with GaInAsP (Soda et al., 1979), followed by room temperature cw operation with GaAs (Koyama et al., 1988). Figure 6.15a shows a typical configuration of a GaAs VCSEL operating at 800 nm, where the GaAs MQW active layer is sandwiched by two distributed Bragg reflectors (DBRs) composed of AlGaAs/AlAs multilayers which form a one-wavelength-long laser cavity vertical to the active layer. The electric current and the optical radiation is confined to a few-μm-diameter area by selective oxidization of an AlAs layer, enabling lowering the threshold current to 225 μA (Huffaker et al., 1994). To overcome the short gain length of only 10s-nm thickness, the DBRs are fabricated with high reflectivity of typically 99.5–99.9% and very low loss (Huffaker & Deppe, 1994), thus leading to high extraction efficiency and high power-conversion efficiency exceeding 50%. The small, low-current VCSELs are suitable as ultra-low-power consumption devices for high-frequency modulation exceeding 10 GHz, required for local area network (LAN), optical interconnection, and optical processing systems (Iga, 2000).

Many VCSELs are fabricated with a monolithic process, where the VCSELs are monolithic-grown and processed on a wafer, electrical contacts are formed and diced from 1-D or 2-D arrays into single devices, as shown schematically in Fig. 6.15b. As a high-power laser emitter, the VCSEL has several advantages: (a) Each VCSEL device

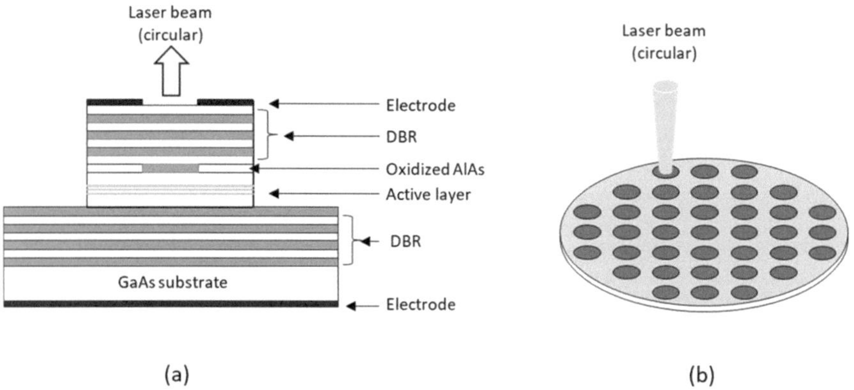

Fig. 6.15 **a** A typical configuration of GaAs VCSEL operating at 800 nm, where a one-wavelength-scale laser cavity is formed by two reflectors above and below a thin active layer, emitting a circular laser beam to the vertical direction. **b** Monolithic fabrication of many VCSELs on a wafer. This figure is not in scale, since the beam diameter is only a few μm whereas the wafer is typically 3 inches

works on the wafer without additional processing, leading to on-wafer screening and low-cost production; (b) Due to short cavity length and monolithic fabrication, the VCSEL operates in a single longitudinal mode and the laser wavelength variation between the VCSEL array elements is less than 1 nm; (c) The VCSEL are not affected by COD, leading to high reliability of less than 10 FIT (failures in time of 10^9 device hours); (d) Circular output beam of ~ 10-degree divergence (full angle) which can be collimated with a circular lens for fiber coupling and direct pumping of solid state lasers; (e) The 2-D VCSEL arrays composed of thousands of low-power single devices driven in parallel can be fabricated as a single package, delivering 400 W quasi-cw power from a 1 × 1 cm module (Data sheet of Princeton Optronics, Inc.).

A chip-scale DPSSL, named Vertical-External-Cavity Surface-Emitting Laser (VECSEL), has been developed by pumping a solid-state laser with VCSEL. In the VECSEL whose configuration is shown in Fig. 6.16, a 0.5-mm-thick Yb:YAG fabricated above a GaAs substrate is pumped by the 940-nm emission from the InGaAs VCSEL. Repetitive Q-switched lasing of the Yb:YAG laser has been obtained at 1030 nm, by forming a laser cavity composed of a high-reflectance mirror and a saturable absorber (Cr:YAG). With this small chip of only 1 mm^3 dimension, Q-switched pulses of 30.4 μJ energy and 450 ps duration (57 kW peak power) have been generated at 1.79 kHz repetition rate with an average power of 54 mW (Yue et al., 2022). The VECSEL, which is a compact, high-peak power laser, can be fabricated in a 2D array like VCSEL. It works as a converter from the VCSEL whose power is limited by the small volume of the active layer to high peak-power laser pulses.

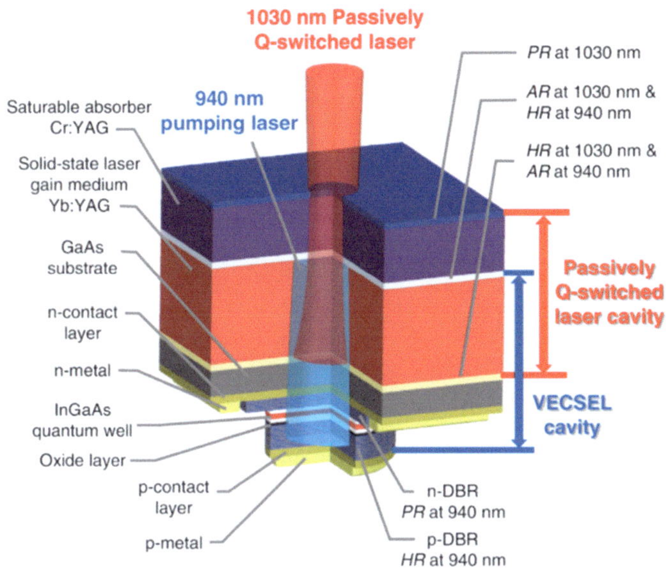

Fig. 6.16 Schematic of a VECSEL (vertical-external-cavity surface-emitting laser), which is a 1-mm^3 size DPSSL. A Yb:YAG, which is pumped by 940-nm emission from a InGaAs quantum well, lases at 1030 nm by Q-switching with a Cr:YAG saturable absorber. Reprinted from Yue et al. (2022). © 2022 Springer Nature

(b) **Photonic-Crystal Surface-Emitting Laser (PCSEL)**

Photonic crystals refer to the wavelength-scale periodic dielectric structures with forbidden gaps to the electromagnetic waves, analogous to the electronic band gap in semiconductors (Joannopoulos et al., 2008; Yablonovitch, 1993). The multilayer dielectric film is an example of a one-dimensional (1D) photonic crystal, where an optical wave of wavelength λ cannot propagate in the $\lambda/4$-stack multilayers at normal incidence (see Sect. 2.2.3.2). Figure 6.17a shows a dispersion curve of an optical wave in the multilayers, where the group velocity of the optical wave $v_g = \partial\omega / \partial k$ becomes smaller as the wavenumber$k$ increases, eventually becoming $v_g = 0$ at $k = \pi/a$ (a is the multilayer period) due to Bragg reflection, corresponding to formation of a bandgap to the optical wave.

Figure 6.17b shows an example of a two-dimensional (2D) photonic crystal comprising triangular array of circular holes in a dielectric slab. The optical wave propagating in the x-direction is diffracted to backward (−180°), ± 60° and ± 120° directions simultaneously, resulting in coupling of the six waves. The photonic bandgaps are created to these waves due to periodic variation of the refractive index. The 2D and 3D photonic crystals are used in micro-optic circuits, for example, for wave guiding, data storage, and data extraction by forming line and point defects (filling the airholes) as marked by the dotted lines in Fig. 6.17b (Noda, 2006).

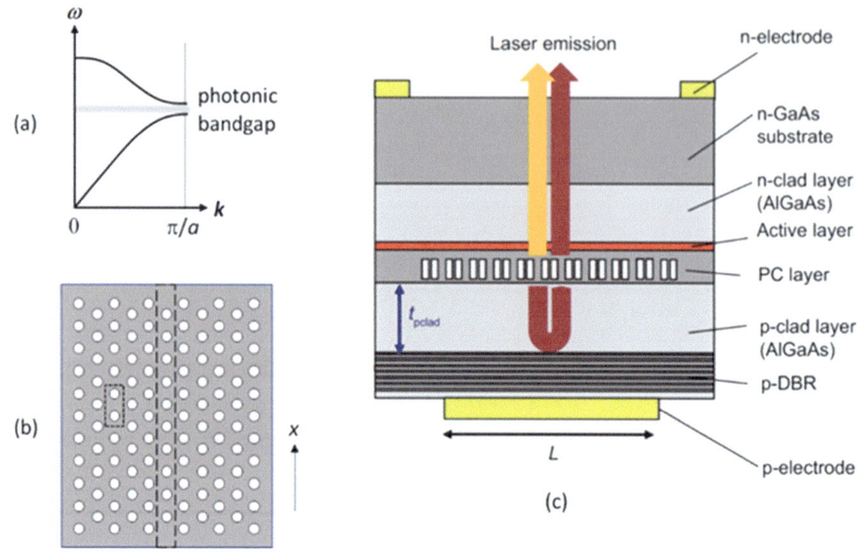

Fig. 6.17 **a** Dispersion curve of a 1D photonic crystal. **b** A 2D photonic crystal composed of a triangular array of circular airholes. **c** Schematic structure of a PCSEL where an MQW GaAs active layer and a photonic crystal (PC) layer is sandwiched between n-clad and p-clad layers. The two holes in the PC layer denote a pair of holes with different shapes. The downward emission is reflected by a reflector (p-DBR) to increase the laser power upward through the n-GaAs substrate. Reprinted from Inoue et al. (2022). © 2022 Springer Nature

Referring to Fig. 6.17c, the 2D photonic crystal is placed adjacent to an active layer with an emission frequency in the photonic bandgap. Then the photonic crystal acts as a 2D Bragg reflector, which suppresses amplification in the x–y plane and allows diffraction out of the plane, yielding stimulated emission vertical to the active layer surface when the gain becomes higher than the loss. The laser emission, which is spatially coherent and single frequency over a circular area of 350 μm in diameter, was first demonstrated by Imada et al. (1999).

To obtain higher power lasing from broader areas, various improvements have been made: array structure, fabrication method, non-circular air-holes, double-lattice photonic crystal, and reflection of the downward diffractive beam. With the 500 μm diameter GaAs PCSEL schematically shown in Fig. 6.17c where a photonic crystal is formed by two adjacent holes (double lattices) to select a single transverse mode while suppressing high-order spatial modes, a circular beam of 10 W peak power has been generated at ~ 940 nm with nearly diffraction-limited divergence of ~ 0.1°, corresponding to a brightness of over 300 MW cm^{-2} sr^{-1} (Yoshida et al., 2021). A design for the PCSEL to achieve 100-W-to-kW-class single mode operation with a water-cooled 3–10 mm-diameter single chip has been shown, based on detailed analyses of ultra-large-area PCSELs for increasing the threshold margin between the fundamental and the higher-order modes and increasing the coupling to the radiative waves emitted in the vertical direction (Inoue et al., 2022). Figure 6.18 shows a 3-mm-diameter PCSEL, designed and fabricated based on this concept, with which single-mode lasing of 50 W CW power with 0.05° divergence and brightness of 1 GW cm^{-2} sr^{-1} has been demonstrated, marking an important milestone toward realization of a single-mode, small-divergence 1-kW semiconductor laser (Yoshida et al., 2023).

Short-pulse lasing has been also achieved by passive Q-switching of the PCSEL by inserting absorption regions generating 35-ps pulses (Morita et al., 2021) and by an innovative scheme, named "self-evolving carrier-photon dynamics", where lasing region moves from one edge to another by chirping the lattice constant, generating 27-ps pulses. (Inoue et al., 2023). It is expected that compact, high-power, high-brightness PCSELs will bring strong impacts to a wide variety of applications such as material processing, laser radar, medicine, and aerospace, replacing various types of conventional high-power lasers. When it is used for pumping other lasers, it will extremely simplify the pumping geometry.

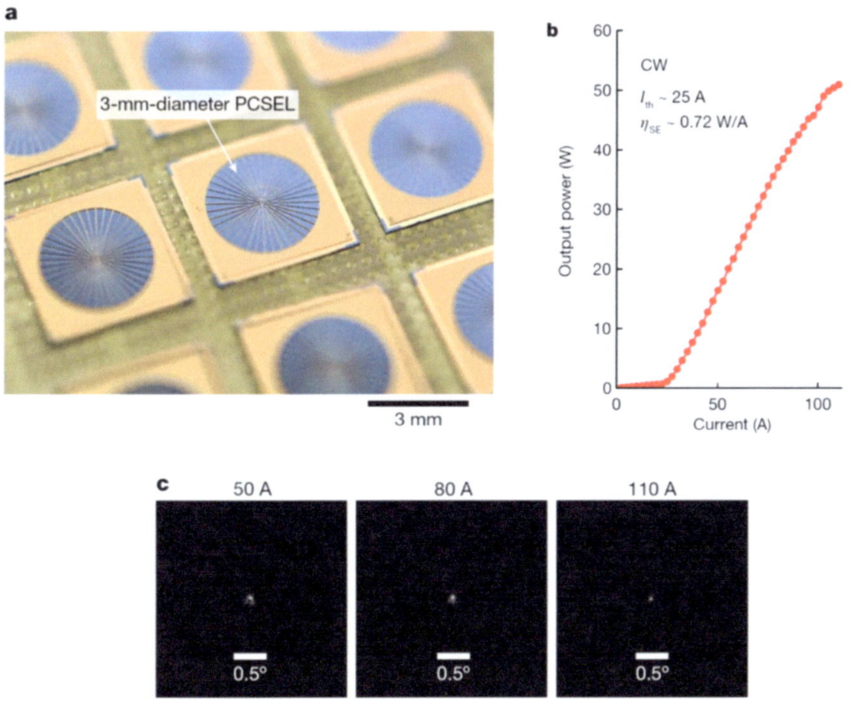

Fig. 6.18 Demonstration of 50-W single-mode CW operation of a 3-mm diameter PCSEL. **a** Photograph of the fabricated 3-mm diameter PCSEL chips. Each chip is mounted on a package via a sub-mount in the p-side down configuration. **b** CW laser output power vs. the injection current. **c** Far field patterns of the laser beam at 50, 80 and 110 A, showing smaller divergence of ~ 0.05° at the higher current. Reprinted from Yoshida et al. (2023). © 2023 Springer Nature

6.3.3.3 Requirements to Laser Diodes as IFE Driver

Let us consider an IFE power plant. Denoting the laser efficiency by η_d, a pellet gain by G, and the plant efficiency (for conversion of the fusion power to the electricity) by η_g, it is required that the product $\eta_d \eta_g G$ exceeds 1 to supply the electricity outside the plant. Assuming $\eta_g = 0.33$ (efficiency of a fission power plant) and $\eta_d = 0.15$ (described in the following), a pellet gain of at least 20 is required for $\eta_d \eta_g G \geq 1$. For order-of-magnitude estimation, let us assume that an IFE reactor is operated at 10 Hz and driven by a laser driver of 1 MJ total energy per pulse. If a pellet gain of 100 is achieved at 10 Hz, 330 MW electrical power is generated. Since 67 MW is required for laser operation, this power station can supply 263 MW electricity.

The 1 MJ laser energy may be supplied by many laser beams, for example, by 100 beams, i.e., 10 kJ per beam. Let us consider a Yb:YAG laser as a candidate for the IFE driver, where 100 J, 10 Hz operation (Mason et al., 2017) and efficient conversion to the third harmonics (Phillips et al., 2021) have been demonstrated and scale-ups to higher energies with various architectures are undertaken as described in Sect. 5.3.2. Based on these developments, it will be possible to form a 10-kJ beam with one hundred 100-J Yb:YAG lasers, or with smaller numbers of higher energies. An overall LD pump-to-UV conversion efficiency of 25% can be achieved, considering the optical–optical efficiency of 36% with the Yb:YAG active mirror (Wang et al., 2020) and the third harmonic conversion efficiency of 68% (Phillips et al., 2021). Including a LD electrical efficiency of ~ 60%, the total electrical efficiency of this laser system is 15%.

Status and perspectives of high-power LDs for IFE has been reviewed by Häfner et al. (2022), where LD pumping of Nd:glass laser is considered. Although the requirements to LDs are different for pumping Yb:YAG and Nd:glass in terms of the pump wavelength (9xx nm vs. 8xx nm) and the pulse duration (1.2 ms vs. 0.3 ms), most aspects in this study are applicable. The most demanding issues are the production cost and the long-term reliability, in addition to requirements for performance improvements such as brightness (W/cm^2 sr). The cost and the production volume are closely related. Although the LD cost has decreased rapidly to less than 1 \$/W with an increase in production volume, the LD cost of less than ~ 0.01 \$/W is required to make an IFE power plant cost competitive. Considering the total LD power of 40 MW for pumping the 1 MJ, 10 Hz driver, which exceeds the present total production volume, significant innovations are required in the design and manufacturing of the high-power LDs, such as implementation of massive automation in the labor-intensive processes in manufacturing, assembly and inspection which weighs large fraction of the production cost. Another issue is the requirement for the very long LD lifetime (mean time to failure): 9.5 Giga-shots for 30-year lifetime of an IFE reactor operating at 10 Hz. The surface emitting LD such as VCSEL may be suitable in this respect due to its lower cost in mass production and absence of COD of the output facet.

6.4 Excimer Laser

Extensive investigations were undertaken on the excimer laser in the twentieth century as a candidate for the laser fusion driver due to its short wavelength and high intrinsic efficiency (Ewing, 1979): for example, at Sandia National Laboratory (Hoffman et al., 1976), Los Alamos National Laboratory (Rosocha et al., 1987), Naval Research Laboratory (Obenschain et al., 1996), Rutherford Appleton Laboratory (Shaw et al., 1993), Keio University (Suda et al., 1986), University of Electro-Communications (Ueda et al., 1987), and Electrotechnical Laboratory (Owadano et al., 1999). In this section, we describe basic aspects of the excimer laser and its future prospect as a fusion driver by referring to investigations at the Naval Research Laboratory (NRL), USA.

The KrF and ArF excimer lasers have unique properties which are advantageous for the direct-drive laser fusion (Obenschain et al., 2015, 2020). Since these lasers operate at short wavelengths (KrF at 249 nm and ArF at 193 nm), the laser light is absorbed at a high plasma density, resulting in high ablation pressure with less probability for the plasma instabilities whose threshold intensities increase at shorter laser wavelengths (Montgomery, 2016). These lasers also have broad spectral bandwidths (KrF of 3 THz and ArF of 10 THz), which are advantageous for smoothing the spatial and temporal irradiation nonuniformities by applying the induced spatial incoherence (Lehmberg et al., 1987) and for suppressing the plasma instabilities, especially the cross beam energy transfer (CBET) (Bates et al., 2018) which has been shown to be a major driver-power loss in direct-drive implosion at 351 nm (Myatt et al., 2017).

The energy level diagram of the KrF excimer laser (which is similar for the ArF excimer laser) is shown in Fig. 6.19, where Kr^+ and F^- form an ionic-bound excited state whereas a weakly bound covalent state is formed as the ground state. Lasing takes place in the bound-free transition, with broad spectral width, from $B^2\sum_{1/2}$ state (KrF*) to $X^2\sum_{1/2}$ state. Since the B-X transition is dipole-allowed, the radiative lifetime is 6.5 ns for KrF* and 2.7 ns for ArF*, requiring temporally continuous energy extraction during the electron beam pumping of over 100 ns.

The KrF* and ArF* states are formed through a neutral channel ($R* + F_2 \rightarrow RF* + F$) and an ion channel $\left(R^+ + F^- + M \rightarrow RF* + M\right)$, where R refers to either Kr or Ar and M stands for a buffer gas which conserves the energy and momentum in the reaction. The optimum conditions for the electron-beam pumped excimer lasers have been determined from detailed analyses of the energy deposition and atomic processes by, for example, Petrov et al. (2002) and Giuliani et al. (2002) for KrF, and Petrova et al. (2022) for ArF.

The schematic of an electron-beam pumped excimer laser is shown in Fig. 6.20. A laser cell filled with $Ar/Kr/F_2$ mixture (~ 90/10/0.2) for KrF laser or Ar/F_2 mixture (~ 100/0.3) for ArF laser with a total pressure of ~ 1 atm is pumped with high-voltage electron beams from both sides of the cell with the power density of ~ 1 MW/cm^3. The electron beams are guided by an external magnetic field to prevent electron beam pinching, and the laser gas is recirculated for repetitive operation.

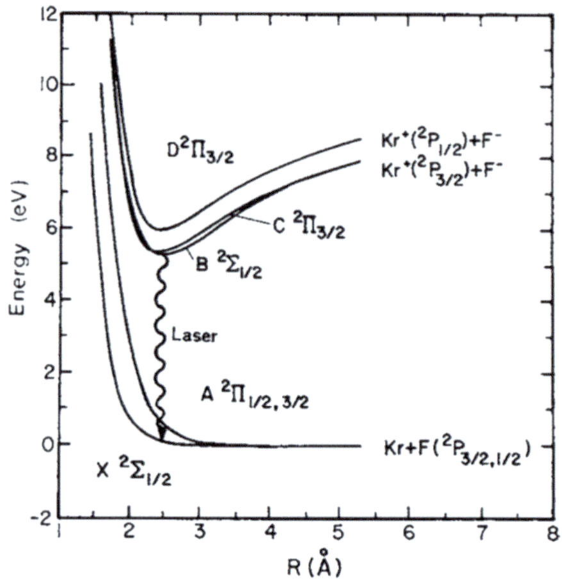

Fig. 6.19 Simplified energy level diagram of KrF laser. The lasing is obtained in the allowed transition from $B^2 \sum_{1/2}$ to $X^2 \sum_{1/2}$

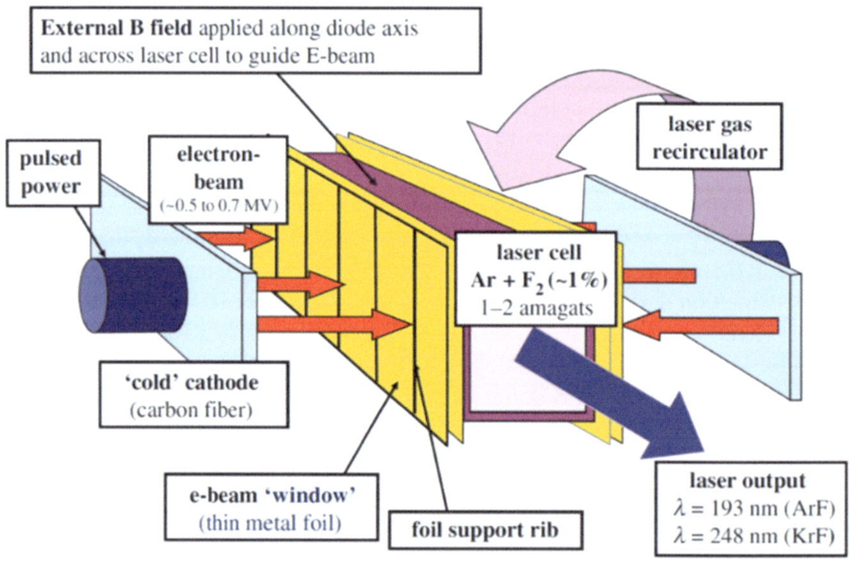

Fig. 6.20 Schematic of an electron-beam pumped excimer laser. Reprinted from Obenschain et al. (2020). © 2020 The Royal Society Publishing

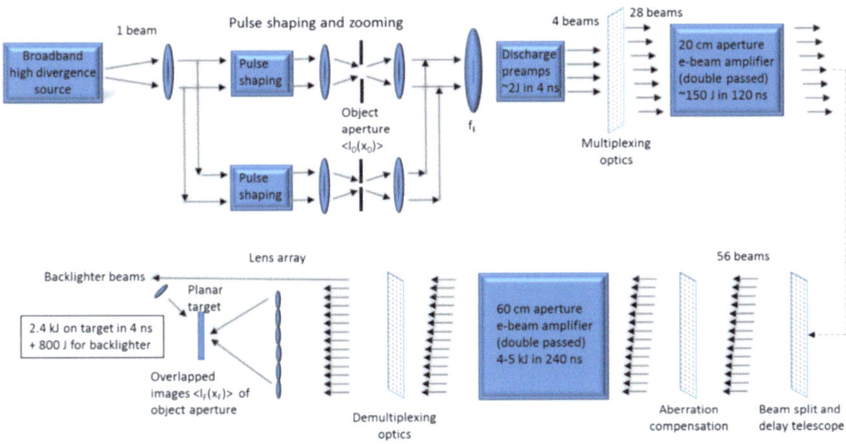

Fig. 6.21 Schematic of an optical system of the excimer laser Nike at NRL. Reprinted from Obenschain et al. (2015). © 2015 Optical Society of America

The schematic of the optical system of the Nike KrF laser at the Naval Research Laboratory is shown in Fig. 6.21. To extract the energy from KrF* which has a short radiative lifetime, a 4-ns laser pulse produced in the front-end is split in several steps with multiplexing optics and finally into 56 beams with sequential delay of 4 ns to extract the energy almost continuously from the 60-cm aperture final amplifier which is pumped by a 240 ns-duration electron beam. These pulses are demultiplexed to form a single 4 ns pulse before target irradiation. An incoherent light source at the front end is relay-imaged during the amplification stages all the way to the final target so that the target is irradiated with incoherent light at high intensity. This laser system was routinely operated at 4 kJ output energy, corresponding to laser fluence of ~ 1.4 J/cm^2 and intensity of ~ 5.6 MW/cm^2 (assuming a circular laser beam of 60-cm diameter).

The performance of the KrF and ArF lasers based on the plasma kinetics simulation, verified by the experimental data, show that the intrinsic efficiency (ratio of the output laser energy to the electron beam energy) and the electric wall-plug efficiency are 12% and 7.2% for KrF and 17% and 10.2% for ArF lasers, respectively (Obenschain et al., 2020; Petrova et al., 2022).

Laser wavelength scaling of various physics processes show that shorter laser wavelength leads to higher hydrodynamic efficiency and reduced laser-plasma instabilities (Schmitt & Obenshcain, 2023a), enabling a target design for an IFE reactor with less power and energy of the laser driver (Schmitt & Obenshcain, 2023b). The laser wavelength dependence of the target energy gain based on one-dimensional hydrocode simulations (Fig. 6.22) shows that the higher target gain is obtained with the ArF laser compared with the KrF laser and the third harmonic of the Nd:glass laser. With the ArF laser, a target gain of over 200 is expected with 1 MJ laser energy for the conventional ignition design and ~ 0.4 MJ with the shock ignition design, respectively (Obenschain et al., 2020).

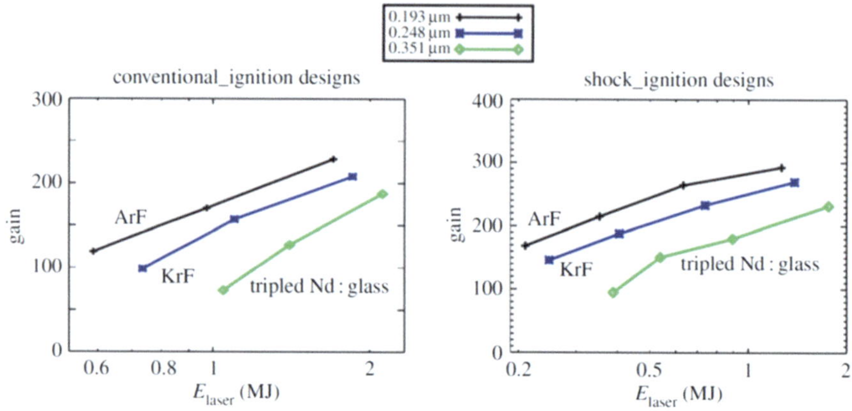

Fig. 6.22 One dimensional hydrocode simulations of target energy gains as a function of laser energy for conventional and shock-ignited targets for the laser wavelengths of 193 nm, 248 nm, and 351 nm, respectively Reprinted from Obenschain et al. (2020). ©2020 The Royal Society Publishing

It is necessary to test the applicability of the excimer laser to the IFE driver, where repetitive operation of ~ 2 Giga shots at 5 Hz over 10 years is required. A repetitively pulsed KrF laser, named Electra, was developed at NRL to test the lifetime of major components of the excimer laser: the pulse-power and pulse transforming lines with high-voltage high-current switches, the electron beam diodes composed of cathodes, anodes and pressure membranes which transmit the electron beam, laser-gas circulators, and the laser cell where fluorine molecules and other chemically activate atoms and ions are contained. Electra has achieved a lifetime of 50,000 shots at 5 pps for the electrical components, with 2% change in the transmission of a fused silica window over 250,000 shots (Obenschain et al., 2015). For long term operation of the ArF laser at high fluence, further developments of the ArF optics such as high quality MgF_2 windows and fluorine-resistant coatings which are used in the lithography industry will be required.

References

Abate, J. A., Lund, L., Brown, D., et al. (1981). Active mirror: A large-aperture medium-repetition rate Nd:Glass amplifier. *Applied Optics, 20*, 351.

Abu-Shawareb, H., et al. (2022). Lawson criterion for ignition exceeded in an inertial fusion experiment. *Physical Review Letters, 129*, Article 075001.

Abu-Shawareb, H., et al. (2024). Achievement of target gain larger than unity in an inertial fusion experiment. *Physical Review Letters, 132*, Article 065102.

Andre, M., Courteille, J. C., Le Gall, J. Y., et al. (1979). OCTAL: A new neodymium glass laser built in the Centre d'Etudes de Limeil for laser fusion experiments. *Proceedings of SPIE, 164*, 8.

Baisden, P. A., Atherton, L. J., Hawley, R. A., et al. (2016). Large optics for the National Ignition Facility. *Fusion Science and Technology, 69*, 295.

Banerjee, S., Mason, P., Ertel, K., et al. (2016). 100-J level nanosecond pulsed diode pumped solid state laser. *Optics Letters, 41*, 2089.
Banerjee, S., Mason, P., Phillips, J., et al. (2020). Pushing the boundaries of diode-pumped solid-state lasers for high-energy applications. *High Power Laser Science and Engineering, 8*, Article e20.
Basov, N. G., Krokhin, O. N., Sklizkov, G. V., et al. (1972). A powerful laser setup and investigation of the efficiency of high temperature heating of a plasma. *Soviet Journal of Experimental and Theoretical Physics, 35*, 109.
Batani, D., Colaïtis, A., Consoli, F., et al. (2023). Future for inertial-fusion energy in Europe: A roadmap. *High Power Laser Science and Engineering, 11*, Article e83.
Bates, J. W., Myatt, J. F., Shaw, J. G., et al. (2018). Mitigation of cross-beam energy transfer in inertial-confinement-fusion plasmas with enhanced laser bandwidth. *Physical Review E, 97*, Article 061202.
Bayramian, A., Armstrong, J., Ault, E., et al. (2007). The mercury project: A high average power, gas cooled laser for inertial fusion energy development. *Fusion Science and Technology, 52*, 383.
Bayramian, A., Armstrong, J., Beer, G., et al. (2008). High-average-power femto-petawatt laser pumped by the Mercury laser facility. *Journal of the Optical Society of America B: Optical Physics, 25*, B57.
Beach, R., Benett, W. J., Freitas, B. L., et al. (1992). Modular microchannel cooled heatsinks for high average power laser diode arrays. *IEEE Journal of Quantum Electronics, QE-28*, 966.
Betti, R., Zhou, C. D., Anderson, K. S., et al. (2007). Shock ignition of thermonuclear fuel with high areal density. *Physical Review Letters, 98*, Article 155001.
Boehly, T. R., Brown, D. L., Craxton, R. S., et al. (1997). Initial performance results of the OMEGA laser system. *Opt. Comm., 133*, 495.
Born, M., & Wolf, E. (1999). *Principles of optics: Electromagnetic theory of propagation, interference and diffraction of light* (7th ed.). Cambridge University Press.
Brons, J., Pervak, V., Fedulova, E., et al. (2014). Energy scaling of Kerr-lens mode-locked thin-disk oscillators. *Optics Letters, 39*, 6442.
Brown, D. C., Kelly, J. H., & Abate, J. A. (1981). "Active-mirror amplifiers: progress and prospects. *IEEE Journal of Quantum Electronics, QE-17*, 1755.
Buresi, E., Coutant, J., Dautrey, R., et al. (1985). Laser program development at CEL-V: Overview of recent experimental results. In *17 European conference on laser interaction with matter, Rome (Italy)*, 18–22 Nov.
Calendron, A.-L., Meier, J., Hemmer, M., et al. (2018). Laser system design for table-top X-ray light source. *High Power Laser Science and Engineering, 6*, Article e12.
Campbell, E. M., Hunt, J. T., Bliss, E. S., et al. (1986). Nova experimental facility. *Review of Scientific Instruments, 57*, 2101.
Chénais, S., Druon, F., Forget, S., et al. (2006). On thermal effects in solid state lasers: The case of ytterbiumu-doped materials. *Progress in Quantum Electronics, 30*, 89.
Chi, H., Baumgarten, C. M., Jankowska, E., et al. (2019). Thermal behavior characterization of a kilowatt-power-level cryogenically cooled Yb:YAG active mirror laser amplifier. *Journal of the Optical Society of America B: Optical Physics, 36*, 1084. https://doi.org/10.1364/JOSAB.36.001084
Craxton, R. S., Anderson, K. S., Boehly, T. R., et al. (2015). Direct-drive inertial confinement fusion: A review. *Physics of Plasmas, 22*, Article 110501.
Denis, V., Nicolaizeau, M., Neauport, L., et al. (2021). LMJ 2021 facility status. In: *SPIE Proceedings, High Power Lasers for Fusion Research VI* (Vol. 11666, p. 1166603).
Divoky, M., Smrz, M., Chyla, M., et al. (2014). Overview of the HiLASE project: High average power pulsed DPSSL systems for research and industry. *High Power Laser Science and Engineering, 2*, Article e14.
Dixit, S. N., Feit, M. D., Perry, M. D., et al. (1996). Designing fully continuous phase screens for tailoring focal-plane irradiance profiles. *Optics Letters, 21*, 1715.

Drake, R. P. (2009). Perspectives of high-energy-density physics. *Physics of Plasmas, 16*, Article 055501.

Erhard, S., Karszewski, M., Stewen, C., et al. (2000). Pumping schemes for multi-kW thin disk lasers. In *Advanced solid state lasers*. OSA Technical Digest Ser. paper MB16.

Ertel, K., Banerjee, S., Mason, P. D., et al. (2011). Optimising the efficiency of pulsed diode pumped Yb:YAG laser amplifiers for ns pulse generation. *Optics Express, 19*, 26610. https://doi.org/10.1364/OE.19.026610

Ewing, J. J. (1979). Excimer lasers. In M. L. Stitch (Ed.), *Laser handbook* (Vol. 3, pp.135–197). North Holland

Garban-Labaune, C., Fabre, E., Max, C. E., et al. (1981). Effect of laser wavelength and pulse duration on laser-light absorption and back reflection. *Physical Review Letters, 48*, 1018.

Giesen, A., Hűgel, H., Voss, A., et al. (1994). Scalable concept for diode-pumped high-power solid-state lasers. *Applied Physics B, 58*, 365.

Giuliani, J. L., Petrov, G. M., & Dasgupta, A. (2002). Electron energy deposition in an electron-beam pumped KrF amplifier: Impact of the gas composition. *Journal of Applied Physics, 92*, 1200.

Gomez, M. R., Slutz, S. A., Jennings, C. A., et al. (2020). Performance scaling in magnetized liner inertial fusion experiments. *Physical Review Letters, 125*, Article 155002.

Häfner, C., Bayramian, A., Betts, S., et al. (2017). High average power, diode pumped petawatt laser systems: a new generation of lasers enabling precision science and commercial applications. *Proceedings of SPIE, 10241*, 1024102-1–1024102-5

Häfner, C., Holly, C., Hoffman, H.-D., et al. (2022). Status and perspective of high-power pump diodes for inertial fusion energy drivers. A white paper prepared for the IFE Science and Technology Community Strategic Planning Workshop, National Ignition & Photon Science, Lawrence Livermore National Laboratory.

Hayashi, I., Panish, M. B., Foy, P. W., et al. (1970). Junction lasers which operate continuously at room temperature. *Applied Physics Letters, 17*, 109.

Herkommer, C., Krotz, P., Jung, R., et al. (2020). Ultrafast thin-disk multipass amplifier with 720 mJ operating at kilohertz repetition rate for applications in atmospheric research. *Optics Express, 28*, 30164.

Hoffman, J. M., Hays, A. K., & Tisone, G. C. (1976). High power UV noble-gas-halide lasers. *Applied Physics Letters, 28*, 538.

Hopps, N. W., Oades, K., Andrew, J., et al. (2015). Comprehensive description of the Orion laser facility. *Plasma Physics and Controlled Fusion, 57*, Article 064002.

Huffaker, D. L., & Deppe, D. G. (1994). Low threshold vertical-cavity surface-emitting lasers based on high contrast distributed Bragg reflectors. *Applied Physics Letters, 70*, 1781.

Huffaker, D. L., Deppe, D. G., Kumar, K., et al. (1994). Native-oxide defined ring contact for low threshold vertical-cavity lasers. *Applied Physics Letters, 65*, 97.

Hüller, S., Raj, G., Luo, M., et al. (2020). Crossed beam energy transfer between optically smoothed laser beams in inhomogeneous plasmas. *Philosophical Transactions of the Royal Society A, 378*, 20200038.

Hurricane, O. A., Callahan, D. A., Casy, D. T., et al. (2016). Inertially confined fusion plasmas dominated by alpha-particle self-heating. *Nature Physics, 12*, 800.

Hurricane, O. A., Callahan, D. A., Casy, D. T., et al. (2024). Energy principles of scientific breakeven in an inertial fusion experiment. *Physical Review Letters, 132*, 065103.

Iga, K. (2000). Surface-emitting laser—Its birth and generation of new optoelectronics field. *IEEE Journal of Selected Topics in Quantum Electronics, 6*, 1201.

Imada, M., Noda, S., Chutinan, A., et al. (1999). Coherent two-dimensional lasing action in surface-emitting laser with triangular-lattice photonic crystal structure. *Applied Physics Letters, 75*, 316.

Inoue, T., Morita, R., Nigo, K., et al. (2023). Self-evolving photonic crystals for ultrafast photonics. *Nature Communications, 14*, 50.

Inoue, T., Yoshida, M., Gelleta, J., et al. (2022). General recipe to realize photonic-crystal surface-emitting lasers with 100-W-to-1-kW single-mode operation. *Nature Communications, 13*, 3262. https://doi.org/10.1038/s41467-022-30910-7

Joannopoulos, J. D., Johnson, S. G., Winn, J. N., et al. (2008). *Photonic crystals—Molding the flow of light* (2nd ed.). Princeton University Press.

Kato, Y., Kuroda, J., Mizumoto, Y., et al. (1978). High power, high intensity, and extremely short light source of glass laser system GEKKO IV. In *Proceedings on 13th International Congress High Speed Photography and Photonics* (p. 233).

Kato, Y., Mima, K., Miyanaga, N., et al. (1984). Random phasing of high-power lasers for uniform target acceleration and plasma-instability suppression. *Physical Review Letters, 53*, 1057.

Kato, Y., Yoshida, K., Kuroda, J., & Yamanaka, C. (1981). 3.4-TW performance of a Nd:phosphate glass laser with output aperture of 20 cm. *Applied Physics Letter, 38*, 72.

Kawanaka, J., Miyanaga, N., Kawashima, T., et al. (2008). New concept for laser fusion energy driver by using cryogenically-cooled Yb:YAG ceramic. *Journal of Physics: Conference Series, 112*, 032058.

Keller, U., Weingarten, K. J., Kartner, F. X., et al. (1996). Semiconductor saturable absorber mirrors (SESAM's) for femtosecond to nanosecond pulse generation in solid-state lasers. *IEEE Journal of Selected Topics in Quantum Electronics, 2*, 435.

Kirillov, G. A., Kochemasov, G. G., Bessarab, A. V., et al. (2000). Status of laser fusion research at VNIIEF (Arzamas-16). *Laser and Particle Beams, 18*, 219.

Kodama, R. (2021). Fast ignition laser fusion energy research in Japan. Fast Ignition Research Consortium, Laser Fusion Strategy Committee of IFE Forum, IAEA-CN-123/45.

Kodama, R., Norreys, P. A., Mima, K., et al. (2001). Fast heating of ultrahigh-density plasma as a step towards laser fusion ignition. *Nature, 412*, 798.

Koechner, W. (2006). *Solid-state laser engineering, 6-th Edn.* Springer.

Koyama, F., Kinoshita, S., & Iga, K. (1988). Room temperature cw operation of GaAs vertical cavity surface emitting laser. *Transactions IEICE, E71*, 1089.

Kritcher, A. L., Zystra, A. B., Weber, C. R., et al. (2024). Design of the first fusion experiment to achieve target energy gain G > 1. *Physical Review E, 109*, Article 025204.

Krupke, W. F. (1989). Solid state laser driver for an ICF reactor. *Fusion Science and Technology, 15*, 377.

Ladish, J. S. (1979). Helios, a 20 TW CO_2 laser fusion facility. In *Laser 79 Opto-Electronics: Proceedings of Fourth Conference* (p. 294). IPC Science and Technology Press

Le Pape, S., Berzak Hopkins, L. F., Divol, L., et al. (2018). Fusion energy output greater than the kinetic energy of an imploding shell at the National Ignition Facility. *Physical Review Letters, 120*, Article 245003.

Lehmberg, R. H., & Obenschain, S. P. (1983). Use of induced spatial incoherence for uniform illumination of laser fusion targets. *Optics Communications, 46*, 27.

Lehmberg, R. H., Schmitt, A. J., & Bodner, S. E. (1987). Theory of induced spatial incoherence. *Journal of Applied Physics, 62*, 2680.

Lindl, J. D. (1995). Development of the indirect-drive approach to inertial confinement fusion and the target physics basis for ignition and gain. *Physics of Plasmas, 2*, 3933.

Lindl, J. D., Amendt, P., Berger, R. L., et al. (2004). The physics basis for ignition using indirect-drive targets on the National Ignition Facility. *Physics of Plasmas, 11*, 339.

Lindl, J. D., Landen, O. L., Edwards, J. et al. (2014). Review of the National Ignition Campaign 2009–2012. *Physics Plasmas, 21*, 020501; "Erratum", *Physics Plasmas,21*, 129902.

Ma, T., & Betti, R. (2022). Report of the 2022 Fusion Energy Sciences Basic Research Needs Workshop. U.S. Department of Energy, Office of Science. https://events.bizzabo.com/IFEBRN 2022/home

Maiman, T. H., Hoskins, R. H., Haenens, I. J. D., et al. (1961). Stimulated optical emission in fluorescent solids, II. Spectroscopy and stimulated emission in ruby. *Physical Review, 123*, 1151.

Mason, P. D., Baneerjee, S., Ertel, K., et al. (2015). DiPOLE100: A 100 J, 10 Hz DPSSL using cryogenic gas cooled Yb:YAG multi slab amplifier technology. *Proceedings of SPIE, 9513*, 951302.

Mason, P., Baneerjee, S., Smith, J., et al. (2018). Development of a 100 J, 10 Hz laser for compression experiments at the high energy density instrument at the European XFEL. *High Power Laser Science and Engineering, 6*, Article e65.

Mason, P., Divoky, M., Ertel, K., et al. (2017). Kilowatt average power 100 J-level diode pumped solid state laser. *Optica, 4*, 438.

McMahon, J. M., Emmett, J. L., Holozrichter, J., et al. (1973). A glass disk laser amplifier. *IEEE Journal of Quantum Electronics, 9*, 992.

Michel, P., Glenzer, S. H., Divol, L., et al. (2010). Symmetry tuning via controlled crossed-beam energy transfer on the National Ignition Facility. *Physics of Plasmas, 17*, Article 056305.

Miquel, J.-L., Lion, C., & Vivini, P. (2016). The laser mega-joule: LMJ & PETAL status and program overview. *Journal of Physics Conference Series, 688*, 012067. https://doi.org/10.1088/1742-6596/688/1/012067

Miyaji, G., Miyanaga, N., Urushihara, S., et al. (2002). Three-directional spectral dispersion for smoothing of a laser irradiance profile. *Optics Letters, 27*, 725.

Montgomery, D. S. (2016). Two decades of progress in understanding and control of laser plasma instabilities in indirect drive inertial fusion. *Physics of Plasmas, 23*, Article 055601.

Morita, R., Inoue, T., De Zoysa, M., et al. (2021). Photonic-crystal lasers with two-dimensionally arranged gain and loss sections for high-peak-power short-pulse operation. *Nature Photonics, 15*, 311.

Moses, E. I., Lindle, J. D., Spaeth, M. L., et al. (2016). Overview: Development of the National Ignition Facility and the transition to a user facility for the ignition campaign and high energy density scientific research. *Fusion Science and Technology, 69*, 1. https://doi.org/10.13182/FST15-128

Myatt, J. F., Follett, R. K., Shaw, J. G., et al. (2017). A wave-based model for cross-beam energy transfer in indirect-drive inertial confinement fusion. *Physics of Plasmas, 24*, Article 056308.

Negel, J.-P., Loescher, A., Voss, A., et al. (2015). Ultrafast thin-disk multipass laser amplifier delivering 1.4 kW (4.7 mJ, 1030 nm) average power converted to 820 W at 515 nm and 234 W at 343 nm. *Optics Express, 23*, 21064.

Noda, S. (2006). Recent progresses and future prospects of two- and three-dimensional photonic crystals. *Journal of Lightwave Technology, 24*, 4554.

Nubbemeyer, T., Kaumanns, M., Ueffing, M., et al. (2017). 1 kW, 200 mJ picosecond thin-disk laser system. *Optics Letters, 42*, 1381.

Nuckolls, J., Wood, L., Thiessen, A., & Zimmerman, G. (1972). Laser compression of matter to super-high densities: Thermonuclear (CTR) applications. *Nature, 239*, 139.

Obenschain, S. P., Bodner, S. E., Colombant, D., et al. (1996). The Nike KrF laser facility: Performance and initial target experiments. *Physics of Plasmas, 3*, 2098.

Obenschain, S., Lehmberg, R., Kehne, D., et al. (2015). High-energy krypton fluoride lasers for inertial fusion. *Applied Optics, 54*, F103. https://doi.org/10.1364/AO.54.00F103

Obenschain, S. P., Schmitt, A. J., Bates, J. W., et al. (2020). Direct drive with the argon fluoride laser as a path to high fusion gain with sub-megajoule laser energy. *Philosophical Transactions of the Royal Society A, 378*, 20200031. https://doi.org/10.1098/rsta.2020.0031

Ogino, J., Tokita, S., Kitajima, S., et al. (2022). 10-J, 100-Hz conduction-cooled active mirror laser. *Optics Continuum, 1*, 1270. https://doi.org/10.1364/OPTCON.456554

Orth, C. D., Payne, S. A., & Kupke, W. F. (1996). A diode-pumped solid-state laser for inertial fusion energy. *Nuclear Fusion, 36*, 75.

Owadano, Y., Okuda, I., Matsumoto, Y., et al. (1999). Overview of 'Super-ASHURA' KrF laser program. *Fusion Engineering and Design, 44*, 91.

Pak, A., Zystra, A. B., Baker, K. L., et al. (2024). Observations and properties of the first laboratory fusion experiment to exceed a target gain of unity. *Phys Rev E, 109*, Article 025203.

Payne, S. A., Bibeau, C., Beach, R. J., et al. (1998). Diode-pumped solid-state lasers for inertial fusion energy. *Journal of Fusion Energy, 17*, 213.

Petrov, G. M., Giuliani, J. L., & Dasgupta, A. (2002). Electron energy deposition in an electron-beam pumped KrF amplifier: Impact of beam power and energy. *Journal of Applied Physics, 91*, 2662.

Petrova, T., Giuliani, J., Dasgupta, A., et al. (2022). Plasma kinetics of the NRL high efficiency electron-beam-pumped argon fluoride laser. NRL/6720/MR-2022/1. Naval Research Laboratory.

Phillips, J. P., Banerjee, S., Mason, P., et al. (2021). Second and third harmonic conversion of a kilowatt average power, 100-J-level diode pumped Yb:YAG laser in large aperture LBO. *Optics Letters, 46*, 1808.

Reagan, B. A., Baumgarten, C., Jankowska, E., et al. (2018). Scaling diode-pumped, high-energy picosecond lasers to kilowatt average powers. *High Power Laser Sci. Eng., 6*, Article e11.

Rhodes, M. A., Woods, B., Deyoreo, J. J., et al. (1995). Performance of large-aperture optical switches for high-energy inertial-confinement fusion lasers. *Applied Optics, 34*, 5312.

Rosocha, L. A., Hanlon, J. A., McLeod, J., et al. (1987). Aurora multikilojoule KrF laser system prototype for inertial confinement fusion. *Fusion Science and Technology, 11*, 497.

Ross, I., White, M., Boon, J., et al. (1981). Vulcan—A versatile high-power glass laser for multiuser experiments. *IEEE Journal of Quantum Electronics, 17*, 1653.

Rothenberg, J. E. (1997). Comparison of beam-smoothing methods for direct-drive inertial confinement fusion. *Journal of the Optical Society of America B: Optical Physics, 14*, 1664.

Rubery, M. S., Rosen, M. D., Aybar, N., et al. (2024). Hohlraum reheating from burning NIF implosions. *Physical Review Letters, 132*, Article 065104.

Saltarelli, F., Graumann, I. J., Lang, L., et al. (2019). Power scaling of ultrafast oscillators: 350 W average-power sub-picosecond thin-disk laser. *Optics Express, 27*, 31465.

Saraceno, C. J., Sutter, D., Metzger, T., et al. (2019). The amazing progress of high-power ultrafast thin-disk lasers. *Journal of the European Optical Society-Rapid Publications, 15*, 15.

Schmidt, B., & Schaefer, M. (2018). Advanced industrial laser systems and applications. *Proceedings of SPIE, 10525*, 1052502.

Schmitt, A. J., & Obenshcain, S. P. (2023a). The importance of laser wavelength for driving inertial confinement fusion targets. I. Basic physics. *Physics of Plasmas, 30*, 012701.

Schmitt, A. J., & Obenshcain, S. P. (2023b). The importance of laser wavelength for driving inertial confinement fusion targets. II. Target design. *Physics of Plasmas, 30*, 012702.

Seidl, P. A., Barnard, J. J., Faltens, A., et al. (2013). Research and development toward heavy ion driven inertial fusion energy. *Physical Review ST Accelerators and Beams, 16*, 024701.

Sekine, T., Kurita, T., Hatano, Y., et al. (2022). 253 J at 0.2 Hz, LD pumped cryogenic helium gas cooled Yb:YAG ceramic laser. *Optics Express, 30*, 44385.

Shaw, M. J., Edwards, B., Hirst, G. J., et al. (1993). Development of high-performance KrF and Raman laser facilities for inertial confinement fusion and other applications. *Laser and Particle Beams, 11*, 331.

Shaw, M. J., House, R. K., Williams, W. H., et al. (2008). Laser Performance Operations Model (LPOM): A computational system that automates the setup and performance analysis of the National Ignition Facility. *Journal of Physics Conference Series, 112*, 032022.

Shaw, M., Williams, W., House, R., et al. (2004). Laser Performance Operations Model (LPOM). *Proceedings of SPIE, 5341*, 73.

Skupsky, S., Short, R. W., Kessler, T., et al. (1989). Improved laser-beam uniformity using the angular dispersion of frequency-modulated light. *Journal of Applied Physics, 66*, 3456.

Soda, H., Iga, K., Kitahara, T., & Suematsu, Y. (1979). GaInAsP/InP surface emitting injection lasers. *Japanese Journal of Applied Physics, 18*, 2329.

Soures, J. M., Hutchinson, R. J., Jacobs, S. D., et al. (1983). OMEGA: A short wavelength laser for fusion experiments. In *Proceedings on 10th IEEE/NPSS Symposium Fusion Engineering, SOFE'83* (p. 1392). IEEE.

Spaeth, M. L., Manes, K. R., Kalanter, D. H., et al. (2016b). Description of the NIF laser. *Fusion Science and Technology, 69*, 25.
Spaeth, M. L., Wegner, P. J., Suratwala, T. I., et al. (2016a). Optics recycle loop strategy for NIF operations above UV laser-induced damage threshold. *Fusion Science and Technology, 69*, 265.
Speck, D. R., Bliss, E. S., Glaze, J. A., et al. (1981). The Shiva laser-fusion facility. *IEEE Journal of Quantum Electronics, 9*, 1599.
Streifer, W., Scifres, D. R., Harnagel, G. L., et al. (1988). Advances in diode laser pumps. *IEEE Journal of Quantum Electronics, 24*, 883.
Suda, A., Obara, M., & Noguchi, A. (1986). Performance characteristics of the ArF excimer laser using a low-pressure argon-rich mixture. *Journal of Applied Physics, 60*, 3791.
Tabak, M., Hammer, J., Glinsky, M. E., et al. (1994). Ignition and high gain with ultrapowerful lasers. *Physics of Plasmas, 1*, 1626.
Tamer, I., Keppler, S., Korner, J., et al. (2019). Modeling of the 3D spatio-temporal thermal profile of joule-class Yb^{3+}-based laser amplifiers. *High Power Laser Science and Engineering, 7*, Article e42.
Thompson, G., & Kirby, P. (1973). (GaAl)As lasers with a heterostructure for optical confinement and additional heterojunctions for extreme carrier confinement. *IEEE Journal of Quantum Electronics, QE-9*, 311.
Tsang, W. T. (1981). Extremely low threshold (AlGa)As modified multiquantum well heterostructure lasers grown by molecular-beam epitaxy. *Applied Physics Letters, 39*, 786.
Ueda, K., Nishioka, & H., Takuma, H. (1987). Characteristics of e-beam pumped KrF laser system. In *Excimer Lasers and Optics* (Vol. 0710). Proceedings on SPIE.
Volodin, B. L., Dolgy, S. V., Melnik, E. D., et al. (2004). Wavelength stabilization and spectrum narrowing of high-power multimode laser diodes and arrays by use of volume Bragg gratings. *Optics Letters, 29*, 1891.
Wang, Y., Chi, H., Baumgarten, C., et al. (2020). 1.1 J Yb:YAG picosecond laser at 1 kHz repetition rate. *Optics Letters, 45*, 6615.
Wisoff, P. J., Bowers, M. W., Erbert, G. V., et al. (2004). NIF injection laser system. *Proceedings of SPIE, 5341*, 146.
Yablonovitch, E. (1993). Photonic band-gap structures. *Journal of the Optical Society of America B: Optical Physics, 10*, 283.
Yamanaka, C., Kato, Y., Izawa, Y., Yoshida, K., et al. (1981). Nd-doped phosphate glass laser systems for laser-fusion research. *IEEE Journal of Quantum Electronics, 17*, 1639.
Yamanaka, C., Nakai, S., Matoba, M., et al. (1981). The LEKKO VIII CO_2 gas laser system. *IEEE Journal of Quantum Electronics, Q-17*, 1678.
Yoshida, M., De Zoysa, M., Ishizaki, K., et al. (2021). Photonic-crystal lasers with high-quality narrow-divergence symmetric beams and their application to LiDAR. *Journal of Physics: Photonics, 3*, Article 022006.
Yoshida, M., Katsuno, S., Inoue, T., et al. (2023). High-brightness scalable continuous-wave single-mode photonic-crystal laser. *Nature, 618*, 727. https://doi.org/10.1038/s41586-023-06059-8
Yue, J., Tanaka, K., Yonezawa, G., et al. (2022). Chip-scale high-peak-power semiconductor/solid-state vertically integrated laser. *Nature Communications, 13*, 5774. https://doi.org/10.1038/s41467-022-33528-x
Zheng, W., Wei, X., Zhu, Q., et al. (2016). Laser performance of the SG-III laser facility. *High Power Laser Science and Engineering, 4*, e21.
Zheng, Y., Kawashima, T., Satoh, N., et al. (2014). Stable-wavelength and narrow-bandwidth high-power external-cavity diode stack. *Appl. Phys. Exp., 7*, Article 092702.
Zhu, J., Zhu, J., Li, X., et al. (2018). Status and development of high-power laser facilities at the NLHPLP. *High Power Laser Science and Engineering, 6*, Article e55.
Zou, J. P., Le Blanc, C., Audebert, P., et al. (2008). Recent progress on LULI high power laser facilities. *IOP Journal of Physics Conference Series, 112*, 032021.
Zylstra, A. B., Kritcher, A., Ross, S. et al. (2021b). Improving implosion energy coupling at the NIF. In *28th IAEA Fusion Energy Conference (FEC 2020)*, Nice, France, May 10–15.

Zylstra, A. B., Hurricane, O. A., Callahan, D. A., et al. (2022). Burning plasma achieved in inertial fusion. *Nature, 601*, 542.

Zylstra, A. B., Kritcher, A. L., Hurricane, O. A., et al. (2021a). Record energetics for an inertial fusion implosion at NIF. *Physical Review Letters, 126*, Article 025001.

Chapter 7
High-Power, Ultrashort-Pulse Lasers

Abstract Gérard Mourou and Donna Strickland have opened "ultrafast optical science" by invention of the chirped pulse amplification (CPA). In parallel, generation of ultrashort-duration optical pulses of only a few-optical-cycles became possible due to development of the broad-band Titanium-doped sapphire laser by Moulton. Based on these inventions, ultra-high peak-power CPA lasers of 10 PW peak power and 10^{23} W/cm^2 focused intensity have been developed, opening a new field called "relativistic laser science". Development of 100 PW ~ EW class lasers are pursued with the Ti:sapphire laser, optical parametric chirped pulse amplification (OPCPA), and new schemes such as post-compression and coherent beam combining. High-energy, short-pulse (~ ps) CPA Nd:glass lasers are also developed for high-field and high-energy density science. In this Chapter, a brief overview of high-power short-pulse lasers is presented in Sect. 7.1. Then we will describe detailed accounts of the high-power ultrashort-duration CPA Ti:sapphire laser (Sect. 7.2), the high-energy CPA Nd:glass laser (Sect. 7.3), and other approaches for ultrahigh-power generation (Sect. 7.4). In the final section, developments of LD-pumped repetitive ultrafast lasers are presented (Sect. 7.5).

Keywords Chirped pulse amplification · CPA · Optical parametric chirped pulse amplification · OPCPA · Post compression · Coherent beam combining

7.1 Introduction—From TW to PW and Toward EW

Strickland and Mourou (1985) have invented chirped pulse amplification (CPA) to generate high-power, ultrashort optical pulses with compact lasers, and established the Center for Ultrafast Optical Science (CUOS) at the University of Michigan. In 1986, P. Moulton at MIT Lincoln Laboratory developed a new Ti-doped sapphire crystal (Ti:Al_2O_3), suitable as a tunable laser covering broad wavelength range in the near-infrared (Moulton, 1986). Generation of 60-fs pulses were demonstrated with a self-mode-locked Ti:sapphire laser by Spence et al. (1991). This work has

Y. Kato et al., *High Power Laser and Plasma Science*, Springer Series on Atomic, Optical, and Plasma Physics 130, https://doi.org/10.1007/978-3-032-08930-4_7

triggered development of 1–10 TW CPA Ti:sapphire lasers for investigation of ultrafast science at Washington State University (Zhou et al., 1995), CUOS (Du et al., 1995), University of California San Diego (Barty et al., 1996; Walker et al., 1999), Laboratoire d'Optiquée Applique (Chambaret et al., 1996), and JAERI Advanced Photon Research Center (Yamakawa et al., 1998).

In parallel with the Ti:sapphire laser development, the CPA concept was applied to the Nd:glass laser, initially on a compact Nd:glass CPA laser at Stanford University (Barty et al., 1988), then to existing high-energy Nd:glass laser facilities to realize 10s-TW laser pulses of ps-duration (Danson et al., 1993; Rouyer et al., 1993; Yamakawa et al., 1991).

The ultrahigh-peak power lasers have given broad impacts to science and applications (Mourou, 1997). Coordinating leading scientists in high-intensity lasers, the OECD Global Science Forum has published the recommendations: (1) systematic support is needed for laser development and applications, (2) large-scale facilities offer advantages at the highest performance level, and (3) co-ordination mechanisms in the community should be strengthened (OECD Global Science Forum, 2001). Following this report, PW-class Ti:sapphire lasers were developed at many laboratories worldwide, including Advanced Photon Research Center (APRC) in Japan (Aoyama et al., 2003; Kiriyama et al., 2010), Institute of Atomic and Molecular Science in Taiwan (Chu et al., 2004), Central Laser Facility in UK (Hooker et al., 2006), Berkeley Lab Laser Accelerator Center (BELLA) in USA (Nakamura et al., 2017), and IBS Center for Relativistic Laser Science (CoReLS) in Korea (Sung et al., 2017). Development of 10 PW Ti:sapphire laser was also started at Laboratoire d' Optique Appliquee (LOA) in France (Giambruno et al., 2011) and Shanghai Institute of Optics and Fine Mechanics (SIOM) in China (Gan et al., 2017).

To investigate high field science at extreme conditions, Extreme Light Infrastructure (ELI) was established as a user facility under the concept described in the ELI Whitebook (Mourou et al., 2011). The ELI, operated by the European Research Infrastructure Consortium (ERIC) of European Commission, is comprised of three facilities: ELI Beamlines (ELI-Beamlines) in Czech Republic, ELI Attosecond Light Pulse Source (ELI-ALPS) in Hungary, and ELI Nuclear Physics (ELI-NP) in Romania. In these ELI laboratories, advanced lasers are developed as user facilities with complementary architectures and performances, emphasising high peak power, ultrashort-duration pulses, and high average-power repetitive operation (https://eli-laser.eu/). The 10 PW operation has been demonstrated for the first time in 2020 with the CPA Ti:sapphire laser at ELI-NP (Lureau et al., 2020).

To further cultivate new regimes of high-field and high-energy science (Bulanov et al., 2015; Mourou et al., 2006; Tajima et al., 2017), next generation lasers with a peak power of exawatt (10^{18} W) to zettawatt (10^{21} W) are pursued with various concepts: chirped beam CPA, OPCPA, post compression, and coherent beam combining. Toward cultivating EW laser, a 100 PW laser based on OPCPA, named SEL-100 PW, is under construction in Shanghai, China (Du et al., 2023). High-energy, ps-duration CPA Nd:glass lasers are also developed for laser fusion, high energy density science, and OPCPA pumping. Another important aspect is the repetitive-operation high-intensity ultrafast lasers. The facilities and R&D activities of the

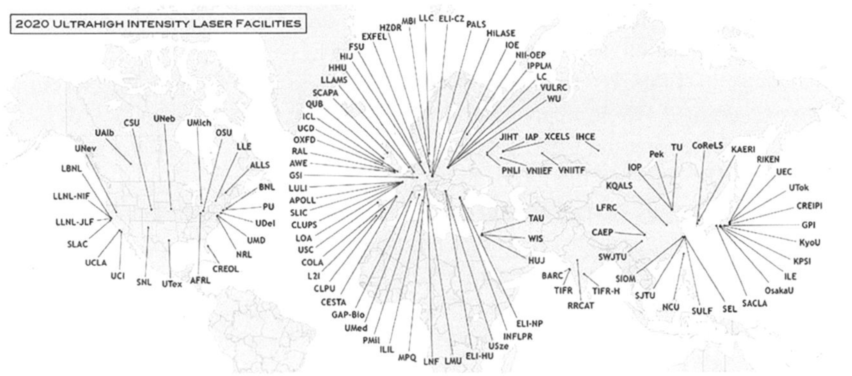

Fig. 7.1 Ultrahigh intensity laser facilities in the world in 2020. Curtesy of ICUIL. https://www.icuil.org/activities/laser-labs.html. (Refer to the ICUIL home page for the full names and countries of the laboratories shown in this map.)

petawatt and exawatt class lasers are reviewed in Bucksbaum et al. (2018) and Danson et al. (2019).

In 2020, there are over 100 ultrahigh intensity laser systems exceeding a peak power of 10 TW or an intensity at focus of 10^{19} W/cm^2, as shown in the ICUIL World Map (Fig. 7.1) with its interactive version providing performance of each facility (https://www.easymapmaker.com/map/ICUIL_World_Map_v3). To support co-ordination in the high-intensity laser community, the International Committee on Ultra-High Intensity Lasers (ICUIL) was established as a Working Group of the International Union for Pure and Applied Physics (IUPAP) in 2005, organizing The International Conference on Ultra-High Intensity Lasers (ICUIL) in every two years. In parallel with ICUIL, Asian Intense Laser Network was started, organizing the Asian Symposium on Intense Laser Science (ASILS) in every 1–2 years in various countries in Asia. Reflecting rapid evolution in high-power ultrafast lasers and their applications, many other conferences are held actively in these fields.

7.2 High-Power CPA Ti:Sapphire Laser

7.2.1 Introduction

In recent years, the 10 PW Ti:sapphire laser systems have been developed (Li et al., 2018, 2022; Lureau et al., 2020; Radier et al., 2022), and the laser intensity at focus of over 10^{23} W/cm^2 has been realized (Yoon et al., 2021). Even higher peak power could be generated with the Ti:sapphire laser by increasing the beam aperture, if limitations accompanied with increasing the beam aperture are overcome, such as growth of high-quality, larger-aperture Ti:sapphire crystals and suppression of spontaneous emission amplification in the large Ti:sapphire crystals.

We show in Table 7.1 the high-power CPA Ti:sapphire lasers with the output power exceeding 1 PW in many institutes in the world, including the 10 PW lasers at ELI-NP (Lureau et al., 2020), Apollon (Papadopoulos et al., 2016) and SIOM (Li et al., 2018). Most of the lasers shown in Table 7.1 are operated with the pulse duration of over ~ 25 fs. Development of a high-power laser with a pulse duration shorter than ~ 25 fs requires thorough control of the spectral phase over a whole spectral bandwidth: 200 nm to generate a 15-fs pulse, for example.

The repetition rate of high-power Ti:sapphire lasers exceeding ~ 1 PW is ~ 0.1 Hz or lower at present, limited by the repetition rate of the pump laser. For example, high-energy flash-lamp pumped Nd:glass lasers delivering over 500 J in total are used to pump the final amplifier of the 10 PW lasers. Due to advances in the LD-pumped high-energy solid-state lasers, a 1-PW, 10 Hz CPA Ti:sapphire laser has been realized at ELI Beamlines (Häfner et al., 2017), and extension to higher rep-rate and higher power is envisioned by utilizing the diode-pumped solid state lasers (DPSSLs) described in Sect. 6.3.

7.2.2 Architecture of High-Power CPA Ti:Sapphire Laser System

A schematic diagram of the CPA concept has been shown in Fig. 2.15. In a high-power Ti:sapphire laser system, a combination of a grating stretcher (Öffner stretcher) and a grating compressor (Treacy compressor) is used for producing a multi-PW pulse by stretching an ultrashort pulse by ~ 10,000 times to 10s-ns duration and recompressing to an ultrashort pulse, while keeping the intensity low to avoid self-focusing, laser damage and saturation during amplification and propagation. To describe the complex high-power CPA Ti:sapphire laser systems in detail, we show in Fig. 7.2 an example of a layout with reference to the 10 PW high power laser system at ELI-NP (Lureau et al., 2020), with addition of a plasma mirror for pre-pulse reduction and focusing optics for target irradiation. The laser system is composed of three major sections: (a) front end, (b) power amplification, and (c) pulse compression and target irradiation.

At the front-end section, ~ 20 ps duration, ~ 10 mJ energy, ~ 100 nm bandwidth pulses with high contrast are stably generated at a 10 Hz repetition rate. In the power amplification section, the laser pulse is stretched in time to ~ 20 ns and amplified with multiple-stage power amplifiers, while gradually expanding the beam diameter, generating output laser pulses of ~ 300 J energy in 130 mm beam diameter at 1 shot/min repetition rate.

In the pulse compression and target irradiation section, the high energy laser pulse is compressed in time to < 23 fs, resulting in 243 J energy and over 10 PW peak power. The compressed high power laser pulse is focused with an off-axis parabolic mirror on a target. A plasma mirror is used after pulse compression in other laser

Table 7.1 High-power Ti:sapphire lasers with the output power of over ~ 1 PW, in operation (op) or under development (dev) in 2023

Country	Institute	Facility	Peak power (PW)	Energy (J)	Pulse width (fs)	Rep rate	Status
UK	RAL	Astra Gemini	0.5 (× 2)	15	30	20 s	op
France	APOLLON	F1	10 (4 in 2022)	150	15	1 min	op
Germany	CALA	ATLAS-3000	3	60	20	1 Hz	op
	HZDR	DRACO	1	30	30	1 Hz	op
Spain	CLPU	VEGA 3	1	30	30	1 Hz	op
Czechia	ELI-BL	L3 HAPLS	1 (0.5 in 2021)	30	< 30	10 Hz	op
Romania	ELI-NP		10 (× 2)	242	22.7	1 min	op
	INFLPR	CETAL-PW	1	25	25	0.1 Hz	op
Hungary	ELI-ALPS	HF-2PW	2	> 34	< 17	10 Hz	dev
USA	LBNL	BELLA	1.4	42	30	1 Hz	op
	SLAC	MEC Upgrade	1	150	150	10 Hz	dev
	U. Nebraska	Diocles	1	30	30	0.1	op
Canada	INRS	ALLS	0.5	10	20	2.5 Hz	op
Japan	KPSI	J-KAREN-P	1	30	30	1 Hz	op
Korea	CoReLS		4.2	83	19.6	1 Hz	op
China	SIOM	SULF-10PW	10	217	21	3 min	op

Two-beam facilities are indicated by "× 2" in the peak power, whereas the laser performance is for a single laser beam

Abbreviations: RAL—Rutherford Appleton Laboratory, CALA—Centre for Advanced Laser Applications, HZDR—Helmholtz-Zentrum Dresden-Rossendorf Laboratory, CLPU—Spanish Pulsed Lasers Center, ELI-BL—ELI Beamlines, ELI-NP—ELI Nuclear Photonics, ELI-ALPS—ELI Attosecond Light Pulse Sources, INFLPR—Center for Advanced Laser Technologies National Institute for Laser, Plasma, and Radiation Physics, LBNL—Lawrence Berkeley National Laboratory, SLAC—Stanford Linear Accelerator Center, INRS—National de la Recherche Scientifique, KPSI—Kansai Photon Science Institute, CoReLS—IBS Center for Relativistic Laser Science, SIOM—Shanghai Institute of Optics and Fine Mechanics

systems to clean residual pre-pulses for irradiation of ultra-thin targets (Thaury et al., 2007).

A small fraction of the compressed pulse is led to the diagnostics station where many laser parameters are monitored, such as energy, near field pattern, far field pattern, wavefront, and spectral phase. The measured wavefront is fed back to a deformable mirror for wavefront correction. The measured spectral phase is fed

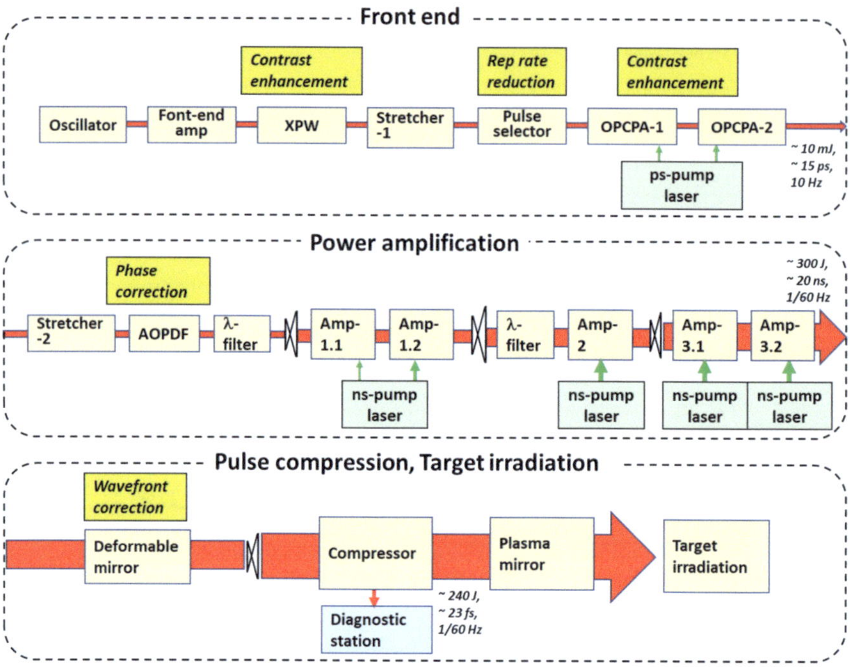

Fig. 7.2 Schematic layout of a high-intensity CPA Ti:sapphire laser system, which is grouped into three sections: front-end, power amplification, and pulse compression and target irradiation. Abbreviations: XPW—cross-polarized wave generation, AOPDF—acousto-optic programmable dispersive filter, λ-filter—spectral filter to pre-compensate the spectral changes in the amplifiers. Detailed accounts of this layout are described in the text

back to an acousto-optic programmable dispersive filter (AOPDF) located before the amplifiers to generate a transform-limited pulse.

7.2.3 Modelling for High-Peak-Power, Ultrashort-Pulse Laser Systems

In Chaps. 4 and 6, we have seen that the computer codes MALAPROP (Simmons et al., 1981) and its outgrowth LPOM (Laser Performance and Operations Model) (Shaw et al., 2004, 2008) took key roles in the design, performance evaluation, and operation of the high energy Nd:glass laser systems at LLNL. Similar codes are necessary for the short-pulse (ps-duration) and ultrashort-pulse (fs-duration) high-peak-power laser systems, such as CPA Nd:glass lasers, CPA Ti:sapphire lasers, and OPCPA lasers.

A numerical code for analyzing propagation of a broad band laser pulse over a whole laser system is required for these laser systems. The Miró software developed by Morice (2003) incorporates various functions required for analyzing the beam propagation in short- and ultrashort-pulse high power laser systems: the paraxial nonlinear Schrödinger approximation and broadband computation including dispersion up to the second order. The code comprises main optical components: a four-level multi-pass amplifier including both saturation and gain narrowing; broadband and noncollinear frequency conversion; the grating, including high-order dispersion effects; and laser damage effects. This code has been validated with 60-kJ CPA Nd:glass laser, Laser Integration Line (LIL), a prototype of LMJ, and is used to design and analyze the Apollon 10-PW laser (Zou et al., 2022).

7.2.4 *Performance of High-Power CPA Ti:Sapphire Laser System*

High power CPA laser systems are designed and operated to generate well controlled high intensity laser pulses. The following are some of the important parameters which characterize the performance of high-power ultrashort-pulse CPA Ti:sapphire laser systems.

(a) high peak power,
(b) transform-limited ultrashort pulse,
(c) high contrast ratio,
(d) diffraction-limited focusing,
(e) pointing stability.

We will describe how these parameters are controlled in the ultrashort CPA Ti:sapphire laser systems.

7.2.4.1 High Peak Power

Generation of a high peak-power laser pulse requires considerations on the broad bandwidth and uniform spatial distribution without being affected by parasitic lasing in the high gain amplifiers.

(a) **Broad Spectral Width**

A Ti:sapphire crystal has a broad (~ 110 nm FWHM) absorption band with a peak at ~ 500 nm and a broad fluorescence (230 nm FWHM) with a peak at 780 nm (Moulton, 1986). During amplification, the spectral distribution changes significantly due to spectral gain narrowing and gain saturation (Barty et al., 1996). As an example of the spectral evolution, Fig. 7.3 shows spectra of the input and amplified spectra, normalized at the peak, of the final three-pass Ti:sapphire amplifier of 100-mm-diameter for the 30-fs, 1.5-PW laser at GIST (Yu et al., 2012). After amplification, the spectral distribution becomes flat due to amplification saturation of the short wavelength components. Due to the flat spectral phase as shown in Fig. 7.3a, a 30

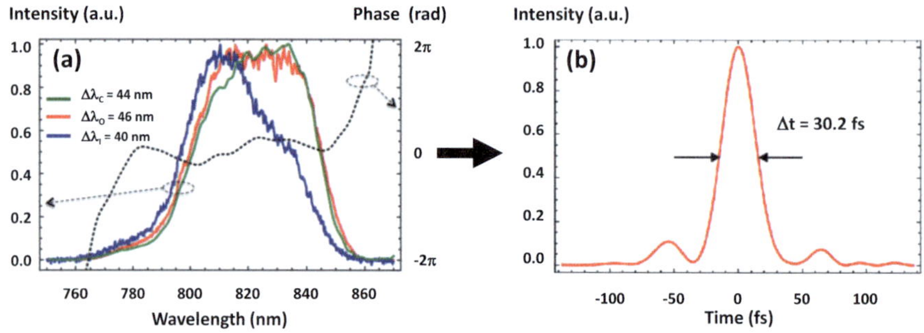

Fig. 7.3 **a** Evolution of the laser spectrum due to amplification in a Ti:sapphire crystal. Spectra of the input (measured in blue line) and amplified (measured in red line and calculated in green line) pulses. Also shown is the measured spectral phase. **b** Temporal pulse profile reconstructed from the spectral distribution of the amplitude and phase of the amplified pulse. Reprinted from Yu et al. (2012). © 2012 Optical Society of America (Note that the characters in the original figure are enlarged for visibility.)

fs-duration laser pulse is obtained after pulse compression, which is close to the Fourier-transform limit of 26 fs.

High power laser beam of broad spectral width has been generated with various methods to control the spectral distribution: spectral filtering of high-gain region with an etalon, resulting in 61 nm FWHM spectral width and 32.9 fs pulse width (Aoyama et al., 2003); strongly suppressing the long wavelength component of the OPCPA output, resulting in 84 nm FWHM width and 18.9 fs pulse width (Sung et al., 2017); two spectral filter mirrors at the entrance of the amplifiers (shown by "λ-filter" in Fig. 7.2), resulting in over 70 nm FWHM width and less than 22.7 fs pulse width (Lureau et al., 2020). In designing the 10 PW Apollon laser, the spectral evolution of a broadband pulse in an amplifier chain has been calculated to design an optimum spectral filter by solving the Frantz–Nodvik equation in small time and frequency steps to generate the output spectrum broad enough to produce a 15 fs output pulse (Giambruno et al., 2011).

(b) **Ti:Sapphire Amplifier**

Since the fluorescence lifetime of Ti:sapphire is 3.2 μs, intense pumping of sub-μs duration is required to create high population inversion density. For amplification of a pulsed laser beam, Ti:sapphire crystals of typically circular shapes are pumped along the axial direction with the second-harmonic (532 nm) of Nd:YAG or Nd:glass laser pulses of tens-ns duration. The ~ 800 nm laser pulse is amplified in the Ti:sapphire crystals in several passes (typically 3–5 passes) in each crystal. The energy of the Ti:sapphire laser is limited partly by the available size of the Ti:sapphire crystal with high optical quality (wavefront distortion and striation). The largest crystals used in the current laser systems are 200 mm in diameter at ELI-NP (Lureau et al., 2020) and 235 mm at SIOM (Li et al., 2018).

(c) **Parasitic Lasing**

The parasitic lasing in the Ti:sapphire crystals are suppressed by index-matched absorbing solid claddings (Aoyama et al., 2003; Sung et al., 2017) or liquids (Lureau et al., 2020). When this approach is not sufficient, especially in large aperture multi-pass amplifiers under strong pumping, the timings between the pumping pulses and the incoming pulses are adjusted to keep the transverse gain low by extracting the stored energy by laser amplification before the next pumping (Li et al., 2018; Lureau et al., 2020).

(d) **Spatial Uniformity**

Generation of a high energy laser pulse with uniform spatial distribution (high filling factor) and broad spectral width at the output of the final amplifier is required to generate a high-peak power, ultra-short laser pulse. Nonuniformities of the laser beam arising from Fresnel diffraction during the beam propagation is alleviated by image-relaying of the laser beam through the whole laser chain using the beam expander/spatial filters as implemented in high-energy glass lasers (see Sect. 3.5). Off-axis parabolic telescopes are used instead of refractive lenses to avoid dispersion and chromatic aberration (Zou et al., 2015). The spatial uniformity of the amplified beam is also controlled by pumping each Ti:sapphire crystal with spatially uniform pumping beams, which are realized by spatially homogenizing the pump beam (Kiriyama et al., 2010), or by image-relaying the uniform output of the pumping lasers on the crystal.

(e) **Laser Fluence**

Noting that the stimulated emission cross section of Ti:sapphire is $\sigma_{||} = 2.8 \times 10^{-19}$ cm^2 at 795 nm for polarization parallel to the crystal axis, the saturation fluence of the Ti:sapphire is $E_s = h\upsilon/\sigma \cong 0.9$ J/cm^2. A high laser fluence of ~ 2 J/cm^2 is required for efficient extraction of the stored energy, especially in the final amplifiers.

(f) **Laser Damage**

Laser damage is a serious limiting factor of the output power of ultrahigh-power lasers. In the configuration shown in Fig. 7.2, approximately 330 J energy is extracted with a circular beam of 130-mm diameter and 900-ps duration, corresponding to the intensity of 2.7 GW/cm^2 and the average fluence of 2.5 J/cm^2. To avoid the laser damage to the down-streams optics, the diameter of the amplified beam is expanded to 180-mm to reduce the fluence to 1.4 J/cm^2 at the deformable mirror and further to 450 mm to reduce the fluence to 0.23 J/cm^2 at the compressor gratings.

7.2.4.2 Transform-Limited Ultrashort Pulse

A Fourier-transform-limited (FTL) pulse is generated when the spectral phase is constant over the spectral bandwidth. Referring to Sect. 2.3, the FTL pulse is generated when the group delay dispersion (GDD), the third order dispersion (TOD), the fourth order dispersion (FOD) and higher order terms are controlled to make the phase flat over the whole spectrum. We will describe various methods to control the spectral phase to generate a high power FTL pulse.

(a) **Front-End**

The requirement to the frontend of a high-power laser system is to stably generate a broadband pulse of ~ 10-mJ energy for subsequent amplification with the Ti:sapphire amplifiers. In the next subsection, we will describe various methods to generate such pulses. Here we assume that an FTL pulse is generated in the frontend.

(b) **Global Dispersion Management**

A FTL pulse generated in the front-end propagates in a laser system which is comprised of a pulse stretcher, multi-stage amplifiers, and a pulse compressor. Dispersion compensation over the whole system is required to generate a high-power FTL pulse at the output. This condition is expressed as

$$\phi^{\text{system}} = \phi^{\text{stretcher}} + \phi^{\text{amplifier}} + \phi^{\text{compressor}} = 0, \tag{7.1}$$

where the three terms on the right-hand side refer to the phases of the stretcher, the amplifier, and the compressor, respectively. (Here "amplifier" includes all optical components which the laser beam passes through.) Expanding the phase in a Taylor series as Eq. (2.142), this condition is met if Eq. (7.1) is satisfied in each order of the expansion:

$$\begin{aligned}
\phi_2^{\text{stretcher}} + \phi_2^{\text{amplifier}} + \phi_2^{\text{compressor}} &= 0\,,\\
\phi_3^{\text{stretcher}} + \phi_3^{\text{amplifier}} + \phi_3^{\text{compressor}} &= 0\,,\\
\phi_4^{\text{stretcher}} + \phi_4^{\text{amplifier}} + \phi_4^{\text{compressor}} &= 0\,,
\end{aligned} \tag{7.2}$$

and in higher order terms.

The second, third and fourth-order dispersions of the parallel gratings for pulse compression are given in Eqs. (2.179), (2.180) and (2.181), respectively. Since the dispersion of an aberration-free stretcher are given by reversing the sign of these equations, dispersions of the stretcher and the compressor cancel each other when the same configurations (groove density, incidence angle and grating separation) are used for the stretcher and the compressor.

With the amplifier materials added in the laser system, the dispersion of the amplifier can be compensated by changing the configurations of the stretcher and/or the compressor (Kane & Squire, 1997a). Using Eq. (7.2) for the second and the third order dispersions and Eq. (2.180) for $\phi_3^{\text{compressor}}$, the second and the third order dispersions of the whole system are compensated simultaneously by adjusting the diffraction angle of the compressor θ_d as

$$\frac{\lambda}{d}\frac{\sin\theta_d}{\cos^2\theta_d} = \frac{-2\pi c}{\lambda}\frac{\phi_3^{\text{stretcher}} + \phi_3^{\text{amplifier}}}{\phi_2^{\text{stretcher}} + \phi_2^{\text{amplifier}}} - 1. \tag{7.3}$$

Although the second and the third order dispersions are compensated for in this method, the fourth-order dispersion introduced in the compressor is not compensated by the stretcher. Kane and Squire (1997b) have shown that it is possible to control the fourth-order dispersion of the laser system by using a grating in the compressor with a larger groove-density than the groove density of the grating in the stretcher and by taking advantage of the material dispersion by multi-passing in the amplifier and other optics. Still the compressed pulse is accompanied with pedestals due to residual spectral dispersion.

Toward global dispersion management of a high-power CPA laser system, we compare in Table 7.2 the dispersion properties (signs of ϕ_2, ϕ_3, ϕ_4, and ϕ_5) of a grating stretcher, a grating compressor, optical materials, a grism pair, and an acousto-optic dispersive filter. The ϕ_2, ϕ_3 and ϕ_4 of transparent materials are derived from the material dispersion as expressed in Eqs. (2.158), (2.161) and (2.162), respectively. As a numerical example, the dispersion terms of ϕ_2, ϕ_3 and ϕ_4 are calculated as 1.2478×10^5 fs^2, 9.9957×10^4 fs^3 and -4.3911×10^4 fs^4, respectively, at 800 nm for a 10 PW CPA Ti:sapphire laser system, which is comprised of a regenerative amplifier, three-stage multi-pass amplifiers, two-stage power amplifiers, and several other optical elements (Li et al., 2017).

(c) **Static Compensation for Material Dispersion**

The material dispersion can be compensated if we can find an optical device which has dispersions with similar magnitudes, but opposite signs compared with the material dispersion in each order. Li et al. (2015) proposed to use a grism pair for compensation of the material dispersions of high-power CPA Ti:sapphire lasers. Noting that the ratio of TOD to GVD (ϕ_3/ϕ_2) is ~ 0.7 for most of optical materials, the grism can be designed to take the similar ϕ_3/ϕ_2 ratio, but with an opposite sign.

A grating-prism (grism) is an optical device composed of a prism and a grating, where an optical beam is incident to a grating after passing through a prism. A transmission grism shown in Fig. 7.4a is used in astronomical instruments as a dispersive optics for spectral imaging (Deen et al., 2017). Tournois (1993) has found that a compressor composed of a pair of transmission grisms can have zero third-order

Table 7.2 Dispersion properties of the material and several types of stretchers and compressors used to control the dispersion of high power, ultrafast laser systems

	ϕ_2	ϕ_3	ϕ_4	ϕ_5
Grating stretcher	+	−	+	−
Grating compressor	−	+	−	+
Material dispersion	+	+	−	
Grism pair	−	−	−	−
A-O dispersive filter	+/−	+/−	+/−	+/−

The signs of the dispersion terms are shown by + , −, or +/−. (The A-O dispersive filter can be used in both plus and minus signs.) In general, the FOD (ϕ_4) of transparent materials is small in magnitude and takes either plus or minus signs depending on the material and the wavelength region dispersion when we consider a diffraction equation for the transmission grism, which is given by $\sin\theta_d = m\lambda/d - n\sin\theta_i$ where θ_i and θ_d are the incidence and diffraction angles, n is the refractive index of the prism, d is the grating period, and m is the diffraction order. Extending this concept, Kane and Squire (1997a) have proposed a new type of stretcher/compressor composed of a pair of transmission grisms which has negative ϕ_2 and negative ϕ_3 with its ϕ_3/ϕ_2 ratio close to that of the material dispersion, thus enabling simultaneous second- and third-order compensation. A more efficient (higher throughput) stretcher based on a pair of reflection grisms was developed (Gibson et al., 2006) and applied to generate sub-10-fs pulses by controlling the phase over the spectral range of 720–1000 nm in an OPCPA system (Tavella et al., 2007).

A layout of a stretcher composed of a pair of reflection grisms is shown schematically in Fig. 7.4b, where an optical path is indicated by the line in red. In this configuration, the dispersion takes place in propagation through the prism, diffraction by the grating, and angular dispersion at the exit of the prism. The reflection grating is separate from the prism with a small space of ~ 1 mm. The laser beam dispersed by the first grism is further dispersed and collimated by the second grism. The collimated beam out of the second grism may be returned to the grism pair with a roof mirror to compensate for the spectral displacement and to further increase the dispersion. The dispersion properties of this device were analyzed in detail by optical raytracing by Zheng and Zacharias (2009), where dependence of the ϕ_3/ϕ_2 ratio on various grism parameters was investigated for application to high-power non-collinear optical chirped pulse amplification.

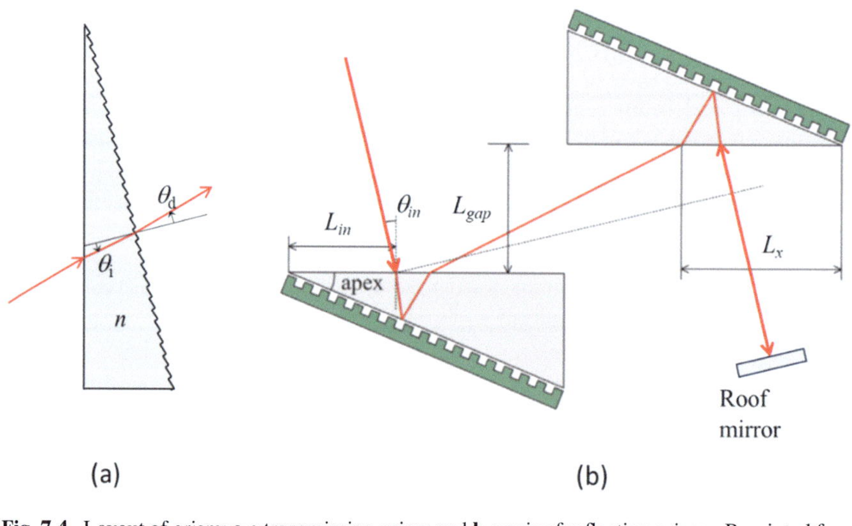

Fig. 7.4 Layout of grism: **a** a transmission grism and **b** a pair of reflection grisms. Reprinted from Li et al. (2015), with permission from Elsevier; permission conveyed through Copyright Clearance Center, Inc. ©2015 Elsevier

In application to high-power OPCPA Ti:sapphire lasers, the grism-pair is useful to compensate for the material dispersion by placing it after the Öffner stretcher. With the grism-pair installed in a laser system, the setting of the compressor grating is now given by the following equation:

$$\frac{\lambda}{d}\frac{\sin\theta_d}{\cos^2\theta_d} = \frac{-2\pi c}{\lambda}\frac{\phi_3^{stretcher} + \phi_3^{grism} + \phi_3^{amplifier}}{\phi_2^{stretcher} + \phi_2^{grism} + \phi_2^{amplifier}} - 1, \tag{7.4}$$

where the superscript *grism* refers to the grism pair. Global optimization to minimize the dispersion of a high-power Ti:sapphire laser system using the grism pair was investigated by Li et al. (2015). In parallel with minimizing GDV and TOD to zero by adjusting the compressor in accordance with Eq. (7.7), the fourth-order dispersion was also minimized by adjusting the grism parameters such as the incidence angle θ_{in}, the gap distance L_{gap}, and the incident point to the second grism L_{x}. Figure 7.5 compares the intensity and the spectral phase of a compressed pulse, where high-order dispersions were compensated by three different approaches: (a) GVD and TOD are compensated by adjusting the compressor (incidence angle and the grating separation) in accordance with Eq. (7.6) (black dashed line marked by TODc), (b) FOD is reduced while leaving some amount of TOD to reduce the pulse width by adjusting the compressor grating (blue dotted line marked by TOD&FOD), and (c) the GOD, TOD and FOD are minimized by adjusting the grism pair and the compressor gratings (red dotted curve marked by FODc). The intensity and the spectral phase of (c) is very close to a FTL pulse, shown in a shaded region, over the wavelength of 765–840 nm with only small deviation due to the residual fifth order dispersion (FiOD).

This approach was applied to a multi-PW CPA Ti:sapphire laser, resulting in a 5 PW output power with a pulse duration of 24 fs and the spectral phase of less than 1.2 rad over output spectral width of 80 nm. Due to the static property of the grism pairs, the pulse duration remained stable over a long duration of time (Li et al., 2017).

(d) **Dynamic Dispersion Control**

In addition to the static compensation of material dispersion, it is necessary to control residual dispersion with a flexible device. Dynamic control of the spectral phase and the spectral intensity is possible with an acousto-optic programmable dispersive filter (AOPDF) to adjust these parameters to the desired values at the amplifier output by forming a feedback loop (Tournois, 1997). As shown in Fig. 7.6, an acoustic wave is propagated in an acousto-optic crystal, in which an optical wave polarized in the x-direction (ordinary axis) is diffracted into an optical wave polarized in the y-direction (extraordinary axis). Taking the propagation direction as z with $z = 0$ at the input of the optical wave, the diffraction takes place at when the following phase matching condition is satisfied:

$$K[z(\omega)] = k_1(\omega) - k_2(\omega) = \frac{\omega}{c}[n_1(\omega) - n_2(\omega)]. \tag{7.5}$$

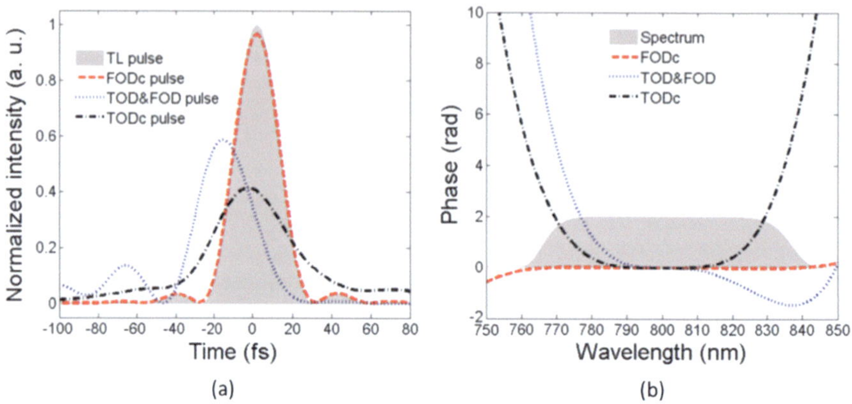

Fig. 7.5 Normized intensity (**a**) and the spectral phase (**b**) of a compressed pulse, where the spectral phase due to material dispersion is compensated for by three different approaches, as described in the text. The red dotted curve (marked by FODc) shows that the spectral phase is controlled to within 0.2 rad over 765–840 nm with a grism pair, resulting in a temporal shape close to an FTL pulse. Reprinted from Li et al. (2015), with permission from Elsevier; permission conveyed through Copyright Clearance Center, Inc. ©2015 Elsevier

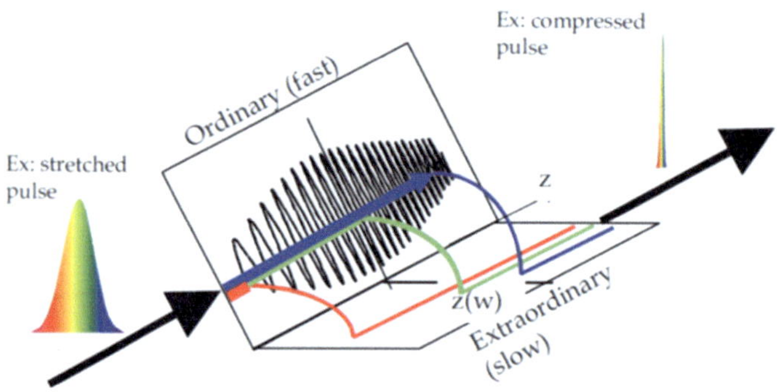

Fig. 7.6 Schematic of an AOPDF, when it is used for pulse compression of a chirped pulse. A stretched pulse of negative chirp polarized in the ordinary axis (x-axis) is diffracted to an optical wave polarized in the extraordinary axis (y-axis) by a chirped acoustic wave in an acousto-optic crystal. The diffracted wave forms a compressed pulse after propagation due to material dispersion. A transform-limited pulse can be generated by properly designing the spectrum of the acoustic wave. Reprinted from Oksenhendler and Forget (2010). © 2010 InTech

Here K, $k_1(\omega)$, and $k_2(\omega)$ are the wavenumbers of the acoustic wave, the incident wave, and the diffracted wave, and $n_1(\omega)$ and $n_2(\omega)$ are the refractive indices for the $k_1(\omega)$ and $k_2(\omega)$ waves, respectively. When an incident wave is composed of broad spectrum, a frequency component ω, is diffracted only at $z = z(\omega)$ where the phase matching condition is satisfied.

Writing the group velocities of the optical waves 1 and 2 as $v_{g1}(\omega)$ and $v_{g2}(\omega)$, respectively, the group delay of the frequency component ω at the output of the crystal of length L is given by

$$\tau(\omega) = \frac{z(\omega)}{v_{g1}(\omega)} + \frac{L - z(\omega)}{v_{g2}(\omega)}. \tag{7.6}$$

When we specify the group delay $\tau(\omega)$ for the frequency ω, the interaction point $z(\omega)$ is determined from this equation. Since the wavenumber of the acoustic wave K at $z(\omega)$ is determined as

$$K[z(\omega)] = \frac{\omega}{\Delta v_p(\omega)}, \tag{7.7}$$

where $\Delta v(\omega) = c/[n_2(\omega) - n_1(\omega)]$ is the differential phase velocity, the spectrum of the acoustic wave is determined by specifying the group delay. The amplitude of the acoustic wave is also determined from the amplitude of the optical wave, since the diffraction efficiency from the wave 1 to the wave 2 is proportional to the amplitude of the acoustic wave. Using these relations, the acoustic transducer can be programmed to generate the acoustic wave of the wavenumber and the amplitude in the crystal suitable for generating the laser pulse with the specified spectral phase and amplitude distributions (Verluise et al., 2000).

In the laser configuration shown in Fig. 7.2, the spectral phase and the amplitude after the pulse compressor are fed to the AOPDF located before the amplifier chain and the output pulse duration is optimized by an iterative process. Figure 7.7 shows (a) the spectral phase and the amplitude measured at the laser output and (b) the temporal pulse shape reconstructed from these distributions of the 10 PW laser at ELI-NP. The spectral phase is flat to within 1.2 rad over 85 nm bandwidth and the pulse duration of 22.7 fs is close to the FTL-pulse duration calculated for the measured spectrum (Lureau et al., 2020).

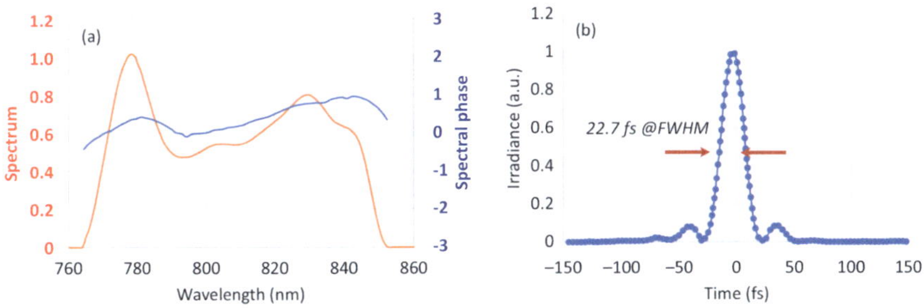

Fig. 7.7 **a** The spectral phase and amplitude of the laser pulse at the output of the 10 PW laser system and **b** a pulse shape reconstructed from these measurements. Reprinted from Lureau et al. (2020). © 2020 Cambridge University Press (Note that the characters in the original figure are enlarged for visibility.)

7.2.4.3 High Contrast Ratio

In Fig. 7.8, we show a schematic of a temporal profile of a high-intensity laser pulse on target, where the intensity is in a logarithmic scale (normalized to ~ 10^{20} W/cm^2) and the time measured from the peak of a main pulse is in an arbitrary scale. On both sides of $t < 0$ and $t > 0$, a high intensity, ultrafast laser pulse is often accompanied with (a) nano-second duration low intensity pedestals, (b) 10s-ps duration pedestal with exponential increase in time to $t = 0$, and (c) short duration pre-pulses and post-pulses (Ziegler et al., 2021). These pedestals and pre/post-pulses are characterized by the "contrast ratio" which is the intensity ratio of the pedestals and pre/post pulses to the peak of the main pulse.

A low-density plasma is formed when a target is pre-ionized by the pedestals and pre-pulses before arrival of the high-intensity main pulse. The preformed plasmas are cold since the intensities of the pedestals and pre-pulses are low in general. The threshold for plasma creation, which corresponds to the damage threshold of the target material, depends on the target material and the pulse duration τ, with the threshold energy scaling as $\tau^{1/2}$ (Stuart et al., 1995). The single-shot damage threshold of a fused silica has been measured to be 10^{11} and 10^{12} W/cm^2 at $\tau = 1$ ns and 10 ps respectively with 800 nm laser (Tien et al., 1999). The laser contrast

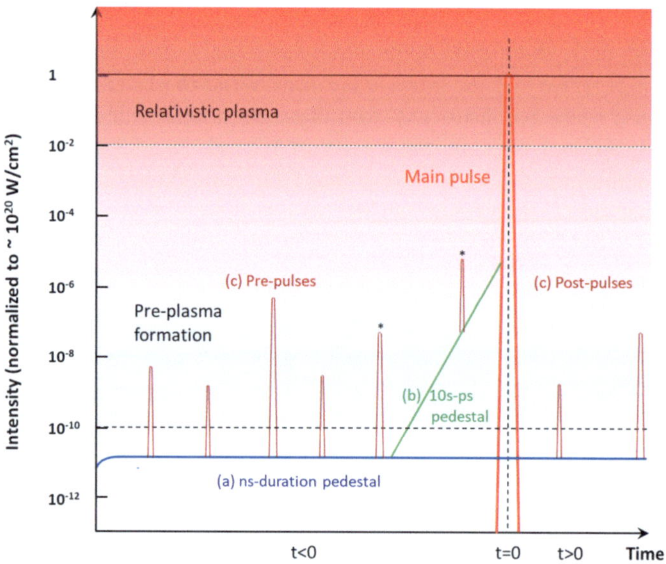

Fig. 7.8 Schematic of a temporal profile of a high-intensity laser pulse on target, where the intensity is normalized to ~ 10^{20} W/cm^2 and the time is in an arbitrary scale. An ultrafast high intensity laser pulse, marked by "main pulse" is accompanied with **a** ns-duration pedestal, **b** 10s-ps duration pedestal with exponential increase in time, and **c** pre-pulses and post-pulses. The pre-pulses marked by the asterisks arise from the post-pulses as described in the text. The pre-plasma is formed at the intensity of ~ 10^{11} W/cm^2 or lower, depending on the target material

of better than 10^{-10} is required in general to implement clear experiments, since the pedestal has longer duration of over 10 ns and many targets such as plastics and metals have damage thresholds lower than that of fused silica. In the following, we describe the origins of the pedestals and pre/post pulses and the approaches to reduce them.

(a) **Nanosecond-Duration Pedestal**

In typical high-power Ti:sapphire CPA laser systems, an ultrafast laser pulse of nJ energy generated in an oscillator is amplified up to ~ 100 μJ with a regenerative (or multi-pass) amplifier located after the oscillator in the front-end of the laser system. The intensity contrast ratio of the output laser pulse from the regenerative amplifier is ~ 10^6–10^7 due to the amplified spontaneous emission (ASE) (Ivanov et al., 2003), and degrades further during downstream amplification due to saturated amplification of the main pulse in the power amplifiers (Veisz, 2010).

In many high-power CPA laser systems, the long-duration pedestal due to ASE is reduced by two methods: cross-polarized wave (XPW) generation and optical parametric chirped pulse amplification (OPCPA). The XPW is a passive nonlinear filter which selectively transmits intense pulses, improving the contrast ratio by 10^{2-3}. When a linearly polarized laser beam propagates in a crystal of cubic or tetragonal symmetry with its crystal axis angled to the laser polarization, an optical wave with a polarization perpendicular to the incident laser beam (XPW) is generated due to the third-order nonlinear susceptibility (Minkovski et al., 2002; Schroeder et al., 1995). Since the XPW generation efficiency is proportional to the square of the laser intensity, the XPW generation is useful to reduce the pedestal and thus to improve the contrast ratio.

A typical layout of XPW generation is shown in Fig. 7.9. A linearly polarized fundamental wave (FW) is weakly focused on a nonlinear crystal to increase the laser intensity required to achieve high XPW generation efficiency. The generated XPW is selected with an analyzer whose direction is orthogonal to the polarizer. Often a BaF_2 crystal is used for XPW generation since it is transparent in visible and ultraviolet regions and high-quality single crystals are available. The crystal is z-cut where the surface is normal to the [001] axis, and its [100] axis is oriented at angle β from the polarization of FW. [Alternatively, a [011]-cut crystal has been used to achieve higher XPW generation efficiency (Canova et al., 2008)].

The XPW generation is one of the third-order nonlinear optical processes which is described as

$$P_i^{(3)}(\omega_i, \boldsymbol{k}_i) = \frac{\varepsilon_0}{4} \sum_{jkl} \chi_{ijkl}^{(3)}(\omega_j, \omega_k, \omega_l) E_j(\omega_j, \boldsymbol{k}_j) E_k(\omega_k, \boldsymbol{k}_k) E_l(\omega_l, \boldsymbol{k}_l), \tag{7.8}$$

where the subscripts denote the crystallographic axes. From the energy and momentum conservations, $\omega_i = \omega_j + \omega_k + \omega_l$ and $k_i = k_j + k_k + k_l$ hold. The XPW is a degenerate process, i.e., $|\omega_j| = |\omega_k| = |\omega_l|$ where one of them takes a negative value. Many terms in Eq. (7.8) are reduced to smaller numbers, since the

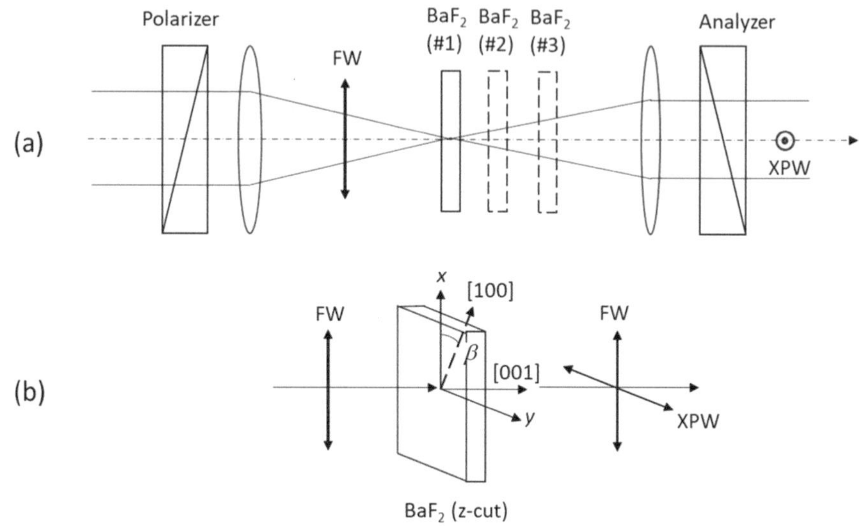

Fig. 7.9 A typical layout of cross-polarized wave (XPW) generation. **a** is an overall layout, where one (or more) BaF_2 crystal(s) is placed near the focus of a weakly-focused, linearly polarized fundamental wave (FW). The generated XPW is selected with an analyzer which is normal to the analyzer. **b** is a three-dimensional view of a z-cut BaF_2 crystal where the laser beam (FW) propagates along the crystal [001] axis which is normal to the surface and [100] axis is oriented at β from the polarization direction of the FW

materials of isotropic and cubic symmetry have only four nonzero third-order susceptibility tensor elements (χ_{iiii}, χ_{iijj}, χ_{ijji}, χ_{ijij}) and the Kleinman symmetry relation is applicable in most cases (see Sect. 5.1.2).

Describing the amplitude of the incident wave by A and the amplitude of the cross-polarized wave by B, propagation of these plane waves is described in slowly varying envelope approximation as (Canova et al., 2008)

$$\begin{aligned}\frac{dA}{dz} &= i\big(\underset{SPM}{\gamma_1 AA*A} + \underset{FWM}{\gamma_2 AAB^*} + \underset{FWM}{2\gamma_2 AA*B} + \underset{XPM}{2\gamma_3 AB*B} + \underset{FWM}{\gamma_3 BBA^*} + \underset{XPW}{\gamma_4 BB*B}\big),\\ \frac{dB}{dz} &= i\big(\gamma_5 BB*B + \gamma_4 BBA^* + 2\gamma_4 ABB* + 2\gamma_3 ABA* + \gamma_3 AAB^* + \gamma_2 AA*A\big).\end{aligned} \tag{7.9}$$

Each term in this equation corresponds to self-phase modulation (SPM), cross-phase modulation (XPM), cross-polarized wave generation (XPW) and parametric four wave mixing (FWM), as shown in Eq. (7.9). Considering a linearly polarized laser beam (fundamental wave FW) propagating along the fourfold axis [001] of a z-cut cubic (and tetragonal) crystal as shown in Fig. 7.9, the nonlinear coupling coefficients γ_i ($i = 1$–5) in Eq. (7.9) are given as

$$\begin{aligned} \gamma_1 &= \gamma_0\left[1 - (\sigma/2)\sin^2(2\beta)\right], \\ \gamma_2 &= -(\gamma_0\sigma/2)\sin(2\beta)\cos(2\beta), \\ \gamma_3 &= (\gamma_0/3)(4-\sigma) - \gamma_1, \\ \gamma_4 &= -\gamma_2, \\ \gamma_5 &= \gamma_1. \end{aligned} \tag{7.10}$$

The constants γ_0 and σ, which correspond to the magnitude and the anisotropy of the $\chi^{(3)}$ tensor respectively, are given by,

$$\begin{aligned} \gamma_0 &= \left(6\pi/8\lambda n\right)\chi^{(3)}_{xxxx}, \\ \sigma &= \left[\chi^{(3)}_{xxxx} - 2\chi^{(3)}_{xyxy} - \chi^{(3)}_{xxyy}\right]\Big/\chi^{(3)}_{xxxx}. \end{aligned} \tag{7.11}$$

With the initial conditions $B(0) = 0$ and $A(0) = A_0$, the solutions of Eq. (7.9) for the propagation distance of $z = L$ at low intensity are (Minkovski et al., 2004)

$$\begin{aligned} A &= A_0\exp\left(i\gamma_1 A_0^2 L\right), \\ B &= A_0(\gamma_2/\gamma_1)\left[\exp\left(i\gamma_1 A_0^2 L\right) - 1\right]. \end{aligned} \tag{7.12}$$

From these relations, the XPW generation efficiency is

$$\eta = \left|\frac{B}{A}\right|^2 \cong 2\left(\frac{\gamma_2}{\gamma_1}\right)^2\left[1 - \cos(\gamma_1)A_0^2 L\right]. \tag{7.13}$$

Here we have assumed the coupling coefficients to be real, since their imaginary parts are much smaller when the photon energy is less than approximately half of the band gap (Schroeder et al., 1995). Since $\gamma_2 \propto \sin(4\beta)$, the conversion efficiency becomes maximum at $\beta = 90°/4 = 22.5°$.

We note that the term $\gamma_1 A_0^2 L \equiv S$ in Eq. (7.12) corresponds to the B-integral, a measure of self-focusing described in Sects. 4.2.5 and 5.4.2. At low intensity of $\gamma_1 A_0^2 L \ll 1$, the conversion efficiency is proportional to the square of the laser intensity $I_0 = (c\varepsilon_0 n/2)|A_0|^2$ as

$$\eta = (\gamma_2)^2 A_0^4 L^2 \propto I_0^2. \tag{7.14}$$

Thus, the XPW generation works as a nonlinear filter to reduce low intensity components in space and time. Under the present approximation, the maximum conversion efficiency is obtained at $\beta = 22.5°$ and $\gamma_{\|}' A_0^2 L = \pi$ as

$$\eta_{\max} = \left(\frac{2\sigma}{4-\sigma}\right)^2, \tag{7.15}$$

Using $\sigma = -1.2$ (DeSalvo et al., 1993), $\eta_{max} = 0.21$.

More detailed analyses are required to evaluate the conversion efficiency, since the FW is self-focused, resulting in changes of the beam diameter, intensity and spatial phase (Gouy phase) of FW during propagation in the crystal. This leads to reductions in the overlap of FW and XPW and the conversion efficiency. Figure 7.10 shows dependence of the XPW conversion efficiency on the intensity parameter S for a 2-mm thick z-cut BaF_2 crystal with different crystal angles β, calculated for the FW of Gaussian spatial and temporal distributions (Jullien et al., 2006). The XPW conversion efficiency saturates at $\eta \approx 23\%$ with the crystal angle of $\beta = 22.5°$ at S ~ 5, which corresponds to I_0 ~ 10^{12} W/cm^2 for a 2-mm-thick crystal, whereas higher conversion efficiencies of $\eta \sim 30\,\%$ are reached with smaller crystal angles and at higher laser intensities.

In the first experiment of the XPW generation by Jullien et al. (2005), an energy conversion efficiency of 10% with the intensity contrast improvement of more than 4 orders of magnitude have been demonstrated. A higher conversion efficiency of 28% was attained in a refined experiment (Jullien et al., 2008), where a 2 mm-thick [011]-cut BaF_2 crystal was irradiated in vacuum (to avoid nonlinear effects in the air) with a 25-fs pulse of up to 800 μJ energy with a spatially flat-top profile at an intensity of approximately 10^{12} W/cm^2. Due to SPM and XPM, the spectral width of the XPW was 100 nm HWHM whereas the spectral width of the incident FW was 40 nm FWHM, resulting in the pulse-width shortening from 25 fs to 11.5 fs with phase variation of within 1 rad over the pulse duration.

To overcome the conversion efficiency limit with a single crystal, Jullien et al. (2006) proposed to use two crystals in the series as shown in Fig. 7.9. The optimum condition for high conversion efficiency was searched by varying the thickness, angles and separation of the two crystals and energy and focal length of the laser beam. With two 2-mm thick BaF_2 crystals of $\beta_1 = \beta_2 = 22.5°$, 20% conversion efficiency was achieved, corresponding to 30% internal efficiency which will be achieved by applying anti-reflection coatings on the optical surfaces. Dinh et al. (2025) refined this

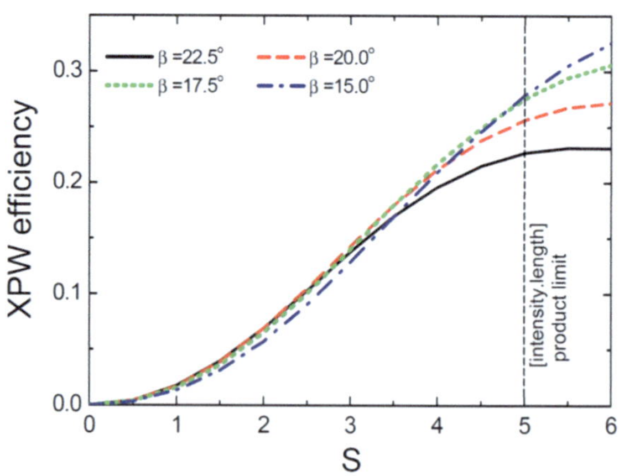

Fig. 7.10 Evolution of the XPW conversion efficiency as a function of $S = \gamma_1 |A_0|^2 L$ for different crystal angles β, calculated with Eq. (7.9) for the incident FW with Gaussian spatial and temporal distributions. Reprinted from Jullien et al. (2006). © 2006 Optical Society of America

approach, where three BaF_2 crystals with the angles $\beta_1 = 22.5°$ and $\beta_2 = \beta_3 = 0°$ were used for XPW generation. In this arrangement, XPW is generated only in the first crystal, whereas the second and the third crystals act as the spatial and temporal filters to select the Gaussian part of the XPW by using the FW for nonlinear filtering. A nearly diffraction-limited XPW ($M^2 = 1.1$) was generated with an energy conversion efficiency of 33% (corresponding to 40% internal efficiency) and 3.2 times pulse-width shortening. Due to careful spatial filtering of the incident FW and the generated XPW, the XPW has long term stabilities of energy (1.5%) and pointing (less than 5 urad). Furthermore, a temporal contrast 10^{-11}–10^{-12} was achieved after further amplification, showing this approach to be suitable for seed injection.

Another approach for pedestal reduction is the optical parametric chirped pulse amplification (OPCPA). An example of the front-end of a high-power ultrafast laser system which incorporates XPW and OPCPA is shown in Fig. 7.11 (Papadopoulos et al., 2017). The optical pulses after the XPW filter are well-controlled transform-limited pulses of ~ 10 fs duration with a high contrast ratio of ~ 10^{-10} and an energy of ~ 10s μJ per pulse. The laser pulses are amplified to ~ 1 mJ with the optical parametric chirped pulse amplification (OPCPA) within a short temporal window of ~ 20 ps centered at the main pulse, thus improving the contrast ratio to over ~ 10^{-13} due to elimination of the pedestals outside of this temporal window. The OPCPA is suitable for this purpose, since amplification takes place only when a nonlinear crystal is pumped by a short duration pumping pulse.

Although not shown in Fig. 7.11 explicitly, the laser pulse after XPW is stretched to ~ 10 ps by chirping with a glass block and goes through a double-pass AOPDF to control the spectral phase of the entire laser chain. After reducing the repetition rate to ~ 10 Hz with a pulse selector (Pockels cell switch), the chirped pulse is amplified by OPCPA with a nonlinear crystal (typically a BBO crystal due to its high conversion efficiency). After amplification with 1 or 2 stage OPCPAs, the pulse is further stretched to ~ 1 ns duration with an Öffner pulse stretcher for amplification with the main amplifiers. The OPCPA gain is controlled below the saturation region

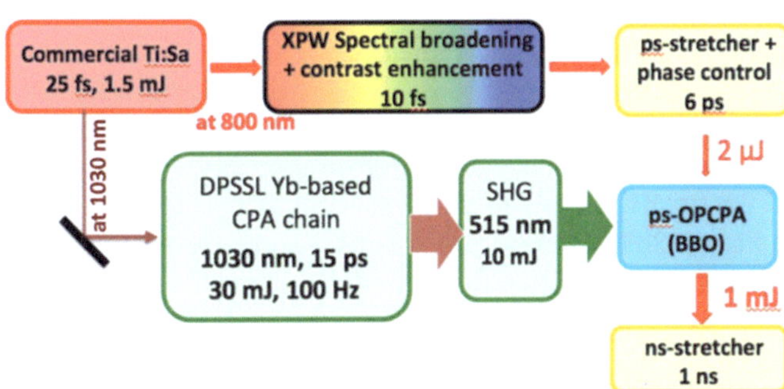

Fig. 7.11 Schematic layout of the front-end of Apollon, a high-power ultrafast laser system. Reprinted from Papadopoulos et al. (2016). © 2016 Cambridge University Press

to limit the level of parametric fluorescence. Typical energy conversion efficiency from the pump beam to the amplified signal beam is ~ 10% under this condition.

The OPCPA pump beam, an important part of this system, is generated by selecting a long-wavelength part (~ 1030 nm) of the spectral component of the Ti:sapphire oscillator pulse to achieve the exact timing between the signal pulse and the pumping pulse for parametric amplification. The selected pulse is stretched to ~ 1 ns, amplified with a Yb-doped fiber amplifier and Yb or Nd-doped crystal amplifiers up to ~ 100 mJ, compressed to ~ 15 ps, and converted to the second harmonic with a nonlinear crystal. The spatial distribution of the pumping beam is controlled close to flat distribution to amplify the signal pulse with high spatial quality.

(b) **10 s-Picosecond-Duration Pedestal**

In a CPA laser, a laser pulse is dispersed into its spectral components in a pulse stretcher and a pulse compressor. When the phase of a spectral component changes, a pedestal is produced since the pulse shape is a Fourier transform of its spectral distribution. The impact of high-frequency spectral phase modulation in the stretcher and compressor to the low-intensity pedestal has been evaluated in Dorrer and Bromage (2008) and Bromage et al. (2012). A general model for this evaluation is shown in Fig. 7.12. A laser beam of the central frequency ω_0, spectral bandwidth $\Delta\omega$ (1/e half-width) and a beam radius W (1/e half-width) is dispersed along the x-direction with the first grating (or a prism) and recombined with the second grating, to generate or compress a frequency chirped beam. The spatial coordinates in the near field before focusing are denoted by (x,y). The spatial coordinates at the far field are denoted by (x',y'), which are related to the spatial frequencies (u,v) as $u = kx'/f$ and $v = ky'/f$ with f the focal length and $k = 2\pi/\lambda$ the wavenumber.

Defining the optical frequency ω as the offset from the central frequency ω_0, the position of the center of the beam $x_0(\omega)$ of frequency ω is related to ω by $x_0(\omega) = \gamma\omega$, where γ is the spatio-spectral coefficient assuming linear relationship between $x_0(\omega)$ and ω as shown in Fig. 7.12b. The spectral modulation arising from the optical components in a stretcher/compressor is modeled with a phase screen which is assumed to generate homogenous, isotropic phase noises $\phi(x, y)$ centered at (x,y) with a correlation length l_c as shown in Fig. 7.12c. The phase noises affect the spatial and temporal structures of the laser beam in the near field and in the far field. From the Fourier transform relation, modulation of an optical wave of frequency ω_0 with the modulation frequency ω generates sub-pulses separated from the main pulse by $t = 2\pi/\omega$. Therefore, the phase noises distributed over broad frequencies as shown in Fig. 7.12c will generate distributed satellite pulses in the time domain. Considering that the modulation frequency ω is related to the space coordinate in the near field as $x_0(\omega) = \gamma\omega$, and the coordinate at the far field is the Fourier transform of the near field, we find that the time of a satellite pulse is directly connected to the coordinate at the far field. This relationship was analyzed quantitatively by Bromage et al. (2012).

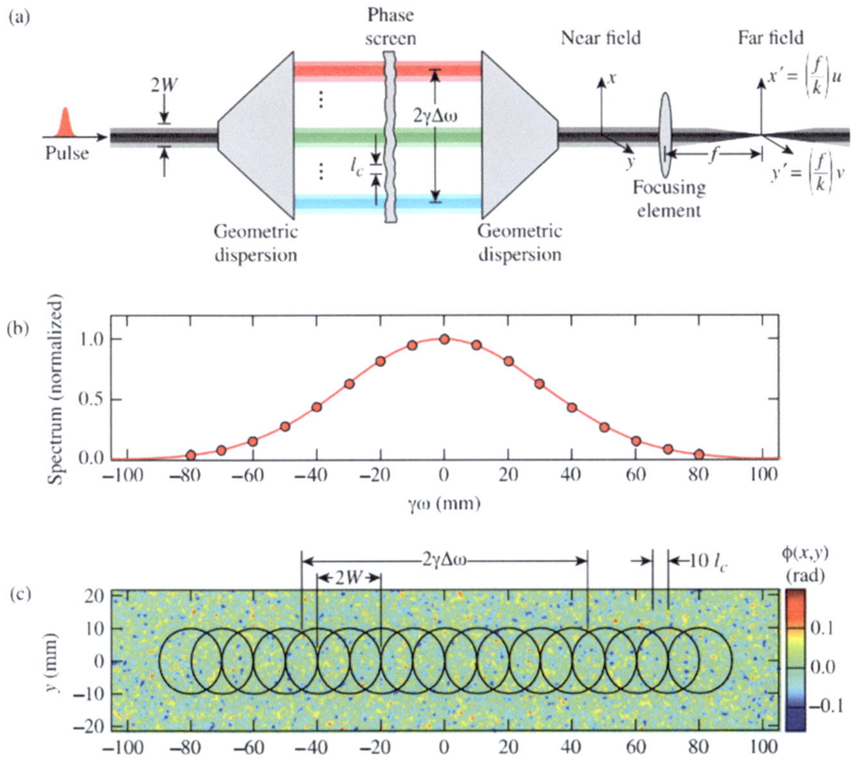

Fig. 7.12 **a** A general model to analyze the impact of phase noises in a stretcher/compressor to the temporal and spatial structures in the near and the far fields. A phase screen represents the phase noises of the optical components in a stretcher/compressor. **b** A Gaussian spectrum expressed in unit of spatial coordinate in the near-field. **c** A phase screen $\phi(x, y)$ used in the simulation. The circles show beam locations for optical frequencies marked in (**b**). This example is based on the following parameters: $\Delta\omega = 100$ THz, corresponding to a 17-fs pulse width, $\gamma = 0.44$ mm/THz, $W = 10$ mm, and $l_c = 0.5$ mm. Reprinted from Bromage et al. (2012), with permission from Optical Society of America. © 2012 Optical Society of America

In the near field, the optical field in the frequency domain is given by

$$\tilde{E}(x, y, \omega) = \tilde{E}_0(x, y, \omega)e^{i\Phi(x,y,\omega)} \cong \tilde{E}_0(x, y, \omega)\left[1 + i\phi(x - \gamma\omega, y)\right] \quad (7.16)$$

where $\tilde{E}_0(x, y, \omega)$ is the optical field without perturbation. The phase screen is represented by the phase function $\phi(x, y, \omega) = f(x - x_0, y) = \phi(x - \gamma\omega, y)$ whose magnitude is assumed to be small. The field in a time domain in the near field is the Fourier transform of Eq. (7.20) as

$$\begin{aligned} E(x, y, t) &= \int_{-\infty}^{\infty} d\omega \tilde{E}(x, y, \omega) e^{-i\omega t} \\ &= E_0(x, y, t) + i \int_{-\infty}^{\infty} d\omega \tilde{E}_0(x, y, \omega) \phi(x - \gamma\omega, y) e^{-i\omega t}. \end{aligned} \tag{7.17}$$

For quantitative evaluation, we express the optical field in the far field, derived from the Fourier transform of $E(x, y, t)$ in (x,y), in terms of the spatial frequencies (u,v) as

$$\begin{aligned} \tilde{E}(u, v, t) &= \int_{-\infty}^{\infty} \int_{-\infty}^{\infty} dxdyE(x, y, t) e^{-iux} e^{-ivy} \\ &= \tilde{E}_0(u, v, t) + i \int_{-\infty}^{\infty} \int_{-\infty}^{\infty} \int_{-\infty}^{\infty} d\omega dxdy \tilde{E}_0(x, y, \omega) \phi(x - \gamma\omega, y) e^{-i(\omega t + ux + vy)}. \end{aligned} \tag{7.18}$$

The intensity at the far field $I(u, v, t) = \left|\tilde{E}(u, v, t)\right|^2$ averaged over an ensemble is expressed in terms of the power spectral density (PSD) of the phase screen as

$$\begin{aligned} \langle I(u, v, t) \rangle &= I_0(u, v, t) + \int_{-\infty}^{\infty} \cdots \int_{-\infty}^{\infty} d\omega d\omega\prime dxdx\prime dydy\prime \\ &\times \tilde{E}_0(x, y, \omega) \tilde{E}_0 * (x\prime, y\prime, \omega\prime) \langle \phi(x - \gamma\omega, y) \cdot \phi * (x\prime - \gamma\omega\prime, y\prime) \rangle \\ &\times e^{-i(\omega-\omega\prime)t} e^{-i(x-x\prime)u} e^{-i(y-y\prime)v} \\ &= I_0(u, v, t) + \int_{-\infty}^{\infty} \int_{-\infty}^{\infty} du\prime dv\prime I_0(u\prime, v\prime, t\prime + \gamma u - \gamma u\prime) PSD(u - u\prime, v - v\prime). \end{aligned} \tag{7.19}$$

Here PSD is the Fourier transform of the correlation function of the phase noise:

$$\begin{aligned} PSD(u, v) &= \int_{-\infty}^{\infty} \int_{-\infty}^{\infty} d\Delta x d\Delta y C(\Delta x, \Delta y) e^{-i\Delta xu} e^{-i\Delta yv} \\ &= \frac{\sigma^2 l_c^2}{2\pi} \exp\left[-\frac{l_c^2(u^2 + v^2)}{2}\right]. \end{aligned} \tag{7.20}$$

In this equation, $C(\Delta x, \Delta y)$ is the correlation function of the phase noise and σ is the root-mean-square (rms) phase noise at a screen in radians:

$$C(\Delta x, \Delta y) = \langle \phi(x, y) \cdot \phi^*(x + \Delta x, y + \Delta y) \rangle = \sigma^2 \exp\left(-\frac{\Delta x^2 + \Delta y^2}{2l_c^2}\right). \tag{7.21}$$

The far field intensity distribution (7.22) is simplified by assuming Gaussian functions for the spatial and spectral intensities as

$$I_0(u, v, t) = I_P \exp\left(-\Delta\omega^2 t^2 - W^2 u^2 - W^2 v^2\right), \tag{7.22}$$

where I_P is the peak intensity at $u = v = t = 0$. Noting that $\gamma \Delta\omega \gg W \gg l_c$ holds in typical cases, the far field intensity becomes,

$$\langle I(u, v, t)\rangle = I_0(u, v, t) + I_P \frac{\sigma^2 l_c}{\sqrt{2}\gamma \Delta\omega}\left(\frac{l_c}{\sqrt{2}W}\right) \exp\left[-W^2\left(u + \frac{t}{\gamma}\right)^2 - \frac{l_c^2 v^2}{2} - \frac{l_c^2 t^2}{2\gamma^2}\right]. \quad (7.23)$$

The magnitude of the pedestal, the second term, is proportional to the variance of the phase noise σ^2 and inversely proportional to the beam radius W since the noise is averaged in larger beam areas. The temporal width is given by $\tau = \gamma/l_c$ which is ~ 5 ps for typical parameters of $\gamma = 0.5$ mm/THz and $l_c = 0.1$ mm. The notable feature which Eq. (7.23) shows is that the lateral position of the pulse peak moves with time as $u = -{}^{t}/_{\gamma}$. Since the instantaneous width is smaller than the time-integrated width in the dispersion direction u, the pedestal at the focus position $u = 0$ becomes small.

This analytic expression was examined with simulation for various optical configurations by Bromage et al. (2012). As an example, Fig. 7.13a shows an Öffner-type pulse stretcher where a grating, a primary mirror, and a retroreflector (denoted by A–A′) are in the near field whereas the secondary mirror is in a far field since each dispersed beam is focused on this mirror. Figure 7.13b shows a time-averaged intensity at the far field in the u-t plane, where the two slanted lines are contributions from the primary mirror and the grating. This spatiotemporal (ST)-coupling was demonstrated in a space- and time-resolved measurement by Ma et al. (2015).

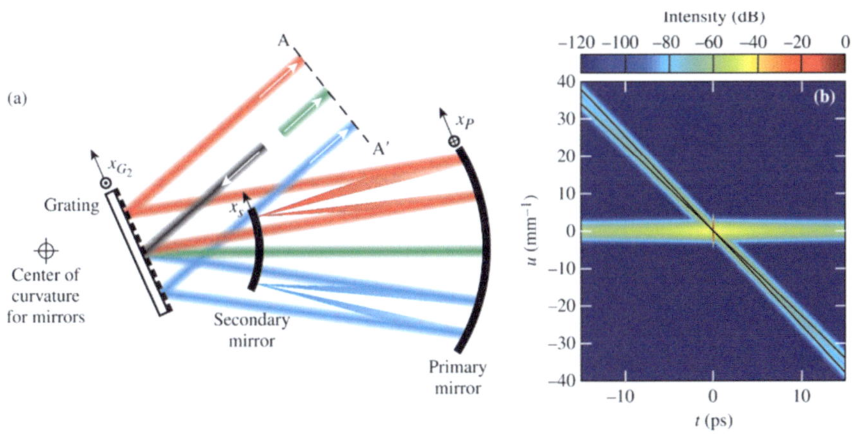

Fig. 7.13 **a** Schematic of an Öffner stretcher to evaluate the effects of the phase noises on the three spectral components to the far field. **b** Far-field average intensity in the u-t plane. The two black lines correspond to u = −t/g for the grating and the first mirror, whereas the bright horizontal line is due to the secondary mirror on which each spectral component is focused. Reprinted from Bromage et al. (2012), with permission from Optical Society of America. © 2012 Optical Society of America

Another notable feature in Fig. 7.13b is the bright horizontal line due to the secondary mirror, since the phase noise on this surface is directly transferred to the far field. This line, which is concentrated at the focus position $u = 0$, produces ps-duration pedestal centered at $t = 0$. The close correlation between the ps-duration pedestal and the surface quality of the optics in a pulse stretcher and a pulse compressor was demonstrated in several works (Bagnoud & Salin, 1999; Hooker et al., 2011; Ranc et al., 2020; Schanz et al., 2019). Among them, Ranc et al. (2020) have demonstrated that the 10s-ps duration pedestal is reduced by ~ two orders of magnitudes by using a high-quality optics with measured RMS roughness of 0.17 nm in a convex mirror of the Öffner stretcher, achieving a contrast ratio of 10^{-12} at 40 ps before the main pulse. These works show that achieving a low pedestal within ~ 100 ps before the main pulse requires using very high-quality optical components, in terms of surface roughness, surface figure and contaminations, in the stretcher and compressors, especially in a convex mirror of the stretcher.

(c) **Short-Duration Pre-pulses and Post-pulses**

Short-duration pre-pulses are almost always observed in addition to the ns-duration pedestal and the 10s-ps duration pedestal. The short-duration pre-pulses have been ascribed to specular reflections at the surfaces of optical components located in the laser beamline such as a regenerative amplifier, a multi-pass amplifier, and a Pockels cell switch.

Didenko et al. (2008) have proposed that some of these pre-pulses may be generated by the post-pulses which propagate after the main laser pulse. In this proposed scheme, shown in Fig. 7.14, the temporal profile of the stretched main pulse is modulated due to interference of the main pulse with a post pulse in a stretcher when the delay time of the post-pulse from the main pulse, t_p, is shorter than the duration of chirped pulse, τ_c, which is typically longer than 100 ps. This intensity modulation is enhanced in the amplifiers due to nonlinear refractive index, which leads to temporal modulation of the spectral phase of the chirped pulse. The pre-pulses may be generated when this spectral modulation is large. Didenko et al. have analyzed this process and found that a few pre-pulses are generated at approximately same temporal separation from the main pulse, i.e., at $t = -t_p, -2t_p$, etc., with an intensity comparable to the intensity of the post-pulse, when the B-integral is of the order of 1 rad. This evaluation was validated in the experiments using regenerative and multi-pass amplifiers.

This concept was tested in a PW laser system (Kiriyama et al., 2020, 2021), where the pedestals and pre-pulses have been eliminated with AOPDF, saturable absorber, and OPCPA in a front-end. Still many pre-pulses are observed, with one-to-one correlation between the pre-pulses and the post-pulses. All these pre-pulses were eliminated by wedging the surfaces of optical components by a few degrees, achieving the intensity contrast ratio of 10^{-12} as shown in Fig. 7.15b. The pedestal closer to the main pulse of within ~ 100 ps was further reduced by reducing the RMS surface roughness of the convex mirror in the Öffner stretcher from 1 to 2 nm (blue) to 0.2 to 0.3 nm (red) (Fig. 7.15c).

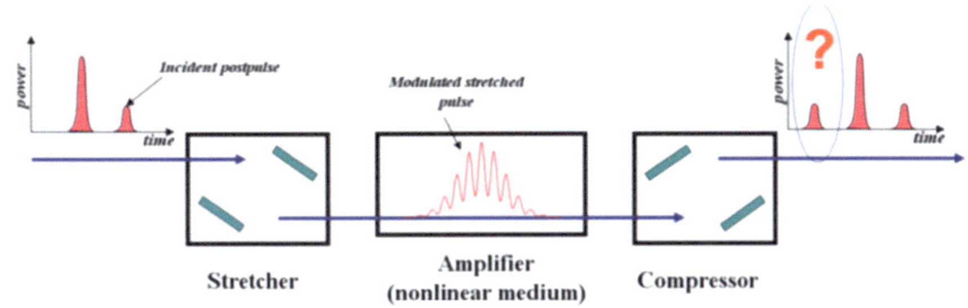

Fig. 7.14 Generation of a pre-pulse due to interference of the main pulse with a post-pulse in a stretcher and modulation enhancement due to nonlinearity in the amplifiers. Reprinted from Didenko et al. (2008). © 2008 Optical Society of America

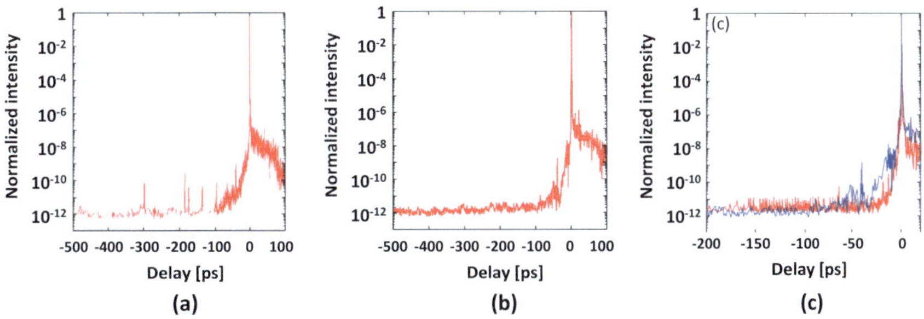

Fig. 7.15 Measurement of pre-pulses of a PW laser: **a** before and **b** after elimination of post-pulses by wedging the surfaces of optical components in the laser system, showing elimination of short-duration pre-pulses. **c** Further reduction of the pedestal within – 50 ps by using a higher quality convex mirror in the Öffner stretcher. Reprinted from Kiriyama et al. (2021). © 2021 Cambridge University Press (Note that the characters in the original figure are enlarged for visibility.)

Further improvement of the temporal contrast of ultrafast laser pulses is required for clean experiments at ultrahigh intensities without being affected by a pre-plasma formed on a target surface. It was found that a plasma mirror (PM) is very effective in reducing pedestals and pre-pulses, especially within a few picoseconds before the main pulse. With a plasma mirror, a laser beam is focused on the anti-reflection coated transparent dielectric surface in vacuum, where most of the pedestals and pre-pulses are transmitted (Kapteyn et al., 1991). As the laser intensity increases in time, a thin plasma layer is formed on the dielectric surface due to nonlinear absorption and heated by inverse bremsstrahlung, resulting in a plasma whose density which exceeds the critical density at the laser wavelength. At this moment, the surface is suddenly turned on to become a "high-reflectance mirror". It was found that the plasma mirror is very robust against changes in the pulse fluence and imperfections of the beam spatial profile (Doumy et al., 2004; Wittmann et al., 2006).

Contrast improvements have been achieved with the plasma mirrors in several 10-TW class Ti:sapphire lasers: 10^4 improvements up to $t = -10$ ps before the main

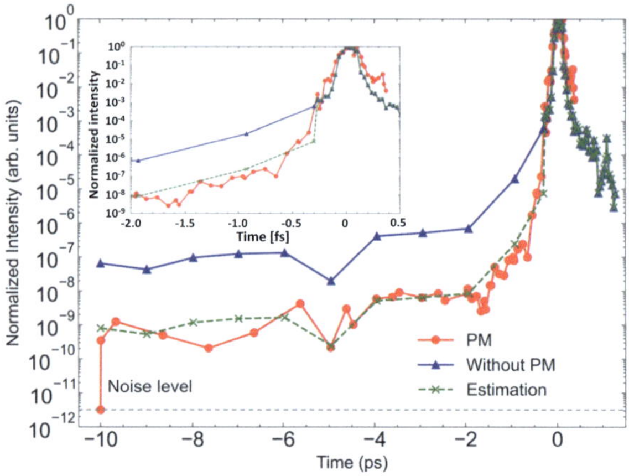

Fig. 7.16 Temporal contrast with (red line) and without (blue line) the plasma mirror. The dotted line is the contrast estimated using the reflectance of the PM (65%) and AR coating (0.8%). The inset is an enlarged plot, showing the PM is triggered from – 0.9 to – 0.3 ps before the pulse peak. Reprinted from Kon et al. (2022). © 2022 Cambridge University Press (Note that the characters in the original figure are enlarged for visibility.)

pulse with a reflectivity of 70% with a double plasma mirror (Thaury et al., 2007), and up to $t = -1$ ps with a reflectivity of 70% with a single plasma mirror (Inoue et al., 2016). By installing a single plasma mirror after the pulse compressor in a 200 TW, 50 fs Ti:sapphire laser, contrast has improved by 10^2 up to $t = -0.5$ ps before the peak of the pulse as shown in Fig. 7.16. The PM reflectivity has exceeded 80% at a fluence of over ~ 10^5 J/cm^2, with stable on-target laser energy and the pulse shape (Kon et al., 2022). Combining PM with other methods, it is now possible to generate ultrashort high-intensity laser pulses exceeding ~ 10^{-12} contrast close to ~ 1 ps before the main pulse.

7.2.4.4 Diffraction-Limited Focusing

Diffraction-limited focusing is required to implement an experiment with a controlled spatial profile on target and to achieve the highest intensity possible with the laser facility. Figure 7.17 shows a layout of the laser system which has achieved the intensity of over 10^{23} W/cm^2 with an f/1.1 focusing optics (Yoon et al., 2021). In this system, the diameter of the laser beam after the final amplifier is expanded from 85 to 280 mm to avoid the damage of the diffraction gratings in the pulse compressor.

The wavefront of the laser beam is first controlled before the compressor using a wavefront sensor (WFS1) and a deformable mirror (DM1) to avoid beam degradation at a far field arising from the space–time coupling of the phase-aberrated beam in the compressor. The wavefront of the focused beam is controlled by another set of

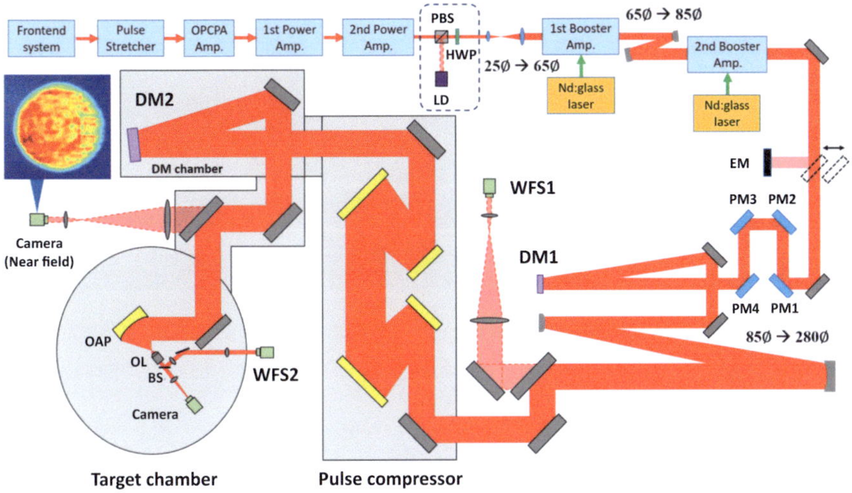

Fig. 7.17 Layout of the CoReLS petawatt laser to achieve the laser intensity of over 10^{23} W/cm^2. The wavefront measured by the wavefront sensors WFS1 and WFS2 are fed back to the deformable mirrors DM1 and DM2 to achieve the diffraction-limited focusing on target. Reprinted from Yoon et al. (2021). © 2021 Optical Society of America (Note that the characters in the original figure are enlarged for visibility.)

the wavefront sensor and the deformable mirror (WFS2 and DM2). Using a low-energy 5 Hz laser beam of the laser system without pumping the booster amplifiers, the wavefront aberration is measured with a Shack-Hartmann wavefront sensor and fed back to a closed-loop system of the deformable mirror to control the wavefront aberration to within 0.30 μm PV and 0.05 μm RMS.

Figure 7.18 shows the far field patterns of the laser beam focused with a f/1.1 off-axis parabolic mirror (OAP) (a) before and (b) after the phase correction where the latter has a circular shape with a spot size of 1.1 μm (FWHM) in diameter. The latter is close to the ideal focal spot image (c). From the measured pulse width of 19.6 fs (FWHM) and the measured on-target energy of 55.6 J, the peak intensity is derived as $I_0 = 1.4 \times 10^{23}$ W/cm^2 for this set of data with a 3-D profile shown in (e).

For systematic evaluation of the stability, the CoReLS laser was operated at 0.1 Hz with all amplifiers activated over 80 consecutive shots. This test has provided the average and the standard deviation (SD) of various laser parameters with the average values as follows: the peak intensity 1.1×10^{23} W/cm^2; the rms wavefront aberration (ϕ_{RMS}) 0.079 μm; the FWHM focal spot size 1.2 μm; the effective focal spot size 1.8 μm; and the Strehl ratio 0.68. Here the effective focal spot is the circular area in which all energy is contained with the diameter defined by $d_{\mathrm{eff}} = 2\sqrt{A_{\mathrm{eff}}/\pi}$ where $A_{eff} = \int i(x, y)dxdy$ is the effective spot area and $i(x, y)$ is the normalized intensity distribution of a measured focal spot. The Strehl ratio, the ratio of measured peak intensity to the peak intensity of an ideal diffraction-limited focal spot, is calculated

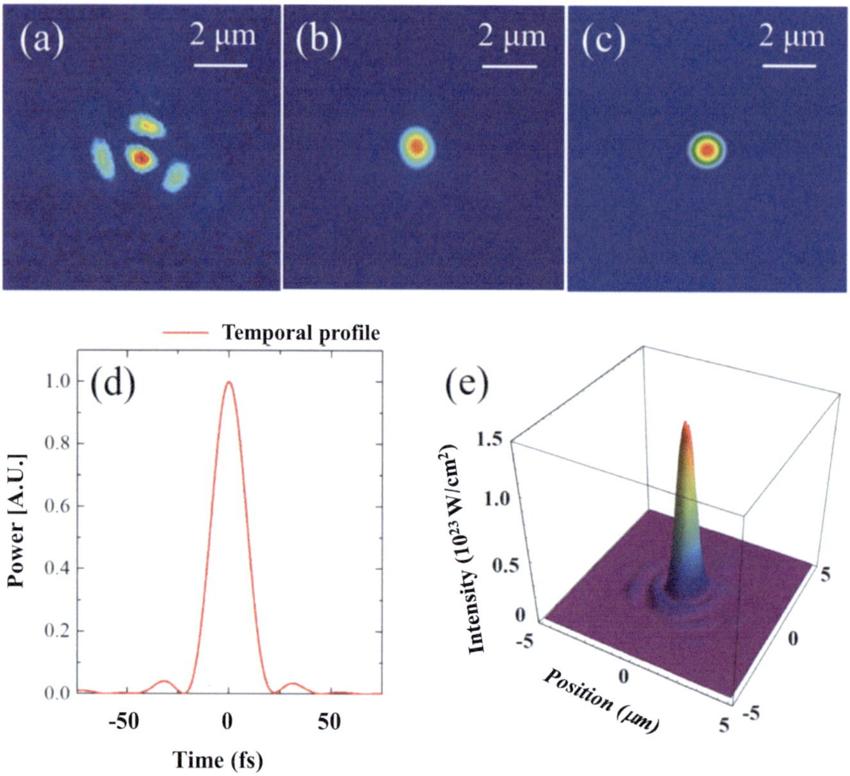

Fig. 7.18 (left) The intensity distributions of the focused laser beam **a** before and **b** after phase correction and **c** the ideal profile. **d** is the measured temporal profile and **e** is the intensity showing the peal intensity exceeding 10^{23} W/cm^2. Reprinted from Yoon et al. (2021). © 2021 Optical Society of America (Note that the characters in the original figure are enlarged for visibility.)

from the measured wavefront aberration ϕ_{RMS} as SR $\simeq \exp\left[-(2\pi\phi_{\mathrm{RMS}}/\lambda)^2\right]$. The 20% variation of the peak intensity arises mainly from the variation of the focal spot profile which is affected by the fluctuation of the wavefront aberration.

7.2.4.5 Pointing Stability

Keeping the pointing stability is essential in using the high intensity lasers. In the laser-driven wake-field electron acceleration for example, a high-intensity laser beam has to interact stably with a gaseous target of a well-controlled density profile to generate a monoenergetic ~ GeV electron beam (Ke et al., 2021), which is injected into undulators to amplify euv radiation (Wang et al., 2021). In the nonlinear Compton scattering, a laser beam of relativistic intensity must collide with a GeV-energy electron beam generated by LWFA with spatial accuracy of a few μm and temporal accuracy of less than 10 fs (Pathak et al., 2021).

In the laser system at CoReLS (Yoon et al., 2021), the pointing stability of the main laser was 3.0 μrad and 3.6 μrad (standard deviations) in the horizontal and vertical directions, respectively, corresponding to the Strehl ration (SR) of 0.94. Since this SR is larger than the measured SR of the PW laser beam of 0.68, this indicates that the pointing stability is affected by some effects other than the laser beam divergence.

The origin of the fluctuation was investigated by propagating a high-quality laser beam through the laser chain using 800-nm laser diode (LD) located before the power amplifiers as shown in Fig. 7.16. The wavefront aberration, measured without firing the amplifiers in 80 consecutive shots with a wavefront sensor (WFS1) located in front of the pulse compressor, resulted in the RMS wavefront aberration of 0.071 ± 0.017 and the Strehl ratio of 0.74 ± 0.08, both close to the measured values of the PW laser beam. This shows that the air turbulence in the beam path before the compressor is the major contributor to the wavefront aberration and the pointing instability of the laser beam. The wavefront aberration and the pointing stability of high-power laser systems can be improved by tightly protecting the laser system from subtle air flow in the laser beam path.

7.3 High-Energy CPA Nd:Glass Laser

The spectral width (FWHM) of the fluorescence of the Nd-doped glasses is 21.8 nm in phosphate and 27.4 nm in silicate glasses (Table 4.1). These widths are broad enough to amplify ~ 500 fs duration laser pulses or ~ 200 fs with the phosphate-silicate mixed glasses by taking advantage of the difference in the fluorescence peak wavelengths: 1054 nm in phosphate and 1062 nm in silicate (Hays et al., 2007). Since large energy is stored in the Nd-doped laser glasses, multi-PW peak-power, sub-ps duration laser pulses can be generated by applying CPA to the Nd:glass lasers.

The PW-class CPA Nd:glass lasers, or high energy petawatt (HEPW) lasers, have been developed worldwide, most of them combined with long-pulse high-energy laser systems or pulsed power facilities to heat high-density plasmas: Nova (Perry et al., 1999) and NIF ARC (Barty et al., 2004) at LLNL, Z-Petawatt at Sandia National Laboratory (Schwarz et al., 2016), and OMEGA EP at LLE (Kelly et al., 2006) in USA; Vulcan Petawatt at RAL (Danson et al., 2004) and Orion at AWE (Parker et al., 2018) in UK; LULI 2000 at LULI (Zou et al., 2008) and PETAL at CEA (Blanchot et al., 2017) in France; PHELIX at GSI (Bagnoud et al., 2010) in Germany; FIREX-I at ILE (Miyanaga et al., 2006) in Japan; and SG-II UP beamline at SIOM NLHPLP (Zhu et al., 2018) and Xingguang-III at Laser Fusion Research Center, China Academy of Engineering Physics (Zhou et al., 2021) in China. More recently, a 10 PW mixed-glass laser system L4 with liquid cooling for rapid operation (every 7 min.) was completed at ELI-Beamlines in Czech Republic (Chériaux et al., 2019). Performances of these laser systems are summarized in Table 7.3.

Table 7.3 Performances of high energy PW laser facilities

Country	Institute	Facility	Amplifier	Peak power (PW) (× beams)	Energy (J) (× beams)	Pulse width (ps)
UK	RAL	Vulcan PW	Nd:P-glass	1	500	0.5
	AWE	Orion	Nd:P-glass	0.4 @526 nm	200	0.5
France	LULI	LULI 2000	Nd:P-glass	0.15	~ 100	1
	LULI	ELFIE	Nd:P/S-glass	0.057 (× 2)	20 (× 2)	0.35
	CESTA	PETAL	Nd:P-glass	1.15	< 1000	0.5
Germany	GSI	PHELIX	Nd:P-glass	0.5	200	0.5
	FSUJ	POLARIS	Yb:FP glass/ CaF_2	0.2	> 17	< 0.1
	HZDR	PEnNELOPE	Yb:CaF_2	1	150	0.15
Czechia	ELI-BL	L4 Aton	Nd:P/S-glass	10	1500	0.15
USA	LLNL	NIF-ARC	Nd:P-glass	1 (× 4)	1000 (× 4)	1
	SNL	Z-Beamlet	Nd:P-glass	0.2	50	0.5
		Z-Petawatt	Nd:P-glass	1	0.4–0.65	0.5
	U Texas	Texas Petawatt	Nd:P/S-glass	1.1	0.18	0.17
	LLE	Omega-EP	Nd:P-glass	0.5 (× 2)	500 (× 2)	1
Japan	ILE	LFEX	Nd:P-glass	2 (× 4)	10,000	1–20
China	NLHPL	SG-II 5 PW	Nd:P-glass	5 (× 2)	5000 (× 2)	1
	LFRC/ CAEP	SG-III ps beam	Nd:P-glass	0.77	370	0.48

Here P, S and FP refer to phosphate, silicate and fluoro-phosphate laser glasses, respectively. The number of laser beams of multiple-beam systems is shown in the parentheses

Similarly to high-power CPA Ti:sapphire laser systems shown in Fig. 7.2, high-energy CPA Nd:glass laser systems are comprised of a front-end, preamplifier, power amplifier, pulse compressor, and target irradiation. Various technical issues must be addressed to generate high energy PW (HEPW) pulses with the Nd:glass lasers (Zuegel et al., 2006). As an example of a HEPW laser, we show in Fig. 7.19 a layout of LFEX-I facility developed for fast ignition of laser fusion at ILE Osaka University, which generates 10 kJ pulse in four beams in a 10-ps trapezoid with a rise time of less than 2 ps (Miyanaga et al., 2006).

At the front end and the preamplifier sections, a frequency-chirped pulse is generated with over 8 nm spectral width to generate a FTL 500 fs pulse after pulse compression and a pulse width of over ~ 2 ns to limit the B-integral by reducing the laser intensity during amplification. Other requirements are high contrast ratio exceeding ~ 10^8 and an output energy of typically ~ 10 mJ for amplification to kJ energy by the main amplifier.

In the configuration shown in Fig. 7.19, short pulses of 150 fs duration, spectral width of 8.2 nm (FWHM) and a single-pulse energy of ~ 1 nJ are generated at 1055 nm

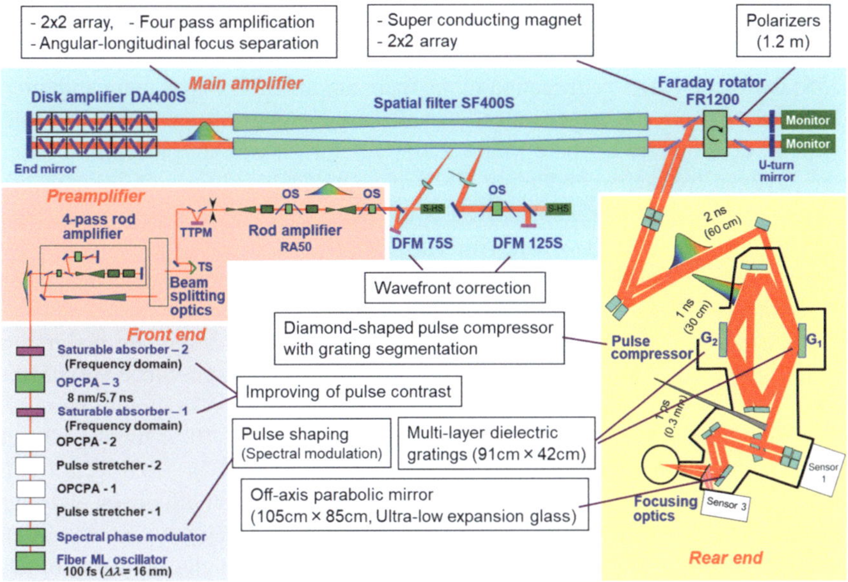

Fig. 7.19 Schematic layout of LFEX-I 10-kJ PW laser system. The laser system is composed of front-end, preamplifier, main amplifier, beam-transport optics, pulse compressor, and focusing optics. Abbreviations: OPCPA—optical parametric chirped pulse amplifier, TS—translation stage, TTPM—tip-tilt/piston mirror, RA—rod amplifier, DFM—deformable mirror, OS—optical shutter (Pockels cell), DA400S—square disk amplifier of 400 mm × 400 mm clear aperture, FR: Faraday rotator, SF: spatial filter, G: multi-layer dielectric grating. Curtesy of Prof. Noriaki Miyanaga

with a LD-pumped Nd:glass oscillator which is mode-locked with a semiconductor saturable absorber. The spectral phase is shaped by combination of a diffraction grating and a spatial light modulator after the oscillator to generate a trapezoid pulse after pulse compression. The nJ pulse is stretched and amplified with two-stage OPCPA to generate a 3 ns, 65 mJ pulse with a pre-pulse ratio of less than 1.5×10^{-8} by precisely overlapping the temporal and spatial profiles of the signal pulse and the pump pulses (Bagnoud et al., 2005; Yoshida et al., 2003). After further reducing pre-pulses with saturable absorbers, the pulse from the front end is further amplified with a 4-pass rod amplifier (RA50), split into four beams, and amplified with two-stage 50-mm diameter rod amplifiers to ~ 1 J per beam. Another approach to amplify ~ 1 nJ pulses of 1-μm wavelength to over 10 mJ is with a LD-pumped regenerative Nd:glass amplifier incorporating a phased mirror to generate a flat-top spatial beam profile and a birefringent filter to supress the spectral gain narrowing (Rybeyre et al., 2003).

The chirped pulse from the preamplifier is amplified with the main amplifier. For efficient energy extraction of the stored energy and high-density spatial packaging, the multi-pass amplifier architecture developed at NIF (Fig. 6.5) is introduced to HEPW facilities with several modifications (Blanchot et al., 2017; Kelly et al., 2006; Miyanaga et al., 2006). The main amplifier of LFEX is a 2 × 2 close-packed array, each composed of an eight-disk amplifier of 40 cm × 40 cm aperture where the laser pulse from the preamplifier is amplified in four-pass configuration. The wavefront distortions are corrected by two deformable mirrors (DFM 75S and DFM 125S) during and after propagation in the main amplifier. The four main amplifiers are protected from the target reflection with one Faraday rotator (FR 1200) with a 1.25-m clear aperture where four Faraday rotator glasses are placed in a large-bore, cryogen-free superconducting 1.4-T magnet developed for medical magnetic resonance imaging (MRI). In other HEPW laser systems, large aperture plasma electrode Pockels cell (PEPC) switches are used for protection. In this laser system, the chirped pulses are compressed with the diamond-shaped compressor with grating segmentation which is shown in Fig. 7.20c.

The major difficulties in the pulse compressor of HEPW lasers are the grating damage, fabrication of large gratings, and the compressor design. Since the spectral width of the HEPW lasers is less than the bandwidth of the Ti:sapphire laser, we can use multilayer dielectric (MLD) gratings which have ~ 10 × higher damage threshold (LDT) than the metal-coated gratings whose LDT is ~ 0.6 J/cm^2. It is possible to optimize the grating structure for high diffraction efficiency and high LDT by calculating the field distribution at and near the grating grooves for various grating structures and laser parameters such as incidence angle, wavelength, and bandwidth (Li, 1993). For example as shown in Fig. 7.20, high laser field is located between the grating grooves to avoid the laser damage at the grating surface (Barty et al., 2004).

In actual HEPW laser systems, a decrease in LDT due to organic contamination of the grating surface has been a serious issue (Ashe et al., 2008; Favrat et al., 2014;

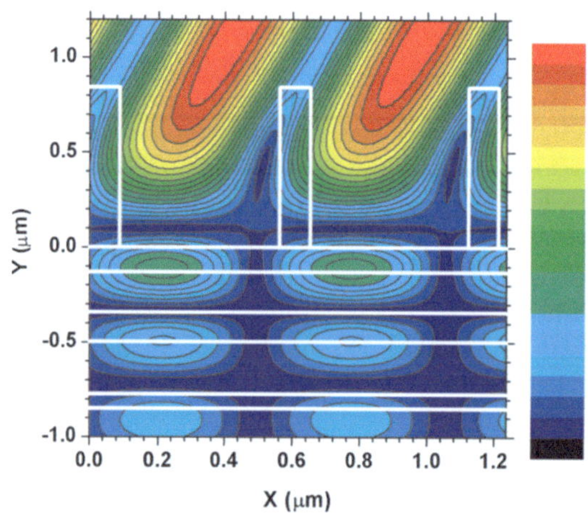

Fig. 7.20 Field distribution in one period of a 1780 lines mm^{-1} grating and the top layers of the MLD stack for an incidence angle of 61°. In this case, the field is distributed between the grooves to avoid laser damage on the grating grooves. The grating has a groove height of 620 nm and a duty cycle (ratio of width and period of the groove) of 0.28. The MLD stack consists of 20 alternating layers of Ta_2O_5 and Al_2O_3. Reprinted from Barty et al. (2004), with permission from IOP Science: permission conveyed through Copyright Clearance Center, Inc. © 2004 IOP Science

Jitsuno et al., 2010). The organic contaminants have been found to originate not only from the chamber wall and the vacuum pumps, but also from the molecular gases in the room air which intrudes into the chamber through the vacuum seals (Jitsuno et al., 2013, 2016). These studies suggest the importance of careful handling of the optics during storage, cleaning, installation, and operation.

A meter-size grating structure is printed with a holographic method where two large-area plane waves are interfered to form a periodic interference pattern on a photoresist coated on a low thermal-expansion substrate (Armstrong & Kessler, 1993). Another method is a scanning-beam interference lithography (SBIL), where small grating image, formed by interference of two small-area phase-locked coherent beams, is raster scanned over a photoresist on a large substrate which is moved with a high-precision interferometer-controlled stage (Schattenburg et al., 2002). The 91 cm × 42 cm gratings for LFEX-I have been fabricated with SBIL (Jitsuno et al., 2008).

Coherent addition of multiple gratings, called tiled grating, is often required for compression of multiple-kJ HEPW lasers, when the beam area on the grating exceeds the maximum fabrication size of the grating. The phasing of multiple gratings has been demonstrated with development of accurate alignment methods (Cotel et al., 2007; Kessler et al., 2004; Qiao et al., 2007, 2008). We note that a combination of one grating and its image formed by a mirror perpendicular to the grating surface works as a self-tiled grating, instead of using two gratings side-by-side (Li et al., 2010, 2015).

Another requirement is to reduce the grating separation distance since the pulse compressor with a standard four-grating design requires a separation distance of ~ 30 m for HEPW laser systems. Since this distance is often too large for placing a vacuum chamber in a limited laboratory space, development of compressor designs to reduce the size of the compressor chamber is required.

Figure 7.21 shows three compressor designs with tiled gratings. In the standard compressor design composed of four gratings shown in (a), the transverse displacement of the spectral components produced by G1, G2 is compensated by G3, G4. The stretched pulse is compressed in two stages: the first pair to half the input pulse width and the second pair to the final FTL width. Larger size gratings are required for G2 and G3 than G1 and G4 due to diffraction of broad spectral components.

In the configuration shown in (b) developed for NIF, the length of the compressor chamber is shortened to ~ half by offsetting the lateral positions of the input and output beams. This is achieved by using mixed gratings, i.e., different groove densities for G1 and G2 (1750 mm^{-1}) and G3 and G4 (1780 mm^{-1}), resulting in shifting G4 upward relative to G1 in this diagram (Barty et al., 2004).

Whereas in (c) developed for LFEX-I, two pairs of tiled gratings (G1, G2) and four plane mirrors (M1 –M4) form a diamond-shaped compressor, which is tolerant to grating-to-grating period variations and to alignment errors (Habara et al., 2010; Rushford et al., 2008). A lower (upper) half of the laser beam, written in blue (red), incident from left follows a path: G1A (G1B)–M1–G2B (G2A)–M2–M3–G2A (G2B)–M4–G1B (G1A) and exits to right. Since the laser beam is incident to each grating once and from the opposite direction in the second path, the angular

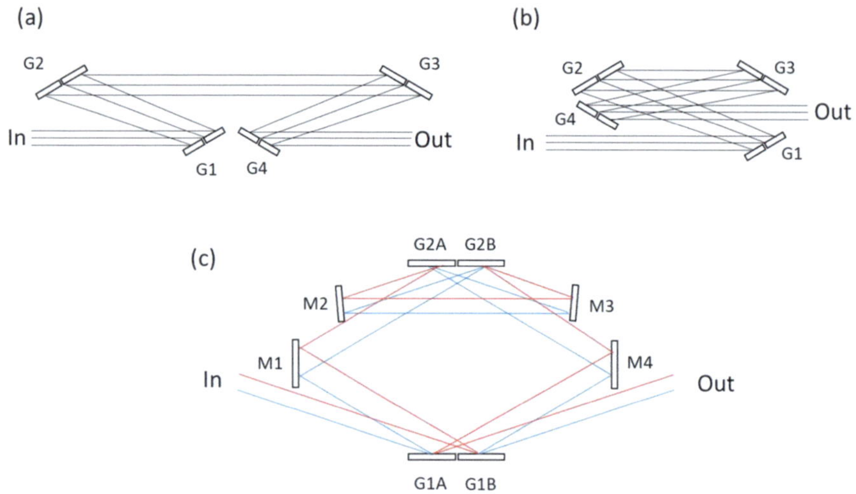

Fig. 7.21 Examples of pulse compressor configurations with tiled gratings for high-energy PW laser systems: **a** standard four-gating compressor, **b** mixed-grating compressor, and **c** diamond-shaped compressor

deviation of the exit beam due to errors in the groove density and the alignment of the gratings does not depend on the beam path in this configuration (Miyanaga et al., 2006).

7.4 Other Approaches to Ultra-High Power Pulse Generation

Following the start of the ELI project (ELI Whitebook, 2011), where 10 PW lasers will be developed by chirped pulse amplification with broadband solid-state lasers, investigation to realize next generation ultra-high power (100 PW—EW) lasers has been pursued in many approaches. In this section, we describe several concepts to extend the peak power to 100 PW ~ exawatt (EW) region: chirped beam CPA, OPCPA, post compression, and coherent beam combining.

7.4.1 Chirped Beam CPA

In the present HEPW Nd:glass lasers described in Sect. 7.2, a few kJ, ~ 500 fs PW pulses are generated by amplification of frequency-chirped pulses of several ns duration. Noting that this energy is only a fraction of the stored energy (25 kJ for a NIF single beam line, for example), a laser pulse of exawatt peak power could be generated if the stored energy of a high-energy beam line is fully extracted. The peak power attainable with the HEPW lasers has been limited by several constraints: insufficient preamplifier bandwidth to support a ~ 100 fs pulse, insufficient pulse stretching required to keep the B-integral small enough to avoid the nonlinear beam break-up, fabrication of large gratings in the pulse compressor, and laser damage in the final optics. To overcome these limitations, a new concept has been proposed to generate an exawatt-scale ultrashort pulse by amplification of a spatially chirped frequency-chirped laser pulse (Barty, 2016; Chesnut & Barty, 2023).

In this concept shown in Fig. 7.22, a frequency-chirped pulse expanded in time to a long duration (~ 20 ns) is amplified to ~ 25 kJ with NIF-equivalent power amplifiers of 37 cm × 37 cm aperture where the B-integral is kept within 6. The input pulse has broad spectrum to support generating a ~ 100-fs pulse after pulse compression by using mixed glasses in pre-amplifiers (Gaul et al., 2010). The input pulse is spatially pre-chirped to reduce the grating distance, thus limiting the size of the gratings G2 and G3 to within affordable 2 m. The 20 ns pulse is compressed with the four gratings (G1–G4) resulting in a pulse with the same spatial chirp and aperture as the input pulse, but with a duration of 3.3 ns not to exceed the damage threshold of G4. The space- and frequency-chirped pulse after G4 is further compressed by the third pair (G5, G6) to generate a 20 kJ, 100 fs, 0.2 EW peak power pulse.

Since the fluence at G6 far exceeds the damage limit of the grating, the 3.3 ns pulse after G4 is split into ~ 20 beams, each compressed with the third compressor, and focused on target as shown in Fig. 7.23a. The intensity greater than 10^{26} W/cm^2 could be achieved by focusing a 200-PW pulse with fast focusing optics. The multiple laser beams are focused from opposite directions for dipole wave generation as shown in Fig. 7.23b, where the field strength can reach close to the theoretical maximum of the electric dipole pulse (Bulanov et al., 2010; Gonoskov et al., 2012, 2013, 2014).

The available size and the damage threshold of the gratings for pulse compression are the critical issues to generate a ~ 100 PW pulse. The damage threshold of gold-coated gratings (which depends critically on the grating morphology) has been

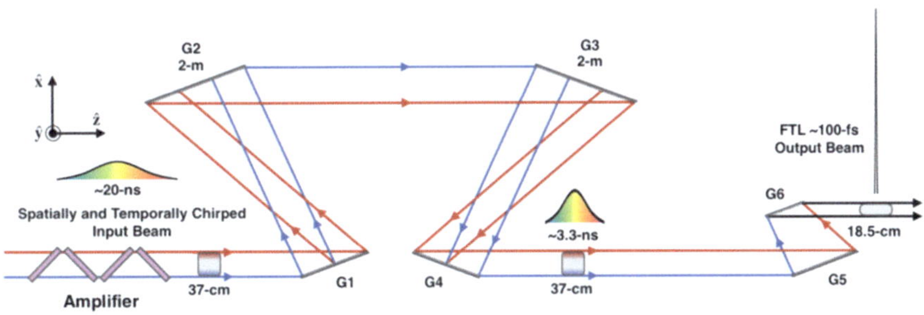

Fig. 7.22 Generation of a 100 fs, 0.2 EW pulse by amplification of a space- and frequency-chirped pulse and pulse compression with a six-grating compressor. Reprinted from Chesnut and Barty (2023). © 2023 Optica Publishing Group

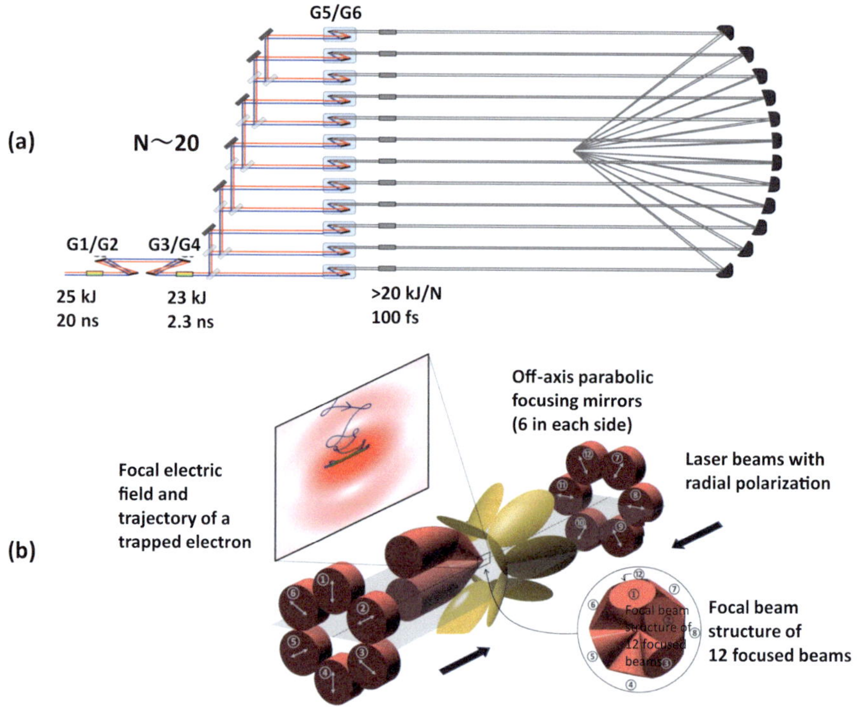

Fig. 7.23 **a** Beam-splitting, pulse compression, and focusing of a 0.2 EW pulse. Reproduced from Barty (2016), with permission from IOP Publishing. © 2016 IOP Publishing. **b** Layout for dipole wave generation where multiple beams with radial polarization are focused symmetrically with fast optics. Reprinted from Gonoskov et al. (2014), with permission from The American Physical Society and the author(s). © 2014 The American Physical Society (Note that the characters in the original figures are enlarged for visibility.)

measured as ~ 0.6 J/cm^2 (Poole et al., 2013; Stuart et al., 1996). Considering these factors, a compressor design based on the chirped beam CPA has been proposed to generate a fs-duration 100 PW pulse (Du et al., 2023; Liu et al., 2021; Wang et al., 2022). In this scheme shown in Fig. 7.24, a frequency-chirped, ~ 4 ns-duration pulse with 200 nm spectral bandwidth is spatially chirped with the prism-pairs (PS-I and PS-II-X) located before and after the OPCPA main amplifier. The lateral shift associated with the chirped beam leads to smoothing of spatial nonuniformities of the laser beam, enabling operating the laser with a higher average fluence on the gratings without causing the grating damage. The space-chirped beam is composed of a central area where full spectral width is covered, and two side areas where the spectrum is not fully covered due to the spatial chirp. The distance between the edge of the central area and the edge of the side area, marked by the black and red (or blue) lines in Fig. 7.24, is defined as the "spatial chirp width".

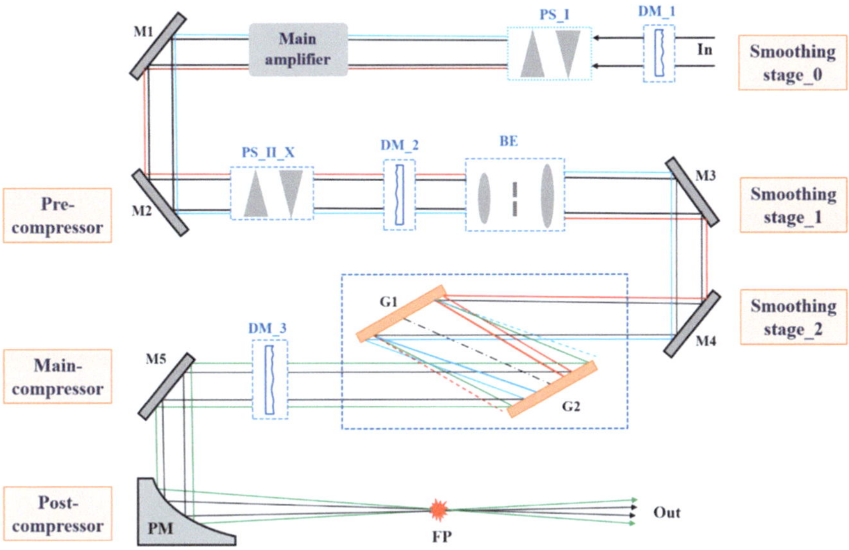

Fig. 7.24 Generation and compression of a space- and frequency-chirped laser pulse. PS-I and PS-II-X—prism pairs for spatial chirping, Main amplifier—OPCPA amplifier, DM-1 –3—deformable mirrors for wavefront correction, BE—beam expander, G1 and G2—singe-pass grating compressor, PM—off-axis parabolic mirror, FP—focal point. Spatial fluence nonuniformities are smoothed by the spatial chirps with the prism pairs and the grating compressor. Reprinted from Du et al. (2023). © 2023 Cambridge University Press

A chirped-beam CPA pulse of 860 mm × 860 mm beam area before the compressor is compressed with a single-pass, single-pair of the gratings (G1 and G2) of 1600 mm × 1000 mm size with long perpendicular distance (~ 2.3 m) sufficient to compensate for the dispersion of a ~ 15 fs pulse. (Compressor design with two grating pairs is given in Liu et al., 2021.) When the width of the chirped beam is 520 mm on the second grating, 7.8% energy loss is introduced due to spectral clipping by the aperture of the second grating. The compressed pulse, which is spatially chirped after compression, is focused with an off-axis parabolic mirror. A slight pulse front tilt introduced by the spatial chirp results in the increase in the pulse duration. Due to spectral clipping and pulse front tilt, it is estimated that the focused intensity decreases by ~ 15% and the pulse duration increases by ~ 11% from 14.3 fs to 15.9 fs (Du et al., 2023). Since the average fluence on G2 is ~ 140 mJ/cm^2 for a 10-fs, 100 PW, 1 kJ pulse of 86 × 86 cm^2 area, which is ~ 25% of the grating damage threshold, it is conceived that ~ 100 PW pulse could be generated with this design.

7.4.2 Ultrahigh-Power OPCPA Laser

As described in Sect. 5.3, optical parametric chirped pulse amplification (OPCPA) has several advantages to generate ultrashort-duration, ultrahigh-power pulses in comparison to the CPA Ti:sapphire laser (Ross et al., 1997). In terms of ultrahigh-power pulse generation, the most attractive feature of OPCPA is its broad bandwidth without gain-narrowing, enabling generation of high-power ultrashort pulses of less than ~ 10 fs duration which is difficult with broadband solid-state lasers. High pulse contrast is achieved by using a pump pulse of ~ ns duration, the pulse width of a chirped signal pulse. The spatial quality of a signal beam is not affected by thermal effects since the pumping energy is not converted to thermal energy in the parametric amplification process. The OPCPA system is composed of fewer amplification stages due to high single-pass gain in nonlinear crystals. The laser beam aperture is limited by the aperture of nonlinear crystals which can be grown larger than the laser crystals. The signal wave is amplified by conserving its spectral and spatial phases, even when the crystal is pumped by multiple beams with different angles, wavelengths, and phases (Marcinkevičius et al., 1998), enabling higher pumping energy and broader gain bandwidth.

Ultra-high-power OPCPA laser requires nonlinear crystals of large nonlinear optical susceptibilities, broad gain bandwidth, and large apertures. Also required are pump lasers of a few ns-duration of rectangular pulse shape to pump the multi-stage optical parametric amplifiers, accurately synchronized to the signal pulse. A pump laser of several kJ energy in green is required at the final stage to generate a 10 fs, 100 PW, 1 kJ pulse in the near infrared.

The nonlinear optical susceptibilities of several crystals, expressed by the contracted *d*-coefficients, are shown in Table 5.1. Among these crystals, BBO and LBO have large nonlinear susceptibilities and can be grown in large sizes: BBO up to ~ 25 mm × 25 mm and LBO up to 130 mm × 130mm (Zeng et al., 2017). For high-power OPCPA laser systems, BBO is often used in the front end whereas LBO is used in the main amplifiers due to availability of larger crystals. Although KDP has smaller nonlinear optical susceptibility compared with LBO, KDP is suitable for ultrahigh power lasers since over 500-mm crystals are available with high damage threshold as described in Sect. 6.2.

High-power OPCPA systems of over 1 PW have been developed or proposed in many laboratories as shown in Table 7.4, many of which utilizing high energy Nd:glass lasers for OPCPA pumping. Either LOB (LiB_3O_5) or partially deuterated KDP (DKDP) crystal is used in the final stages. Bismuth borate (BiBO) and yttrium calcium oxyborate (YCOB) crystals are also studied for ultrabroad band near-infrared OPCPA (Galletti et al., 2020b).

Table 7.4 High power OPCPA facilities, with output power, pulse width, nonlinear crystal and size, center wavelength, bandwidth, and references

Country	Institute	Facility	Crystal (D ratio in %)	Crystal size (mm)	λ_0 (bw) (nm)	Power (PW)	Pulse width (fs)	References
China	SIOM/ SKLHFLP		LBO	100 × 100 × 17	800 (62)	1.02	32	Yu et al. (2015)
	CAEP/ LFRC	SILEX-II	LBO	130 × 130 × 10	790 (65)	4.9	18.6	Zeng et al. (2017)
	SIOM/ NLHPLP	SG-II 5PW	LBO LBO	110 × 17 150 × 150 × 13		1.76 5	21	Zhu et al. (2018)
UK	RAL	Vulcan OPPEL	LBO	100 × 100 × 19	880 (200)	1	30	Musgrave (2021)
Japan	ILE	(design)	LBO	130 × 130	900 (460)	120	8	Li et al. (2021)
Czechia	ELI-BL	L2-DUHA	LBO	> 30	820 (100)	0.1	25	Green et al. (2021)
Japan	ILE	Gekko-EXA (design)	DKDP (65)	800 × 800 × 15	1150 (500)	50	10	Kawanaka et al. (2016)
Russia	IAP	XCELS (proposed)	DKDP (80)	250 × 250 × 43	910 (150)	600 (50 × 12b)	20	Khazanov (2023)
USA	LLE	EP-OPAL	DKDP (>90)	370 × 370	930 (150)	50 (25 × 2b)	20	Bromage et al. (2021)
China	SIOM/ SKLHFLP	SEL-100	DKDP (85)	400 × 400 × 35	920 (200)	100	15	Peng et al. (2021)

Abbreviations: SIOM/SKLHLP—State Key Laboratory for High Field Laser Physics, SIOM; SIOM/NLHPL—National Laboratory for High Power Laser Physics, SIOM; CAEP/LFRC—Laser Fusion Research Center, Chinese Academy of Engineering Physics; RAL—Rutherford Appleton Laboratory; ILE—Institute of Laser Engineering, Osaka University; ELI-BL—ELI Beamlines; IAP—Institute of Applied Physics of the Russian Academy of Sciences; LLE—Laboratory for Laser Energetics, University of Rochester

7.4.2.1 High Power OPCPA with LBO Crystal

Referring to Fig. 5.8 for non-collinear phase matching, the biaxial LBO crystal is suitable for parametric amplification at ~ 800 nm by 526.5 nm pumping with type-I phase matching at $\theta = 90°$ and $\varphi = 13.85°$, noncollinear angle of $\alpha = 1.26°$ and walk-off angle of $\rho = 0.48°$ (Xu et al., 2013). Using a $100 \times 100 \times 17\ mm^3$ LBO crystal, 1.02 PW, 32 fs pulse was generated at SIOM (Yu et al., 2015). A higher-power pulse of 4.9 PW, 18.6 fs, 91.1 J at 800 nm has been generated at CAEP, by pumping two LBO crystals of $130 \times 130 \times 10\ mm^3$ and $165 \times 120 \times 10\ mm^3$ with a 800 J, 3 ns, 526 nm, $105 \times 105\ mm^2$ pulse. An energy transfer efficiency from the pump to the signal beam in the final OPA stage was 25% (Zeng et al., 2017).

Another LBO-based high-power OPCPA is the Vulcan OPPEL (OPCPA PEtawatt Laser) under construction at the Central Laser Facility, Rutherford Appleton Laboratory, with design performance of 30 J energy in 30 fs, i.e., 1 PW peak power (Galletti et al., 2020a). The Vulcan OPPEL is a new beamline added to the Vulcan Nd:glass laser facility. A broadband pulse from a Ti:sapphire oscillator, electronically synchronized to the Vulcan oscillator, is stretched and amplified in three-stages with LBO crystals where the final crystal of $100 \times 100 \times 19\ mm^3$ in size is pumped by Vulcan (Musgrave et al., 2021).

The gain bandwidth of parametric amplification is significantly enhanced by pumping by the laser beams with angular dispersion (Baltuska et al., 2002) and broad spectral width (Zhao et al., 2008). Based on this concept, possibility for generating a single-cycle 100 PW pulse is studied by Li and Kawanaka (2019), where an LBO crystal is pumped with multiple beams with different angles, named wide-angle non-collinear OPCPA (WNOPCPA). In an updated configuration of this concept shown in Fig. 7.25 (Li et al., 2021), a chirped pulse is amplified in LBO crystals, each pumped by two pump beams with different angles thereby amplifying different spectral components for broadening the gain bandwidth and with a time delay to avoid interference between the pump beams. A ns-duration chirped pulse generated in the front end is amplified in five stages, with 130 mm × 130 mm aperture LBO crystals in the last three amplifiers, generating a pulse of ~ 1380 J energy, 400 nm spectral-width after the final amplifier. A 971 J, 8.0 fs, 120 PW pulse is generated after compression with a grating compressor of 70% transmittance. (In the configuration shown in this figure, the amplified pulse is further compressed by post compression, which is described in the next section.).

Repetitively pulsed, highly stable OPCPA lasers pumped by DPSSL (diode-pumped solid-state laser) are developed at ELI Beamlines. A 1 kHz OPCPA laser, L1-Allegra, is now operated as a user facility close to the design performance of 100 mJ energy and 15 fs pulse width with better than 5% energy stability (Bakule et al., 2017). For this laser, accurate synchronization of the pump and signal pulses with 17 fs RMS timing jitter has been achieved (Batysta et al., 2014). A higher power system, L2-DUHA OPCPA, is under construction for developing an x-ray free electron laser driven by high-energy electrons generated by laser wake-field acceleration. In this system, 515-nm pump, 820 nm-centered signal and 2200 nm-centered probe pulses are generated with a common oscillator, for accurate timing of these pulses. A cryogenically cooled, LD pumped Yb:YAG multi-slab amplifier generates 3 ns, 10 J pulses at 10 Hz for pumping BBOs in a front end and LBOs in a main amplification stage. After pulse compression, this system will provide 2.8 J, 25 fs, 120 TW pulses in 10 Hz (Green et al., 2021).

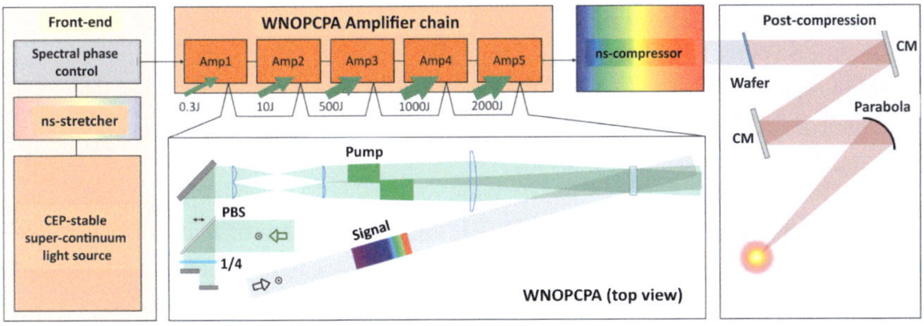

Fig. 7.25 Schematic of wide-angle non-collinear OPCPA (WNOPCPA) where a signal pulse is amplified in five stages with LBO crystals. In each stage, an LBO is pumped by two pump beams with different angles with a time delay. The amplified pulse, compressed with a grating compressor, is further compressed with a post compressor composed of a thin optical plate for spectral broadening and chirped mirrors for final compression. Reprinted from Li et al. (2021). © 2021 Springer Nature (Note that the characters in the original figure are enlarged for visibility.)

7.4.2.2 High Power OPCPA with DKDP Crystal

Partially deuterated KDP crystal (DKDP) is suitable for scaling OPCPA to exawatt power, due to its broad amplification bandwidth and availability of very large, high-quality crystals (see Sect. 5.3.6). The super-broadband parametric amplification by DKDP was discovered by Freidman et al. (2002) and Lozhkarev et al. (2005) and extended to demonstration of up to 1 PW peak power at the Institute of Applied Physics, Russian Academy of Science (Lozhkarev et al., 2006; Shaykin et al., 2009). The optimum deuteration ratio of DKDP for super-broadband amplification was further investigated by Fujioka et al. (2015).

The ultra-high power OPCPA laser with DKDP for super-broadband amplification has been proposed by Dabu (2010), Kawanaka et al. (2016), Shen et al. (2018), Bromage et al. (2019), and Khazanov et al. (2023). In the design study at the Institute of Laser Engineering, Osaka University, a 50 PW, 10 fs pulse with 1150 nm center wavelength is generated using a 65-% deuterated DKDP crystal in the final amplifier (Fujioka et al., 2015; Kawanaka et al., 2016).

At the Laboratory for Laser Energetics, University of Rochester, a DKDP-based OPCPA system, called EP OPAL (OMEGA-EP optical parametric amplifier line), is planned for addition to the OMEGA EP (OMEGA Extended Performance) which is a four-beam high energy Nd:glass laser system with its two beamlines operated in chirped pulse amplification for up to 1 PW output power. The EP OPAL is a two-beam system, where a 370 mm^2 DKDP crystal is pumped by one of the ns-duration beamlines of OMEGA EP, generating a 500-J, 20-fs, 25 PW pulse in each beamline with an expected intensity at focus exceeding 5×10^{23} W/cm^2. This OPCPA system is designed with DKDP of over 90% deuteration fraction to make the angle between the signal and pump beams larger than 1°. As a preliminary study for EP OPAL development, a lower power OPCPA system, called MTW-OPAL (multi-TW optical parametric amplification beamline) has been developed (Bromage et al., 2019). This is a stand-alone system, complete with a front end, a pumping laser, parametric amplification stages, ultra-broadband transport line, a grating pulse compressor, and a vacuum chamber for focusing and characterization. With a 63 mm^2 DKDP crystal in the final amplifier, a 7.3-J, 20-fs, 0.35-TW pulse has been generated in preliminary experiments (Bromage et al., 2021).

A 100 PW OPCPA facility SEL (Station of Extreme Light) is under construction at the Shanghai high repetition rate XFEL and Extreme Light Facility (SHINE) in Pudong District of Shanghai City. The SHINE project is undertaken to investigate quantum electrodynamic physics, including measurement of the vacuum birefringence created under an intense optical field with a linear-polarization X-FEL beam (Shen et al., 2018). Layout of SHINE, which is comprised of XFEL and 100-PW laser (SEL), is shown in Fig. 7.26.

The SEL-100 PW is a full-OPCPA system, comprised of a front-end, three OPCPA stages and the final amplifier with a 400 mm × 400 mm DKDP crystal (~ 85% deuteration ratio) pumped by a 5ns/2 kJ, 526 nm pulse from a Nd:glass laser, with which a 25 J/0.1 Hz pulse centered at 820 nm is amplified to 0.5 kJ with 200nm bandwidth. The amplified pulse is expanded to 800 mm × 800 mm, divided into four beams due to the current size limit of the grating, compressed with four independent grating compressors, and combined at the focus to irradiate a target with a 1.5 kJ, 15 fs, 100 PW pulse. Several key elements of SEL-100 PW are being developed: generation of broadband, carrier-envelop-phase-stabilized seed pulses (Shao et al.,

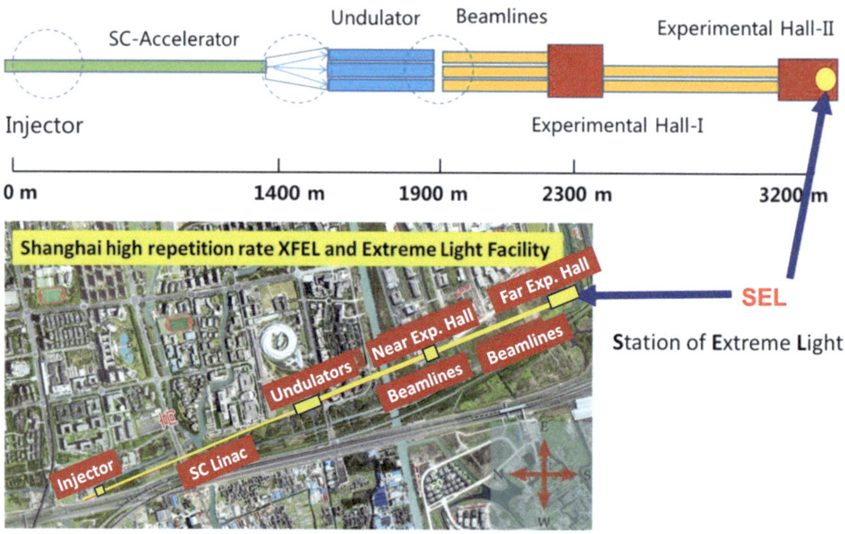

Fig. 7.26 Layout of the SHINE facility under construction. At the Experimental Hall-II located at the end of the 3200-m long XFEL installed in an underground tunnel, the SEL-100 PW laser beam is focused to investigate quantum electrodynamics physics including vacuum birefringence by using both SEL and XFEL beams. The white ring in this photo is Shanghai Synchrotron Radiation Facility SSRF. Curtesy of Prof. Yuxin Leng, © SIOM

2020), 13.4 fs, 0.1 Hz, full-OPCPA front-end (Wang et al., 2022a), and compressor design to compress a large-area laser beam based on the chirped-beam CPA design (Du et al., 2023; Wang et al., 2022b; Wu et al., 2022).

As a candidate for the 4th pillar of the ELI concept (ELI Whitebook, 2011), 600 PW XCELS (eXawatt Center for Extreme Light Studies) system is proposed at the Institute of Applied Physics of the Russian Academy of Sciences. In the XCELS, 12 beams of 15 PW each generated by DKDP-based OPCPA are focused as a converging dipole wave to create an extremely high field of ~ 10^{25} W/cm^2 in a vacuum (Khazanov et al., 2023).

The layout of XCELS is shown in Fig. 7.27. A signal pulse of 910 nm wavelength with 200 nm bandwidth and 3-ns duration generated in the front end is amplified with the intermediate and booster OPCPAs, with the output of 27-cm diameter circular beam of 1250 J energy. From this beam, 12 beams of 5 cm-square shape are split. Each beam is expanded to 25 cm × 25 cm with a reflective telescope and amplified with the final OPCPA to 1250 J, expanded to 55 × 55 cm and compressed with a 4-grating compressor, resulting in a 750 J, 20 fs, 35 PW pulse. The 12 beams are focused with off-axis parabolas in a configuration of two belts near the equatorial plane with which the intensity of $I = cE^2 / 8\pi = 8\pi P / 3\lambda^2$ is reached at the central point, where E is the electric field strength, P is the total power, and λ is the laser wavelength. If the temporal and spatial phases are completely controlled, it is expected that the intensity will reach 32×10^{25} W/cm^2 (Gonoskov et al., 2013). Like other systems, the output power is constrained by the laser damage and the available size of the grating. In the XCELS design, the grating size is 57 cm × 101 cm for G2 and G3, and the fluence on the last grating is 0.174 J/cm^2.

7.4.3 *Post Compression*

Post compression is a method to generate an ultrashort-duration pulse from a short pulse by broadening its spectrum and compressing the spectrum-broadened pulse. Compression of a picosecond laser pulse to a femtosecond pulse has been proposed in the 1960s by broadening its spectral bandwidth using an electrooptic phase modulator (Giordmaine et al., 1968) or nonlinear optical self-phase modulation in liquids (Fisher et al., 1969). This concept was advanced in the 1980s to generate femtosecond laser pulses by spectral broadening with a single-mode fiber by self-phase modulation and pulse compression with a pair of gratings (Grischkowsky & Balant, 1982). Since the energy of the fs pulses generated with the single-mode fiber was limited to a few nanojoules, this concept was extended to generate higher energy fs pulses by increasing the beam area for self-phase modulation with hollow core fibers (HCF) or with bulk materials.

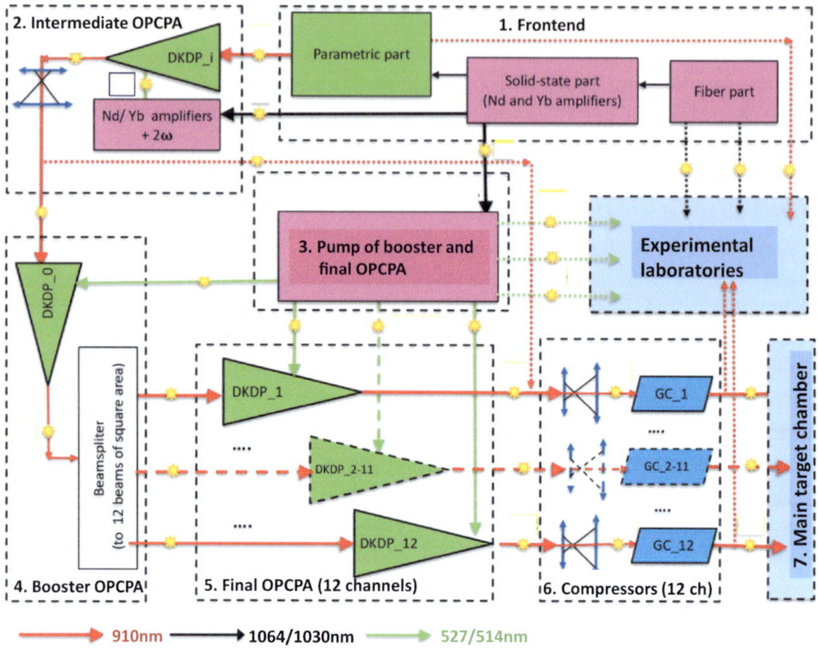

Fig. 7.27 Layout of the 12-beam, 400 PW XCELS system. A signal pulse generated in the frontend is amplified in the intermediate and booster DKDP OPCPAs, split into 12 beams, amplified with the 12 final OPCPAs, compressed with the 12 compressors, and focused in the main target chamber. Reprinted from Khazanov et al. (2023). © 2023 Cambridge University Press (Note that the characters in the original figure are enlarged for visibility.)

7.4.3.1 Post Compression with a Hollow-Core Fiber

By propagating laser pulses from a Ti:sapphire laser through a hollow-core fused silica fiber filled with noble gases at high pressure, a 140 fs/660 μJ laser pulse was converted to a 10 fs/240 μJ pulse, i.e., 14 times pulse compression, with 36% energy conversion efficiency (Nisoli et al., 1996). By refinements of this approach, 3.4 fs (1.5-cycle at 719 nm central wavelength), 2.5 mJ pulses have been generated at 1 kHz with an experimental layout shown in Fig. 7.28 (Ouillé et al., 2020). In this work, the 24 fs-duration pulses from the front-end, converted to circular polarization to mitigate multiphoton ionization and self-focusing, were propagated through a HCF of 2.4 m length and 536 μm inner diameter filled with a helium gas. The output pulse, re-converted to linear polarization, was compressed with 6 pairs of chirped mirrors. The third-order dispersion was compensated by controlling a wedge plate in the beam path to keep the spectral phase to within ~ 2 rad over 600 nm bandwidth (400–1000 nm). The carrier-envelop-phase (CEP) and the pulse energy were stabilized to within 152 mrad and 0.5%, respectively. This laser is used to investigate light waveform effects in laser wake-field electron acceleration at relativistic intensities.

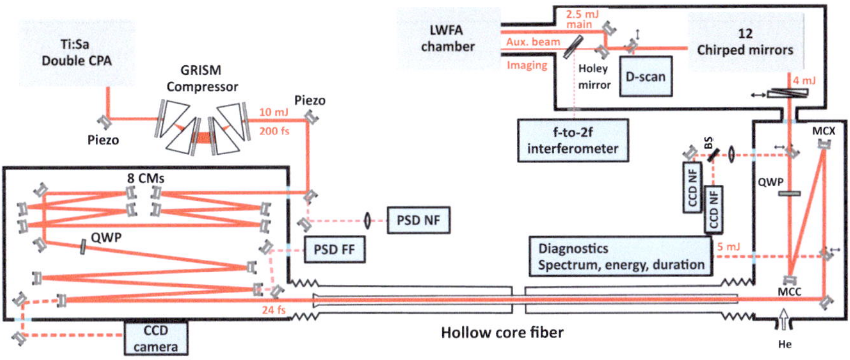

Fig. 7.28 Schematic for generating 1.5-cycle pulses by post compression with a gas-filled hollow core fiber. Abbreviations: CM—chirped mirror, QWP—quarter wave plate, MCX—convex mirror, MCC—concave mirror, PSD—photosensitive detector, NF—near field, FF—far field. Reprinted from Ouillé et al. (2020). © 2020 Springer Nature (Note that the characters in the original figure are enlarged for visibility.)

This approach has been upscaled to generate laser pulses exceeding 1 TW (3.8 fs pulses of 6.1 mJ energy) at 1 kHz, by propagating a laser beam at an intensity slightly above the ionization threshold for efficient spectral broadening through a longer hollow core of 3.75 m (Nagy et al., 2020). As seen in these examples, the post compression with HCF is suitable to generate well-controlled few-cycle pulses with ~ TW peak power.

7.4.3.2 Post Compression with Bulk Materials

(a) Compression After Compressor: CafCA

Post compression of a large-area, high-energy laser beam with a bulk material is suitable to extend this concept to ~ PW scale. In this concept, laser irradiation of the bulk material with uniform intensity distribution over a whole area is required for uniform spectral broadening. As described in Sect. 5.4.3, the spectral broadening $\delta\omega_{\max}$ by self-phase modulation is related to the nonlinear phase delay $\Phi_{NL}^{\max}$ as $\delta\omega_{\max} \simeq \Phi_{NL}^{\max}/\tau_0$, where τ_0 is the laser pulse width. The nonlinear phase delay is expressed as $\Phi_{NL}^{\max} \simeq (2\pi/\lambda_0)L\gamma_{nl}I = B$, where L is the material thickness,γ_{nl} is the nonlinear refractive coefficient, I is the laser intensity, and B is the B-integral which characterizes the self-focusing. From these relations, we find that this method is suitable for generating an ultrashort duration, i.e., broad spectral width, and high-intensity pulse although it is constrained by B-integral.

In an early experiment on post compression, a 19 fs, 11 μJ pulse was generated from a 92 fs, 300 μJ input pulse, i.e., 4.8-fold pulse compression and 4% efficiency, by irradiating a 4.2-mm thick quartz plate with a ~ 1-mm diameter Gaussian beam

(Rolland & Corkum, 1988). This work was followed by Mével et al. (2003), generating 14-fs, 220 μJ pulses with an improved configuration. After these initial works, it was shown theoretically that it is possible to generate sub-exawatt few-cycle pulses, based on analyses of nonlinear propagation of high intensity beam while considering various limiting processes; dispersion, pulse self-steepening, small-scale self-focusing and ionization (Mourou et al., 2014; Voronin et al., 2013). This method, where a spectral-broadened pulse induced by self-phase modulation with a thin dielectric plate or a thin film is compressed with chirped mirrors, is called thin film compression (TFC) (Mourou et al., 2014) or compression after compressor approach (CafCA) (Ginzburg et al., 2019). (Hereafter we call this method CafCA.) A comprehensive review of the post compression with bulk materials is given in Khazanov et al. (2019).

(b) **Controlling the Whole-Beam and Small-Scale Self-focusing**

To increase the spectral width while controlling the self-focusing, several innovative concepts have been introduced in CafCA as reviewed by Khazanov et al. (2019). The first is mitigation of the whole beam self-focusing, where a thin plate acts as a positive lens when the laser intensity is stronger in the center of the beam such as a Gaussian distribution, since the nonlinear phase delay $B = kL\gamma_{nl}I$ is proportional to $I(r)$, where k is the wave number and L is the plate thickness. This phase delay can be cancelled by using a negative lens whose thickness$L(r)$ is inversely proportional to $I(r)$ (Mironov et al., 2014).

Another important concept is the suppression of small scale self-focusing, where a laser beam breaks up into many small hot spots. As described in Sect. 4.4.2, the growth of the lateral intensity perturbation depends on its spatial frequency K. The spatial frequency for the maximum growth is given by $K_m = K_0(2\gamma_{nl}I)^{1/2}$, with the growth rate $g_m = K_0\gamma_{nl}I$ where $K_0 = 2\pi/\lambda_0$ is the wavenumber of the laser beam in a vacuum. In high-energy lasers of nanosecond pulse duration and $\mathrm{l}_0 = 1\ \mu\mathrm{m}$, $(2\gamma_{nl}I)^{1/2} \approx 1.3 \times 10^{-3}$ for $I \approx 3\ \mathrm{GW/cm^2}$ (typical intensity for ns-duration lasers) and $\gamma_{nl} \approx 3 \cdot 10^{-16}\ \mathrm{cm^2/W}$ (valid for phosphate glass and Ti:sapphire crystal). In this case, $K_m \approx 1.3 \times 10^{-3}K_0 \approx (2\pi) \times 13\ \mathrm{cm}^{-1}$ corresponding to a spatial scale of $\Lambda_m = 2\pi \big/ K_m \approx 0.8$ mm and a divergence angle $\theta_\mathrm{m} \sim K_m/K_0 \sim 1.3$ mrad. Growth of these perturbations is suppressed with multiple spatial filters placed in series in an amplifier chain as described in Sect. 3.2.6.

In femtosecond-duration high power lasers, the laser intensity is several TW/cm^2, 3 orders of magnitude higher than the intensity of nanosecond-duration high energy lasers. Since K_m scales as $I^{1/2}$, the spatial perturbation grows at a higher frequency $K_m \approx (2\pi) \cdot 400\ \mathrm{cm}^{-1}$, corresponding to a shorter spatial scale of $\Lambda_\mathrm{m} \approx 25\ \mu\mathrm{m}$ and a larger divergence angle of $\theta_m \approx 40$ mrad. Due to this large divergence angle, the small-scale perturbations are self-filtered by free spatial propagation longer than a distance of $z \approx D/\theta_m$ where D is the beam diameter, without using the long spatial filters (Ginzburg et al., 2018; Mironov et al., 2012). A figure of merit of a nonlinear material for applying CafCA to ultrashort pulses is the ratio $\gamma_{nl} \big/ k_2$, where γ_{nl} characterizes the spectral broadening and k_2, the group velocity dispersion, characterizes the effect of dispersion to the broad-band pulse (Ginzburg et al., 2020).

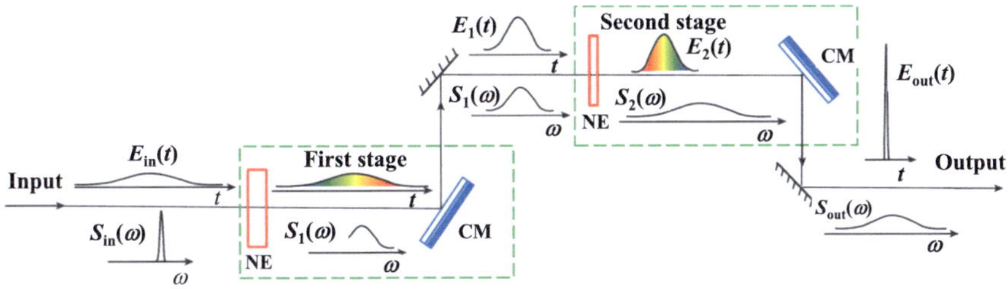

Fig. 7.29 Schematic layout of a two-stage CafCA. In the first stage, an input laser pulse from a high-power laser is spectral broadened with a nonlinear element (NE), propagated in a free space over a distance sufficient for filtering small scale perturbations, and compressed with a chirped mirror (CM). In the second stage, the generated pulse is further compressed to generate an ultrashort output pulse. Reprinted from Ginzburg et al. (2020), with permission from Turpion and the author(s). © 2020 Turpion (Note that the characters in the original figure are enlarged for visibility.)

Very short-duration pulses can be generated with multiple-stage post compression, where a short pulse generated in the first stage is further compressed to an ultrashort pulse in the second stage, with a layout shown in Fig. 7.29 (Ginzburg et al., 2020). To generate an ultrahigh-power laser pulse with this method, very thin nonlinear element of only a few 100 μm thickness is required for keeping the *B*-integral below the breakup limit but large enough for attaining sufficient spectral broadening. As a nonlinear material, a plastic film for spectral broadening has been proposed and tested to overcome the technical difficulty of fabricating a thin transparent high optical-quality plate with large aspect ratio D/L, where D is the beam diameter and L is the plate thickness (Mironov et al., 2015; Mourou et al., 2014).

(c) Propagation of an Ultrashort-Duration, High-Intensity Laser Pulse

In the post compression, it is required to consider propagation of a laser pulse where the pulse length in the propagation direction becomes comparable to the optical wavelength and the intensity becomes close to the ionization threshold of dielectric materials. For this case, the slowly varying envelope approximation (SVEA) which we have assumed in Sects. 2.1.5 and 4.2.6 is not applicable, and nonlinear dispersion should be considered.

As described in Sect. 5.5, it is predicted that the filamentation instability will be stabilized when the pulse duration becomes less than ~ 10 optical cycles (i.e. ~ 25 fs at 800 nm). In this extremely short pulse region, the perturbation growth due to filamentation instability is delayed to the rear side of the pulse due to dependence of the group velocity on the intensity, i.e. the instability becomes convective (Balakin et al., 2016, 2017). If this prediction is valid, although it needs experimental verification, the small scale self-focusing will be suppressed in the propagation of extremely short high-intensity pulses, possibly enabling generation of a few-cycle, extremely high-power laser pulses.

(d) Development of High-Power Post-compression Lasers

The CafCA concept has been examined in various experiments. For example, a compressed pulse of 11 fs duration and 1.5 PW peak power has been generated with a single-stage CafCA (Ginzburg et al., 2021). In this experiment, an input pulse of ~ 60 fs duration was compressed to 11 fs with a 5-mm thick silica plate for self-phase modulation and two chirped mirrors in series for the chirp compensation with total dispersion of $-$ 250 fs^2. This experiment was implemented with an input beam of 54 –74 fs duration, 910 nm central wavelength, 18 cm diameter, and laser energy up to 18 J. This parameter set with 18 J laser energy and 60 fs pulse width corresponds approximately to an input power of 0.3 PW, an average intensity of 1.3 TW/cm^2, average fluence of 70 mJ/cm^2, and B-integral in the silica of $B = 14$. Noting that KDP has larger figure of merit than fused silica, where the inverse of the figure of merit k_2 / γ_{nl} is 1 and 2.5 (in unit of fs^2 PW/cm^3) for silica and KDP respectively, another experiment was implemented with the same configuration but with a 4-mm thick KDP crystal replacing a fused silica plate. An output pulse of 10-fs duration and 1.5 PW peak power has been generated with chirped mirrors of -200 fs^2 dispersion (Shaykin et al., 2021). That is, ~ 6 $\times$ pulse width reduction and ~ 5$\times$ intensity enhancement have been achieved.

Despite the large B-integrals in both experiments, there were no signs of damage on the silica or KDP and the chirped mirrors, indicating suppression of small-scale self-focusing. This suppression was tentatively ascribed to self-filtering of high spatial frequencies, decrease of noise sources at high spatial-frequencies, and possible indication of instability convection proposed by Balakin et al. (2016, 2017).

In another experiment with similar layout, a laser pulse of 820 nm central wavelength, 41 nm spectral width, 22 fs pulse duration, and 7 J energy from a Ti:sapphire laser was passed through two thin (0.4-mm and 0.3-mm thick) glass plates placed at a Brewster angle, resulting in bandwidth broadening to 74 nm capable of pulse compression to 15 fs (Bleotu et al., 2022). In another direction for application to smaller facilities, the post compression with bulk materials was applied to a compact 1 kHz Ti:sapphire laser. In two-stage compression using a 0.5-mm thick fused silica plate in each stage for spectral broadening, compressed pulses of 7 fs duration and 0.7 mJ energy were generated at 1 kHz with close to diffraction-limited focal spot (Stanfield et al., 2021).

These examples show that the post compression with bulk materials is useful to generate sub-10 fs pulses for multi-PW facilities and for high-repetition-rate compact lasers, by simple addition to the existing facilities. When we apply the post compression to generate a 10 fs, 100 PW pulse, a chirped mirror of over 80 cm diameter will be required, considering ~ 0.2 J/cm^2 damage threshold (measured with 1030 nm, 180 fs pulses) of the chirped mirrors designed by electric field reallocation (Melnikus et al., 2017). Compensation of higher order dispersions will be also necessary to achieve the pulse duration shorter than ~ 15 fs.

7.4.4 Coherent Beam Combining

Coherent beam combining (CBC) is an approach to produce a high-power laser beam by coherently combining multiple laser beams by controlling their temporal and spatial phases (Brignon, 2013). In CBC, which is developed under international collaboration (Mourou et al., 2013; Tajima et al., 2013), multiple laser beams, split from a beam generated with a master oscillator, are amplified and their phases are controlled to the same phase with a phase controller for each beam. There are two methods for CBC: tiled-aperture scheme and filled-aperture scheme. In the tiled-aperture scheme, a large-aperture phased-array beam is formed by synchronizing multiple parallel beams (Goodno et al., 2006; Le Dortz et al., 2017). In the filled-aperture scheme, multiple laser beams are superposed collinearly to form one laser beam. The theoretical maximum efficiency of combining multiple beams is 100% for the filled-aperture, whereas it is 68% for the tiled Gaussian beams due to tiling gaps between the beams which lead to far-field side lobes in the composite beam.

An example of the tiled-aperture CBC with fiber lasers is shown in Fig. 7.30 (Fsaifes et al., 2020). The tiled-aperture scheme is suitable for coherent combining of high energy pulsed lasers since this scheme does not require beam combining mirrors which are susceptible to the laser damage. The CBC of 61 fiber amplifiers in a tiled-aperture configuration, shown in Fig. 7.30, has generated at ~ 1 kW average power of 257 fs pulses at 2 MHz repetition rate, i.e., 500 μJ energy per pulse. In this system, 220 fs oscillator pulses are stretched to 2.5 ns with a chirped fiber Bragg grating, followed by a pulse shaper to pre-compensate the dispersion over the whole chain, and split into 64 beams through two stages of 1×8 splitters. The 61 beams individually pass through Delay and Phase Adjustment Modules (DPAM) for independent adjustment of each optical path length for exact match of the phase offset and the delay between all pulses. Each beam is then amplified with a large mode area Yb-doped photonic crystal fiber of 30 μm mode field diameter. The 61 output fibers are accurately arrayed in a hexagonal configuration of 3.2 mm pitch with a laser head, and the output beams are collimated with a hexagonal micro-lens array. The combined and spatially filtered beam is compressed with a Tracy-type pulse compressor. The 61 parallel beams, partially transmitted through a mirror, overlapped with a reference beam generated as one of the 64 beams on a camera for interferometric measurement providing the control signals to DPAM. With feedback control, a residual phase error between two fibers of $\lambda/90\,RMS$ and beam quality of $M^2 = 1.11$ has been achieved.

In the filled-aperture CBC, multiple beams are combined using optical elements such as dielectric polarizers (Müller et al., 2016), diffractive optical elements (Zhou et al., 2018), and intensity beam splitters (Müller et al., 2020). We show in Fig. 7.31 a schematic setup of the filled-aperture CBC of 12 fiber amplifiers, which has generated laser pulses of 254 fs duration of 10.4 kW average power at an 80 MHz repetition rate, i.e., 130 μJ energy per pulse (Müller et al., 2020). The stretched, pulse-shaped and amplified pulses of 150 W average power from the front-end are split into 16 channels with a tree of intensity beam splitters. In each channel, a set of quarter

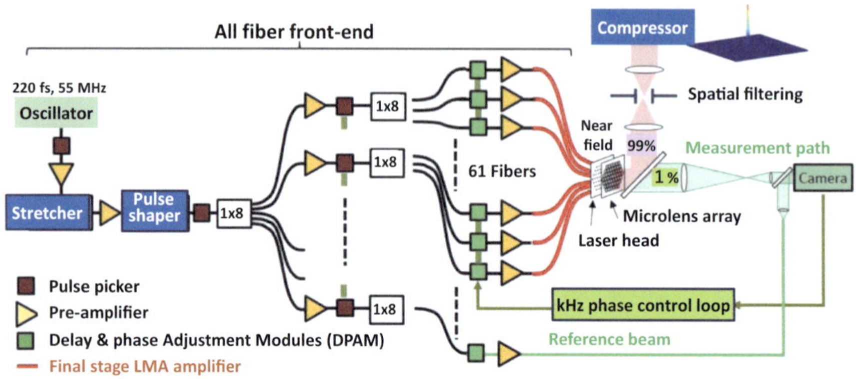

Fig. 7.30 Coherent beam combining with tiled-aperture scheme. Reprinted from Fsaifes et al. (2020). © 2020 Optical Society of America (Note that the characters in the original figure are enlarged for visibility.)

wave plate (WP) and a piezo-actuated mirror (PA) are used for polarization control and phase stabilization, respectively. Each beam is amplified with a 11-m long non-polarization-maintaining ytterbium-doped step-index 20/400 active fiber (20 μm core and 400 μm cladding diameter) coiled to 12 cm diameter, pumped by fiber-coupled LDs up to 1.6 kW per channel. The direction and phase of each amplified beam is controlled by two motorized mirrors (MMs) and a motorized translation stage (MTS), respectively, combined into one beam with the second tree of beam splitters, and compressed with a double-pass Treacy-type compressor.

One of the 16 channels is used as a reference for locking of optical coherence by single-detector electronic frequency-tagging (LOCSET) (Brignon, 2013), where the intensity dithers due to small phase dithers of the piezo-actuated mirror PA at distinct frequency in each channel is detected by a photodiode (PD_1) and fed back to the PA position for constructive interference. Furthermore, offsets of the group delay and the beam direction are detected by the photodiodes PD_2 and PD_3, respectively, and fed to MTS and MMs for locking the 11 beams to the reference beam. All optomechanical components after the beam combination stage are water-cooled to minimize the thermal drift. With this system, an output power of 10.4 kW after compression has been reached with a beam combination efficiency of 96%, a beam quality of $M^2 < 1.2$, and the combined power noise of 0.56% in the frequency span from 1 Hz to 1 MHz. Since the temperature increase of the final beam splitter is only 3 K in this system, it is expected that this method can generate output power of 100 kW level.

Theoretical analyses of the beam-splitter based CBC by Müller et al. (2021) show that average power is limited by the thermo-optical and thermo-mechanical properties of the beam splitter material and its coatings. It is anticipated that 100 kW average power can be obtained with the state-of-the-art intensity beam splitters with 1.4 ppm residual absorption, at only 2% loss of combining efficiency and high beam quality $\left(M^2 \leq 1.1\right)$. Even up to 300 kW level appears feasible by keeping the combination

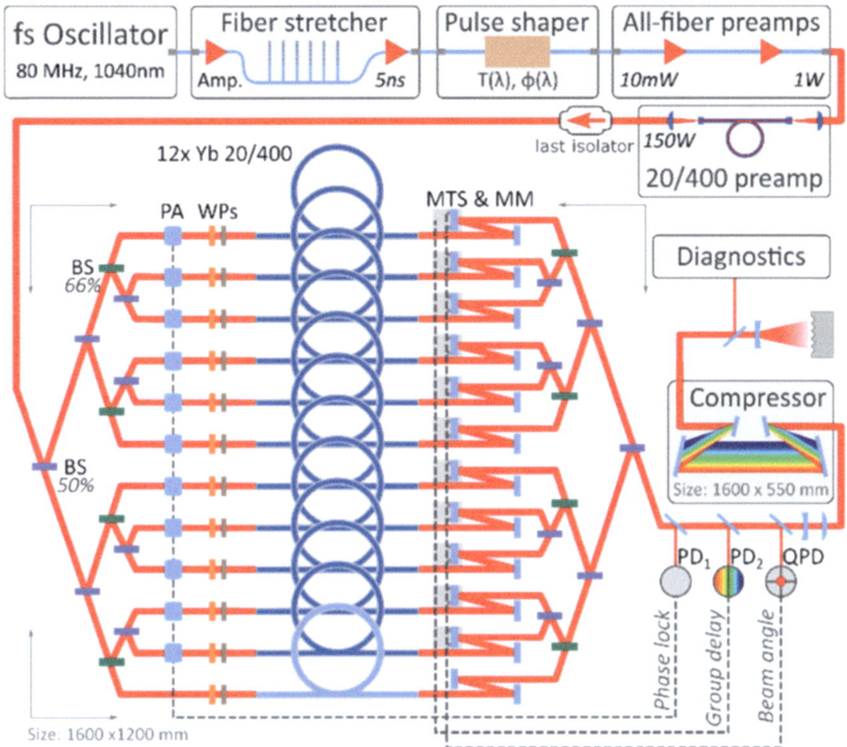

Fig. 7.31 Schematic setup for filled-aperture coherent combining of chirped pulses with 16-channel fiber amplifiers, delivering 10.4 kW average power. Reprinted from Müller et al. (2020), with permission from Optical Society of America. © 2020 Optical Society of America

efficiency of over 90% with M^2 degradation of 0.1. The peak power is limited by the optical damage of the surface in the nanosecond-stretched-pulse regime and/or by nonlinear phase aberration in the ultrashort-pulse regimes. These effects can be mitigated by increasing the beam size, enabling scaling to Joule-level pulse energy. Therefore, the CBC approach will enable realizing demanding applications such as laser wake-field electron accelerators (Leemans, 2011) and space debris removal (Soulard et al., 2014).

7.5 Laser-Diode-Pumped Short-Pulse, High-Power Lasers

Repetitively pulsed ultrashort-duration high-power lasers are required in applications to science and industry. The repetition rate of the high-power Ti:sapphire laser is limited at present by the repetition rate of the pumping laser, which is for example a flashlamp-pumped Nd:glass laser delivering 50–100 J of ~ 20 ns duration at 532 nm. The Ti:sapphire lasers could be operated at 10–100 Hz if they are pumped by high-energy, diode-pumped solid-state lasers (DPSSLs) described in Sect. 7.3.

Laser-diode pumping of broadband gain material is a more direct approach to realize repetitively pulsed, short-duration, high-power lasers with high overall efficiency. For direct LD pumping, laser materials with long gain lifetime are required since the peak power of the LD is limited due to its short upper-level lifetime and internal optical damage. The gain materials of long gain lifetime can store large energies in small volumes, whereas they are accompanied with low small-signal gain and high saturation fluence. Innovative laser architectures are required to fully exploit the potential of these gain materials, while avoiding the laser damage arising from high saturation fluence. In the following we describe rare-earth-doped CaF_2 lasers and Tm:YLF laser, which are developed by taking these aspects into account.

7.5.1 LD-Pumped Rare-Earth-Doped CaF_2 Lasers

High efficiency, repetitive, nanosecond-pulse DPSSLs have been developed with Yb:YAG crystals or ceramics, as described in Sect. 6.3.2. Short pulse DPSSLs can be developed by replacing Yb:YAG in these architectures with other laser materials with broader spectral widths. One of the candidates is the rare-earth (RE)-doped CaF_2. A Yb-doped CaF_2 crystal has broad fluorescence spectrum of ~ 70 nm width, sufficient for amplification of ~ 100 fs-duration ultrashort pulses (Hein et al., 2006). The host material of CaF_2, available as high-quality large size crystals, has broad optical transmission range (0.13–10 μm), small spectral dispersion, and small nonlinear refractive index. In a Yb:CaF_2 crystal, Yb^{3+} substitutes for Ca^{+2}, where excess charge due to Yb^{3+} substitution is compensated for by interstitial F^- ions. This leads to the formation of several fluorescent centers with different symmetries. These fluorescent centers, which are isolated at low doping concentrations (< 0.1 at %), become aggregated at higher concentrations (up to ~ 8%), resulting in broad bands in absorption and emission spectra (Petit et al., 2008).

Spectroscopic and laser properties of Yb:CaF_2 are compared with those of Yb:YAG by Siebold et al. (2009). Due to broader spectral width, Yb:CaF_2 has small emission cross section of 1.7×10^{-21} cm^2 and long radiative lifetime of 2.4 ms, which are 0.085 × and 2.5 ×, respectively, of the corresponding values of Yb:YAG. With the LD-pumped Yb:CaF_2 laser, 99 fs pulses have been generated in mode-locked operation (Friedel et al., 2009).

In the design of a high power Yb:CaF_2 laser system, the laser beam must pass many times through the active crystal to amplify the injected beam and extract the stored energy, due to the small stimulated-emission cross section, high saturation fluence (~ 80 J/cm^2) and long radiative lifetime of Yb:CaF_2. Development of a LD-pumped Yb^{3+}:CaF_2 laser is undertaken at Friedlich-Schiller-University Jena and Helmholtz-Institute Jena (Siebold et al., 2008; Tamer et al., 2019), and at the Helmholtz-Zentrum Dresden-Rossendorf (HZDR) (Albach et al., 2019). The Yb^{3+}:CaF_2 laser system at HZDR, named PENELOPE (Petawatt Energy Efficient Laser for Optical Plasma

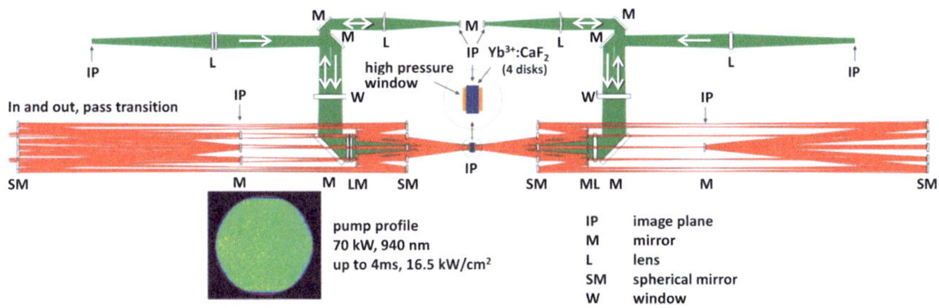

Fig. 7.32 Schematic layout of HEPA I, the first stage of the high-energy power amplifies. The Yb^{3+}:CaF_2 crystal slabs are pumped by high-power LD beams (shown in green) from both faces. A broadband, nanosecond-duration laser beam (shown in red) extracts the stored energy in 12 passes through the crystal. Reprinted from Albach et al. (2019). © 2019 Cambridge University Press (Note that the characters in the original figure are enlarged for visibility.)

Experiment), is planned to deliver 1 PW (150 J, 150 fs) at 1 Hz on target. The PENELOPE consists of a front end, three-stage high-gain broadband amplifiers, and two-stage high-energy power amplifiers (named HEPA I and HEPA II) with amplified energies of 14 J and 200 J with a nanosecond pulse for HEPA I and HEPA II, respectively.

A schematic configuration of HEPA I is shown in Fig. 7.32, where cryogenic gas face-cooled four Yb:CaF_2 disks of 55 mm diameter and 5 mm thickness are pumped from both sides, each with 70 kW LD beams of 940 nm at a pump intensity of 16.5 kW/cm^2. The laser pulse passes 12 times through the amplifier head, or 24 times in a double-pass configuration, before exiting HEPA I. In a benchmark amplification experiment with the double-12 pass configuration at room temperature, an output energy of 12.6 J with 30 mm beam diameter has been demonstrated for a 6 ns-long input pulse of 16 nm spectral width (FWHM) centered at ~ 1030 nm. Since the average fluence of 1.8 J/cm^2 for this case is much lower than the saturation fluence of Yb:CaF_2, there is more room for extracting the stored energy.

Another promising candidate of the LD-pumped short-pulse high-power laser is Nd:CaF_2, which has spectroscopic properties close to those of the Nd-doped phosphate glasses, low nonlinear refractive index coefficient, high thermal conductivity comparable to YAG, and negative thermo-optic coefficient suitable for thermal expansion compensation. Payne et al. (1991) reported fluorescence enhancement in Nd^{3+}:SrF_2 and Nd^{3+}:CaF_2 crystals by co-doping with rare earth buffer ions (RE^{3+}) such as La^{3+} and Y^{3+} to reduce Nd^{3+}–Nd^{3+} clustering. Lasing of LD-pumped Nd^{3+},RE^{3+}:CaF_2 co-doped crystals has been reported in continuous wave (Doualan et al., 2013; Zhu et al., 2015) and mode-locked (Wang et al., 2019) operations. Although these Nd^{3+}, RE^{3+}:CaF_2 crystals are promising, it is difficult to grow high optical quality CaF_2 crystals in large sizes with constant co-doping fraction and uniform doping concentration over the whole volume.

To overcome the difficulties associated with growing large Nd^{3+},RE^{3+}:CaF_2 crystals, Fujioka et al. have fabricated transparent Nd^{3+},La^{3+}:CaF_2 ceramics by the liquid

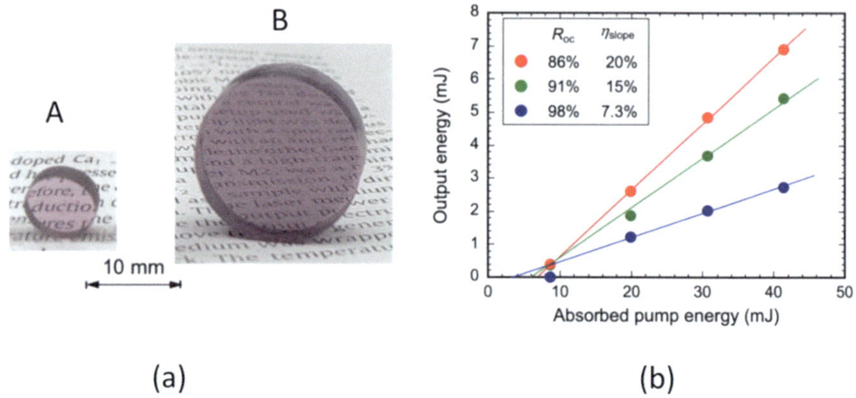

Fig. 7.33 **a** Photographs of Nd,La:CaF_2 transparent ceramics of (A) 3.6 mm-thick and 7.6-mm diameter sample doped with 1.7 mol% Nd and 12.9 mol% La, and (B) 6.3 mm-thick and 17.6-mm diameter sample doped with 1.1 mol% Nd and 9.3 mol% La, respectively. **b** Lasing performance of Nd,La:CaF_2 ceramics of 1.7 mol% Nd doping by LD pumping of 795 nm wavelength and 1 ms duration, showing dependence of the slope efficiency (η_{slope}) on the output coupler reflectivity (R_{oc}). Reprinted from Fujioka et al. (2025). © 2025 Optica Publishing Group

synthesis method (see Sect. 4.3.2.3) with the size of up to 20 mm in diameter. Figure 7.33 shows (a) photographs of two Nd,La:CaF_2 ceramics samples and (b) its lasing performance by LD pumping with diffraction-liming output beam, showing up to 20% slope efficiency (Fujioka et al., 2025). From the spectroscopic measurements and analyses, a Nd,La:CaF_2 ceramic with Nd 1.1 mol% and La 9.3 mol% concentrations has the following properties: fluorescence yield 79%, spectral width 28 nm (22 nm), fluorescence lifetime 494 μs (315 μs), and stimulated emission cross section 2.6×10^{-20} cm^2 (4.2×10^{-20} cm^2), where the numbers in the parentheses correspond to the properties the phosphate laser glass LHG-8. From these properties, we can expect that it is possible to use configurations such as gas-cooled multi-slab amplifiers and active mirrors described in Sect. 5.3.2 to efficiently extract the energies stored in the Nd,La:CaF_2 ceramics with broad spectral widths sufficient to generate high power ultrashort pulses. Since the fabrication method has been established, it will be possible to fabricate the Nd,La:CaF_2 transparent ceramics of large sizes by upgrading the fabrication facilities.

7.5.2 LD-Pumped Tm:YLF Laser

The trivalent thulium ion Tm^{+3} has two un-paired 4f electrons forming the energy levels in a crystal field as shown in Fig. 4.14. In Tm:YLF, Tm^{+3} substitutes for Y^{+3} of $LiYF_4$ (YLF). Excited states in order of increasing energy from the ground state $^3H_6$ are $^3F_4$ (~ 5800 cm^{-1}), $^3H_5$ (~ 8200 cm^{-1}), $^3H_4$ (~ 12,600 cm^{-1}), and higher levels (Walsh et al., 1998). When $^3H_4$ is pumped from the ground state by one photon at 793 nm, two Tm^{3+} ions in the excited state $^3F_4$ are formed simultaneously due to cross relaxation with the ground state $^3H_4 +^3 H_6^3 \rightarrow F_4 +^3 F_4$ with small quantum defect. Due to large Stark splitting of the ground state combined with vibronic coupling, the absorption band of $^3H_4$ is very broad (~ 770–800 nm, peaked near 790 nm), as well as the emission band from $^3F_4$ to the ground state spanning over ~ 350 nm (1650–2000 nm) which is peaked at 1.88 μm. Sub-100 fs pulses have been generated at 2017 nm with a diode-pumped Tm:CLNGG (Li^+-doped $Ca_3Nb_{1.5}Ga_{3.5}O_{12}$; a disordered crystal with smooth spectral profile), by passive mode-locking with single-walled carbon nanotubes as a saturable absorber (Wang et al., 2018).

The $^3F_4$ excited state has a very long radiative lifetime of ~ 15 ms and small emission cross section of ~ 2–3 × 10^{-21} cm^2. The saturation fluence of the quasi-three level Tm laser is 21 J/cm^2, estimated by $E_s = h\nu / \gamma\sigma$ where $\gamma = 1 + g_2 / g_1 = 2$ (g_1 and g_2 are degeneracies of the lower and upper laser levels, respectively). The saturation intensity is 1.4 kW/cm^2, since the upper-level lifetime is 15 ms.

Tm:YLF is suitable for efficient pumping with high power $Al_xGa_{1-x}As$ LDs due to long radiative lifetime of $^3F_4$ level and good spectral match for pumping. However, it is difficult to efficiently extract the stored energy with a seed pulse at the fluence close to the saturation which exceeds the damage threshold (typically less than 10 J/cm^2) and at the intensity exceeding the B-integral-limit (~ 10 GW/cm^2). Sistrunk et al. (2019) have proposed "multi-pulse extraction" (MPE) to alleviate these difficulties. In MPE, the stored energy is extracted by multiple pulses within a radiative lifetime of the excited state, instead of energy extraction in a single pulse, thus reducing the laser fluence of each pulse. In this mode of operation with the seed pulse lower than the saturation fluence, it is expected that the extraction efficiency increases to 73% for 150 pulses per gain lifetime (10 kHz for Tm:YLF), whereas it is only 11% for 10 pulses per gain lifetime. This concept, named Big Aperture Thulium (BAT) laser, is envisioned for driving laser-based electron accelerators at 300 kW average power.

A proof-of-principle experiment was implemented with a configuration shown in Fig. 7.34 (Tamer et al., 2022). The seed pulse from a Tm:YLF oscillator is amplified in 4-passes with an amplifier composed of a 100 mm diameter, 35 mm thick, 6%-at doped Tm:YLF crystal, which is end-pumped on both sides, each with a 793 nm, 20 kW laser diode array (LDA) with its beam profile shaped into a hexagonal profile of 22 mm diameter. With this configuration, a 4 mJ, 20 ns pulse was amplified up to 21.7 J (1.1 GW peak power), i.e. 5,400 times of un-saturated gain. Furthermore, saturated amplification of a ~ 6.8 kHz burst pulse train has been demonstrated, where the Tm:YLF amplifier was pumped with LD up to 18.2 kW and a pulse duration of 25 ms, resulting in the amplified pulse power of 6.4 kW with an optical-to-optical efficiency of 19%.

Although the laser power required to drive the laser-driven electron accelerator is not yet determined, scaling up of this Tm:YLF laser by 10–100 times could be used as the driver of the electron accelerator. The long-wavelength diode-pumped solid-state lasers, such as Er-, Tm- and Ho-doped lasers, are also considered as candidates for the Inertial Fusion Energy (IFE) driver due to diode cost reduction owing to the long radiative lifetime of these active ions (Spinka et al., 2022).

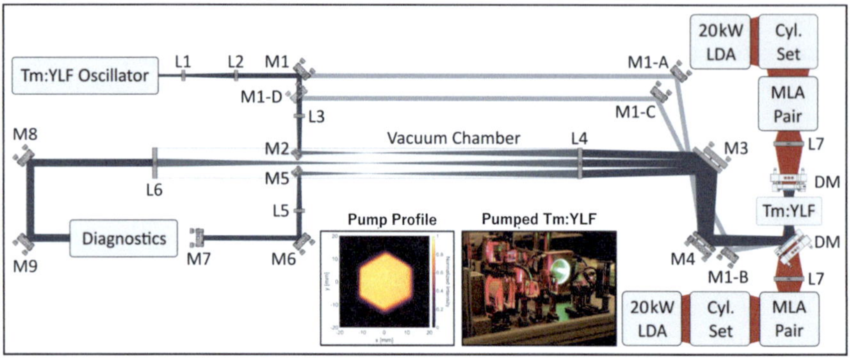

Fig. 7.34 Layout of the Tm:YLF amplifier pumped by 20 kW laser diode arrays (LDA) from both sides. The pump beam profiles, shown in the inset, are homogenized with a pair of microlens arrays (MLA). The seed pulse from the Tm:YLF oscillator is relay-imaged and guided to the amplifier for 4-pass amplification. Reprinted from Tamer et al. (2022). © 2022 Optica Publishing Group (Note that the characters in the original figure are enlarged for visibility.)

References

Albach, D., Loeser, M., Siebold, M., et al. (2019). Performance demonstration of the PENELOPE main amplifier HEPA I using broadband nanosecond pulses. *High Power Laser Science and Engineering, 7*, Article e1.

Aoyama, M., Yamakawa, K., Okahane, Y., et al. (2003). 0.85-PW, 33-fs Ti:sapphire laser. *Optics Letters, 28*, 1594.

Armstrong, J. J., & Kessler, T. J. (1993). Large-aperture, holographic gratings for high-power laser systems. *SPIE Proceedings, 1870.*

Ashe, B., Marshall, K. L., Mastrosimone, D., et al. (2008). Minimizing contamination to multilayer dielectric diffraction gratings within a large vacuum system. *Proceedings of SPIE, 7069*, Article 706902.

Bagnoud, V., Aurand, B., Blazevic, A., et al. (2010). Commissioning and early experiments of the PHELIX facility. *Applied Physics B, 100*, 137.

Bagnoud, V., Begishev, I. A., Guardalben, M. J., et al. (2005). 5 Hz, > 250 mJ optical parametric chirped-pulse amplifier at 1053 nm. *Optics Letters, 30*, 1843.

Bagnoud, V., & Salin, F. (1999). Influence of optical quality on chirped-pulse amplification: Characterization of a 150-nm-bandwidth stretcher. *Journal of the Optical Society of America B: Optical Physics, 16*, 188.

Bakule, P., Antipenkov, R., Green, J. T., et al. (2017). Development of high energy, sub-15 fs OPCPA system operating at 1 kHz repetition rate for ELI-Beamlines facility. *Proceedings of SPIE, 10241*, 102410U.

Balakin, A. A., Kim, A. V., Litvak, A. G., et al. (2016). Extreme self-compression of laser pulses in the self-focusing mode resistant to transverse instability. *Physical Review A, 94*, Article 043812.

Balakin, A. A., Litvak, A. G., Mironov, V. A., et al. (2017). Self-compression of soliton-like laser pulses in the process of self-focusing. *Journal of Optics, 19*, Article 095503. https://doi.org/10.1088/2040-8986/aa792e

Baltuska, A., Fuji, A., & Kobayashi, T. (2002). Controlling the carrier-envelope phase of ultrashort light pulses with optical parametric amplifiers. *Physical Review Letters, 88*, 133901.

Barty, C. P. J. (2016). Nexawatt: A strategy for exawatt peak power lasers based on NIF and NIF-like beam lines. *Journal of Physics: Conference Series, 717*, 012086.

Barty, C. P. J., Guo, T., Le Blanc, C., et al. (1996). "Generation of 18-fs, multiterawatt pulses by regenerative pulse shaping and chirped-pulse amplification. *Optics Letters, 21*, 668.

Barty, C. P. J., Key, M., Britten, J., et al. (2004). An overview of LLNL high-energy short-pulse technology for advanced radiography of laser fusion experiments. *Nuclear Fusion, 44*, S266.

Barty, C. P. J., King, D. A., Yin, G. Y., et al. (1988). 12.8-eV laser in neutral cesium. *Physics Review Letters, 61*, 2201.

Batysta, F., Antipenkov, R., Green, J. T., et al. (2014). Pulse synchronization system for picosecond pulse-pumped OPCPA with femtosecond-level relative timing jitter. *Optics Express, 22*, 30281.

Blanchot, N., Behar, G., Chapuis, J. C., et al. (2017). 115 PW-850 J compressed beam demonstration using the PETAL facility. *Optics Express, 25*, 16957.

Bleotu, P.-G., Wheeler, J., Papadopoulos, D., et al. (2022). Spectral broadening for multi-Joule pulse compression in the APOLLON long focal area facility. *High Power Laser Science and Engineering, 10*, Article e9.

Brignon, A. (Ed.). (2013). *Coherent laser beam combining*. Wiley.

Bromage, J., Bahk, S.-W., Begishev, I. A., et al. (2019). Technology development for ultraintense all-OPCPA systems. *High Power Laser Science and Engineering, 7*, Article e4.

Bromage, J., Bahk, S.-W., Bedzyk, M., et al. (2021). MTW-OPAL: A technology development platform for ultra-intense optical parametric chirped-pulse amplification systems. *High Power Laser Science and Engineering, 9*, Article e63.

Bromage, J., Dorrer, C., & Jungquist, R. K. (2012). Temporal contrast degradation at the focus of ultrafast pulses from high-frequency spectral phase modulation. *Journal of the Optical Society of America B: Optical Physics, 29*, 1125. https://doi.org/10.1364/JOSAB.29.001125

Bucksbaum, P., et al. (2018). *Opportunities in intense ultrafast lasers: Reaching for the brightest light*. The National Academies Press.

Bulanov, S. S., Mur, V. D., Narozhny, N. B., et al. (2010). Multiple colliding electromagnetic pulses: A way to lower the threshold of e^+e^- pair production from vacuum. *Physical Review Letters, 104*, Article 220404.

Bulanov, S. V., Esirkepov, TZh., Kando, M., et al. (2015). On the problems of relativistic laboratory astrophysics and fundamental physics with super powerful lasers. *Plasma Physics Reports, 41*, 1.

Canova, L., Courtev, S., Minkovski, N., et al. (2008). Efficient generation of cross-polarized femtosecond pulses in cubic crystals with holographic cut orientation. *Applied Physics Letters, 92*, Article 231102.

Chambaret, J. P., Le Blanc, C., Cheriaux, G., et al. (1996). Generation of 25-TW, 32-fs pulses at 10 Hz. *Optics Letters, 21*, 1921.

Chériaux, G., Gaul, E., Anipenkov, R., et al. (2019). KJ-10 PW class laser system at 1 shot a minute. *SPIE Proceedings, 10898*, 1089806.

Chesnut, K. D., & Barty, C. P. J. (2023). Ideal spatio-temporal pulse distribution for exawatt-scale lasers based on simultaneous chirped beam and chirped pulse amplification. *Optics Express, 31*, 5687.

Chu, H.-H., Huang, S.-Y., Yang, L.-S., et al. (2004). A versatile 10-TW laser system with robust passive controls to achieve high stability and spatiotemporal quality. *Applied Physics B, 79*, 193.

Cotel, A., Castaing, M., Pichon, P., et al. (2007). Phased-array grating compression for high-energy chirped pulse amplification lasers. *Optics Express, 15*, 2742.

Dabu, R. (2010). Very broad gain bandwidth parametric amplification in nonlinear crystals at critical wavelength degeneracy. *Optics Express, 18*, 11689.

Danson, C. N., Barzanti, L., Chang, Z., et al. (1993). Chirped pulse amplification experiments on the VULCAN Nd:glass laser facility. *Proceedings of SPIE, 1860*, 10.

Danson, C. N., Brummitt, P. A., Clarke, R. J., et al. (2004). Vulcan Petawatt—An ultra-high-intensity interaction facility. *Nuclear Fusion, 44*, S239.

Danson, C. N., Haefner, C., Bromage, J., et al. (2019). Petawatt and exawatt class lasers worldwide. *High Power Laser Science and Engineering, 7*, Article e54.

Deen, C. P., Gully-Santiago, M., Wang, W., et al. (2017). A grism design review and the as-built performance of the silicon grisms for JWST-NIRCam. *Publications of the Astronomical Society of Pacific, 129*, Article 065004.

DeSalvo, R., Sheik-Bahae, M., Said, A. A., et al. (1993). Z-scan measurements of the anisotropy of nonlinear refraction and absorption in crystals. *Optics Letters, 18*, 194.

Didenko, N. V., Konyashchenko, A. V., Lutsenko, A. P., et al. (2008). Contrast degradation in a chirped-pulse amplifier due to generation of prepluses by postpulses. *Optics Express, 16*, 3178. https://doi.org/10.1364/OE.16.003178

Dinh, T.-H., Mori, M., Nagashima, K., et al. (2025). Highly stable and efficient spatiotemporal filtering in cross-polarized wave generation via nonlinearity tailoring in multi-plate crystals. *Optics Express, 33*, 21318.

Dorrer, C., & Bromage, J. (2008). Impact of high-frequency spectral phase modulation on the temporal profile of short optical pulses. *Optics Express, 16*, 3058.

Doualan, J. L., Su, L. B., Brasse, G., et al. (2013). Improvement of infrared laser properties of Nd:CaF_2 crystals via codoping with Y^{3+} and Lu^{3+} buffer ions. *Journal of the Optical Society of America b: Optical Physics, 30*, 3018.

Doumy, G., Quéré, F., Gobert, O., et al. (2004). Complete characterization of a plasma mirror for the production of high-contrast ultraintense laser pulses. *Physical Review E, 69*, Article 026402.

Du, D., Squier, J., Kane, S., et al. (1995). Terawatt Ti:Sapphire laser with a spherical-optic pulse expander. *Optics Letters, 20*, 2114.

Du, S., Shen, X., Liang, W., et al. (2023). A 100-PW compressor based on single-pass single-grating pair. *High Power Laser Science and Engineering, 11*, Article e4.

ELI Whitebook. (2011). Science and technology with ultra-intense lasers. In G. Korn (Ed.). http://www.extreme-light-infrastructure.eu/.

Favrat, O., Sozet, M., Tovena-Pecault, I., et al. (2014). Influence of organic contamination on laser induced damage of multilayer dielectric mirrors by subpicosecond laser pulses. *SPIE, 9237*, Article 923707.

Fisher, R. A., Kelley, P. L., & Gustafson, T. K. (1969). Subpicosecond pulse generation using the optical Kerr effect. *Applied Physics Letters, 14*, 140.

Freidman, G., Andreev, N., Bespalov, V., et al. (2002). Use of KD*P crystals for non-degenerated broadband optical parametric chirped pulse amplification in petawatt lasers. OSA/CLEO, paper CPDA9

Friedel, F., Druon, F., Boudeile, J., et al. (2009). Diode-pumped 99 fs Yb:CaF_2 oscillator. *Optics Letters, 34*, 1474.

Fsaifes, I., Daniault, L., Bellanger, S., et al. (2020). Coherent beam combining of 61 femtosecond fiber amplifiers. *Optics Express, 28*, 20152. https://doi.org/10.1364/OE.394031

Fujioka, K., Fujimoto, Y., Tsubakimoto, K., et al. (2015). Partially deuterated potassium dihydrogen phosphate optimized for ultra-broadband optical parametric amplification. *Journal of Applied Physics, 117*, Article 093103.

Fujioka, K., Matsumoto, Y., Tamaru, Y., et al. (2025). Fabrication, spectroscopic characteristics, and lasing performance of Nd,La:Ca_F2 transparent ceramics. *Optical Materials Express, 15*, 890.

Galletti, M., Oliveira, P., Galimberti, M., et al. (2020a). Ultra-broadband all-OPCPA petawatt facility fully based on LBO. *High Power Laser Science and Engineering, 8*, Article e31.

Galletti, M., Pires, H., Hariton, V., et al. (2020b). Ultra-broadband near-infrared NOPAs based on the nonlinear crystals BiBO and YCOB. *High Power Laser Science and Engineering, 8*, Article e29.

Gan, Z., Yu, L., Li, S., et al. (2017). 200 J high efficiency Ti:sapphire chirped pulse amplifier pumped by temporal dual pulse. *Optics Express, 25*, 5169.

Gaul, E., Martinez, M., Blakeney, J., et al. (2010). Demonstration of a 1.1 petawatt laser based on a hybrid optical parametric chirped pulse amplification/mixed Nd:glass amplifier. *Applied Optics, 49*, 1676.

Giambruno, F., Radier, C., Rey, G., et al. (2011). Design of a 10 PW (150 J/15 fs) peak power laser system with Ti:sapphire medium through spectral control. *Applied Optics, 17*, 2617.

Gibson, E. A., Gaudiosi, D. M., Kapteyn, H. C., et al. (2006). Efficient reflection grisms for pulse compression and dispersion compensation of femtosecond pulses. *Optics Letters, 31*, 3363.

Ginzburg, V. N., Kochetkov, A. A., Potemkin, A. K., et al. (2018). Suppression of small-scale self-focusing of high-power laser beams due to their self-filtration during propagation in free space. *Quantum Electronics, 48*, 325.

Ginzburg, V. N., Yakovlev, I. V., Zuev, A. S., et al. (2019). Compression after compressor: Threefold shortening of 200-TW laser pulses. *Quantum Electronics, 49*, 299.

Ginzburg, V. N., Yakovlev, I. V., Zuev, A. S., et al. (2020). Two-stage nonlinear compression of high-power femtosecond laser pulses. *Quantum Electronics, 50*, 331.

Ginzburg, V., Yakovlev, I., Kochetkov, A., et al. (2021). 11 fs, 1.5 PW laser with nonlinear pulse compression. *Optics Express, 29*, 28297.

Giordmaine, J., Duguay, M., & Hansen, J. (1968). Compression of optical pulses. *IEEE Journal of Quantum Electronics, 4*, 252.

Gonoskov, A., Bashinov, A., Gonoskov, I., et al. (2014). Anomalous radiative trapping in laser fields of extreme intensity. *Physical Review Letters, 113*, Article 014801.

Gonoskov, A., Gonoskov, I., Harvey, C., et al. (2013). Probing nonperturbative QED with optimally focused laser pulses. *Physical Review Letters, 111*, Article 060404.

Gonoskov, I., Aiello, A., Heugel, S., et al. (2012). Dipole pulse theory: Maximizing the field amplitude from 4π focused laser pulses. *Physical Review A, 86*, 053836.

Goodno, G. D., Komine, H., McNaught, S. J., et al. (2006). Coherent combination of high-power, zigzag slab lasers. *Optics Letters, 31*, 1247.

Green, J. T., Antipenkov, R., Bakule, P., et al. (2021). L2-DUHA 100 TW high repetition rate laser system at ELI-beamlines: Key design considerations. *The Review of Laser Engineering, 49*, 106.

Grischkowsky, D., & Balant, A. C. (1982). Optical pulse compression based on enhanced frequency chirping. *Applied Physics Letters, 41*, 1.

Habara, H., Xu, G., Jitsuno, T., et al. (2010). "Pulse compression and beam focusing with segmented diffraction gratings in a high-power chirped-pulse amplification glass laser system. *Optics Letters, 35*, 1783.

Häfner, C., Bayramian, A., Betts, S., et al. (2017). High average power, diode pumped petawatt laser systems: a new generation of lasers enabling precision science and commercial applications. *Proceedings on SPIE, 10241*, 1024102.

Hays, G. R., Gaul, E. W., Martinez, M. D., et al. (2007). Broad-spectrum neodymium-doped laser glasses for high-energy chirped-pulse amplification. *Applied Optics, 46*, 4813.

Hein, J., Kaluza, M., Bödefeld, R., et al. (2006). POLARIS: An all diode-pumped ultrahigh peak power laser for high repetition rates. In H. Schwoerer, B. Beleites, & J. Magill (Eds.), *Lasers and nuclei. Lecture notes in physics 694* (p. 47). Springer.

Hooker, C. J., Collier, J. L., Chekhlov, O., et al. (2006). The Astra Gemini project—A dual-beam petawatt Ti:sapphire laser system. *Journal de Physique IV France, 133*, 673.

Hooker, C., Tang, Y., Chekhlov, O., et al. (2011). Improving coherent contrast of petawatt laser pulses. *Optics Express, 19*, 2193.

Inoue, S., Maeda, K., Tokita, S., et al. (2016). Single plasma mirror providing 10^4 contrast enhancement and 70% reflectivity for intense femtosecond lasers. *Applied Optics, 55*, 5647.

Ivanov, V. V., Maksimchuk, A., & Mourou, G. (2003). Amplified spontaneous emission in a Ti:sapphire regenerative amplifier. *Applied Optics, 42*, 7231.

Jitsuno, T., Motokoshi, S., Okamoto, T., et al. (2008). Development of 91 cm size gratings and mirrors for LFEX laser system. *Journal of Physics: Conference Series, 112*, 032002.

Jitsuno, T., Murakami, H., Kato, K., et al. (2013). Recent progresses on insights of laser damage mechanisms and influence of contamination in optics. *Proceedings of SPIE, 8786*, 87860B.

Jitsuno, T., Murakami, H., Motokoshi, S., et al. (2010). Oil-contamination problem in large-scale pulse-compressor. *Proceedings of SPIE, 7842*, Article 784221.

Jitsuno, T., Murakami, H., Motokoshi, S., et al. (2016). Source of contamination in damage-test sample and vacuum. *Proceedings of SPIE, 9983*, Article 998316.

Jullien, A., Albert, O., Burgy, F., et al. (2005). 10^{-10} temporal contrast for femtosecond ultraintense lasers by cross-polarized wave generation. *Optics Letters, 30*, 920.

Jullien, A., Albert, O., Cheriaux, G., et al. (2006). A two crystal arrangement to fight efficiency saturation in cross-polarized wave generation. *Optics Express, 14*, 2760–2769.

Jullien, A., Rousseau, J.-P., Mercier, B., et al. (2008). Highly efficient nonlinear filter for femtosecond pulse contrast enhancement and pulse shortening. *Optics Letters, 33*, 2353.

Kane, S., & Squire, J. (1997a). Grism-pair stretcher-compressor system for simultaneous second- and third-order dispersion compensation in chirped-pulse amplification. *Journal of the Optical Society of America B: Optical Physics, 14*, 661.

Kane, S., & Squire, J. (1997b). Fourth-order-dispersion limitations of aberration-free chirped-pulse amplification systems. *Journal of the Optical Society of America B: Optical Physics, 14*, 1237.

Kapteyn, H. C., Murnane, M. M., Szoke, A., et al. (1991). Prepulse energy suppression for high-energy ultrashort pulses using self-induced plasma shuttering. *Optics Letters, 16*, 490.

Kawanaka, J., Tsubakimoto, K., Yoshida, H., et al. (2016). Conceptual design of sub-exa-watt system by using optical parametric chirped pulse amplification. *Journal of Physics: Conference Series, 688*, 012044.

Ke, L. T., Feng, K., Wang, W. T., Qin, Z. Y., et al. (2021). Near-GeV electron beam at a few per-mille level from a laser wakefield accelerator via density-tailored plasma. *Physical Review Letters, 126*, Article 214801.

Kelly, J. H., Waxer, L. J., Bagnoud, V., et al. (2006). OMEGA EP: High-energy petawatt capability for the OMEGA laser facility. *Journal de Physique IV France, 133*, 75.

Kessler, T. J., Bunkenburg, J., Hung, H., et al. (2004). Demonstration of coherent addition of multiple gratings for high-energy chirped-pulse-amplified lasers. *Optics Letters, 29*, 635.

Khazanov, E. A., Yu Mironov, S., & Mourou, G. (2019). Nonlinear compression of high-power laser pulses: Compression after compressor approach. *Physics-Uspekhi, 62*, 1096.

Khazanov, E., Shaykin, A., Kostyukov, I., et al. (2023). EXawatt center for extreme light studies. *High Power Laser Science and Engineering, 11*, Article e78.

Kiriyama, H., Miyasaka, Y., Kon, A., et al. (2021). Enhancement of pre-pulse and picosecond pedestal contrast of the petawatt J-KAREN-P laser. *High Power Laser Science and Engineering, 9*, Article e62.

Kiriyama, H., Miyasaka, Y., Sagisaka, A., et al. (2020). Experimental investigation on the temporal contrast of pre-pulses by post-pulses in a petawatt laser facility. *Optics Letters, 45*, 1100.

Kiriyama, H., Mori, M., Nakai, Y., et al. (2010). High-spatiotemporal-quality petawatt-class laser system. *Applied Optics, 49*, 2105.

Kon, A., Nishiuchi, M., Fukuda, Y., et al. (2022). Characterization of the plasma mirror system at the J-KAREN-P facility. *High Power Laser Science and Engineering, 10*, Article e25.

Le Dortz, J., Heilmann, A., Antier, M., et al. (2017). Highly scalable femtosecond coherent beam combining demonstrated with 19 fibers. *Optics Letters, 42*, 1887.

Leemans, W. (2011). White paper of the ICFA-ICUIL joint task force: High power laser technology for accelerators. *ICFA Beam Dynamics Newsletter, 56*, 10–88.

Li, A. X., Qin, C. Y., Zhang, H., et al. (2022). Acceleration of 60 MeV proton beams in the commissioning experiment of the SULF-10 PW laser. *High Power Laser Science and Engineering, 10*, Article e26.

Li, L. (1993). Multilayer modal method for diffraction gratings of arbitrary profile, depth, and permittivity. *Journal of the Optical Society of America a. Optics and Image Science, 10*, 2581.

Li, S., Wang, C., Liu, Y., et al. (2017). High-order dispersion control of 10-petawatt Ti:sapphire laser facility. *Optics Express, 25*, 17488.

Li, W., Gan, Z., Yu, L., et al. (2018). 339 J high-energy Ti:sapphire chirped-pulse amplifier for 10 PW laser facility. *Optics Letters, 43*, 5681.

Li, Z., Kato, Y., & Kawanaka, J. (2021). Simulating an ultra-broadband concept for Exawatt-class lasers. *Scientific Reports, 11*, 151.

Li, Z., & Kawanaka, J. (2019). Possible method for a single-cycle 100 petawatt laser with wide-angle non-collinear optical parametric chirped pulse amplification. *OSA Continuum, 2*, 1125.
Li, Z., Wang, C., Li, S., et al. (2015). Fourth-order dispersion compensation for ultra-high power femtosecond lasers. *Optics Communications, 357*, 71.
Li, Z., Xu, G., Wang, T., et al. (2010). Object–image-grating self-tiling to achieve and maintain stable, near-ideal tiled grating conditions. *Optics Letters, 35*, 2206.
Liu, J., Shen, X., Du, S., et al. (2021). Multistep pulse compressor for 10s to 100s PW lasers. *Optics Express, 29*, 17140.
Lozhkarev, V. V., Freidman, G. I., Ginzburg, V. N., et al. (2005). Study of broadband optical parametric chirped pulse amplification in a DKDP crystal pumped by the second harmonic of a Nd:YLF laser. *Laser Physics, 15*, 1319.
Lozhkarev, V. V., Freidman, G. I., Ginzburg, V. N., et al. (2006). 200 TW 45 fs laser based on optical parametric chirped pulse amplification. *Optics Express, 14*, 446.
Lureau, F., Matras, G., Chalus, O., et al. (2020). High-energy hybrid femtosecond laser system demonstrating 2 × 10 PW capability. *High Power Laser Science and Engineering, 8*, Article e43.
Ma, J., Yuan, P., Wang, J., et al. (2015). Spatiotemporal noise characterization for chirped-pulse amplification systems. *Nature Communications, 6*, 6192.
Marcinkevic˘ius, A., Piskarskas, A., Smilgevic˘ius, V., et al. (1998). Parametric superfluorescence excited in a nonlinear crystal by two uncorrelated pump beams. *Optics Communications, 158*, 101.
Melnikus, S., Tolenis, T., Smalakys, L., et al. (2017). Enhancement of laser-induced damage threshold in chirped mirrors by electric field reallocation. *Optics Express, 25*, 26537.
Mével, E., Tcherbakoff, O., Salin, F., et al. (2003). Extracavity compression technique for high-energy femtosecond pulses. *Journal of the Optical Society of America B, 20*, 105.
Minkovski, N., Saltiel, S. M., Petrov, G. I., et al. (2002). Polarization rotation induced by cascaded third-order processes. *Optics Letters, 27*, 2025.
Minkovski, N., Petrov, G. I., Saltiel, S. M., et al. (2004). Nonlinear polarization rotation and orthogonal polarization generation experienced in a single-beam configuration. *Journal of the Optical Society of America B, 21*, 1659.
Mironov, S. Yu., Ginzburg, V. N., Gacheva, E. I., et al. (2015). Use of polyethylene terephthalate for temporal recompression of intense femtosecond laser pulses. *Laser Physics Letters, 12*, 025301.
Mironov, S. Yu., Lassonde, P., Kieffer, J.-C., et al. (2014). Spatially-uniform temporal recompression of intense femtosecond optical pulses. *The European Physical Journal Special Topics, 223*, 1175.
Mironov, S. Yu., Lozhkarev, V. V., Ginzburg, V. N., et al. (2012). Second-harmonic generation of super powerful femtosecond pulses under strong influence of cubic nonlinearity. *IEEE Journal of Quantum Electronics, 18*, 7.
Miyanaga, N., Azechi, H., Tanaka, K. A., et al. (2006). 10-kJ PW laser for the FIREX-I program. *Journal of Physics IV France, 133*, 81. (Inertial Fusion Sci. Appl. 2005, J.-C. Gauthier, et al.)
Morice, O. (2003). Miró: Complete modeling and software for pulse amplification and propagation in high-power laser systems. *Optical Engineering, 42*, 1530.
Moulton, P. F. (1986). Spectroscopic and laser characteristics of Ti:Al_2O_3. *Journal of the Optical Society of America B: Optical Physics, 3*, 125.
Mourou, G. (1997). The ultrahigh-peak-power laser: Present and future. *Applied Physics B, 65*, 205.
Mourou, G., Brocklesby, B., Tajima, T., & Limpert, J. (2013). The future is fibre accelerators. *Nature Photonics, 7*, 258.
Mourou, G., Korn, G., Sandner, W, & Collier, J. L. (Ed.). (2011). ELI whitebook. In *Science and technology with ultra-intense lasers*. ELI PP Publications.
Mourou, G., Mironov, S., Khazanov, E., et al. (2014). Single cycle thin film compressor opening the door to Zeptosecond-Exawatt physics. *The European Physical Journal Special Topics, 223*, 1181.
Mourou, G., Tajima, T., & Bulanov, S. V. (2006). Optics in the relativistic regime. *Reviews of Modern Physics, 78*, 309.

Müller, M., Aleshire, C., Buldt, J., et al. (2021). Scaling potential of beam-splitter-based coherent beam combination. *Optics Express, 29*, 27900.

Müller, M., Aleshire, C., Klenke, A., et al. (2020). 10.4 kW coherently combined ultrafast fiber laser. *Optics Letters, 45*, 3083.

Müller, M., Kienel, M., Klenke, A., et al. (2016). 1 kW 1 mJ eight-channel ultrafast fiber laser. *Optics Letters, 41*, 3439.

Musgrave, I., Galimberti, M., Galleti, M., et al. (2021). Overview of the OPCPA beamline development for the Vulcan Petta-watt laser facility. *Review of Laser Engineering, 49*, 110.

Nagy, T., Kretschmar, M., Vrakking, M., et al. (2020). Generation of above-terawatt 1.5-cycle visible pulses at 1 kHz by post-compression in a hollow fiber. *Optics Letters, 45*, 3313.

Nakamura, K., Mao, H.-S., Gonsalves, A. J., et al. (2017). Diagnostics, control and performance parameters for the BELLA high repetition rate petawatt class laser. *IEEE Journal of Quantum Electronics, 53*, 4.

Nisoli, M., DeSilvestri, S., & Svelto, O. (1996). Generation of high energy 10 fs pulses by a new pulse compression technique. *Applied Physics Letters, 68*, 2793.

OECD Global Science Forum. (2001). *OECD global science forum workshop on compact high-intensity short-pulse lasers: Future directions and applications*. https://www.oecd.org/sti/inno/2451461.pdf.

Oksenhendler, T., & Forget, N. (2010). Pulse-shaping techniques theory and experimental implementations for femtosecond pulses. In M. Grishin (Ed.), *Advances in solid-state lasers: Development and applications* (Vol. 16, pp. 347–386).

Ouillé, M., Vernier, A., Böhle, F., et al. (2020). Relativistic-intensity near-single-cycle light waveforms at kHz repetition rate. *Light: Science and Applications, 9*, 47.

Papadopoulos, D. N., Ramirez, P., Genevrier, K., et al. (2017). High-contrast 10 fs OPCPA-based front end for multi-PW laser chains. *Optics Letters, 42*, 3530.

Papadopoulos, D. N., Zou, J. P., Le Blanc, C., et al. (2016). The Apollon 10 PW laser: Experimental and theoretical investigation of the temporal characteristics. *High Power Laser Science and Engineering, 4*, e34.

Parker, S., Danson, C., Egan, D., et al. (2018). 400 TW operation of Orion at ultra-high contrast. *High Power Laser Science and Engineering, 6*, Article e47.

Pathak, V. B., Lee, S. K., Pae, K. H., et al. (2021). Strong field physics pursued with petawatt lasers. *AAPPS Bulletin, 31*, 4.

Payne, S. A., Caird, J. A., Chase, L. L., et al. (1991). Spectroscopy and gain measurements of Nd^{3+} in SrF_2 and other fluorite-structure hosts. *Journal of the Optical Society of America B: Optical Physics, 8*, 726.

Peng, Y., Yu, L., Wang, X., et al. (2021). Overview and status of station of extreme light toward 100 PW. *The Review of Laser Engineering, 49*, 93.

Perry, M. D., Pennington, D., Stuart, B. C., et al. (1999). Petawatt laser pulses. *Optics Letters, 24*, 160.

Petit, V., Camy, P., Doualan, J.-L., et al. (2008). Spectroscopy of Yb^{3+}:CaF_2: From isolated centers to clusters. *Physical Review B, 78*, Article 085131.

Poole, P., Trendafilov, S., Shvets, G., et al. (2013). Femtosecond laser damage threshold of pulse compression gratings for petawatt scale laser systems. *Optics Express, 21*, 26341.

Qiao, J., Kalb, A., Guardalben, M. J., et al. (2007). Large-aperture grating tiling by interferometry for petawatt chirped-pulse–amplification systems. *Optics Express, 15*, 9562.

Qiao, J., Kalb, A., Nguyen, T., et al. (2008). Demonstration of large-aperture tiled-grating compressors for high-energy, petawatt-class, chirped-pulse amplification systems. *Optics Letters, 33*, 1684.

Radier, C., Chalus, O., Charbonneau, M., et al. (2022). 10 PW peak power femtosecond laser pulses at ELI-NP. *High Power Laser Science and Engineering, 10*, Article e21.

Ranc, L., Le Blanc, C., Lebas, N., et al. (2020). Improvement in the temporal contrast in the tens of ps range of the multi-PW Apollon laser front-end. *Optics Letters, 45*, 4599.

Rolland, C., & Corkum, P. B. (1988). Compression of high-power optical pulses. *Journal of the Optical Society of America B: Optical Physics, 5*, 641.

Ross, I. N., Matousek, P., Towrie, M., Langley, A. J., & Collier, J. L. (1997). The prospects for ultrashort pulse duration and ultrahigh intensity using optical parametric chirped pulse amplifiers. *Optics Communications, 144*, 125.

Rouyer, C., Mazataud, E., Allais, I., et al. (1993). Generation of 50-TW femtosecond pulses in a Ti:sapphire/Nd:glass chain. *Optics Letters, 18*, 214.

Rushford, M. C., Britten, J. A., Barty, C. P. J., et al. (2008). Split-aperture laser pulse compressor design tolerant to alignment and line-density differences. *Optics Letters, 33*, 1902.

Rybeyre, X., Videau, L., Migus, A., et al. (2003). Nd:glass diode-pumped regenerative amplifier, multimillijoule short-pulse chirped-pulse-amplifier laser. *Optics Letters, 28*, 1374.

Schanz, V. A., Roth, M., & Bagnoud, V. (2019). Picosecond contrast degradation by surface imperfections in chirped-pulse-amplification stretchers. *Journal of the Optical Society of America A. Optics and Image Science, 36*, 1735.

Schattenburg, M.L., Chen, C. G., Heilmann, R. K., et al. (2002). Progress toward a general grating patterning technology using phase-locked scanning beams. *SPIE Proceedings, 4485.*

Schroeder, W. A., McCallum, D. S., Harken, D. R., et al. (1995). Intrinsic and induced anisotropy of nonlinear absorption and refraction in zinc blende semiconductors. *Journal of the Optical Society of America b: Optical Physics, 12*, 401.

Schwarz, J., Rambo, P., Armstrong, D., et al. (2016). Recent laser upgrades at Sandia's Z-backlighter facility in order to accommodate new requirements for magnetized liner inertial fusion on the Z-machine. *High Power Laser Science and Engineering, 4*, Article e36.

Shao, B., Li, Y., Peng, Y., Wang, P., et al. (2020). Broad-bandwidth high-temporal-contrast carrier-envelope-phase-stabilized laser seed for 100 PW lasers. *Optics Letters, 45*, 2215.

Shaw, M., Williams, W., House, R., et al. (2004). Laser performance operations model (LPOM). *Proceedings of SPIE, 5341*, 73.

Shaw, M. J., House, R. K., Williams, W. H., et al. (2008). Laser Performance Operations Model (LPOM): A computational system that automates the setup and performance analysis of the National Ignition Facility. *Journal of Physics: Conference Series, 112*, 032022.

Shaykin, A., Ginzburg, V., Yokovlev, I., et al. (2021). Use of KDP crystal as a Kerr nonlinear medium for compressing PW laser pulses down to 10 fs. *High Power Laser Science and Engineering, 9*, Article e54.

Shaykin, A. A., Freidman, G. I., Garanin, S. G., et al. (2009). 1 petawatt OPCPA laser in Russia: Status and expectations. CLEO Europe.

Shen, B., Bu, Z., Xu, J., et al. (2018). Exploring vacuum birefringence based on a 100 PW laser and an x-ray free electron laser beam. *Plasma Physics and Controlled Fusion, 60*, Article 044002.

Siebold, M., Bock, S., Schramm, U., et al. (2009). Yb:CaF_2—A new old laser crystal. *Applied Physics B, 97*, 47.

Siebold, M., Hornung, M., Boedefeld, R., et al. (2008). Terawatt diode-pumped Yb:CaF_2 laser. *Optics Letters, 33*, 2770.

Simmons, W. W., Hunt, J. T., & Warren, W. E. (1981). Light Propagation through large laser systems. *IEEE Journal of Quantum Electronics, QE-17*, 1727.

Sistrunk, E., Alessi, D. A., Bayramian, A., et al. (2019). Laser technology development for high peak power lasers Achieving kilowatt average power and beyond. *SPEI Proceedings, 11034*, 1103407.

Soulard, R., Quinn, M. N., Tajima, T., & Mourou, G. (2014). ICAN: A novel laser architecture for space debris removal. *Acta Astronautica, 105*, 192.

Spence, D. E., Kean, P. N., & Sibbet, W. (1991). 60-fsec pulse generation from a self-mode-locked Ti:Sapphire laser. *Optics Letters, 16*, 42.

Spinka, T. M., Payne, S. A., Ebbers, C. (2022). Long-wavelength diode-pumped solid-state lasers as inertial fusion energy (IFE) drivers. IFE Science and Technology Community Strategic Planning Workshop, Livermore CA, Feb. 22–24, 2022.

Stanfield, M., Beier, N. F., Hakimi, S., et al. (2021). Millijoule few-cycle pulses from staged compression for strong and high field science. *Optics Express, 29*, 9123.
Strickland, D., & Mourou, G. (1985). Compression of amplified chirped optical pulses. *Optics Communications, 56*, 219.
Stuart, B. C., Feit, M. D., Herman, S., et al. (1996). Optical ablation by high-power short-pulse lasers. *Journal of the Optical Society of America B: Optical Physics, 13*, 459.
Stuart, B. C., Feit, M. D., Rubenchik, A. M., et al. (1995). Laser-induced damage in dielectrics with nanosecond and subpicosecond pulses. *Physical Review Letters, 74*, 2248.
Sung, J. E., Lee, H. W., Yoo, J. Y., et al. (2017). 4.2 PW, 20 fs Ti:sapphire laser at 0.1 Hz. *Optics Letters, 42*, 2058. https://doi.org/10.1364/OL.42.002058
Tajima, T., Brocklesby, W., & Mourou, G. (2013). ICAN: The next laser power house. *Optics and Photonics, 24*(5), 36–43.
Tajima, T., Nakajima, K., & Mourou, G. (2017). Laser accelerators. *Rivista del Nuovo Cimento, 40*, 33.
Tamer, I., Keppler, S., Korner, J., et al. (2019). Modeling of the 3D spatio-temporal thermal profile of joule-class Yb^{3+}-based laser amplifiers. *High Power Laser Science and Engineering, 7*, Article e42.
Tamer, I., Reagan, B. A., Galvin, T., et al. (2022). 1 GW peak power and 100 J pulsed operation of a diode-pumped Tm:YLF laser. *Optics Express, 30*, 46336. https://doi.org/10.1364/OE.475914
Tavella, F., Nomura, Y., Veisz, L., et al. (2007). Dispersion management for a sub-10-fs, 10 TW optical parametric chirped-pulse amplifier. *Optics Letters, 32*, 2227.
Thaury, C., Quere, F., Geindre, J.-P., et al. (2007). Plasma mirrors for ultrahigh intensity optics. *Nature Physics, 3*, 424.
Tien, A.-C., Backus, S., Kapteyn, H., et al. (1999). Short-pulse laser damage in transparent materials as a function of pulse duration. *Physical Review Letters, 82*, 3883.
Tournois, P. (1993). New diffraction grating pair with very linear dispersion for laser pulse compression. *Electronics Letters 29*, 1414.
Tournois, P. (1997). Acousto-optic programmable dispersive filter for adaptive compensation of group delay time dispersion in laser systems. *Optics Communications, 140*, 245.
Veisz, L. (2010). Contrast improvement of relativistic few-cycle light pulses, coherence and ultrashort pulse laser emission. In F. J. Duarte (Ed.), *Coherence and ultrashort pulse laser emission*. Intech Open, ISBN: 978-953-307-242-5.
Verluise, F., Laude, V., Huignard, J.-P., et al. (2000). Arbitrary dispersion control of ultrashort optical pulses with acoustic waves. *Journal of the Optical Society of America B: Optical Physics, 17*, 138.
Voronin, A. A., Zheltikov, A., Ditmire, T., et al. (2013). Subexawatt few-cycle lightwave generation via multipetawatt pulse compression. *Optics Communications, 291*, 299.
Walker, B. C., Toth, C., Fittinghoff, D. N., et al. (1999). A 50-EW/cm^2 Ti:Sapphire laser system for studying relativistic light-matter interactions. *Optics Express, 5*, 196.
Walsh, B. M., Barnes, N. P., & Bartolo, B. D. (1998). Branching ratios, cross sections, and radiative lifetimes of rare earth ions in solids: Application to Tm^{3+} and Ho^{3+} ions in $LiYF_4$. *Journal of Applied Physics, 83*, 2772.
Wang, C., Wang, D., Xu, Y., et al. (2022b). Full-aperture chirped-pulse grating compression with a non-uniform beam. *Optics Communication, 507*, Article 127613.
Wang, S., Jiang, D., Wu, Q., et al. (2019). Spectral properties and highly efficient continuous-wave laser operation in Nd, Gd:CaF_2 crystals. *Journal of Alloys and Compounds, 781*, 629.
Wang, W., Feng, K., Ke, L., Yu, C., et al. (2021). Free-electron lasing at 27 nanometres based on a laser wakefield accelerator. *Nature, 595*, 516.
Wang, X., Liu, X., Lu, X., et al. (2022). 13.4 fs, 0.1 Hz OPCPA front end for the 100 PW-class laser facility. *Ultrafast Science, 2022*, 9894358.
Wang, Y., Zhao, Y., Pan, Z., et al. (2018). 78 fs SWCNT-SA mode-locked Tm:CLNGG disordered garnet crystal laser at 2017 nm. *Optics Letters, 43*, 4286.

Wittmann, T., Geindre, J., Audebert, P., et al. (2006). Towards ultrahigh-contrast ultraintense laser pulses—Complete characterization of a double plasma-mirror pulse cleaner. *Review of Scientific Instruments, 77*, Article 083109.
Wu, F., Hu, J., Liu, X., et al. (2022). Dispersion management for a 100 PW level laser using a mismatched-grating compressor. *High Power Laser Science and Engineering, 10*, Article e38.
Xu, L., Yu, L., Liang, X., et al. (2013). High-energy noncollinear optical parametric-chirped pulse amplification in LBO at 800 nm. *Optics Letters, 38*, 4837.
Yamakawa, K., Shiraga, H., Kato, Y., et al. (1991). Prepulse-free 30-TW, 1-ps Nd:Glass laser. *Optics Letters, 16*, 1593.
Yamakawa, K., Aoyama, M., Matsuoka, S., et al. (1998). Generation of 16-fs pulses at a 1-Hz repetition rate with efficient Ti:Sapphire amplifiers. *Optics Letters, 23*, 525.
Yoon, J. W., Kim, Y. G., Choi, I. W., et al. (2021). Realization of laser intensity over 10^{23} W/cm^2. *Optica, 8*, 630. https://doi.org/10.1364/OPTICA.420520
Yoshida, H., Ishii, E., Kodama, R., et al. (2003). High-power and high-contrast optical parametric chirped pulse amplification in β-BaB_2O_4 crystal. *Optics Letters, 28*, 257.
Yu, T. J., Lee, S. K., Sung, J. H., et al. (2012). Generation of high-contrast, 30 fs, 1.5 PW laser pulses from chirped-pulse amplification Ti:sapphire laser. *Optics Express, 20*, 10807.
Yu, L., Liang, X., Xu, L., et al. (2015). Optimization for high-energy and high-efficiency broadband optical parametric chirped-pulse amplification in LBO near 800 nm. *Optics Express, 40*, 3412.
Zeng, X., Zhou, K., Zuo, Y., et al. (2017). Multi-petawatt laser facility fully based on optical parametric chirped-pulse amplification. *Optics Letters, 42*, 2014.
Zhao, B., Jiang, Y., Sueda, K., et al. (2008). Ultrabroadband noncollinear optical parametric amplification with LBO crystal. *Optics Express, 16*, 18863.
Zheng, J., & Zacharias, H. (2009). Design considerations for a compact grism stretcher for non-collinear optical parametric chirped-pulse amplification. *Applied Physics B, 96*, 445.
Zhou, J., Huang, C.-P., Murnane, M., et al. (1995). Amplification of 26-fs, 2-TW pulses near the gain-narrowing limit in Ti:Sapphire. *Optics Letters, 20*, 64.
Zhou, T., Du, Q., Sano, T., et al. (2018). Two-dimensional combination of eight ultrashort pulsed beams using a diffractive optic pair. *Optics Letters, 43*, 3269.
Zhou, K., Zeng, X., Huang, X., et al. (2021). A comprehensive high-intensity laser platform for string-field physics research. *The Review of Laser Engineering, 49*, 97.
Zhu, J., Zhang, L., Gao, Z., et al. (2015). Diode-pumped femtosecond mode-locked Nd, Y-codoped CaF_2 laser. *Laser Physics Letters, 12*, Article 035801.
Zhu, J., Zhu, J., Li, X., et al. (2018). Status and development of high-power laser facilities at the NLHPLP. *High Power Laser Science and Engineering, 6*, Article e55.
Ziegler, T., Albach, D., Bernert, C., et al. (2021). Proton beam quality enhancement by spectral phase control of a PW-class laser system. *Science and Reports, 11*, 7338.
Zou, J., Coic, H., & Papadopoulos, D. (2022). Spatiotemporal coupling investigations for Ti:sapphire based multi-PW lasers. *High Power Laser Science and Engineering, 10*, Article e5.
Zou, J. P., Le Blanc, C., Audebert, P., et al. (2008). Recent progress on LULI high power laser facilities. *IOP Journal of Physics: Conference Series, 112*, Article 032021.
Zou, J. P., Le Blanc, C., Papadopoulos, D. N., et al. (2015). Design and current progress of the Apollon 10 PW project. *High Power Laser Science and Engineering, 3*, Article e2.
Zuegel, L. D., Borneis, S., Barty, C., et al. (2006). Laser challenges for fast ignition. *Fusion Science and Technology, 49*, 454.

Part III
Laser Plasmas and Inertial Fusion Physics

Chapter 8
Introduction to Plasma Physics

Abstract In the mid-19th century, electrical discharge was investigated in detail by Michael Faraday. In the late 19-th century, William Crookes investigated the electrical discharge in a high-vacuum tube to discover an "electron beam". This work is recognized as the first step of the discovery of the "electron" by Thomson (Philos Mag 44:293, 1897) and Millikan (Philos Mag 6:209, 1910). Research on the physics of ionized gas in the discharge was started by the experiments of I. Langmuir in the 1920's. He named the ionize gas "plasma" in 1928 (Tonks and Lngmuir in Phys Rev 33:195, 1929). In this chapter, characteristics of plasma are briefly described.

Keywords Plasma · Fourth state of matter · Saha equation · Ionization equilibrium · Coupling parameter · Warm dense matter · Hot dense matter · Strongly coupled plasma · Relativistic plasma

8.1 Plasma as the Fourth State of Matter

Plasma is called "the fourth state of matter". The solid phase is the first state. When a solid is heated, it is melted to become liquid: the liquid phase is the second state. When it is further heated, it is evaporated to be gas: the gas phase is the third state. When the gas is heated to a temperature higher than several thousand Kelvin degrees, the gas called plasma is the fourth state of matter.

The four phases of matter and the characteristic temperature are summarized in Table 8.1.

8.2 Saha Equation (Ionization Equilibrium)

When free electrons and bound electrons are in thermal equilibrium, the ionization degree of hydrogen-like gas, namely, the electron density n_e at the temperature T, is related to the total ion number density n_h and the neutral atomic number density n_n

Y. Kato et al., *High Power Laser and Plasma Science*, Springer Series on Atomic, Optical, and Plasma Physics 130, https://doi.org/10.1007/978-3-032-08930-4_8

Table 8.1 Four phases of matter

Phase	Solid state	Liquid state	Gas phase	Plasma
Temperature	0-10^3 °K	10–10^3 °K	10^2–10^4 °K	10^4–higher °K
Particle interactions	Atomic and molecular force	Atomic and molecular force	Van-der-Waals force	Electro-magnetic force

as follows (Saha, 1920):

$$\frac{n_h n_e}{n_n} = 2\left(\frac{2\pi m_e k_B T}{h^2}\right)^{3/2} \frac{g_i}{g_n} \exp\left(-\frac{I}{k_B T}\right) \tag{8.1}$$

where m_e is the electron mass, k_B is the Boltzmann constant, h is the Planck constant, I is the ionization potential of the atom, $(h^2/2\pi m_e k_B T)^{3/2}$ is the cube of the thermal de Broglie wavelength of an electron, and g_n, g_i are the effective numbers of the quantum states of neutral atom and ion, respectively. Note here that only the single ionization is considered.

8.3 Plasmas in the Universe and in Laboratories

Various plasmas in nature and laboratories are shown in the temperature and density parameter space of Fig. 8.1. Properties of plasmas are classified by the two parameters. The one is the coupling parameter Γ and the other is the degree of degeneracy $\epsilon_F/k_B T$ where ϵ_F is the Fermi energy of electron. The coupling parameter Γ is defined by the ratio of average Coulomb interaction energy to the thermal energy, namely $Z^2 e^2/4\pi\varepsilon_0 a k_B T$, where ε_0 is the vacuum permittivity and a is the average inter-ion distance which depends on the density as $(3/4\pi n)^{1/3}$ for the ion density of n, and the ion charge Ze. When $\Gamma > 1$, ions interact with each other strongly and ion motions are strongly correlated. This region is on the right and lower side of the broken line of Fig. 8.1.

On the other hand, when Γ is very large, namely, the matter is in a high density and low temperature state, it is solid/liquid plasma or metal, where ions are strongly correlated, and electrons are degenerate. When the degree of degeneracy is greater than 1, electrons are no longer classical particles. Electron wave functions are extended over the inter-electron distance and the electron–electron and electron-ion interactions should be treated by quantum mechanics.

When an intense laser light irradiates a solid target, the solid surface is melted and evaporated, and dense hot plasmas are generated. Thus, the surface of the laser irradiated solid target is covered by dense plasmas. The laser plasma density and temperature on the surface of the solid target are in the region indicated by the yellow area and the circled area above the yellow area in Fig. 8.1. Note here that the laser plasma parameters extend over the five areas in the density and temperature

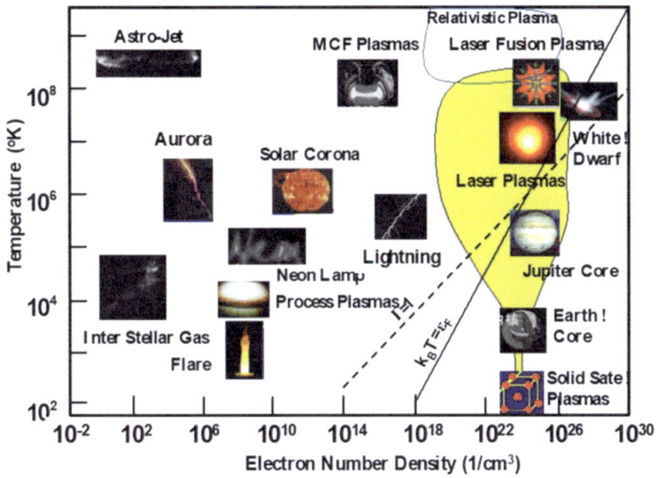

Fig. 8.1. Various plasmas in temperature-density space. The yellow area indicates the laser plasmas which are produced on the surface of a solid target irradiated by intense lasers. The circled area above the yellow area indicates the relativistic plasma produced by ultra-intense lasers.

space: high temperature and low density (ideal) plasmas, high temperature and dense plasma (hot dense matter), high density and low temperature (strongly coupled) plasmas, degenerate electron plasmas (warm dense matter), and relativistic plasmas (the circled area above the laser plasma area). In Part III, the non-relativistic basic plasma physics is described.

References

Millikan, R. A. (1910). A new modification of the cloud method of determining the elementary electrical charge and the most probable value of that charge. *Philosophical Magazine, 6*, 209.

Saha, M. N. (1920). LIII. Ionization in the solar chromosphere. *Philosophical Magazine Series, 6*(40), 472. https://doi.org/10.1080/14786441008636148

Thomson, J. J. (1897). Cathode rays. *Philosophical Magazine, 44*, 293.

Tonks, L., & Langmuir, I. (1929). Oscillations in ionized gases. *Physical Review, 33*, 195.

Chapter 9
Plasma Kinetic Theory

Abstract Plasma is the physical system in which charged particles interact with each other through electromagnetic fields determined by the Maxwell equations. The current density and the charge density included in the Maxwell equations are determined by the motions of the charged particles. The self-consistent coupling between the electromagnetic field and the charged particle motion is schematically shown in Fig. 9.1. The kinetic theory treating plasma as the many body charged particle system is described in this chapter. Since laser plasmas are in high density states where the collisional effects are not negligible in some cases, we derive the kinetic equation including the collisional effects from the Liouville equation.

Keywords Liouville theorem · Liouville equation · Kinetic equation · Vlasov equation · Collision operator · Coulomb log · Plasma fluid dynamics · Magneto-hydrodynamic equation

9.1 Liouville Equation

The kinetics of the interacting charged particle system is derived from the N-body distribution function which satisfies the Liouville theorem (Fig. 9.1). As an introduction to the theorem, we consider the phase space for one dimensional motion of a single particle. In a small time-step, Δt, a particle which starts from area A moves to area B in Fig. 9.2. Let the phase space area of A to be s_A and the area of B to be s_B, where $s_A = dqdp$. The area s_B is estimated as follows:

$$s_B = dQdP = dqdp + \left(\frac{\partial \frac{dq}{dt}}{\partial q} + \frac{\partial \frac{dp}{dt}}{\partial p} \right) \Delta t dqdp + O(\Delta t^2), \tag{9.1}$$

because

$$Q(q + dq, p) = Q(q, p) + dq + \frac{\partial \frac{dq}{dt}}{\partial q} \Delta t dq$$

Y. Kato et al., *High Power Laser and Plasma Science*, Springer Series on Atomic, Optical, and Plasma Physics 130, https://doi.org/10.1007/978-3-032-08930-4_9

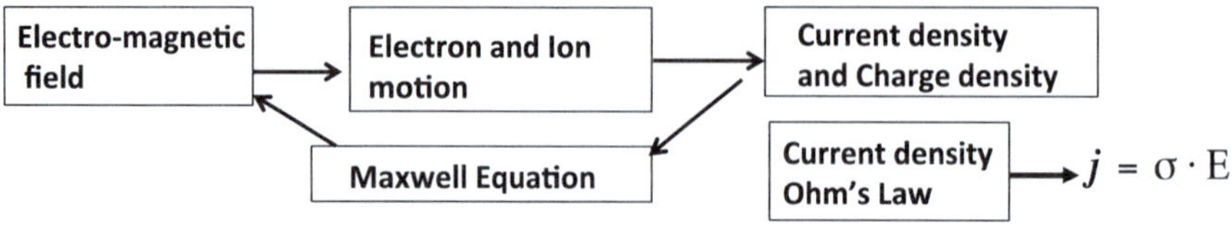

Fig. 9.1. Coupling between the electron and ion motions and the electromagnetic fields.

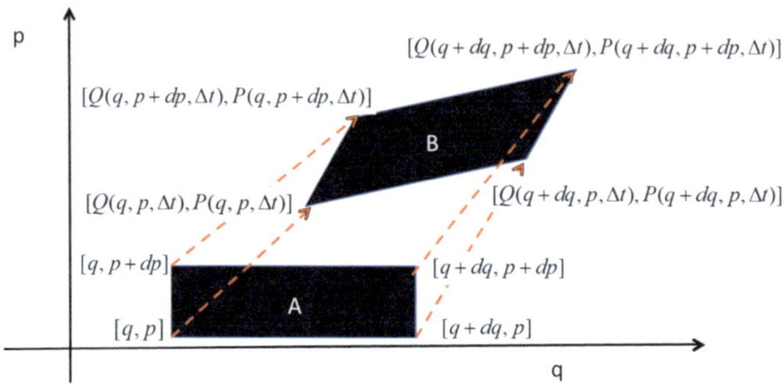

Fig. 9.2. Particle flow in the phase space of 1D motion.

and

$$P(q, p + dp) = P(q, p) + dp + \frac{\partial \frac{dp}{dt}}{\partial p} \Delta t dp.$$

The Hamilton equations of motion for the Hamiltonian $\mathcal{H}(q, p)$ is given by

$$\frac{dq}{dt} = \frac{\partial \mathcal{H}(p, q)}{\partial p} \quad \text{and} \quad \frac{dp}{dt} = -\frac{\partial \mathcal{H}(p, q)}{\partial q} \tag{9.2}$$

Therefore,

$$\frac{\partial \frac{dq}{dt}}{\partial q} + \frac{\partial \frac{dp}{dt}}{\partial p} = 0. \tag{9.3}$$

Equations (9.1) and (9.3) yield

$$\lim_{\Delta t \to 0} \frac{s_B - s_A}{\Delta t} = 0. \tag{9.4}$$

Namely, the phase space area which moves according to the equation of motion is constant. Therefore, we find that the Jacobian of the transform from the variables (q, p) to variables (Q, P) is equal to unity,

$$\frac{\partial(Q,P)}{\partial(q,p)}=\frac{s_B}{s_A}=1. \tag{9.5}$$

The invariance of the phase space volume of the one-dimensional single particle motion can be extended to the multi-dimensional and multi particle case, even when the particles interact with each other if the system is non-dissipative. This is called the Liouville theorem, which says that the Jacobian of transform from $(\boldsymbol{q},\boldsymbol{p})$ to variables $(\boldsymbol{Q},\boldsymbol{P})$ is equal to unity,

$$\frac{\partial(\boldsymbol{Q}_1,\boldsymbol{Q}_2,\ldots,\boldsymbol{Q}_N,\boldsymbol{P}_1,\boldsymbol{P}_2,\ldots,\boldsymbol{P}_N)}{\partial(\boldsymbol{q}_1,\boldsymbol{q}_2,\ldots,\boldsymbol{q}_N,\boldsymbol{p}_1,\boldsymbol{p}_2,\ldots,\boldsymbol{p}_N)}=1. \tag{9.6}$$

Let us give the particle number dN in a small phase space volume $d\tau_1, d\tau_2, d\tau_3, \ldots d\tau_N$ as

$$dN=F(\boldsymbol{q}_1,\boldsymbol{q}_2,\ldots,\boldsymbol{q}_N,\boldsymbol{p}_1,\boldsymbol{p}_2,\ldots,\boldsymbol{p}_N)d\tau_1,d\tau_2,d\tau_3,\ldots,d\tau_N$$

for the i-th particle position $\boldsymbol{q}_i$ and momentum $\boldsymbol{p}_i$, where $i=1,2,3,\ldots,N$. Here $F(\boldsymbol{q}_1,\boldsymbol{q}_2,\ldots,\boldsymbol{q}_N,\boldsymbol{p}_1,\boldsymbol{p}_2,\ldots,\boldsymbol{p}_N)$ is the N-particle distribution function and $d\tau = d\boldsymbol{q}d\boldsymbol{p}$.

Since the phase space volume element $d\tau_1, d\tau_2, d\tau_3, \ldots, d\tau_N$ is constant according to the Liouville theorem, the particle number conservation $dN/dt=0$ requires

$$\frac{dF(\boldsymbol{q}_1,\boldsymbol{q}_2,\ldots,\boldsymbol{q}_N,\boldsymbol{p}_1,\boldsymbol{p}_2,\ldots,\boldsymbol{p}_N)}{dt}=0. \tag{9.7}$$

From this expression, considering the Hamilton equations, it follows that

$$\frac{\partial F}{\partial t}+\sum_{i=1}^{N}\left(\dot{\boldsymbol{q}}_i\frac{\partial F}{\partial \boldsymbol{q}_i}+\dot{\boldsymbol{p}}_i\frac{\partial F}{\partial \boldsymbol{p}_i}\right)=0. \tag{9.8}$$

Here and below, we use Newton's notation for a dot which stands for the derivative with respect to time. This equation can be rewritten for the charged particle system in the external electromagnetic field $(\boldsymbol{E},\boldsymbol{B})$ in the form,

$$\frac{\partial F}{\partial t}+\sum_{i=1}^{N}\frac{\boldsymbol{p}_i}{m}\frac{\partial F}{\partial \boldsymbol{r}_i}-\sum_{j\neq i,1}^{N}\frac{Z_iZ_je^2}{4\pi\varepsilon_0}\left(\frac{\partial}{\partial \boldsymbol{r}_i}\frac{1}{|\boldsymbol{r}_i-\boldsymbol{r}_j|}\right)\frac{\partial F}{\partial \boldsymbol{r}_i}$$
$$+\sum_{i=1}^{N}Z_ie\left(\boldsymbol{E}(\boldsymbol{r}_i,t)+\frac{\boldsymbol{p}_i}{m}\times\boldsymbol{B}(\boldsymbol{r}_i,t)\right)\frac{\partial F}{\partial \boldsymbol{p}_{ii}}=0 \tag{9.9}$$

where $\boldsymbol{p}_i/m=\boldsymbol{v}_i=\dot{\boldsymbol{r}}_i$. From here and on we change the notation from the coordinate $\boldsymbol{q}_i$ to $\boldsymbol{r}_i$. The Eq. (9.9) combined with the Maxwell equations are the exact kinetic equation of plasmas with the given initial condition of N body distribution function F. Here, we assume that the particle velocities are much slower than the speed of light and the particle-particle interactions through magnetic fields are negligible.

This distribution function includes $6N$ variables where N is huge. So, it is impossible to solve the coupled equations directly. The Eq. (9.9) is reduced to one-body distribution function when the particle interaction is weak as described in the following. In this case, we assume that the individual particles move almost independently on a short time scale, much shorter than the collision-mean-free time. We now reduce the Liouville equation (9.9) (N-body kinetic equation) to the equation of the one-body distribution function f_1 by integrating the Liouville equation for the N-1 particle position and momentum to obtain,

$$\frac{\partial f_1}{\partial t} + \frac{1}{m}\boldsymbol{p}\frac{\partial f_1}{\partial \boldsymbol{r}} + Ze\left(\boldsymbol{E} + \frac{\boldsymbol{p}}{m} \times \boldsymbol{B}\right)\frac{\partial f_1}{\partial \boldsymbol{p}} = C(f), \tag{9.10}$$

where $C(f)$ is the collision integral equal to

$$C(f) = \sum_{l=e,i} \iint d\boldsymbol{r}_l d\boldsymbol{p}_l \frac{ZZ_l e^2}{4\pi\varepsilon_0}\left(\frac{\partial}{\partial \boldsymbol{r}}\frac{1}{|\boldsymbol{r}-\boldsymbol{r}_l|}\right)\frac{\partial f_2(\boldsymbol{r},\boldsymbol{p},\boldsymbol{r}_l,\boldsymbol{p}_l)}{\partial \boldsymbol{p}}. \tag{9.11}$$

In Eq. (9.10), $\boldsymbol{E}$ and $\boldsymbol{B}$ are now the self-consistent electro-magnetic field. Note also here that the right-hand-side of Eq. (9.11) is considered as the two particle collisional effects and $f_2(\boldsymbol{r},\boldsymbol{p},\boldsymbol{r}_l,\boldsymbol{p}_l)$ is the two-body distribution function. When the plasma is hot and relatively low density, the two particles are approximately uncorrelated since the average Coulomb interaction energy normalized by the thermal kinetic energy (the coupling parameter $\Gamma = e^2/4\pi\varepsilon_0 a k_B T$ given in Sect. 8.3) is much smaller than unity.

Therefore, the two-particle distribution function $f_2(\boldsymbol{r},\boldsymbol{p},\boldsymbol{r}_l,\boldsymbol{p}_l)$ in the r.h.s. of Eq. (9.11) can be approximated by $f_2(\boldsymbol{r},\boldsymbol{p},\boldsymbol{r}_l,\boldsymbol{p}_l) = f_1(\boldsymbol{r},\boldsymbol{p})f_1(\boldsymbol{r}_l,\boldsymbol{p}_l)$ (Lifshitz & Pitaevskii, 1981). Note that the coupling parameter $\Gamma \approx (1/4\sqrt{3}\pi n\lambda_D^2)^{2/3} \ll 1$ is small compared to unity for the Debye shielding distance λ_D which will be introduced in Chap. 10. The condition of the weak-coupling $\Gamma \ll 1$ is equivalent to the condition corresponding to the large number of particle in the Debye sphere, $n\lambda_D^3 \gg 1$. When the two systems of particles are uncorrelated, the collision integral $C(f)$ in the r.h.s. of Eq. (9.11) vanishes because of the spherical symmetry of the distribution function $f_1(\boldsymbol{r}_l,\boldsymbol{p}_l)$. Then Eq. (9.10) is reduced to

$$\frac{\partial f_1}{\partial t} + \frac{1}{m}\boldsymbol{p}\frac{\partial f_1}{\partial \boldsymbol{r}} + Ze\left(\boldsymbol{E} + \frac{\boldsymbol{p}}{m} \times \boldsymbol{B}\right)\frac{\partial f_1}{\partial \boldsymbol{p}} = 0. \tag{9.12}$$

Here, the fields $\boldsymbol{E}$ and $\boldsymbol{B}$ now include collective fields induced by the plasma electric charge and the electric current.

Equation (9.12) is the kinetic equation of collision-less plasma which is called the Vlasov equation (Vlasov, 1968). The Vlasov equation together with the Maxwell equations is called the system of Vlasov-Maxwell equations, which determines the self-consistent electromagnetic fields in collision-less plasmas and plasma dynamics.

9.2 Collision Operator

When the coupling parameter Γ is finite, it is necessary to take it into account the correlation in the two-body distribution $f_2(\boldsymbol{r}, \boldsymbol{p}, \boldsymbol{r}_l, \boldsymbol{p}_l)$ by solving the Bogolyubov–Born–Green–Kirkwood–Yvon (BBGKY) hierarchy (sometimes called Bogolyubov hierarchy) which is a set of equations describing the dynamics of a system of a large number of interacting particles (see Vereshchagin & Aksenov, 2017). Namely, collision integral $C(f)$ in the r.h.s. of Eq. (9.11) remains finite when the particles are correlated. In the following, we determine the two-body distribution function $f_2(\boldsymbol{r}, \boldsymbol{p}, \boldsymbol{r}_l, \boldsymbol{p}_l)$, (it depends on the coordinates and momenta $\boldsymbol{r}, \boldsymbol{p}, \boldsymbol{r}_l, \boldsymbol{p}_l$ of two particles) in the r.h.s. of Eq. (9.11) approximately by assuming smallness of the parameter Γ.

The two-body distribution function satisfies the equation,

$$\frac{\partial f_2}{\partial t} + \frac{1}{m}\left(\boldsymbol{p}_1 \frac{\partial}{\partial \boldsymbol{r}_1} + \boldsymbol{p}_2 \frac{\partial}{\partial \boldsymbol{r}_2}\right) f_2 - \frac{Z_1 Z_2 e^2}{4\pi\varepsilon_0}\left(\frac{\partial}{\partial \boldsymbol{r}_1}\frac{1}{|\boldsymbol{r}_1 - \boldsymbol{r}_2|}\frac{\partial}{\partial \boldsymbol{p}_1} + \frac{\partial}{\partial \boldsymbol{r}_2}\frac{1}{|\boldsymbol{r}_1 - \boldsymbol{r}_2|}\frac{\partial}{\partial \boldsymbol{p}_2}\right) f_2 = \sum_{l=1}^{2} \frac{Z_l Z_3 e^2}{4\pi\varepsilon_0} \iint \frac{\partial}{\partial \boldsymbol{r}_l}\frac{1}{|\boldsymbol{r}_l - \boldsymbol{r}_3|}\frac{\partial}{\partial \boldsymbol{p}_3} f_3 d\boldsymbol{r}_3 d\boldsymbol{p}_3. \tag{9.13}$$

Let us assume here that the correlation effects, namely the deviation of the three-particle distribution $f_3(\boldsymbol{r}_1, \boldsymbol{p}_1, \boldsymbol{r}_2, \boldsymbol{p}_2, \boldsymbol{r}_l, \boldsymbol{p}_l)$ from $f_2(\boldsymbol{r}, \boldsymbol{p}, \boldsymbol{r}_{2l}, \boldsymbol{p}_2) f_1(\boldsymbol{r}_l, \boldsymbol{p}_l)$, is finite only when $|\boldsymbol{r} - \boldsymbol{r}_l| \approx b \ll a$, where b is approximately the impact parameter of 90° scattering and a is the average inter-particle distance of the plasma. Therefore, the r.h.s. of Eq. (9.13) is much smaller than the left-hand side (l.h.s.) by the factor $(b/a)^3$ and the r.h.s. of Eq. (9.13) can be neglected. Then, Eqs. (9.10) and (9.13) make a closed system. We solve Eq. (9.13) to express the collision operator of Eq. (9.12) in terms of $f_1(\boldsymbol{r}_l, \boldsymbol{p}_l)$, according to the analysis presented in the textbook by Lifshitz and Pitaevskii (1981).

The r.h.s. of Eq. (9.13) is the total derivative df/dt, where the variables $\boldsymbol{r}, \boldsymbol{p}, \boldsymbol{r}_l, \boldsymbol{p}_l$ are regarded as functions of time which satisfy the equation of motion for the two body Hamiltonian,

$$\mathcal{H} = \frac{\boldsymbol{p}_1^2}{2m} + \frac{\boldsymbol{p}_2^2}{2m} + \frac{Z_1 Z_2 e^2}{4\pi\varepsilon_0 |\boldsymbol{r}_1 - \boldsymbol{r}_2|}$$

So, we rewrite Eq. (9.13) as follows,

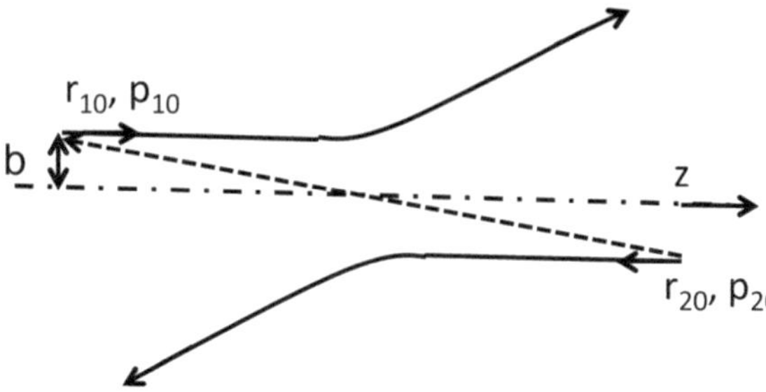

Fig. 9.3. The orbits of two colliding particle orbits with an impact parameter b in the center of mass frame. Initial positions are $\boldsymbol{r}_{1,0}$ and $\boldsymbol{r}_{2,0}$ at t_0 and distance is $|z_{1,0}-z_{2,0}| = |\boldsymbol{r}_{1,0}-\boldsymbol{r}_{2,0}|$ where the direction of z is parallel or anti-parallel to $\boldsymbol{v}_{1,0}-\boldsymbol{v}_{2,0} = \boldsymbol{p}_{1,0}/m - \boldsymbol{p}_{2,0}/m$.

$$\frac{df_2(\boldsymbol{r},\boldsymbol{p},\boldsymbol{r}_2,\boldsymbol{p}_2)}{dt} = 0. \tag{9.14}$$

Let us consider two colliding particles of which positions are initially at $\boldsymbol{r}_{1,0}$ and $\boldsymbol{r}_{2,0}$ at time t_0 and we assume that $|\boldsymbol{r}_{1,0}-\boldsymbol{r}_{2,0}|$ is much longer than the interaction force range as shown in Fig. 9.3. Then, the two particles are assumed to be uncorrelated, and the two-body distribution function is assumed to be

$$f_2(\boldsymbol{r}_{1,0},\boldsymbol{p}_{1,0},\boldsymbol{r}_{2,0},\boldsymbol{p}_{2,0}) = f_1(\boldsymbol{r}_{1,0},\boldsymbol{p}_{1,0})f_1(\boldsymbol{r}_{2,0},\boldsymbol{p}_{2,0}). \tag{9.15}$$

When the time scale of the external force is much longer than the duration of the collision, the explicit time derivative of Eq. (9.14) can be neglected. Therefore, Eq. (9.14) can be reduced to

$$\frac{1}{m}\left(\boldsymbol{p}_1\frac{\partial}{\partial \boldsymbol{r}_1} + \boldsymbol{p}_2\frac{\partial}{\partial \boldsymbol{r}_2}\right)f_2 - \frac{Z_1Z_2e^2}{4\pi\varepsilon_0}\left(\frac{\partial}{\partial \boldsymbol{r}_1}\frac{1}{|\boldsymbol{r}_1-\boldsymbol{r}_2|}\frac{\partial}{\partial \boldsymbol{p}_1} + \frac{\partial}{\partial \boldsymbol{r}_2}\frac{1}{|\boldsymbol{r}_1-\boldsymbol{r}_2|}\frac{\partial}{\partial \boldsymbol{p}_2}\right)f_2 = 0. \tag{9.16}$$

The derivative with respect to the momentum $\boldsymbol{p}_1$ in Eq. (9.11) can be expressed by the derivatives with respect to the variables $\boldsymbol{r}, \boldsymbol{r}_2, \boldsymbol{p}_2$ by using Eq. (9.16). Namely, Eq. (9.11) is rewritten as,

$$C(f) = \sum_{l=e,i}\iint d\boldsymbol{r}_2 d\boldsymbol{p}_2\left[\frac{\boldsymbol{p}}{m}\frac{\partial}{\partial \boldsymbol{r}} + \frac{\boldsymbol{p}_2}{m}\frac{\partial}{\partial \boldsymbol{r}_2} - \frac{ZZ_2e^2}{4\pi\varepsilon_0}\left(\frac{\partial}{\partial \boldsymbol{r}_2}\frac{1}{|\boldsymbol{r}-\boldsymbol{r}_2|}\right)\frac{\partial}{\partial \boldsymbol{p}_2}\right]f_1(\boldsymbol{r}_{1,0},\boldsymbol{p}_{1,0})f_1(\boldsymbol{r}_{2,0},\boldsymbol{p}_{2,0}). \tag{9.17}$$

The derivative with respect to $\boldsymbol{p}_2$ in the third term of the r.h.s. of Eq. (9.17) disappears by integration over $\boldsymbol{p}_2$. When plasma non-uniformity is neglected, the function $f_1(\boldsymbol{r}_{1,0}, \boldsymbol{p}_{1,0}) f_1(\boldsymbol{r}_{2,0}, \boldsymbol{p}_{2,0})$ depends upon the relative coordinate $\boldsymbol{r}_{rel} = \boldsymbol{r} - \boldsymbol{r}_2$. As a result, Eq. (9.17) can be rewritten as,

$$C(f) = \iint d\boldsymbol{r}_{rel} d\boldsymbol{p}_2 \boldsymbol{v}_{rel} \frac{\partial}{\partial \boldsymbol{r}_{rel}} f_1(\boldsymbol{r}_{1,0}, \boldsymbol{p}_{1,0}) f_1(\boldsymbol{r}_{2,0}, \boldsymbol{p}_{2,0}), \tag{9.18}$$

where $\boldsymbol{v}_{rel} = \boldsymbol{v} - \boldsymbol{v}_2$. Considering the center of mass frame for the colliding two particles and introducing the cylindrical coordinates z, ρ, φ with the axis along $\boldsymbol{v}_{rel}$, Eq. (9.18) can be rewritten as,

$$C(f) = \iint dz \rho d\rho d\boldsymbol{p}_2 \boldsymbol{v}_{rel} \frac{\partial}{\partial \boldsymbol{r}_{rel}} f_1(\boldsymbol{r}_{1,0}, \boldsymbol{p}_{1,0}) f_1(\boldsymbol{r}_{2,0}, \boldsymbol{p}_{2,0}).$$

By integration with respect to z, we get

$$C(f) = \iint \rho d\rho d\boldsymbol{p}_2 \boldsymbol{v}_{rel} f_1(\boldsymbol{r}_{1,0}, \boldsymbol{p}_{1,0}) f_1(\boldsymbol{r}_{2,0}, \boldsymbol{p}_{2,0}) \Big|_{z=-\infty}^{z=+\infty}. \tag{9.19}$$

Since $z = -\infty$ means before the collision, the momenta are $\boldsymbol{p}_1 = \boldsymbol{p}_{1,0}$ and $\boldsymbol{p}_2 = \boldsymbol{p}_{2,0}$, respectively. Since $z = +\infty$ corresponds to the case after the collision on the other hand, $\boldsymbol{p}_1 = \boldsymbol{p}_{1,0} + \Delta \boldsymbol{p}$ and $\boldsymbol{p}_2 = \boldsymbol{p}_{2,0} - \Delta \boldsymbol{p}$. Namely, the momentum $\Delta \boldsymbol{p}$ is transferred from particle 2 to 1 by the collision. Therefore, for the uniform plasma, which is applicable for the case of slowly varying space dependent distribution function in comparison with the interaction range, Eq. (9.19) can be rewritten as follows.

$$C(f_1(\boldsymbol{p})) = \iint \rho d\rho d\varphi d\vec{p}_2 v_{rel} \big[f_1(\boldsymbol{p}_1 + \Delta \boldsymbol{p}) f_1(\boldsymbol{p}_2 - \Delta \boldsymbol{p}) - f_1(\boldsymbol{p}_1) f_1(\boldsymbol{p}_2) \big]. \tag{9.20}$$

The transferred momentum $\Delta \boldsymbol{p}$ is related to the impact parameter ρ. Then, the cross section $d\sigma(q, \varphi)$ for the transferred momentum $\Delta \boldsymbol{p}$ can be introduced as $d\sigma(q, \varphi) = \rho d\rho d\varphi$. As a result, we obtain,

$$C(f_1(\boldsymbol{p})) = \iint v_{rel} \big[f_1(\boldsymbol{p} + \Delta \boldsymbol{p}) f_1(\boldsymbol{p}' - \Delta \boldsymbol{p}) - f_1(\boldsymbol{p}) f_1(\boldsymbol{p}') \big] d\sigma(q, \varphi) d\boldsymbol{p}'. \tag{9.21}$$

For the collision due to the long-range Coulomb force, the cross section is divergent in the limit of small, transferred momentum $\Delta \boldsymbol{p}$. In other words, the collision is dominated by small angle scattering. The collisions result in the diffusion in momentum space as described in the "Physical Kinetics" textbook by Lifshitz and Pitaevskii (1981). Therefore, we can rewrite Eq. (9.21) in the form,

$$C(f) = -\frac{\partial}{\partial \boldsymbol{p}} \boldsymbol{S}_p. \tag{9.22}$$

Let us introduce the probability $w(\boldsymbol{p} + \Delta \boldsymbol{p}/2, \boldsymbol{p}' - \Delta \boldsymbol{p}/2 : \Delta \boldsymbol{p})$ for the transition:

$$(\boldsymbol{p}, \boldsymbol{p}') \rightarrow (\boldsymbol{p} + \Delta\boldsymbol{p}, \boldsymbol{p}' - \Delta\boldsymbol{p})$$

with the small change of momentum equal to $\Delta\boldsymbol{p}$. Here $\boldsymbol{p} + \Delta\boldsymbol{p}/2$ and $\boldsymbol{p}' - \Delta\boldsymbol{p}/2$ are the momenta at the middle of the collision. On the other hand, for the transition

$$(\boldsymbol{p} + \Delta\boldsymbol{p}, \boldsymbol{p}' - \Delta\boldsymbol{p}) \rightarrow (\boldsymbol{p}, \boldsymbol{p}'),$$

the transition probability is $w(\boldsymbol{p} + \Delta\boldsymbol{p}/2, \boldsymbol{p}' - \Delta\boldsymbol{p}/2 : -\Delta\boldsymbol{p})$. Noting here that the transition probability w is symmetric for the reverse process, we obtain

$$w(\boldsymbol{p} + \Delta\boldsymbol{p}/2, \boldsymbol{p}' - \Delta\boldsymbol{p}/2 : \Delta\boldsymbol{p}) = w(\boldsymbol{p} + \Delta\boldsymbol{p}/2, \boldsymbol{p}' - \Delta\boldsymbol{p}/2 : -\Delta\boldsymbol{p}). \tag{9.23}$$

Therefore, we can reduce Eq. (9.21) and Eq. (9.22) to

$$\frac{\partial}{\partial \boldsymbol{p}}\boldsymbol{S}_p = \iint \Big[f_1(\boldsymbol{p} + \Delta\boldsymbol{p}) f_1(\boldsymbol{p}' - \Delta\boldsymbol{p}) - f_1(\boldsymbol{p}) f_1(\boldsymbol{p}') \\ w\left(\boldsymbol{p} + \frac{\Delta\boldsymbol{p}}{2}, \boldsymbol{p}' - \frac{\Delta\boldsymbol{p}}{2} : \Delta\boldsymbol{p}\right) d\Delta\boldsymbol{p} d\boldsymbol{p}'\Big] \tag{9.24}$$

Considering the smallness of the momentum change $\Delta\boldsymbol{p}$ for the dominant Coulomb collision, we rewrite Eq. (9.24) as follows,

$$\frac{\partial}{\partial \boldsymbol{p}}\boldsymbol{S}_p = \iint \left[f_1(\boldsymbol{p})\left(\Delta\boldsymbol{p} \cdot \frac{\partial f_1(\boldsymbol{p}')}{\partial \boldsymbol{p}'}\right) - \left(\Delta\boldsymbol{p} \cdot \frac{\partial f_1(\boldsymbol{p})}{\partial \boldsymbol{p}}\right) f_1(\boldsymbol{p}')\right] \\ w(\boldsymbol{p}, \boldsymbol{p}' : \Delta\boldsymbol{p}) d\Delta\boldsymbol{p} d\boldsymbol{p}' \tag{9.25}$$

Note that the integration over the momentum $\Delta\boldsymbol{p}$ should be performed over a half of space since using Eq. (9.23) we added two equivalent collisions to derive Eq. (9.24). Using the relationship $w\Delta\boldsymbol{p} = (\boldsymbol{v} - \boldsymbol{v}')d\sigma$, we obtain from Eq. (9.25) that $\boldsymbol{S}_p$ is equal to

$$\boldsymbol{S}_p = \iint \left[f_1(\boldsymbol{p})\frac{\partial f_1(\boldsymbol{p}')}{\partial \boldsymbol{p}'} - \frac{\partial f_1(\boldsymbol{p})}{\partial \boldsymbol{p}} f_1(\boldsymbol{p}')\right] \overleftrightarrow{\boldsymbol{B}} d\boldsymbol{p}'. \tag{9.26}$$

Here, the tensor $\overleftrightarrow{\boldsymbol{B}}$ is defined as

$$\overleftrightarrow{\boldsymbol{B}} = \frac{1}{2}\int \Delta\boldsymbol{p}\Delta\boldsymbol{p}|\boldsymbol{v} - \boldsymbol{v}'| d\sigma. \tag{9.27}$$

This expression can be written in the form

$$\overleftrightarrow{\boldsymbol{B}} = \frac{1}{2}B\left\{\overleftrightarrow{\boldsymbol{I}} - \frac{(\boldsymbol{v} - \boldsymbol{v}')(\boldsymbol{v} - \boldsymbol{v}')}{|\boldsymbol{v} - \boldsymbol{v}'|^2}\right\} \text{ with } B = \frac{1}{2}\int (\Delta\boldsymbol{p})^2 |\boldsymbol{v} - \boldsymbol{v}'| d\sigma \tag{9.28}$$

and $\overleftrightarrow{I}$ is a unit tensor. For the reduced mass of two particles μ, the transferred momentum equals $\Delta \boldsymbol{p} = \mu(\boldsymbol{v} - \boldsymbol{v}')$. Hence,

$$B = \frac{1}{2}\mu^2|\vec{v} - \vec{v}'|^3 \int \chi^2 d\sigma. \tag{9.29}$$

Here, χ is the scattering angle and the differential cross section $d\sigma$ depends upon χ and relative velocity $v_{rel} = |\boldsymbol{v} - \boldsymbol{v}'|$. The cross section $d\sigma$ depends on the interaction force between particles.

For the Coulomb interaction, we use the Rutherford scattering cross-section to obtain,

$$d\sigma = \frac{8\pi}{\mu^2|\boldsymbol{v} - \boldsymbol{v}'|^4}\frac{d\chi}{\chi^3}. \tag{9.30}$$

Therefore, we have

$$B = \frac{2\pi(ee')^2}{\mu^2|\boldsymbol{v} - \boldsymbol{v}'|}\int_{\chi_{\min}}^{\chi_{max}} \frac{d\chi}{\chi} = \frac{2\pi(ee')^2}{\mu^2|\boldsymbol{v} - \boldsymbol{v}'|}\ln\left(\frac{\chi_{\max}}{\chi_{\min}}\right). \tag{9.31}$$

Here, we assumed the upper limit and the lower limit of the scattering angle. The lower limit is related to the long-distance scattering, and it is determined by the shielding of the Coulomb field by surrounding charged particles. The expression $L = \ln(\chi_{\max}/\chi_{\min})$ is called the "Coulomb logarithm". The lower integration limit $\chi_{\min}$ corresponds to the scattering angle for the collision when the impact parameter is equal to either the Debye length or de Broglie wavelength (it is equal to $\hbar/p$). On the other hand, $\chi_{\max}$ corresponds to the large angle collision limit. It is usually set equal to $\pi/2$, if the corresponding impact parameter is longer than the de-Broglie wavelength.

Combining Eqs. (9.10), (9.11), (9.22), and (9.26), we obtain the kinetic equation in the Fokker Planck form,

$$\frac{\partial f}{\partial t} + \frac{1}{m}\boldsymbol{p}\frac{\partial f}{\partial \boldsymbol{r}} + Ze\left(\boldsymbol{E} + \frac{\boldsymbol{p}}{m} \times \boldsymbol{B}\right)\frac{\partial f}{\partial \boldsymbol{p}}$$
$$= \sum_i \frac{\Gamma_i}{2}\frac{\partial}{\partial \boldsymbol{u}}\int d\boldsymbol{p}'\frac{1}{|\boldsymbol{u}|}\left(\overleftrightarrow{I} - \frac{\boldsymbol{uu}}{|\boldsymbol{u}|^2}\right)\left(\frac{\partial}{\partial \boldsymbol{p}} - \frac{\partial}{\partial \boldsymbol{p}'}\right)f(\boldsymbol{p})f(\boldsymbol{p}'). \tag{9.32}$$

where $\boldsymbol{u} = \boldsymbol{v} - \boldsymbol{v}'$ and $\Gamma_i = 4\pi Z^2 Z_i^2 e^4 \ln(\chi_{\max}/\chi_{\min})$. In the r.h.s. of this equation the collision integral is written in Landau form (see Landau, 1936) (Fig. 9.4).

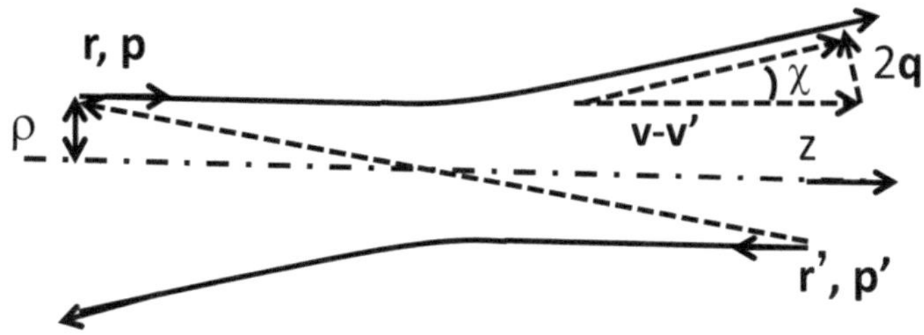

Fig. 9.4. Small angle scattering. $\Delta \boldsymbol{p}$ is perpendicular to $\vec{v} - \vec{v}'$.

9.3 Basic Equations of Plasma Fluid Dynamics

Many plasma phenomena can be described by the system of the electron-ion-hydrodynamics and the Maxwell equations. The hydrodynamics equations are derived by taking the moments of the kinetic equation [Eq. (9.10)]. Integrating the kinetic equation with respect to the variable $\boldsymbol{p}$ we obtain the continuity equation for the electron/ion density,

$$\frac{\partial n_j}{\partial t} + \nabla\left(n_j \boldsymbol{v}_j\right) = 0, \tag{9.33}$$

where $j = e, i$ denotes the species, and the velocity of the electron or ion fluid element equals $\boldsymbol{v}_j = \boldsymbol{p}_j / m_j$.

Multiplying the kinetic equation on $\boldsymbol{p}$ and integrating with respect to $\boldsymbol{p}$ we obtain the equation of motion,

$$\frac{\partial \boldsymbol{v}_j}{\partial t} + \left(\boldsymbol{v}_j \cdot \nabla\right)\boldsymbol{v}_j = \frac{q_j}{m_j}\left(\boldsymbol{E} + \boldsymbol{v}_j \times \boldsymbol{B}\right) - \frac{1}{n_j m_j}\nabla P_j \\ - \sum_k \nu_{jk}\left(\boldsymbol{v}_j - \boldsymbol{v}_k\right). \tag{9.34}$$

Here j and k refer to the electron and the ion, respectively. The charge q_j equals $-e$ and Ze for the electron and the ion, respectively. Note here that the collision term (the last term of r.h.s. of Eq. (9.34)) should be included if the plasma density is high or the temperature is low. The collisional term in Eq. (9.32) comes from the first moment of the r.h.s. of Eq. (9.32).

Since the dynamics of the electron/ion fluid components are determined by the electromagnetic field in Eq. (9.34), it is necessary to find the self-consistent field in plasmas. The electric charge density and the electric current density in plasmas depend on the densities and the velocities of the electron and the ion, respectively, as follows:

$$\rho_e = Zen_i - en_e, \tag{9.35}$$

and

$$\boldsymbol{j} = Zen_i\boldsymbol{v}_i - en_e\boldsymbol{v}_e. \tag{9.36}$$

The following Maxwell equations relate the electric field and the magnetic fields to the charge density and the current density. They are the Ampere' law

$$\frac{1}{\mu_0}\nabla \times \boldsymbol{B} = \boldsymbol{j} + \varepsilon_0\frac{\partial \boldsymbol{E}}{\partial t}, \tag{9.37}$$

Faraday's law (with μ_0 being the vacuum permeability)

$$\frac{\partial \boldsymbol{B}}{\partial t} = -\nabla \times \boldsymbol{E}, \tag{9.38}$$

Poisson equation

$$\nabla \cdot \varepsilon_0 \boldsymbol{E} = \rho_e, \tag{9.39}$$

and

$$\nabla \cdot \boldsymbol{B} = 0, \tag{9.40}$$

which shows that the magnetic field is a divergence-free vector field.

Finally, we need an equation of state (EOS) to close the above set of equations. Namely, pressure P_j is related to the temperature T_j and the density n_j. It is determined by the energy equation. When the plasma is near the thermal equilibrium, the internal energy equation can be written in the form,

$$\frac{\partial}{\partial t}\left(\frac{1}{\gamma_j - 1}P_j\right) + \nabla \cdot \left(\frac{1}{\gamma_j - 1}P_j\boldsymbol{v}_j\right) + P_j\nabla \cdot \boldsymbol{v}_j + \nabla \cdot \boldsymbol{q}_j = 0, \tag{9.41}$$

where γ_j is the adiabatic index (ratio of specific heats) and $\boldsymbol{q}_j$ is the heat flux. When the thermal conductivity is high enough in the given space and time scale, the internal energy equation (9.41) is reduced to the isothermal process due to requirement of the no-temperature gradient $\nabla T_j = 0$. On the other hand, if the thermal conductivities are small enough and the viscous term is neglected, then Eq. (9.41) yields

$$\left(\frac{\partial}{\partial t} + \boldsymbol{v}_j\nabla\right)\left(\frac{P_j}{n_j^{\gamma_j}}\right) = 0. \tag{9.42}$$

This equation is equivalent to the condition of the adiabatic regime in which the pressure of the j species is proportional to $n_j^{\gamma_j}$. In general, to determine the pressure, it is necessary to determine the heat flux and the viscosity which are determined by the higher order momentum equations of the kinetic equations.

9.4 Magnetohydrodynamics Equations

When the space and time scales are much longer than the Debye length and the electron plasma wave oscillation period defined in the next section, respectively, the plasma is assumed to be quasi-neutral. In other words, plasma is regarded as a neutral conductive fluid. Then the Eqs. (10.33) and (10.34) for electron and ion are reduced to the one component fluid equations, which are

$$\frac{\partial \rho}{\partial t} + \nabla(\rho \boldsymbol{v}) = 0, \tag{9.43}$$

$$\rho\left[\frac{\partial \boldsymbol{v}}{\partial t} + (\boldsymbol{v}\cdot\nabla)\right]\boldsymbol{v} = \boldsymbol{j}\times\boldsymbol{B} - \nabla P + \mu\nabla^2\boldsymbol{v} + \frac{\mu}{3}\nabla(\nabla\cdot\boldsymbol{v}). \tag{9.44}$$

Here, the plasma density equals $\rho = m_i n_i + m_e n_e \simeq m_i n_i$. The plasma velocity $\boldsymbol{v} = \frac{m_i \boldsymbol{v}_i + Z m_e \boldsymbol{v}_e}{m_i + m_e} \approx \boldsymbol{v}_i$ is approximately equal to the velocity of the ion component. The total pressure is a sum of the electron and ion pressure $P = P_i + P_e$. The electric current density $\boldsymbol{j} = Zen_i\boldsymbol{v}_i - en_e\boldsymbol{v}_e$ is related to the electric and magnetic field via the generalized Ohm's law:

$$\boldsymbol{j} = \sigma(\boldsymbol{E} + \boldsymbol{v}\times\boldsymbol{B}), \tag{9.45}$$

where the electron inertia and the thermo-electric effects are neglected.

Combining Eqs. (9.37), (9.38), and (9.45) and neglecting the displacement current effects in Eq. (9.37) (this implies nonrelativistic plasma motion) we obtain the induction equation for the magnetic field. It reads

$$\frac{\partial \boldsymbol{B}}{\partial t} = \nabla\times(\boldsymbol{v}\times\boldsymbol{B}) + \nabla\times(\upsilon_m\nabla\times\boldsymbol{B}), \tag{9.46}$$

where $\upsilon_m = 1/\mu_0\sigma$ is magnetic diffusivity.

The force $\boldsymbol{j}\times\boldsymbol{B}$ in the r.h.s. of Eq. (9.44) can be expanded in the form of two terms

$$\boldsymbol{j}\times\boldsymbol{B} = \frac{1}{\mu_0}(\boldsymbol{B}\cdot\nabla)\boldsymbol{B} - \nabla\left(\frac{\boldsymbol{B}^2}{2\mu_0}\right), \tag{9.47}$$

where the first is the magnetic tension force, which occurs because of the curvature of the magnetic field lines, and the second is the magnetic pressure due to the magnetic field inhomogeneity, respectively. The set of above formulated equations is called the Magneto-Hydrodynamics (MHD) equation.

From the induction equation (9.46) in the limit $\upsilon_m \to 0$ corresponding to the ideal (dissipation-less) plasma dynamics when $\boldsymbol{E} + \boldsymbol{v}\times\boldsymbol{B} = 0$, the magnetic field is described by the Eqs. (9.40) and

$$\frac{\partial \boldsymbol{B}}{\partial t} = \nabla \times (\boldsymbol{v} \times \boldsymbol{B}). \tag{9.48}$$

From these equations it follows the conservation of the magnetic flux through the surface S encircled by the contour moving together with the plasma

$$\Phi = \int_S \boldsymbol{B} \cdot d\boldsymbol{S} , \tag{9.49}$$

i.e.

$$\frac{d\Phi}{dt} = 0. \tag{9.50}$$

This is a property of the magnetic field lines to be frozen in the plasma. In other words, the magnetic field topology is preserved in the limit $\nu_m \rightarrow 0$. It is known as the Alfven theorem.

For nonvanishing magnetic diffusivity $\nu_m \neq 0$ the magnetic field is not frozen into plasma and the magnetic topology changes. In this case, one says about the reconnection of the magnetic field lines, which plays fundamental role in space and laboratory plasmas. In the laser plasma the magnetic reconnection foreseen by Alskar'yan et al (1995) has attracted high attention (e.g. see Nilson et al., 2006). The finite resistivity effects are characterized by the dimensionless parameter which is known as the Lundquist number S. It is called also the magnetic Reynolds number. The dimensionless parameter S is equal to the ratio of two characteristic time scales: the magnetic diffusion time $\tau_\sigma = l^2/\nu_m$ and a typical time $\tau_A = l/v_A$ for an Alfvén wave to propagate along the distance l; $S = \tau_\sigma/\tau_A$. Here $v_A = |\boldsymbol{B}|/\sqrt{4\pi n m_i}$ is the Alfvén wave velocity. Note here that in laser plasmas, the generalized Ohm's law should be replaced by the electron force balance equation to consider the thermoelectric effects, since temperature gradient is very large.

References

Askar'yan, G. A., Bulanov, S. V., Pegoraro, F., & Pukhov, A. M. (1995). Magnetic interaction of self-focused channels and magnetic wake excitation in high intense laser pulses. *Comments on Plasma Physics and Controlled Fusion, 17*, 35.

Landau, L. D. (1936). Kinetic equation for the case of coulomb interaction. *Physikalische Zeitschrift der Sowjetunion., 10*, 154.

Lifshitz, E. M., & Pitaevskij, L. P. (1981). *Physical kinetics.* Pergamon.

Nilson, P. M., Willingale, L., Kaluza, M. C., et al. (2006). Magnetic reconnection and plasma dynamics in two-beam laser-solid interactions. *Phys. Rev. Lett., 97*, Article 255001.

Vereshchagin, G. V., & Aksenov, A. G. (2017). *Relativistic kinetic theory with applications in astrophysics and cosmology*. Cambridge University Press.

Vlasov, A. A. (1968). The vibrational properties of an electron gas. *Sov. Phys. Uspekhi, 10*, 721.

Chapter 10
Fundamental Processes in Laser Plasmas

Abstract In this chapter, fundamental processes in laser plasmas such as field ionization, laser propagation, classical laser absorption, and so on are described. When intense laser light irradiates a gas or solid target, plasmas are produced at the focus of the laser. Namely, field ionization takes place by laser interaction with the gas or solid surface and various laser plasma processes follow. Free electrons are heated by the laser and the electron impact ionization starts to produce hot dense plasmas. Electrons heated by the laser carry the energy to evaporate the solid surface which is called "laser ablation". By evaporation of the solid, ablation plasmas called "corona plasmas" are produced. In laser interaction with the corona plasmas, various linear and nonlinear wave phenomena take place. We describe field ionization process, and plasma waves related to the laser plasma interactions. Then we describe the laser propagation and the collisional absorption in the corona plasmas. The collective absorption of laser and related wave phenomena are described in Chap. 11.

Keywords Field ionization · Tunnel ionization · Volkov state · Keldysh theory · Classical absorption · Critical density · Electron impact ionization · Laser ablation · Corona plasma · Collisional absorption

10.1 Laser Plasma Initiation—Field Ionization

At the beginning of intense laser interaction with a solid, the solid surface is evaporated and ionization of gases produced by the evaporation takes place. When the laser field is comparable to the Coulomb field acting on bound electrons of an atom, the bound electrons could obtain energy to escape from the bound state to the free state. This model is called the "tunneling ionization". As shown in Fig. 10.1a, the external electric force $F = eE$ equals to the atomic Coulomb force at the radius $r_0 = \sqrt{e^2/4\pi\varepsilon_0 F}$. The total electric field potential acting on the bound electron is described by a green curve. For a hydrogen atom, the bound electron tunnels from the turning point r_1 near the nucleus to $r_2 = R_c$ as shown in Fig. 10.1b.

Y. Kato et al., *High Power Laser and Plasma Science*, Springer Series on Atomic, Optical, and Plasma Physics 130, https://doi.org/10.1007/978-3-032-08930-4_10

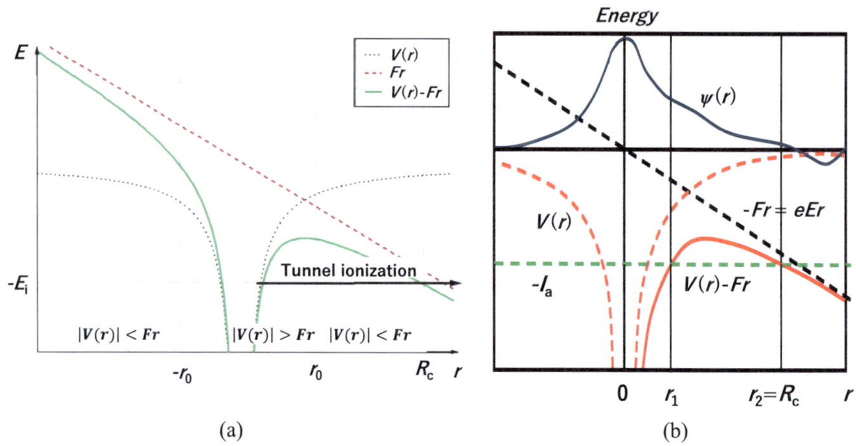

Fig. 10.1 **a** Coulomb potential of a proton (dotted black line) $V(r)$ and the external electric field potential (red broken line)$-Fr$. **b** $\psi(r)$ is the electron wave function. For $r_1 < r < r_2$, the wave function is evanescent

When the external force F is much smaller than the atomic Coulomb force at the Bohr radius $a_B = 4\pi\varepsilon_0\hbar^2/m_e e^2 \approx 5.292 \times 10^{-11}$m, we can apply the WKB approximation for the tunneling electron wave function when $r \approx r_1$ or $r > r_1$. Namely, the inner turning point r_1 and the outer turning point $r_2 = R_c$ in Fig. 10.1b can be determined by the condition

$$\frac{(\hbar k)^2}{2m_e} = Fr - I_a + \frac{c^2}{4\pi\varepsilon_0 r} = 0. \tag{10.1}$$

Here, $(\hbar k)^2/2m_e$ is the electron kinetic energy, $F = |eE|$, I_a is the hydrogen ionization potential, and the ionization energy $\hbar\omega_a$ is

$$\hbar\omega_a = I_a = \frac{m_e e^4}{(4\pi\varepsilon_0)^2\hbar^2}. \tag{10.2}$$

Note that in the region $r_1 < r < r_2$ the electron wave function decays exponentially as

$$\psi(r) \propto \exp\left(-\int_{r_1}^{r} |k(r')|dr'\right),$$

where

$$\frac{(\hbar k(r))^2}{2m_e} = Fr - I_a + \frac{e^2}{4\pi\varepsilon_0 r} < 0. \tag{10.3}$$

Here

$$\begin{pmatrix} r_1 \\ r_2 \end{pmatrix} = \frac{\hbar\omega_a}{4F}\left(1 \mp \sqrt{1 - \frac{4e^2F}{\pi\varepsilon_0(\hbar\omega_a)^2}}\right) = \frac{\hbar\omega_a}{4eE}\left(1 \mp \sqrt{1 - \frac{16E}{E_a}}\right), \tag{10.4}$$

where

$$E_a = \frac{m_e^2 e^5}{(4\pi\varepsilon_0)^3\hbar^4} \tag{10.5}$$

is the electric field of the nuclear e/r^2 at the radius equal to the Bohr radius $a_B = 4\pi\varepsilon_0\hbar^2/c^2m_e = 5.29 \times 10^{-11}$m. The atomic electric field equals $E_a = 5.14 \times 10^{11}$V/m. The ionization potential is $I_a = m_e Ze^2/2a_B = (Z/\nu)13.6$ eV. The ionization time scale is equal to $m_e e^4/\hbar^3 = 4.13 \times 10^{-16}$ s.

Then the ratio between electron probability density inside and outside of the atom, $|\psi(r_1)/\psi(r_2)|^2$, can be estimated as

$$w_{ion} \approx \exp\left(-\int_{r_1}^{r}\sqrt{\frac{2m_e cE}{r'}(r'-r_1)(r_2-r')dr'}\right). \tag{10.6}$$

If the external electric field is substantially weaker that the atomic electric field given by Eq. (10.5), $E \ll E_a$, then the coordinate of the inner turning point is small compared to the outer turning point coordinate, $r_1 \ll r_2$ and $r_2 = \hbar\omega_a/eE$. Integrating of Eq. (10.6) we find that in the weak electric field limit the ionization rate is exponentially small

$$w_{ion} \approx \exp\left(-\frac{2E_a}{3|E|}\right) \tag{10.7}$$

with the normalized electric field. The approximate expression for the ionization rate of the 1 s ground state of hydrogen atom is (e.g. see Landau & Lifshitz, 1965)

$$w_{ion} \approx 4\omega_a\frac{E_a}{|E|}\exp\left(-\frac{2E_a}{3|E|}\right). \tag{10.8}$$

The field ionization by high frequency electric field has been obtained by Keldysh (1965). When the electromagnetic field is given by the vector potential $\boldsymbol{A}(t)$, the free space wave function for the electron with the momentum $\boldsymbol{p}$ in the nonrelativistic limit satisfies the Schrödinger equation,

$$\left\{i\frac{\partial}{\partial t} + \frac{1}{2m_e}[\hbar\nabla + i\boldsymbol{A}(t)]^2\right\}\psi_{\boldsymbol{p}}(\boldsymbol{r}, t) = 0. \tag{10.9}$$

Then, the wave function of a free electron in the electromagnetic field is given by

$$\psi = \frac{1}{(2\pi\hbar)^{3/2}} \exp\left\{ \frac{i}{\hbar} \left[\boldsymbol{p}\cdot\boldsymbol{r} - \frac{m_e}{2} \int_{-\infty}^{t} \left(\boldsymbol{V}_p(t')\right)^2 dt' \right] \right\}, \tag{10.10}$$

where $\boldsymbol{V}_p(t) = (\boldsymbol{p} - e\boldsymbol{A}(t))/m_e$ (Wolkow, 1934). This is so-called the Volkov state, which is also discussed below in Chap. 20.

By using the Volkov state, L. V. Keldysh extended the field ionization theory to the case of oscillating field, which corresponds to the case of the atom ionization by the laser field (Karnakov et al., 2015; Keldysh, 1965; Popov, 2004). The ionization rate is determined by the transition rate from the bound state to the Volkov state. The ionization rate R_K according to the Keldysh theory is given by

$$R_K = \int \lim_{\tau\to\infty} \left(\frac{|M_K(\boldsymbol{p},\tau)|}{\tau} \right) d\boldsymbol{p}. \tag{10.11}$$

In this expression

$$M_K(\boldsymbol{p},\tau) = -\frac{i}{\hbar} \int_0^{\tau} \left\langle \psi_p(\boldsymbol{r},t) | V_{\text{int}}(t) |^2 \psi_0(\boldsymbol{r},t) \right\rangle dt \tag{10.12}$$

is the transition matrix from the bound state to the free state for the electron with the momentum $\boldsymbol{p}$. Here the interaction energy $V_{\text{int}}(t)$ is given by

$$V_{\text{int}}(t) = \frac{ie\hbar}{m_e} \boldsymbol{A}(t)\nabla + \frac{e^2}{2m_e} (\boldsymbol{A}(t))^2. \tag{10.13}$$

The electron wave function $\psi_0(\boldsymbol{r},t)$ is the wave function of the bound state without an external electromagnetic field. Then, the ionization rate R_K can be calculated according to Eq. (10.12). The Keldysh theory has been extended by Perelomov et al. (1966), which is called the PPT model where the electromagnetic field effects on the bound state are taken into account.

The ionization regimes are determined by the Keldysh parameter γ_K, which equals to the ratio of the laser frequency ω to the above introduced frequency ω_a [see Eq. (10.2)] of the electron tunneling through the potential barrier. The Keldysh parameter is given by an expression

$$\gamma_K = \frac{\omega}{\omega_a} = \frac{\omega\sqrt{2m_e I_a}}{eE}. \tag{10.14}$$

In the limit of small parameter when $\gamma_K \ll 1$, we have the tunneling ionization. The multiphoton ionization regime requires $\gamma_K \gg 1$. In the adiabatic limit ($\gamma_K \ll 1$) the angle distribution of the photoelectrons has a maximum along the direction of the laser electric field: $p_\parallel \approx p_\perp/\gamma_K$, where $p_\perp \approx m_e c\sqrt{E/E_a}$. In the opposite limit when $\gamma_K \gg 1$ the distribution is isotropic with $p_\parallel \approx p_\perp \approx 1/\sqrt{K_0 \ln \gamma_K}$. Here $K_0 = I_a/\hbar\omega$.

The ionization rate, depending on whether the Keldysh parameter is smaller or larger than unity, is approximately equal to

$$w(E/E_a,\omega) \approx \begin{cases} \exp\left\{-\frac{2E_a}{3E}\left[1-\frac{1}{10}\left(1-\frac{1}{3}\xi^2\right)\gamma_K^2\right]\right\}, & \gamma_K \ll 1 \\ (K_0E/E_a)^{2K_0} \approx I^{K_0}, & \gamma_K \gg 1 \end{cases} \tag{10.15}$$

In this expression ξ determines the ellipticity of the electromagnetic wave. In the circularly polarized wave $\xi = 1$ and for the linearly polarized wave $\xi = 0$ (For details, see the review by Popov (2004) and the literatures cited therein).

The ionization rate based upon the PPT model (Ammosov et al., 1986; see also Cianppina et al., 2019), called ADK model, is,

$$w_{PPT}(\nu,l,m) = |C_{\nu,l}|^2\sqrt{\frac{6}{\pi}}f_{lm}\mathcal{E}_i\mathcal{F}(t)^{-2\nu+|m|+3/2}\left(1+\gamma_K^2\right)^{|m/2|+3/4}$$
$$A_m(\omega,\gamma_K)\exp\left(-\frac{2}{3}\frac{g(\gamma_K)}{\mathcal{F}(t)}\right). \tag{10.16}$$

In Eq. (10.16), ν, l, m are the principal, angular, and magnetic quantum numbers of the bound electron, respectively, with $\nu = Z\sqrt{I_a/\mathcal{E}_i}$ for the nuclear charge Z,

$$l = \nu - 1, \tag{10.17}$$

$$\mathcal{F}(t) = \frac{|E(t)|I_a^{2/3}}{(2\mathcal{E}_i)^{2/3}E_a}, \tag{10.18}$$

$$f_{lm} = \frac{(2l+1)(l+|m|)!}{2^{|m|}|m|(l-|m|)!}, \tag{10.19}$$

$$g(\gamma_K) = \frac{3}{2\gamma_K}\left[\left(1+\frac{1}{2\gamma_K^2}\right)\mathrm{ArcSinh}(\gamma_K) - \frac{\sqrt{1+\gamma_K^2}}{2\gamma_K}\right], \tag{10.20}$$

$$\left|C_{\nu,l}\right|^2 = \frac{2^{2\nu-2}}{\nu\Gamma(\nu+l+1)\Gamma(\nu-1)}, \tag{10.21}$$

and

$$A_m(\omega,\gamma_K) = \frac{4}{\sqrt{3\pi}}\frac{1}{|m|!}\frac{\gamma_K^2}{1+\gamma_K^2}\sum_{\nu>\nu}^{\infty}\exp(-(\nu-\nu)a(\gamma_K))$$
$$w_m\left(\sqrt{\frac{2\gamma_K}{\sqrt{1+\gamma_K^2}}(\nu-\nu)}\right) \quad . \tag{10.22}$$

Here the functions and parameters are

$$w_m(x) = \exp(-x^2) \int_0^x (x^2 - y^2)^m \exp(y^2) dy, \tag{10.23}$$

$$a(\gamma_K) = 2\left[\mathrm{ArcSinh}(\gamma_K) - \frac{\gamma_K}{\sqrt{1+\gamma_K^2}} \right], \tag{10.24}$$

$$\nu = \frac{\mathcal{E}_i}{\omega}\left(1 + \frac{1}{2\gamma_K^2}\right). \tag{10.25}$$

Note that $\mathcal{F}$, $\mathcal{E}_i$, and ω are the electric field, the electron binding energy, and the laser frequency respectively, normalized by the atomic unit $2I_a/ea_B$ and I_a. $\mathcal{E}_i$ and E_a are the ionization energy and the electric field at the Bohr radius a_B of the hydrogen atom, respectively.

When the Keldysh parameter is small, $\gamma_K \ll 1$, the ionization Eq. (10.16) is reduced, in the quasistatic (QST) regime of tunneling, to

$$w_{QST}(\nu, l, m) = |C_{\nu,l}|^2 \sqrt{\frac{6}{\pi}} f_{lm} \mathcal{E}_i \mathcal{F}(t)^{-2\nu+|m|+3/2} \left(1 + \gamma_K^2\right)^{|m/2|+3/4} \exp\left(-\frac{2}{3\mathcal{F}(t)}\right). \tag{10.26}$$

This equation becomes the Keldysh formula for $\nu = 1,\ l^* = 0,\ l = m = 0,\ Z = 1$ with the ionization rate w_{QST} normalized by $\omega_a = m_e e^4/(4\pi\varepsilon_0)^2\hbar^3$ [see Eq. (10.2)].

At the relativistic laser intensity, the relativistic Volkov state [described in Chap. 20 (Eq. 20.81)] should be used in the above theory, which refers to the results obtained by Popov (2004), Mendonca and Serbeto (2011) and Cianppina et al. (2019).

Once free electrons are generated on a solid target surface, the free electrons absorb the laser and heated to a high temperature. The free electron density further increases by the electron impact ionization and the laser plasma is formed. In the following sections, the laser propagation in plasmas and the laser plasma interactions are discussed.

10.2 Laser Propagation in Plasmas and Classical Absorption

The electromagnetic wave propagates in plasmas according to the Maxwell equations in the conductive media, where the current density $\boldsymbol{j}$ in Eq. (9.36) is determined by the equation of motion of the electrons Eq. (9.34). Here, the ions are assumed to be immobile to the high frequency electromagnetic waves. Let us introduce the vector potential for the electromagnetic fields $\boldsymbol{A}$ as follows,

$$\boldsymbol{E} = -\frac{\partial \boldsymbol{A}}{\partial t}, \quad \boldsymbol{B} = \nabla \times \boldsymbol{A}, \quad \nabla \cdot \boldsymbol{A} = 0, \tag{10.27}$$

where the Coulomb gauge (transverse gauge) is applied. Then, linearizing the system of electro-hydrodynamics—Maxwell equations, we find the electromagnetic wave equation for $\boldsymbol{A}$. It reads

$$\varepsilon_0 \frac{\partial^2 \boldsymbol{A}}{\partial t^2} - \frac{1}{\mu_0} \nabla^2 \boldsymbol{A} = \boldsymbol{j}. \tag{10.28}$$

When the wave amplitude is small enough, thus the nonlinear effects on the electron motion are neglected, and there is no external magnetic field, the first velocity moment of Eq. (9.34) yields the equation of electron motion

$$\frac{\partial \boldsymbol{v}}{\partial t} = \frac{e}{m_e} \frac{\partial \boldsymbol{A}}{\partial t} - \nu_{ei} \boldsymbol{v}, \tag{10.29}$$

where ν_{ei} is the electron–ion collision frequency. In the case of negligibly small collision effects, the electric current density can be shown to be equal to

$$\boldsymbol{j} = -en_0 \, \boldsymbol{v} = -\frac{n_0 e^2}{m_e} \boldsymbol{A}. \tag{10.30}$$

Combining Eqs. (10.28) and (10.30), we obtain the electromagnetic wave equation describing the electromagnetic wave propagation in plasma. It can be written as

$$\frac{\partial^2 \boldsymbol{A}}{\partial t^2} - c^2 \nabla^2 \boldsymbol{A} = -\omega_{pe}^2 \boldsymbol{A}. \tag{10.31}$$

By setting $\boldsymbol{A} = \boldsymbol{A}_0 \exp(i(\boldsymbol{k} \cdot \boldsymbol{r} - \omega t))$, we obtain the dispersion equation for the electromagnetic wave in a uniform plasma in terms of frequency ω and wavevector $\boldsymbol{k}$. Substituting this expression to Eq. (10.31), we obtain

$$\omega^2 = k^2 c^2 + \omega_{pe}^2. \tag{10.32}$$

Here,

$$k^2 = k_x^2 + k_y^2 + k_z^2 \ \text{ and } \ \omega_{pe} = \sqrt{\frac{n_0 e^2}{\varepsilon_0 m_e}} \tag{10.33}$$

are the square of the wave vector and the electron plasma frequency, respectively. Equation (10.32) yields the refractive index of the electromagnetic wave, $N = kc/\omega$,

$$N = \sqrt{\varepsilon(\omega)} = \sqrt{1 - \frac{\omega_{pe}^2}{\omega^2}} = \sqrt{1 - \frac{n}{n_c}}, \tag{10.34}$$

where $\varepsilon(\omega)$ is the dielectric constant of the plasma and

$$n_c = \varepsilon_0 m_e \omega^2 / e^2 \tag{10.35}$$

is the cutoff density (it is also called the "critical density") for the electromagnetic wave of frequency ω. The critical plasma density is expressed as $n_c = 1.1 \times 10^{21} (1\ \mu\text{m}/\lambda)^2 \text{cm}^{-3}$. The electromagnetic wave can propagate in plasma only when the plasma density is below the critical density, $n < n_c$. The electromagnetic wave cannot propagate in plasma and becomes evanescent, when $n > n_c$, because the refractive index is imaginary in the over dense plasma.

When the electron collision with the ions and/or atoms is not small, the phase of the electron oscillations is shifted with respect to the phase for $\nu_{ei} = 0$, causing the laser energy depletion. The Fourier transform of the electron velocity becomes a complex function of the wave frequency:

$$\boldsymbol{v} = \frac{e}{m_e \omega} \boldsymbol{A} (\omega - i\nu_{ei}), \tag{10.36}$$

which leads to the change of the dispersion Eq. (10.32). It takes the form

$$\omega^2 = k^2 c^2 + \omega_{pe}^2 \left(1 - i \frac{\nu_{ei}}{\omega}\right). \tag{10.37}$$

Therefore, the dumping rate κ_L of the laser field is given by

$$\kappa_L = \text{Im}\{\omega\} = \frac{n}{2n_c} \nu_{ei}. \tag{10.38}$$

In this case, while a laser pulse of the energy density $W_L(\boldsymbol{r}, t)$ propagates in the plasma, the function $W_L(\boldsymbol{r}, t)$ obeys the equation

$$\dot{W}_L = \frac{\partial W_L}{\partial t} + \nabla (\boldsymbol{v}_g \cdot W_L) = -2\kappa_L W_L. \tag{10.39}$$

Here $\boldsymbol{v}_g(\boldsymbol{r}, t) = \partial \omega / \partial \boldsymbol{k}$ is the group velocity of the electromagnetic wave.

In plasma of inhomogeneous density which increases above the critical density, the laser pulse is reflected in the high plasma density region as illustrated in Fig. 10.2. Here the plasma density is assumed to be homogeneous in the y direction and has the gradient in x as $n(x) = n_c x / L$. The cutoff coordinate where the plasma density is equal to the critical density is at $x = x_c = L$. However, the electromagnetic wave obliquely incident on the plasma slab is reflected at the turning point x_t where the density is lower than the cutoff density as explained below.

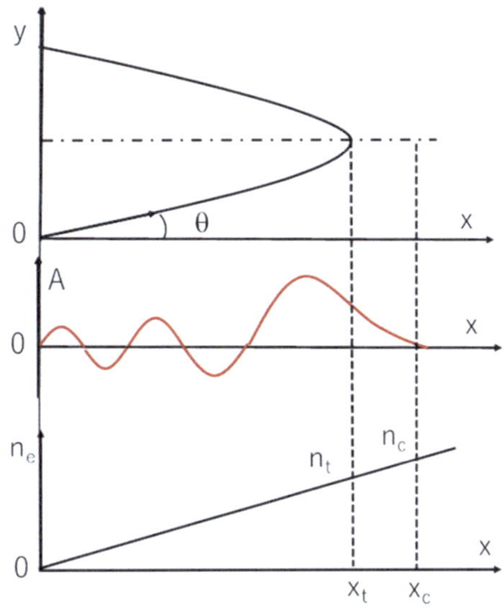

Fig. 10.2 Density profile (bottom), the laser ray trace (top), and the amplitude (middle). The laser is evanescent at $x > x_t$ and the wavelength in the x direction becomes longer in the higher density region of $x > x_t$ (middle)

Let us consider the electromagnetic wave with the electric field along the z-direction (it is so-called the s-polarized electromagnetic wave) incident on the plasma whose density profile is given by $n(x) = n_c x/L$. In this case, the plasma frequency depends on the x coordinate as $\omega_{pe}(x) = \omega\sqrt{n_c x/L}$. The propagating electromagnetic wave is given by

$$\boldsymbol{A} = A_z(x)\boldsymbol{e}_z \exp(-i\omega t + ik_y y),$$

where the y-component of the wave vector is $k_y = k \sin\theta = (\omega/c)\sin\theta$. Substituting this expression to the wave equation (10.31) we obtain

$$\frac{d^2 A_z}{dx^2} + \frac{\omega^2}{c^2}(\varepsilon - \sin^2\theta)A_z = 0. \tag{10.40}$$

As we see, the electromagnetic wave is reflected at the point where $\varepsilon = \sin^2\theta$. For the plasma density of a linear profile, the reflection coordinate is $x_t = L\cos^2\theta$.

In the framework of geometric optics approximation, which is valid when the electromagnetic wave wavelength is substantially smaller than the plasma inhomogeneity scale length, $kL \gg 1$ (Landau & Lifshitz 1980), the coordinates and the components of the wave vector of laser wave packet satisfy the system of equations,

$$\dot{x} = k_x c^2/\omega, \quad \dot{y} = k_y c^2/\omega, \quad \dot{k}_x = -\partial\omega/\partial x, \quad \dot{k}_y = -\partial\omega/\partial y \tag{10.41}$$

The group velocity $\boldsymbol{v}_g(\boldsymbol{r}, t) = \partial\omega/\partial\boldsymbol{k} = \boldsymbol{k}c^2/\omega$ for dispersion Eq. (10.32) is given by $\boldsymbol{v}_g = \boldsymbol{k}c^2/\sqrt{k^2c^2 + \omega_{pe}^2(\boldsymbol{r})}$. Initial conditions at $t = 0$ for Eq. (10.41) are $x = 0$,

$y = y_0, k_x = k\cos\theta$, and $k_y = k\sin\theta$. Since the plasma parameters are stationery and the plasma is homogeneous in the y-direction, the electromagnetic wave frequency ω and the y-component of the wave vector $k_y = k\sin\theta = (\omega/c)\sin\theta$ are constant. From Eq. (10.41) it is easy to obtain the equation of the wave packet trajectory in the (x, y) plane:

$$\frac{dy}{dx} = \frac{\sin\theta}{\sqrt{\cos^2\theta - x/L}}. \tag{10.42}$$

The solution to this equation,

$$y = \pm 2L\sin\theta\sqrt{\cos^2\theta - x/L}, \tag{10.43}$$

shows that the electromagnetic pulse is reflected at the coordinate $x_t = L\cos^2\theta$.

During propagation of the laser in plasmas, its energy is depleted via absorption by electron–ion collision and/or via collective phenomena. The absorption through the electron–ion collision is called the "classical absorption" or "inverse bremsstrahlung". For stationary injections of the laser, the integration of Eq. (10.39) gives for the laser energy flux in the x direction I_x as

$$I_x(x) = I_x(0)\exp(-\Lambda), \tag{10.44}$$

with

$$\Lambda = 2\int_0^{x_t} \frac{2\kappa_L}{v_{gx}} dx = \frac{32}{15}\frac{\nu_{ec}L}{c}\cos^5\theta. \tag{10.45}$$

Here ν_{ec} is the electron–ion collision frequency at the cutoff density, which is approximately proportional to the cutoff density n_c, the ion charge Z, and the plasma temperature in the $-3/2$ power $T^{-3/2}$ (Trubunikov, 1965). According to Eq. (10.44), the classical absorption efficiency η_{ab} is (Kruer, 1988)

$$\eta_{ab} = 1 - \exp(-\Lambda) \tag{10.46}$$

From Eq. (10.44) it is seen that the classical absorption efficiency η_{ab} is higher for the shorter wavelength lasers and lower temperature plasmas. The classical absorption is dominant when the laser wavelength is short like the third harmonic of the Nd glass lasers, which are currently used for implosion experiments in the studies of the laser thermonuclear fusion and when the laser intensity is not very high. For example, with Eq. (10.46), the value of Λ is greater than unity for plastic targets when the scale length L is longer than 10 μm and the plasma electron temperature T_e is lower than a few keV for 0.35 μm laser wavelength. This corresponds to the laser intensity less than 10^{15} W/cm^2, since the required balance of the energy flux

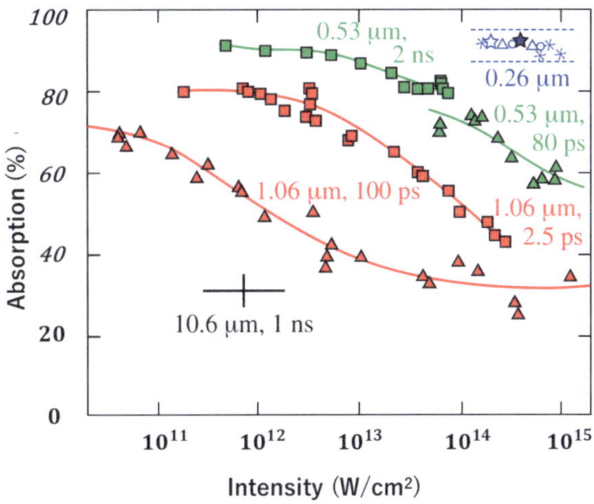

Fig. 10.3 Intensity dependence of the absorption rate for the laser wavelengths of 10.6 μm, 1.06 μm, 0.53 μm and 0.26 μm with various pulse widths. Reprinted from Garban-Labaune et al. (1982), with permission from The American Physical Society and the author(s). © 1982 The American Physical Society

of about $4T_e n_e c_S$ as discussed later with the absorbed laser power of 10^{15} W/cm^2 predicts the temperature of a few keV.

Dependence of the absorption on the laser intensity has been explored in many experiments. In Fig. 10.3 we present the summary of the experimental results, showing the absorption rate for the laser radiation of various wavelengths, ranging from 10.6 μm to 0.25 μm, over broad intensities (Garban-Labaune et al., 1982). The absorption efficiency is lower for longer wavelength lasers and at higher plasma temperatures, because

$$\eta_{ab} \sim \frac{n_c}{T_e^{3/2}} \sim \frac{n_c}{\lambda^2 T_e^{3/2}}.$$

Since the plasma temperature increases with the laser intensity, the classical absorption decreases as shown in Fig. 10.3. For the green laser of $\lambda = 0.53$ μm, the absorption efficiency starts to decrease at the laser intensity I_L higher than 10^{14} W/cm^2.

In summary, the laser classical absorption is weak in general at high intensities and long wavelengths. However, the absorption does not decrease above 10^{14} W/cm^2 for the 1.06 μm laser as shown in Fig. 10.3. This is an indication of the collective absorption, which is described in the next Chapter.

References

Ammosov, M. V., Delone, N. B., & Krainov, V. P. (1986). Tunneling ionization of complex atoms and atomic ions in an alternating electro-magnetic field. *Soviet Physics JETP, 64*, 1191.

Ciappina, M. F., Popruzhenko, S. V., Bulanov, S. V., Ditmire, et al. (2019). Progress toward atomic diagnostics of ultrahigh laser intensities. *Physical Review A, 99*, 043405.

Garban-Labaune, C., Fabre, E., Max, C. E., et al. (1982). Effect of laser wavelength and pulse duration on laser-light absorption and back reflection. *Physical Review Letters, 48*, 1018.

Karnakov, B. M., Mur, V. D., Popruzhenko, S. V., & Popov, V. S. (2015). Current progress in developing the nonlinear ionization theory of atoms and ions. *Physics Uspekhi, 58*, 3.

Keldysh, L. V. (1965). Ionization in the field of a strong electromagnetic wave. *Soviet Physics JETP, 20*, 2354.

Kruer, W. L. (1988). The physics of laser plasma interactions. In *Frontiers of physics series* (Vol. 73). Addison Wesley Publication Company.

Landau, L. D., & Lifshitz, E. M. (1965). *Quantum mechanics* (2nd ed., p. 275). Pergamon.

Landau, L. D., & Lifshitz, E. M. (1980). *The classical theory of fields*. Pergamon.

Mendonca, J. T., & Serbeto, A. (2011). Volkov solutions for relativistic quantum plasmas. *Physical Review E, 83*, 0264.

Perelemov, A. M., Popov, V. S., & Terent'ev, M. V. (1966). Ionization of atoms in an alternating electric field. *Soviet Physics JETP, 23*, 924.

Popov, V. S. (2004). Tunnel and multiphoton ionization of atoms and ions in a strong laser field (Keldysh theory). *Physics Uspekhi, 47*, 855.

Trubunikov, B. A. (1965). Particle interactions in a fully ionized plasmas. *Reviews of Modern Plasma Physics, 1*, 105.

Wolkow, D. M. (1934). Über eine Klasse von Lösungen der Diracschen Gleichung. *Zeitschrift fur Physik, 94*, 250.

Chapter 11
Plasma Waves and Collective Laser Absorption

Abstract This chapter describes theory, simulation, and experiments on propagations of the electron plasma wave and the ion acoustic wave, the laser absorption by resonance mode conversion, and the Brunel absorption on a sharp plasma surface.

Keywords Electron plasma wave · Parametric instability · Phonon-plasmon decay · Two plasmon decay · Landau damping · Ion acoustic wave · Collective laser-plasma interaction · Resonance absorption · Hot electrons · Brunel absorption

11.1 Plasma Waves

This section describes an electrostatic electron wave and an ion wave which are excited in laser plasma interactions. When an intense laser interacts with unmagnetized plasmas, electrostatic waves and scattered electromagnetic waves are generated through linear mode conversion and/or nonlinear mode coupling between the laser and the plasma waves. For example, in the parametric decay described in Chap. 12, the electromagnetic wave (laser) decays into a high frequency plasma wave and a low frequency plasma wave. The high frequency wave is the electron plasma wave, and the low frequency wave is the ion wave. The linear dispersion relations of the electron plasma wave and the ion acoustic wave are derived in this Chapter.

11.1.1 Electron Plasma Wave

Electron plasma wave, called Langmuir wave, is a longitudinal high frequency wave in unmagnetized plasmas. It is assumed that the ions are immobile since their mass is much heavier than that of the electrons. We consider the case, within the framework of 1D approximation, when the electron distribution function $f(x, p, t)$ and the electric field $E(x, t)$ are perturbed around the unperturbed distribution function $f_0(p)$ as

Y. Kato et al., *High Power Laser and Plasma Science*, Springer Series on Atomic, Optical, and Plasma Physics 130, https://doi.org/10.1007/978-3-032-08930-4_11

$$f(x, p, t) = f_0(p) + \delta f(x, p, t) \text{ and } E(x, t) = E_0 + \delta E(x, t) \tag{11.1}$$

with $\delta f(x, p, t) << f_0(p)$ and $E_0 = 0$. Assuming that the plasma is uniform and stationary, the electron-plasma-wave propagates along the x-direction, and the electron momentum distribution function in the y and z directions is not modulated. Therefore, p in Eq. (11.1) is the x-component of the electron momentum. The electric field in x and the electrostatic potential are related as,

$$E(x, t) = -\frac{\partial \Phi}{\partial x} \tag{11.2}$$

When the collisional effects are neglected, the linearized Vlasov equation (9.12) is written as

$$\frac{\partial \delta f(x, p, t)}{\partial t} + v\frac{\partial \delta f(x, p, t)}{\partial x} + e\frac{\partial \delta \phi}{\partial x}\frac{\partial f_0(p)}{\partial p} = 0 \tag{11.3}$$

where $v = p/m_e$ and the electric field perturbation is given by $\delta E = -\partial \delta \phi/\partial x$. The Laplace–Fourier transformation with respect to time and space coordinates of Eq. (11.3) yields

$$-i(\omega - kv)\delta f_{\omega k}(p) + ike\phi_{\omega k}\frac{\partial f_0(p)}{\partial p} = \delta f_k(p, 0). \tag{11.4}$$

Here $\delta f_k(p, 0)$ is the Fourier transform of the initial perturbation of the distribution function, and

$$\delta f_{\omega k}(p) = \int_0^{\infty} dt \int_{-\infty}^{+\infty} dx \delta f(x, p, t) \exp(i\omega t - ikx). \tag{11.5}$$

The inverse Laplace–Fourier transform is

$$\delta f(x, p, t) = \frac{1}{2\pi} \int_{-\infty+i\delta}^{+\infty+i\delta} d\omega \int_{-\infty}^{+\infty} dk f_{\omega k}(p, t) \exp(-i\omega t + ikx). \tag{11.6}$$

Note here that the integration contour for the inverse Laplace transformation lies in the upper half of the complex ω plane due to causality.

The electron density perturbation $\delta n_{\omega k}$ obtained from Eq. (11.3) is

$$\delta n_{\omega k}(p) = \int_{-\infty}^{+\infty} dp \frac{ek\phi_{\omega\kappa}}{\omega - kv}\frac{\partial f_0(p)}{\partial p} + S(\omega, k). \tag{11.7}$$

Here

$$S(\omega, k) = i \int_{-\infty}^{+\infty} dp \frac{\delta f_k(p, 0)}{\omega - kv} \tag{11.8}$$

is the initial value of the electron density perturbation of the wave number k. Considering the Poisson equation (9.39), we find the relationship between the electron density and the electrostatic potential perturbations:

$$e\delta n_{\omega k} = -\varepsilon_0 k^2 \delta\phi_{\omega k}. \tag{11.9}$$

Let us normalize the unperturbed distribution function by the background plasma density n_0 as

$$\overline{f}_0(p) = \frac{f_0(p)}{n_0} \quad \text{which leads to} \quad \int_{-\infty}^{+\infty} dp \overline{f}_0(p) = 1 \tag{11.10}$$

Equations (11.7)–(11.10) yield the temporal evolution of the perturbations of the electrostatic potential $\delta\phi_{\omega k}$ related to the initial perturbation $S(\omega, k)$,

$$\delta\phi_k(t) = -\frac{1}{2\pi i} \int_{-\infty+i\delta}^{+\infty+i\delta} d\omega \frac{eS_{\omega,k}}{\varepsilon_0 k^2 \varepsilon(\omega, k)} \exp(i\omega t). \tag{11.11}$$

Here $\varepsilon(\omega, k)$ is the longitudinal dielectric function normalized by the vacuum dielectric constant, given as

$$\varepsilon(\omega, k) = 1 - \frac{\omega_{pe}^2}{k^2} \int_{-\infty}^{+\infty} dp \frac{m_e}{v - \omega/k} \frac{\partial \overline{f}_0(p)}{\partial p}. \tag{11.12}$$

In taking the inverse Laplace transform of Eq. (11.11), the integration path should be in the upper half of the complex ω-plane due to the causality condition (which comes from the requirement for vanishing the perturbations at $t < 0$) so that the singularities of the integrand are below the integral path, namely assumption of no perturbation. The singularities come from the functions $S(\omega, k)$ and $\varepsilon(\omega, k)$, e.g., the poles where $\varepsilon(\omega, k) = 0$. For example, when a cold beam is included in the initial distribution $f_k(p, 0)$ corresponding to the mono-energetic electron beam with the velocity $v_b = p_b/m_e$, $S(\omega, k)$ has the singularity at $\omega = kv_b$, which is called a beam mode or the Van Kampen (1955) mode.

On the other hand, the singularity $\varepsilon(\omega, k) = 0$ gives the dispersion equation for the relationship between the frequency and wavenumber of the electron plasma wave, which is a longitudinal electron mode. When plasma is non-relativistic and

Maxwellian and the second term in the r.h.s. of Eq. (11.12) is modified accordingly, the dispersion equation takes the form

$$1 - \frac{\omega_{pe}^2}{k^2} \int_{-\infty}^{+\infty} dv \frac{m_e}{v - \omega/k} \frac{\partial \bar{f}_0(v)}{\partial v} = 0, \tag{11.13}$$

where there is a pole at $v = \omega/k = v_{ph}$. Landau (1946) (see also Dawson, 1961) discovered that the integral near the pole corresponds to the wave particle interaction resulting in the wave dumping. This is the so-called Landau damping. Since the integrand of the above equation is assumed to be analytic near the real axis of the complex v-plane, except at $v = v_{ph}$, we could deform the integration path as shown in Fig. 11.1. Namely, the integration path is separated into the small half circle around the pole and the straight part which is called the principal part of integration. Then, Eq. (11.13) can be rewritten as follows,

$$\varepsilon(\omega, k) = 1 - \frac{\omega_{pe}^2}{k^2} \text{p.v.} \left[\int_{-\infty}^{+\infty} dv \frac{m_e}{v - \omega/k} \frac{\partial \bar{f}_0(v)}{\partial v} \right] - i\pi \frac{\omega_{pe}^2}{k^2} \frac{\partial \bar{f}_0(v)}{\partial v} \bigg|_{v=v_{ph}} . \tag{11.14}$$

Here and below "p.v." means the Cauchy principal value of the integral. Note here that the last term of the r.h.s. of Eq. (11.14) is the imaginary part of the dielectric function (permittivity) which is related to the wave dumping. Since it is related to the electrons of the velocity near the phase velocity, those electrons exchange energy with the electron plasma wave.

When the phase velocity is significantly higher than the thermal velocity, the second term of the r.h.s. of Eq. (11.14) is approximately equal to $-(\omega_{pe}/\omega)^2 (1 + 3k^2 v_{Th}^2/\omega^2)$ and the third relatively small term is $i\pi (\omega_{pe}/k)^2 \partial \bar{f}_0(v)/\partial v \big|_{v=v_{ph}}$. Here $v_{Th} = \sqrt{T_e/m_e}$ is the electron thermal velocity for

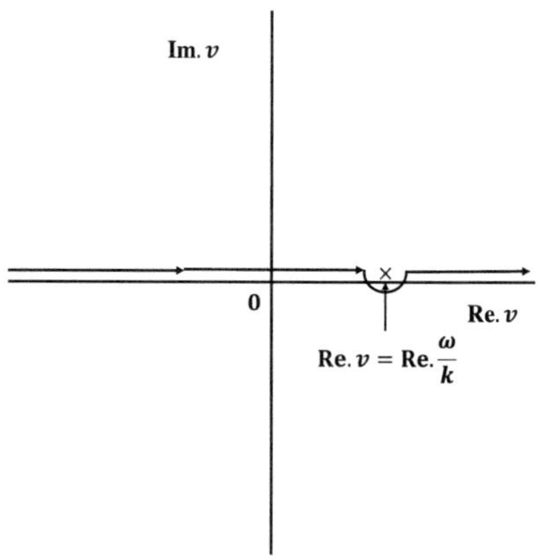

Fig. 11.1 Integrating path for the velocity integral in the dielectric function

the electron temperature T_e in the energy unit. Then the dispersion equation relates the frequency to the wave number approximately as follows,

$$\omega_e(k) = \omega_{pe}\left[\left(1 + \frac{3k^2}{2k_{De}^2}\right) + i\frac{\pi}{2}v_{ph}^2 \left.\frac{\partial \overline{f}_0(v)}{\partial v}\right|_{v=v_{ph}}\right], \tag{11.15}$$

where $k_{De} = \omega_{pe}/v_{e,th}$ is the Debye wave number, which is the inverse to the electron Debye length $\lambda_{De} = \sqrt{\varepsilon_0 T_e/ne^2}$. The imaginary part of the frequency is called the Landau damping decrement when $\partial\overline{f}_0(v)/\partial v\Big|_{v=v_{ph}} < 0$, or the "bump on tail" instability growth rate when $\partial\overline{f}_0(v)/\partial v\Big|_{v=v_{ph}} > 0$.

When plasma is in thermal equilibrium, the Landau damping decrement γ_L is equal to (Chen, 1984),

$$\gamma_L = -\omega_{pe}\sqrt{\frac{\pi}{8}}\left(\frac{v_{ph}}{v_{e,th}}\right)^3 \exp\left(-\frac{1}{2}\frac{v_{ph}^2}{v_{e,th}^2}\right), \tag{11.16}$$

Dawson (1961) re-derived the Landau damping decrement by calculating the energy transfer from the electron plasma wave to the electron kinetic energy. Equation (11.16) indicates that the damping rate is much smaller than the electron plasma frequency when the phase velocity of the electron plasma wave is 3 times higher than the thermal velocity; namely, the wave number is smaller than 1/3 of the Debye wave number. Dependence of the spatial damping rate on the temperature and the phase velocity, i.e., γ_L/v_g where $v_g = \partial\omega_e/\partial k$ of the electron plasma wave, was measured by Malmberg and Wharton (1966). They have shown that the experimental results for the dependence of damping on the ratio of the phase velocity to the thermal velocity agree well with the theoretical prediction.

In the limit of high phase velocity, the following wave equation of the density fluctuation for the electron plasma wave is derived from the electron fluid equation with the Poisson equation as follows,

$$\frac{\partial^2 \tilde{n}}{\partial t^2} + 2\gamma_L\frac{\partial \tilde{n}}{\partial t} - 3v_{Th}^2\frac{\partial^2 \tilde{n}}{\partial x^2} + \omega_{pe}^2\tilde{n} = S, \tag{11.17}$$

where S is the external source term for exciting the plasma wave. The Landau damping term is implemented phenomenologically. As it will be described in the following section, the external source term can include the nonlinear laser plasma effects.

11.1.2 Ion Acoustic Wave

We have assumed above that the ions are at rest in the electron plasma wave. As far as the ion wave is concerned, its frequency is much lower than the frequency of the electron plasma wave. As a result, electron inertia can be neglected and electron density can be approximated by the Boltzmann distribution assuming that the electron component of the plasma is in thermal equilibrium, i.e. the electron density is described by the expression

$$n_e = n_{e0} \exp\left(\frac{e\phi}{T_e}\right) \underset{e\phi << T_e}{\approx} n_{e0}\left(1 + \frac{e\phi}{T_e}\right), \tag{11.18}$$

if we assume $|e\phi/T_e| << 1$.

The fluctuation of the ion velocity distribution function is derived by the same way for the electron in the electron plasma wave [see Eq. (11.4)]. As a result, we obtain

$$-i(\omega - kv)\delta f_{i,\omega k}(v) + i\frac{kZe\phi_{\omega k}}{m_i}\frac{\partial f_{i,0}(v)}{\partial v} = \delta f_{i,k}(p, 0). \tag{11.19}$$

where Ze is the ion charge and m_i is the ion mass. Hence the ion density perturbation is related to the perturbation of the electrostatic potential as

$$n_i(k, \omega) = -\int_{-\infty}^{+\infty} \frac{dv}{m_i}\frac{Zek\phi(k, \omega)}{(\omega - kv)}\frac{\partial f_{i,0}(v)}{\partial v} + S_i(k, \omega), \tag{11.20}$$

where

$$S_i(k, \omega) = \int_{-\infty}^{+\infty} \frac{dv}{m_i}\frac{f_{i,k}(v, 0)}{(\omega - kv)}\frac{\partial f_{i,0}(v)}{\partial v}.$$

Equations (11.18), (11.19) and (11.20) with the Poisson equation provide,

$$\varepsilon_0 k^2 \phi_{\omega k} = -e^2 n_{e0}\frac{\phi_{\omega k}}{T_e} - i\frac{(Ze)^2 n_{i0}}{m_i}\int_{-\infty}^{+\infty} dv \frac{\phi_{\omega k}}{\omega - kv}\frac{\partial f_{i,0}(v)}{\partial v} + ZeS_{i,\omega k}. \tag{11.21}$$

Therefore, the dielectric function of low frequency fluctuation is defined by

$$\varepsilon(k, \omega) = 1 + \frac{k_D^2}{k^2} - \frac{\omega_{pi}^2}{k^2}\int_{-\infty}^{+\infty} dv \frac{1}{v - \omega/k}\frac{\partial \bar{f}_{i,0}(v)}{\partial v}. \tag{11.22}$$

This yields an expression for electrostatic potential as a function of time,

$$\phi(t) = \frac{1}{2\pi\varepsilon_0 i} \int_{-\infty+i\delta}^{+\infty+i\delta} d\omega \frac{ZeS_{i,k\omega}}{k^2\varepsilon(k,\omega)} \exp(-i\omega). \tag{11.23}$$

Here,

$$\omega_{pi} = \sqrt{\frac{Z^2e^2n_{i0}}{\varepsilon_0 m_i}}$$

is the ion plasma frequency. The dispersion equation obtained from a pole of the integrand of Eq. (11.23), $\varepsilon(k,\omega)$, has a form

$$\omega^2 = \frac{k^2c_S^2}{1+k^2/k_{De}^2} \int_{-\infty}^{+\infty} dv \frac{\omega^2}{(\omega-kv)^2} \frac{\partial \bar{f}_{i,0}(v)}{\partial v}. \tag{11.24}$$

This expression can be rewritten as

$$\omega^2 = \frac{k^2c_S^2}{1+k^2/k_{De}^2}\left[P\int_{-\infty}^{+\infty} dv \frac{\omega^2}{(\omega-kv)^2} \frac{\partial \bar{f}_{i,0}(v)}{\partial v} + i\pi\frac{\omega^2}{k^2}\left.\frac{\partial \bar{f}_{i,0}(v)}{\partial v}\right|_{v=v_{ph}}\right]. \tag{11.25}$$

Here,

$$c_S = \sqrt{\frac{ZT_e}{m_i}}$$

is the sound velocity, which is approximately equal to the phase velocity of the long wavelength ion acoustic wave when the ion temperature is lower than the electron temperature. Then, the ratio of the phase velocity to the ion thermal velocity, $c_S/v_{i,th} = \sqrt{ZT_e/T_i}$ can be large for high Z and/or high electron temperature. In this case, the ion acoustic dispersion equation is reduced to

$$\omega = kc_S\sqrt{\frac{1+(3T_i/ZT_e)\left(1+k^2/k_{De}^2\right)}{1+k^2/k_{De}^2}}$$
$$\left[1+i\frac{\pi}{2}\frac{v_{ph}^2}{1+(3T_i/ZT_e)\left(1+k^2/k_{De}^2\right)}\left.\frac{\partial \bar{f}_{i,0}(v)}{\partial v}\right|_{v=v_{ph}}\right]. \tag{11.26}$$

The Landau damping rate of the ion acoustic wave for the Maxwellian plasma is given by the expression

$$\mathrm{Im}\{\omega\} = -\omega_{ac}\sqrt{\frac{\pi}{8}}\left(\frac{v_{ph}}{v_{i,th}}\right)^3 \exp\left(-\frac{v_{ph}^2}{2v_{i,th}^2}\right),$$

where

$$\omega_{ac} = kc_S\sqrt{\frac{1+(3T_i/ZT_e)\left(1+k^2/k_{De}^2\right)}{1+k^2/k_{De}^2}} \tag{11.27}$$

is a real part of the frequency, and

$$\frac{v_{ph}}{v_{i,th}} = \frac{\omega_{ac}}{kv_{i,th}} = \sqrt{\frac{ZT_e\left[1+(3T_i/ZT_e)\left(1+k^2/k_{De}^2\right)\right]}{T_i\left(1+k^2/k_{De}^2\right)}}. \tag{11.28}$$

Therefore, the ion acoustic wave is a weakly damped mode for long wavelength modes whose wave number is small, $k < k_{De}$ with $ZT/T_e > 5$. In this case, the ratio of the damping rate to the wave frequency is approximately equal to $|\mathrm{Im}\{\omega\}|/\omega_{ac} \leq 0.2$.

11.1.3 Summary

As we have discussed above, there are two electrostatic modes in un-magnetized plasmas. They are the electron plasma wave of high frequency and the ion acoustic wave of low frequency. The electron plasma wave is an electric charge separation wave, in which electrons and ions oscillate in opposite phase with a negligible impact to the ion motion. The frequency of the electron wave is finite in the long wavelength limit as shown by a solid line in Fig. 9.10. The decrement due to Landau damping is small provided the wavelength is large enough compared with the Debye length.

On the other hand, electrons and ions oscillate in phase for the ion acoustic wave which is the mass density wave driven by the plasma pressure. In the long wavelength limit, the frequency of the ion acoustic wave is proportional to the wave number as shown by a broken line in Fig. 11.2. The damping of the ion acoustic wave is small only in high Z plasmas with high enough electron to ion temperature ratio.

11.2 Collective Laser Plasma Interactions

When the effect of the collision is negligibly weak, the energy of the incident laser radiation decays or scatters through exciting electrostatic waves, which are the electron plasma wave and the ion acoustic wave. The laser-wave interaction processes are described schematically in Fig. 11.3.

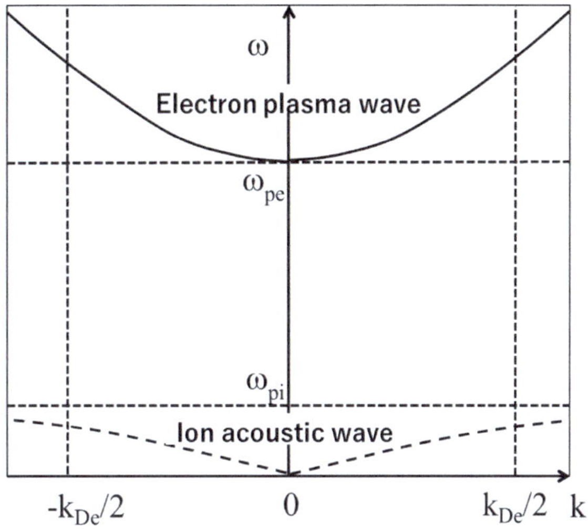

Fig. 11.2 Dispersion relations of electron plasma wave and ion acoustic wave in the (k, ω) plane

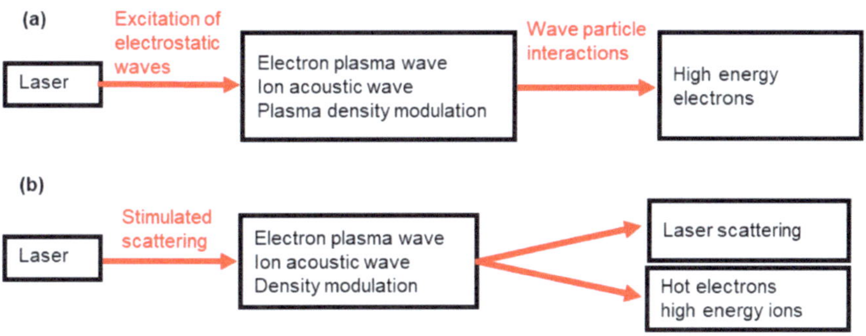

Fig. 11.3 **a** Collective absorption; **b** stimulated scattering processes

In the collective absorption of laser radiation interacting with plasmas shown in Fig. 11.3a, important absorption mechanisms are resonance absorption and the Brunel or "vacuum heating" absorption. Through these processes, the laser energy is mainly converted into high energy electrons. For development of the parametric instabilities (Fig. 11.3a) and the nonlinear laser scattering (Fig. 11.3b), frequencies and wave numbers of the involved waves should satisfy the conditions as follows.

Frequency matching: $\omega_L = \omega_{e1} + \omega_{e2}$ or $\omega_L = \omega_e + \omega_S$;

Wave vector matching: $\boldsymbol{k}_L = \boldsymbol{k}_{e1} + \boldsymbol{k}_{e2}$ or $\boldsymbol{k}_L = \boldsymbol{k}_e + \boldsymbol{k}_S$.

Here, ω_L, ω_e and ω_S are the frequencies of the incident laser, the electron plasma wave and the scattered electromagnetic wave, respectively, and $\boldsymbol{k}_L$, $\boldsymbol{k}_e$ and $\boldsymbol{k}_S$ are the corresponding wave vectors. Note that ω_{e1}, ω_{e2} are the frequencies of two electron plasma waves for the two plasmon decay and ω_e, ω_S are the frequences of the electron plasma wave and the ion acoustic wave.

11.3 Resonance Absorption

The resonance absorption of the electromagnetic wave interacting with an inhomogeneous density plasma is one of the fundamental processes in the plasma electrodynamics (Denisov, 1957; Eastabrook et al., 1975; Forslund et al., 1975; Ginzburg, 1964; Uenoyama et al., 1981; Takabe & Mulser, 1982).

We consider that a p-polarized electromagnetic wave of frequency ω is incident into a non-uniform plasma as shown in Figs. 10.2 and 11.4. Near the critical density surface, where the plasma density can be approximated locally by the function $n(x) = n_c x/L$, the electric field $\boldsymbol{E}$ of the wave is related to the z-component of magnetic field $\boldsymbol{B}$ by Ampere's law,

$$\nabla \times (\boldsymbol{e}_z B_z) = -i\frac{\omega}{c^2}\varepsilon(\omega)\boldsymbol{E}, \tag{11.29}$$

where $\boldsymbol{e}_z$ is a unit vector in the z-direction and

$$\varepsilon(\omega) = 1 + \frac{\omega_{pe}^2}{\omega(\omega + i\nu)} \tag{11.30}$$

is the dielectric function of plasma with a finite dissipation ν, which is the electron collision frequency or the effective damping rate of the electron plasma wave. The p-polarization means that the laser electric field is in the incidence plane, (x, y), and the magnetic field is parallel to the z-direction. The longitudinal component of electric field is given by

$$E_x = c\frac{\sin\theta}{\varepsilon(\omega)} B_z|_{x=0} = \frac{E_d}{\varepsilon(\omega)} \tag{11.31}$$

with the driver field is defined as $E_d = c\sin\theta\, B_z|_{x=0}$. Note here that $B_z|_{x=0}$ is the laser magnetic field tunneled from the turning point to the point where the plasma

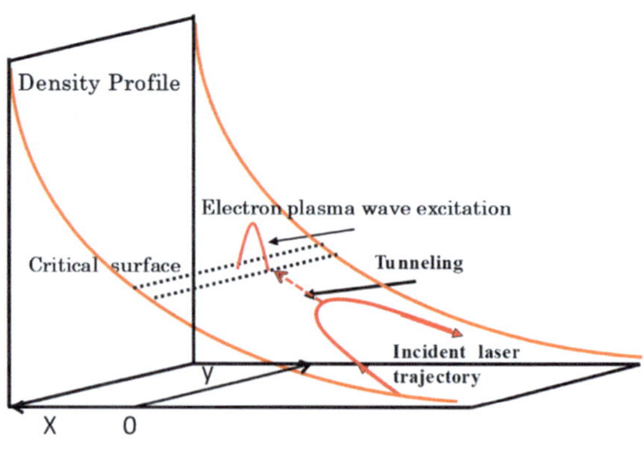

Fig. 11.4 Geometry of resonance absorption in a laser produced corona plasma. Plasma density profile, laser propagation and excited longitudinal field profile are shown

density is equal to the critical density n_c. The dielectric function of the plasma $\varepsilon(\omega)$ becomes zero near the critical density, resulting in the resonantly intensified longitudinal component of the electric field.

The z-component of laser magnetic field B_z satisfies the following wave equation

$$-\omega^2 B_z - c^2 \nabla^2 B_z + \omega_{pe}^2 B_z = 0. \tag{11.32}$$

When the plasma is inhomogeneous as shown in Fig. 11.4 and the laser incident angle is θ, then Eq. (11.32) is written as

$$\frac{d^2 B_z}{dx^2} - k_0^2 \left(\frac{x}{L} + \sin^2 \theta \right) B_z = 0, \tag{11.33}$$

where $k_0 = \omega/c$. The plasma density profile is assumed to depend on x as $n = n_c(1 + x/L)$. The coordinate of the turning point of the electromagnetic wave is $x_t = L \sin^2 \theta$. The solution of Eq. (11.33) expressed in terms of the Airy function is

$$B_z = B_z(0) \mathrm{Ai}\left[-(k_0 L)^{2/3} \left(\frac{x - x_t}{L} \right) \right]. \tag{11.34}$$

The Airy function is defined as

$$\mathrm{Ai}(\xi) = \frac{1}{\pi} \int_0^\infty \exp\left(\xi q + \frac{q^3}{3} \right) dq. \tag{11.35}$$

Assuming that the electromagnetic wave amplitude is $B_{z0} = B_z(-L)$ at the plasma boundary of $x = -L$, we find that

$$B_z = B_{z0} \frac{\mathrm{Ai}\left[-(k_0 L)^{2/3} \left(\frac{x - x_t}{L} \right) \right]}{\mathrm{Ai}\left[-(k_0 L)^{2/3} \cos^2 \theta \right]}. \tag{11.36}$$

Using the asymptotic expansion of the Airy function in the limit $k_0 L >> 1$, we obtain from Eqs. (11.34) and (11.36) the amplitude of the laser field at the critical density surface ($x = 0$) as (Ginzburg, 1964),

$$\begin{aligned} B_z\left((k_0 L)^{2/3} \sin^2 \theta \right) &= B_{z0} \frac{\mathrm{Ai}\left[(k_0 L)^{2/3} \sin^2 \theta \right]}{\mathrm{Ai}\left[-(k_0 L)^{2/3} \cos^2 \theta \right]} \\ &\approx \Phi\left(\xi_c^{1/2} \right) \frac{B_{z0}}{\sqrt{2\pi k_0 L} \sin \theta}, \end{aligned} \tag{11.37}$$

where $\xi_c = (k_0 L)^{2/3} \sin^2 \theta$ and

$$\Phi(\tau) = 2.3\tau \exp\left(-\frac{2}{3}\tau^2\right) \tag{11.38}$$

with $\tau = (k_0 L)^{1/3} \sin\theta$. As a result, we can write the expression giving the relationship between the laser amplitude in vacuum B_{z0} and the electric field driving the resonance at the critical density surface. It reads

$$E_d = \Phi(\tau) \frac{cB_{z0}}{\sqrt{2\pi k_0 L}} \tag{11.39}$$

The conversion efficiency from the laser to the electron plasma wave is determined by the function $\Phi(\tau)$ which is shown in Fig. 11.5. The incidence angle corresponding to the maximum conversion efficiency corresponds to the maximum of $\Phi(\tau)$. As a function of the plasma density scale length, it is

$$\sin\theta_{\max} \approx 0.8 \times (k_0 L)^{-1/3}. \tag{11.40}$$

For example, the optimum incident angle is $\theta_{\max} \approx 0.8$ rad for $L = 10\lambda$. Near the cut-off density point, the frequency and the wave number matching conditions are met: here the electron plasma frequency ω_{pe} is equal to the laser frequency ω and the laser wavenumber k_y to the electrostatic field wave number. As shown in Fig. 11.4, the laser is reflected at the turning point where the density is $n_t = n_c \cos\theta$. The laser field tunnels into the cut-off density region where the electron plasma wave frequency is close to the laser frequency. This leads to the resonant conversion of the incident laser radiation into the electron plasma wave which dissipates through the

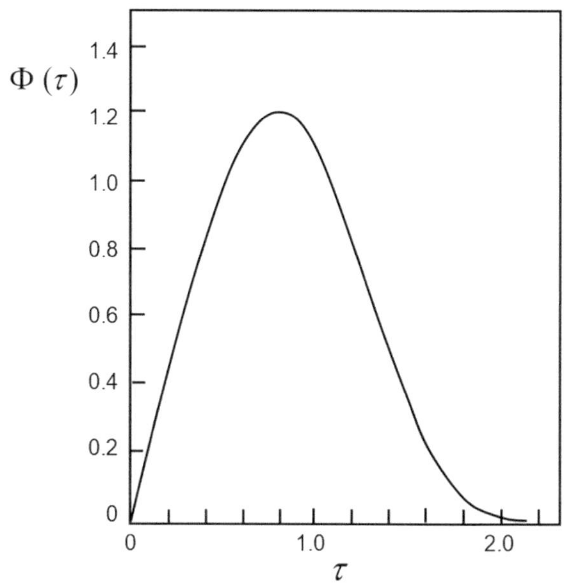

Fig. 11.5 A plot of the function $\Phi(\tau)$, which characterizes the efficiency of the resonance absorption

collisional and collision-less processes. In Part IV, we continue the discussion of the plasma resonance physics, addressing in detail the linear and nonlinear mechanisms of limitations of electric field amplitude as well as charged particle acceleration and generation of attosecond electromagnetic pulses.

Resonance absorption has been widely investigated by experiments, theory and simulation. Figure 11.6 shows an example of the experimental result, where a laser pulse from a Nd-Glass laser of 1.06 μm wavelength, 0.2 J energy and 25 ps duration irradiated a copper target with the focal spot diameter of 35 μm, at the laser intensity of about $5 \times 10^{14}\ \mathrm{W/cm^2}$. X-ray emission intensity was measured at the energies of 12 keV and 26 keV. Figure 11.6 shows that the X-ray emissions are maximum at the incident angle of ~ 15° for the p-polarization, but not for the s-polarization. From the ion velocity spectra, the expansion velocity was estimated to be about $2 \times 10^7\ \mathrm{cm/sec}$. Therefore, the plasma scale length was 5 μm. So, the optimum angle given by Eq. (11.40) is about 14° which reasonably agrees with Fig. 11.6. Namely, in this experiment, it is concluded that the dominant absorption process is the resonance absorption.

When the plasma is hot and collisions are weak, the electron plasma wave excited at the critical surface propagates down to the lower density plasma. In the under dense plasma, where $\omega_{pe} < \omega$, the phase velocity of the plasma wave can be comparable to the electron thermal velocity. Then the electron plasma wave is dissipated by wave-particle interaction, i.e., through the Landau damping.

The absorption intensity due to the resonance mode conversion is estimated as follows. Since the absorption power density is given by $\boldsymbol{j} \cdot \boldsymbol{E} = j_x E_x$, the absorption power I_{abs} is

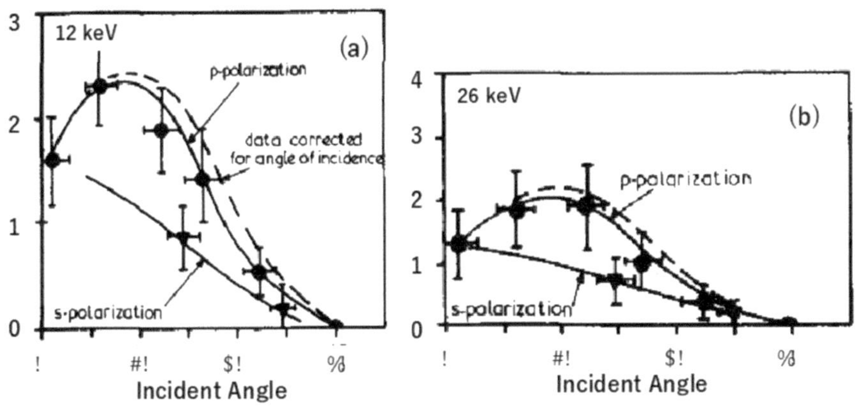

Fig. 11.6 Angle and polarization dependences of the 12 keV and 26 keV X-ray signals. The vertical axis is the X-ray intensity in a relative scale and the horizontal axis is the incidence angle in degrees. Reprinted from Luther-Davis (1978), with permission from AIP Publishing: permission conveyed through Copyright Clearance Center, Inc. © 1978 AIP Publishing

$$I_{abs} = \int_{-L}^{\infty} j_x E_x dx \approx 2\nu \int_{-L}^{\infty} \frac{\varepsilon_0 E_d^2(x)}{|\varepsilon(\omega, x)|^2} dx. \tag{11.41}$$

Here, the incident laser power is

$$I_L = 2c \frac{|B_{z0}|^2}{\mu_0}, \tag{11.42}$$

$$E_d^2(x) = \frac{|cB_{z0}\Phi(\tau)|^2}{2\pi k_0 L}, \tag{11.43}$$

$$|\varepsilon(\omega, x)|^2 \approx (x/L)^2 + (\nu/\omega)^2. \tag{11.44}$$

Integration of Eq. (11.41) yields

$$I_{abs} = c \frac{|B_{z0}\Phi(\tau)|^2}{\mu_0}. \tag{11.45}$$

The resonance absorption efficiency is obtained by dividing the absorbed power given by Eq. (11.45) by the incident laser power (11.42) as

$$\eta_{abs} = \frac{\Phi^2(\tau)}{2}. \tag{11.46}$$

This indicates that the resonance absorption efficiency could be substantially high for the optimum incident angle given by Eq. (11.40).

Laser absorption due to the resonance absorption was investigated by the 2D PIC simulations by Estabrook et al. (1975) for the following conditions: initial plasma scale length $L = 3\lambda$ with the density changing from n = 0 to 1.7 n_c, the laser intensity 10^{16} W/cm^2, the simulation time $140\pi/\omega$ corresponding 0.2 ps fo 1 μm wavelength laser, and the ion to electron mass ratio of 100. Figures 11.6 and 11.7 show the experimental and simulation results of the angular dependence of absorption efficiency. For the scale length $L = 3\lambda$, the optimum incidence angle is approximately equal to 22°, and the absorption efficiency is 70%. The angular dependence is similar to that of the theoretical prediction. However, in the simulations the maximum absorption angle is smaller by 5°. In addition, the absorption is finite (1/4 of the maximum absorption) for the normal incidence. These results indicate the importance of other processes, such as plasma density profile modification, plasma surface rippling and other absorption processes such as the parametric decay instabilities near the critical density region and the two plasmon decay at the quarter critical region. These processes are described below.

In actual situation, the laser absorption by the resonance mode conversion is limited due to various reasons. Let us discuss here the effects of the plasma density profile steepening. When the laser intensity is strong, the laser radiation pressure

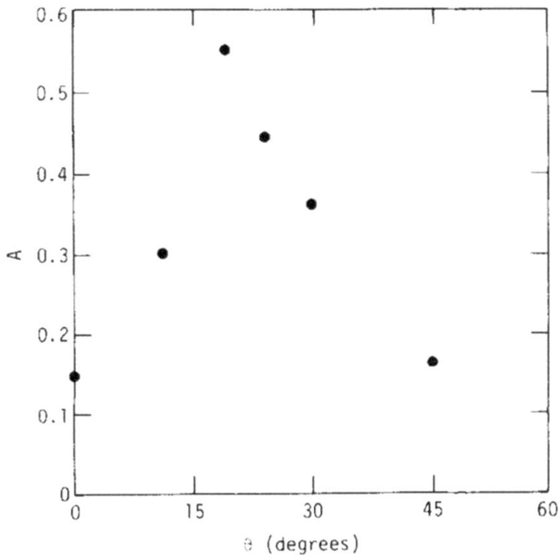

Fig. 11.7 Simulation result of the laser absorption efficiency (here $A = \eta_{abs}$) for various incidence angles. Reproduced from Estabrook et al. (1975), with permission from AIP Publishing: permission conveyed through Copyright Clearance Center, Inc. © 1975 AIP Publishing

and/or the ponderomotive force of the longitudinal fields can modulate the density profile. The density profile is steepened near the critical density region. An example of the self-consistent density profile is shown in Fig. 11.8. In the simulation, once the density steepening occurs, the resonance absorption is strongly modified (Estabrook et al., 1975). The electron plasma wave is excited near the critical surface. Therefore, the plasma is pushed out by the ponderomotive force, and the cavity is formed as seen in Fig. 11.8b. Since the electron plasma wave trapped in the cavity is continued to be amplified, the wave breaking occurs to generate hot electrons. The hot electron pressure breaks the cavity as shown in Fig. 11.8c. So, the density profile modification causes the wave breaking of the electron plasma wave to produce hot electrons of a few tens keV, as it was observed by Luther Davis (1978). The hot electron energy spectrum of the simulation is shown in Fig. 11.9.

Theoretical analysis on the nonlinear behavior of electron plasma waves excited by the resonant mode conversion was presented by Uenoyama et al., (1981) and Takabe and Mulser (1982). Since the Maxwellian tail electrons are accelerated by large amplitude electron plasma waves, "hot electrons" (also called supra-thermal electrons) are generated. The slope temperature of the hot electrons, T_h, is empirically scaled to laser intensity as follows (Estabrook et al., 1975; Forslund et al., 1975a),

$$T_h \approx 13 T_e^{1/4} \left(I_L \lambda_L^2\right)^{0.39} \ \text{keV}, \tag{11.47}$$

$$T_h \approx 7 T_e^{1/3} \left(I_L \lambda_L^2\right)^{0.39} \ \text{keV}. \tag{11.48}$$

Here T_e is in keV and $I_L \lambda_L^2$ is in 10^{15} Wμm^2/cm^2. When T_e is higher than several tens of keV, the ablatively accelerated fuel layer is preheated by hot electrons and

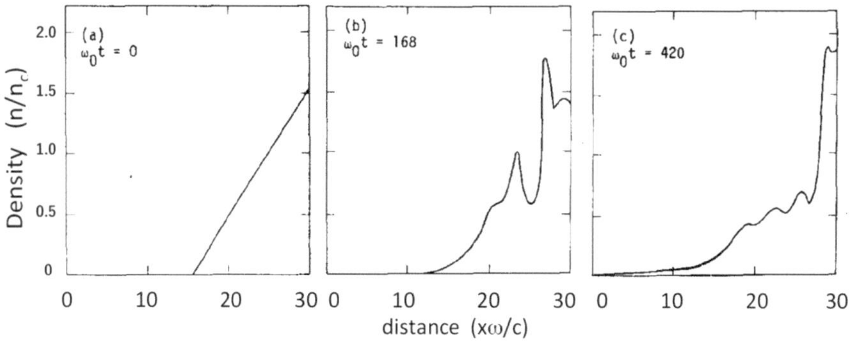

Fig. 11.8 Evolution of density profile steepening for a short scale-length under-dense plasma. **a** Initial density profile, **b** cavity formation near the critical density, and **c** steepening of density profile at the critical density. Reprinted from by Estabrook et al. (1975), with permission from permission from AIP Publishing: permission conveyed through Copyright Clearance Center, Inc. © 1975 AIP Publishing

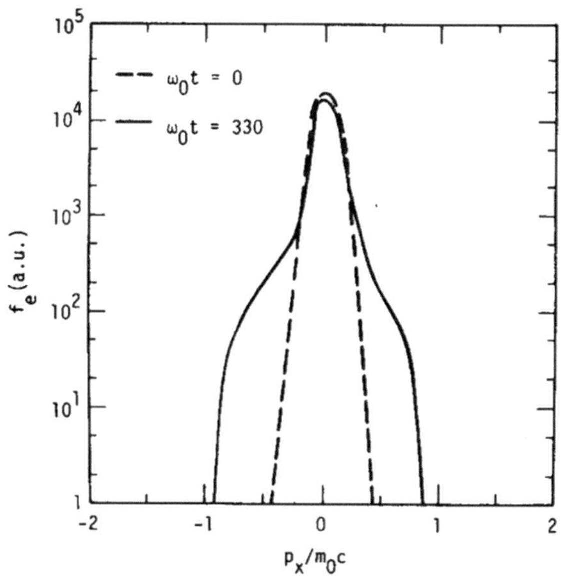

Fig. 11.9 Hot electron spectrum after electron plasma wave is fully developed. The slope temperature reaches 60 keV. Reprinted from by Estabrook et al. (1975), with permission from AIP Publishing: permission conveyed through Copyright Clearance Center, Inc. © 1975 AIP Publishing Publishing

the preheating will prevent the compression. So, the value of $I_L\lambda_L^2$ is required to be lower than 0.5×10^{15} W μm^2/cm^2.

11.4 Brunel Absorption (Not-So Resonant Resonance Absorption)

Density gradient is very sharp in laser plasma interaction of short pulses (picosecond and sub-picosecond pulses). Then, the electrons in the target can be pulled out into vacuum by the laser directly. Once the electrons are pulled out, strong electrostatic fields are generated on the surface and the electrons in vacuum are injected into the plasma with high energy. This process proposed by Brunel (1987) is the dominant absorption mechanism for the short pulse laser obliquely incident on a steep gradient plasma-vacuum interface. In the case of normal incidence or s-polarization, the second harmonic oscillation of the ponderomotive force causes the processes similar to the Brunel-absorption (Sanz et al., 2012a).

The absorbed energy flux is estimated by the average kinetic energy of the hot electrons multiplied on the electron flux. Let us assume that the average injecting electron kinetic energy is $\eta m_e v_{osc}^2/2$. The total number of electrons extracted into vacuum, N, is estimated by the total charge which screens the external electric field. It can be found according to the Poisson equation to be equal to $N = \varepsilon_0 E_0/e$. As a result, the absorbed energy flux is equal to

$$I_{abs} = \eta \frac{\omega}{4\pi} \frac{\varepsilon_0 E_0}{e} m_e v_{osc}^2 = \frac{\eta}{2\pi} v_{osc} \frac{\varepsilon_0 E_0^2}{2}, \tag{11.49}$$

where the numerical coefficient η is approximately equal to 1.57 (Brunel, 1987). Since the incident laser intensity is given by $I_L = c\eta\varepsilon_0 E_0^2/A^2$, the absorption efficiency for the Brunel process is

$$\eta_{abs} = \eta \frac{v_{osc} A^2}{2\pi c}, \tag{11.50}$$

where A is the enhancement factor due to field swelling near the critical density. The factor A depends upon the density profile near the critical density region, being greater than unity. From the relationships obtained above, it follows that the absorption efficiency can exceed 10% for the laser intensity of $I_L \lambda_L^2 = 10^{16} (\mu\text{m})^2 \text{W/cm}^2$.

Note here that a similar process takes place in the plasma interacting with an obliquely incident intense short laser pulse. When a sub-picosecond laser with relativistic intensity irradiates a solid target, the laser pulse penetrates the over dense plasma, where the electron motion in the surface layer is similar to the Brunel process, although the electron motion is relativistic and the magnetic field effects are important. In this process, multi-MeV electrons are generated when the normalized amplitude $a_0 = eA/m_e c^2$ is higher than unity. This process has been analyzed in Sanz et al. (2012a, 2012b).

Analytical description of dynamics of electrons and ions near the plasma-vacuum interface irradiated by high-intensity laser pulse is presented in this chapter of Part IV.

References

Brunel, F. (1987). Not-so-resonant, resonant absorption. *Physical Review Letters, 59*, 52.

Chen, F. F. (1984). *Introduction to plasma physics and controlled fusion*. Plenum Press.

Dawson, J. M. (1961). On Landau damping. *Physics of Fluids, 4*, 869.

Denisov, N. G. (1957). On a singularity of the field of an electromagnetic wave propagated in an inhomogeneous plasma. *JETP, 4*, 544.

Estabrook, K., Valeo, E. J., & Kruer, W. L. (1975). Two-dimensional relativistic simulations of resonance absorption. *Physics of Fluids, 18*, 1151.

Forslund, D. W., Kindel, J. M., Lindman, E. L., et al. (1975). Theory and simulation of resonant absorption in a hot plasma. *Physical Review A, 11*, 679.

Ginzburg, V. L. (1964). *Propagation of electromagnetic waves in plasmas*. Pergamon New York.

Landau, L. D. (1946). Oscillations of an electron plasma. *Journal of Physics (USSR), 10*, 25.

Luther Davis, B. (1978). Evidence of resonance absorption in laser-produced plasmas from the polarization and angular dependence of high-energy X-ray bremsstrahlung emission. *Applied Physics Letters, 32*, 209.

Malmberg, J. H., & Wharton, C. B. (1966). Dispersion of electron plasma waves. *Physical Review Letters, 17*, 175.

Sanz, J. R., Debayie, A., & Mima, K. (2012a). Model for ultra-intense laser-plasma interaction at normal incidence. *Physical Review E, 85*, Article 046411.

Sanz, J. R., Debayie, A., & Mima, K. (2012b). Ultra-intense laser/plasma interaction at normal incidence: Relativistic mirror effects, higher harmonic generation and absorption. *Comptes Rendus Mecanique, 340*, 894.

Takabe, H., & Mulser, P. (1982). Self-consistent treatment of resonance absorption in a streaming plasma. *Physics of Fluids, 25*, 2304.

Uenoyama, T., Mima, K., & Watanabe, T. (1981). Nonlinear effects in collective absorption. *Journal of the Physical Society of Japan, 50*, 3107.

Van Kampen, N. G. (1955). On the theory of stationary waves in plasmas. *Physica, 21*, 949.

Chapter 12
Nonlinear Wave Phenomena in Laser Plasma Interaction

Abstract In the laser fusion, targets are irradiated by laser pulses with typical intensity and pulse duration of $I_L = 10^{15}\ \mathrm{W/cm^2}$ and a few nano-second, respectively. When the intense laser pulse propagates in a long-scale, non-uniform, hot corona plasma, various nonlinear wave phenomena take place in various density regions. In this chapter, the nonlinear wave phenomena of laser-plasma interaction are described.

Keywords Nonlinear wave interaction · Corona plasma · Stimulated Raman scattering · Stimulated Brillouin scattering · Cross beam energy transfer · Parametric decay instability · Filamentation instability · Self-focusing · Two-plasmon decay · Oscillating two-stream instability

12.1 Nonlinear Wave Interactions in Laser Plasmas

As described in Chap. 11, the intense laser propagates and interacts with a corona plasma as shown in Fig. 11.4, where the laser is injected from the right-hand side and interacts with corona plasmas through various linear and nonlinear processes which depend on the plasma density and temperature. When the laser intensity is high enough and the electron quiver velocity in the laser field is comparable to the electron thermal velocity, the laser induces density fluctuations and excites the electron plasma wave as well as the ion acoustic wave through nonlinear laser-plasma interaction. The nonlinear excitation of various waves in laser plasma, depending on their type, is sometimes called the "parametric instability", "stimulated scattering", or "resonant three wave coupling". As a result of developing the parametric instabilities near the critical plasma density region and in the vicinity of the quarter critical density region, the laser electromagnetic energy is converted into the plasma waves. These processes together with the resonant mode conversion can be called the "collective absorption", which is an additional channel of the laser energy absorption other than the inverse bremsstrahlung due electron–ion collisions.

A sketch presented in Fig. 12.1 shows the plasma density regions in which specific nonlinear processes take place. The parametric instability in region (a) of Fig. 12.1

Y. Kato et al., *High Power Laser and Plasma Science*, Springer Series on Atomic, Optical, and Plasma Physics 130, https://doi.org/10.1007/978-3-032-08930-4_12

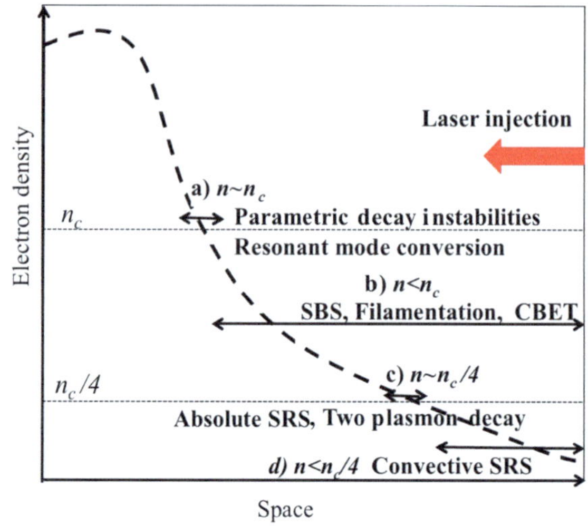

Fig. 12.1 Nonlinear laser plasma interaction in a pre-plasma formed in front of a solid target surface

near the critical density surface excites both the ion acoustic wave and the electron plasma wave. On the other hand, the parametric instability in region (b) near quarter of the critical plasma density excites two electron plasma waves via the nonlinear scattering of laser radiation, which is known as the stimulated Raman scattering (SRS). The SRS in region (c) is called absolute SRS, since the group velocities of the excited electron plasma and electromagnetic waves are approximately equal to zero. The convective stimulated Raman scattering (SRS) develops in region (d), where the density is lower than a quarter of the critical density $n_c/4$.

Stimulated Brillouin scattering (SBS), filamentation, and cross beam energy transfer (CBET) take place in the plasma up to the cut-off density region ($\omega \leq \omega_{pe}$), region (b). Note here that the CBET is the seeded SBS in which two laser beams with slightly different frequencies transfer energy from the higher frequency laser beam to the lower frequency beam through the SBS process, where the frequency difference matches with the ion acoustic wave frequency (Igumenshchev et al., 2010, 2012; Kruer et al., 1996; Michel et al., 2009). The convective SRS takes place also in region (d).

12.2 Nonlinear Wave Equations

The lowest order nonlinearities are proportional to the square of amplitudes of waves. One of these nonlinearities, dominant in un-magnetized plasmas, is the ponderomotive force which is proportional to the square of oscillation velocity (Molares & Lee, 1974). Let us write the equation of motion of the electron component:

$$E = \frac{\partial \boldsymbol{v}_e}{\partial t} + (\boldsymbol{v}_e \cdot \nabla)\boldsymbol{v}_e = -\frac{\nabla P}{n_e m_e} - \frac{e}{m_e}(\boldsymbol{E} + \boldsymbol{v}_e \times \boldsymbol{B}) - \nu_l \boldsymbol{v}_e. \tag{12.1}$$

Here $\boldsymbol{v}_e$ is the electron velocity and the electric and magnetic field are $\boldsymbol{E} = -\partial \boldsymbol{A}/\partial t - \nabla\phi$ and $\boldsymbol{B} = \nabla \times \boldsymbol{A}$ respectively where $\boldsymbol{A}$ and ϕ are vector and scalar potentials of electromagnetic field, P is the plasma pressure, and ν is the collision frequency. Then, in a uniform and stationary plasma, the small amplitude oscillation velocity $\boldsymbol{v}_1$ of the first order smallness satisfies the following equation,

$$\frac{\partial \boldsymbol{v}_1}{\partial t} = \frac{e}{m_e}\left(\nabla\phi + \frac{\partial \boldsymbol{A}}{\partial t}\right) - \frac{1}{n_0 m_e}\nabla P - \nu_l \boldsymbol{v}_1. \tag{12.2}$$

When the dissipation term $\nu_l \boldsymbol{v}_1$ is negligibly small, the transverse component of oscillation velocity driven by the laser field is

$$\boldsymbol{v}_t = \frac{e}{m_e}\boldsymbol{A}.$$

The equation of motion for the longitudinal component of the electron velocity $\boldsymbol{v}_l$ including the second order nonlinearity with respect to the laser field amplitude can be obtained after some algebra. It can be written as (Kruer, 1988)

$$\frac{\partial \boldsymbol{v}_l}{\partial t} = \frac{e}{m_e}\nabla\phi - \frac{1}{n_0 m_e}\nabla P - \nu_l \boldsymbol{v}_l + \frac{1}{m_e}\boldsymbol{F}_{NL}, \tag{12.3}$$

where ν_l is the effective damping rate of longitudinal wave. The force in in the r.h.s. of this equation is given by

$$\boldsymbol{F}_{NL} = -\nabla\left[\frac{m_e}{2}\left(\boldsymbol{v}_t - \frac{e}{m_e}\boldsymbol{A}\right)^2\right] = -\nabla\Phi_p \tag{12.4}$$

with Φ_p being the "ponderomotive potential", which occurs from the electron quiver motion in the electromagnetic wave.

The equation for the electron density modulations $\tilde{n}$ follows from Eq. (12.3). It can be cast in the form

$$\frac{\partial^2 \tilde{n}}{\partial t^2} + \nu_l \frac{\partial \tilde{n}}{\partial t} - 3v_e^2 \nabla^2 \tilde{n} + \omega_{pe}^2 \tilde{n} = -n_0 \frac{\partial \nabla \cdot \boldsymbol{v}_l}{\partial x} - \frac{n_0}{m_e}\nabla \cdot \boldsymbol{F}_{NL}, \tag{12.5}$$

Here, we neglected the nonlinear terms due to the electron density modulations in the longitudinal direction. The Poisson equation

$$\nabla^2\phi = \frac{e}{\varepsilon_0}\tilde{n} = \frac{m_e}{e n_0}\omega_{pe}^2 \tilde{n}$$

is also used to derive Eq. (12.5).

In the second order nonlinearity, the equation for the transverse component of the electron velocity $\boldsymbol{v}_t$ is given by,

$$\frac{\partial \boldsymbol{v}_t}{\partial t} = \frac{e}{m_e}\frac{\partial \boldsymbol{A}}{\partial t} - \nu \boldsymbol{v}_t + \frac{1}{m_e}\boldsymbol{F}_{t,NL}. \tag{12.6}$$

Note here that the curl of the function $\boldsymbol{F}_{t,NL}$ is equal to zero,

$$\nabla \times \boldsymbol{F}_{t,NL} = \nabla \times \left[m_e \boldsymbol{v}_1 \times (\nabla \times \boldsymbol{v}_1) - e\boldsymbol{v}_1 \times (\nabla \times A) - \frac{m_e}{2}\nabla(\boldsymbol{v}_1)^2 \right] = 0$$

since $\nabla \times \boldsymbol{v}_1 = 0$. As a result, we can neglect $\boldsymbol{F}_{t,NL}$ in Eq. (12.6) within the framework of the present approximation.

Now we consider the electromagnetic field in the form of superposition of two electromagnetic waves,

$$\boldsymbol{A} = \frac{1}{2}(\boldsymbol{A}_0 \exp(i\boldsymbol{k}_0 \cdot \boldsymbol{r} - i\omega_0 t) + \boldsymbol{A}_1 \exp(i\boldsymbol{k}_1 \cdot \boldsymbol{r} - i\omega_1 t + i\psi) + c.c.), \tag{12.7}$$

in which the first of them with amplitude $|\boldsymbol{A}_0|$ corresponds to the incident laser radiation and the second wave with amplitude $|\boldsymbol{A}_1|$ represents the scattered light. In this expression "c.c." denotes complex conjugate, and ψ is the phase of the scattered electromagnetic wave. The ponderomotive potential defined by Eq. (12.4) can be written as

$$\begin{aligned}\Phi_p = \frac{e^2}{2m_e}\{&\boldsymbol{A}_0 \cdot \boldsymbol{A}_1[\cos((\boldsymbol{k}_0 - \boldsymbol{k}_1)\cdot \boldsymbol{r} - (\omega_0 - \omega_1)t - \psi) \\ &+ \cos((\boldsymbol{k}_0 + \boldsymbol{k}_1)\cdot \boldsymbol{r} - (\omega_0 + \omega_1)t - \psi)] \\ &+ \boldsymbol{A}_0^2 \cos^2(\boldsymbol{k}_0 \cdot \boldsymbol{r} - \omega_0 t - \psi) + \boldsymbol{A}_1^2 \cos^2(\boldsymbol{k}_1 \cdot \boldsymbol{r} - \omega_1 t)\}\end{aligned} \tag{12.8}$$

Since the frequency of the longitudinal plasma waves in the underdense plasma is smaller than the electromagnetic wave frequency, the first term of the ponderomotive force in Eq. (12.4) can be in resonance with a longitudinal wave. When the matching conditions

$$\boldsymbol{k}_l = \boldsymbol{k}_0 - \boldsymbol{k}_1, \qquad \omega_l = \omega_0 - \omega_1, \tag{12.9}$$

are met, the longitudinal wave with the wavevector and frequency $(\boldsymbol{k}_l, \omega_l)$ is resonantly excited.

In summary, the nonlinear coupling of the incident laser wave with other modes occurs via the nonlinear electric current

$$\boldsymbol{j}_{NL} = -\frac{e^2 \tilde{n}}{m_e}\boldsymbol{A}_0. \tag{12.10}$$

Then, the scattered wave equation is obtained by the Maxwell equation. It reads

$$\frac{\nabla \times \nabla \times \boldsymbol{A}_2}{\mu_0} = \dot{\boldsymbol{j}}_{NL} - \varepsilon_0 \frac{\partial^2 \boldsymbol{A}_2}{\partial t^2}.$$

As a result, the wave equation for the scattered electromagnetic wave has a form

$$\frac{\partial^2 \boldsymbol{A}_2}{\partial t^2} - c^2 \nabla^2 \boldsymbol{A}_2 = \frac{\dot{\boldsymbol{j}}_{NL}}{\varepsilon_0}, \tag{12.11}$$

Note that Eqs. (12.4), (12.5), and (12.11) are the system of equations describing the coupled longitudinal and electromagnetic waves.

12.3 Stimulated Raman Scattering

The stimulated Raman scattering (SRS) is the process of transformation of the incident laser radiation into a scattered transverse electromagnetic wave and an electron plasma wave. The scattering process is schematically shown in Fig. 12.2, where the frequency and wave number matching conditions are given as follows,

$$\boldsymbol{k}_0 = \boldsymbol{k}_s + \boldsymbol{k}_e, \qquad \omega_0 = \omega_s + \omega_e. \tag{12.12}$$

The electron density fluctuation of the electron plasma wave is excited by the ponderomotive force. This process is described by the following equation which is obtained from Eq. (12.5) and the ponderomotive potential given by Eq. (12.8):

$$\begin{aligned} \frac{\partial^2 \tilde{n}}{\partial t^2} + 2\gamma_L \frac{\partial \tilde{n}}{\partial t} - 3v_e^2 \nabla^2 \tilde{n} + \omega_{pe}^2 \tilde{n} = \frac{n_0 e^2}{2m_e^2} \nabla^2 \{ &\boldsymbol{A}_0 \cdot \boldsymbol{A}_s [\cos((\boldsymbol{k}_0 - \boldsymbol{k}_s) \cdot \boldsymbol{r} \\ &-(\omega_0 - \omega_s)t - \psi)]\}. \end{aligned} \tag{12.13}$$

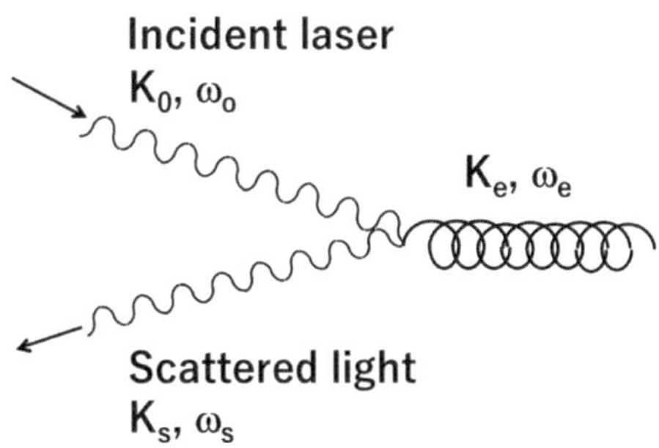

Fig. 12.2 Three wave resonance of the stimulated Raman scattering, where propagations of the electromagnetic waves and the electron plasma wave are shown by wavy and circulating lines, respectively

Here $\boldsymbol{A}_0$, $\boldsymbol{k}_0$, and ω_0 and $\boldsymbol{A}_s$, $\boldsymbol{k}_s$, ω_s are the amplitude, wavevector and frequency of the incident and scattered electromagnetic waves, respectively, and γ_L is the Landau damping of the electron plasma wave. The wave equation of the scattered laser light includes the nonlinear current $\boldsymbol{j}_{NL}$ given by Eq. (12.10), which originates from the incident laser EM field. Note that one of the nonlinear currents in Eq. (12.10) (it is the term in the r.h.s. of this equation) is related to the density modulation associated with the electron plasma wave. The scattered electromagnetic wave is described by the following wave equation,

$$\frac{\partial^2 \boldsymbol{A}}{\partial t^2} + 2\nu_{EM}\frac{\partial \boldsymbol{A}}{\partial t} - c^2\nabla^2\boldsymbol{A} + \omega_{pe}^2\boldsymbol{A} = \frac{\boldsymbol{j}_{NL}}{\varepsilon_0}. \tag{12.14}$$

Here, we set

$$\boldsymbol{A} = \frac{1}{2}\left[\boldsymbol{A}_0\exp(i\boldsymbol{k}_0\cdot\boldsymbol{r} - i\omega_0 t) + \boldsymbol{A}_s\exp(i\boldsymbol{k}_s\cdot\boldsymbol{r} - i\omega_s t) + c.c.\right],$$

and ν_{EM} is the damping rate of the electromagnetic wave. The nonlinear current $\boldsymbol{j}_{NL}$ is related to the density modulation in the electron plasma wave and to the modulation of the incident laser amplitude, when we retain only linear terms with respect to $\tilde{n}$, $\boldsymbol{v}_l$, and $\boldsymbol{A}_s$.

For the slowly varying amplitude and phase of interacting waves, we can obtain the coupled mode equations describing the excited plasma wave during development of the Raman back scattering instability.

$$\frac{\partial A_s}{\partial t} \pm v_{gs}\frac{\partial A_s}{\partial x} + i\frac{\omega_{pe}^2}{2\omega_s}\frac{x}{L}A_s + \nu_s A_s = -i\frac{\omega_{pe}^2}{4\omega_s n_0}A_0\tilde{N}^*, \tag{12.15}$$

$$\frac{\partial \tilde{N}}{\partial t} + v_{gs}\frac{\partial \tilde{N}}{\partial x} + i\frac{\omega_{pe}^2}{2\omega_{ek}}\tilde{N} + \gamma_L\tilde{N} = -i\frac{(\boldsymbol{k}_0 - \boldsymbol{k}_s)^2}{4\omega_{ek}}\frac{n_0 e^2}{m_e^2}A_0A_s^*. \tag{12.16}$$

In these equations, the electromagnetic waves and the electron plasma wave propagate along the x direction, the incident laser and the scattered electromagnetic wave are polarized in the y direction, and $\tilde{N}$ is the amplitude of $\tilde{n}(\boldsymbol{r}, t)$,

$$\tilde{n}(\boldsymbol{r}, t) = \frac{1}{2}\tilde{N}\left[\exp(i\boldsymbol{k}_e\cdot\boldsymbol{r} - i\omega_{ek}t) + c.c.\right], \tag{12.17}$$

$$\omega_{ek} = \omega_0 - \omega_s, \quad \text{and} \quad \boldsymbol{k}_e = \boldsymbol{k}_0 - \boldsymbol{k}_s = (k_0 - k_s)\boldsymbol{e}_x, \tag{12.18}$$

$$\omega_{ek} = \sqrt{\omega_{pe}^2 3k_e^2 v_{th,e}^2}, \quad \omega_0 = \sqrt{k_0^2c^2 + \omega_{pe}^2} \text{ and } \omega_s = \sqrt{k_s^2c^2 + \omega_{pe}^2}. \tag{12.19}$$

Equations (12.15) and (12.16) yield the growth rate of the SRS instability including the damping effects (Drake, 1974; Forslund et al., 1975; Kruer, 1988)

$$\Gamma_{SRS} = \frac{k_e v_0}{4\gamma_L}\sqrt{\frac{\omega_{pe}^2}{\omega_{ek}\omega_s}}, \tag{12.20}$$

where $v_0 = |eA_0/m_e|$. The growth rate is maximum at the quarter critical density, where $\omega_{pe} = \omega_0/2$. The maximum growth rate is equal to

$$\Gamma_{SRS,\max} = \sqrt{\frac{3}{2}}\frac{v_0}{4c}\omega_0. \tag{12.21}$$

As we see, the SRS mode is localized in the region near the quarter critical surface. The two plasmon decay instability and the SRS occur simultaneously in this region. Although the threshold of the two plasmon decay instability becomes lower for the lower electron temperature, the SRS instability dominates if the electron temperature is higher than several keV since the electron plasma waves generated by the two plasmon decay instability become highly dumped modes.

The threshold of the SRS instability depends upon the wave dissipation or plasma non-uniformity. To obtain the instability threshold, which corresponds to the zero growth rate, we neglect the convection terms in Eqs. (12.15) and (12.16) assuming the plasma density to be homogeneous, and find the condition

$$\Gamma_{SRS,\max} > \sqrt{\nu_s\gamma_L}. \tag{12.22}$$

From this condition it follows that the threshold depends upon the Landau damping of the electron plasma wave and collisional damping of the scattered light. This is known as the convective instability threshold. When the plasma density is inhomogeneous, the threshold depends upon the density gradient. According to the analysis by Rosenbluth (1972), Liu et al. (1973), and Pesme et al. (1973), the amplification factor A showing the increase of the amplitude of the unstable modes is given by

$$A = \exp\left(\frac{\Gamma_{CRS}^2 L^2}{v_{gs}v_{ge}}\right). \tag{12.23}$$

In an inhomogeneous plasma the instability threshold is given by

$$\Gamma_{SRS,\max} > \frac{\sqrt{v_{gs}v_{ge}}}{L}, \tag{12.24}$$

according to Mima and Nishikawa (1984).

Excitation of large amplitude electron plasma waves during development of the SRS instability leads to generation of high energy electrons. This process can result in instability saturation as seen in the results obtained with the PIC computer simulation carried out by Estabrook et al. (1980). The electron energy spectrum obtained in this paper has high energy tail as shown in Fig. 12.3, which increases the dumping of the electron plasma wave.

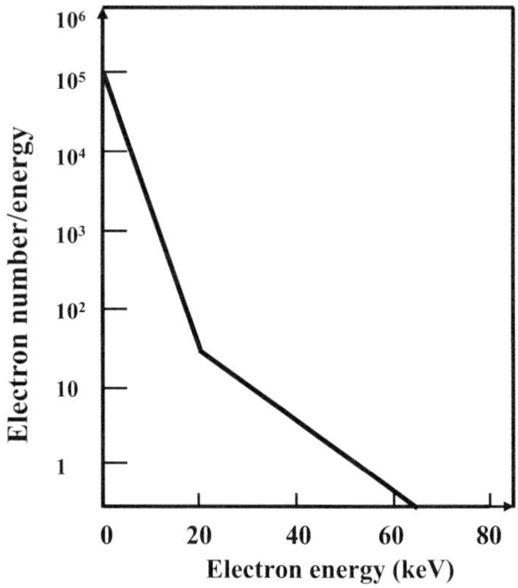

Fig. 12.3 Hot electron spectrum obtained with the PIC simulation by Estabrook et al. (1980)

The convective Stimulated Raman Scattering process becomes important for large scale plasmas like cavity plasmas in a hohlraum (Froula et al., 2009). The resonance conditions given by Eqs. (12.18) and (12.19) for the back scattered SRS yield the phase velocity of the resonant plasma wave as follows,

$$\frac{v_{ph,e}}{c} \cong \frac{\sqrt{n}}{\sqrt{n_c - n} - 2\sqrt{n}}.$$

This relation indicates that the resonant electron velocity is a few times the electron thermal velocity when the plasma temperature is a few keV and $n/n_c < 0.1$. On the other hand, the resonant velocity is significantly higher than the thermal velocity for $n/n_c > 0.1$.

The scattered light intensity depends strongly on the plasma density as shown in Figs. 12.4 and 12.5. Figure 12.4 shows experimental spectra of the SRS back scattered radiation. In these experiments, the OMEGA laser beam (University of Rochester) of $\lambda = 0.35$ μm wavelength irradiated the large-scale under-dense plasma. The spectrum of the scattered radiation clearly demonstrates time development of the SRS instability for the plasma densities of (a) $0.101\, n_c$, (b) $0.113\, n_c$, and (c) $0.13\, n_c$.

Froula et al. (2009) have shown that the observed stimulated Raman scattering in a large-scale high-temperature plasma scales strongly with the plasma density, increasing by an order of magnitude when the electron density is increased by 20%. As shown in Fig. 12.5, the SRS reflectivity depends strongly on the plasma density. This is consistent with the linear theory, including pump depletion in a uniform plasma.

The Landau damping of the resonant electron plasma wave due to SRS, which is significant to reduce the SRS, is smaller for $n/n_c < 0.1$. The reflectivity of the SRS rapidly increases with the increase in the plasma density from $0.1\, n_c$ to

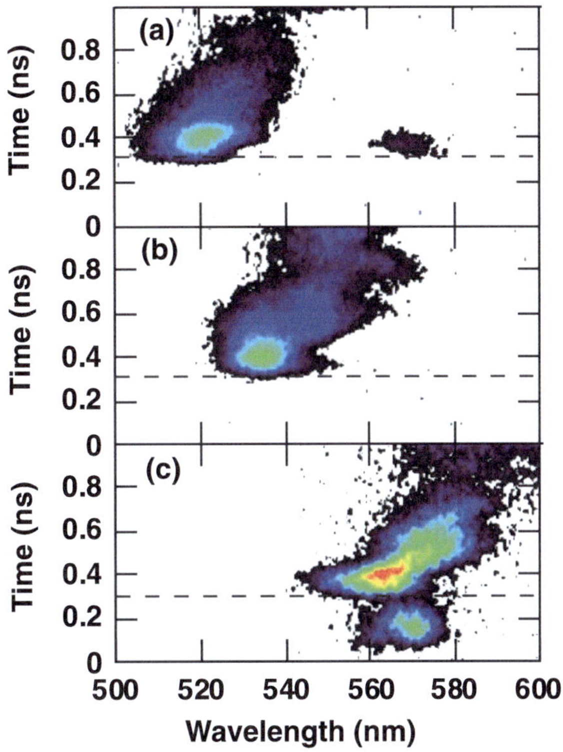

Fig. 12.4 Typical experimental spectra of the back scattered SRS radiation. The $\lambda = 0.35$ μm laser (OMEGA at the University of Rochester) irradiated an under-dense plasma target. The mode spectrum versus time for the plasma densities of (a) $0.101\,n_c$, (b) $0.113\,n_c$, and (c) $0.13\,n_c$. Reprinted from Froula et al., (2009), with permission from The American Physical Society and the author(s). © 2009 The American Physical Society

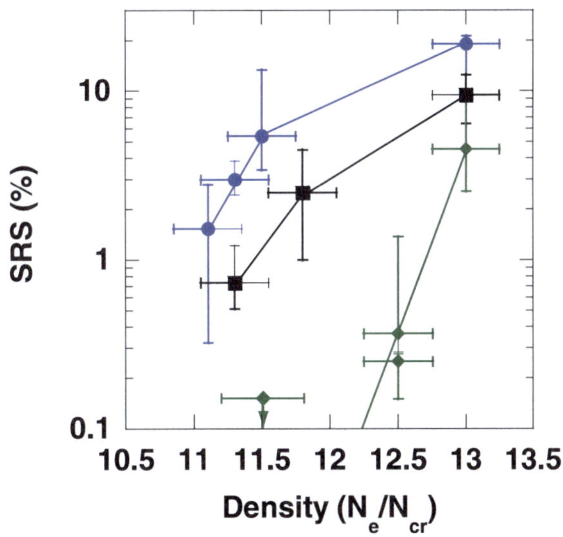

Fig. 12.5 Dependence of the SRS reflectivity on the plasma density and the laser intensities of 9×10^{14} W/cm^2 (blue circles), 12.2×10^{14} W/cm^2 (black squares), and 3.9×10^{14} W/cm^2 (green diamonds). The density scale is in unit of percent. Reprinted from Froula et al., (2009), with permission from The American Physical Society and the author(s). © 2009 The American Physical Society

$0.13\,n_c$(Fig. 12.5) and the scattered light frequency approaches to half of the laser frequency (Fig. 12.4).

12.4 Absolute Raman Instability at Quarter Critical Density and Forward Raman Scattering

The stimulated Raman scattering instability developing at the quarter critical density becomes an absolute instability since the group velocity of the scattered wave is very small. In an inhomogeneous plasma, the threshold for the absolute instability is (Rosenbluth, 1972)

$$\Gamma_{SRS} = \frac{\omega_0}{4}\left(\pi \frac{c}{0.9\omega_0 L}\right)^{2/3},$$

where L is the scale-length of the plasma inhomogeneity. We note that the growth rate calculated for homogeneous plasma is $\sqrt{3/32}\omega_0 v_0/c$ according to Eq. (12.21). The amplification factor A showing the increase of the amplitude of the unstable modes of the forward Raman scattering (which is the convective instability) is given by

$$A = \exp\left(2\pi \frac{\Gamma^2 L}{v_{gs}\omega_{pe}}\right). \tag{12.25}$$

12.5 Two Plasmon Decay Instability

Two plasmon decay instability represents a resonant decay of a light wave into two electron plasma waves as shown in Fig. 12.6. This three-wave interaction is called the "two plasmon decay (TPD)" (Goldman & DuBois, 1966; Liu et al., 1973; Simon et al., 1983). Since the frequency of the electron plasma wave is approximately equal to ω_{pe}, the resonance conditions given by Eq. (12.18) require $\omega_{pe} = 2\omega_0$. Therefore, this TPD instability develops in the region near the quarter critical density surface where the electron density is $n = n_c/4$. The TPD mode is localized in a narrow region, wherein the localized electric field and local heating cause the density profile steepening, which then suppress the instability. As a result, the TPD instability occurs in the form of bursts. Although the instability is intermittent, a significant number of hot electrons, typically several tens of keV energy, is generated causing preheating of an imploding shell.

The coupled mode equations for the TPD are derived from Eqs. (12.2), (12.3), (12.8), (12.10), and (12.11), yielding

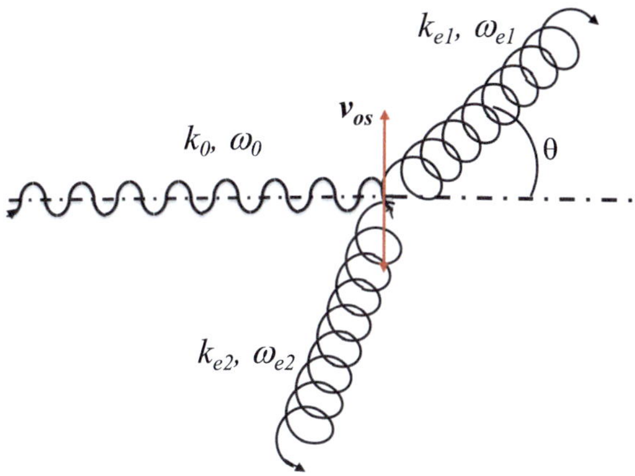

Fig. 12.6 Coupling geometry of the two plasmon decay, where propagations of the electromagnetic wave and the electron plasma wave are shown by the wavy and the circulating lines, respectively

$$\frac{\partial^2 \tilde{n}}{\partial t^2} + 2\gamma_L \frac{\partial \tilde{n}}{\partial t} - 3v_{th,e}^2 \nabla^2 \tilde{n} + \omega_{pe}^2 \tilde{n} = n_0 \nabla^2 (\boldsymbol{v}_l \cdot \boldsymbol{v}_0) - \frac{\partial}{\partial t} \nabla\nabla \cdot (\tilde{n}\boldsymbol{v}_0) \tag{12.26}$$

Let us write the expression for the density modulation in the coupling of two electron plasma waves $\tilde{n}$ in the form

$$\tilde{n}(\boldsymbol{r}, t) = \frac{1}{2}\left[N_1 \exp(i\boldsymbol{k}_{e1} \cdot \boldsymbol{r} - i\omega_{e1} t) + N_2 \exp(i\boldsymbol{k}_{e2} \cdot \boldsymbol{r} - i\omega_{e2} t) + c.c.\right], \tag{12.27}$$

where N_1 and N_2 are slowly varying functions of time t and coordinate $\boldsymbol{r}$. Using these, we obtain the equations for the amplitudes of the coupled modes:

$$\frac{\partial N_1}{\partial t} + \boldsymbol{v}_{g1} \cdot \nabla N_1 + \gamma_L N_1 + \omega_{pe}^2 N_1 = -\frac{i}{2}(\boldsymbol{k} - \boldsymbol{k}_0) \cdot \boldsymbol{v}_0 N_2^* \left(1 - \frac{k^2 \omega_{pe}^2}{(\boldsymbol{k} - \boldsymbol{k}_0)^2 \omega_{e1}\omega_{e2}}\right), \tag{12.28}$$

$$\frac{\partial N_2}{\partial t} + \boldsymbol{v}_{g2} \cdot \nabla N_2 + \gamma_L N_2 + \omega_{pe}^2 N_2 = \frac{i}{2}\boldsymbol{k} \cdot \boldsymbol{v}_0 N_1^* \left(1 - \frac{(\boldsymbol{k} - \boldsymbol{k}_0)^2 \omega_{pe}^2}{\boldsymbol{k}^2 \omega_{e1}\omega_{e2}}\right), \tag{12.29}$$

where $\boldsymbol{k} = \boldsymbol{k}_{e1}$ and $\boldsymbol{k}_{e1}\boldsymbol{k} - \boldsymbol{k}_0 = \boldsymbol{k}_{e2}$. When the plasma temperature is not very high, the mode frequencies are approximately equal to the Langmuir frequency, $\omega_{e1} \approx \omega_{e2} \approx \omega_{pe}$.

In the case of homogeneous plasma density and laser pulse intensity, Eqs. (12.28) and (12.29) yield the growth rate of the TPD instability as

$$\Gamma_{TPD} = -\gamma_L + \left|\frac{\boldsymbol{k} \cdot \boldsymbol{v}_0 k^2 \omega_{pe}^2\left[(\boldsymbol{k} - \boldsymbol{k}_0)^2 - \boldsymbol{k}^2\right]}{2k|\boldsymbol{k} - \boldsymbol{k}_0|}\right|. \tag{12.30}$$

The instability growth rate is maximum for $k >> k_0$ and $\boldsymbol{k} \cdot \boldsymbol{k}_0 = \pm kk_0 \cos 45°$. In this case, it is equal to

$$\Gamma_{TPD,\max} = \frac{k_0 v_0}{\sqrt{2}}. \tag{12.31}$$

From Eqs. (12.31) and (12.32), it follows that the threshold of the instability development corresponds to the laser amplitude given by

$$\frac{v_0}{c} = \frac{\sqrt{2}\gamma_L}{k_0 c}. \tag{12.32}$$

When plasma inhomogeneity is strong, the convection loss determines the threshold. To analyze that limit, we replace the terms $\boldsymbol{v}_{g1} \cdot \nabla N_1$ and $\boldsymbol{v}_{g2} \cdot \nabla N_2$ with the terms $i\left(3kv_{th,e}^2/L\omega_{pe}\right)N_1$ and $-i\left(3kv_{th,e}^2/L\omega_{pe}\right)N_2$, respectively in Eqs. (12.28) and (12.29) to model the mode convection, and obtain the instability threshold,

$$\frac{v_0^2}{c^2} = \frac{\left(k_0 v_{th,e}\right)^2}{L^2}. \tag{12.33}$$

Here L is the density scale-length at the quarter critical density region. This threshold is lower than that for the stimulated Raman scattering instability at the quarter critical density unless the plasma is quite hot (Kruer, 1988).

Experimental results of the TPD instability have been published since the early 1980s in several papers (e.g. Short et al., 1984). Two plasmon decay and three-halves harmonic generation have been also observed together with the filamentation instability (Seka et al., 2009). (Filamentation instability is described in Sect. 12.8.) The two-plasmon-decay in direct drive implosion experiments and multi-beam effects on fast-electron generation associated with the TPD were investigated by Stoeckl et al. (2003) in the spherical implosion experiments with OMEGA laser of 0.35 μm wavelength, 23 kJ laser energy and 1 ns flat top pulse. According to the LILAC hydrodynamic simulation, the plasma scale length in this experiment was 100–150 μm for the target diameter of 900–1100 μm at the peak of the laser pulse. The experimental results show the hard X-ray intensity increases by a factor of 10 and the 3/2 harmonic intensity by a factor of 5 when the laser intensity is increased by only 30%. This indicates that the TPD instability builds up in this parameter regime and should be considered in future large-scale direct implosion.

12.6 Stimulated Brillouin Scattering

In the under-dense plasma, the incident laser radiation is also scattered by the electron/ion density fluctuations associated with the ion acoustic waves. In this process, which is known as Stimulated Brillouin Scattering (SBS), the incident electromagnetic

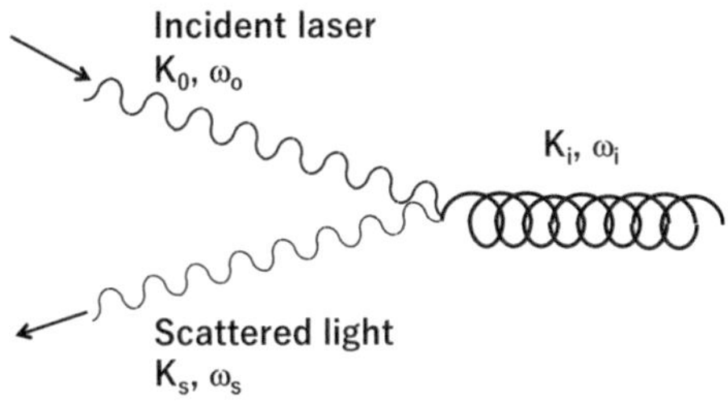

Fig. 12.7 Resonant three wave coupling diagram for the parametric decay in the critical density region, where propagation of the electromagnetic wave is shown by the wavy line, and the electron plasma wave and the ion acoustic wave are shown by the circulating lines, respectively (Drake et al., 1974)

wave decays into a scattered electromagnetic wave ($\omega_s, \boldsymbol{k}_s$) and an ion acoustic wave ($\omega_i, \boldsymbol{k}_i$). A schematic of the SBS process is shown in Fig. 12.7. The frequency and wave vector matching conditions for the resonant decay are,

$$\omega_0 = \omega_s + \omega_i, \quad \text{and} \quad \boldsymbol{k}_0 = \boldsymbol{k}_s + \boldsymbol{k}_i. \tag{12.34}$$

The coupled mode equations for the electromagnetic and ion-acoustic waves are

$$\frac{\partial^2 \boldsymbol{A}}{\partial t^2} + 2\nu_{EM}\frac{\partial \boldsymbol{A}}{\partial t} - c^2\nabla^2\boldsymbol{A} + \omega_{pe}^2\boldsymbol{A} = -\,\omega_{pe}^2\frac{\tilde{n}_l}{n_0}\boldsymbol{A}, \tag{12.35}$$

$$\frac{\partial^2 \tilde{n}_l}{\partial t^2} + 2\nu_a\frac{\partial \tilde{n}}{\partial t} + \omega_{pi}^2\lambda_e^2\nabla^2\tilde{n} = n_0\frac{Z}{m_i}\nabla^2\Phi_p, \tag{12.36}$$

where Φ_p is the ponderomotive potential given by Eq. (12.8) and ν_a is the dumping rate of the ion acoustic wave. When dumping is negligible, the growth rate is maximum for the back scattering of SBS and is given by

$$\Gamma_{SBS} = \frac{k_0 v_0 \omega_{pi}}{\sqrt{8k_0 c_S \omega_0}} \tag{12.37}$$

for the weak coupling case, which requires $\Gamma_{SBS} < \omega_{pi}$. This condition corresponds to the requirement to the laser intensity of $I_L < 10^{14}$ W/cm^2 in 1 keV low-Z plasmas.

In the strong coupling limit of $v_0/v_e >> 8\sqrt{Zm_e/m_i}$, the growth rate of the SBS instability is

$$\Gamma_{SBS} = \frac{\sqrt{3}}{2}\left(\frac{k_a^2 v_0^2}{4\omega_s}\omega_{pi}^2\right)^{1/3}, \tag{12.38}$$

where v_0, k_a and ω_s are the electron quiver velocity, ion acoustic wave number, and the frequency of the scattered electromagnetic wave, respectively.

12.7 Parametric Electron–Ion Decay and Oscillating Two Stream Instability

For developing the parametric instabilities, the matching conditions for the frequencies and wave-vectors are

$$\omega_0 = \omega_e + \omega_i, \quad \text{and} \quad \boldsymbol{k}_0 = \boldsymbol{k}_e \pm \boldsymbol{k}_i, \tag{12.39}$$

respectively. Referring to Fig. 12.8, the incident electromagnetic wave with frequency and wave-vector $(\omega_0, \boldsymbol{k}_0)$ decays into an electron plasma wave $(\omega_e, \boldsymbol{k}_e)$ and an ion acoustic wave $(\omega_i, \boldsymbol{k}_i)$. Since the electromagnetic wave frequency is about the electron plasma wave frequency, $\omega_0 \approx \omega_e$, and both are substantially larger than of the frequency of the ion-acoustic wave, $\{\omega_0, \omega_e\} >> \omega_i$, this instability develops in the region where the plasma density equals the critical density. Through this instability, the electron plasma-wave turbulence heats electrons by the wave particle resonance and produces high energy electrons whose energy is several 10 keV (Dubois & Goldman, 1965; Kruer, 1988; Mima & Nishikawa, 1984; Nishikawa, 1968a, 1968b; Silin, 1965; Valeo et al., 1972; Yamanaka et al., 1972).

Parametric instability is suppressed due to the density profile steepening, since this parametric instability is convective. The threshold of instability derived by Perkins et al. (1971) is

$$\Gamma_P = \frac{\pi}{2} \frac{\sqrt{v_{ge} c_S}}{L}. \tag{12.40}$$

Here, v_{ge} and c_S are the electron plasma wave group velocity and ion acoustic wave velocity. The maximum growth rate of instability is given by (Kruer, 1988)

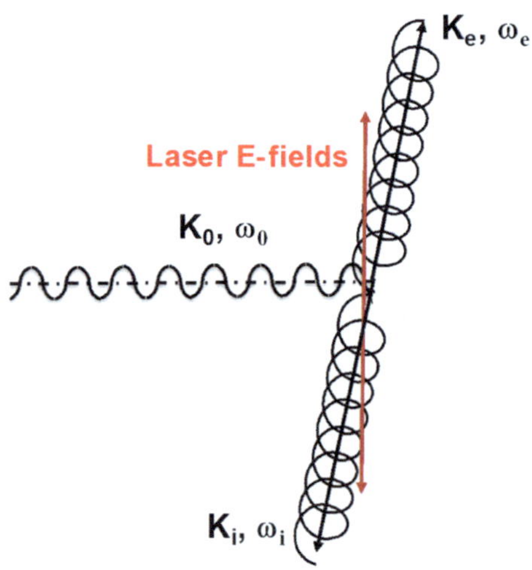

Fig. 12.8 Resonant three wave coupling diagram for the parametric decay in the critical density region, where propagation of the electromagnetic wave is shown by the wavy line, and the electron plasma wave and the ion acoustic wave are shown by the circulating lines, respectively

$$\Gamma_{P,\max} = \frac{\sqrt{3}}{2}\omega_L \left[\frac{Zm_e}{m_i} \frac{(kv_0)^2}{16\omega_L^2} \right]^{1/3}. \tag{12.41}$$

This expression has been obtained assuming that the intensity of the laser light is high enough and the growth rate is higher than the ion wave frequency. Here, v_0 is the electron quiver velocity in the laser field, m_e, m_i, and Z are the electron mass, the ion mass and the charge number, and k is the wave number of the unstable mode. The threshold of the parametric decay instability is

$$\left(\frac{v_0}{v_{th,e}} \right)^2 > \sqrt{\frac{k_D}{k}} \left(\frac{m_i}{Zm_e} \right)^{1/4} \left(\frac{2\pi}{k_0 L} \right)^3, \tag{12.42}$$

where $k_D = \omega_{pe}/v_{th,e}$ is the Debye wave number and $v_{th,e}$ is the electron thermal velocity. This indicates that the instability threshold depends strongly on the scale-length of the plasma density inhomogeneity. For example, if the electron temperature is $T_e = 5$ keV and the intensity and the wavelength of the laser are 10^{15} W/cm^2 and 0.35 μm respectively, we have $\left(v_0/v_{th,e}\right)^2 \sim 0.01$. Under these conditions, $k_D L$ is required to be greater than 200. This condition is not satisfied when density steepening occurs.

In the slightly over-dense region, it is known that two electron-plasma-waves propagating oppositely are coupled through laser electric field and zero frequency ion density fluctuation becomes unstable. This is called the oscillating two stream instability (Mima & Nishikawa, 1984; Nishikawa, 1968b). Since four waves are coupled in this process, it is also called a four-wave process.

As a result, in the process of target implosion, the parametric instabilities near the cut-off density region are very weak and the effect of those instabilities is small in comparison with the resonance absorption, and two plasmon decay instability as well as stimulated Raman scattering, which are discussed in the following section.

12.8 Self-focusing and Filamentation Instability

The self-focusing (SF) of the laser beam in plasma has a character of instability. This is the process of the laser intensity modulation in transverse direction due to modulation of the plasma refractive index (Askar'yan, 1962; for details see e.g. Cohen et al., 1991; Kaw et al., 1973; Schmitt, 1988). The mechanism of this instability is illustrated in Fig. 12.9.

This instability plays an important role in achieving uniform irradiation of the target both in the indirect and direct drive implosion. In the SF instability, four waves, incident laser, two scattered light, and electron density fluctuation, are coupled. This instability is a kind of modulational instability. The coupled wave equations can be obtained by extending Eqs. (12.13) and (12.14) to the relativistic case. We find

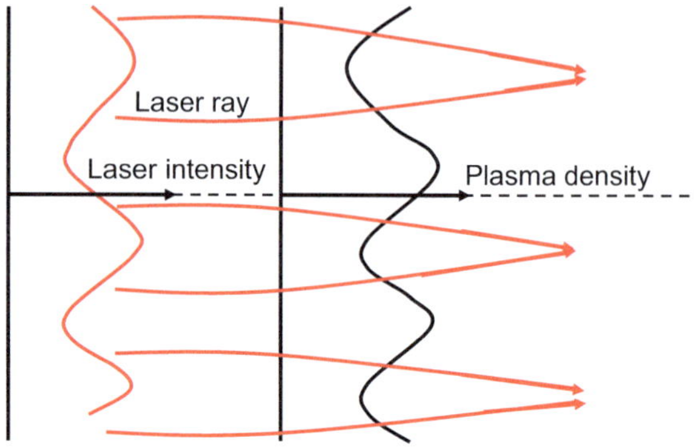

Fig. 12.9 Self-focusing/filamentation instability of the laser light in plasma

$$-3v_{e,th}^2\nabla^2\tilde{n}_e + \omega_{pe}^2(\tilde{n}_e - \tilde{n}_i) = n_0 c^2 \nabla^2 \gamma_e, \tag{12.43}$$

$$\frac{\partial^2 \boldsymbol{A}}{\partial t^2} + 2\nu_{EM}\frac{\partial \boldsymbol{A}}{\partial t} - c^2\nabla^2\boldsymbol{A} + \frac{\omega_{pe}^2}{\gamma_e}\boldsymbol{A} = -\omega_{pe}^2\frac{\tilde{n}_e}{\gamma_0 n_0}\boldsymbol{A}_0, \tag{12.44}$$

with

$$\boldsymbol{A} = \frac{1}{2}\left[\boldsymbol{A}_0 + \sum_s \boldsymbol{A}_s \exp(i\boldsymbol{k}_{s\perp}\cdot\boldsymbol{r}_\perp)\right]\exp(ik_0 z - i\omega_0 t) + c.c. \tag{12.45}$$

Here, $\boldsymbol{A}_0$ is the vector potential of the incident laser with wavenumber and frequency k_0, ω_0 and $\vec{A}_s(z,t)$ is the slowly varying amplitude of the vector potential of the intensity modulation in the perpendicular direction $\boldsymbol{r}_\perp = (x, y)$. When the electron–ion charge neutrality $(\tilde{n}_e - \tilde{n}_i) = 0$ is met and the laser modulation amplitudes are small in comparison with $\boldsymbol{A}_0$, the above equations are linearized to $\boldsymbol{A}_s(z,t)$, leading to the following coupled mode equations:

$$\frac{\partial A_+}{\partial t} + v_g\frac{\partial A_+}{\partial z} + i\frac{K^2c^2}{2\omega_0}A_+ + \nu A_+ = i\frac{\omega_{pe}^2}{2\omega_0}\frac{\tilde{n}_K}{n_0}A_0, \tag{12.46}$$

$$\frac{\partial A_-}{\partial t} + v_g\frac{\partial A_-}{\partial z} - i\frac{K^2c^2}{2\omega_0}A_- + \nu A_- = -i\frac{\omega_{pe}^2}{2\omega_0}\frac{\tilde{n}_{-K}}{n_0}A_0. \tag{12.47}$$

Here we assume that the modulation is in x-direction and the polarizations of the electromagnetic waves is in y-direction for the two obliquely propagating electromagnetic waves:

$$\sum_s \boldsymbol{A}_s(z,t)\exp(ik_{sx}x) = \left[A_+(z,t)\exp(iKx) + A_-(z,t)\exp(-iKx)\right]. \tag{12.48}$$

At non-relativistic laser intensity, $\boldsymbol{A}_s(z,t) < 1$, the linearization of Eq. (12.43) yields

$$\tilde{n}_K(z,t) = \frac{n_0 e^2}{4T_e^2}\left(A_0^* A_+ + A_0 A_-^*\right)\exp(iKx) \tag{12.49}$$

Equations (12.46), (12.47), and (12.44) yield

$$\frac{\partial A_+}{\partial t} + v_g \frac{\partial A_+}{\partial z} + i\frac{K^2c^2}{2\omega_0}A_+ + \nu A_+ = i\frac{\omega_{pe}^2}{8\omega_0}\frac{e^2}{T_e^2}\left(A_0^* A_+ + A_0 A_-^*\right)A_0, \tag{12.50}$$

$$\frac{\partial A_-^*}{\partial t} + v_g \frac{\partial A_-^*}{\partial z} - i\frac{K^2c^2}{2\omega_0}A_-^* + \nu A_-^* = -i\frac{\omega_{pe}^2}{8\omega_0}\frac{e^2}{T_e^2}\left(A_0 A_-^* + A_0^* A_+\right)A_0^*. \tag{12.51}$$

Assuming $\left\{A_+, A_-^*\right\} \propto \exp(iQz - i\Omega t)$ in Eqs. (12.50) and (12.51), we obtain the dispersion equation for Q and Ω,

$$\left(\Omega + v_g Q + i\nu\right)^2 - \left(\frac{K^2c^2}{2\omega_0} - i\frac{\omega_{pe}^2}{8\omega_0}\frac{e^2}{T_e^2}|A_0|^2\right)^2 = -\frac{\omega_{pe}^4}{64\omega_0^2}\frac{e^4}{T_e^4}|A_0|^4. \tag{12.52}$$

The maximum growth rate of the filamentation instability $\mathrm{Im}\{\Omega\}$ obtained from the above equation is

$$\mathrm{Im}\{\Omega\} = \omega_0 \frac{n_0}{8n_c}\left(\frac{eA_0}{mcv_{th,e}}\right)^2 - \nu \tag{12.53}$$

for the transverse modulation wave number $K = k_0\left(eA_0/2m_e c^2\right)$. In otherwise, the spatial growth rate is given by $Q = \mathrm{Im}\{\Omega\}/v_g$.

Therefore, for the moderate laser intensity, the filament size 1/K is relatively large and the filamentation instability is stabilized when the scale-length of the plasma inhomogeneity is small compared with the instability wavelength, i.e. $L < 1/Q$. As depicted in Fig. 12.9, the laser intensity is modulated in a transverse direction by the plasma density modulation due to the action of the ponderomotive force.

References

Askar'yan, G. A. (1962). Effects of the gradient of a strong electromagnetic beam on electrons and atoms. *Soviet Physics—JETP, 15*, 1088.

Cohen, B. F., Lasinski, A. B., Langdon, B., & Cummings, J. C. (1991). Dynamics of ponderomotive self-focusing in plasmas. *Physics of Fluids B, 3*, 766.

Drake, J. F., Kaw, P. K., & Lee, Y. C. (1974). Parametric instabilities of electromagnetic waves in plasmas. *Physics of Fluids, 17*, 778.

DuBois, D. F., & Goldman, M. V. (1965). Radiation-induced instability of electron plasma oscillations. *Physical Review Letters, 14*, 544.
Estabrook, K., Kruer, W. L., & Lasinski, B. F. (1980). Heating by Raman backscatter and forward scatter. *Physical Review Letters, 45*, 1399.
Forslund, D. W., Kindel, J. M., & Lindman, E. L. (1975). Theory of stimulated scattering processes in laser-irradiated plasmas. *Physics of Fluids, 18*, 1002.
Froula, D. H., Divol, L., London, R. A., et al. (2009). Observation of the density threshold behavior for the onset of stimulated Raman scattering in high-temperature hohlraum plasmas. *Physical Review Letters, 103*, 0450006.
Goldman, M. V., & Dubois, D. F. (1966). Parametric plasmon-photon interactions: Part I. *Annals of Physics (New York), 38*, 117.
Igumenshchev, I. V., Edgell, D. H., Goncharov, V. N., et al. (2010). Crossed-beam energy transfer in implosion experiments on OMEGA. *Physics of Plasmas, 17*, 122708.
Igumenshchev, I. V., Seka, W., Edgell, D. H., et al. (2012). Crossed-beam energy transfer in direct-drive implosions. *Physics of Plasmas, 19*, 056314.
Kaw, P. K., Schmidt, G., & Wilkocx, T. (1973). Filamentation and trapping of electromagnetic radiation in plasma. *Physics of Fluids, 16*, 1522.
Kruer, W. L. (1988). *The physics of laser plasma interactions*. In *Frontiers of Physics Series* (Vol. 73). Addison Wesley Publishing Company.
Kruer, W. L., Wilks, S. C., et al. (1996). Energy transfer between crossing laser beams. *Physics of Plasmas, 3*, 382.
Liu, C. S., Rosenbluth, M., & White, R. B. (1973). Parametric scattering instability in inhomogeneous plasmas. *Physical Review Letters, 31*, 697.
Michel, P., Divol, L., Williams, E. A., et al. (2009). Tuning the implosion symmetry of ICF targets via controlled cross-beam-energy-transfer. *Nuclear Fusion, 22*, 561.
Mima, K., & Nishikawa, K. (1984). *Basic plasma physics II*. In A. A. Galeev & R. N. Sudan (Chap. 6.5, p. 452).
Morales, G. J., & Lee, Y. C. (1974). Ponderomotive-force effects in a nonuniform plasma. *Physical Review Letters, 33*, 1016.
Nishikawa, K. (1968a). Parametric excitation of coupled waves I. General formulation. *Journal of the Physical Society of Japan, 24*, 916.
Nishikawa, K. (1968b). Parametric excitation of coupled waves. II. Parametric plasmon-phonon interaction. *Journal of the Physical Society of Japan, 24*, 1152.
Pesme, D., Laval, G., & Pellat, R. (1973). Parametric instabilities in bounded plasmas. *Physical Review Letters, 31*, 203.
Perkins, F. W. & Flick, J. (1971). Parametric instabilities in inhomogeneous plasmas. *Physics of Fluids, 14*, 2012.
Rosenbluth, M. N. (1972). Parametric instabilities in inhomogeneous media. *Physical Review Letters, 29*, 565.
Schmitt, A. J. (1988). The effects of optical smoothing techniques on filamentation in laser plasmas. *Physics of Fluids, 31*, 3079.
Seka, W., Edgell, D. H., Myatt, J. F., et al. (2009). Two-plasmon-decay instability in direct-drive confinement fusion experiments. *Physics of Plasmas, 16*, 052701.
Short, R. W., Seka, W., & Tanaka, K. A. (1984). Two-plasmon decay and three-halves harmonic generation in filamentations in a laser-produced plasma. *Physical Review Letters, 52*, 1496.
Silin, V. P. (1965). Parametric resonance in a plasma. *JETP, 21*, 1127.
Simon, A., Short, R. W., Williams, E. A., & Dewandre, T. (1983). On the inhomogeneous two-plasmon instability. *Physics of Fluids, 26*, 3107.
Stoeckl, C., Bahr, R. E., Yaakobi, B., et al. (2003). Multibeam effects on fast-electron generation from two-plasmon-decay instability. *Physical Review Letters, 90*, 235002.
Valeo, E., Oberman, C., & Perkins, F. W. (1972). Saturation of the decay instability for comparable electronic and ion temperatures. *Physical Review Letters, 28*, 340.
Yamanaka, C., Yamanaka, T., Sasaki, T., Yoshida, K., Waki, M., & Kang, H. B. (1972). Anomalous heating of a plasma by a laser. *Physical Review A, 6*, 2335.

Chapter 13
Heat Transport and Laser Ablation

Abstract In this chapter, we describe transport of the deposited laser energy from the corona-plasma to the solid surface, ablation of the solid target surface, hydrodynamics of plasma expansion, and the self-generated magnetic fields.

Keywords Electron heat transport · Nonlocal heat transport · Spitzer–Harm heat flux · Radiation heat transport · Ablation pressure · Heat wave · Shock wave · Rocket model · Self-generated magnetic field

13.1 Introduction

When an intense pulse laser irradiates a solid target as shown in Fig. 13.1, the laser is absorbed through various processes: atomic excitation including multi-photon absorption, optical phonon excitation, multi-photon ionization, free-free transition (collisional absorption), and so on. Then, the solid surface is heated to the temperature of the order of 10 eV or higher and the plasma layer is formed, with the laser intensity reaching higher than 10^{11} W/cm^2. Note here that the free streaming electron heat flux $n_e v_{th,e} T_e$ is the order 10^{11} W/cm^2 for $T_e = 10$ eV and $n_e \approx n_c$ (the electron density is about the critical plasma density).

Once a plasma layer is formed on the solid surface, the laser is absorbed near the critical density of the plasma mainly through inverse bremsstrahlung as described in Sect. 3.4. When the laser intensity increases to 10^{11} W/cm^2 or higher, the plasma is further heated up to a few keV electron temperature, the inverse bremsstrahlung is lowered, and the collective absorption starts as described in Chap. 4. When the pulse duration of the laser is longer than a few 10 picoseconds, the plasma hydrodynamic motion starts. Namely, electrons and radiation carry the absorbed laser energy to the solid surface and then the surface is ablated.

The ablated plasma expands into vacuum and the ablation reaction (ablation pressure) pushes the target. As shown in Fig. 13.1, the ablation pressure compresses the target to launch a shock wave into the solid. The above processes are summarized

Y. Kato et al., *High Power Laser and Plasma Science*, Springer Series on Atomic, Optical, and Plasma Physics 130, https://doi.org/10.1007/978-3-032-08930-4_13

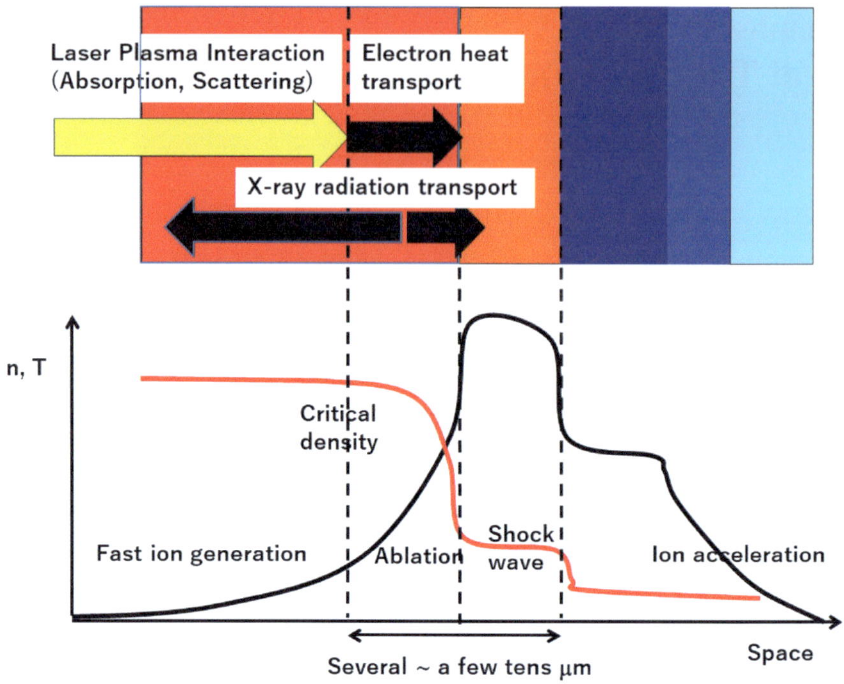

Fig. 13.1 Profiles of the plasma density (black line) and the temperature (red line) of a laser-heated solid foil

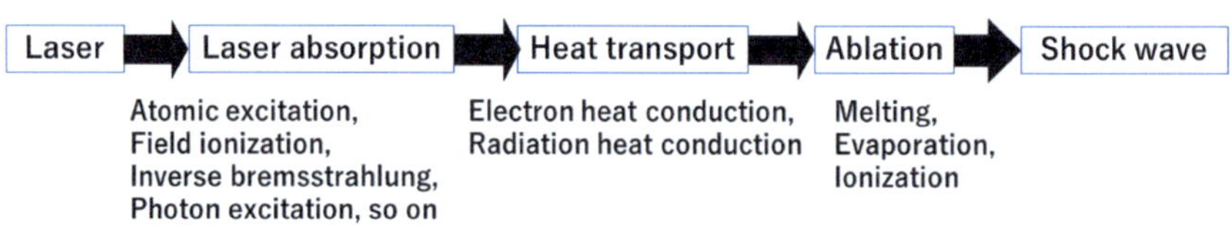

Fig. 13.2 Processes related to laser ablation

in Fig. 13.2. Heat conduction by electron and radiation and ablation hydrodynamics are described in the following Section.

13.2 Electron Heat Transport

Linear and nonlinear electron heat transport depends upon the temperature scale-length relative to the electron mean free path. In this section, Spitzer–Harm heat conduction and nonlocal heat conduction are described.

13.2.1 Classical Theory of Heat Transport

The absorbed laser energy is carried into the solid target by electron heat conduction. Time evolution of the electron temperature is described by an energy conservation equation. When the spatial scale length of the electron temperature gradient is much longer than the electron-mean-free-path l_e, the heat flux is proportional to the temperature gradient according to the Fourier's law. So, the electron heat flux $\boldsymbol{q}_e$ is given by

$$\boldsymbol{q}_e = -\kappa_e \nabla T_e \tag{13.1}$$

The plasma heat conductivity κ_e has been derived by Braginskii (1958, 1965). For a fully ionized hydrogen plasma,

$$\kappa_e = 3.2 k_B l_e v_{th,e} n_e \tag{13.2}$$

The mean-free-path is defined by $l_e = v_{th,e}/\nu_e = v_{th,e}\tau_e$, where ν_e, l_e, τ_e, and $v_{th,e}$ are the electron collision frequency, mean-free-path, and the electron thermal velocity respectively. $k_B = 1.380 \times 10^{-23}$ J/K or $k_B = 8.617 \times 10^{-5}$ eV/K is Boltzmann constant. For hydrogen plasmas, the electron–ion collision frequency ν_e is given by

$$\nu_e = \frac{4\sqrt{2\pi} n_i e^4 \ln \Lambda}{3\sqrt{m_e}(k_B T_e)^{3/2}(4\pi\varepsilon_0)^2} \tag{13.3}$$

according to the Spitzer–Harm model (Spitzer, 1956). Here, $\ln \Lambda$ is the Coulomb logarithm, which is approximately equal to

$$\ln \Lambda = 24 - \ln\left(n_e^{0.5} T_e^{-1}\right) \tag{13.4}$$

for the electron–electron collision when $T_e > 10$ eV,

$$\ln \Lambda = 23 - \ln\left(n_e^{0.5} Z T_e^{-1.5}\right) \tag{13.5}$$

for $T_e < 10Z^2$ eV, and

$$\ln \Lambda = 24 - \ln\left(n_e^{0.5} T_e^{-1}\right) \tag{13.6}$$

when $T_e > 10Z^2$ eV (Huba, 2000). In these expressions, the electron density n_e is measured in cm^{-3} and the electron temperature T_e in eV.

Then, the space–time evolution of the electron temperature is subject to the following electron heat conduction equation when the plasma pressure is uniform,

$$\frac{\partial(\rho c_P T_e)}{\partial t} = -\nabla \cdot \boldsymbol{q}_e = \nabla \cdot \kappa_e \nabla T_e \tag{13.7}$$

where c_P is the heat capacity for constant pressure. Note here that when the plasma pressure is inhomogeneous, the ambipolar electric field exists and the electric field effect on the heat conduction must be considered. In this case, there is a dependence between the heat flux $\boldsymbol{q}_e$ and the electron temperature T_e on one hand and the electric field $\boldsymbol{E}$ and the electric current density $\boldsymbol{j}$ on the other hand. This relationship is determined by the system of equations (Spitzer, 1953):

$$\frac{\boldsymbol{j}}{\sigma_\perp} = \boldsymbol{E} + \frac{\alpha}{\sigma_\perp} \nabla \boldsymbol{T_e} \tag{13.8}$$

$$\boldsymbol{q} = -\beta \boldsymbol{E} - \kappa_\perp \nabla \boldsymbol{T_e} \tag{13.9}$$

These equations are the generalized Ohm's law, where $\sigma_\perp$ and $\kappa_\perp$ are the electric and thermal conductivities of plasma in the direction perpendicular to the magnetic field, and the coefficients α and β represent the thermo-electric effects related to the Nernst-Ettinghausen effect. The ratio $\alpha/\beta = -S$ where S is the Seebeck coefficient and the term $-\beta \boldsymbol{E}$ is related to the Peltier effect (Spitzer, 1953, 1956).

13.2.2 Non-local Heat Transport

The diffusion approximation of the electron heat transport breaks down when the electron mean-free path is finite in comparison with the temperature gradient. The finite temperature gradient effects are included to analyze the experimental results by introducing a flux limiter, which describes the saturation of the heat conductivity when the heat flux is higher than a fraction of the free streaming limit,

$$f\, n_e \sqrt{\frac{k_B T_e}{m_e}} k_B T$$

where $f \approx 0.1-0.3$ is so-called the flux limit factor.

The concept of the flux limiter was introduced by Bickerton (1973). The kinetic theory of the electron heat transport in steep heat transport regime has been investigated in many works (e.g. Bell et al., 1981; Nishiguchi et al., 1992; Craxton et al., 2015). When there is a temperature gradient, the electron distribution function becomes non-isotropic. According to Eq. (9.30), the electron distribution function satisfies the equation in non-uniform unmagnetized plasmas,

$$\frac{\partial f}{\partial t} + \frac{1}{m} \boldsymbol{p} \frac{\partial f}{\partial \boldsymbol{r}} + Ze\left(\boldsymbol{E} + \frac{\boldsymbol{p}}{m} \times \boldsymbol{B}\right) \frac{\partial f}{\partial \boldsymbol{p}}$$

$$= \sum_i \frac{\Gamma_i}{2} \frac{\partial}{\partial \boldsymbol{u}} \int d\boldsymbol{p}' \frac{1}{|\boldsymbol{u}|} \left(\overleftrightarrow{I} - \frac{\boldsymbol{u}\boldsymbol{u}}{|\boldsymbol{u}|^2} \right) \left(\frac{\partial}{\partial \boldsymbol{p}} - \frac{\partial}{\partial \boldsymbol{p}'} \right) f(\boldsymbol{p}) f(\boldsymbol{p}') = C(f) \tag{13.10}$$

where $\boldsymbol{u} = \boldsymbol{v} - \boldsymbol{v}'$ and $\Gamma_i = \left(Z^2 Z_i^2 e^4/\varepsilon_0\right) \ln(\chi_{\max}/\chi_{\min})$. In the r.h.s. of this equation the collision integral is written in Landau form (see Landau, 1936).

For the case when the electron temperature is inhomogeneous with the temperature gradient in the x-direction, the velocity distribution function $f_e(\boldsymbol{v},x)$ is anisotropic and given by the series of spherical harmonics,

$$f_e(v,x) = f_0(v,x) P_0(\xi) + f_1(v,x) P_1(\xi) + f_2(v,x) P_2(\xi) + \cdots \tag{13.11}$$

Here $v = \sqrt{v_x^2 + v_y^2 + v_z^2}$, $\xi = v_x/v$, and $P_0(\xi) = 1, P_1(\xi) = \xi, P_2(\xi) = (3\xi^2 - 1)/2$ are the Legendre Polynomials which are normalized as

$$\frac{1}{2} \int_{-1}^{1} d\xi \, (P_l(\xi))^2 = 1 \text{ and } \frac{1}{2} \int_{-1}^{1} d\xi P_l(\xi) P_m(\xi) = 0 \text{ if } l \neq m$$

For simplicity to understand the kinetic effects in the electron thermal transport, we consider only the electron–ion collision and assume that the electron-to-ion mass ratio is zero by neglecting the ion thermal velocity effects. This leads to reducing the integrand in r.h.s. of Eq. (13.10) to

$$\frac{n_i \Gamma_i}{2 m_e^2} \frac{\partial}{\partial \boldsymbol{v}} \left(\frac{1}{|\boldsymbol{v}|} \left(\overleftrightarrow{I} - \frac{\boldsymbol{v}\boldsymbol{v}}{|\boldsymbol{v}|^2} \right) \frac{\partial}{\partial \boldsymbol{v}} f_e(\boldsymbol{v}) \right) \tag{13.12}$$

Substituting the expression for the distribution function given by Eqs. (13.11)–(13.12) we obtain

$$C(f) = \sum_{j=1}^{\infty} \frac{n_i \Gamma_i}{2 m_e^2} f_j(v,x) \frac{\partial}{\partial \boldsymbol{v}} \left(\frac{1}{|\boldsymbol{v}|} \left(\overleftrightarrow{I} - \frac{\boldsymbol{v}\boldsymbol{v}}{|\boldsymbol{v}|^2} \right) \frac{\partial}{\partial \boldsymbol{v}} P_j(\xi) \right) \tag{13.13}$$

Note here that there is no energy transfer in the electron–ion collisions within the framework of the present approximation, so that the action of collision operator on the function of the absolute value $v = |\boldsymbol{v}|$ of velocity $\boldsymbol{v}$ is equal to zero, and it is finite only for the non-isotropic part as follows from Eq. (13.13).

When the temperature gradient is small and the plasma is stationary, we retain the terms of $j = 0$, 1 in Eqs. (13.10) and (13.13) only, to obtain the equation giving the relationship between the functions $f_0(v,x)$ and $f_1(v,x)$,

$$\xi v \frac{\partial f_0}{\partial x} - \frac{e}{m_e} \xi E \frac{\partial f_0}{\partial v} = -\sqrt{3} \frac{n_i \Gamma_i}{m_e^2 v^3} f_1 \xi \tag{13.14}$$

Then, we obtain the electron heat flux in the x direction,

$$q_x = \int d^3\mathbf{v}\frac{m_e v^2}{2} v_x f_1(v,x) P_1\left(\frac{v_x}{v}\right) = -\int d^3\mathbf{v}\frac{m_e^3 v^6}{2n_i\Gamma_i}\xi^2 v\left(v\frac{\partial f_0}{\partial x} - \frac{e}{m_e}E_x\frac{\partial f_0}{\partial v}\right) \quad (13.15)$$

Equation (13.15) shows that the heat flux is the higher-order velocity moment of the electron velocity distribution function. The main contribution to the integral in Eq. (13.15) comes from the point where $v \approx 3v_{th,e}$. In other words, the heat flux is very sensitive to the form of the high energy tail of the distribution function. In realistic situations which could be realized in laser heated plasmas, the form of the high energy tail could be distorted and not described by the Maxwellian distribution as can be seen in Fig. 13.3.

When the isotropic part of the distribution function $f_0(v)$ is assumed to be Maxwellian locally, then, we obtain the thermal conductivity κ and the Peltier coefficient β (see r.h.s. of Eq. (13.9)) by carrying out the integration of Eq. (13.15):

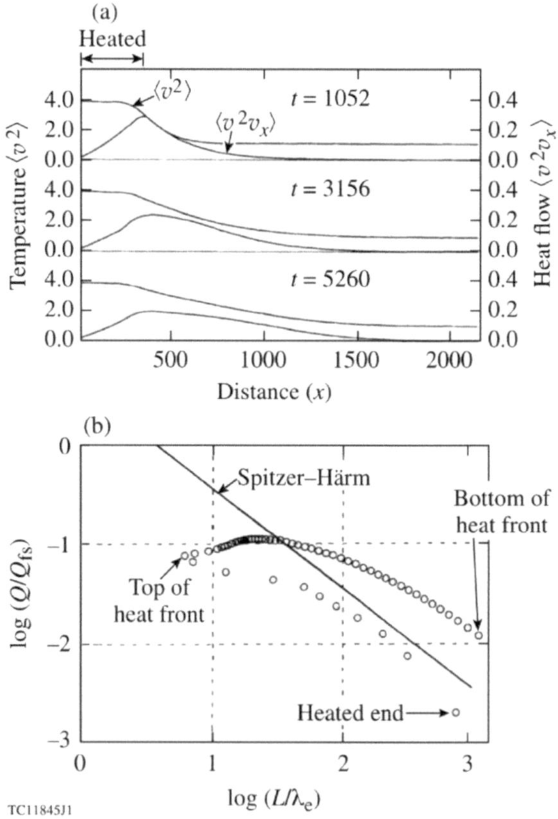

Fig. 13.3 The Fokker-Planck simulation of the non-local heat transport. **a)** Temperature profile and heat flow normalized by free streaming heat flow in the x-direction which is normalized by the Debye length. **b)** Heat flux versus temperature scale length normalized by the electron collision mean-free path length. Reprinted from Bell et al. (1981), with permission from The American Physical Society and the author(s). © 1981 The American Physical Society

$$\kappa = \frac{k_B n_e m_e^2}{6\sqrt{2\pi} n_i \Gamma_i v_{th,e}^7} \int_0^\infty dv v^{11} \exp\left(-\frac{v^2}{2v_{th,e}^2}\right) = \frac{640 k_B n_e m_e^2 v_{th,e}^5}{\sqrt{2\pi} n_i \Gamma_i} \tag{13.16}$$

$$\beta = \frac{e n_e m_e^2}{3\sqrt{2\pi} n_i \Gamma_i v_{th,e}^5} \int_0^\infty dv v^9 \exp\left(-\frac{v^2}{2v_{th,e}^2}\right) = \frac{128\, e n_e m_e^2 v_{th,e}^5}{\sqrt{2\pi} n_i \Gamma_i} \tag{13.17}$$

Here we use the fact that the integral in the r.h.s. of these equations is expressed in terms of the Euler gamma function $\Gamma(x)$:

$$\int_0^\infty dv v^a \exp\left(-\frac{v^2}{2w^2}\right) = \frac{1}{2}\left(\sqrt{2}w\right)^{1+a} \Gamma\left(\frac{1+a}{2}\right)$$

The heat conductivity given by Eq. (13.16) can be compared with the Braginskii heat conductivity given by Eq. (13.2). For the electron–ion collisions where the collision frequency ν_e is given by Eq. (13.3), Eq. (13.16) takes the form

$$\kappa = \frac{5.4}{Z^2} k_B l_e v_{th,e} n_e \tag{13.18}$$

The difference between the expressions of Eqs. (13.16) and (13.2) for the hydrogen plasma is a factor 5.2/3.2, which is because the electron–electron collision effects are not included in Eq. (13.16). But for high Z plasmas, Eqs. (13.16) and (13.17) are good approximations.

When the plasma density is homogeneous and there is no electron current, the electric field is determined by the current neutrality condition

$$\int d^3\mathbf{v} \frac{m_e^3 v^4}{2n_i \Gamma_i} \xi^2 v \left(v \frac{\partial f_0}{\partial x} - \frac{e}{m_e} E_x \frac{\partial f_0}{\partial v}\right) = 0 \tag{13.19}$$

which yields

$$E_x = -\frac{3k_B}{e} \frac{dT_e}{dx}$$

Therefore, effective heat conductivity is reduced by a factor of 0.4.

When the scale length of the temperature inhomogeneity is finite, the above expression for the thermal conductivity is modified because of the break-down of the local thermal equilibrium. The energy is mainly carried by electrons with the velocity $v \approx 3v_{th,e}$ whose mean free path is almost 100 times longer than the average electron collision mean-free path length. Namely, the perturbations are strong, $|f_1(v,x)| > |f_0(v,x)|$ for $v \approx 3v_{th,e}$, and the distribution function $f_0(v,x)$ is no more Maxwellian when the temperature scale length is less than 100 times the average electron mean free path. Therefore, the electrons with the velocity higher

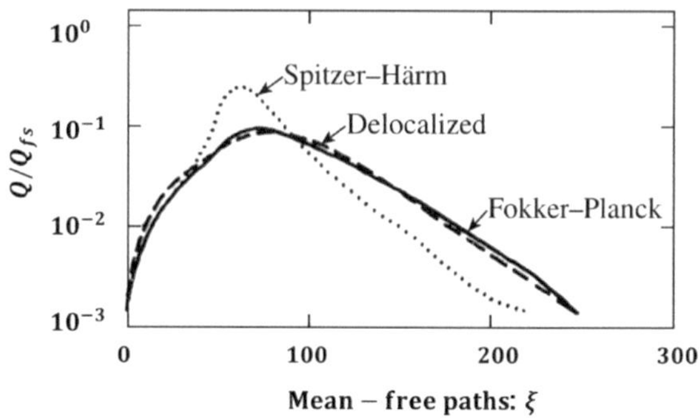

Fig. 13.4 Comparison of the non-local heat flux model by Luciani with the Spitzer–Harm and the Fokker–Planck simulation result. Reprinted from Luciani et al. (1983), with permission from The American Physical Society and the author(s). © 1983 The American Physical Society

than 3 times of the thermal velocity could be mixed between the regions with high and low temperatures. This affects the high-energy tail of the distribution function, which is strengthened in the low-temperature region and reduced in the high-temperature region. Thus, the heat flux is strengthened in the low-temperature region and reduced in the high-temperature region as shown in Fig. 13.3. It has been recognized that non-local heat transport plays an important role in the laser plasma interaction and the laser ablation processes.

Luciani et al. (1983) introduced a delocalization model suggested by Bell et al. (1981), wherein the heat flux at one point is given by a weighted average (convolution integral) of the Spitzer–Harm heat flux at all other points, appropriately weighted according to the electron mean free path. As shown in Fig. 13.4, the results they obtained are consistent with those obtained using the one-dimensional Fokker–Planck code for the test problem. Their model can describe the preheating effects beyond the base of the heat front and has the merit of being far easier to implement in a hydrodynamics code than a Fokker–Planck model. The non-local nature of heat conductivity has been implemented in most of recent implosion simulation codes.

13.3 Laser Ablation

When an intense pulse laser irradiates a solid surface, the deposited laser energy is transported into the solid. Then, the solid surface is ablated, and a hot dense plasma is generated on the target surface. The plasma pressure pushes the solid target to generate a shock wave when the heat front propagation becomes subsonic. The plasma pressure behind the sub-sonic heat wave is called the "ablation pressure" (Mima et al., 1994). In this section, the laser-driven heat wave and the ablation process are described.

13.3.1 Heat Wave

The heat front propagation velocity is determined by the heat conduction Eq. (13.7). The self-similar solution $T(\xi)$ of this equation in partial derivatives satisfies the following ordinary differential equation:

$$-\xi \frac{d(c_p \rho T)}{d\xi} = \frac{d}{d\xi}\left(2\kappa(T)\frac{dT}{d\xi}\right) \tag{13.20}$$

In this approach, we use the self-similar variable $\xi = \delta x/\sqrt{t}$. Note that although we have used a notation ξ for the electromagnetic wave polarization ellipticity as well as for normalized velocity variable, here its usage will not cause any ambiguity.

For an ideal plasma, $c_p = 5Zk_B/2m_i$, Z is the average ion charge, and the thermal conductivity coefficient is proportional to $T^{5/2}$, namely, $\kappa = \kappa_0(T/T_0)^{5/2}$, according to Eq. (13.16). As a result, we find that $\delta = c_p\rho/2\kappa_0$. On the other hand, keeping $\delta = 1$ in the definition of independent variable ξ, we can choose variable $T(\xi)/T_0(2\kappa_0/\rho c_p)^{2/5}$ instead the $T(\xi)$ and reduce Eq. (13.20) to

$$\frac{d}{d\xi}\left(T^{5/2}\frac{dT}{d\xi}\right) + \xi\frac{dT}{d\xi} = 0$$

By implementing these relationships and solving numerically this equation, we obtain a self-similar solution,

$$T = T_0 f(\xi) \tag{13.21}$$

When the temperature is T_0 at the point $x = 0$, which is the coordinate where the laser energy is deposited, the initial conditions for the self-similar function is as follows: $f(0) = 1$, $f'(0) = 0$. The solution of Eq. (13.20) is shown in Fig. 13.5. The position of the heat front $\xi = \xi_f$ is determined by the condition $f(\xi_f) = 0$. Near the point $\xi = \xi_f$ the self-similar solution has approximately the form

$$f(\xi) = \left(1 - \frac{\xi}{\xi_f}\right)^{2/5} \tag{13.22}$$

where

$$\xi_f = \sqrt{\frac{44.8 l v_{e,th}}{15}} \tag{13.23}$$

It is easy to find that the heat wave front propagates as $x_f(t) = \sqrt{44.8 l v_{e,th} t/15}$. So, the heat wave-front velocity decreases with time according to

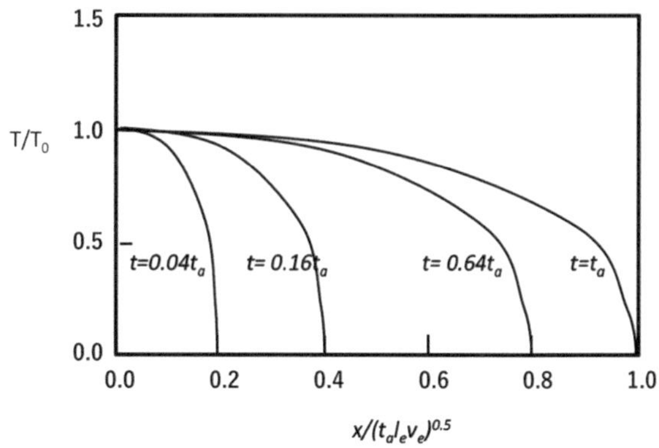

Fig. 13.5 Heat wave propagation from the boundary of fixed temperature

$$\dot{x}_f(t) = \sqrt{\frac{11.2\, l v_{e,th}}{15\, t}} \tag{13.24}$$

with time-development of the temperature profile shown in Fig. 13.5.

Let us assume that the heat front velocity is lower than the sound velocity, when the time is less than $t < t_a$ which can be estimated to be $t_a = (11.2 m_e / 15 m_i)\tau_e$, where $\tau_e = 1/\nu_e$ is the mean free time. When the temperature is equal to 1 keV and the plasma is of the solid hydrogen density, we estimate $t_a \sim 10$ ps. So, after this time, the ablative acceleration of the target starts. The ablative acceleration of the spherical shell to the center has been widely applied for laser fusion as discussed in the following sections.

13.3.2 *Stationary Ablation*

In stationary ablation, mass flow, momentum flow, and energy flux are conserved across the ablation front, as illustrated in Fig. 13.6. In the frame moving with the ablation front, the following conservation relations between the down-stream (laser heating region: laser cut-off) and the up-stream (the shocked solid region) are satisfied according to the equation of continuity, the equation of motion, and the equation of enthalpy conservation as follows:

$$\rho_c v_c = \rho_A v_A \tag{13.25}$$

$$\rho_c v_c^2 + P_c = \rho_A v_A^2 + P_A \approx P_A \tag{13.26}$$

$$\left(\frac{\gamma_c P_c}{\gamma_c - 1} + \frac{1}{2}\rho_c v_c^2 \right) v_c - I_{abs} = \left(\frac{\gamma_A P_A}{\gamma_A - 1} + \frac{1}{2}\rho_A v_A^2 \right) v_A \approx 0 \tag{13.27}$$

Fig. 13.6 Density profiles of the ablation and the shock wave due to laser irradiation of a target surface

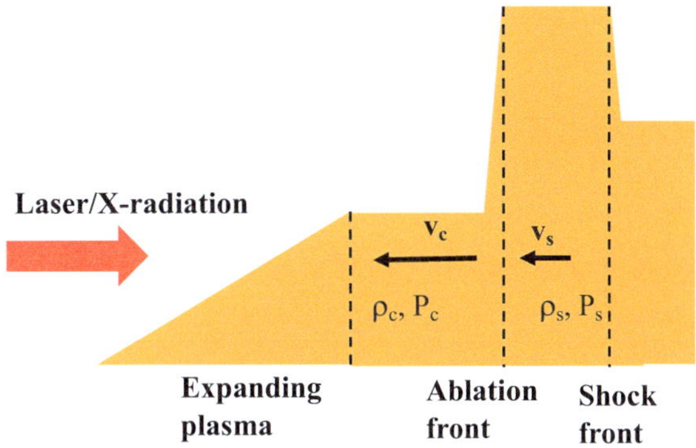

when $v_c \gg v_A$. Here, ρ_c, ρ_A, v_c, v_A, P_c, P_A, γ_c, γ_A, and I_{abs} are mass densities at the laser cut-off and the ablation front, flow velocities at the cut-off and the ablation front, pressures at the cut-off and the ablation front, adiabatic constants at the cut-off and the ablation front, and the absorbed laser intensity, respectively. For $\rho_c \gg \rho_A$ (which is satisfied for the laser-heated solid target), $v_c \gg v_A$ according to Eq. (13.25). So, the last conditions of Eqs. (13.26) and (13.27) are satisfied.

13.3.3 Laser Driven Ablation Pressure

When $\rho_A \gg \rho_c$, which is met with laser heated solid target, we assume $v_c \gg v_A \cong 0$ and that the flow velocity at the cut-off is the isothermal sound velocity, $v_c^2 = P_c/\rho_c$. We also assume the ideal equation of state (EOS) of the cut-off plasma, i.e. $\gamma_c = 5/3$. Then Eq. (13.27) is reduced to

$$2.5\sqrt{P_c^3/\rho_c} = I_{abs} \text{ which gives } P_c = \left(\frac{I_{abs}}{2.5}\right)^{2/3} \rho_c^{1/3} \tag{13.28}$$

Therefore, Eqs. (13.27) and (13.28) yield the ablation pressure P_A and the mass ablation rate $\dot{m}$ for a film target of the area mass $m(t)$ as follows

$$P_A = 2\left(\frac{I_{abs}}{2.5}\right)^{2/3} \rho_c^{1/3} \tag{13.29}$$

$$\dot{m} = \rho_c v_c = \sqrt{\rho_c P_c} = \left(\frac{I_{abs}}{2.5}\right)^{1/3} \rho_c^{2/3} \tag{13.30}$$

Since the cut-off density is proportional to $1/\lambda_L^2$ for the laser wavelength λ_L, Eqs. (13.29) and (13.30) can be rewritten as

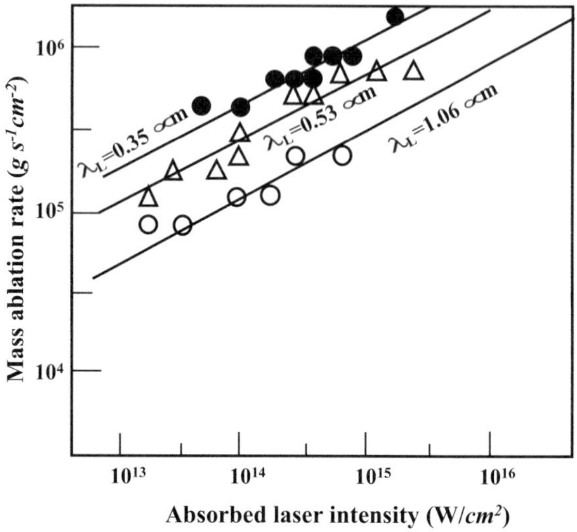

Fig. 13.7 Laser intensity and wavelength dependences of the mass ablation rate, where solid circles, open triangles, and open circles are for the laser wavelengths of 0.35 μm, 0.53 μm and 1.06 μm, respectively. Reprinted from Key et al. (1983), with permission from The American Institute of Physics Publishing: permission conveyed through Copyright Clearance Center, Inc. © 1983 The American Institute of Physics

$$P_A = 2\left(\frac{I_{abs}}{2.5}\right)^{2/3}\left(\frac{\rho_0}{\lambda_L^2}\right)^{1/3} \tag{13.31}$$

$$\dot{m} = \rho_c v_c = \sqrt{\rho_c P_c} = \left(\frac{I_{abs}}{2.5}\right)^{1/3}\left(\frac{\rho_0}{\lambda_L^2}\right)^{2/3} \tag{13.32}$$

Here ρ_0 is the cut-off mass density for a $\lambda_L = 1$ μm laser and the laser wavelength λ_L is measured in the unit of 1 μm. The ablation pressure and the mass ablation ratio are higher for shorter wavelength lasers, while the absorbed laser intensity is the same.

Dependences of the ablation scaling laws Eqs. (13.31) and (13.32) on the absorbed laser intensity and the laser wavelength have been verified in several laser ablation experiments. In Fig. 13.7, the scaling law of the mass ablation rate is compared with the experimental results obtained at the Rutherford Appleton Laboratory (Key et al., 1983). The agreement is better in the low intensity region. The experimental ablation mass rates are relatively smaller than the theoretical prediction in the higher intensity region. This may be due to two- and three-dimention (2&3D) effects and nonlinear laser plasma interactions.

13.3.4 Radiation Driven Ablation Pressure

According to Eq. (13.27), the expanding plasma energy flux is related to the X-ray flux as follows,

$$q_{ex} = 2.5 P_c v_c = 2.5 \rho_c v_c^3 \tag{13.33}$$

Since the ablated plasma temperature is equal to the radiation temperature T_r, the rarefaction iso-thermal expansion velocity is

$$v_c = \sqrt{\frac{(1+Z)}{Am_p}T_r} \tag{13.34}$$

Therefore, the ablation plasma density and the ablation pressure are given by combining Eqs. (13.33) and (13.34), yielding

$$\rho_c = \frac{q_{ex}}{2.5}\left(\frac{Am_p}{(1+Z)T_r}\right)^{3/2} \tag{13.35}$$

$$P_A = 2\rho_c v_c^2 = \frac{2q_{ex}}{2.5}\left(\frac{Am_p}{(1+Z)T_r}\right)^{1/2} \tag{13.36}$$

When the temperature T_r is measured in units of eV, then the black body radiation flux is given by

$$q_{ex} = \sigma_B\left(1.16 \times 10^4 T_r\right)^4 \tag{13.37}$$

where σ_B is Stefan-Boltzmann constant. Equations (13.35), (13.36), and (13.37) yield the expressions for the radiation driven ablation pressure and mass ablation rate:

$$P_A = 2\rho_c v_c^2 = \frac{2\sigma_B}{2.5}\left(\frac{Am_p}{1+Z}\right)^{1/2} T_r^{3.5}\left(1.16 \times 10^4\right)^4 \tag{13.38}$$

$$\dot{m} = \rho_c v_c = \frac{\sigma_B}{2.5}\frac{Am_p}{(1+Z)}T_r^3\left(1.16 \times 10^4\right)^4 \tag{13.39}$$

For a Beryllium ablator of $Z = 4$, $A = 9$ and T_r measured in eV, Eqs. (13.38) and (13.39) take a form

$$P_A \simeq 1.1\, T_r^{3.5} \text{ bar} \tag{13.40}$$

$$\dot{m} \simeq 0.7\, T_r^3 \text{ g/cm}^2\text{ s} \tag{13.41}$$

In Fig. 13.8, this scaling is compared with a more sophisticated scaling obtained by John Lindl (1995).

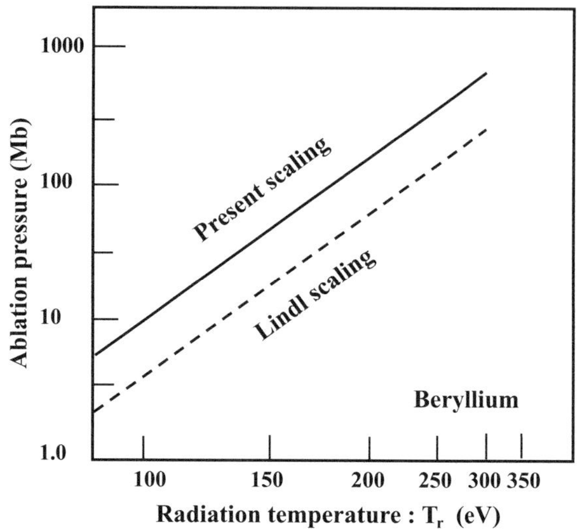

Fig. 13.8 Ablation pressure scaling laws. Solid-line presents the scaling of Eq. (13.40), and the dash-line is the scaling by Lindl

13.3.5 Rocket Model of Ablative Acceleration

In the laser fusion, it is important to accelerate a spherical shell radially to implode a fuel pellet. The important figure of merit of the implosion is the energy conversion efficiency from the absorbed laser energy to the kinetic energy of an imploding shell. This efficiency is called the "hydrodynamic efficiency". It is calculated using the rocket model described below.

Let us assume that a solid plate target of the initial area density $m(t = 0) = m_0$ is accelerated by the laser ablation pressure. Time dependence of the target velocity $v(t)$ is described by the equation of motion of the plate,

$$\frac{d(m(t)v(t))}{dt} = P_A(t) + v(t)\frac{dm(t)}{dt} \tag{13.42}$$

According to Eqs. (13.31) and (13.32) the ablation pressure in the r.h.s. of this equation equals $P_A(t) = -2v_c\dot{m}(t)$. Equation (13.42) is called the Rocket equation which was derived by Moore (1810). This equation can be rewritten as

$$m(t)\frac{dv(t)}{dt} = -2v_c\frac{dm(t)}{dt} \tag{13.43}$$

Integrating this equation, we find the velocity of accelerating target,

$$v(t) = 2v_c \ln\left(\frac{m_0}{m(t)}\right) \tag{13.44}$$

Therefore, the final kinetic energy of the accelerated film is

$$\frac{m_f v_f^2}{2} = 2m_f v_c^2 \ln^2\left(\frac{m_0}{m_f}\right) \tag{13.45}$$

Here, m_f and v_f are the final areal mass density and the final velocity of the target.

If we assume that $v_c^2 = P_c/\rho_c$ and the ideal gas EOS, i.e. $\gamma_c = 5/3$, we find the relationship between the laser absorption intensity given by Eq. (13.27) and the mass ablation rate. It reads

$$I_{abs} = -3v_c^2 \dot{m} \tag{13.46}$$

Integrating this equation, we obtain

$$3v_c^2(m_0 - m_f) = \int_0^{t_f} I_{abs} dt = \text{absorbed laser energy} \tag{13.47}$$

Equations (13.45) and (13.47) yield energy conversion efficiency

$$\eta_h = \frac{m_f v_f^2}{2\int_0^{t_f} I_{abs} dt} = \frac{2}{3}\psi \frac{(\ln \psi)^2}{1-\psi}, \tag{13.48}$$

where $\psi = m_f/m_0$ is the normalized mass remaining after the end of the ablative acceleration.

This is called the "hydrodynamic efficiency" for the ablative acceleration, which becomes maximum of approximately 0.3 at about $\psi = 1/3$. Hydrodynamic efficiency is one of the key issues in the laser fusion.

13.4 Self-generated Magnetic Field via Thermo-electric Effects

Associated with the electron flow driven by the laser heating, the magnetic fields are generated on the target surface, which affects the electron heat transport. In this section, the thermoelectric effects relevant to the magnetic field generation are discussed.

13.4.1 Introduction

Strong magnetic field generation is one of the important processes in the space–time evolution of the laser plasma. The magnetic field generation and the plasma dynamics are closely coupled to laser heating, electron heat transport, and hydrodynamics.

In the plasma under the Local Thermodynamical Equilibrium (LTE), the density and temperature gradients, the hydrodynamic motion and the heat flux are coupled with the electro-magnetic field through the Biermann Battery effects, the Nernst-Ettinghausen effect, the magnetic field advection by plasma flow, the dynamo process, and so on.

When gradients of the temperature and density are steep and the plasma electron energy distribution is no more Maxwellian, the thermo-electric instabilities, such as the Weibel Instability and the heat flux inhibition become important, leading to magnetic field reconnection, current filamentation, plasma jet formation, formation of magnetic vortex and so on. In this section, the magnetic field generation and advection mechanisms are described.

13.4.2 Magnetic Field Generation in LTE Non-uniform Plasmas

When a solid target is irradiated locally with an intense laser, the laser produced plasma has both the temperature gradient and the density gradient. When the laser intensity is moderate and the pulse duration and the laser spot size are not very small, the electron kinetic effects are limited, and the local thermal equilibrium is retained. Then, the electron dynamics is described by the electron fluid equation, the heat conduction equation, and the Maxwell equations with neglection of the displacement current as follows,

$$m_i\left(\frac{\partial \boldsymbol{v}}{\partial t} + (\boldsymbol{v}\cdot\nabla)\boldsymbol{v}\right) = -e(\boldsymbol{E} + \boldsymbol{V}\times\boldsymbol{B}) + \frac{1}{n}\boldsymbol{j}\times\boldsymbol{B} - \frac{1}{n}\nabla P + \boldsymbol{R_T} + \boldsymbol{R_u} \tag{13.49}$$

$$\frac{3}{2}\left(\frac{\partial(nk_BT)}{\partial t} + \nabla\cdot(nk_BT\boldsymbol{v})\right) + nk_BT\nabla\cdot\boldsymbol{v} + \frac{\boldsymbol{j}}{en}\cdot\nabla(nk_BT) = -\nabla\cdot\boldsymbol{q} \tag{13.50}$$

$$\nabla\times\boldsymbol{B} = \frac{\boldsymbol{j}}{\mu_0} \tag{13.51}$$

$$\frac{\partial\boldsymbol{B}}{\partial t} = -\nabla\times\boldsymbol{E} \tag{13.52}$$

Here, $\boldsymbol{v}$ and $\boldsymbol{V}$ are the electron and the ion flow velocity, respectively, $\boldsymbol{q} = -\overset{\leftrightarrow}{\kappa}\cdot\nabla T$ is the heat flux, $\overset{\leftrightarrow}{\kappa}$ is the heat conductivity tensor in the magnetized plasma, $\boldsymbol{R_T}$ is the thermo-electric force, and $\boldsymbol{R_u}$ is the resistive (or the electron–ion friction) force due to electron–ion collision:

$$\boldsymbol{R_T} = -\beta_0 n\nabla T \quad \text{and} \quad \boldsymbol{R_u} = \frac{\alpha_0}{\tau_{ei}ne}m\boldsymbol{j}$$

when the magnetic field effects are negligible. Here $\alpha_0 = 0.3-0.5$ and $\beta_0 = 0.7-1.5$ for Z = 1–90, and the mean free time is (Braginskii, 1958)

$$\tau_{ei} = \frac{3(4\pi\varepsilon_0)^2 T^{1.5}\sqrt{m}}{4\sqrt{2\pi} n_i Z^2 e^4 \ln(\chi_{\max}/\chi_{\min})}$$

The first order correction to $\boldsymbol{R_T}$ due to the magnetic field is

$$\boldsymbol{R_{TN}} = -n\frac{\beta_0}{\delta_0}\tau_{ei}\frac{e}{m}\boldsymbol{B}\times\nabla T \tag{13.53}$$

where $\beta_0/\delta_0 \simeq 0.8-0.9$ for Z = 1–90 (Braginskii, 1965).

Neglecting the inertial terms in the left-hand side of Eq. (13.49) and taking the curl, we obtain,

$$\nabla\times(\boldsymbol{E}+\boldsymbol{V}\times\boldsymbol{B}) = \nabla\times\left(\frac{1}{en}\boldsymbol{j}\times\boldsymbol{B}\right) - \nabla\times\left(\frac{1}{en}\nabla P\right) + \nabla\times(\eta\boldsymbol{j}) + \frac{1}{e}\nabla\times\boldsymbol{R_T} \tag{13.54}$$

where $\eta = \alpha_0 m/\tau_{ei} n e^2 = \alpha_0/\varepsilon_0\,\tau_{ei}\omega_{pe}^2$ is the resistivity. Combining Eqs. (13.52), (13.50) and (13.51), we obtain

$$\begin{aligned}\frac{\partial B}{\partial t} &= \nabla\times(\boldsymbol{v}\times\boldsymbol{B}) + \nabla\times\left(\frac{1}{en}\nabla P\right) - (\nabla\eta)\\ &\times\boldsymbol{j} - \frac{c^2\alpha_0}{\tau_{ei}\omega_{pe}^2}\nabla\times\nabla\times\boldsymbol{B} + \frac{1}{e}\nabla\times\boldsymbol{R_T}\end{aligned} \tag{13.55}$$

Equation (13.55) describes the spatial and temporal evolution of the quasi-static magnetic field in plasmas. The first term of the r.h.s. of Eq. (13.55) is the magnetic convection associated with the electron flow, the second term is the Bierman battery effect due to non-parallel density and pressure gradients, and the third term is the generation of magnetic field fluctuations by the transverse resistivity gradient when the external current is injected, the fourth term is the magnetic field diffusion, and the last term is the thermal force associated with the heat flux. Note here that the last term can be rewritten in the form

$$-n\frac{\beta_0}{\delta_0}\tau_{ei}\frac{1}{m}\nabla\times(\boldsymbol{B}\times\nabla T) \propto \nabla\times(\boldsymbol{q}\times\boldsymbol{B})$$

This means that the magnetic field is advected by electron heat flux, which is called the Nernst effect.

The Bierman battery effect has been observed around the laser spot on a plane target. As shown in Fig. 13.9, there is a large density gradient in the target normal direction and the temperature gradient in the lateral direction due to the local laser heating. The Bierman effect was observed first by Stamper and Ripin (1975).

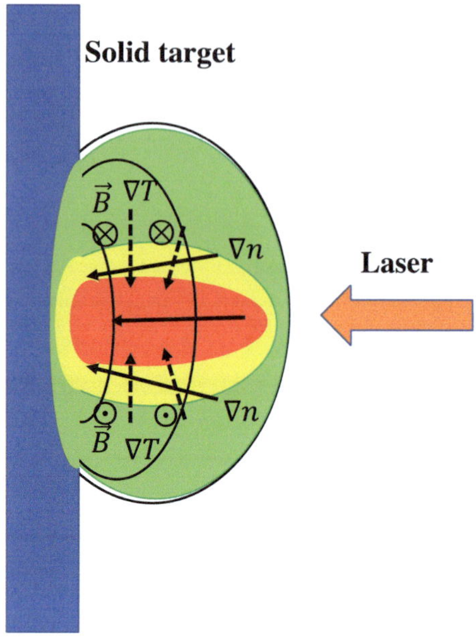

Fig. 13.9 Schematic view the temperature and density profiles and the self-generated magnetic field in a laser plasma

The magnitude of the generated magnetic field is estimated as follows. The magnetic field generated by the Bierman battery effect is carried out by the plasma expansion and by the Nernst effect, where both are directed towards vacuum and inside of the target. Let us balance the generation rate and the loss by the plasma expansion with sound speed in Eq. (12.64). For the temperature $k_B T = 1\,\text{keV}$, the sound velocity $c_s \approx 10^7\,\text{m/s}$, and the scale length of the plasma density inhomogeneity is $L = 100\,\mu\text{m}$, we obtain

$$B \approx \frac{k_B T}{e c_s L} \approx 100\,\text{Tesla} \tag{13.56}$$

In this case, the self-generated magnetic field is estimated to be the order of mega Gauss. The Mega Gauss magnetic fields have been observed by the Faraday rotation (Stamper & Ripin, 1975; Stamper et al., 1978) and by the proton radiography by Li et al. (2007) whose results are shown in Fig. 13.10.

Another unique phenomenon is the heat flux filamentation instability due to coupling of the Hall heat flux and the Biermann battery effects. This instability is called the Tidman-Shanny instability (T-S instability) (Hirano & Ogasawara, 1980; Tidman & Shanny, 1974) which takes place when the heat flows into the lower density plasmas. This may happen in the precursor of a strong shock front or when the shock breaks out in the rear side of a foil target.

The mechanism of the T-S instability is described as follows. When the temperature is modulated in the x-direction, the magnetic field is generated by the Bierman battery effect in the y-direction as shown in Fig. 13.11. Then, the transverse heat flux

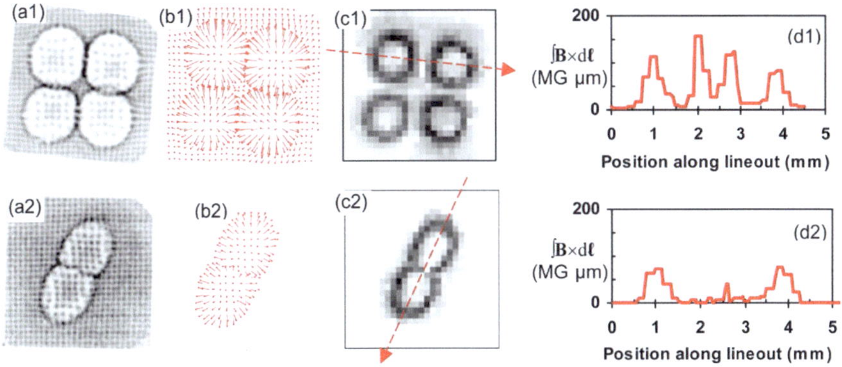

Fig. 13.10 Magnetic field generated by laser irradiation on a plastic foil with either 4 (upper figures) or 2 (lower figures) beams, observed by backlighting of a 14 MeV proton beam. (a1,2) radiography images generated by proton deflection by the magnetic field, (b1,2) proton displacement deduced from the images, (c1,2) magnetic field map deduced from the displacements, and (d1,2): lineouts along the red arrows. The colliding magnetic fields are merging, and reconnection takes place in (c2). The magnetic field strength is of the order of 100 MG as shown in (d1) and (d2). Reprinted from Li et al. (2007), with permission from The American Physical Society and the author(s). © 2007 The American Physical Society

in the x-direction is induced, which is proportional to $q_x \boldsymbol{e}_x \propto B_y \boldsymbol{e}_y \times q_{z0} \boldsymbol{e}_z$. Since $B_y \propto (dT/dx)\boldsymbol{e}_x \times (dn/dz)\boldsymbol{e}_z$, and $q_{z0} = -\kappa(dT/dz)$ these result in

$$q_x \propto \frac{dT}{dx}\frac{dn}{dz}\frac{dT}{dz} \tag{13.57}$$

Equation (13.57) indicates that the transverse heat flux (in the x-direction) is directed toward the higher temperature region and the temperature modulation increases when $\frac{dn}{dz}\frac{dT}{dz} > 0$, leading to the thermal instability development. As shown in Fig. 13.11, the laser energy is mainly deposited near the critical density surface and then the density gradient and the temperature gradient are in the same direction (z-direction) in the under-dense region. Therefore, the T-S instability could happen in the under-dense plasma. However, when the corona plasma temperature is of the order of keV, thermal smoothing will smear out the temperature modulation and stabilize the T-S instability. On the other hand, when the laser intensity is low and the plasma temperature is lower than 100 eV, this instability will affect the laser plasma homogeneity.

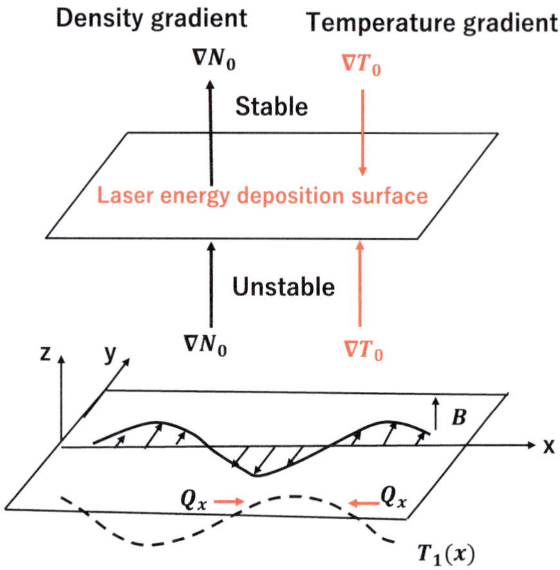

Fig. 13.11 Mechanism of the Tidman-Shanny thermal instability

References

Bell, A. R., Evans, R. G., & Nicholas, D. J. (1981). Electron energy transport in steep temperature gradients in laser produced plasmas. *Physical Review Letters, 46*, 243.

Bickerton, R. J. (1973). Thermal conduction limitations in laser fusion. *Nuclear Fusion, 13*, 457.

Braginskii, S. I. (1958). Transport phenomena in a completely ionized two-temperature plasma. *Soviet Physics—Journal of Experimental and Theoretical Physics, 6*, 358.

Braginskii, S. I. (1965). *Review of plasma physics* (Vol. 1, pp. 205–311) (A. M. A. Leontovich, Eds.). Consultant Bureau.

Craxton, R. S., Anderson, K. S., Boehly, T. R., et al. (2015). Direct-drive inertial confinement fusion: A review. *Physics of Plasmas, 22*, Article 110501.

Hirano, A., & Ogasawara, M. (1980). Magnetic field generating thermal instability including the Nernst effects. *The Physical Society of Japan, 50*, 668.

Huba, J. D. (2000). *NRL plasma formulary* (pp. 34–35). Naval Research Laboratory.

Key, M. K., Toner W. T., Goldsack T. J., et al. (1983). A study of ablation by laser irradiation of plane targets at wavelengths 1.05, 0.53, and 0.35 μm. *Physics of Fluid, 26*.

Landau, L. D. (1936). Kinetic equation for the case of coulomb interaction. *Phys. z. Sowjetunion., 10*, 154.

Li, C. K., Seguin, F. H., Frenje, J. A., et al. (2007). Observation of Mega-gauss-field topology changes due to magnetic reconnection in laser produced plasmas. *Physical Review Letters, 99*, Article 055001.

Lindl, J. (1995). Development of the indirect-drive approach to inertial confinement fusion and the target physics basis for ignition and gain. *Physics of Plasmas, 2*, 3933.

Luciani, J. F., Mora, P., & Virmont, J. (1983). Nonlocal heat transport due to steep temperature gradients. *Physical Review Letters, 51*, 1664.

Mima, K., Kitagawa, Y., Yamanaka, M., et al. (1994). System analysis of the laser fusion reactor KOYO driven by laser diode pumped solid state laser. In *IAEA Conference on Plasma Physics and Controlled Nuclear Fusion* (Vol. 2, IAEA-CN-60/F-6, p. 663).

Moore, W. (1810). Theory of the motion of Rockets. *A Journal of Natural Philosophy, Chemistry and the Arts, XXVII*, Article IV.

Nishiguchi, A., Mima, K., Azechi, H., et al. (1992). Kinetic effects of electron thermal conduction on implosion hydrodynamics. *Physics of Fluids B, 4*, 417.

Spitzer, L., Jr. (1953). Transport phenomena in a completely ionized gas. *Physical Review, 89*, 977.

Spitzer, L., Jr. (1956). *Physics of fully ionized gases*. Inter-science.

Stamper, J. A., & Ripin, B. H. (1975). Faraday-rotation measurements of mega-gauss magnetic fields in laser-produced plasmas. *Physical Review Letters, 34*, 138.

Stamper, J. A., McLean, E. A., & Ripin, B. H. (1978). Studies of spontaneous magnetic fields in laser-produced plasmas by Faraday rotation. *Physical Review, 40*, 1177.

Tidman, D. A., & Shanny, R. A. (1974). Field-generating thermal instability in laser-heated plasmas. *Physics of Fluids, 17*, 1207.

Chapter 14
Laser Fusion

Abstract The laser-driven ablation processes and the ablation pressure for implosion have been described in Chap. 13. In this Chapter, fusion gain, ablative acceleration, direct-drive target physics, indirect-drive- target physics, and alternative concepts (electron and ion fast ignition and shock ignition) are described, together with recent achievements of integrated laser fusion experiments. The laser fusion concept and the related issues are also presented. Uniform irradiation technology and the hydrodynamic instabilities related to the laser fusion are described in the following Chapters.

Keywords Laser fusion · Direct-drive · Indirect-drive · Uniform irradiation · Hohlraum · Central ignition · Fast ignition · Shock ignition · Isobaric ignition · Isochoric ignition · Fusion gain · Richtmyer-Meshkov instability · Rayleigh-Taylor instability · Turbulent mixing · Inertial fusion reactor

14.1 History of Laser Fusion Research

In 1917, Albert Einstein suggested the theory of stimulated emission of light that led to the laser development. The first laser oscillation, based on the Einstein theory, was demonstrated by T. H. Maiman in 1960 (Maiman et al., 1960). For the laser invention and developments, N. Basov, A. Prokhorov, and C. H. Towns received the Nobel Prize in Physics in 1963. On the other hand, it has been recognized that nuclear fusion energy is the energy source of our universe. It is the origin of the energy of our Sun (Atkinson & Houtermans, 1929) and of the stars (Bethe, 1937, 1939; Bethe & Peierls, 1934; Gamow & Teller, 1938).

Right after demonstration of the laser, it was suggested by John Nuckolls, Edward Teller and Stirling Colgate in the USA and Andrei Sakharov in the USSR, to use nuclear fusion induced by lasers to solve the energy problem. Following the suggestions, the pioneering works for heating plasmas to the thermonuclear temperature with high power lasers have been published by Basov and Krokhin (1964), Dawson (1964), Hora (1964), Krokhin (1969), Floux et al. (1969), and Floux (1971). A new

Y. Kato et al., *High Power Laser and Plasma Science*, Springer Series on Atomic, Optical, and Plasma Physics 130, https://doi.org/10.1007/978-3-032-08930-4_14

concept of fusion ignition and burn by laser implosion was proposed by Nuckolls (1972), which has extended the spherically imploding shock concept discovered by Guderley (1942) to the laser fusion. Since then, laser fusion research has been undertaken all over the world and many high power laser facilities have been developed for the fusion research: GEKKO I, GEKKO II, GEKKO IV, GEKKO MII, GEKKO XII, and LFEX at ILE Osaka University in Japan; JANUS, CYCLOPS, ARUGUS, SHIVA, NOVA, and NIF at LLNL, OMEGA at LLE Univ. Rochester, HELIOS at LANL, and PHAROS and NIKE at NRL in USA; PHEBUS at Limeil CEA, GLECO at LULI Ecole Polytechnic, and LMJ at Bordeaux in France; ASTERIX Iodine laser at Garching in Germany; Shengan II at Shanghai Institute of Optics and Fine Mechanics and Shenguan III at Myanyang in China; and VULCAN at Rutherford Appleton Laboratory in UK.

At present, ignition and burning of the imploded fusion fuel have been achieved by the National Ignition Facility (NIF) at LLNL (Moses, 2009), USA. However, commercial applications will be still more than two decades away.

14.2 Introduction

Laser fusion, which is the one of the inertial confinement fusion (ICF) schemes, is described in this section as an application of high-power laser and plasma science. The inertial fusion energy (IFE) will be achieved with multiple choices of energy drivers and schemes of ignition and burn scenario like implosion and hot spark formation as shown in Figs. 14.1 and 14.2 (Mima, 2012). The direct drive is the scenario where many high intensity laser beams irradiate uniformly a spherical shell containing nuclear fusion fuel. By stagnation at the center of the spherical pellet, the nuclear fusion fuel is highly compressed and heated to induce the nuclear fusion. The indirect drive scenario starts with the conversion of the laser or heavy ion beams into soft x-rays in a cylindrical cavity of a few mm size. A pellet containing nuclear fusion fuel is imploded by the x-ray radiation in the cavity. In both schemes, the imploding spherical pellet must be stable to hydrodynamic instabilities, such as the Rayleigh–Taylor (RT) instability, to achieve the high compression necessary for ignition. The hydrodynamic instabilities are initiated by various spherical symmetry imperfections of the fuel pellet and the irradiation non-uniformity on the pellet surface.

The severe requirement of hydrodynamic stability could be relaxed by the advanced concepts like fast ignition (Tabak, 1994), shock ignition (Betti et al., 2007; Ribeyre et al., 2009), and so on. In these cases, compression is achieved by the direct drive or indirect drive at first. Then, the imploded core plasma is heated by an intense spherical shock wave which is driven by another intense short laser pulse to produce a hot spot. In the following, critical issues of science and technology toward realization of IFE are described. The final goal is to obtain an economical electric power plant based on laser fusion.

The research was started in 1968 when production of the laser fusion neutron from lithium-deuterium targets was reported (Basov, 1968) for the first time. The

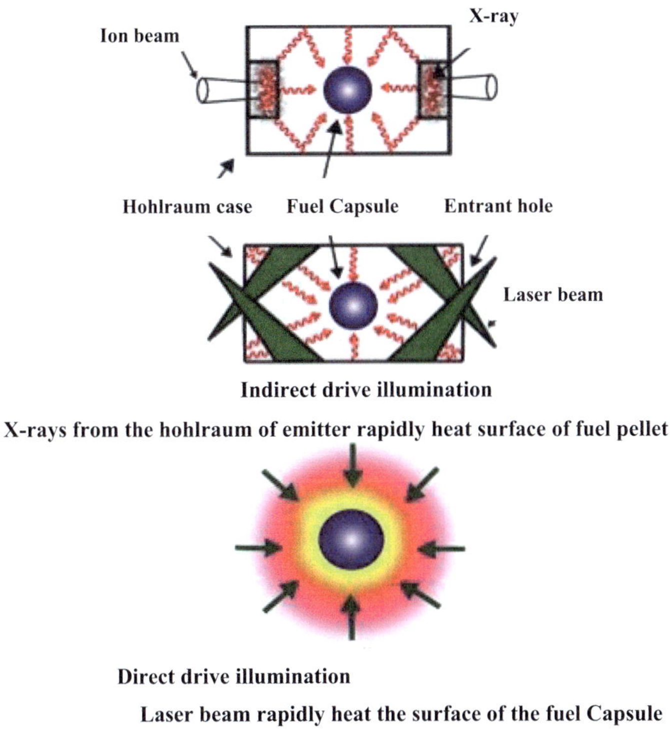

Fig. 14.1 Implosion schemes of IFE driven by laser or ion beams

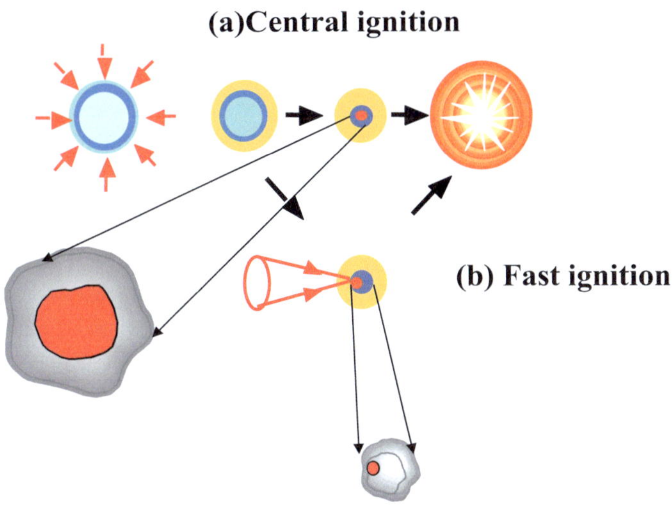

Fig. 14.2 Two ignition schemes: **a** central hot spark ignition (isobaric ignition) and **b** fast ignition (iso-choric ignition)

national research center for laser fusion was established in the USA at the Lawrence Livermore Laboratory and in other countries in the early 1970s. In 1972, J. Nuckolls (Nuckolls et al., 1972) proposed that it is necessary to compress the fusion fuel to the density higher than 1000-times solid-density to achieve ignition with a reasonable laser pulse energy. There are two main schemes for pellet implosion, i.e., direct and indirect drive. These concepts are schematically shown in Fig. 14.1. In the direct drive implosion, the surface of a fuel pellet is directly and uniformly irradiated by many laser beams. In contrast, in the indirect-drive implosion, the driver energy is converted into soft x-rays in a cavity which is filled with quasi-black body X-ray radiation. The soft x-rays are then absorbed on the surface of the fuel pellet to generate ablation pressure to drive the implosion. There is another scheme called hybrid implosion (He et al., 2016; Nishimura et al. 2000), which combines the direct and indirect implosion concepts.

As for ignition of the imploded fuel, there are basically two concepts, i.e., the central hot spark ignition (isobaric ignition, Fig. 14.4a) (Nuckolls, 1972), and isochoric ignition (Fig. 14.4b) which is valid for fast ignition (Tabak et al., 1994) or shock ignition (Betti et al., 2007). The corresponding ignition processes are illustrated in Fig. 14.2 (Eliezer & Mima, 2009). When the implosion is spherically symmetric, the central hot spark is formed as shown in Fig. 14.2a, where the temperature of the plasma at the center is lower than the outer shell. Then, the central plasma is adiabatically compressed and heated till the pressure of the outer shell and the central plasma becomes equal. So, it is called "isobaric ignition". The ignition condition for the isobaric model is estimated by Meyer-ter-Vehn (Meyer-ter-Vehn, 1982), as described in Sect. 14.3.

Even if the implosion produces a compressed fuel of uniform density but fails to produce a central hot spark, we can still ignite it by introducing an intense short pulse laser where the pulse duration is shorter than the expansion time of the compressed core. It is called "fast ignition" or "shock ignition". These schemes are also called "isochoric ignition", since the plasma density of the hot spark is same as the density of the cold fuel. The hot spark is formed at the edge of a high-density core plasma for fast ignition as shown in Fig. 14.2b. So, the fast ignition is also called "off-center ignition". The plasma density of the hot spark in the fast ignition and the shock ignition is expected to be several times higher than that for the central hot spark ignition (Tajima et al., 2000). Therefore, the hot spark size indicated by the red area in Fig. 14.2b is much smaller than that for the isobaric ignition. Namely, the required laser energy for ignition and high gain could be smaller by several times.

The red circle in Fig. 14.3 shows the fusion gain of the recent NIF indirect-drive experiment (Ma et al., 2021), and the point design for direct drive with NIF is shown by the blue circle (Moses et al., 2009). Figure 14.3 also shows the gains required in laser fusion reactors KOYO and KOYO-F which are the conceptual designs for the central ignition and the fast ignition, respectively (Johzaki et al., 2007; Norimatsu et al., 2006; Tomabechi et al., 2006). The green and yellow areas in Fig. 14.3 show possible gains depending on laser pulse energy for direct drive (Atzeni & Meyer-ter-Vehn, 2004) and fast ignition (Johzaki et al., 2007), respectively. The credibility of the gain by the direct-drive central ignition target is discussed in the next section.

It is believed that uniform irradiation configuration of NIF 192 beams will be able to achieve fusion ignition and burning, and energy gain of higher than 10. If fast ignition works well, we can achieve a higher gain of 100–200 with less than 1 MJ laser energy as shown by the yellow area of Fig. 14.4. The recent progress in fast ignition and its credibility and uncertainty are described in Sect. 14.4.

In inertial fusion, the key process is fusion fuel compression by the implosion. The mass M_f of fusion fuel is

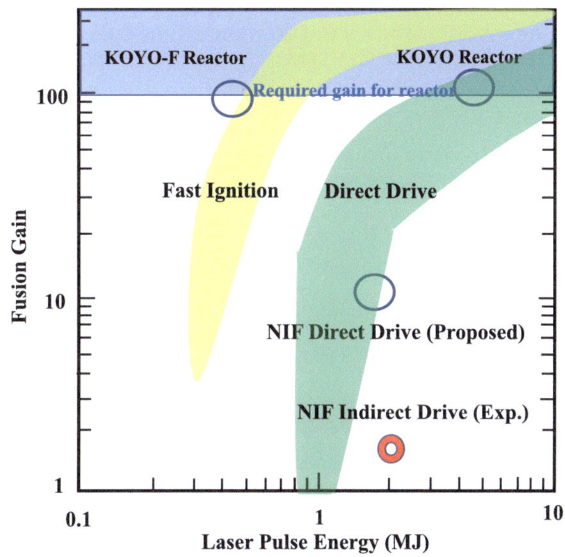

Fig. 14.3 Predicted fusion gain for central ignition and fast ignition

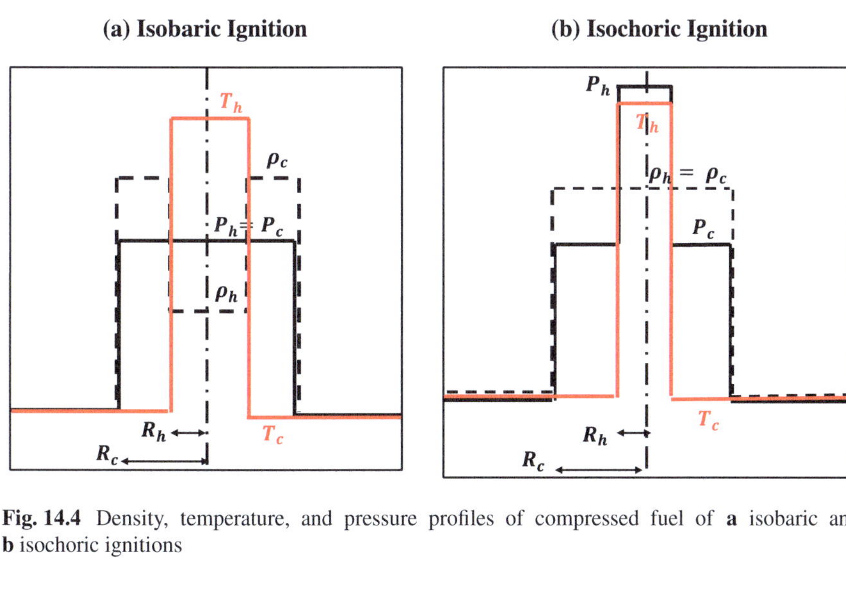

Fig. 14.4 Density, temperature, and pressure profiles of compressed fuel of **a** isobaric and **b** isochoric ignitions

$$M_f = \frac{4\pi\rho R^3}{3} = 4.19\frac{(\rho R)^3}{\rho^2} \approx 113/\left(\rho\,[\mathrm{g/cm^2}]\right)^2$$

for the fuel radius R, where $\rho = 3\,\mathrm{g/cm^2}$ is the areal mass density required for efficient fusion burning, as described in Sect. 14.3. Without compression, the DT fuel mass density is $\rho = 0.2\,\mathrm{g/cm^2}$. So, the above requirement implies the total DT fuel mass of about 2.8 kg, which is an undesired value for the controlled fusion reactor. However, when it is compressed to 2000 times the DT solid density $\rho = 400\,\mathrm{g/cm^2}$, the total mass is 1.0 mg of the DT fuel pellet. The driver energy $\mathcal{E}_d$ which is necessary to compress the DT fuel to $400\,\mathrm{g/cm^2}$ of the sphere radius R is,

$$\eta_a \eta_h \mathcal{E}_d = 3nT\left(\frac{4\pi R^3}{3}\right) \tag{14.1}$$

where n is the D, T number density. From Eq. (14.1) follows,

$$\mathcal{E}_d = \frac{4\pi T(\rho R)^3}{2.5 m_p \rho_0^2 \eta_a \eta_h}\left(\frac{\rho_0}{\rho}\right)^2 \tag{14.2}$$

where the ideal gas EOS has been used, m_p is the proton mass, $\rho = 0.2\,\mathrm{g/cm^2}$ is the initial solid DT density, and the temperature T is measured in energy units. Here, η_a and η_h are the laser absorption efficiency and the hydrodynamic efficiency, which is the coupling efficiency of absorbed laser energy to the compressed plasma internal energy. Taking laser energy $\mathcal{E}_d$ that heats the plasma to the threshold temperature of 4 keV (where fusion energy equals the bremsstrahlung losses) and inducing a 30% burn of the nuclear fuel (namely $\rho = 3\,\mathrm{g/cm^2}$), we get

$$\mathcal{E}_d[J] = \frac{1.3 \times 10^{12}}{\eta_a \eta_h}\left(\frac{\rho_0}{\rho}\right)^2\left(\frac{T}{4\ \mathrm{keV}}\right). \tag{14.3}$$

Taking typical values of $\eta_a = 0.8$ and $\eta_h = 0.1$ the inertial fusion ignition and nuclear burn require a pulsed (few nanoseconds) laser system with an energy

$$\mathcal{E}_d[J] = 1.4 \times 10^{13}\left(\frac{\rho_0}{\rho}\right)^2\left(\frac{T}{4\ \mathrm{keV}}\right). \tag{14.4}$$

It is clear from this result that without compression one requires an unrealistic pulse laser with energy of 16 TJ for ignition, while a 3000 compression (i.e. $600\,\mathrm{g/cm^3}$) reduces the required laser energy to 1.8 MJ. Note that the compression of a plastic shell with an initial density of about $1\,\mathrm{g/cm^3}$ to the density of $600\,\mathrm{g/cm^3}$ has been demonstrated experimentally at Osaka University (Azechi et al., 1991) and the density up to 1300 times solid density of DT has been achieved at NIF (Park et al., 2014). The Mega-Joule laser system at Livermore (NIF) achieved the ignition with the energy of about 2 MJ and the fusion energy output became higher than the input laser energy, even though the coupling efficiency is low for the indirect drive (Abu-Shawareb et al., 2022; Hurricane et al., 2019; Ma & Betti, 2023). The best record of the burning plasma is measured to be

$$T_h = 10.9\,\text{keV}, \quad P_h = 589\,\text{Gbar}, \quad \rho_h R_h = 0.44\,\text{g/cm}^2.$$

Since Nuckolls et al. published the concept in Nature (Nuckolls et al., 1972), the laser fusion research has successfully realized the plasma compression and ignition by implosion. However, it is still necessary to control further the laser plasma interaction, and the hydrodynamic instabilities, like Rayleigh–Taylor and Richtmyer-Meshkov instabilities, are still critical issues toward the better implosion.

14.3 Fusion Ignition Condition

The fusion ignition condition has been proposed by Lawson (1955), which is called the "Lawson Criterion". In inertial fusion, when the hot spark satisfies the ignition condition, the surrounding plasma is heated up by alpha-particles produced by fusion reaction and then the fusion propagation burn takes place if the ignition condition is met. We discuss the ignition condition in this section. The typical imploded plasma structures with a hot spark are shown in Fig. 14.4. The structure of the isobaric ignition is shown in Fig. 14.4a and that of the isochoric ignition is shown in Fig. 14.4b. The density and pressure of the cold fuel and of the hot spark are described by ρ_c, P_c, ρ_h, P_h, respectively. The effective temperature of the cold fuel and the hot spark are taken to be $T_c = m_{DT}P_c/2\rho_c$ and $T_h = m_{DT}P_h/2\rho_h$, respectively. Note that $P_c = P_h = P$ for isobaric ignition and $\rho_c = \rho_h = \rho$ for isochoric ignition. The spherical radius of the cold fuel and the hot spark are R_c and R_h, respectively.

The ignition condition is given by the condition where the energy density of the hot spark increases by the fusion reaction heating. Namely, the internal energy density of the hot spark $\mathcal{E}_h$ satisfies the following equation

$$\frac{d\mathcal{E}_h}{dt} = W_\alpha - W_r - W_e - W_m. \tag{14.7}$$

Here W_α is the alpha particle energy deposition, and W_r, W_e, W_m are radiation loss, thermal conduction loss, and mechanical work, respectively. When the ideal gas EOS is assumed for the hot spark, $\mathcal{E}_h = 3P_h/2$ with the hot spark pressure P_h. When we neglect the neutron heating, the alpha particle heating rate is given by

$$W_\alpha = \frac{1}{5}A_{DT}\rho_h^2\langle\sigma v\rangle f_\alpha, \tag{14.8}$$

where $\langle\sigma v\rangle$ is the fusion reaction rate, $A_{DT} = 4 \times 10^{39}\ kg^{-1}$, and f_α is the deposition-fraction of alpha particle energy. Here we can approximate it as $f_\alpha \approx R_h/l_\alpha$ for

$R_h/l_\alpha < 1/2$ and $f_\alpha \approx 1 - l_\alpha/4R_h$ when $R_h/l_\alpha \gg 1/2$ for the alpha particle mean free path (Atzeni & Meyer-Ter-Vehn, 2004).

As described in the previous section, compression and hot spark formation are the crucial issues in the IFE research. In particular, efficient uniform plasma compression by implosion is essential. The central area is the hot spark, and the outer area is the cold dense shell shown in Fig. 14.4. The quality of the implosion depends upon the laser absorption efficiency η_a, hydrodynamics η_h, isentrope, and implosion velocity (laser irradiation uniformity). Namely, the important plasma physics related to high quality implosion is the laser plasma interaction, heat conduction and radiation hydrodynamics. In the following, the ignition condition is derived for the ideal case.

The ignition criterion is determined by the energy balance between the fusion self-heating and the energy losses, namely the right-hand side of Eq. (14.7) is positive. So,

$$W_\alpha - W_r - W_e - W_m > 0. \tag{14.9}$$

Here, $W_e = Sq/V \approx S\kappa T_h/VR_h = 3\kappa T_h/R_h^2$, radiation energy loss is $W_r = A_b\rho_h^2 T_h^{0.5}$ for $A_b = 3.05 \times 10^{17}\ \mathrm{J\,cm^3\,g^{-2}\,s^{-1}\,keV^{-0.5}}$ and the mechanical work is $W_m = SP_h u/V = 3P_h u/R_{ph}$ for the velocity of the hot spark expansion $u = S\sqrt{(P_h - P_c)/4\rho_c}$. As a result, Eq. (14.9) can be written as

$$(\rho_h R_h)^2\left(\frac{1}{5}\langle\sigma v\rangle f_a - A_b T_h^{0.5}\right) - 3\kappa T_h - 3R_h P_h u > 0. \tag{14.10}$$

According to Atzeni and Meyer-Ter-Vehn (2004), Eq. (14.10) is reduced to the following ignition criterion for both the isobaric and isochoric ignitions:

$$3\kappa R_h T_h \geq 6\sqrt{\frac{\rho_h}{\rho_c}}\ \mathrm{g\,cm^2\,keV} \tag{14.11}$$

for the temperature range of $5\,\mathrm{keV} < \mathrm{T_h} < 15\,\mathrm{keV}$.

14.4 Gain Scaling of Laser Fusion

Figure 14.3 shows the possible gain areas for the indirect drive proposed by Lindl (1998), the direct drive with isobaric ignition predicted by Atzeni and Meyer-ter-Vehn (2004), and the fast ignition proposed by Johzaki et al. (2007) for the fuel density equal to $\rho = 300\,\mathrm{g/cm^3}$. In this section, we describe the fusion gain scaling for the central ignition according to Atzeni and Meyer-ter-Vehn (2004).

The fusion gain G is defined as

$$G = \frac{Q_{DT} M_f \phi}{\mathcal{E}_L}, \tag{14.12}$$

where ϕ is the fraction of fusion burn and Q_{DT} is the DT fusion energy per unit mass. The DT fuel-mass density during the burning decreases according to the following equation,

$$\frac{d\rho}{dt} = -\rho \frac{\rho}{m_{DT}} \langle \sigma(v) v \rangle, \tag{14.13}$$

where $\sigma(v)$ is the cross section of DT fusion reaction at the relative velocity v, $\langle \cdots \rangle$ is the average over the Maxwellian at the burning plasma temperature, and m_{DT} is the average nuclear mass. Equation (14.13) is integrated for a fixed plasma temperature to yield

$$\frac{1}{\rho} - \frac{1}{\rho_0} = \frac{\langle \sigma(v) v \rangle}{m_{DT}} t,$$

where ρ_0 is the initial mass density and t is the burning time. Since the plasma expands with the sound velocity c_s the burning time is approximately equal to $t \approx R_f / c_s$. Then the burning fraction is defined as

$$\phi = 1 - \frac{\rho}{\rho_0} = \frac{H_f}{H_F + H_f}. \tag{14.14}$$

where

$$H_f = \rho_c (R_f - R_h) + \rho_h R_h \tag{14.15}$$

is the fuel areal mass density, and

$$H_f = \frac{m_{DT} c_s}{\langle \sigma(v) v \rangle} \tag{14.16}$$

The required total internal energy of the fuel plasma $\mathcal{E}_L = \eta \mathcal{E}_L$ is proportional to the sum of the hot spark internal energy $M_h e_h$ and the cold fuel internal energy $M_c e_c$

$$\eta \mathcal{E}_L = M_c e_c + M_f e_f \tag{14.17}$$

Here, η is the laser-plasma coupling efficiency, hot spark specific energy is $e_h = 1.4 \times 10^{11} T_h$ J/(kg keV), and the cold fuel specific energy is given by $e_c = 3P_c/2\rho_c$, where P_c is the DT cold fuel pressure. Since exact EOS is complicated, we shall assume the ideal gas EOS for the ions and the Fermi–Dirac EOS for the electrons (Eliezer et al., 2002; Lee & More, 1984) with a multiplier (isentrope parameter) α. Namely, the pressure $P_c = \alpha P_{FD}$ (in mega bars) and the fuel specific Fermi energy e_F (in J/g) are given in terms of the density ρ [g/cm^3] as

$$P_{FD}\,[\text{Mbar}] = A_{\text{deg}}\rho^{5/3} \quad \text{with} \quad A_{\text{deg}} = 2.17\,[\text{Mbar}\,\text{g}^{5/3}\,\text{cm}^{-5}] \tag{14.18}$$

$$e_F = \alpha e_{FD} \quad \text{with} \quad e_{FD} = 3.3 \times 10^5 r^{2/3}\,[\text{J/g}]. \tag{14.19}$$

Therefore, we obtain $e_c = (3/2)\alpha A_{\text{deg}}\rho^{5/3}$. For the structure of Fig. 14.4, we estimate the cold fuel mass as,

$$M_c = \frac{4\pi}{3}\rho_c\left(R_f^3 - R_h^3\right) = \frac{4\pi}{3}\rho_c R_f^3\left(1 - \chi^3\right), \tag{14.20}$$

where $\chi = R_h/R_f$. Note that

$$M_c \approx \frac{4\pi}{3}\rho_c R_f^3 = \frac{4\pi}{3}\frac{H_f^3}{\rho_c^2} \propto \mathcal{E}_L$$

and the burning fraction is proportional to H_f when $\chi \ll 1$. From Eqs. (14.12) and (14.14) it follows that the fusion gain is proportional to the cubic root of the driver energy $\mathcal{E}_L^{1/3}$ for $H_f < H_F$ and given compressed fuel density. In the relationships written above and below, ΔR is the main fuel shell thickness and R_h is the hot spark radius, as shown in the density profile of Fig. 14.4. Since the burning fraction is proportional to $\rho\Delta R + \rho_h R_h \approx \rho\Delta R$ when $H_f \gg \rho_h R_h$, the gain is proportional to $\rho\Delta R$. Namely the gain (which is called the limited gain) increases with the 1/3 power of the input laser energy for large driver energy, as shown by the broken line in Fig. 14.5.

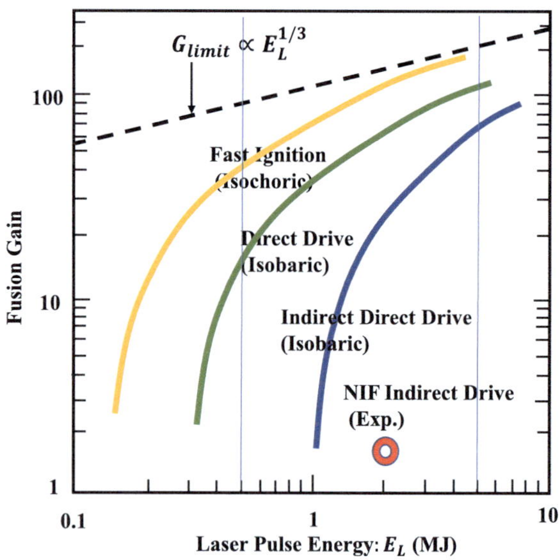

Fig. 14.5 Gain curves for indirect drive, direct drive, and fast ignition. The fusion gain of the NIF experiment is shown by a red circle

The gain curve in Fig. 14.5 is determined for a given coupling efficiency to produce a hot spark of 10 keV temperature. Then the main fuel total mass and the areal mass density increase is proportional to $(\mathcal{E}_L - \mathcal{E}_{Lh})^{1/3}$, where $\mathcal{E}_{Lh}$ is the driver energy to ignite the fuel by producing the hot spark. In Fig. 14.5, the energy $\mathcal{E}_{Lh}$ is assumed to be 1.0 MJ, 0.3 MJ, and 0.15 MJ for indirect drive, direct drive and fast ignition/shock ignition, respectively. The isentrope α is assumed to be 3, and the coupling efficiency η is assumed to be of 5%, 15% and 15% for indirect drive, direct drive, and fast ignition/shock ignition, respectively. As for the NIF experimental gain shown in Fig. 14.5, it was achieved with the input laser energy of 2.15 MJ, resulting in the fusion output of 3.5 MJ (Ma & Betti, 2023).

14.5 Implosion Physics

When multiple laser beams irradiate a pellet surface which, for example, can be a plastic capsule containing DT ice layer as shown in Fig. 14.6. The pellet surface is ablated and a corona plasma is formed by laser absorption and heating, where field ionization and electron impact ionization take place. Laser beams penetrate the corona plasma up to the cut-off region and deposit the energy or partly reflected. The laser beams are absorbed or scattered in the under-dense plasma by the classical process and various non-linear collective processes. In those processes, the absorption efficiency, heating uniformity, and hot electron mitigation are important from the standpoint of implosion efficiency. Through those processes, the corona plasma is heated to a few keV, and the pellet surface is further ablated by the electron and radiation heat conduction from the cut-off region to the pellet surface. Then the ablation pressure pushes and implodes the pellet shell as described in Chap. 13 (Fig. 13.6).

In the laser plasma interaction, resonance absorption, self-focusing, parametric instabilities, stimulated Raman scattering, and stimulated Brillouin scattering, which are the plasma instabilities described in Sect. 12.5, are the important processes. As for the heat conduction, the kinetic effects (non-local heat transport) described in Sect. 12.6 and the thermal smoothing have been investigated widely. Pressure of the hot dense ablation plasma drives a shock wave and accelerates the fuel shell toward the center of the pellet. This is called "the laser driven implosion".

If the implosion is spherically uniform and the accelerated fuel shell shrinks and stagnates at the center, then a strong spherical shock wave (Guderley shock) is generated to form a hot spark at the center of an ultra-dense plasma. However, the implosion becomes non-spherical due to hydrodynamic instabilities. The hydrodynamic instabilities grow from initial perturbations on the target and non-uniformity of laser irradiation and leads to breakup of the shell and quenching of the hot spark at the center as schematically shown in Fig. 14.7.

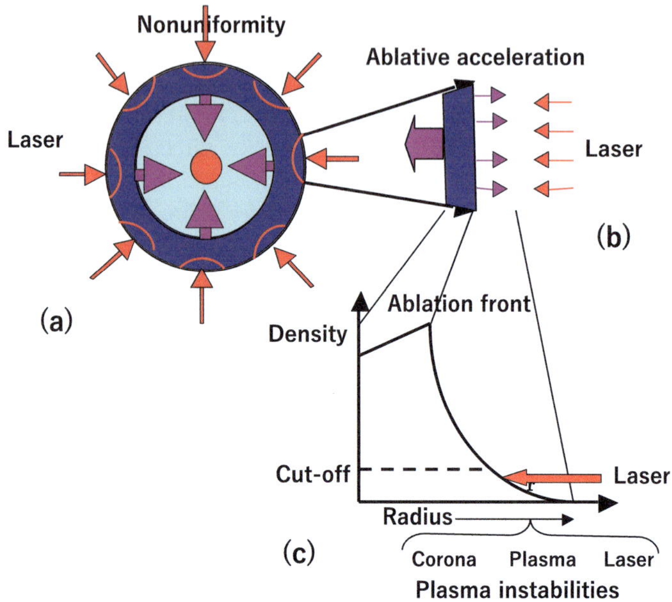

Fig. 14.6 Laser plasma interaction physics and hydrodynamics in the directly driven implosion

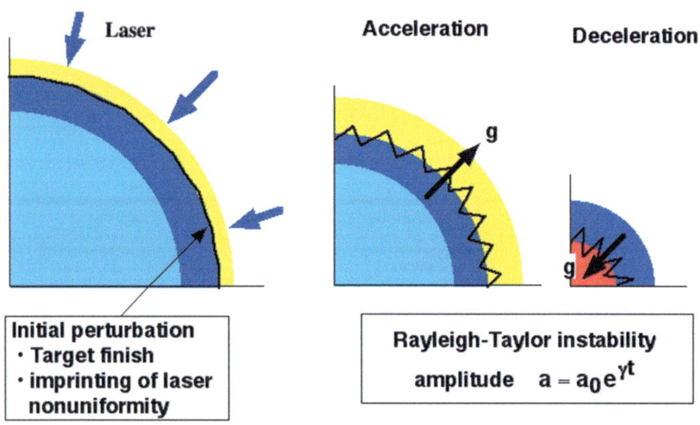

Fig. 14.7 Inhomogeneity and instability in the laser-driven implosion

14.6 Critical Issues of Implosion

14.6.1 Implosion Physics Processes

In the implosion physics, the critical issues are laser absorption, plasma instability, ablative acceleration of the fuel shell, and stability of the implosion (Bodner, 2000). The laser absorption and laser plasma instability have been described in Chaps. 10, 11 and 12, and the ablative acceleration of fuel shell is described in Chap. 13. Here we consider implosion processes including the laser plasma interactions and the radiation hydrodynamics. The processes related to the implosion physics are schematically shown in Fig. 14.8.

In the initial phase of implosion, separation of the absorption region (cut-off region) from the ablation front on a solid target is small as shown in Fig. 14.8. Therefore, the non-uniformity of the laser irradiation is strongly imprinted on the target surface (initial imprint). Then, the shock wave driven by the main pulse can excite the Richtmyer–Meshkov (RM) instability, which is coupled with the original target roughness and the initial imprint by laser irradiation no-uniformity. The ablation pressure non-uniformity is relaxed from the laser non-uniformity in the later time due to thermal smoothing during the energy transport from the absorption region to the ablation surface. In the main pulse acceleration phase, the Reyleigh–Taylor (RT) instability grows. However, the shorter wavelength perturbation is stabilized due to the ablative stabilization. If the in-flight aspect ratio is too large, the accelerated shell breaks easily (shell break).

Even if the non-uniformity of the accelerated shell is small, mixing of the hot spark with the surrounding dense cold fuel will take place due to explosive growth of the RT instability/RM instability at the interface between the hot spark and the cold

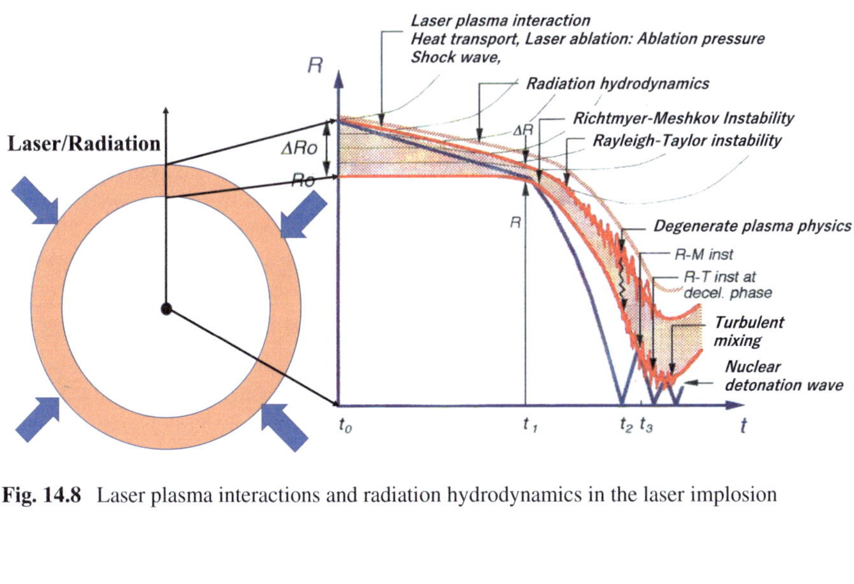

Fig. 14.8 Laser plasma interactions and radiation hydrodynamics in the laser implosion

fuel in the stagnation phase. In general, the RT instability is not generated as a single mode, but many modes grow simultaneously, finally inducing the turbulent mixing. The growth rate of the mixing layer thickness would be smaller than the linear growth rate due to nonlinear saturation and mode–mode coupling in the nonlinear phase.

When the spherical non-uniformity is large, the hot spark plasma is mixed with the high-density cold plasma surrounding the spark. Then, the hot spark plasma is cooled down and the fusion ignition does not take place. Therefore, the mixing layer thickness at the interface between the outer cold and dense shell and the hot spark should be sufficiently smaller than the hot spark radius. If we can manage to keep the spherical symmetry and ignite the fuel, then, the fire polishing by alpha particle heating can smooth the burning wave front even with some level of non-uniform structure in the main fuel, and the fusion gain will be obtained as expected in Sect. 14.4.

The elementary processes mentioned above such as initial imprint, R–M instability, thermal smoothing, R–T instability, nonlinear saturation, and turbulent mixing have been investigated independently and separately with adequately modeled experiments on plane targets of various structures and irradiation conditions of lasers as discussed in Sect. 14.7. In direct drive implosion by laser, the most critical physical issue is the instability of the implosion (Bodner, 2000) as discussed above.

14.6.2 Implosion Experiments

Concerning the implosion stability, experimental data on various parameters of the compressed core are compared with a one-dimensional simulation. Typical examples are shown in Fig. 14.9 (Azechi et al., 1991; Nakai, 1990; Nakai & Takabe, 1996). In the series of implosion experiments with the Gekko XII laser at Osaka University (Azechi et al., 1991), plastic shell targets of CDT doped with Si were imploded to measure both the areal density and the neutron yield. The areal density is inferred from the yield of Si activated by DT neutrons. Namely, the areal density is deduced from the neutron yield and the Si activation cross section. The experimental areal density and the average density of the imploded plastic shell are shown in Figs. 14.9a and 14.10, respectively. The experimental neutron yield is normalized by the neutron yield of one-dimensional simulation in Fig. 14.9b. The experimental result by OMEGA-60 laser at the University of Rochester is also shown in Fig. 14.9b.

The measured areal densities agree with the one-dimensional prediction up to a convergence ratio (CR; initial target radius divided by the final radius) of about 30, when the irradiation uniformity is improved by using a random phase plate (RPP) (Kato et al., 1984). The neutron yield, however, decreased faster as the CR increased, as shown in Fig. 14.9b, to about 10^{-3} of the one-dimensional calculation at CR of 30 where the high-density compression was achieved. The estimated radius of the hot spark with one-dimensional simulations was $\sim$ 5 μm at the hot-spark CR of $\sim$ 50. Possible explanation for the discrepancy in the neutron yield would be that the remaining irradiation and target non-uniformities might cause a collapse of the hot

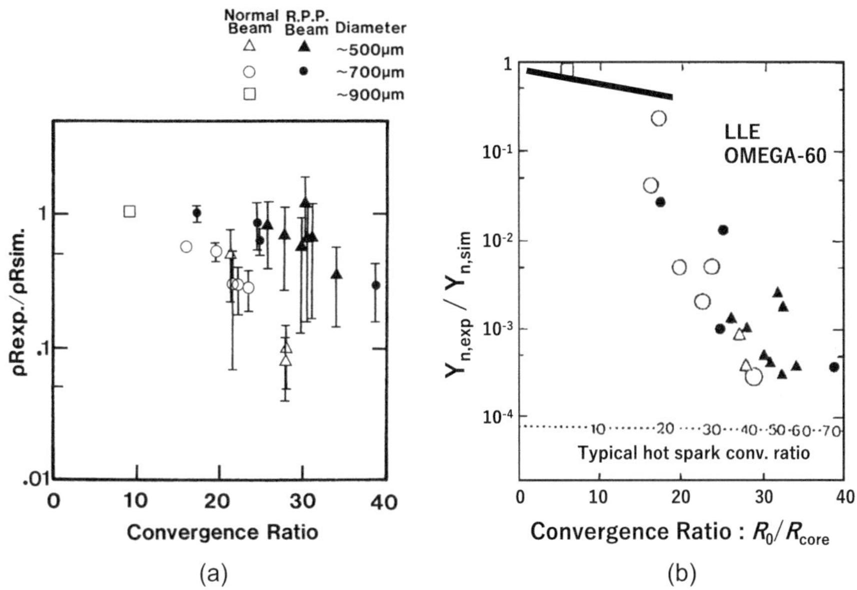

Fig. 14.9 High density shell implosion and degradation of implosion. The ratios of the experiment results to the one-dimensional simulations are potted versus the conversion ratio (CR) for **a** areal density and **b** neutron yield. Courtesy of Prof. H. Azechi

spark region, from where most of the neutrons are generated in the one-dimensional simulation.

14.6.3 Stability of Implosion

As described above, one of the key issues of direct drive implosion is the RT/RM instabilities, which are related to non-uniformities of laser irradiation and fuel pellet. It has been a serious problem with the laser fusion ignition because they cause the mixing of the hot spark and the cold main fuel. The growth rate of RT instability at the ablation/contact surfaces during the implosion can be reduced by enlarging the density scale length of the ablation front and the hot spark surface. To realize the stable implosion, it has been explored to increase the ablation velocity for reducing the RT growth rate according to Eq. (16.17), called "Takabe-Bodner formula", as described in Sect. 16.3. They could be achieved by (a) shock preheating, (b) radiative heating, and (c) nonlocal heat transport by hot electrons. Together with the progress in fuel pellet fabrication and uniform laser irradiation technologies, it has been expected that fusion ignition, burn and energy gain will be achieved with multi-beam mega joule laser, if the implosion processes are properly controlled.

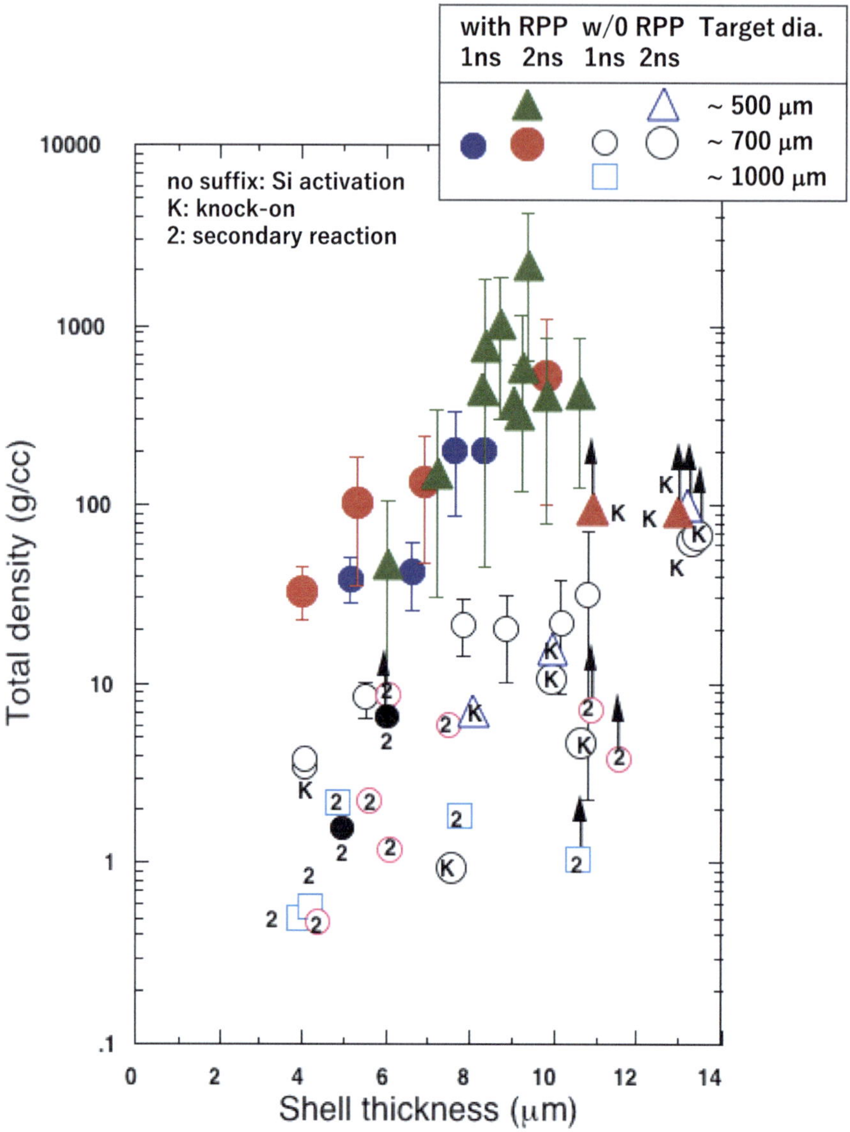

Fig. 14.10 Density of the plastic plasma measured by the Si activation for various shell thicknesses and shell diameters. Solid circles and triangles are for the GekkoXII laser with RPP, and open circles, triangles, and squares are without RPP. Open circles, triangles, and squares with "K" and "2" are the densities measured by neutron knock-on proton and neutron yield of the secondary DT reaction of CD shells, respectively. Courtesy of Prof. N. Miyanaga

Recently, significant progress in direct-drive implosions has been achieved at the LLE, University of Rochester (McCrory et al., 2002). The LLE experiments were conducted on the OMEGA 60-beam, 30 kJ, UV laser system. The improvements in the irradiation uniformity have significantly improved the performance of implosions. A single-beam non-uniformity of 3% (averaged over 300 ps) was achieved with full implementation of 1-THz bandwidth two-dimensional smoothing by spectral dispersion (SSD) (Skupsky, 1989) and polarization smoothing with birefringence wedges (Regan, 2000) which are described later in Sect. 14.8. This corresponds to an on-target non-uniformity of less than 1% rms due to beam overlap. The beam-to-beam power imbalance has been reduced to below 5% rms. With these conditions, moderate CR targets ($\sim$ 15) produce $\sim$ 30% of the neutron yield predicted by one-dimensional prediction and nearly 100% of the predicted fuel and shell areal densities. At predicted CR values close to 40, the primary neutron yield was $\sim$ 20% of the one-dimensional prediction. These results are also plotted in Fig. 14.7b. The remarkable progress of implosions shows the validation of direct-drive inertial fusion for heading towards high-gain laser IFE for power plants.

Other than the laser smoothing techniques, there are several stabilization techniques as indicated in the above. They can be categorized as (a) shock preheating of the shell with a tailored pulse such as a picket-fence pre-pulse or a sharp rise front edge before the main pulse for compression (McKenty et al., 2001), (b) radiation preheating with a high Z doped foam coat or high Z thin layer coat on the fuel pellet (Bodner et al., 2000), and (c) multi collar irradiation (Otani et al., 2007) using nonlocal heat transport by hot electrons.

14.7 Present Status and Critical Issues of Direct and Indirect Drive

14.7.1 Direct Drive

The high compression of a small fuel contained in a capsule is one of the major challenges in IFE. Compression is achieved by the pressure of the high temperature ablated plasma produced by the laser irradiation on the surface of a spherical DT target. With laser intensities $\approx 2 \times 10^{15}$ W/cm^2 of 0.35 μm wavelength, the ablation pressure is of the order of 100 Mbar which accelerates the target shell to the velocity of about 300 km/s within a few nanoseconds. Although the efficiency of the ablative acceleration is of the order of 10–15%, the energy required for the compression is relatively low, about 30–50 kJ. More energy is needed for the DT heating to achieve ignition. The most critical issue of IFE is the heating of a small portion of the compressed fuel, less than 1%, to create a hot spark, that will trigger the burning wave propagation in the remaining cold fuel. The size of the hot spot is determined by the mean free path of the α particle, which corresponds to the areal mass of about 0.2 g/cm^2 (Tikhonchuk, 2011).

In conventional IFE schemes, which are the direct- and indirect-drive implosions, the compression and heating are achieved by specially choosing the laser pulse shape and the target structure as shown in Fig. 14.11 for the direct drive. The target has the form of a shell with an external ablator layer (CH), a solid DT fuel layer (DT ice), and a DT saturated vapor inside (DT gas). The laser pulse has three parts: the pre-pulse launches a shock through the solid shell and compresses and moves the shell inward in the following steps: (1) the rising part accelerates the shell to the velocity that will enable the fuel assembly, (2) the high intensity part launches the shock through the shell and (3) heats the hot spot to the ignition temperature at the moment of stagnation.

In the direct drive, the target is irradiated directly by multiple laser beams. The direct drive approach is potentially more efficient, but it requires very uniform laser irradiation with many beams. It also requires controlling the laser plasma interactions which may cause the instabilities related to small-scale modulations of the laser intensity or imperfections of the shell surface. Moreover, the energetic electrons created in the laser plasma interaction may preheat the central part of the target and reduce the compression (Atzeni & Meyer-Ter-Vehn, 2004).

In the history of the direct drive implosion experiments, high-density compression has been demonstrated with the multi-beam and multi-10 kJ laser facilities at ILE, Osaka University and LLE, University of Rochester. In 1988, the CDT plastic capsules were imploded with 10 kJ, 0.53 μm wavelength laser pulse at Gekko XII laser of ILE to achieve 1000 times compression as shown in Fig. 14.10. As the next steps of the direct drive approach, the MJ-class multi-beam implosion and ignition experiments are expected. This could be done with NIF in USA and LMJ in France.

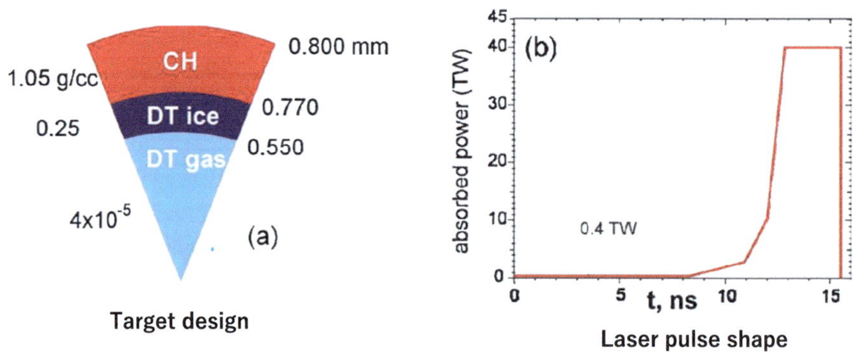

Fig. 14.11 Ignition target design for direct drive laser fusion: **a** fuel shell target, **b** laser pulse shape for a point ignition design

14.7.2 Indirect Drive

In the indirect drive (ID), the laser light is converted into thermal X-ray radiation inside a cavity containing a fuel pellet. The physical processes in the cavity, called a hohlraum, are shown in Fig. 14.12. Many laser beams are injected into the cavity through laser entrant holes. The geometry of the laser beams, the cavity and the capsule, and laser plasma processes are shown in Fig. 14.12. In the cavity, the laser beams interact with the cavity plasma, X-rays are generated in the cavity plasma, and the radiation are confined through absorption and re-emission on the cavity wall. The ablated plasma from the cavity wall interacts with laser beams and the capsule. The radiation confinement, the radiation ablation and the radiation driven implosion are described in this Section.

The cavity used at NIF is cylindrical as shown in Fig. 14.13 (Lindl, 2010; Moses et al., 2009). The high-Z cylindrical cavity contains a fuel capsule inside. There are many other proposals such as rugby ball cavity, 6-hole octahedral cavity, and so on, which are tested by experiments and simulations (Lan et al., 2016; Vandenboomgaerde et al., 2007).

The small-scale irradiation uniformity on the fuel capsule is much better in the ID. However, the ID approach is less efficient because of conversion of the laser light to the X-rays (60–80% efficiency) and the fact that part of the thermal radiation is spent to produce cavity plasma and misses the target. In the NIF experiments, approximately 30 kJ out of 1.8 MJ is converted into the kinetic energy of the imploding

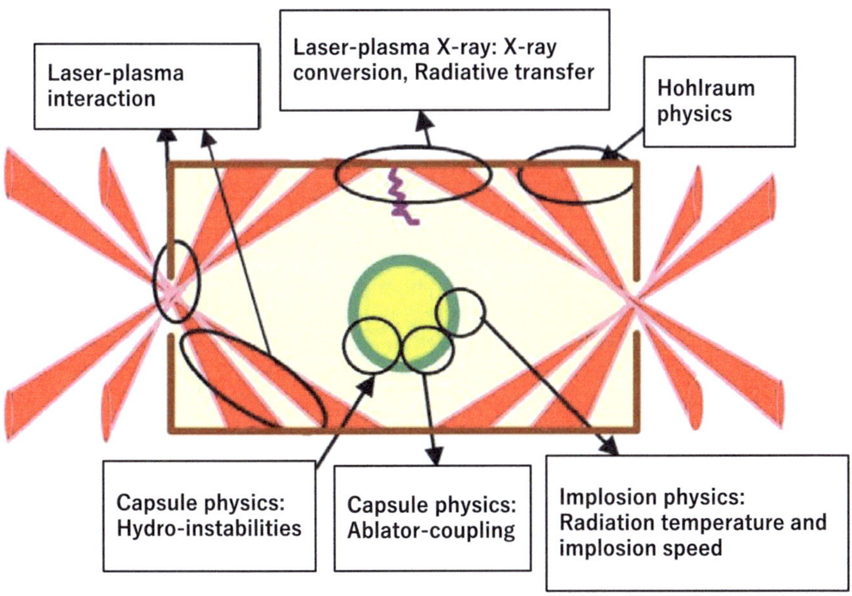

Fig. 14.12 A typical geometry and physical processes of the indirect drive laser fusion

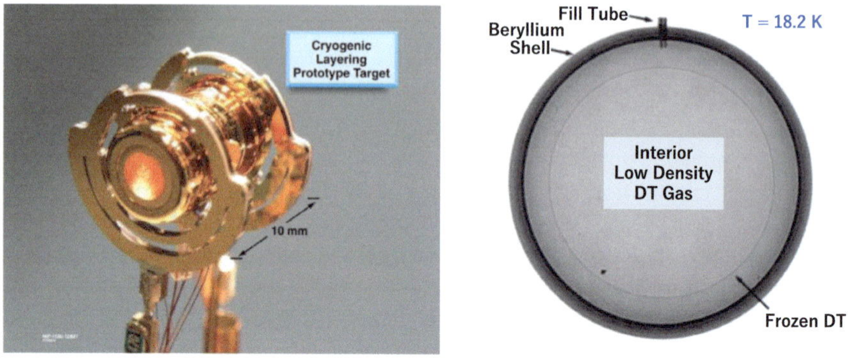

Fig. 14.13 Cylindrical hohlraum and cryogenic fuel target at LLNL. Reprinted from Lindl (2010), with permission from IOP Publishing: permission conveyed through Copyright Clearance Center, Inc. © 2010 IOP Publishing

shell. Optimization of the target design, the laser pulse shape and the irradiation geometry have been implemented by the theory, simulation and experiment. The implosion symmetry has been the critical issue of the NIF implosion. As shown in Fig. 14.14, the wavelengths of each laser beams are carefully selected to mitigate the beam energy transfer by the SBS process. Owing to optimization for uniform irradiation, the shape of the imploded core plasma becomes symmetric as shown in the bottom right of Fig. 14.14.

Extensive numerical simulations have shown that ignition condition will be achieved with the laser energy of 1.0–1.8 MJ (Ebrard & Chaput, 2010; Lindl, 1995, 1998, 2010; Moses et al., 2009). However, these predictions have not been realized because of the implosion non-uniformity. The Rayleigh–Taylor instability and the Richtmyer–Meshkov instability (described in Sect. 14.9) at the interface between the hot spark and the dense cold fuel are strongly unstable during the stagnation. Therefore, the adiabatic compression and heating of the hot spark are significantly reduced.

Recent progress of the ignition experiments at NIF, LLNL was reported in the 2020 IAEA fusion energy conference (Patel et al., 2021). In the experiments since 2013, several new implosion designs have been proposed by machine learning to combine previous experimental data and the simulation results. In 2013, O. Hurricane, et al. introduced a new pulse shape which has higher foot pulse to stabilize the implosion and stagnation to increase the hot spot internal energy. As a result, higher fusion yield was achieved up to 30 kJ. Then, in 2019, a new capsule made by diamond-like carbon called High Density Carbon (HDC) was introduced and the fusion yield reached 60 kJ, where the implosion processes were further improved to increase the energy coupling to the hot spark.

Further improvements were achieved with the two implosion designs, called HDC HyE (High Density Carbon Capsule Hybrid-E) (Zylstra, 2021) and HDC Iraum implosion. In HDC HyE shown in Fig. 14.15a, the low mode asymmetry due to

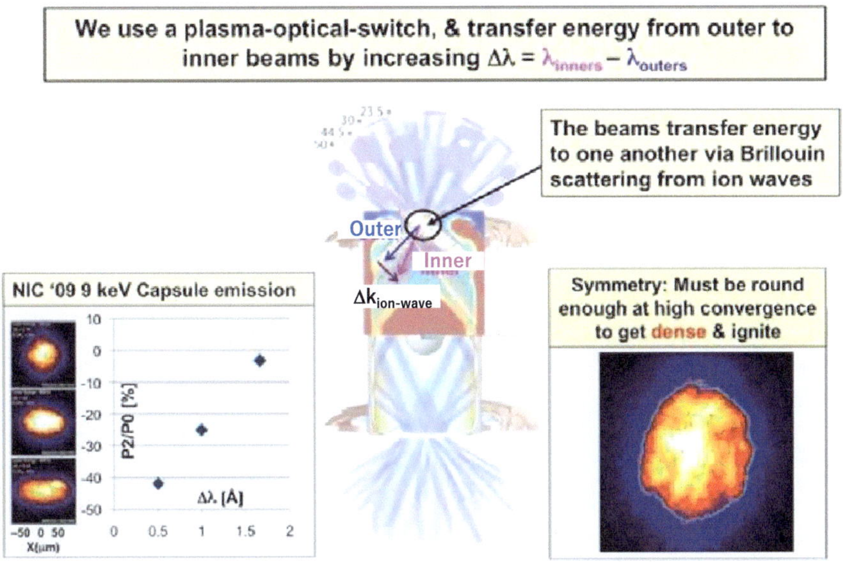

Fig. 14.14 Irradiation geometry of NIF laser beams and control of the wavelength of each beam for achieving the uniform irradiation on the inner surface of the cavity. Reprinted from Lindl (2010), with permission from IOP Publishing: permission conveyed through Copyright Clearance Center, Inc. © 2010 IOP Publishing

deflection of the inner cone beams by the plasma generated by the outer cone beams shown by yellow clouds is mitigated by making the inner cone beam wavelength longer for enhancing the Cross Beam Energy Transfer (CBET). On the other hand, in HDC Iraum shown in Fig. 14.15b, the I-shape cavity is introduced to suppress expansion of the plasmas produced by the outer cone beams to mitigate the deflection of the inner cone beams.

The HDC HyE and HDC Iraum implosions were conducted on the National Ignition Facility (NIF) with larger capsules, where thicker fuel layers are introduced to mitigate the fuel-ablator mix (Baker et al., 2020; Le Pape et al., 2018). In the HDC HyE, the symmetry is improved by increasing the crossbeam energy transfer (CBET) by introducing larger red shift of the inner cone beams. The enhancement of the CBET compensates for the reduction of the inner cone laser intensity near the middle of the cavity as shown in Fig. 14.15a. Another improvement in the symmetry is introduced in the HDC Iraum which stands for "High Density Carbon Capsule in I-shaped hohlraum". In the HDC Iraum, the cylindrical cavity wall is deformed by adding two concave rings which reduce expansion of the plasmas ablated by the outer cone beams as shown in Fig. 14.15b. With these approaches, the low mode asymmetry has been significantly reduced even for smaller CCR (Cavity to Capsule ratio), namely, higher energy coupling of the cavity radiation to the capsule has been achieved with larger capsules.

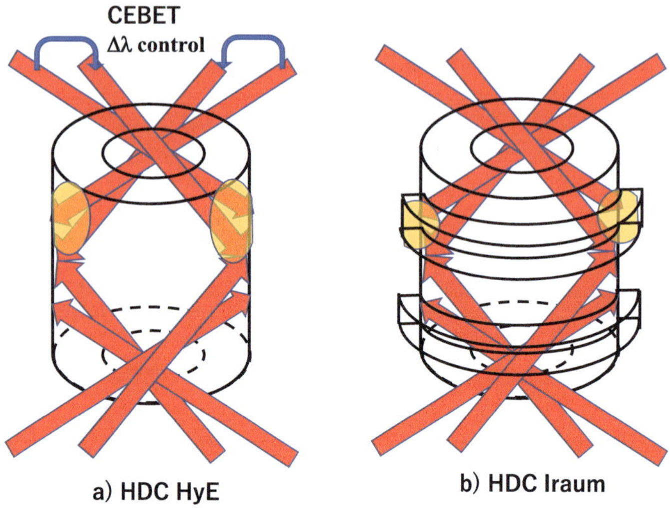

Fig. 14.15 Improved target designs. **a** Reduction of laser intensity of the inner cone beams is supplemented by the crossbeam energy transfer (CBET). **b** The plasmas are confined in the concave rings to reduce the deflection of the inner cone beams

These experiments have achieved record high values of the implosion energetics as well as the fusion yield, as shown in Fig. 14.16. The fusion output reached higher than 100 kJ and the amplification factor of the fusion yield by the alpha self-heating has increased to 3–6 times. In the summer of 2021, the breakthrough in the NIF experiments has been achieved, reaching the fusion yield of 1.35 MJ with 1.9 MJ laser input as shown by a purple square in the right-top corner in Fig. 14.16, where a high-quality target with a bigger capsule and a cavity was selected (Abu-Shawareb, 2022). In December 2022, the breakeven has been achieved, namely the yield reached 3.5 MJ for 2.15 MJ input laser energy. These experiments are historical achievements in the laser fusion research because the proof of principle of the central hot spark concept has been demonstrated (Ma & Betti, 2023).

14.8 Alternative Ignition Concepts

14.8.1 Electron Driven Fast Ignition

Alternative schemes that separate the fuel assembly and the hot spot ignition open possibilities for more efficient use of the laser energy, responding better to the reactor requirements. Instead of the isobaric ignition in the standard approach, the hot spot is created in an already compressed fuel at a much higher pressure. However, a very

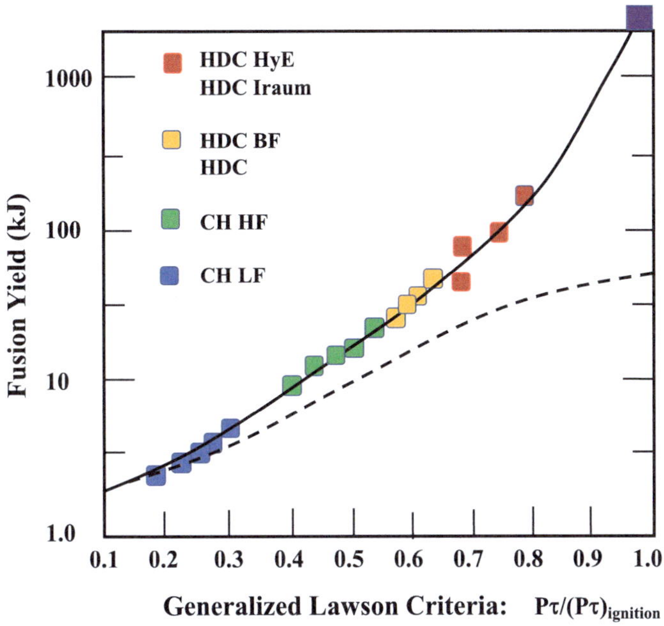

Fig. 14.16 Fusion yield versus the generalized Lawson criteria. The purple square is the 2021.8.8 NIF shot result

short hot spot lifetime of the order of 10 ps imposes strong requirements on the laser power that exceeds a petawatt level.

In the electron driven fast ignition (EFI) scheme (Tabak et al., 1994), the off-center hot spot is created by a laser-accelerated energetic electron beam carrying 50–100 kJ energy. It proceeds in two steps shown in Fig. 14.17b. First, a laser pulse with ~ 10^{19}–10^{20} W/cm^2 intensity and ~ 10 PW power creates a hole in a plasma around the compressed core to the density of the order of a few tens g/cm^3, thus preparing the conditions for efficient electron acceleration close to the core. The lifetime of 30 ps channel provides a launching window for the second, more powerful ignition pulse. This pulse with a power exceeding 10 PW accelerates electrons to ~ 1–5 MeV energy that are stopped in the core creating the ignition conditions in a small spot of 30–100 μm size.

Channel formation is the less developed part of the fast ignition scheme. The formation of high-quality transparent channels in dense plasmas has not been demonstrated yet. Alternative versions of the fast ignition scheme rely either on providing a shorter path to the core for the ignition pulse, or on energy transport with ions. Since the ions are suffering less from instabilities, their transport could be more easily controlled, and they can be generated at a larger distance from the core from a separate auxiliary target. However, the ion acceleration with lasers could be efficient only at very high intensities exceeding 10^{20} W/cm^2. Such laser parameters are rather

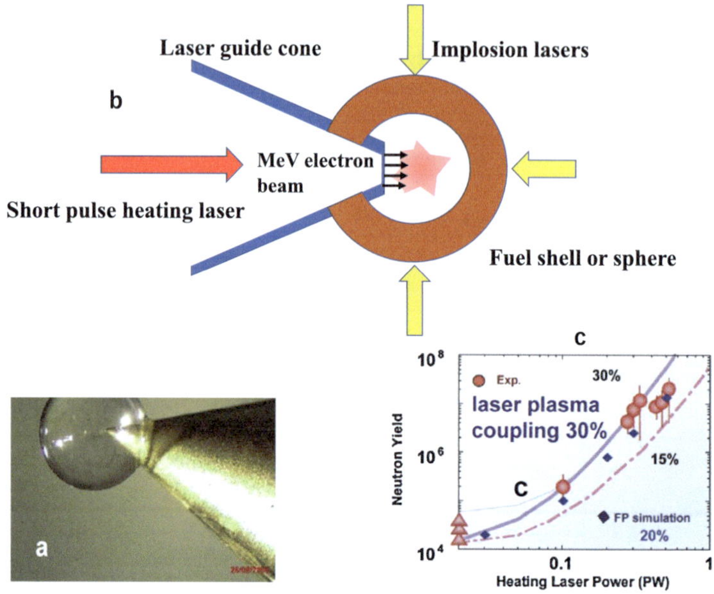

Fig. 14.17 An example of laser-generated-electron driven fast ignition (EFI), where a hollow gold cone is attached to a thin deuterated polystyrene shell. **a** a photo of a fuel capsule with a cone, **b** schematic figure of the EFI concept, and **c** neutron yield versus heating laser power

demanding, and the proof of principle experiments are unlikely to take place in the near future.

The most significant breakthrough in the fast ignition was achieved in a UK-Japan joint experiment (Kodama et al., 2002; Mima et al., 2012). A thin deuterated polystyrene shell with a hollow gold cone (Fig. 14.17a) was imploded with the 9 ns-duration beams of GEKKO XII laser system with a total energy of 2.5 kJ (Azechi et al., 2009). The ignition pulse with the maximum power of 0.4 PW was injected into the cone's interior at the moment of stagnation. Energetic electrons generated near the tip of the cone heated the imploded plasma and increased the D-D neutron yield by a factor of thousand. The time window of the efficient neutron emission was consistent with the core stagnation time of 50 ps, and 20–30% of the 500 J ignition pulse energy was transferred to the core.

This very successful experiment served as a base for two ongoing IFE projects. The project FIREX (Azechi et al., 2009) at the Institute of Laser Engineering, Osaka University aims to demonstrate fast ignition of the DT fuel by using the cone-in-shell target assembly (Kodama, 2001). The first step consists in repeating the experiment (Kodama et al., 2002) at a higher injected energy of a few kJ. The LFEX laser with 10 kJ energy and 1.5–5 ps pulse duration has been constructed. Figure 14.18 shows an X ray image of a tip of the cone and a compressed plasma when the LFEX laser pulse was injected. The contrast of the x-ray image will be improved since the laser pedestal was high in this experiment. The second stage to upgrade the LFEX and the

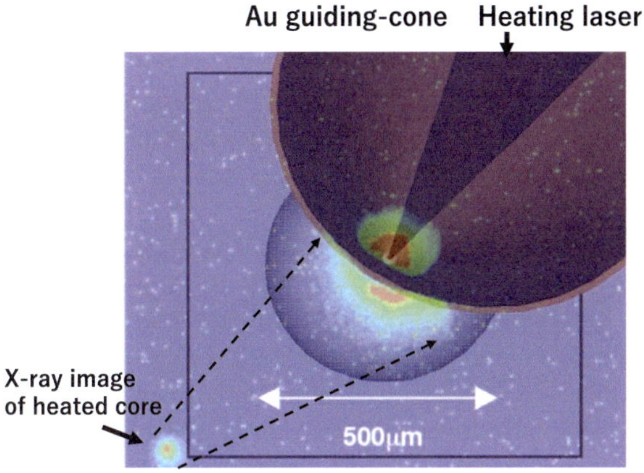

Fig. 14.18 Self-emitted X-ray intensity distribution of the imploded plasma heated by a short pulse laser in the cone-shell target experiment

compression laser system GEKKO XII to 50 and 100 kJ levels, respectively, will be endorsed when the first milestone is achieved.

In Europe, the HiPER project was proposed in 2007 (Dunne, 2007). This is the first international project in IFE and its initial research and development phase is supported by the European Commission. The project aims on building a facility that will demonstrate the technical feasibility of an IFE power plant by operating about 100 high gain shots with the repetition rate of a few hertz in burst mode. The major issues are: (i) construction of an efficient high repetition rate laser system; (ii) mass target fabrication, target acceleration and laser beam steering; (iii) development of the reactor technology including the first wall and final optics; (iv) target design that provides the required gain and technologically acceptable fabrication. Two target design concepts are under development. The first concept, corresponding to the fast ignition scheme, is rather like the FIREX project, implying the electron beam injection in the compressed core through a hollow cone. The second concept is the recently developed scheme of the shock ignition.

Efficient transformation of the laser energy to hot electrons has been demonstrated with sufficiently short laser pulses of less than 1 ps duration. The major issues for the hot electron source are the energy spectrum and the angular divergence, which are closely related to the plasma density profile seen by the intense laser pulse (Honrubia et al., 2010). In the case of a cone guided target, the density profile is sensitive to the laser pulse preceding the main short pulse (pre-pulse or pedestal). It was found in numerical simulations and experiments that the laser pedestal level should be very low to avoid plasma production and expansion of the inner wall of the cone. The effect of the laser pre-pulses needs to be verified in the coming experiments with laser pulses of longer duration up to 10 ps.

The physics of electron acceleration with solid targets and the role of cone environment were studied. The efficiency of the laser pulse coupling to the electrons increases with the laser intensity, achieving 40–50% at the intensities exceeding 10^{19}–10^{20} W/cm^2. The mean electron energy scales approximately with a cubic root of the laser intensity, following the Beg's law (Beg et al., 1997; Hains et al., 2009).

The electron distribution measured right behind the acceleration zone is strongly divergent with a characteristic opening angle of the order of 20–30°. This is due to small scale modulation of the acceleration surface by the laser ponderomotive force and due to the Weibel instability in the target right behind the acceleration zone. The magnetic focusing of the electrons in indispensable for creation of a hot spot of 20–30 μm size in the core.

The electron beam transport is the major problem of the fast ignition scheme. The current density of the electron beam exceeds 10^{14} A/cm^2. The total current is higher than 1 GA, which is many orders of magnitude higher than the Alfvén limit. The return current of the ambient plasma neutralizes the beam current. But it produces an energy dissipation, and it is also a source of instabilities that increase the beam divergence (Honrubia et al., 2010). It is not known whether the beam can be transported to 100–200 μm distance in a dense plasma without strong divergence and losses. One promising possibility is a guiding effect of the resistivity gradient that was demonstrated recently (Ramakrishna, 2010).

A significant progress in understanding the physics of the electron beam ejection from the cone has been achieved in recent years. The efficient electron acceleration requires no pre-plasma in the cone, which implies a high contrast ignition laser beam. It is not known whether the appropriate interaction conditions can be maintained during 10–15 ps since the actual experiments have been limited to less than 1 ps pulse duration.

Longer pulse experiments are planned on the LFEX and Omega EP facilities. To guide the accelerated electrons toward the cone tip, a double cone geometry has been proposed (Cai et al., 2009). An inner thinner cone serves to guide and accelerate electrons along the surface, while the external cone protects the path for injection of the ignition laser pulse. Figure 14.19 shows that high energy electrons are effectively blocked by the vacuum gap in the double cone by self-generated electro-magnetic fields.

The experimental studies of the fast ignition scheme are conducted at the University of Rochester and Osaka University with the installations OMEGA EP and LFEX.

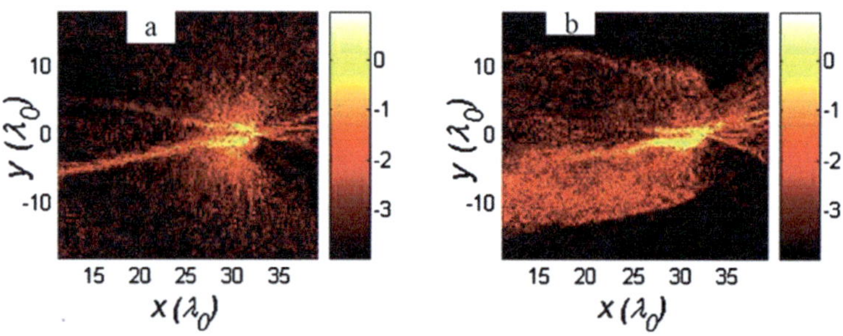

Fig. 14.19 Hot electron confinement by a double wall cone target. Reprinted from Cai et al. (2009). © 2009 The American Physical Society

Further full-scale experiments could be conducted on NIF and LMJ when these facilities are equipped with high intensity short pulse PW beams.

During 2016–2020, efficient hot electron guiding experiments were carried out at ILE to demonstrate significant heating of the imploded plasma. Figure 14.20 shows a schematic of the laser driven capacitor coil for generating strong magnetic field to guide the hot electrons, and the experimental results of the electron temperature of the heated dense plasma (Fujioka et al., 2016; Matsuo et al., 2020; Sakata et al., 2018).

In addition to guiding of the relativistic electrons, another important issue in the fast ignition is to understand the heating mechanism of the cold dense plasma. There are three major mechanisms for isochoric heating (Kemp et al., 2006): drag heating where the core plasma is heated directly by the MeV electrons, resistive heating by the electrons of the return current generated to maintain current neutrality in a plasma, and diffusive heating where thermal electrons transport their energy from the laser heated region at the critical density to the high density cold region. Among these mechanisms, it was found by experiment and simulation that the thermal diffusive heating plays an essential role to heat the cold dense plasma of > 10 g/cm^3 to over keV in the laser heating of multi-picosecond duration (Higashi et al., 2020; Matsuo et al., 2020; Sawada et al., 2019).

It was shown by Kidder (1974) that high density fuel will be generated by spherical homogenous isentropic compression of a solid target with a properly programmed laser pulse. Since we can separate high-density compression and heating with the fast ignition scheme, it is possible to heat the high-density, low-temperature fuel produced by implosion of a spherical solid target in place of compressing a thin shell target which is susceptible to hydrodynamic instabilities as described in Sect. 14.6.1. In recent fast ignition experiments with the external magnetic field at ILE (Fujioka et al., 2016; Matsuo et al., 2020; Sakata et al., 2018), spherical solid ball targets are

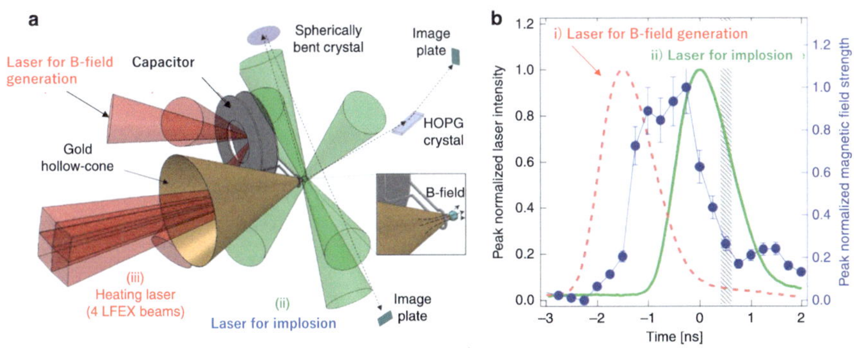

Fig. 14.20 **a** Schematic view of the target of the electron-driven fast ignition (EFI) with a laser driven capacitor coil for applying strong magnetic field on an imploded plasma, **b** temporal profiles of the laser pulse driving the capacitor coil (red dotted line), magnetic field strength (blue solid circles), and implosion laser pulse (green line). Reprinted from Sakata et al. (2018). © 2018 Springer Nature

used for implosion. When a simpler layout of core heating is developed, this approach may be preferable for fusion power plant design due to ease of target fabrication.

14.8.2 Ion Beam Driven Fast Ignition

Although the electron beam driven fast ignition (EFI) has many advantages, the EFI has some drawbacks such as the spreading and long stopping range of the electron beam. As an alternative scheme to the fast ignition by relativistic electrons, the ion driven fast ignition (IFI) has been proposed by Roth et al. (2001), Roth et al. (2009), Honrubia et al. (2009), Fernandes et al. (2014), and Atzeni et al. (2002). A typical geometry of the ion acceleration and the imploded plasma is shown in Fig. 14.21.

The ion beam is much more collimated in comparison with the electron beam. However, the laser to ion energy conversion efficiency is relatively low in comparison with the electron beam. Conversion efficiency is about 6–12% according to the experiments by Snavely et al. (2000) and Robinson et al. (2008). The ion beam interacts with the imploded plasma classically and can deposit the energy locally, since it is relatively easy to control the ion transport. Proton fast ignition (PFI) is an advanced option because of the high transfer efficiency from the beam to the hot spark, which is the critical issue for the EFI. However, it is not yet solved to find a proper geometry of connecting the efficient laser ion acceleration system to the implosion plasma. Novel target geometries such as flat-top cone targets seem to be appropriate to increase these efficiencies (Flippo et al., 2008).

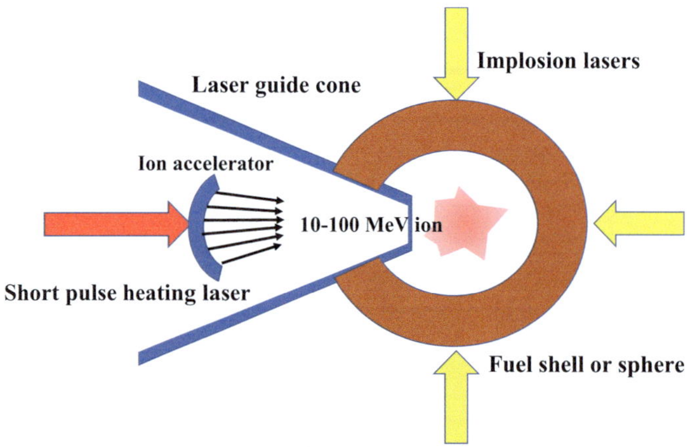

Fig. 14.21 Schematic view of the IFI concept

14.8.3 Shock Ignition

Shock ignition is an alternative scheme that combines a more stable and secure shell implosion with a central hot spot ignition initiated with an additional laser pulse (Betti et al., 2007; Guskov et al., 2017; Ribeyre et al., 2009). A spherical shell is imploded by multiple laser beams as in the direct drive scheme. A relatively low implosion velocity in the range of 220—290 km/s provides the density of a few hundred of g/cm^3 at stagnation, but with a low hot spot temperature. The latter is boosted by launching a strong spherical shock at the end of the free coasting phase of the implosion. This ignition shock is further amplified due to spherical convergence and due to collision with the main shock reflected from the target center. The ignition occurs under the condition where the hot spot pressure is significantly larger than the pressure of the surrounding dense fuel. Such a non-isobaric fuel assembly requires less energy to ignite than the conventional DD or ID schemes. The ignition laser power is smaller by order of magnitude than in the fast ignition scheme because the convergence and collision of shocks provide power amplification.

The parameter that defines the efficiency of the shock ignition is the ratio of the hot spot pressure to the pressure of the shell. Figure 14.22 shows the trajectories of the converging shell and the shock fronts. Enhancement of the ignition shock amplitude is expected due to its convergence (from S_5 to S_6 in Fig. 14.22a) and collision with the main shock reflected from the center (at S_6, S_7, S_8, and S_9 in Fig. 14.22a and b). As shown in Fig. 14.22b, the last collision increases the hot spot pressure by a factor of five compared to the shell pressure. These selective heating of the hot spot and additional shell compression strongly reduce the ignition threshold, by a factor of 20 for the pressure ratio of 5 (Lafon, 2010). Numerical simulations for the baseline HiPER target (similar to that shown by the isochoric gain curve in Fig. 14.5) predict the gain of 90 for the 180 kJ, 50 TW of the compression pulse and 100 kJ, 200 TW ignition pulse with the launching window of the ignition pulse of 250 ps. These parameters are within the specifications of the NIF and LMJ facilities, thus promising a full-scale verification of the present scheme.

However, the NIF and LMJ are not designed for the direct-drive experiments. The positions of the laser beams around the interaction chamber are such that irradiation of polar zones is much stronger than in the equator. Implementation of specially designed Polar Direct Drive (PDD) implosion scheme (Skupsky, 2004) will be tested on the NIF. It is expected that it will improve the implosion symmetry by redirecting and deviating the beams. A combination of the PDD implosion and shock ignition is now under investigation. It might happen that the ignition shock does not need very high symmetry. High intensity of the ignition pulse implies high plasma temperatures and enhanced lateral smoothing of the pressure between the absorption and the ablation surfaces. Moreover, smaller shell acceleration and multiple crossings of the inner shell boundary by shocks play stabilization roles for the shell implosion.

The shock ignition approach has renewed interest in the studies of laser plasma interaction. In the standard DD scheme, the laser intensities are limited to 10^{15} W/cm^2 at the third harmonic (the wavelength of 351 nm) to enable the collisional laser

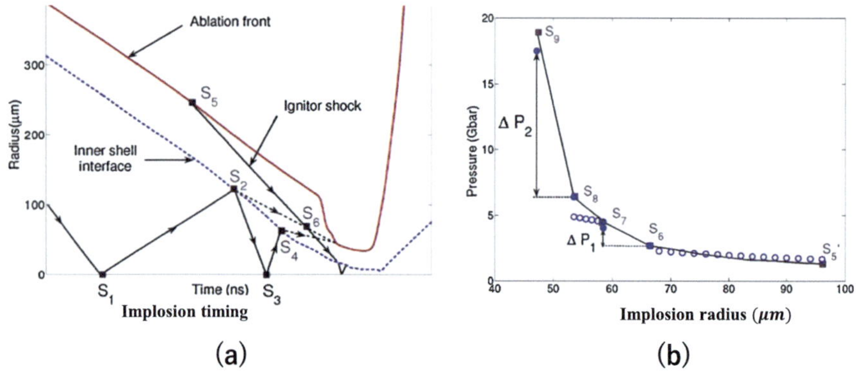

Fig. 14.22 Propagation of the ignition shock generated by the tailored pulse

absorption and not to enter in the domain where the parametric instabilities operate. Lower implosion velocities provide better security for the laser plasma interaction at the shell acceleration stage. However, the risk of excitation of stimulated scattering processes at the ignition phase is much higher. These effects were studied recently in massive kinetic simulations of the laser plasma interaction (Klimo et al., 2010) and the experiments are underway. The simulation shows competition between the stimulated Brillouin scattering (SBS) and stimulated Raman scattering (SRS) processes. The SBS is initially very strong, but this situation changes as soon as the SRS near the quarter of critical density starts to grow. This is an absolute parametric instability that transforms up to half of incident laser energy into plasma waves and consequently in hot electrons. Moreover, the scattered SRS wave at half of the laser frequency undergoes the secondary SRS instability in the vicinity of its own quarter critical density. This process is manifested by the reflected signal at the quarter of the laser frequency and by the absorption increasing to 70%. Consequently, the SBS-SRS competition enables efficient transformation of the incident laser energy into hot electrons with the characteristic energy of 30–40 keV as shown in Fig. 14.23. The SRS is considered to be an undesirable process in the standard DD scheme as it produces energetic electrons with a mean free path comparable to or larger than the shell thickness, which strongly increases the entropy of the imploding shell. It is different, however, for the shock ignition. The ignition pulse arrives on the target much later when the shell is approximately half through on its way to the center. Consequently, its areal density is about 10 times higher, and the SRS-generated electrons are stopped in the outer part of the shell thus increasing further the ablation pressure. This beneficial role of the SRS instability in the context of shock ignition needs to be verified in experiments.

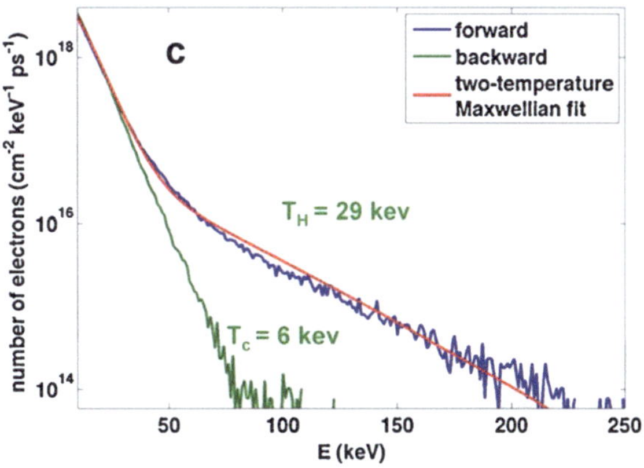

Fig. 14.23 Bulk and hot electron energy spectra in the shock ignition. Reprinted from Klimo et al. (2010), with permission from IOP Publishing: permission conveyed through Copyright Clearance Center, Inc. © 2010 IOP Publishing

14.9 Comments on Inertial Fusion Energy (IFE) Reactor

Sustainable energy production presents many serious challenges for the IFE. Assuming the laser efficiency of 10% and the thermal cycle efficiency of 40%, the power plant with the recirculating power of less than 20% requires the fusion gain higher than 100. Moreover, to produce the electric power of 1 GW with the released energy in each shot limited to 250 MJ, the reactor should operate with the repetition rate of 10 Hz and with the laser pulse energy less than 2 MJ. The challenge for the driver is to construct a compact laser facility with the average power of 20 MW delivering 10 pulses per second with a controlled temporal shape. The technology of diode pumping of solid-state lasers is a promising approach. The major challenges are the diodes cost, robustness, and the temporal and spatial profiles of the laser pulse (Ma et al., 2023; Haefner et al., 2023).

Moreover, the existing ignition schemes rely on the conversion of the laser light into the third harmonic (350 nm), which imposes very severe limits on the damage of the output optic elements. The target fabrication and the target engagement in the interaction vessel constitute another specific domain of the ICF technology. The typical time of fabrication of a present-day target is about a few days and its cost is more than several tens of thousands of dollars. The major factors are a complicated structure, a very high precision of the assembly and a high quality of the surface. The power plant targets need to cost less than 20 cents and be fabricated much faster (keeping in mind that about one million targets per day are needed). The existing technologies, in principle, permit to fulfill these requirements, although the targets need to be sufficiently simple and robust. Launching the target into the target center with a velocity of more than 100 m/s with a precision of 50 μm is the problem that is not yet resolved.

The science and technology of the reactor first wall are partly like those for the magnetic fusion energy reactor, regarding neutron damage, tritium generation and retention, average heat flux, etc. However, high-energy ion bombardment, pulsed X-ray flux, and repetitive heat pulse effects are very different from the conditions expected in the MFE reactor. Some issues such as tritium breeding and low vacuum requirement are simpler because of the high modularity of the inertial fusion reactor and the possibility for replacement of damaged parts in a relatively short time of a few weeks.

14.10 Conclusion

Inertial confinement fusion is crossing a crucial and decisive point in its history. Expected ignition demonstration on the NIF facility called for three projects for energy production that are developed in parallel. The LIFE project in the USA aims on the power plant prototype based on the indirect drive scheme. The European and Japanese projects are more conservative, aiming on the fusion energy production either in a single shot or in a burst mode by using the alternative schemes based on the fast ignition of a target with a cone or on the shock ignition of a spherical target. The alternative ignition schemes are more attractive in many aspects regarding laser energy, target fabrication, and reactor operation. Many physical and technological problems require their solutions. Reliable operation in the next coming years of the NIF and LMJ facilities will allow us to test both options in full scale.

References

Abu-Shawareb, H., et al. (Indirect Drive ICF Collaboration). (2022). Lawson criterion for ignition exceeded in inertial fusion experiment. *Physical Review Letters, 129*, 075001.

Atkinson, R., & Houtermans, F. (1929). "Aufbaumöglichkeit in Sternen", Zur Frage der Aufbaumöglichkeit der Elemente in Sternen. *Zeitschrift Für Physik, 54*, 656.

Atzeni, S., Temporal, M., & Honrubia, J. (2002). A first analysis of fast ignition of precompressed ICF fuel by laser- accelerated protons. *Nuclear Fusion, 42*, L1–L4.

Atzeni, S., & Meyer-ter-Vehn, J. (2004). *The physics of inertial fusion.* Claredon Press.

Azechi, H., Jituno, T., Kanabe, K., et al. (1991). High-density compression experiments at ILE, Osaka. *Laser and Particle Beams, 9*, 193.

Azechi, H., Mima, K., Fujimoto, Y., et al. (2009). Plasma physics and laser development for the fast-ignition realization experiment (FIREX) project. *Nuclear Fusion, 49*, Article 104024.

Baker, K. l., Thomas, C.A., Casey, D.T., Hohenberger, M., et al. (2020). Hot-spot parameter scaling with velocity and yield for high-adiabat layered implosions at the National Ignition Facility. *Physical Review E, 102*, 023210.

Basov, N. G., & Krokhin, O. N. (1964). The conditions of plasma heating by the optical generator radiation. In P. Grivet & N. Bloembergen (Eds.), *Proceedings of the Third International Symposium on Quantum Electronics*, Paris, 1963 (p. 1373). Columbia University Press.

Basov, N., Kriukov, P., Zakharov, S. D., Senaski, Yu., et al. (1968). Experiments on the observation of neutron emission at a focus of high-power laser radiation on a deuteride surface. *IEEE Journal of Quantum Electronics, QE4*, 864.

Beg, F. N., Bell, A. R., Dangor, A. E., et al. (1997). A study of picosecond laser-solid interactions up to 10^{19} W cm^{-2}. *Physics of Plasmas, 4*, 447.

Bethe, H. A. (1939). Energy production in stars. *Physical Review, 55*, 434.

Bethe, H. A., & Peierls, R. (1934). The neutrino in nature. *Nature, 133*, 532.

Bethe, H. A. (1937). Nuclear physics B. Nuclear dynamics, theoretical. *Reviews of Modern Physics, 9*, 69.
Betti, R., Zhou, C. D., Anderson, K. S., Perkins, W., et al. (2007). Shock Ignition of thermonuclear fuel with high areal density. *Physical Review Letters, 98*, Article 155001.
Bodner, S. E., et al. (2000). High-gain direct drive target design for laser fusion. *Physics of Plasmas, 7*, 2298.
Cai, H., Mima, K., Zhou, W., et al. (2009). Enhancing the number of high-energy electrons deposited to a compressed pellet via double cones in fast ignition. *Physical Review Letters, 102*, Article 245001.
Dawson, J. M. (1964). On the production of plasma by giant pulse lasers. *Physics of Fluids, 7*, 981.
Dunne, M., et al. (2007). Technical background and conceptual design report 2007. HiPER Website www.hiper-laser.org
Ebrard, J., & Chaput, J. M. (2010). LMJ on its way to fusion. *Journal of Physics: Conference Series, 244*, Article 032017.
Eliezer, S., & Mima, K. (2009). *Application of laser plasma interactions*. CRC Press.
Eliezer, S., et al. (2002). Fundamentals of equations of state. *World Scientific*. https://doi.org/10.1142/4872
Fernandez, J. C., Albright, S. J., Beg, F. N., et al. (2014). Fast ignition with laser-driven proton and ion beams. *Nuclear Fusion, 54*, Article 054006.
Flippo, K. A., d'Humi'eres, E., Gaillard, S. A., et al. (2008). Increased efficiency of short-pulse laser-generated proton beams from novel flat-top cone targets. *Physics of Plasmas, 15*, 056709.
Floux, F. (1971). Nuclear fusion by laser radiation. *Nuclear Fusion, 11*, 635.
Floux, F., Cognard, D., Bobin, J. L., et al. (1969). Nuclear fusion by laser radiation. *Academy of Sciences, 269*, 697.
Fujioka, S., Arikawa, Y., Kojima, S., et al. (2016). Fast ignition realization experiment with high-contrast kilo-joule peta-watt LFEX laser and strong external magnetic field. *Physics of Plasmas, 23*, Article 056308.
Gamow, G., & Teller, E. (1938). The rate of selective thermonuclear reactions. *Physical Review, 53*, 608.
Guderley, G. (1942). Starke kugelige und zylindrische verdichtungsstosse in der nahe des kugelmitterpunktes bnw. der zylinderachse. *Luftfahrt-Forsch., 19*, 302.
Gus'kov, S. Yu., Zmitreenko, N. V., & Rahimly, O. R. (2017). A converging shock wave for ignition of a pre-compressed target of laser thermonuclear fusion. https://www.mathnet.ru/links/c32fca8a09f3a5699f0d8b655443627d/ipmp2310.pdf (in Russian).
Haefner, C. L., Alexander, N., & Ricard, B., et al. (2023). Memorandum of laser fusion energy. https://www.bmbf.de/SharedDocs/Downloads/de/2023/230522-memorandum-laser-inertial-fusion-energy.pdf.
Hains, M. G., Wei, M. S., Beg, F. N., et al. (2009). Hot-electron temperature and laser-light absorption in fast ignition. *Physical Review Letters, 102*, Article 045008.
He, X. T., et al. (2016). A hybrid-drive non-isobaric-ignition scheme for inertial confinement fusion. *Physics of Plasmas, 23*, Article 082706.
Honrubia, J. J., Fernandes, J. C., Temporal, M., et al. (2009). Fast ignition of inertial fusion target by laser-driven carbon beams. *Physics of Plasmas, 16*, Article 102701.
Honrubia, J. J., Fernández, J. C., Temporal, M., Hegelich, B. M., Meyer-ter-Vehn, J. (2010). Fast ignition by laser-driven carbon beams. *Journal of Physics: Conference Series*, *244*, 022038.
Hora, H. (1964). Nonlinear effect of expansion of laser produced plasmas. IPP 6/27 Institut für Plasmaphysik – Garching.
Hurricane, O. A., et al. (2019). High performance indirect-drive cryogenic implosions at high adiabat on the National Ignition Facility. *Plasma Physics and Controlled Fusion, 61*, Article 014033.
Johzaki, T., Mima, K., & Nakano, Y. (2007). Simulation study of fast ignition and burn characteristics of fast ignition DT targets. *Plasma and Fusion Research, 2*, 041.
Kato, Y., Mima, K., Miyanaga. N., et al. (1984). Random phasing of high-power lasers for uniform target acceleration and plasma-instability suppression. *Physical Review Letters*, *53*, 1057.

Kemp, A. J., Sentoku, Y., Sotnikov, V., & Wilks, S. C. (2006). Collisional relaxation of superthermal electrons generated by relativistic laser pulses in dense plasma. *Physical Review Letters, 97*, Article 235001.

Kidder, R. E. (1974). Theory of homogeneous isentropic compression and its application to laser fusion. *Nuclear Fusion, 14*, 53.

Klimo, O., et al. (2010). Particle-in-cell simulations of laser–plasma interaction for the shock ignition scenario. *Plasma Physics and Controlled Fusion, 52*, Article 055013.

Kodama, R., Norreys, P., & Mima, K. (2001). Fast heating of ultrahigh-density plasma as a step towards laser fusion ignition. *Nature, 412*, 798.

Kodama, R., et al. (2002). Fast ignition scalable to laser fusion ignition. *Nature, 418*, 933.

Krokhin, O. N. (1969). *Proceedings of the International School of Physics—Enrico Fermi (Varenna), 48th Course.*

Lafon, M., Ribeyre, X., & Schurtz, G. (2010). Gain curves and hydrodynamic modeling for shock ignition. *Physics of Plasmas, 17*, Article 052704.

Lan, K., Liu, J., Li, Z., et al. (2016). Progress on octahedral spherical hohlraum study. *Matter and Radiation at Extremes, 1*, 8.

Lawson, J. D. (1955). Some criteria for a power producing thermonuclear reactor. *Proceedings of the Physical Society, Section B. 70*, 6.

Le Pape, S., Berzak Hopkins, L. F., Divo, L., et al. (2018). Fusion energy output greater than the kinetic energy of an imploding shell at the National Ignition facility. *Physical Review Letters, 120*, Article 245003.

Lee, Y. T., & More, R. M. (1984). An electron conductivity model for dense plasmas. *Physics of Fluids, 27*, 1273.

Lindl, J. (1995). Development of the indirect-drive approach to inertial confinement fusion and the target physics basis for ignition and gain. *Physics of Plasmas, 2*, 3933.

Lindl, J. (1998). *Inertial confinement fusion-the quest for ignition and energy gain using indirect drive*. Springer.

Lindl, J. D., et al. (2010). Progress toward Ignition on the National Ignition Facility. *Nuclear Fusion, 51*, 094024. *Proceedings of the 23rd Fusion Energy Conference*, Daejeon, Korea (2010), IAEA, Vienna. https://doi.org/10.1088/0029-5515/51/9/094024.

Ma, T., & Betti, R. (2023). Inertial Fusion Energy. Report of the 2022 Fusion Energy Sciences: Basic Research Needs Workshop. Office of Science, US-DOE. https://science.osti.gov/-/media/fes/pdf/workshop-reports/2023/IFE-Basic-Research-Needs-Final-Report.pdf

Ma, X., Qiu, L., & Wang, Y. (2021). A measurement method of curvature radius and shell thickness of laser fusion capsule by combining computer vision and virtual laser differential confocal technology. *iSPIE Future Sensing Technologies, 11914*, 119140V.

Maiman, T. H., et al. (1960). Stimulated optical radiation in ruby. *Nature, 187*, 493.

Matsuo, K., Higashi, N., Iwata, N., et al. (2020). Peta-pascal pressure driven by fast isochoric heating with a multi-pico-second intense laser pulse. *Physical Review Letters, 124*, Article 035001.

McCrory, R. L., et al. (2002). Direct-drive inertial confinement fusion research at the Laboratory for Laser Energetics: Charting the path to thermonuclear ignition. In *Proceedings of the 19th IAEA Fusion Energy Conference* (Lyon, France, 14–19, October 2002), IAEA-CN-94/1F-1.

McKenty, P. W., et al. (2001). Analysis of a direct-drive ignition capsule designed for National Ignition Facility. *Physics of Plasmas, 8*, 2315.

Meyer-ter-Vehn, J. (1982). On the energy gain of fusion targets: The model of Kidder and Bodner improved. *Nuclear Fusin, 22*, 561.

Mima, K. (2012). *Fusion physics* (Kikuchi, M., Lackner, K., & Tran, M. Q., Eds.) (Chap. 10, p. 1043). IAEA. ISBN 978-92-0-1-130410-0.

Moses, E. I., et al. (2009). Ignition on the National Ignition Facility: A path towards inertial fusion energy. *Nuclear Fusion, 49*, Article 104022.

Nakai, S., et al. (1990). X-ray and particle diagnostics of a high-density plasma by laser implosion (invited). *Review of Scientific Instruments, 61*, 3235; *Proceedings of the 13th IAEA Conference*

on Plasma Physics and Controlled Fusion Research (1–6 October 1990, Washington D.C.) IAEA-CN-53/B-1-3.
Nakai, S., & Takabe, H. (1996). Principles of inertial confinement fusion-physics of implosion and concept of inertial fusion energy. *Reports on Progress in Physics, 59*, 1071.
Nishimura, H., et al. (2000). Indirect-direct hybrid target experiments with the GEKKO XII laser. *Nuclear Fusion, 40*, 547.
Norimatsu, T., Johzaki, T., Azechi, H., Mima, K., et al. (2006). Conceptual design of laser fusion reactor KOYO-fast-target design and the fueling system. *Journal De Physique IV, 133*, 841.
Nuckolls, J., et al. (1972). Laser compression of matter to super-high densities. *Nature, 239*, 139.
Otani, K., Shigemori, K., Sakaiya, T., Fujioka, S., et al. (2007). Reduction of the Tayleigh-Taylor instability growth with cocktail color irradiation. *Physics of Plasmas, 14*, Article 122702.
Park, H. S., Hurricane, O. A., Callahan, D. A., et al. (2014). High-adiabat high-foot inertial confinement fusion implosion experiments on the National Ignition Facility. *Physical Review Letters, 112*, Article 055001.
Patel, P. K., et al. (2021). Progress in the US inertial confinement fusion program. In *2021 IAEA-FEC (Fusion Energy Conference).*
Ramakrishna, B., Kar, S., Robinson, A. P. L., et al. (2010). Laser-driven fast electron collimation in targets with resistive boundary. *Physical Review Letters, 105*, Article 135001.
Regan, S., Marozas, J. A., Kelly, J. H., et al. (2000). Experimental investigation of smoothing by spectral dispersion. *Journal of the Optical Society of America b: Optical Physics, 17*, 1483.
Ribeyre, X., Schurtz, G., Lafon, M., et al. (2009). Shock ignition: An alternative scheme for HiPER. *Plasma Physics and Controlled Fusion, 51*, Article 015013.
Robinson, A. P. L., Zepf, M., Kar, S., et al. (2008). Radiation pressure acceleration of thin foils with circularly polarized laser pulses. *New Journal of Physics, 10*, Article 013021.
Roth, M., Cowan, T. E., Key, M. H., et al. (2001). Fast ignition by intense laser-accelerated proton beams. *Physical Review Letters, 86*, 436.
Roth, M., et al. (2009). Review on the current status and prospects of fast ignition in fusion targets driven by intense laser generated proton beams. *Plasma Physics and Controlled Fusion, 51*, Article 014004.
Sakata, S., Lee, S., Morita, H., et al. (2018). Magnetized fast isochoric laser heating for efficient creation of ultra-high-energy-density sates. *Nature Communications, 9*, 3937.
Sawada, H., Sentoku, Y., Yabuuchi, T., et al. (2019). Monochromatic 2D Kα emission images revealing short-pulse laser isochoric heating mechanism. *Physical Review Letters, 122*, Article 155002.
Skupsky, S., Short, R. W., Kessler, T., Craxton, R. S., et al. (1989). Improved laser-beam uniformity using the angular dispersion of frequency-modulated light. *Journal of Applied Physics, 66*, 3456.
Skupsky, S., et al. (2004). Polar direct drive on the National Ignition Facility. *PhysIcs of Plasmas, 11*, 2763.
Snavely, R. A., Key, M. H., Hatchett, S. P., Cowan, T. E., et al. (2000). Intense high-energy proton beams from petawatt-laser irradiation of solid. *Physical Review Letters, 85*, 2945.
Tabak, M., et al. (1994). Ignition and high gain with ultrapowerful lasers. *Physics of Plasmas, 1*, 1626.
Tajima, T., Mima, K., & Baldis, H. (2000). *High field science*. Kluwer Academic/Plenum Publishers.
Tikhonchuk, V., & Mima, K. (2011). Alternative schemes for the inertial fusion energy. *Fusion Engineering and Design, 86*, 504.
Tomabechi, K., Kozaki, Y., Miyanaga, N., Norimatsu, T., et al. (2006). Conceptual design of laser fusion reactor KOYO-fast. *Journal De Physique IV, 133*, 837.
Vandenboomgaerde, M., Bastian, J., Casner, A., et al. (2007). Prolate-Spheroid (Rugby-Shaped) hohlraum for inertial confinement fusion. *Physical Review Letters, 99*, Article 065004.
Zylstra, A. S., Kritcher, A. L., Hurricane, Q. A., et al. (2021). Record energetics for inertial fusion implosion. *Physical Review Letters, 126*, Article 025001.

Chapter 15
Uniform Laser Irradiation

Abstract Laser irradiation uniformity is an important issue to realize spherically symmetric implosion not only for the direct drive but also for the indirect drive. The optical technologies to introduce random phase in space and time have been successful for uniform irradiation. In this chapter, the smoothing technologies are described.

Keywords Random phase plate · Induced spatial incoherence · Smoothing by spectral dispersion

15.1 Random Phase Plate

One of the critical issues for spherically symmetric laser implosion is the uniformity of laser irradiation on a target. Since a large aperture high power laser beam is usually accompanied by the phase distortion and is affected by diffraction at the aperture, the laser intensity on target is non-uniform. Kato et al. proposed and demonstrated mitigating the irradiation non-uniformity by introducing the random phase plate (RPP), which is illustrated schematically in Fig. 15.1a, b where a randomly distributed phase plate is placed before the focusing lens (Kato et al., 1982, 1984). Initially, a two-level phase plate of square elements with phases of either 0 or π in random distribution across the beam was introduced. The element size was chosen so that the diffraction-limited focal spot size associated with each element is adjusted to the target size. Namely, the laser beam is broken into separate beam-lets which overlap with each other on the target.

Phase-plate elements are typically much smaller than the spatial scale of the phase variation of the laser beam. So, the irradiation pattern becomes independent of the phase variations. The envelope of the intensity pattern on target $f(x, y)$ in Fig. 15.1b is described by a sinc-squared function; namely,

$$f(x, y) = \frac{\sin^2 X}{X^2} \frac{\sin^2 Y}{Y^2}, \tag{15.1}$$

Y. Kato et al., *High Power Laser and Plasma Science*, Springer Series on Atomic, Optical, and Plasma Physics 130, https://doi.org/10.1007/978-3-032-08930-4_15

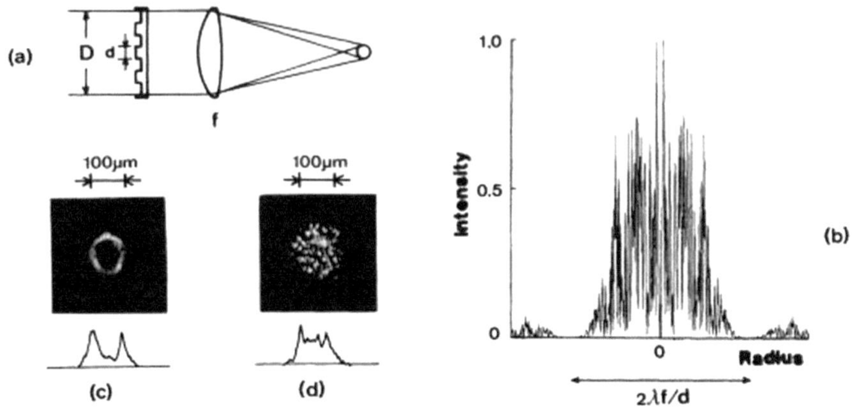

Fig. 15.1 Random phasing of a laser beam for uniform target irradiation. **a** Schematic of a random phase plate (RPP), **b** calculated beam intensity profile at a target position with RPP, observed focal patterns of the laser beam **c** without and **d** with the RPP. Reprinted from Kato et al. (1984). © 1984 The American Physical Society

where $X = dx/2F\lambda_0$, and $Y = dy/2F\lambda_0$ with the focal length F, the laser wavelength λ_0, and the phase element size dx and dy in the x and y directions.

After the proposal by Kato et al., LLE and LLNL continued to improve the phase plates. LLE employed two-level phase plates with a hexagonal pattern, since 20% of the laser energy is diffracted outside the first zero of the sinc-squared envelopes. Furthermore, the two-level phase plate was replaced by the continuous phase plate in which almost 100% of the beam energy is focused on target. The phase-plate was further developed at LLNL, where large, 80-cm-diameter phase plates were applied to the Nova laser by Dixit et al. (1993, 1994). The "kino-form" phase plates, like the continuous random phase plates, were used on the NIF. An alternative approach to improve the original two-level phase-plate concept was proposed by Deng et al. (1986) in the form of a lens array.

15.2 Induced Spatial Incoherence

For mitigating the speckles associated with the RPP-generated pattern, a novel technique known as induced spatial incoherence (ISI) was developed at NRL (Lehmberg & Obenschain, 1983; Lehmberg et al., 1987), where the speckles are smoothed out by temporal movement of the speckle pattern. An echelon is placed before the focus lens, as shown schematically in Fig. 15.2 for one dimension. The steps in the echelon delay different transverse sections of the beam by a laser coherence time or larger, which is related to the bandwidth of the laser. As with a phase plate, the echelon breaks up the beam into beam-lets, each of which forms a well-defined intensity envelope on target.

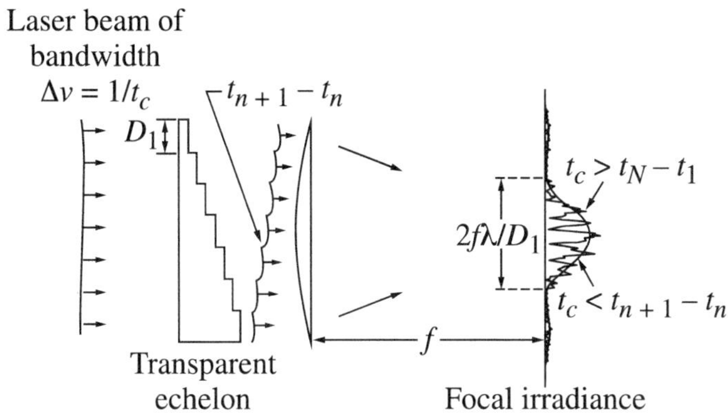

Fig. 15.2 Concept of the ISI: a broad band laser beam is broken up into many beamlets with a transparent echelon, where each beamlet has a different time delay longer than the coherent time of the laser beam. Reprinted from Lehmberg and Obenschain (1983) with permission from Elsevier: permission conveyed through Copyright Clearance Center, Inc. © 1983 Elsevier

Instantaneously, there is a high-modulation interference pattern (speckle) on target, but the speckle fluctuates randomly in time since contributions to the interference pattern from different echelon steps are temporally incoherent. When averaged over many coherence times, the speckle structure becomes smoothed, with the nonuniformity decreasing in time. The laser bandwidth must be sufficiently large (with a coherence time of 2 ps which is achievable with Nd:glass lasers) to smooth the beam before significant imprinting of the nonuniformity onto the target.

15.3 Smoothing by Spectral Dispersion (SSD)

15.3.1 SSD with Coherent Light

Smoothing by spectral dispersion (SSD) was developed by Skupsky et al. (1989) at LLE to provide the temporal smoothing of the speckles due to the phase plates and meet the damage constraints imposed on the glass lasers such as OMEGA and NIF. Like ISI, this approach uses broadband light and time delays to create temporally varying phase-front variations across the beam. The basic SSD concept is shown in Fig. 15.3, where phase modulated bandwidth is imposed upon the laser beam by an electro-optic modulator and dispersed by a grating to result in color variations across the beam. The change of the phase distribution across the beam combined with the use of a phase plate at the focus lens results in temporal smoothing of the beam at focus.

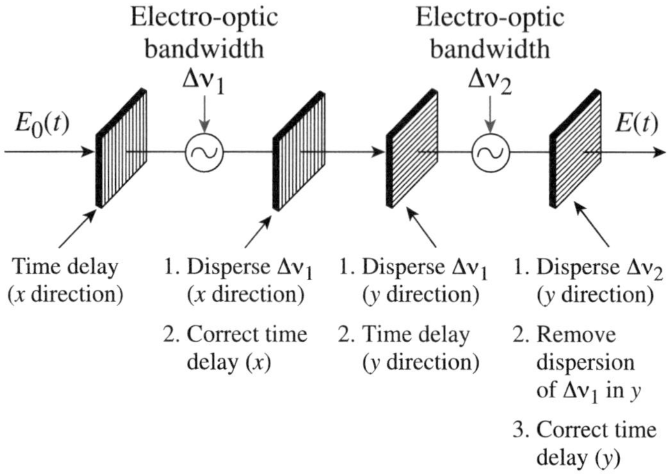

Fig. 15.3 Concept of the two-dimensional SSD. Reprinted from Skupsky and Craxton (1999) with permission from AIP Publishing: permission conveyed through Copyright Clearance Center, Inc. © 1999 AIP Publishing

15.3.2 SSD of Amplified Spontaneous Emission

Further improvement of the laser irradiation uniformity has been proposed at ILE, Osaka University (Miyaji et al., 2002; Nakano et al., 1993). It is an application of SSD with a space–time incoherent laser pulse. Since in-coherent light is scanned in space, the intensity profile of the focused laser is extremely smooth as shown in Fig. 15.4, where intensity profiles at focus are compared with the GEKKO XII laser, from left to right, for coherent light without and with RPP and incoherent light generated as amplified stimulated emission (ASE) without and with RPP.

However, the beam intensity filling factor of the optics is extremely small for the ASE beam. So, the laser optics can be damaged if the average fluence of the beam is not limited to a very low level. Therefore, the ASE beam with SSD has not been used except for very precise intensity control.

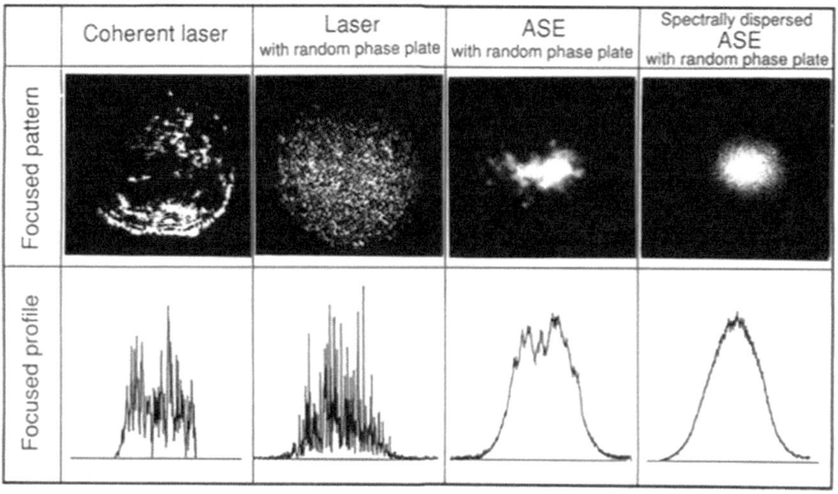

Fig. 15.4 Focused laser intensity profiles of the Gekko XII laser: from left to right, coherent laser, coherent laser with RPP, ASE with RPP, and ASE with SSD and RPP, respectively. Courtesy of Prof. Noriaki Miyanaga

References

Deng, X., Liang, X., Chen, Z., Yu, W., & Ma, R. (1986). Uniform illumination of large targets using a lens array. *Applied Optics, 25*, 377.

Dixit, S. N., Lawson, J. K., Manes, K. R., et al. (1994). Kinoform phase plates for focal plane irradiance profile control. *Optics Letters, 19*, 417.

Dixit, S. N., Thomas, I. M., Woods, B. W., et al. (1993). Random phase plates for beam smoothing on the Nova laser. *Applied Optics, 32*, 2543.

Kato, Y., & Mima, K. (1982). Random phase shifting of laser beam for absorption profile smoothing and instability suppression in laser produced plasmas. *Applied Physics B, 29*, 186.

Kato, Y., Mima, K., Miyanaga, N., et al. (1984). Random phasing of high-power lasers for uniform target acceleration and plasma-instability suppression. *Physical Review Letters, 53*, 1057.

Lehmberg, R. H., & Obenschain, S. P. (1983). Use of induced spatial incoherence for uniform illumination of laser fusion targets. *Optics Communication, 46*, 27.

Lehmberg, R. H., Schmitt, A. J., & Bodner, S. E. (1987). Theory of induced spatial incoherence. *Journal of Applied Physics, 62*, 2680.

Miyaji, G., Miyanaga, N., Urushihara, S., Suzuki, K., Matsuoka, S., Nakatsuka, M. (2002). Three-directional spectral dispersion for smoothing a laser irradiance profile. *Optics Letters, 27*, 725.

Nakano, H., Miyanaga, N., Yagi, K., Tsubakimoto, K., et al. (1993). Partially coherent light generated by using single and multimode optical fibers in a high-power Nd:glass laser system. *Applied Physics Letters, 63*, 580.

Skupsky, S., & Craxton, R. S. (1999). Irradiation uniformity for high compression laser-fusion experiments. *Physics of Plasmas, 6*, 2157.

Skupsky, S., Short, R. W., Kessler, T., Craxton, R. S., et al. (1989). Improved laser-beam uniformity using the angular dispersion of frequency-modulated light. *Journal of Applied Physics, 66*, 3456.

Chapter 16
Hydrodynamic Instabilities

Abstract The implosion spherical symmetry is broken by the plasma instabilities. In particular, the hydrodynamic instabilities and the electro-magnetic thermal effects in imploding plasmas are important. In this chapter, relevant hydrodynamic instabilities and thermo-electric effects are described.

Keywords Hydrodynamic instabilities · Richtmyer-Meshkov instability · Rayleigh-Taylor instability

16.1 Introduction

Non-uniformities of the laser irradiation and the pellet (surface profile and density distribution) initiate the implosion non-uniformities, which cause the turbulent mixing between the cold main fuel and the hot spark at the final stage of the implosion. When a non-uniform shock wave passes through the contact surface with density jump, the contact surface deforms, and the non-uniformity is amplified by the Richtmyer-Meshkov (R-M) Instability in the initial stage of implosion. On the other hand, in the final stage of implosion (beginning of the stagnation phase), the shock wave reflected from the pellet center hits the interface between the main fuel and the hot spark, where the interface distortion is amplified again by the Richtmyer-Meshkov Instability driven, leading to the turbulent mixing of target material.

On the other hand, when the pellet shell is accelerated inward by the ablation pressure, the ablation front will be hydro-dynamically unstable due to the Rayleigh-Taylor (R-T) instability. The amplified non-uniformities of the imploding shell are the seeds for the R-M instability at the stagnation phase. In this Chapter, these hydrodynamic instabilities are described.

Y. Kato et al., *High Power Laser and Plasma Science*, Springer Series on Atomic, Optical, and Plasma Physics 130, https://doi.org/10.1007/978-3-032-08930-4_16

16.2 Richtmyer-Meshkov Instability

The processes of the implosion non-uniformity are schematically shown in Fig. 7.8. The non-uniformity of the laser irradiation generates non-uniform ablation pressure, and the implosion asymmetry increases due to the hydrodynamic instability. At the initial phase of irradiation, separation of the absorption region (cutoff region) from the ablation front is small. Therefore, the non-uniformity of the laser irradiation can be strongly imprinted on the target surface (initial imprint). The shock wave driven by the main pulse amplifies the surface rippling on the target through the R-M instability as described in the following. The surface rippling is due to the original target roughness and the initial imprint. The coupling of the laser irradiation non-uniformity to the ablation pressure non-uniformity is weak in a later time due to thermal smoothing in the energy transport from the absorption region to the ablation surface.

The R-M instability was found in the pioneering works of Robert D. Richtmyer at Los Alamos National Laboratory (theoretical work in 1960) and Evgeny E. Meshkov at the Russian Federal Nuclear Center-All-Russian Research Institute of Experimental Physics (RFNC-VNIIEF) (the first experiments to study this instability in 1969).

When a shock wave passes through a corrugated interface from the low-density side to the high-density side, non-uniform fluid flow appears as shown in Fig. 16.1a, which is a manifestation of the R-M instability. The schematic development of the M-R instability is illustrated in Fig. 16.1b. When the shock passes through the interface from a lighter layer (with density ρ_1) to a higher layer (with density ρ_2), the impulse acceleration occurs at the interface. Namely, the "impulse gravity" is applied in the direction from the lighter layer to the heavier layer. In such a case, the R-M instability can be called the "impulse Reyleigh-Taylor instability". As shown in Fig. 16.1(b), when a shock wave propagates through the density gradient, the shear flow is generated. Let us assume that the pressure jump at the shock front is P_s the shock front is inclined by an angle θ relative to the contact surface, and the density gradient is $(\rho_1 - \rho_2)/d$, where d is the width of the density jump as shown in Fig. 16.2.

The equation of motion of a fluid element is given by,

$$\frac{D\boldsymbol{v}}{Dt} = -\frac{\nabla P}{\rho}. \tag{16.1}$$

Here the operator in the l.h.s. of this equation is

$$\frac{D}{Dt} = \frac{\partial}{\partial t} + (\boldsymbol{v} \cdot \nabla).$$

Taking the y component (directed along the contact surface) of Fig. 16.1, we obtain

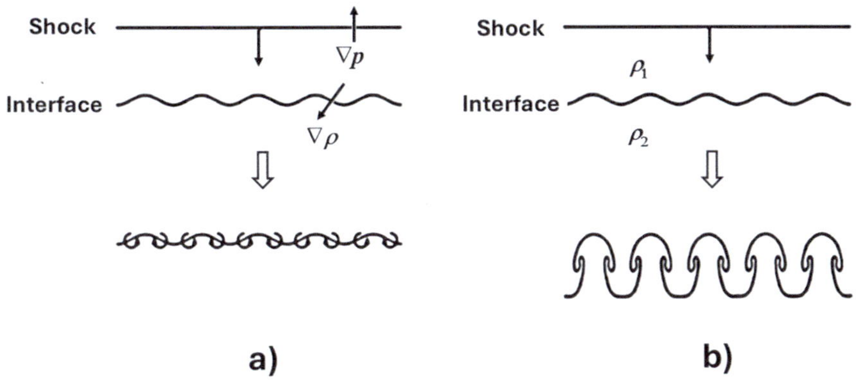

Fig. 16.1 **a** Geometry of the R-M instability, and **b** mechanisms of the instability

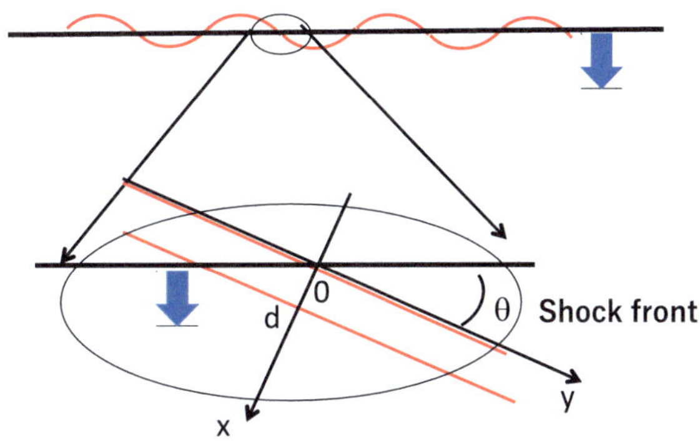

Fig. 16.2 A shock propagates in the x-direction with an angle θ with respect to the y-axis. The density jumps from r_1 to r_2 from $x = 0$ to d

$$\frac{Dv_y}{Dt} = -\frac{1}{\rho(x)}\frac{\partial P}{\partial y} \approx \frac{\sin\theta}{\rho(x)}\frac{\partial P}{\partial x}. \tag{16.2}$$

Let us consider the density profile given by the function

$$\rho(x) = \rho_1 + \frac{\rho_2 - \rho_1}{d}x, \quad (0 < x < d).$$

Integrating Eq. (16.2) over time t in the shock wave reference frame, where $x = x_0 + U_s t$ with U_s being the effective shock velocity at the contact surface, we obtain,

$$v_y(x) = \frac{\Delta P_s \sin\theta}{\rho(x)U_s}\Theta(U_s t - x). \tag{16.3}$$

Here, $\Theta(x)$ is the Heaviside step function, which is equal to unity, when the argument is positive, and is equal to zero for $x < 0$. This means that the shear flow is generated by the interaction of the shock wave obliquely incident on the density discontinuity. Note here that the flow is incompressible, i.e. the derivatives of the shear flow in Eq. (16.3) are negligible. As a result, the vorticity $\boldsymbol{\omega} = \omega \boldsymbol{e}_z$ (here $\boldsymbol{e}_z$ is a unit vector in the z-direction) can be found. It is equal to

$$\omega = \frac{\partial v_y(x)}{\partial x} = -\frac{(\rho_1 - \rho_2)\Delta P_s \sin\theta}{\rho d U_s}. \tag{16.4}$$

The vorticity generation can be also described by taking the curl of Eq. (16.1),

$$\frac{D\boldsymbol{\omega}}{Dt} = \frac{\nabla\rho \times \nabla P}{\rho^2}. \tag{16.5}$$

Therefore, the vorticity oscillates along the ripple of the interface as shown in Fig. 16.1b. Note here that the vorticity in the z direction is negative when $\rho_2 > \rho_1$, i.e. $Att > 0$, where

$$Att = \frac{\rho_2 - \rho_1}{\rho_2 + \rho_1} \tag{16.6}$$

is the Attwood number. The vortex flow deforms the contact surface as shown in the bottom of Fig. 16.1a.

Richtmyer's linear instability was first verified by the experiment of Meshkov. It was performed by shock tube experiments, in which a perturbed interface of different gases was separated by a sinusoidally-shaped thin membrane. The flow was visualized by Schlieren photography. Subsequently, the R-M instability was investigated for stronger shocks, three-dimensional perturbations, and reacceleration by the reflected shock wave (Aleshin et al., 1988; Benjamin, 1992; Brouillette & Sturtevant, 1994; Sadot et al., 1998; Vasilenko, 1992; Vetter & Sturtevant, 1995).

More recently, a technique for generating corrugated interface without membrane by using opposed gas flows was developed (Collins & Jacobs, 2002; Jacobs & Krivets, 2005; Jones et al., 1997). These experiments have enabled repeatable 2D and 3D single-mode perturbation experiments. The gas flows are visualized by the planar laser-induced fluorescence (PLIF). An example of the experimental results is shown in Fig. 16.3.

16.3 Rayleigh-Taylor Instability

The key element that governs the implosion stability has been identified as the Rayleigh-Taylor (R-T) instability, leading to requirements for the laser and X-ray irradiation uniformity and the fuel pellet sphericity. A schematic geometry of the R-T instability mode is shown in Fig. 16.4. In the case of ablative acceleration of a shell, the ablated low-density plasma (fluid 1) pushes the high-density shell (fluid 2), and the shell is accelerated. Since the effective gravity acts in the z-direction on the acceleration reference frame, the interface is distorted as shown in Fig. 16.4b.

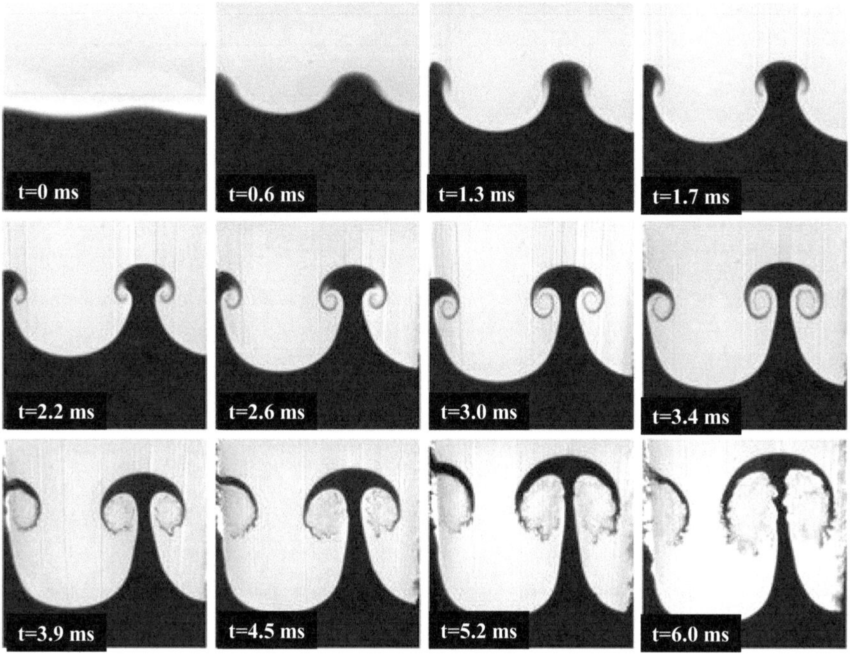

Fig. 16.3 Shock tube experiment of the temporal evolution of the R-M instability driven by the vortex, in which a shock wave accelerates an air/SF_6 interface. Reprinted from Jacobs and Krivets (2005) with permission from AIP Publishing: permission conveyed through Copyright Clearance Center, Inc. © 2005 AIP Publishing

Since the distortion is a surface wave, the mode structure is expected as shown in Fig. 16.4c.

The two-dimensional incompressible fluid motion for the linear analysis of the R-T instability is described by the system of hydrodynamics equations,

$$\rho \frac{\partial v_x}{\partial t} = -\frac{\partial \delta P}{\partial x}, \tag{16.7}$$

$$\rho \frac{\partial v_z}{\partial t} = -\frac{\partial \delta P}{\partial z} - g\delta\rho, \tag{16.8}$$

with the incompressibility condition:

$$\frac{\partial v_x}{\partial x} + \frac{\partial v_z}{\partial z} = 0. \tag{16.9}$$

Since $\rho(z)$ is the profile of the ambient plasma, the equation of continuity yields

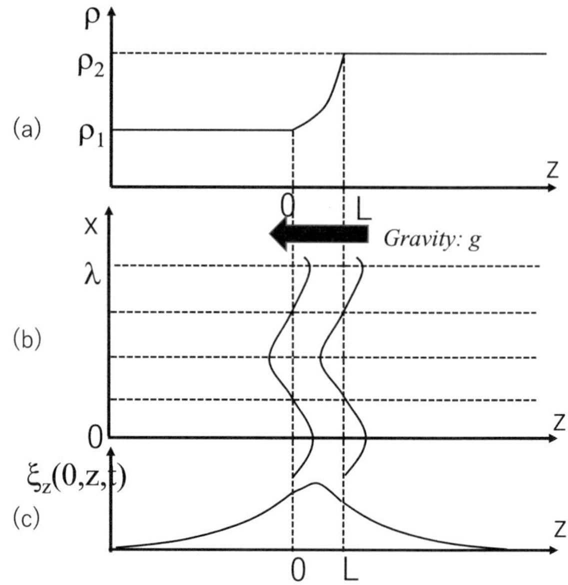

Fig. 16.4 Schematic of the Rayleigh-Taylor instability of the two-fluid interface. **a** Density profile between fluid 1 and fluid 2 with transition layer between z = 0 and z = L, **b** the sinusoidal distortion of the interface at z = 0 and z = L, and **c** the R-T mode structure for the displacement of the fluid element

$$\frac{\partial \delta\rho}{\partial t} + v_z \frac{\partial \rho(z)}{\partial z} = 0. \tag{16.10}$$

To analyze the RT instability, we look for the solution of Eqs. (16.7)–(16.10) representing the perturbations of the fluid velocity and density as a sinusoidal mode along the interface between the regions of the fluid (along the x-direction), For the mode having a wave number k, and depending on time as $\exp(\gamma t)$ we represent the velocity, the density and the pressure perturbations as follows:

$$\begin{pmatrix} v_x(x,z,t) \\ v_z(x,z,t) \\ \delta\rho(x,z,t) \\ \delta P(x,z,t) \end{pmatrix} = \begin{pmatrix} \tilde{v}_x(z) \\ \tilde{v}_z(z) \\ \tilde{\rho}(z) \\ \tilde{P}(z) \end{pmatrix} \exp(ikx + \gamma t) \tag{16.11}$$

Substituting these functions to Eqs. (16.7)–(16.10) we obtain the system of two ordinary differential equations for $\tilde{v}_z(z)$ and $\tilde{\rho}(z)$:

$$\gamma\rho \frac{d\tilde{v}_z}{dz} + k^2 \tilde{p} = 0, \tag{16.12}$$

$$\frac{d\tilde{P}}{dz} + \left(\gamma\rho - \frac{g}{\gamma}\frac{d\rho}{dz}\right)\tilde{v}_z = 0. \tag{16.13}$$

Equations (16.12) and (16.13) yield

$$\frac{d}{dz}\left(\rho\frac{d\tilde{v}_z}{dz}\right)+k^2\left(\rho-\frac{g}{\gamma^2}\frac{d\rho}{dz}\right)\tilde{v}_z=0 \tag{16.14}$$

The dispersion equation giving the relationship between the wave number k and the RT instability growth rate γ follows from Eq. (16.14) for the surface mode with the boundary condition $\tilde{v}_{z\to\pm\infty}\to 0$.

Describing the ambient density profile shown in Fig. 16.4 as

$$\rho(z)=\begin{cases}\rho_1 & \text{for } z\le 0\\ \rho_1\exp\left[\frac{z}{L}\ln\left(\frac{\rho_2}{\rho_1}\right)\right] & \text{for } 0\le z\le L\,,\\ \rho_2 & \text{for } z\ge L\end{cases} \tag{16.15}$$

we obtain the following dispersion equation

$$\left(k^2+A-\frac{K^2}{4}\right)\left(\exp\left(2L\sqrt{A}\right)-1\right)+2k\sqrt{A}\left(\exp\left(2L\sqrt{A}\right)+1\right)=0, \tag{16.16}$$

Here

$$A=k^2\left(1-\frac{Kg}{\gamma^2}\right)+\frac{K^2}{4},\quad \text{and}\quad K=\frac{1}{L}\ln\left(\frac{\rho_2}{\rho_1}\right). \tag{16.17}$$

For a discontinuous density jump $L\to 0$ Eq. (16.16) gives the well-known growth rate of the R-T instability,

$$\gamma=\sqrt{Att\,k\,g}. \tag{16.18}$$

with Att equal to Attwood number defined by Eq. (16.6).

The growth rate of the R-T instability at the ablation front is determined by the ablating convection and the finite density scale length as follows,

$$\gamma=\alpha\sqrt{\frac{kg}{1+kL}}+\beta k v_a, \tag{16.19}$$

where k is the wave number of the growing mode, g is the acceleration, v_a is the ablation velocity, and L is the density scale length at the ablation front. Empirical values of $\alpha=0.9$ and $\beta=3-4$ have been found using the comparison of numerical and experimental results for direct-drive implosions (Betti et al., 1998; Lindl, 1995; Takabe et al., 1983, 1985).

The concepts of stabilizing the implosion aim to enlarge the characteristic scale length L and to increase the ablation velocity v_a with the reduced density of the ablation front ρ_a and keeping the ablation pressure $\rho_a v_a$ constant. These are (a) shock preheating, (b) radiation energy transport, and (c) nonlocal heat transport by hot electrons, as previously mentioned. The gain estimations in recent numerical simulations of direct-drive implosions include consideration of these instability processes and the stabilizing mechanism. Together with the progress in fuel pellet fabrication and laser beam control technology, it can be reasonably expected that fusion ignition, burn, and energy gain will be demonstrated by direct-drive implosions with irradiation of multi-beam mega-joule lasers if the beams are properly arranged for the direct drive. The Rayleigh-Taylor instability is described more precisely in Atzeni and Meyer-Ter-Vehn (2004).

References

Aleshin, A. N., Gamalli, E. G., Zaitsev, S. G., Lazareva, E. V., et al. (1988). Nonlinear and transitional stages in the onset of the Richtmyer-Meshkov instability. *Soviet Technical Physics Letters, 14*, 466.

Atzeni, S., & Meyer-ter-Vehn, J. (2004). *The physics of inertial fusion*. Claredon Press.

Benjamin, R. F. (1992). Experimental observations of shock stability and shock-induced turbulence. In W. P. Dannevik, A. C. Buckingham, & C. E. Leith (Eds.), *Advances in compressible turbulent mixing*.

Betti, R., Goncharov, V. N., McCrory, R. L., & Verdon, C. P. (1998). Growth rates of the ablative Rayleigh-Taylor instability in inertial confinement fusion. *Physics of Plasma, 5*, 1446.

Brouillette, M., & Sturtevant, B. (1994). Experiments on the Richtmyer-Meshkov instability: Single-scale perturbations on a continuous interface. *Journal of Fluid Mechanics, 263*, 271.

Collins, B. D., & Jacobs, J. W. (2002). PLIF flow visualization and measurements of the Richtmyer-Meshkov instability of an air/SF interface. *Journal of Fluid Mechanics, 464*, 113.

Jacobs, J. W., & Krivets, V. V. (2005). Experiments on the late-time development of single-mode Richtmyer- Meshkov instability. *Physics of Fluids, 17*, Article 034105.

Jones, M. A., & Jacobs, J. W. (1997). A membrane-less experiment for the study of Richtmyer-Meshkov instability of a shock-accelerated gas interface. *Physics of Fluids, 9*, 3078.

Lindl, J. (1995). Development of the indirect-drive approach to inertial confinement fusion and the target physics basis for ignition and gain. *Physics of Plasmas, 2*, 3933.

Sadot, O., Erez, L., Alon, U., et al. (1998). Study of nonlinear evolution of single-mode and two bubble interaction under Richtmyer-Meshkov instability. *Physical Review Letters, 80*, 1654.

Takabe, H., Montierth, L., & Morse, R. L. (1983). Self-consistent eigenvalue analysis of Rayleigh-Taylor instability in an ablating plasma. *Physics of Fluids, 26*, 2299.

Takabe, H., Mima, K., et al. (1985). Self-consistent growth rate of the Rayleigh-Taylor instability in an ablatively accelerating plasma. *Physics of Fluids, 28*, 3679.

Vasilenko, A. M., Buryakov, O. V., Kuropatenko, V. F., et al. (1992). Experimental research of gravitational instability and turbulization of flow at the noble gases interface. In W. P. Dannevik, A. C. Buckingham, & C. E. Leith (Eds.), *Advances in compressible turbulent mixing* (pp. 581–606). Lawrence Livermore National Laboratory Report Conf-8810234.

Vetter, M., & Sturtevant, B. (1995). Experiments on the Richtmyer-Meshkov instability of an air/SF6 interface. *Shock Waves, 4*, 247.

Chapter 17
Generation and Transport of Radiation in Laser Plasmas

Abstract Emission and absorption of x-ray radiation play key roles in hydrodynamic properties of laser produced plasmas. In this chapter, we overview basic aspects of radiation hydrodynamics and atomic models in laser-produced plasmas. Application of the atomic models to plasma x-ray lasers, plasma diagnostics and EUV lithography are briefly described.

Keywords Radiation hydrodynamics · Radiation transport · Collisional radiative model · Atomic processes · Atomic models

17.1 Introduction

When a solid material is irradiated with ps-ns duration pulsed laser, a plasma is created where the density and temperature change rapidly in space and time; many orders in the electron density (10^{19}–10^{24} cm^{-3}) and the temperature (100 eV–a few keV). X-rays are emitted from a limited region of plasma which satisfies the optimum condition for x-ray emission. The emitted x-rays are partly absorbed in the plasma, contributing to plasma heating and reducing the x-ray emission outside the target.

Hydrodynamic properties of the laser produced plasmas are affected by radiation, since a large fraction of the absorbed laser energy is emitted by x-rays; 70–80% for high-Z materials like Au and several % for low-Z materials like plastics, where Z is the atomic number. Generation and transport of radiation in laser-produced plasmas is analyzed with the radiation hydrodynamics where hydrodynamics, radiation transport and atomic physics are analyzed self-consistently.

Y. Kato et al., *High Power Laser and Plasma Science*, Springer Series on Atomic, Optical, and Plasma Physics 130, https://doi.org/10.1007/978-3-032-08930-4_17

17.2 Radiation Hydrodynamics

17.2.1 Hydrodynamic Equations

In general, fluid dynamics is described by the basic equations (17.1), (17.2) and (17.3) which correspond to mass conservation, momentum conservation and energy conservation, respectively:

$$\frac{\partial \rho}{\partial t} + \nabla \cdot (\rho \boldsymbol{u}) = 0, \tag{17.1}$$

$$\frac{\partial (\rho \boldsymbol{u})}{\partial t} + \nabla \cdot (\rho \boldsymbol{u}\boldsymbol{u} + p) = 0, \tag{17.2}$$

$$\frac{\partial (\varepsilon_{kin} + \varepsilon_{\text{int}})}{\partial t} + \nabla \cdot \left[(\varepsilon_{kin} + p_{kin})\boldsymbol{u} \right] = 0. \tag{17.3}$$

Here, ρ, $\boldsymbol{u}$, and p_{kin} are mass density, velocity, and pressure, respectively. With the equation of states of ideal gases, energy density ε_m and pressure p_m of a material are described as

$$\begin{aligned} \varepsilon_m &= \varepsilon_{kin} + \varepsilon_{\text{int}} = \frac{3}{2} n_e k T_e + \frac{3}{2} n_i k T_i + \sum_i I_p n_p \\ p_m &= n_e k T_e + n_i k T_i, \end{aligned} \tag{17.4}$$

where n_e, n_i are the number densities and T_e, T_i are the temperatures of the electrons and the ions, respectively. The internal energy density is $\varepsilon_{\text{int}} = \sum_p I_p n_p$ where I_p is the ionization potential and n_p is the atomic population of the p-th state of an ion where the summation is taken for all ions. Contribution from the radiation pressure, which is important in astronomical objects, is not included in these equations since it is small (~ 1%) compared to the plasma pressure in laser-produced plasmas.

17.2.2 Radiation Transport

The x-rays generated in laser-produced plasmas are either absorbed in the plasma contributing to plasma heating or emitted outside the plasma contributing to plasma cooling. Evolution of radiation intensity during propagation is described by a radiation transport equation,

$$\frac{1}{c} \frac{\partial I_\nu}{\partial t} + \boldsymbol{\Omega} \cdot \nabla I_\nu = \eta_\nu - \chi_\nu I_\nu \equiv -\chi_\nu (I_\nu - S_\nu), \tag{17.5}$$

where I_ν is the spectral intensity, η_ν is the emission coefficient, χ_ν is the absorption coefficient, and $\mathbf{\Omega}$ is a unit vector in the propagation direction (Lokke & Grasberger, 1977). We have omitted the terms arising from scatterings such as Compton scattering and other processes in the right-hand side since these terms are small in laser-produced plasmas. We define the "source function" S_ν in (17.5) as

$$S_\nu = \frac{\eta_\nu}{\chi_\nu}. \tag{17.6}$$

In thermal equilibrium $\eta_\nu = \chi_\nu B_\nu$ where B_ν is the Planck function, since the emission and absorption processes are balanced in detail, leading to $S_\nu = B_\nu$. With the absorption coefficient χ_ν, we can derive the radiation mean free path $1/\chi_\nu$ and the optical depth $\tau_\nu = \chi_\nu l$ where l is a length of the plasma.

The material energy density ε_m is related to the radiation intensity as

$$\frac{d\varepsilon_m}{dt} = 4\pi \int (\chi_\nu J_\nu - \eta_\nu) d\nu + Q, \tag{17.7}$$

where J_ν is the angle-averaged intensity

$$J_\nu = \frac{1}{4\pi} \int I_\nu d\Omega, \tag{17.8}$$

and Q represents any additional energy sources.

17.2.3 Atomic Processes

The electronic states are excited, deexcited, or ionized by collisional and radiative processes. We list typical processes with the following notations. E_p: bound states with $p = 1$ for the ground state, e_{free}: free electron generated by ionization, e, e': free electrons before and after interaction, p, p': photons before and after interaction, and E_pE_q: doubly excited state. The photo emission is due to the spontaneous emission, whereas the stimulated emission is included in the photo-absorption as the reduced absorption expressed by the population density difference between the lower and upper states. The last terms on the right-hand side of these processes denote the rate coefficients used in the rate equation (17.15).

(a1) Collisional excitation	$E_p + e \rightarrow E_q + e'$	C^e_{pq}
(a2) Collisional deexcitation	$E_q + e \rightarrow E_p + e'$	C^d_{qp}
(b1) Collisional ionization	$E_p + e \rightarrow e_{free} + e'$	C^i_p
(b2) Three body recombination	$E_p + e_1 + e_2 \rightarrow E_qE_r + e_2 \rightarrow E_1E_r + e'$	C^r_p
(c1) Photo excitation	$E_p + p \rightarrow E_q$	R^e_{pq}

(c2) Photo deexcitation	$E_q \rightarrow E_p + p'$	R^d_{qp}
(d1) Photo ionization	$E_p + p \rightarrow e_{free}$	R^i_p
(d2) Radiative recombination	$e \rightarrow E_p + p'$	R^r_p

In Fig. 17.1, we show schematic diagrams of these processes with the following notations: electronic transitions by vertical solid lines; incoming and outgoing particles (electrons in black and photons in red) by tilted downward and upward lines respectively; Coulomb interaction between electrons by vertical dashed lines in green; and electromagnetic interaction between radiation and electron by vertical dash-dotted lines in red. Each atomic process is encircled by a blue shaded square, with initial and final states on left and right, respectively. In the three-body recombination (b2), the first incoming electron e_1 is captured and forms an unstable state, which is subsequently stabilized by collision with the second incoming electron e_2 and resulting in ejection of an electron e'.

In addition to these processes, there are other processes to populate the electronic states. When a highly excited state whose energy is above the ionization threshold is formed in a Z^{Q+} ion, for example, by excitation of an inner shell electron or simultaneous excitation of two outer-shell electrons, this highly excited state stabilizes automatically by ejecting a free electron and forming a stable electronic state in a $Z^{(Q+1)+}$ ion. This process is called *autoionization*. Its reverse process is the *electron*

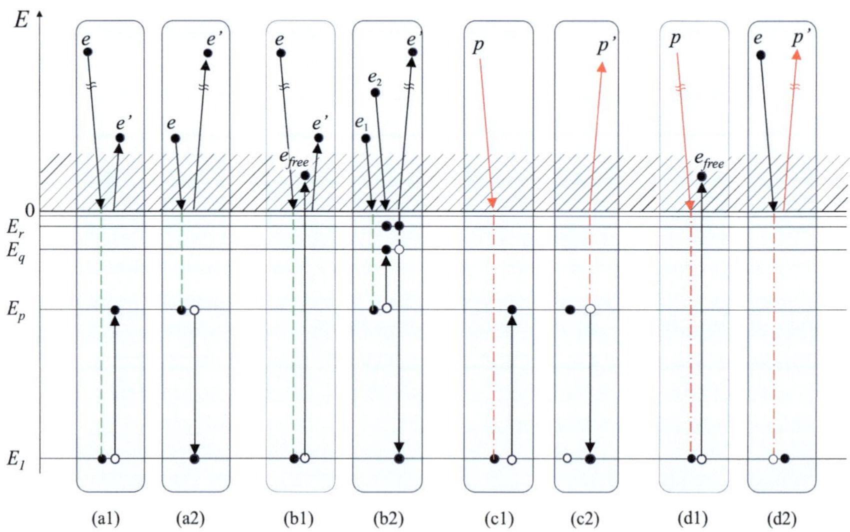

Fig. 17.1 Schematic of atomic processes, where (a1)–(d2) correspond to the processes described in the text. The vertical scale is energy, where 0 is the ionization threshold. Incoming and outgoing (generated) particles, *e* for free electrons and *p* for photons, are shown by tilted downward and upward lines, respectively

capture, where an electron is captured by a $Z^{(Q+1)+}$ ion and a highly excited state is formed in a Z^{Q+} ion.

Explicit forms of the collisional and radiative rates are given, for example, in (Chung et al., 2005, 2016). The electron collisional excitation rate coefficient C^e_{pq} for transition from p to q with transition energy $E_{pq} = E_q - E_p > 0$ is obtained by integrating the collisional cross section $\sigma_c^{pq}(\varepsilon)$ over the electron distribution $f_e(\varepsilon) = n_e 4\pi(\pi kT_e)^{-3/2}\varepsilon^{1/2}e^{-\varepsilon/kT_e}$ as

$$C^e_{pq} = n_e \int_{E_{pq}}^{\infty} v(\varepsilon)\sigma_c^{pq}(\varepsilon)f_e(\varepsilon)d\varepsilon$$

$$= 1.578 \times 10^{-5}\frac{\overline{g}(E_{pq}/T_e)\exp(-E_{pq}/T_e)}{E_{pq}\sqrt{T_e}}f_{pq} \tag{17.9}$$

where $v(\varepsilon)$ is the electron velocity, $\varepsilon = (1/2)mv^2$ is the electron energy, and $\overline{g}(E_{pq}/T_e)$ is the average gaunt factor. From the principle of detailed balance, which states that in equilibrium any process is exactly counterbalanced by its inverse process, the rate for collisional deexcitation $q \to p$ (C^d_{qp}) is obtained from C^e_{pq} as $n_p C^e_{pq} = n_q C^d_{qp}$ where n_p and n_q are the population densities in thermal equilibrium of the p and q states, respectively, leading to $n_p/n_q = (g_p/g_q)\exp(E_{pq}/T_e)$ where g_p is the statistical weight of the p level.

The collisional ionization rate C^i_p from level p is derived, for example, from the ionization rate from Z^{Q+} to $Z^{(Q+1)+}$ ion (Lotz, 1968, 1969)

$$\frac{C_{Q,Q+1}}{n_e} = 2.97 \times 10^{-6}\left(\frac{\xi}{I_p\sqrt{T_e}}\right)E_1(U) \tag{17.10}$$

where ξ is the number of electrons in the outer shell of the ion which are ionized, I_p is the ionization potential and $E_1(U)$ is the exponential integral of the first kind with argument $U = I_p/T_e$. The rate for the three-body collisional recombination C^r_p is obtained from C^i_p by the detailed balance.

The photo-excitation rate for $p \to q$ is

$$R^e_{pq} = 4\pi \int_0^{\infty} \sigma_v^{pq} J_v \frac{dv}{hv}. \tag{17.11}$$

Here $\sigma_v^{pq} = (\pi e^2/mc)f_{pq}\phi(v)$ is the absorption cross section where f_{pq} is the absorption oscillator strength, $\phi(v)$ is the normalized line profile function, and $J_v = (1/4\pi)\int I_v d\Omega$ is the angle-averaged photon intensity. The radiative rates of the stimulated and spontaneous emission of the $q \to p$ transition are given by the first and second terms on the right-hand side of the following equation respectively,

$$R^d_{qp} = 4\pi \left(\frac{n_p}{n_q}\right) \int_0^\infty \sigma_\nu^{pq} \left(J_\nu + \frac{2h\nu^3}{c^2}\right) e^{-h\nu/kT} \frac{d\nu}{h\nu}. \tag{17.12}$$

The photoionization rate coefficient R^i_0 is derived from the photoionization cross section $\sigma_{Q,Q+1}(E)$ at photon energy E

$$R^i_o = 4\pi \int \sigma_{Q,Q+1}(E) J(\nu) \frac{d\nu}{h\nu}. \tag{17.13}$$

The rate of its inverse process of radiative recombination R^r_0 is, for a Maxwellian electron distribution,

$$R^r_0 = \left(\frac{n_Q}{n_{Q+1}}\right)^{\mathrm{LTE}} 4\pi \int \sigma_{Q,Q+1}(E) \left(J(\nu) + \frac{2h\nu^3}{c^2}\right) e^{-h\nu/kT_e} \frac{d\nu}{h\nu}. \tag{17.14}$$

Here $\left(n_Q/n_{Q+1}\right)^{\mathrm{LTE}}$ is the density ratio of the Z^{Q+} and $Z^{(Q+1)+}$ ions at the local thermodynamic equilibrium.

So far, we have assumed that ions are isolated, and relevant parameters such as energy levels, transition rates, etc. are calculated with atomic models for free ions. However, ions in finite density plasmas are affected by "high density effects" due to electrostatic potentials of free electrons and neighboring ions. For example, electrons in high lying states with small binding energies will be ionized. The ionization threshold energy is lowered due to this ionization, which is referred to as ionization population depression (IPD), pressure ionization, or continuum lowering (Stewart & Pyatt, 1966). This effect is treated in collisional radiative models by limiting the maximum quantum number of the bound states in the rate equations. Another example of the high-density effect is the electron degeneracy. At high electron densities or low temperatures, the electron distribution is described by the Fermi–Dirac distribution instead of the Maxwellian distribution.

17.2.4 Population Kinetics

Atomic populations in the electronic states of various charged ions are derived by solving the coupled rate equations for the collisional and radiative processes (Chung et al., 2005; McWhirter et al., 1963; Scott, 2016),

$$\frac{d}{dt}n(p) = \left[\sum_{q<p} C^e_{qp} n(q) - \sum_{q>p} C^e_{pq} n(p)\right] - \left[\sum_{q<p} C^d_{pq} n(p) - \sum_{q>p} C^d_{qp} n(q)\right]$$

$$+\left[\sum_{q<p} R^{e}_{qp} n(q) - \sum_{q>p} R^{e}_{pq} n(p)\right] - \left[\sum_{q<p} R^{d}_{pq} n(p) - \sum_{q>p} R^{d}_{qp} n(q)\right]$$
$$-\left(C^{i}_{p} - C^{r}_{p}\right) - \left(R^{i}_{p} - R^{r}_{p}\right). \tag{17.15}$$

Here C and R denote the collisional and radiative processes, e and d in the superscript denote the excitation and deexcitation processes, and p and q in the subscript denote the electronic states, respectively. In the right-hand side, the terms in the first and the second parentheses correspond to the collisional excitation and deexcitation processes, the third and the fourth to the radiative excitation and deexcitation processes, and the fifth and the sixth to the collisional and radiative ionization and recombination processes, respectively. There are also autoionization and electron capture processes as described in Sect. 17.2.3.

17.2.5 General Features of Population Distribution

When a solid target is irradiated with a pulsed laser, the material temperature increases rapidly in time, leading to an increase in the ionization charge. The plasma reaches a quasi-steady state at and near the peak of the laser pulse. After the peak of the laser pulse, the plasma temperature decreases gradually, leading to recombination of ions with electrons as the temperature decreases in time. These phases of the plasma are called ionizing phase, quasi-steady state phase, and recombining phase, respectively. The mechanism and population distribution changes in time reflecting the change of the plasma phases.

17.2.5.1 Population Distribution in LTE Plasmas

When plasma is in local thermodynamic equilibrium (LTE), distribution of the population density $n(i)$ of the bound state i is described by the Boltzmann distribution

$$\frac{n(p)}{n(q)} = \frac{g_p}{g_q} e^{-(E_p - E_q)/kT_e}, \tag{17.16}$$

where g_p and E_p are the statistical weight and energy of level p and T_e is the electron temperature. In LTE, the ratio of the number density of the $(Q+1)$-charged ion n_{Q+1} multiplied by the electron density n_e to the density of the Q-charged ion n_Q is expressed as

$$\frac{n_{Q+1}}{n_Q} n_e = \frac{U_{Q+1}}{U_Q} e^{-(E+I_Q)/\,kT_e} \cdot 2\left(\frac{4\pi p^2}{h^3}\right) dp. \tag{17.17}$$

Here E is the kinetic energy and $2(4\pi p^2/h^3)dp$ is the density of states of free electrons with momentum p, and I_Q is the ionization potential of the Q-charged ion. U_Q is the partition function,

$$U_Q = \sum_{p \in [Q]} g_p e^{-(E_p - E_0)/kT_e} \tag{17.18}$$

where E_0 is the energy of the ground state, and summation is taken over all bound states p. Integrating (17.17) with E using the relation $E = p^2/2m_e$, we obtain

$$\frac{n_{Q+1}}{n_Q} n_e = 2\left(\frac{2\pi m_e kT_e}{h^2}\right)^{3/2} \frac{U_{Q+1}(T_e)}{U_Q(T_e)} e^{-I_Q/kT_e}, \tag{17.19}$$

This relation is called Saha equation (Zel'dovich & Raizer, 1966). The distribution of the electronic states and the ion in the thermal equilibrium is described by the Saha–Boltzmann equation, which is the combined relation of (17.17) and (17.19).

17.2.5.2 Population Distribution in Non-LTE Plasmas with Hydrogenic Model

When plasma is not in thermal equilibrium, which is often the case for laser-produced plasmas, population distribution is derived by solving the rate equation (17.15). T. Fujimoto has shown general features of the population distribution in various plasma conditions, based on hydrogen-like ions, by deriving $n(p)$ assuming quasi-steady-state $dn(p)/dt = 0$ for the excited states $p = 2, 3, \ldots$ (Fujimoto, 1979a, 1979b, 1980a, 1980b, 1985). This hydrogen atom model scales to hydrogenic ions with a nuclear charge z by n_e/z^7 for the electron density and T_e/z^2 for the electron temperature. Based on this study, we briefly describe population dynamics in various plasma phases.

Figure 17.2 shows schematically characteristics of the population density of excited levels, in a form of "phase diagram", including population mechanism and population distribution in the insets (Fujimoto, 1980b). Figure 17.2a is the diagram of an ionizing plasma, where the population distribution $n(p)$ is evaluated for the processes proportional to the ground state density $n(1)$, assuming density of the ionized state $n(\text{ion}) = 0$. In this case, the atoms are ionized by the ionization flow of electrons through the excited levels between the ground and continuum states. The ground state population $n(1)$ decreases in time as

$$\frac{d}{dt} n(1) = -S_{\text{CR}} n(1) n_e, \tag{17.20}$$

where S_{CR} is the collisional-radiative ionization rate coefficient. The population $n(p)$ is determined from $n(1)$ as $n(p) = [Z(p)/Z(1)] r_1(p) n(1)$ for $i = 2, 3, \ldots$, where $r_1(p)$ is the population coefficient for the ionization process, and

$$Z(p) = \frac{g_p}{2U_Q}\left(\frac{h^2}{2\pi mkT_e}\right)^{3/2} e^{I(p)/kT_e}. \tag{17.21}$$

Here U_Q is the partition function of the Q-charged ion.

Figure 17.2b is the diagram of a recombining plasma, where $n(p)$ is evaluated for the processes proportional to $n(\text{ion})$, assuming $n(1) = 0$. In this case, the ions are recombined by the recombination flow through the excited levels to the ground state. The ion population $n(\text{ion})$ decreases as

$$\frac{d}{dt}n(\text{ion}) = -\alpha_{\text{CR}}\text{n}(\text{ion})\text{n}_\text{e}, \tag{17.22}$$

where α_{CR} is the collisional-radiative recombination rate coefficient. The population $n(p)$ is determined from $n(\text{ion})$ as $n(p) = Z(p)r_0(p)n(\text{ion})\text{n}_\text{e}$, where $r_0(i)$ is the population coefficient for the recombination process. The population coefficients $r_0(p)$ and $r_0(p)$, which are composed of the collisional and radiative rate coefficients described in (17.15), are functions of n_e and T_e.

The ionizing and recombining plasmas shown in Fig. 17.2a and b correspond to the idealized plasmas in the ionization phase and the recombination phase, respectively. In actual plasmas both of $n(1)$ and $n(\text{ion})$ are populated. In this case, the ionization-recombination in plasma is expressed as superposition of Eqs. (17.20) and (17.22). However, since it is found in many cases that one of the two components dominate over the other, the plasma is characterized as either "predominantly ionizing plasma" or "predominantly recombining plasma". Therefore, it is useful to examine Fig. 17.2a and b separately.

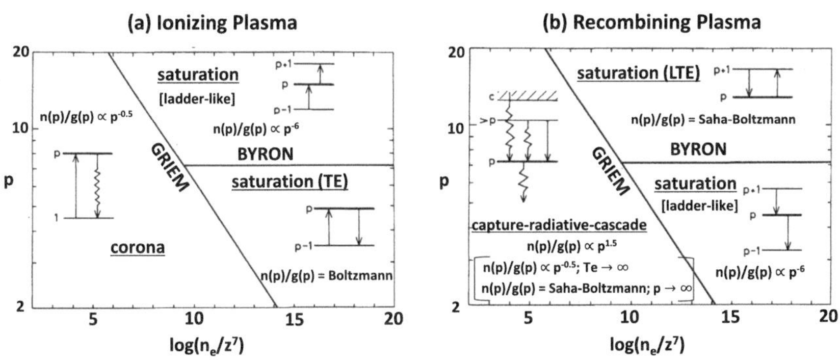

Fig. 17.2 Schematic phase diagram of excited levels of **a** ionizing and **b** recombining plasmas. For each phase, phase name, population mechanism, and population density distributions are shown. The collisional are radiative transitions are marked by the arrowed straight and zig-zag lines, respectively. Griem's boundary and Byron's boundary are marked as GRIEM and BYRON, respectively. Reprinted from Fujimoto (1980b), with permission from The Japan Physical Society and the author. © 1980 The Japan Physical Society (Note that the characters in the original figure are enlarged for visibility.)

17.2.5.3 Population Distribution in Ionizing and Recombining Plasmas

The dynamics and population distribution change drastically with density and temperature of the plasma. Figure 17.2a shows several phases in a 2-D space of the electronic level and the electron density, where phase boundaries are marked by GRIEM and BYRON. Considering that the collisional rate coefficients are proportional to the electron density, the Griem's boundary divides the n_e regions into low- and high-density at the density where the collisional depopulating rate (C^e, C^d and C^i) become comparable to the radiative decay rate R^d (Griem, 1964). Considering an ionizing plasma, in the low-density region, called corona phase, most electrons are populated at the ground level, since the excitation rate to excited levels is small. The excited levels are populated by electron collision from the ground level. The population $n(p)$ of an excited level is balanced by the radiative decay, as shown schematically in the inset, resulting in the distribution described by $n(p)/g(p) \propto p^{-0.5}$.

With increase in n_e, the collisional processes become gradually dominant over the radiative processes. Taking level $p = 4$ as an example, this level is in the corona phase below $n_e/z \leq 10^{11.5}$. However, the collisional excitation/deexcitation processes become dominant at higher densities, reaching Boltzmann distribution at very high density.

The excited levels in the higher density region are further divided by Byron's boundary (Byron et al., 1962), where the collisional excitation and deexcitation rate to an adjacent level become comparable, i.e., $C^d_{p(p-1)} \approx C^e_{p(p+1)}$. In the region above this boundary, ladder-like excitation takes place, resulting in the distribution $n(p)/g(p) \propto p^{-6}$ which has small dependence on p at high-lying levels, i.e., the distribution is saturated. The principal quantum number of the level satisfying condition is $p_B \approx \left(z^2\chi_H/3kT_e\right)^{1/2}$, where $\chi_H = 13.6\,\text{eV}$ is the Rydberg energy of hydrogen. This situation arises from Eq. (17.9) where the collisional excitation rate is larger for the smaller transition energy E_{pq}, corresponding to smaller energy gap in higher excited levels.

In a recombining plasma, the population kinetics in the low-density region is described by the capture-radiative-cascade (CRC) scheme, where the excited levels are populated by the direct radiative recombination supplemented by cascade from the higher-lying levels. The depopulating process is radiative decay to lower-lying levels, resulting in the distribution $n(p)/g(p) \propto p^{1.5}$ since the radiative decay rate is larger for lower excited levels.

At high densities, the distribution in high-lying levels above the Byron's boundary are in "partial LTE" due to the balance between the collisional excitation and de-excitation processes, resulting in the Boltzmann distribution. In the levels below the Byron's boundary, the collisional de-excitation prevails over the collisional excitation, leading to ladder-like de-excitation with the distribution given by $n(p)/g(p) \propto p^{1.5}$.

The validity criteria for the excited states to be in the partial LTE are examined for the ionizing, recombining, and ionization-balance plasmas by Fujimoto and McWhirter (1990).

17.2.5.4 Population Dynamics in Plasma X-Ray Lasers

We can compare population distribution evaluated by theoretical models with experimental observation of the emission spectra. Another approach is observation of x-ray lasers, which are very sensitive to population distribution, population kinetics, and hydrodynamics. When population is inverted, i.e., the population of a higher-level u is larger than that of a lower-level l, the radiation emitted in the transition from u to l is amplified due to stimulated emission with a gain coefficient given by $\gamma(\omega) = \left(n_u/g_u - n_l/g_l\right)\left(\pi^2c^2/\omega_0^2\right)g(\omega)A_{ul}$ (see Part II, Eq. (3.89)). Here $n_{u,l}$ and $g_{u,l}$ are population densities and degeneracies of the levels u and l, respectively, $\hbar\omega_0 = E_u - E_l$ is the transition energy, $g(\omega)$ is the normalized line shape function, and A_{ul} is the spontaneous emission rate from u to l. When the gain exceeds the losses (due to absorption, scattering, refraction etc.), the radiation intensity increases exponentially with the amplifier length, reaching an intensity many orders of magnitudes higher than the spontaneous emission intensity. Since the transition energy scales as z^2 with the ion charge z, x-ray amplification is obtained in plasmas composed of high-z ions. However, high pumping rate is required to create large inversion for ray amplification, since A_{ul} is proportional to ω_0^3.

Population inversion is created in a low-density recombining plasma, where the population in higher levels are larger than the lower levels as $n(p)/g(p) \propto p^{1.5}$. Figure 17.3a shows a simplified energy level diagram of a z-charged hydrogen-like ion. The largest inversion is obtained in the Balmer-α (H_α) line due to $3d$-$2p$ transition since the $2p$ level decays very quickly to $1s$ by Lyman—α emission. This scheme requires rapid plasma expansion and cooling, where the optimum region for high inversion density is limited in time and space (Pert, 1976, 1979). Amplification of the H_α line at 18.2 nm in hydrogen-like carbon (C VI) has been achieved with a gain $g = 12.5\ \text{cm}^{-1}$ by irradiation of a 7-mm diameter and 5-mm long carbon fiber with a 2-ps-duration laser pulse (Zhang et al., 1995). The H_α amplification was extended to shorter wavelengths, resulting in $g \sim 8.6\ \text{cm}^{-1}$ for Na XI H_α at 5.42 nm (Azuma et al., 1990).

Amplification has been observed also in a lithium-like Al-ion recombination x-ray laser with a gain of $g \sim 2$–$3\ \text{cm}^{-1}$ for the $4f$-$3d$ line at 15.4 nm (Calliron et al., 1990; Hara et al., 1989; Jaegle et al., 1987). Ionization and population dynamics of the Li-like recombination laser show good agreement with the experiment for the $4f$-$3d$ line, but with discrepancy for the $5f$-$3d$ line (Klisnick et al., 1990). The latter issue has been investigated by considering the deexcitation and recombination processes through doubly excited states in a dense Li-like recombining plasma (Kawachi & Fujimoto, 1997).

The soft x-ray amplification has been realized also in ionizing plasmas of non-hydrogenic ions. Figure 17.3b shows a schematic energy level diagram of a neon-like ion (ground state configuration $1s^22s^22p^6$), which has simple excited state configurations since $2p^6$ is a closed shell. The $2p^6$ state is collisional-excited to the $2p^53p$ state, which is composed of 5 levels. Among these levels, excitation rate to the $J = 0$ state by electron monopole excitation is the highest, whereas other levels of $2p^53p$ are collisional-excited, and populated by the collisional decay from the higher lying

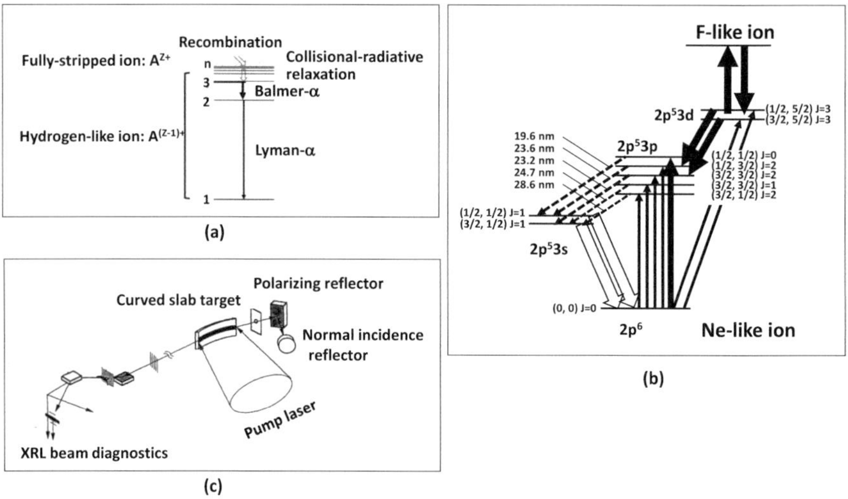

Fig. 17.3 **a** Balmer-α laser in recombination-phase hydrogenic z-charged ions. **b** Schematic energy level diagram of Ne-like ions. Collisional and radiative processes are shown by solid and open arrows respectively, with the widths qualitatively corresponding to the transition rates. Laser amplification is observed in 5 lines due to transitions from $2p^5 3p$ to $2p^5 3s$ levels, where lasing wavelengths of Ne-Ge are shown. The levels are written in $(j_1, j_2)J$ notation where, for example, (1/2, 3/2) J = 2 denotes $2p^5{}_{1/2}3p_{3/2}$ $(J = 2)$. **c** Soft x-ray amplification of Ge $J = 0$–0 line at 19.6 nm with a curved slab target. Linearly polarized x-ray laser is generated using a half cavity composed of a multilayer polarizer and a reflector. The x-ray laser beam is measured with a flat-field grating spectrometer

$2p^5 3d$ level and by the dielectronic recombination from the F-like ions. Population inversion is created between $2p^5 3p$ and $2p^5 3s$, since $2p^5 3s$ decays in less than 1 ps to $2p^6$ by radiative transition (Rosen et al., 1985).

In the first successful demonstration of the collisional excitation x-ray laser implemented with Ne-like Se using an exploding foil target, intense lasing was obtained in the two $J = 2$–1 lines at 20.6 nm and 20.9 nm, whereas lasing of the $J = 0$–1 line at 18.3 nm which was predicted to be the strongest was not observed (Matthews et al., 1985). This "mystery" has been resolved later with Ne-like Ge, where the $J = 0$–1 line lased stronger than the $J = 2$–1 lines in the ionizing phase with a curved slab target, shown in Fig. 17.2c, to compensate for the x-ray laser beam bending caused by plasma density gradient (Daido et al., 1995; Murai et al., 1994a, 1994b).

The collisional excitation x-ray laser scheme has been extended to shorter wavelengths near the carbon K-edge of 4.48 nm with nickel-like W (ground state configuration $3d^{10}$) (MacGowan et al., 1990; Maxon et al., 1988), and to the table-top-size compactification by transient-gain, travelling-wave short-pulse pumping (Kawachi et al., 2004; Nickles et al., 1997).

17.3 Atomic Models

17.3.1 Introduction

As described in previous sections, population distribution over the atomic states in various charged ions is required to determine the radiative and hydrodynamic properties of the laser-produced plasmas. For example, the internal energy of a plasma is composed of the thermal energy $(3/2)(n_e kT_e + n_i kT_i)$ and the ionization energy stored in the ions as $\varepsilon_{ion} = \sum_{q,n} n_{q,n} E_{q,n}$ where $n_{q,n}$ and $E_{q,n}$ are the population and ionization energy of the n-th level of a q-charged ion, respectively. The population distribution is derived by solving the rate equations, where proper atomic models are required to incorporate the atomic states, whose number easily exceeds millions when treated in detail.

In analyzing the spectra, energies of the electronic states and transition rates are calculated with wavefunctions of electronic configurations based on various atomic models. Although there are atomic models where electron configurations are described in detail, the computer code based on these models is not sometimes suitable to run with radiation hydrodynamic code due to limitations in memory capacity and computation time. Simplification of the atomic model has been pursued to make the atomic code compatible with other calculations and yet maintain essential features of the phenomena (Fontes et al., 2016; Hansen, 2016). In this section, we introduce typical atomic models developed for analyses of radiation hydrodynamics of laser produced plasmas.

17.3.2 Detailed Configuration Accounting

To analyze and quantitatively predict the dynamics of laser produced plasmas, precise and detailed atomic parameters such as energy, oscillator strength, and radiative and nonradiative transition rates are required. Among various codes to calculate atomic data of multi-electron atoms, GRASP (General Purpose Relativistic Atomic Structure Program) is based on the Multi-Configuration Dirac-Fock (MCDF) formulation (Dyall et al., 1989; Grant et al., 1980; Parpia et al., 1996).

Corresponding to recent advances in plasma physics, atomic codes with detailed configuration accounting like GRASP are required to provide high-precision, full-scale data. As an example, we show in Fig. 17.4 a comparison between an experiment and a theoretical calculation of the x-ray transmission through an aluminum foil heated uniformly with thermal x-ray radiation generated by laser irradiation (Iglesias et al., 1995). The electronic states considered in the calculation in the L-shell transitions are $1s^2 2l^m \rightarrow 1s^2 2l^{m-1} n'l'$ and K-shell excited state $1s^1 2l^{m+1} \rightarrow 1s^1 2l^m n'l'$ for $3 \leq n' \leq 7$ and $l' < n'$ for each ionization stage. Using densities and temperatures derived from a hydrodynamic code, line energies and oscillator strengths are calculated with MCDF and the opacities with OPAL code (Rogers, 1992). In Fig. 17.4, we

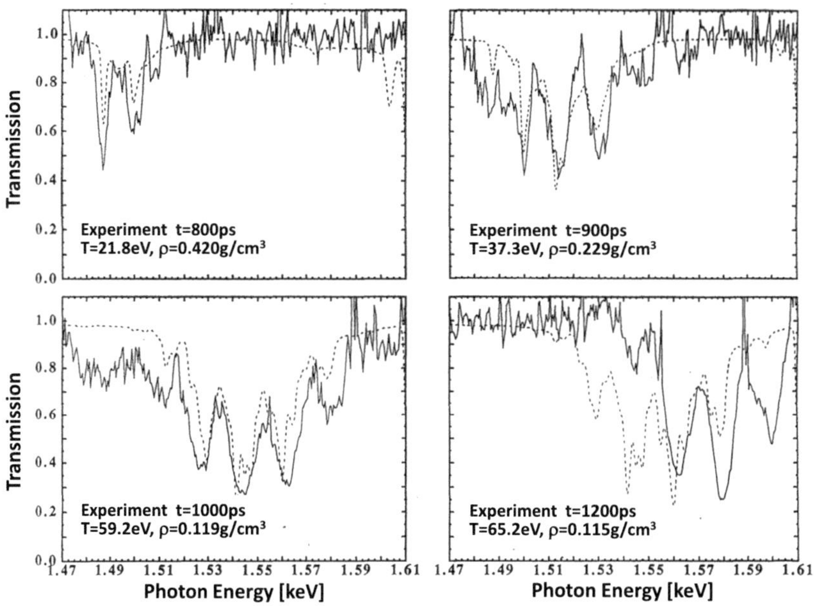

Fig. 17.4 Transmission through an aluminum foil, irradiated with laser-generated x-rays at different times after the start of the heating. The experimental spectra (solid lines) are compared with the OPAL/MCDF model (dotted lines) evaluated at the material temperature T and the density r shown in each figure. Reprinted from Iglesias et al. (1995), with permission from Elsevier: permission conveyed through Copyright Clearance Center, Inc. © 1995 Elsevier (Note that the characters in the original figure are enlarged for visibility.)

find good agreement between the experiment and the transmission model in terms of line energies and opacities, except for some differences at $t = 1200$ ps.

The HULLAC (Hebrew University Livermore Laboratory Atomic Physics Code) is another relativistic detailed code, with which computation time of the detailed codes such as MCDF is reduced by orders of magnitude (Bar-Shalom et al., 1989, 2001). In this code, calculation of the interaction potential $1/r_{ij}$ where r_{ij} is the distance between two electrons, one of the most time consuming part of MCDF, is replaced by an average potential $V(r_i)$ which depends only on the radial coordinate r_i. Focused on simplicity to use, HULLAC provides various atomic data, such as the state and transition energies, spontaneous emission rates, collisional ionization cross sections, auto-ionization rates, radiation ionization cross sections, and radiative recombination rates. A set of programs named MICCRON has been developed as an extension of HULLAC to include several ionization stages (Klapisch & Bar-Shalom, 1997).

17.3.3 Unresolved Transition Array

Many spectral lines emitted from laser produced plasmas often overlap in spectral observations due to various broadening effects and instrumental resolutions. These spectra arise from transitions between two electronic configurations, each composed of many states. The unresolved spectral bands make it difficult to interpret the spectra and identify the ionization stages. A general model named Unresolved Transition Arrays (UTA) has been developed to analyze the unresolved spectra (Bauche et al., 1988, 1996).

The frequency distribution of the observed spectral lines is determined as the average of the transition frequencies weighted by the radiative transition rates, not by the statistical weights. Considering the radiative transition from configuration *B* to configuration *A*, which are comprised of states *b* and *a* respectively, the *n*-th moment of the weighted distribution of the line wavenumbers is written as

$$\mu_n = \sum_{a,b} (E_b - E_a)^n w_{ab}/W. \tag{17.23}$$

Here E_a and E_b are the energies of the *a* and *b* states respectively. The weight of the transition is the line strength $w_{ab} = |\langle a|\boldsymbol{D}|b\rangle|^2$ where $\boldsymbol{D} = -\sum_i e\boldsymbol{r}_i$ is the electric dipole and $W = \sum_{a,b} w_{ab}$. The use of this weight implies that the level populations are assumed to be proportional to the statistical weight $2J + 1$ and the Boltzmann factor is constant within a configuration. The average wavenumber of the transition arrays is expressed by μ_1, and its root-mean square deviation σ, which corresponds to the spectral width of the band, is given by the square root of the variance $\sigma^2 = \mu_2 - (\mu_1)^2$.

The formulae and coefficients required to calculate μ_1 and μ_2 have been tabulated for quick computation of various transition arrays which are often observed in the spectra of laser-produced plasmas: simple arrays of $l^N l' - l^N l''$ and $l^{N+1} - l^N l'$, general arrays involving more than two subshells $l_1^{N+1} l_2^{N_2} l_3^{N_3} l_4^{N_4}$ o.s. $- \mathrm{l}_1^{N} \mathrm{l}_2^{N_2+1} \mathrm{l}_3^{N_3} \mathrm{l}_4^{N_4}$ o.s., asymmetric arrays composed of several subarrays, and the spectrum composed of subarrays due to large spin–orbit interactions. Here the symbols l, l', l_1 etc. stand for $nl, n'l', n_1l_1$, etc., and o.s. refers to open spectator subshells.

As an example of the UTA calculation, Fig. 17.5 shows the transition array of Pr XVI $4d^8$-$4d^7 4f$ which is composed of two subarrays (Bauche et al., 1988). Thin vertical lines show the transitions, each with a height proportional to its strength, except many weak lines which are represented with 3% of the highest line. The dotted curve **a** is the envelope calculated by adding two gaussian curves for each subarrays, curve **b** is the Gaussian using μ_1 and μ_2, and curve **c** is the skewed Gaussian using μ_1, μ_2 and μ_3, where μ_3 is the third-moment which describes the asymmetric distribution.

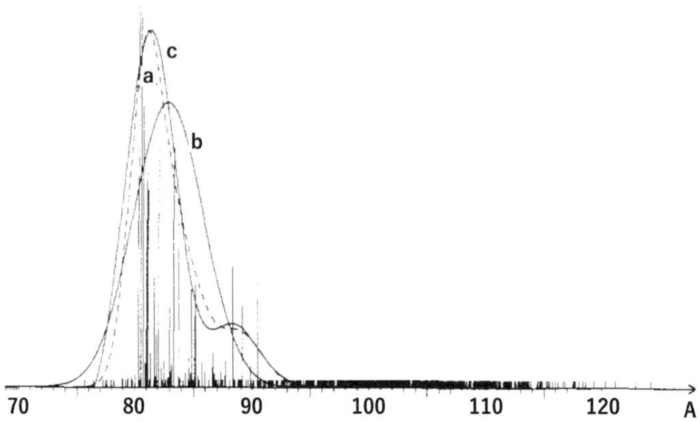

Fig. 17.5 Transition array of Pr XVI $4d^8$-$4d^7 4f$, which is composed of two subarrays. See text for the notations. Reprinted from Bauche et al. (1988), with permission from The Institute of Physics Publishing: permission conveyed through Copyright Clearance Center, Inc. © 1988 The Institute of Physics Publishing

17.3.4 Collisional Radiative Model with Super-Configurations

In analyzing complicated spectra such as UTA, solving the rate equations for all atomic configurations involved in the atomic processes is impractical due to enormous number of configurations, especially in highly excited states of high-Z elements. To reduce the computation complexity, Bar-Shalom et al. introduced the concepts of *super-transition arrays* (STA) and *super-configurations* (SC) (Bar-Shalom et al., 1989). Collisional radiative models based on the SCs have been developed in two codes: SCROLL (Bar-Shalom et al., 1997a, 1997b) and AVERROĒS/TRANSPEC (Peyrusse, 2000, 2001).

Figure 17.6 shows a schematic diagram of the collisional radiative model based on SCs. Considering a $+Q$-charged ion, a SC Ξ_i is formed by grouping numerous configurations of similar energies, named *supershells* σ_i, which are unions of energetically adjacent ordinally atomic *subshells* $s \in \sigma$, $s \equiv \mathbf{j}_s \equiv n_s l_s j_s$. Excitation and deexcitation between SCs take place by collisional and radiative processes. STAs are formed by radiative transitions between the bound states in SCs. A SC above the ionization limit composed of free electrons is formed by ionization. Some of the free electrons recombine with $+Q$ ion, whereas other electrons autoionizes to form SC Ξ' of the $+(Q+1)$ ion.

A SC of a q-electron ion, where $q = Z - Q$, is defined by a set of supershell occupation numbers q_σ where $\sum_\sigma q_\sigma = q$. The q_σ electrons are distributed among the subshells $s \in \sigma$ in all possible ways subject to $\sum_{s\in\sigma} q_s = q_\sigma$. The partition function (PF) of a q-electron ion is expressed by the summation of binomial distributions of the q_s electrons in the subshell s, and multiplied by the Boltzmann factor,

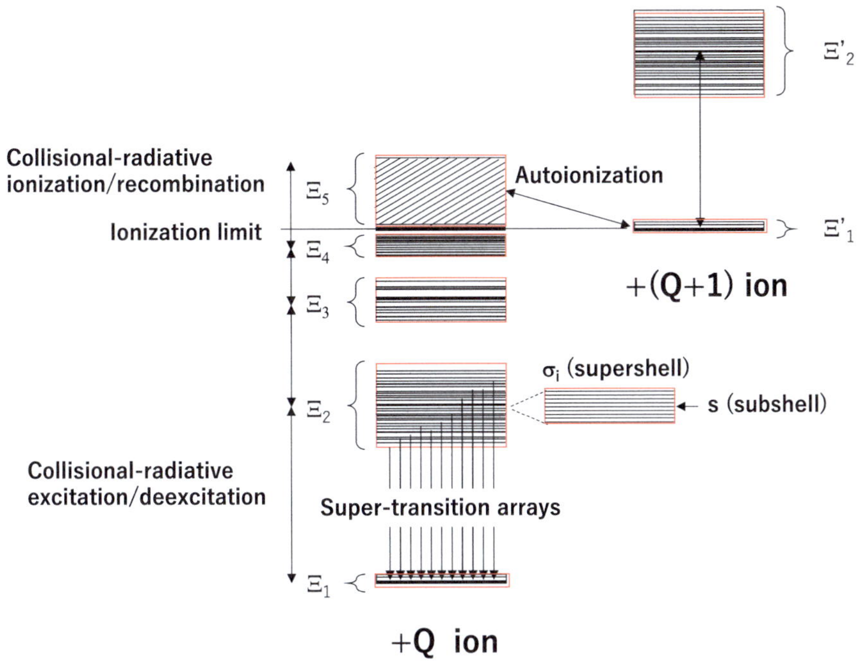

Fig. 17.6 Schematic diagram of non-LTE super-configuration collisional radiative model (SCROLL). Super-configurations are denoted by Ξ, where Ξ_1 corresponds to the ground state and Ξ_5 to the continuum state above the ionization limit. Each SC is an assembly of supershells σ, each composed of the atomic subshells s. With the collisional radiative processes, a SC is coupled to other SCs of the $+Q$-ion, and to SCs of the $+(Q+1)$-ions. Super-transition arrays are observed in radiative emissions between the SCs

$$U_q = \sum_{\sum_s q_s = q} \prod_s \begin{bmatrix} g_s \\ q_s \end{bmatrix} X_s^{q_s}, \quad X_s = e^{-(\varepsilon_s - \mu)/kT}, \tag{17.24}$$

where g_s and ε_s are the statistical weight and energy of s respectively, and μ is the chemical potential.

The collisional radiative rate equations for the SCs are

$$\frac{\mathrm{dN}_\Xi}{\mathrm{dt}} = -N_\Xi \sum_{\Xi'} R_{\Xi\Xi'} + \sum_{\Xi'} N_{\Xi'} R_{\Xi'\Xi} \tag{17.25}$$

where N_Ξ is the SC population density and $R_{\Xi\Xi'}$ is the SC transition rate for all the atomic processes connecting Ξ and Ξ'. The SC transition rate is decomposed into the transition rates between the supershells and subshells as

$$R_{\Xi\Xi'} = \frac{1}{U_\Xi(g)} \sum_{\sigma\in\Xi, \sigma'\in\Xi'} U_\sigma R_{\sigma\sigma'}, \quad R_{\sigma\sigma'} = \frac{1}{g_\sigma} \sum_{s\in\sigma, s'\in\sigma'} g_s R_{ss'} \tag{17.26}$$

Here $U_{\Xi}(g)$ is the partition function of Ξ. The notation (g) stands for the set of statistical weights $\{g_s\}$ of various shells. The transition rates $R_{\sigma\sigma'}$ are calculated by configuration-averaging the state-to-state transition rates $R_{ss'}$ for the radiative and collisional excitation/deexcitation and ionization/recombination processes. The occupation numbers of an active electron in the initial state and an active hole in the final state are calculated by the configuration occupation polynomial. Calculation of many PFs in this process is simplified by an efficient recursion algorithm with which a PF of some charge state is derived by using other PFs which had already been calculated in previous steps.

In the SCROLL code, population distribution is determined in iterative two-step loop calculations. Initially large supershells are defined, SCs are constructed over these supershells, average transition rates between SCs are computed, and collisional radiative rate equations are solved to determine the SC populations. The *ionization temperature* defined by Busquet (described in Sect. 17.3.7) is initially adopted as an effective temperature for the Boltzmann distribution. The computation is repeated by gradually splitting large supershells into smaller supershells, until convergence is reached on the SC populations. When the average charge determined by this SC population is different from the initial value, second loop computation is implemented with a revised effective temperature, until convergence is reached on the population, the average charge, and the effective temperature. The emission and absorption coefficients are determined from the population distribution. For spectroscopic applications, the super- and sub-shells may be further divided to meet the required resolution. An algorithm for systematic splitting of a large supershell into small supershells in minimum numbers is described in Bar-Shalom et al. (2000).

In the SCROLL it is assumed that population distribution is described by the Boltzmann distribution. However, it is not obvious in what conditions this distribution is established and how the effective temperature is determined. J. Bauche et al. have investigated the relations between the population and the excitation energy (a) in each super-configuration and (b) between the super-configurations (Bauche & Bauche-Arnoult, 2000; Bauche et al., 2004, 2006). The semi-logarithmic plots of these relations are well fitted by straight lines, with their slopes corresponding to the effective temperatures: "super-configuration temperatures" for (a), and "ionic excitation temperatures" for (b), respectively.

As an example, Fig. 17.7 shows plots of the four super-configurations, marked by (a) ~ (d) respectively, of the Xe^{25+} (Cu-like) ion at $T_e = 450\,\text{eV}$ and $n_e = 1.2 \times 10^{20}\,\text{cm}^{-3}$. Each point in this diagram shows the average energy (in abscissa) and the reduced population (in ordinate) of a configuration of a supershell C_i calculated by a level-by-level collisional radiative model for many levels which comprise C_i. The reduced population $n(C_i)/g(C_i)$ of C_i is expressed as

$$\frac{n(C_i)}{g(C_i)} = n(SC)\exp\left[-\frac{E(C_i) - E_{av}(SC)}{kT(SC)}\right]. \tag{17.27}$$

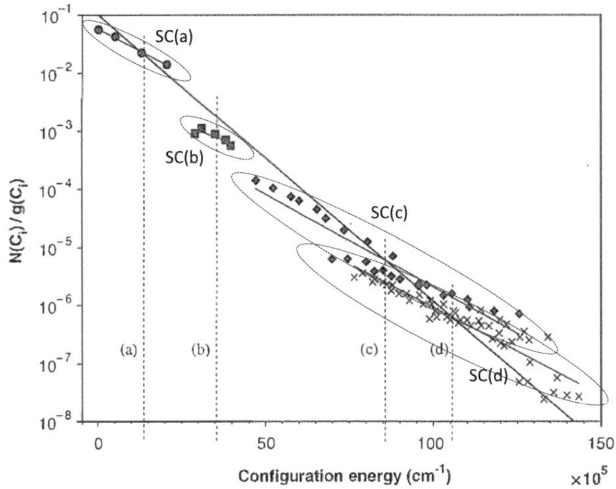

Fig. 17.7 Determination of the super-configuration temperatures $T(SC)$ and the ionic excitation temperature $T(\mathbf{I})$ for Xe^{25+} (Cu-like) ions at $T_e = 450$ eV and $n_e = 1.2 \times 10^{20}$ cm^{-3} from the plot of the reduced population (ordinate) versus energy (abscissa) of the configurations belonging to four low-energy super-configurations $3^{18}4^1$, $3^{18}5^1$, $3^{17}4^2$ and $3^{17}4^15^1$, which are encircled and marked by (a)–(d) respectively. (Here, for example, $3^{18}4^1$ denotes a configuration where 18 and 1 electrons are in $n = 3$ and 4 shells respectively.) $T(SC)$ and $T(\mathbf{I})$ are determined from the slopes of the points in each SC and the line connecting the SCs, respectively. Reprinted from Bauche et al. (2006), with permission from Elsevier: permission conveyed through Copyright Clearance Center, Inc. © 2006 Elsevier

Here $n(C_i)$, $g(C_i)$, and $E(C_i)$ are the population, degeneracy, and energy of the configuration C_i respectively, $n(SC)$ and $E_{av}(SC)$ are the average population and degeneracy-weighted average energy of SC respectively, and $T(SC)$ is the *super-configuration temperature*. The $T(SC)$, determined from the slope of the straight line fitted to the points in each SC, are 183, 300, 165 and 178 eV for (a)–(d) respectively.

Similarly, the super-configuration population $n(SC)$ of an ion **I** is described by

$$\frac{n(SC)}{g(SC)} = n(\mathbf{I}) \exp\left[-\frac{E(SC) - E(SC_1)}{kT(\mathbf{I})}\right]. \tag{17.28}$$

Here $g(SC)$ and $E(SC)$ are the degeneracy and average energy of the super-configuration, $E(SC_1)$ is the energy of the ground state of the super-configurations of **I**, $T(\mathbf{I})$ is the *ionic excitation temperature*, and $n(\mathbf{I})$ is the average-state population of the ion with degeneracy 1. From the long straight line in Fig. 17.7 connecting the ground state (top-left) and the average points of the super-configurations (a)–(d), $T(\mathbf{I})$ of the Xe^{25+} ion is determined to be 76 eV. This example shows that the super-configuration populations are described by decreasing-exponential functions, characterized by effective temperatures $T(SC)$ and $T(\mathbf{I})$ which are, respectively, ~ 2 and 6 times lower than the electron temperature.

The decreasing exponential population was further analyzed by Bauche et al. (2006) and Busquet (2006). Due to the principle of detailed balance for the collisional and radiative excitation/deexcitation processes where the ratio of the excitation and deexcitation rates is proportional to $\exp(-\Delta E/kT_e)$, the population becomes close to the thermal equilibrium characterized by T_e. However, the effective temperatures within a super-configuration $T(SC)$ and between the super-configurations $T(\mathbf{I})$ become lower due to population reduction of the upper states arising from the spontaneous emission whose rate is proportional to $(\Delta E)^3$ and for which a counter process does not exist.

17.3.5 Screened Hydrogenic Model

In detailed configuration accounting, populations of the electronic states are calculated for each electron configuration of various charge-stage ions. This calculation becomes complex due to an extremely large number of electronic configurations, especially for mid–large Z ions with many electrons partially filling the valence shell. For preliminary data analyses and evaluation of new experimental plans, for example, it is desirable to use atomic models with which a wide range of plasma conditions are accessed quickly with reasonable accuracy.

The screened hydrogenic model was developed to evaluate the radiative emission and absorption properties of plasmas in non-LTE conditions (Lokke & Grasberger, 1977), with improvements to adapt to high density plasmas (Marchand et al., 1990; More, 1982). In this model, the energy levels, which are characterized by the principal quantum number n and the l-splitting (Perrot, 1989), are constructed with the relativistic hydrogenic formulas including screening of the nuclear charge by the electrons. The screened charge q_n is a function of the screening constants $\sigma(n, m)$ and the occupation numbers N_n and N_m of the n-th and m-th hydrogenic shells, respectively, as

$$q_n = Z - \sum_{m<n} \sigma(n, m) N_m - 0.5\sigma(n, n)(N_n - 1). \tag{17.29}$$

On the right-hand side of this equation, Z is the nuclear charge, and the second and the third terms correspond to screening by the inner-shell and the n-th shell electrons, respectively. Based on this atomic model, the population distribution over n-states is calculated by solving the rate equations using the collisional and radiative rate constants derived for this model.

As an example of the atomic code with a screened hydrogenic model, the FLYCHK is a compact and versatile code used as a module for macroscopic codes. Detailed atomic data files are generated in the post processer FLYSPECTRA to calculate the emission and absorption spectra from the population distribution derived from FLYCHK (Chung et al., 2005).

17.3.6 Average Atom Model

In the average atom model, atomic processes are calculated with the collisional and radiative rate equations for a virtual average ion with a non-integer charge, which represents the ions of various charge states (Post et al., 1977). Kinetic equations for this ion are solved to determine the mean-state occupations and free electron densities with the rates calculated using the average wave functions. The charge of the average ion is determined by iterating these calculations starting from an assumed distribution such as in LTE until reaching to the stationary distributions of the bound and free states. The thermodynamic self-consistency required for the equation of state in hydrodynamic simulations is assured with this model. This model does not require preparation of extensive atomic databases. The average atom model was applied to analyze the pressure ionization in laser-fusion target simulation (Zimmermann & More, 1980) and the opacities of non-LTE high density plasmas (Rozsnyai, 1977).

As an example, we describe the application of the average atom model together with the screened hydrogenic model to analyze the soft x-ray emission spectra of gold (Nishikawa et al., 1993; Takabe & Nishikawa, 1994), irradiated with a 0.35 mm laser at an intensity of 3×10^{14} W/cm^2. Radiation hydrodynamic simulation ILESTA (Takabe et al., 1988) coupled with collisional-radiative rate equations on this experimental condition shows that the soft x-rays are emitted mainly from a flat-density region between the ablation front and the expanding plasma of electron temperature $T_e \sim 200$ eV and plasma density $r \sim 0.1$ g/cm^3, at which the effective charge of the ion Au^{+Q*} is evaluated as $Q* \sim 25$.

Figure 17.8a shows experimental data (Nishimura et al., 1991) and simulation result (Takabe & Nishikawa, 1994) of the x-ray emission spectrum of gold. The observed spectrum has characteristic features: a strong peak at ~ 200 eV and relatively flat region above 400 eV, whereas the simulation spectrum has a peak at ~ 800 eV instead of the flat region in the experimental spectrum. A similar spectrum is reported with the LASNEX simulation (Goldston et al., 1987). Figure 17.8b shows the spectrum with the improved simulation which better agrees with the experimental spectrum, as described in the following.

Figure 17.8c shows the energy levels of gold as a function of the number of bound electrons, calculated with the screened hydrogenic model: upper traces with the principal quantum number n only, i.e. l-states are degenerate (More, 1982), and lower traces with l-splitting by including the angular quantum number l (Perrot, 1989). The horizontal axis in these graphs is the number of the bound electrons ($Z - Q*$), which is ~ 54 at 200 eV and 0.1 g/cm^3. The electron binding energy becomes stronger as the number of bound electrons decreases, since screening of the nuclear charge is reduced. In addition, the energies of the electronic states are affected by various factors; hundreds of eV by l-splitting, tens of eV by difference in the charge state, tens of eV by spin–orbit interaction, and ~ eV by natural broadening and thermal fluctuations.

The emissivity of gold evaluated with the average atom model without and with l-splitting are shown in the upper and lower traces in Fig. 17.8d, respectively. The

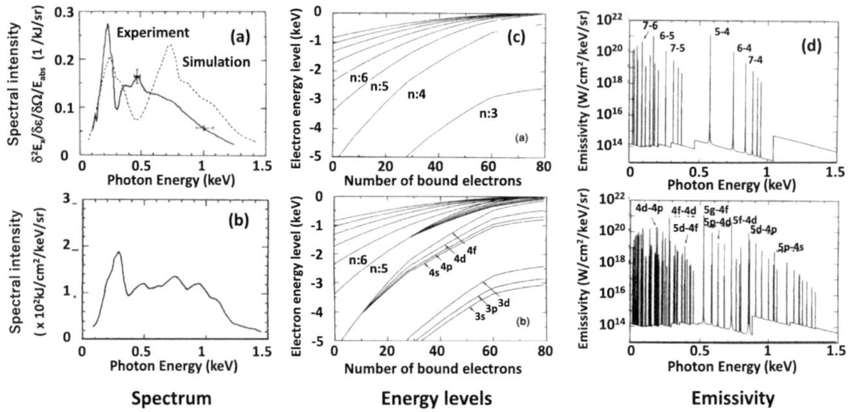

Fig. 17.8 The x-ray emission of gold. **a** Experimental (solid line) and simulated (dotted line) x-ray emission spectra. **b** Simulated x-ray spectrum by the average ion model including the l-splitting and other broadening effects. **c** Energy level of gold plotted against the number of bound electrons, calculated with the average-ion screened hydrogenic model, without (upper traces) and with (lower traces) the l-splitting. **d** Emissivity of gold calculated without (upper traces) and with (lower traces) the l-splitting. Courtesy of Prof. Takeshi Nishikawa

spectrum with l-splitting is quite different from the spectrum with n only, since the intense lines arising from the transitions between different n-states in the upper traces become weaker due to spitting into l-components. In the lower traces, we find that the lines below 500 eV are mostly due to transitions between the l-split levels of $n = 4$ such as $4d$-$4p$ and $4f$-$4d$, whereas the lines above 500 eV are due to transitions between $n = 4$ and $n = 5$ levels such as $4f$-$5g$, $4d$-$5f$, and $4p$-$5d$. In addition, each line in Fig. 17.8d with l-splitting distributes over ~ 10 eV due to different charge states. The x-ray spectrum with revised hydrodynamic simulation incorporating the l-splitting and broadenings due to other effects is shown in Fig. 17.6b, which agrees with the experimental spectrum. This example shows that the average atom model with inclusion of l-splitting is effective to analyze the x-ray spectra of mid- to high-Z materials.

The average atom model has been applied also to EUV lithography (Nishihara et al., 2006) to find the optimum condition of the EUV source, where high conversion efficiency is required at 13.5 nm with a 2% bandwidth (in-band emission), corresponding to the reflection band of the Mo/Si multilayers. The candidate materials for efficient 13.5 nm generation are Sn (Z = 50) (Shimada et al., 2005; Spitzer et al., 1996) and Xe (Z = 54) (Endo et al., 2004). The average atom model with l-splitting and incorporating the screened charge Q_{nl} was used in the analyses (Nishikawa et al., 2004).

Since accurate emission wavelength is required for this application, the transition energies were calculated with HULLAC which includes the energy shift due to configuration interactions. Figure 17.9a shows calculated wavelengths of the 5-4 and 4-4 transitions in Sn (upper traces) and Xe (lower traces) at various charge states

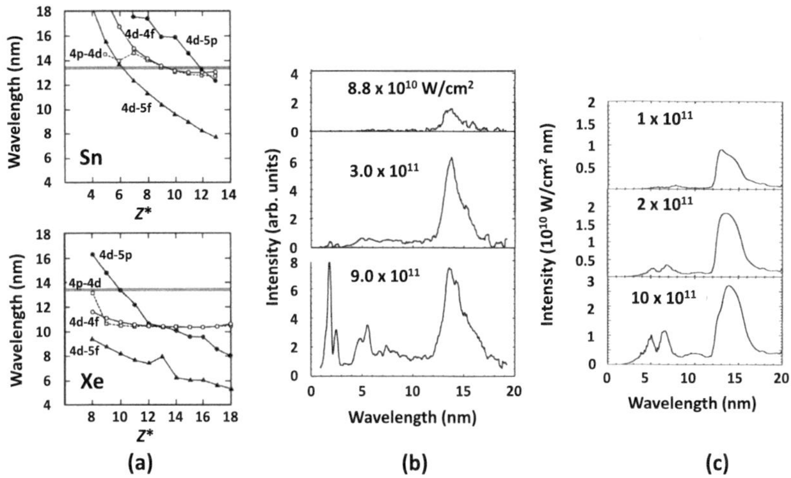

Fig. 17.9 **a** Emission wavelengths of Sn (upper diagram) and Xe (lower diagram) for $\delta n = 0$ and 1 transitions. The in-band wavelength at 13.5 nm is shown by the horizontal lines. Observed (**b**) and calculated (**c**) emission spectra of Sn at irradiation intensities of ~ 10^{11}–10^{12} W/cm^2. Courtesy of Prof. Takeshi Nishikawa

$Z*$. The $4p$-$4d$ and $4d$-$4f$ lines match well to the 13.5 nm band over broad charge states of $Sn^{8+\text{ to }12+}$ due to the configuration interaction. However, these lines shift to shorter wavelengths in Xe, although Xe is attractive as a debris-free x-ray source.

Figure 17.9b and c show the observed and calculated emission spectra of Sn, respectively, with 1-mm laser irradiation at the intensities of ~ 10^{11}–10^{12} W/cm^2. The emission band at 12–17 nm is due to the $\Delta n = 0$ transitions, whereas the short wavelength band of 4-8 nm is due to the $\Delta n = 1$ and 2 (5-4 and 6-4) transitions with higher charge states of $Sn^{+14\text{ to }+21}$. Both from the experiment and the simulation, the conversion efficiency from the laser energy to the in-band x-rays is evaluated as ~ 3% for the laser wavelength of 1 micrometer.

Dependence of the conversion efficiency on the laser wavelength was investigated by Nishihara et al. (2006). In general, laser-to-x-ray conversion efficiency is expected to be higher with shorter wavelength lasers, since they are absorbed at higher plasma density with higher absorption rate. Figure 17.10 shows spatial profiles of the ion density n_i, electron temperature T_e, average ionization state Z^* and in-band emission intensity S^*_{EUV} and S_{euv} of the Sn plasma irradiated with (a) 0.26 mm and (b) 10.6 mm lasers, respectively. The irradiation intensity is 10^{12} W/cm^2 at 0.26 mm and 10^{10} W/cm^2 at 10.6 mm where the highest in-band conversion efficiency is obtained for each wavelength. The laser pulse duration is 10 ns in both cases. Denoting the emissivity, absorption coefficient and emission energy density by η^ν, χ^ν, and E^ν, respectively, the in-band emission intensity over all directions at each position in the plasma is described by $S^*_{\text{EUV}} = (4\pi\eta^\nu)_{\text{EUV}}$, whereas the effective in-band emission intensity which is reduced from S^*_{EUV} due to absorption during propagation in the plasma is described by $S_{\text{EUV}} = (4\pi\eta^\nu - c\chi^\nu E^\nu)_{\text{EUV}}$.

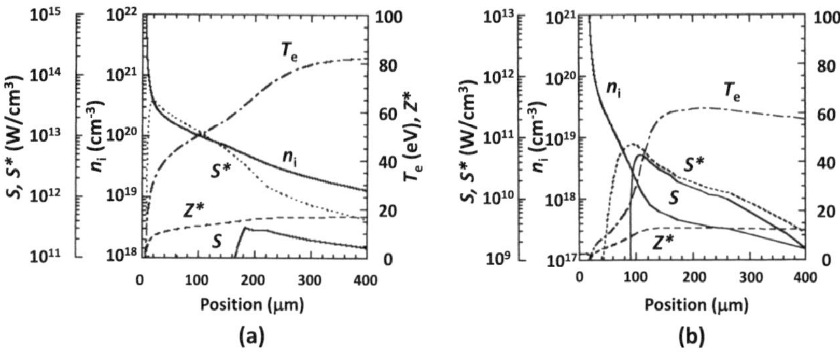

Fig. 17.10 Spatial profiles of the ion density n_i, electron temperature T_e, effective charge Z^*, in-band emission intensity S^*, and effective in-band emission intensity S, for irradiation of **a** 0.26-mm and **b** 10.6-mm wavelength lasers. Note that the vertical scales of n_i, S^* and S are different in (**a**) and (**b**). Courtesy of Prof. Takeshi Nishikawa (Note that the characters in the original figure are enlarged for visibility.)

The laser energy is deposited mainly at the critical density of $n_{cr}(\mathrm{cm}^{-3}) = (1.12 \times 10^{21})/\lambda_0^2(\mu\mathrm{m})$. With 0.26 mm laser irradiation, laser energy is deposited near its critical density of $n_{cr} = 1.7 \times 10^{22}\mathrm{cm}^{-3}$, or $n_\mathrm{i} \simeq 2 \times 10^{21}\mathrm{cm}^{-3}$ for $Z^* \simeq 10$, which is located at 20 mm from the initial target surface. In this region, the x-rays are emitted with high S^*_{EUV}. However, the emitted x-rays are absorbed in the high-density LTE plasma of long scale-length (~ 245 mm), resulting in the effective intensity S_{EUV} with 2-orders of magnitudes smaller than S^*_{EUV}. With 10.6 mm laser irradiation, the ion density corresponding to the critical density is $n_\mathrm{i} \sim 2 \times 10^{18}\mathrm{cm}^{-3}$. At this low density, x-ray absorption is small since LTE plasma is not formed, resulting in S_{EUV} closely located to the peak of S^*_{EUV} with similar magnitudes. This study has shown that a high conversion efficiency exceeding 6% is expected with a 10.6-mm CO_2 laser.

In a prototype EUV lithography system, liquid Sn droplets are irradiated with a CO_2 laser with a pre-pulse to smash the droplet creating a lower-density, larger size Sn particles which are irradiated by a main laser pulse, resulting in the in-band conversion efficiency of 2.3% (Mizoguchi et al., 2010). In another system with refinements, close to 5% maximum conversion efficiency (3.3% in daily operation) has been achieved (Banine et al., 2013). In this industrial lithography system, where a CO_2 laser of 80-kHz repetition rate and 47 kW average power is used for EUV generation, 22-nm resolution semiconductor devices are produced with a production rate of ~ 100 wafers/h. As a next step of the EUV lithography, ~ 3 nm resolution lithography system is conceived using a ~ 6.8 nm EUV source, reflection band of La/B multilayers (Makhotkin et al., 2012), with Gd as a candidate of the source material (Li et al., 2011).

17.3.7 Radiation-Dependent Ionization Model

A compact atomic physics code is required to couple the atomic physics and the hydrodynamics online. Busquet has developed a non-LTE atomic model RADIOM (radiation-dependent ionization model), with which the average charge is evaluated for non-LTE plasmas with a simple model (Busquet, 1982, 1993).

In thermally equilibrium plasmas, the population of the electronic states and the ionic distribution, determined uniquely by the electron temperature T_e, are described by the Boltzmann and Saha equations, respectively. It is known that the distribution of the ionic charges is narrow, comprising a sharp peak at n_{max} and small broadening around this peak. The average charge of the ions $\langle Q\rangle$ represents the ionization balance, and the electron density n_e is derived from $\langle Q\rangle$ as $n_e = \left(\rho N_A / Z\right)\langle Q\rangle$. Here n_Q and n are the number densities of the Q-charged and total ions respectively, ρ is the density, Z is the atomic number, and N_A is the Avogadro number. In thermal equilibrium, the average charge is (Zel'dovich & Raizer, 1966)

$$\langle Q\rangle = \frac{n_e}{n} = \frac{AT_e^{3/2}}{n} e^{-\bar{I}_{<Q>+1/2}/kT_e}, \tag{17.30}$$

where $A = 2\left(2\pi mk / h^2\right) = 6 \times 10^{21} \mathrm{cm}^{-3}\mathrm{eV}^{-3/2}$ and T_e is the electron temperature. $\bar{I}_{\langle Q\rangle+1/2}$ is the ionization potential of a virtual $\langle Q\rangle$-charged ion, which is approximated by an average of the ionization potential of Q- and $(Q+1)$-charged ions. The optimum values of $\langle Q\rangle$ and $\bar{I}_{\langle Q\rangle+1/2}$, which are mutually related, are found in a few iterations.

In non-LTE plasmas, populations of high-lying states are reduced due to radiative cooling by the spontaneous emission, leading to reduction of the effective temperature to characterize the Boltzmann distribution as described in Sect. 17.3.4, and due to reduction of the average charge estimated by LTE at T_e. To derive the average charge of a non-LTE plasma, Busquet has defined an "ionization temperature" T_z, with which the average charge of a non-LTE plasma is derived from a standard LTE equation

$$\langle q\rangle^{\mathrm{non-LTE}}(T_e) = \langle q\rangle^{\mathrm{LTE}}(T_z). \tag{17.31}$$

In LTE plasmas, n_{Q+1}/n_Q is expressed by the Saha equation (17.19). Busquet has found that the maximum abundance $n^{\max}$ is attained at $n_{Q+1}/n_Q = 1$. Since $\langle Q\rangle$ is proportional to $n^{\max}$, T_z satisfying Eq. (17.31) will be found if an "extended Saha equation" applicable to non-LTE plasmas is derived.

The charge distribution of a non-LTE plasma is evaluated in RADIOM with a simplified atomic model shown in Fig. 17.11. Noting that energies of the excited states from the ground state are larger than half of the ionization energy in most ions, all excited states are grouped into one state named a "superlevel" or a "thermal band". The superlevel and the ground state of the + Q-ion are denoted by (Q,c) and (Q,g), respectively, in Fig. 17.11. The unbound states above the ionization limit and the ground state of the (Q + 1)-ion denoted by (Q + 1,g) are also included in (Q,c). It is assumed that these states are in thermal equilibrium at high density due to collisional processes and their populations are described by the Boltzmann distribution. The isolated levels between these levels are not included in this model since their contribution to the ionization balance is small compared to higher levels. Therefore, in this model, (Q,g), ($Q + 1$,g) and ($Q + 2$,g) is a part of ($Q - 1$,c), (Q,c) and ($Q + 1$,c), respectively.

Populations of (Q,g), (Q,c) and their neighboring states are calculated by solving the rate equations between these states using the rate coefficients of the collisional and radiative processes and the partition functions evaluated for these states. The ratio of the reduced number densities of the ($Q + 1$)-ion and the Q-ion, which corresponds to the extended Saha equation for non-LTE plasma, is given as,

$$R_{Q+1/2} \equiv \frac{n_{Q+1}/U_{Q+1}}{n_Q/U_Q} = \frac{U_e}{n_e} e^{-(I_Q/kT_e)} \times \frac{1 + \beta u^3/(\gamma \langle E/B \rangle) + [\mathrm{Aug/Col}]}{1 + \beta u^3/\gamma + [\mathrm{Aug/Col}]}. \tag{17.32}$$

In this equation, $u = I_Q \big/ kT_e$ is the ratio of the ionization potential I_Q to the electron temperature; $\beta = b \cdot T_e^{7/2}/n_e$ is the collisional parameter to describe the

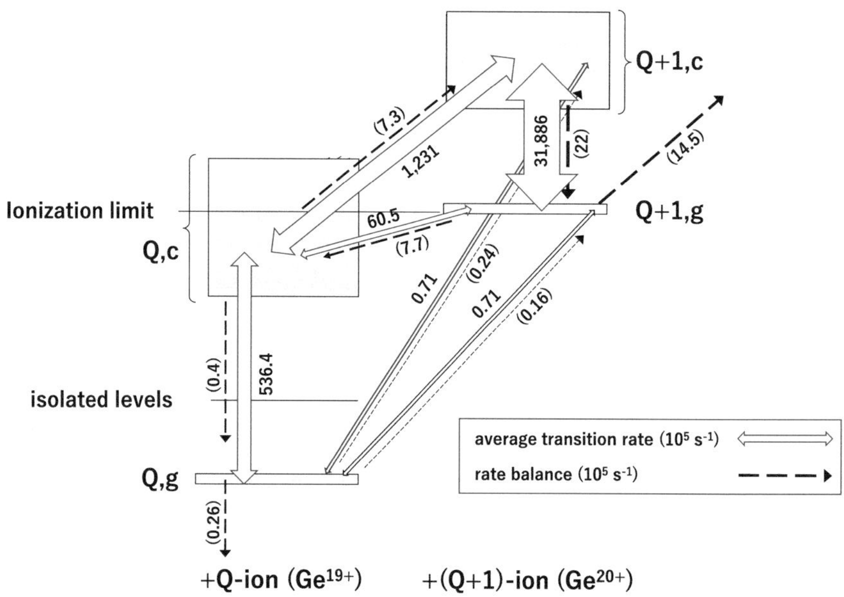

Fig. 17.11 Schematic diagram of the energy levels used in RADIOM. The energy levels of Q-charged ion are grouped into the ground state (Q,g) and the excited states (Q,c) which includes the free state above the ionization limit and ($Q + 1$,g). The average transition rate and the rate balance for Ge^{19+} and Ge^{20+} at $n_e = 10^{21}$ cm^{-3} and $T_e = 250$ eV are shown by the open and dotted arrows, respectively. Reprinted from Busquet (2006), with permission from Elsevier: permission conveyed through Copyright Clearance Center, Inc. © 2006 Elsevier

temperature and density dependence of the collisional to radiative rate ratio ($b = 1.34 \times 10^{13}$ cm^{-3}eV$^{-7/2}$); $\langle E/B \rangle$ is the radiative energy density divided by its LTE value estimated at the average transition energy of this charge state; $\gamma = \langle \overline{g}/\Delta E^3 \rangle I_Q^3$ is the parameter merging the effective Gaunt factor $\overline{\gamma}$ and the effective transition energy ΔE; $U_e = aT_e^{3/2}$ is the free electron partition function ($a = 6 \times 10^{21}$ cm^{-3}eV$^{-3/2}$); and U_Q is the non-LTE partition function at the effective temperature T_q. The term $\left[\text{Aug/Col}\right]$ is added to account for the Auger transition and the inverse dielectronic capture processes which are important in high Z materials (Busquet et al., 2009).

By equating $(n_{Q+1}/U_{Q+1})\big/(n_Q/U_Q) = 1$ in (17.19) for LTE and in (17.32) for non-LTE, the following result is obtained,

$$\frac{U_e}{n_e}e^{-(I_Q/kT_z)} = \frac{U_e}{n_e}e^{-(I_Q/kT_e)} \times \frac{1 + \beta u^3/(\gamma \langle E/B \rangle) + \left[\text{Aug/Col}\right]}{1 + \beta u^3/\gamma + \left[\text{Aug/Col}\right]}. \tag{17.33}$$

This leads to

$$\frac{T_z}{T_e} = \left\{1 - \frac{1}{u}\ln\left[\frac{1 + u\beta^3\big/(\gamma \langle E/B \rangle) + \left[\text{Aug/Col}\right]}{1 + u\beta^3\big/\gamma + \left[\text{Aug/Col}\right]}\right]\right\}^{-1} \tag{17.34}$$

With (17.34), the ionization temperature T_z/T_e is determined as a function of u, b, $\langle E/B \rangle$, and g. Thus, the average charge of a non-LTE plasma is determined by Eq. (17.31) using T_z in place of T_e. It was confirmed with detailed collisional radiative models that RADIOM yields very good values of the average charge and T_z when the plasma conditions are not too far from LTE (Klapisch & Bar-Shalom, 1997). Although it is expected that the ionization temperature T_z is close to the effective temperature which characterizes the Boltzmann distribution, for example the ionic excitation temperature $T(\mathbf{I})$ described in Sect. 17.3.4, detailed models such as the harmonic model (Busquet, 2006) are required to include various processes which affect the population distribution.

The opacity of non-LTE plasma is evaluated by a modified source function S_ν,

$$S_\nu = \frac{\eta_\nu}{\chi_\nu} \approx B_\nu \times \frac{1 + \beta u^3/(\gamma \langle E/B \rangle) + \left[\text{Aug/Col}\right]}{1 + \beta u^3/\gamma + \left[\text{Aug/Col}\right]}, \tag{17.35}$$

where B_ν is the Planck function, and η_ν and χ_ν are the emission and absorption coefficients, respectively. The emissivity of non-LET plasmas is reduced compared to LTE due to reduction of the population of high lying states.

Figure 17.11 also shows the population flux between the sets of Ge^{19+} and Ge^{20+} at $n_e = 10^{21}$ cm^{-3} and $T_e = 250$ eV. The excitation (upward) and deexcitation (downward) fluxes are shown by the unshaded arrows with their widths qualitatively corresponding to the transition rates. The average of the excitation and deexcitation rates are shown near these arrows. The rate balances, the difference of these rates, are shown in the parentheses with their directions corresponding to the flow direction. We note that the population flow between neighboring charge ions by ionization and recombination is larger than the flow inside each superlevel, with a characteristic transition time (inverse of the transition rate) of shorter than 1 ns. Under this plasma condition, the major flow is from Ge^{20+} to Ge^{21+}. This example shows that the population of an atomic level of an ion in high density plasmas must be determined including the atomic processes connecting the neighboring ions.

References

Azuma, H., Kato, Y., Yamakawa, K., et al. (1990). Short-pulse pumping of a recombination Balmer-α laser of hydrogenic sodium. *Optics Letters, 15*, 1011.

Banine, V., et al. (2013). EUV lithography: Status, future requirements and challenges. In *International Workshop on EUV and Soft X-ray Sources*, Dublin.

Bar-Shalom, A., Oreg, J., Goldstein, W. H., et al. (1989). Super-transition arrays: A model for the spectral analysis of hot, dense plasma. *Physical Review A, 40*, 3183.

Bar-Shalom, A., Oreg, J., & Klapisch, M. (1997a). Non-LTE superconfigurational collisional radiative model. *Journal of Quantitative Spectroscopy & Radiative Transfer, 58*, 427.

Bar-Shalom, A., Oreg, J., & Klapisch, M. (1997b). Collisional radiative model for heavy atoms in hot non-local thermodynamical-equilibrium plasmas. *Physical Review E, 56*, R70.

Bar-Shalom, A., Oreg, J., & Klapisch, M. (2000). Recent developments in the SCROLL model. *Journal of Quantitative Spectroscopy & Radiative Transfer, 65*, 43.

Bar-Shalom, A., Klapisch, M., & Oreg, J. (2001). HULLAC, an integrated computer package for atomic processes in plasmas. *Journal of Quantitative Spectroscopy & Radiative Transfer, 71*, 169.

Bauche, J., & Bauche-Arnoult, C. (1996). Recent progress in the global description of atomic transitions. *Physica Scripta, T65*, 99.

Bauche, J., Bauche-Arnoult, C., & Klapisch, M. (1988). Unresolved transition arrays. *Physica Scripta, 37*, 659.

Bauche, J., Bauche-Arnoult, C., & Fournier, K. B. (2004). Model for computing superconfiguration temperatures in nonlocal-thermodynamic-equilibrium hot plasmas. *Physical Review E, 69*, Article 026403.

Bauche, J., Bauche-Arnoult, C., & Peyrusse, O. (2006). Effective temperatures in hot dense plasmas. *Journal of Quantitative Spectroscopy & Radiative Transfer, 99*, 55.

Bauche, J., & Bauche-Arnoult, C. (2000). Temperatures in non-LTE hot plasmas. *Journal of Physics B: Atomic, Molecular and Optical Physics, 33*, L283.

Busquet, M. (1982). Mixed model: Non-local-thermodynamic equilibrium, non-coronal-equilibrium simple ionization model for laser-created plasmas. *Physical Review A, 21*, 2302.

Busquet, M. (1993). Radiation-dependent ionization model for laser-created plasmas. *Physics of Fluids, 85*, 4191.

Busquet, M. (2006). Onset of pseudo-thermal equilibrium within configurations and super-configurations. *Journal of Quantitative Spectroscopy & Radiative Transfer, 99*, 131.

Busquet, M., Colombant, D., Klapisch, M., et al. (2009). Improvements to the RADIOM non-LTE model. *High Energy Density Physics, 5*, 270.

Byron, S., Stabler, R. C., & Bortz, P. I. (1962). Electron-ion recombination by collisional and radiative processes. *Physical Review Letters, 8*, 376.

Carillon, A., Edwards, M. J., Grande, M., et al. (1990). Soft X-ray amplification in aluminium recombining plasma produced from a thin coated fibre. *Journal of Physics B: Atomic, Molecular and Optical Physics, 23*, 147.

Chung, H.-K., Chen, M. H., Morgan, W. L., et al. (2005). FLYCHK: Generalized population kinetics and spectral model for rapid spectroscopic analysis for all elements. *High Energy Density Physics, 1*, 3.

Chung, H. K., Hansen, S. B., Scott, H. A., (2016) "Generalized collisional radiative model using screened hydrogenic levels," in Modern Methods in Collisional Radiative Modeling of Plasmas, Y. Ralchenko (ed.), Springer, Switzerland.

Daido, H., Kodama, R., Murai, K., et al. (1995). Significant improvement in the efficiency and brightness of the J = 0–1 19.6-nm line of the germanium laser by use of double-pulse pumping. *Optics Letters, 20*, 61.

Dyall, K. G., Grant, I. P., Johnston, C. T., et al. (1989). GRASP: A general-purpose relativistic atomic structure program. *Computer Physics Communications, 55*, 425.

Endo, A., Abe, T., Suganuma, T., et al. (2004). Design of high-average-power clean EUV light source based on laser-produced Xenon plasma. *Proceedings of SPIE, 5196*, 256.

Fontes, C. J., Cogan, J., Abdallar, Jr, J. (2016). Self-consistent large-scale collisional-radiative modeling. In Y. Ralchenko (Ed.), *Modern methods in collisional radiative modeling of plasmas*. Springer.

Fujimoto, T., & McWhirter, R. W. P. (1990). Validity criteria for local thermodynamic equilibrium in plasma spectroscopy. *Physical Review A, 42*, 6588.

Fujimoto, T. (1979a). Kinetics of ionization-recombination of a plasma and population density of excited ions. I. Equilibrium plasma. *Journal of the Physical Society of Japan, 47*, 265.

Fujimoto, T. (1979b). Kinetics of ionization-recombination of a plasma and population density of excited ions. II. Ionizing plasma. *Journal of the Physical Society of Japan, 47*, 273.

Fujimoto, T. (1980a). Kinetics of ionization-recombination of a plasma and population density of excited ions. III. Recombining plasma—High temperature case. *Journal of the Physical Society of Japan, 49*, 1561.

Fujimoto, T. (1980b). Kinetics of ionization-recombination of a plasma and population density of excited ions. IV. Recombining plasma—Low temperature case. *Journal of the Physical Society of Japan, 49*, 1569.

Fujimoto, T. (1985). Kinetics of ionization-recombination of a plasma and population density of excited ions. V. Ionization-recombination and equilibrium plasma. *Journal of the Physical Society of Japan, 54*, 2905.

Goldstone, P. D., Goldman, S. R., Mead, W. C., et al. (1987). Dynamics of high-Z plasmas produced by a short-wavelength laser. *Physical Review Letters, 59*, 56.

Grant, I. P., McKenzie, Norrigan, P. H., et al. (1980). Atomic multiconfigurational Dirac-Fock package. *Computer Physics Communications, 21*, 207.

Griem, H. R. (1964). *Plasma spectroscopy*. McGraw-Hill.

Hansen, S. B. (2016). Balancing detail and completeness in collisional-radiative models. In Y. Ralchenko (Ed.), *Modern methods in collisional radiative modeling of plasmas*. Springer.

Hara, T., Ando, K., Kusakabe, N., et al. (1989). Soft X-ray lasing in an Al plasma produced by a 6 J laser. *Japanese Journal of Applied Physics, 28*, L1010.

Iglesias, C. A., Chen, M. H., McWilliams, D. L., et al. (1995). Observations on radiation transfer experiments using *K*-shell absorption spectra. *Journal of Quantitative Spectroscopy & Radiative Transfer, 54*, 185.

Jaegle, P., Jamelot, G., Calliron, A., et al. (1987). Soft-x-ray amplification by lithiumlike ions in recombining hot plasmas. *Journal of the Optical Society of America B: Optical Physics, 4*, 563.

Kawachi, T., Sasaki, A., Tanaka, M., et al. (2004). Observation of strong soft-X-ray amplification at 8.8 nm in the transient collisional-excitation scheme. *Physical Review A, 69*, 033805.

Kawachi, T., & Fujimoto, T. (1997). Deexcitation and recombination of excited ions through doubly excited levels in a dense recombining plasma: Lithiumlike recombining plasma laser. *Physical Review E, 55*, 1836.

Klapisch, M., & Bar-Shalom, A. (1997). Checking the validity of Busquet's ionization temperature with detailed collisional radiative models. *Journal of Quantitative Spectroscopy & Radiative Transfer*, *58*, 687.

Klisnick, A., Sureau, A., Guennou, H., et al. (1990). Effective rates for Li-like ions; calculated XUV gains in Al10+. *Applied Physics B, 50*, 153.
Li, B., Dunne, P., Higashiguchi, T., et al. (2011). Gd plasma source modeling at 6.7 nm for future lithography. *Applied Physics Letters, 99*, 231502.
Lokke, W. A., & Grasberger, W. H. (1977). *XNSQ-U a Non-LTE emission and absorption coefficient subroutine*. UCRL-52276. Lawrence Livermore National Lab.
Lotz, W. (1968). Electron-impact ionization cross-sections and ionization rate coefficients for atoms and ions from hydrogen to calcium. *Zeitschrift Für Physik, 216*, 241.
Lotz, W. (1969). Electron-impact ionization cross-sections and ionization rate coefficients for atoms and ions from scandium to zinc. *Zeitschrift Für Physik, 220*, 466.
MacGowan, B. J., Maxon, S., Da Silva, L. B., et al. (1990). Demonstration of X-ray amplifiers near the carbon K edge. *Physical Review Letters, 65*, 420.
Makhotkin, I. A., Zoethout, E., Louis, E., et al. (2012). Spectral properties of La/B-based multilayer mirrors near the boron K absorption edge. *Optics Express, 20*, 11778.
Marchand, R., Caillie, S., & Lee, Y. T. (1990). Improved screening coefficients for the hydrogenic ion model. *Journal of Quantitative Spectroscopy & Radiative Transfer, 43*, 149.
Matthews, D. L., Hagelstein, P. L., Rosen, M. D., et al. (1985). Demonstration of a soft X-ray amplifier. *Physical Review Letters, 54*, 110.
Maxon, S., Hagelstein, P., MacGowan, B., et al. (1988). Calculation and design of a Ni-like Eu soft-x-ray laser. *Physical Review A, 37*, 2227.
McWhirter, R. W. P., & Hearn, A. G. (1963). A calculation of the instantaneous population densities of the excited levels of hydrogen-like ions in a plasma. *Proceedings of the Physical Society, 82*, 641.
Mizoguchi, H., Abe, T., Watanabe, Y., et al. (2010). 1st generation laser-produced plasma source system for HVM EUV lithography. *Proceedings of SPIE, 7636*, 763608.
More, R. M. (1982). Electronic energy-levels in dense plasmas. *Journal of Quantitative Spectroscopy & Radiative Transfer, 27*, 345.
Murai, K., Shiraga, H., Yuan, G., et al. (1994a). Lasing properties of the J = 0–1 and the J = 2–1 lines of a neonlike germanium soft-x-ray laser. *Journal of the Optical Society of America B: Optical Physics, 11*, 2287.
Murai, K., Yuan, G., Kodama, R., et al. (1994b). Double-pass amplification in Ge soft x-ray laser with a polarizing half cavity. *Japanese Journal of Applied Physics, 33*, L600.
Nickles, P. V., Shlyaptsev, V. N., Kalachnikov, M., et al. (1997). Short pulse X-ray laser at 32.6 nm based on transient gain in Ne-like titanium. *Physical Review Letters, 78*, 2748.
Nishihara, K., Sasaki, A., Sunahara, A., & Nishikawa, T. (2006). Conversion efficiency of LPP sources. In V. Bakshi (Ed.), *EUV sources for lithography* (Chap. 11, pp. 339–370). SPIE Press.
Nishikawa, T., Takabe, H., & Mima, K. (1993). Line profile modeling for non-LTE partially ionized plasmas based on average atom model with *l*-splitting. *Laser and Particle Beams, 11*, 81.
Nishikawa, T., Sasaki, A., Furukawa, H., et al. (2004). Atomic models and hydrodynamic simulation for laser-plasma EUV source. In *EUV Source Workshop*, Santa Clara, CA, February 2004.
Nishimura, H., Takabe, H., Kondo, K., et al. (1991). X-ray emission and transport in gold plasmas generated by 351-nm laser irradiation. *Physical Review A, 43*, 3073.
Parpia, F. A., Fisher, C. F., & Grant, I. P. (1996). GRASP92: A package for large-scale relativistic atomic structure calculations. *Computer Physics Communications, 94*, 249.
Perrot, F. (1989). Fast calculation of electronic structure in plasmas: The screened hydrogenic model with L-splitting. *Physica Scripta, 39*, 332.
Pert, G. J. (1976). Model calculations of XUV gain in rapidly expanding cylindrical plasmas. *Journal of Physics B: Atomic, Molecular and Optical Physics, 9*, 3301.
Pert, G. J. (1979). Model calculations of XUV gain in rapidly expanding cylindrical plasmas. II. *Journal of Physics B: Atomic, Molecular and Optical Physics, 12*, 2067.
Peyrusse, O. (2000). A superconfigration model for broadband spectroscopy of non-LTE plasmas. *Journal of Physics B: Atomic, Molecular and Optical Physics, 33*, 4303.

Peyrusse, O. (2001). On the superconfiguration approach to model NLTE plasma emission. *Journal of Quantitative Spectroscopy & Radiative Transfer, 71*, 571.

Post, D. E., Jensen, R. V., Tarter, C. B., et al. (1977). Steady-state radiative cooling rates for low-density, high-temperature plasmas. *Atomic Data and Nuclear Data Tables, 20*, 397.

Rogers, F. J., & Iglesias, C. A. (1992). OPAL opacities. *The Astrophysical Journal Supplement Series, 79*, 291.

Rosen, M. D., Hagelstein, P. L., Matthews, D. L., et al. (1985). Exploding-foil technique for achieving a soft X-ray laser. *Physical Review Letters, 54*, 106.

Rozsnyai, B. F. (1977). A collisional-radiative average atom model for hot plasmas. *Physical Review B, 55*, 7507.

Scott, H. A. (2016). Collisional-radiative modeling for radiation hydrodynamics codes. In Y. Ralchenko (Ed.), *Modern methods in collisional radiative modeling of plasmas*. Springer.

Shimada, Y., Nishimura, H., Nakai, M., et al. (2005). Characterization of extreme ultraviolet emission from laser-produced spherical tin plasma generated with multiple laser beams. *Applied Physics Letters, 86*, Article 051501.

Spitzer, R. C., Orzechowski, T. J., Phillion, D. W., et al. (1996). Conversion efficiencies from laser-produced plasmas in the extreme ultraviolet regime. *Journal of Applied Physics, 79*, 2251.

Stewart, J. C., & Pyatt, K. D. (1966). Lowering of ionization potentials in plasmas. *The Astrophysical Journal, 144*, 1203.

Takabe, H., & Nishikawa, T. (1994). Computational model for non-LTE atomic process in laser produced plasmas. *Journal of Quantitative Spectroscopy & Radiative Transfer, 51*, 379.

Takabe, H., Yamanaka, M., Mima, K., et al. (1988). Scalings of implosion experiments for high neutron yield. *Physics of Fluids, 31*, 2884.

Zel'dovich, Ya. B., & Raizer, Yu. (1966). *Physics of shock waves and high temperature hydrodynamic phenomena*. Academic Press.

Zhang, J., Key, M. H., Norreys, P. A., et al. (1995). Demonstration of high gain in a recombination XUV laser at 18.2 nm driven by a 20 J, 2 ps glass laser. *Physical Review Letters, 74*, 1335.

Zimmerman, G. R., & More, R. (1980)0 Pressure ionization in laser fusion target simulation. *Journal of Quantitative Spectroscopy & Radiative Transfer, 23*, 522.

Chapter 18
Laser Driven Neutron Source

Abstract The neutron sources have been widely applied to nuclear science, material diagnostics, biology, medical application, infra-structure check, security check, and so on. The conventional neutron sources are fission reactors and accelerator-driven sources. Recently, the laser-driven neutron source has emerged. In this chapter, research and development of the laser-driven neutron source is reviewed.

Keywords Laser-driven neutron source · Accelerator-driven neutron source · Spallation nuclear reaction · Photo-nuclear reaction · Low-energy nuclear reaction · Fusion neutron source · Target normal sheath acceleration · Neutron time-of-flight · Nuclear resonance absorption

18.1 Introduction

Demand is growing in the neutron sources for scientific and industrial applications, such as deeply penetrating probe in the isotope identification, nuclear materials damage test studies, the cancer therapy of Boron Neutron Cancer Therapy (BNCT), and so on (Alvarez et al., 2014; Brenner et al., 2016, 2019; Mima et al., 2018; Passoni et al., 2020; Roth, 2018; Roth et al., 2013). The neutron generation processes driven by accelerators and lasers are summarized in Fig. 18.1. The high flux neutron facilities (spallation neutron sources) consist with high energy proton beams of multi-GeV energies. The multi-GeV proton beam is injected into heavy metal targets to produce neutrons by spallation nuclear reactions as shown in the top-left of Fig. 18.1.

A multi-GeV proton collides with many heavy nuclei to produce several hundreds of neutrons, where the neutron generation efficiency is the highest, of the order of $10^{11} n/\mathrm{J}$ (beam energy). As a compact neutron facility, a compact proton accelerator of several MeV is used as shown in the top-right of Fig. 18.1. For several MeV protons, energy loss is dominated by electron scattering so that the neutron generation efficiency is lower than that for the spallation nuclear reactions of the multi-GeV proton.

Y. Kato et al., *High Power Laser and Plasma Science*, Springer Series on Atomic, Optical, and Plasma Physics 130, https://doi.org/10.1007/978-3-032-08930-4_18

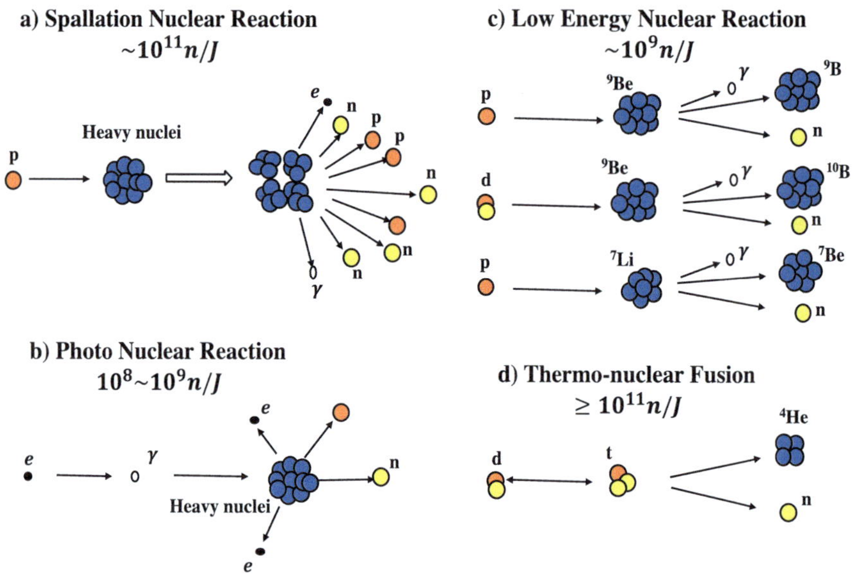

Fig. 18.1 Energy efficiencies (n/J) of accelerator driven neutron and laser driven neutron sources. For accelerator driven neutron, n/J = (neutron number)/(particle beam pulse energy) and for laser, n/J = (neutron number)/(laser pulse energy) (Mima et al., 2018). **a** Spallation neutron source driven by GeV range proton accelerator, **b** Photo nuclear neutron source driven by electron accelerator or ultra-intense laser, **c** Neutron source driven by several MeV light particles from ion accelerators or ultra-intense laser ion source, **d** Fusion neutron source with ICF or MCF

The neutron generation efficiency in the case of MeV proton beams is of the order of $10^9 n/\text{J}$ (beam energy) where the stripping nuclear reaction and/or nuclear fusion are relevant as shown in Fig. 18.2. Note that Li(d,n) and Be(d,n) reactions are higher for the neutron generation than the Li(p,n) and Be(p,n) reactions. However, in the case of accelerator driven neutron source, a deuteron accelerator is not feasible because the hallow of the deuteron beam hits the accelerator wall to generate neutrons and the accelerator area will be contaminated. On the other hand, in the laser driven neutron source, deuteron can be accelerated by short pulse high intensity laser and applied to generate neutrons, since the laser driven acceleration area is very small, so that neutron generation in the acceleration area is not a problem.

As for laser fusion neutron sources (Fig. 18.1d), cluster fusion and laser implosion thermonuclear fusion have been explored. In the laser implosion fusion, a capsule containing D and T mixture is imploded to produce hot dense plasma as described in Chap. 6. This neutron source size is of the order of $100\,\mu\text{m}$ and the pulse duration is sub-nano second. The total neutron yield has increased up to 10^{18}/pulse or more, although the repetition rate is very low (less than 0.001 Hz). This may provide the opportunity of opening a new neutron science. Another process is the photo-nuclear reactions (Fig. 18.1, left hand bottom), where the gamma rays higher than 10 MeV interact with heavy metal targets. The gamma rays are generated by

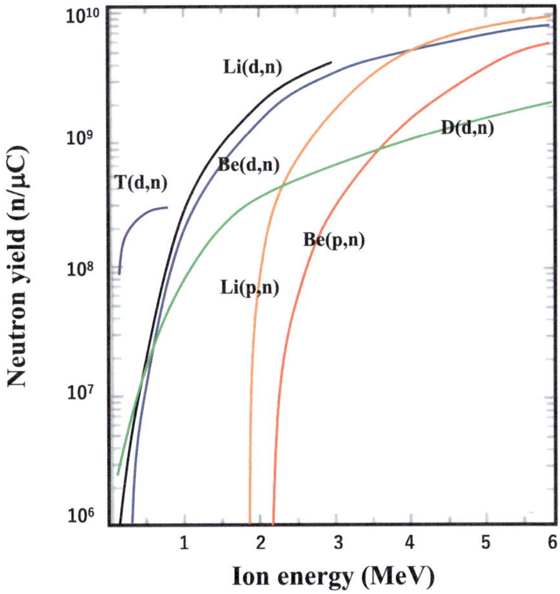

Fig. 18.2 Ion energy dependences of the neutron yield per μC (beam energy) for several reactions with MeV-energy ion beams

high energy electrons accelerated by electron linear accelerator or ultra-intense laser plasma interactions.

Since the high-power laser technology is rapidly evolving (Danson et al., 2015, 2019) in terms of repetition rate and peak power, the laser-driven short-pulse neutron sources could significantly increase the global capacity to meet the demands in various industrial, scientific and healthcare sectors. Namely, it is expected to apply the laser driven neutron sources (LDNS) to material diagnostics, nuclear physics, diagnostics of infrastructure soundness, and inspection of nuclear materials by the combination of laser-driven x-rays.

In Sects. 18.2 and 18.3, we overview R&D of Laser Driven Neutron Sources (LDNS) and present status in Japan, Germany, UK, and USA. In Japan, the peta-watt laser facility LFEX has been used for characterizing the laser driven neutron source. In Germany, the efficient neutron sources have been investigated by the FELIX laser at GSI. In the UK, the LDNS has been investigated by using the Vulcan laser at the Science and Technology Facilities Council (STFC) Rutherford Appleton Laboratory (Brenner et al., 2016). In the USA, for example, repetitive production of the laser driven neutron has been explored at the University of Michigan (Hah et al., 2016, 2018).

18.2 Overview of Research on Laser Driven Neutron Source

18.2.1 Laser Ion Acceleration for Neutron Source

When intense lasers interact with matter, extremely high electro-magnetic fields are generated as described in Part IV of this book. In high energy density plasmas, various high energy particle beams and radiation are generated. There are many processes to accelerate particles to high energies in laser plasmas. When the intense short pulse laser interacts with a solid target, strong oscillating electric fields are generated in the target. Then, electrons oscillate with relativistic velocity in plasma and oscillating electrostatic fields are generated. The electrostatic field generations are due to: (1) the charge separation by the electron quiver motion in the steep density gradient on the solid target surface, (2) resonant excitation of electron plasma waves near the critical density region, (3) longitudinal electron motion caused by the $\boldsymbol{E} \times \boldsymbol{B}$ force in the relativistic intensity laser field, (4) electron plasma wave excitation by the Stimulated Raman scattering and/or the self-modulation of the laser beam, and (5) stochastic electron acceleration by the crossing laser beams, and so on.

Once high density and high energy electrons with MeV are generated (Beg et al., 1997; Haines, 2019) and injected into a solid and/or expanded into vacuum, high-energy bremsstrahlung x-rays and high energy ions due to the strong sheath fields are generated. The strong sheath field reaches teravolt/m which delivers protons and ions with broad band energy distribution in the MeV range (Borghesi, 2014). The accelerated ion bunches can then be used for neutron generation (Alejo et al., 2016) by inserting a converter foil behind the accelerator foil or choosing a target material in which neutron generation occurs in tandem with the ion acceleration. These micro-scale particle beam sources are an interesting innovation in accelerator technology. They offer a new versatile source that is highly flexible and responsive in the delivery of a choice of neutron beams.

There are many publications on various acceleration mechanisms of the laser-driven ion beam (Bolton et al., 2018): (1) target normal sheath acceleration (TNSA) (Borghesi et al., 2006; Clark et al., 2000; Cowan et al., 2004; Gitomer et al., 1986; Higginson et al., 2018; Logan et al., 2006; Roth et al., 2002, 2009; Snavely et al., 2000; Yogo et al., 2018), (2) laser breakout afterburner (BOA) (Albright et al., 2007; Hegelich et al., 2002; Yin et al., 2007), (3) Coulomb explosion (Bulanov et al., 2002; Ditmire et al., 2000; Nishihara et al., 2001; Sakabe et al., 2004), (4) radiation pressure acceleration (RPA) (Esirkepov et al., 2004; Henig et al., 2009; Robinson et al., 2004), and so on. See also Part IV.

As shown in Fig. 18.3a, the TNSA has been the appropriate pitcher for the laser driven neutron source, because of more relaxed experimental requirements. The image of the TNSA driven neutron source is shown in Fig. 18.3a. Here, the intense short pulse laser is focused onto the front side of a thin foil with high intensity ($I_L > 10^{18}$ W/cm^2) to generate MeV electrons. The MeV electrons expand into both sides of the foil to generate the sheath fields, as described in Chap. 4.

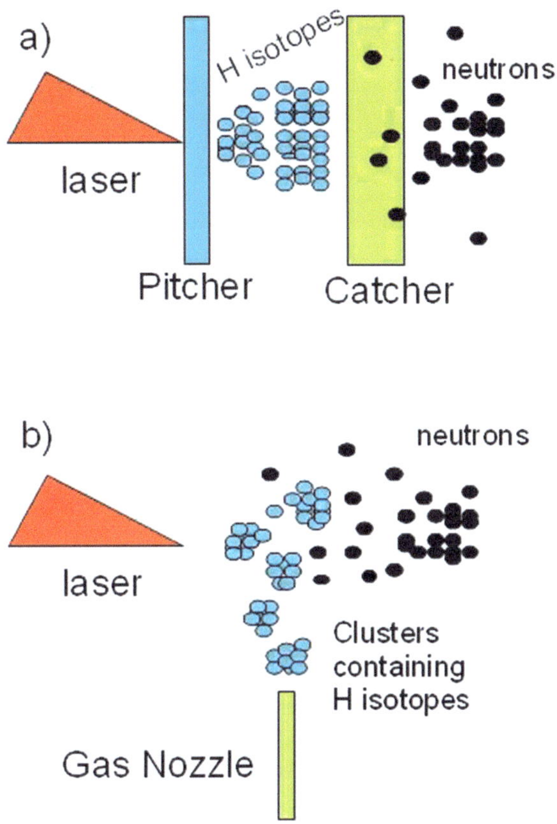

Fig. 18.3 Schematic of **a** Pitcher-Catcher neutron source with TNSA and **b** cluster fusion

The strong sheath electrostatic fields (TV/m) on both front and rear surface accelerate ions on the target surface. Ions accelerated from the rear surface are more preferred due to their better beam properties in terms of directionality and bunch density. The duration of the ion bunch is related to the duration of the electrostatic field at the rear side which depends on the duration of the laser pulse and the motion of the electrons. Thus, the ion beam pulse duration can be as short as a few picoseconds near the pitcher surface. The TNSA ions have broad energy spectra extended up to several tens of MeV, where the peak ion energy and the particle flux depend on various laser and target parameters (Zeil, et al., 2010).

18.2.2 Neutron Sources Driven by Laser Driven Ion Beam

There are numerous examples of laser ion beam driven fusion and neutron generation. For example, the reader is referred to the reviews (Alejo et al., 2016, 2017; Ledingham & Galster, 2010). The most representative examples rely on D samples:

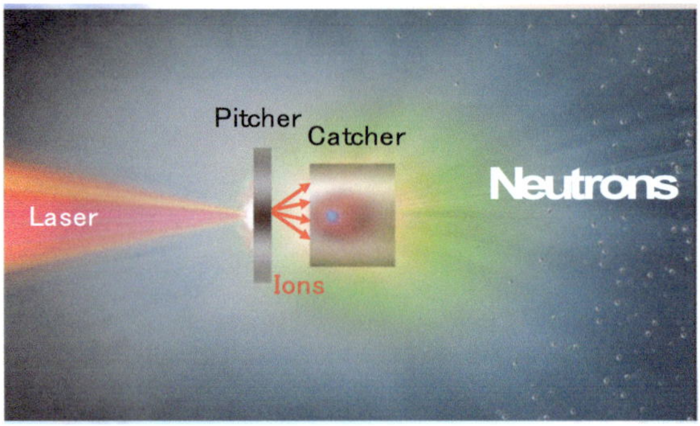

Fig. 18.4 Pitcher-Catcher scheme of laser driven neutron source

solid CD_2 target (Youssef et al., 2005), deuterated plastic targets (Izumi et al., 2002), and heavy water (D_2O) sprayed target (Ter-Avetisyan et al., 2005). The H or D ion beams accelerated by TNSA impact on a catcher target, where the nuclear reaction and neutron production take place. This scheme, called a "pitcher-catcher set-up" (see Fig. 18.4), can sometimes be substituted by a single thicker target in which both the ion acceleration and nuclear reactions take place in the same material but at different depths (Willingale et al., 2011). The intense laser driven neutron sources by the pitcher-catcher scheme have been demonstrated by Roth et al. (2013), Ker et al. (2016), etc.

For gaseous or cluster targets with D content, Coulomb explosion plays a dominant role in acceleration of ions which trigger the neutron generation by interaction with other D ions and neutral atoms from neighboring clusters (see Fig. 18.3b) (Ditmire et al., 2000; Madison et al., 2004; Parks et al., 2001). Experimentally, neutron production by laser irradiation of clusters has reached 5.5×10^6/shot neutrons (Lu et al., 2009).

Since the TNSA ion beam is well collimated in the forward direction, the ion beam irradiates a small spot of the order of millimeter diameter on the surface of the catcher. The catcher is a few millimeter block of Li, Be, CD_2, D_2O, deuterium ice, and so on for sub-10 MeV ion beams. By attaching a moderator of a few cm to the rear side of the catcher target, the catcher works also as a compact moderator. Furthermore, the TNSA ion beam pulse duration is sub-nano second in the front of the catcher. So, the source size of the laser neutron source could be not only small but also the neutron pulse duration near the moderator could be short. The short pulse and small size are the most important and most attractive points of the laser driven neutron source, since it is possible to realize high resolution neutron spectroscopy with a relatively short beam line.

18.2.3 Applications of Laser Driven Neutron Source

In the large-scale spallation neutron source, the pulse duration of the thermal neutrons is of the order of 100 ms. As a result, approximately equal to 100 m the TOF distance is required for the 0.1% energy resolution. On the other hand, the pulse duration of the thermal neutron driven by laser is of the order of 1 ms. So, the TOF distance is required to be order of 1 m. This means that the required neutron flux is reduced by four orders of magnitude.

A typical application of the high-resolution neutron spectroscopy is the neutron reflectometry. As shown in Fig. 18.5, the nano-meter scale structure of the surface can be measured by the reflection angle which depends upon the neutron energy as

$$\lambda_z = \frac{\lambda}{\sin\theta}, \tag{18.1}$$

where $\lambda = h/\sqrt{2m\epsilon_n}$ is the neutron de Broglie wavelength for the energy equal to ϵ_n, θ is the reflection angle, and h is the Planck constant. For a fixed angle θ the de Broglie wavelength depends upon the neutron energy according to Eq. (18.1). So, the reflected neutron spectrum is related to the depth distribution of the surface structure. As an example, the neutron reflectometry is applied to the Li-ion battery as shown in Fig. 18.6. Here, the Li density concentration distribution on the interface between electrodes and electrolytes can be measured by the reflectometry. The energy dependent transmission image and the neutron depth profiling with the neutron-Li nuclear reaction,

$$n + {}^6_3\text{Li} \rightarrow {}^4_2\text{He}(2.055\ \text{keV}) + {}^3_1\text{He}(2.727\ \text{keV}),$$

are also applied for the Li-ion battery characterization.

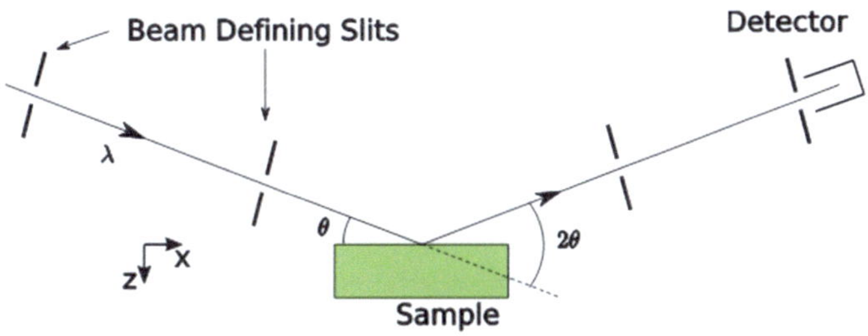

Fig. 18.5 Schematic layout of the neutron reflectometry

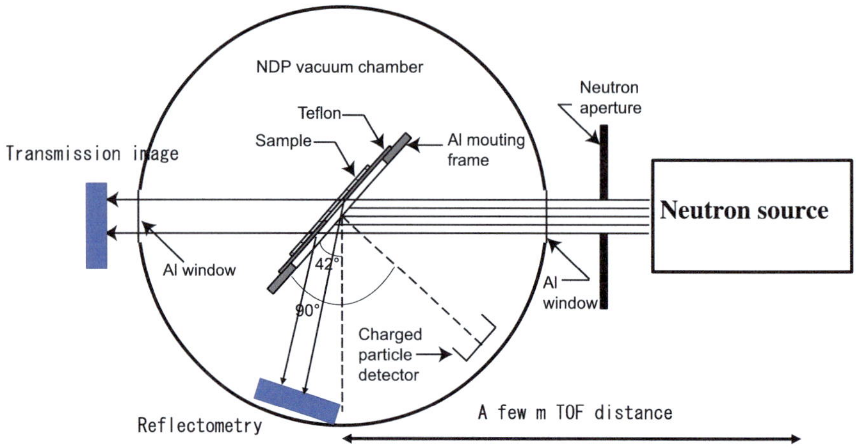

Fig. 18.6 Application of compact laser driven neutron source for Li-ion battery diagnostics

18.3 Present Status of Research on Laser Neutron Source

18.3.1 Projects in UK

In the last 5 years in the UK, the PLATINUM project has been explored to develop the key technology required for application of laser driven neutron source (Brenner et al., 2019). In this project, the Vulcan laser at the Science and Technology Facilities Council (STFC) Rutherford Appleton Laboratory has been used in collaboration with industry to assess the feasibility of using laser-driven neutron source for isotope inspection in nuclear waste barrels along with the in-line imaging capability using laser-driven x-rays (Musgrave et al., 2013).

The experiments have been also carried out by the DiPole laser which is operational at 10 Hz pulse rates. In the experiments, highly anisotropic, beam-like neutron emission with peak flux of the order of 10^9 n/sr was obtained from light nuclei reactions in a pitcher–catcher scenario, by employing MeV ions driven by a sub-petawatt laser (Kar et al., 2016). The spatial profile of the neutron beam, fully captured for the first time by employing a CR39 nuclear track detector, shows a FWHM divergence angle of ~ 70°, with a peak flux nearly an order of magnitude higher than the isotropic component elsewhere. The observed beamed flux of neutrons is highly favorable for a wide range of applications, and for further transport and moderation to thermal energies. A systematic study employing various combinations of pitcher–catcher materials indicates that the dominant reactions are $d(p, n + p)\,^1_1H$ and $d(d, n)\,^3_2He$. The observed anisotropy in the neutrons has been analyzed by numerical simulation. The spatial and spectral profiles are most likely related to the directionality and high energy of the projectile ions.

In the PLATINUM project, significant effort is dedicated to characterizing and optimizing the laser-driven concept of neutron source generation with a view to their use in scientific and industrial applications. Brenner et al. (2019) describe the overview of research in this field carried out in the last 5 years within the UK and the highlight key developments in enabling technology required for the exploitation of these beams for applications. This paper also introduces the activities of the PLATINUM project on collaboration with industry aiming to assess the feasibility of using laser-driven, beamed neutrons for isotope inspection in nuclear waste barrels along with the in-line imaging capability using laser-driven x-rays. With recent delivery of the high-peak-power laser technology operational at 10 Hz pulse rates, a laser-driven, multi-modal nuclear waste assay beamline installation within the waste management environment is envisioned.

18.3.2 Research in Germany

The group led by M. Roth of Technical University Darmstadt has been working on efficient high energy neutron generation. In 2012, experiments at the Trident laser at Los Alamos National Laboratory demonstrated the neutron yield of 10^{10}n/shot obtained by using 80 J pulses of 0.6 ps duration focused on 3 μm diameter spot (Jung et al., 2013; Roth et al., 2013). In 2013, the non-isotropic forward yield was further optimized with a cylindrical tungsten reflector around the beryllium converter as shown in Fig. 18.7. In 2015, the experiments at the PHELIX laser system at GSI in Darmstadt demonstrated the neutron yield of 2×10^{11}n/shot, where the laser pulse energy and the duration are 80 J and 0.45 ps, respectively (Roth et al., 2018).

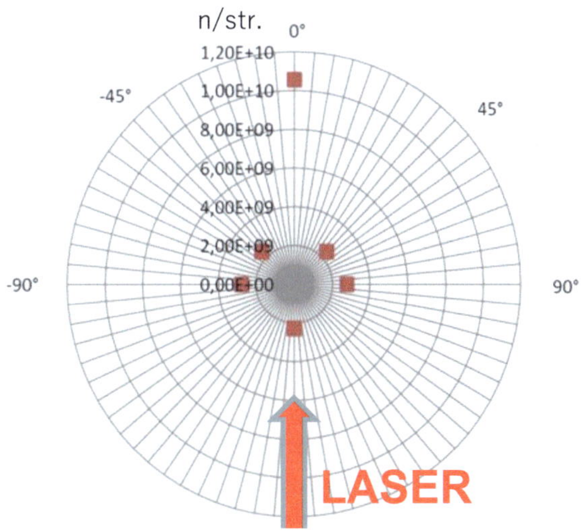

Fig. 18.7 The angular dependent neutron flux of the experiments by the Trident laser at Los Alamos National Laboratory USA. Reprinted from Roth et al. (2013), with permission from The American Physical Society and the author(s). © 2013 The American Physical Society

18.3.3 Research in the USA

The laser driven neutron sources have been explored in the USA at the University of Texas, the University of Michigan, and others. Experiments on the $^{7}Li(p,n)^{7}Be$ reaction with 60 J, 180 fs laser pulses have been carried out with the Texas Petawatt Laser Facility in Austin, Texas (Storm et al., 2013). Protons are accelerated by the target normal sheath acceleration mechanism. The neutrons were generated in the nuclear reactions caused by the subsequent proton bombardment of a pure lithium foil of natural isotopic abundance. The neutron energy ranged up to 2.9 MeV. The total yield was estimated to be 1.6×10^7 neutrons per steradian. An EUV light camera, used to image the target rear surface, correlated variations in the proton yield and the peak energy to target rear surface ablation. Calculations using the hydrodynamic code *FLASH* indicated that the ablation resulted from a laser pre-pulse of prolonged intensity. The ablation severely limited the proton acceleration and neutron yield.

At the University of Michigan, a heavy water stream was irradiated by a compact femto-second laser with high repetition rate as shown in the experimental set up of Fig. 18.8. The λ^3 laser system at the University of Michigan operates at a 0.5 kHz repetition rate and delivers up to 18 mJ energy per pulse on heavy water stream targets. The laser was focused to a 1.3 μm full width at half maximum focal spot, with the maximum peak intensity of 3×10^{19} W/cm^2. A 30 μm diameter stream of room temperature D_2O was delivered through a 20 μm diameter capillary tube. The deuterium nuclei from the vapor in the chamber also stopped deuterons laser-accelerated from the laser interaction, increasing the overall neutron yield (Hah et al., 2016, 2018). The deuteron energy spectra were measured with mylar filter stacks. The deuteron energies were measured up to 0.9 MeV, and the total number of deuterons above 0.4 MeV was comparable to the yield by a 2D PIC simulation (Hah et al., 2018).

The production of fast neutrons via deuteron acceleration and their subsequent d(d, n)^{3}He fusion reactions at a laser intensity of 3×10^{19} Wcm^{-2} were also demonstrated at the University of Michigan using the λ^3 laser system. Neutron fluxes were measured by the bubble detector shown in Fig. 18.8 at different incident laser energies. From the number of bubbles, the neutron flux was estimated and plotted in Fig. 18.9. At each laser pulse energy, 3–5 measurements were performed with different laser pulse energies from 6 to 12 mJ. The calibrated neutron flux varies from the order of 10^4 to 10^5 neutrons/s. Without an extra pitcher-catcher arrangement, accelerated deuterons collide with the deuterium nuclei in the heavy water stream ("prompt") and the vapor in the chamber ('delayed'), yielding a 2.45 ± 0.32 MeV energy neutron flux of 2.2×10^5 per second. The neutron average flux is comparable to other laser-based neutron sources (Ditmire et al., 1997, 1999; Giulietti et al., 2017; Hah et al., 2016, 2018; Lancaster et al., 2004; Yang et al., 2004; Zagar et al., 2005), although the neutron source presented here requires only a few mJ of pulse energy, which can be achieved with a compact system.

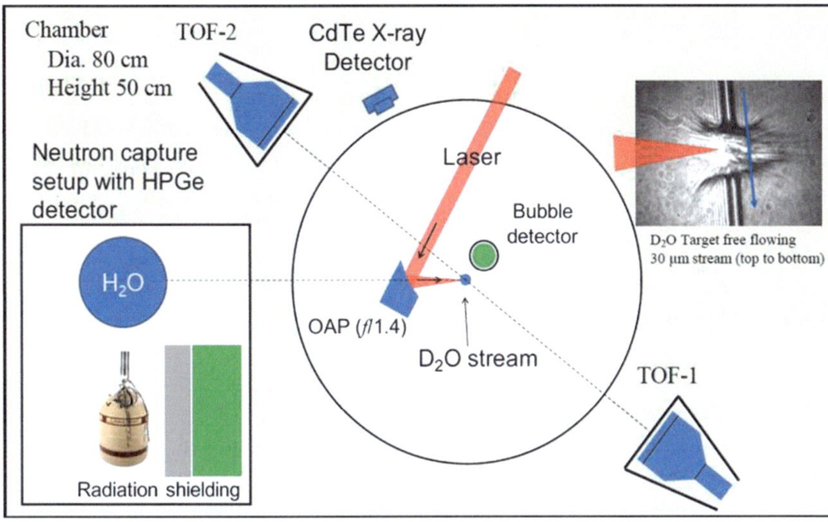

Fig. 18.8 Layout of the chamber and the detector for irradiation of a D_2O stream with the λ^3 laser system at the University of Michigan. Reprinted from Hah et al. (2018), with permission from IOP Publishing: permission conveyed through Copyright Clearance Center, Inc. © 2018 IOP Publishing

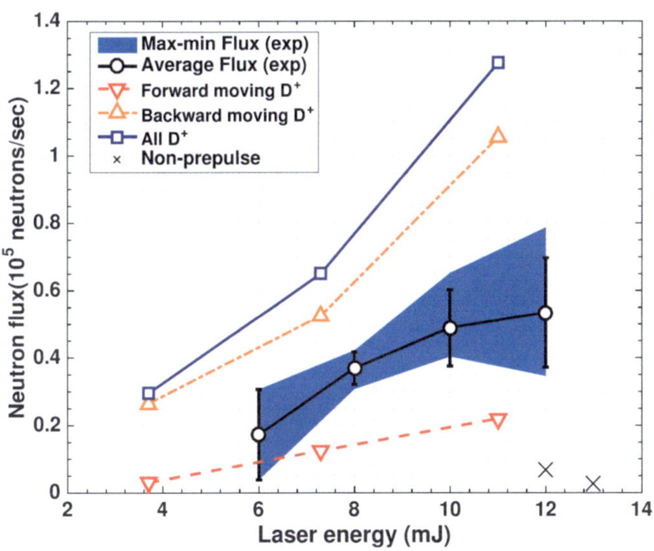

Fig. 18.9 The laser pulse energy dependence of the average neutron flux with the λ^3 laser system at the University of Michigan. Experimental data are shown in open circles, with their variations with blue shades. The neutron flux calculated with PIC simulation is shown for the neutrons generated with forward moving D^+ in the heavy water (red triangles) and with backward D^+ mainly in low-density vapor in the chamber (yellow triangles), with their sums with blue open squares. Reprinted from Hah et al. (2016). © 2016 AIP Publishing

18.3.4 Research in Japan

Research and development of laser driven neutron source was carried out for 5 years since 2016 as a joint project of the Institute of Laser Engineering (ILE) Osaka University, the Graduate School for the Creation of New Photonics Industries (in Hamamatsu City), Kyoto University, and Tohoku University under the support of Japan Science and Technology Agency.

At ILE, neutron generation and diagnostics were explored with the kJ, PW LFEX laser. A typical set-up of the experiment is shown in Fig. 18.10 (Mirfayzi et al., 2020). Thin film CD_2 polyethylene targets are irradiated with a 1.5 ps, 300 J laser pulse to generate deuteron and proton beam. A typical energy spectrum of the accelerated ions is shown in Fig. 18.11. Protons are from the contaminant layer on the rear side of the film, whose energy per nucleon is higher than twice the deuteron energy. The proton and deuteron beams are injected onto a Be catcher target. Although proton energy is higher than deuteron energy, the main contribution to neutron generation comes from the deuteron beam. The total neutron flux is higher than 10^{10} n/sr in average.

A compact cryogenic moderator was introduced to moderate the neutron energy (Mirfayzi et al., 2020). A solid hydrogen moderator was set near the rear of the catcher target as shown in Fig. 18.10. Here, the cold neutrons are detected by a ^{3}He proportional counter which is set in the laser irradiation direction. Figures 18.12a and b show the TOF signal measured at 8.2 m from the moderator and the energy spectrum of the thermal neutrons, respectively. Whereas Figs. 18.12c and d show the TOF measured with a He^3 proportional counter at 20 cm from the moderator and the energy spectrum of cold neutrons, respectively. In this work, the thermal and cold neutrons were generated as a small-spot and short-duration pulse by combining the

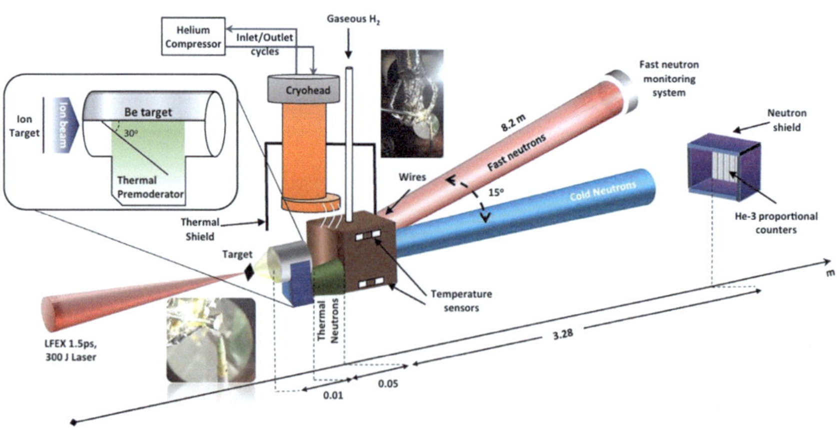

Fig. 18.10 Experimental set up for generation and measurement of the thermal and cold neutrons with LFEX laser at ILE, Osaka University. Reprinted from Mirfayzi et al. (2020). © 2020 Springer Nature

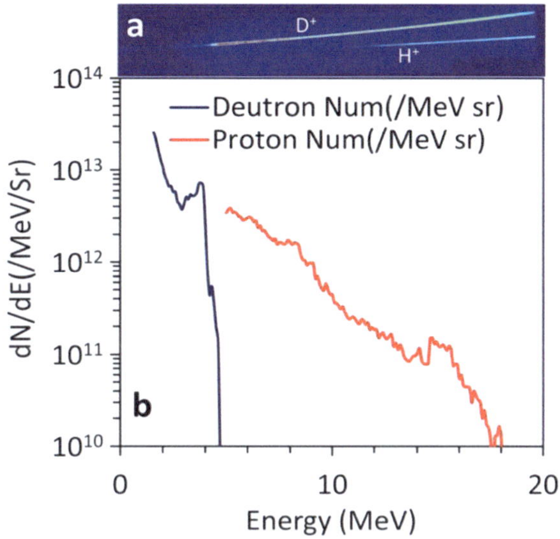

Fig. 18.11 Energy spectra of deuterons and protons generated from a thin CD_2 film irradiated with in a 300J/1/5 ps laser pulse. Reprinted from Mirfayzi et al. (2020). © 2020 Springer Nature

laser driven neutron generation and the compact cryogenic moderator. Such characteristic features of the thermal/cold laser driven neutron pulse will be applicable for compact high resolution neutron spectroscopy, such as neutron nuclear resonance measurement.

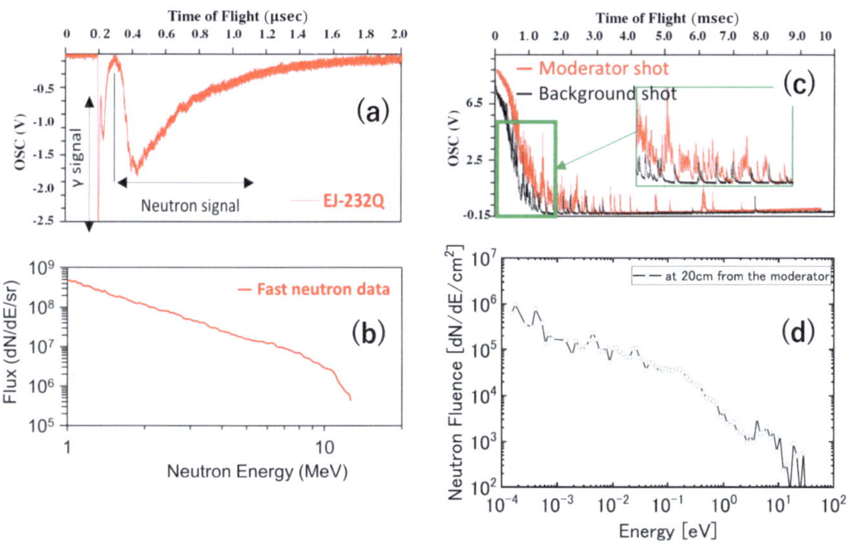

Fig. 18.12 Time of flight (TOF) and energy spectra of the thermal neutrons (**a** and **b**) and the cold neutrons (**c** and **d**), respectively. Reprinted from Mirfayzi et al. (2020). © 2020 Springer Nature

As a new field of material imaging, the neutron nuclear resonance diagnosis has attracted much attention in recent years. As shown in Fig. 18.13, the neutron energy dependent cross sections of the nuclear resonance absorption (NRA) have very narrow line widths and unique resonance energies for different materials. Therefore, it is possible by the neutron nuclear resonance absorption imaging to provide images of elemental distributions as well as to diagnose materials in heavy metal boxes, for example, for nuclear safeguards and security applications (Paradela, et al., 2017). Furthermore, the material dependent temperature in a heavy metal box could be measured by the Doppler broadening of the resonance absorption line shapes.

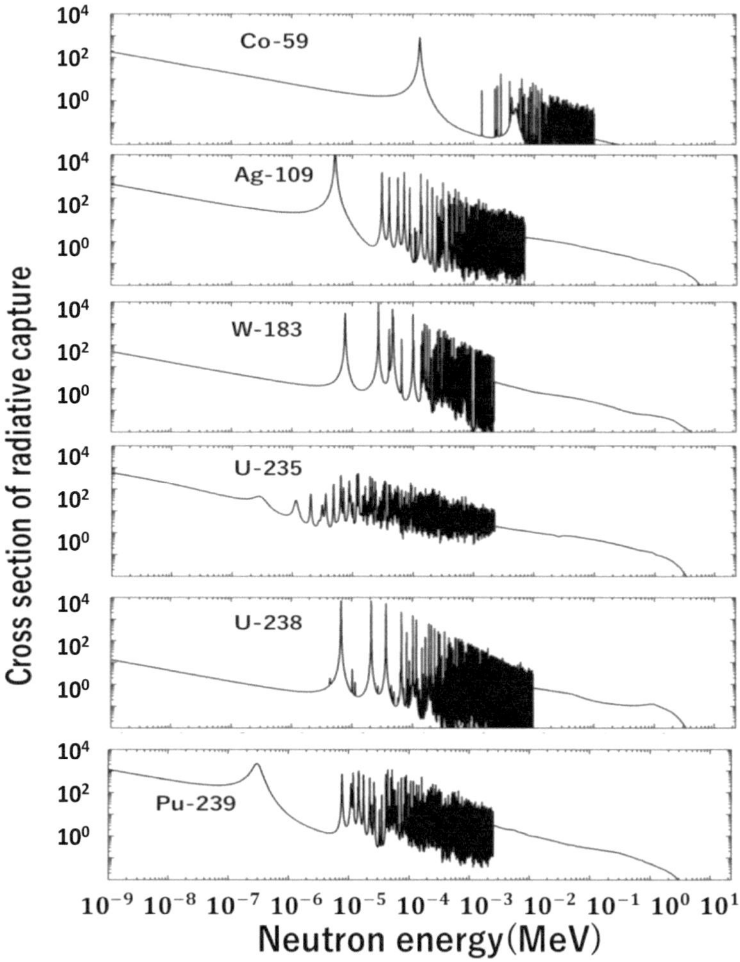

Fig. 18.13 Cross sections of neutron absorption of various nuclei. Reprinted from Paradela et al. (2017). © 2017 EDP Sciences

Although the accelerator driven neutron sources (ADNS) have already been applied to the neutron resonance absorption (NRA) measurements, it requires long n-TOF beam line because of the long pulse duration of ADNS due to large source size and long duration of the driver beam. On the other hand, the neutron pulse is much shorter in LDNS. In recent experiments at Osaka university, the pulse width of the epi-thermal neutron was measured to be a few microseconds.

Recently, measurement of the neutron nuclear resonance lines has been demonstrated with a relatively short TOF diagnostic system, less than 2 m for the thermal neutrons (Yogo et al., 2021; Mima et al., 2022; Yogo et al., 2023). The layout of the experiment is shown in Fig. 18.14. Here, a three-layered high Z sample made of In, Ag, and Ta each with 0.1 mm-thickness was set at1.8 m from the 3 cm compact moderator. Behind the sample, the time dependent neutron flux is measured by a Li-6 glass scintillator with PMT. With this arrangement, the NRA peaks at 5.19 eV of Ag, 4.28 eV of Ta, and 3.19 eV of In have been clearly measured. The n-TOF's of these resonance neutrons were about 57 μs, 62.5 μs, and 65.5 μs for Ag, Ta, and In, respectively.

The FWHM width of NRA lines can be determined by the n-TOF absorption-line-measurements. For example, the measured Ta n-TOF line extends from 61.5 μs to 63.5 μs as shown in Fig. 18.12 for the neutron-resonance energy of 4.28 eV. From this, the full width of the n-TOF signals is estimated to be about $\Delta\tau = 2.0\,\mu$s. This means that the energy spread of the neutrons at the sample is $\Delta\mathcal{E}/\mathcal{E} = 2.0/62.5 = 0.03$ at 4.28 eV. Note that the intrinsic resonance line width of NRA, namely the width of the relevant nuclear energy levels, is much narrower than this energy width. This

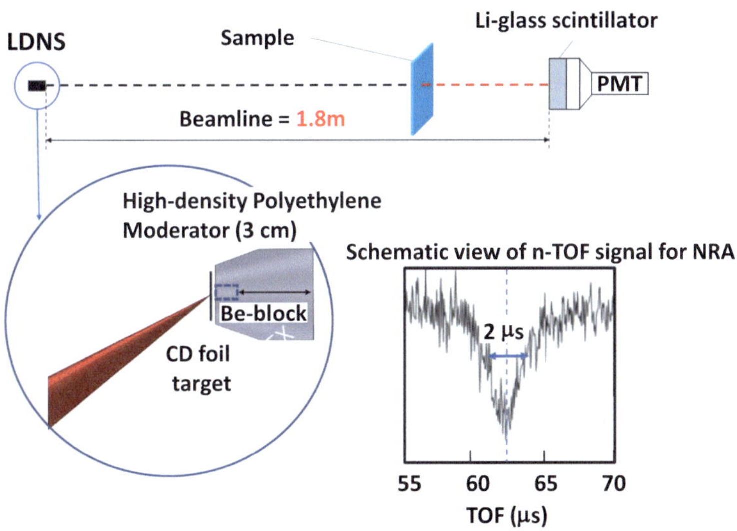

Fig. 18.14 Schematic layout of the NRA imaging experiment by LDNS and an illustration of the n-TOF signal

energy width is the neutron energy spread at a given TOF, which depends not only on the moderator size but also on the neutron collimation of the transport system.

The neutron pulse width at the moderator surface $\Delta\tau$ is estimated as

$$\Delta\tau = \sqrt{\tau_n^2 + (\Delta L/v_n)^2},$$

where τ_n is of the order of 1 ns for the fast neutron pulse entering the moderator of ΔL is the size of the moderator flying with velocity v_n. Since the epithermal neutron velocity is $v_n \approx 3 \times 10^{14}$ m/s, the moderator size is $\Delta L \approx 0.03$, and $\Delta\tau \approx \Delta L/v_n \approx$ 1 ms. This width is approximately equal to the measured width but smaller by a factor of two. We conclude that the epithermal neutron pulse duration is shorter than 2 ms at the moderator surface and the epithermal neutron energy resolution $\Delta\mathcal{E}/\mathcal{E}$ at the sample located at 1.8 m from the source is $v_n\tau/L \approx 3\%$ in the present system. From Doppler broadening of neutron resonance absorption, temperature profiling of materials in a shielded environment has been demonstrated in a single laser shot (Lan et al., 2024). Note here that the NRA experiments with LDNS were also reported from the GSI group at the Nuclear Photonics Conference NP2021.

References

Albright, B. J., Yin, L., Bowers, K. J., Hegelich, B. M., et al. (2007). Relativistic Buneman instability in the laser breakout afterburner. *Physics of Plasmas, 14*, Article 094502.

Alejo, A., Ahmed, H., Green, A., Mirfayzi, S. R., et al. (2016). Recent advances of laser-driven neutron sources. *Il Nuovo Cimento C, 38C*, 188.

Alejo, A., Krygier, A. G., Ahmed, H., Morrison, J. T., et al. (2017). High flux, beamed neutron sources employing deuteron-rich ion beams from D_2O-ice layered targets. *Plasma Physics and Controlled Fusionn, 59*, Article 064004.

Alvarez, J., Fernandez-Tobias, J., Mima, K., et al. (2014). Laser driven neutron sources: Characteristics, applications and prospects. *Physics Procedia, 60*, 29–38.

Beg, F. N., Bell, A. R., Dangor, A. E., et al. (1997). A study of picosecond laser-solid interactions up to 10^{19} W cm^{-2}. *Physics of Plasmas, 4*, 447.

Bolton, P. R., Pardoni, K., & Schreiber, J. (2018). *Part I: Laser-driven particle acceleration and associated energetic photon and neutron generation: Current understanding and basic capability, Part II: Applications of laser-driven particle acceleration.* CRC Press, Taylor & Francis Group. http://www.crcpress.com

Borghesi, M. (2014). Laser driven ion acceleration: State of art. *Nuclear Instruments and Methods in Physics Research, 740*, 6.

Borghesi, M., Fuchs, J., Bulanov, S. V., Mackinnon, A. J., et al. (2006). Fast ion generation by high-intensity laser irradiation of solid targets and applications. *Fusion Science and Technology, 49*, 412.

Brenner, C. M., Mirfayzi, S. R., Rusby, D. R., et al. (2016). Laser-driven x-ray and neutron source development for industrial applications of plasma accelerators. *Plasma Physics and Controlled Fusion, 58*, Article 014039.

Brenner, C. M., Mirfayzi, S. T., Alejo, A., et al. (2019). A review of recent UK activities on laser-driven neutrons for industrial applications. *The Review of Laser Engineering, 46*, 665.

Bulanov, S., Esirkepov, T., Khoroshkov, V. S., Kuznetsov, A. V., & Pegoraro, F. (2002). Oncological hadron therapy with laser ion accelerators. *Physics Letters A, 299*, Article 240247.

Clark, E. L., et al. (2000). Measurements of energetic proton transport through magnetized plasma from intense laser interactions with solids. *Physical Review Letters, 84*, 670.

Cowan, T., et al. (2004). Ultralow emittance, multi-MeV proton beams from a laser virtual-cathode plasma accelerator. *Physical Review Letters, 92*, Article 204801.

Danson, C., Hillier, D., Hopps, N., & Neely, D. (2015). Petawatt class lasers worldwide. *High Power Laser Science and Engineering, 3*, E3.

Danson, C., Haefner, C., Bromage, J., et al. (2019). Peta watt and exawatt class lasers worldwide. *High Power Laser Science and Engineering, 7*, E54.

Ditmire, T., Tisch, J. W. G., Springate, E., et al. (1997). High-energy ions produced in explosions of superheated atomic clusters. *Nature, 386*, 54.

Ditmire, T., Zweiback, J., Yanovsky, V., et al. (1999). Nuclear fusion from explosions of femtosecond laser-heated deuterium clusters. *Nature, 398*, 489.

Ditmire, T., Donnelly, T., Rubenchik, A. M., et al. (2000). Nuclear fusion in gases of deuterium cluster with a femtosecond laser. *Physics of Plasmas, 7*, 1993.

Esirkepov, T. Zh., Borgeshi, M., Bulanov, S. V., Mourou, G., & Tajima, T. (2004). Highly efficient relativistic-ion generation in the laser-piston regime. *Physical Review Letters, 92*, Article 175003.

Gitomer, S. J., et al. (1986). Fast ions and hot electrons in the laser–plasma interaction. *Physics of Fluids, 29*, 2679.

Giulietti, D., Boutoux, G., Ache, M., et al. (2017). D+D fusion reactions in 10^{18} W/cm^2 intensity and repetitive laser-plasma interactions. *Europhysics Letters, 119*, 65001.

Hah, J., Petrov, G. M., Nees, J. A., et al. (2016). High repetition-rate neutron generation by several-mJ, 35 fs pulses interacting with free-flowing D_2O. *Applied Physics Letters, 109*, Article 144102.

Hah, J., Nees, J. A., Hammig, M. D., et al. (2018). Characterization of a high repetition-rate laser-driven short-pulsed neutron source. *Plasma Physics and Controlled Fusion, 60*, Article 054011.

Haines, B. M., Olson, R. E., Sweet, W., et al. (2019). Robustness to hydrodynamic instabilities in indirectly driven layered capsule implosions. *Physics of Plasmas*, *26*, 012707.

Hegelich, M., Karsch, S., Pretzler, G., et al. (2002). MeV ion jets from short-pulse-laser interaction with thin foils. *Physical Review Letters, 89*, Article 085002.

Henig, A., Steinke, S., Schnrer, M., et al. (2009). Radiation-pressure acceleration of ion beams driven by circularly polarized laser pulses. *Physical Review Letters, 103*, Article 245003.

Higginson, A., Gray, R. J., King, M., et al. (2018). Near-100 MeV protons via a laser-driven transparency enhanced hybrid acceleration scheme. *Nature Communications, 9*, 724.

Izumi, N., Sentoku, Y., Habara, H., et al. (2002). Observation of neutron spectrum produced by fast deuterons via ultra-intense laser plasma interactions. *Physical Review E, 65*, Article 036413.

Jung, D., Falk, K., Guler, N., et al. (2013). Characterization of a novel, short pulse laser-driven neutron source. *Physics of Plasmas, 20*, Article 056706.

Kar, S., Green, A., Ahmed, H., et al. (2016). Beamed neutron emission driven by laser accelerated light ions. *New Journal of Physics, 18*, Article 053002.

Lan, Z., Arikawa, Y., Mirfayzi, S. R., et al. (2024). Single-shot laser-driven neutron resonance spectroscopy for temperature profiling. *Nature Communications, 15*, 5365. https://doi.org/10.1038/s41467-024-49142-y

Lancaster, K. J., Karsch, S., Habara, H., et al. (2004). Characterization of 7Li(p,n)7Be neutron yields from laser produced ion beams for fast neutron radiography. *Physics of Plasmas, 11*, 3404.

Ledingham, K. W. D., & Galster, W. (2010). Laser-driven particle and photon beams and some applications. *New Journal of Physics, 12*, Article 045005.

Logan, B. G., et al. (2006). Assessment of potential for ion-driven fast ignition. *Fusion Science and Technology, 49*, 399.

Lu, H. Y., Liu, J. S., Wang, C., et al. (2009). Efficient fusion neutron generation from heteronuclear clusters in intense femtosecond laser fields. *Physical Review A*, *80*, 051201(R).

Madison, K. W., Patel, P. K., Price, D., Edens, A., Allen, M., Cowan, T. E., Zweiback, J., & Ditmire, T. (2004), Fusion neutron and ion emission from deuterium and deuterated methane cluster plasmas. *Physics of Plasmas*, *11*, 270.

Mima, K., Nakai, S., Kato, Y., & Kiyanagi, Y. (2018a). Preface to the special issue on status of research and development of laser driven neutron sources. *Review of Laser Engineering, 46*, 558.

Mima, K., Yogo, A., Mirfayzi, S. R., et al. (2022). Laser-driven neutron source and nuclear resonance absorption imaging at ILE, Osaka University: Review. *Applied Optics, 81*, 2398.

Mirfayzi, S. R., Yogo, A., Lan, Z., Mima, K., et al. (2020). Proof-of-Principle experiment for laser-driven cold neutron source. *Scientific Reports, 10*, 20157.

Musgrave, I., Boyle, A., Carroll, D., et al. (2013). Recent developments on the Vulcan high power laser facility. *Proceedings of SPIE, 8780*, 878003.

Nishihara, K., Amitani, H., Murakami, M., et al. (2001). High energy ions generated by laser driven Coulomb explosion of cluster. *Nuclear Instruments and Methods in Physics Research Section A, 464*, 98.

Paradela, C., Heyse, J., Kopecky, S., et al. (2017). Neutron resonance analysis for nuclear safeguards and security applications. *EPJ Web of Conferences, 146*, 09002. https://doi.org/10.1051/epjconf/201714609002

Parks, P. B., Cowan, T. E., Stephens, R. B., & Campbell, E. M. (2001). Model of neutron-production rates from femtosecond-laser–cluster interactions. *Physical Review A, 63*, 063203.

Passoni, M., Arioli, F. M., Cialfi, L. C., et al. (2020). Advanced laser-driven neutron sources and their applications in materials and nuclear science. *Plasma Physics and Controlled Fusion, 62*, Article 014022.

Robinson, A. P. L., Gibbon, P., Zepf, M., Kar, S., et al. (2004). Relativistically correct hole-boring and ion acceleration by circularly polarized laser pulses. *Plasma Physics and Controlled Fusion, 51*, Article 024004.

Roth, M., Blazevic, A., Geissel, M., et al. (2002). Energetic ions generated by laser pulses: A detail study on target properties. *Physical Review Accelerators and Beams, 5*, Article 061301.

Roth, M., et al. (2009). Review on the current status and prospects of fast ignition in fusion targets driven by intense laser generated proton beams. *Plasma Physics and Controlled Fusion, 51*, Article 014004.

Roth, M., Jung, D., Falk, K., et al. (2013). Bright laser driven neutron source based on the relativistic transparency of solid. *Physical Review Letters, 110*, Article 044802.

Roth, M. (2018). *Applications of laser-driven particle acceleration* (Bolton, P., Parodi, K., & Schreiber, J., Eds.) (Chap. 6, p. 93). CRC Press.

Sakabe, S., Shimizu, S., Hashida, M., et al. (2004). Generation of high-energy protons from the Coulomb explosion of hydrogen clusters by intense femtosecond laser pulses. *Physical Review A, 69*, Article 023203.

Snavely, R. A., Key, M. H., Hatchett, S. P., Cowan, T. E., et al. (2000). Intense high-energy proton beams from petawatt-laser irradiation of solid. *Physical Review Letters, 85*, 2945.

Storm, M., Jiang, S., Wertepny, D., Orban, C., Morrison, J., Willis, C., McCary, E., Belancourt, P., Snyder, J., Chowdhury, E., Bang, W., Gaul, E., Dyer, G., Ditmire, T., Freeman, R. R., Akli, K. (2013). Fast neutron production from lithium converters and laser driven protons. *Physics of Plasmas, 20*, 053106. https://doi.org/10.1063/1.4803648

Tel-Avetisyan, S., Schnuerer, M., Hilscher, U., et al. (2005). Fusion neutron yield from a laser-irradiated heavy-water spray. *Physics of Plasmas, 12*, Article 012702.

Willingale, L., Petrov, G. M., Maksimchuk, A., et al. (2011). Comparison of bulk and pitcher-catcher targets for laser-driven neutron production. *Physics of Plasmas, 18*, Article 083106.

Yang, J. M., McKenna, P., Ledingham, K. W. D., et al. (2004). Neutron production by fast protons from ultraintense laser-plasma interactions. *Journal of Applied Physics, 96*, 6912.

Yin, L., Albright, B. J., Hegelich, B. M., et al. (2007). Monoenergetic and GeV ion acceleration from the laser breakout afterburner using ultrathin targets. *Physics of Plasmas, 14*, Article 056706.

Yogo, A., Hanayama, R., et al. (2018). Developments of laser-driven neutron source based on "pitcher-catcher" method. *Review of Laser Engineering, 46*, 582.

Yogo, A., Mirfayzi, S. R., Arikawa, Y., et al. (2021). Single shot radiography by a bright source of laser-driven thermal neutrons and x-rays. *Applied Physics Express, 14*, 106001.

Yogo, A., Lan, Z., Arikawa, Y., et al. (2023). Laser-driven neutron generation realizing single-shot spectroscopy. *Physical Review X, 13*, Article 011011.

Youssef, A., Habara, H., Tanaka, K. A., et al. (2005). Broad-range neutron spectra identification in ultra-intense laser interactions with carbon-deuterated plasmas. *Physics of Plasmas, 12*, Article 110703.

Zagar, T. J., Galy, J., Magill, J., & Kellett, M. (2005). Laser-generated nanosecond pulse neutron sources: scaling form VULCAN to table-top. *New Journal of Physics, 7*, 253.

Zeil, K., Bock, S., Bussmann, M., et al. (2010). The scaling of proton energies in ultrashort pulse laser plasma acceleration. *New Journal of Physics, 12*, Article 045015.

Part IV
High Field Science with Lasers

Chapter 19
Relativistic Plasmas: Dimensionless Parameters

Abstract Various regimes of the high-power laser plasma interaction with matter leading to formation of relativistic plasma are described by dimensionless parameters which are presented below.

Keywords Relativistic laser plasma interactions · Key parameters · Electron in electromagnetic wave · Dimensionless parameters · Principle of qualitative scaling

Physical systems obey scaling laws, which can also be presented as similarity rules. In the theory of similarity and modelling, the key role is played by dimensionless parameters, which characterize the phenomena we consider.

The intensity of an electromagnetic wave is defined by its electric field through the expression: $I = (c/4\pi)E^2$, where E is the electric (and B is the magnetic field) field in the wave. The intensity is equal to the value of the Poynting vector

$$\boldsymbol{S} = \frac{c}{4\pi}\boldsymbol{E} \times \boldsymbol{B}. \tag{19.1}$$

The power of an electromagnetic wave is equal to the Poynting vector integral over the wave beam transverse cross section $\sum$,

$$\mathcal{P}_{las} = \frac{c}{4\pi}\oint_{\Sigma} ([\boldsymbol{E} \times \boldsymbol{B}] \cdot \boldsymbol{n})\, d\boldsymbol{\Sigma} \tag{19.2}$$

where $\boldsymbol{n}$ is a unit vector normal to the surface $\sum$. The time integral of the power over the interval equal to the laser pulse duration gives the electromagnetic (laser) pulse energy, $\mathcal{E}_{las} = \int_{-\infty}^{+\infty} \mathcal{P}_{las} dt = \overline{\mathcal{P}}_{las}\tau_{las}$. Here $\overline{\mathcal{P}}_{las}$ and τ_{las}, are the average power and pulse duration. Other important parameters are the pulse frequency, ω, which is related to its wavelength, $\lambda = 2\pi c/\omega$, and the electromagnetic pulse polarization.

Y. Kato et al., *High Power Laser and Plasma Science*, Springer Series on Atomic, Optical, and Plasma Physics 130, https://doi.org/10.1007/978-3-032-08930-4_19

19.1 Charged Particle Dynamics

The probabilities of the processes involving extremely high intensity electromagnetic field interaction with electrons, positrons and photons are determined by several parameters.

The first of the dimensionless parameters, which characterizes the electromagnetic wave packet, is the ratio of the pulse length, $l_{las} = c\tau_{las}$, to the radiation wavelength, λ. We shall denote this ratio as

$$N_p = l_{las}/\lambda. \tag{19.3}$$

It is equal to the number of wavelengths per pulse and is Lorentz invariant.

When the normalized dimensionless electromagnetic wave amplitude

$$a_0 = \frac{eE_0}{m_e \omega c} = \frac{eE_0 \lambda}{2\pi m_e \omega c} \tag{19.4}$$

exceeds unity, the energy of the electron quivering in the field of the wave, $eE_0\lambda/2\pi$, becomes relativistic (Mourou et al., 2006; Marklund & Shukla, 2006; Di Piazza et al., 2012; King & Heinzl, 2016; Gonoskov et al., 2022; Yu et al., 2024). The electromagnetic wave intensity I_{las} expressed via the normalized amplitude, a_0, is given by

$$I_{las} = \frac{\pi}{2}\frac{a_0^2}{\lambda^2}\frac{m_e c^3}{r_e} \approx 1.37 \times 10^{18} a_0^2 \left(\frac{1\,\mu\text{m}}{\lambda}\right)^2 \frac{\text{W}}{\text{cm}^2}, \tag{19.5}$$

where $r_e = e^2/m_e c^2 \approx 2.8 \times 10^{-13}$cm is the classical electron radius. If the wave is focused into a one wavelength spot, this intensity corresponds to the laser power

$$\mathcal{P}_{las} = a_0^2 \frac{m_e^2 c^5}{2e^2} \approx a_0^2 \times 21 \text{ GW}. \tag{19.6}$$

In a plane electromagnetic wave the parameter is related to the Lorentz and gauge invariant, which being expressed via the 4-potential of the electromagnetic field, A^μ, is equal to $a_0 = e\sqrt{A^\mu A_\mu}/m_e c^2$. Here and below, a summation over repeating indices, $\mu = 0, 1, 2, 3$, is assumed.

The relativistic electron interacting with an electromagnetic wave emits high-energy photons. Here, for the sake of simplicity, we analyse the dynamics of a radiating electron in a homogeneous rotating electric field. The rotating electric field can correspond to the nodes of two colliding electromagnetic waves, where the wave magnetic field vanishes, and/or to the electron interaction with an electromagnetic wave in near-critical plasmas in the frame of reference moving with the group velocity of the wave.

19.2 Radiation

For the ultrarelativistic electron rotating with the frequency ω the power emitted is proportional $\gamma_e = a_0$ to the fourth power of its energy, $\mathcal{E}_e$, (see Landau & Lifshitz, 1980)

$$\mathcal{P}_\gamma = \varepsilon_{rad} m_e c^2 \omega \gamma_e^4. \tag{19.7}$$

This case, in particular, corresponds to the regime of nonlinear Thomson scattering. In Eq. (19.7) the dimensionless parameter,

$$\varepsilon_{rad} = \frac{4\pi r_e}{3\lambda} = 1.17 \times 10^{-8} \left(\frac{1\mu m}{\lambda}\right), \tag{19.8}$$

proportional to the ratio of the classical electron radius r_e and the electromagnetic wave wavelength λ, characterizes the role of radiation losses.

The maximal rate at which an electron can acquire the energy from the electromagnetic field is approximately equal to $m_e c^2 \omega a_0$. The condition of the balance between the acquired and lost energy for the electron Lorentz factor equal to $\gamma_e = a_0$ shows that the radiation friction effects become dominant at

$$a_0 > a_{rad} = \varepsilon_{rad}^{-1/3}, \tag{19.9}$$

i.e., at the electromagnetic wave intensity above (Zel'dovich and Illarionov, 1972)

$$I_R = \left(\frac{3}{2}\right)^{2/3} \frac{m_e^{8/3} c^5 \omega^{4/3}}{4\pi e^{10/3}} = 2.65 \times 10^{23} \left(\frac{1\,\mu\text{m}}{\lambda}\right)^{4/3} \frac{\text{W}}{\text{cm}^2}. \tag{19.10}$$

The characteristic frequency of the emitted radiation is proportional to the cube of the electron energy,

$$\omega_c \approx 0.3 \omega \gamma_e^3. \tag{19.11}$$

19.3 Quantum Effects

Using expressions (19.7) and (19.11) for total power emitted by the electron $\mathcal{P}_\gamma$ and for the emitted radiation frequency ω_c we can find the number of photons emitted during one period of the wave. It is approximately equal to the ratio of the power $\mathcal{P}_\gamma$ multiplied to the wave period $2\pi/\omega_0$ to the photon energy $\hbar\omega_c$, i.e.

$$N_\gamma = \frac{8\pi}{9} \alpha \gamma_e, \tag{19.12}$$

where $\alpha = r_e/\lambda_C = e^2/\hbar c \approx 1/137$ is the fine structure constant with $\lambda_C = \hbar/m_e c = r_e/\alpha \approx 3.86 \times 10^{-11}$ cm being the Compton scattering wavelength.

Quantum Electrodynamics (QED) effects become important, when the energy of the photon generated by Thomson (Compton) scattering is of the order of the electron energy,

$$\hbar\omega_c \approx m_e c^2 \gamma_e, \tag{19.13}$$

i.e. when the particle normalized energy is equal to $\gamma_e > \sqrt{m_e c^2/\hbar\omega}$. In the case of one micron wavelength laser, it is approximately equal to 10^3, i.e. the electron energy is of the order of 500 MeV. For $\gamma_e \approx a_0$ Eqs. (19.11) and (19.13) yield the quantum electrodynamics limit on the electromagnetic field amplitude,

$$a_0 > \sqrt{a_S}. \tag{19.14}$$

Here, the dimensionless parameter

$$a_S = \frac{cE_S\lambda}{2\pi m_e c^2} = \frac{m_e c^2}{\hbar\omega} = \frac{\lambda}{\lambda_C} = 4.2 \times 10^5 \left(\frac{\lambda}{1\mu m}\right) \tag{19.15}$$

is the normalized critical electric field of quantum electrodynamics,

$$E_S = \frac{m_e^2 c^3}{e\hbar}. \tag{19.16}$$

It is also known as the Sauter or Schwinger electric field (see Sauter, 1931). The electric field equal to E_S produces the work on the electron over the distance of the Compton wavelength, λ_C, equal to $m_e c^2$. Its value is equal to

$$E_S \approx 1.3 \times 10^{18} \mathrm{V/cm}. \tag{19.17}$$

The critical magnetic field, $B_S = m_e^2 c^2/e\hbar$, equals

$$B_S \approx 4.4 \times 10^9 \mathrm{T}. \tag{19.18}$$

The electromagnetic radiation intensity of the electromagnetic wave with the amplitude E_S is

$$I_S = \frac{m_e^4 c^7}{4\pi e^2 \hbar^2} = 2.36 \times 10^{29} \frac{\mathrm{W}}{\mathrm{cm}^2}. \tag{19.19}$$

The above obtained quantum electrodynamics limit given by Eq. (19.14), $a_0^2 > a_S$, corresponds to the condition $\chi_e > 1$, where χ_e is the relativistic and gauge invariant parameter given by

$$\chi_e = \frac{\sqrt{-(F^{\mu\nu}p_\nu)^2}}{E_S m_e c} \approx 2\frac{a_0}{a_S}\gamma_e. \tag{19.20}$$

It characterizes the probability of the gamma-photon emission by the electron with 4-momentum in the field of the electromagnetic wave. The 4-tensor of the electromagnetic field is defined as

$$F_{\mu\nu} = \partial_\mu A_\nu - \partial_\nu A_\mu, \tag{19.21}$$

where $A^\mu = (\varphi, \boldsymbol{A})$ is the four-vector potential with scalar φ and vector $\boldsymbol{A}$ potentials.

The quantum electrodynamics limit is reached for the electromagnetic wave intensity of the order of

$$I_Q = \frac{m_e^3 c^5 \omega}{8\pi e^2 \hbar} = 5.75 \times 10^{23} \left(\frac{1\,\mu\text{m}}{\lambda}\right) \frac{\text{W}}{\text{cm}^2}. \tag{19.22}$$

The Lorentz invariant dimensionless parameter

$$\chi_\gamma = \frac{\hbar\sqrt{-(F^{\mu\nu}k_\nu)^2}}{E_S m_e c} \approx a_0 \frac{\hbar^2 \omega\,\omega_\gamma}{m_e^2 c^4}. \tag{19.23}$$

determines the probability of the electron–positron pair creation by the photon with the energy $\hbar\omega_\gamma$ in the electromagnetic field given by $F_{\mu\nu}$ via the Breit–Wheeler process (Breit & Wheeler, 1934). The Breit–Wheeler pair production mechanism corresponding to the reaction

$$\gamma' + n\gamma = e^+ + e^- \tag{19.24}$$

is one of the most simple processes of the electron-positron pair generation in collision of the gamma-photon with the electromagnetic wave in the multi-photon process ($n \gg 1$).

In contrast with classical physics, in quantum theory the vacuum is not a void substance. It is filled with emerging and disappearing virtual particle-antiparticle pairs. The respond of virtual electron-positron pairs on the electromagnetic field action results in the electromagnetic field amplitude dependent refraction index, i.e. to the vacuum polarization, birefringence and dichroism. The electric field with the amplitude approaching the critical electric field of quantum electrodynamics given by Eq. (19.16) can create electron-positron plasma from vacuum.

Different regimes of electromagnetic interaction are realized depending on the parameters E/E_S, χ_γ, and χ_e. In the limit $E/E_S \to 1$ the electromagnetic waves can create electron–positron pairs from vacuum, the electromagnetic wave can interact via photon–photon scattering and vacuum polarization. When the parameter χ_e becomes large, the multi-photon Compton scattering results in the high-energy photon generation. For $\chi_\gamma > 1$ the multi-photon Breit–Wheeler, process leads to electron–positron pair generation via the gamma-photon interaction with the strong electromagnetic field. The nonlinear Thomson scattering regime is realized for $a_0 >> 1$ with the scattering cross section depending on the electromagnetic field amplitude. If $a_0 > (4\pi r_e/3\lambda)^{-1/3}$ radiation friction effects play a key role.

As one can see two dimensionless parameters, a_0 and χ_e, can be used to subdivide the (a_0, χ_e) plane into four domains shown in Fig. 19.1 where we express the parameter $\chi_e \approx (a_0/a_S)\gamma_e$ in the form $\chi_e \approx a_0^2/a_S$, i.e. assuming that the normalized electron energy equals $\gamma_e = a_0$. The $\chi_e = 1$ line divides the plane into the radiation reaction description of the interaction domain ($\chi_e < 1$) and quantum electrodynamics description of interaction domain ($\chi_e > 1$). The $a_0\varepsilon_{rad}^{1/3} = 1$ line divides the plane into radiation dominated ($a_0 > \varepsilon_{rad}^{-1/3}$) and non-dissipative ($a_0 < \varepsilon_{rad}^{-1/3}$) regimes of interaction domains. We note that the $a_0 = \varepsilon_{rad}^{-1/3}$ threshold comes from the requirement for an electron to emit the amount of energy per electromagnetic wave period equal to the energy gain from the electromagnetic wave during the wave period.

The comparison of expressions given by Eqs. (19.10) and (19.22) shows that for the electromagnetic radiation wavelength equal to $\lambda \approx 0.8\ \mu$m the intensities I_R and I_Q are of the same order of magnitude, corresponding to the intersection point of the lines $a_0\varepsilon_{rad}^{1/3} = 1$ and $\chi_e = 1$ in Fig. 19.1. They are equal at the frequency

$$\omega^* = \frac{e^4 m_e}{\hbar^3} = \frac{m_e c^2}{\hbar \alpha^2} \tag{19.25}$$

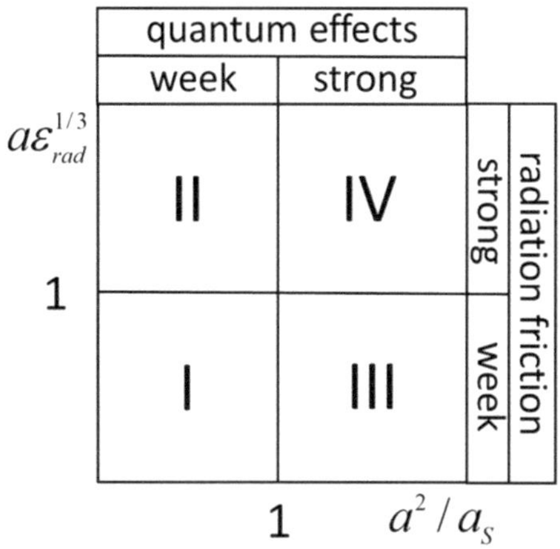

Fig. 19.1 Regimes of electromagnetic field interaction with matter on the plane of parameters: the normalized electromagnetic wave amplitude $a\varepsilon_{rad}^{1/3}$ and the parameter $a^2/a_S = \chi_e$. 4 domains correspond to (I) relativistic electron—electromagnetic field interaction with neither radiation friction nor quantum electrodynamics effects; (II) electron—electromagnetic wave interaction is dominated by radiation friction; (III) quantum electrodynamics effects are important with insignificant radiation friction effects; (IV) both quantum electrodynamics and radiation friction determine radiating charged particle dynamics in electromagnetic field

for normalized amplitude

$$a^* = \left(\frac{3c}{2r_e\omega^*}\right)^{1/3} = \frac{\hbar c}{e^2} = \frac{1}{\alpha}, \tag{19.26}$$

i.e. $a^* = 137$, for which the radiation intensity equals $I^* = 2.6 \times 10^{22}$ W/cm^2.

Here we would like to attract attention to the relationship between the well-known critical electric field of classical electrodynamics E_{cr}, the critical electric field of quantum electrodynamics E_S and the electric field E^*. They can be written as $E_{cr} = e/r_e^2$, $E_S = e/(r_e\lambda_C)$, and $E^* = e/\lambda_C^2$, respectively. In other words, we have $E_S = E_{cr}\alpha$, and $E^* = E_{cr}\alpha^2$.

When the electric field becomes of the order of the critical field of quantum electrodynamics the number of electron-positron pairs created in vacuum can be estimated considering the work produced by the electric field over the distance $\Delta x = \hbar/\Delta p \approx \hbar c/\Delta\mathcal{E}$ as follows from the Heisenberg uncertainty principle. It is equal to

$$\Delta\mathcal{E} = 2\mathcal{N}_{ep}m_ec^2 = \sqrt{eE\hbar c}$$

For the number of created electron-positron pairs we obtain $\mathcal{N}_{ep} = \Delta\mathcal{E}/2m_ec^2 = \sqrt{E/4E_S}$.

When the electric field approaches critical electric field of quantum electrodynamics, $E/E_S \to 1$, the electron-positron pair creation can be considered as a tunneling process from the "negative frequency" state with the energy

$$\mathcal{E}_- = -\sqrt{m_e^2c^4 + p^2(z)c^2} + eEz \tag{19.27}$$

to the "positive frequency" state with the energy

$$\mathcal{E}_+ = +\sqrt{m_e^2c^4 + p^2(z)c^2} + eEz \tag{19.28}$$

as it is illustrated in Fig. 19.2. For relatively week electric field $E/E_S < 1$ the sub-barrier tunneling occurs in the quasi-classical regime with the probability (Berestetskii, Lifshitz, and Pitaevskii, 1982)

$$w \approx \exp\left(-2\int_{z_2}^{z_1} |p(z)|dz\right) = \exp\left(-2\frac{m_e^2c^3}{e\hbar E}\int_{-1}^{1}\sqrt{1-\zeta^2}d\zeta\right), \tag{19.29}$$

where the barrier border coordinates z_1 and z_2 are determined by the condition $p(z) = 0$. This yields (we can see an analogy with the ionization process, Keldysh, 1965)

$$w \approx \exp\left(-\pi\frac{E_S}{E}\right). \tag{19.30}$$

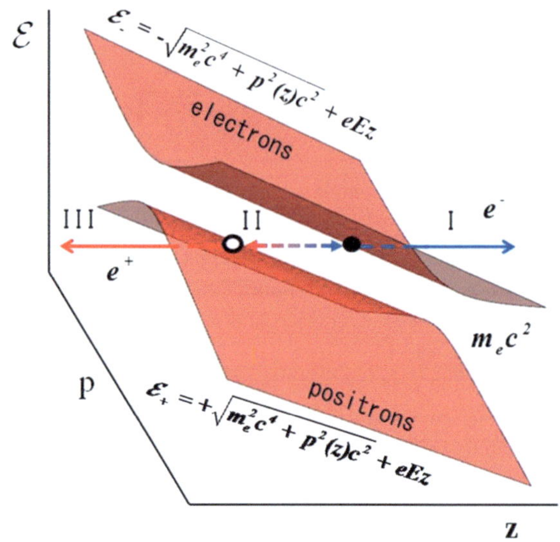

Fig. 19.2 Electron-positron pair creation from vacuum as a tunneling process in strong electric field

Creation of electron-positron pairs plays a very important role in a vast variety of processes in physics and astrophysics (see Dittrich and Gies (2000), and review article by Ruffini et al. (2009) and references therein).

19.4 Electromagnetic Waves in the Plasma

The electromagnetic wave behavior in plasma differs from its behavior in vacuum and depends on the electron density. In a plasma with a density n, a displacement of the electrons with respect to the ions generates the electric field. Its ratio to the laser electric field is

$$\frac{E}{E_0} = \frac{4\pi n e^2}{m_e \omega^2} = \left(\frac{\omega_{pe}}{\omega}\right)^2 = \frac{n}{n_{cr}}, \tag{19.31}$$

where

$$\omega_{pe} = \sqrt{\frac{4\pi n e^2}{m_e}}, \tag{19.32}$$

is the Langmuir frequency and

$$n_{cr} = \frac{m_e \omega^2}{4\pi e^2}, \tag{19.33}$$

is the critical electron density. The low amplitude electromagnetic wave does not propagate in the plasma with the density above critical.

The dimensionless parameter

$$\frac{\omega_{pe}}{\omega} = \sqrt{\frac{n}{n_{cr}}} = \sqrt{\frac{4\pi n e^2}{m_e \omega^2}}, \tag{19.34}$$

is a measure of the plasma collective response to a periodic electromagnetic field.

When an electromagnetic wave propagates through a plasma, its group velocity, $v_{gr} = \partial\omega/\partial k$, and phase velocity, $v_{ph} = \omega/k$, are not equal to each other in contrast to the case of vacuum, where $v_{gr} = v_{ph}$, and are related as $v_{gr} v_{ph} = c^2$. While in vacuum the dispersion equation for the frequency, ω, and wave vector $\boldsymbol{k}$, takes the form $\omega^2 = k^2 c^2$, in a plasma it becomes

$$\omega^2 = k^2 c^2 + \omega_{pe}^2. \tag{19.35}$$

This dispersion equation can be rewritten as becomes

$$kc = \sqrt{\omega^2 - \omega_{pe}^2}, \tag{19.36}$$

which shows that an electromagnetic wave with a frequency below the Langmuir frequency cannot propagate through the plasma, i.e. an overdense plasma with the electron density higher than the critical density is not transparent. The electromagnetic field evanescence length in a high-density plasma is of the order of the collisionless skin depth, $d_e = c/\omega_{pe}$. For a finite amplitude, circularly polarized electromagnetic wave the dispersion equation takes the form (Akhiezer & Polovin, 1956):

$$\omega^2 = k^2 c^2 + \frac{\omega_{pe}^2}{\sqrt{1 + a_0^2}} \tag{19.37}$$

showing that the effective critical density $n_{cr} = m_e \omega^2 (1 + a_0^2)/4\pi e^2$ increases as the electromagnetic wave amplitude grows, i.e., the plasma is more transparent to high intensity electromagnetic radiation. The dependence of the wave frequency on the amplitude results in a broad variety of such nonlinear processes accompanying the electromagnetic wave interaction with plasmas as the parametric instabilities, self-modulation and self-focusing and to the generation of various nonlinear entities as the relativistic solitons, vortices and wake waves.

The collective response of plasma, in addition to the transverse electromagnetic mode, exhibits longitudinal plasma oscillations, known as the Langmuir waves. The electric field in a Langmuir wave oscillates with frequency equal to ω_{pe} (see Eq. (19.32)). The group velocity of the Langmuir waves vanishes, $v_{gr} = \partial\omega/\partial k = 0$, and their phase velocity is determined by the wave number being equal to $v_{ph} = \omega_{pe}/k$.

Relativistic effects change the dispersion equation due to the dependence of the Langmuir frequency on the wave amplitude. The frequency of a longitudinal wave depends on its amplitude $a_0 = eE_0/m_e\omega_{pe}c$ as $\omega \approx \omega_{pe}(1 - 3a_0^2/4)$ for $a_L \ll 1$ and $\omega \approx \omega_{pe}/\sqrt{8a_0}$ as in the case of relativistically intense electromagnetic wave when $a_0 \gg 1$.

Finite amplitude laser radiation in a collisionless plasma is subject of several instabilities (Kruer, 1988). The stimulated Raman scattering instability develops on the electron time scale. Stimulated backward scattering results in the erosion of the laser pulse front and to anomalous laser pulse energy depletion (Bulanov et al., 1992). Stimulated forward Raman scattering leads to the self-modulation of the relatively long laser pulse with a modulation length of the order of the Langmuir wavelength, $2\pi c/\omega_{pe}$. If the electromagnetic wave packet length is of the order of or shorter than the Langmuir wave wavelength, it excites the wake plasma wave non-adiabatically, which is of great importance for the Laser Wake Field Acceleration (LWFA) paradigm formulated by Tajima and Dawson (1979) [for details see review article by Esarey et al., (2009)], within the framework of which the electron acceleration up to high energy occurs over the acceleration length substantially small compared the size of conventional accelerators.

The phenomenon of self-focusing (see the paper by Askar'yan (1962) as well as Part III and Chap. 9), which appears owing to the dependence of the refractive index of the medium on the intensity of the electromagnetic wave, is of fundamental importance for physics of nonlinear processes in the laser plasma. In the relativistic regime (Litvak, 1969), a laser pulse is self-focused because both of the relativistic dependence of the mass of the electron on its energy and of the redistribution of electrons under the action of the ponderomotive pressure of the electromagnetic pulse. The relativistic self-focusing threshold is determined by the critical power of radiation and the electron density of the plasma (Sun et al., 1987):

$$\mathcal{P}_{\text{cr}} = \overline{\mathcal{P}}(\omega/\omega_{pe})^2, \tag{19.38}$$

where

$$\overline{\mathcal{P}} = 2m_e^2c^5/e^2 \approx 17 \text{ GW} \tag{19.39}$$

is the characteristic power. If the power of laser radiation $\mathcal{P}_{\text{las}}$ is significantly higher than the self-focusing threshold, $\mathcal{P}_{\text{las}} \gg \mathcal{P}_{\text{cr}}$, the pulse can decay into several filaments. The corresponding dimensionless parameter is $\mathcal{P}_{\text{las}}/\mathcal{P}_{\text{cr}}$.

The electrons accelerated in relativistic laser plasmas produce the electric current, which generates the quasi-static magnetic field. Since for the relativistic electrons the ratio of the velocity to speed of light, v/c, is of the order of unity, the magnetic part of the force acting on charged particles is approximately equal to that part, which is due the electric field. This causes significant effects of magnetic interaction in laser plasmas, including the coalescence of self-focusing channels (Askar'yan et al.,

1994), the electron vortex system formation (Bulanov et al., 1996a, 1996b) described by the Hasegawa-Mima theory (Hasegawa & Mima, 1978), the ion acceleration by the magnetic vortex mechanism (Bulanov & Esirkepov, 2007; Park, et al., 2019), and the magnetic field line reconnection foreseen by Askar'yan et al. (1995) [see also Ref. (Bulanov, 2017)]. The later phenomenon is of paramount importance for modelling the processes in astrophysical objects under the conditions of terrestrial laboratories as a part of the so-called laboratory astrophysics (Remington et al., 2006; Bulanov et al., 2009).

The counter-relationship between the quasi-static magnetic field and the electron vortices is a consequence of the Maxwell equations, i.e. of

$$\nabla \times \boldsymbol{B} = -\frac{4\pi en}{c}\boldsymbol{v}, \tag{19.40}$$

which yields for the electron fluid vorticity,

$$\nabla \times \boldsymbol{v} = \frac{c}{4\pi en}\Delta \boldsymbol{B}.$$

The quasi-static magnetic field and the associated with the magnetic field electron vortices stay in the plasma for a long time due to their low propagation velocity. At the plasma-vacuum interface, the self-generated magnetic field sustains the electric charge separation thus enhancing the ion acceleration (Bulanov & Esirkepov, 2007).

The quasistatic magnetic field generation in relativistic laser plasmas occurs due to the fast electron beam interaction with the background plasma. It can be understood in terms of the Weibel instability (Weibel, 1959) or in the generic case, in terms of the electromagnetic filamentation instability. When the fast electron beam propagates in plasma, its electric current is compensated by the current carried by the plasma electrons. A repulsion of the oppositely directed electric currents results in the electron beam filamentation and in the generation of a strong magnetic field (Califano et al., 1997). An electromagnetic filamentation instability leads to the generation of a quasistatic magnetic field and is associated with many small-scale current filaments. Each filament consists of a direct and of a return electric current which repel each other. This produces a strong electric field, which accelerates the ions in a radial direction. In the long-term evolution, the successive coalescence of the small-scale current filaments forms a large-scale magnetic structure. This process is accompanied by the reconnection of the magnetic field lines, by the formation of current sheets, and by strong ion acceleration inside these sheets (Higashimori et al., 2012; Sakai et al., 2002). This process is illustrated in Fig. 19.3.

The filamentation phenomena are of great interest for the explanation of the quasi-static magnetic field origin in laser plasmas irradiated by relativistically strong electromagnetic wave (Askar'yan et al.,). Counter-streaming electric current configurations naturally appear in space at the fronts of colliding electron-positron and electron-ion plasma clouds (Kazimura et al., 1998) as in the cases of the Gamma Ray Bursts (GRB) and in shock waves in supernova remnants. The filamentation

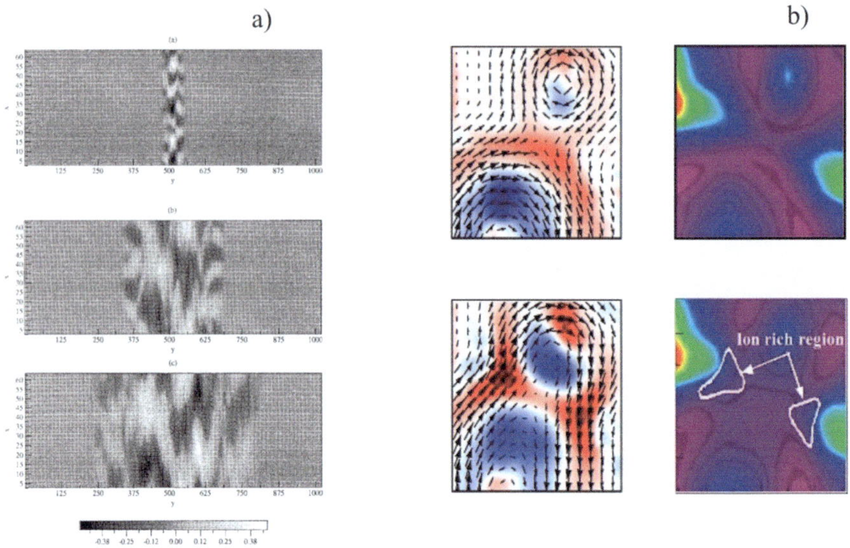

Fig. 19.3 **a** Time development of the self-generated magnetic fieldeld at the front of colliding two electron-positron clouds Kazimura et al., (1998). **b** Vector plots of the magnetic fields and spatial distribution of the charge density, spatial distribution of the electron density. The arrows show the motion of the electrons. (Reproduced from Sakai, et al (2002), with the permission of AIP Publishing, see also Hoshino (2012))

instability generates the magnetic field required by the theory of the synchrotron afterglow in gamma ray bursts.

The filamentation instability developing in the vicinity of shock wave fronts together with other types of instabilities plays the role of the source of strong electromagnetic turbulence invoked in the theoretical models of the Fermi acceleration of cosmic rays (Matsumoto et al., 2013; Medvedev & Loeb, 1999; Silva et al., 2003; Sironi et al., 2013). A realization of the Fermi acceleration mechanism of Type A at the shock wave front is discussed in (Berezinskii et al., 1990). Large, stable electromagnetic structures have been observed in laser-produced counter-streaming plasmas (Kugland et al., 2012).

A low frequency wave can propagate through the plasma if the plasma electrons do not screen the electric field of the wave. The condition for wave propagation implies that the convection electric current $-env$ is smaller than the displacement current $\partial_t E/4\pi$ in the wave, i.e.

$$env \leq \frac{\omega E}{4\pi}. \tag{19.41}$$

In the nonrelativistic limit the electron quiver velocity approximately equals $v \approx eE/m_e\omega$, and the condition of transparency is equivalent to $\omega > \omega_{pe}$. In the ultrarelativistic limit the electron velocity is of the order of speed of light in vacuum,

$v \approx c$, the plasma becomes transparent if the ratio of the laser frequency to the relativistic Langmuir frequency is above unity,

$$\frac{\omega(1 + a_0^2)^{1/4}}{\omega_{pe}} > 1, \tag{19.42}$$

which can be considered as a dimensionless parameter characterizing the relativistic transparency/opaqueness of plasmas. In other words, plasma with the density

$$n < n_{cr}\sqrt{1 + a_0^2} \tag{19.43}$$

approximately a_0 times higher than the critical n_{cr} (see definition of critical density by Eq. (19.33)) is transparent for electromagnetic radiation of the amplitude a_0.

In the opposite case when the plasma is overdense with the density above critical, $n > n_{cr}\sqrt{1 + a_0^2}$, it cannot be considered as homogeneous. The laser radiation penetrates the plasma over the evanescence length of the order of $c(1 + a_0^2)^{1/4}/\omega_{pe}$. At the sharp plasma vacuum interface, the super-intense laser radiation extracts electrons from the plasma and accelerates those outwards. This anomalous electron heating mechanism is known as "the vacuum electron heating" (Brunel, 1987). It plays an important role in enhancing the ion acceleration during relatively long pulse laser interaction with solid targets (Yogo et al., 2017).

At a steep plasma-vacuum interface, the laser radiation forces the electron fluid oscillate back and force. The light reflection at the "oscillating relativistic mirrors" results in generation of high order harmonics (Bulanov et al., 1994; Teubner & Gibbon, 2009; Lamac et al., 2023). In space, the reflected electromagnetic radiation has a form of a train of ultra-short (attosecond) pulses (Naumova et al., 2004).

19.5 Principle of Qualitative Scaling in Modelling Astrophysical Phenomena

During the laser–matter interaction, the high-power electromagnetic radiation impact on the target generates shock waves; the converging gas motion is accompanied by the Rayleigh–Taylor (RT) and Richtmayer–Meshkov instability development; collimated plasma jets are formed; the matter equation of state acquires new properties under extreme pressure, density and temperature conditions, which of paramount importance for the high energy density state of matter physics. In the limit of relativistically intense electromagnetic field, the laser plasma emits high-energy charged particle beams and high- and low-frequency powerful electromagnetic radiation. Gathering of these facts which are principal for both space and laboratory physics has initiated works in the so-called "laboratory astrophysics" with the aim to model the processes of key importance for the space objects under laboratory conditions.

In contemporary relativistic plasma astrophysics (it is also called "astro-particle physics"), central are the questions on the mechanisms of charged particle acceleration (Berezinskii et al., 1990) and high energy gamma photon generation (Aharonian, 2004). In space plasmas basic mechanisms of cosmic ray acceleration are the reconnection of magnetic field lines, which is accompanied by the strong and regular electric field generation (it occurs in the planet magnetospheres, in binary stellar systems, in accretion disks, in the magnetar magnetospheres, etc, see Philippov & Kramer, 2022) and with collisionless shock waves, at the fronts of which the charged particle acceleration occurs (this happens in interplanetary space, during supernova explosions, in colliding galaxies, etc.).

The most energetic particles observed in the Universe are the ultra-high energy cosmic rays (UHECR) whose maximal energy corresponds to the Greizen-Zatsepin-Kuz'min (GZK) cutoff of approximately 5×10^{19}eV (Greizen, 1966; Zatsepin & Kuz'min, 1966). This limit is set by the energy loss effect of interactions of the protons with the microwave background radiation over cosmological distances.

In the discussion of the origine of high-energy particles widely used is the so-called Hillas diagram (Hillas, 1984). Underlying idea of the Hillas approach is based on the condition that charged particle of the energy $\mathcal{E}_a = m_a c^2 \gamma_a$ cannot have the Larmor radius (it is equal to $r_{Ba} = p_a c / Z_\varepsilon eB \approx m_a c^2 \gamma_a / Z_a eB$) larger than the size of the region where it is accelerated, R. This relationship can be written in a similar form $\mathcal{E}_a < \eta Z_a eBR$ with $\eta = v/c$, where $Z_a e$, m_a are the particle electric charge and mass, respectively, and B is the magnetic field value in the cosmic accelerator and is the velocity with which the magnetic field evolves. It can be found from consideration of the acceleration rate $\dot{\mathcal{E}}_a^+$ given by the integral over the particle trajectory

$$\dot{\mathcal{E}}_a^+ = Z_a e \int \boldsymbol{E} \cdot d\boldsymbol{v} \tag{19.44}$$

The electric field in this expression, $\boldsymbol{E}$, is approximately equal to $E \approx (v/c)B$. This yield for the charged particle energy $\mathcal{E}_a \approx \eta Z_a eBR$, i.e. logarithm of the accelerator size and logarithm of the magnetic field magnitude are related as (Fig. 19.4)

$$\log B = \log(\mathcal{E}_a / \eta Z_a e) - \log R. \tag{19.45}$$

Implementation of the radiation loss effects determined by the radiation friction force

$$\boldsymbol{f}_{rad} = -\frac{2G_a}{3c} r_a^2 \boldsymbol{v} \gamma_a^2 \left[\left(\boldsymbol{E} + \frac{1}{c} \boldsymbol{v} \times \boldsymbol{B} \right)^2 - \left(\frac{1}{c} \boldsymbol{v} \cdot \boldsymbol{E} \right)^2 \right] \propto \gamma_a^2 B^2 \tag{19.46}$$

with the Gaunt factor $G_e = \mathrm{I_{QED}}/\mathrm{I_C}$ giving the impact of quantum effects and $r_a = Z_a^2 e^2 / m_a c^2$ yields $\dot{\mathcal{E}}_a = \dot{\mathcal{E}}_a^+ - \dot{\overline{\mathcal{E}}}_a$. It has been considered by Medvedev (2003), Ptitsyna and Troitsky (2010) and Sotirov (2023). Here the energy loss equals

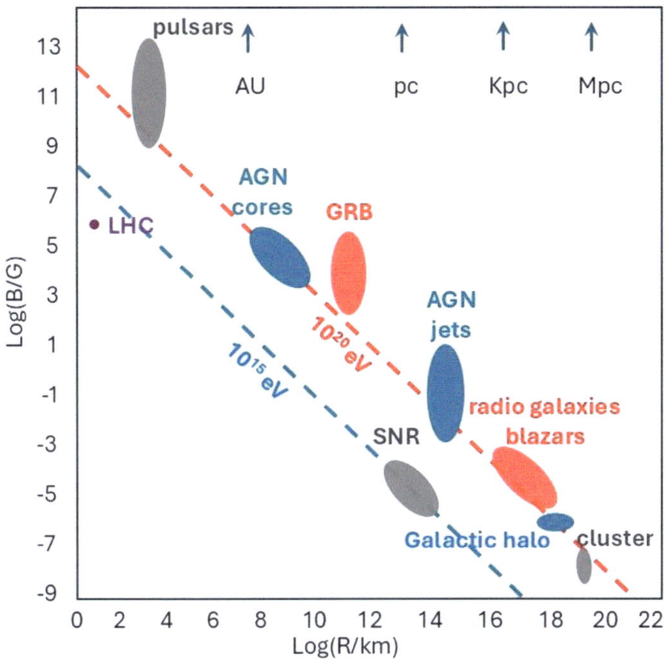

Fig. 19.4 Logarithm of the magnetic field strength versus the logarithm of the size of various cosmic ray accelerators

$$\dot{\mathcal{E}}_a = f_{rad} \cdot v.$$

This results in the radiation friction cutoff

$$\mathcal{E}_{\text{rad}} = m_a c^2 \left(\frac{2R}{3r_a} \right)^{1/3} \tag{19.47}$$

If we trace a relationship between astrophysics and laser physics, we can see many publications devoted to the laboratory modeling of astrophysical processes (e.g. review articles Remington et al., 2006; Bulanov et al., 2009; Esirkepov & Bulanov, 2012 and references therein). As is known there has been an interest in modeling space physics with laboratory experiments for many years. The first modeling of processes fundamental for space physics in terrestrial laboratories has been done by Kristian Birkeland, who more than 100 years ago conducted first experiments on studying the auroral regions in the Earth's magnetosphere (Birkeland, 1908). Later, progress has been achieved in the laboratory modeling of various processes (Alfvén & Falthammar, 1963; Fortov, 2007), including the magnetic field reconnection (Frank, 2010, Yamada et al., 2010), and collisionless shock waves (Romagnani et al., 2008), which provide mechanisms for charged particle acceleration under various astrophysical conditions.

Below, we briefly discuss the ways toward modeling of astrophysical processes and extreme field regimes with super-power lasers. As far as it concerns the astrophysics, the main attention should be paid to the concept of *limited similarity* in using the dimensionless parameters characterizing the processes in the laser and astrophysical plasmas. Here, we address two key-important problems relevant to the problem of ultrarelativistic particle acceleration under astrophysical conditions: magnetic reconnection and collisionless shock waves. In the extreme field limits we consider the regimes of dominant radiation reaction, changing the electromagnetic wave-matter interaction. These regimes are also studied in connection to high energy charged particle acceleration in space plasmas. In particular, magnetic reconnection in the relativistic limit when the effects of displacement current (Syrovatskii, 1966) can be studied in underdense laser plasma (Gu & Bulanov, 2021). The ultra-intense laser radiation interaction with matter will result in a new powerful source of ultra-high brightness gamma-rays and will make possible the electron–positron pair creation in vacuum in a multiphoton process. This will allow modeling under terrestrial laboratory conditions the processes in astrophysical objects.

In an ideal, but not realistic case the laboratory modeling requires absolute similarity with the same dimensionless quantities under the experiment conditions and of the modeled phenomenon (Barenblatt, 1996; Sedov, 1959). However, since these restrictive requirements can hardly be fulfilled, modeling is usually conducted within the framework of approximate (incomplete) similarity when only a few basic parameters are the same. Laboratory experiments for studying astrophysical phenomena are of two types (Falthammar, 1974). The first type of experiments can be referred to as configuration modeling, which is aiming at simulating the actual configuration of a system or in other words reproducing the global geometry (see, e.g., Zhong et al. (2010), where the results of the laser-plasma experiments on the modelling of the global structure of the coronal mass ejections are presented). The second type of experiment corresponds to *process simulation,* or it is aimed at local properties of physical processes. In this case the properties of physical processes relevant to astrophysical phenomena are studied.

In one of the first works on the charged particle acceleration by strong electromagnetic waves in astrophysical plasmas, pulsars (Beskin et al., 1993; Gunn & Ostriker, 1969) have been considered as sources of ultra-intense radiation. Pulsars were considered as oblique rotators with non-parallel rotation and magnetic dipole axes, as illustrated in Fig. 19.5 (left frame). The power of magneto-dipole radiation emitted by rotating magnetic dipole is given by the expression

$$W = \frac{2}{3}\frac{\mu_p^2(\sin\theta)^2\Omega_p^4}{c^3}, \tag{19.48}$$

where μ_p is the magnetic momentum, θ is an angle between the rotational and magnetic dipole axes and Ω_p is the pulsar rotation frequency. Although the plasma effects modify the pulsar electrodynamics (Beskin et al., 1993; Philippov & Kramer, 2022), the expression (19.48) gives the pulsar electromagnetic energy losses, as follows from the theoretical model of the pulsar magnetosphere. The magnetic

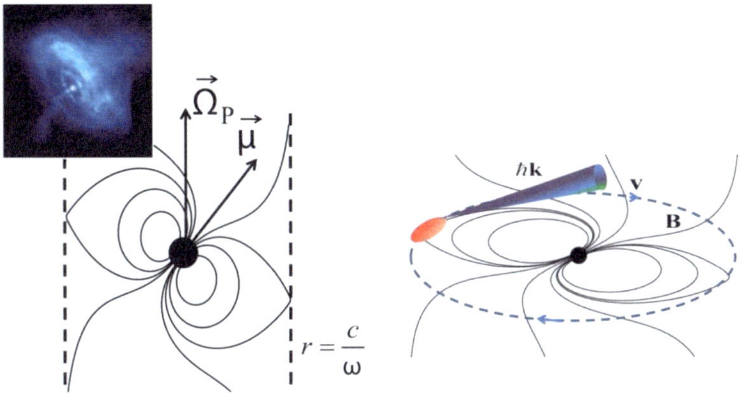

Fig. 19.5 Left: Oblique rotator. The inset: the Crab pulsar Hester (2008). Right: Schematic pulsar magnetosphere, according to reference Ginzburg and Zheleznyakov (1975). A rotating relativistic electron lump emits electromagnetic radiation by the antenna mechanism

moment is related to the pulsar magnetic field and radius as $\mu \approx B R_p^3$. For typical values of $R_p = 10^6$ cm and $B = 10^{12}$ G we obtain $\mu_p = 10^{30}\,\mathrm{G\,cm}^3$. The electromagnetic wave intensity at the distance r is equal to $I = W/4\pi r^2$. At the wave zone boundary, $r = c/\Omega_p$, the dimensionless amplitude of the electromagnetic field is

$$a_p = \frac{e\mu_p\Omega_p^2}{m_e c^4}, \tag{19.49}$$

For the Crab pulsar with the rotation frequency $\Omega_p = 200\,s^{-1}$ we find $a_p = 2 \times 10^{10}$.

In the limit of high radiation intensity, the effects of the radiation damping should be incorporated into the theory of the electromagnetic wave interaction with plasmas. A dimensionless parameter,

$$\varepsilon_{rad} = \frac{2e^2\Omega_p}{3m_e c^3}, \tag{19.50}$$

gives a value of the wave amplitude, $a_{rad} = \varepsilon_{rad}^{-1/3}$, above which the radiation damping cannot be neglected. For $\Omega_p = 200\,s^{-1}$ this yields $a_{rad} = 10^7$, which is substantially less than above found value of $a_p = 2 \times 10^{10}$.

The antenna mechanism of the pulsar radiation emission has been proposed has been proposed by Ginzburg and Zheleznyakov (1975). According to this mechanism in the pulsar magnetosphere, which is a rotating magnetic dipole, the magnetic dipole interaction with a plasma at the magnetosphere periphery induces strong modulations of the electron density, an electron density lump. The phase velocity of the electron lump can be arbitrarily close to the speed of light in vacuum. It is directed along a circle as depicted in Fig. 19.5 (right frame). As a result of the curvilinear acceleration, the electron lump emits radiation, whose properties are similar to the synchrotron radiation (Ginzburg, 1979).

In the context of relativistic laboratory astrophysics, it is remarkable that the relativistic rotating dipole can naturally be formed in the laser plasma. Laser-plasma interactions provide an opportunity to reproduce nonlinear electrodynamics effects under astrophysical conditions in the laboratory. In reference (Esirkepov et al., 2004) it is demonstrated that high-power coherent synchrotron-like radiation can be generated by the relativistic charge density wave rotating self-consistently inside a relativistic electromagnetic-dipole solitary wave, dwelling in a laser plasma. The soliton emits the radiation which is emanated from the electron density hump rotating in the wall of the soliton cavity, similar to coherent synchrotron-like emission. The emission of the spiral wave correlates to the rotation of the electron density hump in the cavity wall, and it leads to the spiral modulations of the electron density. The density hump gyrates in a circle, and the period of revolution is exactly equal to the soliton period. The polarization of the spiral wave corresponds to the well-known synchrotron radiation and the density hump emission is coherent.

References

Aharonian, F. A. (2004). *Very high energy cosmic gamma radiation: A crucial window on the extreme universe* (World Scientific).

Akhiezer, A. I., & Polovin, R. V. (1956). Theory of wave motion of an electron plasma. *Soviet Physics—JETP, 3*, 696.

Alfvén, H., & Falthammar, C. G. (1963). *Cosmic electrodynamics*. Clarendon.

Askar'yan, G. A. (1962). Effects of the gradient of a strong electromagnetic beam on electrons and atoms. *Soviet Physics—JETP, 15*, 1088

Askar'yan, G. A., Bulanov, S. V., Pegoraro, F., & Pukhov, A. M. (1994). Magnetic interaction of self-focusing channels and the electromagnetic radiation: Their merging and energy accumulation. *JETP Letter, 60*, 251.

Askar'yan, G. A., Bulanov, S. V., Pegoraro, F., & Pukhov, A. M. (1995). Magnetic interaction of self-focused channels and magnetic wake excitation in high intense laser pulse. *Comments on Plasma Physics and Controlled Fusion, 17*, 35.

Barenblatt, G. I. (1996). *Scaling, self-similarity, and intermediate asymptotics: dimensional analysis and intermediate asymptotics*. Cambridge University Press.

Berestetskii, V. B., Lifshitz, E. M., & Pitaevskii, L. P. (1982). *Course of theoretical physics, Vol. 4: Quantum electrodynamics*. Pergamon.

Berezinskii, V. S., Bulanov, S. V., Ginzburg, V. L., Dogiel, V. A., & Ptuskin, V. S. (1990). *Astrophysics of cosmic rays*. North Holland/Elsevier.

Beskin, V. S., Gurevich, A. V., & Istomin, Ya. N. (1993). *Physics of the pulsar magnetosphere*. Cambridge Univ. Press.

Birkeland, K. (1908). *The Norwegian aurora Polaris expedition, 1902–1903*. Aschenhoug.

Breit, G., & Wheeler, J. A. (1934). Collision of two light quanta. *Physical Review, 46*, 1087.

Brunel, F. (1987). Not-so-resonant, resonant absorption. *Physical Review Letters, 59*, 52.

Bulanov, S. V., Inovenkov, I. N., Kirsanov, V. I., et al. (1992). Nonlinear depletion of ultrashort and relativistically strong laser pulses in an underdense plasma. *Physics of Fluids B, 4*, 1935.

Bulanov, S. V., Naumova, N. M., & Pegoraro, F. (1994). Interaction of an ultrashort, relativistically strong laser pulse with an overdense plasma. *Physics of Plasmas, 1*, 745.

Bulanov, S. V., Lontano, M., Esirkepov, T. Zh., Pegoraro, F., & Pukhov, A. M. (1996a). Electron vortices produced by ultraintense laser pulses. *Physical Review Letters, 76*, 3562.
Bulanov, S. V., Lontano, M., Esirkepov, T. Zh., Pegoraro, F., & Pukhov, A. M. (1996b). Magnetic interaction and magnetic wake of high intensity laser pulses in plasmas. *Physica Scripta T, 63*, 280.
Bulanov, S. V., & Esirkepov, T. Zh. (2007). Comment on "Collimated multi-MeV ion beams from high-intensity laser interactions with underdense plasma." *Physical Review Letters, 98*, Article 049503.
Bulanov, S. V., Esirkepov, T. Zh., Habs, D., Pegoraro, F., & Tajima, T. (2009). Relativistic laser-matter interaction and relativistic laboratory astrophysics. *The European Physical Journal D, 55*, 483.
Bulanov, S. V. (2017). Magnetic reconnection: From MHD to QED. *Plasma Physics and Controlled Fusion, 59*, Article 014029.
Califano, F., Pegoraro, F., & Bulanov, S. V. (1997). Spatial structure and time evolution of the Weibel instability in collisionless inhomogeneous plasmas. *Physical Review E, 56*, 963.
Dittrich, W., & Gies, H. (2000). *Probing the quantum vacuum. Perturbative effective action approach in quantum electrodynamics and its application.* Springer tracts in modern physics (Vol. 166). Springer.
Esarey, E., Schroeder, C. B., & Leemans, W. P. (2009). Physics of laser-driven plasma-based electron accelerators. *Reviews of Modern Physics, 81*, 1229.
Esirkepov, T. Zh., Bulanov, S. V., Nishihara, K., & Tajima, T. (2004). Soliton synchrotron afterglow in a laser plasma. *Physical Review Letters, 92*, Article 255001.
Esirkepov, T. Zh., & Bulanov, S. V. (2012). Fundamental physics and relativistic laboratory astrophysics with extreme power lasers. *European Astronomical Society Publications Series, 58*, 7.
Falthammar, C. G. (1974). Laboratory experiments of magnetospheric interest. *Space Science Reviews, 15*, 803.
Fortov, V. E. (2007). Intense shock waves and extreme states of matter. *Physics-Uspekhi, 50*, 333.
Frank, A. G. (2010). Dynamics of current sheets underlying flare-type events in magnetized plasmas. *Physics-Uspekhi, 53*, 941.
Ginzburg, V. L., & Zheleznyakov, V. V. (1975). On the pulsar emission mechanisms. *Annual Review of Astronomy and Astrophysics, 13*, 511.
Ginzburg, V. L. (1979). *Theoretical physics and astrophysics.* Pergamon Press.
Gonoskov, A., Blackburn, T. G., Marklund, M., & Bulanov, S. S. (2022). Charged particle motion and radiation in strong electromagnetic fields. *Reviews of Modern Physics, 94*, Article 045001.
Greisen, K. (1966). End to the cosmic-ray spectrum? *Physical Review Letters, 16*, 748.
Gu, Y. J., & Bulanov, S. V. (2021). Magnetic field annihilation and charged particle acceleration in ultra-relativistic laser plasmas. *High Power Laser Science and Engineering, 9*, Article e2.
Gunn, J. E., & Ostriker, J. P. (1969). Acceleration of high-energy cosmic rays by pulsars. *Physical Review Letters, 22*, 728.
Hasegawa, A., & Mima, K. (1978). Pseudo-three-dimensional turbulence in magnetized nonuniform plasma. *Physics of Fluids, 21*, 87.
Hester, J. (2008). The crab nebula: An astrophysical chimera. *Annual Review of Astronomy and Astrophysics, 46*, 127.
Hillas, A. M. (1984). The origin of ultra-high-energy cosmic rays. *Annual Review of Astronomy and Astrophysics, 22*, 425.
Higashimori, K., Yokoi, N., & Hoshino, M. (2012). Explosive turbulent magnetic reconnection. *Physical Review Letters, 110*, Article 255001.
Hoshino, M. (2012). Stochastic particle acceleration in multiple magnetic islands during reconnection. *Physical Review Letters, 108*, Article 135003.
Kazimura, Y., Sakai, J.-I., Neubert, T., & Bulanov, S. V. (1998). Generation of a small-scale quasi-static magnetic field and fast particles during the collision of electron-positron plasma clouds. *ApJ, 498*, L183.

Kugland, N. L., Ryutov, D. D., Chang, P.-Y., et al. (2012). Self-organized electromagnetic field structures in laser-produced counter-streaming plasmas. *Nature Physics, 11*, 809.
Keldysh, L. V. (1965). Ionization in the field of a strong electromagnetic wave. *Soviet Physics JETP, 20*, 2354.
King, B., & Heinzl, T. (2016). Measuring vacuum polarization with high-power lasers. *High Power Laser Science and Engineering, 4*, 1.
Kruer, W. L. (1988). *The physics of laser plasma interactions.* Addison-Wesley Publishing Co.
Lamač, M., Chaulagain, U., Nejdl, J., & Bulanov, S. V. (2023). Generation of intense magnetic wakes by relativistic laser pulses in plasma. *Scientific Reports, 13*, 170113.
Landau, L. D., & Lifshitz, E. M. (1980). *Course of theoretical physics, Vol. 2: The classical theory of fields.* Pergamon.
Litvak, A. G. (1969). Finite-amplitude wave beams in a magnetoactive plasma. *Soviet Physics—JETP, 30*, 344.
Marklund, M., & Shukla, P. (2006). Nonlinear collective effects in photon-photon and photon-plasma interactions. *Reviews of Modern Physics, 78*, 591.
Matsumoto, Y., Amano, T., & Hoshino, M. (2013). Electron acceleration in a nonrelativistic shock with very high Alfvén Mach Number. *Physical Review Letters, 111*, Article 215003.
Medvedev, M. V., & Loeb, A. (1999). Generation of magnetic fields in the relativistic shock of gamma-ray burst sources. *ApJ, 526*, 697.
Medvedev, M. V. (2003). Constraint on electromagnetic acceleration of highest energy cosmic rays. *Physical Review E, 67*, 045401(R).
Mourou, G., Tajima, T., & Bulanov, S. V. (2006). Optics in the relativistic regime. *Reviews of Modern Physics, 78*, 309.
Naumova, N. M., Nees, J. A., Sokolov, I. V., Hou, B., & Mourou, G. A. (2004). Relativistic generation of isolated attosecond pulses in a λ3 focal volume. *Physical Review Letters, 92*, Article 063902.
Park, J., Bulanov, S. S., Bin, J., Ji, Q., Steinke, S., Vay, J. L., Geddes, C. G. R., Schroeder, C. B., Leemans, W. P., Schenkel, T., & Esarey, E. (2019). Ion acceleration in laser generated megatesla magnetic vortex. *Physics of Plasmas, 26*, Article 103108.
Philippov, A., & Kramer, M. (2022). Pulsar magnetospheres and their radiation. *Annual Review of Astronomy and Astrophysics, 60*, 495.
Di Piazza, A., Muller, C., Hatsagortsyan, K. Z., & Keitel, C. H. (2012). Extremely high-intensity laser interactions with fundamental quantum systems. *Reviews of Modern Physics, 84*, 1177.
Ptitsyna, K. V., & Troitsky, S. V. (2010). Physical conditions in potential accelerators of ultra-high-energy cosmic rays: updated Hillas plot and radiation-loss constraints. *Physics-Uspekhi, 53*, 691.
Remington, B., Drake, R., & Ryutov, D. (2006). Experimental astrophysics with high power lasers and Z pinches. *Reviews of Modern Physics, 78*, 755.
Romagnani, L., Bulanov, S. V., Borghesi, M., Audebert, P., Gauthier, J. C., Löwenbrück, K., MacKinnon, A. J., Patel, P., Pretzler, G., Toncian, T., & Willi, O. (2008). Observation of collisionless shocks in laser-plasma experiments. *Physical Review Letters, 101*, Article 025004.
Ruffini, R., Vereshchagin, G., & Xue, S. S. (2009). Electron-positron pairs in physics and astrophysics: From heavy nuclei to black holes. *Physics Reports, 487*, 1.
Sauter, F. (1931). On the behavior of an electron in a homogeneous electric field in Dirac's relativistic theory. *Zeitschrift Für Physik, 69*, 742.
Sakai, J. I., Saito, S., Mae, H., Farina, D., Lontano, M., Califano, F., Pegoraro, F., & Bulanov, S. V. (2002). Ion acceleration, magnetic field line reconnection, and multiple current filament coalescence of a relativistic electron beam in a plasma. *Physics of Plasmas, 9*, 2959.
Sedov, L. I. (1959). *Similarity and dimensional methods in mechanics.* Academic Press.
Silva, L. O., Fonseca, R. A., Tonge, J. W., Dawson, J. M., Mori, W. B., & Medvedev, M. V. (2003). Interpenetrating plasma shells: Near-equipartition magnetic field generation and non-thermal particle acceleration. *ApJ, 596*, L121.
Sironi, L., Spitkovsky, A., & Arons, J. (2013). The maximum energy of accelerated particles in relativistic collisionless shocks. *ApJ, 771*, 54.

Sotirov, S. (2023). General constraints on sources of high-energy cosmic rays from interaction losses. *Physical Review D, 107*, Article 123018.
Sun, G.-Z., Ott, E., Lee, Y. C., & Guzdar, P. (1987). Self-focusing of short intense pulses in plasmas. *Physics of Fluids, 30*, 526.
Syrovatskii, S. I. (1966). Dynamic dissipation of a magnetic field and charged particle acceleration. *Soviet Astronomy, 10*, 270.
Tajima, T., & Dawson, J. M. (1979). Laser electron accelerator. *Physical Review Letters, 43*, 267.
Teubner, U., & Gibbon, P. (2009). High-order harmonics from laser-irradiated plasma surfaces. *Reviews of Modern Physics, 81*, 445.
Weibel, E. S. (1959). Spontaneously growing transverse waves in a plasma due to an anisotropic velocity distribution. *Physical Review Letters, 2*, 83.
Yamada, M., Kulsrud, R., & Ji, H. (2010). Magnetic reconnection. *Reviews of Modern Physics, 82*, 603.
Yogo, A., Mima, K., Iwata, N., Tosaki, S., Morace, A., Arikawa, Y., Fujioka, S., Johzaki, T., Sentoku, Y., Nishimura, H., & Sagisaka, A. (2017). Boosting laser-ion acceleration with multi-picosecond pulses. *Scientific Reports, 7*, 42451.
Yu, T.-P., Liu, K., Zhao, J., Zhu, X. L., Lu, Y., Cao, Y., Zhang, H., Shao, F.-Q., & Sheng, Z.-M. (2024). Bright X/gamma-ray emission and lepton pair production by strong laser fields: a review. *Reviews of Modern Plasma Physics, 8*, 24.
Zatsepin, G. T., & Kuz'min, V. A. (1966). Upper limit of the spectrum of cosmic rays. *JETP Letter, 4*, 78.
Zel'dovich, Ya. B., & Illarionov, A. F. (1972). Scattering of a strong wave by an electron in a magnetic field. *Soviet Physics—JETP, 34*, 467.
Zhong, J., Li, Y., Wang, X., et al. (2010). Modelling loop-top X-ray source and reconnection outflows in solar flares with intense lasers. *Nature Physics, 6*, 984.

Chapter 20
Electron Interaction with Strong Electromagnetic and Plasma Waves

Abstract Here we discuss the basic properties of electron interaction with strong electromagnetic and potential waves in vacuum and plasma in classical and quantum regimes.

Keywords Electron-EM wave interaction · Lawson-Woodward theorem · Wolkow State · Charged particle interaction with strong electromagnetic wave · Radiation reaction

20.1 Description Within the Approximation of Classical Mechanics and Electrodynamics

Analytical description and computer simulation of the laser radiation interaction with plasmas within the framework of the classical electrodynamics approximation is based on the system of the Maxwell equations for electromagnetic field and the Minkovskii equations for the charged particle motion (Barut, 2010; Landau & Lifshitz, 1980). The Maxwell equations can be written in the covariant form as

$$\varepsilon^{\alpha\beta\gamma\delta}\partial_\beta F_{\gamma\delta} = 0, \tag{20.1}$$

$$\partial_\beta F^{\alpha\beta} = -\frac{4\pi}{c} j^\alpha, \tag{20.2}$$

where $\partial_\alpha = \partial/\partial x_\alpha$ is a partial derivative with respect to variable $x_\alpha = (t, \boldsymbol{r})$, and $F_{\gamma\delta}$ is the 4-tensor of the electromagnetic field. Here t and $\boldsymbol{r} = (x, y, z)$ are time and coordinates, respectively. On the right-hand side of Eq. (20.2) $j^\alpha = (\rho, \boldsymbol{j})$ is the four-vector of the electric current density. Expressing the tensor $F_{\alpha\beta}$ in terms of the electric and magnetic fields, which are equal to

$$\boldsymbol{E} = -\frac{1}{c}\partial_t \boldsymbol{A}, \quad \boldsymbol{B} = \nabla \times \boldsymbol{A}, \tag{20.3}$$

Y. Kato et al., *High Power Laser and Plasma Science*, Springer Series on Atomic, Optical, and Plasma Physics 130, https://doi.org/10.1007/978-3-032-08930-4_20

respectively, we can write the electromagnetic tensor $F_{\alpha\beta}$ and the dual tensor ${}^*F^{\alpha\beta} = (1/2)\varepsilon^{\alpha\beta\gamma\delta}F_{\gamma\delta}$, where $\varepsilon^{\alpha\beta\gamma\delta}$ is a four-dimensional fully antisymmetric unit tensor, as:

$$F_{\alpha\beta} = (\boldsymbol{E}, \boldsymbol{B}), \quad F^{\alpha\beta} = (-\boldsymbol{E}, \boldsymbol{B}), \quad {}^*F^{\alpha\beta} = (-\boldsymbol{B}, -\boldsymbol{E}). \tag{20.4}$$

The 4×4 matrices $(\boldsymbol{E}, \boldsymbol{B})$ and $(-\boldsymbol{E}, \boldsymbol{B})$ are given by expressions

$$(\boldsymbol{E}, \boldsymbol{B}) = \begin{pmatrix} 0 & E_x & E_y & E_z \\ -E_x & 0 & -B_z & B_y \\ -E_y & B_z & 0 & -B_x \\ -E_z & -B_y & B_x & 0 \end{pmatrix}, \quad (-\boldsymbol{E}, \boldsymbol{B}) = \begin{pmatrix} 0 & -E_x & -E_y & -E_z \\ E_x & 0 & -B_z & B_y \\ E_y & B_z & 0 & -B_x \\ E_z & -B_y & B_x & 0 \end{pmatrix} \tag{20.5}$$

and the matrix $(-\boldsymbol{B}, -\boldsymbol{E})$ has the form

$$(-\boldsymbol{E}, -\boldsymbol{B}) = \begin{pmatrix} 0 & -B_x & -B_y & -B_z \\ B_x & 0 & E_z & E_y \\ B_y & -E_z & 0 & E_x \\ B_z & E_y & -E_x & 0 \end{pmatrix}. \tag{20.6}$$

In 3D notation the Maxwell equations (20.1) and (20.3) take the form, where the first pair of equations is

$$\nabla \times \boldsymbol{E} = -\frac{1}{c}\partial_t \boldsymbol{B}, \tag{20.7}$$

$$\nabla \cdot \boldsymbol{B} = 0, \tag{20.8}$$

and the second pair is

$$\nabla \times \boldsymbol{B} = \frac{4\pi}{c}\boldsymbol{j} + \frac{1}{c}\partial_t \boldsymbol{E}, \tag{20.9}$$

$$\nabla \cdot \boldsymbol{E} = 4\pi\rho. \tag{20.10}$$

From the components of the electromagnetic field tensor $F_{\alpha\beta}$ one can make two invariants:

$$F_{\alpha\beta}F^{\alpha\beta} = 2(\boldsymbol{B}^2 - \boldsymbol{E}^2) = \text{inv}, \tag{20.11}$$

and

$$\varepsilon^{\alpha\beta\gamma\delta}F_{\alpha\beta}F_{\gamma\delta} = -4(\boldsymbol{E} \cdot \boldsymbol{B}) = \text{inv}. \tag{20.12}$$

They are known as the Poincare invariants of the electromagnetic field.

The equations of motion of charged particle the species a in the electromagnetic field, written in Minkowskii's form, are

$$\frac{dp_a^\alpha}{ds} = \frac{e_a}{c} F^\alpha_\beta u_a^\beta, \tag{20.13}$$

where the particle velocity four-vector $u_a^\alpha = (c, \boldsymbol{v}_a)$ is equal to

$$u_a^\alpha = \frac{dx_a^\alpha}{ds} \tag{20.14}$$

with $x_a^\alpha = (ct, \boldsymbol{r}_a)$ being the four-coordinate of the charged particle. The proper time s is defined as follows:

$$s = \int \sqrt{1 - v_a^2/c^2} dt. \tag{20.15}$$

The index, a, labels the different species of the particles.

The electric current density four-vector is $j^\alpha = \sum_a (\rho_a c, \rho_a \boldsymbol{v}_a)$. The electric charge density $\rho(x)$ and the density of the electric current $\boldsymbol{j}(x)$ can be represented in terms of the Dirac delta function:

$$\rho(\boldsymbol{r}, t) = \sum_a e_a \delta(\boldsymbol{r} - \boldsymbol{r}_a(t)), \tag{20.16}$$

and

$$\boldsymbol{j}(\boldsymbol{r}, t) = \sum_a e_a \dot{\boldsymbol{r}} \delta(\boldsymbol{r} - \boldsymbol{r}_a(t)), \tag{20.17}$$

where $\boldsymbol{r}_a(t)$ is the radius-vector, which corresponds to the position of the a-th particle at time t. The Maxwell equations are inherently relativistic.

The equations of charged particle motion (20.13) written in a three dimensional form are

$$\frac{d\boldsymbol{p}_a}{dt} = e_a \boldsymbol{E} + \frac{e_a}{c} \boldsymbol{v}_a \times \boldsymbol{B} \tag{20.18}$$

and

$$\frac{d\boldsymbol{r}_a}{dt} = c \frac{\boldsymbol{p}_a}{\sqrt{m_a^2 c^2 + |\boldsymbol{p}_a|^2}}. \tag{20.19}$$

The electromagnetic field in vacuum for vanishing electric charge density $\rho(\boldsymbol{r}, t) = 0$ and the electric current density $\boldsymbol{j}(\boldsymbol{r}, t) = 0$ obeys Eqs. (20.18) and (20.19), which can be written for vector and scalar potentials in a three dimensional form as follows:

$$\Delta \boldsymbol{A} - \frac{1}{c^2}\partial_{tt}\boldsymbol{A} = 0, \tag{20.20}$$

$$\Delta \varphi - \frac{1}{c^2}\partial_{tt}\varphi = 0. \tag{20.21}$$

Here we have used the Lorentz gauge condition, which requires

$$\partial_\alpha A_\alpha = 0. \tag{20.22}$$

In the case when the electromagnetic potential $\boldsymbol{A}$ depends on one coordinate x and time t only, the wave equation (20.20) describes a plane electromagnetic wave $\boldsymbol{A} = A_k(x, t)\boldsymbol{e}_k$ with $A_k(x, t)$ given by

$$A_k(x, t) = f_k(t - x/c) + g_k(t + x/c), \tag{20.23}$$

where $\boldsymbol{e}_k$ with $k = 2, 3$ are unit vectors perpendicular to the wave propagation direction, and f_k and g_k are functions determined by the initial conditions. The forms of electromagnetic waves in two- and tree-dimensional configurations are discussed in detail by Whitham (1974), Konopinski (1981), and Sokolov and Ternov (1986), Bulanov (2021).

In the electromagnetic wave propagating along the x axis in the positive direction, with $g_k = 0$ in Eq. (20.23), the electric and magnetic field are

$$\boldsymbol{E} = -\frac{1}{c}\boldsymbol{A}', \quad \boldsymbol{B} = -\frac{1}{c}\boldsymbol{e}_1 \times \boldsymbol{A}', \tag{20.24}$$

being perpendicular to each other. Here a prime denotes the differentiation with respect to the variable $t - x/c$ and $\boldsymbol{e}_1$ is the unit vector in the direction of the wave propagation.

Substituting expressions (20.24) for the electric and magnetic field in the right hand side of Eq. (20.18) and changing an independent variable from t to

$$u = \omega(t - x/c) \tag{20.25}$$

with ω being equal to the inverse time scale, we can rewrite this equation in the components as follows:

$$\frac{d\boldsymbol{p}_\perp}{du} = \frac{e\,\omega}{c}\frac{d\boldsymbol{A}}{du}, \tag{20.26}$$

$$\left(1 - \frac{p_\parallel}{m_e c\,\gamma}\right)\frac{dp_\parallel}{du} = \frac{e}{\gamma\, m_e c^2}\left(\boldsymbol{p}_\perp \cdot \frac{d\boldsymbol{A}}{du}\right) \tag{20.27}$$

where $\boldsymbol{p}_\perp$ and $p_\parallel$ are perpendicular and parallel to the wave propagation direction components of the particle momentum and $\gamma = \sqrt{1 + p_\perp^2/m_e^2c^2 + p_\parallel^2/m_e^2c^2}$ is the relativistic Lorentz factor, respectively. Here for the sake of brevity we drop the subscript "a".

Integrating Eq. (20.26) we obtain

$$\boldsymbol{p}_\perp = \boldsymbol{a}_\perp + \boldsymbol{\pi}_\perp. \tag{20.28}$$

This corresponds to conservation of the transverse component of the particle generalized momentum, which is a consequence of the homogeneity along the transverse coordinates of the problem under consideration. From Eq. (20.27) follows that

$$\left[1 + (\boldsymbol{a}_\perp + \boldsymbol{\pi}_\perp)^2 + p_\parallel^2\right]^{1/2} - p_\parallel = h, \tag{20.29}$$

whose conservation is a consequence of the electromagnetic field dependence on the coordinate x and time t in the form $t - x/c$ and the Noether theorem. The constants $\boldsymbol{\pi}_\perp$ and h on the right-hand sides of Eqs. (20.28) and (20.29) are determined by initial conditions. Here we have introduced normalized dependent variables: $\boldsymbol{\pi}_\perp = \boldsymbol{p}_\perp/m_ec$, $p_\parallel/m_ec$, and $\boldsymbol{a}_\perp = e\mathbf{A}/m_ec^2$.

Solution of Eq. (20.29) with respect to $p_\parallel$ and γ gives for the longitudinal component of the particle momentum

$$p_\parallel = \frac{(\boldsymbol{a}_\perp + \boldsymbol{\pi}_\perp)^2 + 1 - h^2}{2h} \tag{20.30}$$

and for the particle energy

$$\gamma = \frac{(\boldsymbol{a}_\perp + \boldsymbol{\pi}_\perp)^2 + 1 + h^2}{2h}, \tag{20.31}$$

respectively.

Equation (20.19) yields the relationships between the coordinate x, time t and the phase $u = t - x/c$ as follows

$$x = c\left(\frac{1 - h^2}{2h^2}\right)u + \frac{c}{2h^2}\int (\boldsymbol{a}_\perp + \boldsymbol{\pi}_\perp)^2 du, \tag{20.32}$$

$$\boldsymbol{r}_\perp = \frac{c}{h}\int (\boldsymbol{a}_\perp + \boldsymbol{\pi}_\perp)\, du, \tag{20.33}$$

$$t = \left(\frac{1 + h^2}{2h^2}\right)u + \frac{1}{2h^2}\int (\boldsymbol{a}_\perp + \boldsymbol{\pi}_\perp)^2 du. \tag{20.34}$$

Here $\boldsymbol{r}_\perp = (y, z)$ is the transverse coordinate. Expressions (20.28)–(20.34) describe the particle trajectory in the parametric form.

In the frame of reference, where the particle was at rest before the interaction with the electromagnetic wave, the constants on the right hand sides of Eqs. (20.28) and (20.29) are equal to $\boldsymbol{\pi}_\perp = 0$ and $h = 1$, respectively. In this frame of reference the (normalized by mc^2) particle kinetic energy $K = \gamma - 1$ and longitudinal and transverse components of the momentum $\boldsymbol{p}$ are given by expressions

$$K = \frac{\boldsymbol{a}_\perp^2}{2}, \quad p_\parallel = \frac{\boldsymbol{a}_\perp^2}{2}, \quad \boldsymbol{p}_\perp = \boldsymbol{a}_\perp. \tag{20.35}$$

We see that for $|\boldsymbol{a}_\perp| > 2$ the particle acquires a relativistic energy, $K > 1$, and the value of the longitudinal component of its momentum $p_\parallel$ is larger than that of the transverse component $|\boldsymbol{p}_\perp|$.

For the case of the particle having the momentum $\boldsymbol{p}_0 = p_0 \boldsymbol{e}_1$ before interaction Eqs. (20.30) and (20.31) give for the longitudinal momentum and for the particle energy

$$p_\parallel = p_0 + \frac{a_\perp^2}{2h}, \quad \gamma = \gamma_0 + \frac{a_\perp^2}{2h}, \tag{20.36}$$

respectively. Here the constants h and γ_0 are equal to $h = \gamma_0 - p_0$ and $\gamma_0 = \sqrt{1 + p_0^2}$.

In the case of counter-propagating electromagnetic pulse and ultrarelativistic charged particle, when $h \approx 2p_0$, the energy is by a factor $1/2p_0$ smaller than that has been given by Eq. (20.35) for $p_0 = 0$. In the configuration of the co-propagating electromagnetic pulse and charged particle, we have $h \approx 1/2p_0$. This gives the particle longitudinal momentum and energy

$$p_\parallel \approx p_0 a_\perp^2, \quad \gamma = p_0 a_\perp^2. \tag{20.37}$$

As we can see, the energy is by a factor $2p_0$ larger than in the case of $p_0 = 0$.

We can use expressions (20.28)–(20.34) for describing the particle trajectory in the parametric form in the frame of reference, where the particle is at rest on average. In this frame of reference the average particle momentum vanishes, i.e. $\boldsymbol{\pi}_\perp = 0$ and for the value of the constant h we find $h = \sqrt{1 + \overline{a_\perp^2}}$. As a result, we obtain

$$x = \frac{c}{2h^2} \int (\boldsymbol{a}_\perp^2 - \overline{\boldsymbol{a}_\perp^2})\, du, \quad \boldsymbol{r}_\perp = \frac{c}{h} \int \boldsymbol{a}_\perp du, \quad t = u + \frac{1}{2h^2} \int (\boldsymbol{a}_\perp^2 - \overline{\boldsymbol{a}_\perp^2})\, du, \tag{20.38}$$

$$p_\parallel = \frac{1}{2h} (\boldsymbol{a}_\perp^2 - \overline{\boldsymbol{a}_\perp^2}), \quad \boldsymbol{p}_\perp = \boldsymbol{a}_\perp. \tag{20.39}$$

The electromagnetic wave with $|\boldsymbol{a}_\perp| \gg 1$ is called superintense or relativistically strong. The normalized amplitude of laser radiation can be expressed, via the radiation intensity $I = c|\boldsymbol{E}|^2/4\pi$ and the wavelength λ, as:

$$|\boldsymbol{a}_\perp| = \sqrt{\frac{I}{1.37 \times 10^{18}\mathrm{W/cm}^2}\left(\frac{1\,\mu\mathrm{m}}{\lambda}\right)^2}. \tag{20.40}$$

As an example, we consider a circularly polarized monochromatic electromagnetic wave whose electromagnetic potential $\boldsymbol{A} = A_k(x, t)\boldsymbol{e}_k$ is equal to

$$\boldsymbol{A} = A_0\left\{\boldsymbol{e}_y \cos\left[\omega\left(t - \frac{x}{c}\right)\right] + \boldsymbol{e}_z \sin\left[\omega\left(t - \frac{x}{c}\right)\right]\right\}. \tag{20.41}$$

Assuming that the particle was at rest before the interaction with the electromagnetic wave, we obtain for the components of the particle momentum

$$p_\parallel = \frac{a_0^2}{2}, \quad \boldsymbol{p}_\perp = a_0\left\{\boldsymbol{e}_y \cos\left[\omega\left(t - \frac{x}{c}\right)\right] + \boldsymbol{e}_z \sin\left[\omega\left(t - \frac{x}{c}\right)\right]\right\}. \tag{20.42}$$

The particle trajectory is given by

$$x = \frac{a_0^2}{2 + a_0^2}ct, \quad \boldsymbol{r}_\perp = a_0\frac{c}{\omega}\left[\boldsymbol{e}_y \cos\left(\frac{2\omega t}{2 + a_0^2}\right) + \boldsymbol{e}_z \sin\left(\frac{2\omega t}{2 + a_0^2}\right)\right] \tag{20.43}$$

As we can see, during the time interval, over which the wave phase changes on 2π (it is equal to $\Delta t = \pi(2+a_0^2)/\omega$), the particle displacement along the wave propagation direction is equal to $\Delta x = \pi a_0^2 c/\omega$ and the particle excursion in the transverse direction is $\Delta y = \pi a_0 c/\omega$, i.e. they are proportional to a_0 and a_0^2, respectively.

According to the written above relationships, the interacting with the electromagnetic wave particle can gain a substantially large amount of kinetic energy, provided that the wave period is sufficiently long and the wave intensity is sufficiently high, i.e. $|a_0| \gg 1$. However, as a result of the charged particle interaction with the finite width electromagnetic pulse consisting of many oscillations, in which the amplitude changes from zero to a maximum and then tends to zero again, an originally stationary particle is left stationary after the pulse has passed the particle. This statement is known as the Lawson-Woodward theorem (Lawson, 1979; Woodward, 1947).

As an example illustrating charged particle dynamics in strong electromagnetic field, we consider the charged particle motion in the crossed field electromagnetic wave. In the crossed-field electromagnetic wave, the electric and magnetic field are constant, i.e. the vector potential is given by

$$\boldsymbol{a}_\perp(u) = a_0 u \boldsymbol{e}_y. \tag{20.44}$$

In the frame of reference, where the particle was at rest before the interaction with the electromagnetic wave, Eqs. (20.28)–(20.34) yield:

$$p_{\parallel} = \frac{a_0^2 u^2}{2}, \quad \boldsymbol{p}_{\perp} = a_0 u \boldsymbol{e}_y, \quad t = \frac{u}{\omega}\left(1 + \frac{a_0^2 u^2}{6}\right), \quad x = \frac{a_0^2 u^3 c}{6\omega}, \quad y = \frac{a_0^2 u^2 c}{2\omega} \tag{20.45}$$

In the limit $t \to \infty$, the kinetic energy of the particle grows as

$$K \approx \left(\frac{m_e c^2}{2^{1/3}}\right)(3a_0^2 \omega t)^{2/3}. \tag{20.46}$$

To reveal the limits of applicability of the Lawson-Woodward theorem we consider here the charged particle interaction with the linearly polarized electromagnetic pulse whose envelope is given by a Gaussian function (see also Lai, 1980):

$$\boldsymbol{E}(x, t) = E_0 \exp\left[-\frac{\omega^2 (t - x/c)^2}{\omega^2 \tau^2}\right] \cos\left[\omega\left(t - \frac{x}{c}\right) + \phi\right] \boldsymbol{e}_y, \tag{20.47}$$

where ϕ is the wave phase. According to Eq. (20.24) the vector potential is related to the wave electric field by

$$\boldsymbol{A}(t - x/c) = -c \int^{t-x/c} \boldsymbol{E}(u) du. \tag{20.48}$$

In Eq. (20.47) the wave amplitude rises from zero at $u = \omega(t - x/c) \to -\infty$ and vanishes at $u \to +\infty$. However, the associated vector potential left behind for $u \to +\infty$ does not necessarily vanish. The final value of the vector potential seen by the particle equals

$$\boldsymbol{A}(u \to +\infty) = -\frac{c}{\omega} \int_{-\infty}^{+\infty} \boldsymbol{E}(u) du. \tag{20.49}$$

The final particle momentum and kinetic energy can be found by using Eqs. (20.38) and (20.45). For the electric field given by Eq. (20.47) the integral (20.49) is the (cosine) Fourier transform of the envelope function. The integral (20.49) for the integrand given by Eq. (20.47) is equal to

$$\boldsymbol{A}(u \to +\infty) = -E_0 c \tau \exp\left[-\frac{(\omega\tau)^2}{4}\right] \cos(\phi) \boldsymbol{e}_y. \tag{20.50}$$

For the wide enough envelope width, $\omega\tau \gg 1$, when the pulse contains many oscillations, the integral (20.50) is exponentially small, i.e. the energy finally gained by the particle, as it follows from (20.35), is also exponentially small. This is equivalent to the Lawson-Woodward theorem.

Moreover, if the pulse has only a few waves, the energy gain can be significant, provided $\cos(\phi) \neq 0$. If the phase is such that $\cos(\phi) = 0$, the work produced by the laser field along the particle trajectory vanishes. In the limiting case of zero frequency, $\omega = 0$, the final value of the vector potential is

$$\boldsymbol{A}(u \to +\infty) = -E_0 c\tau \cos(\phi)\boldsymbol{e}_y, \tag{20.51}$$

which gives the crossed electromagnetic field. This case corresponds to the crossed electromagnetic field configuration analysed above. As we see, as longer the pulse as larger energy gain is.

The Lawson-Woodward theorem can be violated, in addition, due to non-one-dimensional electromagnetic field configuration and due the radiation damping effects, which violate the conservation laws.

20.2 The Case of Arbitrary Speed of Propagation of an Electromagnetic Wave

Here we analyse the particle motion in the electromagnetic wave, which propagates with the velocity, $c\beta_{ph}$, not equal to the speed of light in vacuum. In the plane wave this is the phase velocity. In particular, the phase velocity of the electromagnetic wave propagating in a plasma is superluminal with $\beta_{ph} > 1$. We use the Hamilton-Jacobi equation formalism with the Hamilton-Jacobi equation describing motion of a particle with a mass m_e and charge e (i.e. we consider the electron motion) in the electromagnetic field given by the four-vector potential A_α. In geometry determined by the metric tensor

$$g_{\alpha\beta} = g^{\alpha\beta} = \mathrm{diag}(1, -1, -1, -1)$$

the Hamilton-Jacobi equation for the action S has a form (Landau & Lifshitz, 1980)

$$g^{\alpha\beta}\left(\partial_\alpha S + \frac{e}{c}A_\alpha\right)\left(\partial_\beta S + \frac{e}{c}A_\beta\right) = m_e^2 c^2. \tag{20.52}$$

For A_α depending on one independent variable $u = \omega(t - x/c\beta_{ph})$ we represent the solution of Hamilton-Jacobi equation as

$$S = -p_\alpha^{(0)} x^\alpha + \frac{m_e c^2}{\omega} F(u) \tag{20.53}$$

with $p_\alpha^{(0)} = (\gamma^{(0)} m_e c, \boldsymbol{p}^{(0)})$, for which we have $p_\alpha^{(0)} p^{(0)\alpha} = m_e^2 c^2$. For the four-vector $p_\alpha^{(0)}$ we have $p_\alpha^{(0)} p^{(0)\alpha} = m_e^2 c^2$. Substituting the function S from Eqs. (20.53) to (20.52) we obtain

$$F'^2 + \frac{2\beta_{ph}^2 h_0}{\beta_{ph}^2 - 1} F' - \frac{\beta_{ph}^2}{\beta_{ph}^2 - 1}\left(\frac{p_\alpha a^\alpha}{m_e c} - a_\alpha a^\alpha\right) = 0, \tag{20.54}$$

where $h_0 = (p_0 - p_1/\beta_{ph})/m_e c$, $a_\alpha = eA_\alpha/mc^2$ and a prime denotes a differentiation with respect to the variable u. In the Lorentz gauge (20.22) we can choose the 4-potential in the form

$a_\alpha = (0, 0, \boldsymbol{a}_\perp)$, with $\boldsymbol{a}_\perp = (a_2, a_3)$, $p_\alpha a^\alpha = \boldsymbol{p}_\perp \cdot \boldsymbol{a}_\perp$ and $a_\alpha a^\alpha = a_\perp^2$.

As a result, we obtain

$$S = -p_\alpha^{(0)} x^\alpha + \frac{m_e c^2}{\omega} \int \frac{\sqrt{\beta_{ph}^4 h_0^2 + \beta_{ph}^2\left(\beta_{ph}^2 - 1\right)\left(\frac{\boldsymbol{p}_\perp \cdot \boldsymbol{a}_\perp}{m_e c} + a_\perp^2\right)} - \beta_{ph}^2 h_0}{\beta_{ph}^2 - 1} du. \tag{20.55}$$

Differentiation of the action (20.55) with respect to the coordinates and time gives the expressions for the momentum components and energy, which have the form

$$p_1 = p_1^{(0)} + m_e c \frac{\sqrt{\beta_{ph}^4 h_0^2 + \beta_{ph}^2\left(\beta_{ph}^2 - 1\right)\left(\frac{\boldsymbol{p}_\perp \cdot \boldsymbol{a}_\perp}{m_e c} + a_\perp^2\right)} - \beta_{ph}^2 h_0}{\beta_{ph}^2 - 1}, \tag{20.56}$$

$$\boldsymbol{p}_\perp = \boldsymbol{p}_\perp^{(0)} + m_e c \boldsymbol{a}_\perp, \tag{20.57}$$

$$\gamma = \frac{p_1}{\beta_{ph} m_e c} + h_0. \tag{20.58}$$

The particle trajectory can be found in the parametric form by differentiating (20.55) with respect to the $p_\alpha^{(0)}$ components. As a result, we find

$$x_1 = \frac{1}{\omega m_e h_0} \int p_1 du, \quad \boldsymbol{r}_\perp = \frac{1}{\omega m_e h_0} \int (\boldsymbol{p}_\perp^{(0)} + m_e c \boldsymbol{a}_\perp) du, \quad t = \frac{u}{\omega} + \frac{x}{\beta_{ph} c}. \tag{20.59}$$

Here we have assumed that $\beta_{ph} \geq 1$, which corresponds to the electromagnetic wave propagating in an underdense plasma. For relatively weak electromagnetic pulse, or/and the phase velocity close to the speed of light in vacuum, $a_\perp^2 \ll h_0^2/(\beta_{ph}^2 - 1)$, the particle energy and momentum scale according to expressions (20.30)–(20.35). In the high intensity limit $a_\perp^2 \gg h_0^2/\left(\beta_{ph}^2 - 1\right)$, we have $p_1 \approx m_e c a_\perp/(\beta_{ph}^2 - 1)^{1/2}$.

The charged particle motion in the field of an electromagnetic wave whose phase velocity is not equal to the speed of light in vacuum can also be described with the Hamiltonian

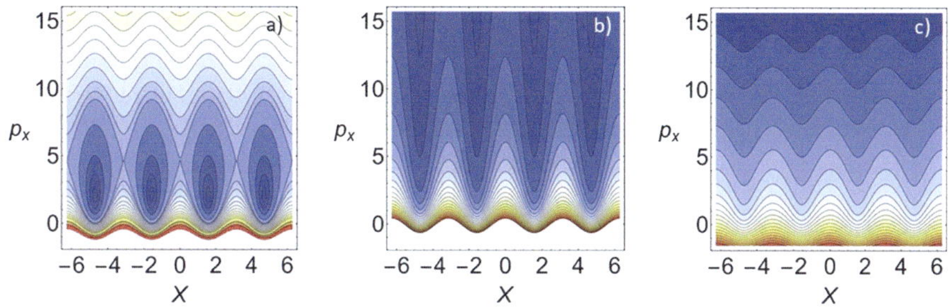

Fig. 20.1 Contours of constant level in the (X, p_x) plane of the Hamiltonian given by Eq. (20.60) for $a_0 = 2$: **a** subluminal phase velocity $\beta_w = 0.9$; **b** luminal phase velocity $\beta_w = 1$; **c** superluminal phase velocity $\beta_w = 1.1$

$$\left[1 + (\boldsymbol{a}_\perp + \boldsymbol{\pi}_\perp)^2 + p_x^2\right]^{1/2} - \beta_{ph} p_x = h. \tag{20.60}$$

Equation (20.29) describes the case when the electromagnetic wave propagates with luminal phase velocity for which $\beta_{ph} = 1$.

The contours of constant levels of the Hamiltonian, given by Eq. (20.60) elucidate the charged particle dynamics when the phase velocity of the electromagnetic wave β_{ph} changes. We assume that $|\boldsymbol{a}_\perp + \boldsymbol{\pi}_\perp| = a_0 \cos X$. Figure 20.1 shows isocontours for the Hamiltonian (20.60) for $a_0 = 2$: (a) subluminal phase velocity $\beta_{ph} = 0.9$; (b) luminal phase velocity $\beta_{ph} = 1$; (c) superluminal phase velocity $\beta_{ph} = 1.1$.

We can see in Fig. 20.1a, in the case of the electromagnetic wave wave propagating with a subluminal phase velocity ($\beta_w = 0.9$) on the phase plane there is a separatrix that encloses the trajectories of trapped charged particles. Below and above the separatrix the trajectories correspond to untrapped (transient) charged particles. In Fig. 20.1b, we have a phase plot of the luminal electromagnetic wave ($\beta_w = 1$). While the phase velocity of the wave approaches the luminal case the separartix extends to infinity. The phase plane that corresponds to the interaction of the charged particle with the electromagnetic wave with superluminal phase velocity ($\beta_w = 1.1$) does not have any separatrix. All trajectories are the trajectories of the transient charged particles, whose energy periodically changes.

20.3 Resonant Acceleration of Electrons Trapped by Potential Wave in a Magnetic Field

It is known that a quasi-static magnetic field can modify the wave–particle interaction in a plasma. Under certain conditions, the magnetic field can create a regime of so-called "unlimited" (without phase slippage) particle acceleration (surfatron acceleration) as shown by Katsouleas and Dawson (1983). The situation in which the magnetic field strongly affects the acceleration of particles trapped by a potential wave is typical of space and laboratory plasmas (Artemyev et al., 2018; Berezinskii et al., 1990; Bulanov et al., 2013; Hosokai et al., 2010; Reynolds, 2008). This circumstance gives rise to the wide scope of scientific problems discussed in literature. The acceleration of charged particles in a magnetic field (which is also referred to as the $\boldsymbol{v} \times \boldsymbol{B}$ acceleration) is a fundamental phenomenon.

The basic idea of the acceleration mechanism can be described as follows. In the frame of reference moving with the phase velocity of the wave V_w, the electric field directed along the wave front appears. The induced electric field is determined by a standard formula of the Lorentz transformation:

$$E = \frac{\beta_w B_{||}}{\sqrt{1-\beta_w^2}}. \tag{20.61}$$

where $\beta_w = V_w/c$ and $B_{||}$ is the magnetic field component parallel to the wave front. The necessary condition for acceleration is that the particle must be trapped by the wave. It is easy to show that, for the particle to be trapped (confined), it is necessary that the component of the electric field perpendicular to the wave front, $E_\perp$, be large enough:

$$E_\perp > B_{||}\frac{\sqrt{v^2/c^2-\beta_w^2}}{1-\beta_w^2}. \tag{20.62}$$

Here where v is the particle velocity. The condition for unlimited acceleration ($v \to c$) is

$$E_\perp > \frac{B_{||}}{\sqrt{1-\beta_w^2}}. \tag{20.63}$$

In this case, the energy of particle grows as $\mathcal{E} > eB_{||}vt/\sqrt{1-\beta_w^2}$.

In a real situation, the front of the wave is always curved and its curvature changes with time. This refers equally to the wake plasma wave behind a short laser pulse, shock waves generated at the boundary of the self-focusing channel, and shock waves in space plasmas. The effects of the shock wave front curvature have been studied Bulanov and Sakharov (1986) and (2000). We assume that the wave front is a smooth surface whose shape and position are determined by a given function of time $S(\boldsymbol{r}, t)$. This function is related to the structure of the electric field that affects the accelerated particle. We also assume that the wave propagates in an inhomogeneous magnetic field. To describe the charged particle motion, we use the methods of Lagrange mechanics. The Lagrange function of a particle moving with relativistic velocity in the electric and magnetic fields, which can be expressed in terms of the electrostatic potential $\Phi(\boldsymbol{r}, t)$ and vector potential $\boldsymbol{A}(\boldsymbol{r}, t)$, has the form:

$$\mathcal{L} = -m_e c^2\left(1-\frac{v^2}{c^2}\right)^{1/2} + \frac{e}{c}(\boldsymbol{A}\cdot\boldsymbol{v}) - e\Phi, \tag{20.64}$$

We introduce the curvilinear coordinates with q_j, and $j = 1, 2, 3$, where q_1 is the coordinate along the normal to the surface and q_2 and q_3 are the coordinates along the surface respectively.

For the particle trapped by the wave, the coordinate $q_1(t)$, which is a given function of time, determines the position of the wave front. The motion of a trapped particle along the front is described by the Lagrangian

$$\mathcal{L}^*(q_2, q_3, \dot{q}_2, \dot{q}_3) = \mathcal{L}^*(q_1, \dot{q}_1, q_2, q_3, \dot{q}_2, \dot{q}_3). \tag{20.65}$$

The corresponding system of Lagrange equations is

$$\frac{d}{dt}\frac{\partial \mathcal{L}^*}{\partial q_i} = \frac{\partial \mathcal{L}^*}{\partial q_i} \qquad i = 2, 3. \tag{20.66}$$

In other words, we are dealing with a nonautonomous Lagrange system with holonomic constraints dependent on time (see Arnold, 1978).

As an example, we consider the case when the position of the front is determined by the equation

$$S(x, y, t) = x - v_{ph}t + \frac{\alpha y^2}{2}, \tag{20.67}$$

where v_{ph} is the propagation velocity of the wave and is the front curvature $\alpha = 1/R$ with R being the radius of curvature in the vicinity of $y = 0$. Further, we assume that the curvature is small so that $\alpha y \ll 1$. The magnetic field is assumed to be uniform: $\boldsymbol{B} = B_0\boldsymbol{e}_y$. Then, the Lagrange function takes the form

$$\mathcal{L}^*(y, z, \dot{y}, \dot{z}, t) = -\frac{eB_0}{c}\left(v_{ph}t - \frac{\alpha y^2}{2}\right)\dot{z} - m_e c^2\left(1 - \beta^2\left(1 - \frac{2\alpha y\dot{y}}{v_{ph}}\right) + (1 + \alpha^2 y^2)\frac{\dot{y}^2}{v_{ph}^2}\right)^{1/2}. \tag{20.68}$$

For the initial conditions $y = 0$ and $\dot{y} = 0$, equations of motion (20.66) with the Lagrange function (20.68) have the solution

$$y^{(0)}(t) = 0, \quad \dot{y}^{(0)}(t) = 0, \tag{20.69}$$

$$z^{(0)}(t) = c\frac{(1 - \beta^2)\beta\,\Omega_c t}{(1 + \beta^2\Omega_c^2 t^2)^{1/2}}, \tag{20.70}$$

$$\gamma^{(0)}(t) = \left(\frac{1 + \beta^2\Omega_c^2 t^2}{1 - \beta^2}\right)^{1/2}, \tag{20.71}$$

where $\Omega_c = eB_0/m_e c$ is the electron cyclotron frequency calculated for the magnetic field B_0 and $\gamma = \mathcal{E}/m_e c^2$ with $\mathcal{E}$ being the electron energy. The particle remains at the parabola apex and is accelerated along the z-axis. In the limit $\beta\,\Omega_c t \to \infty$, the particle energy increases as $\mathcal{E} = m_e c^2 \beta\,\Omega_c t/(1-\beta^2)^{1/2}$.

20.4 Quantum Regime of the Electron Interaction with the Electromagnetic Wave

The quantum electrodynamics Lagrangian describing an electron (positron) dynamic in the electromagnetic field is a sum of local products of the operators ψ and A_μ. The operator $\psi(x)$ is a positron creation and an electron annihilation operator, and $\overline{\psi}(x)$ is an electron creation and a positron annihilation operator, respectively. As far it concerned the electromagnetic field the operator $A_\mu(x)$ creates or annihilates the photon at the point $x = (t, \boldsymbol{r})$. The Lagrangian has a form:

$$\mathcal{L}(x) = \overline{\psi}(x)\big[\gamma^\mu(i\partial_\mu - eA_\mu(x)) - m_e\big]\psi(x) - \frac{1}{4}F_{\mu\nu}(x)F^{\mu\nu}(x), \tag{20.72}$$

where $F_{\mu\nu} = \partial_\mu A_\nu - \partial_\nu A_\mu$ is the electromagnetic field operator, and γ_μ are the Dirac 4-matrices. Here we use the quantum electrodynamics notations, assuming that the Planck constant and speed of light in vacuum are equal to unity, $\hbar = c = 1$. The Lagrangian (20.60) can be written as

$$\mathcal{L}(x) = \overline{\psi}(x)(i\gamma^\mu\partial_\mu - m_e)\psi(x) - \frac{1}{4}F_{\mu\nu}(x)F^{\mu\nu}(x) - e\overline{\psi}(x)\gamma^\mu\psi(x)A_\mu(x), \tag{20.73}$$

where the first term corresponds to free electron, the second term is for photon and the third term corresponds to electron-photon interaction.

If $\psi(x)$ is the particle wave function (bi-spinor) and $A_\mu(x)$ is the vector potential of the external electromagnetic field, a variation of the Lagrangian (20.73) gives the Dirac equation (see Berestetskii et al., 1982), which is a relativistic wave equation for a spin-1/2 massive particle describing the electron interaction with the electromagnetic field:

$$\big[\gamma_\mu(i\partial_\mu - eA_\mu) + m_e\big]\psi = 0. \tag{20.74}$$

The electron motion in the plane electromagnetic wave is described by the Volkov solution of the Dirac equation (Wolkow, 1935). The Volkov state has been discussed above in Chap. 10, but from here on we will not be afraid of repetitions when they are justified by the context and the need to present other aspects of the problems under consideration. To find this solution, it is convenient to rewrite the Dirac equation in the form of the second order equation acting on Eq. (20.74) by the operator

$\gamma_\mu(i\partial_\mu - eA_\mu) + m$. This operation results in

$$\left[(i\partial_\mu - eA_\mu)^2 - m_e^2 - \frac{i}{2} e F_{\mu\nu}\sigma^{\mu\nu}\right]\psi = 0. \tag{20.75}$$

Here the matrix $\sigma^{\mu\nu}$ is related to the Dirac matrices γ_μ as

$$\sigma^{\mu\nu} = \frac{1}{2}(\gamma^\mu\gamma^\nu - \gamma^\nu\gamma^\mu) = (\boldsymbol{\alpha}, i\boldsymbol{\Sigma}). \tag{20.76}$$

Using representations $A_\mu = (\varphi, \boldsymbol{A})$ and $F^{\mu\nu} = (-\boldsymbol{E}, \boldsymbol{B})$ we can rewrite Eq. (20.75) in the three-dimensional form. In normal units it is

$$\left[\left(\frac{i\hbar}{c}\partial_t - \frac{e}{c}\varphi\right)^2 - \left(i\hbar\nabla + \frac{e}{c}\boldsymbol{A}\right)^2 - m_e^2 c^2 + \frac{e\hbar}{c}\boldsymbol{\Sigma B} - \frac{ie\hbar}{c}\boldsymbol{\alpha E}\right]\psi = 0. \tag{20.77}$$

The last two terms in the square brackets on the left-hand side of this equation appear due to the particle spin.

In the plane electromagnetic wave with the wave 4-vector $k_\alpha = (\omega, \boldsymbol{k})$ $(k_\alpha k^\alpha = 0)$ the four-potential A_μ depends on the variable u, which can be written as $u = k_\alpha x^\alpha$ (see also Eq. (20.25)). Using Eq. (20.77) written as

$$\left[-\partial^2 - 2ie(A\partial) + e^2A^2 - m_e^2 - ie(\gamma k)(\gamma A')\right]\psi = 0 \tag{20.78}$$

with $\partial^2 = \partial_\alpha\partial^\alpha$ we seek its solution in the form

$$\psi = \exp(-ipx)F(u), \tag{20.79}$$

where p is a constant 4-vector, on which we can impose a condition, $p^2 = m^2$. For a free particle, when $A_\mu = 0$, the vector p_μ is the four-momentum. Equations (20.78) and (20.79) yield for the function $F(u)$ an ordinary differential equation

$$2i(kp)F' + \left[-2e(pA) + e^2A^2 - ie(\gamma k)(\gamma A')\right]F = 0. \tag{20.80}$$

Integration of this equation gives

$$\psi = \frac{w}{(2p_0)^{1/2}}\exp(iS)\left[1 + \frac{e(\gamma k)(\gamma A)}{2(kp)}\right], \tag{20.81}$$

where

$$S = -px - \int_0^{kx}\left[\frac{e}{(kp)}(pA) - \frac{e^2}{2(kp)}A^2\right]du, \tag{20.82}$$

is the same function as the classical action for the charged particle moving in the field of the plane electromagnetic wave (see Eq. (20.4) in the limit $\beta_{ph} = 1$). This is the Volkov solution of the Dirac equation for the electron interacting with the electromagnetic wave (Wolkow, 1935). The bispinor w describes the polarization (spin) effects. It obeys the Dirac equation for a free particle,

$$(\gamma^\mu p_\mu - m_e)w = 0, \tag{20.83}$$

with additional constraints imposed:

$$w\overline{w} = 2m_e, \quad \overline{w}\gamma^\mu w = 2p^\mu. \tag{20.84}$$

The current density, $j^\mu = \overline{\psi}\gamma^\mu\psi$, by using Eqs. (20.82), (20.83), and (20.84), can be obtained as follows:

$$j^\mu = \frac{1}{p_0}\left\{p^\mu - eA^\mu + k^\mu\left[\frac{e}{(kp)}(pA) - \frac{e^2}{2(kp)}A^2\right]\right\}. \tag{20.85}$$

20.5 Electron Interaction with Plasma Wave

Here we discuss the electron dynamics in the field of the plasma wave, limited to the one-dimensional approximation (Esirkepov et al., 2006). The electron motion is described by the Hamilton equations with the Hamiltonian

$$\mathcal{H} = \sqrt{1 + p_x^2} - \phi(x, t). \tag{20.86}$$

For the electrostatic potential depending only on the variable $X = x - \beta_w t$, the Hamiltonian (20.86) has a symmetry implying conservation of the integral of motion

$$\sqrt{1 + p_x^2} - \beta_w p_x - \phi(X) = h_0. \tag{20.87}$$

The solution of this equation gives for the normalized electron energy $\gamma = \sqrt{1 + p_x^2}$ dependence on the variable X:

$$\gamma(X) = \gamma_w^2\left[\phi(X) + h_0 \pm \sqrt{(\phi(X) + h_0)^2 - 1/\gamma_w^2}\right]. \tag{20.88}$$

Here the plus sign in the front of the square root corresponds to the variable X increasing with time and the minis sign is for the decreasing variable X. The constant h_0 is determined by the initial conditions at $X = X_0$:

$$h_0 = \sqrt{1 + p_0^2} - \beta_w p_0 - \phi_0, \tag{20.89}$$

with p_0 and ϕ_0 being the electron momentum and electrostatic potential at $X = X_0$, respectively. For ultrarelativistic electron injected in the accelerating phase of the wake field, $\gamma_0 = \sqrt{1 + p_{x0}^2} \gg 1$, the integral value given by Eq. (20.89) is approximately equal to

$$h_0 \approx p_0(1 - \beta_w) - \phi_0, \tag{20.90}$$

which gives for the electron energy

$$\gamma(X) \approx \gamma_0 + \frac{\phi(X) - \phi_0}{1 - \beta_w}. \tag{20.91}$$

As we may see from Eq. (20.91), the energy achieved by the electron interacting with the plasma wave propagating with the relativistic energy is proportional to the electrostatic potential in the wave and to $2\gamma_w = 2/\sqrt{1 - \beta_w^2}$, which in the limit when β_w approaches unity grows in both cases of subluminal ($\beta_w < 1$) and superluminal ($\beta_w > 1$) phase velocity. In the luminal case with $\beta_w = 1$, Eq. (20.91) predicts an infinite energy for the electron. However, it will take an infinite time (infinite acceleration length) to reach this energy.

To illustrate the various regimes of interaction of an electron with an electrostatic wave, we present here patterns of phase planes that correspond to the contours of constant levels of the Hamiltonian, given by Eqs. (20.86, 20.87) when the phase velocity β_w changes. For simplicity we assume that the potential $\phi(X)$ has the form $\phi(X) = \phi_0 \cos X$. Figure 20.2 shows isocontours for the Hamiltonian given by Eq. (20.86) for $\phi_0 = 0.1$: (a) subluminal phase velocity $\beta_w = 0.95$; (b) luminal phase velocity $\beta_w = 1$; (c) superluminal phase velocity $\beta_w = 1.1$.

As seen in Fig. 20.2a, in the case of the electrostatic wave propagating with a subluminal phase velocity ($\beta_w = 0.95$) on the phase plane there is a separatrix that

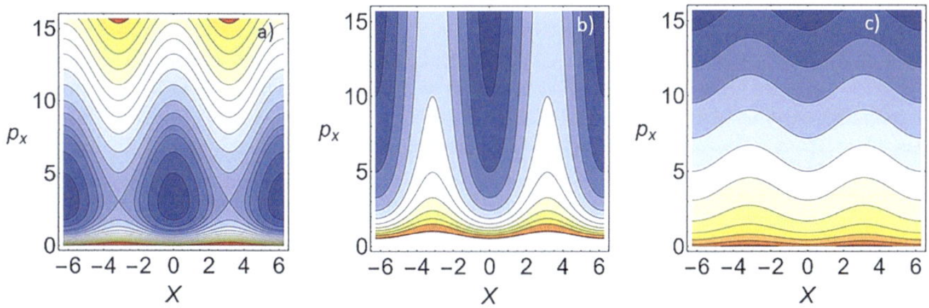

Fig. 20.2 Contours of constant level in the (X, p_x) plane of the Hamiltonian given by Eq. (20.86) for $\phi_0 = 0.1$: **a** subluminal phase velocity $\beta_w = 0.95$; **b** luminal phase velocity $\beta_w = 1$; **c** superluminal phase velocity $\beta_w = 1.1$

encloses the trajectories of trapped charged particles. Below and above the separatrix the trajectories correspond to untrapped charged particles. In Fig. 20.2b, we have a phase plot of the luminal electrostatic wave ($\beta_w = 1$). While the phase velocity of the wave approaches the luminal case the separartix blows up to infinity. The phase plane that corresponds to the interaction of the charged particle with the electrostatic wave with superluminal phase velocity ($\beta_w = 1.1$) does not have any separatrix. All trajectories are the trajectories of the transient charged particles, whose energy periodically changes.

We note that the QED consideration of the electron interaction with electrostatic wave is presented in the paper by Mendonça and Serbeto (2011).

References

Arnold, V. I. (1978). *Mathematical methods of classical mechanics*. Springer.

Artemyev, A. V., Neishtadt, A. I., Vainchtein, D. L., Vasiliev, A. A., Vasko, I. Y., & Zelenyi, L. M. (2018). Trapping (capture) into resonance and scattering on resonance: Summary of results for space plasma systems. *Communications in Nonlinear Science and Numerical Simulation, 65*, 111.

Barut, A. O. (2010). *Electrodynamics and classical theory of fields and particles*. Dover Books on Physics.

Berestetskii, V. B., Lifshitz, E. M., & Pitaevskii, L. P. (1982). *Course of theoretical physics, Vol. 4: Quantum electrodynamics*. Pergamon.

Berezinskii, V. S., Bulanov, S. V., Ginzburg, V. L., Dogiel, V. A., & Ptuskin, V. S. (1990). *Astrophysics of cosmic rays*. North Holland/Elsevier.

Bulanov, S. V. (2021). Electron dynamics in the field of strong plasma and electromagnetic waves: A review. *Physics of Wave Phenomena, 29*, 1.

Bulanov, S. V., Esirkepov, T. Zh., Kando, M., Pirozhkov, A. S., & Rosanov, N. N. (2013). Relativistic mirrors in plasmas–novel results and perspectives. *Physics Uspekhi, 56*, 429.

Bulanov, S. V., & Sakharov, A. S. (1986). Acceleration of particles captured by a strong potential wave with a curved wave front in a magnetic field. *JETP Letters, 44*, 543.

Bulanov, S. V., & Sakharov, A. S. (2000). Effect of the magnetic field on the resonant particle acceleration. *Plasma Physics Reports, 26*, 1005.

Esirkepov, T. Zh., Bulanov, S. V., Yamagiwa, M., & Tajima, T. (2006). Electron, positron, and photon Wakefield acceleration: Trapping, wake overtaking, and ponderomotive acceleration. *Physical Review Letters, 96*, Article 014803.

Hosokai, T., Zhidkov, A., Yamazaki, A., Mizuta, Y., Uesaka, M., & Kodama, R. (2010). Electron energy boosting in laser-wake-field acceleration with external magnetic field B$\sim$ 1 T and laser prepulses. *Applied Physics Letters, 96*, 121501.

Katsouleas, T., & Dawson, J. M. (1983). Unlimited electron acceleration in laser-driven plasma waves. *Physical Review Letters, 51*, 392.

Konopinski, E. J. (1981). *Electromagnetic fields and relativistic particles*. McGrow Hill Book Company.

Lai, H. M. (1980). Particle acceleration by an intense solitary electromagnetic pulse. *Physics of Fluids, 23*, 2373.

Landau, L. D., & Lifshitz, E. M. (1980). *Course of theoretical physics, Vol. 2: The classical theory of fields*. Pergamon.

Lawson, J. D. (1979). Lasers and accelerators. *IEEE Transactions on Nuclear Science, 26*, 4217.

Mendonça, J. T., & Serbeto, A. (2011). Volkov solutions for relativistic quantum plasmas. *Physical Review E, 83*.

Reynolds, S. P. (2008). Supernova remnants at high energy. *Annual Review of Astronomy and Astrophysics, 46*, 89.
Sokolov, A. A., & Ternov, I. M. (1986). *Radiation from relativistic electrons.* AIP.
Whitham, G. B. (1974). *Linear and nonlinear waves.* Wiley.
Wolkow, D. M. (1935). Über eine Klasse von Lösungen der Diracschen Gleichung. *Zeitschrift fur Physik, 94*, 250.
Woodward, P. M. (1947). A method for calculating the field over a plane aperture required to produce a given polar diagram. *Journal of IEEE, 93*(Part III A), 1554.

Chapter 21
Transverse and Longitudinal Waves in a Plasma

Abstract Here in this section, we discuss the properties of finite amplitude electromagnetic and plasma waves propagating in collisionless plasmas. Small-amplitude electromagnetic and plasma waves can be described in the linear approximation framework as waves propagating in the plasma without interacting with each other and with their frequencies independent of the wave amplitude. Finite amplitude, electromagnetic waves in the relativistic limit are described by nonlinear equations in partial derivatives. Nonlinear interaction leads to coupling of transverse electromagnetic waves with longitudinal plasma waves resulting, in particular, in the generation of the wake plasma waves left in the plasma behind short laser pulse. High enough amplitude plasma waves undergo the wave steepening and breaking leading to formation of the electron density singularity. Due to nonlinear evolution of the electromagnetic wave the relativistic electromagnetic solitons occur in collisionless plasma, and the electron vortices are formed.

Keywords Relativistically intense transverse and longitudinal waves in plasmas · Relativistic solitons and vortices

21.1 Finite Amplitude Electromagnetic Waves

The interest towards nonlinear plasma wave studies stems from the problem of electron acceleration by wake field, which has been proposed (Tajima & Dawson, 1979) (see also review articles Esarey et al., 2009; Mourou et al., 2006, 2011; Tajima & Malka, 2020; Tajima et al., 2020). It is related to the development of a compact accelerator of high-energy electrons and positrons (Leemans & Esarey, 2009). As has been demonstrated theoretically and experimentally (see review article Bulanov et al., 2013), wake plasma waves can also be used as mirrors moving with a relativistic velocity; in reflecting the photons from relativistic mirror their frequency can be substantially upshifted and the intensity of reflected radiation can be increased.

For finite-amplitude waves, an exact solution describing wave propagation in cold collisionless plasmas was found by Akhiezer and Polovin (1956). The use

Y. Kato et al., *High Power Laser and Plasma Science*, Springer Series on Atomic, Optical, and Plasma Physics 130, https://doi.org/10.1007/978-3-032-08930-4_21

of hydrodynamics approximation for describing the electron and ion component dynamics and Maxwell equations for a self-consistent electromagnetic field allowed deriving the system of equations for coupled electromagnetic and plasma waves. This system of equations has the form (Farina & Bulanov, 2001a, 2001b; Kozlov et al., 1979)

$$\partial_{tt}\boldsymbol{A} - \Delta\boldsymbol{A} + \partial_t \nabla\phi = n_a \boldsymbol{v}_a - n_e \boldsymbol{v}_e, \tag{21.1}$$

$$\Delta\phi = n_e - n_a, \tag{21.2}$$

$$\partial_t n_b + \nabla \cdot (n_b \boldsymbol{v}_b) = 0, \tag{21.3}$$

$$\partial_t \boldsymbol{P}_b = -\nabla(\rho_b \phi + \gamma_b) + \boldsymbol{v}_b \times \nabla \times \boldsymbol{P}_b, \tag{21.4}$$

where the indices e and a denote the electron and ion components, and the index b can be either e or a. The generalized momentum $\boldsymbol{P}_b$ and the relativistic gamma-factor γ_b are related to the particle momentum $\boldsymbol{p}_b$ and the vector potential $\boldsymbol{A}$ as

$$\boldsymbol{P}_b = \boldsymbol{p}_b + \rho_b \boldsymbol{A}, \tag{21.5}$$

and

$$\gamma_b = \sqrt{1 + |\boldsymbol{p}_b|^2}. \tag{21.6}$$

The parameter ρ_b is equal to $em_b/e_b m_e$. The velocity of an electron (ion) fluid element is

$$\boldsymbol{v}_b = \boldsymbol{p}_b/\gamma_b.$$

In this system of equations, the variables are normalized as follows. The units of length and time are c/ω_{pe} and ω_{pe}^{-1}, velocity is normalized to c, momentum to $m_e c/e$, the vector and electrostatic potential to $m_e c/e$, and density to n_0.

In an ansatz corresponding to one-dimensional geometry, the dependence of all functions on the time t and coordinate x is assumed to be of the form

$$X = x - v_g t \text{ and } \tau = t - v_g x, \tag{21.7}$$

which corresponds to a plane wave propagating with constant velocity. For the sake of simplicity, we assume below that $e_\alpha = |e|$, i.e., $\rho_\beta \equiv -\rho = -m_e/m_\alpha$. Equations for the electromagnetic and electrostatic potentials,

$$A_y + iA_z = a(X)\exp(i\omega\tau) \text{ and } \phi = \phi(X), \tag{21.8}$$

can be rewritten in the form (Farina & Bulanov, 2001a, 2001b)

$$\phi'' = \frac{\beta_g}{1-\beta_g^2}\left(\frac{\psi_e}{R_e} - \frac{\psi_a}{R_a}\right), \tag{21.9}$$

$$a'' + \omega^2 a = a\frac{\beta_g}{1-\beta_g^2}\left(\frac{1}{R_e} - \frac{\rho}{R_a}\right), \tag{21.10}$$

where the prime denotes differentiation with respect to X and $\beta_g = v_g/c$ is the group velocity of the electromagnetic wave normalized to the speed of light in the vacuum c, equal to the phase velocity of the plasma wave. We have also introduced the functions

$$\psi_e = \Gamma_e + \phi, \tag{21.11}$$

$$\psi_a = \Gamma_a - \rho_a \phi, \tag{21.12}$$

$$R_e = \sqrt{\psi_e^2 - (1-\beta_g^2)(1+a^2)}, \tag{21.13}$$

$$R_\alpha = \sqrt{\psi_\alpha^2 - (1-\beta_g^2)(1+\rho_a^2 a^2)}, \tag{21.14}$$

where the constants Γ_e and Γ_a are determined by the boundary conditions. For example, if the field has a constant amplitude ($a = a_0$, $\phi = 0$), and the plasma is at rest at $x \to \pm\infty$, then $\Gamma_e = \sqrt{1 + a_0^2}$ and $\Gamma_a = \sqrt{1 + \rho_a^2 a_0^2}$.

The density and the relativistic Lorentz factors of the electron and ion components $b = e, \alpha$ are equal to

$$n_b = \beta_b \frac{(\psi_b - \beta_b R_b)}{R_b(1-\beta_g^2)}, \tag{21.15}$$

and

$$\gamma_b = \frac{\psi_b - \beta_g R_b}{1-\beta_g^2}, \tag{21.16}$$

respectively.

System of equations (21.9) and (21.10) has the integral

$$\frac{1-\beta_g^2}{2}(a'^2 + \omega^2 a^2) + \frac{1}{2}\phi'^2 + \frac{\beta_g}{1-\beta_g^2}\left(R_e + \frac{R_a}{\rho_a}\right) = \text{constant}. \tag{21.17}$$

21.2 Transverse Electromagnetic Waves

21.2.1 The Case of Circular Polarization

In the case of a transverse electromagnetic wave with a circular polarization, Eqs. (21.9) and (21.10) possess a solution describing a constant-amplitude wave with $a(X) = a_0$ and $\phi = 0$, i.e.,

$$A_y = a_0 \cos(\omega\tau) \, \text{and} \, A_z = a_0 \sin(\omega\tau). \tag{21.18}$$

The dependence of the wave frequency on the amplitude follows from Eq. (21.10). It reads

$$\omega^2 = \gamma_g^2 \left(\frac{1}{\Gamma_e} + \frac{\rho_a}{\Gamma_a} \right), \tag{21.19}$$

where $\gamma_g = 1/\sqrt{1 - \beta_g^2}$ is the gamma-factor calculated for the wave group velocity. Taking into account the relation between the phase and group velocity of electromagnetic wave, $v_{ph} v_g = c^2$, and introducing the wave number k, for which the phase velocity equals $v_{ph} = \omega/k$, we can write the dispersion equation for the frequency and wave number of the relativistically intense electromagnetic wave (it is written here in non-normalized form):

$$\omega^2 = k^2 c^2 + \omega_{pe}^2 \left(\frac{1}{\sqrt{1 + a_0^2}} + \frac{\rho_a}{\sqrt{1 + \rho_a^2 a_0^2}} \right). \tag{21.20}$$

As we see, the relativistic effects change the wave frequency, but for circularly polarized wave they do not lead to the appearance of higher-order harmonics in the frequency spectrum. In the field of a circularly polarized electromagnetic wave, electrons rotate with frequency given by Eq. (21.19). The electron energy in the wave is equal to $m_e c^2 \sqrt{1 + a_0^2}$. The longitudinal component of the electron momentum vanishes, and the transverse component is $m_e c a_0$.

In a linearly polarized electromagnetic wave, the transverse and longitudinal components of the electromagnetic field and of the electron momentum are coupled to each other (Akhiezer & Polovin, 1956; Chian, 1981). In a wave of a small but finite amplitude, the transverse component of the electric field oscillates with the frequency

$$\omega \approx kc + \frac{\omega_{pe}^2}{2kc} \left(1 - \frac{a_0^2}{2} \right). \tag{21.21}$$

The longitudinal component oscillates with the doubled frequency, having an amplitude approximately equal to a_0^2.

Equations describing a plane electromagnetic wave take a particularly simple form if we make the Lorentz transformation to the frame of reference moving with the group velocity. In this boosted frame of reference, the wave group velocity vanishes and the phase velocity $\overline{v}_{ph} = \overline{\omega}/\overline{k}$, by virtue of the relation $v_{ph}v_g = c^2$, becomes equal to infinity, and hence the wave number also vanishes. This means that in the boosted frame of reference all the values depend only on time (Chian, 1981). Harmonic oscillations with the frequency

$$\overline{\omega} = \omega_{pe}\sqrt{\frac{1}{\sqrt{1+a_0^2}} + \frac{\rho_a}{\sqrt{1+\rho_a^2 a_0^2}}} \tag{21.22}$$

correspond to a circular-polarization wave. Here, the bar denotes the values in the boosted frame of reference.

A linearly polarized electromagnetic wave is described by a system of equations of coupled nonlinear oscillators. Oscillations in the longitudinal direction occur at odd harmonics of the fundamental frequency and the frequency spectrum of transverse oscillations comprises the even harmonics.

The circular and linear polarizations are isolated cases of general electromagnetic mode: there is no change in the polarization and there is no rotation of the polarization plane. In a relativistically strong electromagnetic wave with elliptical polarization, which also has a longitudinal component of the electric field, rotation of the polarization plane occurs due to the nonlinearly induced birefringence (Chakraborty et al., 1983; Smetanin et al., 2004). This process is similar to the Faraday rotation of the polarization plane of the electromagnetic wave propagating through a magnetized medium.

21.3 Longitudinal Plasma Waves

For a given function $a(X)$, Eq. (21.9) describes longitudinal waves. They can be excited by a finite length electromagnetic pulse. In particular, if $a = a_0 = 0$, this equation describes free Langmuir oscillations. If for the sake of simplicity we neglect the ion motion effects, the dependence of the relativistic Lorentz factor γ_e of the electrons in the wave on the coordinate, which we redefine for plasma waves as $X = x - v_{ph}t$, because their group velocity is equal to zero, can be represented implicitly as

$$\frac{1}{\sqrt{2}}X = \beta_{ph}\Big[(\gamma_m+1)^{1/2}\mathrm{E}(\Psi|\kappa) - (\gamma_m+1)^{-1/2}\mathrm{F}(\Psi|\kappa)\Big] - (\gamma_m-\gamma_e)^{1/2}. \tag{21.23}$$

Here

$$\mathrm{F}(\Psi \mid \kappa) = \int_0^{\Psi} \left(1 - \kappa \sin^2 \theta\right)^{-1/2} d\theta \quad \text{and} \quad \mathrm{E}(\Psi \mid \kappa) = \int_0^{\Psi} \left(1 - \kappa \sin^2 \theta\right)^{1/2} d\theta \tag{21.24}$$

are the elliptic integrals of the first and second kinds, respectively (Olver et al., 2010).

In Eq. (21.23)

$$\Psi = \pm \mathrm{Arcsin}\left[\sqrt{(\gamma_m - \gamma_e)/(\gamma_m - 1)}\right] \tag{21.25}$$

is the amplitude,

$$\kappa = (\gamma_m - 1)/(\gamma_m + 1) \tag{21.26}$$

is the modulus, and γ_m is the maximum gamma-factor value of electrons in the wave.

The wavelength of longitudinal plasma waves is given by

$$\lambda_p = \frac{2^{5/2} v_{ph}}{\omega_{pe}}\left[(\gamma_m + 1)^{1/2}\mathrm{E}\left(\frac{\gamma_\mathrm{m} - 1}{\gamma_\mathrm{m} + 1}\right) - (\gamma_\mathrm{m} + 1)^{-1/2}\mathrm{K}\left(\frac{\gamma_\mathrm{m} - 1}{\gamma_\mathrm{m} + 1}\right)\right]. \tag{21.27}$$

where $\mathrm{K}(\kappa) = \mathrm{F}(\pi/2 \mid \kappa)$ and $\mathrm{E}(\kappa) = \mathrm{E}(\pi/2 \mid \kappa)$ are the complete elliptic integrals of the first and second kinds (Olver et al., 2010).

In the limit of a large amplitude, the wavelength dependence on the parameters is given by

$$\lambda_p \underset{\gamma_m \to \infty}{=} \frac{2^{5/2}\gamma_m^{1/2} c}{\omega_{pe}}, \tag{21.28}$$

which is equivalent to the dependence of the frequency of plasma oscillations

$$\omega_p \underset{\gamma_m \to \infty}{=} \pi \frac{\omega_{pe}}{2^{3/2}\gamma_m^{1/2}} \text{ and } \omega_p \underset{\gamma_m \to 1}{=} \omega_{pe}\left[1 - 3\left(\frac{p_m}{m_e c}\right)^2\right], \tag{21.29}$$

in the limits of large and small values of the amplitudes. Here, the maximum value of the momentum of the electrons in the wave, p_m, is related to the maximum gamma-factor γ_m as $p_m = m_e c\sqrt{\gamma_m^2 - 1}$.

21.4 Coupling of Transverse and Longitudinal Motion

In the linearly polarized waves in plasma, the transverse and longitudinal motions of electrons are always coupled as was shown by Akhiezer and Polovin (1956), Kaw and Dawson (1970), Chian (1981), Smetanin et al. (2004) and Pesch and Kull (2007). To demonstrate typical properties of the coupled transverse and the longitudinal waves, we use the property of the plane electromagnetic wave to have zero wave number, $k' = 0$, in the frame of reference moving with the group velocity, $v_g = c\beta_g$. Here we use the approach developed by Chian (1981), Bulanov et al. (2005), Bulanov (2021).

In the boosted frame of reference moving with the velocity V, the wave frequency ω' and wave number k' are given by expressions, respectively,

$$\omega' = \frac{\omega - kV}{\sqrt{1 - V^2/c^2}} \text{ and } k' = \frac{k - \omega V/c^2}{\sqrt{1 - V^2/c^2}}. \tag{21.30}$$

If the velocity V is equal to $V = kc^2/\omega$, i.e., it is equal to the wave group velocity, the wave number k' vanishes. Using the dispersion equation for the electromagnetic wave frequency and wave number, we find that the wave frequency in the boosted frame of reference equals the Langmuir frequency,

$$\omega' = \omega_{pe} \text{ and } k' = 0. \tag{21.31}$$

Since $k' = 0$, the plane electromagnetic wave here is the plasma oscillations with homogeneous amplitude. It is easy to show that in the boosted frame of reference the wave magnetic field vanishes. In this frame of reference, Eqs. (21.1)–(21.4) can be cast in the form

$$\boldsymbol{p}'_\perp - m_e c \boldsymbol{a}'_\perp = \text{constant}, \tag{21.32}$$

$$\partial_{t'} n' = 0, \tag{21.33}$$

$$\partial_{t'} p'_x = -eE'_x, \tag{21.34}$$

$$\partial_{t'} E'_x = \frac{4\pi e}{m_e}\left(n' \frac{p'_x}{\sqrt{1 + (p'_x/m_e c)^2 + a_\perp^2}} - n'_0 \frac{p_{0x}}{\sqrt{1 + (p_{0x}/m_e c)^2}} \right), \tag{21.35}$$

$$\partial_{t't'} \boldsymbol{a}'_\perp + \frac{4\pi n' e^2}{m_e} \frac{\boldsymbol{a}'_\perp}{\sqrt{1 + (p'_x/m_e c)^2 + a_\perp^2}} = 0, \tag{21.36}$$

where a prime denotes the variables in the boosted frame of reference,

$$\boldsymbol{a}'_\perp = a'_y\boldsymbol{e}_y + a'_z\boldsymbol{e}_z,\ \boldsymbol{p}'_\perp = p'_y\boldsymbol{e}_y + p'_z\boldsymbol{e}_z,\ p_{0x} = -m_e v_g/\sqrt{1-v_g^2/c^2},$$

and n'_0 is the density of the ambient ions in the boosted frame of reference, which is higher than the ion density in the laboratory frame, n_0, by a factor $\gamma_g = 1/\sqrt{1-v_g^2/c^2}$. Further, where it will not cause an ambiguity, we drop a prime notation and use normalized variables:

$$p_x/m_e c \to p_x,\ \boldsymbol{p}'_\perp/m_e c \to \boldsymbol{p}_\perp,\ n'/n_0 \to \gamma_g,\ eE'/m_e\omega_{pe} \to E,\ \omega_{pe}t' \to t. \tag{21.37}$$

As a result, the system of equations (21.32)–(21.36) takes the form

$$\partial_{tt}p_x = \frac{p_{0x}}{\sqrt{1+p_{0x}^2}} - \frac{p_x}{\sqrt{1+p_x^2+a_\perp^2}}, \tag{21.38}$$

$$\partial_{tt}\boldsymbol{a}_\perp + \frac{\boldsymbol{a}_\perp}{\sqrt{1+p_x^2+a_\perp^2}} = 0. \tag{21.39}$$

In particular cases of $\boldsymbol{a}_\perp = 0$, Eqs. (21.38) and (21.39) have the solution in the form of longitudinal Langmuir oscillations, and if $p_x = p_{0x}\sqrt{1+a_0^2}$, they describe transverse circularly polarized oscillations with the amplitude a_0. In general case, the system of equations (21.38) and (21.37) describes the coupled transverse and longitudinal oscillations whose oscillation frequencies, ω_x, ω_y, and ω_z, are different and depend on the oscillation amplitudes.

The system of equations (21.32), (21.38) and (21.39) can be cast to the Hamiltonian form with the Hamiltonian given by

$$\mathcal{H}(\boldsymbol{E},\boldsymbol{p}) = \frac{1}{2}\left(E_x^2+E_y^2+E_z^2\right) + \sqrt{1+p_x^2+p_y^2+p_z^2} - \frac{p_{0x}}{\sqrt{1+p_{0x}^2}}p_x. \tag{21.40}$$

The trajecrtories of the system (21.32), (21.38) and (21.39) lie on the surface of constant value of the Hamiltonian. Surface of equal value of the Hamiltonian taken in the form

$$\mathcal{H}(E,p_x,p_\perp) = E^2/2 + \sqrt{1+p_x^2+p_\perp^2} - v_0 p_x = 1.75$$

is shown in Fig. 21.1 for $v_0 = 0.8$. It is seen that the motion is periodic with the frequencies depending on the oscillation amplitudes.

Performing the Lorentz trasform to the laboratory frame of reference, we find the coupled transverse and longitudinal waves whose dependence on time and coordinate is given by $\omega_j\gamma_g(t - x\beta_g)$ with $j = x, y, z$, i.e., they propagate with superluminal phase velocity equal to $1/\beta_g$.

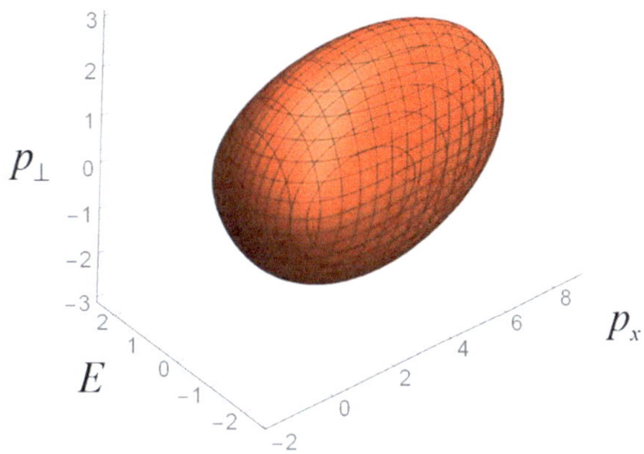

Fig. 21.1 Surface of equal value of the Hamiltonian $\mathcal{H}(E, p_x, p_\perp) = E^2/2 + \sqrt{1 + p_x^2 + p_\perp^2} - v_0 p_x = 1.75$ for $v_0 = 0.8$

In Figs. 21.2 and 21.3, we present the results of numerical integration of the system of Eqs. (21.30)–(21.37) in the case of linearly polarized waves with $p_z = 0$ and $E_z = 0$ for initial, at $t = 0$, conditions $\beta_g = 0.98$, $p_0 = 5$, $p_{x0} = p_0$, $p_{y0} = -2$, $E_{x0} = 1$, and $E_{y0} = -2$. Figure 22.2 shows the dependence of longitudinal and transverse momenta and of the electron density in the laboratory frame of reference on the coordinate $\tau = \gamma_g(t - x\beta_g)$. Figure 22.3 presents the dependence of longitudinal and transverse momenta and of the electron density in the laboratory frame of reference on the variable τ. As is seen, the period of oscillations of the longitudinal momentum p_x and electric field E_x and electron density n is approximately larger than the oscillation period of the transverse components of the momentum and electric field, p_y and E_y, by a factor $\gamma_g = 1/\sqrt{1 - \beta_g^2}$. In Fig. 21.2, we see that the electron density is modulated with the period of transverse oscillations. In its turn, the transverse oscillations, p_y and E_y, are modulated with the period of longitudinal oscillations. The period and amplitude of transverse oscillations are modulated by longitudinal oscillations too.

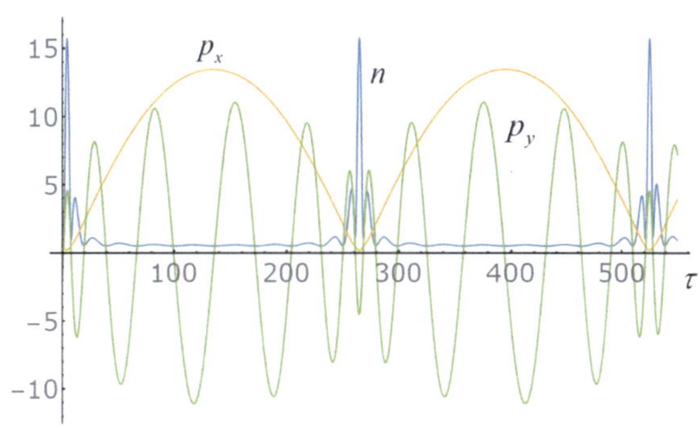

Fig. 21.2 Coupled transverse and longitudinal waves: the longitudinal momentum p_x devided by $2p_0$, transverse momentum $p_y\beta_g/\sqrt{1 - \beta_g^2}$, and the electron density $n\sqrt{1 - \beta_g^2}$ versus the coordinate τ for $\beta_g = 0.98$, $p_0 = 5$, $p_{x0} = p_0$, $p_{y0} = -2$, $E_{x0} = 1$, and $E_{y0} = -2$

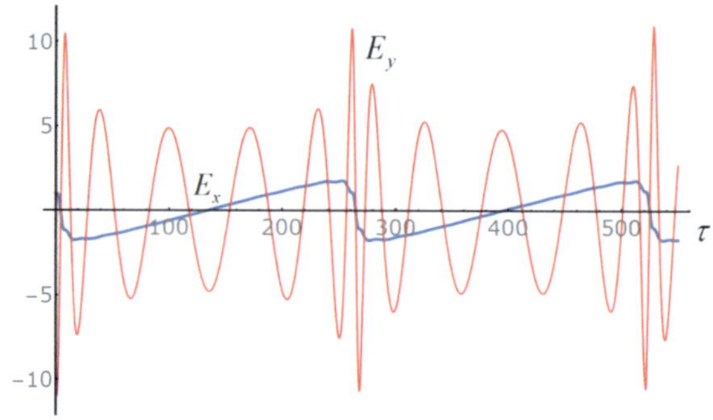

Fig. 21.3 Coupled transverse and longitudinal waves: the x- and y-components of the electric field, E_x and E_y, versus the coordinate τ for the same parameters as in Fig. 22.1

21.5 The Breaking of Plasma Wave

21.5.1 Maximum Amplitude of Langmuir Wave

The amplitude of steady-state plasma waves for a given value of the phase velocity v_{ph} cannot be arbitrarily large. It follows from Eq. (21.21) that the maximum value of the Lorentz factor γ_m, associated with the maximum momentum of the electrons in the wave p_m satisfies the condition

$$\gamma_m = \sqrt{1 + \frac{p_m^2}{m_e^2 c^2}} \leq \gamma_{ph} = \frac{1}{\sqrt{1 - v_{ph}^2/c^2}}. \tag{21.41}$$

This means that the velocity of the particles cannot exceed the phase velocity of the wave. When $\gamma_m \to \gamma_{ph}$, a singularity occurs at which the particle density becomes infinite. The integral (21.17) conservation implies the relation

$$E^2 + 2\gamma_e \left(\frac{m_e \omega_{pe} c}{e}\right)^2 = \text{constant} \tag{21.42}$$

between the electric field and the electron energy, which in turn imposes a constraint on the electric field in a nonlinear plasma wave:

$$E_{AP} = \frac{m_e \omega_{pe} c}{e} \sqrt{2(\gamma_{ph} - 1)}, \tag{21.43}$$

which is often referred to as the Akhiezer and Polovin limit value of the field (Akhiezer & Polovin, 1956).

Because strongly nonlinear wake waves play a key role in laser-matter interaction, we discuss the wave breaking effect below in more detail.

21.5.2 Breaking of Plasma Waves Due to the Phase Mixing

Nonlinear plasma waves break according to a completely different scenario than finite-amplitude gravity waves on the surface of water, which is described by Witham (1974). For the sake of clarity, we assume that the amplitude of the plasma wave is nonrelativistic and the ion component is immobile. Equations (21.2), (21.3) and (21.4) within the framework of the one-dimensional approximation result in the equations for the electron velocity and the self-consistent electric field,

$$\partial_t v + v\partial_x v = -\frac{e}{m_e}E, \tag{21.44}$$

$$\partial_t E + v\partial_x E = 4\pi\, en(\varepsilon x)v. \tag{21.45}$$

In these equations, the density of immobile ions $n(\varepsilon x)$ is assumed to be weakly inhomogeneous with the inverse inhomogeneity scale length proportional to a small parameter $\varepsilon \ll 1$. Although the transformation to the Lagrangian variables gives a solution of this system in quadratures for arbitrarily large amplitude waves, we solve it by using the perturbation method, in order to find whether harmonics of different orders that appear in higher orders of the perturbation theory are in resonance with each other.

Expanding $v(x, t)$, $E(x, t)$, and $n(\varepsilon x)$ in power series in the small parameter ε, we obtain

$$v = \varepsilon v^{(1)} + \varepsilon^2 v^{(2)} + \cdots, \tag{21.46}$$

$$E = \varepsilon E^{(1)} + \varepsilon^2 E^{(2)} + \cdots, \tag{21.47}$$

$$n = n_0 + \varepsilon n_0' x + \cdots, \tag{21.48}$$

where a prime denotes differentiation with respect to the coordinate.

In the first order in the small parameter, the solution of Eqs. (21.44) and (21.45) is given by

$$v^{(1)} = v_m \sin(\omega_{pe} t - kx), \tag{21.49}$$

$$E^{(1)} = -\frac{m_e v_m \omega_{pe}}{e} \cos(\omega_{pe} t - kx), \tag{21.50}$$

where v_m is the maximum value of the electron velocity in the wave.

In the second order in the small parameter ε, the equations for $v^{(2)}$ and $E^{(2)}$ that follow from Eqs. (21.44) and (21.45) are

$$\partial_t v^{(2)} + e\frac{E^{(2)}}{m_e} = \frac{v_m^2 k}{2}\sin\big(2(\omega_{pe}t - kx)\big), \tag{21.51}$$

$$\begin{aligned}\partial_t E^{(2)} - 4\pi e n_0 v^{(2)} &= \frac{m_e v_m^2 \omega_{pe} k}{2e}\{1 - \cos\big[2(\omega_{pe}t - kx)\big]\} \\ &+ 4\pi e n_0' x v_m \sin(\omega_{pe}t - kx)\end{aligned} \tag{21.52}$$

They have the solution

$$\begin{aligned}v^{(2)} &= \frac{n_0' x v_m}{4n_0}\big[2\omega_{pe}t\cos(\omega_{pe}t - kx) - \sin(\omega_{pe}t - kx)\big] \\ &- \frac{v_m^2 k}{2\omega_{pe}}\{1 - \cos\big[2(\omega_{pe}t - kx)\big]\},\end{aligned} \tag{21.53}$$

$$\begin{aligned}E^{(2)} &= \frac{\pi e n_0' x v_m}{\omega_{pe}}\big[2\omega_{pe}t\sin(\omega_{pe}t - kx) - \cos(\omega_{pe}t - kx)\big] \\ &- \frac{m_e v_m^2 k}{2e}\sin\big[2(\omega_{pe}t - kx)\big]\end{aligned} \tag{21.54}$$

As can be seen, the plasma wave harmonics are not in a resonance in a uniform plasma with $n = \text{const}$, which means that $n_0' = 0$. This follows from the fact that the group and phase velocities of the plasma wave are not equal. The group velocity of the plasma wave in cold plasma vanishes, while the phase velocity is finite and equal to the ratio $v_{ph} = \omega_{pe}/k$. In a homogeneous plasma, the plasma wave breaks if its amplitude v_m is so large that the inequality $v_m > v_{ph}$ is satisfied. In an inhomogeneous plasma where $n_0' \neq 0$, the plasma wave, being a wave with a continuous spectrum, breaks for an arbitrarily small but finite amplitude due to phase mixing, leading to an increase in the wave number with time. Formally, the increase in the wave number is described by the first terms in the right-hand sides of Eqs. (21.53) and (21.54), containing the terms proportional to the time, which corresponds to the resonance between the first and second harmonics. It is easy to show that the subsequent harmonics are also in resonance. This leads to nonlinear steepening of the wave profile and eventually to its breaking.

Phase mixing has a simple explanation. In general, the wave dependence on time and coordinate can be represented as

$$v = v_m \exp(i\psi),$$

where $\psi(x, t)$ is its phase or eikonal. The wave number and frequency are respectively equal to the derivative of the eikonal with respect to the coordinate ($k = \partial_x\psi$) and to minus the derivative with respect to time ($\omega = -\partial_t\psi$). The equality of mixed partial derivatives $\partial_{tx}\psi = \partial_{xt}\psi$ implies the equation

$$\partial_t k = -\partial_x \omega. \tag{21.55}$$

In a stationary but inhomogeneous medium, the solution of this equation is given by the expression that describes the growth of the wave number

$$k = k_0 - \omega' t, \tag{21.56}$$

where k_0 is its initial value. Because plasma waves can be represented as an ensemble of non-interacting oscillators, the oscillators have different frequencies in a medium with inhomogeneous parameters, which lead to the growth of the phase difference between them. This phenomenon is called phase mixing. As a result of phase mixing, the wave profile is modulated such that the wavelength asymptotically decreases to the limit where the effects of dissipation, dispersion, or nonlinearity become significant.

Nonlinear effects become important when the displacement of electrons in the wave approaches the wavelength value. In a collisionless plasma, neighboring electrons overtake each other, which corresponds to wave breaking, also referred to as the particle trajectory self-intersection. Phase mixing occurs not only due to the inhomogeneity of the plasma but also because of the non-one-dimensional geometry (Dawson, 1959); in the limit of relativistic energies of the electron oscillations, phase mixing occurs due to the plasma wave amplitude inhomogeneity (Drake et al., 1976).

Here, we analyze the breaking of an arbitrarily large-amplitude plasma wave in 1-D, 2-D and 3-D geometry. In the hydrodynamic approximation, the plasma waves are described by the equations (Bulanov et al., 2012; Dawson, 1959)

$$\partial_t n + r^{-s} \partial_r (r^s n v) = 0, \tag{21.57}$$

$$\partial_t p + v \partial_r p = -E, \tag{21.58}$$

$$\partial_t E + v r^{-s} \partial_r (r^s E) = v, \tag{21.59}$$

for the density, the electron momentum, and the electric field. Here, the index s takes the values 1, 2, 3 for planar, cylindrical, and spherical waves, respectively. The velocity is related to the electron momentum as $v = p/\sqrt{1+p^2}$. It is normalized to the speed of light in vacuum c, and the momentum unit is $m_e c$. The electron density n is normalized to the density of immobile ions n_0. The time and coordinate units are respectively ω_{pe}^{-1} and c/ω_{pe}.

To solve the problem under consideration it is convenient to use the Lagrange variables. We perform transformation from the Euler variables (r, t) to the Lagrange variables (r_0, t), with the relationship between derivatives with respect to time and coordinate,

$$(\partial_t)_L = (\partial_t + v \partial_r)_E, \tag{21.60}$$

$$((\partial_{r_0} r)^{-1} \partial_{r_0})_L = (\partial_r)_E, \tag{21.61}$$

where the indices L and E denote the Lagrange and the Euler variables. Application of this transformation to Eqs. (21.57)–(21.59) yields

$$\partial_t(r^s n) + r^s n\,(\partial_{r_0} r)^{-1}\partial_{r_0} v = 0, \tag{21.62}$$

$$\partial_t p = -E, \tag{21.63}$$

$$\partial_t(r^s E) = r^s v, \tag{21.64}$$

where the velocity of the electron fluid element is $v = \partial_t \xi$ and $\partial_{r_0} r = 1 + \partial_{r_0}\xi$. Here and below, $\xi(r_0, t)$ is a displacement of the electron fluid element from its initial position r_0. It determines the relationship between the Lagrange and the Euler variables,

$$r = r_0 + \xi(r_0, t') \text{ and } t = t'. \tag{21.65}$$

The solution of Eq. (21.62) yields for the electron density

$$n = \frac{r_0^s}{(r_0 + \xi)^s(1 + \partial_{r_0}\xi)}. \tag{21.66}$$

From Eq. (21.64) we obtain for the electric field,

$$E = \frac{1}{(s+1)}\left(r_0 + \xi - \frac{r_0^{s+1}}{(r_0 + \xi)^s}\right). \tag{21.67}$$

Using expressions (21.66) and (21.67), the equation of electron motion (21.58) can be rewritten in Hamiltonian form with the Hamiltonian,

$$\mathcal{H}(p, \xi) = \sqrt{1 + p^2} + \Pi(\xi), \tag{21.68}$$

where the potential function $\Pi(\xi)$ is equal to $(r_0 + \xi)^2/2$ for $s = 0$, i.e. for plane waves, it is

$$\Pi(\xi) = (r_0 + \xi)^2/4 + (r_0^2/2)\ln(r_0 + \xi) \text{ for } s = 1,$$

which corresponds to cylindrical waves, and

$$\Pi(\xi) = \left((r_0 + \xi)^2/2 + r_0^3/(r_0 + \xi)\right)/3, \text{ when } s = 2,$$

in the case of spherical waves.

From the relations found above and the fact that the Hamiltonian given by Eq. (21.68) is independent of time, $\mathcal{H} = h$, it follows that the solution of

Eqs. (21.62)–(21.63) can be written in quadrature:

$$t = \int_{\xi_1}^{\xi} \frac{[h - \Pi(s)]ds}{\sqrt{[h - \Pi(s)]^2 - 1}}. \tag{21.69}$$

To find the period of nonlinear oscillations, one should integrate the right-hand side of Eq. (21.69) over a whole cycle. In the case of a spherical wave, for which $s = 2$, it is shown by Bulanov et al. (2012) that the oscillation period in the limit $p_{\mathrm{m}} \gg 1$ is equal to $T = 2\sqrt{6p_{\mathrm{m}}}$. In general, the period, and hence the frequency, depends on the coordinate r_0, which means that the plasma oscillations are waves with a continuous spectrum.

The wave breaking of nonlinear plasma waves corresponds to the vanishing of the Jacobian of the transformation from the Euler to the Lagrange coordinates, which is equal to

$$J = \frac{(r_0 + \xi)^s}{r_0^s}(1 + \partial_{r_0}\xi). \tag{21.70}$$

The vanishing condition is $\partial_{r_0}\xi = -1$. As follows from Eq. (21.66), the electron density tends to infinity when $J \to 0$.

To estimate the characteristic time of wave breaking, we note that each element of the electron fluid undergoes periodic oscillations, and represent the periodic function describing the electron displacement in the form of a Fourier series:

$$\xi(r_0, t) = \sum_{j=1}^{\infty} \xi_j(r_0) \exp(ij\Omega(r_0)t), \tag{21.71}$$

with the frequency Ω that depends on the coordinate r_0. The dependence of the oscillation frequency $\Omega(r_0)$, as mentioned above, can be due to the plasma inhomogeneity and/or inhomogeneity of the wave amplitude. Differentiating expression (21.71) with respect to the coordinate r_0 and time t, we obtain for the displacement derivatives (Brantov et al., 2008)

$$\partial_{r_0}\xi(r_0, t) = \sum_{j=1}^{\infty} \left[\partial_{r_0}\xi_j(r_0) + it\partial_{r_0}\Omega(r_0)j\xi_j(r_0)\right] \exp(ij\Omega(r_0)t), \tag{21.72}$$

$$\partial_t\xi(r_0, t) = \sum_{j=1}^{\infty} \Omega(r_0)j\xi_j(r_0) \exp(ij\Omega(r_0)t). \tag{21.73}$$

In the limit $t \to \infty$, these equations yield an asymptotic relation between the space and time derivatives of ξ:

$$\partial_{r_0}\xi(r_0, t) = t\partial_{r_0}\ln(\Omega(r_0))\partial_t\xi(r_0, t). \tag{21.74}$$

From this expression it follows that the gradient of the displacement $\partial_{r_0}\xi$ increases with time. This phenomenon is the phase mixing discussed above. In relativistically strong plasma waves, the electron velocity $\partial_t\xi$ changes periodically from $+c$ in one half-cycle, to $-c$ in another, and remains roughly constant during each half cycle. It hence follows that the dependence of the displacement on time has the form of a saw-tooth oscillation: the displacement varies from $-\xi_{\max}$ to $+\xi_{\max}$, being a linear function of time during the half-cycle, $\xi(t) \approx \pm ct$. This gives the relation

$$\Omega = \pi c/2\xi_{\max} \tag{21.75}$$

between the amplitude of the oscillations and their frequency. As a result, we find the onset of wave breaking corresponding to the condition $\partial_{r_0}\xi = -1$:

$$t_{br} \approx \frac{1}{c\,\partial_{r_0}\ln(\Omega(r_0))}. \tag{21.76}$$

The weaker the inhomogeneity of the wave parameters is, the longer the wave-breaking time.

21.6 Relativistic Electromagnetic Solitons

Among the rich variety of nonlinear effects, solitons are of fundamental importance for nonlinear wave theory (Dodd et al., 1982; Whitham, 1974; Zakharov et al., 1984). The solitons are the nonlinear waves that can propagate in various media over a large distance without changing their shape. Relativistic electromagnetic solitons propagating in collisionless plasma have been extensively studied theoretically (Esirkepov et al., 1998; Farina & Bulanov, 2001a, 2001b; Farina et al., 2000; Kaw et al., 1992; Kozlov et al., 1979; Lehmann et al., 2006; Lontano et al., 2002; Poornakala et al., 2002), with computer simulations (Bulanov et al., 1992; Bulanov et al., 1999; Bulanov et al., 2001; Esirkepov et al., 2002a, 2002b; Esirkepov et al., 2004; Kim et al., 2004; Liu et al., 2015; Mancic et al., 2006; Naumova et al., 2001a, 2001b; Sanchez-Arriaga & Lefebvre, 2011a; Sanchez-Arriaga & Lefebvre, 2011b; Sanchez-Arriaga et al., 2011; Saxena et al., 2013; Sentoku et al., 1999; Wu et al., 2019), and in the experiments on the laser-plasma interaction (Borghesi et al., 2002; Chen et al., 2007; Kando et al., 2007, 2009; Pirozhkov et al., 2007; Romagnani et al., 2010; Sarri et al., 2011). Typically, solitons in relativistic plasmas can be regarded as pulses of electromagnetic radiation trapped inside the cavities formed in the plasma electron density by the pulse ponderomotive pressure. In the limit of small but finite soliton amplitude they are described by the nonlinear Schrodinger equation (NSE). For the properties of the NSE solitons see Refs. (Akhmediev & Ankiewicz, 1997;

Dodd et al., 1982; Whitham, 1974; Zakharov et al., 1984). Relativistic electromagnetic solitons can provide one of the ways of anomalous absorption of laser energy by transforming it into the energy of fast particles and into energy of high and low frequency electromagnetic radiation.

21.6.1 Circularly Polarized Electromagnetic Soliton

The system of equations (21.7–21.17) admits solutions in the form of the solitons. In the case of immovable ions, it convenient to rewrite it for the functions

$$A = A_y + iA_z = a(X)\exp(i\omega\tau) \text{ and } p_{||} = \beta_g b(X), \tag{21.77}$$

where X and τ are defined by Eq. (21.7) and $p_{||}$ is the x-component of the electron momentum. As a result, we obtain (Esirkepov et al., 1998)

$$a'' + \omega^2 a = \frac{\gamma_g^2 a}{\gamma - b}, \tag{21.78}$$

$$(\gamma - \beta^2 b)'' = \frac{b}{\gamma - b}, \tag{21.79}$$

where $\gamma = \sqrt{1 + a^2 + \beta_g^2 b^2}$, $\gamma_g = 1/\sqrt{1 - \beta_g^2}$, and a prime denotes a differentiation with respect to the variable X. This system of equations with boundary conditions $a(\pm\infty) = b(\pm\infty) = 0$ describes a one-dimensional electromagnetic soliton propagating through a cold collisionless plasma. In this system of equations, the variables are normalized as follows. The units of length and time are c/ωpe and ωpe−1, velocity is normalized to c, momentum to mec/e, the vector and electrostatic potential to mec/e, and density to n0. The soliton speed and frequency are less than the speed of light in vacuum, $\beta_g < 1$, and the electron plasma frequency, $\omega < 1$.

In the limit of small but finite wave amplitude, the soliton solution is given by hyperbolic secant expression (Kaw et al., 1992; Kozlov et al., 1979)

$$A(X, \tau) = a_m \frac{\exp(i\omega\tau)}{\cosh(\xi)}, \tag{21.80}$$

with the soliton amplitude $a_m = 2\sqrt{1 - \omega^2/\gamma_g^2}$, frequency $\omega \approx \gamma_g(1 - a_m^2/8)$, and $\xi = X\sqrt{1 - \omega^2/\gamma_g^2}$. This expression describes an isolated envelope soliton. It corresponds to the class of the NSE solitons.

Another exact solution can be obtained in the limit of a soliton with zero propagation velocity, when $\beta_g = 1$ and $b = 0$. Equation (21.76) takes the form

$$a'' + [\omega^2 - (1 + \gamma'')/\gamma]a = 0. \tag{21.81}$$

Using the substitution $a = \sinh u$, $\gamma = \cosh u$ we can cast this equation into the form

$$u'' = \sinh u(1 - \omega^2 \cosh u). \tag{21.82}$$

Its solution describing the localized soliton is (Esirkepov et al., 1998)

$$A(\xi, \tau) = \frac{2\sqrt{1 - \omega^2} \cosh(\xi) \exp[i(\omega\tau)]}{\cosh^2(\xi) + \omega^2 - 1}. \tag{21.83}$$

The relationship between the soliton amplitude a_m and the soliton frequency is given by

$$a_m = \frac{2\sqrt{1 - \omega^2}}{\omega^2}. \tag{21.84}$$

The electron density, $n = 1 + \gamma''$, has a form of cavity with thc dcnsity minimum equal to

$$n(0) = 1 - \frac{4(1 - \omega^2)^2}{\omega^4}. \tag{21.85}$$

At the soliton amplitude $a_m = \sqrt{3}$ (the soliton frequency is $\omega = \sqrt{2/3}$) the density in the soliton center vanishes. The electromagnetic field trapped inside the soliton comprises of rotating electric and magnetic field. In three-dimension geometry the electromagnetic field topology is non-trivial (Esirkepov et al., 2002a, 2002b). The soliton consists of oscillating electrostatic and electromagnetic fields confined in a prolate cavity of the electron density. The cavity size is of the order of few laser wavelengths. The soliton field resembles an oscillating electromagnetic dipole. The oscillating toroidal magnetic field and poloidal electric field correspond to the TM (transverse magnetic field structure). In the TE-transverse electric configuration the soliton has poloidal magnetic field and toroidal electric field.

The theoretical description of these solitons implies that the ions are immovable because the time of the soliton formation is much shorter than the ion response time so that ions can be assumed to be at rest. Ions can be assumed to be at rest during approximately $\sqrt{m_i/m_e}$ oscillation periods of the electromagnetic field trapped inside the soliton. For longer times the ponderomotive pressure of the electromagnetic field inside the soliton starts to dig a hole in the ion density, and the parameters of the soliton change. A low frequency electromagnetic wave packet remains in the plasma, being well confined inside the slowly expanding plasma cavity. This nonlinear electromagnetic wave packet is called the “postsoliton”. They have been observed in computer simulations by Naumova et al. (2001a, 2001b) and in the laser-plasma interaction experiments by Borghesi et al. (2002). In inhomogeneous plasma target the postsolitons being close to the plasma-vacuum interface can emit low frequency electromagnetic radiation in the terahertz frequency range (Kando et al., 2009; Sentoku et al., 1999).

21.6.2 Effect of Plasma Inhomogeneity

In an inhomogeneous density plasma, an electromagnetic wave pulse moves with not-constant group velocity. Within the framework of the geometric optics approximation the wave packet is governed by the Hamilton equations (Witham, 1974)

$$\dot{\boldsymbol{k}} = -\partial_{\boldsymbol{x}} \mathcal{H}, \quad \dot{\boldsymbol{x}} = \partial_{\boldsymbol{k}} \mathcal{H} \tag{21.86}$$

where the Hamiltonian is the wave frequency.

Using Eq. (21.86) we can find the acceleration of the wave packet. It reads

$$\ddot{x}_i = \frac{\partial^2 \mathcal{H}}{\partial k_i \partial x_j} \frac{\partial \mathcal{H}}{\partial k_j} - \frac{\partial^2 \mathcal{H}}{\partial k_i \partial k_j} \frac{\partial \mathcal{H}}{\partial x_j}. \tag{21.87}$$

In collisionless plasma the wave frequency is equal to $\omega = \sqrt{k^2c^2 + \omega_{pe}^2(x)}$, i.e. the Hamiltonian dependence on the wave vector and coordinate is given by

$$\mathcal{H} = \sqrt{k^2c^2 + \omega_{pe}^2(x)}.$$

In the case when the plasma density dependence on the coordinate is given by

$$n = n_0(1 + x/L_x + y^2/L_y^2),$$

which corresponds to the plasma channel of the width L_y in inhomogeneous plasma, from Eq. (21.87) it follows that

$$\ddot{x} = -\frac{c^2}{2L_x}, \quad \ddot{y} + \frac{c^2}{L_y^2} y = 0, \tag{21.88}$$

i.e. the wave packet moves along the x axis as $x = x_0 + v_g t - wt^2/2$ with the acceleration $w = -c^2/2L_x$ directed towards the low density side and oscillates in the transverse direction with the frequency c/L_y. Such soliton behavior has been observed in simulation results presented by Sentoku et al. (1999).

The propagation in inhomogeneous media of solitons described by the nonlinear Schrödinger equation was discussed by Chen and Liu (1976) (see also Liu et al., 2019) who showed that this problem admits multi-soliton solutions that were obtained with the inverse scattering method (Dodd et al., 1982; Gardner et al., 1967; Whitham, 1974;

Zakharov et al., 1984). These exact solutions show that the solitons are accelerated toward the plasma–vacuum interface with an acceleration proportional to the gradient of the plasma density as in Eq. (21.87). Taking into account the fact that the nonlinear Schrödinger equation giving the soliton in homogeneous density plasma (21.78), with so-called the "lens-transform",

$$\tau \to t - w(x - v_g t + 2wt^2/3), \quad X \to x - v_g t + wt^2, \tag{21.89}$$

is transformed to the nonlinear Schrödinger equation describing nonlinear waves in inhomogeneous plasma with the density linearly dependent on the coordinate, $n = 1+wx$. From Eq. (21.89) we see that the time derivative of the soliton coordinate vanishes at $t^* = v_g^2/2w$. For the time $t < t^*$ the soliton propagates along the x-axis in a positive direction. At time $t = t^*$ the soliton bounces back and propagates down towards low density region.

The physical mechanism of these solitons' formation is related to the laser pulse frequency downshifting while it interacts with the plasma. The laser pulse loses its energy as it generates the plasma wake waves. The plasma frequency in the under-sense plasma is much lower than the carrier frequency of the laser pulse resulting in the adiabatic laser-plasma interaction when the number of photons of the electromagnetic wave is adiabatically conserved (Bulanov et al., 1992). Because of this constraint, the decrease of the laser pulse energy must be accompanied by the lowering of the carrier frequency and decreasing of the laser pulse group velocity. The group velocity tends to zero at $\omega \to \omega_{pe}$. The laser pulse energy depletion length is approximately equal to $l_{dep} = l_{las}(\omega/\omega_{pe})^2$. This leads to trapping the depleted portion of the laser pulse in the cavity behind the pulse in the form of the soliton with almost zero group velocity.

21.7 Long-Wavelength Limit

21.7.1 Laser Pulse Energy Depletion

Considering that the electromagnetic wave and electrostatic potential amplitude are slow functions of the coordinates and time we can cast Eqs. (21.1)–(21.6) in the form (Bulanov et al., 1992)

$$\partial_X \phi + \frac{k_p^2}{2}\left(\frac{1+a^2}{(1+\phi)^2} - 1\right) = 0, \tag{21.90}$$

$$2\omega_0 \partial_t a + \frac{\omega_{pe}^2}{\omega_0^2} c^2 \partial_{XX}^2 a + 2v_g \partial_{Xt}^2 a = -\omega_{pe}^2 \frac{\phi}{1+\phi} a, \tag{21.91}$$

where $X \to x - v_g t$, the group velocity of the electromagnetic wave is $v_g = c^2 k_0/\omega_0$, and the electromagnetic wave is given by $a(X,t)\exp(-\omega_0 t + k_0 x)$ and $\phi(X,t)$ with normalized electromagnetic $a = eE/m_e\omega_0 c$ and electrostatic $\phi = e\varphi/m_e c^2$ potentials.

For a short and strong enough the laser pulse with the duration $\tau_{las} < \sqrt{2}a/\omega_{pe}$ and $|a| \gg 1$ the plasma wave inside the laser pulse has

$$\phi(X,t) = \frac{k_p^2}{2}\int_{\infty}^{X} dX' \int_{\infty}^{X'} dX''|a^2(X'',t)|. \tag{21.92}$$

Matching this solution with the solution for the free wave behind the laser pulse we obtain the amplitude of the wake wave left in a plasma behind the laser pulse,

$$\Delta\phi = \phi_{\max} - \phi_{\min} = \sqrt{A(A+4)} \text{ with } A = \left(\int k_p|a^2(X,t)dX\right)^2. \tag{21.93}$$

The maximal electric field for the optimal laser pulse duration equals $E_w = (\sqrt{2}m_e\omega_0 c/e)|a|$. We note that more detailed discussion of the wake wave generation by the laser pulse will be presented below in Chap. 24.

To address the laser pulse evolution, we pay attention to the fact that Eq. (21.91) has the first integral

$$\int dX\left[|a^2| - i\frac{v_g}{\omega_0}\left(a\partial_X a^* - a^*\partial_X a\right)\right] = \text{constant}, \tag{21.94}$$

where a^* is the complex conjugate amplitude of the laser pulse.

Taking $a(X,t)$ in the form $|a(X,t)|\exp(i\psi)$ with the phase $\psi(X,t)$ we find local wave frequency equal to time derivative of the eikonal function

$$\omega = -\partial_t\big(k_0 x - \omega_0 t + \psi(x - v_g t, t)\big) \simeq \omega_0 + v_g\partial_X\psi. \tag{21.95}$$

Using this relationship we rewrite Eq. (21.94) in the form

$$\int dX\,|a^2|\omega = \text{constant}. \tag{21.96}$$

Since $|a^2| = (eE/m_e\omega c)^2$ the integrand in Eq. (21.96), $|a^2|\omega$, is proportional to the photon number density $N = E^2/8\pi\hbar\omega$.

In addition, from Eq. (21.91) we can obtain the equation

$$\partial_t\int dX\left(i\frac{v_g}{\omega_0}\partial_X a + a*\right)\left(i\frac{v_g}{\omega_0}\partial_X a * -a\right) = \frac{1}{2}v_g\frac{\omega_{pe}^2}{\omega_0^2}\int dX\,|a^2|\partial_X\left(\frac{1}{1+\phi}\right). \tag{21.97}$$

Rearranging this equation, we obtain a relation showing the rate of energy loss of the laser pulse

$$\partial_t \int dX \left(\frac{E^2 + B^2}{4\pi} \right) = -\frac{1}{\pi} \left(\frac{m_e \omega_{pe} c}{4e} \right) \int dX \, |a^2| \partial_X \left(\frac{1}{1+\phi} \right). \tag{21.98}$$

The right-hand side of this equation is equal to the rate of the laser energy transfer into the plasma wake wave (taken with opposite sign).

Assuming that a short laser pulse (the pulse duration is of $\tau_{las} < 1/||a|\omega_{pe})$ is characterized by average frequency $\langle \omega(t) \rangle$ and by the integral $Q(t) = \int dX \, |a|^2$, one can obtain from the obtained above equations that

$$Q(t) = (1 - t/t_{nl})^{-1/3} Q(0), \tag{21.99}$$

$$\langle \omega(t) \rangle = (1 - t/t_{nl})^{-1/3} \langle \omega(0) \rangle, \tag{21.100}$$

where the time of the nonlinear depletion of the laser energy is equal to

$$t_{nl} = \frac{16}{3k_p Q(0)} \frac{\omega_0^2}{\omega_{pe}^3}, \tag{21.101}$$

From the solution (21.99–21.100) follows that the decrease of the laser energy and the laser pulse frequency downshift are accompanied by the growth of the normalized laser pulse amplitude and as a consequence of Eqs. (21.92) and (21.93) by the increase of the wake wave amplitude.

21.7.2 *Laser Pulse Instability in Collisionless Plasma. Relativistic Self-focusing*

For adiabatically slow rise of the laser pulse amplitude the electron mass increases accordingly. As a result, as farther from the pulse front is as higher the group velocity of the electromagnetic wave is which leads to compression of the laser pulse and to the pulse front steepening is fastest where $|a| \simeq 1$. The characteristic time of the pulse front steepening is approximately equal to

$$t_{st} = \frac{\omega_0^2}{c\,\omega_{pe}^2} \frac{1}{\partial_X |a|} \bigg|_{|a|=1}. \tag{21.102}$$

Super-intense laser radiation in a plasma is subject to a host of instabilities (see also Part III where non-relativistic limit is discussed). The fastest is the so-called stimulated Raman scattering (SRS) instability, which develops on the electron timescale. In the SRS an electromagnetic pump wave $(\omega_0, \boldsymbol{k}_0)$ is scattered by an electron density wave ($\omega_{pe}, \boldsymbol{k}_{pe}$), which in turn is excited by the beating of the pump and of the scattered ($\omega_1, \boldsymbol{k}_1$) waves. Thus, there is a feedback loop, which results in the set of instability. The frequencies and wave numbers of the waves involved satisfy the matching (decay) conditions:

$$\omega_0 - \omega_1 = \omega_{pe}, \tag{21.103}$$

$$\boldsymbol{k}_0 - \boldsymbol{k}_1 = \boldsymbol{k}_{pe}. \tag{21.104}$$

The case when the vectors $\boldsymbol{k}_1$ and $\boldsymbol{k}_0$ are nearly parallel ($\boldsymbol{k}_1 \approx \boldsymbol{k}_0$ and $k_{pe} = c/\omega_{pe} \ll k_0$) corresponds to the stimulated forward Raman scattering (SFRS). It leads to self-modulation of the laser pulse with modulation length of the order of $2\pi/k_{pe} = 2\pi c/\omega_{pe}$ (Antonsen & Mora, 1992; Mori et al., 1994; Sprangle et al., 1992). The growth rate of the SFRS instability for small but finite amplitude of the laser pulse is given by the expression

$$\gamma_{SFRS} = \omega_0 \left(\frac{\omega_{pe}}{\omega_0}\right)^2 \frac{|a|}{2\sqrt{2}}. \tag{21.105}$$

The case when the vectors $\boldsymbol{k}_1$ and $\boldsymbol{k}_0$ are antiparallel ($\boldsymbol{k}_1 \approx -\boldsymbol{k}_0$ and $\boldsymbol{k}_{pe} \approx 2\boldsymbol{k}_0$) corresponds to stimulated backward Raman scattering (SBRS). For ultrashort super-intense laser pulses, this instability is the first to develop. In contrast to SFRS instability, the SBRS instability is convective. The perturbations inside the pulse arise from the thermal noise of the electron plasma density ahead of the pulse. For a time interval of the order of the pulse duration, a steady-state solution is established inside the pulse. The backward stimulated Raman scattering has a maximum growth rate in the region where $|a| \simeq 2$. It can be expressed as

$$\gamma_{SBRS} = \omega_0 \left(\frac{\omega_{pe}}{\omega_0}\right)^{2/3} \frac{1}{2^{2/3}\sqrt{3}}. \tag{21.106}$$

The relations between pump and scattered wave vectors are illustrated in Fig. 21.4 for SFRS and SBRS cases.

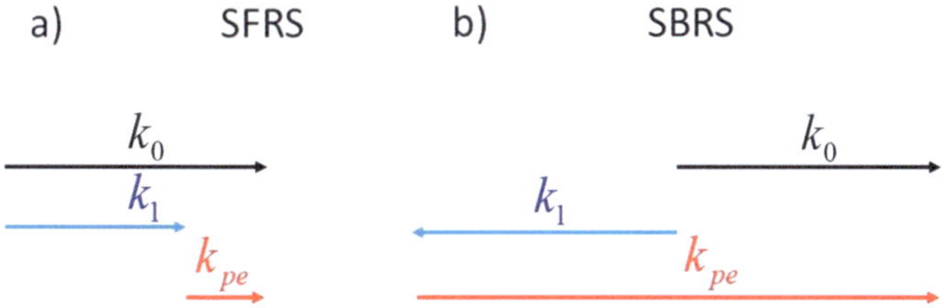

Fig. 21.4 The pump and scattered wave vectors in the case of **a** stimulated forward Raman scattering and **b** stimulated backward Raman scattering

Large-scale modulation instability can also affect the laser pulse. Considering the long-wavelength perturbations with the wave number $q \ll k_0$ and using Eqs. (21.90) and (21.91) we can find the maximum growth rate of this instability is achieved for $|a| \simeq \sqrt{2}$ being equal to

$$\gamma_{SM} = \frac{q}{4}\left(\frac{\omega_{pe}}{\omega_0}\right)^2. \tag{21.107}$$

Among other instabilities of the laser radiation in an underdense plasma, probably, the most impressive phenomenon is the self-focusing of the electromagnetic wave. The self-focusing, discovered by Askar'yan in 1962, appears due to the nonlinear change of the refraction index of the medium in the region where a high intensity electromagnetic wave propagates, e. g. in the case of relativistically intense electromagnetic wave due to a counter-play of the ponderomotive force effects and relativistic change of the electron mass (Bespalov & Talanov, 1966; Borisov et al., 1992; Sun et al., 1987).

A standard approach for theoretical description of the self-focusing is based on using the Nonlinear Schrödinger Equation (NSE) for the envelope of the laser pulse $a(\boldsymbol{r}_\perp, t)$, which can be obtained in the long-wavelength limit similar to Eq. (21.91) (e.g. see Landau & Lifshitz, 1984; Sulem & Sulem, 2007). Here $\boldsymbol{r}_\perp$ is the coordinate transverse with respect to the direction of the laser pulse propagation. In the case of relativistic self-focusing of the laser pulse in the collisionless plasma whose density corresponds to the plasma frequency ω_{pe}, the NSE equation can be written as

$$2i\omega_0\partial_t a + c^2\Delta_\perp a + \frac{\omega_{pe}^2}{4}|a|^2 a = 0, \tag{21.108}$$

where $\Delta_\perp = \partial_{yy} + \partial_{zz}$.

Writing the electromagnetic wave envelope as

$$a = R\exp(i\Phi)$$

and substituting this function in Eq. (21.108) we can represent the NSE equation in the form of the Madelung equations:

$$\omega_0\partial_t R^2 + c^2\nabla_\perp(R^2\nabla_\perp\Phi) = 0,$$
$$2\omega_0\partial_t\Phi + c^2(\nabla_\perp\Phi)^2 - \frac{c^2}{R}\Delta_\perp R - \frac{\omega_{pe}^2}{4}R^2 = 0.$$

The Madelung form is also known as the "hydrodynamic form" of the NSE equation. These equations have an exact solution describing constant amplitude $R = R_0$ plane wave whose phase is a linear function of time depending nonlinearly on the wave amplitude,

$$\Phi_0 = \omega_{pe}^2 R_0^2 t / 8\omega_0,$$

i.e.

$$a_0 = R_0 \exp\left(i \frac{\omega_{pe}^2 R_0^2}{8\omega_0} t \right).$$

We can rewrite this system of equations introducing the "density", $\rho = R^2$, and "velocity" $\boldsymbol{v}_\perp = \nabla_\perp \Phi$ and casting it in the form

$$\omega_0 \partial_t \rho + c^2 \nabla_\perp (\rho \boldsymbol{v}_\perp) = 0, \tag{21.109}$$

$$2\omega_0 \partial_t \boldsymbol{v}_\perp + c^2 (\boldsymbol{v}_\perp \nabla_\perp) \boldsymbol{v}_\perp = \frac{c^2}{\sqrt{\rho}} \nabla_\perp \Delta_\perp \sqrt{\rho} + \frac{\omega_{pe}^2}{4} \nabla_\perp \rho. \tag{21.110}$$

Linearizing the system of equations (21.109)–(21.110) around the equilibrium solution, i.e. considering the functions ρ and $\boldsymbol{v}_\perp$ to be in the form

$$\rho = \rho_0 + \rho_1 \quad \text{and} \quad \boldsymbol{v}_\perp = \boldsymbol{v}_{\perp 0} + \boldsymbol{v}_{\perp 1} \tag{21.111}$$

with small perturbations ρ_1 and $\boldsymbol{v}_{\perp 1}$, we obtain from Eqs. (21.109) and (21.110) the system of linear equations in partial derivatives,

$$\omega_0 \partial_t \rho_1 + c^2 \rho_0 \nabla_\perp \cdot \boldsymbol{v}_{\perp,1} = 0,$$
$$2\omega_0 \partial_t \boldsymbol{v}_{\perp 1} + \frac{c^2}{2\rho_0} \nabla_\perp \Delta_\perp \rho_1 + \frac{\omega_{pe}^2}{4} \nabla_\perp \rho_1 = 0.$$

Seeking for dependence of perturbations representing the functions $\rho_1(\boldsymbol{r}_\perp, t)$ and $\boldsymbol{v}_{\perp 1}$ as

$$\rho_1(\boldsymbol{r}_\perp, t) = \rho_1 \exp(-i\omega t + i\boldsymbol{k}_\perp \cdot \boldsymbol{r}_\perp) \text{ and } \boldsymbol{v}_{\perp 1}(\boldsymbol{r}_\perp, t) = \boldsymbol{v}_{\perp 1} \exp(-i\omega t + i\boldsymbol{k}_\perp \cdot \boldsymbol{r}_\perp)$$

and substituting these to the system of linearized equations above we obtain two algebraic equations

$$i\omega_0 \omega \rho_1 - c^2 k_\perp \rho_0 d_1 = 0, \tag{21.112}$$

$$4i\omega_0 \omega d_1 - \left(k_\perp^2 c^2 - \frac{\omega_{pe}^2}{2} R_0^2 \right) \rho_1 = 0, \tag{21.113}$$

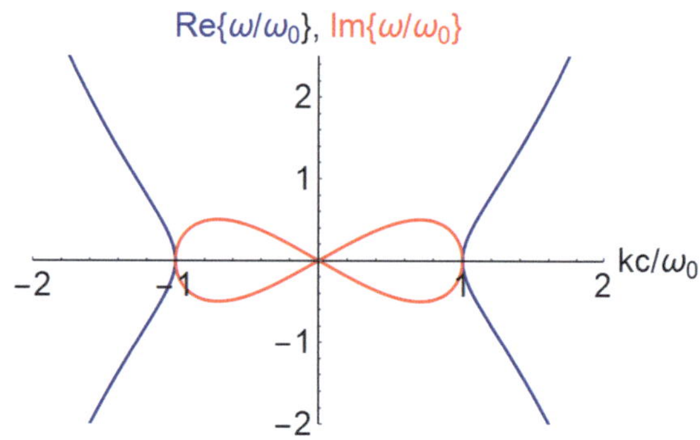

Fig. 21.5 Real and imaginary parts of the frequency ω on the plane $(k_\perp c/\omega_0, \omega/\omega_0)$ for $\omega_{pe}R_0/\omega_0 = 2$ in the case of self-focusing instability

where $d_1 = (\boldsymbol{k}_\perp \cdot \boldsymbol{v}_\perp)$ and $k_\perp^2 = k_y^2 + k_z^2$. The condition of equality of their determinant to zero gives the dispersion equation for the frequency dependence on the wave number.

A condition that the determinant of this system is equal to zero gives the growth rate of the self-focusing instability equal to imaginary part of the frequency:

$$\gamma = \pm \frac{k_\perp c}{\omega_0}\sqrt{k_\perp^2 c^2 - \frac{\omega_{pe}^2}{4}R_0^2}. \tag{21.114}$$

In Fig. 21.5 we plot on the plane $(k_\perp c/\omega_0, \omega/\omega_0)$ the dependence of real and imaginary parts of the frequency ω.

According to Eq. (21.114) the instability develops for perturbation with the wave vector in the interval $k_\perp \in (0, \omega_{pe}R_0/2c)$. From this condition one can find the self-focusing threshold written in terms of the laser pulse power

$$\mathcal{P} > \overline{\mathcal{P}}\frac{n_{cr}}{n_e}, \tag{21.115}$$

where

$$\overline{\mathcal{P}} = \frac{2m_e c^3}{r_e} = 17.5\,\mathrm{GW},$$

$r_e = e^2/m_e c^2$ is the classical electron radius, and $n_{cr} = m_e\omega_0^2/4\pi e^2$ is the critical plasma density.

When the laser power is well above the self-focusing threshold the laser pulse can break up into multiple filaments, which can then interact to each other. The nonlinear modification of the refraction index leads to calescence of the filaments (see Lezhnin et al., 2024). Another mechanism of the self-focusing filament merging (this is the magnetic interaction) has been proposed by Askar'yan et al. (1994). In this case the filaments are attracted to each other due to attraction of the electric current filaments inside the self-focusing channels formed by the accelerated electron beams. This

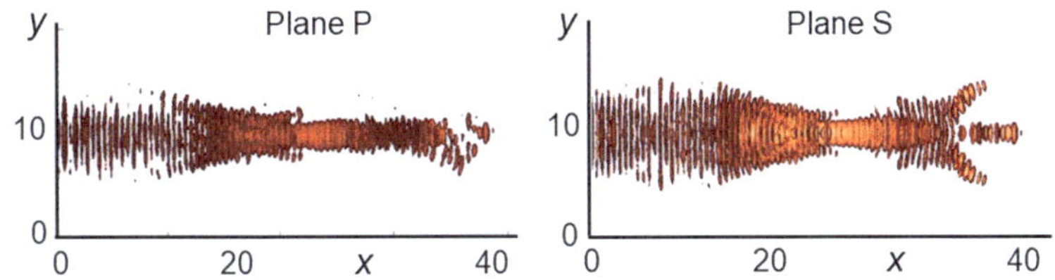

Fig. 21.6 Relativistic self-focusing of the semi-infinite linearly polarized laser pulse (the electric field is parallel to the y-axis) of the amplitude $a_0 = 4.2$ and laser width $12\lambda_0$ in underdense plasma whose density corresponds to the Langmuir frequency equal to $\omega_{pe} = 0.45\omega_0$. The results of 3D PIC simulations are shown for time $t = (2\pi/\omega_0) \times 60$. Reproduced from Naumova et al. (2002), with the permission of AIP Publishing

effect is illustrated in Fig. 21.6, the results of 3D PIC simulations by Naumova et al. (2002) are presented. In the leading part of the laser pulse propagating in an underdense plasma we see three filaments formed in the plane perpendicular to the electric field of the linearly polarized pulse, which then merge leading to a single self-focusing channel.

21.7.3 Electromagnetic-Electron Rings

The process of relativistic self-focusing for high-power laser pulses leads to the formation of multifilament and, in particular, the ring-shaped transverse structures as reported in numerous theoretical and experimental works (Borisov et al., 1992; Kovalev & Bychenkov, 2019; Mori et al., 1988; Naseri et al., 2016; Zhao et al., 2016). In the paper by Valenta et al. (2021), the authors present a study focused on the formation of the electromagnetic rings in the regime of high-power, ultrashort laser pulses, where the longitudinal charge separation and the generation of Langmuir waves cannot be neglected.

Using the results presented in Sect. 21.3 we can find the distribution of the electron density in the wake wave left behind the laser pulse driver. It can be written as

$$n_e(X, r) = \frac{n_0 \beta_{ph}}{\beta_{ph} - \beta_e(X, r)}, \tag{21.116}$$

where β_e is the normalized electron velocity in the wake wave. It can be approximated as

$$\beta_e = \frac{a_0^2}{2} \exp\left(-\frac{r^2}{R^2}\right) \cos\left(k_w X - \frac{r^2}{R^2}\right). \tag{21.117}$$

Here $\beta_{ph} = v_{ph}/c$ is the normalized phase velocity of the wake wave (it is equal to the group velocity of the driver laser pulse), a_0 is the normalized laser pulse amplitude,

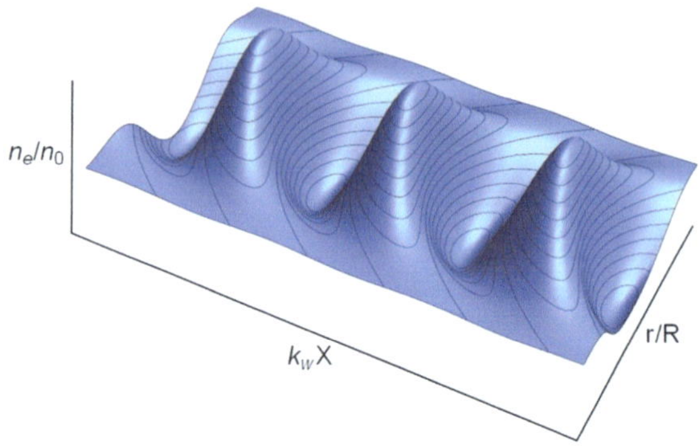

Fig. 21.7 Distribution of the normalized electron number density, n_e/n_0, as the function of the X and r coordinates being formed in a plasma under the action of the electromagnetic wave packet as described by Eqs. (21.116) and (21.117) for $\beta_{ph} = 0.99$, $a_0 = 0.5$, $k_w = 0.2$ and $R = 2$

R is the transverse size of the driver, and $k_w = 2\pi/\lambda_w$ with λ_w is the wavelength of the plasma wake wave. The electron density here has a form of paraboloids moving with relativistic velocity (see Bulanov & Sakharov, 1991; Matlis et al., 2006) (Fig. 21.7).

The propagation of the electromagnetic wave packet (it can be, e.g., a rear part of the driver pulse or a separate "witness" pulse trailing behind the driver, which corresponds to the Photon Accelerator paradigm according to Wilks et al. (1989) and Mendonca (2001, 2024) in the plasma modulated by the Langmuir wave can be described within the framework of the geometric optics approximation. The trajectories of particles ("photons"), each having coordinates $\boldsymbol{x} = (X, r)$ and momentum $\boldsymbol{k} = (k_X, k_r)$ are determined by the Hamilton system of equations with the Hamiltonian given by

$$\Omega(\boldsymbol{x}, \boldsymbol{k}) = \left[\boldsymbol{k}^2 + n(\boldsymbol{x})\right]^{1/2} - k_X \beta_{ph}. \tag{21.118}$$

Here, k_X and k_r are measured in the units of c/ω with ω equal to the initial frequency of the electromagnetic wave and $n(X, r)$ is electron density normalized by the critical plasma density $n_{cr} = m_e\omega^2/4\pi e^2$, which also equal to $\left(\omega_{pe}(X, r)/\omega\right)^2$.

A numerical solution of the system of the Hamilton equations,

$$\begin{aligned} \dot{\boldsymbol{x}} &= \partial_{\boldsymbol{k}}\Omega, \\ \dot{\boldsymbol{k}} &= -\partial_{\boldsymbol{x}}\Omega, \end{aligned} \tag{21.119}$$

for photons with several different initial values of r is displayed in Fig. 21.8. The photons initially propagate parallel to the Langmuir wave propagation direction with $k_X = 10$, whereas the Langmuir wave is described by Eqs. (21.17) and (21.18) with $a_0 = 0.6$, $L = \lambda_p/2$, $n_0 = 0.01$, and $\beta_{ph} = 0.95$. The illustration shows that the trajectories of photons interacting with the Langmuir wave structures are curved toward and outward the propagation axis due to the changes in the plasma refractive index. In regions of electron density depressions (i.e., where $n_e/n_0 < 1$), the plasma enhances focusing. On the other hand, in regions of electron density peaks (i.e., where $n_e/n_0 > 1$), the plasma enhances defocusing. The focusing and defocusing of photons are, thus, induced by the excitation of the Langmuir wave. In addition, in the presence of plasma density gradients, the photons undergo energy (and, thus, the frequency) changes. (Esirkepov et al., 2006; Wilks et al., 1989). Since the plasma density profile and the initial photon distribution in Fig. 21.8 are axially symmetric, the defocused photons form a halo, which may in the high intensity limit evolve into an electromagnetic ring well separated from the axial part of the pulse. An analysis on how the transverse inhomogeneity affects the evolution of an electromagnetic wave as it interacts with a Langmuir wave can be found in Bulanov and Sakharov (1991) publication, where the authors consider the induced focusing.

The electromagnetic ring seen in 3D PIC simulations during the driver pulse propagation in plasma is shown in Fig. 21.9. In the simulations the driver laser pulse has the normalized amplitude $a_0 = 16$. The driver has Gaussian temporal and spatial profiles with the full-with-at half-maximum duration $\tau_0 = 12\,T_0$, where $T_0 = c/\lambda_0$ is the cycle period of the driver, and the beam waist $w_0 = \lambda_0$. The driver pulse propagates in a homogeneous hydrogen plasma with the electron number density $5 \times 10^{-3} n_c$.

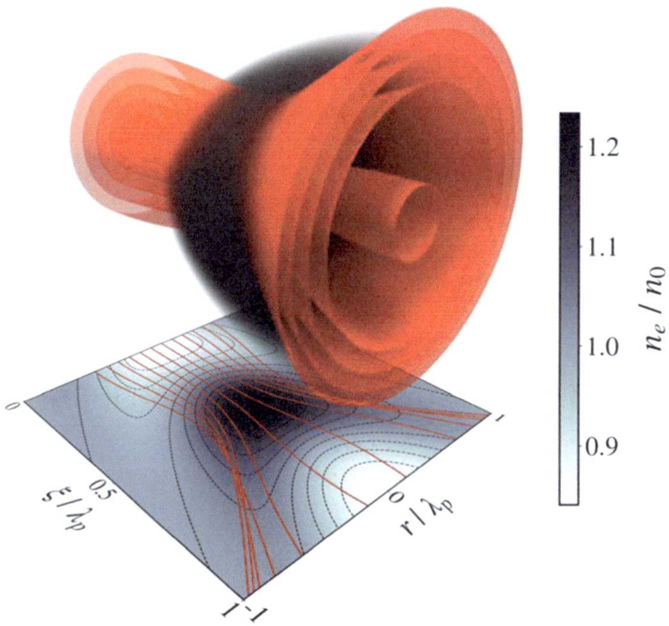

Fig. 21.8 Numerical solution of system of equations (21.120) showing the changes in the distribution of photons (reds) during their interaction with the Langmuir wave (grays). The photons, having several different initial values of r, initially propagate parallel to the Langmuir wave propagation direction (from left to right). The projection at the bottom shows the photon trajectories (solid lines) and the normalized electron number density, with its iso-contours (grays and dashed lines) in the (X, r)-plane. Reproduced from Valenta et al. (2021), with the permission of AIP Publishing

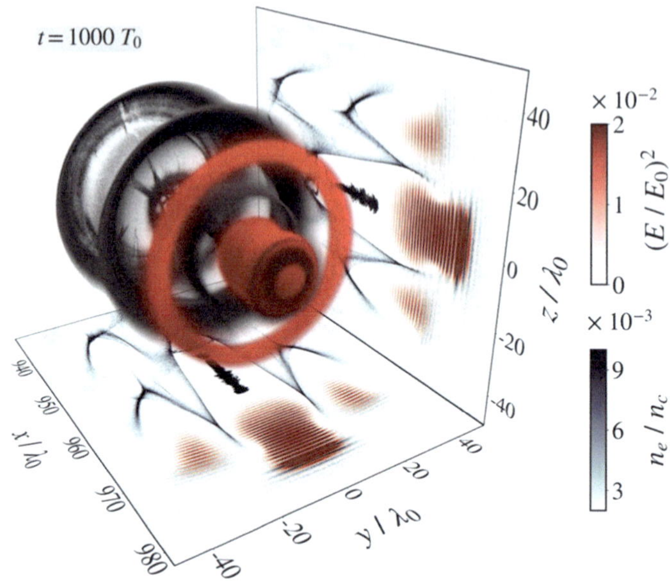

Fig. 21.9 3D representation of the normalized driver field intensity, $(E/E_0)^2$, and the normalized electron number density, n_e/n_c, at the time $t = 1000T_0$ obtained from the PIC simulation. Projections show the cross sections of the driver field intensity and the electron density in the xy- and xz-planes. Reproduced from Valenta et al. (2021), with the permission of AIP Publishing

The electromagnetic rings in laser interaction with an underdense plasma arise due to the interplay between of the relativistic self-focusing, Langmuir wave generation and the laser pulse defocusing induced by the Langmuir waves.

21.8 Electron Vortices (Quasi-static Magnetic Field)

A high-power laser pulse due to the self-focusing can propagate in underdense plasmas as a short and narrow "bullet" having finite size in transverse and longitudinal directions. The interaction of the laser field with plasma electrons results in the generation of fast electrons. Laser accelerated electrons moving along the self-focusing channel, carry the electric current producing the quasi-static magnetic field (Askar'yan et al., 1994) when the corresponding motion of the plasma electron component takes the form of the vortex row (Bulanov et al., 1996a, 1996b; Lamač et al., 2023; Yue et al., 2021). Near the laser pulse the vortex row is symmetrical. Then due to the bending instability development, which is similar to renowned von Karman vortex row instability in hydrodynamics (Lamb, 1932), it evolves into an antisymmetric configuration.

21.8.1 Stationary Electron Vortex in Collisionless Plasma

Electron vortical motion forms azimuthal electric current generates the magnetic field directed along the vortex axis. Radial distribution of the electron density, n_e, magnetic field, B_z, electric field arrizing due to the electric charge separation, E_r, and electron velocity, $v_{e\theta}$, inside the vortex found as a solution of the electron hydrodynamics equations and of the Maxwell equations, which can be written in the form (Lezhnin et al., 2016)

$$\partial_r B_z = \frac{4\pi e n_e v_{e\vartheta}}{c}, \tag{21.120}$$

$$\frac{1}{r}\partial_r(rE_r) = 4\pi e(n_i - n_e), \tag{21.121}$$

$$E_r + \frac{v_{e\vartheta}}{c} B_z = \frac{m_e v_{e\vartheta}^2}{r\sqrt{1 - v_{e\vartheta}^2/c^2}}. \tag{21.122}$$

This is the system of three equations for four variables. There could be an infinite number of vortex shapes, however, not all of them are stable. Here, we assume the dependence of the electric field on radius as

$$E_r = r\exp(-\alpha r^2), \tag{21.123}$$

for which the electric field vanishes at the axis and at $r \to 0$ and $r \to \infty$. In Fig. 21.10 we show radial distribution of the electron density, n_e, magnetic field, B_z, electric field, E_r, and electron velocity, $v_{e\theta}$, inside the vortex given by the solution of Eqs. (21.120)–(21.123). There the electric and magnetic field are normalized on $m_e\omega c/e$, the electron velocity $v_{e\vartheta}$ is normalized on speed of light in vacuum, the electron density is normalized on the ion density n_i, and the radius is measured in units of c/ω.

21.8.2 Von Karman like Instability of the Vortex Row in Dispersive Media

In contrast to the Euler hydrodynamics in collisionless plasmas due to the dispersion effects the dynamics of the vortices acquire novel properties. This follows from the freezing of the generalized vorticity (it is the curl of the generalized momentum)

$$\boldsymbol{\Omega} = \nabla \times [\boldsymbol{p} - (e/c)\boldsymbol{A}]$$

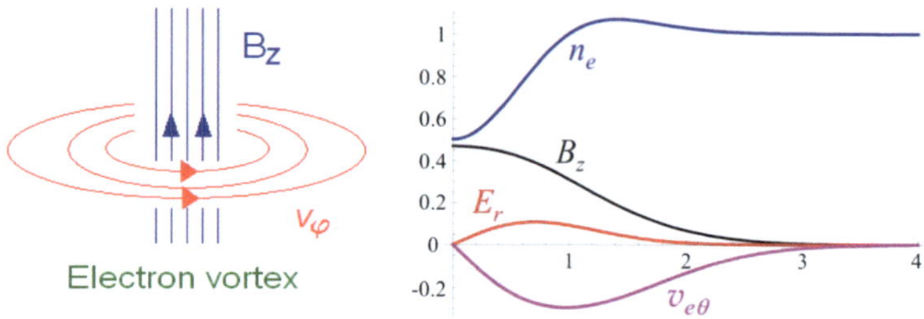

Fig. 21.10 Left: Electron vortex. Right: Radial distribution of the electron density, n_e, magnetic field, B_z, electric field arrizing due to the electric charge separation, E_r, and electron velocity, $v_{e\theta}$, inside the vortex for $E_0 = 1/4$ and $\alpha = 1$. Reproduced from Lezhnin et al. (2016), with the permission of AIP Publishing

in the electron fluid. The property of the vector field (in the present case, this field is $\boldsymbol{\Omega}$) to be frozen to the medium moving with velocity $\boldsymbol{v}$ is described by the equation

$$\partial_t(\nabla \times \boldsymbol{\Omega}) = \nabla \times [\boldsymbol{v} \times (\nabla \times \boldsymbol{\Omega})]. \tag{21.124}$$

We consider here the case when the electron fluid motion is slow enough to be regarded as incompressible and nonrelativistic with unmovable ion component of plasma. This leads to the following relationship between the electron velocity and the magnetic field

$$\boldsymbol{v} = -\frac{c}{4\pi e n}\nabla \times \boldsymbol{B}, \tag{21.125}$$

which follows from the Maxwell equations where the displacement current (it is equal to $c^{-1}\partial_t \mathrm{E}$) effects are negligibly week Taking the magnetic field to be directed along the z-axis, and considering the motion of the electron component in the (x, y) plane ($\boldsymbol{B} = B\boldsymbol{e}_z, \boldsymbol{p} = m_e\boldsymbol{v}$ and n = constant), we obtain

$$(\partial_t + \boldsymbol{e}_z \times \nabla B \cdot \nabla)(\Delta B - B) = 0. \tag{21.126}$$

Here $\nabla = \boldsymbol{e}_x\partial_x + \boldsymbol{e}_y\partial_y$, $\Delta = \partial_{xx} + \partial_{yy}$, and the time and space coordinates are normalized on $\omega_{Be}^{-1} = a^{-1}(\omega/\omega_{pe}^2)$ and c/ω_{pe}, respectively. This equation is known as the Hasegawa-Mima equation in the limit of zero drift velocity (Hasegawa & Mima, 1978) or/and the Charney-Obukhov equation in the limit of zero Rossby velocity related to the β-plane approximation (see Charney, 1947; Pedlosky, 1987; Dolzhansky, 2012).

According to Horton and Hasegawa (1994), Eq. (21.89) has a discrete vortex solution for which the generalized vorticity, $\boldsymbol{\Omega} = \boldsymbol{e}_z\Omega$, is localized at the point $\boldsymbol{r} = \boldsymbol{r}_j$:

$$\Omega = \Delta B - B = \sum_j \Gamma_j \delta[\boldsymbol{r} - \boldsymbol{r}_j], \tag{21.127}$$

where Γ_j are the constants giving the vortex amplitudes and the vortex coordinate is $\boldsymbol{r_j} = (x_j, y_j)$. The solution of Eq. (21.127) for the magnetic field (Abrikosov, 1957) is given by

$$B = \sum_j B_j, \quad B_j[\boldsymbol{r}, \boldsymbol{r}_j(t)] = -\frac{\Gamma_j}{2\pi} K_0\big[\left|\boldsymbol{r} - \boldsymbol{r}_j(t)\right|\big], \tag{21.128}$$

with $K_n(x)$ being modified Bessel functions of the n-th kind (Olver et al., 2010). The modified Bessel function of the 0-th kind has an integral representation

$$K_0(x) = \int_0^\infty \cos(x \sinh t)\, dt;$$

in the limit $x \to 0$ we have $K_0(x) \approx \ln\left(\frac{2}{x}\right) - 0.5772$,
and for $x \to \infty$ we have $K_0(x) \approx \sqrt{\frac{\pi}{2}} \frac{\exp(-x)}{\sqrt{x}}$.

The limit $x \to 0$ corresponds to the distance much smaller than c/ω_{pe}, which corresponds to the Euler hydrodynamics. The dynamics of interacting point vortices in hydrodynamics have been discussed by Lamb (1932) and Sedov (1971).

The vortex dynamics in dispersive medium is determined by the functions $\boldsymbol{r_j}(t)$ which should be found from the equations of characteristics of Eq. (21.127):

$$\dot{\boldsymbol{r}}_j = \boldsymbol{e}_z \times \nabla \cdot \sum_{k \neq j} B_k[\boldsymbol{r}_j(t), \boldsymbol{r}_k(t)]. \tag{21.129}$$

Using these expressions, we find the equations of the vortex motion

$$\dot{x}_j = -\frac{1}{2\pi} \sum_{k \neq j} \Gamma_k \frac{y_j - y_k}{|\boldsymbol{r}_j - \boldsymbol{r}_k|} K_1[|\boldsymbol{r}_j - \boldsymbol{r}_k|], \tag{21.130}$$

$$\dot{y}_j = \frac{1}{2\pi} \sum_{k \neq j} \Gamma_k \frac{x_j - x_k}{|\boldsymbol{r}_j - \boldsymbol{r}_k|} K_1[|\boldsymbol{r}_j - \boldsymbol{r}_k|], \tag{21.131}$$

with $|\boldsymbol{r}_j - \boldsymbol{r}_k| = \sqrt{(x_j - x_k)^2 + (y_j - y_k)^2}$. This system has a Hamiltonian form with the Hamiltonian

$$\mathcal{H}(x_j, y_j) = \frac{1}{2\pi} \sum_{k \neq j} \Gamma_k K_0[|\boldsymbol{r}_j - \boldsymbol{r}_k|], \tag{21.132}$$

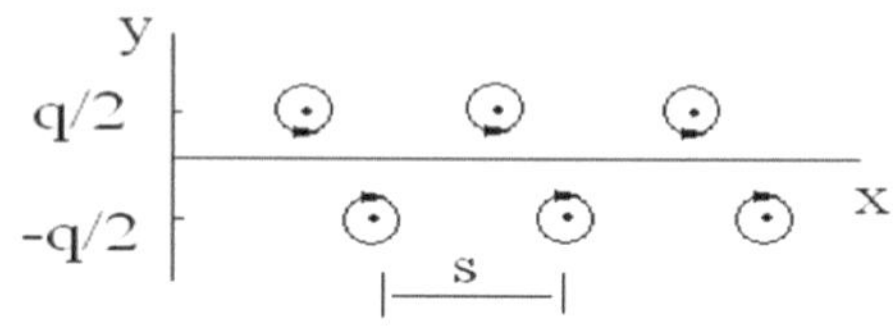

Fig. 21.11 Antisymmetric vortex row

In 2D configurations the vortex systems have typically a double chain of vortices with opposite signs. In Euler hydrodynamics a symmetric double chain row is unstable while antisymmetric row is stable for a specific set of the parameters (Lamb, 1932). The analysis of the stability of the single and double vortex chains in collisionless plasma described by the Hasegawa-Mima equations is done by Bulanov et al. (1996a, 1996b), Bulanov et al. (1997) (Fig. 21.11).

In a double chain the opposite sign vortices have coordinates and amplitudes equal to

$$x_j^{(0)} = js + Ut, \quad y_j^{(0)} = q/2, \quad -\infty < j < +\infty, \quad \Gamma_j = -1, \tag{21.133}$$

for the upper chain, and

$$x_k^{(0)} = (k+\sigma)s + Ut, \quad y_j^{(0)} = -q/2, \quad -\infty < k < +\infty, \quad \Gamma_k = 1, \tag{21.134}$$

for the lower chain. The distance between neighboring vortices in a chain is s, the distance between the chains in the y–direction is q, and the lower chain is shifted along the x-direction by σs. The shifts $\sigma s = 0$ and $\sigma s = s/2$ correspond to the symmetric and antisymmetric rows respectively. In Eqs. (21.133) and (21.134)

$$U = \frac{q}{\pi} \sum_{k=0}^{\infty} \frac{K_1(\rho_k')}{\rho_k'}, \qquad \rho_k' = \sqrt{(k+\sigma)^2 s^2 + q^2}, \tag{21.135}$$

is the velocity of the vortex row in the x-direction. For $s \ll 1$ and $q \ll 1$ from Eq. (21.135) follows the known result valid for Euler hydrodynamics (Lamb, 1932):

$$U = \frac{1}{2s} \coth\left(\frac{\pi q}{s}\right) \quad \text{for} \quad \sigma = 0 \tag{21.136}$$

and

$$U = \frac{1}{2s} \tanh\left(\frac{\pi q}{s}\right) \quad \text{for} \quad \sigma = 1/2. \tag{21.137}$$

Far from the vortex row the magnetic field and the electron velocity decrease exponentially.

From Eqs. (21.130) and (21.131) we can obtain the linearized equations of motion of the vortices representing the solutions in the form

$$\begin{pmatrix} x_j \\ y_j \end{pmatrix} = \begin{pmatrix} X \\ Y \end{pmatrix} \exp(\gamma t + ij\varphi), \quad \begin{pmatrix} x'_k \\ y'_k \end{pmatrix} = \begin{pmatrix} X' \\ Y' \end{pmatrix} \exp(\gamma t + ik\varphi), \tag{21.138}$$

for the upper and lower chains. The dispersion equation giving the relationship between the real and imaginary parts of γ and φ has a form

$$\begin{aligned} \gamma = i\frac{\varsigma}{\pi} \sum_{k=0}^{\infty} \frac{(k+\sigma)}{\rho_k'^2} K_2(\rho'_k) \sin[(k+\sigma)\varphi] \\ \mp \frac{1}{\pi} \left\{ \sum_{j=0}^{\infty} \frac{K_1\left(\rho'_j\right)}{\rho'_j} [1-\cos(j\varphi)] \right. \\ \left. - \sum_{k=0}^{\infty} \left[\frac{K_1(\rho'_k)}{\rho'_k} - q^2 \frac{K_2(\rho'_k)}{\rho_k'^2} \right] [1 + \varsigma \cos((k+\sigma)\varphi)] \right\} \\ \times \left\{ \sum_{j=0}^{\infty} \left[\frac{K_1\left(\rho'_j\right)}{\rho'_j} - K_2\left(\rho'_j\right) \right] [1-\cos(j\varphi)] \right. \\ \left. + \sum_{k=0}^{\infty} \left[\left(s\frac{k+\sigma}{\rho'} \right)^2 K_2(\rho'_k) - \frac{K_1(\rho'_k)}{\rho'_k} \right] [1 - \varsigma \cos((k+\sigma)\varphi)] \right\}, \end{aligned} \tag{21.139}$$

where $\varsigma = \pm 1$ depends on the parity of the perturbation. The symmetrical with $\sigma = 0$ vortex row is always unstable. In the limit $s \ll q \ll 1$ and $q \ll 2\pi s/\varphi \ll 1$ we reproduce Rayleigh's result for the growth rate

$$\mathrm{Re}\{\gamma\} = \kappa U \sqrt{\kappa q}, \tag{21.140}$$

where $\kappa = \varphi/s$, of the bending instability of a finite width fluid stream (Lamb, 1932). When the perturbation wavelength is larger than one and $q < 1 < 2\pi s/\varphi$, the instability growth rate is

$$\mathrm{Re}\{\gamma\} = \kappa^2 U \sqrt{\kappa q} \tag{21.141}$$

If the distance between point vortices is larger than one, $s > 1$, the growth rate is exponentially small.

In the case of the antisymmetric vortex row with $\sigma = 1/2$ the perturbations behave in more complicated way than in the symmetric vortex system. As noted in Lamb's monograph (Lamb, 1932), in the Euler hydrodynamics the antisymmetric von Karman's vortex row is stable for

$$q = 0.281\ s. \tag{21.142}$$

In the Euler hydrodynamics a point vortex is described by expression

$$-\frac{\Gamma_j}{2\pi}\ln[|\boldsymbol{r}-\boldsymbol{r}_j(t)|],$$

which can be obtained the limit $(\omega_{pe}/c)|\boldsymbol{r}-\boldsymbol{r}_j(t)| \to 0$ from expression (21.128). Here we use dimensional variables. In the opposite limit, when $(\omega_{pe}/c)|\boldsymbol{r}-\boldsymbol{r}_j(t)| \to \infty$, the magnetic field (21.129) is proportional to

$$\exp\left[-\frac{\omega_{pe}}{c}|\boldsymbol{r}-\boldsymbol{r}_j(t)|\right].$$

Such the vortex properties change the behavior of the vortex row instability. By direct inspection of Eq. (21.129) we find that for large distance between neighboring vortices the antisymmetric row is stable when

$$s/2 < q < 3s^2/4. \qquad (21.143)$$

The dependence of the growth rate on s and φ is shown in Fig. 21.12.

In inhomogeneous plasmas the vortices move in the direction perpendicular to the plasma density gradient (Lezhnin et al., 2018; Mandal et al., 2020). In the 3D case, the vortices have a toroidal form with the ring-shape magnetic field (Yue et al., 2021). This vortex property plays an important role in the laser ion acceleration in the regime of the magnetic vortex acceleration mechanism (Bulanov & Esirkepov, 2007; Fukuda et al., 2009; Kuznetsov et al., 2001; Nakamura et al., 2010; Park et al., 2019; Psikal, 2021).

As we see in collisionless plasma the von Karman vortex row can be stable in a wider range of parameters than in Euler hydrodynamics. The nonlinear stage of the vortex row instability has been studied numerically in Bulanov et al. (1997), and Califano et al. (1999). It has been shown that the row in that range of parameter is stable according to the Lyapunov criterion of stability.

21.8.3 *Mechanism of the Formation of Electromagnetic Solitons and Electron Vortices*

The physical mechanism that produces the solitons and vortices is different from the standard process where the nonlinear steepening of the wave is counterbalanced by the effect of dispersion. The solitons occur due the frequency downshift of the laser pulse. This downshift appears because of the energy depletion of the laser pulse as has been discussed above. Interacting with the underdense plasma the laser pulse loses its energy as it generates electrostatic and magnetostatic wake fields behind it. The frequency of these fields is much lower than the carrier frequency of the laser pulse and the laser-plasma interaction is adiabatic. In this case the number of quanta

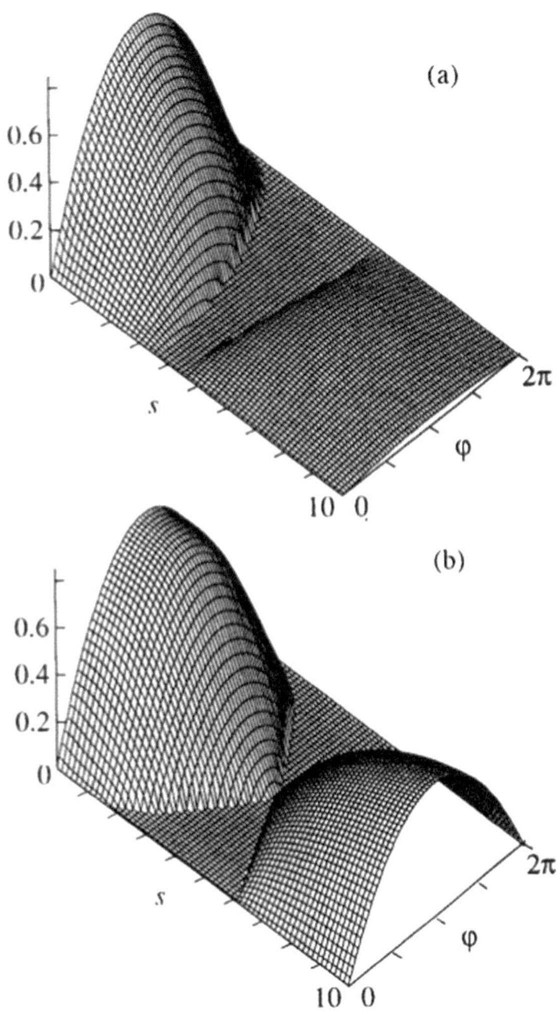

Fig. 21.12 Dependence on s and φ of the growth rate of the instability for the antisymmetric vortex row in the case described by the Hasegawa-Mima equation (**a**) and within the framework of the Euler hydrodynamics (**b**) (Bulanov et al., 1997)

of the electromagnetic field is adiabatically conserved. Because of this constraint, the decrease of the radiation energy must be accompanied by the downshift of the carrier frequency. The pulse depletion length is $l_{depl} = l_{las}(\omega_0/\omega_{pe})^2$. On the other hand the group velocity of the laser pulse decreases as the carrier frequency is downshifted, and the pulse group velocity tends to zero at $\langle\omega\rangle \to \omega_{pe}$. Thus, the depleted portions of the pulse, e.g. inside the lateral filaments, lag behind and convert their energy into solitons or vortices with almost zero group velocity. As far as it is concerned electron vortex they can be formed due the laser frequency downshift at the end of the laser pulse evolution in plasma as has been shown by Naumova et al. (2001) and Lezhnin et al. (2018). In other cases, as for the vortices formed inside the self-focusing channel or at the plasma-vacuum interface, they are produced by the electric current of fast electrons accelerated by the laser pulse (Bulanov & Esirkepov, 2007; Bulanov et al., 1996a, 1996b; Lamač et al., 2023; Park et al., 2019; Yue et al., 2021).

References

Abrikosov, A. A. (1957). The magnetic properties of superconducting alloys. *Journal of Physics and Chemistry of Solids, 2*, 199; On the magnetic properties of superconductors of the second group. *Soviet Physics JETP, 5*, 1174.

Akhiezer, A. I., & Polovin, R. V. (1956). Theory of wave motion of an electron plasma. *Soviet Physics—JETP, 3*, 696.

Akhmediev, N. N., & Ankiewicz, A. (1997). *Nonlinear pulses and beams*, Springer.

Antonsen, T. M., & Mora, P. (1992). Self-focusing and Raman scattering of laser pulses in tenuous plasmas. *Physical Review Letters, 69*, 2204.

Askar'yan, G. A., Bulanov, S. V., Pegoraro, F., & Pukhov, A. M. (1994). Magnetic interaction of self-focusing channels and the electromagnetic radiation: their merging and energy accumulation. *JETP Letters, 60*, 251.

Askar'yan, G. A. (1962). Effects of the gradient of a strong electromagnetic beam on electrons and atoms. *Soviet Physics—JETP, 15*, 1088.

Bespalov, V. I., & Talanov, V. I. (1966). Filamentary structure of light beams in nonlinear liquids. *JETP Letters, 3*, 307.

Borghesi, M., Bulanov, S. V., Campbell, D. H., et al. (2002). Macroscopic evidence of soliton formation in multiterawatt laser plasma interaction. *Physical Review Letters, 88*, Article 135002.

Borisov, A. B., Borovskiy, A. V., Korobkin, V. V., Prokhorov, A. M., Shiryaev, O. B., Shi, X. M., Luk, T. S., McPherson, A., Solem, J. C., Boyer, K., & Rhodes, C. K. (1992). Observation of relativistic and charge-displacement self-channeling of intense subpicosecond ultraviolet (248 nm) radiation in plasmas. *Physical Review Letters, 68*, 2309.

Brantov, A. V., Esirkepov, T. Zh., Kando, M., et al. (2008). Controlled electron injection into the wake wave using plasma density inhomogeneity. *Physics of Plasmas, 15*, 073111.

Bulanov, S. V. (2021). Electron dynamics in the field of strong plasma and electromagnetic waves: A review. *Physics of Wave Phenomena, 29*, 1.

Bulanov, S. V., & Sakharov, A. S. (1991). Induced focusing of the electromagnetic radiation in the wake-field plasma wave. *JETP Letters, 54*, 203.

Bulanov, S. V., Inovenkov, I. N., Kirsanov, V. I., et al. (1992). Nonlinear depletion of ultrashort and relativistically strong laser pulses in an underdense plasma. *Physics of Fluids B, 4*, 1935.

Bulanov, S. S., Fedotov, A. M., & Pegoraro, F. (2005). Damping of electromagnetic waves due to electron-positron pair production. *Physical Review E, 71*, 016404.

Bulanov, S. S., Maksimchuk, A., Schroeder, C. B., et al. (2012). Relativistic spherical plasma waves. *Physics of Plasmas, 19*, 020702.

Bulanov, S. S., Schroeder, C. B., Esarey, E., & Leemans, W. P. (2013). Electromagnetic cascade in high-energy electron, positron, and photon interactions with intense laser pulses. *Physical Review A, 87*, Article 062110.

Bulanov, S. V., & Esirkepov, T. Zh. (2007). Comment on "Collimated multi-MeV ion beams from high-intensity laser interactions with underdense plasma. *Physical Review Letters, 98*, 049503.

Bulanov, S. V., Lontano, M., Esirkepov, T. Zh., Pegoraro, F., & Pukhov, A. M. (1996). Electron vortices produced by ultraintense laser pulses. *Physical Review Letters, 76*, 3562.

Bulanov, S. V., Lontano, M., Esirkepov, T. Zh., Pegoraro, F., & Pukhov, A. M. (1996). Magnetic interaction and magnetic wake of high intensity laser pulses in plasmas. *Physica Scripta T, 63*, 280.

Bulanov, S. V., Esirkepov, T. Zh., Lontano, M., & Pegoraro, F. (1997). Stability of vortex streets and vortex rows in the framework of the Hasegawa-Mima equation. *Plasma Physics Reports, 23*, 660.
Bulanov, S. V., Naumova, N. M., Vshivkov, V. A., Dudnikova, G. I., Liseikina, T. V., Esirkepov, T. Zh., Kamenets, F. F., Califano, F., & Pegoraro, F. (1999). Interaction of petawatt laser pulses with underdense plasmas. *Plasma Physics Reports, 25*, 701.
Bulanov, S. V., Califano, F., Dudnikova, G. I., et al. (2001). Relativistic interaction of laser pulses with plasmas. In V. D. Shafranov (Ed.), *Reviews of plasma physics* (Vol. 22, p. 227). Kluwer Academic/Plenum Publishers.
Califano, F., Prandi, R., Pegoraro, F., & Bulanov, S. V. (1999). Two-dimensional electron-magnetohydrodynamic instabilities. *Physics of Plasmas, 6*, 2332.
Chakraborty, B., Khan, M., & Bhattacharya, B. (1983). Nonlinearly evolved Faraday rotation due to strong radiation in a plasma. *Physical Review A, 28*, 1047.
Charney, J. (1947). Dynamics of long waves in a baroclinic westerly current. *Journal of the Atmospheric Sciences, 4*, 135.
Chen, L. M., Kotaki, H., Nakajima, K., Bulanov, S. V., Tajima, T., Gu, Y. Q., Peng, H. S., Hua, J. F., An, W. M., Tang, C. X., & Lin, Y. Z. (2007). Self-guiding of 100 TW femtosecond laser pulses in centimeter-scale underdense plasma. *Physics of Plasmas, 14*, 040703.
Chen, H. H., & Liu, C.-S. (1976). Solitons in nonuniform media. *Physical Review Letters, 37*, 693.
Chian, A.C.-L. (1981). Relativistically strong-coupled transverse-longitudinal waves in an electron-ion plasma. *Physical Review A, 24*, 2773.
Dawson, J. M. (1959). Nonlinear electron oscillations in a cold plasma. *Physical Review, 113*, 383.
Dodd, R. K., Eilbeck, J. C., Gibbon, J. D., & Morris, H. C. (1982). *Solitons and nonlinear wave equations*. Academic Press.
Dolzhansky, F. V. (2012). *Fundamentals of geophysical hydrodynamics*. Springer.
Drake, J. F., Lee, Y. C., Nishikawa, K., & Tsintsadze, N. L. (1976). Breaking of large-amplitude waves as a result of relativistic electron-mass variation. *Physical Review Letters, 36*, 196.
Esarey, E., Schroeder, C. B., & Leemans, W. P. (2009). Physics of laser-driven plasma-based electron accelerators. *Reviews of Modern Physics, 81*, 1229.
Esirkepov, T. Zh., Bulanov, S. V., Nishihara, K., & Tajima, T. (2004). Soliton synchrotron afterglow in a laser plasma. *Physical Review Letters, 92*, 255001.
Esirkepov, T. Zh., Kamenets, F., Bulanov, S. V., & Naumova, N. M. (1998). *Low-frequency relativistic electromagnetic solitons in collisionless plasmas. JETP Letters, 68*, 36.
Esirkepov, T. Zh., Bulanov, S. V., Daido, H., Kato, Y., Khoroshkov, V. S., Kitagawa, Y., Mima, K., Nagai, K., Nishihara, K., Sakabe, S., Pegoraro, F., & Tajima, T. (2002). Proposed double-layer target for the generation of high-quality laser-accelerated ion beams. *Physical Review Letters, 89*, 175003.
Esirkepov, T. Zh., Nishihara, K., Bulanov, S. V., & Pegoraro, F. (2002). Three-dimensional relativistic electromagnetic sub-cycle solitons. *Physical Review Letters, 89*, 275002.
Esirkepov, T. Zh., Bulanov, S. V., Yamagiwa, M., & Tajima, T. (2006). Electron, positron, and photon wakefield acceleration: Trapping, wake overtaking, and ponderomotive acceleration. *Physical Review Letters, 96*, 014803.
Farina, D., & Bulanov, S. V. (2001a). Relativistic electromagnetic solitons in the electron-ion plasma. *Physical Review Letters, 86*, 5289.
Farina, D., & Bulanov, S. V. (2001b). Dark solitons in electron–positron plasmas. *Physical Review E, 64*, Article 066401.
Farina, D., Lontano, M., & Bulanov, S. V. (2000). Electromagnetic solitons in magnetized plasmas. *Physical Review E, 62*, 4146.
Fukuda, Y., Faenov, A. Y., Tampo, M., et al. (2009). Energy increase in multi-MeV ion acceleration in the interaction of a short pulse laser with a cluster-gas target. *Physical Review Letters, 103*, Article 165002.
Gardner, C. S., Greene, J. M., Kruskal, M. D., & Miura, R. M. (1967). Method for solving the Korteweg-deVries equation. *Physical Review Letters, 19*, 1095.

Hasegawa, A., & Mima, K. (1978). Pseudo-three-dimensional turbulence in magnetized nonuniform plasma. *Physics of Fluids, 21*, 87.
Horton, W., & Hasegawa, A. (1994). Quasi-two-dimensional dynamics of plasmas and fluids. *Chaos, 4*, 227.
Kando, M., Fukuda, Y., Pirozhkov, A. S., Ma, J., Daito, I., Chen, L.-M., Esirkepov, T. Z., Ogura, K., Homma, T., Hayashi, Y., Kotaki, H., Sagisaka, A., Mori, M., Koga, J. K., Daido, H., Bulanov, S. V., Kimura, T., Kato, Y., & Tajima, T. (2007). Demonstration of Laser-frequency upshift by electron-density modulations in a plasma wakefield. *Physical Review Letters, 99*, Article 135001.
Kando, M., Pirozhkov, A. S., Kawase, K., Esirkepov, T. Z., Fukuda, Y., Kiriyama, H., Okada, H., Daito, I., Kameshima, T., Hayashi, Y., Kotaki, H., Mori, M., Koga, J. K., Daido, H., Faenov, A. Y., Pikuz, T., Ma, J., Chen, L.-M., Ragozin, E. N., … Bulanov, S. V. (2009). Enhancement of photon number reflected by the relativistic flying mirror. *Physical Review Letters, 103*, Article 235003.
Kaw, P., & Dawson, J. (1970). Relativistic nonlinear propagation of laser beams in cold overdense plasmas. *Physics of Fluids, 13*, 472.
Kaw, P. K., Sen, A., & Katsouleas, T. (1992). Nonlinear 1D laser pulse solitons in a plasma. *Physical Review Letters, 68*, 3172.
Kim, J. B., Bulanov, S. V., Suk, H., & Ko, I. S. (2004). Kinetic relativistic solitons in electron-positronion plasmas. *Physics Letters A, 329*, 464.
Kovalev, V. F., & Bychenkov, V. Yu. (2019). Analytic theory of relativistic self-focusing for a Gaussian light beam entering a plasma: Renormalization-group approach. *Physical Review E, 99*, 043201.
Kozlov, V. A., Litvak, A. G., & Suvorov, E. V. (1979). Envelope solitons of relativistic strong electromagnetic waves. *Soviet Physics—JETP, 49*, 75.
Kuznetsov, A. V., Esirkepov, T. Zh., Kamenets, F. F., & Bulanov, S. V. (2001). On the efficiency of fast ion acceleration by relativistically strong laser pulses in an underdense plasma. *Plasma Physics Reports, 27*, 211.
Lamač, M., Chaulagain, U., Nejdl, J., & Bulanov, S. V. (2023). Generation of intense magnetic wakes by relativistic laser pulses in plasma. *Scientific Reports, 13*, Article 170113.
Lamb, H. (1932). *Hydrodynamics*. Cambridge University Press.
Landau, L. D., & Lifshitz, E. M. (1984). Course of theoretical physics, Vol. 8: Electrodynamics of continuous media. Pergamon Press.
Leemans, W. P., & Esarey, E. (2009). Laser-driven plasma-wave electron accelerators. *Physics Today, 62*, 44.
Lehmann, G., E., Laedke, W., & Spatschek, K. H. (2006). Stability and evolution of one-dimensional relativistic solitons on the ion time scale. *Physics of Plasmas, 13*, 092302.
Lezhnin, K. V., Qu, K., Fisch, N. J., & Bulanov, S. V. (2024). On parallel laser beam merger in plasmas. *Physics of Plasmas, 31*, 032114.
Lezhnin, K. V., Kamenets, F. F., Esirkepov, T. Zh., Bulanov, S. V., Gu, Y. J., Weber, S., & Korn, G. (2016). Explosion of relativistic electron vortices in laser plasmas. *Physics of Plasmas, 23*, 093116.
Lezhnin, K. V., Kamenets, F. F., Esirkepov, T. Zh., & Bulanov, S. V. (2018). On annihilation of the relativistic electron vortex pair in collisionless plasmas. *Journal of Plasma Physics, 84*, 905840610.
Liu, C. S., Tripathi, V. K., & Eliasson, B. (2019). *High-power laser-plasma interaction*. Cambridge University Press.
Lontano, M., Bulanov, S., Koga, J., Passoni, M., & Tajima, T. (2002). A kinetic model for the one-dimensional electromagnetic solitons in an isothermal plasma. *Physics of Plasmas, 9*, 2562.
Mancic, A., Hadzievski, L., & Skoric, M. (2006). Dynamics of electromagnetic solitons in a relativistic plasma. *Physics of Plasmas, 13*, Article 052309.

Mandal, D., Vashistha, A., & Das, A. (2020). Spontaneous formation of coherent structures by an intense laser pulse interacting with overdense plasma. *Journal of Plasma Physics, 86*, 905860606.

Matlis, N. H., Reed, S., Bulanov, S. S., Chvykov, V., Kalintchenko, G., Matsuoka, T., Rousseau, P., Yanovsky, V., Maksimchuk, A., Kalmykov, S., Gennady, S., & Downer, M. C. (2006). Snapshots of laser wakefields. *Nature Physics, 2*, 749.

Mendonça, J. T. (2001). *Theory of photon acceleration*. Institute of Physics Publishing.

Mendonça, J. T. (2024). Photon acceleration by superluminal ionization fronts. *Symmetry, 16*, 112.

Mori, W. B., Joshi, C., Dawson, J. M., Forslund, D. W., & Kindel, J. M. (1988). Evolution of self-focusing of intense electromagnetic waves in plasma. *Physical Review Letters, 60*, 1298.

Mori, W. B., Decker, C. D., Hinkel, D. E., & Katsouleas, T. (1994). Raman forward scattering of short pulse high-intensity laser. *Physical Review Letters, 72*, 1482.

Mourou, G., Tajima, T., & Bulanov, S. V. (2006). Optics in the relativistic regime. *Reviews of Modern Physics, 78*, 309.

Mourou, G. A., Korn, G., Sandner, W., & Collier, J. L. (eds.). (2011). *Eli-extreme light infrastructure: Science and technology with ultra-intense lasers, whitebook*. THOSS Media GmbH.

Nakamura, T., Bulanov, S. V., Esirkepov, T. Zh., & Kando, M. (2010). High-energy ions from near-critical density plasmas via magnetic vortex acceleration. *Physical Review Letters, 105*, 135002.

Naseri, N., Rozmus, W., & Pesme, D. (2016). Self-channeling of intense laser pulses in underdense plasma and stability analysis. *Physics of Plasmas, 23*, Article 113101.

Naumova, N. M., Bulanov, S. V., Esirkepov, TZh., Farina, D., Nishihara, K., Pegoraro, F., Ruhl, H., & Sakharov, A. S. (2001). Formation of electromagnetic post-solitons in plasmas. *Physical Review Letters, 87*, 185004.

Naumova, N. M., Bulanov, S. V., Nishihara, K., Esirkepov, TZh., & Pegoraro, F. (2002). Polarization effects and anisotropy in three-dimensional relativistic self-focusing. *Physical Review E, 65*, 045402.

Naumova, N. M., Bulanov, S. V., Esirkepov, T. Zh., Farina, D., Nishihara, K., Pegoraro, F., Ruhl, H., & Sakharov, A. S. (2001). Formation of electromagnetic post-solitons in plasmas. *Physical Review Letters, 87*, 185004.

Naumova, N. M., Koga, J., Nakajima, K., Tajima, T., Esirkepov, T. Zh., Bulanov, S. V., & Pegoraro, F. (2001). Polarization, hosing and long-time evolution of relativistic laser pulses. *Physics of Plasmas, 8*, 4149.

Olver, F. W. J., Lozier, D. W., Boisvert, R. F., & Clark, C. W. (2010). *NIST handbook of mathematical functions*. Cambridge University Press.

Park, J., Bulanov, S. S., Bin, J., Ji, Q., Steinke, S., Vay, J. L., Geddes, C. G. R., Schroeder, C. B., Leemans, W. P., Schenkel, T., & Esarey, E. (2019). Ion acceleration in laser generated megatesla magnetic vortex. *Physics of Plasmas, 26*, Article 103108.

Pedlosky, J. (1987). *Geophysical fluid dynamics*. Springer.

Pesch, T. C., & Kull, H.-J. (2007). Linearly polarized waves with constant phase velocity in relativistic plasmas. *Physics of Plasmas, 14*, 083103.

Pirozhkov, A. S., Ma, J., Kando, M., Esirkepov, T. Zh., Fukuda, Y., Chen, L.-M., Daito, I., Ogura, K., Homma, T., Hayashi, Y., Kotaki, H., Sagisaka, A., Mori, M., Koga, J. K., Kawachi, T., Daido, H., Bulanov, S. V., Kimura, T., Kato, Y., & Tajima, T. (2007). Frequency multiplication of light back-reflected from a relativistic wake wave. *Physics of Plasmas, 14*, 123106.

Poornakala, S., Das, A., Kaw, P. K., Sen, A., Sheng, Z. M., Sentoku, Y., Mima, K., & Nishikawa, K. (2002). Weakly relativistic one-dimensional laser pulse envelope solitons in a warm plasma. *Physics of Plasmas, 9*, 3802.

Psikal, J. (2021). Laser-driven ion acceleration from near-critical Gaussian plasma density profile. *Plasma Physics and Controlled Fusion, 63*, Article 064002.

Romagnani, L., Bigongiari, A., Kar, S., Bulanov, S. V., Cecchetti, C. A., Esirkepov, T. Zh., Galimberti, M., Jung, R., Liseykina, T. V., Macchi, A., Osterholz, J., Pegoraro, F., Willi, O., & Borghesi,

M. (2010). Observation of magnetized soliton remnants in the wake of intense laser pulse propagation through plasmas. *Physical Review Letters, 105*, 175002.
Sanchez-Arriaga, G., Siminos, E., & Lefebvre, E. (2011). Relativistic solitary waves with phase modulation embedded in long laser pulses in plasmas. *Physics of Plasmas, 18*, Article 082304.
Sanchez-Arriaga, G., & Lefebvre, E. (2011a). Two-dimensional s-polarized solitary waves in relativistic plasmas. I. The fluid plasma model. *Physical Review E, 84*, 036404.
Sanchez-Arriaga, G., & Lefebvre, E. (2011b). Two-dimensional s-polarized solitary waves in plasmas. II. Stability, collisions, electromagnetic bursts, and post-soliton evolution. *Physical Review E, 84*, 036404.
Sarri, G., Kar, S., Romagnani, L., Bulanov, S. V., Cecchetti, C. A., Galimberti, M., Gizzi, L. A., Heathcote, R., Jung, R., Kourakis, I., Osterholz, J., Schiavi, A., Willi, O., & Borghesi, M. (2011). Observation of plasma density dependence of electromagnetic soliton excitation by an intense laser pulse. *Physics of Plasmas, 18*, Article 080704.
Saxena, V., Kourakis, I., Sanchez-Arriaga, G., & Siminos, E. (2013). Interaction of spatially overlapping standing electromagnetic solitons in plasmas. *Physics Letters A, 377*, 473.
Sedov, L. I. (1971). *A course in continuum mechanics. Volumes. I-IV*. Wolters-Noordhoff Publishing.
Sentoku, Y., Esirkepov, T. Zh., Mima, K., Nishihara, K., Califano, F., et al. (1999). Bursts of super reflected laser light from inhomogeneous plasmas due to the generation of relativistic solitary waves. *Physical Review Letters, 83*, 3434.
Smetanin, I. V., Esirkepov, T. Zh., Farina, D., Koga, et al. (2004). Evolution of an intense elliptically polarized electromagnetic wave in underdense plasmas. *Physics Letters A, 320*, 438.
Sprangle, P., Esarey, E., Krall, J., & Joyce, G. (1992). Propagation and guiding of intense laser pulses in plasmas. *Physical Review Letters, 69*, 2200.
Sulem, C., & Sulem, P. L. (2007). *The nonlinear Schrodinger equation: Self-focusing and wave collapse*. Springer Science & Business Media
Sun, G.-Z., Ott, E., Lee, Y. C., & Guzdar, P. (1987). Self-focusing of short intense pulses in plasmas. *Physics of Fluids, 30*, 526.
Tajima, T., & Dawson, J. M. (1979). Laser electron accelerator. *Physical Review Letters, 43*, 267.
Tajima, T., & Malka, V. (2020). Laser plasma accelerators. *Plasma Physics and Controlled Fusion, 62*, 034004.
Tajima, T., Yan, X. Q., & Ebisuzaki, T. (2020). Wakefield acceleration. *Reviews of Modern Plasma Physics, 4*, 1.
Valenta, P., Grittani, G. M., Lazzarini, C. M., Klimo, O., & Bulanov, S. V. (2021). On the electromagnetic-electron rings originating from the interaction of high-power short-pulse laser and underdense plasma. *Physics of Plasmas, 28*, Article 122104.
Whitham, G. B. (1974). *Linear and nonlinear waves*. Wiley.
Wilks, S. C., Dawson, J. M., Mori, W. B., Katsouleas, T., & Jones, M. E. (1989). Photon accelerator. *Physical Review Letters, 62*, 2600.
Wu, D., Yu, W., Fritzsche, S., Zheng, C. Y., & He, X. T. (2019). Formation of relativistic electromagnetic solitons in overdense plasmas. *Physics of Plasmas, 26*, Article 063107.
Yue, D. N., Chen, M., Geng, P. F., et al. (2021). Dynamics of moving electron vortices and magnetic ring in laser plasma interaction. *Physics of Plasmas, 28*, Article 042303.
Zakharov, V. E., Manakov, S. V., Novikov, S. P., & Pitaevskii, L. P. (1984). *Theory of solitons*. Consultants Bureau.
Zhao, T. Z., Behm, K., Dong, C. F., Davoine, X., Kalmykov, S. Y., Petrov, V., Chvykov, V., Cummings, P., Hou, B., Maksimchuk, A., Nees, J. A., Yanovsky, V., Thomas, A. G. R., & Krushelnick, K. (2016). High-flux femtosecond X-ray emission from controlled generation of annular electron beams in a laser wakefield accelerator. *Physical Review Letters, 117*, Article 094801.

Chapter 22
Laser Interaction with an Overdense Plasma

Abstract When the relativistically strong laser pulse interacts with an underdense plasma the laser absorption depends on the laser amplitude, incidence angle, laser polarization and plasma density profile. For moderate laser intensity and weal plasma density inhomogeneity the laser pulse absorption happens at the plasma resonance region at the critical plasma density surface. This process is accompanied by the charged particle acceleration. In the case of sharp vacuum-plasma interface, the laser loses the energy via the "vacuum heating" process. It also results in the charged particle acceleration and generation of high order harmonics via the relativistic oscillating mirror mechanism as well as in the generation of the surface waves.

Keywords Oblique incidence of EM wave on overdense plasma slab · Plasma resonance · Electron and ion acceleration · Surface waves

22.1 Physical Processes Determining the Laser Interaction with an Overdense Plasma

The interaction between the laser pulse and plasma targets depends critically on the ratio of the pulse carrier frequency ω to the Langmuir frequency ω_{pe}. In the case of a transparent plasma considered in previous Sect. 22.3, where the laser frequency is larger than the Langmuir frequency, $\omega \gg \omega_{pe}$, the main absorption mechanisms of an ultra-short, relativistically intense pulse arise either from the losses due to the radiation of wake plasma waves behind the pulse or from its modulation due to forward and backward Raman scattering. Instead, in the case of a laser pulse interacting with a collisionless overdense plasma, $\omega \ll \omega_{pe}$, the main absorption mechanisms are determined by the ratio of the electron quiver amplitude in the electromagnetic field of the amplitude E oscillating with the frequency ω, equal to

$$r_E = \frac{eE}{m_e\omega^2}$$

to the plasma non-uniformity scale length

Y. Kato et al., *High Power Laser and Plasma Science*, Springer Series on Atomic, Optical, and Plasma Physics 130, https://doi.org/10.1007/978-3-032-08930-4_22

$$L = \left.(\partial_x \ln n)^{-1}\right|_{n=n_{cr}}$$

and by the pulse polarization and incidence angle. Here n_{cr} is the critical value of plasma density for which $\omega = \sqrt{4\pi n_{cr} e^2/m_e}$. For $\eta = r_E/L \ll 1$ the main absorption mechanism for p-polarized pulses is due to non-linear processes, occurring in the plasma resonance region (Denisov, 1957; Gershman et al., 1957; Ginzburg, 1962; Gildenburg, 1964; Landau & Lifshitz, 1984; Kruer, 1988; Metelskii et al., 2024). We remind here that for the p-polarization (from the German word parallel) of the obliquely incident on the target pulse the electric field lies in the incidence plane while for the s-polarized (from the German word senkrecht) laser pulse the electric field is perpendicular to the incidence plane. On the contrary, if the oscillation amplitude r_E is larger than the typical non-uniformity scale length $\eta \gg 1$, the laser energy is absorbed through the mechanism of electron "vacuum heating" (Brunel, 1987; Gibbon & Bell, 1992). It is also known that the interaction between matter and extremely powerful laser pulses is accompanied by the generation of high harmonics of electromagnetic radiation and by the production of fast particles (Bulanov et al., 1994; Metelskii et al., 2024; Teubner & Gibbon, 2009).

22.2 Oblique Incidence of the Electromagnetic Wave on an Inhomogeneous Plasma

22.2.1 The Lorentz Transform to the Boosted Frame of Reference

Here we consider an electromagnetic wave with carrier frequency ω, wave vector $\boldsymbol{k}$ and normalized amplitude $a_0 = eE/m_e\omega c$. The wave is obliquely incident on a plasma (located in the region $x > 0$) and forms an angle θ with respect to the normal to the plasma surface. At large distances from the plasma boundary, $x \to \infty$, the plasma is overdense, $\omega \gg \omega_{pe} = \sqrt{4\pi n e^2/m_e}$. In the vacuum region ($x < 0$) the frequency ω and wave vector $\boldsymbol{k}$ are connected through the dispersion equation $\omega^2 = k_x^2 c^2 + k_y^2 c^2$ where $k_x = k\cos\theta$, $k_y = k\sin\theta$. Following Bourdier (1983), we perform a Lorentz transformation to the frame of reference frame $\mathcal{M}$ moving in the y direction along the plasma target surface with velocity V_M. In the $\mathcal{M}$ frame of reference, the frequency ω' and the components of the wave vector $\boldsymbol{k}'$ are

$$\omega' = \frac{\omega - k_y \beta_M c}{\sqrt{1-\beta_M^2}}, \quad k_x' = k_x, \quad k_y' = \frac{k_y - \omega\beta_M/c}{\sqrt{1-\beta_M^2}}, \tag{22.1}$$

where $\beta_M = V_M/c$. For

$$V_M = k_y c^2/\omega = c\sin\theta$$

the y-component k'_y of the wave vector in the boosted frame of reference $\mathcal{M}$ is equal to zero. Thus, in this frame of reference, the electromagnetic wave propagates only in the x-direction, and the problem under consideration is reduced to that of normal incidence of the wave on the plasma, where both ions and electrons move with velocity V_M along the y-direction. We note that the boost velocity $V_M = k_y c^2/\omega$ equals the y-component of the wave group velocity in the laboratory frame of reference $\mathcal{L}$.

In the boosted frame of reference $\mathcal{M}$, the frequency is $\omega' = \omega \cos\theta$ and wave vector has only one non-zero component

$$\boldsymbol{k}' = (k'_x, 0, 0) = (k\cos\theta, 0, 0).$$

In the case of p-polarized electromagnetic waves, the electric and magnetic fields in the boosted frame of reference are

$$\boldsymbol{E}' = (0, B_0\cos\theta, 0) \text{ and } \boldsymbol{B}' = (0, 0, B_0\cos\theta),$$

respectively.

For s-polarized electromagnetic waves, we have

$$\boldsymbol{E}' = (0, 0, E_0\cos\theta) \text{ and } \boldsymbol{B}' = (0, E_0\cos\theta, 0).$$

The unperturbed plasma density n' in the boosted frame of reference $\mathcal{M}$ is

$$n' = n\gamma_M = n/\cos\theta,$$

where the relativistic factor γ_M depends on $\beta_M = \sin\theta$ as $\gamma_M = 1/\sqrt{1-\beta_M^2} = 1/\cos\theta$. Electrons and ions move in the boosted frame of reference with equal velocities, $-\boldsymbol{V}_M$. Their unperturbed momentum and energy are given by $m_a c\beta_M\gamma_M = m_a c\tan\theta$ and $m_a c^2\gamma_M = m_a c^2/\cos\theta$, where $a = i, e$. The number of oscillations in the laser pulse $N_{las} = \omega t_{las}/2\pi$, where t_{las} is the laser pulse duration, as well as the Langmuir frequency ω_{pe} are Lorentz invariant. In the $\mathcal{M}$ frame of reference all quantities are taken to depend on the time t' and the coordinate x' only. A sketch of the electromagnetic wave electric and magnetic fields as well as of the wave vectors in the $\mathcal{L}$ and in the $\mathcal{M}$ frames is shown in Fig. 22.1.

While the Lorentz transformation to the boosted frame of reference $\mathcal{M}$ simplifies the incidence problem from oblique to normal, several paradoxes remain to be understood. First comes the question of what is the driving force acting on the particles in the case of p-polarized waves, since in the $\mathcal{M}$ frame reference the x-component of the electric field of the laser pulse vanishes. A second question is related to the location of the plasma resonance. In the $\mathcal{M}$ frame of reference the plasma resonance is located at the point $x = x_p$, where the frequency ω of the electromagnetic wave equals the local value of the Langmuir frequency. In the $\mathcal{M}$ frame of reference the value of the Langmuir frequency ω_{pe} is unchanged, while the frequency ω of the laser pulse decreases, $\omega' = \omega\cos\theta$. To elucidate these and other problems we shall first

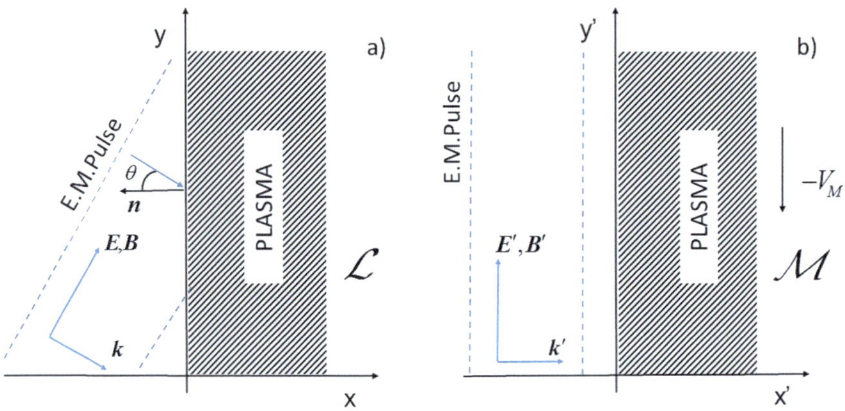

Fig. 22.1 Electromagnetic wave incident on an inhomogeneous plasma in the laboratory frame of reference $\mathcal{L}$, **a** and in the boosted frame of reference $\mathcal{M}$ (**b**). The plasma is located in the $x > 0$ half plane (it is shown by shaded area). In the $\mathcal{L}$ frame of reference, the incidence angle is θ. In the case of s-polarized (p-polarized) electromagnetic wave the magnetic (electric) field vector in the (x, y) plane. In the boosted frame of reference $\mathcal{M}$ the plasma moves in the y-direction with the velocity equal to $-V_M = -c \sin\theta$

analyze the limit of a small amplitude electromagnetic wave in the boosted frame of reference $\mathcal{M}$.

22.2.2 *Propagation of Electromagnetic Waves in Inhomogeneous Plasma*

In the boosted frame of reference $\mathcal{M}$, by linearizing the cold plasma fluid equations and Maxwell's equations, Eqs. (22.1)–(22.4), we obtain the equations for the electric and magnetic field components whose dependence on time and coordinate is assumed to be in the form

$$\boldsymbol{E}'(x', t') = \boldsymbol{E}'(x') \exp(-i\omega' t') \quad \text{and} \quad \boldsymbol{B}'(x', t') = \boldsymbol{B}'(x') \exp(-i\omega' t'), \tag{22.2}$$

is

$$\frac{d}{dx'}\left[\frac{1}{\varepsilon'(\omega', x')}\frac{d}{dx'}\left(B'_z - \beta_M E'_x\right)\right] + \frac{\omega'^2 - \omega'^2_{pe}(x')}{c^2 \varepsilon'(\omega', x')}\left(B'_z - \beta_M E'_x\right) = 0, \tag{22.3}$$

$$\left(E'_x - \beta_M B'_z\right) + \frac{\beta_M}{\varepsilon'(\omega', x')}\left(B'_z - \beta_M E'_x\right) = 0 \tag{22.4}$$

for p-polarized waves, and is

$$\frac{d^2}{dx'^2}E'_z + \frac{\omega'^2 - \omega'^2_{pe}(x')}{c^2}E'_z = 0 \tag{22.5}$$

for s-polarized waves.

In Eqs. (22.3) and (22.4), the dielectric constant $\varepsilon'(\omega', x')$ is equal to

$$\varepsilon'(\omega', x') = 1 - \frac{\omega'^2_{pe}(x')}{\omega'^2}\left(1 - \beta_M^2\right) \tag{22.6}$$

Choosing a plasma density profile described by a linear function of the x' coordinate for which $\omega'^2_{pe}(x') = \omega^2_{pe}x'/L$ we write Eq. (22.5) for s-polarized wave as

$$\frac{d^2}{dx'^2}E'_z + \left(\frac{\omega'^2}{c^2} - \frac{\omega^2_{pe}}{c^2}\frac{x'}{L}\right)E'_z = 0, \tag{22.7}$$

whose solution can be expressed in terms of Airy functions (Olver et al., 2010):

$$E'_z(x') = E_0\mathrm{Ai}\left(-\left(\frac{\omega_{pe}L}{c}\right)^{2/3}\left(\frac{x'}{L} - \frac{\omega'^2}{\omega^2_{pe}}\right)\right), \tag{22.8}$$

where E_0 is constant. This expression shows that the s-polarized wave reflects at the coordinate $x'_r = L(\omega'/\omega_{pe})^2$, i.e. at the point $x'_r = L(\omega/\omega_{pe})^2\cos^2\theta$. Using asymptotic representation of the Airy function we find from Eq. (22.8) that for $x' \gg x'_r$ the wave amplitude decreases exponentially as

$$E'_z(x') = \frac{E_0}{2\sqrt{\pi}\,\varsigma^{1/4}}\exp(-\varsigma) + \cdots \tag{22.9}$$

where $\varsigma = -(\omega_{pe}L/c)^{2/3}\left(\omega'^2/\omega^2_{pe} - x'/L\right)$. We note that in the vacuum region, where $\varepsilon'(\omega', x') = 1$ the x-component of the electric field vanishes, $E'_x = 0$.

Now we consider the case of p-polarized electromagnetic wave. From Eqs. (22.2)–(22.5) we see that the reflection point of the electromagnetic waves occurs where $\omega'^2 - \omega'^2_{pe}(x')$ as in the case of s-polarized wave. In the case of p-polarized waves, the plasma resonance point x'_p occurs where $\varepsilon'(\omega', x'_p) = 0$, which corresponds to singular point for Eqs. (22.3) and (22.4), i.e. where $\omega'^2 - \omega'^2_{pe}\left(x'_p\right)\left(1 - \beta_M^2\right)$.

Near the plasma resonance point, were

$$\varepsilon'(\omega', x') \approx \frac{x' - x'_p}{L}, \tag{22.10}$$

the field $\left(B'_z - \beta_M E'_x\right)$ has a logarithmic singularity

$$B'_z - \beta_M E'_x \approx B_m \left[1 + \frac{\beta_M^2 \omega'^2 (x' - x'_p)^2}{2c^2(1 - \beta_M^2)} \ln \left(\frac{\beta_M \omega' (x' - x'_p)}{c\sqrt{1 - \beta_M^2}} \right) \right]. \tag{22.11}$$

From Eq. (22.4), it follows that at the plasma resonance point, for $x' \to x'_p$, the field $E'_x - \beta_M B'_z$ becomes infinite

$$E'_x - \beta_M B'_z \approx \frac{B_m \beta_M L}{x' - x'_p}, \tag{22.12}$$

The constant B_m is approximately equal to B_0/γ_M if the wavelength of electromagnetic radiation of of the order of the plasma inhomogeneity. In general, it can be found by asymptotic matching expression, given by Eq. (22.11), to the solution of Eq. (22.3) far from the plasma resonance region (Ginzburg, 1962).

In the boosted frame of reference $\mathcal{M}$, Langmuir oscillations correspond to oscillations with a frequency $\omega' = \omega_{pe}\sqrt{1 - \beta_M^2} = \omega_{pe}\cos\theta$. They have a magnetic field in the z-direction and an electric field in the x-direction related by $B'_z = E'_x \sin\theta$. This relationship must be satisfied for the oscillations to be potential in the $\mathcal{L}$ laboratory frame of reference. These expressions clarify the first two paradoxes formulated above. Indeed, in the $\mathcal{M}$ frame of reference the driving force is the $e\boldsymbol{v} \times \boldsymbol{B}/c$ force. The plasma resonance occurs at the same x-location both in the $\mathcal{L}$ and in the $\mathcal{M}$ frame of reference: in the $\mathcal{M}$ frame of reference the mode frequency changes while the Langmuir frequency remains invariant, but the expression of the dielectric constant $\varepsilon'(\omega', x'_p)$ changes so that the resonance condition is now given by $\omega' = \omega_{pe}\sqrt{1 - \beta_M^2}$.

Summarizing the main results obtained above, the oblique incidence of the p-polarized electromagnetic wave on the inhomogeneous density plasma slab shows different pattern compare with the case of the s-polarized electromagnetic wave. The s-polarized wave reflects at the point $x = x_r$, where $x_r = L(\omega/\omega_{pe})^2\cos^2\theta$. At $x \gg x_r$ the wave amplitude decreases exponentially. I the of p-polarization, wave also reflects at the point $x = x_r$ and exponentially decays for $x \gg x_r$, but in the vicinity of the points, where $\varepsilon(\omega, x) = 0$ the electric field formally grows to infinity. Its maximal amplitude is determined by dissipation, transformation to the Langmuir waves or by nonlinear effects (see Ginzburg, 1962; Kruer, 1988; Bulanov et al., 1990; Metelskii et al., 2024 and literature cited therein). Figure 22.2 illustrates a difference in the behavior of the s- and p-polarized electromagnetic waves in inhomogeneous plasma.

In the close vicinity of the plasma resonance point the electron component dynamics can be considered as the Langmuir wave excitation by the driver electric field E_d parallel to the gradient of the plasma density. Using expressions given by Eqs. (22.8) and (22.9) we can find for the driver field (see also Part III, Eq. (22.39))

$$E_d \approx 2.3 \frac{E_0}{\sqrt{2\pi}} \left(\frac{\omega L}{c} \right)^{1/6} \sin\theta \, \exp\left(-\frac{2}{3} \left(\frac{\omega L}{c} \right)^{2/3} \sin^2\theta \right), \tag{22.13}$$

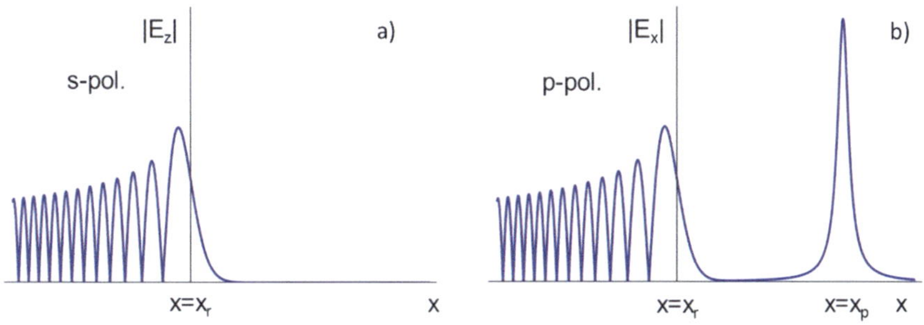

Fig. 22.2 Absolute value of the electric field of the wave obliquely incident on an inhomogeneous plasma slab. **a** s-polarized electromagnetic wave; **b** *p*-polarized electromagnetic wave, in the $\mathcal{L}$ frame of reference

which vanishes for grazing incidence of the wave ($\theta \to 0$) and tends to zero for $\omega L/c \to \infty$. It has a maximum at $\theta_m = \text{ArcSin}[(c/2\omega L)^{1/3}]$ equal to

$$E_d \approx 0.23 E_0 (\omega L/c)^{5/6}. \tag{22.14}$$

22.3 Nonlinear Langmuir Waves in the Region of Critical Plasma Density

In order to obtain analytical estimates of the physical processes that are involved into the plasma dynamics in the region near the surface of critical plasma density, we will adopt a simplified one-fluid, cold electron model. Instead of the Eulerian coordinates x, t' we use the Lagrangian variablesx_0 and τ introduced in Sect. 22.3. The equation

$$x = x_0 + \xi(x_0, \tau), \quad v_x = \partial_\tau \xi \tag{22.15}$$

gives the relationship between the Eulerian and Lagrangian variables. Here $\xi(x_0, \tau)$ is the displacement of the elementary volume of the electron fluid from the initial position x_0 and v_x is the electron velocity.

Using the results obtained in Sect. 22.3, we write the solution of the continuity equation for the electron density n can be written as a function of these variables in the form

$$n'(x_0, \tau) = \frac{n'(x_0, 0)}{1 + \partial_{x_0} \xi}. \tag{22.16}$$

The solution of Maxwell's equations for the longitudinal electric field E_x is

$$E_x' = 4\pi e \int_0^{\xi(x_0,\tau)} n_i(x_0 + s)ds, \tag{22.17}$$

with $n_i(x_0) = n'(x_0, 0)$ being the ion density distribution, which in the present case, is assumed to be given.

If we consider a linear density profile $n_i(x_0) = n_{cr}x_0/L$, where n_{cr} is the critical plasma density: the scale-length L of the plasma inhomogeneity is chosen such that that the plasma resonance $\omega' = \omega_{pe}\sqrt{1 - \beta_M^2}$ occurs at $x_p = L$, E_x' is given by

$$E_x' = 4\pi e n_{cr}\left(\frac{x_0\xi}{L} + \frac{\xi^2}{2L}\right), \tag{22.18}$$

for $x_0 + \xi > 0$, that is inside the region occupied by the plasma, and by

$$E_x' = -4\pi e n_{cr}\left(\frac{\xi^2}{2L}\right), \tag{22.19}$$

for $x_0 + \xi < 0$, that is outside the plasma. In the case of a step-like density profile for which $n' = n_0'$ for $x_0 > 0$ and $n' = 0$ for $x_0 < 0$, we obtain

$$E_x' = 4\pi e n_0'\xi, \tag{22.20}$$

for $x_0 + \xi > 0$, and

$$E_x' = -4\pi e n_0' x_0, \tag{22.21}$$

for $x_0 + \xi < 0$.

Here we consider a linear density profile. Then, we obtain for the displacement $\xi(x_0, \tau)$ driven by an electromagnetic wave with a weakly relativistic amplitude, for which $a = eE_d/m_e\omega c < 1$, the equations

$$\partial_{\tau\tau}\xi + \omega'^2\left(\frac{x_0\xi}{L} + \frac{\xi^2}{2L}\right) + \partial_{\tau\tau}\xi(\partial_\tau\xi)^2\frac{3\gamma_M}{2c^2} = \frac{eE_d}{m_e}\sin\omega'\tau, \tag{22.22}$$

for $x_0 + \xi > 0$, and

$$\partial_{\tau\tau}\xi - \omega'^2\frac{x_0^2}{2L} + \partial_{\tau\tau}\xi(\partial_\tau\xi)^2\frac{3\gamma_M}{2c^2} = \frac{eE_d}{m_e}\sin\omega'\tau, \tag{22.23}$$

in the vacuum region where $x_0 + \xi < 0$. In these equations the Langmuir frequency depends on the Lagrange coordinate x_0, and the non-linear terms arise from the plasma non-uniformity, $\omega'^2(\xi^2/2L)$, and from relativistic effects, $\partial_{\tau\tau}\xi(\partial_\tau\xi)^2(3\gamma_M/2c^2)$. To obtain these equations we have assumed that the function $(B'_z - \beta_M E'_x)$ is given and is of the order of E_d. Furthermore, we have neglected the change in the electron velocity in the y-direction due to the E_y component of the electric field.

22.3.1 Plasma Resonance

For $\eta = r_E/L \ll 1$, that is for $\xi/L \ll 1$, the plasma resonance occurs near the point where $\omega' = \omega_{pe}\sqrt{1-\beta_M^2}$. (This problem is also discussed in Chap. 11). In this region strong Langmuir oscillations are excited, with a large amplification of the x-component of the electric field E'_x, up to values for which resonance saturation effects become important. These effects can be characterized in terms of a dimensionless parameter s, small compared to unity. The maximum value of the amplitude of the electric field, the width of the region, where the strong electric field is localized, and the characteristic time of the resonance saturation are approximately equal to

$$E'_{\max} = \frac{E_d}{s}, \quad \Delta x = sL, \quad \Delta t' = \frac{1}{s\omega'}, \tag{22.24}$$

respectively.

In the case of an ultra-short laser pulse of weakly relativistic intensity, the main mechanisms of resonance saturation are the relativistic detuning of the oscillations, the self-intersection of the electron trajectories and the finite duration of the laser pulse. At the initial stage of the growth, when the amplitude of the electric field is still small, we can neglect the non-linear terms on the left-hand side of Eq. (22.22). Then, we find for the displacement $\xi(x_0, \tau)$ as a function x_0, τ

$$\xi(x_0, \tau) \approx -\frac{r_E L}{x_0 - L}\left(\frac{\omega' \sin\Omega(x_0)\tau}{\Omega(x_0)} - \sin\omega'\tau\right), \tag{22.25}$$

where $\Omega(x_0)$ is a function of the Lagrangian coordinate x_0

$$\Omega(x_0) = \omega'\left(\frac{x_0}{L}\right)^{1/2}. \tag{22.26}$$

At the resonance point, $x_0 = x_p = L$, where $\Omega(x_p) = \omega'$, the amplitude of the oscillations grows linearly with time as

$$\xi(x_p, \tau) \approx -\frac{r_E}{2}\omega'\tau\cos\omega'\tau, \tag{22.27}$$

while the width of the resonance region decreases with time as $1/\omega'\tau$. At time $t \approx 2/(\omega'\sqrt{\eta})$, the amplitude of the oscillations becomes equal to the resonance width $\Delta x = L\sqrt{\eta}$, the Jacobian of transformation from the Euler to Lagrange variables,

$$\partial_{x_0} x = 1 + \partial_{x_0} \xi$$

vanishes and the electron trajectories start to self-intersect. The maximum oscillation amplitude and the dimensionless parameter determining the amplitude and width of the plasma rsonance are approximately equal to (Koch & Albritton, 1974; Bulanov et al., 1977)

$$\xi_{\max} \approx \Delta x \approx \sqrt{r_E L}, \quad \text{and} \quad s_N \approx \sqrt{\frac{r_E}{L}}, \tag{22.28}$$

i.e. $s_N \approx \sqrt{\eta}$, respectively.

Relativistic non-linearity leads to a frequency detuning

$$\delta\omega = -3\gamma_M^2 \omega'^3 \xi_0^2 / 8c^2,$$

depending on the oscillation amplitude ξ_0. The dependence of this amplitude in the plasma resonance region as a function of the Lagrangian coordinate x_0 can be found from the algebraic equation (for the properties of nonlinear oscillator see Landau & Lifshitz, 1976)

$$\xi_0 \left(\frac{\omega'}{2} \frac{x_0 - L}{L} - \frac{3\gamma_M^2 \omega'^3}{16c^2} \xi_0^2 \right) = \frac{r_E \omega'}{2}, \tag{22.29}$$

We see that according to Bulanov and Kovrizhnykh (1976) for $\eta < a_0^{4/3}$, where a_0 is the dimensionless amplitude of the pulse, the maximum value of the oscillation amplitude is given by $\xi_{\max} \approx r_E / a_0^{2/3}$, the width of the resonance is approximately equal to $\Delta x = a_0^{2/3} L$ and the resonance saturation time is $\Delta t' \approx 1/\delta\omega' \approx 1/(\omega' a_0^{2/3})$. We also see that in this regime the nonlinearity arising from plasma inhomogeneity can be neglected.

For an ultrashort laser pulse, if its duration $\Delta t' \approx 2\pi N_p/\omega'$ (here N_p is the number of the wave periods in the pulse) is sufficiently short, the maximum oscillation amplitude is $\xi_{\max} \approx r_E/(\omega' \Delta t') = r_E \pi N_p$, and the width of the resonance region is $\Delta x \approx L/(\pi N_p)$.

Summarizing, we can see that depending on the values of the laser pulse properties under consideration, the dimensionless parameter s is given by the largest among

$$\left\{ s_N = \eta^{1/2}, \quad s_R = a_0^{2/3}, \quad s_T = (\pi N_p)^{-1} \right\}. \tag{22.30}$$

Here $s_N = \sqrt{r_E/L}$ is related to the self-intersection of the electron trajectories (to the breaking of the Langmuir oscillation in the plasma resonance region),

$s_R = (2\pi r_E/\lambda)^{2/3}$ to the relativistic detuning of the frequency and s_T is due to the finite duration of the pulse.

In the case of rather short pulses, with $s_T = (\pi N_p)^{-1} > \{s_N, s_R\}$, the amplitude of the Langmuir oscillations stops growing after the pulse ends. These oscillations remain localized around the resonance point since in cold plasma their group velocity, $v_g \approx v_{the}(\lambda_{De}/\Delta x)$ with v_{the} the electron thermal velocity and λ_{De} their Debye length, is small. However, their effective wave number increases and their phase velocity decreases according to the formula

$$\partial_{t'} k = -(1 - \beta_M^2)\partial_{x'}\omega_{pe}, \tag{22.31}$$

which describes the change of the wave vector k in an inhomogeneous plasma in the geometrical optics approximation. This formula corresponds to the phase mixing of Langmuir oscillations discussed in the previous Sect. 22.3. The solution of Eq. (22.32) yields for the wave number

$$k(t') = k(0) - \text{sign}(\partial_x \omega_{pe}) \left[\frac{\omega_{pe}(1 - \beta_M^2)^{1/2}}{L} \right] t', \tag{22.32}$$

which depending on the sign of the density gradient tends to plus or minus infinity when $t' \to \infty$. The increase of the wave number can also be seen directly, without the assumption $kL \gg 1$ from the expression of ξ, as a function of x_0 and of τ for $\tau > 2\pi N_p/\omega'$:

$$\xi(x_0, \tau) = r_E \frac{L\omega'}{(x_0 - L)\Omega} [\sin(\Omega\tau) - \sin(\Omega\tau - \psi_0)]. \tag{22.33}$$

Here we have supposed that the magnetic field in the pulse has the form $E_d(\tau) = E_{d0}\sin(\omega'\tau)$ for $0 < \tau < 2\pi N_p/\omega'$ and $E_d(\tau) = 0$ for $\tau > 2\pi N_p/\omega'$. The phase ψ_0 is given by expression

$$\psi_0 = \frac{2\pi N_p \Omega}{\omega'} = 2\pi N_p \left(1 - \frac{x_0 - L}{L}\right). \tag{22.34}$$

From Eq. (22.33) we see that in the plasma resonance region the maximum electric field is $E_{\max} = \pi E_d N_p$ and that the field is localized in a region of width $\Delta x = L/(\pi N_p)$. In this region

$$\xi(x_0, \tau) = r_E \frac{L\omega'}{(x_0 - L)\Omega} \sin\left(\frac{\pi N_p (x_0 - L)}{L}\right) \cos(\Omega\tau). \tag{22.35}$$

In this equation we may disregard, in the dependence of Ω on the Lagrangian variable x_0, the difference between the Eulerian variable x_0 and x_0 is substantially small compared to the plasma inhomogeneity scale-length. This is consistent with the fact that the non-linearity arising from the inhomogeneity of plasma density have been neglected. Then we find that the points of constant phase $\phi = \Omega(x_0)\tau$ move with the wave phase velocity

$$v_{ph} = \partial_\tau x_0|_{\phi=\text{constant}} = -\frac{\Omega}{\tau\partial_{x_0}\Omega} = -\frac{\Omega^2}{\phi\partial_{x_0}\Omega}, \tag{22.36}$$

which is negative for positive values of the gradient of Ω. Differentiation of Eq. (22.35) with respect to x_0 shows that for large τ the gradient of the oscillation amplitude has a term that is a growing function of time:

$$\partial_{x_0}\xi(x_0, \tau) \approx -\xi_{\max} r_E \partial_{x_0}\Omega\, \tau \sin(\Omega\tau). \tag{22.37}$$

where

$$\xi_{\max} = r_E \frac{L\omega'}{(x_0 - L)\Omega} \sin\left(\frac{\pi N_p(x_0 - L)}{L}\right) \approx r_E \pi N_p. \tag{22.38}$$

The breaking of Langmuir oscillations occurs when the Jacobean of the transformation from the Euler to Langrange coordinates vanishes, i.e. when $\partial_{x_0}\xi = -1$. This corresponds to the time $\tau_{br} = (\xi_{\max}\partial_{x_0}\Omega)^{-1}$ and to the value of the wave phase $\phi = \Omega\tau_{br} = \Omega/(\xi_{\max}\partial_{x_0}\Omega)$ such that $\sin\phi = 1$. Following Isichenko and Yankov (1986) we can describe the structure of the wave near the wave-breaking point $x = x_{br}$ and look for the dependence of $v = dx/dt'$ on x in phase space. From Eq. (22.15) it follows that in the vicinity of the point x_{br}, we have for $\delta x = x - x_{br}$ and $\delta x_0 = x_0 - x_{br}$

$$\delta x = \delta x_0 + \partial_{x_0}\xi\; \delta x_0 + \frac{1}{2}\partial_{x_0 x_0}\xi\; (\delta x_0)^2 + \frac{1}{6}\partial_{x_0 x_0 x_0}\xi\; (\delta x_0)^3 + \cdots. \tag{22.39}$$

Due to the conditions $\partial_{x_0}\xi = -1$ and $\sin(\Omega\tau_{br}) = 1$, the first three terms in Eq. (22.39) vanish and

$$\delta x = \frac{1}{6}\partial_{x_0 x_0 x_0}\xi\; (\delta x_0)^3 \approx \left(\partial_{x_0}\Omega\; \tau\right)^2 \frac{(\delta x_0)^3}{6}. \tag{22.40}$$

The velocity v of the electrons in the vicinity of the wave breaking point can be written as

$$v = \partial_\tau\xi = -\xi_{\max}\Omega \sin[(\Omega + \partial_{x_0}\Omega\delta x_0)_{x=x_{br}}\tau], \tag{22.41}$$

which gives

$$v = -\xi_{\max}\Omega\sin(\Omega\tau)\cos(\partial_{x_0}\Omega\delta x_0\tau) = -r_E\pi N_p\Omega\left[1 - \frac{1}{2}\left(\tau\delta x_0\partial_{x_0}\Omega\right)^2\right]. \tag{22.42}$$

Using Eq. (22.40) we can express the velocity v as a function of δx_0 in the form

$$v = -r_E \pi N_p \Omega \left[1 - \frac{\left(6\tau \partial_{x_0} \Omega\right)^{2/3}}{2} (\delta x_0)^{2/3} \right]. \tag{22.43}$$

We see from this equation that the spatial dependence of the electron velocity has a cusp at the wave breaking point.

22.3.2 Laser Energy Absorption at the Plasma Resonance

Following Gildenburg (1964), Silin (1985), and Kruer (1988), see also discussion in Part III, we can calculate the energy absorption in the plasma resonance region. Using a simple model for the electric field at the plasma resonance region

$$E_x = \frac{E_d}{\varepsilon'(\omega', x')} \tag{22.44}$$

we implement in the dielectric constant the wave damping, ν':

$$\varepsilon'(\omega', x') = 1 - \frac{\omega_{pe}^2(x')}{\omega'(\omega' + i\nu'(x'))}. \tag{22.45}$$

Within the framework of this approximation the maximal electric field and the resonance width at the plasma resonance equal E_d/s_ν and $\Delta x = s_\nu L$, respectively, with $s_\nu = \nu'(x_p)/\omega'$. In general, the parameter s_ν can be considered as one of the parameters in expression (22.30) and/or due to energy loss via the electron–ion collisions.

The absorbed energy power is

$$P_{abs} = \int_{-\infty}^{+\infty} \nu \frac{E_x^2}{8\pi} dx = \frac{\nu}{8\pi} \int_{-\infty}^{+\infty} \frac{E_d^2}{|\varepsilon|^2} dx. \tag{22.46}$$

For a linear profile of the plasma density in the vicinity of the plasma resonance point thwe dielectric constant can be approximated as

$$\varepsilon = \frac{x_p - x'}{L} + i s_\nu. \tag{22.47}$$

Substituting expression (22.47) to Eq. (22.46) and approximating the driver field E_d by its value at the plasma resonance point $x = x_p$ we obtain

$$P_{abs} = \frac{\nu E_d^2(x_p)L}{8\pi} \int_{-\infty}^{+\infty} \frac{du}{u^2 + s_\nu^2}. \tag{22.48}$$

Since the integral is equal to π/s_ν the absorption power is

$$P_{abs} = \frac{\omega' E_d^2(x_p)L}{8}. \tag{22.49}$$

It does not depend on the particular absorption mechanism similarly to the case of the power absorbed by a resonantly driven oscillator (Landau & Lifshitz, 1976).

22.3.3 Electron Acceleration in the Plasma Resonance Region

Using expressions (22.24) and (22.30) for the value of the electric field and the width of the plasma resonance for approximate estimates of the energy acquired by the electron during the passage of the plasma resonance, we find that the energy $\mathcal{E} = eE_x\Delta x = eE_dL$ does not depend on the dissipation/nonlinearity that determines the resonance parameters. However, a more detailed examination leads to the conclusion that rigorous description should take into account the spatial and temporal dependence of Langmuir oscillations in the resonance region.

Implementing Eq. (22.17) we can write the equation for the electric field excited in plasma resonance in the case of linear profile of the plasma density

$$\ddot{E}_x + \nu\dot{E}_x + \omega^2\frac{x_0}{L}E_x = \omega^2 E_d\cos(\omega t). \tag{22.50}$$

Here we neglect the effects of nonlinearity due to the plasma inhomogeneity assuming that the electron displacement is much less than the inhomogeneity scale length, $\xi \ll L$.

Solution of Eq. (22.50) describing stationary oscillations (in the limit $t \gg 1/\nu$) gives

$$E_x = \mathrm{Re}\left\{\frac{E_d\exp(i\omega t)}{1 - x_0/L + i\nu/\omega}\right\}. \tag{22.51}$$

Contours of equal value of the electric field given by this solution in the plasma resonance region in the (x, t) plane showing the down ramp directed (i.e. it is directed towards low plasma density) phase velocity of the Langmuir oscillations. Due to that the electrons are accelerated when they co-propagate along with the Langmuir oscillations. The phase velocity is approximately equal to $v_{ph} = \omega\Delta x = s\omega L$, where the parameter s is given by Eqs. (22.24) and (22.30). Within some limits the phase

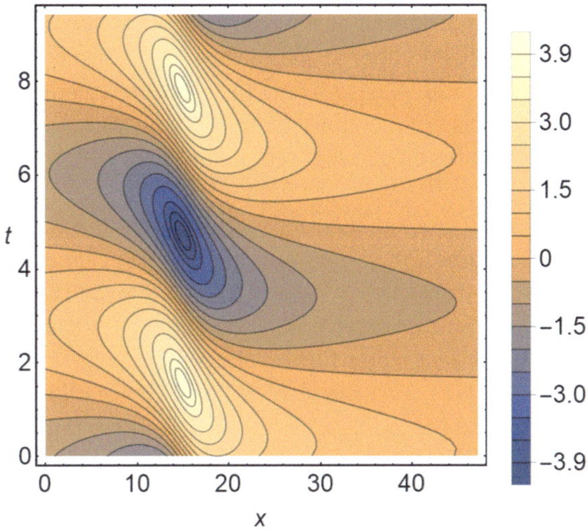

Fig. 22.3 Contours of equal value of the electric field in the plasma resonance region in the (x, t) plane showing the downward directed phase velocity of the Langmuir oscillations

velocity can be arbitrarily large or small. If $2\pi sL/\lambda > 1$ the phase velocity of the Langmuir oscillations is superluminal, i.e. $v_{ph}/c > 1$. In the case of the ultra-short driver laser pulse, the width of the region, where the Langmuir oscillations are localized, is $\Delta x = L/\pi N_p$. For $N_p < 2L/\lambda$ the phase velocity $v_{ph} = 2cL/\lambda N_p$ is superluminal. Then according to Eq. (22.31) the oscillation wave number changes due to the phase mixing effect which asymptotically results in the slowing down of the phase velocity Fig. 22.3.

The energy of the electron passing the plasma resonance can be found by calculating the integral along the electron trajectory

$$\mathcal{E}_e = e\int_{-\infty}^{+\infty} E_x dx = e\int_{-\infty}^{+\infty} \mathrm{Re}\left\{\frac{E_d \exp(i\omega t)|_{t=x/c}}{1 - x/L + i\nu/\omega}\right\} dx = 2\pi e E_d L \exp\left(-\frac{\nu L}{c}\right). \tag{22.52}$$

22.3.4 *Ion Acceleration in the Localized High-Frequency Electric Field*

The nature of the acceleration of the ions and of that of the electrons is different due to their large mass difference. Electrons gain their energy approximately in a period of the laser radiation due to the mechanism of "vacuum heating" (Brunel, 1987), or as a result of the wave-particle interaction with the strong Langmuir wave excited in the plasma resonance region, as it has been discussed above. Because of their large mass, ions only react to the slowly varying electric field produced by charge separation.

This field is produced by the ponderomotive force that arises from the high frequency fields localized either at the plasma resonance region or in the vicinity of a sharp boundary in the plasma where the electromagnetic wave becomes evanescent.

In the case of electrons, the ponderomotive force is compensated by the electric field due to the charge separation. Ions instead can gain energy from their interaction with this localized electric field.

In order to show this, let us calculate the motion in the $\mathcal{M}$ frame of reference of a test particle with mass m_a and electric charge $Z_a e$ in a high-frequency electric field $E'_x(x_0, \tau)$ which is assumed to be a given function of the Lagrangian variable x_0 and of time, as given e.g. at the plasma resonance, by Eq. (22.18). We introduce Langrangian variables for the electrons $x = x_0 + \xi(x_0, \tau)$ and for the test particles $x = x_a + \xi_a(x_a, \tau)$ (see also Eqs. (22.15)). Following Landau and Lifshitz (1976) we separate the motion of the test particles into a slowly varying part X and a fast-oscillating part $\tilde{\eta}$ so that

$$x = \tilde{\eta} + X = x_a + \xi_a(x_a, \tau) \equiv x_0 + \xi(x_0, \tau). \tag{22.53}$$

Then we have

$$x_0 = X + (\tilde{\eta} - \xi). \tag{22.54}$$

Inserting these expressions into the equation of motion of the ions in the electric field $E'_x(x_0, \tau)$ we obtain

$$\frac{d^2X}{d\tau^2} + \frac{d^2\tilde{\eta}}{d\tau^2} = \frac{Z_a e}{m_a \gamma_M^3}\left(E'_x(X, \tau) + (\tilde{\eta} - \xi)\frac{dE'_x(X, \tau)}{dX}\right). \tag{22.55}$$

By averaging this equation over the high frequency, we obtain for slowly varying part

$$\frac{d^2X}{d\tau^2} = \frac{Z_a e}{m_a \gamma_M^3}\left\langle(\tilde{\eta} - \xi)\frac{dE'_x(X, \tau)}{dX}\right\rangle, \tag{22.56}$$

where $\langle\cdots\rangle$ stand for averaging over fast oscillations. Using the approximation of small but finite amplitude oscillations, $|\tilde{\eta} - \xi| \ll X$, we find that $E'_x(x_0, \tau) = E'_x(X, \tau)$. We obtain for the fast oscillating $\xi(X, \tau)$ and $\tilde{\eta}(X, \tau)$ the equations

$$\frac{d^2\xi}{d\tau^2} = \frac{e}{m_e \gamma_M^3} E'_x(X, \tau) \text{ and } \frac{d^2\tilde{\eta}}{d\tau^2} = \frac{Z_a e}{m_a \gamma_M^3} E'_x(X, \tau), \tag{22.57}$$

which give

$$\xi = \frac{e}{m_e \gamma_M^3 \omega_{pe}^2} E'_x(X, \tau) \text{ and } \tilde{\eta} = -\frac{Z_a e}{m_a \gamma_M^3 \omega_{pe}^2} E'_x(X, \tau). \tag{22.58}$$

As we see $\tilde{\eta} = -(Z_\alpha m_e/m_\alpha)\xi$. So that the equation for the slowly varying part is

$$\frac{d^2X}{d\tau^2} = -\frac{Z_\mathrm{a}c}{2m_\mathrm{a}\gamma_M^3}\left(1 + \frac{Z_\mathrm{a}m_\varepsilon}{m_\mathrm{a}}\right)\frac{d\langle E_s'^2(X,\tau)\rangle}{dX}, \tag{22.59}$$

The force term on the right-hand side of Eq. (22.59) is a slowly varying function of time. If the test particles are electrons ($Z_a = -1$ and $m_a = m_e$) this force vanishes, since for the electrons the ponderomotive force is compensated for by the electric field due to the charge separation. On the contrary, Eq. (22.59) indicates that in the case of the ions, they can acquire energy. Let us assume that the initial velocity of the ions along the x-direction vanishes, that is that the characteristic time of the high frequency field excitation is much shorter than the time scale of the ion motion. Then from Eq. (22.59) we obtain that the change of the ion kinetic energy $\Delta\mathcal{E}_a'$ in the $\mathcal{M}$ frame of reference due to the ponderomotive force is given by

$$\Delta\mathcal{E}_\mathrm{a}' = \frac{m_\mathrm{a}\gamma_M}{2}\left(\frac{dX}{d\tau}\right)^2 = \frac{Z_\mathrm{a}e}{2m_\mathrm{a}\gamma_M^3}\left(1 + \frac{Z_\mathrm{a}m_\varepsilon}{m_\mathrm{a}}\right)\left(\langle E_s'^2(X_\mathrm{a})\rangle - \langle E_s'^2(X,\tau)\rangle\right), \tag{22.60}$$

where X_a is the initial coordinate of the particle. Under the action of the electric field the ions leave the region where the high frequency field is localized and their final energy depends on their initial position and is given by

$$\Delta\mathcal{E}_\mathrm{a}' = \frac{Z_\mathrm{a}c}{2m_\mathrm{a}\gamma_M^3}\left(1 + \frac{Z_\mathrm{a}m_\varepsilon}{m_\mathrm{a}}\right)\langle E_x'^2(X_\mathrm{a})\rangle. \tag{22.61}$$

Typical values of the ion energy are $\Delta\mathcal{E}_a' \approx Z_a\mathcal{E}_e'$ with $\mathcal{E}_e'$ the electron quiver energy. The energy spectrum $N_a'(\Delta\mathcal{E}_a')$ of the fast ions, i.e., the distribution of the number of fast ions accelerated up to the energy $\Delta\mathcal{E}_a'$, can be found from the flux conservation in velocity space

$$\frac{dN_a'}{d\Delta\mathcal{E}_a'} = n_a'\left|\frac{dX_a}{d\Delta\mathcal{E}_a'}\right|, \tag{22.62}$$

where n_a' is the density of the ions of the a species.

If the high frequency field is localized in the vicinity of the plasma boundary and decays exponentially with evanescence length c/ω_{pe}, the energy spectrum of the fast ions can be written as

$$\frac{dN_a'}{d\Delta\mathcal{E}_a'} = \frac{2n_a'}{\omega_{pe}\Delta\mathcal{E}_a'}, \tag{22.63}$$

In the neighborhood of the plasma resonance region the high-frequency field distribution has the form

$$\frac{E_{\max}\Delta x}{\sqrt{(x-x_p)^2+\Delta x^2}}, \tag{22.64}$$

with Δx the width of the plasma resonance region. Then the energy of the fast ions depends on Δx and on $X_a - x_p$ according to

$$\Delta\mathcal{E}_a' = \frac{Z_a\mathcal{E}_e'\Delta x^2}{(X_a-x_p)^2+\Delta x^2}, \tag{22.65}$$

and the energy spectrum can be written as

$$\frac{dN_a'}{d\Delta\mathcal{E}_a'} = \frac{n_a\Delta x}{2}\frac{Z_a\mathcal{E}_s'}{\Delta\mathcal{E}_a^{13/2}\left(Z_a\mathcal{E}_e'-\Delta\mathcal{E}_a'\right)^{1/2}}. \tag{22.66}$$

22.4 "Vacuum Heating" of Electrons

The term "vacuum heating" after Brunel (1987) (see also Gibbon & Bell, 1992) is used for the case of laser interaction with the plasma-vacuum interface accompanied by the generation of a copious amount of high energy electrons. The "vacuum heating" regime requires the parameter $\eta = r_E/L$ to be much larger than unity, i.e. the quiver radius of the electrons to be substantially larger than the scale of the plasma density inhomogeneity. In this limit, the electrons in the region near the plasma boundary are expelled from the plasma into the vacuum region. In our analysis of the motion of these electrons we will disregard the effect of the driving term in the electron hydrodynamic equation, which will only appear through the initial value of the electron velocity. After one half period of the electron motion, when the electrons come back into the plasma, their trajectories self-intersect, and we can no longer use the cold fluid equation employed. Nevertheless, the solution we will present can be useful to obtain estimates of the typical time scales of the process, the values of the electric field and that of the electron density.

22.4.1 Electron Dynamics Near Plasma-Vacuum Interface

We assume that the electrons are initially located inside the target (within the interval $-l/2 < x < l/2$). Under the action of a short laser pulse, the electrons gain a momentum $p_{e,0}$ directed along the x axis. The expansion of the electron cloud into vacuum is determined by the self-consistent electric field, which can also accelerate light ions (Bulanov et al., 1994). This acceleration mechanism will be considered below. It can be regarded as an extension of the well-known mechanism of ion acceleration during the expansion of plasma into vacuum (Gurevich et al., 1966) to the case in which plasma quasineutrality is strongly violated and the particles have relativistic energies.

To describe the expansion of an electron cloud into a vacuum, we will use one-dimensional equations of electron hydrodynamics:

$$\partial_t n_e + \partial_x (n_e v_e) = 0, \tag{22.67}$$

$$\partial_t p_e + v_e \partial_x p_e = -E, \tag{22.68}$$

$$\partial_x E = 4\pi e (Z_a n_a - n_e), \tag{22.69}$$

where n_e is the electron density, the electron velocity v_e and momentum p_e are related to each other by the expression

$$v_e = c \frac{p_e}{(m_e^2 c^2 + p_e^2)^{1/2}}, \tag{22.70}$$

and the ion density $n_i(x)$ is assumed to be uniform inside the interval $-l/ < x < l/2$ and equal to zero outside of it. To find the solution to Eqs. (22.67)–(22.70), we switch from the Euler variables x, t and t to the Lagrange variables x_0 and t according to the relationship given by Eq. (22.15): $x = x_0 + \xi_e(x_0, t)$. Here, x_0 is the coordinate of an element of the electron fluid at the initial instant $t = 0$ the Lagrange coordinate of electrons is $x_0 \in \{-l/2, l/2\}$. The electron displacement from initial position $\xi_e(x_0, t)$ is equal to the distance by which an element of the electron fluid is displaced from its initial position over time t. At the initial instant, we have $\xi_e(x_0, 0) = 0$. In the Lagrange variables, Eqs. (22.67)–(22.69) become (see Akhiezer & Lyubarskii, 1951; Dawson, 1959; Bulanov et al., 1990)

$$\partial_t p_e = -4\pi Z_i e^2 \int\limits_0^{\xi_e(x_0,t)} n_i(x_0 + s) ds, \tag{22.71}$$

$$\partial_t \xi_e = c \frac{p_e}{(m_e^2 c^2 + p_e^2)^{1/2}}. \tag{22.72}$$

In our case, the density profile of heavy ions $n_i(x)$ is given by expression $n_i(x) = n_0 \theta(l^2/4 - x^2)$, where $\theta(x)$ is the Heaviside step function: $\theta(z) = 0$ at $z < 0$ and $\theta(z) = 1$ at $z > 0$. As a result, the right-hand side of Eq. (22.71) takes the form

$$eE(x_0,t) = -4\pi n_0 Z_i e^2 \begin{cases} l/2 - x_0, & l/2 < x_0 + \xi_e(x_0,t) \\ \xi_e(x_0,t), & -l/2 < x_0 + \xi_e(x_0,t) < l/2 \\ -l/2 - x_0, & x_0 + \xi_e(x_0,t) < -l/2 \end{cases} \tag{22.73}$$

Here we consider the dependence of the electric field on the coordinate and time near the front of the electron cloud. It follows from Eq. (22.73) that, in the Lagrange variables, the electric field has the form $E(x_0,t) = -4\pi n_0 Z_i e(l/2 - x_0)$. To find the electric field profile as a function of the Euler coordinate x, it is necessary to use the solutions to Eqs. (22.71) and (22.72):

$$p_e(x_0,t) = p_{e,0}(x_0) - 2\pi n_0 Z_i e^2 (l - 2x_0)t, \tag{22.74}$$

$$\xi_e(x_0,t) = \frac{c}{\kappa(x_0)}\left\{\left[m_e^2c^2 + p_{e,0}^2(x_0)\right]^{1/2} - \left[m_e^2c^2 + \left(p_{e,0}(x_0) - \kappa(x_0)t\right)^2\right]^{1/2}\right\}, \tag{22.75}$$

where the function $p_e(x_0)$ describes the initial distribution of the initial distribution of the electron momentum in the cloud expanding to vacuum and

$$\kappa(x_0) = 2\pi n_0 Z_i e^2 (l - 2x_0)t. \tag{22.76}$$

It can be seen that the displacement of an electron from its initial position ($\xi_e(x_0,0) = 0$) first increases as $\xi_e(x_0,0) \approx v_{e,0}t$, where $v_{e,0} = cp_{e,0}/(m_e^2c^2 + p_{e,0}^2)^{1/2}$ is the initial electron velocity. In a time of $t_m = p_{e,0}(x_0)/(\pi n_0 Z_i e(2l - x_0))$, the displacement reaches its maximum value

$$\xi_{e,m}(x_0,t) = \frac{c}{\kappa(x_0)}\left\{\left[m_e^2c^2 + p_{e,0}^2(x_0)\right]^{1/2} - m_e c\right\}, \tag{22.77}$$

and then decreases. It follows from the expressions obtained that the closer the Lagrange coordinate x_0 of an element of the electron fluid to the boundary of the plasma layer, the longer the time required for the element to return to its initial position ($\approx t_m$). For $x_0 \to l/2$ (which corresponds to the particles located at the front of the electron cloud), the return time tends to be infinite. Near the front of the electron cloud, where $\kappa(x_0) \to 0$, we have

$$\xi_e(x_0,t) = c\frac{p_{e,0}(x_0)t}{\left[m_e^2c^2 + p_{e,0}^2(x_0)\right]^{1/2}} - 2\pi n_0 Z_i e^2\left(\frac{l}{2} - x_0\right)\frac{m_e^2c^3t^2}{\left[m_e^2c^2 + p_{e,0}^2(x_0)\right]^{3/2}}. \tag{22.78}$$

If we assume that, near the front of the cloud, the initial distribution of the electron momentum $p_{e,0}(x_0)$ is uniform; i.e., $p_{e,0}$ does not depend on the coordinate x_0, then, sig Eqs. $x = x_0 + \xi_e(x_0,t)$ and (22.76), we find the relation between the Lagrange and Euler coordinates:

$$x_0 = \frac{4(x - v_{e,0}t) + blt^2}{2(2 + bt^2)}, \tag{22.79}$$

where

$$b = \frac{4\pi n_0 Z_i e^2 m_e^2 c^3}{\left[m_e^2 c^2 + p_{e,0}^2(x_0)\right]^{3/2}} \tag{22.80}$$

and $v_{e,0} = cp_{e,0}/\left(m_e^2c^2 + p_{e,0}^2\right)^{1/2}$.

It follows from this that the electric field profile near the front of the electron cloud is described by the expression

$$E(x,t) = -4\pi n_0 Z_i e \frac{2(x - v_{e,0}t) - l}{2 + bt^2}\theta(v_{e,0}t + l/2 - x). \tag{22.81}$$

It can be seen that the electric field vanishes at the front of the cloud (i.e., at $x = v_{e,0}t + l/2$). The gradient of the electric field is proportional to the electron density and depends on time as

$$\partial_x E = -\frac{8\pi n_0 Z_i e}{2 + bt^2}. \tag{22.82}$$

22.4.2 Acceleration of Ions at the Front of the Electron Cloud

Here we consider the acceleration of ions with a mass m_a and charge $Z_a e$ in the electric field described by expression (22.81). The equation of motion of an ion in this field has the form

$$\dot{p}_\alpha = -4\pi n_0 Z_i Z_a e^2 \frac{(x - v_{e,0}t) - l/2}{1 + bt^2/2}\theta\left(v_{e,0}t + \frac{l}{2}\right), \tag{22.83}$$

$$\dot{x}_a = c\frac{p_a}{\left(m_a^2c^2 + p_a^2\right)^{1/2}}, \tag{22.84}$$

where the superior dot stands for the time derivative and p_a and x_a are the momentum and coordinate of an ion, respectively, and $\theta(\cdot)$ is the unit step Heaviside theta function.

Nonrelativistic limit. In the case of non-relativistic ion energies, Eqs. (22.83) and (22.84) can be transformed into one equation for x_a:

$$\ddot{x}_a + \left(\frac{\mu b}{1 + bt^2/2}\right)x_a = \mu b\frac{v_{e,0}t + l/2}{1 + bt^2/2}, \tag{22.85}$$

where

$$\mu = \frac{Z_a m_e}{m_a}\left[1 + \left(\frac{p_{e,0}}{m_e c}\right)^2\right]^{3/2}. \tag{22.86}$$

The solution to Eq. (22.85) for given initial coordinate and velocity of an ion, $x_{a,0}$ and $\dot{x}_{a,0}$, respectively can be expressed in terms of hypergeometric functions:

$$\begin{aligned} x_a(t) &= \frac{l}{2} + \left(x_{a,0} - \frac{l}{2}\right){}_2F_1\left(a, \beta; \gamma; -\frac{bt^2}{2}\right) \\ &+ t\left[v_{e,0} + (\dot{x}_{a,0} - v_{e,0}){}_2F_1\left(-\beta, -a; \gamma + 1; -\frac{bt^2}{2}\right)\right]. \end{aligned} \tag{22.87}$$

Here, ${}_2F_1(\alpha, \beta; \gamma; z)$ is the hypergeometric function (Olver et al., 2010) determined as

$${}_2F_1(\alpha, \beta; \gamma; z) = \sum_{n=0}^{\infty} \frac{(\alpha)_n(\beta)_n}{(\gamma)_n}\frac{z^n}{n!} = 1 + \frac{\alpha\beta}{\gamma}\frac{z}{1!} + \frac{\alpha(\alpha+1)\beta(\beta+1)}{\gamma(\gamma+1)}\frac{z^2}{2!} + \ldots,$$

where $(a)_n$ is the Pochhammer symbol equal to

$$(x)_n = \frac{\Gamma(x+n)}{\Gamma(x)} = x(x+1)\ldots(x+n-1)$$

with $\Gamma(z)$ being the Euler gamma-function. The parameters α, β, and γ in Eq. (22.87) are related to μ given by Eq. (22.86) by the expressions

$$\alpha = -\frac{1}{4} - i\frac{(8\mu - 1)^{1/2}}{4}, \quad \beta = -\frac{1}{4} + i\frac{(8\mu - 1)^{1/2}}{4}, \quad \gamma = \frac{1}{2}. \tag{22.88}$$

If at the initial instant of time, $t = 0$, an ion be at rest $\dot{x}_{a,0} = 0$ and located at the point $x_{a,0} = l/2$ (i.e., at the front of the electron cloud) it then follows from Eq. (22.87) that the ion coordinate depends on time as

$$x_a(t) = \frac{l}{2} + v_{e,0}t\left[1 - {}_2F_1\left(\frac{1-(1-8\mu)^{1/2}}{4}, \frac{1+(1-8\mu)^{1/2}}{4}; \frac{3}{2}; -\frac{bt^2}{2}\right)\right]. \tag{22.89}$$

It can be seen that the second term on the right-hand side of this expression increases with time. At the instant $t = t^*$, which is determined by the equation

$${}_2F_1\left(\frac{1-(1-S\mu)^{1/2}}{4}, \frac{1+(1-S\mu)^{1/2}}{4}; \frac{3}{2}; -\frac{bt^{*2}}{2}\right) = 0, \tag{22.90}$$

the ion coordinate becomes $x_\alpha(t^*) = l/2 + v_{e,0}t^*$ i.e., in the course of acceleration, the ion crosses the front of the electron cloud and leaves it behind. As a result, the ion gains the velocity

$$\dot{x}_a(t^*) = v_{e,0}\left[1 + \frac{\mu b t^{*2}}{24}\,{}_2F_1\left(\frac{5-(1-\delta\mu)^{1/2}}{4}, \frac{5+(1-\delta\mu)^{1/2}}{4}; \frac{5}{2}; -\frac{bt^{*2}}{2}\right)\right]. \tag{22.91}$$

Since, after crossing the front of the electron cloud, the ion velocity becomes higher than the maximum electron velocity, which is, in turn, assumed to be nonrelativistic, the ion energy should be no higher than $m_a c^2$. Using the asymptotic of the hypergeometric function at $z \to \infty$,

$${}_2F_1(\alpha, \beta; \gamma; z) \approx \frac{\Gamma(\gamma)\Gamma(\beta-\alpha)}{\Gamma(\beta)\Gamma(\gamma-\alpha)}(-z)^{-\alpha} + \frac{\Gamma(\gamma)\Gamma(\alpha-\beta)}{\Gamma(\alpha)\Gamma(\gamma-\beta)}(-z)^{-\beta}, \tag{22.92}$$

where $\Gamma(z)$ is the Euler gamma-function (Olver et al., 2010), we find from Eq. (22.91) that, in the limit $t^* \to \infty$, the ion gains the velocity

$$\dot{x}_a(t^*) \approx v_{e,0}\left[1 + \frac{\pi^{1/2}2^{1/4}\mu}{8}\left(\frac{2}{b}\right)^{(1-\sqrt{1-8\mu})/4} \frac{\Gamma(\sqrt{1-8\mu}/2)}{\Gamma(5+\sqrt{1-8\mu}/4)^2} t^{*(\sqrt{1-8\mu}-1)/2}\right]. \tag{22.93}$$

It can be seen that, at $t^* \to \infty$, the ion velocity is approximately equal to the expansion velocity of the electron cloud. Therefore, in the nonrelativistic limit, the characteristic energy of fast ions is equal to

$$\mathcal{E}_a \approx \frac{m_a v_{e,0}^2}{2} \equiv \frac{m_a}{m_e}\mathcal{E}_e. \tag{22.94}$$

As in the case considered by Gurevich et al. (1966), this energy does not depend on the ion charge; this fact, in both cases, reflects the kinematic relationship corresponding to a situation in which the ion overtakes the front of the electron cloud.

We note that the limit of relativistic ion energy is analyzed in Bulanov et al. (2004a, 2004b).

22.5 Surface Waves in Cold Plasma

The electromagnetic waves in the vicinity of the boundary between two plasma regions can be analyzed within two approaches. In the first of them the so-called Fresnel problem is solved for finding the reflection/transmission of the electromagnetic wave incident on the two-plasma. In the second case, one solves the problem corresponding to obtaining the conditions of existence of the localized electromagnetic mode, i.e. of the surface wave. The Fresnel problem has been addressed in the beginning of the present Chapter in regard with the oblique incidence of the electromagnetic wave on the inhomogeneous plasma slab.

Here, following Aleksandrov and Rukhadze (1999) and Landau and Lifshitz (1984), we consider the surface waves, which are localized at the interface between two plasmas (in a particular case they are localized at the plasma-vacuum interface). Within the framework of the cold plasma approximation the dielectric tensor $\varepsilon_{ij}(\omega, \boldsymbol{x})$ in the frame of reference $\mathcal{L}$ can be written as to

$$\varepsilon_{ij}(\omega, \boldsymbol{x}) = \varepsilon(\omega, \boldsymbol{x})\delta_{ij} \text{ with } \varepsilon(\omega, \boldsymbol{x}) = 1 - \sum_a \frac{\omega_{pa}^2}{\omega^2}, \tag{22.95}$$

where δ_{ij} is the Kronecker delta,

$$\omega_{pa} = (4\pi n_a(\boldsymbol{x}) Z_a e^2/m_a)^{1/2}$$

and $a = e, i$ stands for electron, ion species.

Regarding the plasma-vacuum interface we assume that the density $n_\alpha(\boldsymbol{x})$ is equal to $n_a(\boldsymbol{x}) = n_{0a}\theta(x)$, i.e. $n_a = n_0$ for $x > 0$ and $n_a = 0$, i.e. $\varepsilon(\omega, \boldsymbol{x}) = 1$, if $x < 0$.

In the frame of reference $\mathcal{L}$ Eqs. (22.2)–(22.5) for the electric and magnetic field,

$$\boldsymbol{E}(x, z, t) = \boldsymbol{E}(x)\exp(-i\omega t + k_z z) \text{ and } \boldsymbol{B}(x, z, t) = \boldsymbol{B}(x)\exp(-i\omega t + k_z z).$$

take the form (the z-axis is directed along the plasma surface)

$$\frac{d}{dx}\left(\frac{1}{\varepsilon(\omega, x)}\frac{dB_z}{dx}\right) - \left(\frac{k_z^2}{\varepsilon(\omega, x)} - \frac{\omega^2}{c^2}\right)B_z = 0, \tag{22.96}$$

this is the equation for the B-wave, which in the case of the Fresnel problem, is the p-polarized wave, and the equation

$$\frac{d^2 E_z}{dx^2} - \left(k_z^2 - \frac{\omega^2}{c^2}\varepsilon(\omega, x)\right)E_z = 0, \tag{22.97}$$

for the E-wave (it the s-polarized wave in the case of the Fresnel problem).

For the components E_x, E_y, B_x, and B_y we have

$$E_x = -i\frac{k_z}{k_z^2 - \varepsilon\omega^2/c^2}\frac{dE_z}{dx}, \quad B_y = -i\frac{\omega}{c(k_z^2 - \varepsilon\omega^2/c^2)}\frac{dE_z}{dx}, \tag{22.98}$$

$$E_y = i\frac{\omega}{c(k_z^2 - \varepsilon\omega^2/c^2)}\frac{dB_z}{dx}, \quad B_x = -i\frac{k_z}{k_z^2 - \varepsilon\omega^2/c^2}\frac{dB_z}{dx}. \tag{22.99}$$

In addition, Eqs. (22.96)–(22.99) should be solved taking into account the boundary conditions at $x > 0$:

$$\{E_y\}_{x=0} = \{E_z\}_{x=0} = \{B_y\}_{x=0} = \{B_z\}_{x=0} = 0, \tag{22.100}$$

where a notation $\{f\}_{x=0}$ stands for the discontinuity of the function $f(x)$ at the interface $x = 0$, i.e.

$$\{f\}_{x=0} = f(x+\delta) - f(x-\delta) \quad \text{for} \quad \delta \to 0.$$

We seek the solution of Eq. (22.96) for the B-wave in the form of an evanescent wave in the regions $x < 0$ and $x > 0$

$$B_{z1} = B_0 \exp(ik_z z + \kappa_1 x), \; \kappa_1 = (k_z^2 - \omega^2/c^2)^{1/2} \, \text{for}\, x < 0, \tag{22.101}$$

$$B_{z2} = B_0 \exp(ik_z z - \kappa_2 x), \; \kappa_2 = (k_z^2 + |\varepsilon|\omega^2/c^2)^{1/2} \, \text{for}\, x > 0, \tag{22.102}$$

where the boundary condition $\{B_z\}_{x=0} = 0$ is already satisfied. The second boundary condition $\{\varepsilon^{-1}\partial_x B_z\}_{x=0} = 0$ gives

$$\frac{dB_{z1}}{dx}|_{x=0} = \frac{1}{\varepsilon}\frac{dB_{z2}}{dx}|_{x=0}, \tag{22.103}$$

or

$$\kappa_1 = \frac{\kappa_2}{|\varepsilon|}, \tag{22.104}$$

to satisfy which it should be $|\varepsilon| > 1$ and $\varepsilon < 0$. From Eq. (22.104) it follows the dispersion equation for the frequency ω and wave number k_z of the surface wave,

$$k_z^2 c^2 = \frac{\omega^2(\omega_{pe}^2 - \omega^2)}{\omega_{pe}^2 - 2\omega^2}, \tag{22.105}$$

or

$$\omega = \frac{\sqrt{k^2c^2 + \omega_{pe}^2 - \sqrt{k^4c^4 + \omega_{pe}^4}}}{\sqrt{2}}. \tag{22.106}$$

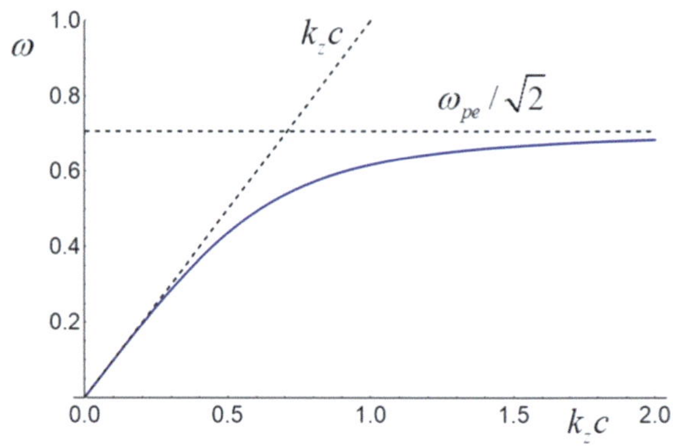

Fig. 22.4 Dependence of the frequency of the surface wave on the z component of the wave vector for $\omega_{pe} = 1$

We assumed that the dielectric constant equals $\varepsilon = 1 - \omega_{pe}^2/\omega^2$.

The dependence of the frequency of the surface wave on the z component of the wave vector is plotted in Fig. 22.4 for $\omega_{pe} = 1$. As we see in the long-wavelength limit when $k_z c \to 0$ the phase and group velocity ($v_{ph} = \omega/k_z$, $v_g = \partial_{k_z}\omega$) tend to speed of light in vacuum. In the short-wavelength limit when $k_z c \to \infty$ the wave frequency is approximately equal to $\omega \approx \omega_{pe}/\sqrt{2}$. The group velocity of the wave vanishes as $v_g = o(1/k_z^3)$ and the phase velocity equals $\omega_{pe}/\sqrt{2}k_z$.

The surface wave excited at the surface of the solid density target irradiated by the relativistically intense laser pulse plays a crucial role in generating high frequency electromagnetic radiation as demonstrated by Lamac et al. (2023) and in exciting the hosing mode of the laser pulse in collisionless plasma (Naumova et al., 2001).

References

Akhiezer, A. I., & Lyubarskii, G. Y. (1951). On the nonlinear theory of electron plasma oscillations. *Doklady Akad. Nauk SSSR, 80*, 193.

Aleksandrov, A. F., & Rukhadze, A. A. (1999). Lectures on the electrodynamics of plasma-like media. M: Ed. Moscow State University

Bourdier, A. (1983). Oblique incidence of a strong electromagnetic wave on a cold inhomogeneous electron plasma. Relativistic effects. *Physics Fluids, 26*, 1804

Brunel, F. (1987). Not-so-resonant, resonant absorption. *Physical Review Letters, 59*, 52.

Bulanov, S. V., Kovrizhnykh, L. M., & Sakharov, A. S. (1990). Regular mechanisms of the electron and the ion acceleration in the interaction of strong electromagnetic waves with an inhomogeneous plasma. *Physics Reports, 186*, 1.

Bulanov, S. V., Esirkepov, T. Zh., Koga, J., Tajima, T., Farina, D. (2004a). Interaction of electromagnetic waves with plasma in the radiation-dominated regime. *Plasma Physics Reports, 30*, 196.

Bulanov, S. V., Esirkepov, T. Zh., Koga, J., & Tajima, T. (2004b). Concerning the maximum energy of ions accelerated at the front of a relativistic electron cloud expanding into vacuum. *Plasma Physics Reports, 30*, 21.

Bulanov, S. V., & Kovrizhnykh, L. M. (1976). Maximum amplitude of the HF-field in the plasma resonance region. *Soviet Journal of Plasma Physics, 2*, 58.

Bulanov, S. V., Kovrizhnykh, L. M., & Sakharov, A. S. (1977). Finite amplitude Langmuir oscillations in the plasma resonance region. *Soviet Physics—JETP, 45*, 949.

Bulanov, S. V., Naumova, N. M., & Pegoraro, F. (1994). Interaction of an ultrashort, relativistically strong laser pulse with an overdense plasma. *Physics of Plasmas, 1*, 745.

Dawson, J. M. (1959). Nonlinear electron oscillations in a cold plasma. *Physical Review, 113*, 383.

Denisov, N. G. (1957). On a singularity of the field of an electromagnetic wave propagated in an lnhomogeneous plasma. *SOVIET PHYSICS JETP-USSR, 4*, 544.

Gershman, B. N., Ginzburg, V. L., & Denisov, N. G. (1957). Propagation of electromagnetic waves in plasma (ionosphere). *Uspekhi Fizicheskikh Nauk, 61*, 561.

Gibbon, P., & Bell, A. R. (1992). Collisionless absorption in sharp-edged plasmas. *Physical Review Letters, 68*, 1535.

Gildenburg, V. B. (1964). Nonlinear effects in an inhomogeneous plasma. *Soviet Physics JETP, 19*, 1456.

Ginzburg, V. L. (1962). *Propagation of electromagnetic waves in plasma.* Gordon and Breach Science Publishers Inc.

Gurevich, A. V., Pariiskaya, L. V., & Pitaevskii, L. P. (1966). Self-similar motion of rarefied plasma. *Soviet Physics—JETP, 22*, 449.

Isichenko, M. B., & Yan'kov, V. V. (1986). Evolution of Langmuir waves in a cold plasma. *Soviet Journal of Plasma Physics, 12*, 98.

Koch, P., & Albritton, J. (1974). Electron and ion heating through resonant plasma oscillations. *Physical Review Letters, 32*, 1420.

Kruer, W. L. (1988). *The physics of laser plasma interactions.* Addison-Wesley Publishing Co.

Lamač, M., Mima, K., Nejdl, J., Chaulagain, U., & Bulanov, S. V. (2023). Anomalous relativistic emission from self-modulated lasma mirrors. *Physical Review Letters, 131*, 205001.

Landau, L. D., & Lifshitz, E. M. (1976). *Course of Theoretical Physics, Vol. 1. Mechanics.* Pergamon.

Landau, L. D., & Lifshitz, E. M. (1984). *Course of theoretical physics, Vol. 8: Electrodynamics of continuous media.* Pergamon Press.

Metelskii, I. I., Kovalev, V. F., & Bychenkov, V. Yu. (2024). Relativistic-nonlinear resonant absorption and generation of harmonics of electromagnetic radiation in an inhomogeneous plasma. *Physics-Uspekhi, 194*, 457.

Naumova, N. M., Koga, J., Nakajima, K., Tajima, T., Esirkepov, T. Zh., Bulanov, S. V., & Pegoraro, F. (2001). Polarization, hosing and long-time evolution of relativistic laser pulses. *Physics of Plasmas, 8*, 4149.

Olver, F. W. J., Lozier, D. W., Boisvert, R. F., & Clark, C. W. (2010). *NIST handbook of mathematical functions.* Cambridge University Press.

Silin, V. P. (1985). Absorption of radiation by turbulent laser plasmas. *Soviet Physics Uspekhi, 28*, 136.

Teubner, U., & Gibbon, P. (2009). High-order harmonics from laser-irradiated plasma surfaces. *Reviews of Modern Physics, 81*, 445.

Chapter 23
Electromagnetic Radiation Emission by the Electron Interacting with an Electromagnetic Wave

Abstract Radiative cooling of electron beams interacting with counter-propagating electromagnetic waves is analyzed, taking into account the quantum modification of the radiation friction force. Central attention is paid to the evolution of the energy spectrum of electrons accelerated by the laser wake field acceleration mechanism. As an electron beam loses energy to radiation, the mean energy decreases, and the form of energy distribution also changes due to quantum–mechanical spectral broadening.

Keywords Thomson scattering · Radiation friction force · High-energy electron interaction with EM wave · Quantum effects on the radiation friction force

23.1 Thomson Scattering

Here we derive expressions for the energy of electromagnetic field and formulate the Pointing theorem. Multiplying the first of the Maxwell Eqs. (20.7) on the vector $\boldsymbol{E}/4\pi$ and the second of the Maxwell Eqs. (20.9) on $\boldsymbol{B}/4\pi$ and adding them we obtain

$$\partial_t\left(\frac{E^2+B^2}{8\pi}\right)+c(\boldsymbol{j}\cdot\boldsymbol{E})=-\frac{c}{4\pi}\nabla\cdot(\boldsymbol{E}\times\boldsymbol{B}) \tag{23.1}$$

Integrating this expression over the volume H enclosed by the surface Σ and using the Gauss theorem we find

$$\partial_t\int\limits_{\mathrm{H}}\left(\frac{E^2+B^2}{8\pi}\right)d^3x+c\int\limits_{\mathrm{H}}(\boldsymbol{j}\cdot\boldsymbol{E})\,d^3x=-\frac{c}{4\pi}\oint\limits_{\Sigma}[(\boldsymbol{E}\times\boldsymbol{B})\cdot\boldsymbol{n}]d\Sigma \tag{23.2}$$

where $\boldsymbol{n}$ is the unit vector normal to the surface Σ. The first term on the left-hand side of this equation is a time derivative of a full energy of the electromagnetic field in the volume H with $\left(E^2+B^2\right)/8\pi$ equal to the electromagnetic energy density. The second term on the left-hand side of Eq. (23.2) is equal to the energy losses. The

Y. Kato et al., *High Power Laser and Plasma Science*, Springer Series on Atomic, Optical, and Plasma Physics 130, https://doi.org/10.1007/978-3-032-08930-4_23

term on the right-hand side corresponds to the energy losses due to the radiation,

$$\oint_{\Sigma} (\mathcal{S} \cdot \boldsymbol{n}) d\Sigma \tag{23.3}$$

with the Pointing vector $\mathcal{S}$ given by Eq. (19.1).

The electric and magnetic fields of a point charge (electron) moving with the velocity $\boldsymbol{v}$ and acceleration $\dot{\boldsymbol{v}}$ observed at the distance $\boldsymbol{R}$ are given by expressions (Landau & Lifshitz, 1980)

$$\boldsymbol{E} = \frac{e\left(1 - v^2/c^2\right)}{(R - \boldsymbol{R} \cdot \boldsymbol{v}/c)^3}\left(\boldsymbol{R} - \frac{\boldsymbol{v}}{c}R\right) + \frac{e}{c^2(R - \boldsymbol{R} \cdot \boldsymbol{v}/c)^3}\left\{\boldsymbol{R} \times \left[\left(\boldsymbol{R} - \frac{\boldsymbol{v}}{c}R\right) \times \dot{\boldsymbol{v}}\right]\right\} \tag{23.4}$$

$$\boldsymbol{B} = \frac{1}{R}\boldsymbol{R} \times \boldsymbol{E} \tag{23.5}$$

Here all the functions on the right-hand sites are taken at the retarded time t' determined by the equation

$$t' + \frac{R\left(t'\right)}{c} = t \tag{23.6}$$

The first term on the right-hand site of Eq. (23.4) corresponds to the field from the particle moving with the velocity $\boldsymbol{v}$. It is the near-zone field, which is at large distance proportional to $1/R^2$. The second term depends on the particle acceleration $\dot{\boldsymbol{v}}$. It is related to the electromagnetic wave emitted by the charged particle. At large distance this term is proportional to $1/R$. It can be written as

$$\boldsymbol{E} = \frac{e}{c^2 R}\frac{e}{(1 - \boldsymbol{n} \cdot \boldsymbol{v}/c)^3}\left\{\boldsymbol{n} \times \left[\left(\boldsymbol{n} - \frac{\boldsymbol{v}}{c}\right) \times \dot{\boldsymbol{v}}\right]\right\} \tag{23.7}$$

with the magnetic field equal to

$$\boldsymbol{B} = \boldsymbol{n} \times \boldsymbol{E} \tag{23.8}$$

where $\boldsymbol{n}$ is the unit vector in the direction of radiation.

The radiation intensity emitted in the solid angle do equal to

$$dI = \frac{c}{4\pi}E^2R^2do \tag{23.9}$$

Using Eq. (23.7) it can be written as

$$dI = \frac{e^2}{4\pi c^3}\left\{\frac{2(\boldsymbol{n}\cdot\dot{\boldsymbol{v}})(\boldsymbol{v}\cdot\dot{\boldsymbol{v}})}{c(1-\boldsymbol{v}\cdot\boldsymbol{n}/c)^5} + \frac{\dot{\boldsymbol{v}}^2}{(1-\boldsymbol{v}\cdot\boldsymbol{n}/c)^4} + \frac{\left(1-v^2/c^2\right)(\boldsymbol{n}\cdot\dot{\boldsymbol{v}})^2}{(1-\boldsymbol{v}\cdot\boldsymbol{n}/c)^6}\right\} do \quad (23.10)$$

If the charged particle interacts with the circularly polarized electromagnetic wave its velocity and acceleration are perpendicular to each other. In this case (see Landau & Lifshitz, 1980) the intensity of the electromagnetic wave emitted in the solid angle do equals

$$dI = \frac{e^2 w^2}{4\pi c^3}\left\{\frac{1}{(1-(v/c)\cos\theta)^4} - \frac{\left(1-v^2/c^2\right)\sin^2\theta\cos^2\varphi}{(1-(v/c)\cos\theta)^6}\right\} do \quad (23.11)$$

where

$$w = \frac{4\omega a_0 c}{\left(2+a_0^2\right)^2}, \quad v = \frac{a_0 c\sqrt{8+a_0^2}}{2+a_0^2} \quad (23.12)$$

and θ is an angle between $\boldsymbol{n}$ and $\boldsymbol{v}$ and φ is an azimuthal angle of the vector $\boldsymbol{n}$ with the $\boldsymbol{v}$, $\dot{\boldsymbol{v}}$ plane.

In the rest frame of reference the total power of radiation, $\mathcal{P}_\gamma = d\mathcal{E}_\gamma/dt$, emitted by the charged particle equals

$$\mathcal{P}_\gamma = \frac{2e^2 w^2}{3c^3}, \quad (23.13)$$

where w is the particle acceleration. In the boosted frame of reference for the 4-momentum of emitted radiation we have

$$dp_\gamma^\mu = -\frac{2e^2}{3c}\frac{du^\nu}{ds}\frac{du_\nu}{ds}dx^\mu \quad (23.14)$$

Here we have used that in the rest frame of reference the space components of the 4-velocity u^μ vanish and

$$\frac{du^\nu}{ds}\frac{du_\nu}{ds} = \frac{w^2}{c^2} \quad (23.15)$$

Using equations of particle motion in the form

$$\frac{du^\mu}{ds} = \frac{e}{m_e c^2}F^{\mu\nu}u_\nu \quad (23.16)$$

we can rewrite Eq. (23.14) as

$$dp_\gamma^\mu = -\frac{2e^4}{3m_e^2c^5}\left(F^{\lambda\nu}u_\nu\right)\left(F_{\lambda\kappa}u^\kappa\right)dx^\mu \tag{23.17}$$

Expressing 4-tensor $F^{\mu\nu}$ via the external electromagnetic field we find the total energy $\mathcal{E}_\gamma$ of the radiation emitted by charged particle

$$\mathcal{E}_\gamma = \frac{2e^4}{3m_e^2c^5}\int_{-\infty}^{+\infty}\gamma_e^2\left[\left(\boldsymbol{E}+\frac{1}{c}\boldsymbol{v}\times\boldsymbol{B}\right)^2-\left(\frac{1}{c}\boldsymbol{v}\cdot\boldsymbol{E}\right)^2\right]dt \tag{23.18}$$

with $\gamma_e = 1/\sqrt{1-v^2/c^2}$. The power emitted equals

$$\mathcal{P}_\gamma = \frac{2e^4}{3m_e^2c^5}\int_{-\infty}^{+\infty}\boldsymbol{v}\,\gamma_e^2\left[\left(\boldsymbol{E}+\frac{1}{c}\boldsymbol{v}\times\boldsymbol{B}\right)^2-\left(\frac{1}{c}\boldsymbol{v}\cdot\boldsymbol{E}\right)^2\right]dt \tag{23.19}$$

As an example, we consider the power of the radiation emitted by the charged particle moving in the field of circularly polarized electromagnetic wave in the frame of reference where the particle on average is at rest. The particle trajectory, in this case, has been obtained above. This yields

$$\mathcal{P}_\gamma = \frac{2e^4E_0^2}{3m_e^2c^3}\gamma_e^2 = \frac{2e^4E_0^2}{3m_e^2c^3}\left[1+\left(\frac{eE_0}{m_e\omega c}\right)^2\right] = \frac{2}{3}m_ec^2\omega\frac{r_e}{\lambda}a_0^2\left(1+a_0^2\right) \tag{23.20}$$

This expression can be rewritten as

$$\mathcal{P}_\gamma = \frac{2}{3}\frac{r_e}{\lambda}m_ec^2\omega\gamma_e^2\left(\gamma_e^2-1\right) \tag{23.21}$$

which shows that the radiation losses of the ultrarelativistic particle with the gamma factor much larger than unity, $\gamma_e \gg 1$, are proportional to the fourth power of its energy.

If the particle moves in homogeneous oscillating electric field, e. g. which corresponds to the field in the antinodes of standing linearly polarized electromagnetic wave, the averaged over the oscillation period power is equal to

$$\mathcal{P}_\gamma = \frac{r_e}{3\lambda}m_ec^2\omega a_0^2 \tag{23.22}$$

i.e. it is proportional to the second power of particle energy.

As far is concerning the frequency spectrum of the radiation emitted by the electron (here we consider the electron rotating along the circular trajectory, i.e. the electron in the frame of reference, where it is on average at the rest), it is given by the expression

$$\frac{dI_\gamma(\omega_\gamma, \mathcal{E})}{d\omega_\gamma} = \frac{\sqrt{3}}{2\pi}\frac{e^2}{c}\left(\frac{\mathcal{E}}{m_e c^2}\right)u(\omega_\gamma, \mathcal{E}) \int\limits_{u(\omega_\gamma, \mathcal{E})}^{\infty} K_{5/3}(x)dx \tag{23.23}$$

where $K_\nu(x)$ is the modified Bessel function. The electron energy is assumed to be ultrarelativistic, i.e. $\mathcal{E} \gg m_e c^2$. The function $u(\omega_\gamma, \mathcal{E}) = \omega_\gamma/\omega_c$ is the ratio of the frequency of the radiation emitted ω_γ to the critical frequency ω_c equal to

$$\omega_c = \frac{3c}{2R}\left(\frac{\mathcal{E}}{m_e c^2}\right)^3 \tag{23.24}$$

where R is the curvature radius of the electron trajectory. In the limit $u(\omega_\gamma, \mathcal{E}) \ll 1$ expression (23.23) yields

$$\frac{dI_\gamma(\omega_\gamma, \mathcal{E})}{d\omega_\gamma} \approx \frac{e^2}{c}\left(\frac{\omega_\gamma R}{c}\right)^{1/3}. \tag{23.25}$$

For $u(\omega_\gamma, \mathcal{E}) \ll 1$ we have

$$\frac{dI_\gamma(\omega_\gamma, \mathcal{E})}{d\omega_\gamma} \approx \sqrt{\frac{3\pi}{2}\frac{e^2}{c}}\left(\frac{\mathcal{E}}{m_e c^2}\right)u^2(\omega_\gamma, \mathcal{E}) \exp(-u(\omega_\gamma, \mathcal{E})) \tag{23.26}$$

The characteristic energy of a photon emitted via nonlinear Thomson scattering of the circularly polarized electromagnetic wave scales with the electron quiver energy $\mathcal{E}$ as

$$\omega_c = \frac{3}{2}\omega_0\left(\frac{\mathcal{E}}{m_e c^2}\right)^3 \tag{23.27}$$

Here ω_0 is the laser frequency. The energy of the electron rotating under the action of an electromagnetic wave with an amplitude of $a_0 = eE/m_e\omega_0 c$ approximately equals $\mathcal{E} = a_0 m_e c^2$. For a laser frequency of the order of $10^{15}\mathrm{s}^{-1}$ the emitted photon energy is in the gamma-ray range provided $a_0 > 4 \times 10^2$ which corresponds to a laser intensity higher than $10^{22}\mathrm{W/cm}^2$.

23.2 Radiation Friction Effects

In the limit of extremely high wave intensity the radiation friction effects on the charged particle dynamics become dominant (Zel'dovich, 1975; Zhidkov et al., 2002; Bulanov et al., 2004; Blackburn, 2020; Zhang, 2013; Zhidkov et al., 2014). At these limits the electron dynamics become dissipative with fast conversion of the electromagnetic wave energy to hard electromagnetic radiation, which for typical laser parameters is in the gamma-ray range. For laser radiation with 1 μm wavelength the radiation friction force changes the scenario of the electromagnetic wave interaction with matter at the intensity approximately equal to $I_R = 10^{23}\mathrm{W/cm^2}$. For the laser intensity close to I_R the abundant electron–positron pair creation comes into play. In this regime, the electron (positron) interaction with the electromagnetic field is principally determined by a counter-play between the radiation friction and quantum effects. The quantum electrodynamics effects weaken the electromagnetic emission by the relativistic electron resulting in the lowering of the radiation friction. In an extremely high intensity electromagnetic field vacuum loses its property to be an empty substance showing such various nonlinear quantum electrodynamics processes as vacuum polarization, electron–positron pair plasma creation, and other properties depending on the electromagnetic field amplitude.

23.2.1 Dynamics of Radiating Electron

In order to describe the relativistic electron dynamics in the electromagnetic field we shall use the equations of electron motion with the radiation friction force in the Landau–Lifshitz form (Pomeranchuk, 1940; Landau & Lifshitz, 1980) with a form-factor taking into account the quantum electrodynamics weakening of the radiation friction (Sokolov et al., 2010). The equations of electron motion, $a = e$ and $e_a = -e$, with the implemented radiation friction force read

$$\frac{d\boldsymbol{p}}{dt} = -e\left(\boldsymbol{E} + \frac{1}{c}\boldsymbol{v} \times \boldsymbol{B}\right) + \boldsymbol{f}_{rad} \tag{23.28}$$

The radiation friction force in the Landau-Lifshitz form written in the covariant form is given by

$$f^{\mu}_{rad} = \frac{2e^3}{3m_e c^3}\left\{\partial_\lambda F^{\mu\nu} u_\nu u_\lambda - \frac{e}{m_e c^2}\left[F^{\mu\lambda} F_{\nu\lambda} u^{\nu} - \left(F_{\nu\lambda} u^{\lambda}\right)\left(F^{\nu\kappa} u_\kappa\right) u^{\mu}\right]\right\} \tag{23.29}$$

The 3D form reads

$$\begin{aligned}
\boldsymbol{f}_{rad} &= \frac{2e^3}{3m_e c^3}\gamma\left\{(\partial_t + \boldsymbol{v}\cdot\nabla)\boldsymbol{E} + \frac{1}{c}\boldsymbol{v} \times [(\partial_t + \boldsymbol{v}\cdot\nabla)\boldsymbol{B}]\right\} \\
&+ \frac{2e^4}{3m_e^2 c^4}\left\{\boldsymbol{E} \times \boldsymbol{B} + \frac{1}{c}\boldsymbol{B} \times (\boldsymbol{B} \times \boldsymbol{v}) + \frac{1}{c}\boldsymbol{E}(\boldsymbol{v}\cdot\boldsymbol{E})\right\} \\
&- \frac{2e^4}{3m_e^2 c^5}\gamma^2 \boldsymbol{v}\left\{\left(\boldsymbol{E} + \frac{\boldsymbol{v}}{c} \times \boldsymbol{B}\right)^2 - \frac{1}{c^2}(\boldsymbol{v}\cdot\boldsymbol{E})^2\right\}.
\end{aligned} \tag{23.30}$$

The first term on the right-hand side is of the order of $\approx \varepsilon_{rad}\gamma a_0$. For the second term on the right-hand side we see that it is of the order of $\approx \varepsilon_{rad} a_0^2$. The third term is of the order of $\approx \varepsilon_{rad}\gamma^2 a_0^2$.

Retaining highest order term, the radiation friction force can be written as

$$\boldsymbol{f}_{rad} = -\frac{2}{3} r_e^2 G_e(\chi_e) \boldsymbol{p} \sqrt{1 + \frac{p^2}{m_e^2 c^2}} \left[\left(\boldsymbol{E} + \frac{1}{c} \boldsymbol{v} \times \boldsymbol{B} \right)^2 - \left(\frac{1}{c} \boldsymbol{v} \cdot \boldsymbol{E} \right)^2 \right] \tag{23.31}$$

It assumes the ultra-relativistic electron energy. On the right hand side, the form factor $G_e(\chi_e)$ depending on the dimensionless parameter χ_e describes the radiation friction reduction due to quantum effects. As noticed above, quantum electrodynamics effects weaken the radiation friction because in the quantum regime the charged particle emits less radiation. In the quantum electrodynamics regime, the recoil due to photon emission becomes important. The total radiated intensity is reduced by a factor depending on the quantum parameter χ_e given by Eq. (19.28). In Eq. (23.31), we take into account the quantum electrodynamics effects by using the form-factor, $G(\chi_e)$, equal to the ratio of the full radiation intensity, I, to the intensity emitted by a classical electron, I_{cl}, with the same energy. It is also called the Gaunt factor. Using the results of (Sokolov et al., 2010), we can write form-factor $G(\chi_e)$ as

$$G(\chi_e) = -\frac{3}{4} \int_0^\infty \left[\frac{4 + 5\chi_e x^{3/2} + 4\chi_e^2 x^3}{\left(1 + \chi_e x^{3/2}\right)^4} \right] \Phi'(x) x dx \tag{23.32}$$

where $\Phi(x)$ is the Airy function. Here, we neglect the effects of the discrete nature of the photon emission in quantum electrodynamics.

In the limit $\chi_e \ll 1$ the form-factor $G(\chi_e)$ tends to unity as

$$G(\chi_e) = 1 - \frac{55\sqrt{3}}{16} \chi_e + \cdots \approx 1 - 5.95\chi_e + \cdots \tag{23.33}$$

For $\chi_e \gg 1$ it tends to zero as

$$G(\chi_e) = \frac{32\pi}{3^{5/6} 27\, \Gamma(1/3)\, \chi_e^{4/3}} - \cdots \approx \frac{0.56}{\chi_e^{4/3}} - \cdots \tag{23.34}$$

where $\Gamma(x)$ is the Gamma function. Although the form-factor $G(\chi_e)$ tends to zero at $\chi_e \gg 1$ the radiation losses grow proportionally to $\chi_e^{2/3}$.

Expression (23.32) and asymptotical dependences (23.33) and (23.34) are not convenient for implementing them in computer codes. For the sake of calculation simplicity, we shall use the approximation

$$G(\chi_e) = \frac{1}{\left(1 + 8.9\chi_e + 2.41\chi_e^2\right)^{2/3}} \tag{23.35}$$

Within the interval $0 < \chi_e < 10$ the accuracy of approximation is better than 1%.

The parameter χ_e expressed via the electric and magnetic field reads

$$\chi_e = \frac{\gamma_e}{E_S}\sqrt{\left(\boldsymbol{E} + \frac{1}{c}\boldsymbol{v} \times \boldsymbol{B}\right)^2 - \left(\frac{1}{c}\boldsymbol{v} \cdot \boldsymbol{E}\right)^2} \tag{23.36}$$

We note that the leading term in the Landau-Lifshitz equation with a quantum correction appears from the Fokker–Planck equation obtained by Niel et al. (2018).

23.3 Coupled Longitudinal and Transverse Electromagnetic Modes

We introduce the normalized variables

$$\tau = \omega t, \quad \boldsymbol{q} = \frac{\boldsymbol{p}}{m_e c}, \quad \boldsymbol{a} = \frac{e\boldsymbol{E}}{m_e \omega c} \tag{23.37}$$

In the case of a plane electromagnetic wave propagating in low density plasma we can write the equations of electron component motion with the radiation friction force in the form given by Eq. (23.31) and electromagnetic field evolution taken in the frame of reference moving with wave group velocity as (Bulanov, 2021)

$$\dot{\boldsymbol{q}} = -\boldsymbol{a} - \frac{\varepsilon_{rad}}{\gamma_e} G(\chi_e)\left\{\gamma_e^2 \dot{\boldsymbol{a}} - \boldsymbol{a}(\boldsymbol{q} \cdot \boldsymbol{a}) + \boldsymbol{q}\left[(\gamma \boldsymbol{a})^2 - (\boldsymbol{q} \cdot \boldsymbol{a})^2\right]\right\} \tag{23.38}$$

$$\dot{\boldsymbol{a}} = \boldsymbol{\beta} - \boldsymbol{\beta}_0 \tag{23.39}$$

Here $\boldsymbol{\beta}$, is the electron velocity normalized on c. The normalized velocity $\boldsymbol{\beta}_0$ is the velocity of the boost. In the boosted frame of reference the wave magnetic field vanishes. A dot denotes a differentiation with respect to normalized time, $\omega_{pe} t$. The dimensionless parameter ε_{rad} is equal to

$$\varepsilon_{rad} = \frac{4\pi r_e}{3\lambda_p} \tag{23.40}$$

where $r_e = e^2/m_e c^2$ is the classical electron radius and $\lambda_p = 2\pi c/\omega_{pe}$.

In Figs. 23.1 and 23.2 we show the results of numerical integration of the system (23.38,23.39) in the case like that presented in Figs. 21.1 and 21.2, but with the radiation reaction effects taken into account. The coupled electromagnetic-electrostatic wave is linearly polarized with $p_z = 0$ and $E_z = 0$ for initial, at $t = 0$, conditions for $\beta_g = 0.96$, $p_0 = 3.5$, $p_{x0} = p_0$, $p_{y0} = -3$, $E_{x0} = 2$, and $E_{y0} = -5$. The dimensionless parameter ε_{rad} given by Eq. (23.40) equals 10^{-3}. Figure 23.1 shows the dependence of the x- and y-components of the electric field, E_x and E_y, and the electron density $n\sqrt{1-\beta_g^2}$ versus the coordinate $\tau = \gamma_g(t - x\beta_g)$. As it can be seen the electromagnetic wave decays fast: during a time interval approximately equal to the wave period the wave amplitude decreases down to the value below $\varepsilon_{rad}^{-1/3}$. Then the wave amplitude decreases gradually. This wave evolution due to the radiation friction is also seen in Fig. 23.2 showing the trajectories of coupled longitudinal and transverse oscillations in the planes (p_x, E_x) and (p_y, E_y), respectively, for the same parameters as in Fig. 23.1

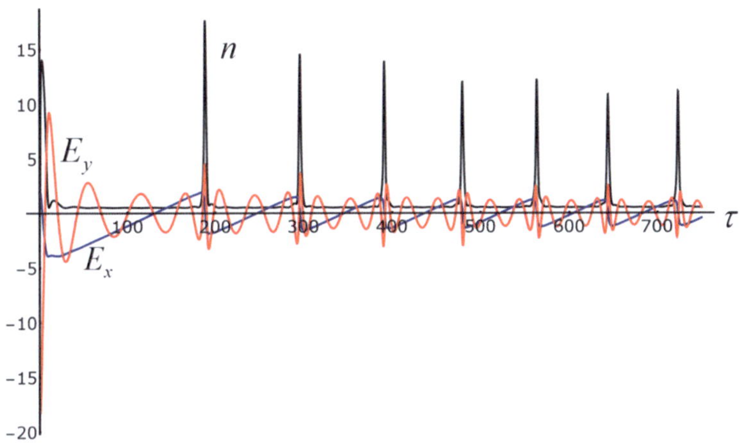

Fig. 23.1 Coupled transverse and longitudinal waves:, the x- and y-components of the electric field, E_x and E_y, and the electron density $n\sqrt{1-\beta_g^2}$ versus the coordinate τ for $\beta_g = 0.96$, $p_0 = 3.5$, $p_{x0} = p_0$, $p_{y0} = -3$,$E_{x0} = 2$, and $E_{y0} = -5$. The dimensionless parameter ε_{rad} equals 10^{-3}

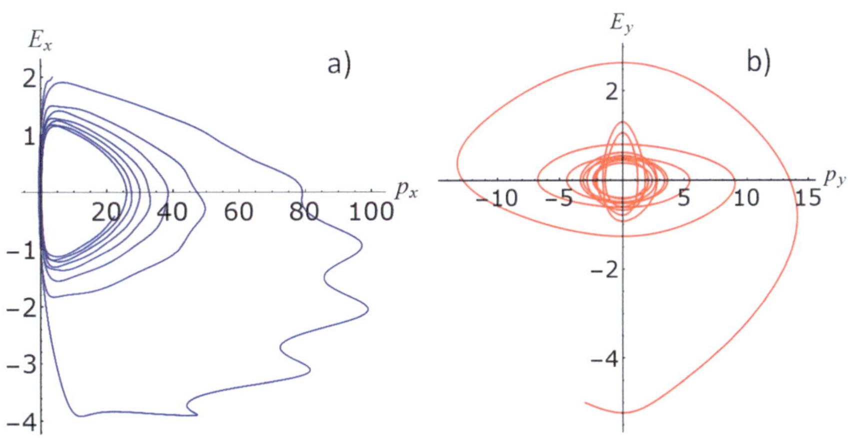

Fig. 23.2 Trajectories of coupled longitudinal and transverse oscillations in the planes (p_x, E_x) and (p_y, E_y), respectively, for the same parameters as in Fig. 23.1

Very fast decay of the electromagnetic wave in the near critical density plasma due to the radiation loss effects provides an underlying mechanism of the high efficiency gamma flash generation by the multi-petawatt power lasers (Hadjisolomou et al., 2021, 2022; Kolenatý et al., 2022; Lezhnin et al., 2018; Nakamura et al., 2012; Ridgers et al., 2012).

23.4 Radiating Electron Dynamics in a Rotating Electric Field

As an example, illustrating radiation friction effects on the charged particle dynamics, we consider the case when the electromagnetic wave effects on the radiating electron is modeled by a rotating electric field. As noted above, it can correspond either to the electric field seen by the electron in the boosted frame of reference moving with the group velocity of the circularly polarized electromagnetic wave propagating in the plasma, or to the electric field in the antinodes of standing circularly polarized electromagnetic wave. The rotating electric field is given by

$$\boldsymbol{a} = a_0(\boldsymbol{e}_2 \cos\tau + \boldsymbol{e}_3 \sin\tau) \tag{23.41}$$

where $\boldsymbol{e}_2$ and $\boldsymbol{e}_3$ are unit vectors along the coordinate axis in the plane perpendicular to the direction of the electromagnetic wave propagation. Here we introduce the normalized variables,

$$\tau = \omega t, \quad \boldsymbol{q} = \frac{\boldsymbol{p}}{m_e c}, \quad \boldsymbol{a} = \frac{e\boldsymbol{E}}{m_e \omega c} \tag{23.42}$$

We represent the electron momentum as

$$\begin{pmatrix} q_1 \\ q_\parallel \\ q_\perp \end{pmatrix} = \begin{pmatrix} 1 & 0 & 0 \\ 0 & \cos\tau & \sin\tau \\ 0 & -\sin\tau & \cos\tau \end{pmatrix} \begin{pmatrix} q_1 \\ q_2 \\ q_3 \end{pmatrix} \tag{23.43}$$

Here $q_\parallel$ and $q_\perp$ are the components of the electron momentum parallel and perpendicular to the electric field (see Fig. 23.3).

In the case of a rotating electric field, Eq. (23.36) yields for the quantum electrodynamics parameter

$$\chi_e = \frac{a_0}{a_S}\sqrt{1 + q_1^2 + q_\perp^2} \tag{23.44}$$

For the rotating electric field the equations of the electron motion (23.28) can be written as

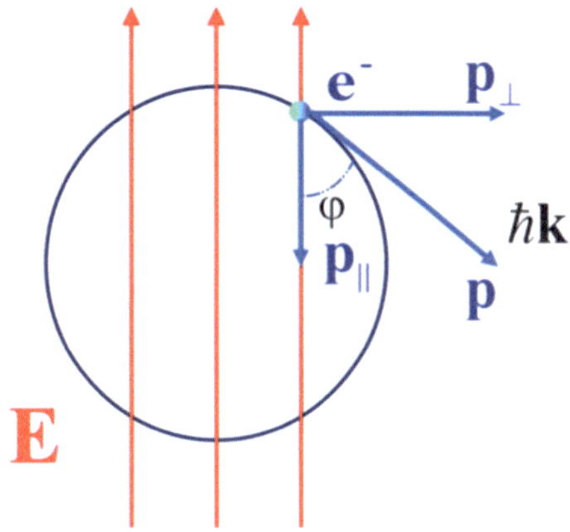

Fig. 23.3 Electron in rotating electric field

$$\frac{d\boldsymbol{q}}{d\tau}=\boldsymbol{a}-\varepsilon_{rad}\frac{G(\chi_e)}{\gamma_e}\left\{\gamma_e^2\frac{d\boldsymbol{a}}{d\tau}-\boldsymbol{a}(\boldsymbol{a}\cdot\boldsymbol{q})+\boldsymbol{q}\left[(\gamma_e\boldsymbol{a})^2-(\boldsymbol{a}\cdot\boldsymbol{q})^2\right]\right\} \tag{23.45}$$

with the electron Lorentz factor equal to $\gamma_e=\sqrt{1+q_1^2+q_2^2+q_3^2}$.

Substituting expressions (23.47) into the equations of the electron motion (23.49), we obtain

$$\frac{dq_\perp}{d\tau}-q_{||}=-\varepsilon_{\rm rad}G(\chi_e)\left[\gamma_e a_0+a_0^2\frac{q_\perp}{\gamma_e}\left(1+q_\perp^2\right)\right] \tag{23.46}$$

$$\frac{dq_{||}}{d\tau}+q_\perp=a_0-\varepsilon_{\rm rad}G_e(\chi_e)a_0^2q_{\|}\frac{\left(1+q_\perp^2\right)}{\gamma_e} \tag{23.47}$$

In this case, the quantum electrodynamics parameter χ_e is equal to

$$\chi_e=\frac{a_0}{a_S}\sqrt{1+q_\perp^2} \tag{23.48}$$

In this example, we assume that the longitudinal component of the electron momentum, q_1, is much less than the transverse momentum component $\sqrt{q_{||}^2+q_\perp^2}$, which corresponds to the case of the laser pulse interacting with a near-critical density plasma, when $n\approx m_e\omega_0^2/4\pi e^2$ and $q_1^{(0)}\leq 1$, the change of the momentum component is negligible provided that the laser pulse duration is small enough.

Multiplying Eq. (23.46) by $q_\perp/\gamma_e$ and Eq. (23.47) by $q_{\|}/\gamma_e$ and taking the sum, we find

$$\frac{d\gamma_e}{d\tau}=a_0\frac{q_{||}}{\gamma_e}-\varepsilon_{\rm rad}G(\chi_e)\left[a_0q_\perp+a_0^2\frac{\left(\gamma_e^2-1\right)}{\gamma_e^2}\left(1+q_\perp^2\right)\right] \tag{23.49}$$

which shows how the electron acquires energy from the electromagnetic wave and loses it due to radiation friction.

This equation shows how the electron acquires energy from the electromagnetic wave and loses it due to radiation friction. The stationary solutions of Eqs. (23.41) in the limit $\chi_e \ll 1$, i.e. $G(\chi_e) = 1$, give relationships between the momentum components $q_\parallel$ and $q_\perp$, and the amplitude of the electromagnetic field a as follows. If the EM field amplitude is relatively small, that is, $1 \ll a_0 \ll \varepsilon_{rad}^{-1/3}$, Eqs. (23.45) yield for the components of the electron momentum perpendicular and parallel to the electric field (see also Bulanov et al., 2011),

$$q_\perp \approx a_0 \tag{23.50}$$

$$q_\parallel \approx \varepsilon_{rad} a_0^4 \tag{23.51}$$

In the opposite limit of strong field, when $a_0 \gg \varepsilon_{rad}^{-1/3}$, the momentum components are given by

$$q_\perp \approx \frac{1}{\sqrt{\varepsilon_{rad} a_0}} \tag{23.52}$$

and

$$q_\parallel \approx \left(\frac{a_0}{\varepsilon_{rad}}\right)^{1/4} \tag{23.53}$$

Typical solutions of this system of equations show that for $a_0 < \varepsilon_{rad}^{-1/3}$ the electron oscillations in the $(q_\perp, q_\parallel)$ plane decay slowly while for the laser pulse amplitude values equal to or above $\varepsilon_{rad}^{-1/3}$, the electron oscillations in the rotating coordinate system decay during a time of the order of or less than the wave period. In Fig. 23.4 we depict these dependences for different electromagnetic wave frequency.

The dependence on time of the quantum electrodynamics parameters χ_e and $G(\chi_e)$ demonstrates that for $a_0 = 0.25\varepsilon_{rad}^{-1/3}$, i.e., at the laser intensity of the order of $10^{21}\,\mathrm{W/cm^2}$ for 1 μm wavelength laser radiation, the quantum reducing of the radiation friction force is negligibly small. Asymptotically at $t \to \infty$ we have $\chi_e = 0.1$ and $G^{1/3}(\chi_e) = 0.95$. When $a_0 = 0.75\varepsilon_{rad}^{-1/3}$, the quantum correction of the radiation friction force becomes more significant with $\chi_e = 0.3$ and $G^{1/3}(\chi_e) = 0.75$ in the limit $t \to \infty$.

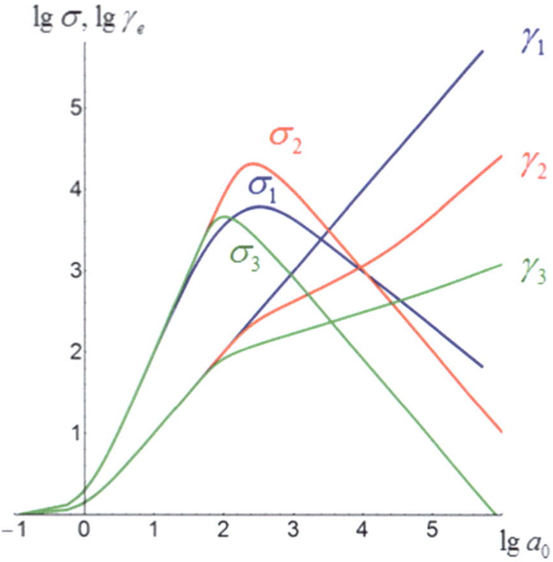

Fig. 23.4 Dependences of $\log_{10}(\sigma/\sigma_T)$ and $\log_{10}(\gamma_e)$ on $\log_{10}(a)$ for (1) $\omega = \omega_1/12.5$, (2) $\omega = \omega_1$, and (3) $\omega = 12.5\,\omega_1$

23.5 Integral Scattering Cross Section

According to Eqs. (23.45) and (23.49) the energy flux reemitted by the electron is equal to

$$\frac{e(\boldsymbol{v}\cdot\boldsymbol{E})}{m_e c^2\omega} \approx \varepsilon_{rad} G(\chi_e)\gamma_e\big(a_0 q_\perp + a_0^2 q_\perp^2\big) \tag{23.54}$$

This expression is a condition of the balance between the rate of energy acquired by the electron from the electromagnetic field and the rate of the radiation losses. The integral scattering cross section by definition (Landau & Lifshitz, 1980) equals the ratio of the reemitted energy flux to the Poynting vector amplitude, $cE^2/4\pi$. This yields an expression for the cross section dependence on the electron momentum,

$$\sigma = \sigma_T G(\chi_e)\left(\frac{q_\perp}{a_0} + q_\perp^2\right) \tag{23.55}$$

Here, σ_T is the Thomson scattering cross section, $\sigma_T = 8\pi r_e^2/3 = 6.65\times10^{-25}\mathrm{cm}^2$. Expression (23.55) can also be approximated by

$$\sigma = \sigma_T\left(\frac{\gamma_e^2 G(\chi_e)}{1+\varepsilon_{rad}^2\gamma_e^6 G^2(\chi_e)}\right) \tag{23.56}$$

In the range of the wave amplitudes of $1 \ll a_0 \ll (\varepsilon_{rad} G(\chi_e))^{-1/3}$, the transverse component of the particle momentum (it is perpendicular to the instant direction of the electric field) is substantially larger than the longitudinal component (it is perpendicular to the electric field direction) $q_\perp \gg q_\parallel$. In this regime, the integral scattering cross section grows as

$$\sigma \approx \sigma_T \left(1 + a_0^2\right) \tag{23.57}$$

For $\chi_e \ll 1$, i.e.

$$1 \ll \varepsilon_{rad}^{-1/3} \ll a_0 \ll \varepsilon_{rad} a_S^2 \tag{23.58}$$

it reaches a maximum of $\sigma \approx 0.53\, \sigma_T \varepsilon_{rad}^{-2/3}$ at $a_0 \approx 1.1\, \varepsilon_{rad}^{-1/3}$. Then for $a_0 \gg \varepsilon_{rad}^{-1/3}$ the scattering cross section decreases as

$$\sigma \approx \sigma_T / a_0 \varepsilon_{rad} \tag{23.59}$$

as seen in Fig. 23.4. This dependence is a consequence of the fact that the maximal power reemitted by the electron cannot exceed the level of $e\,E\,c$. This puts a constraint on the scattering cross section:

$$\sigma < 4\pi e/E \tag{23.60}$$

equivalent to (23.55). In the limit

$$\varepsilon_{rad}^{-1/3} \ll a_0 \ll \varepsilon_{rad} a_S^2$$

the electron momentum components the momentum components are given by Eqs. (23.52) and (23.53) i.e. the electron energy scales as $\gamma_e \approx (a_0/\varepsilon_{rad})^{1/4}$. In this limit the perpendicular component of the electron momentum is much less than the longitudinal, $q_\perp \ll q_\parallel$, in contrast to the case $a_0 \ll \varepsilon_{rad}^{-1/3}$, when $q_\perp \gg q_\parallel$ and the electron energy scales as $\gamma_e \approx a_0$.

Dependences of the scattering cross section normalized to the Thomson scattering cross section σ_T and the normalized to $m_e c^2$ electron energy γ_e on the laser pulse amplitude are shown in Fig. 23.4. The dimensionless parameters $\varepsilon_{rad} = 1.75 \times 10^{-8}$ and $a_S = 2.8 \times 10^5$ (curves 2) correspond to the electromagnetic wave frequency equal to. For $\omega = \omega_1/12.5$ (curves 1) the electromagnetic wavelength is about $\lambda = 10\ \mu\text{m}$, which is typical for the CO_2 laser wavelength range. Case 3 with $\omega = 12.5\,\omega_1$ corresponds to the parameters in the domain where the radiation friction lowered by quantum effects.

23.6 Radiating Electron Interaction with Multiple Electromagnetic Waves

The electron motion in the standing electromagnetic wave generated by several colliding focused electromagnetic pulses exhibits remarkable features. The multiple colliding laser pulse concept formulated by Bulanov et al. (2010a) is beneficial for achieving an extremely high amplitude of coherent electromagnetic field. Since the topology of electric and magnetic fields oscillating in time of multiple colliding laser pulses is far from trivial and the radiation friction effects are significant in the high field limit, the dynamics of charged particles interacting with the multiple colliding laser pulses demonstrates remarkable features corresponding to random walk trajectories, limit circles, attractors, regular patterns and Levy flights. Under extremely high intensity conditions, the nonlinear dissipation mechanism stabilizes the particle motion resulting in the charged particle trajectory being located within narrow regions and in the occurrence of a new class of regular patterns made by the particle ensembles (Esirkepov & Bulanov, 2017; Gonoskov et al., 2014, 2022; Jirka et al., 2016).

In Fig. 23.5 we show the cross section on the electromagnetic wave amplitude and parameter ε_{rad}.

Within the framework of the multiple colliding laser pulses (MCLP) concept the laser radiation with given energy $\mathcal{E}_{las}$ is subdivided into several beams each of them having $1/N$ of the laser energy, where N is the number of the beams. If the beams interfere in the focus in a constructive way, i.e. their electric fields are summed, the resulting electric field and the laser intensity are equal to $E_N = \sqrt{N} E_{las}$ and to

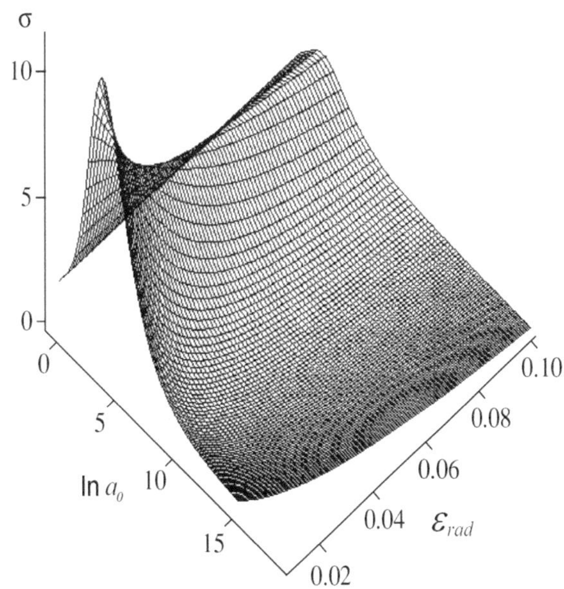

Fig. 23.5 Scattering cross section versus the electromagnetic wave amplitude and parameter ε_{rad}

$I_N = NI_{las}$, respectively. Here E_{las} and I_{las} are the electric field and the intensity of the laser light. For a large number of beams there is a diffraction constraint on the electric field amplitude in the focus region. In the limit $N \to \infty$ the electromagnetic field can be approximated by the 3D dipole configurations (Bulanov et al., 2010a) for which the electric field maximum is given by

$$E_m = 8\pi \sqrt{\frac{\mathcal{P}_{las}}{3c\lambda^2}} \tag{23.61}$$

where $\mathcal{P}_{las}$ is the laser power.

Since the radiation friction and quantum electrodynamics processes both depend on the particle's momentum, the strength of the present electromagnetic field, and on their mutual orientation, it is crucial in understanding the dynamics of charged particles in the electromagnetic field in the regime of radiation dominance. Even in the simplest multiple colliding laser pulses case, two counter-propagating plane waves, the particle behavior in the standing wave is quite complicated. It demonstrates regular and chaotic motion, random walk, limit circles and strange attractors. As is well known, the standing wave configuration is widely used in classical electrodynamics and in quantum electrodynamics theory. This is because in the planes where the magnetic field vanishes, the charged particle interact with an oscillating pure electric field. This provides great simplification of the theoretical description.

According to expressions (23.31) and (23.36) the radiation friction force and the parameter χ_e, characterizing the role of quantum effects, are proportional to

$$\sqrt{\left(\gamma_e \boldsymbol{E} + \frac{\boldsymbol{p}}{m_e c} \times \boldsymbol{B}\right)^2 - \left(\frac{\boldsymbol{p}}{m_e c} \cdot \boldsymbol{E}\right)^2} \tag{23.62}$$

In the case of an electron interaction with a plane electromagnetic wave propagating along the x-axis with phase and group velocity equal to speed of light in vacuum the function (23.63) is equal to

$$E(\gamma_e - p_x/m_e c) \tag{23.63}$$

For an electron co-propagating with the electromagnetic wave the linear combination of the electron energy and momentum, $h_e = \gamma_e - p_x/m_e c$ is of the order of $1/2\gamma_e$, i.e. the effects of radiation friction and the role of quantum effects are not significant. On the contrary, the function (23.63) can be enhanced to approximately $2\gamma_e$, when the electron interacts with a counter-propagating laser pulse. Therefore, the head-on collision configuration has an apparent advantage for strengthening the electron- electromagnetic -wave interaction and, in particular, for enhancing the gamma-ray production due to nonlinear Thomson or/and Compton scattering. The head-on collision is apparently realized in the case of multiple electromagnetic wave configuration.

23.7 Electron Beam Slowing Down Due to the Nonlinear Thomson Scattering

23.7.1 Ultrarelativistic Electron Beam Slowing Down

We consider a head-on collision of an ultrarelativistic electron with a laser pulse. The laser ponderomotive pressure pushes the electron sideways with respect to the pulse propagation and also changes the longitudinal component of the electron momentum. In the electron rest frame (see below) the laser pulse duration is $\overline{\tau}_{las} \approx \tau_{las}\sqrt{1+a_0^2/2\gamma_e}$. The electron is not scattered aside by the laser ponderomotive force provided its energy is large enough,

$$\gamma_e > c\tau_{las}a_0/2w_\perp \tag{23.64}$$

where $w_\perp$ is the focus width. As we see, for a one-micron, 1 PW, i.e. $a = 10_0^2$, 30 fs duration laser pulse focused to an one-wavelength spot this condition requires $\gamma_e > 500$, i.e. the electron energy above 250 MeV. Further we assume that this condition is respected.

When a counter-propagating electron collides with the laser pulse its longitudinal momentum decreases as

$$p_x = p_0 + m_e c \frac{a_0^2 m_e c}{2\left[\left(m_e^2c^2 + p_0^2\right)^{1/2} - p_0\right]} \tag{23.65}$$

In the boosted frame of reference where the electron is at the rest the laser frequency is related to the laser frequency in the laboratory frame, ω_0, as

$$\overline{\omega}_0 = \frac{\omega_0}{\sqrt{1+a_0^2}}\sqrt{\frac{1-\beta_0}{1+\beta_0}} \tag{23.66}$$

where $\beta_0 = p_0/\sqrt{m_e^2c^2 + p_0^2}$. The bar "$^-$" denotes the frequency value in the frame of reference where the electron on average is at rest. The characteristic frequency of the radiation emitted by the electron in this frame of reference is equal to $\overline{\omega}_m = 0.3\,\overline{\omega}_0 a^3$, Koga et al. (2006). The photon energy, $\hbar\overline{\omega}_m = 0.3\,\hbar\overline{\omega}_0 a_0^3$, satisfies condition $\hbar\overline{\omega}_m = m_e c^2\gamma_e$ for $a_0 = \sqrt{m_e c^2/0.3\,\hbar\overline{\omega}_0}$, which corresponds to the parameters when the quantum electrodynamics description should be used. For lower laser amplitudes, the photon generation can be described as the nonlinear Thomson scattering process.

Dividing the radiation power $\varepsilon_{rad} m_e c^2 \overline{\omega}_0 \gamma^4$ with $\gamma \approx a_0$ by the typical energy of the emitted photon, $\hbar\overline{\omega}_m = 0.3\,\hbar\overline{\omega}_0 a_0^3$, and multiplying by $2\pi/\overline{\omega}_0$ we find the number of photons emitted by the electron during one wave period, $N_\gamma \approx (9\pi/4)\alpha a_0$.

The photon energy in the laboratory frame of reference is

$$\hbar\omega_m = 0.3\hbar\omega_0 \frac{a_0^3}{1+a_0^2}\frac{1-\beta_0}{1+\beta_0} \approx 1.2\hbar\omega_0 a_0 \gamma_0^2 \tag{23.67}$$

We note that in the interaction of a GeV electron with a petawatt power laser pulse the radiation friction effects must be incorporated. Here we shall consider it as a perturbation, i.e. in the Landau-Lifshitz form (Landau & Lifshitz, 1980), which limits of applicability have been discussed by Bulanov et al., (2011a, 2011b).

In order to estimate ω_γ we should find the electron beam energy taking into account the radiation losses. One can obtain the equation for the x-component of the electron momentum

$$\frac{dp_x}{dt} = -4\varepsilon_{rad}\omega_0 a^2(2t)\frac{p_x^2}{m_e c} \tag{23.68}$$

Its solution is given by

$$p_x(t) = \frac{p_x(0)m_e c}{m_e c + 4\varepsilon_{rad}\omega_0 p_x(0)\int_0^t a^2(2s)ds} \tag{23.69}$$

If we assume an $a(t)$ dependence of the form $a(t) = a_0 \exp\left(-t^2/2\tau^2\right)$, Eq. (23.66) can be rewritten as

$$p_x(t) = \frac{p_x(0)m_e c}{m_e c + 4\varepsilon_{rad}\omega_0 p_x(0)\tau a_0^2\sqrt{\pi/32}\left[\mathrm{erf}\left(\sqrt{2}t/\tau\right) - 1\right]} \tag{23.70}$$

Here erf(x) is the error function equal to $\mathrm{erf}(x) = \left(\sqrt{\pi}/2\right)\int_0^x \exp\left(-t^2\right)dt$. Equation (23.70) shows that for large enough $p_x(0)\tau a_0^2$ the electron momentum tends to the limit of

$$p_x(t) \to_{t\to\infty} \sqrt{\frac{\pi}{8}}\left(\frac{m_e c}{4\varepsilon_{rad}\omega_0\tau a_0^2}\right) \tag{23.71}$$

in accordance with the theory formulated in Landau and Lifshitz (1980). For $a_0 = 10^2$ and $\omega_0\tau = 6$ the electron momentum is about $p_x(\infty)/m_e c = 500$.

In the limit when quantum corrections weaken the radiation friction, i.e. when according to Eq. (23.31) the Gaunt factor is approximately equal to $G(\chi_e) \approx 0.56/\chi_e^{4/3}$, the equation of the electron motion with the radiation friction force given by Eq. (23.34) can be written in the form

$$\frac{dp_x}{dt} = -\eta\,\varepsilon_{rad}\omega_0 m_e c a^{2/3}(2t) a_S^{4/3}\left(\frac{p_x}{m_e c}\right)^{2/3} \tag{23.72}$$

with the dimensionless coefficient approximately equal to $\eta \approx 0.888$. The solution of this equation is

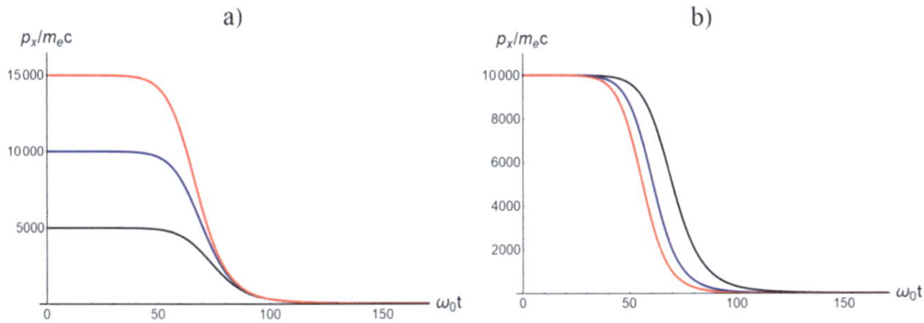

Fig. 23.6 Time dependence of the electron momentum according to Eqs. (23.28) and (23.31) for the electron interacting with Gaussian electromagnetic pulse $a(t) = a_0 \exp\left(-((t - t_m)/(t_m/6))^2\right)$ with $t_m = 40(2\pi/\omega_0)$. **a** Dependence of the electron momentum on time for different initial momentum: $p_x(0)/m_e c = 15000$, $p_x(0)/m_e c = 10000$ and $p_x(0)/m_e c = 5000$. The electromagnetic pulse amplitude equals $a_0 = 100$. **b** The same as in the frame **a** for initial momentum: $p_x(0)/m_e c = 10000$ and varying laser pulse amplitude: $a_0 = 100$, $a_0 = 200$, and $a_0 = 300$

$$3\left(\frac{p_x}{m_e c}\right)^{1/3} = 3\left(\frac{p_x(0)}{m_e c}\right)^{1/3} - \eta\, \varepsilon_{rad}\, \omega_0 m_e c a_S^{4/3} \int_0^t a^{2/3}(2t)dt \tag{23.73}$$

Figure 23.6 illustrates the electron dynamics under the radiation friction while it interacts with strong electromagnetic pulse. The frame Fig. 23.6 (a) shows the time dependence of the electron momentum according to Eqs. (23.28) and (23.31) in the case of interaction with the electromagnetic pulse of Gaussian form, $a(t) = a_0 \exp(-((t - t_m)/(t_m/6))^2)$, where $t_m = 40(2\pi/\omega_0)$. The initial electron momentum equals $p_x(0)/m_e c = 15000$, $p_x(0)/m_e c = 10000$ and $p_x(0)/m_e c = 5000$, respectively. The electromagnetic pulse amplitude equals $a_0 = 100$. (b) The same as in frame Fig. 23.6. (a) for the initial momentum equal to $p_x(0)/m_e c = 10000$ and varying laser pulse amplitude: $a_0 = 100$, $a_0 = 200$, and $a_0 = 300$.

23.8 Change of the Energy Spectrum of the Electron Beam

The radiation losses lead to the change of the energy spectrum of the electrons interacting with strong localized electromagnetic field (see Bulanov et al., 2024). This process has been considered by Telnov (1997) for laser cooling of electron beams for linear colliders. The evolution of the electron energy distribution can be described by the kinetic equation

$$\partial_t f + \partial_{p_x}(\dot{p}_x f) = 0 \tag{23.74}$$

where the radiation friction force, $\dot{p}_x$, is given by either

$$\frac{dp_x}{dt} = -4\varepsilon_{rad}\omega_0 a^2(2t)\frac{p_x^2}{m_e c} \tag{23.75}$$

in the limit $\chi_e \ll 1$ or by

$$\frac{dp_x}{dt} = -\eta\,\varepsilon_{rad}\omega_0 m_e c a^{2/3}(2t) a_S^{4/3}\left(\frac{p_x}{m_e c}\right)^{2/3} \tag{23.76}$$

when $\chi_e > 1$, respectively. The solution of Eq. (23.75) is the function $f(t, p)$ which is constant on the characteristics (e.g. see Ibragimov, 2010). The equations for the characteristics of Eq. (23.75) are

$$dt = \frac{dp}{\dot{p}_x} = \frac{df}{f\,\partial_{p_x}\dot{p}_x} \tag{23.77}$$

In the cases of the radiation friction force given by Eqs. (23.76) and (23.77), the radiation friction force can be represented in the form

$$\dot{p}_x = \upsilon(t)\varpi(p_x) \tag{23.78}$$

Here the functions $\upsilon(t)$ and $\varpi(p_x)$ are

$$\upsilon(t) = -4\varepsilon_{rad}\omega_0 m_e c a^2(2t) \text{ and } \varpi(p_x) = \left(\frac{p_x}{m_e c}\right)^2, \tag{23.79}$$

in the case corresponding to Eq. (23.76), and

$$\upsilon(t) = -\eta\,\varepsilon_{rad}\omega_0 m_e c a^{2/3}(2t) a_S^{4/3} \text{ and } \varpi(p_x) = \left(\frac{p_x}{m_e c}\right)^{2/3}, \tag{23.80}$$

for the radiation friction force given by Eq. (23.77), respectively. Introducing the function $F = \varpi f$ and changing the variables and to

$$s = -\int^{t} \upsilon(\tau)d\tau, \quad \text{and} \quad w = \int^{p_x} \frac{dp}{\varpi(p)} \tag{23.81}$$

we rewrite Eq. (23.75) as

$$\partial_s F - \partial_w F = 0 \tag{23.82}$$

The initial problem solution of this equation is

$$F(s, w) = F_0(w - s) \tag{23.83}$$

with $F_0(w)$ determined by the initial conditions. It yields for the distribution function $f(t, p_x)$ the expression

$$f(t,p_x) = \frac{\varpi\left(p_{x,0}\right)}{\varpi\left(p_x\right)} f_0\left(p_{x,0}\right), \tag{23.84}$$

where $f_0\left(p_{x,0}\right)$ is the distribution function at $t = 0$.

As an example, we consider the electron distribution function before the electron beam interaction with the electromagnetic field of the super-Gaussian form

$$f_0\left(p_{x,0}\right) = \frac{1}{\Gamma(1+1/q)\Delta p_0} \exp\left[-\left(\frac{p_{x,0}-p_{b,0}}{\Delta p_0}\right)^q\right] \tag{23.85}$$

with positive index q. It describes the electron beam with the average momentum $p_{b,0}$ and the width in the momentum space equal to Δp_0. Here we use the electron momentum normalization to

$$\Pi_1 = -\frac{m_e c}{4\varepsilon_{rad} a_0^2} \tag{23.86}$$

Using relationships obtained above we obtain in the case when $\dot{p}_x$ is given by Eq. (23.76)

$$f_1(t,p_x) = \frac{1}{\Gamma(1+1/q)\Pi_1 \Delta p_0 (1-p_x s_1 t)^2} \exp\left\{-\left[\frac{p_x - p_{b,0}(1-p_x s_1 t)}{\Delta p_0 (1-p_x s_1 t)}\right]^q\right\}, \tag{23.87}$$

where $s_1 = 4\varepsilon_{rad}\omega_0 m_e c a_0^2 = \omega_0/\Pi_1$. According to this expression, the average electron momentum and the distribution width decrease as

$$p_b(t) = \frac{p_{b,0}}{1+p_{b,0}s_1 t} \quad \text{and} \quad \Delta p(t) \approx \frac{\Delta p_0}{1+p_{b,0}s_1 t} \tag{23.88}$$

Figure 23.7 shows the electron beam distribution functions for $q = 8$, (a) $f_1(t, p_x)$ and (b) $f_2(t, p_x)$, at $(s_1 t, s_2 t) = 0.0, 0.1, 0.2, 0.3, 0.4, 0.5$. As is clearly seen, in both cases the radiation friction effects result in the decreasing of the average electron momentum and to the narrowing the distribution. In the case of distribution function $f_1(t, p_x)$ shown in Fig. 23.7a, the spectrum narrowing is relatively faster than in the case of the function $f_2(t, p_x)$ presented in Fig. 23.7b.

In the case, when the radiation friction force is given by Eqs. (23.77) the electron distribution function is

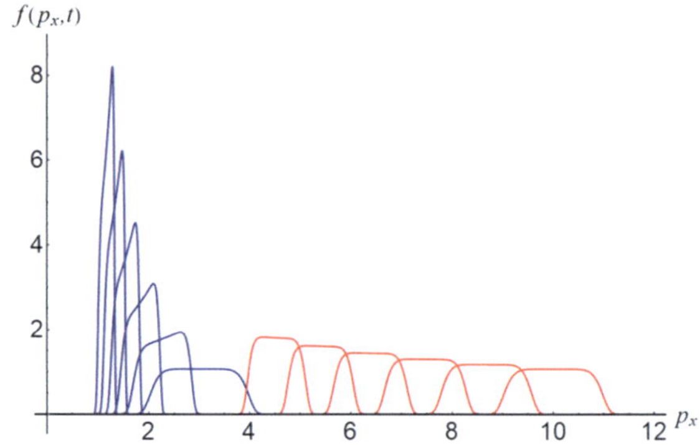

Fig. 23.7 Electron beam distribution functions versus time for $q = 8, f_1(t, p_x)$ (shown by blue color) and $f_2(t, p_x)$ (shown by red color), at $(s_1 t, s_2 t) =$ 0.0, 0.1, 0.2, 0.3, 0.4, 0.5

$$f_2(t, p_x) = \frac{1}{\Gamma(1 + 1/q)\Pi_2 \Delta p_0}\left(1 + \frac{s_2 t}{p_x^{1/3}}\right)^2 \exp\left\{-\left[\frac{\left(p_x^{1/3} + s_2 t\right)^3 - p_{b,0}}{\Delta p_0}\right]^q\right\} \tag{23.89}$$

with $s_2 = \eta\, \varepsilon_{rad}\omega_0 m_e c a_S^{4/3} a_0^{2/3} = \omega_0/\Pi_2$. For the sake of simplicity here we assume that the normalized wave amplitude is constant being equal to a_0, which can correspond to the circularly polarized radiation.

23.9 Electron Energy Spectrum Broadening Due to Quantum Mechanical Stochasticity

As we discussed above, within the framework of the approximation based on classical electrodynamics, the expressions for the radiation friction force, including the case when the quantum effects are taken into account with the Gaunt factor $G(\chi_e)$, radiation losses for electrons interacting with the electromagnetic wave result in a drift of the electron distribution function towards lower electron momentum with additional narrowing of the momentum distribution function. The quantum mechanical stochasticity leads to the appearance of diffusion in the momentum space. The Fokker–Planck equation implementing energy drift and diffusion takes the form (Neitz & Di Piazza, 2013)

$$\partial_t f = \partial_{p_x}\left\{A(p_x) f + \frac{1}{2}\partial_{p_x}\left[B(p_x) f\right]\right\} \tag{23.90}$$

where the energy drift and diffusion coefficients are

$$A(p_x) = \frac{2\alpha m_e^2 c^3}{3\hbar}\beta\chi_e^2 \quad \text{and} \quad B(p_x) = \frac{55}{8\sqrt{3}}\frac{\alpha m_e^3 c^4}{\hbar}\gamma\chi_e^3 \tag{23.91}$$

Estimating characteristic energy drift and diffusion time as

$$\tau_{drift} = p_x/A(p_x) \text{ and } \tau_{diff} = 2p_x^2/B(p_x) \tag{23.92}$$

we find that the energy drift evolves faster than the energy diffusion, $\tau_{drift} \ll \tau_{diff}$, at relatively low electron energy: $\gamma < (8/55\sqrt{3})(a_S/a_0) \approx 8 \times 10^5/a_0$ (see also discussion by Neitz and Di Piazza (2013), Vranic et al. (2016); Ridgers et al. (2017)). If the normalized field amplitude equals $a_0 = 10^3$, which can be reached with multi-petawatt lasers, electron energy $8 \times 10^2 m_e c^2$ corresponds to 400 MeV. Taking into account that the QED parameter equals $\chi_e = (a_0/a_S)\gamma$, we find that the energy diffusion is faster than the energy drift for $\chi_e > 1/3$.

Introducing normalized time and momentum, $t/T = \tau$ and $p_x/\Pi = p$, with

$$T = \frac{3^{3/2}55}{128}\frac{\hbar}{\alpha m_e c^2}\left(\frac{a_S}{a_0}\right) \text{ and } \Pi = \frac{16}{55\sqrt{3}} m_e c \left(\frac{a_S}{a_0}\right) \tag{23.93}$$

we can rewrite Eq. (23.87) in the form

$$\partial_\tau f = \partial_p \left\{ (p+\overline{\gamma})^2 \frac{p}{\overline{\gamma}} f + \partial_p \left[\overline{\gamma}(p+\overline{\gamma})^3 f \right] \right\} \tag{23.94}$$

with

$$\Pi = \sqrt{\left(\frac{m_e c}{\Pi}\right)^2 + p^3} \tag{23.95}$$

Figure 23.8 shows the results of the electron beam evolution for the initial distribution function given by Eq. (23.82) with $q = 2, p_0 = 5, \Delta p_0 = 1$. Frame (a) shows the function $f(\tau, p)$. In Frame (b) we plot equal value contours of $f(\tau, p)$ on the plane (τ, p). We observe a systematic drift of the energy spectrum towards low energies due to radiation losses. Quantum diffusion, which occurs due to the recoil effects, results in a smoothing of the distribution in the region of vicinity of the maximum electron momentum and broadening of the whole momentum distribution.

Effects of strong radiation reaction, including quantum mechanical effects on electron dynamics, have been observed in experiments involving collisions of laser-accelerated high-energy electrons with a counter-propagating laser pulse by Cole et al., (2018), Poder et al., (2018), Los et al. (2025).

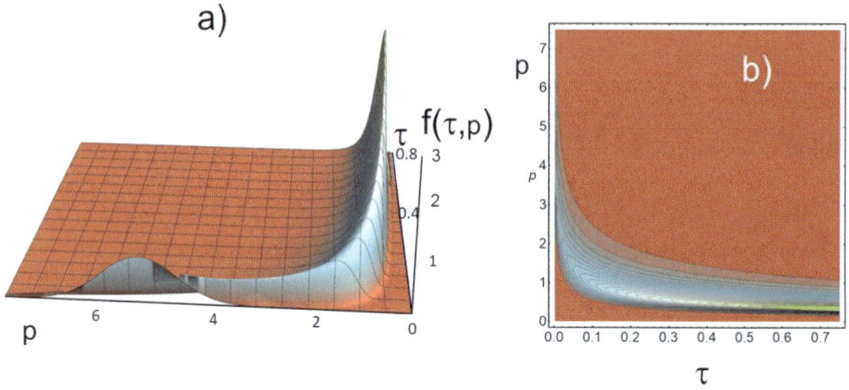

Fig. 23.8 Electron beam evolution for initial distribution function given by Eq. (23.82) with $q = 2$, $p_0 = 5$, $\Delta p_0 = 1$; **a** distribution function $f(\tau, p)$; **b** equal value contours of $f(\tau, p)$ on the plane (τ, p). The electron momentum is normalized on Π given by Eq. (23.90)

23.10 Ultra-Relativistic Electron Slowing Down in the 3D Dipole Electromagnetic Field

The3D dipole electromagnetic field configuration Bulanov et al. (2010b) have proposed as the approximation of the field in the focus of multiple colliding laser pulses. Than this configuration has been actively used in a number of publications (e.g., see Gonoskov et al., 2014, 2017, 2022; Bulanov et al., 2015; Magnusson et al., 2019; Jirka et al., 2022).

23.10.1 3D Electromagnetic Wave

If following Artyukov et al. (2018) we represent the electromagnetic potential in the form

$$\boldsymbol{A} = \boldsymbol{l} \times \nabla u, \tag{23.96}$$

with $\boldsymbol{l}$ being the unit vector independent on time and coordinates, we can see that $\boldsymbol{A}$ satisfies the Maxwell equations provided the function $u(\boldsymbol{r}, t)$ obeys the scalar wave equation

$$\Delta u - \frac{1}{c^2}\partial_{tt} u = 0 \tag{23.97}$$

Considering the spherically symmetric solution of this equation,

$$u(r,t) = \frac{f(r-ct) + g(r+ct)}{r} \tag{23.98}$$

we write a regular at the center solution (at $r \to 0$) as

$$u(r,t) = \frac{f(ct+r) - f(ct-r)}{r} \tag{23.99}$$

where f is an arbitrary function of its argument and $r = \sqrt{x^2+y^2+z^2}$.

Expression (23.96) yields for the electromagnetic potential, electric and magnetic fields

$$\boldsymbol{A} = \boldsymbol{l} \times \boldsymbol{n}\partial_r u, \tag{23.100}$$

$$\boldsymbol{E} = -\frac{1}{c}\boldsymbol{l} \times \boldsymbol{n}\partial_{rt} u, \tag{23.101}$$

$$\boldsymbol{B} = \left(\partial_{rr}u + \frac{1}{r}\partial_r u\right)\boldsymbol{l} \times \boldsymbol{n} - \left(r\partial_r \frac{1}{r}\partial_r u\right)(\boldsymbol{l}\cdot\boldsymbol{n})\boldsymbol{l} \tag{23.102}$$

where $\boldsymbol{n}$ is a vector along the $\boldsymbol{r}$ direction, i.e. $\boldsymbol{n} = \boldsymbol{r}/|\boldsymbol{r}|$.

Taking the function f in the form $f(x) = \cos kx/k$ we obtain for $u(r,t)$ in Eq. (23.99)

$$u(r,t) = \frac{2\sin\omega t \sin kr}{kr} \tag{23.103}$$

with $\omega = kc$. Assumption on $\boldsymbol{l} = \boldsymbol{e}_z$ and calculation of $\partial_r u$ and its substitution to Eqs. (23.100)–(23.102) gives for

$$\boldsymbol{A} = 2\sin\omega t \frac{\rho(kr\cos kr - \sin kr)}{kr^3}\boldsymbol{e}_\varphi, \tag{23.104}$$

$$\boldsymbol{E} = -2\cos\omega t \frac{\rho(kr\cos kr - \sin kr)}{r^3}\boldsymbol{e}_\varphi, \tag{23.105}$$

$$\begin{aligned}\boldsymbol{B} = {} & 2\sin\omega t \frac{z\rho\left(3kr\cos kr + \left(k^2r^2 - 3\right)\sin kr\right)}{kr^5}\boldsymbol{e}_\rho \\ & + 2\sin\omega t \frac{\left(\left(2z^2-\rho^2\right)kr\cos kr + \left(2z^2-\rho^2\right)\sin kr\right)}{kr^5}\boldsymbol{e}_z\end{aligned} \tag{23.106}$$

where $\rho = \sqrt{x^2+y^2}$ and $\boldsymbol{e}_\varphi$, $\boldsymbol{e}_\rho$, and $\boldsymbol{e}_z$ are unit vectors in φ, ρ, z directions. Equations (23.104)–(23.106) describe the TE mode (it is known as the transverse electric mode) of electromagnetic field with toroidal electric and poloidal magnetic field. The Transverse Magnetic (TM) mode can be obtaining by changing $\boldsymbol{E}$ and $\boldsymbol{B}$ with $\boldsymbol{B}$ and $\boldsymbol{E}$, i.e. $\boldsymbol{E}_\varphi \to \boldsymbol{B}_\varphi, \boldsymbol{B}_\rho \to \boldsymbol{E}_\rho, \boldsymbol{B}_\rho \to \boldsymbol{E}_\rho$.

When an ultrarelativistic electron crosses a region of strong electromagnetic fields it loses the energy due to radiation friction. To reveal the characteristic features of the electron slowing down during interaction with the 3D electromagnetic field, we consider the case of a strong field ($a_0 \gg 1$).

As is noted above Eqs. (23.104)–(23.106) describe the TE mode. To obtain the TM mode one should change $\boldsymbol{E}$ and $\boldsymbol{B}$ with $\boldsymbol{B}$ and $\boldsymbol{E}$, i.e. change $E_\varphi \to B_\varphi$ and $B_\rho \to E_\rho$. We rewrite the 3D dipole field equations as follows. For toroidal magnetic and electric field components we have

$$\begin{pmatrix} B_\phi \\ E_\phi \end{pmatrix} = \begin{pmatrix} a_{TM} \\ a_{TE} \end{pmatrix} \sqrt{\frac{2}{\pi}} \sin(t)\rho \left[\frac{\sin R - R\cos R}{R^3} \right] \tag{23.107}$$

with $\rho = \sqrt{x^2 + y^2}$ and $R = \sqrt{\rho^2 + z^2}$. The poloidal components of the electric and magnetic fields are given by

$$\begin{pmatrix} E_\rho \\ B_\rho \end{pmatrix} = \begin{pmatrix} a_{TM} \\ a_{TE} \end{pmatrix} \sqrt{\frac{2}{\pi}} \cos(t)\rho z \left[\frac{3R\cos R + \left(R^2 - 3\right)\sin R}{R^5} \right] \tag{23.108}$$

and

$$\begin{pmatrix} E_\rho \\ B_\rho \end{pmatrix} = \begin{pmatrix} a_{TM} \\ a_{TE} \end{pmatrix} \sqrt{\frac{2}{\pi}} \cos(t)\rho z \left[\frac{\left(2z^2 - \rho^2\right)R\cos R - \left(2z^2 - \rho^2 + R^2 - \rho^2 R^2\right)\sin R}{R^5} \right] \tag{23.109}$$

We assume that the field is a superposition of the TE and TM modes $a_{TM} = a_{TE} = a_0$ with no phase difference. The peak field amplitude is related to the total laser power according to

$$a_m = \sqrt{\frac{8}{9\pi}} a_0 \approx 550 \sqrt{\frac{\mathcal{P}}{1\ \text{PW}}} \tag{23.110}$$

Figure 23.9 illustrates the dipole electromagnetic field configuration.

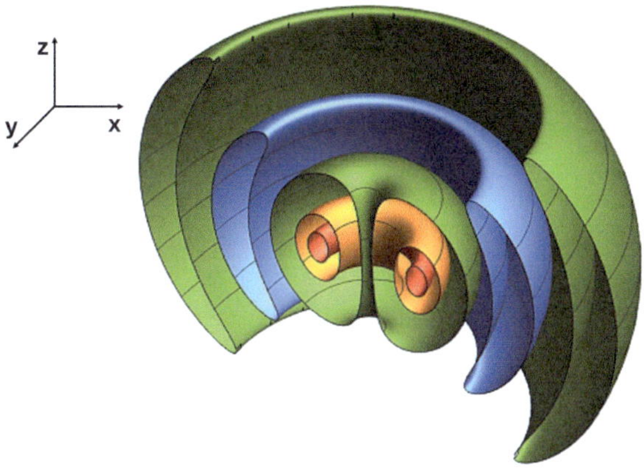

Fig. 23.9 Dipole electromagnetic field

23.10.2 *Electron Interaction with the Dipole Electromagnetic Field*

The longitudinal component of the momentum of the electron interacting with the electromagnetic field dominates over the transverse component acquired in the electromagnetic field meaning $p_z \gg a_0$. For an electron moving parallel to the symmetry axis (z axis) of the 3D dipole electromagnetic configuration. the electron motion equation takes the form

$$p_z' = -\varepsilon_{rad} p_z^2 K_z(\rho, z) \tag{23.111}$$

where the primed momentum denotes differentiation with respect to the coordinate z and the function (see Magnusson et al., 2019)

$$K_z(\rho, z) = \left(E_\rho - B_\phi\right)^2 + \left(E_\phi + B_\rho\right)^2 \tag{23.112}$$

The function $K_z(\rho, z)$ is equal to zero on the z axis, has a ρ^2 scaling for small radii $\rho \ll 1$, and reaches its maximum of $K_{z,m} \approx 0.24 a_0^2$ at $\rho \approx 2$ and $z = 0$. The maximum value of the invariant quantum parameter χ_e that can be achieved by an electron moving parallel to the z axis is $0.49\gamma_e a_0/a_S$. For an electron with initial momentum $p_{z,0}$ and initial coordinate $\rho = \rho_0$, the solution to Eq. (23.111) is then given by

$$p_z(z) = \frac{p_{z,0}}{1 + (2/\pi m_e c)\varepsilon_{rad} p_{z,0} a_0^2 \Phi(\rho_0, z)} \tag{23.113}$$

where the function $\Phi(\rho_0, z)$ is

$$\Phi(\rho, z) = \frac{\pi}{2a_0^2 \int_{-\infty}^{z} K_z(\rho, s) ds} \tag{23.114}$$

The dependence of the electron momentum on the coordinate z according to Eq. (23.113) shows a rapid depletion and then the saturation. At $(2/\pi m_e c)\varepsilon_{rad} p_{z,0} a_0^2 \Phi(\rho_0, z) \gg 1$ the electron momentum $p_z(z)$ becomes independent on its initial value, $p_{z,0}$. The final momentum, at $z \to \infty$, tends to

$$p_{z,\infty} = \frac{\pi m_e c}{2\varepsilon_{rad} a_0^2 \Phi(\rho_0, \infty)} \tag{23.115}$$

References

Artyukov, I. A., Vinogradov, A. V., D'yachkov, N. V., & Feshchenko, R. M. (2018). Energy density in a collapsing electromagnetic wave. *Quantum Electronics, 48*, 1073.

Blackburn, T. G. (2020). Radiation reaction in electron–beam interactions with high-intensity lasers. *Reviews of Modern Plasma Physics, 4*, 1.

Bulanov, S. V., Esirkepov, T. Zh., Koga, J., & Tajima, T. (2004). Interaction of electromagnetic waves with plasma in the radiation-dominated regime. *Plasma Physics Reports, 30*, 196.

Bulanov, S. V., Echkina, E. Y., Esirkepov, T. Zh., Inovenkov, I. N., Kando, M., Pegoraro, F., & Korn, G. (2010a). Unlimited ion acceleration by radiation pressure. *Physical Review Letters, 104*, 135003.

Bulanov, S. S., Mur, V. D., Narozhny, N. B., Nees, J., & Popov, V. S. (2010b). Multiple colliding electromagnetic pulses: A way to lower the threshold of pair production from vacuum. *Physical Review Letters, 104*, 220404.

Bulanov, S. V., Esirkepov, T. Z., Kando, M., Koga, J. K., Pirozhkov, A. S., Rosanov, N. N., & Zhidkov, A. G. (2011a). Relativistic tennis with photons: Frequency up-shifting, light intensification and ion acceleration with flying mirrors. *AIP Conference Proceedings, 1320*, 235.

Bulanov, S. V., Esirkepov, T. Z., Kando, M., Koga, J. K., & Bulanov, S. S. (2011b). Lorentz-Abraham-Dirac versus Landau-Lifshitz radiation friction force in the ultrarelativistic electron interaction with electromagnetic wave (exact solutions). *Physical Review E, 84*, 056605.

Bulanov, S. V., Yogo, A., Esirkepov, T. Zh., Koga, J. K., Bulanov, S. S., Kondo, K., & Kando, M. (2015). Stochastic regimes in the driven oscillator with a step-like nonlinearity. *Physics of Plasmas, 22*, 063108.

Bulanov, S. V. (2021). Electron dynamics in the field of strong plasma and electromagnetic waves: A review. *Physics of Wave Phenomena, 29*, 1.

Bulanov, S. V., Grittani, G. M., Shaisultanov, R., Esirkepov, T. Z., Ridgers, C. P., Bulanov, S. S., Russell, B. K., & Thomas, A. G. R. (2024). On the energy spectrum evolution of electrons undergoing radiation cooling. *Fundamental Plasma Physics, 9*, Article 100036.

Cole, J. M., Behm, K. T., Gerstmayr, E., Blackburn, T. G., Wood, J. C., & Baird, C. D. (2018). Experimental evidence of radiation reaction in the collision of a high-intensity laser pulse with a laser-wakefield accelerated electron beam. *Physical Review X, 8*, Article 011020.

Esirkepov, T. Zh., & Bulanov, S. V. (2017). Paradoxical stabilization of forced oscillations by strong nonlinear friction. *Physics Letters A, 381*, 2559.

Gonoskov, A., Bashinov, A., Bastrakov, S., Efimenko, E., Ilderton, A., Kim, A., Marklund, M., Meyerov, A., Mironov, I., & Sergeev, A. M. (2017). Ultrabright GeV photon source via controlled electromagnetic cascades in laser-dipole waves. *Physical Review X, 7*, Article 041003.

Gonoskov, A., Bashinov, A., Gonoskov, I., Harvey, C., Ilderton, A., Kim, A., Marklund, M., Mourou, G., & Sergeev, A. M. (2014). Anomalous radiative trapping in laser fields of extreme intensity. *Physical Review Letters, 113*, Article 014801.

Gonoskov, A., Blackburn, T. G., Marklund, M., & Bulanov, S. S. (2022). Charged particle motion and radiation in strong electromagnetic fields. *Reviews of Modern Physics, 94*, Article 045001.
Hadjisolomou, P., Jeong, T. M., Valenta, P., Korn, G., & Bulanov, S. V. (2021). Gamma-ray flash generation in irradiating a thin foil target by a single-cycle tightly focused extreme power laser pulse. *Physical Review E, 104*, 015203.
Hadjisolomou, P., Jeong, T. M., Valenta, P., Kolenatý, D., Versaci, R., Olšovcová, V., & Bulanov, S. V. (2022). Gamma-ray flash in the interaction of a tightly focused single-cycle ultra-intense laser pulse with a solid target. *Journal of Plasma Physics, 88*, 905880104.
Ibragimov, N. H. (2010). *A practical course in differential equations and mathematical modelling: Classical and new methods, nonlinear mathematical models, symmetry and invariance principles.* World Sci.
Jirka, M., Klimo, O., Bulanov, S. V., Esirkepov, T. Zh., Gelfer, E., Bulanov, S. S., Weber, S., & Korn. (2016). Electron dynamics and γ and e−e+ production by colliding laser pulses. *Physical Review E, 93*, 023207.
Jirka, M., Sasorov, P. V., & Bulanov, S. V. (2022). On the new and old physics in the interaction of a radiating electron with the extreme electromagnetic field. *Physical Review D, 105*, Article 113004.
Koga, J., Esirkepov, T. Zh., & Bulanov, S. V. (2006). Nonlinear Thomson scattering in the strong radiation damping regime. *Physics of Plasmas, 12*, 093106.
Kolenatý, D., Hadjisolomou, P., Versaci, R., Jeong, T. M., Valenta, P., Olšovcová, V., & Bulanov, S. V. (2022). Electron-positron pairs and radioactive nuclei production by irradiation of high-Z target with gamma-photon flash generated by an ultra-intense laser in the lambda-cube regime. *Physical Review Research, 4*, Article 023124. https://doi.org/10.1103/PhysRevResearch.4.023124
Landau, L. D., & Lifshitz, E. M. (1980). *Course of theoretical physics, Vol. 2: The classical theory of fields.* Pergamon.
Lezhnin, K. V., Kamenets, F. F., Esirkepov, T. Zh., & Bulanov, S. V. (2018). On annihilation of the relativistic electron vortex pair in collisionless plasmas. *Journal of Plasma Physics, 84*, 905840610.
Los E., Arran, C., Gerstmayr, E., et al. (2025). A Bayesian framework to investigate radiation reaction in strong fields. *High Power Laser Science and Engineering, 13*, e25.
Magnusson, J., Gonoskov, A., Marklund, M., Esirkepov, T. Z., Koga, J. K., Kondo, K., Kando, M., Bulanov, S. V., Korn, G. & Bulanov, S. S. (2019). Laser-particle collider for multi-GeV photon production. *Physical Review Letters, 122*, 254801.
Nakamura, T., Koga, J. K., Esirkepov, T. Z., Kando, M., Korn, G., & Bulanov, S. V. (2012). High-power γ-ray flash generation in ultraintense laser-plasma interactions. *Physical Review Letters, 108*, Article 195001.
Neitz, N., & Di Piazza, A.. (2013). Stochasticity effects in quantum radiation reaction. *Physical Review Letters, 111*, 054802.
Niel, F., Riconda, C., Amiranoff, F., Duclous, R., & Grech, M. (2018). From quantum to classical modeling of radiation reaction: a focus on stochasticity effects. *Physical Review E, 97*, 043209.
Poder, K., Tamburini, M., Sarri, G., Di Piazza, A., Kuschel, S., Baird, C. D., Behm, K., Bohlen, S., Cole, J. M., Corvan, D. J., Duff, M., Gerstmayr, E., Keitel, C. H., Krushelnick, K., Mangles, S. P. D., McKenna, P., Murphy, C. D., Najmudin, Z., Ridgers, C. P., … Zepf, M. (2018). Experimental signatures of the quantum nature of radiation reaction in the field of an ultraintense laser. *Physical Review X, 8*, Article 031004.
Pomeranchuk, I. (1940). On the maximum energy which the primary electrons of cosmic rays can have on the earth's surface due to radiation in the earth's magnetic field. *Journal of Physics, 2*, 65.
Ridgers, C. P., Brady, C. S., Duclous, R., Kirk, J. G., Bennett, K., Arber, T. D., Robinson, A. P. L., & Bell, A. R. (2012). Dense electron-positron plasmas and ultraintense γ rays from laser-irradiated solids. *Physical Review Letters, 108*, Article 165006.
Ridgers, C. P., Blackburn, T. G., Del Sorbo, D., Bradley, L. E., Slade-Lowther, C., Baird, C. D., Mangles, S. P. D., McKenna, P., Marklund, M., Murphy, C. D., & Thomas, A. G. R. (2017).

Signatures of quantum effects on radiation reaction in laser-electron-beam collisions. *Journal of Plasma Physics, 83*, 715830502.

Sokolov, I. V., Naumova, N. M., Nees, J. A., & Mourou, G. (2010). Pair creation in QED-strong pulsed laser fields interacting with electron beams. *Physical Review Letters, 105*, Article 195005.

Telnov, V. (1997). Laser cooling of electron beams for linear colliders. *Physical Review Letters, 78*, 4757.

Vranic, M., Grismayer, T., Fonseca, R. A., & Silva, L. O. (2016). Electron–positron cascades in multiple-laser optical traps. *Plasma Physics and Controlled Fusion, 59*, Article 014040.

Zel'dovich, Ya. B. (1975). Interaction of free electrons with electromagnetic radiation. *Soviet Physics Uspekhi, 18*, 79.

Zhang, S. (2013). Pre-acceleration from Landau–Lifshitz series. *Progress of Theoretical and Experimental Physics, 2013*, 123A01.

Zhidkov, A. G., Koga, J., Sasaki, A., & Uesaka, M. (2002). Radiation damping effects on the interaction of ultraintense laser pulses with an overdense plasma. *Physical Review Letters, 88*, Article 185002.

Zhidkov, A. G., Masuda, S., Bulanov, S. S., Koga, J., Hosokai, T., & Kodama, R. (2014). Radiation reaction effects in cascade scattering of intense, tightly focused laser pulses by relativistic electrons: Classical approach. *Physical Review STAB, 17*, Article 054001.

Chapter 24
Laser Electron Accelerators

Abstract Here we discuss the electron acceleration by the laser pulse interacting with underdense plasma target via generation of plasma wake waves propagating with relativistic velocity.

Keywords Laser wake field · LWFA · Electron injection to accelerating phase of wake wave · Energy scaling

The development of collective methods for the acceleration of charged particles is among the most attractive fields concerning the relativistic laser plasma. The advantages of collective methods for acceleration are known since the mid-1950s, when were proposed several schemes of plasma accelerators of particles. It is known that the electric field in standard accelerators of charged particles is limited by the threshold of the electrical breakdown developed on the walls of a chamber; for this reason, accelerators should be large (the length of the Large Hadron Collider at CERN is 27 km). Such a limitation is absent for collective electric fields in the plasma; consequently, their dimension can be many orders of magnitude smaller.

In the laser methods of electron acceleration, the mechanism of acceleration involving the electric field of wake plasma waves proposed by Tajima and Dawson (1979) (see also review Esarey et al., 2009) is the most developed. In this method, a short laser pulse excites a wake plasma wave in the plasma whose phase velocity $v_{ph} = v_w$ is equal to the group velocity v_g of the electromagnetic pulse. The development of the theory and experiment has resulted in demonstration of the 8 GeV electrons accelerated with 1 PW short pulse laser guided by the capillary discharge channel over the distance of 20 cm (Gonsalves et al., 2019), and 9.2 GeV electrons accelerated by the laser guided inside the plasma channel of 30 cm length (Picksley et al., 2024). Another approach is based on using the relativistic electron beam for driving the plasma wake wave accelerating the witness electron bunch (Chen et al., 1985). The progress in charged particles acceleration resulted in the occurrence of the scientific direction aimed at the providing compact free electron laser based on the using of the electrons accelerated by the plasma waves excited either by the laser

Y. Kato et al., *High Power Laser and Plasma Science*, Springer Series on Atomic, Optical, and Plasma Physics 130, https://doi.org/10.1007/978-3-032-08930-4_24

or by the electron beam-driver or by both the means (see Grüner et al., (2007) and review article Galetti et al., (2024) and references there in).

24.1 Basic Parameters of the Wake Field

Although the wake wave generation by a finite length laser pulse characterized by the function $a(X)$, can be described by equation obtained above, for the sake of simplicity it is convenient to re-derive this result. Assuming that the ions are at rest and the electron temperature is equal to zero in the one-dimensional approximation the wake wave driven by a given electromagnetic pulse can be written as a system of equations in partial derivatives:

$$\partial_t n + \partial_x (nv) = 0, \tag{24.1}$$

$$\partial_t p + v\partial_x p = -E - \frac{1}{2\gamma}\partial_x |\boldsymbol{a}|^2, \tag{24.2}$$

$$\partial_x E = 1 - n. \tag{24.3}$$

Here we use the normalized variables with the wake wave electric field $\boldsymbol{E} = E\boldsymbol{e}_x$ measured in the units of $m_e\omega_{pe}c/e$. The transverse electromagnetic pulse is characterized by its vector potential $\boldsymbol{A}(x,t)$; being normalized $m_e c^2/e$ it is $\boldsymbol{a}(x,t)$. In the 1D geometry due to the homogeneity of the problem along the transverse directions the transverse component of the generalized electron momentum, $\boldsymbol{p}_\perp - \boldsymbol{a} =$ constant. Using the generalized transverse momentum conservation, we obtain that the electron relativistic Lorentz factor is given by

$$\gamma = \sqrt{1 + |\boldsymbol{a}|^2 + p^2}. \tag{24.4}$$

The electrostatic potential φ (it is normalized by $m_e c^2/e$) and the electric field E are related to each other by $E = -\partial_x \varphi$. For the sake of simplicity here and below we assume that the electromagnetic wave is circularly polarized with $|\boldsymbol{a}| = a_0$.

In the case of the driver pulse and wake wave propagating with a constant phase velocity $v_w = \beta_w$ all the functions in Eqs. (24.1)–(24.3) depend on the variable

$$X = x - \beta_w t. \tag{24.5}$$

As a result Eqs. (24.1)–(24.3) are reduced to ordinary differential equations:

$$(\gamma - \beta_w p)' = -E, \tag{24.6}$$

$$E' = \frac{v}{\beta_w - v}. \tag{24.7}$$

Here a prime denotes a differentiation with respect to the variable X defined by Eq. (24.5).

Expressing the electric field via the electrostatic potential φ as $E = -\varphi'$ and using Eqs. (24.4), (24.6) and (24.7), we find the equation for the electrostatic potential, which has the form

$$\varphi'' = \beta_w \gamma_w^2 \left(\frac{1+\varphi}{\sqrt{(1+\varphi)^2 - (1+a_0^2)\gamma_w^{-2}}} - \frac{1}{\beta_w} \right), \tag{24.8}$$

where $\gamma_w = 1/\sqrt{1-\beta_w^2}$.

We assume that the driver pulse has a constant amplitude, $|a_0| =$ constant for $X < 0$ and its amplitude is equal to zero for $X > 0$. Multiplying Eq. (24.8) by φ' and integrating over X we obtain the integral

$$\frac{1}{2}(\varphi')^2 = \beta_w \gamma_w^2 \left(\sqrt{(1+\varphi)^2 - (1+a_0^2)\gamma_w^{-2}} - \frac{\varphi}{\beta_w} \right) + C \tag{24.9}$$

Here C is a constant determined by the boundary conditions. If at the driver pulse front $X = 0$ the potential $\varphi(0)$ and the electric field, $E(0) = -\varphi'(0)$, are equal to zero, the integration constant is given by

$$C = -\beta_w \gamma_w^2 \sqrt{1 - (1+a_0^2)\gamma_w^{-2}}. \tag{24.10}$$

The electric field and electrostatic potential dependence on the coordinate X are shown in Fig. 24.1a for the wake wave generated by a flat top driver pulse with the amplitude $a_0 = 5$ in a plasma with $\beta_w = 0.9999$ ($\gamma_w = 70$). The electrostatic potential has a maximum at the point where the $\varphi' = 0$, i. e. where the electric field vanishes. Equations (24.9) and (24.10) yield for the maximum potential

$$\varphi_{\max} = 2\beta_w \gamma_w^2 \left(\beta_w - \sqrt{1 - (1+a_0^2)\gamma_w^{-2}} \right), \tag{24.11}$$

which, in the limit $a_0 \ll 1$, is given by $\varphi_{\max} = a_0^2/\beta_w$ and cannot exceed the value $2\beta_w^2\gamma_w^2$ reached for the driver laser amplitude equal to

$$|a_0| = \beta_w \gamma_w. \tag{24.12}$$

At the minimum the electrostatic potential vanishes, $\varphi_{\min} = 0$.

If the driver pulse amplitude is above $\beta_w \gamma_w$, the wake wave breaks. Above the wave breaking threshold the wave cannot be described within the framework of the stationary wave approximation. The electric field maximum corresponds to the

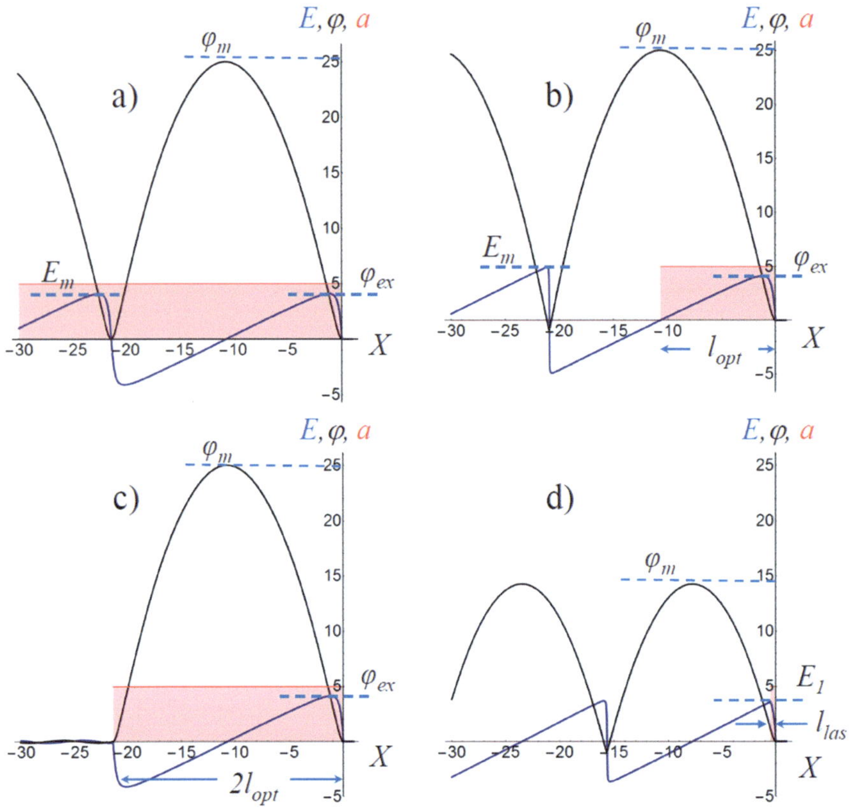

Fig. 24.1 Electric field and electrostatic potential of the wake wave generated **a** by a semi-infinite flat top driver pulse with the amplitude $a_0 = 5$ in a plasma with $\beta_w = 0.9999$ ($\gamma_w = 70$); **b** by a flat top driver pulse (indicated as $a(X)$) of the optimal length l_{opt} with the amplitude $a_0 = 5$ in a plasma with β_w=1, ($l_{las} = 10.6969$); **c** by a wakeless driver pulse (indicated as $a(X)$) of the double optimal length, $l_{las} = 2l_{opt}$ with the amplitude $a_0 = 5$ in a plasma with β_w=1, ($l_{las} = 21.3939$); **d** by an ultra-short driver pulse (indicated as $a(X)$) with the amplitude $a = 5$ and length $l_{las} = 0.5$, i.e. $l_{las} = 0.04674\ l_{opt}$, in a plasma with β_w=1 (Bulanov et al., 2016)

extremum of the second derivative of electrostatic potential, i. e. when the right-hand side of Eq. (24.8) vanishes. This condition gives an expression for the electrostatic potential at the electric field extremum:

$$\varphi_{ex} = \pm\sqrt{1 + a_0^2} - 1. \tag{24.13}$$

Substitution of this expression into Eq. (24.9) with the constant C given by Eq. (24.10) results in

$$|E_\mathrm{m}| = \sqrt{\gamma_w^2 \mp \sqrt{1 + a_0^2} - \beta_w\gamma_w\sqrt{\gamma_w^2 - (1 + a_0^2)}}. \tag{24.14}$$

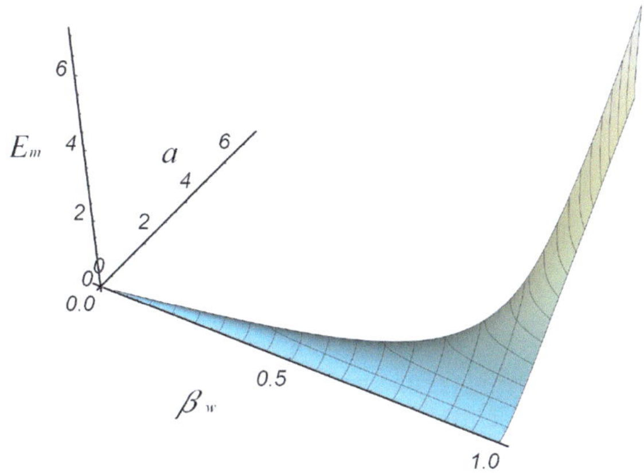

Fig. 24.2 Maximum electric field dependence on the driver pulse amplitude a_0 and wake wave phase velocity β_w (Bulanov et al., 2016)

For small laser pulse amplitude, when $a_0 \ll 1$, the maximum electric field is proportional to the square of the driver pulse amplitude:

$$|E_{\rm m}| \approx a_0^2/2\beta_w. \tag{24.15}$$

Figure 24.2 shows the maximum electric field dependence on the driver pulse amplitude a and the wake wave phase velocity β_w. For a given phase velocity of the wake wave the stationary wave can exist provided the driver amplitude is less than $a_{\rm lim} = \sqrt{\gamma_w^2 - 1}$.

The upper limit on the electric field in a stationary wake wave inside the laser pulse corresponds to the condition given by Eq. (24.12). This yields

$$\max[|E_{\rm m}|] = \gamma_w\sqrt{1 - \gamma_w^{-1}}. \tag{24.16}$$

As for the wake wave left in a plasma the amplitude of the laser driver pulse with optimal duration cannot exceed the Akhiezer-Polovin wave breaking limit (Akhiezer & Polovin, 1956)

$$E_{A-P} = \sqrt{2(\gamma_w - 1)}. \tag{24.17}$$

24.2 Optimal Laser Pulse Length

In the case of a cold plasma, when the front of the driver pulse propagates with the velocity equal to the speed of light in vacuum, which corresponds to the limit $\beta_w \to 1$ (i.e. $\gamma_w \to \infty$), Eq. (24.8) takes the form

$$\varphi'' = \frac{1}{2}\left(\frac{1+a_0^2}{(1+\varphi)^2} - 1\right). \tag{24.18}$$

The integral (24.9), in the case of constant amplitude driver pulse, for the boundary conditions $\varphi(0) = 0$ and $\varphi'(0) = 0$ has the form

$$(\varphi')^2 = \frac{\varphi\left(a_0^2 - \varphi\right)}{1+\varphi}. \tag{24.19}$$

We see that the electrostatic potential value varies between the minimum equal to 0 and the maximum, which equals a_0^2. The maximal value of the electric field is reached at the point of the electrostatic potential extremum where its second derivative vanishes. Using Eq. (24.18) we find that at the electric field maximum the electrostatic potential is

$$\varphi_{ex} = \sqrt{1+a_0^2} - 1. \tag{24.20}$$

Using this expression we obtain for the electric field maximum,

$$|E_{\mathrm{m}}| = \sqrt{1+a_0^2} - 1. \tag{24.21}$$

Solution of Eq. (24.18) for the potential can be expressed in terms of elliptic functions. The dependence of the electrostatic potential φ on the coordinate X can be written in the implicit form:

$$X = 2\left[\sqrt{\frac{(a_0^2 - \varphi)(1+\varphi)}{\varphi}} + \mathrm{E}\left(\sin\frac{1}{\sqrt{\varphi}}\middle| -a_0^2\right)\right]. \tag{24.22}$$

Here $\mathrm{E}(\Psi|\kappa)$ is the elliptic integrals of the second kind.

The laser pulse is optimal for the excitation of the large amplitude wake field left behind it, if at the rear side of the electrostatic potential is maximal. From Eq. (24.19) follows the expression for the optimal pulse length, l_{opt}:

$$l_{opt} = \int_0^{a_0^2} \sqrt{\frac{1+\varphi}{\varphi\left(a_0^2 - \varphi\right)}}\, d\varphi = 2\,\mathrm{E}(-a_0^2), \tag{24.23}$$

where $\mathrm{E}(\kappa)$ is the complete elliptical integral of the second kind.

The wake wave wavelength equals $\lambda_w = 4l_{opt}$ in the limit of small laser amplitude $a_0 \ll 1$ and it is about $2l_{opt}$ for $a_0 \gg 1$.

In the small amplitude limit, $a_0 \ll 1$, the optimal laser length is

$$l_{opt} = \pi + \frac{\pi}{4} a_0^2 - \frac{3\pi}{64} a_0^4 + O\left[a_0^2\right]^3, \tag{24.24}$$

and the wake wave wavelength in dimensional units equals $\lambda_w = 2\pi c/\omega_{pe}$.

If the driver amplitude equals unity, $a_0 = 1$, the optimal laser length is 3.8202. For large laser amplitude, $a_0 \gg 1$, we have

$$l_{opt} = 2a_0 + \left(\frac{1}{2} + 2\ln 2 - \ln a_0\right)\frac{1}{a_0} + O\left[\frac{1}{a_0}\right]^3 \tag{24.25}$$

with the wake wave wavelength in dimensional units equal to $\lambda_w = 4ca_0/\omega_{pe}$.

In Fig. 24.3 we plot the dependence of the optimal laser length on the pulse amplitude.

The electric field and electrostatic potential dependence on the coordinate X are shown in Fig. 24.1b, for the wake wave generated by a flat top driver pulse of the optimal length l_{opt}. The amplitude is $a_0 = 5$. Equation (24.23) for this laser amplitude gives $l_{opt} = 10.6969$. The plasma is characterized by $\gamma_w = 70$.

The wake wave left behind the driver laser pulse is described by Eq. (24.18) with $a_0^2 = 0$. For the boundary conditions $\varphi(-l_{opt}) = a_0^2$ and $\varphi'(-l_{opt}) = 0$ the equation for the electric field, $E = -\varphi'$, yields

$$E^2 = a_0^2 - \varphi - \frac{a_0^2 - \varphi}{(1 + a_0^2)(1 + \varphi)}. \tag{24.26}$$

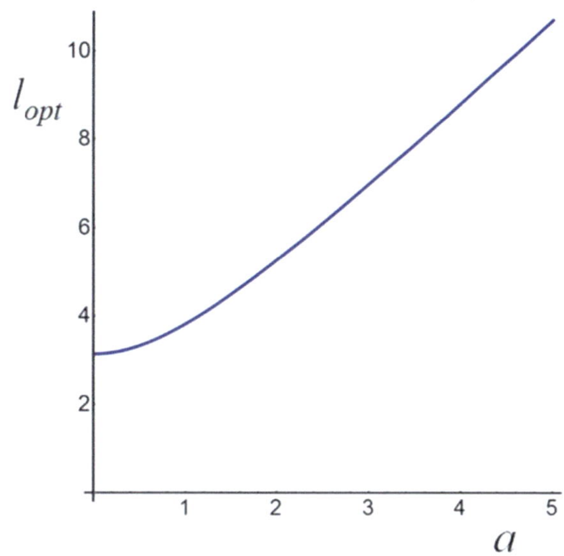

Fig. 24.3 Optimal laser length l_{opt} versus the driver pulse amplitude a_0

From this equation it follows that the electrostatic potential in the wake wave varies between maximal and minimal values at the points where $E = -\varphi' = 0$:

$$-\frac{a_0^2}{1+a_0^2} < \varphi < a_0^2. \tag{24.27}$$

The electric field maximum is determined by the condition $\varphi'' = 0$, which gives the extremum of the electrostatic potential derivative reached at the points where $\varphi = 0$. From Eq. (24.26) it follows that the electric field maximum equals

$$E_m = \frac{a_0^2}{\sqrt{1+a_0^2}}. \tag{24.28}$$

It is larger than the maximum of the wake field inside the driver laser pulse given by Eq. (24.21) as it is also seen in Fig. 24.1b.

If the laser pulse has double the optimal length, $l_{las} = 2\, l_{opt}$, then the wake wave left behind the laser pulse vanishes with the electric field and electrostatic potential localized only inside the laser pulse as illustrated in Fig. 24.1c, where the electric field and electrostatic potential of the wake wave are generated by a wakeless driver pulse.

Formally a laser pulse of double the optimal length does not lose energy in the wake wave generation because the energy lost at the front returns back at the rear side of the pulse. Actually, the laser pulse etching (see consideration of this effect below) and the laser pulse self-modulation will result in the change of the laser pulse length, amplitude and form, eventually leading to appearance of a final amplitude wake-wave in the plasma behind the laser pulse and to the laser pulse energy depletion.

24.3 Propagation Velocity of the Laser Pulse Front

In the expressions obtained above the phase velocity of the wake wave, β_w, is equal to the propagation velocity of the laser pulse front. Apart from its dependence on the group velocity of the electromagnetic wave determined by the plasma density, it is also determined by the process of the laser pulse energy depletion, which results in the change of the laser pulse form manifesting itself in the formation of the shock wave like front and in the carrier frequency down-shift. The laser energy depletion associated with the so-called etching of the pulse, makes the pulse front velocity β_D become smaller than the group velocity of the laser radiation β_D. This process is illustrated in Fig. 24.4.

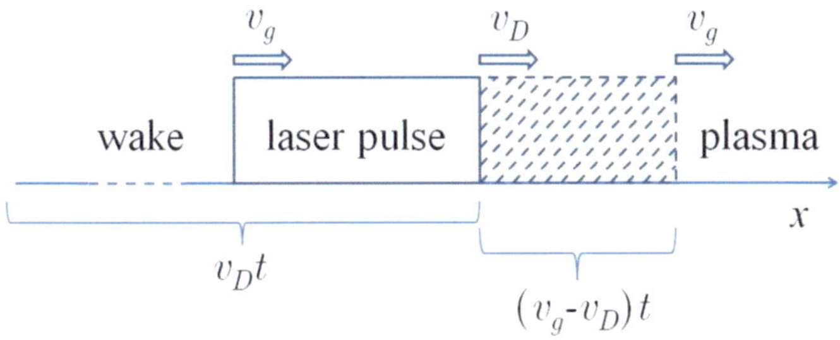

Fig. 24.4 Laser energy depletion leads to the laser pulse shortening $l_{las} = l_{las,0} - (v_g - v_D)t$ and to slower propagation of the pulse front with the velocity $v_D < v_g$. The shaded region shows the damped part of the pulse (Bulanov et al., 2016)

In order to find the laser pulse front velocity we should take into account the balance between the lost laser energy

$$\mathcal{E}_{las} = \frac{m_e^2 \omega^2 c^2 a_0^2}{4\pi e^2} (v_g - v_D)t \tag{24.29}$$

and the wake wave energy

$$\mathcal{E}_w = m_e c^2 n a_0^2 v_D t. \tag{24.30}$$

This gives the relationship between the group velocity of the laser pulse and the velocity of propagation of the pulse front, which in normalized units reads:

$$\beta_D = \frac{\beta_g}{1 + (\omega_{pe}/\omega)^2}. \tag{24.31}$$

The group velocity β_g of the relativistically intense electromagnetic wave depends on its amplitude: associated with the group velocity Lorentz factor in the case of a circularly polarized wave

$$\gamma_g = \frac{1}{\sqrt{1 - \beta_g^2}} = \left(\frac{\omega}{\omega_{pe}}\right)\left(1 + a_0^2\right)^{1/4}. \tag{24.32}$$

In the limit $a \gg 1$ it is proportional to the square root of the wave amplitude. This dependence has important implications for determining the wake wave breaking threshold, which in its turn determines the threshold of the electron self-injection to the wake field. Due to the laser energy depletion the laser pulse length decreases as

$$l_{las} = l_{las,0} - (v_g - v_D)t, \tag{24.33}$$

where $l_{las,0}$ is the initial pulse length. From this expression and Eqs. (24.30) and (24.31) follows that the depletion length is equal to

$$l_{dep} = \frac{l_{las,0}}{\beta_g - \beta_D} \approx l_{las,0}\left(\frac{\omega}{\omega_{pe}}\right)^2. \tag{24.34}$$

We note that the laser pulse with the length $l_{las,0}$ substantially larger than the wavelength of the wake wave is unstable against various instabilities. As a result the self-modulated wake wave regime occurs, which has been studied in experiments on short but not too short laser pulse interaction with underdense plasma targets. In the self-modulated wake field regime, the depletion length is shorter than the length given by expression (24.23), in which we should instead $l_{las,0}$ use the wavelength of the wake wave $4\,\mathrm{E}(-a_0^2)$.

24.4 Electron Trapping into a Wake Wave

In the laser wake-field accelerator concept introduced by Tajima and Dawson (1979), a long-living wake-field, induced by a short intense laser pulse in its wake in a low-density collisionless plasma, accelerates duly injected electrons. To provide electrons, one must use an externally pre-accelerated electron bunch or exploit the effect of self-injection due to a longitudinal Langmuir wave break in homogeneous and/or in the inhomogeneous plasma targets and/or a transverse wave break (Brantov et al., 2008; Buck et al., 2013; Bulanov et al., 1992, 1997a, 1997b, 1998; Geddes et al., 2008; Suk et al., 2001). The optical injection mechanism can also provide pre-accelerated electron beams with the required properties (Faure et al., 2006; Kotaki et al., 2004, 2009; Umstadter et al., 1996). The two-color ionization injection scheme (Yu et al., 2014) and an external static magnetic field (Hosokai et al., 2006) enable the control of the properties of the injected/accelerated electron beams.

Electrons immersed in the wake field oscillate in a regular way. In a plasma with the electron density well below the critical one $n_e \ll n_{cr} = m_e\omega^2/4\pi e^2$, the wake wave phase velocity $v_w = c\beta_w$ is close to the speed of light in vacuum, which corresponds to a large value of the relativistic Lorentz factor, γ_w. The group velocity of the wake wave is equal to zero and the phase velocity v_w is equal to the velocity at which the laser pulse front propagates. The relation between the electromagnetic pulse wavelength λ and the wavelength λ_w of the wake field is given by

$$\lambda_w = \lambda\frac{\sqrt{\gamma_w}}{\pi}\,\mathrm{E}(-a_0^2). \tag{24.35}$$

24.5 Effect of Plasma Inhomogeneity

Plasma inhomogeneity effects can lower the phase velocity of the wake wave left behind the driver laser pulse thus lowering the injection energy threshold. Within the framework of the approximation assuming that the wake wave is below the wave breaking threshold and assuming, in particular, that the electron displacement inside the wake is substantially less than the scale length of the plasma density inhomogeneity, we can write the phase of the wake wave in the form

$$\Psi(x,t) = \omega_{pe}(x)\left[\int_0^x \frac{dx}{v_g(x)} - t\right] = \omega_{pe}(x)\int_{x_{las}(t)}^x \frac{dx}{v_g(x)}. \tag{24.36}$$

Here $\omega_{pe}(x)$ and $v_g(x)$ are the local values of the Langmuir frequency and the laser pulse group velocity, respectively. The group velocity is equal to

$$v_g(x) = \frac{kc^2}{\sqrt{k^2c^2 + \omega_{pe}^2(x)}}, \tag{24.37}$$

The function $x_{las}(t)$ is the laser pulse coordinate at time t given by the equation

$$\int_{x_{las}(t)}^x \frac{dx}{v_g(x)} = t. \tag{24.38}$$

In a homogeneous plasma the phase given by Eq. (24.36) depends on the variable X as

$$\Psi(x,t) = -\frac{\omega_{pe}X}{v_g}. \tag{24.39}$$

In an inhomogeneous plasma the local wave number of the wake wave is the gradient of the phase function, $k_w = \partial_x\Psi$, and the local frequency is proportional to its rate of change, $\omega_w = -\partial_t\Psi$. Differentiating the phase Ψ, given by Eq. (24.36), with respect to time and coordinate we find the frequency and wave number of the wake wave:

$$\omega_w = \omega_{pe}(x) \tag{24.40}$$

and

$$k_w = \frac{\omega_{pe}(x)}{v_g(x)} + \partial_x\omega_{pe}(x)\int_x^{x_{las}(t)} \frac{dx}{v_g(x)}, \tag{24.41}$$

respectively.

In a down-ramp density plasma, where the coordinate derivative of the Langmuir frequency is negative, the wave number $k_w(x)$ increases with the growth of the distance between the laser pulse and the point with the coordinate x. This corresponds to decreasing of the wake wave phase velocity, defined as x

$$v_w = \frac{\omega_w}{k_w} = \frac{v_g(x)}{1 + v_g(x)\frac{\partial_x \omega_{pe}(x)}{\omega_{pe}(x)} \int\limits_x^{x_{las}(t)} \frac{dx}{v_g(x)}}, \tag{24.42}$$

i.e. lowering the injection threshold. Eventually the wake wave. The change of the wake wave wavelength, $\lambda_w(x) = 2\pi/k_w(x)$, leads to a change of the wake wave crest position, where the electron injection occurs. The position of the wake wave crest is marked by the value of the phase Ψ. Using Eqs. (24.36) and (24.42) we can find the velocity of the point of constant phase, v_Ψ, and to calculate the corresponding gamma factor $\gamma_\Psi = 1/\sqrt{1-\beta_\Psi^2}$ with $\beta_\Psi = v_\Psi/c$, which can be written in the limit of low density plasma, $n \ll n_{cr}$, as x

$$\gamma_\Psi = \frac{1}{\sqrt{\left(\frac{\omega_{pe}}{\omega}\right)^2 + \Psi \frac{2c}{\omega_{pe} L}}}. \tag{24.43}$$

Here we make use of the fact that the local value of the laser pulse group velocity is given by $v_g = c\sqrt{1-\omega_{pe}^2(x)/\omega^2}$ and introduce the plasma inhomogeneity scale-length $L = |\omega_{pe}/\partial_x \omega_{pe}|$.

As we can see, the plasma inhomogeneity effects are important provided the plasma inhomogeneity scale length is small enough and/or the wake wave phase is sufficiently large: $L < 2\Psi c\omega^2/\omega_{pe}^3$.

Figure 24.5 illustrates the above discussed effects. Plots are drawn in arbitrary units. Figure 24.4a shows the electron density profile approximated by the expression

$$n_e(x) = \frac{n_1 + n_2}{2} - \frac{n_1 - n_2}{2} \tanh\left(\frac{x}{L}\right), \tag{24.44}$$

where n_1 and n_2 are values of the electron density at $x \to \pm\infty$, respectively. In the plot (a) $n_1 = 1.5$, $n_2 = 1$, and $L = 5$. Figure 24.5b presents the trajectories of the constant phase points in the (x, t) plane. As we see, the distance between the trajectories in the region of the plasma inhomogeneity becomes smaller the larger the time t is. In Fig. 24.5c we see the phase dependence on the coordinate x at $t = t_m$ for $t_m = 125$ with decreasing scale of the inhomogeneity in the region in the vicinity of the coordinate $x = 0$. Figure 24.5d presents the profile of the wake wave phase velocity v_w at $t = 0,\ 0.125t_m,\ 0.25t_m,\ 0.5t_m,\ 0.75t_m,\ t_m$. Within the region of the plasma inhomogeneity the wake wave phase velocity at first for $x < 0$ gradually decreases and then, where $x > 0$, gradually increases. Within the interval where

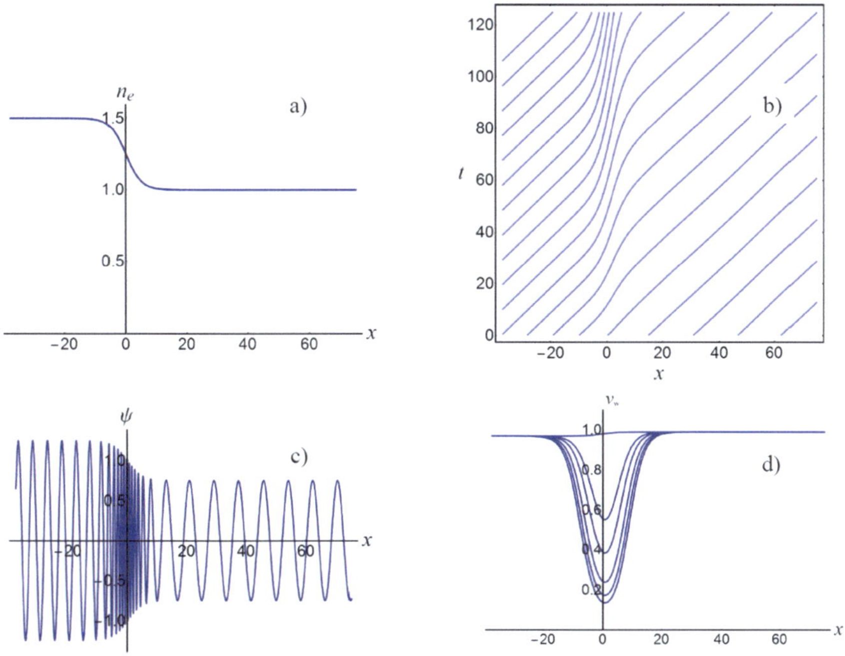

Fig. 24.5 Plasma inhomogeneity effects on the phase velocity of the wake wave. **a** Electron density profile given by Eq. (24.44) with $n_1 = 1.5$, $n_2 = 1$, and $L = 5$. **b** The trajectories of the constant phase points in the (x, t) plane. **c** The phase dependence on the coordinate x at $t = t_m$ for $t_m = 125$. **d** The wake wave phase velocity v_w versus coordinate x at $t = 0,\ 0.125t_m,\ 0.25t_m,\ 0.5t_m,\ 0.75t_m,\ t_m$. Plots are drawn in arbitrary units (Bulanov et al., 2016)

the phase velocity becomes lower than the electron quiver velocity the wake wave breaks resulting in the electron trapping. Then the electrons moving together with the wave appear in the region of high phase velocity. During the injection process, the electrons remain trapped by the wave.

24.6 Electron Injection at the Interface Between Two Plasma Regions

When the plasma density changes abruptly at the interface between two plasmas the injection mechanism is different from the one considered above. We consider the plasma configuration shown in Fig. 24.6 with the plasma density changing from n_1 at $x \to -\infty$ to n_2 for $x \to +\infty$. At the border between these two plasmas the electrons with the momentum equal to p_1 appear inside the wake wave generated by the laser pulse in the lower plasma density region. The momentum p_1 equals either

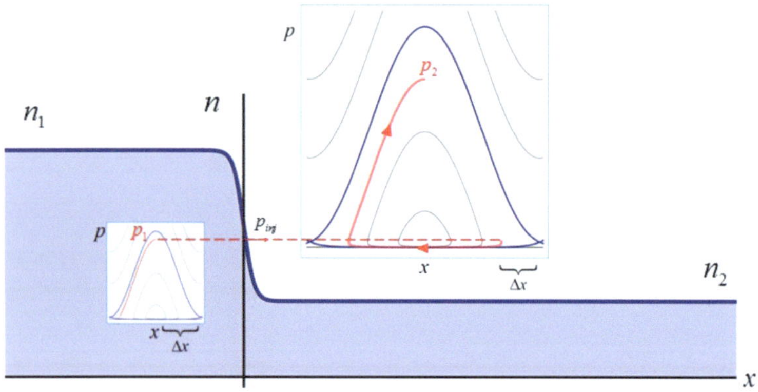

Fig. 24.6 The electron bunch injection at the interface between two regions with different plasma density. In the plasma region on the left-hand side with the density n_1 the LWFA electron bunch is pre-accelerated to the momentum p_1 reached at the interface between the two plasmas. In the plasma region on the right-hand side the density is substantially lower than in the region on the left-hand side, $n_2 \ll n_1$. Here the electron bunch occurs in the region in the phase plane (x, p) within the separatrix (provided $p_1 = p_{inj} = p_{w,2}$, with $p_{w,2} = m_e v_{w,2}/\sqrt{1 - v_{w,2}^2/c^2}$) close to the front of the wake wave bucket. Then the electron bunch is accelerated moving along the trajectory (red curve) close to the separatrix (Bulanov et al., 2016)

$\gamma_{w,1} = \omega/\omega_{pe,1}$ if they are pre-injected into the wake-field or it is about the quiver momentum determined by the wake-wave amplitude, if no pre-injection occurs, as in the case of the injection mechanism formulated by Bulanov et al. (1998). Here $\omega_{pe,1} = \sqrt{4\pi n_1 e^2/m_e}$.

For the sake of brevity we assume that the electrons are pre-injected, the laser pulse driver of optimal length has the amplitude equal to unity, and the length of the left-hand side plasma region is of the order of the acceleration length.

In the case under consideration, the electrons with maximum momentum are localized at the top of the separatrix in the phase plane (x, p) as shown in Fig. 24.6. The distance between the electron bunch and the front side of the wake wave δx is of the order of the laser pulse driver of optimal length given by Eq. (24.23). Since both the relativistic electron bunch and the laser pulse move with the velocity close to the speed of light, at the time, when they enter the plasma region on right hand side, the distance between them remains almost unchanged. As a result the electron bunch in the plasma region on left hand side appears to be relatively close to the front of the wake wave bucket. In order to be further accelerated the electron bunch should be inside the region in the phase plane enclosed by the separatrix. This is equivalent to the requirements $n_2 \ll n_1$ and $p_1 \approx p_{w,2}$, where the injection momentum $p_{inj} \approx p_{w,2}$ corresponds to the electrons with the energy equal to $\gamma_{w,2}$ calculated for the phase velocity of the wake-wave in the right hand side region.

If the requirements formulated above are fulfilled, the electron bunch being trapped inside the separatrix close the wake-wave front then moves along the trajectory in the phase plane shown by the red curve on the upper right side in Fig. 24.6. In the case, when $n_2 \ll n_1$ the energy of accelerated electrons can be of the order of $\gamma_{w,2}^2 \approx n_1/n_2$.

24.7 Wake Wave at the Threshold of Breaking

Equations (24.6) and (24.7) result in an equation for the longitudinal component of the electron momentum, p,

$$(\gamma_e - p\beta_w)'' = \frac{p}{\gamma_e\beta_w - p}, \tag{24.45}$$

where the electron gamma–factor, γ_e, is given by Eq. (24.4). The solution of this equation becomes singular when the denominator on the right-hand side vanishes. The singularity corresponds to the case, where the electron velocity p/γ_e becomes equal to the wave phase velocity β_w, which means wave breaking. In a stationary wave, the singularity occurs at the maximum of the electron velocity $v_m = p_m/\gamma_m$. We let X_m denote the wave-breaking coordinate.

To find the structure of the singularity, we expand the electron momentum and the gamma-factor near the singularity at $\delta X = X - X_m \to 0$, assuming $\delta p = p_m - p$ to be small. We write the electron momentum in the form

$$p = p_m + \delta p + O(\delta p^2), \tag{24.46}$$

where $p_m = \beta_w\sqrt{(1 + a_m^2)/(1 - \beta_w^2)}$, $\delta p/|p_m| \ll 1$. Here $a_m = a(X_m)$ is the vector potential value at the wave-breaking point.

Keeping the leading terms on both sides of Eq. (24.45), we obtain

$$\left(\delta p^2\right)'' = -\eta^2 \frac{1}{\delta p} \tag{24.47}$$

with $\eta^2 = 2\beta_w\gamma_w^2(1 + a_m^2)$. Multiplying the left- and right-hand sides of the equation by $\left(\delta p^2\right)'$ and integrating over X, we find

$$\delta p\left[2\,\delta p\left(\delta p'\right)^2 + \eta^2\right] = f^2, \tag{24.48}$$

where f is the integration constant. If the product $\delta p \delta p'$ tends to zero as $\delta p \to 0$, which requires the vanishing of the right-hand side of Eq. (24.48), then this equation with $f = 0$ implies that

$$\delta p = -\left(\frac{3}{2^{3/2}}\eta\delta X\right)^{2/3} = -\beta_w\gamma_w^3\left(\frac{3\sqrt{1+a_m^2}}{2\beta_w}\delta X\right)^{2/3}. \tag{24.49}$$

For the electron velocity, this then yields

$$v = \beta_w - \frac{\beta_w}{\gamma_w}\left[\frac{3\left(1+a_m^2\right)^{1/4}}{2\beta_w}\delta X\right]^{2/3}. \tag{24.50}$$

In catastrophe theory, this type of singularity is called the 'cusp catastrophe' [3]. In local coordinates, it can be written as the mapping

$$y_1 = x_1, \quad y_2 = x_2^3 + x_1 x_2. \tag{24.51}$$

Although the electron density tends to infinity in the vicinity of the singularity as

$$n = \frac{\beta_w}{\beta_w - v} \approx \gamma_w\left[\frac{3\left(1+a_m^2\right)^{1/4}\beta_w}{2\delta X}\right]^{2/3}, \tag{24.52}$$

the singularity is integrable, and hence the breaking plasma wave contains a finite number of particles.

24.8 Transverse Wake Wave Breaking

The wave breaking regimes considered above were found within the framework of the one-dimensional approximation, under the condition that the transverse inhomogeneity scale substantially exceeds both the wavelength of the wake wave and the value of the electron displacement in the wake-field. Nonlinear wave breaking acquires new features in the three-dimensional geometry, where the relativistically strong laser pulse is relatively narrow in the transverse direction or/and in the case of a wake wave excited inside a plasma channel, e.g., inside a plasma-filled capillary. Spatially inhomogeneous wake waves have the shape of paraboloids. Paraboloidal structures enhance the accelerated charged particle focusing and can be used for the electromagnetic radiation focusing. The curvature of constant-phase surfaces, $1/R$, increases with the distance from the driver laser pulse. The curvature radius R decreases to the value at which it becomes equal to the value of the electron displacement in the wake-field, ξ_w. After that, electron trajectory self-intersection occurs, which is equivalent to wave breaking (Bulanov et al., 1997a, 1997b). This process results in the injection of part of the electrons into the accelerating phase of the wake-field.

Transverse inhomogeneity of a wake wave appears due to a transverse inhomogeneity of the wake wave frequency ω_w, which is due to its dependence on the amplitude of a relativistically strong wave. In turn it is determined by a transverse inhomogeneity of the driver laser pulse and/or the transverse inhomogeneity of the plasma density when the laser pulse propagates inside a plasma channel. The coordinate dependence of the wake wave frequency near the axis can be approximated by the expression

$$\omega_w \approx \omega_w(0) + \delta\omega_w(r/S)^2, \tag{24.53}$$

where $\delta\omega_w$ is the difference between the values of the frequency on the channel axis and outside. It is equal to $\delta\omega_w \approx \omega_w(0)$, when the wakewave is generated by a high-intensity laser pulse with the amplitude $a \gg 1$. The dependence of the wave phase on coordinates and time is

$$x - v_w t + \psi\,(x, r, t)\frac{v_w}{\omega_w(0)} = \frac{r^2}{2R}. \tag{24.54}$$

Actual localization of constant-phase surfaces in a nonlinear wake wave compared to those in a linear wave differs by the value of the electron displacement ξ_w. We can conclude that when the curvature radius R becomes equal to the displacement amplitude ξ_w, the wave breaks. This corresponds to the scenario of transverse wave breaking (Bulanov et al., 1997a, 1997b). From Eq. (24.54), it is easy to see that the curvature radius decreases with the distance from the laser pulse. Taking this fact into account, we find that the distance between the laser pulse and the position where the first wave breaking occurs is equal to

$$\frac{\omega_w(0)S^2}{2\delta\omega_w\xi_w}. \tag{24.55}$$

The number of cycles of a regular wake wave is

$$N_w \approx \frac{\omega_w^2(0)S^2}{4\pi c\,\delta\omega_w\xi_w}. \tag{24.56}$$

Figure 24.7 schematically illustrates the electron trajectory self-intersection in the course of transverse wake wave breaking. The ponderomotive pressure of ultrashort laser pulses pushes the electrons away in the forward direction and transversely, forming a cavity in the electron density (Bulanov et al., 1997a, 1997b; Pukhov & Meyer-ter-Vehn, 2002). The attracting force of the electric field, appearing due to the electric charge separation, pulls the electrons into the cavity. The dashed curves show the position of constant-phase surfaces calculated in the linear approximation.

To describe the transverse wake wave breaking we assume that the electron displacement direction is perpendicular to the constant-phase surfaces obtained in the linear approximation. Constant-phase surfaces are given by equations $\mathbf{M}_0(y, z) =$ constant and $\mathbf{M}(y, z) =$ constant respectively. Mapping the surface $\mathbf{M}_0$ to the surface $\mathbf{M}$ is given by

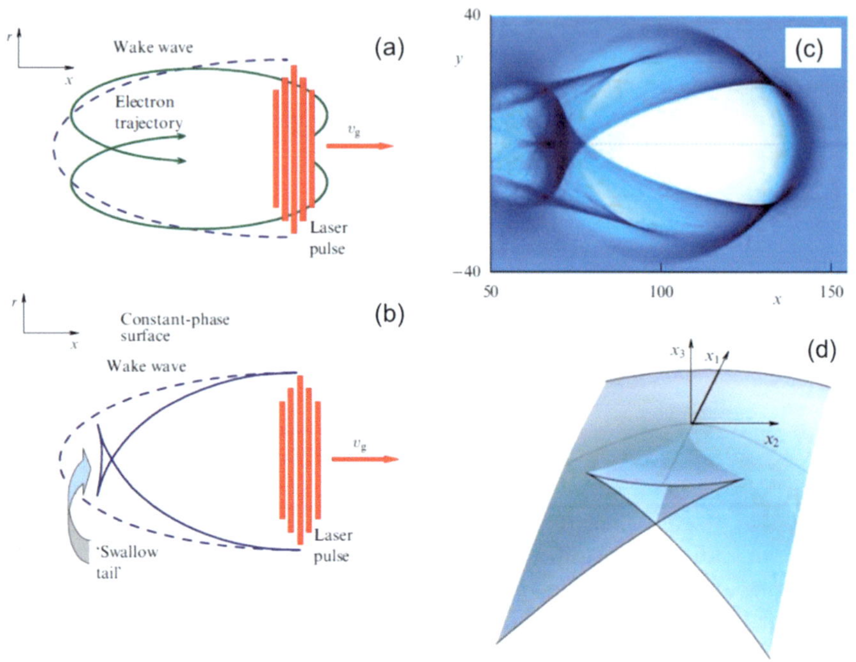

Fig. 24.7 Wake wave breaking in the first cycle of a wave behind the driver laser pulse. **a** Ponderomotive pressure pushes the electrons in the transverse direction and forward. Under the attracting force of the electric field, which occurs due to the electric charge separation inside the cavity, the electrons far enough behind the laser pulse move into the cavity. The dashed curve shows a constant-phase surface calculated in the linear approximation. **b** Nonlinear distortion of the wake wave form corresponding to the singularity called the 'swallow tail' in catastrophe theory. **c** Distribution of the electron density in the (x, y) plane for $a = 3$ and $n_e = 0.001n_{cr}$ at the time $t = 155(2\pi/\omega)$. **d** The swallow-tail catastrophe. Reproduced from Bulanov et al. (2013) @ Uspekhi Fizicheskikh Nauk 2013

$$\boldsymbol{M}(y_0, z_0) = \boldsymbol{M}_0(y_0, z_0) + \xi \boldsymbol{n}(y_0, z_0), \tag{24.57}$$

with coordinates y_0 and z_0. Here $\boldsymbol{n}$ is the vector normal to the surface $\mathbf{M}_0(y, z) =$ constant equal to

$$\boldsymbol{n} = \frac{\partial_{y_0}\boldsymbol{M} \times \partial_{z_0}\boldsymbol{M}}{\left|\partial_{y_0}\boldsymbol{M} \times \partial_{z_0}\boldsymbol{M}\right|}, \tag{24.58}$$

This gives for the coordinates of a new position of the surface of constant phase in the (x, r) plane

$$x = x_0 + \xi_w(r_0)\frac{R}{\sqrt{R^2 + r_0^2}}, \tag{24.59}$$

$$r = r_0 - \xi_w(r_0)\frac{r_0}{\sqrt{R^2 + r_0^2}}, \tag{24.60}$$

where $x_0 = r_0^2/2R$ and r_0 are the constant-phase surfaces obtained in the linear approximation, which are assumed to have the shape of paraboloids of revolution.

Different scenarios of transverse wave breaking can be realized depending on the particular form of the function $\xi_w(r_0)$. For generic form $\xi_w(r_0)$, the singularities formed during the wave break are structurally stable because they have topological properties corresponding to fundamental catastrophes (Poston & Stewart, 1978). In the case of a homogeneous displacement $\xi_w(r_0)$, the mapping $\{(x_0, r_0) \to (x, r)\}$, given by Eqs. (24.59) and (24.60), transforms the paraboloid of revolution into a surface 'parallel' to it.

For $\xi_w \geq R$, the cross section of this surface has a singularity corresponding to the 'swallow tail' catastrophe, which is shown in Fig. 24.7d. In the three-dimensional space with local coordinates (x_1, x_2, x_3), a catastrophe of that type is related to the surface on which the polynomial

$$y^4 + x_1 y^2 + x_2 y + x_3 \tag{24.61}$$

has multiple roots.

In Fig. 24.7, we also present the results of computer simulations of the electromagnetic pulse interaction with a plasma target, which illustrates the formation of the discussed singularity. Here and below, the simulations are performed using the relativistic electromagnetic code REMP [6] based on the particle-in-cell (PIC) method. In the simulations, a laser pulse with the amplitude $a = 3$ corresponding to the radiation intensity $1.93 \times 10^{19} \times (0.8\,\mu\text{m}/\lambda)^2\,\text{W/cm}^2$ and $15\lambda \times 50\lambda$ in size propagates in a plasma slab with a thickness of $145\,\lambda$ and the density $n = 3 \times 10^{17} \times (0.8\,\mu\text{m}/\lambda)^2\,\text{cm}^{-3}$. Figure 24.7c shows the electron density distribution in the (x, y) plane. We see a region where the swallow tail singularity is formed in the cavity behind the laser pulse. Also, the initial stage of the electron bunch injection into the wake-field is shown.

24.9 Bow Wave

Wake plasma wave generation by a laser pulse provides one of the examples of realization of a general physical phenomenon well known in other fields of physics. For example, a moving ship leaves wake waves behind it, known also as Kelvin waves (Lamb, 1932). In addition to these waves, the front part of the ship emits a bow wave, which, in particular, determines the outer boundary of the ship wake wave. Bow waves are well known in observations of fluid motion under conditions on Earth and in space plasma.

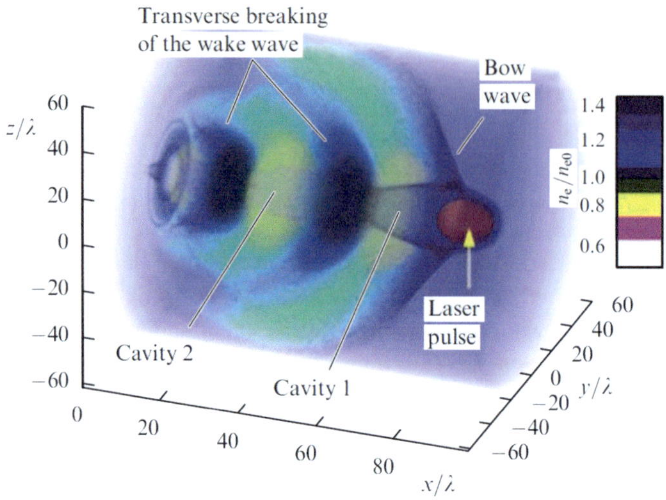

Fig. 24.8 Electron density distribution in the wake behind a laser pulse. The laser pulse is shown by a surface of constant electromagnetic energy density $E^2 + B^2 = (2e/m_e\omega c)^2$ for the time $t = 100(2\pi/\omega)$. Reprinted figure with APS permission from Esirkepov et al. (2008)

Continuing the analogy between the processes specific to the dynamics of continuous media and the nonlinear interaction of intense electromagnetic radiation with plasma, it would be natural to find an analogue of the bow wave forming at the front of a laser pulse (Esirkepov et al., 2008). The clearest bow wave of this kind can be seen in the interaction of a strong narrow electromagnetic pulse with collisionless low-density plasma.

In Fig. 24.8, we present the results of PIC simulations of bow wave formation in the three-dimensional geometry. A linearly polarized laser pulse with the amplitude $a = 6.62$ and size $10\lambda \times 10\lambda \times 10\lambda$ interacts with plasma with the electron density equal to $n = 1.14 \times 10^{18} \times (1\,\mu\text{m}/\lambda)^2\ \text{cm}^{-3}$.

Excitation of a high amplitude wake wave by a laser pulse is accompanied by the formation of a cavity in the electron density in the first cycle (see Figs 24.8 and 24.9).

As can be seen in Fig. 24.9, the longitudinal electric field E_x is localized in the region with a size much larger than the waist of the laser pulse and that of the cavity. As a result, the electrostatic potential value φ is larger than in the case where it corresponds to only the cavity size. We also note that the cavity length is much larger than its width. Transverse wake wave breaking leads to the emission of electron bunches in the radial directions in other cycles of the wake wave, enhancing the electrostatic potential there.

The form of the cavity in the electron density behind the laser pulse can be found from the fact that in a strong bow wave, all electrons in the way of the laser pulse are pushed aside, forming a multi-stream configuration in the collisionless plasma. The plasma inside the cavity is positively charged. It attracts those electrons from side regions that have not experienced the ponderomotive force action. The transverse

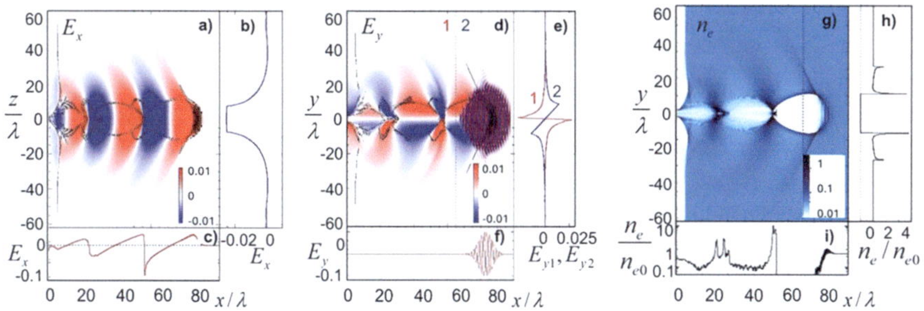

Fig. 24.9 Two-dimensional slices of the electric field distribution, normalized to $m_e\omega c/e$: **a** E_x, **d** E_y, and **g** of the electron density at $t = 100(2\pi/\omega)$. **b**, **e**, **h** Vertical and **c**, **f**, **i** horizontal one-dimensional slices are shown by dashed lines. In (**a**, **d**), black lines show the electron density for $n_e/n_0 = 0.75,\ 1.5$. Reprinted figure with APS permission from Esirkepov et al. (2008)

radial component of the electric field near the axis is linearly proportional to the radius, $E_\perp = 2\pi n_0 e r$. Under the action of this field, electrons from the outer region are drawn into the cavity. It is easy to show that for a laser pulse with a radius of the order of or smaller than $2\pi ca/\omega_{pe}$, the electrons moving inwards remain nonrelativistic. The dependence of their coordinate on time is given by

$$r(t) = r(0)\cos(\omega_{pe}t/\sqrt{2}).$$

For the time equal to $t = \pi/\sqrt{2}\omega_{pe}$, the electrons reach the cavity axis. As a result, the cavity boundary is determined by the surface

$$r = w_0 \cos\left[\frac{\omega_{pe}}{\sqrt{2}}\left(t - \frac{x}{v_w}\right)\right], \tag{24.62}$$

where w_0 is the laser pulse waist.

A bow wave in the laser plasma is a complex nonlinear electromagnetic object. The location where the bow wave is detached from the cavity boundary, clearly seen in Figs. 24.8 and 24.9g, deserves more attention. The (x, r) cross section of this region shows a strong modulation of the electron density, whose shape corresponds to a cusp-like singularity (Poston & Stewart, 1978). The bow wave physics has important implications for understanding mechanisms of the high order harmonics generation in relativistic laser plasmas (Pirozhkov et al., 2012).

24.10 Nonlinear Plasma Wave in Finite-Temperature Plasma

Here we consider properties of nonlinear plasma waves in finite-temperature plasma (Bulanov et al., 2012a, 2012b). The laser plasma temperature in the case of ultra-short, femtosecond laser pulses is determined by the parameters of the prepulse and pedestal interaction with the irradiated target during the time before the arrival of the main pulse. In the case typical for multi-terawatt laser radiation, plasma is formed in the process of photoionization. In such plasmas, the electron temperature is of the order of the quiver energy of electrons moving in the ionizing radiation, i.e., it does not exceed the KeV level, which is much lower than the energy of electrons moving in the field of the main pulse, which is above the MeV level. The electron energy distribution does not correspond to the Maxwell distribution and can be approximated in the framework of the 'water bag' model, which, under circumstances different from those stated above, would be considered too artificial.

In the water bag model, the distribution function $f_e(p, x, t)$ is constant in the phase-plane domain bounded by the curves $p_+(x, t)$ and $p_-(x, t)$, and vanishes outside this domain. The constant is proportional to the electron density and inversely proportional to the width of the distribution in the momentum space, which is related to the electron temperature.

The distribution function evolution is described by the Vlasov equation

$$\partial_t f_e + v\partial_x f_e - E\partial_p f_e = 0, \tag{24.63}$$

with the electric field determined by

$$\partial_x E = 1 - n_e, \tag{24.64}$$

where all the variables are written in the dimensionless form using standard normalization. Equation (24.63) describes an incompressible fluid flow in the phase space.

Taking the first moment, we obtain the electron density

$$n_e(x, t) = g_e[p_+(x, t) - p_-(x, t)]. \tag{24.65}$$

Here, the constant

$$g_e = \frac{n_0}{p_{+,0} - p_{-,0}} \tag{24.66}$$

is the ratio of the initial density value to the initial width of the distribution, which is assumed to be uniform.

It follows from Eqs. (24.63)–(24.65) that the functions $p_+(x, t)$, $p_-(x, t)$ and $E(x, t)$ satisfy the system of equations

$$\partial_t p_+ + \frac{p_+}{\sqrt{1+p_+^2}} \partial_x p_+ = -E, \tag{24.67}$$

$$\partial_t p_- + \frac{p_-}{\sqrt{1+p_-^2}} \partial_x p_- = -E, \tag{24.68}$$

$$\partial_x E = 1 - g_e(p_+ - p_-), \tag{24.69}$$

For waves propagating with a constant velocity β_w, i.e., for the solutions of Eqs. (24.67)–(24.69) in terms of the independent variable $X = x - \beta_w t$, this system can be reduced to

$$h_+' = -E, \tag{24.70}$$

$$E' = 1 - \frac{p_+(h_+) - p_-(h_+ - \beta_w \Delta p_0)}{\Delta p_0}, \tag{24.71}$$

where the prime denotes the differentiation with respect to the variable X,

$$h_+ = \sqrt{1+p_+^2} - \beta_w p_+ \tag{24.72}$$

and

$$\Delta p_0 = p_{+,0} - p_{-,0}. \tag{24.73}$$

Equations (24.70) and (24.71) can be written in the Hamiltonian form. For a symmetric initial distribution with $p_{+,0} = -p_{-,0}$ the Hamiltonian function is given by

$$\mathcal{H}(E, h_+) = \frac{E^2}{2} + \gamma_w^2 h_+ - \frac{W(\gamma_w h_+) - W(\gamma_+(h_+ + \beta_w \Delta p_0))}{2\Delta p_0}, \tag{24.74}$$

where

$$W(z) = z\sqrt{z^2 - 1} - \ln(z + \sqrt{z^2 - 1}). \tag{24.75}$$

Its asymptotic behavior in the limits $z \to 1$ and $z \to \infty$ is

$$W(z) = \begin{cases} \left(4\sqrt{2}/3\right)(z-1)^{3/2} & \text{for } z \to 1 \\ z^2 - \ln(2z) & \text{for } z \to \infty \end{cases}. \tag{24.76}$$

In Fig. 24.10, we plot constant-value contours of Hamiltonian (24.71) in the (E, h_+) plane for $\beta_w = 0.8$ and $\Delta p_0 = 0.1$.

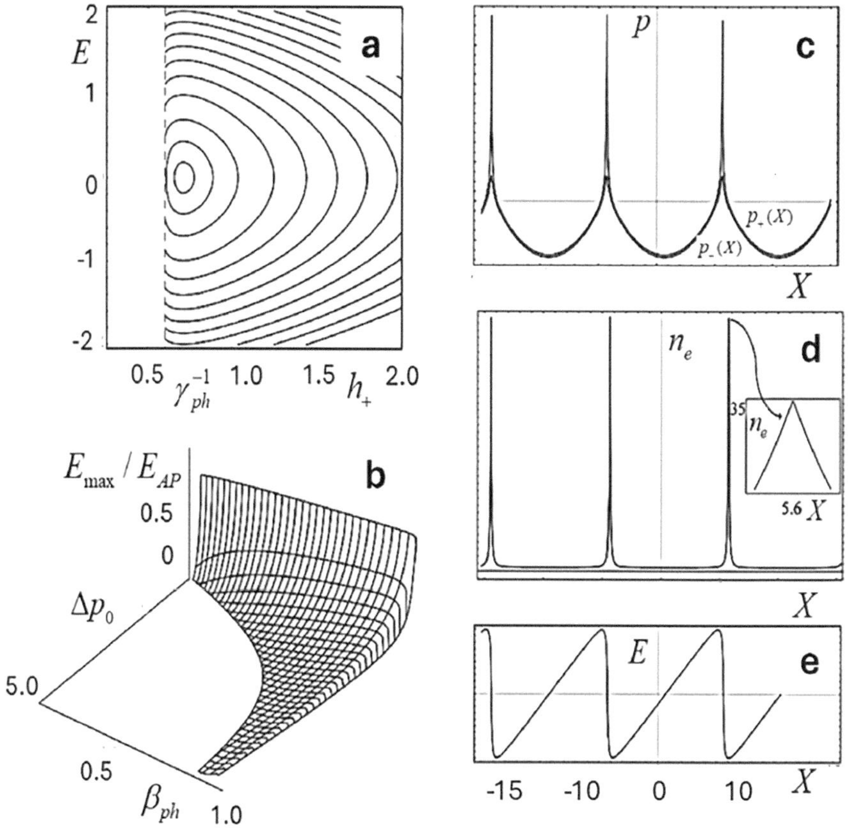

Fig. 24.10 **a** Constant-value contours for Hamiltonian (24.74) on the (E, h_+) plane for $\beta_w = 0.8$ and $\Delta p_0 = 0.1$. **b** Dependence of the maximum of the electric field E_m normalized to E_{AP} on the phase velocity β_w and the width of the distribution function Δp_0. **c** Electron phase plane. **d** Electron density versus the coordinate X (the inset shows the density dependence on the coordinate at the wave crest). **e** Electric field versus the coordinate X. Reproduced from Bulanov et al. (2012a), with the permission of AIP Publishing

The solutions of (24.70) and (24.71) in the limit $h_+ \to \gamma_w^{-1}$ become singular, as follows from expression (24.69) and from the behavior of the Hamiltonian system trajectories seen in Fig. 24.10a. At the singularity, the electron momentum on the upper bounding curve tends to

$$p_+ \to p_{+,m} = \frac{\beta_w}{\sqrt{1-\beta_w^2}}, \tag{24.77}$$

which corresponds to wave breaking when the velocity of the electrons with the momentum p_+ becomes equal to the phase velocity of the wave. At the point of wave breaking, the electron momentum on the lower bounding curve is equal to

$$p_{-,m} = p_{+,m} - \beta_w \gamma_w^2 \Delta h_0 - \sqrt{\gamma_w^2 \Delta h_0^2 - 2\gamma_w^3 \Delta h_0}, \tag{24.78}$$

where

$$\Delta h_0 = \sqrt{1 + p_{+,0}^2} - \sqrt{1 + p_{-,0}^2} - \beta_w \big(p_{+,0} - p_{-,0}\big). \tag{24.79}$$

The effects of a finite temperature of the plasma decrease the value of the maximum electric field in a stationary wave compared with the Akhiezer-Polovin electric field value E_{AP} given by Eq. (22.30). For $\Delta p_0 \ll 1/\beta_w \gamma_w$, the maximum electric field is equal to

$$E_m = E_{AP} - \frac{2(\beta_w \gamma_w)^{3/2}}{3\sqrt{\gamma_w - 1}} \sqrt{\Delta p_0}, \tag{24.80}$$

In the limit $\Delta p_0 \to 2\beta_{ph}\gamma_{ph}$, it vanishes (Fig. 24.10b).

In Figs 24.10c–e, we show the nonlinear wave structure: the phase plane, the electron density, and the electric field for $\beta_w = 0.995$ and $\Delta p_0 = 0.25$.

In contrast to the wave breaking in cold plasma, the density at the wave-breaking point remains finite in a finite-temperature plasma,

$$n_{e,m} = \gamma_w^2 \beta_w \left(\sqrt{1 + \frac{2}{\beta_w \gamma_w \Delta p_0}} - \beta_w \right), \tag{24.81}$$

In the vicinity of the wave-breaking point $X = X_m$, the density dependence on the coordinate can be represented in the form

$$n_e(X) \underset{X \to X_m}{=} n_{e,m} - \left(\frac{\sqrt{n_{e,m} \gamma_{ph}^3}}{\Delta p_0} \right) |\delta X| \tag{24.82}$$

with $\delta X = X - X_m$. This form of the wave corresponds to nonlinear waves called ‘peakons’ (see Fig. 24.10d). Among the best-known peakons are Stockes waves on a water surface (Whitham, 1974).

24.11 Laser Wake-Field Accelerator

In the laser wake-field accelerator (LWFA) concept introduced by Tajima and Dawson (1979), a long-living strong wake-field, induced by a short intense laser pulse in its wake in a low-density collisionless plasma, accelerates duly injected electrons. Within the framework of the laser wake field accelerator paradigm a regular and strong electric field having the form of a wave propagating with phase velocity close to speed of light in vacuum is excited in underdense plasmas by short pulse relativistically intense laser radiation. The electrons injected into the wake-field acceleration phase with the initial energy corresponding to the velocity equal to the wake wave phase velocity are then accelerated up to significantly higher energy. The achievable electron energy is determined by several processes among which the first is the electron slippage with respect to the accelerating phase of the wake wave and the second is the laser energy depletion. The dephasing and depletion lengths, being inversely proportional to the plasma density are of the same order.

24.12 Electron Interaction with Wake Waves

Assuming that the characteristic change in the electron density in the wake wave is of the order of the plasma density and considering the weakly nonlinear wave that is of interest to the discussed concept of a laser electron-positron collider (Leemans & Esarey, 2009), we can estimate the amplitude of the electrostatic potential in the wave as φ_w (see Eqs. (24.27) and (24.28)). The energy of electrons accelerated by the wake wave is $(1-\beta_w)^{-1}$ times greater than φ_w, i.e., the electron gamma-factor is

$$\gamma_e = 2\gamma_w^2. \tag{24.83}$$

The electron acceleration length is

$$l_{acc} \approx \frac{\lambda_w}{2(1-\beta_w)}, \tag{24.84}$$

which is equivalent to the expression

$$l_{acc} \approx \lambda\gamma_w^3. \tag{24.85}$$

Using this we obtain the relation between the acceleration length, the laser wavelength, and the energy of the fast electrons:

$$l_{acc} \approx \lambda\gamma_e^{3/2}. \tag{24.86}$$

For the laser wavelength $\lambda = 1\,\mu\text{m}$ and the electron energy 1 TeV, i.e., for $\gamma_e = 2\times10^6$, we find that the acceleration length (the accelerator size) should be of the order 1 km. In the case of 4 GeV electron energy the acceleration length is about 10 cm. In experiments on laser wake-field acceleration by laser pulses of a multi-terawatt power, the achieved energy of electrons is of the order of 4 GeV (Leemans et al., 2014), 4.2 GeV (Kim et al., 2013), and 8 GeV (Gonsalves et al., 2019) with the several centimeters plasma size of a few centimeters. In the experiments the high-power laser pulse guiding is provided by the plasma channel formed in the capillary discharge plasma. For details of the capillary discharge physics see publications by Ehrlich et al. (1996), Hosokai et al. (2000), Bobrova et al. (2002), Bagdasarov et al. (2021, 2022).

Here we describe the electron dynamics in the wake wave, limited to the one-dimensional approximation (Esirkepov et al., 2006), for the geometry where all functions depend on the time t and one coordinate x. Within the framework of classical electrodynamics, one-dimensional motion of relativistic electrons in the fields of an electromagnetic and wake plasma wave is described by the Hamiltonian

$$\mathcal{H} = \sqrt{m_e^2 c^4 + c^2 P_\parallel^2 + [cP_\perp + e A_\perp(x,t)]^2} - e\,\varphi\,(x,t), \tag{24.87}$$

where $P_\parallel$ and $P_\perp$ are the components of the generalized momentum, $A_\perp$ is the vector potential of the laser pulse, and φ is the wake-field scalar potential.

Neglecting the effects of dispersion on the propagation of electromagnetic waves, we assume that the vector and scalar potentials, $A_\perp$ and φ, depend only on the variable $X = x - v_g t$. It is assumed here that the group velocity of the laser pulse v_g satisfies $0 < v_g < c$. The Hamiltonian in Eq. (24.87) has a symmetry corresponding to the Lie group with the generators

$$v_g \partial_x + \partial_t, \quad \partial_y, \quad \text{and} \quad \partial_z. \tag{24.88}$$

The Noether theorem implies that there are integrals of motion

$$\mathcal{H} - p v_g = m_e c^2 h_0, \quad P_\perp = P_{\perp,0}, \tag{24.89}$$

where constants h_0 and $P_{\perp,0}$ are determined by the initial conditions. Using the dimensionless variables introduced above, $\beta_{ph} = v_g/c$, $\Phi(X) = e\varphi(X)/m_e c^2$, $a(X) = eA_\perp(X)/m_e c^2$, $p_x = P_\parallel/m_e c$, we write the first integral of motion as

$$h(p_x, X) \stackrel{\text{def}}{=} \sqrt{1 + p_x^2 + a^2(X)} - \varphi(X) - \beta_{ph} p_x = h_0. \tag{24.90}$$

It then follows that the kinetic energy acquired by an electron on the trajectory interval from X_0 to X is

$$\mathcal{E} = \gamma_{ph}^2 \left\{ \varphi(X) - h_0 \pm \beta_{ph} \sqrt{[\varphi(X) - h_0]^2 - \gamma_{ph}^{-2}\left[1 + a^2(X)\right]} \right\} - 1. \tag{24.91}$$

where the plus sign corresponds to the coordinate X increasing with time and the minus sign corresponds to decreasing X.

The behavior of the system with the Hamiltonian given by Eq. (24.87) is illustrated in Fig. 24.11, which presents the phase portrait for electrons with $P_{\perp,0} = 0$. The wake wave is excited by a short electromagnetic pulse of circular polarization with the envelope given by

$$a(X) = a_0\left[\exp(-4\ln(2)X^2/l_{las}^2) - 1/16\right]\theta(l_{las} - |X|), \tag{24.92}$$

where $a_0 = 2$, $l_{las} = 10\,\lambda$, and $\theta(x)$ is the Heaviside function, equal to 1 for $x \geq 0$ and 0 for $x < 0$.

Each electron trajectory $\{X(t), p_x(t)\}$ in the (X, p_x) plane corresponds to a constant level line of the function $h(X, p_x)$ in (24.90). The phase plane consists of the domains containing the trajectories of particles trapped in the wake-field and two domains

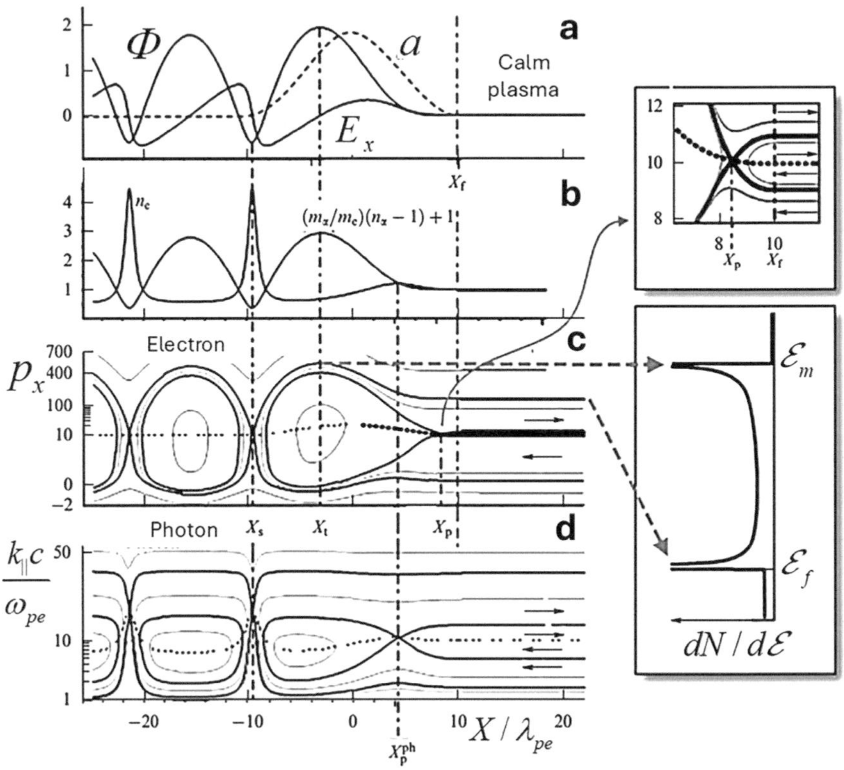

Fig. 24.11 **a** Wake wave excited by a short laser pulse. Dependences of the dimensionless quantities Φ, a, and E_x on the coordinate X; X_f is the coordinate of the laser pulse front. **b** Normalized density of electrons and ions. **c** Electron phase plane. X_S, X_t and X_p are the respective coordinates of a singular point of the separatrix, its crest, and the point of intersection of the separatrix branches. The top right inset shows an enlarged fragment of **c**, the domain where the ponderomotive force of the laser pulse acts on electrons. The bottom inset shows the energy spectrum of the electron beam $dN/d\mathcal{E}$, which consists of the contributions from the particles at the separatrix crest and from the particles that have overtaken the laser pulse; $\mathcal{E}_m$ is the maximum energy of an electron accelerated at the front of the laser pulse. **d** Photon phase plane. The bold lines represent the separatrix and the thin lines show other trajectories in **c**, **d**. Curves in **c** and the upper inset correspond to the dependence $p_x = \beta_{ph}\gamma_{ph}\sqrt{1 + a^2(X)}$, and in **d**, to the dependence $k_x c = \beta_{ph}\gamma_{ph}\omega_{pe}(X)$. Reprinted figure with APS permission from Esirkepov et al. (2006)

corresponding to the transition particles. These areas are separated by singular trajectories, called separatrices, which merge at the singular points, which are located on the curve

$$p_x(X) = \beta_{ph}\gamma_{ph}\sqrt{1 + a^2(X)}. \tag{24.93}$$

On this curve, the radicand in the right-hand side of Eq. (24.91) vanishes, i. e.

$$[\varphi(X) - h_0]^2 = \gamma_{ph}^{-2}\left[1 + a^2(X)\right]. \tag{24.94}$$

We consider an electron moving along the separatrix, whose trajectory originates at a singular point X_S. Its kinetic energy reaches a maximum $\mathcal{E}_m$ when the electron position is at the crest of the separatrix at the point X_t. For a sufficiently short laser pulse with $l_{las} < \lambda_w/2$, the points X_S and X_t in the first cycle of the wake wave correspond to a local minimum and a local maximum of the potential: $\varphi_{\min} = \varphi(X_s)$ and $\varphi_{\max} = \varphi(X_t)$. Hence, the maximum value of the energy acquired by the electron moving along the separatrix is

$$\mathcal{E}_m = \gamma_{ph}^2\left\{(\varphi_{\max} - \varphi_{\min}) + \beta_{ph}\sqrt{(\varphi_{\max} - \varphi_{\min})^2 + 2\gamma_{ph}^{-2}(\varphi_{\max} - \varphi_{\min})}\right\} + \mathcal{E}_{inj}, \tag{24.95}$$

where $\mathcal{E}_{inj} = \gamma_{ph} - 1$ is the electron energy at the point X_S. In the limit $\gamma_{ph} \gg 1$, the electron energy is given by

$$\mathcal{E}_m \approx 2\gamma_{ph}^2(\varphi_{\max} - \varphi_{\min}) + \gamma_{ph} - 1. \tag{24.96}$$

The lowest value of the scalar potential is at the point of the maximum of the electron density in the wake wave with $\varphi_{\min} \geq -1 + 1/\gamma_{ph}$.

The value

$$h(X, p_x) = 1/\gamma_{ph} - \varphi_{\min}$$

corresponds to the separatrix separating the first-cycle trajectories of the electrons trapped by the wake-field from the electron trajectories reflected by the potential at the point X_S. As a result of reflection, the electron acquires a speed greater than the phase velocity of the wake wave and overruns the wake wave and the laser pulse. Asymptotically as $X \to +\infty$, the electron energy tends to

$$\mathcal{E}_f = \gamma_{ph}^2\left\{|\varphi_{\min}| + \beta_{ph}\sqrt{|\varphi_{\min}|^2 + 2\gamma_{ph}^{-2}|\varphi_{\min}|}\right\} + \mathcal{E}_{inj}. \tag{24.97}$$

If the wake-field is excited by a strong laser pulse with $a \gg 1$, then the scalar potential $\varphi_{\min}$ tends to its lowest value $-1 + 1/\gamma_{ph}$, which gives the electron energy

$$\mathcal{E}_{f,m} = 2(\gamma_{ph}^2 - 1). \tag{24.98}$$

We conclude that in a weakly nonlinear wake-field (it corresponds to $a = 1$) the maximum energy of the accelerated electrons is related to the electron density in the plasma by

$$\mathcal{E}_m = \gamma_e m_e c^2 = 2m_e c^2 \gamma_{ph}^2, \tag{24.99}$$

where the relativistic gamma-factor equals $\gamma_{ph} = \omega_0/\omega_{pe} \equiv \lambda_w/\lambda_0 = (n_{cr}/n_0)^{1/2}$. The acceleration length is $l_{acc} = \lambda_w \gamma_{ph}^2$, i.e. being expressed via the laser wavelength it equals $l_{acc} = \lambda \gamma_{ph}^3$. It is related to the electron energy as $l_{acc} = 2\lambda_0 \gamma_e^{3/2}$, which can be represented as

$$\mathcal{E}_m = m_e c^2 \left(\frac{l_{acc}}{2\lambda}\right)^{2/3} = 0.3 \left[\left(\frac{l_{acc}}{1\ \text{km}}\right)\left(\frac{1\,\mu\text{m}}{\lambda}\right)\right]^{2/3} \text{TeV}. \tag{24.100}$$

As well known, in the mid-1950s, Enrico Fermi formulated the question about the highest particle energy that can be achieved with standard accelerators, considering an accelerator size of the order of the circumference of Earth's equator, and found the limit 1 PeV (10^{15} eV). Substituting $l_{acc} = 40\,000$ km for the acceleration length in expression (24.100), we obtain an energy of the same order. The use of a multistage wake-field accelerator (see the discussion below) may increase the electron energy by several orders of magnitude.

24.13 Energy Spectrum of Fast Electrons

Here, we discuss the shape of the energy spectrum of electrons accelerated by the wake-field (Bulanov & Tajima, 2005; Esirkepov et al., 2006). We assume that the electron beam is injected with the parameters that provide the conditions for entering a neighborhood of the point X_S in the phase plane. Then the electrons move along the separatrix with the distribution of their density given by a function $\mathcal{N}(X)$, which leads to a broadening of the spectrum from the initial energy $\mathcal{E}_{inj}$ to the maximum energy $\mathcal{E}_m$, corresponding to the top of the separatrix. In addition, the electron energy spectrum has a singularity at the energy $\mathcal{E}_f$ (see the lower inset on the right in Fig. 24.11). Near the top of the separatrix, the dependence of the energy of the beam particles on the coordinate X can be approximated by a parabola:

$$\mathcal{E}(X) = \mathcal{E}_m \left[1 - \frac{(X - X_t)^2}{l_{acc}^2}\right], \tag{24.101}$$

where $l_{acc} = \lambda_w \gamma_w^2$ is the acceleration length. This implies that the spectrum of the electrons collected in the detector, equal to the integral over time, is given by

$$\left.\frac{dN}{d\mathcal{E}}\right|_{\mathcal{E}\to\mathcal{E}_m-0} = \frac{\mathcal{N}(X)}{|d\mathcal{E}_m/dX|} \approx \frac{\mathcal{N}(X_t)l_{acc}}{2\sqrt{\mathcal{E}_m(\mathcal{E}_m-\mathcal{E})}}, \tag{24.102}$$

where it is assumed that $\mathcal{E}_m > \mathcal{E}$.

If electrons are injected in the vicinity of a singular point on the separatrix in other cycles of the wake wave, then the resulting energy spectrum contains a superposition of peaks given by Eq. (24.97) with different values of the maximum energy in general.

The contribution of electrons overtaking the laser pulse can be found from Eq. (24.91). It is easy to find that the energy spectrum is then described by the expression

$$\left.\frac{dN}{d\mathcal{E}}\right|_{\mathcal{E}\to\mathcal{E}_f+0} \approx \frac{\mathcal{N}(X_f)X_f}{\mathcal{E}-\mathcal{E}_f}, \tag{24.103}$$

where the energy $\mathcal{E}_f$ is determined by Eq. (24.94).

24.14 Energy Scaling

Due to relativistic self-focusing the laser pulse amplitude and waist are different from their values in vacuum. The self-focusing results in the laser pulse channeling inside a finite radius channel. The channel radius r_c can be found from the fact that the ponderomotive force of the laser radiation acting on the plasma electrons in the radial direction is balanced by the force due to charge separation electric field. In other words, the pondemotive potential $m_e c^2(\sqrt{1+a_0^2}-1) \underset{a_0^2>>1}{\approx} m_e c^2 a_0$ is equal to the electrostatic potential $\approx 2\pi n e r_c^2$. This condition gives for the channel radius

$$r_c \approx c\sqrt{a_0}/\omega_{pe}$$

(in dimensionless units it is $r_c = \sqrt{a}$). Although while the laser pulse propagates in plasma, its amplitude and waist change its power $\mathcal{P}$ remains the same. Using this we find a relationship between the electromagnetic wave amplitude and power $\mathcal{P}$,

$$a_0^3 = \frac{\mathcal{P}}{\overline{\mathcal{P}}}\gamma_{ph}^{-2}, \tag{24.104}$$

where

$$\overline{\mathcal{P}} = 2m_e^2c^5/e^2 = 17\,\text{GW}$$

is the characteristic power in the theory of relativistic self-focusing as given by Eq. (19.39). The criterion for self-focusing is given by Eq. (19.38). Note here that formula (24.104) and some other formulas must contain a numerical coefficient approximately equal to one, the value of which depends on the specific geometry. Often, we will not specify the value of these coefficients.

A finite waist laser pulse generates the wake field having the form called the "cavity", "blowout", and/or "bubble" (see Pukhov & Meyer-ter-Vehn, 2002). The transverse size of the cavity is of the order of $r_w = c\sqrt{a}/\omega_{pe}$. The form of the cavity in the electron density behind the laser pulse can be found, as in the case of the self-focusing channel, from the fact that all electrons in the way of the laser pulse are pushed aside, forming a multi-stream configuration in the collisionless plasma. The plasma inside the cavity is positively charged. It attracts those electrons from the side regions that have not experienced pondermotive force action.

The electrostatic potential inside the cavity is about $\varphi_c = a_0$ and the electric field is of the order of $\sqrt{a}$. Using these expressions and Eq. (24.104) we can find the accelerated electron energy expressed in terms of the laser power and plasma density (via the ratio $\omega_{pe}/\omega = \gamma_{ph}^{-1}$) we find

$$\mathcal{E}_e = m_e c^2 \left(\frac{\mathcal{P}}{\overline{\mathcal{P}} \gamma_{ph}^2} \right)^{2/3}, \tag{24.105}$$

where we have taken into account the dependence of the laser group velocity and the pulse etching velocity on the laser amplitude (see Lu et al., 2007). As we see, the fast electron energy is for given laser power higher for lower plasma density, i.e. as high as the ratio γ_{ph}^{-1} is.

24.14.1 Threshold of Relativistic Self-focusing

Considering the case when the driver laser pulse is at the threshold of relativistic self-focusing, we find that the self-focusing condition (19.38) imposes a constraint on the maximum of this ratio $\mathcal{P} = \overline{\mathcal{P}} \gamma_{ph}^2$ when the maximal electron energy scales as (Bulanov et al., 2016)

$$\mathcal{E}_m = 2 m_e c^2 \left(\frac{\mathcal{P}}{\overline{\mathcal{P}}} \right). \tag{24.106}$$

According to Eq. (24.106) a PW class laser can provide electrons with the energy of the order of 60 GeV. With a 200 TW laser the achievable electron energy is about 12 GeV. The acceleration length

$$l_{acc} \approx \lambda \left(\frac{\mathcal{P}}{\overline{\mathcal{P}}} \right)^{3/2} \tag{24.107}$$

is equal to 15 m, in the first case, and is equal to 37.5 cm, in the second.

The laser energy $\mathcal{E}_{las}$ required for acceleration of electrons up to the energy given by Eq. (24.106) can be found from the condition on the laser pulse length l_{las} to be equal to the optimal length $l_{opt}(a)$ given by Eq. (24.23). Under this condition, the laser energy is

$$\mathcal{E}_{las} = 2\overline{\mathcal{E}}\left(\frac{\mathcal{P}}{\overline{\mathcal{P}}}\right)\gamma_{ph}E(-a_0^2). \tag{24.108}$$

with characteristic energy $\overline{\mathcal{E}}$ equal to

$$\overline{\mathcal{E}} = \frac{2\pi\overline{\mathcal{P}}}{\omega} = \frac{4\pi m_e^2 c^5}{e^2\omega} = 54\left(\frac{\lambda}{1\,\mu\text{m}}\right)\mu\text{J}. \tag{24.109}$$

In the case of the maximum of the accelerated electron energy given by Eq. (24.108) corresponding to the condition on the laser amplitude $a_0 = 1$, for which $E(-a_0^2) = 1.9101$, the laser energy and power are related to each other as

$$\mathcal{E}_{las} = 3.8\,\overline{\mathcal{E}}\left(\frac{\mathcal{P}}{\overline{\mathcal{P}}}\right)^{3/2}. \tag{24.110}$$

Combining relationships given by Eqs. (24.106) and (24.110) we find electron energy scaling versus the laser energy. It reads

$$\mathcal{E}_m = 2m_e c^2\left(\frac{\mathcal{E}_{las}}{7.6\,\overline{\mathcal{E}}}\right)^{2/3}. \tag{24.111}$$

For example, 1.5 kJ laser can accelerate electrons up to 30 GeV. The required electron density, $n = n_{cr}m_e c^2/\mathcal{E}_m$ with $n_{cr} = m_e\omega^2/4\pi e^2 = 1.74 \times 10^{21}\,\text{cm}^{-3}$ for laser wavelength $\lambda = 800$ nm, is approximately equal to $3 \times 10^{15}\,\text{cm}^{-3}$. If the dimensionless amplitude of the laser pulse is greater than one (slightly), say it is equal to $a = \overline{a}$, then the expression for the electron energy (24.111) changes as follows: $\mathcal{E}_m = 2m_e c^2\left(\overline{a}^2\mathcal{E}_{las}/\overline{\mathcal{E}}\right)^{2/3}$. For $\overline{a} = 2$ the electron energy calculated above is approximately 2.5 times higher. Further consideration of the energy scaling of the LWFA accelerated electrons can be found in Valenta et al. (2025) and Valenta and Bulanov (2025). On other electron beam parameters see Lundh et al. (2011).

Here we did not consider radiation damping, which is weak. Schroeder et al. (2010) and Golovanov et al. (2022) have analyzed the radiation friction effects on the electron acceleration within the framework of the LWFA paradigm. The LWFA generated electron beams in the high repetition rate regime are studied in Lazzarini et al. (2024).

24.15 Multi-stage Accelerator Configurations

A multi-stage LWFA accelerator consists of several succeeding accelerating stages inside of which the wake waves are generated by a single or by multiple laser pulses. The using of multi-stage LWFA configurations can provide the conditions for the electron acceleration over the distance substantially above the energy depletion length for a single laser pulse. Typical multi-stage accelerator consists of the injector and boosters.

The electron injection can be provided either in the nonlinear wake wave breaking in homogeneous or inhomogeneous plasmas, or by optical, or by ionization injection mechanisms and from the clusters in the gas-cluster targets.

The second and subsequent stages are used for enhancing the energy of the accelerated electron bunch and for improving its quality. Enhancing the bunch quality in a two stage LWFA accelerator has been demonstrated in experiments by Steinke et al. (2016). An electron bunch having been accelerated in the previous (injector) stage has the length determined by the process of the wake wave breaking and by the length of the wake wavelength. In the subsequent stage, if the plasma density there is substantially lower than in the injector stage, it appears to have a length significantly shorter than the wake wave wavelength. Under these conditions, the bunch rotation in the phase space can lead to the electron energy narrowing provided the bunch leaves the stage being at the top of the separatrix.

Here we discuss the injector stage. Assuming the wake wave breaking mechanism of the electron injection, we shall consider the first stage(s) of the multistage accelerator as the injector. The condition of the wake wave breaking corresponds to the case when the wake field given by Eq. (24.28) exceeds the wave-breaking threshold (24.17). In the limit $\gamma_w \gg 1$ this condition yields a relationship between the amplitude of the driver laser pulse and γ_w:

$$1 + a_0^2 \geq \gamma_w, \tag{24.112}$$

which is equivalent to the condition

$$a_0^{3/2} \geq \omega/\omega_{pe}. \tag{24.113}$$

Using the relationship between the driver laser power, the laser amplitude and the plasma density given by Eq. (24.113) we can write the wake wave breaking condition in the form

$$\mathcal{P} > \overline{\mathcal{P}}\left(\frac{\omega}{\omega_{pe}}\right)^4, \tag{24.114}$$

which is well above the threshold of the relativistic self-focusing given by Eq. (1.50). The driver length optimal for the wake wave generation under the condition $a \approx 1$ for the driver power at the wave-breaking threshold (24.114), equals

$$l_{opt} = \frac{\pi c}{\omega_{pe}}. \tag{24.115}$$

This results in the requirement on the laser pulse energy $\mathcal{E}_{las} = \mathcal{P} l_{opt}/c$:

$$\mathcal{E}_{las} = \overline{\mathcal{E}} \left(\frac{\omega}{\omega_{pe}} \right)^3. \tag{24.116}$$

From Eqs. (24.113) and (24.115) follows that the energy of the laser-injector can also be expressed in terms of the laser power:

$$\mathcal{E}_{las} = \overline{\mathcal{E}} \left(\frac{\mathcal{P}}{\overline{\overline{\mathcal{P}}}} \right)^{3/2}. \tag{24.117}$$

For example, the energy of a 2 PW laser injector should be of the order of 30 J, and the energy of a 200 TW laser is approximately 30 times lower.

As we see, the 1D approximation implies too high requirements for the laser pulse power and energy. These requirements can be mitigated in the 3D regime when the laser pulse undergoing the relativistic self-focusing forms a cavity (bubble) in an underdense plasma. Inside the cavity, the wake wave breaking develops in the transverse-wave-breaking regime described above. The electron energy $\approx m_e c^2 a$ is of the order of or above the injection threshold.

In the plasma with inhomogeneous density the phase mixing phenomenon leads to the wake wave breaking with the electron injection even for the initial electron energy determined by the wake wave amplitude well below the injection threshold in homogeneous plasmas. However, in the substantially low wave amplitude this regime requires a good enough quality wake wave to be formed and it may result in the electron injection into a bucket located far behind the laser pulse driver.

In the case of a multi-stage configuration with equal stages (its setup is illustrated in Fig. 24.12) the energy gain in each subsequent stage is given by

$$\mathcal{E} = \mathcal{E}_{inj} + m_e c^2 \frac{a^4 + 2a^2}{(1 - \beta_w)(1 + a_0^2)} \tag{24.118}$$

provided the electron has been injected at the local minimum of the electrostatic potential as it is illustrated in Fig. 24.12. Having interacted with J stages, the electron injected in the injector with the energy $\mathcal{E}_{inj}$ equal to $m_e c^2 \gamma_{inj}$ with $\gamma_{inj} = \gamma_w$ achieves the final energy $\mathcal{E}_J = m_e c^2 \gamma_J$, where

$$\gamma_J = \gamma_{inj} + \frac{J\left(a_0^4 + 2a_0^2\right)}{(1 - \beta_w)(1 + a_0^2)}. \tag{24.119}$$

As seen, for $a \approx 1$ the final energy is of the order of $\gamma_J \approx 3J\gamma_w^2$. The length of one stage is equal to the dephasing length, $l_{acc} \approx \lambda \gamma_w^3$. Thus the total length of the

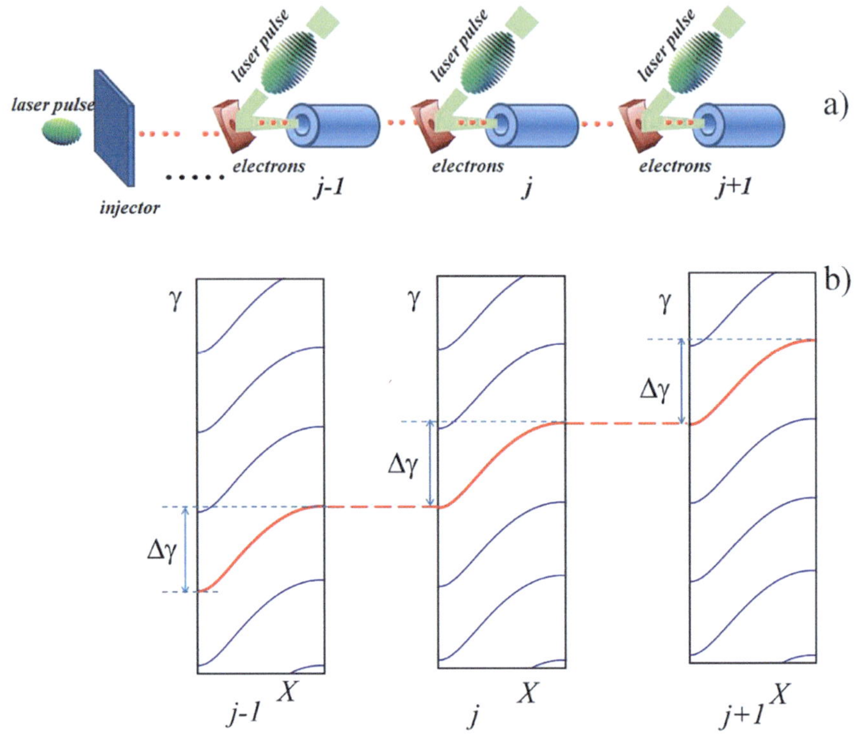

Fig. 24.12 A multi-even-stage accelerator comprising the injector and equal stages. **a** Schematic of the multi-equal stage LWFA accelerator. **b** Phase plots (X, γ) corresponding to the $j-1$-th, j-th, and $j+1$-th stages of the accelerator (Bulanov et al., 2016)

multi-stage accelerator with equal length stages is

$$L_{acc} = J\lambda\gamma_w^3 = \lambda\left(\frac{\gamma_J}{3}\right)^{3/2} J^{-1/2} \tag{24.120}$$

i.e. for fixed final energy of the accelerated electrons, γ_J, the acceleration length it is shorter for a larger the number of stages J. Apparently, the number of the stages cannot be arbitrarily large, because the length of one stage cannot be smaller than $l_{st} = \lambda(l_{las}/\pi\lambda)^3$ under the conditions requiring the laser pulse length to be optimal for the wake wave excitation.

An additional constraint on the number of the stages J stems from the matching requirement between the electron bunch accelerated in the $j-1$-th stage with the wake field in the j-th stage. Under more or less realistic conditions, when the matching is not ideal, the unavoidable deterioration of the electron beam quality may become unacceptably high for a substantially large number of stages.

The number of stages can be made smaller in a multi-uneven-stage LWFA accelerator, whose schematic view is presented in Fig. 24.13. In this configuration, the plasma density in the succeeding stage is lower than the plasma density in the preceding stage, and the length of the succeeding stage is larger than that of the preceding.

The energy gained by the electrons accelerated in the preceding stage

$$\gamma_{j-1} \approx \gamma_{w.j-1}^2 \tag{24.121}$$

(here we assume that the laser driver pulse normalized amplitude is of the order of unity, $a_{j-1} \approx 1$) is equal to the injection energy for the succeeding stage, i.e.

$$\gamma_{w.j-1}^2 \approx \gamma_{w,j}. \tag{24.122}$$

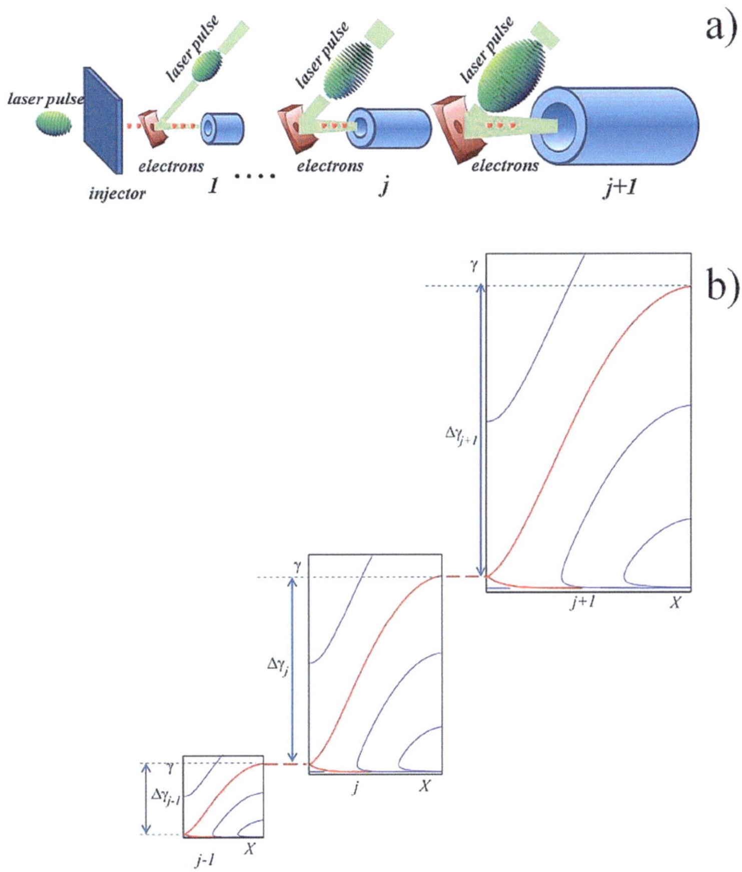

Fig. 24.13 Multi-uneven-stage accelerator comprising the injector and unequal stages. **a** Schematic of the milti unequal stage LWFA accelerator. **b** Phase plots (X, γ) corresponding to the $j-1$-th, j-th, and $j+1$-th stages of the accelerator (Bulanov et al., 2016)

It is assumed that the electrons enter the wake field in succeeding stage in the phase corresponding to the singular point, X-point, on the separatrix as shown in Fig. 24.13. Equation (24.122) gives a relationship between the parameters of the $j-1$-th and j-th stages. From Eq. (24.122) it follows that for known plasma density in the first stage, n_1, determined by the matching conditions with the injector stage, the density in the j-th stage is equal to

$$n_j = n_{cr}\left(\frac{n_1}{n_{cr}}\right)^{2j}. \tag{24.123}$$

The j-th stage length depends on the stage number as

$$l_j = \lambda\left(\frac{n_{cr}}{n_1}\right)^{3j} = \lambda\left(\frac{l_1}{\lambda}\right)^{3j}, \tag{24.124}$$

where $l_1 = \lambda(n_{cr}/n_1)^3$ is the length of the first stage. The energy of the electrons achieved as a result of interacting with J stages equals

$$\gamma_J = \sum_{j=1}^{J}\frac{n_{cr}}{n_j} = \sum_{j=1}^{J}\left(\frac{n_{cr}}{n_1}\right)^{2j} = \frac{(n_{cr}/n_1)^{2J}-1}{1-(n_1/n_{cr})^2} \approx \left(\frac{n_{cr}}{n_1}\right)^{2J}. \tag{24.125}$$

As we see, the accelerated electron energy grows exponentially with the number of stages in the limit $J \to \infty$:

$$\gamma_J \approx \exp\left(J \ln\frac{n_{cr}}{n_1}\right). \tag{24.126}$$

In the relationships obtained above the electron density inside the first stage has remained undetermined. It can be found from the condition imposed on the required electric charge of the accelerated electron bunch. The ratio n_1/n_{cr} can also be chosen for desired electron energy γ_J and the finite number of the stages, J. The required laser power and energy are determined by the parameters of the last, J-th stage.

24.16 Injection Mechanisms

The energy spread of the LWFA accelerated electrons is determined by the injection mechanism. The basic injection mechanisms are as follows. The first of them is the so-called "spontaneous injection" which occurs when nonlinear wake wave breaks in uncontrolled way (Bulanov et al., 1992). This is a robust injection regime. However, it results in relatively broad electron energy distribution. The narrow energy distribution can be realized by using the density down-ramp injection in the tailored target (Bulanov et al., 1998; Suk et al., 2001) or the optical injection mechanism (Umstadter et al., 1996). Another robust injection mechanism is based on the ionization of the dopped gas (Chen et al., 2006) or on the use of the cluster-plasma and nano-particle targets (Aniculaesei et al., 2024; Fukuda et al., 2007). The narrow electron energy spectrum has been demonstrated experimentally in the case of the down-ramp density injection (Faure et al., 2010; Geddes et al., 2008). High quality LWFA electrons can be generated via the optimal beam loading as has been realized by using the AI approach (Kirchen et al., 2021).

Substantial attention has been paid to the generation of the high-quality electron beams by combining the LWFA and PWFA regimes when the LWFA accelerated electrons are used as the electron beam driver (see e.g. experimental results in the paper by Masson-Laborde et al. 2014) and theoretical consideration by Pousa et al. (2019).

The fulfillment of the high directionality of the accelerated electron beams requires considering the transverse laser-target form optimization or/and the cutting the accelerated electron beam into the beamlets with needed parameters e.g. by using active lensing techniques.

Due to the multi-parametric nature of the processes related to the electron beam generation the active use of the AI methods is necessary (see Irshad et al., 2024).

24.17 Direct Laser Acceleration

Apart from the above discussed well-established laser-wakefield acceleration (LWFA), electrons can be also accelerated by the so-called direct laser acceleration (Pukhov et al., 1999). The mechanism of DLA takes the advantage of a resonant process between the long-range "betatron-like" oscillations induced by self-generated radial electric and azimuthal magnetic fields in plasma and Doppler-shifted oscillations within a laser field. Numerous theoretical (Arefiev et al. 2012; Jirka et al., 2020; Valenta et al., 2024) as well as experimental (Gahn et al., 1999; Willingale et al., 2013; Singh et al., 2022) studies have already proven the feasibility and provided understanding of this method.

To find the energy of the electrons accelerated via the DLA we consider the electromagnetic wave propagating inside the infinite length waveguide assuming that there is no a plasma inside. As known the electromagnetic waves propagating in the waveguide are of two types (Landau & Lifshitz, 1984; Rayleigh, 1897; Strutt, 1897): they are the TM (so-called E-waves with transverse magnetic field) and TE (the B-waves with transverse electric field) modes. The relationship between the wave frequency ω and the component k_z of the wave vector along the direction of the wave propagation are given by equation

$$\omega^2 = (k_z^2 + \kappa^2)c^2, \tag{24.127}$$

where κ is the eigen-number determined by the form of the waveguide cross section.

The group velocity of the wave with the wave number k_z along the wave-guide direction is given by the derivative

$$v_{gz} = \frac{\partial \omega}{\partial k_z} = \frac{c k_z}{\sqrt{k_z^2 + \kappa^2}} = \frac{c^2 k_z}{\omega}. \tag{24.128}$$

For given eigen-number κ the group velocity changes from 0 to speed of light c when the wave number k_z changes from 0 to ∞. Since the phase velocity equals $v_{ph\,z} = \omega / k_z$ we have the relationship $v_{g\,z} v_{ph\,z} = c^2$.

For the sake of simplicity, we consider the lowest order TE mode propagating along rectangular waveguide of the size equal to l_x and l_y in the x and y directions, respectively. The TE mode is determined by the z component of magnetic field

$$B_z = B_0 \exp[i(k_z z - \omega t)] \cos(k_x x) \cos(k_y y)$$

with the x and y components of the wave-vector

$$k_x = n_1 \pi / l_x \quad \text{and} \quad k_y = n_2 \pi / l_y,$$

where to n_1 and n_2 are integer. The eigen-number κ equals

$$\kappa = \sqrt{k_x^2 + k_y^2} = \pi \sqrt{\left(\frac{n_1}{l_x}\right)^2 + \left(\frac{n_2}{l_y}\right)^2}.$$

In the lowest κ mode TE_{10}, for which the numbers are $n_1 = 1$ and $n_2 = 0$, i.e. $k_y = 0$, and $\kappa = \pi / l_x$ the group velocity is

$$v_{g\,z} = \frac{c k_z}{\sqrt{k_z^2 + \pi^2 / l_x^2}}. \tag{24.129}$$

Performing the Lorentz transformation into the frame of reference moving with the group velocity (24.129) we find that in the boosted frame of reference the k_z' becomes equal to zero, the wave frequency is $\omega' = \pi c / l_x$, and the magnetic field amplitude does not change:

$$B_z' = B_0 \cos(\pi c t' / l_x) \cos(\pi x' / l_x); \tag{24.130}$$

and the electric field has the y component equal to

$$E_y' = B_0 \sin(\pi c t' / l_x) \sin(\pi x' / l_x). \tag{24.131}$$

Below in this section we will drop a prime denoting the variables in the boosted frame of reference.

Numerical integration of equations of relativistic electron motion in the electric and magnetic field given by Eqs. (24.130) and (24.131) for initial conditions corresponding to the case when the electron in the laboratory frame of reference is at rest, shows that the electron energy is linearly proportional to the field amplitude:

$$a_{DLA} = \frac{eB_0 l_x}{\pi m_e c^2}.$$

In the laboratory frame of reference, the electron energy is approximately equal to

$$\mathcal{E}_e = m_e c^2(\gamma_e - 1) = m_e c^2\left[\gamma_g(\sqrt{1 + a_{DLA}^2 + p_0^2} - p_0\beta_g) - 1\right]. \tag{24.132}$$

Here are the normalized group velocity $\beta_g = v_{gz}/c = k_z/\sqrt{k_z^2 + \pi^2/l_x^2}$, the corresponding gamma-factor $\gamma_g = 1/\sqrt{1 - \beta_g^2} = \sqrt{1 + (k_z l_x/\pi)^2}$, and the initial electron momentum in the boosted frame of reference $p_0 = \beta_g/\sqrt{1 - \beta_g^2} = k_z l_x/\pi$. In the limit $a_{DLA}^2 \gg p_0^2$, i.e. $eB_0/m_e k_z c^2 \gg 1$ we have for the electron energy

$$\mathcal{E}_e = m_e c^2(\gamma_e - 1) \approx m_e c^2 \gamma_g a_{DLA}. \tag{24.133}$$

In the opposite limit when $eB_0/m_e k_z c^2 \ll 1$ the energy is approximately equal to

$$\mathcal{E}_e = m_e c^2(\gamma_e - 1) \approx m_e c^2 a_{DLA}^2/2. \tag{24.134}$$

In Fig. 24.14a we show normalized electron energy in the laboratory frame of reference vs. time $\omega t/2\pi$ for the wave amplitude $a_{DLA} = 30$ and initial electron coordinate equal to $x(0) = -\pi/12$. We can see clearly that electron oscillations are nonlinear. Figure 24.14b presents the dependence of normalized electron energy on the wave amplitude a_{DLA} confirming dependences given by expressions (24.133) and (24.134).

Now we find the energy scaling of electrons accelerated by the DLA mechanism when the laser pulse propagates in underdense plasma being self-focused, i.e. the laser power is above the self-focusing threshold

$$\mathcal{P} > \overline{\mathcal{P}}\gamma_g^2.$$

The laser pulse propagates along the self-focusing channel of the radius

$$l_x = c\sqrt{a_{DLA}}/\omega_{pe}.$$

The relationship between the wave amplitude and laser power for given plasma density, i.e. for given

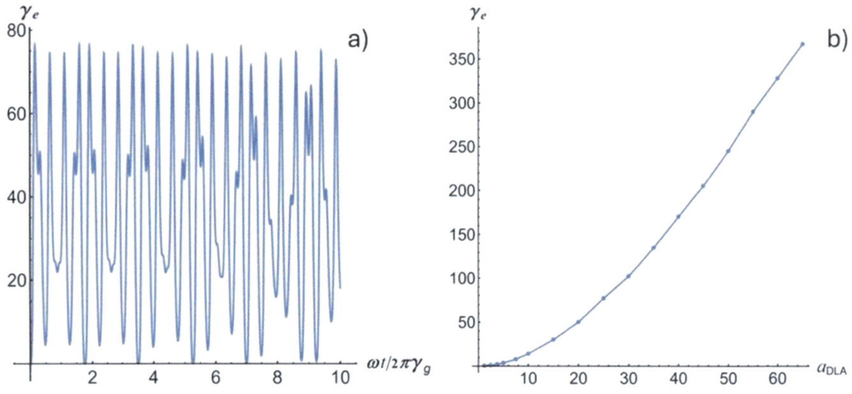

Fig. 24.14 **a** Normalized electron energy versus time for $a_{DLA} = 30$ and $x(0) = -\pi/12$; **b** Normalized electron energy versus wave amplitude a_{DLA}

$$\gamma_g = \sqrt{n_{cr}/n},$$

is given by Eq. (24.104), which we rewrite as

$$a_{DLA} = \left(\mathcal{P}/\overline{\mathcal{P}}\gamma_g^2\right)^{1/3}.$$

Using these relationships and Eq. (24.133) we obtain dependence of the accelerated electron energy on the laser pulse power and γ_g determined by the plasma density

$$\mathcal{E}_e = m_e c^2(\gamma_e - 1) \approx m_e c^2 \gamma_g \sqrt{\frac{\mathcal{P}}{\overline{\overline{\mathcal{P}}}}}. \tag{24.135}$$

For example, if 10 PW laser propagates in a plasma with the density of $n = 10^{-2} n_{cr}$ the electron energy is approximately equal to 1.5 GeV.

24.18 Capillary Discharges for Optical Guiding and for Optics of Charged Particle Beams

24.18.1 Types of Capillary Discharges. MHD Model

The capillary discharge is a versatile plasma device that is being explored for several important applications, including as a source of extreme ultraviolet (EUV) radiation (Andreic et al., 2003), as an active medium for soft X-ray lasers (Rocca et al., 1994), as a plasma waveguide for transporting high-power laser beams over long distances for laser wake-field accelerators (Ehrlich et al., 1996; Hosokai et al., 2000), and for focusing high-energy charged particle beams in a process known as active lensing (Chiadroni et al., 2018, Van Tilborg et al., 2015). We note that plasma lenses (Chen et al., 1987) are also used to focus high energy charged particle beams. As the name suggests, the capillary discharge in its principal scheme is simple: the base element is a narrow channel of sub-millimeter radius and the length on the order of ten centimeters made in an insulator material (metallic capillary is also used in the studies of the EUV sources; see Chan et al., 2014). The channel is ended in a longitudinal direction by the two electrodes. An external low inductance forming line is used to apply a voltage between the electrodes and to form a short pulse with a typical duration on the order of hundred nanosecond of the electric current through the cross section of the channel. The capillary discharge has been operated in two modes: ablative mode and gas-filled mode.

In the ablative capillary mode, the plasma is produced by initiating the discharge along the wall of material and the ablation of the capillary wall material acts as the fuel for the plasma. When the main breakdown occurs, a high current pulse is generated, and it heats up and ionizes the ablated wall material rapidly to form the plasma within the capillary. In the gas-pre-filled capillary discharge mode, gas is used as fuel for producing the plasma. In both cases, the gas is ionized by the electric current excited by the applied strong enough voltage, which results in the formation of plasma inside the channel. The ablative capillary discharge plasma waveguides were applied for the LWFA electron acceleration by Hosokai et al. (2000) and Kameshima et al. (2009).

The applications mentioned above require capillary discharges with different parameters depending on specific case.

For theoretical and computer simulation description of the capillary plasma dynamics the one-fluid magnetohydrodynamics model is used. The MHD system of equations is as follows (see also Part III where the MHD equations were derived).

The continuity equation for the mass density ρ is

$$\partial_t \rho + \nabla(\rho \boldsymbol{v}) = 0. \tag{24.136}$$

Here $\boldsymbol{v}$ is the velocity of the ion component. It obeys the equation of motion

$$\partial_t \boldsymbol{v} + (\boldsymbol{v}\nabla)\boldsymbol{v} = -\frac{1}{\rho}\nabla(p_e + p_i) + \frac{1}{c\rho}\mathbf{j}\times\mathbf{B}, \tag{24.137}$$

where p_e and p_i are the electron and ion thermal pressure, respectively, $\mathbf{j}$ is the electric current density and $\mathbf{B}$ is the magnetic field. Here we neglect the viscousity effects which are not significant for the capillary dynamics in the regimes corresponding to the waveguide formation. The Ohm law

$$\mathbf{E} = -\frac{1}{c}\boldsymbol{v}\times\mathbf{B} + \frac{\mathbf{j}}{\sigma} \tag{24.138}$$

gives a relationship between the electric field **E**, magnetic field **B**, and the electric current density **j** as it follows from the equations of the electron component motion for given plasma velocity $\boldsymbol{v}$ and electric conductivity σ. The heat transport equations for electron and ion components can be written as

$$\rho\ \partial_t \varepsilon_e + (\boldsymbol{v}\nabla)\varepsilon_e + p_e \nabla \cdot \boldsymbol{v} = \nabla(\kappa_e \nabla T_e) + \frac{\mathbf{j}^2}{\sigma} - Q_{ie}, \tag{24.139}$$

$$\rho\ \partial_t \varepsilon_i + (\boldsymbol{v}\nabla)\varepsilon_i + p_i \nabla \cdot \boldsymbol{v} = \nabla(\kappa_i \nabla T_i) + Q_{ie}, \tag{24.140}$$

with ε_e and ε_i being the specific energy, κ_e and κ_i thermal conductivity coefficients of the electron and ion components. The reduced Maxwell equations where the effect of the displacement current is neglected for non-relativistic plasma motion are

$$\mathbf{j} = \frac{c}{4\pi} \nabla \times \mathbf{B}, \tag{24.141}$$

$$\partial_t \mathbf{B} = -c\nabla \times \mathbf{E}, \tag{24.142}$$

and

$$\nabla \cdot \mathbf{B} = 0. \tag{24.143}$$

The electric quasi-neutrality equation for electron n_e and ion density n_i and the ion charge Z_i is

$$n_e - Z_i n_i = 0. \tag{24.144}$$

Discussion of the ion viscosity effects as well as the Nernst-Ettinghausen effect in the generalized Ohm's law can be found in Ref. Bobrova et al. (2002).

The above written system of MHD equations has been actively used for computer simulations of various regimes of the capillary discharge development aimed at modeling active media of X-Ray lasers (Bobrova et al., 1996), at modeling waveguides for ultra-short laser pulse transport over large distance for LWFA electron acceleration (Bagdasarov et al., 2021; Bobrova et al., 2002; Gonsalves et al., 2019), at modelling active lenses (Bagdasarov et al., 2017).

24.18.2 Capillary Discharges of Pinch and Dissipative Types

In the plasma dynamics in the MHD regime the key role is played by Amper's $\mathbf{j} \times \mathbf{B}$ force in the r.h.s. of Eq. (24.137) by the magnetic diffusion, which appears due to finite electric conductivity of plasma. The later process is characterized by magnetic diffusivity

$$\nu_m = \frac{c^2}{4\pi\sigma}.$$

The magnetic field dissipation also results in the plasma heating described by the term $\mathbf{j}^2/\sigma$ in the r.h.s. of Eq. (24.139). Amper's force causes the pinching of the electric current carrying plasma. Strong resistive dissipation of the magnetic field mitigates the pinch effect. Whether or not plasma dynamics is dissipative is determined by the value of so-called magnetic Reynolds number. It is named in the analogy with the Reynolds number in hydrodynamics characterizing the role of fluid viscosity. The magnetic Reynolds number is also known as the Lundquist number, defined as

$$\mathrm{Re}_m = \frac{v_A l}{\nu_m}. \tag{24.145}$$

Here $v_A = B/\sqrt{4\pi\rho}$ is the Alfvén velocity and l is the typical space-scale of the plasma configuration. Alfven velocity is a speed with which the waves propagate through magnetized plasma. Depending on the value of the magnetic Reynolds number there are two types of capillary discharges (see Sasorov, 2018).

In the case of high electric conductivity, when the magnetic Reynolds number is large,

$$\mathrm{Re}_m = \frac{v_A l}{\nu_m} \gg 1$$

capillary discharge plasma undergoes the pinch effect: the electric current at the capillary walls is screened because of the smallness of the skin layer thickness compared to the capillary radius. Plasma is compressed by the magnetic pressure towards the capillary axis. This process develops via the radially converging shock waves. They heat the plasma and form dynamically a high-density hot filament at the channel axis as required for realizing conditions needed for an active medium for soft X-ray lasers.

In the limit of relatively low electric conductivity of capillary plasma corresponding to dissipative regime of the capillary discharge, for the magnetic Reynolds number below unity

$$\mathrm{Re}_m = \frac{v_A l}{\nu_m} \ll 1,$$

the plasma pinching due to Amper's force has relatively small effect; the plasma is Joule heated by the electric current which is smoothly distributed across the capillary; the Joule heating is balanced by the thermal conduction towards the capillary walls; thermal pressure is almost uniform with the plasma is confined by the capillary walls. The plasma waveguides formed in the capillary discharge development in the dissipative regime are mainly used in the studies of the LWFA electron acceleration (Leemans et al., 2014).

24.18.3 Plasma Waveguide

The results of the MHD simulation of the dissipative capillary discharge presented in Bobrova et al. (2002) show that the pinch effect is insignificant under the conditions of the gas-filled capillary discharge waveguide. Furthermore, there is no screening of the axial electric field, and consequently it is uniform across the diameter of the capillary. The magnetic field pressure is much less than the plasma pressure, and the plasma pressure is constant across the capillary, there is no difference between electron and ion temperatures ($T_e = T_i = T$). As a result, the distribution of plasma inside the capillary is determined by the balance between Ohmic heating and cooling due to electron heat conduction. This allows one to formulate a simple model of plasma under such equilibrium conditions.

The radial profile of the plasma temperature is determined by the equation for heat flow

$$\frac{1}{r}\frac{d}{dr}\left(r\kappa_e\frac{dT}{dr}\right)+\sigma E^2=0. \tag{24.146}$$

Here κ_e is the coefficient of thermal conductivity and σ is the electric conductivity of plasma. They are given by equations

$$\kappa_e=\frac{n_e T_e}{m_e\nu_{ei}}\Gamma_1(x_e,w) \tag{24.147}$$

and

$$\sigma=\frac{e^2 n_e}{m_e\nu_{ei}}[1-\Gamma_5(x_e,w)]^{-1}. \tag{24.148}$$

The parameter

$$x_e=\frac{\omega_{Be}}{\nu_{ei}}$$

with $\omega_{Be}=eB/m_e c$ being the electron Larmore frequency is large for magnetized electrons and small in the limit of small magnetic field. The Lorentz parameter

$$w=\frac{\Lambda_{ee}}{\sqrt{2}z\Lambda_{ei}}$$

characterizes the relative role of the electron-electron and electron-ion collisions. The functions $\Gamma_j(x_e,w)$ are defined by Bobrova and Sasorov (1993) and Bobrova et al. (1995). The electron-ion and ion-ion collision frequencies are equal to

$$\nu_{ei} = \frac{4\sqrt{2\pi}e^4 z^2 n_i \Lambda_{ei}}{3\sqrt{m_e}T_e^{3/2}} \quad \text{and} \quad \nu_{ii} = \frac{4\sqrt{\pi}e^4 z^4 n_i \Lambda_{ii}}{3\sqrt{Am_A}T_i^{3/2}}, \tag{24.149}$$

where z is the ion mean electric charge and Λ_{ei} and Λ_{ii} are the Coulomb logarithms for electron-ion and ion-ion collisions.

Equation (24.146) is a nonlinear ordinary differential equation, which should be solved with the boundary conditions at the capillary axis and wall:

$$\left.\frac{dT}{dr}\right|_{r=0} = 0 \quad \text{and} \quad T|_{r=R_0} = 0, \tag{24.150}$$

i.e. we consider the case when the plasma temperature at the capillary wall is negligibly small compared to the temperature in the center of the capillary.

Using expressions (24.147)–(24.149) for the coefficient of thermal conductivity $\kappa_e = \kappa_0 T^{5/2}$ and for the electric conductivity $\sigma = \sigma_0 T^{3/2}$ of plasma, where κ_0 and σ_0 are constant, we introduce normalized radius $\xi = r/R_0$ and the variable $u(\xi)$ related to the plasma temperature by

$$u(\xi) = (T(\xi)/A)^{7/2}, \tag{24.151}$$

with

$$A = (7\sigma_0 R_0^2 E^2/2\kappa_0)^{1/2},$$

we cast Eq. (24.146) in the form

$$\frac{1}{\xi}\frac{d}{d\xi}\left(\xi\frac{du}{d\xi}\right) + u^{3/7} = 0 \tag{24.152}$$

which we should solve with the boundary conditions

$$\left.\frac{du}{d\xi}\right|_{\xi=0} = 0 \quad \text{and} \quad u|_{\xi=1} = 0. \tag{24.153}$$

Bobrova et al. (2002) solved numerically the boundary problem defined by Eqs. (24.152) and (24.153) and found that the plasma density and temperature near the capillary axis have parabolic dependences on the radius with the temperature maximum and density minimum, respectively. The electron density profile near the axis is described by

$$\frac{n_e(r)}{n_e(0)} \approx 1 + 0.33\frac{r^2}{R_0^2} \tag{24.154}$$

with the density on the axis $n_e(0)$ determined by the mean gas density and effective charge. This density profile is what is needed to guide the laser pulse.

Integrating of the wave equation describing the electromagnetic pulse propagation within the waveguide with the parabolic plasma density profile in transverse direction,

$$n_e(r) = n_e(0) + n_e''(0)\, r^2/R_0^2,$$

one can find that a Gaussian laser beam can propagate through a plasma channel having a constant spot size w_0 equal to

$$w_0 = (\lambda R_0)^{1/2} \left(\frac{n_{cr}}{2\pi^2 n(0)} \right)^{1/4} \left(\frac{n_e(0)}{n_e''(0) R_0^2} \right)^{1/4}. \tag{24.155}$$

This expression can be rewritten in the form

$$w_0 = \left(\frac{2}{\pi r_e n_e''(0)} \right)^{1/4},$$

where $r_e = e^2/m_e c^2$. It shows that the matched spot size does not depend on the laser wavelength λ. Using relationships found above the laser spot size can be written as

$$w_0 \approx 1.05 \times 10^5 \frac{\sqrt{R\,[\mu\mathrm{m}]}}{(z n_{0i}\,[\mathrm{cm}^{-3}])^{1/4}}.$$

Taking as an example the capillary radius $R = 150\mu$m and an initial hydrogen pressure of 67 mbar, we find that laser spot size is of $w_0 = 30\,\mu$m.

In Fig. 24.15 we present a comparison of the simulated and measured electron density profiles. As seen, the simulations based on the integration of the system of MHD equations (24.136)–(24.144) show formation of parabolic profile of the electron density as required for the laser pulse guiding.

24.18.4 Large Aspect Ratio Capillary

It is instructive to present here a model describing the equilibrium of plasma inside a flat capillary, that is, a capillary in which all variables depend on the coordinate y (see theory in Bagdasarov et al., 2017, and experiment in Drobniak et al., 2025). This configuration approximates large aspect ratio rectangular capillaries. In Fig. 24.16 we show the distribution of the electron density and temperature inside the capillary obtained with 2D MHD simulation.

Flat capillaries might be used to transport electron beams with high emittance asymmetries, as well as assist in forming spatially flat electron bunches for final focusing before the interaction point. A symmetric emittance of about 1 mm mrad

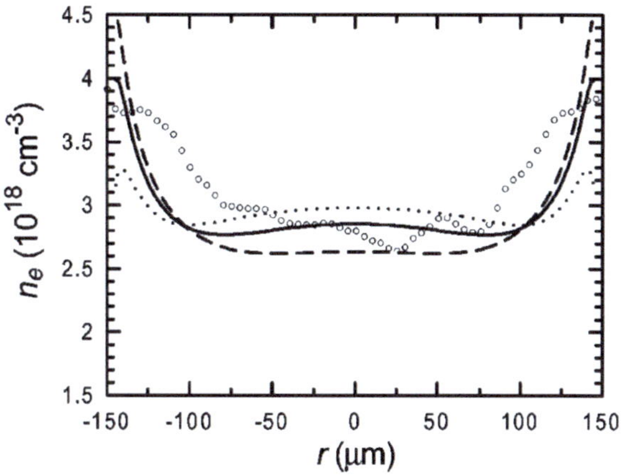

Fig. 24.15 Comparison of the simulated and measured electron density profiles. Open circles correspond to experimental results. The dotted line shows the electron density radial profile for $t = 555$ ns, the solid line for $t = 560$ ns, and the dashed line for $t = 565$ ns. Reprinted figure with APS permission from Bobrova et al. (2002)

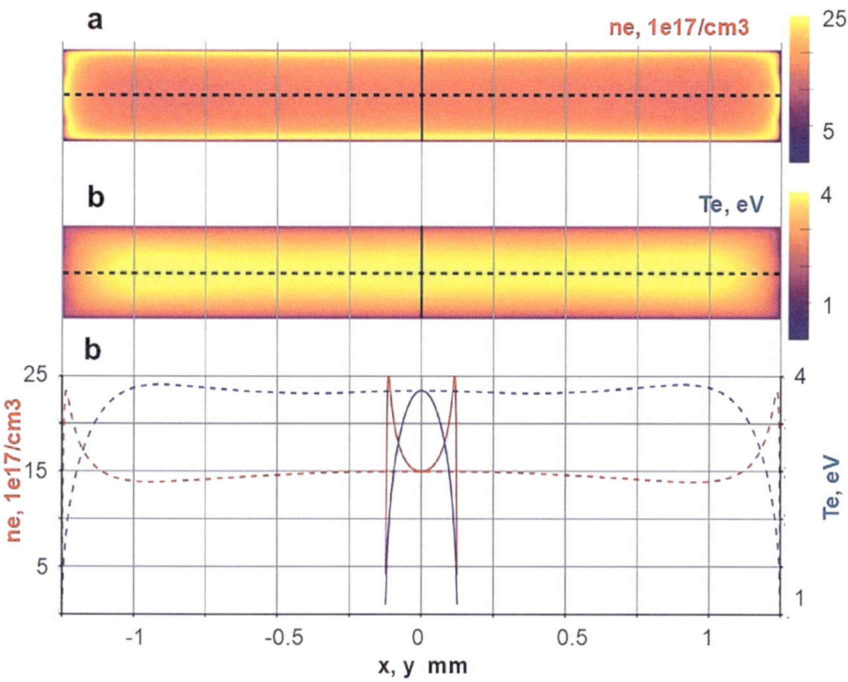

Fig. 24.16 Distribution of the electron: **a** density and **b** temperature at t = 240 ns in the (x, y)-plane. **c** electron density distributions along the line y = 0 (dashed red line) and along the x = 0 (solid red line) and electron temperature versus x along the y = 0 line (dashed blue line) and versus the coordinate y along the x = 0 (solid blue line). Reproduced from Bagdasarov et al. (2017), with the permission of AIP Publishing

is needed for coherent light sources to reach X-ray wavelengths of the order of 1 Å. Linear electron–positron colliders need asymmetric emittances and polarized electrons, with the smaller vertical emittance required to be on the order of 0.1 mm mrad. Asymmetric emittances are needed to reduce the beam-induced synchrotron radiation (beam-strahlung) in the interaction, and electron polarization is required because the effective luminosity can be up to 2 orders of magnitude larger (depending on the process as discussed by Kando et al., 2007).

In the case of the capillary of large aspect ratio cross-section, i.e., with one side considerably larger than the other, we can assume that the parameters of the discharge at the equilibrium are almost independent of the x coordinate, provided we exclude regions of the size of the order of L (it equals the size of the capillary along the short side) in the vicinities of the shorter sides of the rectangle. Hence, all parameters of plasma and magnetic field depend only on the y coordinate. In this case, the dependence of the plasma temperature on the y coordinate is described by the heat transport equation

$$\frac{d^2u}{d\eta^2} + u^{3/7} = 0, \tag{24.156}$$

where $\eta = y/L$ and $u(\eta) = (T(\eta)/A)^{7/2}$ with $A = (7\sigma_0 L^2 E^2/2\kappa_0)^{1/2}$. The boundary conditions for this equation are

$$\left.\frac{du}{d\eta}\right|_{\eta=0} = 0 \quad \text{and} \quad u|_{\eta=\pm 1} = 0. \tag{24.157}$$

Multiplying both sides of Eq. (24.156) by $du/d\eta$ and integrating over η, we obtain the integral

$$\left(\frac{du}{d\eta}\right)^2 = \frac{7}{5}\left(u_0^{10/7} - u^{10/7}\right), \tag{24.158}$$

where u_0 is equal to the value of the function $u(\eta)$ at the capillary center, $\eta = 0$. The boundary conditions (24.157) have been used. Using the integral (24.158), we can write the solution of the boundary problem in quadrature

$$\int_0^{u/u_0} \frac{ds}{\sqrt{1 - s^{10/7}}} = \frac{1}{u_0^{2/7}}\sqrt{\frac{7}{5}}(1 - |\eta|), \tag{24.159}$$

For the constant u_0, we have

$$u_0 = \left[\sqrt{\frac{5}{7}}\int_0^1 \frac{ds}{\sqrt{1 - s^{10/7}}}\right]^{-7/2} = \left[\sqrt{\frac{7}{5\pi}}\frac{\Gamma(6/5)}{\Gamma(17/10)}\right]^{7/2} \approx 0.25, \tag{24.160}$$

where $\Gamma(x)$ is the Euler gamma function.

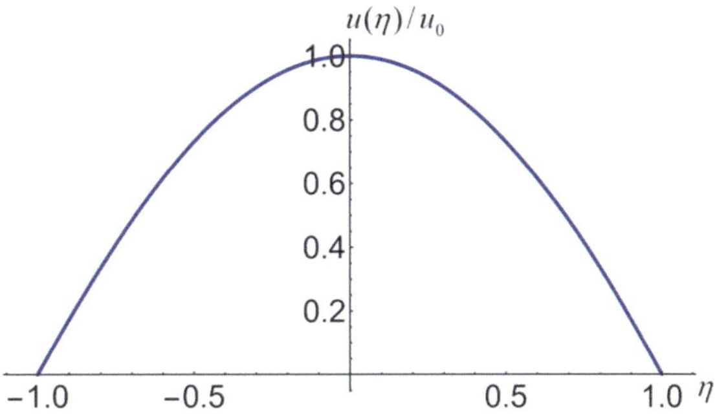

Fig. 24.17 Dependence of the function $u(\eta)/u_0$ on the coordinate η within the interval $[-1, 1]$

The integral on the left-hand side of the Eq. (24.159) can be expressed via hypergeometric functions, ${}_2F_1(\alpha, \beta, \gamma; z)$. As a result, the solution of the boundary problem (24.156) and (24.157) can be written in the implicit form

$$u(\eta)_2F_1\left[\frac{1}{2}, \frac{7}{10}, \frac{17}{10}; \left(\frac{u(\eta)}{u_0}\right)^{10/7}\right] = u_0^{5/7}\sqrt{\frac{7}{5}}(1 - |\eta|). \tag{24.161}$$

Figure 24.17 shows the dependence of the function $u(\eta)/u_0$ on the coordinate η. Near the maximum, the function $u(\eta)$ can be approximated by the parabola. It is easy to find that for $\eta \ll 1$.

$$u(\eta) = u_0\left(1 - \frac{1}{2u_0^{4/7}}\eta^2 + \frac{1}{56u_0^{8/7}}\eta^4 - \cdots\right) \approx 0.25(1 - 1.10\eta^2 + 0.09\eta^4 - \cdots). \tag{24.162}$$

The electron density distribution along the coordinate η can be found from the condition of constant pressure ($p = p_e + p_i$) across the capillary. In the case of full ionization $z = 1$, we have

$$p = 2n(\eta)T(\eta) = 2n(0)T(0). \tag{24.163}$$

so that the electron density profile is given by

$$n_e(\eta) = \frac{p}{2T(\eta)} = n_e(0)\frac{T(0)}{T(\eta)} = n_e(0)\left(\frac{u_0}{u(\eta)}\right)^{2/7}. \tag{24.164}$$

The electron density at the center $n_e(0)$ is equal to

$$n_e(0) = \langle n_e\rangle \frac{10\,\Gamma(17/10)}{7\sqrt{\pi}\,\Gamma(6/5)} \approx 0.8\langle n_e\rangle, \tag{24.165}$$

where

$$\langle n_e \rangle = \frac{1}{2L} \int_{-L}^{L} n_e(y) dy, \tag{24.166}$$

is a mean electron density. Near the axis $\eta = 0$ the electron density distribution is described by the formula

$$\begin{aligned} n_e(\eta) &= n_e(0)\left(1 + \frac{1}{7u_0^{4/7}}\eta^2 + \frac{1}{49u_0^{8/7}}\eta^4 + \cdots\right) \\ &\approx n_e(0)\left(1 + 0.31\,\eta^2 + 0.20\,\eta^4 + \cdots\right). \end{aligned} \tag{24.167}$$

As we see, this density profile is also suitable for guiding the laser pulse.

Now we can find the magnetic field inside the capillary created by the electric current whose density equals $\boldsymbol{j} = j\boldsymbol{e}_z = \sigma E \boldsymbol{e}_z$. The magnetic field is directed along the x direction: $\boldsymbol{B} = B(\eta)\boldsymbol{e}_x$. It should be found from the Eq. (24.141), which takes the form

$$\frac{dB}{d\eta} = \frac{4\pi j}{c} = \frac{4\pi \sigma_0 T^{3/2} E L}{c} = h_1 \left(\frac{u(\eta)}{u_0}\right)^{3/7}, \tag{24.168}$$

where h_1 the magnetic field gradient is equal to

$$h_1 = \left(\frac{7^6 2^5 \pi}{5^3}\right)^{1/4} \left(\frac{\Gamma(6/5)}{\Gamma(17/10)}\right)^{3/2} \frac{\sigma_0^{7/4} L^{5/2} E^{5/2}}{c\,\kappa_0^{3/4}} = 5.94 \times 10^{-10} \frac{\sigma_0^{7/4} L^{5/2} E^{5/2}}{\kappa_0^{3/4}}. \tag{24.169}$$

Using the integral given by Eqs. (24.158) and (24.168) we can find dependence of the magnetic field on the coordinate η. It reads

$$B(\eta) = \sqrt{\frac{7}{5}} h_1 u_0^{2/7} \sqrt{1 - \left(\frac{u(\eta)}{u_0}\right)^{10/7}}. \tag{24.170}$$

This dependence can be written as

$$B(\eta) = -\frac{h_1}{u_0^{3/7}} \frac{du}{d\eta}.$$

Expanding the expression (24.170) near $u = u_0$, i.e., for $\eta \ll 1$, we obtain the following expression for dependence of the magnetic field on the coordinate near the central plane

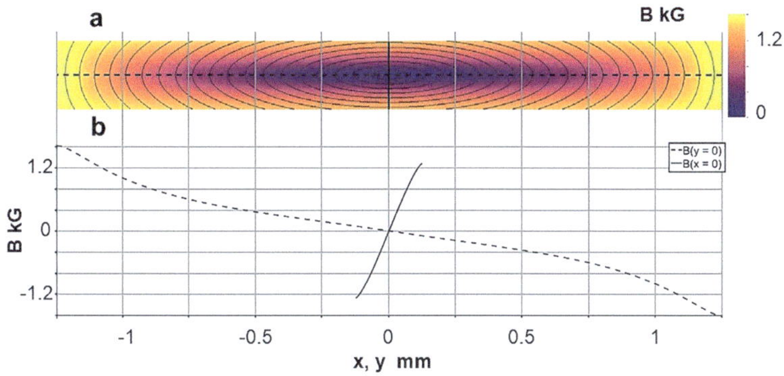

Fig. 24.18 Distribution of the magnetic field at $t = 240$ ns inside the capillary. **a** Magnetic field lines in the (x, y)-plane. **b** B_x versus y along the line $x = 0$ (solid black line) and B_y versus coordinate x along the line $y = 0$ (dashed black line). Reproduced from Bagdasarov et al. (2017), with the permission of AIP Publishing

$$B(\eta) = h_1\eta - h_3\eta^3 + \cdots = h_1\eta(1 - 0.16\eta^3 + \cdots) = h_1\left(\frac{y}{L} - 0.16\left(\frac{y}{L}\right)^3 + \cdots\right), \tag{24.171}$$

with

$$h_3 = \frac{h_1}{14u_0^{4/7}}.$$

The magnetic field maximum is reached at the capillary boundary, $\eta = 1$, where $u(1) = 0$, and is equal to

$$B_m = h_1\sqrt{\frac{7}{5}u_0^{2/7}} \approx 0.796\, h_1. \tag{24.172}$$

In Fig. 24.18 we present the distribution of the magnetic field inside the capillary obtained with 2D MHD computer simulation. It shows the magnetic field dependence along the y coordinate which can be approximated by Eq. (24.171).

From the expressions written above, the electric current inside the capillary is proportional to the electric field in the 5/2 power as in the case of circular capillary discharges (Bobrova et al., 2002).

24.18.5 Active Lenses for Charged Particle Beam Focusing

Typical gradient of the magnetic field created by the electric current inside the capillary is on the order of kT/m. This makes the capillary discharge based compact

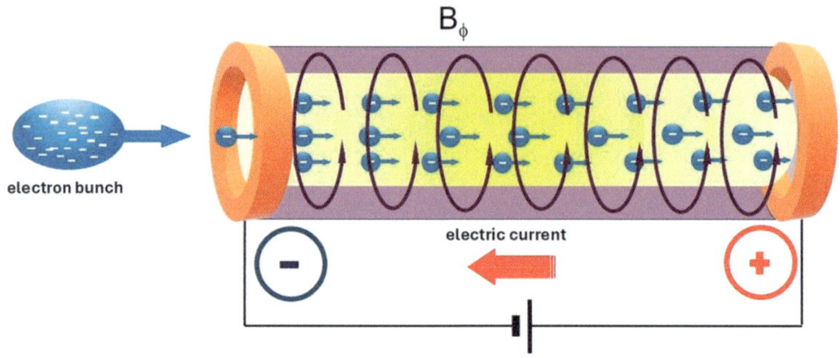

Fig. 24.19 Schematic of active lensing of charged particle beam moving through the capillary magnetic field (e.g. see Chiadroni et al. 2018)

devices applicable for focusing charged particle beams. The first idea of using the magnetic field driven by axial electric current to focus the proton beam was conceived by Panofsky and Baker in 1950. Focusing is due to plasma discharge current either in gas-filled capillary or in an evacuated channel by the azimuthal magnetic field (see schematic of active lensing in Fig. 24.19 as well as the magnetic field in the flat capillary, which can be used for preparing flat electron (positron) beams. Because of the external discharge current, this plasma lens has been named as active. In the case of LWFA accelerated electrons the active lensing has been studied theoretically and realized experimentally in many recently published papers (see Bagdasarov et al., 2017; Chiadroni et al., 2018; Lindstrøm, 2021; Lindstrøm et al., 2018; Röckemann et al., 2018; Sjobak et al., 2021; Steinke et al., 2016; Van Tilborg et al., 2015).

In what follows, we illustrate the active lensing solving the problem of a motion of charged particle in focusing magnetic field given by Eq. (24.171). We consider the motion of an electron in the magnetic field of the rectangular capillary discharge in the framework of a simplified one-dimensional model. This approach is valid for the electron dynamics in azimuthal magnetic field for the particles with zero angular momentum. The trajectory of a charged particle moving in a constant one-dimensional magnetic field can be found from the conservation of the particle energy and of the z-component of the generalized momentum

$$\sqrt{1 + p_x^2 + p_y^2 + p_z^2} = \gamma_0 \quad \text{and} \quad p_z + A_z(\eta) = p_{0z} \tag{24.173}$$

Here, $\sqrt{1 + p_x^2 + p_y^2 + p_z^2} = \gamma$ is the relativistic Lorentz factor and $A_z(\eta)$ is the z-component of the vector potential $\boldsymbol{A} = A_z(\eta)\boldsymbol{e}_z$. The x component of the momentum, $p_x = p_{0x}$, is constant. These integrals of the particle motion follow from the independence of the magnetic field on time and z-coordinate. The normalizations of the momentum on $m_e c$ and of the electromagnetic potential on $m_e c^2/e$ are used. The magnetic field is equal to the curl of the vector potential, $\boldsymbol{B} = \nabla \times \boldsymbol{A}$ i.e. $B = (dA_z/d\eta)(m_e c^2/eL)$.

Equations (24.158) and (24.170) yield the relationship between the z-component of the electromagnetic potential and the function $u(\eta)$

$$A_z(\eta) = A_0 u(\eta), \tag{24.174}$$

where

$$A_0 = -\frac{e h_1 L}{m_e c^2 u_0^{3/7}}.$$

Using equations for the particle coordinates,

$$\dot{\eta} = \frac{p_y}{\gamma} \quad \text{and} \quad \dot{\zeta} = \frac{p_z}{\gamma}$$

with $\eta = y/L$ and $\zeta = z/L$, where a dot stands for the time derivative, and Eq. (24.158), we can obtain the dependence of the particle normalized ζ coordinate on the η-coordinate in quadrature

$$\zeta(\eta) = \zeta_0 + \frac{1}{u_0^{5/7}} \sqrt{\frac{5}{7}} \int_{u_0}^{u(\eta)} \frac{(p_{0,z} - A_0 u) du}{\sqrt{[1 - (u/u_0)^{10/7}][\gamma_0^2 - 1 - p_{0,x}^2 - (p_{0,z} - A_0 u)^2]}}. \tag{24.175}$$

Since in the region near the z-axis for $\eta << 1$ the magnetic field can be approximated by a linear function of the η-coordinate, the aberration in the focusing of high energy electrons can be made weak. This implies the smallness of two parameters

$$\delta_1 = \frac{e h_1 Y_0^2}{p_{0,z} m_e c^2 L} \quad \text{and} \quad \delta_3 = \frac{h_3 Y_0^2}{h_1 L^2}, \tag{24.176}$$

where the amplitude of oscillations Y_0 is approximately equal to the initial coordinate y_0. The first small parameter shows that in the limit $\delta_1 \ll 1$, the electron with a large enough momentum along the z-axis, $p_{0,z}$ or/and small initial position y_0 moves inside the nonadiabatic region, where the particle is not magnetized. The electron oscillates in the y-direction around the position $y = 0$ and moves with approximately constant velocity $v_z \approx c$ along the z-axis. In addition, the smallness of the parameter δ_3 indicates the role of oscillation anharmonism due to the nonlinear dependence of the z-component of the electron momentum on the coordinate even in the case of the magnetic field linearly dependent on y. The second parameter δ_3 should be small to make insignificant the effects of the nonlinear term in the magnetic field dependence on the coordinate given by Eq. (24.171).

Small but finite anharmonism results in the dependence of the oscillation frequency on the amplitude, which leads to the intersection of the particle trajectory with the $y = 0$ axis at different coordinates z. This leads to the aberration of the magnetic lens. The electrons being focused do not meet after the lens in one focal point. The further the trajectories initially are from the axis, the closer to the lens they intersect the axis. It is convenient to illustrate this by considering the electron motion near the axis in the magnetic field given by Eq. (24.171). Using the smallness of the electron displacement from the axis and retaining the terms of the order not higher than y^3, the equations of electron motion can be written in the form of an anharmonic oscillator equation

$$\frac{d^2y}{dt^2} + \Omega_A^2 y = \mathrm{K}_A y^3, \qquad (24.177)$$

with the oscillation frequency and the parameter of nonlinearity equal to

$$\Omega_A = \sqrt{\frac{eh_1 p_{0,z}}{m_e L \gamma_0^2}} \quad \text{and} \quad \mathrm{K}_A = \frac{e}{m_e L^3 \gamma_0^2}\left(\frac{eh_1^2 L}{2m_e c^2} + h_3 p_{0,z}\right), \qquad (24.178)$$

respectively. The nonlinearity parameter can also be written as

$$\mathrm{K}_A = \frac{\Omega_A^2}{Y_0^2}\left(\frac{\delta_1}{2} + \delta_3\right)$$

with δ_1 and δ_3 given by Eq. (24.176).

We consider the electron focusing by two magnetic configurations, by a "long focusing system" and by a "short focusing system". In the first case, the particles are focused at or near the capillary end. In the case of the short focusing system, the focus is located far from the capillary end. In the long focusing system, the particle trajectory $y|_{t=z/c}$ is given by the solution of Eq. (24.177). It can be expressed in terms of elliptic functions. In the case of the small but finite oscillation amplitude, which to the third order of y is (e.g. see Landau & Lifshitz, 1976)

$$y(t) = Y_0 \cos\left(\Omega_A t - \frac{3K_A}{8\Omega_A} Y_0^2 t\right) + \frac{Y_0^3 K_A}{32\Omega_A^2} \cos(3\Omega_A t). \qquad (24.179)$$

It yields the relationship between the amplitude of linear oscillations Y_0 and initial coordinate y_0, which reads

$$Y_0 \approx y_0 - \frac{y_0}{32}\left(\frac{\delta_1}{2} + \delta_3\right).$$

According to Eq. (24.179), the time at which the trajectory intersects the axis $y = 0$ is approximately equal to

$$t_A \approx \frac{\pi}{2\Omega_A} + \frac{3\pi K_A}{16\Omega_A^2} y_0^2. \qquad (24.180)$$

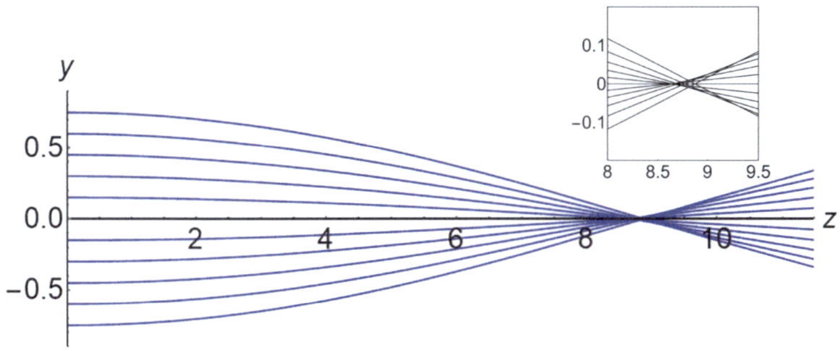

Fig. 24.20 Electron trajectories in the magnetic field (24.171). The inset shows the close-up of the focus region

It depends on the initial coordinate y_0 and nonlinearity, whose effect is characterized by the parameter $K_A \neq 0$, which leads to the finite focus length equal to

$$\Delta z_A = \frac{3\pi K_A c}{16\Omega_A^2} \Delta y_0^2. \tag{24.181}$$

where Δy_0 is the half-width of the particle beam.

Figure 24.20 shows the typical trajectories of the electrons, moving in the magnetic field whose dependence of the coordinate is given by Eq. (24.171). The inset shows the close-up of the focus region. Initial momentum components are equal to $p_{0,y} = 0$ and $p_{0,z} = 30$. The y and z coordinates at $t = 0$ are $y_0 \in [-0.75, 0.75]$ and $z_0 = 0$. The oscillation frequency is $\Omega_A = \pi/10$, and the parameter of nonlinearity K_A approximately equals $0.157\,\Omega_A^2$. The farther the particles are from the axis at $t = 0$, at the larger distance along the z-direction they intersect the axis. This is an example of positive spherical aberration.

In the short focusing system case, the capillary length is substantially smaller than the focus length. As a result of the movement in the magnetic field, the particle acquires the transverse velocity

$$v_0 = -\frac{Z}{c}\left(\Omega_A^2 y_0 - K_A y_0^3\right). \tag{24.182}$$

Here Z is the capillary length and y_0 is the particle coordinate at $t = 0$. Since the particle trajectory in the region outside of the capillary is given by

$$y = y_0 + v_0 \frac{z}{c}$$

we can find the coordinate where the electron intersects the axis. It is equal to

$$z_f = -c\frac{y_0}{v_0} \approx \frac{c^2}{Z\Omega_A^2}\left(1 + \frac{K_A}{\Omega_A^2}y_0^2\right), \tag{24.183}$$

i.e., the focus length is

$$\Delta z_A = \frac{K_A c^2}{Z\Omega_A^4}(\Delta y_0)^2. \tag{24.184}$$

24.19 Applications

The LWFA accelerated electron beams are considered for a waste range of applications (see Albert & Thomas, 2016; Esarey et al., 2009) as the electron-positron collider (Leemans & Esarey, 2009), radiation therapy (Brummer et al., 2020; Svendsen et al., 2021), compact free electron lasers (Assmann et al., 2020; Couprie, 2018; Galetti et al., 2024; Wang et al., 2021), compact source of high brightness hard photons (Corde et al., 2013).

References

Akhiezer, A. I., & Polovin, R. V. (1956). Theory of wave motion of an electron plasma. *Soviet Physics—JETP, 3*, 696.

Albert, F., & Thomas, A. G. R. (2016). Applications of laser wakefield accelerator-based light sources. *Plasma Physics and Controlled Fusion, 58*, Article 103001.

Andreić, Ž, Ellwi, S., Pleslić, S., & Kunze, H.-J. (2003). In-band EUV radiation of ablative capillary discharges in PVC. *Vacuum, 71*, 229.

Aniculaesei, A., Ha, T., Ioffe, S., et al. (2024). Aacceleration of a high-charge electron bunch to 10 GeV in a 10-cm nanoparticle-assisted wakefield accelerator. *Matter and Radiation at Extremes, 9*, Article 014001.

Arefiev, A. V., Breizman, B. N., Schollmeier, M., & Khudik, V. N. (2012). Parametric Amplification of Laser-Driven Electron Acceleration in Underdense Plasma. *Physical Review Letters, 108*, 145004.

Assmann, R. W., et al. (2020). EuPRAXIA conceptual design report. *The European Physical Journal Special Topics, 229*, 3675.

Bagdasarov, G. A., Bobrova, N. A., Boldarev, A. S., et al. (2017). On production and asymmetric focusing of flat electron beams using rectangular capillary discharge plasmas. *Physics of Plasmas, 24*, Article 123120.

Bagdasarov, G. A., Bobrova, N. A., Olkhovskaya, O. G., et al. (2021). Creation of an axially uniform plasma channel in a laser-assisted capillary discharge. *Physics of Plasmas, 28*, Article 053104.

Bagdasarov, G. A., Kruchinin, K. O., Molodozhentsev, A. Yu., et al. (2022). Discharge plasma formation in square capillary. *Physical Review Research, 4*, Article 013063.

Bobrova, N. A., & Sasorov, P. V. (1993). MHD equations for a fully ionized plasma of complex composition. *Plasma Physics Reports, 19*, 409.

Bobrova, N. A., Lazzaro, E., & Sasorov, P. V. (1995). Magnetohydrodynamic two-temperature equations for multicomponent plasma. *Physics of Plasmas, 12*, 022105.

Bobrova, N. A., Bulanov, S. V., Razinkova, T. L., & Sasorov, P. V. (1996). Dynamics of capillary discharges. *Plasma Physics Reports, 22*, 387.

Bobrova, N. A., Esaulov, A. A., Sakai, J.-I., Sasorov, P. V., Spence, D. J., Butler, A., Hooker, S. M., & Bulanov, S. V. (2002). Simulations of a hydrogen-filled capillary discharge waveguide. *Physical Review E, 65*, 016407.

Brantov, A. V., Esirkepov, T. Zh., Kando, M., et al. (2008). Controlled electron injection into the wake wave using plasma density inhomogeneity. *Physics of Plasmas, 15*, Article 073111.

Brummer, T., Debus, A., Pausch, V., Osterhoff, J., & Gruner, F. (2020). Design study for a compact laser-driven source for medical x-ray fluorescence imaging. *Physical Review Special Topics—Accelerators and Beams, 23*, 031601.

Buck, A., Wenz, J., Xu, J., Khrennikov, K., Schmid, K., Heigoldt, M., Mikhajlova, J. M., Geisler, M., Shen, B., Krausz, F., et al. (2013). Shock-front injector for high-quality laser–plasma acceleration. *Physical Review Letters, 110*, Article 185006.

Bulanov, S. V., & Tajima, T. (2005). On the quasi-monoenergetic electron beam generation in the laser wakefield acceleration. *Journal of the Particle Accelerator Society of Japan, 2*, 35.

Bulanov, S. V., Inovenkov, I. N., Kirsanov, V. I., et al. (1992). Nonlinear depletion of ultrashort and relativistically strong laser pulses in an underdense plasma. *Physics of Fluids B, 4*, 1935.

Bulanov, S. V., Pegoraro, F., Pukhov, A. M., & Sakharov, A. S. (1997a). Transverse-wake wave breaking. *Physical Review Letters, 78*, 4205.

Bulanov, S. V., Esirkepov, TZh., Lontano, M., & Pegoraro, F. (1997b). Stability of vortex streets and vortex rows in the framework of the Hasegawa-Mima equation. *Plasma Physics Reports, 23*, 660.

Bulanov, S. V., Naumova, N. M., Pegoraro, F., & Sakai, J.-I. (1998). Particle injection into the acceleration phase of the wake field behind the laser pulse. *Physical Review E, 58*, 5257.

Bulanov, S. S., Schroeder, C. B., Esarey, E., & Leemans, W. P. (2013). Electromagnetic cascade in high-energy electron, positron, and photon interactions with intense laser pulses. *Physical Review A, 87*, 062110.

Bulanov, S. V., Esirkepov, T. Zh., Kando, M., & Koga, J. (2016). Relativistic mirrors in laser plasmas (analytical methods). *Plasma Sources Science and Technology, 25*, Article 053001.

Bulanov, S. V., Esirkepov, T. Zh., Hayashi, Y., Kiriyama, H., Koga, J. K., Kotaki, H., Mori, M., & Kando, M. (2016). On some theoretical problems of laser wake-field accelerators. *Journal of Plasma Physics, 82*, 905820308.

Bulanov, S. V., Esirkepov, T. Zh., Kando, M., Koga, J., Pirozhkov, A. S., Nakamura, T., Bulanov, S. S., Schroeder, C. B., Esarey, E., Califano, F., & Pegoraro, F. (2012a). On the breaking of a plasma wave in a thermal plasma. I. The structure of the density singularity. *Physics of Plasmas, 19*, 113102.

Bulanov, S. V., Esirkepov, T. Zh., Kando, M., Koga, J., Pirozhkov, A. S., Nakamura, T., Bulanov, S. S., Schroeder, C. B., Esarey, E., Califano, F., & Pegoraro, F. (2012b). *On the breaking of a plasma wave in a thermal plasma. II. Electromagnetic wave interaction with the breaking plasma wave. Physics of Plasmas, 19*, 113103.

Chan, L. S., Tan, D., Saboohi, S., Yap, S. L., & Wong, C. S. (2014). Operation of an electron beam initiated metallic plasma capillary discharge. *Vacuum, 103*, 38.

Chen, P., Dawson, J. M., Huff, R. W., & Katsouleas, T. (1985). Acceleration of electrons by the interaction of a bunched electron beam with a plasma. *Physical Review Letters, 54*, 693.

Chen, P., Su, J. J., Katsouleas, T., Wilks, S., & Dawson, J. M. (1987). Plasma focusing for high-energy beams. *IEEE Transactions on Plasma Science, 15*, 218.

Chen, M., Sheng, Z.-M., Ma, Y., & Zhang, J. (2006). Electron injection and trapping in a laser wakefield by field ionization to high-charge states of gases. *Journal of Applied Physics, 99*, Article 056109.

Chiadroni, E., Anania, M. P., Bellaveglia, M., et al. (2018). Overview of plasma lens experiments and recent results at SPARC_LAB. *Nuclear Instruments and Methods in Physics Research Section A, 909*, 16.

Corde, S., Ta Phuoc, K., Lambert, G., Fitour, R., Malka, V., Rousse, A., Beck, A., & Lefebvre, E. (2013). Femtosecond x rays from laser-plasma accelerators. *Reviews of Modern Physics, 85*, 1.

Couprie, M. E. (2018). Towards compact free electron–laser based on laser plasma accelerators. *Nuclear Instruments and Methods in Physics Research, Section A, 909*, 5.

Drobniak, P., Adli, E., Bergravf-Anderson, H., et al. (2025). Development of a nonlinear plasma lens for achromatic beam transport. *Nuclear Instruments and Methods in Physics Research—Section A, 1072*, Article 170223.

Ehrlich, Y., Cohen, C., Zigler, A., Krall, J., Sprangle, P., & Esarey, E. (1996). Guiding of high intensity laser pulses in straight and curved plasma channel experiments. *Physical Review Letters, 77*, 4186.

Esarey, E., Schroeder, C. B., & Leemans, W. P. (2009). Physics of laser-driven plasma-based electron accelerators. *Reviews of Modern Physics, 81*, 1229.

Esirkepov, T. Zh., Bulanov, S. V., Yamagiwa, M., & Tajima, T. (2006). Electron, positron, and photon wakefield acceleration: Trapping, wake overtaking, and ponderomotive acceleration. *Physical Review Letters, 96*, Article 014803.

Esirkepov, T. Zh., Kato, Y., & Bulanov, S. V. (2008). Bow wave from ultraintense electromagnetic pulses in plasmas. *Physical Review Letters, 101*, Article 265001.

Faure, J., et al. (2006). Controlled injection and acceleration of electrons in plasma wakefields by colliding laser pulses. *Nature (London), 444*, 737.

Faure, J., Rechatin, C., Lundh, O., Ammoura, L., & Malka, V. (2010). Injection and acceleration of quasimonoenergetic relativistic electron beams using density gradients at the edges of a plasma channel. *Physics of Plasmas, 18*, Article 083107.

Fukuda, Y., Akahane, Y., Aoyama, M., Hayashi, Y., Homma, T., Inoue, N., Kando, M., Kanazawa, S., Kiriyama, H., Kondo, S., Kotaki, H., Masuda, S., Mori, M., Yamazaki, A., Yamakawa, K., Echkina, E. Yu., Inovenkov, I. N., Koga, J., & Bulanov, S. V. (2007). Ultrarelativistic electron generation during the intense, ultrashort laser pulse interaction with clusters, *Physics Letters A, 363*, 130.

Gahn, C., Tsakiris, G. D., Pukhov, A., Meyer-ter-Vehn, J., Pretzler, G., Thirolf, P., Habs, D., & Witte, K. J. (1999). Multi-MeV Electron Beam Generation by Direct Laser Acceleration in High-Density Plasma Channels. *Physical Review Letters, 83*, 4772.

Galetti, M., Assmann, R., Couprie, M. E., et al. (2024). Prospects for free-electron lasers powered by plasma-wakefield-accelerated beams. *Nature Photonics, 18*, 780.

Geddes, C. G. R., Nakamura, K., Plateau, G. R., Toth, C. S., Cormier-Michel, E., Esarey, E., Schroeder, C. B., Cary, J. R., & Leemans, W. P. (2008). Plasma-density-gradient injection of low absolute-momentum-spread electron bunches. *Physical Review Letters, 100*, Article 215004.

Golovanov, A. A., Nerush, E. N., & Kostyukov, I. Yu. (2022). Radiation reaction-dominated regime of wakefield acceleration. *New Journal of Physics, 24*, 033011.

Gonsalves, A. J., Nakamura, K., Daniels, J., Benedetti, C., et al. (2019). Petawatt laser guiding and electron beam acceleration to 8 GeV in a laser-heated capillary discharge waveguide. *Physical Review Letters, 122*, Article 084801.

Grüner, F., Becker, S., Schramm, U., et al. (2007). Design considerations for table-top laser-based VUV and X ray free electron lasers. *Applied Physics B, 86*, 431.

Hosokai, T., Kando, M., Dewa, H., Kotaki, H., Kondo, S., Hasegawa, N., Nakajima, K., & Horioka, K. (2000). Optical guidance of terrawatt laser pulses by the implosion phase of a fast z-pinch discharge in a gas-filled capillary. *Optics Letters, 25*, 10.

Hosokai, T., Kinoshita, K., Zhidkov, A., Maekawa, A., Yamazaki, A., & Uesaka, M. (2006). Effect of external static magnetic field on the emittance and total charge of electron beams generated by laser-wakefield acceleration. *Physical Review Letters, 97*, Article 075004.

Irshad, F., Eberle, C., & Foerster, F. M. (2024). Pareto optimization and tuning of a laser wakefield accelerator. *Physical Review Letters, 133*, 085001.

Jirka, M., Vranic, M., Grismayer, T., & Silva, L. O. (2020). Scaling laws for direct laser acceleration in a radiation-reaction dominated regime. *New Journal of Physics*, *22*, 083058.

Kameshima, T., Kotaki, H., Kando, M., et al. (2009). Laser pulse guiding and electron acceleration in the ablative capillary discharge plasma. *Physics of Plasmas, 16*, Article 093101.

Kando, M., Fukuda, Y., Pirozhkov, A. S., Ma, J., Daito, I., Chen, L.-M., Esirkepov, T. Z., Ogura, K., Homma, T., Hayashi, Y., Kotaki, H., Sagisaka, A., Mori, M., Koga, J. K., Daido, H., Bulanov, S. V., Kimura, T., Kato, Y., & Tajima, T. (2007). Demonstration of Laser-frequency upshift by electron-density modulations in a plasma wakefield. *Physical Review Letters, 99*, Article 135001.

Kim, H. T., Pae, K. H., Cha, H. J., Kim, I. J., Yu, T. J., Sung, J. H., et al. (2013). Enhancement of electron energy to the multi-GeV regime by a dual-stage laser-wakefield accelerator pumped by petawatt laser pulses. *Physical Review Letters, 111*, Article 165002.

Kirchen, M., Jalas, Messner, P., et al. (2021). Optimal beam loading in a laser-plasma accelerator. *Physical Review Letters, 126*, 174801.

Kotaki, H., Masuda, S., Kando, M., Koga, J. K., et al. (2004). Head-on injection of a high quality electron beam by the interaction of two laser pulses. *Physics of Plasmas, 11*, 3296.

Kotaki, H., Daito, I., Kando, M., Hayashi, Y., Kawase, K., Kameshima, T., Fukuda, Y., Homma, T., Ma, J., Chen, L.-M., Esirkepov, TZh., Pirozhkov, A. S., Koga, J. K., Faenov, A., Pikuz, T., Kiriyama, H., Okada, H., Shimomura, T., Nakai, Y., … Bulanov, S. V. (2009). Electron optical injection with head-on and counter-crossing colliding laser pulses. *Physical Review Letters, 103*, Article 194803.

Lamb, H. (1932). *Hydrodynamics*. Cambridge University Press.

Landau, L. D., & Lifshitz, E. M. (1976). *Course of theoretical physics, Vol. 1. Mechanics*. Pergamon.

Landau, L. D., & Lifshitz, E. M. (1984). *Course of theoretical physics, Vol. 8: Electrodynamics of continuous media*. Pergamon Press.

Lazzarini, C. M., Grittani, G. M., Valenta, P., Zymak, I., Antipenkov, R., Chaulagain, U., Goncalves, L. V. N., Grenfell, A., Lamac, M., Lorenz, S., Nevrkla, M., Spacek, A., Sobr, V., Szuba, W., Bakule, P., Korn, G., & Bulanov, S. V. (2024). Ultrarelativistic electron beams accelerated by terawatt scalable kHz laser. *Physics of Plasmas, 31*, Article 030703.

Leemans, W. P., & Esarey, E. (2009). Laser-driven plasma-wave electron accelerators. *Physics Today, 62*, 44.

Leemans, W. P., Gonsalves, A. J., Mao, H.-S., Nakamura, K., Benedetti, C., Schroeder, C. B., Cs Tóth, J., Daniels, D. E., Mittelberger, S. S. B., Vay, J.-L., Geddes, C. G. R., & Esarey, E. (2014). Multi-GeV electron beams from capillary-discharge-guided sub-petawatt laser pulses in the self-trapping regime. *Physical Review Letters, 113*, Article 245002.

Lindstrøm, C. A., Adli, E., Boyle, G., et al. (2018). Emittance preservation in an aberration-free active plasma lens. *Physical Review Letters, 121*, Article 194801.

Lindstrøm, C. A. (2021). Staging of plasma-wakefield accelerators. *Physical Review Special Topics—Accelerators and Beams, 24*.

Lu, W., Tzoufras, M., Joshi, C., Tsung, F. S., Mori, W. B., Vieira, J., Fonseca, R. A., & Silva, L. O. (2007). Generating multi-gev electron bunches using single stage laser wakefield acceleration in a 3D nonlinear regime. *Physical Review Accelerators and Beams, 10*, Article 061301.

Lundh, O., Lim, J., Rechatin, C., et al. (2011). Few femtosecond, few kiloampere electron bunch produced by a laser–plasma accelerator. *Nature Physics, 7*, 219. https://doi.org/10.1038/nphys1872

Masson-Laborde, P. E., Mo, M. Z., Ali, A., et al. (2014). Giga-electronvolt electrons due to a transition from laser wakefield acceleration to plasma wakefield acceleration. *Physics of Plasmas, 21*, 123113.

Panofsky, W. K. H., & Baker, W. (1950). A focusing device for the external 350-MeV proton beam of the 184-inch cyclotron at Berkeley. *Review of Scientific Instruments, 21*, 445.

Picksley, A., Stackhouse, J., Benedetti, C., Nakamura, K., Tsai, H. E., Li, R., Miao, B., Shrock, J. E., Rockafellow, E., Milchberg, H. M., Schroeder, C. B., van Tilborg, J., Esarey, E., Geddes, C. G. R., & Gonsalves, A. J. (2024). Matched guiding and controlled injection in dark-current-free, 10-GeV-class, channel-guided laser-plasma accelerators. *Physical Review Letters, 133*, Article 255001.

Pirozhkov, A. S., Kando, M., Esirkepov, TZh., Gallegos, P., Ahmed, H., Ragozin, E. N., Faenov, A. Y., Pikuz, T. A., Kawachi, T., Sagisaka, A., Koga, J. K., Coury, M., Green, J., Foster, P., Brenner, C., Dromey, B., Symes, D. R., Mori, M., Kawase, K., … Bulanov, S. V. (2012). Soft-X-ray harmonic comb from relativistic electron spikes. *Physical Review Letters, 108*, Article 135004.

Poston, T., & Stewart, I. (1978). *Catastrophe theory and its applications*. Pitman.

Pousa, F., Martinez de la Ossa, A., Brinkmann, A., & Assmann, R. W. (2019). Compact multistage plasma-based accelerator design for correlated energy spread compensation. *Physical Review Letters, 123*, Article 054801.

Pukhov, A., & Meyer-ter-Vehn, J. (2002). Laser wake field acceleration: The highly non-linear broken-wave regime. *Applied Physics B, 74*, 355.

Pukhov, A., Sheng, Z.-M., & Meyer-ter-Vehn, J. (1999). Particle acceleration in relativistic laser channels. *Physics of Plasmas, 6*, 2847.

Rocca, J. J., Shlyaptsev, V., Tomasel, F. G., et al. (1994). Demonstration of a discharge pumped table-top soft-X-ray laser. *Physical Review Letters, 73*, 2192.

Röckemann, J.-H., Schaper, L., Barber, S., Bobrova, N., Boyle, G., Bulanov, S., Delbos, N., Floettmann, K., Kube, G., Lauth, W., Leemans, W., Libov, V., Maier, A., Meisel, M., Messner, P., Sasorov, P., Schroeder, C., van Tilborg, J., Wesch, S., & Osterhoff, J. (2018). Direct measurement of focusing fields in active plasma lenses. *Physical Review Special Topics—Accelerators and Beams, 21*, Article 122801.

Sasorov, P. V. (2018). Plasma dynamics in capillary discharges, X-ray lasers 2016. In T. Kawachi, S. V. Bulanov, H. Daido, Y. Kato (Eds.,) *Proceedings of the 15th International Conference on X-Ray Laserts. Springer Proceedings in Physics* (Vol. 202, p. 45). Springer International Publishing AG.

Schroeder, C. B., Esarey, E., Geddes, C. G. R., Benedetti, C., & Leemans, W. P. (2010). Physics considerations for laser-plasma linear colliders. *Physical Review Special Topics—Accelerators and Beams, 13*, Article 101301.

Singh, P. K., Li, F.-Y., Huang, C.-K. et al. (2022). Vacuum laser acceleration of super-ponderomotive electrons using relativistic transparency injection. *Nature Communications, 13*, 54.

Sjobak, K. N., Adli, E., Corsini, R., et al. (2021). Strong focusing gradient in a linear active plasma lens. *Physical Review Accelerators and Beams, 24*, 121306.

Steinke, S., Van Tilborg, J., Benedetti, C., Geddes, C. G. R., Schroeder, C. B., Daniels, J., Swanson, K. K., Gonsalves, A. J., Nakamura, K., Matlis, N. H., Shaw, B. H., Esarey, E., & Leemans, W. P. (2016). Multistage coupling of independent laser-plasma accelerators. *Nature, 530*(7589), 190.

Strutt, W. (Lord Rayleigh). (1897). On the passage of electric waves through tubes, or the vibrations of dielectric cylinders. *Philosophical Magazine, 43*, 125.

Suk, H., Barov, N., Rosenzweig, J. B., & Esarey, E. (2001). Plasma electron trapping and acceleration in a plasma wake field using a density transition. *Physical Review Letters, 86*, 1011.

Svendsen, K., Guenot, D., Svensson, J. B., Petersson, K., Persson, A., & Lundh, O. (2021). A focused very high energy electron beam for fractionated stereotactic radiotherapy. *Science and Reports, 11*, 5844.

Tajima, T., & Dawson, J. M. (1979). Laser electron accelerator. *Physical Review Letters, 43*, 267.

Umstadter, D., Kim, J. K., & Dodd, E. (1996). Laser injection of ultrashort electron pulses into wakefield plasma waves. *Physical Review Letters, 76*, 2073.

Valenta, P., Maslarova, D., Babjak, R., Martinez, B., Bulanov, S. V., & Vranic, M. (2024). Direct laser acceleration: A model for the electron injection from the walls of a cylindrical guiding structure. *Physical Review E* 109, 065204.

Valenta, P., & Bulanov, S. V. (2025). Optimized scaling for reaching maximum electron energy from laser wakefield accelerator. *Proceedings of SPIE, 13534*, Laser Acceleration of Electrons, Protons, and Ions VIII, 1353406 (5 June 2025). https://doi.org/10.1117/12.3058376.

Valenta, P., Esirkepov, T. Z., Ludwig, J. D., Wilks, S. C., & Bulanov, S. V. (2025). Bayesian optimization of electron energy from laser wakefield accelerator. *Physical Review Accelerators and Beams,* 28, 094601.

Van Tilborg, J., Steinke, S., Geddes, C. G. R., et al. (2015). Active plasma lensing for relativistic laser-plasma-accelerated electron beams. *Physical Review Letters, 115*, Article 184802.

Wang, W., Feng, K., Ke, L., et al. (2021). Free-electron lasing at 27 nanometres based on a laser wakefield accelerator. *Nature, 595*, 516.

Whitham, G. B. (1974). *Linear and nonlinear waves.* Wiley.

Willingale, L., Thomas, A. G. R., Nilson, P. M., et al. (2013). Surface waves and electron acceleration from high-power, kilojoule-class laser interactions with underdense plasma. *New Journal of Physics, 15*, 025023.

Yu, L.-L., Esarey, E., Schroeder, C. B., et al. (2014). Two-color laser-ionization injection. *Physical Review Letters, 112*, Article 125001.

Chapter 25
Laser Ion Accelerators

Abstract There are four basic ion acceleration mechanisms in relativistic laser plasma. they are the Coulomb explosion, when almost all electrons are blowed out from the target by the laser radiation, the Target Normal Sheath Acceleration mechanism the electric charge separation electric field generated on the target surface by hot electrons, the Magnetic Vortex acceleration mechanism arizes due to the self-generated magnetic field, and the most efficient mechanism of the ion acceleration by radiation pressure of the laser light.

Keywords Ion acceleration in laser plasma · Coulomb explosion · Target normal sheath acceleration mechanism · Ion acceleration by light radiation pressure

25.1 Mechanisms of the Laser Ion Acceleration

That ions can be generated in the interaction of laser radiation with various targets, it was found in experiments in the early 1960-ies. Under the conditions of the experiments in those years the ion source was the plasma formed at the surface of the relatively large targets heated by the laser radiation. Beams of fast ions ejected from the front target surface were directed towards the laser pulse. The laser power growth, the transition to shorter pulses and the increase of the radiation contrast has led to the realization of the regimes, when the ions are accelerated in the direction of the laser radiation propagation. Their source is localized at the rear side of the target with respect to the laser beam. One of the first theoretical works predicted such a mode of acceleration, have been published by Denavit (1992).

As the theory of charged particle acceleration by an intense electromagnetic radiation predicts, with the usage of existing and/or planned lasers it is possible to obtain the ion beams of energies in the range from hundreds of KeV to GeV, which corresponds to the needs of controlled thermonuclear fusion (Bychenkov et al., 2001; Roth et al., 2001), of hadron therapy with laser accelerated ions (Bulanov & Khoroshkov, 2002; Bulanov et al., 2002, 2014) and fundamental physics (Esirkepov et al., 2004). The choice of lasers able to generate the femtosecond pulses is more preferable.

Y. Kato et al., *High Power Laser and Plasma Science*, Springer Series on Atomic, Optical, and Plasma Physics 130, https://doi.org/10.1007/978-3-032-08930-4_25

This conclusion follows from the steeper dependence of the ion energy on the laser radiation intensity in the case of ultra-short pulses in comparison with the long pulse lasers.

Unlike the discussed above laser electron acceleration, based primarily on the acceleration by the plasma wake waves in the low-density plasma, the high energy ions can be generated by various mechanisms.

The simplest mechanism of acceleration, but less efficient in the case of ion acceleration, is based on the property of the sufficiently large amplitude electromagnetic wave to accelerate charged particles in vacuum by the direct acceleration mechanism. In a planar electromagnetic wave, a particle with the charge $Z_a e$ and mass $m_a = A_a m_p$ achieves the energy

$$\mathcal{E}_a = m_a c^2 \left[1 + \left(\frac{Z_a m_e}{m_a} \right)^2 \frac{a_0^2}{2} \right], \tag{25.1}$$

where the normalized wave amplitude is given by $a_0 = eE_0/m_e \omega c$ (see Eq. (20.36)). In order to accelerate a proton to the energy of 200 MeV by the sharply focused electromagnetic pulse, the required power of the laser radiation must be of the order of one petawatt. To accelerate fully ionized carbon ions up to 400 MeV/nucleon energy, the radiation power should be equal to 8 PW, which imposes too high requirements for laser systems. For this reason, main attention in the theory and the experiment is paid to collective methods of acceleration, because in the process of interaction of electromagnetic waves with plasma the long-lived static or low-frequency collective electromagnetic field can be formed, which may accelerate charged particles to the energies significantly greater than the value given by the expression (25.1).

The basic idea of laser ion accelerators is based on the high efficiency of laser energy conversion into the energy of fast ions in the process of its interaction with the plasma in the limit, when the laser power reaches the petawatt level. Collimated beams of fast ions are recorded in experiments on the interaction of laser pulses with solid and gas targets (see review articles, Borghesi et al., 2006; Daido et al., 2012; Macchi et al., 2013; Mourou et al., 2006). The number of particles in the beam and the conversion efficiency of laser energy into the energy of fast ions can reach 10^{13} and 10%, respectively.

There are several basic mechanisms of ion acceleration in the interaction of laser radiation with solid and gas targets. The fundamental difference between the solid targets from the gas targets is that the electron density in the solid targets is so high that such targets are not transparent to laser radiation. In other words, the electron density exceeds the critical density $m_e \omega^2 / 4\pi e^2$ for the radiation with the frequency ω. In the gas target, the ionization of a gas by laser radiation leads to plasma formation with subcritical density, that is, the target is transparent to the laser pulses. The plasma with a near-critical density can be formed by irradiating a solid target by the prepulse that precedes the main laser pulse, with many orders of magnitude smaller amplitude, but having many orders of magnitude greater duration (Esirkepov et al., 2014; Matsukado et al., 2003). Due to its relatively long duration the formation of a plasma corona occurs with the size substantially larger than the target thickness that in its turn has a significant impact on the whole process of interaction. In particular, the plasma corona may absorb a substantial part of the laser pulse energy, with the result of the plasma electrons heating whose energy then can be transferred to the ions.

In the process of the solid target irradiation by powerful laser, the ion acceleration occurs in the main following regimes: by the Coulomb explosion, as the ion acceleration in the layer at the target surface in the electric field of charge separation, and by the radiation pressure of a strong electromagnetic wave.

25.2 Coulomb Explosion

Apparently, from the point of view of the underlying mechanism the Coulomb explosion is the simplest process of ion acceleration (Last et al., 1997; Nishihara et al., 2001).

In this regime, almost all of the electrons are removed from the focus area of the irradiated target under the action of ponderomotive pressure of the laser radiation. The remaining ion core expands (explodes) due to the Coulomb repulsion of positive electric charges. In the most pure form the Coulomb explosion regime can be realized in the interaction of laser radiation with the clusters (Last et al., 1997; Nishihara et al., 2001; Kovalev & Bychenkov, 2003). The clusters are clumps of matter of solid-state density and micron size, Krainov and Smirnov (2002). Usually, they are formed during the gas expansion to vacuum. A typical distance between the individual clusters is equal to several micrometres.

As a result of the cluster irradiation, the electrostatic potential is created when electrons are ejected under the action of ponderomotive force of the laser radiation. Its maximum value equals to the electrostatic potential at the surface of the charged sphere with the radius R_{cl} and the density of charged particles $n_{0,\alpha}$:

$$\varphi_{\max} = 2\pi n_{0,\alpha} e R_{cl}^2/3. \tag{25.2}$$

Assuming that the ion temperature is equal to zero and that the ions move along the radius, we can find the ratio between the kinetic energy of ions

$$\mathcal{E} = \sqrt{m_a^2 c^4 + p_r^2 c^2} - m_a c^2 \tag{25.3}$$

and their potential energy

$$\Pi(r_0, t) = 4\pi e^2 Q(r_0)\left(\frac{1}{r_0 + \xi_\alpha(r_0, t)} - \frac{1}{r_0}\right), \tag{25.4}$$

which corresponds to the energy integral $\mathcal{E} - \Pi(r_0, t) = \text{constant}$. Here r_0 and t are the Lagrange variables, where r_0 is the ion initial coordinate, $\xi_\alpha(r_0, t)$ is the ion displacement at time t from the initial position, and the number of particles $Q(r_0)$ within the sphere of the radius r_0:

$$Q(r_0) = \int_0^{r_0} n_{0,\alpha}(r_0) r_0^2 dr_0. \tag{25.5}$$

In the process of the ion cloud expansion the kinetic energy increases, reaching the value $4\pi e^2 Q(r_0)/r_0$ in the limit $\xi_\alpha \to \infty$.

The ion energy is determined by the coordinate of its initial position inside the cloud. Assuming that the ion density $n_{0,\alpha}$ inside the cloud is homogeneous, we obtain for the final ion energy the value

$$\mathcal{E}_\alpha = 2\pi e^2 n_{0,\alpha} r_0^2/3,$$

which cannot exceed $\mathcal{E}_{\alpha,\max} = 2\pi e^2 n_{0,\alpha} R_{cl}^2/3$. Using the fact that the ion energy is proportional to r_0^2, we can find the ion energy spectrum $\mathcal{N}_\alpha(\mathcal{E}) = df_\alpha/d\mathcal{E}$, which by virtue of the flux conservation in the phase space is equal to $4\pi r_0^2 dr_0/d\mathcal{E}$. From this, it follows that

$$\mathcal{N}_\alpha(\mathcal{E}) = \frac{3R}{2Z_\alpha^2 e^2}\sqrt{\frac{\mathcal{E}}{\mathcal{E}_{\alpha,\max}}}\theta\left(\mathcal{E}_{\alpha,\max} - \mathcal{E}\right), \tag{25.6}$$

where $\theta(x)$ is the defined above the unit step Heaviside function. The energy spectrum of this form has been demonstrated in the experiments Sakabe et al., (2004) with the cluster targets irradiated by the high intense laser radiation.

In the limit of relatively low ion energy, $\mathcal{E}_\alpha < m_\alpha c^2$, we can use a non-relativistic approach. Within the framework of this approximation the solution of the equations of motion can be written in the Lagrange variables r_0, t as

$$\frac{1}{2}\ln\left(\frac{2\xi_\alpha + r_0 + 2\sqrt{\xi_\alpha^2 + r_0\xi_\alpha}}{r_0}\right) + \frac{\sqrt{\xi_\alpha^2 + r_0\xi_\alpha}}{r_0} = \sqrt{\frac{2}{3}}\omega_{p\alpha} t. \tag{25.7}$$

On the right-hand side of Eq. (25.7) $\omega_{p\alpha} = \sqrt{4\pi n_{0,\alpha} Z_\alpha e^2/m_\alpha}$ is equal to the ion plasma frequency. When the displacement is small, $\xi_\alpha \ll r_0$, the ions move with constant acceleration, when $\xi_\alpha \approx r_0(\omega_{p\alpha} t)^2/6$, while in the limit $\xi_\alpha \to \infty$ we have $\xi_\alpha \approx \sqrt{2/3} r_0 \omega_{p\alpha} t$. In the latter case, the ion velocity is constant. The characteristic time of the ion cloud expansion approximately equal to the inverse value of the ion plasma frequency $\omega_{p\alpha}^{-1}$.

Now we consider the ion acceleration in multi-species cluster, assuming that the concentration of the species ions is relatively small: $Z_\beta n_\beta \ll Z_\alpha n_\alpha$. Here $Z_\beta e$ and n_β are equal to the electric charge and density of the minor species ions, respectively. Within the framework of this approximation, the dynamics of impurity ions can be described as the motion of a test particle in a given electric field, generated by the spatial distribution of the electric charge of the main component ions. Solving the problem of the ion motion in this electric field, we can show, that the energy spectrum of the impurity ions has the form

$$\mathcal{N}_\beta(\mathcal{E}) = \frac{3R\sqrt{\mathcal{E}_{\beta,\max} - \mathcal{E}}}{\sqrt{2\mathcal{E}_{\beta,\max}}Z_\alpha Z_\beta e^2}\left(\frac{n_{0,\beta}}{n_{0,\alpha}}\right)\theta\left(\mathcal{E}_{\beta,\max} - \mathcal{E}\right)\theta\left(\mathcal{E} - \mathcal{E}_{\beta,\min}\right), \tag{25.8}$$

where the maximum and minimum energies are equal to

$$\mathcal{E}_{\beta,\max} = \frac{8\pi Z_\alpha Z_\beta e^2 n_{0,\alpha} R^2}{3} \tag{25.9}$$

and

$$\mathcal{E}_{\beta,\min} = \frac{4\pi Z_\beta e^2 R^2}{3}\left(Z_\alpha n_{0,\alpha} - Z_\beta n_{0,\beta}\right). \tag{25.10}$$

As an illustration of ion acceleration in the process of Coulomb explosion of the cluster, we present in Fig. 25.1 the results of the 3D PIC computer simulations of the laser radiation interaction with the cluster. The linearly polarized with the electric field parallel to the z-axis laser pulse propagates in the x direction. Its normalized amplitude is equal to 10, which corresponds to the radiation intensity $I = 1.37 \times 10^{20}$ W/cm^2. The cluster comprising of the electrons and protons, $m_\alpha/m_e = 1836$, has a diameter equal to 0.2 μm. The distribution of the plasma density, equal to 100 n_{cr}, at $t = 0$, is homogeneous inside the cluster. The simulation has been carried out on the mesh of the size $1024 \times 1024 \times 1024$. The total particle number is 3×10^6. The top row shows the spatial distribution of electrons and protons in consecutive moments of time. It is seen that the electrons are blown off by the electromagnetic wave in the direction of the wave propagation. The remaining proton expands with a small deviation from spherical symmetry. In the lower part of the figure, we show the energy spectrum of protons at time $t = 40$ fs. The dashed line shows the spectrum described by formula (25.6).

In the case when the density distribution within a cluster is not uniform, the Coulomb explosion leads at some moment of time to the formation of a singularity in the ionic component. The singularity corresponds to the gradient catastrophe of specific type. Because of the singularity development, the velocity gradient and the density of ions become infinite at some coordinate as shown by Kaplan et al. (2003).

Figure 25.2 shows the ion density distribution and the profile of their velocity for time instants before and during formation of the singularity in the density of ions. In the initial configuration at $t = 0$ the ion density dependence on the radius in the spherically symmetric cluster is given by expression $n_\alpha(r_0) = (4/\pi^{1/2}R^3)\exp(-r_0^2/R^2)$. Initial radius of the cluster is equal to $R = 0.5\,\mu m$. As it is seen in Fig. 25.2a and b the ion cloud expands with the velocity proportional to the radius in the region close the cluster centre. At some distance from the cluster centre the ion velocity

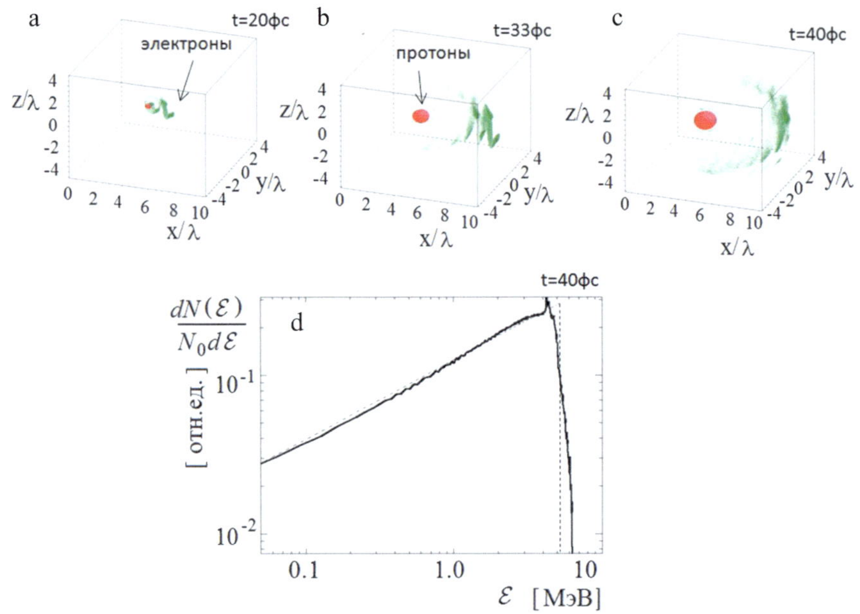

Fig. 25.1 Interaction of laser radiation with a cluster. Top row: **a–c** Distributions of electrons and protons at consecutive instants of time. Bottom row: **d** Energy spectrum of protons at time $t = 40$ fs. The dashed line shows the spectrum described by formula (25.6). Reproduced from Bulanov et al., (2014) @ Uspekhi Fizicheskikh Nauk 2014

reaches the maximum and then decreases. Therefore the internal ion layers move faster than the ions at the periphery of the cluster, leading to the steepening of the ion velocity profile and to formation at some instant of time the singularity in the density distribution of the ions. This singularity is integrable: although the density there tends to infinity, in its neighbourhood the finite number of particles remains.

The singularity formation, with corresponding maximum of the ion velocity has as a consequence the quasi-mono-energetic ion energy spectrum. The energy spectrum of the ions has the form $\mathcal{N}_\alpha(\mathcal{E}) \propto 1/\sqrt{\mathcal{E}_{\alpha,\max} - \mathcal{E}}$.

The ion acceleration in the Coulomb explosion mechanism is realized under the condition that the electric field of the laser radiation exceeds the value of the electric field on the surface of the cluster: $E_{las} > 4\pi n_{0,\alpha} e R_{cl}/3$. It can be rewritten as $a > \omega_{pe}^2 R_{cl} \lambda / 6\pi c^2$. The maximal energy of accelerated ions is given by the expression $\mathcal{E}_\alpha = m_e c^2 \omega_{pe}^2 R_{cl}^2 / 6c^2$. As for applications in hadron therapy the proton energy of about 200 MeV is required, for solid-state density cluster, $n_{0,\alpha} \approx 10^{23}$ см$^{-3}$, we can find the cluster size equal to $R_{cl} \approx \lambda \approx 1$ μm, the number of accelerated particles per one cluster: $N \approx 4 \times 10^{12}$. For the required normalized laser field amplitude this yields $a > 200$, i.e. the radiation intensity should be of the order of 5×10^{22} W/cm^2, which corresponds to the petawatt power class laser.

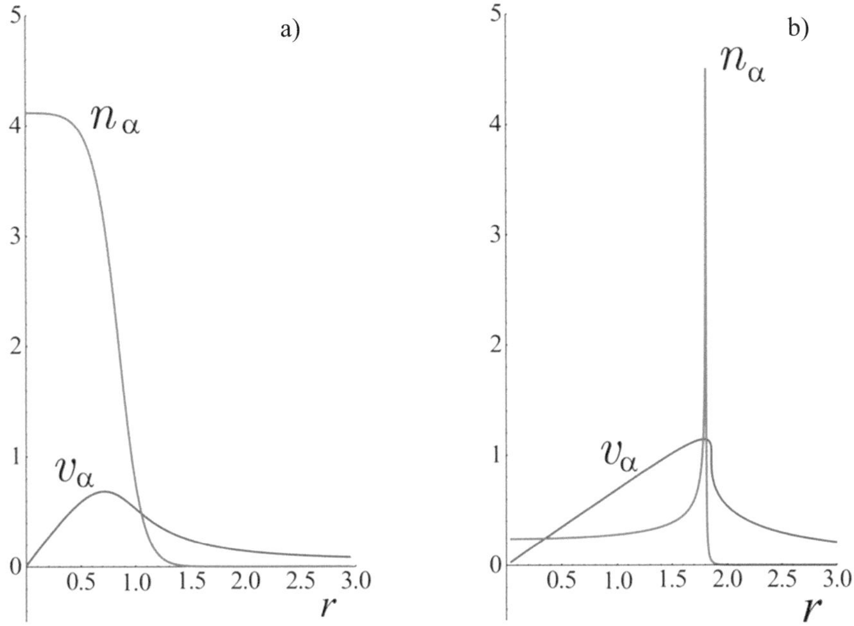

Fig. 25.2 Radial density distribution of ions and the profile of ion velocity (in arb. units) inside a cluster for time, **a** t = 0.5 and **b** t = 1.5. Reproduced from Bulanov et al., (2014) @ Uspekhi Fizicheskikh Nauk 2014

25.3 Target Normal Sheath Acceleration Mechanism

Next acceleration mechanism corresponds to the case when the ions are accelerated by the charge separation electric field at the front of the plasma cloud expanding to vacuum. A strong electric field is formed inside the layer generated by hot electrons at the target rear side with respect to the direction of the laser pulse propagation, Wilks et al., (2001). This is the so-called Target Normal Sheath Acceleration (TNSA) mechanism, which is typically realized in the experiments in the process of the moderate intensity laser radiation interaction with thin metallic or/and plastic foil targets. Usually in the experiments the fast protons are detected, whose appearance in the case of metal foils is explained by the presence of the nanometre thickness contamination water layer on the target surface.

A schematic description of this acceleration mechanism is presented in Fig. 25.3. The mechanism can be realized both in static and in dynamic modes.

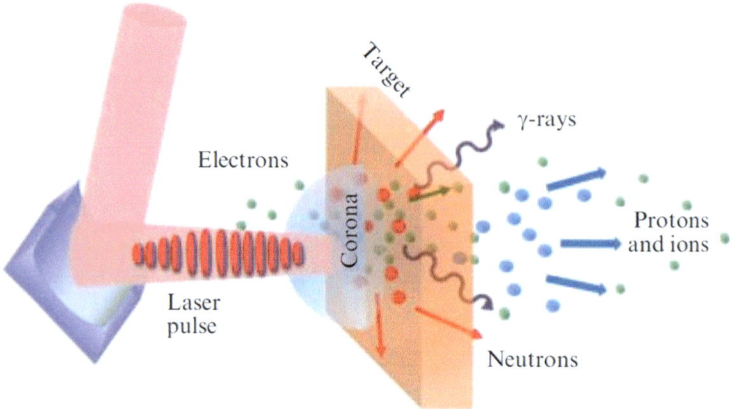

Fig. 25.3 Ion acceleration in the target near-surface layer. A relatively long, small amplitude prepulse creates a plasma corona on the target's front side. The main laser pulse generates in the plasma corona high-energy electrons which penetrate through the dense target. The acceleration of ions in the near-surface layer on the target's rear side occurs in the electric field which is generated by fast electrons. The interaction of laser radiation with solid targets is also accompanied by the generation of gamma-rays, neutrons, positrons, and other debris from nuclear reactions. Reproduced from Bulanov et al., (2014) @ Uspekhi Fizicheskikh Nauk 2014

25.3.1 Static Mode of the Ion Acceleration

In the process of laser radiation interaction with a thin target, the electrons are heated at the front target surface relatively to the direction of the laser pulse propagation. Fast electrons propagate through the target leaving it at the rear side of the target, where they form the layer with high electric field, which accelerates the ions, as shown in Fig. 25.3. In order to find the electric field distribution in the near-surface layer, it is necessary to solve the Poisson equation

$$\frac{d^2\phi}{dx^2} = n_{e,h}(\phi) - \theta(-x), \tag{25.11}$$

where $n_{e,h}$ is the density of hot electrons. This equation uses the dimensionless variables. The electrostatic potential ϕ is normalized on $T_{e,h}/e$, where $T_{e,h}$ is the temperature of fast electrons. The hot electron density and the density of positive electric charge proportional to the difference between the ion concentration and the concentration of cold electrons, are normalized by $\delta n_{0,\alpha}$. The spatial coordinate is measured in units of the Debye radius $r_{D,e} = \sqrt{T_{e,h}/4\pi\delta n_{0,e}e^2}$ with characteristic electron density $\delta n_{0,e}$, which in the limit $x \to -\infty$ is equal to the density of the positive electric charge. The unit step Heaviside function, $\theta(x)$ describes the profile the positive electric charge. The ions are assumed to be at rest. For the sake of simplicity, we neglect the influence of the cold background electrons on the electrostatic potential distribution. The dependence of electron density from the electrostatic potential in the right hand side of Eq. (25.11) is determined by the electron distribution function.

Widely accepted approximation used in the publications devoted to the ion acceleration by the charge separation electric field assumes that the energy distribution of electrons obeys the Boltzmann law. Within the framework of the one-dimensional geometry approximation the Boltzmann distribution function of the electrons leads to the infinite energy of accelerated ions. The reason for that is the presence in the electron distribution of the electrons with formally infinite energy, which results in the electrostatic potential dependence on the coordinate described by the function which tends to infinity logarithmically in the limit $x \to +\infty$. In the case of conditions more realistic for the interaction of short laser pulses with various targets, the electron distribution is terminated at some maximum energy, which is in the order of magnitude equal to the energy of the electron quivering in the laser radiation field. Such the distribution functions lead to the dependence of fast electron density from the potential $n_e(\phi)$ of the form corresponding to the so-called Kappa-function

$$n_{e,h}(\phi) = \left(1 + \frac{\kappa - 1}{\kappa}\phi\right)^{\frac{1}{\kappa-1}}, \tag{25.12}$$

where the index κ is determined by the specific form of the electron distribution function. Formally the relationship (25.12) corresponds to polytrophic dependence of the electron pressure from the density $p = p_0(n/n_0)^\kappa$. In the particular case, when the index equals $\kappa = 3$, the electron energy distribution is described by the function, which takes a constant value for energies smaller than the maximum $\mathcal{E}_{e,\max} = (\kappa/(\kappa - 1))T_{e,h}$ and vanishes at $\mathcal{E}_e > \mathcal{E}_{e,\max}$.

In Fig. 25.4 we present the results of numerical solution of Eq. (25.11) for the boundary conditions $\phi \to 0$ at $x \to -\infty$ and $d\phi/dx \to 0$ in the limit $x \to +\infty$. Figure 25.4a shows the dependence of the positive electric charge in the target $\delta n_{0,\alpha}(x)$ (dashed curve), of the hot electron density $n_{e,h}(x)$, of the electric field $E(x)$ and of the electrostatic potential $\phi(x)$ for the polytope index equal to $\kappa = 3$. In the region $x \geq 0$ the electron density and the electric field vanish at the finite distance from the target surface. Here the electrostatic potential takes the value $\phi_{\max} = -\mathcal{E}_{a,\max}/Z_a e = -(\kappa/(\kappa - 1))T_{e,h}/Z_a e$.

By describing the ion acceleration of a species β within the framework of the test particle approximation, i.e., by considering the ion motion in a given electric field, allows to find the energy spectrum of ions $\mathcal{N}_\beta(\mathcal{E})$. It has the form

$$\mathcal{N}_\beta(\mathcal{E}) \propto \frac{1}{(\mathcal{E}_{\beta,\max} - \mathcal{E})^s} \tag{25.13}$$

with the index $s = \kappa/[2(\kappa - 1)]$ (see Fig. 25.4b). If $\kappa = 3$, than the index s is equal to 3/4. The ion energy spectrum has an integrable singularity at $\mathcal{E} = \mathcal{E}_{\beta,\max}$.

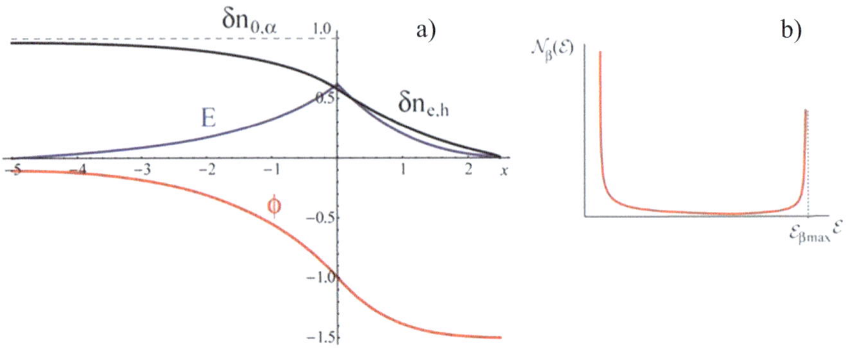

Fig. 25.4 **a** Dependence of the positive charge distribution inside the target, $\delta n_{0,\alpha}(x)$; a (dashed line), the hot electron density $n_{e,h}(x)$, the electric field $E(x)$, and electrostatic potential $\phi(x)$ on the coordinate x for the power index equal to $\kappa = 3$. **b** Energy spectrum of fast ions. Reproduced from Bulanov et al., (2014) @ Uspekhi Fizicheskikh Nauk 2014

We note here that in the case of the Maxwell distribution of the hot electrons, which lead to the dependence $n_e(\phi) = \exp(-\phi)$ in the right hand side of Eq. (25.11), which is a limit of the function (25.12) at $\kappa \to 1$, the ion energy spectrum is exponential.

25.3.2 Dynamic Mode of the Ion Acceleration

In the dynamic mode the ions are accelerated by an electric field formed due to the electric charge separation in the layer at the front expanding into vacuum plasma cloud Gurevich et al. (1966). Within the framework of the approximation, assuming the plasma charge quasi-neutrality, when $n_\alpha = n_{e,h} = n$, the ion motion is described by the system of equations

$$\partial_t n + \partial_x (nv) = 0, \tag{25.14}$$

$$\partial_t v + v\partial_x v = \partial_n \phi \; \partial_x n. \tag{25.15}$$

In the right had side of Eq. (25.15) the derivative of the potential with respect to the density is equal to $\partial_n \phi = dn/d\phi$ with the function $n(\phi)$ given by Eq. (25.12). The ion velocity v is normalized by the ion sound speed $c_s = \sqrt{T_{e,h}/m_\alpha}$.

The general approach to solving such systems of hydrodynamic type equations is described in Landau and Lifshitz (1987). The self-similar solution of Eqs. (25.14) and (25.15), in which the functions n depend v on the variable $\zeta = x/c_s t$, obeys the system of ordinary differential equations

$$(v - \xi)n' = nv', \tag{25.16}$$

$$(v-\xi)v' = \partial_n\phi\, n', \tag{25.17}$$

where a prime denotes the differentiation with respect to ζ. The condition for the existence of nontrivial solutions of a system of linear algebraic Eqs. (25.16) and (25.17) is that the determinant vanishes,

$$(v-\zeta)v' - \partial_n\phi\, n' = 0. \tag{25.18}$$

This condition gives the relationships

$$v-\xi = \sqrt{n\partial_n\phi} \;\; and \;\; dn/dv = \sqrt{n/\partial_n\phi}. \tag{25.19}$$

Using them for the case of $\kappa = 3$ it is possible to represent the solution of Eqs. (25.16) and (25.17) as

$$v = \frac{c_s}{2}\left(\frac{x}{c_s t} + \sqrt{3}\right), \tag{25.20}$$

$$n = \frac{n_0}{2}\left(1 - \frac{1}{\sqrt{3}}\frac{x}{c_s t}\right). \tag{25.21}$$

For known electron density the electrostatic potential is given by expression $\phi = (3/2)(n^2-1)$.

For self-similar variable which takes the value $\zeta = \sqrt{3}$, the ion density vanishes. Here the fast ion energy has the maximum $\mathcal{E}_{b,\max} = (3/2)T_{e,h}$. In the vicinity of the maximal energy the energy spectrum of fast ions has the form

$$\mathcal{N}_b(\mathcal{E}) \propto \frac{2\sqrt{2}}{\sqrt{\mathcal{E}_{b,\max} - \mathcal{E}}}. \tag{25.22}$$

In this expression the singularity $\mathcal{E} \to \mathcal{E}_{b,\max}$ is also integrable.

It is often assumed that in the aforementioned expressions the electron temperature $T_{e,h}$ is related to the EM wave amplitude a_0 as equal to ponderomotive energy (Wilks et al., 1992) $T_{e,h} = m_e c^2(\sqrt{1+a_0^2} - 1)$. However, the effect of the electron recirculation near the plasma-vacuum interface (Bulanov et al., 2015; Kemp & Divol, 2012a, 2012b) can substantially increase the electron energy resulting to the laser accelerated ion energy enhancement (Yogo et al., 2016, 2017).

25.4 The Ion Acceleration by the Radiation Pressure

Among other mechanisms in the high laser intensity limit the greatest efficiency has the mechanism of the ion acceleration by radiation pressure Esirkepov et al., (2004), when an irradiated thin target moves as a whole under the action of the electromagnetic wave radiation pressure.

This acceleration mechanism has attracted increasing attention both in theoretical and experimental works, owing to its prediction to generate the high-quality ion beams with high efficiency of laser energy conversion into the energy of accelerated ions.

In Fig. 25.5, in order to illustrate the acceleration of ions in the regime of domination of the radiation pressure of light, we show the results of PIC computer simulation of the radiation acceleration of ions in a three-dimensional configuration (Esirkepov et al., 2004). As can be seen, in the course of interaction with a thin plasma layer, the laser pulse forms a thin shell, enclosing the laser pulse like a cocoon. The thin plasma shell holds the electromagnetic radiation inside, preventing its spreading. Ions in the front part of the shell are accelerated to an energy exceeding GeV and have a quasi-mono-energetic spectrum (Fig. 25.5c).

The force acting on the target is equal to the momentum flux of an electromagnetic wave, which is proportional to the Poynting vector: $\boldsymbol{S} = c\,\boldsymbol{E} \times \boldsymbol{B}/4\pi$. For the sake of simplicity we assume that the electromagnetic wave is circularly polarized, it has the frequency ω and wave vector $\boldsymbol{k}$, and propagates along the x axis. The wave is given by the vector potential

$$\boldsymbol{A} = A_0\left[\cos(\omega t - kx)\boldsymbol{e}_y + \sin(\omega t - kx)\boldsymbol{e}_z\right] \tag{25.23}$$

where $\boldsymbol{e}_y$ and $\boldsymbol{e}_y$ are unit vectors along the y and z axis, and $k = |\boldsymbol{k}|$ is the wave number. Calculating the electric $\boldsymbol{E} = -\partial_t \boldsymbol{A}/c$ and magnetic field $\boldsymbol{B} = \nabla \times \boldsymbol{A}$, we

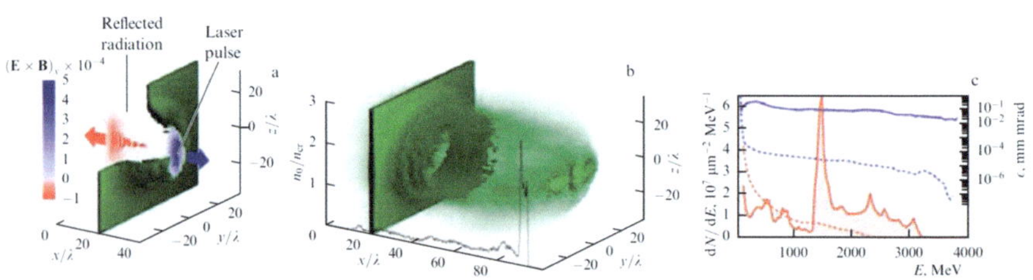

Fig. 25.5 Results of a three-dimensional PIC simulation of radiation pressure acceleration. An electromagnetic pulse forms a thin plasma shell that holds it. **a** The ion density distribution and the x-component of the Poynting vector $\boldsymbol{S} = c\,\boldsymbol{E} \times \boldsymbol{B}/4\pi$ in the (x, y) plane at z = 0 for $t = 40 \times 2\pi/\omega$. **b** Isosurface of the ion density for $t = 100 \times 2\pi/\omega$; the black curve shows the ion density distribution on the axis. **c** The ion energy spectrum (red curve) and transverse emittance of electrons (blue dashed curve) and ions (blue solid curve). Reprinted figure with APS permission from Esirkepov et al. (2004) as follows: Esirkepov et al. (2004). https://doi.org/10.1103/PhysRevLett.92.175003. Copyright (2004) by the American Physical Society

find the Pointing vector

$$S = c\,\omega k A_0^2 e_x. \tag{25.24}$$

It is proportional to the product of the frequency ω and the wave number k. In the frame of reference moving with the velocity of the target $v = \beta c$, the product of the frequency ω and the wave number is given by the expression

$$\overline{\omega k} = \omega k \frac{1+\beta^2}{1-\beta^2} - \left(\omega^2 + k^2\right)\frac{\beta}{1-\beta^2} = \omega^2 \frac{(\beta_g - \beta)(1 - \beta_g \beta)}{1-\beta^2}. \tag{25.25}$$

In order to derive this expression, we have used the relationship between the frequency, wave number and group velocity of an electromagnetic wave, $v_g = c\beta_g = kc^2/\omega$.

Summing the values of the Poynting vector in the incident and reflected waves and subtracting the value of the Poynting vector in the transmitted wave, we find the force acting on the target. It is given by the expression

$$F = \left(1 + |\rho|^2 - |\tau|^2\right) S, \tag{25.26}$$

where $|\rho|^2$ and $|\tau|^2$ are the reflection and transmission суть коэффициенты coefficients of the electromagnetic wave. They are connected by the relationship

$$|\rho|^2 + |\tau|^2 + |\alpha|^2 = 1. \tag{25.27}$$

Here $|\alpha|^2$ is the absorption coefficient. Using expressions (25.25)–(25.27), we find that the equation of motion of the surface element of a thin target can be written in the form

$$\frac{1}{\left(1-\beta_\alpha^2\right)^{3/2}} \frac{d\beta_\alpha}{dt} = \left(\frac{K\,E^2}{4\pi \sigma_0 m_\alpha c}\right) \frac{(\beta_g - \beta_\alpha)(1 - \beta_g \beta_\alpha)}{\left(1 - \beta_\alpha^2\right)}, \tag{25.28}$$

where

$$K = 2|\rho|^2 + |\alpha|^2, \tag{25.29}$$

and $\sigma_0 = n_0 l_0$ is equal to the surface density of the target with the thickness l_0. In Eq. (25.27)

$$\beta_\alpha = p_{\alpha,x} \Big/ \left(m_\alpha^2 c^2 + p_{\alpha,x}^2\right)^{1/2} \tag{25.30}$$

is the x component of the normalized velocity of the surface element of the target, $\beta_g =$ is the normalized group velocity, and $E^2 = \left(\omega A/c\right)^2$. As we see, the radiation pressure vanishes in the limit $\beta_\alpha \to \beta$. This constraint must be taken into account for a sharply focused laser pulses. In the case of complete absorption of radiation by the target that corresponds to the equalities $|\rho|^2 = 0$ and $|\alpha|^2 = 1$, the radiation pressure on the target is two times less than in the case of a perfectly reflective target, when $|\rho|^2 = 1$ and $|\alpha|^2 = 0$.

In the case of the constant amplitude laser pulse (E = constant), the solution of Eq. (25.28) can be written in the form

$$\ln \frac{1 - \beta\beta_g + \beta\left(1 - \beta_g^2\right)^{1/2}\left(1 - \beta^2\right)^{1/2}}{\left(\beta_g - \beta\right)\left[1 + \left(1 - \beta_g^2\right)^{1/2}\right]} - \beta_g \left[\arctan \frac{\left(1 - \beta_g^2\right)^{1/2}\left(1 - \beta^2\right)^{1/2}}{\left(\beta_g - \beta\right)} - \arccos \beta_g\right] = \beta_g\left(1 - \beta_g^2\right)^{3/2}\left(\frac{K\, E^2}{4\pi\sigma_0 m_\alpha c}\right)t\,. \tag{25.31}$$

Above we have considered the radiation pressure of the laser pulse acting on the flat foil target. Here we describe the laser pulse interaction with thin deformable target. For the sake of simplicity we assume that the foil target is opaque for incident radiation and the foil target is considered an ideally reflecting mirror.

In the laboratory frame of reference, the equation of motion of a surface element of thin foil target is

$$\frac{d\boldsymbol{p}}{dt} = \mathcal{P}\frac{\boldsymbol{\nu}}{\sigma}, \tag{25.32}$$

where $\boldsymbol{p}$ is the momentum of the surface element, $\boldsymbol{\nu}$ is a unit vector directed along the normal to the mirror surface, $\mathcal{P}$ is the relativistically invariant pressure of light, and $\sigma = nl$ is the surface area density (n and l are respectively the plasma density and the foil thickness). We assume that the foil acceleration and curvature are small compared with the values corresponding to the electromagnetic wave frequency and wavelength. The laser-foil interaction geometry is illustrated in Fig. 25.6.

It is convenient to introduce the coordinates η and ζ related to the independent variables x, y, z as

$$x = x(\eta, \zeta, t), y = y(\eta, \zeta, t), z = z(\eta, \zeta, t). \tag{25.33}$$

The coordinates η and ζ are markers of the surface element of the shell. We consider the surface element Δs containing the number $\Delta N = \sigma \Delta s$ of particles. This number remains constant during the motion of the shell. The position and shape of the shell are given by the equation of the surface,

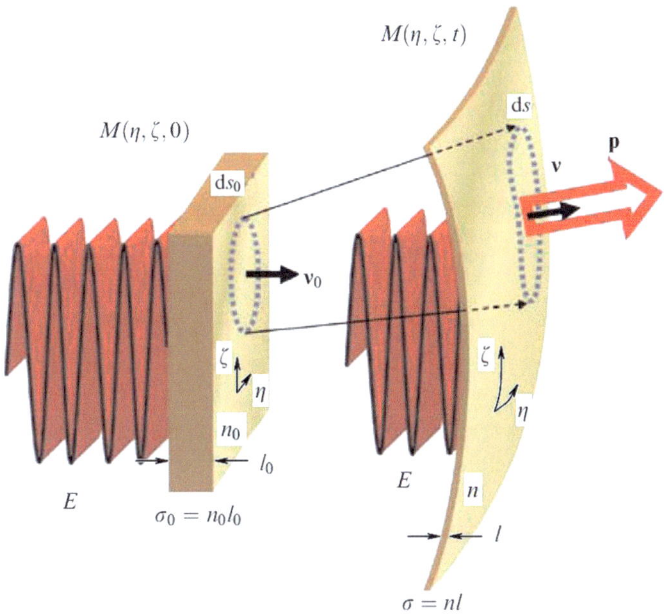

Fig. 25.6 Thin shell evolution under the action of the electromagnetic wave radiation pressure. Reproduced from Bulanov et al. (2013) @ Uspekhi Fizicheskikh Nauk 2013

$$\boldsymbol{M} = \boldsymbol{M}(\eta, \zeta, t) = \{x(\eta, \zeta, t), y(\eta, \zeta, t), z(\eta, \zeta, t)\}, y = y(\eta, \zeta, t), z = z(\eta, \zeta, t). \tag{25.34}$$

Following the rules of differential geometry, we find that at a regular point of the surface, the surface element Δs and the unit normal vector $\boldsymbol{\nu}$ are given by

$$\boldsymbol{\nu}\Delta s = \partial_\eta \boldsymbol{M} \times \partial_\zeta \boldsymbol{M}\, d\eta d\varsigma \tag{25.35}$$

and

$$\boldsymbol{\nu} = \frac{\partial_\eta \boldsymbol{M} \times \partial_\zeta \boldsymbol{M}}{\left|\partial_\eta \boldsymbol{M} \times \partial_\zeta \boldsymbol{M}\right|}. \tag{25.36}$$

The particle number conservation gives $\sigma \Delta s = \sigma_0 \Delta s_0$, where the initial surface plasma density is $\sigma_0 = n_0 l_0$. Hence, it follows that the surface density is given by

$$\sigma = \frac{\sigma_0}{\left|\partial_\eta \boldsymbol{M} \times \partial_\zeta \boldsymbol{M}\right|}. \tag{25.37}$$

Using these relations, we can write equations of motion (25.32) in the form

$$\sigma_0 \partial_t p_i = \mathcal{P} \varepsilon_{ijk} \partial_\eta x_j \partial_\varsigma x_k, \tag{25.38}$$

$$\partial_t x_i = c \frac{p_i}{(m^2c^2 + p_k p_k)^{1/2}}, \tag{25.39}$$

where ε_{ijk} is the unit totally antisymmetric tensor (the Levi–Civita symbol), $i = 1, 2, 3$, and summation over repeated indices is assumed. Here we drop the index α.

In the particular case of the shell irradiating circularly polarized electromagnetic wave propagating along the x axis with the speed of light in vacuum of the amplitude $E = E(x - t/c)$, the radiation pressure $\mathcal{P}$ is

$$\mathcal{P} = K \frac{E^2}{2\pi} \left[\frac{(m_\alpha^2 c^2 + p_x^2)^{1/2} - p_x}{(m_\alpha^2 c^2 + p_x^2)^{1/2} + p_x} \right]. \tag{25.40}$$

The parameter K defined by Eq. (25.29) is less than unity.

Assuming that the target moves along the x axis, with the initial conditions determining a flat plasma slab uniform in the y, z directions, and $y(0) = \eta$ and $z(0) = \zeta$, we can write the equation for the x-component of momentum as

$$\frac{d}{dt} \left(\frac{\beta_x^{(0)}}{\sqrt{1 - \beta_x^{(0)}}} \right) = K \frac{E^2}{2\pi m_\alpha \sigma_0} \left[\frac{1 - \beta_x^{(0)}}{1 + \beta_x^{(0)}} \right]. \tag{25.41}$$

For normalized slab velocity $\beta_x^{(0)}$ we have $\beta_x^{(0)} = p_x^{(0)} / \sqrt{m_\alpha^2 c^2 + p_x^{(0)2}}$. It and the momentum $p_x^{(0)}$ depend on time only.

The solution of Eq. (25.41) is formally equivalent to the solution to the problem of the motion of a charged particle pushed by the radiation pressure of an electromagnetic wave Landau and Lifshitz (1987). For the electromagnetic wave with a constant amplitude $E = E_0 =$ constant and $K = 1$ and for the initial value of the particle momentum equal to zero, the momentum dependence on time can be written in implicit form as (see also Esirkepov et al., 2004).

$$\frac{3p_x^{(0)}}{2m_\alpha c} + \left(\frac{p_x^{(0)}}{m_\alpha c} \right)^3 + \left[1 + \left(\frac{p_x^{(0)}}{m_\alpha c} \right)^2 \right]^{3/2} - 1 = \tau. \tag{25.42}$$

The normalized time τ is equal to $\tau = E_0^2 t / 2\pi \sigma_0 m_\alpha c$. In the limit $\tau \to \infty$ Eq. (25.42) yields

$$p_x^{(0)} = m_\alpha c \left[\left(\frac{3\tau}{4} \right)^{1/3} - \left(\frac{6}{\tau} \right) \right] + \dots. \tag{25.43}$$

In the case of a finite-length electromagnetic pulse, the electric field E_0 on a moving shell, i.e., at $x = x(t)$, depends on time as $E_0(x - x(t)/c)$. In this expression, the function $x(t)$ should be found from Eq. (25.39). We introduce the new variable

$$\psi = \omega_0(t - x^{(0)}(t)/c, \tag{25.44}$$

equal to the phase of the electromagnetic wave at $x = x^{(0)}(t)$. Differentiation of ψ with respect to time yields

$$\dot{\psi} = \omega_0(1 - v^{(0)}(t)/c) = \omega_0 \frac{m_\alpha c\gamma^{(0)} - p_x^{(0)}}{m_\alpha c\gamma^{(0)}}. \tag{25.45}$$

Changing the variable from time t to phase ψ and using the fluence of the electromagnetic wave

$$w(\psi) = \int\limits_{-\infty}^{\psi} \frac{KE_0^2(\psi')}{4m_\alpha\sigma_0\omega^2\lambda_0} d\psi' \equiv \int\limits_{-\infty}^{\psi} \frac{R(\psi')}{2\lambda_0} d\psi', \tag{25.46}$$

where $R(\psi) = KE_0^2(\psi)/2m_\alpha\sigma_0\omega^2$, allows us to represent the solution of Eq. (25.41) in the form

$$p_x^{(0)}(\psi) = m_\alpha c \frac{2w(\psi)(w(\psi)+1)}{2w(\psi)+1}. \tag{25.47}$$

It follows from expression (25.45) that the relation between time t to phase ψ is

$$\psi + \int\limits_0^{\psi} (2w(\psi') + w^2(\psi'))d\psi' = \omega_0 t, \tag{25.48}$$

In the case of a constant-amplitude electromagnetic wave with

$$R = R_0 = KE_0^2/2m_\alpha\sigma_0\omega^2,$$

expressions (25.46) and (25.48) yield

$$w(\psi) = (R_0/2\lambda_0)\psi$$

and

$$3\psi + 3(R_0/2\lambda_0)\psi^2 + 2(R_0/2\lambda_0)^2\psi^3 = 3\omega_0 t, \tag{25.49}$$

It follows that in the limit $\omega_0 t \gg \lambda_0/R_0$, the x-component of the momentum of the surface element, $p_x^{(0)}$, depends on time as

$$p_x^{(0)}(t) \approx m_\alpha c \left(\frac{3R_0}{\lambda_0}\omega_0 t\right)^{1/3}, \tag{25.50}$$

in agreement with

$$3\psi + 3(R_0/2\lambda_0)\psi^2 + 2(R_0/2\lambda_0)^2\psi^3 = 3\omega_0 t, \tag{25.51}$$

From Eqs. (25.47) we can find that the energy of the accelerated proton is equal to

$$\mathcal{E}_p = m_p c^2\left(1 + \frac{2w^2}{1+2w}\right). \tag{25.52}$$

In other words, in the nonrelativistic limit ($w << 1$) the kinetic energy of accelerated ions for a fixed pulse length is proportional to the square of the radiation intensity.

The relationship (25.52) between the ion energy $\mathcal{E}_p$ and the fluence w is a consequence of the energy–momentum conservation. We can represent the laser pulse as a cloud of photons having the energy $\mathcal{N}_\gamma \mathcal{E}_\gamma = \hbar\omega\mathcal{N}_\gamma$ and momentum $\mathcal{N}_\gamma \boldsymbol{p}_\gamma = \hbar \boldsymbol{k}_0 \mathcal{N}_\gamma$, with $\mathcal{N}_\gamma$ equal to the number of photons. When the photon cloud hits the ion shell which has initial energy $\mathcal{E}_{0,\alpha}\mathcal{N}_\alpha = \mathcal{E}_{las}$ and momentum $\boldsymbol{p}_{0,\alpha}\mathcal{N}_\alpha$, where $\mathcal{N}_p$ is the total number of accelerated protons and

$$\mathcal{E}_{las} = \int (E^2/4\pi)dV,$$

the energy momentum conservation gives

$$\mathcal{N}_\gamma \mathcal{E}_{0,\gamma} + \mathcal{N}_\alpha \mathcal{E}_{0,\alpha} = \mathcal{N}_\gamma \mathcal{E}_\gamma + \mathcal{N}_\alpha \mathcal{E}_\alpha \tag{25.53}$$

and

$$\mathcal{N}_\gamma \boldsymbol{p}_{0,\gamma} + \mathcal{N}_\alpha \boldsymbol{p}_{0,\alpha} = \mathcal{N}_\gamma \boldsymbol{p}_\gamma + \mathcal{N}_\alpha \boldsymbol{p}_\alpha, \tag{25.54}$$

where

$\mathcal{E}_{0,\alpha} = \sqrt{m_\alpha^2 c^4 + \left|\boldsymbol{p}_{0,\alpha}\right|^2 c^2}$ and $\mathcal{E}_\alpha = \sqrt{m_\alpha^2 c^4 + \left|\boldsymbol{p}_\alpha\right|^2 c^2}$.

Assuming normal incidence of the laser pulse ($\boldsymbol{k}_0 = (k_0, 0, 0)$ and $\boldsymbol{k} = (k, 0, 0)$) on the receding target ($\boldsymbol{p}_{0,\alpha} = (p_{0,\alpha}, 0, 0)$ and $\boldsymbol{p}_\alpha = (p_\alpha, 0, 0)$), we find from Eqs. (25.53) and (25.54) that

$$p_\gamma = -p_{0,\gamma}\frac{(\mathcal{E}_{0,\alpha} - p_{0,\alpha}c)}{(\mathcal{E}_{0,\alpha} + p_{0,\alpha}c) + 2\mathcal{N}_\gamma \mathcal{E}_{0,\gamma}/\mathcal{N}_\alpha} \tag{25.55}$$

for the momentum of reflected photon. If the photon cloud energy is relatively small compared to the ion energy, i. e. $(\mathcal{E}_{0,\alpha} + p_{0,\alpha}) >> 2\mathcal{N}_\gamma \mathcal{E}_{0,\gamma}/\mathcal{N}_\alpha$, according to Eq. (25.53) the frequency of reflected photon equals $\omega = \omega_0(1 - \beta_0)/(1 + \beta_0)$ with the mirror normalized velocity $\beta_0 = v_{0,\alpha}/c = p_{0,\alpha}c/\mathcal{E}_{0,\alpha}$. The last term in the denominator in the right-hand side of Eq. (25.55) describes the recoil effect which leads to further downshift of the reflected photon frequency.

In the case of the laser pulse interaction with the target which is initially at the rest the reflected light frequency is equal to

$$\omega = \frac{\omega_0}{1 + 2\mathcal{N}_\gamma \mathcal{E}_{0,\gamma}/m_\alpha \mathcal{N}_\alpha m_\alpha c^2} \tag{25.56}$$

and the energy of accelerated ion is given by

$$\mathcal{E}_\alpha = m_\alpha c^2 \frac{2\left(\mathcal{N}_\gamma \mathcal{E}_{0,\gamma}/\mathcal{N}_\alpha m_\alpha c^2\right)^2}{1 + 2\mathcal{N}_\gamma \mathcal{E}_{0,\gamma}/\mathcal{N}_\alpha m_\alpha c^2} \tag{25.57}$$

It corresponds to expression (25.52) with $w = \mathcal{N}_\gamma \mathcal{E}_{0,\gamma}/\mathcal{N}_\alpha m_\alpha c^2$.

The conversion efficiency of laser energy into the energy of fast ions, $\kappa_{eff} = \mathcal{E}_{las}/\mathcal{N}_p \mathcal{E}_p$, where $\mathcal{N}_p$ is the total number of accelerated particles, is equal to

$$\kappa_{eff} = \frac{2w}{1 + 2w}. \tag{25.58}$$

Using the relationships (25.52) and (25.53) we can show, that the generating $\mathcal{N}_p = 5 \times 10^{11}$ protons per second (it is one of the requirements of hadron therapy to the parameters of the laser accelerator), with the particle energy equal to 250 MeV rrequires the lasers with the repetition rate of 1 Hz and with energy per one pulse equal to $\mathcal{E}_{las} = 40$ J. If the pulse duration equals 30 fs, than for the laser power we obtain the value of the order of 1 PW. The acceleration efficiency in this case is $\kappa_{eff} = 0.5$.

To implement the optimal acceleration mode, we should choose the thickness of the target to be on the opacity threshold for laser radiation Vshivkov et al., (1998), which assumes that the laser pulse amplitude satisfies the condition

$$a_0 \leq \varepsilon_p. \tag{25.59}$$

Here the amplitude a_0 is related to the vector potential value A_0 by $a_0 = eA_0/m_e c^2$. The dimensionless parameter

$$\varepsilon_p = \frac{2\pi n e^2 l_0}{m_e \omega c}, \tag{25.60}$$

has been introduced in the paper Vshivkov et al., (1998). It characterizes the relativistic transparency threshold of a thin plasma layer. For example, if the petawatt laser pulse is focused to the 3 μm region, i.e. the laser intensity reaches the level of $I = 10^{21}$ W/cm^2, the thickness of the solid density foil target with the electron density about $n_e = 10^{23}$ cm^{-3} should be equal to 0.2 μm.

The relationship can be rewritten in the form $E_{las} \leq 2\pi n e l_0$, the r. h. s. of which being the Lorentz invariant does not change during the target acceleration. The l. h. s. equal to the electric field on the target in the rest frame of the reference depends on time which may result in weakening of the acceleration mechanism. By choosing a tailored profile of the laser pulse the ion acceleration by the laser radiation pressure can be optimized, Mackenroth and Bulanov (2019).

Additional constraint on the target and laser parameters stem from the requirement of a very high quality of accelerated ion beam (in more detail the issues related to the ion beam quality are discussed below). The beam must have a small energy width. The transverse inhomogeneity of the laser radiation intensity and, therefore, of the fluence inhomogeneity determines the dependence of the accelerated ion energy from the transverse coordinates. We note that as shown by Klimo et al., (2008) in the case of relatively low laser intensity the use of circular polarized laser pulses can enhance the accelerated ion beam quality.

Considering the beam laminarity, it is possible to select a group of ions in the paraxial region with small energy spread. In accordance with the theory and simulation predictions the ion beam has the form of a thin shell, whose element velocity is maximal on the axis. In the paraxial region, the dependence of the ion energy from the radius can be approximated by the function

$$\mathcal{E}_\alpha(r) = \mathcal{E}_{\alpha,\max}\left(1 - r^m/R_\perp^m\right), \tag{25.61}$$

with the index m equal to 2 for a laser pulse with a Gaussian profile, and with $m > 2$ for pulses described by a super-Gaussian function. It is assumed that the laser pulse has a radius $R_\perp$.

For a high-quality beam shaping we can use the screen with a small aperture. The radius of the hole Δr is related to the ion beam energy width $\Delta\mathcal{E}_a$ as

$$\Delta r = R_\perp\left(\Delta\mathcal{E}_a/\mathcal{E}_{a,\max}\right)^{1/m}. \tag{25.62}$$

It follows that in order to obtain the beam with the narrow energy spectrum, $\Delta\mathcal{E}_a/\mathcal{E}_{a,\max} \leq 2\%$, we need to have $\Delta r/R_\perp \approx 0.14$, if the index equals $m = 2$, and ≈ 0.38, for $m = 3$. Using Eqs. (25.34) and (25.35), we find that the required energy and power of the laser radiation must be equal to 75 J and 2.5 PW in the case of a Gaussian pulse, and to 20 J and 700 TW, if $m = 3$.

It is convenient to present the dependences obtained above in dimensional form. The ion energy dependence on the laser pulse parameters in the nonrelativistic limit $\mathcal{E}_a << m_a c^2$ is given by

$$\mathcal{E}_a = 10\left(\frac{10^{11}}{N_{tot}}\right)^2 \frac{m_p}{m_\alpha}\left(\frac{\mathcal{E}_{las}}{1J}\right)^2 [\text{MeV}]. \tag{25.63}$$

where m_p is the mass of the proton; in the ultrarelativistic case, where $\mathcal{E}_a >> m_a c^2$, we have

$$\mathcal{E}_a = 100 \frac{10^{11}}{N_{tot}} \frac{\mathcal{E}_{las}}{1\text{kJ}} [\text{GeV}]. \tag{25.64}$$

The acceleration length in the limit of an ultrarelativistic ion energy, $\mathcal{E}_a/m_a c^2 \to \infty$, is

$$l_{acc} = \frac{l_{las}}{1-\beta_a} \approx 2 l_{las} \left(\frac{\mathcal{E}_a}{m_a c^2} \right)^2. \tag{25.65}$$

It depends on the laser pulse duration l_{las}/c and on the final energy of accelerated ions $\mathcal{E}_a$.

Considering the ion beam accelerator for applications in high-energy physics, we conclude that its parameters must satisfy several conditions, in addition to the requirement of achieving the maximum energy of the particles. In particular, an important parameter such as the luminosity of the beam characterizes the number of particles produced in reactions by colliding beams in the accelerator. We note that controlling the transverse shape of colliding ion beams allows enhancing the beam luminosity value. The luminosity of the beam of ions accelerated in the radiation-pressure-dominated regime is given by

$$\mathcal{L} = 10^{34} \frac{f}{1\,\text{kHz}} \left(\frac{\mathcal{N}_{tot}}{10^{12}} \right)^2 \left(\frac{10^{-4}}{\sigma_\perp} \right)^2 [\text{cm}^{-2}\text{s}^{-1}]. \tag{25.66}$$

Here, $\mathcal{N}_{tot}$ is the number of particles in each beam, $\sigma_\perp$ is the transverse size of the beam, and f is the laser repetition rate. The product of the luminosity and the reaction cross section gives the reaction rate. As we can see, the luminosity can be increased by increasing the number of particles per bunch N_{tot} and/or by increasing the laser repetition rate f, or by decreasing the transverse size of the ion beam $\sigma_\perp$.

First indication on the radiation pressure effects in laser plasma have been obtained in the experiments by Kar et al., (2008). Later on, the results obtained in experiments with high-power lasers interacting with thin foil targets demonstrating the ion acceleration by the laser light pressure have been published as Kar et al. (2012) and Kim et al. (2016). Multiple laser pulse ion acceleration is discussed by Wan et al. (2022).

25.5 Rayleigh–Taylor Like Instability of the Accelerated Shell

The accelerated thin shell is unstable with respect to the development of a mode corresponding to the Rayleigh–Taylor instability (see Chandrasekhar, 1961 and Part III for the theory of RT instability in incompressible fluid). This instability plays a fundamental role in space plasmas and in applications to problems of inertial confinement fusion. In space plasmas and in laser plasma, the Rayleigh–Taylor instability affects the scenario of the interaction of intense radiation with matter. In the case of an accelerated thin shell, it can cause its breakup into individual clumps, Ott (1972), Manheimer et al. (1984).

The thin-shell stability has been analyzed by Pegoraro and Bulanov (2007) and Bulanov et al. (2013) by using system of Eqs. (25.38) and (25.39). Linearizing these equations around the unperturbed solution given by Eqs. (25.42) and (25.47), we obtain

$$\partial_\psi (p_x^{(0)} \partial_\psi \xi_x^{(1)}) = \frac{R(\psi)}{2\pi} (\partial_\eta \xi_y^{(1)} + \partial_\zeta \xi_z^{(1)}), \tag{25.67}$$

$$\partial_\psi \left(\frac{1}{p_x^{(0)}} \partial_\psi \xi_y^{(1)} \right) = -\frac{R(\psi)}{2\pi} \partial_\eta \xi_x^{(1)}, \tag{25.68}$$

$$\partial_\psi \left(\frac{1}{p_x^{(0)}} \partial_\psi \xi_z^{(1)} \right) = -\frac{R(\psi)}{2\pi} \partial_\zeta \xi_x^{(1)}, \tag{25.69}$$

where the perturbations $\xi_x^{(1)}(\eta, \zeta, \psi)$, $\xi_y^{(1)}(\eta, \zeta, \psi)$, and $\xi_z^{(1)}(\eta, \zeta, \psi)$, leading to shell deformation in the transverse direction, are assumed to be small. We seek the solution within the framework of the WKB approximation. We represent the functions $\xi_i^{(1)}(\eta, \zeta, \psi)$ in the form

$$\xi_x^{(1)} \propto \exp\left[-i(q\eta + r\zeta) + \int_0^\psi \Gamma(\psi') d\psi' \right], \tag{25.70}$$

under the assumption of a slow dependence of the growth rate $\Gamma(\psi)$ on ψ:

$$\partial_\psi \Gamma(\psi) / \Gamma^2(\psi) \ll 1, \tag{25.71}$$

Substituting dependence (25.71) in system (25.67)–(25.68) and solving the algebraic dispersion equation, we obtain the instability growth rate

$$\Gamma(\psi) = (q^2 + r^2)^{1/4} \left(\frac{R(\psi)}{2\pi} \right)^{1/2}, \tag{25.72}$$

Using expression (25.49), we find that for a constant-amplitude electromagnetic wave, the perturbations depend on time as

$$\xi_x^{(1)} \propto \exp\left[-i(q\eta + r\zeta) + \left(\frac{t}{\tau_r} \right)^{1/3} \right], \tag{25.73}$$

where the characteristic time of the instability development τ_r in the ultrarelativistic limit is

$$\tau_r = \omega_0^{-1}(2\pi)^{3/2}\frac{R_0^{3/2}}{6(q^2+r^2)^{1/4}\lambda_0^2}, \tag{25.74}$$

Taking into account that $R_0 = KE_0^2(\psi)/2m_\alpha\sigma_0\omega^2$, we see that the characteristic time of instability is proportional to the square root of the ratio of the radiation pressure to the ion mass. In other words, the higher the ion mass is, the faster the instability develops, and the higher the radiation pressure is, the slower the perturbation grows.

The expansion of plasma in the transverse direction along the surface of the shell can lead to a slowdown in the nonlinear stage of instability, characterized by the formation of cusp singularities and of high-density plasmoids.

The experimental evidence for a Rayleigh–Taylor-like instability driven by radiation pressure of an ultra-intense laser pulse has been reported by Palmer et al. (2012).

25.6 'Unlimited' Ion Acceleration by Radiation Pressure

The expansion of the plasma along the surface of the shell (it may be a result of the instability discussed above) can lead to a substantial increase in the energy of ions accelerated by the radiation pressure (Bulanov et al., 2010) because, due to the shell expansion, the number of ions in the region of interaction with the electromagnetic pulse is reduced, which leads to an increase in their energy. To show this, we consider the case where the shell moves in the longitudinal direction with a momentum $p_x \gg m_a c$, with relatively small transverse components of the momentum corresponding to expansion of the shell, $p_y << m_a c$ and $p_z \ll m_a c$. Using these conditions, we seek the solution of Eqs. (25.38) and (25.39) assuming that the transverse coordinates and momenta are linear functions of the Lagrangian coordinates η and ζ:

$$y = \Lambda_y(t)\eta,\ z = \Lambda_z(t)\zeta, \tag{25.75}$$

$$p_y = \pi_y^{(0)}(t)\eta,\ z = \Lambda_z(t)\zeta. \tag{25.76}$$

Relations (25.75) and (25.76) with coordinate independent $\pi_y^{(0)}$ and $\pi_z^{(0)}$ describe the expansion of the shell with uniform deformation and correspond to the local approximation of solutions in the near-axial region. It follows from Eq. (25.39) that the functions Λ_y and Λ_z satisfy the equations

$$\frac{d\Lambda_y}{dt} = \frac{\pi_y^{(0)}}{m_a\gamma}, \tag{25.77}$$

$$\frac{d\Lambda_z}{dt} = \frac{\pi_z^{(0)}}{m_a\gamma} \tag{25.78}$$

with the initial conditions $\Lambda_y(0) = 1$ and $\Lambda_z(0) = 1$. Here, the ion gamma-factor is $\gamma = \sqrt{1 + (p_x/m_a c)^2}$. For simplicity, we assume that $\Lambda_y = \Lambda_z = \Lambda$ and $\pi_y^{(0)} = \pi_z^{(0)} = \pi^{(0)}$. The surface density changes according to the relation $nl = n_0 l_0/\Lambda^2$. Substituting these expressions in Eq. (25.39) for the longitudinal component of the momentum and changing the variable to ψ, we obtain the equation for the deformation Λ. It reads

$$\frac{d^2\Lambda}{d\psi^2} = h^2\Lambda^2, \tag{25.79}$$

where $h = \sqrt{KE_0^2\pi^{(0)}/2\pi m_a n_0 l_0 \omega^2 c}$.

We assume that the laser pulse with a duration t_{las} has a constant amplitude equal to E_0 in the interval $0 < \psi < \omega_0 t_{las}$. cm and zero outside it, i.e., the coefficient h in Eq. (25.74) is constant inside this interval. The solution of Eq. (25.79) is then given by the Weierstrass elliptic function $\wp(u, \{g_2, g_3\})$ Olver et al. (2010):

$$\Lambda(\psi) = \wp\big(\tilde{\psi}_* - \tilde{\psi}, \{g_2, g_3\}\big), \tag{25.80}$$

where we set $\tilde{\psi} = 6^{1/2}\psi/6$ and g_2 and g_3 are constant, g_2 and $g_3 = 1 - \Lambda'(0)$. The value $\tilde{\psi}_*$ is determined by the minimal positive solution of the equation $\wp\big(\tilde{\psi}_*, \{g_2, g_3\}\big) = 1$.

The equation with the Weierstrass elliptic function $\wp(u, \{g_2, g_3\}) = z$ gives the value of z for which

$$u = \int_{-\infty}^{z} \frac{dt}{\sqrt{4t^3 - tg_2 - g_3}}. \tag{25.81}$$

For known solution of Eq. (25.80), $\Lambda(\psi)$, we can find the longitudinal component of the momentum by integrating the equation

$$\frac{dp_x}{d\psi} = \frac{m_a v_E^2}{\omega_0 l_0} \frac{m_a c\gamma}{m_a c\gamma - p_x}\Lambda^2 \tag{25.82}$$

with $v_E^2 = E_0^2/2\pi n m_\alpha$.

We discuss two limit cases. For the shell that is not expanding, i.e., for $\Lambda = 1$, it follows from Eq. (25.82) that asymptotically as $t \to \infty$, the ion momentum depends on time as

$$p_x \approx m_a c\left(\frac{t}{\tau_{1/3}}\right)^{1/3} \tag{25.83}$$

which corresponds to expression (25.50). Here, $\tau_{1/3} = 4l_0 c/3v_E^2$. It can be seen that the ion momentum depends on time as

$$p_x \approx m_a c \left(\frac{t}{\tau_{3/5}} \right)^{3/5}, \tag{25.84}$$

where $\tau_{3/5} = \left[48 m_a l_0 c / 125 \left(\pi^{(0)} v_E \right)^2 \right]^{1/3}$. It can be seen that the surface density decreases as $nl \propto t^{4/5}$, and the longitudinal component of the momentum, $p_x \propto t^{3/5}$, increases more rapidly than in the case of a not expanding shell.

The value of the phase ψ (it is the coordinate of the ions relative to the electromagnetic pulse) is equal to the integral

$$\psi = \omega_0 \left(t - \frac{x}{c} \right) = \int (1 - \beta_x(t')) dt' \tag{25.85}$$

with the normalized velocity of the shell $\beta_x(t) = p_x(t)/m_a c \gamma(t)$. Substituting the dependence of the momentum on time in the integrand in the form $p_x \approx m_\alpha c (t/\tau_k)^k$, which for $k = 1/3,\ 3/5$ corresponds to the relations found above, we obtain

$$\psi = \omega_0 t - \omega_0 \tau_k \frac{(t/\tau_k)^{1-2k}}{k+1} {}_2F_1 \left(\frac{k+1}{2k}, \frac{1}{2}, \frac{3k+1}{2k}; -\left(\frac{t}{\tau_k} \right)^{2k} \right). \tag{25.86}$$

Here where ${}_2F_1(a, b, c; z)$ is the Gauss hypergeometric function (Olver et al., 2010). Asymptotically, in the limit as $t \to \infty$, it follows from this expression that

$$\psi \to \omega_0 \tau_k \frac{(t/\tau_k)^{1-2k}}{1-2k} + \frac{\omega_0 \tau_k}{\pi^{1/2}} \Gamma\left(\frac{2k-1}{2k} \right) \Gamma\left(\frac{2k+1}{2k} \right) \tag{25.87}$$

i.e. for $k > 1/2$ the phase is finite. The phase shift between the position of the ions and the laser pulse does not increase. In a long laser pulse, the ions become trapped, and their energy increases formally without a bound due to the decreasing number of accelerated particles.

We note here that the examples of unlimited acceleration are well known for electrons accelerated by an electromagnetic wave propagating along a magnetic field in the autoresonant regime (Milant'ev, 1997), when an electron is trapped in an electrostatic wave propagating perpendicularly to the magnetic field (Katsouleas & Dawson, 1983; Bulanov & Sakharov, 1986), and in the case of an electron acceleration by a wake wave in an inhomogeneous plasma (Bulanov et al., 2001; Mourou et al., 2006).

Figure 25.7 shows the results of a computer simulation of the interaction of strong laser radiation with an ellipsoidal target $10\lambda_0 \times 7.5\lambda_0$ in size. The initial plasma density corresponds to the ratio $\omega_{pe}/\omega_0 = 6$. A laser pulse $25\lambda_0 \times 25\lambda_0$ in size with the amplitude $a_0 = 125$ compresses the target in the longitudinal direction. The target expands transversely, and, as a result, the radiation pressure of the laser pulse accelerates the target in accordance with the scenario described above. The

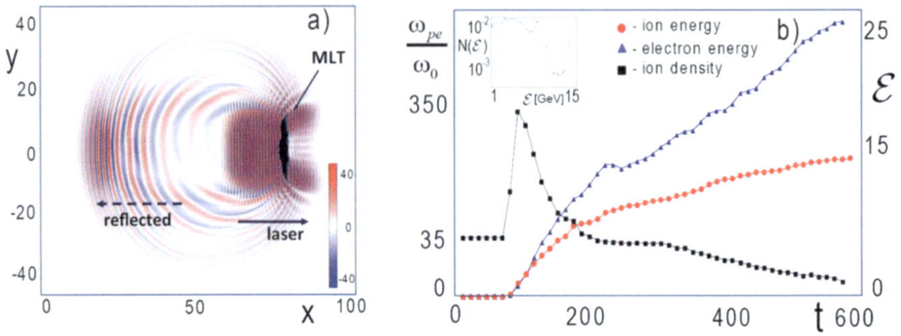

Fig. 25.7 **a** The laser pulse, the reflected wave, and the accelerated plasma target are shown as a superposition of the distribution of the z-component of the electric field and of the ion density in the plane (x, y) for $t = 112.5$. **b** Dependence of the ion and electron energy and of the ion density n/n_{cr} on time. The inset shows the energy spectrum of ions at $t = 600$. (Reproduced from Bulanov et al., 2010). Reprinted figure with APS permission from (Bulanov et al., (2010) as follows: Bulanov et al. (2010). https://doi.org/10.1103/PhysRevLett.104.135003. Copyright (2010) by the American Physical Society

wavelength of the reflected light increases with time, as it should for a wave reflected from a receding mirror that moves with acceleration. At the initial stage of the interaction, the plasma density increases due to the longitudinal compression of the target, and then it decreases monotonically. For the time $t = 600$, the ion energy reaches 14 GeV, and the electron energy becomes equal to 27 GeV. For a target not extending in the transverse direction, the ion energy would be equal to 3 GeV. In the inset in Fig. 25.7b, which shows the energy spectrum of ions at $t = 600$, the spectrum is seen to have a quasi-mono-energetic form.

25.7 Ion Acceleration in an Extended Plasma

In the case of the electromagnetic wave interaction with an extended plasma, the ions are accelerated under the conditions when the radiation pressure drives the plasma ahead, acting as a piston. This regime is called "hole-boring" (Wilks et al., 1992). When a constant amplitude, E_{las}, electromagnetic pulse irradiates a homogeneous plasma, a stationary solution exists, which describes the plasma-vacuum interface moving with constant velocity $d\beta/dt = 0$. In the frame of reference moving with the velocity $c\beta$, the surface position is determined by the balance between the flux of electromagnetic wave momentum and the ion momentum flux. This yields for the velocity β where (Bulanov et al., 2012 a)

$$\beta = \frac{\sqrt{4\beta_g \mathrm{B}_E^2 + \left(1-\beta_g^2\right)^2 \mathrm{B}_E^4} - \mathrm{B}_E^2\left(1+\beta_g^2\right)}{2\left(1-\beta_g \mathrm{B}_E^2\right)} \tag{25.88}$$

with

$$\mathrm{B}_E = \frac{E_{las}}{\sqrt{2\pi n_0 m_a c^2}}. \tag{25.89}$$

The electric field, E_{las}, is taken in the laboratory frame of reference. Here we assume that $|\rho|^2 = 1$ and $|\alpha|^2 = 0$, i.e., $K = 2$, and take into account the fact that in the rest frame of the plasma-vacuum interface the plasma density is equal to $n = n_0/\sqrt{1-\beta^2}$. Apparently, the plasma-vacuum interface velocity cannot exceed the group velocity of the electromagnetic wave, $\beta \le \beta_g$.

In the boosted frame of reference, $\mathcal{M}$, moving with velocity $c\beta$, the ions bounce at the laser pulse front. Their momentum changes from $-m_\alpha c/\sqrt{1-\beta^2}$ to $m_\alpha c/\sqrt{1-\beta^2}$, as illustrated in Fig. 25.8. In the laboratory frame of reference, $\mathcal{L}$, the momentum and the energy of the fast ions are equal to gives

$$p_\alpha = 2m_a c\left(\frac{\beta}{1-\beta^2}\right) \text{ and } \mathcal{E}_a = m_a c^2\left(\frac{1+\beta^2}{1-\beta^2}\right) \tag{25.90}$$

with β given by Eq. (25.88).

In the case when $\beta_g = 1$, we recover the result obtained by Naumova et al., (2009) and Robinson et al., (2012), which gives

$$\beta = \frac{\mathrm{B}_E}{1+\mathrm{B}_E}. \tag{25.91}$$

This yields for the ion kinetic energy

$$\mathcal{E}_\alpha = 2m_\alpha c^2 \frac{\mathrm{B}_E^2}{1+2\mathrm{B}_E}, \quad \beta = \frac{\mathrm{B}_E}{1+\mathrm{B}_E} \tag{25.92}$$

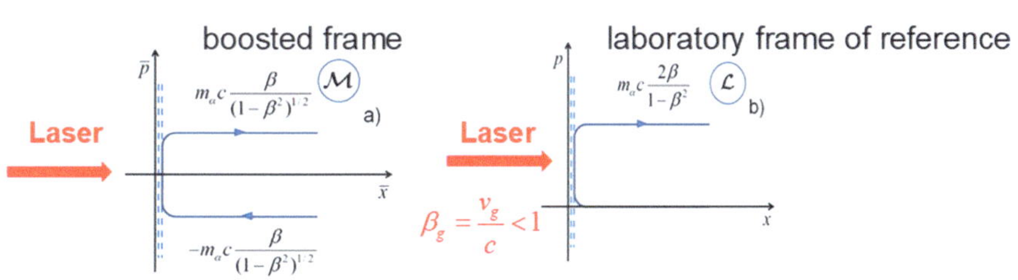

Fig. 25.8 Phase plane of the ions accelerated at the front of the laser pulse interacting with a homogeneous plasma. **a** In the boosted frame of reference, $\mathcal{M}$, the ion momentum changes from $-m_a c/\sqrt{1-\beta^2}$ to $m_a c/\sqrt{1-\beta^2}$. **b** In the laboratory frame of reference, $\mathcal{L}$, the ion momentum changes from zero to $2m_a c\beta/(1-\beta^2)$. Reproduced from Bulanov et al., (2012 a), with the permission of AIP Publishing

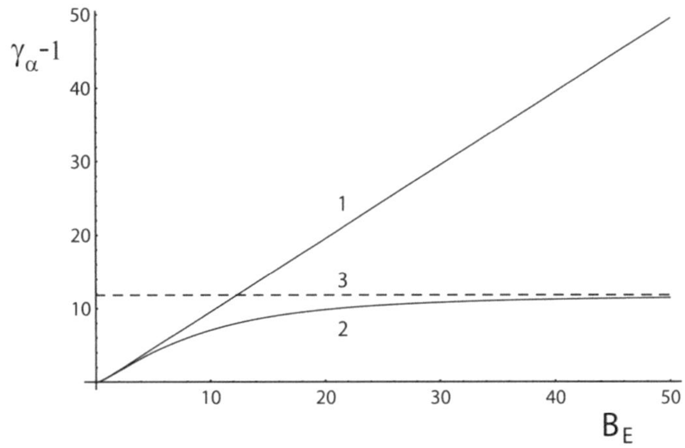

Fig. 25.9 Normalized ion kinetic energy, $(\mathcal{E}_\alpha - m_\alpha c^2)/m_\alpha c^2$, versus the normalized laser pulse amplitude, B_E: (1) for $\beta_g = 1$ and (2) for $\beta_g = 0.925$; (3) The dashed line corresponds to $2\beta_g^2/(1 - 2\beta_g^2)$

Figure 25.9 shows the normalized ion kinetic energy, $(\mathcal{E}_\alpha - m_\alpha c^2)/m_\alpha c^2$, as a function of the normalized laser pulse amplitude, B_E, for $\beta_g = 1$ and $\beta_g < 1$.

In the limit $B_E << 1$ we have the energy scaling with the laser pulse amplitude and plasma density

$$\mathcal{E}_\alpha \approx \left(\frac{I}{2.5 \times 10^{21} W/cm^2}\right)\left(\frac{10^{21} cm^{-3}}{n_0}\right) \text{GeV}. \tag{25.93}$$

When $B_E >> 1$ the energy scaling has the form

$$\mathcal{E}_\alpha \approx \left(\frac{m_\alpha}{m_p}\right)^{1/2}\left(\frac{I}{5 \times 10^{21} W/cm^2}\right)^{1/2}\left(\frac{10^{21} cm^{-3}}{n_0}\right)^{1/2} \text{GeV}. \tag{25.94}$$

These for 10 PW laser pulse with the amplitude equal to $a_0 \approx 30$ and for the plasma density approximately equal to the critical density n_{cr} corresponds to the ion energy in the 30 GeV range.

The region of counter-penetrating beams of accelerated ions is unstable against Buneman-like and Weibel-like instabilities which results in the broadening of the ion energy spectrum as shown by Bulanov et al., (2012 a).

25.8 Parameters Corresponding to Different Ion Acceleration Mechanisms

The basic mechanisms of acceleration are (i) Target Normal Sheath Acceleration (TNSA) (Wilks et al., 2001), (ii) Coulomb Explosion (CE) (Last et al., 1997; Nishihara et al. 2001), (iii) Radiation Pressure Acceleration (RPA) (Bulanov et al., 2013; Esirkepov et al., 2004), and (iv) Magnetic Vortex Acceleration (MVA) (Bulanov & Esirkepov, 2007; Bulanov et al., 2010; Fukuda et al., 2009; Hakimi et al., 2022; Kuznetsov et al., 2001; Nakamura et al., 2010; Park et al., 2019), (v) Hole-boring mechanism of the ion acceleration in extended plasma (Bulanov et al., 2012; Naumova et al., 2009; Wilks et al., 1992). There are also several composite mechanisms, which are either the combinations of basic ones or somehow the enhancement of the basic ones, e. g. the Directed Coulomb Explosion (Bulanov et al., 2008).

As was shown above, the ion acceleration mode in the process of Coulomb explosion can be realized if the laser pulse has a sufficiently large intensity, provided the surface density of the target, $\sigma_0 = n_0 l_0$, is not too large. In the opposite limit of high target surface density and a relatively small electromagnetic wave intensity, a slow heating of the electrons occurs. In this case, the mechanism of ion acceleration at the front of the plasma cloud becomes valid. The boundary between these two regimes in the parameter plane 'radiation amplitude-target surface density' is given by condition (25.54), which can be represented in the form $a_0 = \sigma_0 \lambda_0 r_e$. Here, $r_e = e^2/m_e c^2$ is the classical electron radius. Condition (25.54), as was noted above, also determines the region of parameters where the radiative acceleration of ions comes into play.

All these mechanisms can be parametrized in terms of two dimensionless parameters, α_a and δ_a (Bulanov et al., 2016), which are defined as

$$\alpha_a = \frac{a_0}{\varepsilon_p} \text{ and } \delta_a = \frac{d_e}{\lambda\!\!\!^-} \sqrt{a_0}, \tag{25.95}$$

where we assume $a_0 \gg 1$. The first parameter α_a is the ratio of normalized laser EM field strength, a_0 to ε_p, which is a normalized surface density determined by Eq. (25.60).

The second parameter δ_a is equal to the collisionless skin-depth $d_e\sqrt{a_0}$, with $d_e = c/\omega_{pe}$, normalized on $\lambda\!\!\!^-$. We take also into account the relativistic dependence of the plasma frequency $\omega_{pe}/\sqrt{a_0}$ on the laser pulse amplitude assuming $a_0 \gg 1$. The parameter α_a can be written in the form $\alpha_a = E_0/E_m$, where $E_m = 2\pi e n_e l$ is the maximum electric field, which can be produced by a charge slab of the density n_e and thickness l.

The regions in the parameter plane $\alpha_a - \delta_a$, corresponding to the acceleration of ions in the interaction of laser radiation with plasma targets, are shown schematically in Fig. 25.10.

The target can be considered as a thin foil, provided its thickness l is substantially less than the laser wavelength $\lambda = 2\pi\lambda\!\!\!^-$. In Fig. 25.10, the dashed parabolic curve

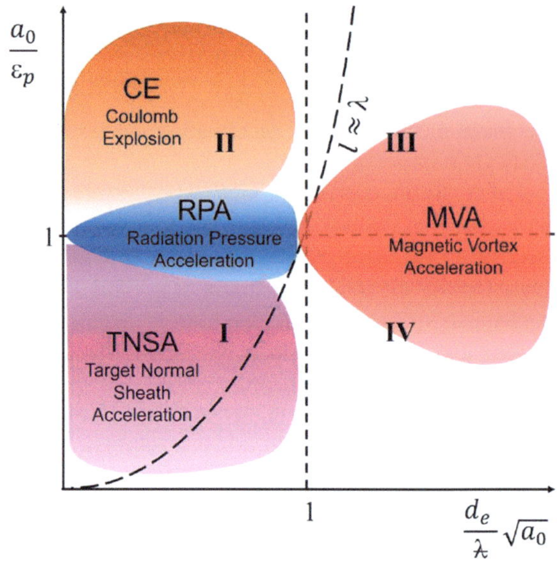

Fig. 25.10 The basic laser ion acceleration mechanisms in the plane of dimensionless parameters, a_0/ε_p and $d_e\sqrt{a_0}/\lambda\!\!\!^-$, characterizing the laser amplitude and the target transparency. Reproduced from Bulanov et al., (2016), with the permission of AIP Publishing

$l \approx \lambda$ subdivides the parameter plane in two sub-domains: a thin foil target for $l < \lambda$ and an extended plasma for $l > \lambda$.

The condition $a_0 = \varepsilon_p$, i.e., $\alpha_a = 1$, marks the transparency threshold for laser pulses' interactions with a thin foil target $l < \lambda$. In Fig. 25.9, domains (I) and (II) correspond to overdense opaque and transparent slab, respectively. If the laser radiation interacts with an opaque target in the parameter domain (I), a relatively small portion of hot electrons can leave the target, forming a sheath with an electric charge separation. The corresponding electric field accelerates ions in the TNSA regime. Above the line $a_0/\varepsilon_p = 1$, i.e. $\alpha_a = 1$, in domain (II), the laser radiation is so intense that it blows out almost all electrons from the irradiated region of the target. The remaining ions undergo fast expansion, known as Coulomb explosion, due to the repelling of non-compensated positive electric charges. At the opaqueness/transparency threshold, in the vicinity of the $a_0/\varepsilon_p = 1$ line in Fig. 25.10, the conditions for ion acceleration in the RPA regime are realized.

In regions (III) and (IV), in Fig. 25.10, where $d_e\sqrt{a_0}/\lambda\!\!\!^- > 1$, i.e., $\delta_a > 1$, the plasma is underdense. We consider here the case of near-critical density plasma when the high power laser pulse undergoes the relativistic self-focusing. The laser pulse loses its energy with the energy depletion length equal to $l_{dep} = a_0\left(\omega/\omega_{pe}\right)^2 l_{las}$, as we have discussed above. For an ultra-short (several wavelength) laser pulse, $l_{las} \approx \lambda$, the choice of the plasma target thickness l equal to the energy depletion length, $l \approx l_{dep}$, provides for maximum laser-target coupling. This condition takes the form $l = a_0\left(c\omega/\omega_{pe}^2\right)$, i.e., equivalent to the condition $a_0/\varepsilon_p = 1$ at the border between domains (III) and (IV) in Fig. 25.9. Under this condition, the laser pulse deposits almost all its energy in the plasma producing many fast electrons in the vicinity of the target rear surface. This results in quasi-static magnetic field generation at the plasma vacuum interface by the electric current carried by fast electrons, which is crucially important for realization of the magnetic vortex acceleration mechanism.

We note that experimentally the energy of the laser accelerated protons according to Ziegler et al., (2024) is approximately equal to 150 MeV.

25.9 Ion Acceleration via 'Nonlinear Vacuum Heating' Mechanism

As we can see, the ion acceleration with charge separation fields generated on the thin foil target surfaces, including target normal sheath acceleration, Wilks et al., (2001). Although, increasing attention is paid to different mechanisms, the TNSA mechanism remains crucial to understanding the physics of ion acceleration at moderate laser intensity regimes. The generation of an ion accelerating charge separation field is governed by the absorption mechanism of laser energy into electrons. In the case of interactions between an obliquely incident laser and overdense plasma, one of the most predominant absorbed mechanisms is the Brunel effect (Brunel, 1987). In this model, the electric field of the laser directly accelerates electrons at the plasma-vacuum interface to momentum $p_{\max} = 2m_e c a_0$ at maximum resulting in the efficient electron temperature equal to (Wilks et al., 2001)

$$T_e = m_e c^2 \left(\sqrt{1 + a_0^2} - 1 \right). \tag{25.96}$$

To explain the acceleration beyond $2m_e c a_0$, a nonlinear effect should be implemented into the analysis. Paradkar et al. (2012), Krasheninnikov (2014), Kemp and Divol (2012a, 2012b), Liseikina et al. (2015), Bulanov et al., (2015 b) theoretically showed stochastic heating of electrons by assuming a model dependent on the coordinate of the electric target. The heating mechanism of electrons driven by $\boldsymbol{j} \times \boldsymbol{B}$ force was precisely discussed by Mulser et al. (2007), and Debayle et al., (2013). Using particle-in-cell simulations, Taguchi et al. (2004) found that the heating of electrons transiting through a small size target can be higher than that by electron quivering and called this phenomenon 'nonlinear resonance absorption'.

The nonlinear vacuum heating of electrons depends on the laser pulse polarization. For obliquely incident on the target laser pulse having the p-polarization (i.e. the electric field of the electromagnetic wave is in the incidence plane), the electron heating is more efficient than in the case of the s-polarized electromagnetic wave, when the wave electric field is perpendicular to the incidence plane. The experiments by Yogo et al., (2016) on the laser ion acceleration with the thin foil target have demonstrated this effect as well as the accelerated ion energy above the electron temperature given by Eq. (25.91). See also the experiment results on the ion acceleration with varying laser pulse length in Yogo et al., (2017) where the indication on the effect of the electron recirculation on the ion acceleration is clearly observed.

25.9.1 Dynamics of Circulating Electrons

Here we consider a plane electromagnetic wave interacting with the thin foil located in the plane $x = 0$. The laser angle of incidence equals φ (see Fig. 25.11). The electromagnetic wave is linearly polarised with the electric field lying in the incidence plane. To simplify the consideration, we perform a Lorentz transformation from the laboratory frame K to the frame of reference K' moving with velocity $V_M = c \sin \varphi$ along the target surface (along the y direction). In the K' frame of reference, the electromagnetic wave is incident normally onto the surface (Bourdier, 1983; Bulanov et al., 1994; Vshivkov et al., 1998), and the gamma factor associated with the velocity V_M is given by

$$\gamma_M = \frac{1}{\sqrt{1 - V_M^2/c^2}} = \frac{1}{\cos \varphi}. \tag{25.97}$$

When the electromagnetic wave interacts with the plasma with a steep gradient assuming that the plasma surface can be treated as an ideal conductor, we obtain the boundary condition for a tangential component of the electric field $\boldsymbol{E}_t = 0$. Then the electric and magnetic field are written as

$$\boldsymbol{E}'(x', t') = \boldsymbol{E}_0\left(e^{i\omega'(t'-x'/c)} - e^{i\omega'(t'+x'/c)}\right), \tag{25.98}$$

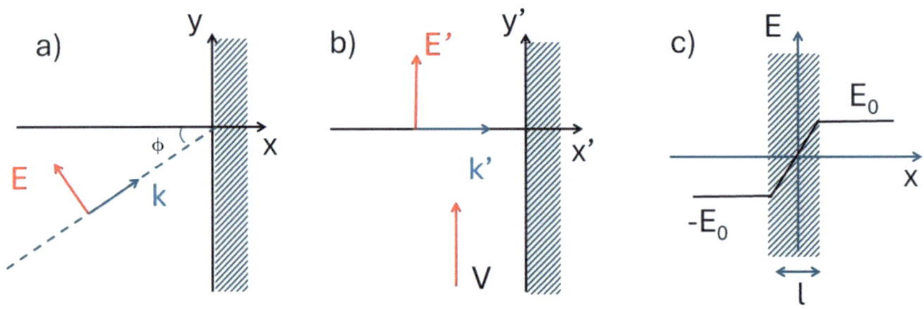

Fig. 25.11 Electromagnetic wave obliquely incident on the thin foil target. **a** The wave in the laboratory frame of reference $\mathcal{L}$. **b** The wave in the boosted frame of reference $\mathcal{M}$. For the p-polarisation wave, the electric field vector is in the x, y plane. In the boosted frame of reference, the plasma electrons move along the target with a velocity $V = -c \sin \varphi$ in the y' direction. **c** A charge separation electric field on the x axis. (see Yogo et al., 2016)

$$\boldsymbol{B}'(x', t') = \boldsymbol{E}_0\Big(e^{i\omega'(t'-x'/c)} + e^{i\omega'(t'+x'/c)}\Big), \tag{25.99}$$

respectively. Therefore, the real part of the magnetic field at the surface ($x' = 0$) is given by

$$\boldsymbol{B}'(x' = 0, t') = \big(0, -2E_0 \sin\theta \cos\omega' t', 2E_0 \cos\theta \cos\omega' t'\big). \tag{25.100}$$

In the boosted frame of reference, K', the acceleration of electrons in the x' direction is driven by the Lorentz force equal to $-(e/c)\boldsymbol{v} \times \boldsymbol{B}$, where the magnetic field is taken at the $x' = 0$ plane and the velocity of the electrons is equal to $\boldsymbol{v} = (0, -V, 0)$. As a result, the equation of motion for electrons yield for the x component of the electron momentum

$$\frac{dp'_x}{dt'} = 2m_e V \omega' a_0 \cos\theta \cos\omega' t'. \tag{25.101}$$

Hence, the x'-component of the electron momentum depends on time as

$$p' = 2m_e V a_0 \cos\theta \sin\omega' t', \tag{25.102}$$

where $V = c \sin\varphi$ (see Eq. (25.97)).

Within the framework of the nonlinear vacuum heating mechanism, the electrons quivered by the laser field are kicked out by the steep-like electric field at the plasma boundary. As a result, the electromagnetic field energy is irreversibly converted into the electron kinetic energy, thus enhancing the electron heating. To describe the nonlinear motion of electrons, we assume that a step-like charge separation electric field is formed near the ion layer with a positive electric charge.

Hence, the equations of electron motion can be written in the form

$$\dot{p} + \varepsilon_p \text{sign}(x) = a_0 \cos t, \tag{25.103}$$

$$\dot{x} = \frac{p}{\sqrt{1 + p^2}}, \tag{25.104}$$

Here the sign function is equal to $\text{sign}(x) = 1$ for $x > 0$ and $\text{sign}(x) = -1$ for $x < 0$, and a dot denotes a differentiation with respect to time, t. The electron momentum is normalized on $m_e c$, the time is measured in units ω^{-1} and the coordinate is normalized on c/ω.

The parameter

$$\varepsilon_p = \frac{2\pi n_0 e^2 l}{m_e \omega c},$$

introduced by Vshivkov et al., (1998), is proportional to the charge separation electric field, $E_l = 2\pi e n_0 l$, normalised on $m_e \omega c/e$, for a thin foil with an electric charge surface density equal to $e n_0 l$. See also Eq. (25.60).

The formulated model of a driven oscillator with a step-like nonlinearity (see also Bulanov et al., 2015) helps to determine and elucidate the conditions when the stochastic vacuum heating takes place. The one-dimensional model used bears the key features of the problem addressed in Sheng et al. (2002), Paradkar et al., (2012), Krasheninnikov (2014), but in a significantly simpler mathematical approximation, because a full description of the charged particle interaction with electromagnetic waves implies a consideration of a seven-dimensional dynamical system.

While the electron trajectory crosses the $x = 0$ plane, the sign of the static electric field changes abruptly. This can be considered as a 'collision', during which the oscillating driver electric field produces the work changing the electron energy. Depending on the phase, when the 'collision' happens the work can be either positive or negative, resulting in the electron energy increasing or decreasing. The enhancement of the energy transfer from the driver electric field to the electron is expected to occur when a resonant condition takes place:

$$a_0 = \frac{\pi}{2} \varepsilon_p, \tag{25.105}$$

which can be equivalent to the relationship of $E_l = E_0$. The expression for the momentum maximum is found for two different nonlinear regimes: nonlinear resonance and stochastic heating (for details see Bulanov et al., 2015).

In a nonlinear resonance regime, the electrons gain a momentum $\Delta p = a_0$ during the passage of the plane $x = 0$. Then, the momentum grows proportionally to the square root of time, written in

$$p_m = \sqrt{a_0 \varepsilon_p t/2} = a_0 \sqrt{t/\pi}, \tag{25.106}$$

for the regime of nonlinear resonance.

In the case of stochastic heating, the momentum dependence on time has a character of diffusion with the diffusion coefficient

$$D_{pp} = \frac{2\Delta p \Delta p}{\tau} = \frac{a_0^2 \varepsilon_p}{2 p_m}, \tag{25.107}$$

where we used the fact that the oscillation cycle duration equals $\tau = 4 p_m/\varepsilon_p$. This yields for the time dependence of the maximum electron momentum

$$p_m = \left(\frac{a_0^2 \varepsilon_p t}{2} \right)^{1/3} = a_0 \left(\frac{t}{\pi} \right)^{1/3}, \tag{25.108}$$

To analyse in more detail the properties of nonlinear oscillations we have integrated Eqs. (25.102) and (25.103) numerically. In the time duration of the laser

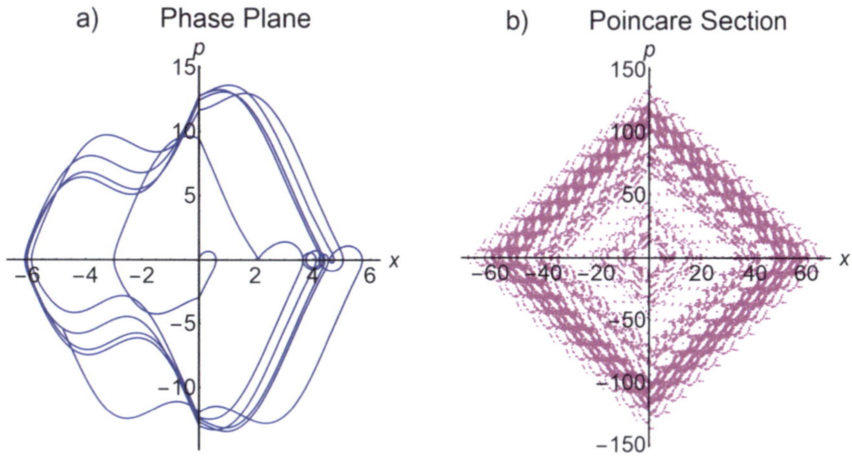

Fig. 25.12 Driven oscillations obtained for $a_0 = 3.14$ and $\varepsilon_p = 2$. **a** The phase plane (x, p). **b** The Poincaré section showing the particle positions on the phase plane at the discrete time with the time step equal to the period of the driving force, 2π

pulse, 60 fs (t = 125 in Fig. 25.12), the momentum reaches $8m_e c$. The oscillation amplitude is limited both in the momentum and coordinate directions, as shown in Fig. 25.12a. The trajectory tightly fills the domain in the phase plane demonstrating the ergodic property of the system under consideration. In Fig. 25.12b we plot a Poincaré section with the form of a finite thickness web, which is broadened due to the stochastic nature of the particle motion. The property of the particle to migrate on the phase plane along the web is typical for minimal chaos regimes (Chernikov et al., 1987). In other words, chaotic behaviour has begun to emerge but does not evolve completely under these conditions.

25.9.2 The Typical Energy Gain of Fast Protons

The nonlinear vacuum heating conjecture is equivalent to the assumption that the energy conversion from the electromagnetic field to electron energy depends on time according to the nonlinear mechanism introduced above. The average square of the x-component of the electron momentum $p_{nvh,x}$ scales with a laser pulse duration proportional to the number of wave periods N for the two different regimes of nonlinear vacuum heating as:

$$< p_{nvh,x}^2 > \propto \begin{cases} N < p_{x'}^2 > \text{ (nonlinear resonance)} \\ N^{2/3} < p_{x'}^2 > \text{ (stochastic heating)} \end{cases}, \tag{25.109}$$

To estimate the typical energy gain of fast protons, we use the step-like charge separation electric field equal to $E_0 \tanh(x/l)$ in the interval $-l/2 - \Delta x \ll x \ll l/2 + \Delta x$, where Δx is the electron displacement and l is the thickness of the positively charged layer of the foil target (Fig. 25.10c). When the target thickness is thin enough to satisfy the condition $l \approx \Delta x$, the electric field E0 can be expressed as $E_0 = 2\pi e n_0 l \approx 2\pi e n_0 \Delta x$ with the ion density equal to n_0. Assuming that the protons can be treated as test particles initially located on the surface of the layer, the proton energy gain $\mathcal{E}_p$ can be estimated to be of the order of the electron kinetic energy, $\mathcal{E}_e$ corresponding to the electron momentum given by Eq. (25.109).

25.10 Formation of High-Quality Ion Beams

For majority of applications of laser accelerators, including injectors (Krushelnick et al., 2005), controlled laser nuclear fusion (Atzeni & Meyer-ter-Vehn, 2004), and hadron therapy (Bulanov et al., 2014), the quality of the proton beam is of key importance because, if no additional measures are taken, fast protons with a wide energy spectrum are generated. For this reason, the question arises of how one can provide beams with narrow enough energy distribution, whose width ratio to the characteristic energy, $\Delta\mathcal{E}_a/\mathcal{E}_a$, is also small enough.

From the typical experiments on the laser ion acceleration it follows that the energy distribution in the range of energies smaller than the maximal $\mathcal{E}_{\min} \ll \mathcal{E}_a \ll \mathcal{E}_{\max}$ has the quasi-thermal form, with an efficient temperature several times less than $\mathcal{E}_{\max}$. For example, this spectrum is unacceptable for the medical applications, because it does not provide the optimal shape of the Bragg curve and, consequently, a substantial excess of radiation dose in the target area, which will lead to unacceptably high damage of healthy tissues.

An approach to control laser accelerated ion beams is associated with the ability to implement specially shaped targets. It was proposed by Bulanov and Khoroshkov (2002) (see also Esirkepov et al., 2002) to make use of a double-layer target for generating beams with a controllable quality. In this setup, the target consists of two layers: the first of which irradiated by the laser radiation consists of partially ionized heavy ions with charge $Z_a e$ and mass m_a, and the second layer, more distant with respect to the laser beam, consists of lighter ions, e.g., from ionized hydrogen, i.e., protons and electrons. In the initial configuration, the electric charge of positive ions is compensated by the negative charge of the electrons. The number of electrons in the first layer is assumed to be much larger than in the second. The second layer must have a sufficiently small thickness (the dimension along the direction of propagation of the laser pulse) and a sufficiently small width (the size in the transverse direction).

Under the action of powerful radiation, the target material becomes almost instantaneously ionized. As a result, the target is transformed into a thin plasma layer with the above-critical density. Then, the ponderomotive force of the laser radiation pushes the electrons away from the focus area. Since the characteristic time of the hydrodynamic expansion of a micron thick plasma is on the order of nanoseconds, it significantly exceeds the femtosecond duration of the laser pulse. Under these conditions, the heavy ions remain at rest during the time of pulse exposure to the target. This leads to the formation of a positively charged ion layer with the lateral size on the order of the laser focus. However, after a period of time around the inverse plasma frequency of heavy ions, equal to

$$\frac{1}{\omega_{pa}} = \sqrt{\frac{m_a}{4\pi n_a Z_a e^2}},$$

where $\omega_{p\alpha}$ is the plasma frequency of heavy ions, the ion layer starts to expand under the action of Coulomb repulsion of like charges, similar to that discussed above with regard to Coulomb explosion.

If the ratio $\mu^{1/2}/Z_a$, where $\mu = m_a/m_p$, is large enough, the characteristic time of proton motion is much less than the time of the Coulomb explosion of the heavy ion layer. Under these conditions, the thin proton layer of the double-layer target is accelerated in a given electric field.

In order to estimate the characteristic energy of accelerated ions, we assume that all electrons produced in the ionization process in the focus region of the laser beam are ejected from this region under the action of the ponderomotive radiation pressure. In this mode, the electric field in the vicinity of the positively charged layer is $E = 2\pi n_{0,a} e Z_a l$. Here, $n_{0\alpha}$ is the ion number density inside the target, and l is the target thickness. The region of the strong electric field localization has a lateral size on the order of the size of the focal area, $R_\perp$. As a result, we can estimate the energy acquired by the proton:

$$\mathcal{E}_{\max} = 2\pi n_{0a} e^2 Z_a l R_\perp.$$

Knowing the size of a focal region, $R_\perp$, we can evaluate the required power of the laser pulse, and then, having assigned a pulse duration, find its energy.

The electric field configuration can be approximated by the electrostatic field induced by an oblate charged ellipsoid with a major semi-axis equal to the focal spot size $R_\perp$ and a minor semi-axis (half the thickness) l. The expression giving the electric field of a charged disk can be found in book by Landau and Lifshitz (1984). Using this solution, we can write the electric field dependence on the coordinates outside of the target:

$$E_x = \frac{4\pi n_{0a} e Z_a l R_\perp^2}{3} \frac{1}{R_\xi} \partial_x \xi, \tag{25.110}$$

$$E_\rho = \frac{4\pi n_{0a} e Z_a l R_\perp^2}{3} \frac{1}{R_\xi} \partial_\rho \xi, \tag{25.111}$$

where x and $\rho = \sqrt{y^2 + z^2}$ are the polar coordinates, and the function R_ξ and the variable ξ are expressed as

$$R_\xi = (\xi + R_\perp^2)\sqrt{\xi + l}, \tag{25.112}$$

$$\xi = \frac{1}{4}\left[\sqrt{\left(\rho - \sqrt{R_\perp^2 - l^2}\right)^2 + x^2} + \sqrt{\left(\rho + \sqrt{R_\perp^2 - l^2}\right)^2 + x^2}\right]^2 - R_\perp^2. \tag{25.113}$$

From formulas (25.110) and (25.111) follow that the electric field is at its maximum on the target surface and decreases rapidly outside the region on the order of $R_\perp$ in size.

The proton energy spectrum can be found, assuming that the ions move in the paraxial region, which, in turn, corresponds to the structure of our double-layer model. On the axis, the radial component of the electric field vanishes, and the longitudinal component takes the form

$$E_x = \frac{8\pi n_{0a} e Z_a l R_\perp^2}{3} \frac{1}{R_\perp^2 - l^2 + x^2}. \tag{25.114}$$

To find the proton energy distribution one should solve the kinetic equation for the proton distribution function $f_p(x, v, t)$. As is well known, the solution of the kinetic equation for the distribution function can be represented as the relationship

$$f_p(x, v, t) = f_{0p}(x_0, v_0), \tag{25.115}$$

where $f_{0p}(x_0, v_0)$ is the distribution function at the initial instant of time $t = 0$, while $x(t)$ and $v(t)$ are the coordinate and velocity along the particle trajectory (the characteristic) originating at the point (x_0, v_0) in the phase space.

We consider the case when, at the initial instant of time, all the particles are at rest and the spatial distribution of their density is given by the function $n_{0p}(x_0)$. These initial conditions are satisfied by the distribution function of the form

$$f_{0p}(x_0, v_0) = n_{0p}(x_0)\delta(v_0),$$

where $\delta(v_0)$ is the Dirac delta function. The number of particles in the volume $dxdv$ of the phase space equals

$$dn = f_p dx dv = f_p v dv dt = f_p \frac{d\mathcal{E} dt}{m_p}. \tag{25.116}$$

Integration of this expression integrates over time yields the energy spectrum of accelerated particles in the form

$$\mathcal{N}_p(\varepsilon) d\varepsilon = d\varepsilon \int n_{0p}(x_0)\delta(v_0)\frac{dt}{m_p} = \frac{n_{0p}(x_0)}{m_p}\left|\frac{dt}{dv}\right|_{v=v_0} d\varepsilon. \tag{25.117}$$

Here, the Lagrange coordinate x_0 and the Jacobean $|dt/dv|_{v=v_0}$ are functions of the particle energy $\mathcal{E}$. The coordinate x_0 dependence on energy $\mathcal{E}$ is implicitly defined by the integral of motion

$$\mathcal{E}(x, x_0) = \mathcal{E}_0 + e(\varphi(x) - \varphi(x_0)), \tag{25.118}$$

where $\varphi(x)$ is the electrostatic potential. For the electric field given by Eq. (25.114), one can obtain

$$\varphi(x) = -\frac{4\pi n_{0a} e Z_a l R_\perp^2}{3\sqrt{R_\perp^2 - l^2}} \arctan\left(\frac{\sqrt{R_\perp^2 - l^2}}{x}\right), \tag{25.119}$$

where x is the running coordinate of the particle. The Jacobean $|dt/dv|_{v=v_0}$ is equal to the inverse value of the particle acceleration at the initial instant of time, i.e., $1/|eE(x_0)|$, where the dependence $E(x)$ is given by Eq. (25.114). On the other hand, the function $1/|eE(x_0)|$ equals $|dx_0/d\mathcal{E}|$. Therefore, we obtain the following expression for the energy spectrum of particles:

$$\mathcal{N}_p(\mathcal{E}) d\mathcal{E} = n_{0p}(x_0) \left|\frac{dx_0}{d\mathcal{E}}\right|_{x_0 = x_0(\mathcal{E})} d\varepsilon. \tag{25.120}$$

As follows from formula (25.114), the electric field is locally homogeneous in the near-axis region of the target. Due to this, the energy spectrum is determined by the form of the dependence $n_{0p}(\varphi^{-1}(\mathcal{E}/e))$.

We may conclude that a unified approach to producing an ion beam with a small energy spectrum width is based on the acceleration of the proton (ion) layers at the initial instant of time, which have a sufficiently small thickness Δx_0. In this case, the energy spread in the proton beam is proportional to the thickness of the layer Δx_0.

The parameter characterizing the property of the beam to be compressed in the longitudinal direction is the longitudinal emittance $\varepsilon_\parallel$ defined as the product of the energy width $\Delta\mathcal{E}$ and the beam duration Δt. Using the above expressions, we find

$$\varepsilon_\parallel = \Delta\mathcal{E}\Delta t \approx \frac{\Delta x_0}{R_\perp}\sqrt{\frac{m_p \mathcal{E}}{2}}. \tag{25.121}$$

For $\Delta\mathcal{E} = 100$ MeV, $\Delta x_0 = 0.1\,\mu$m, and $R_\perp = 5\,\mu$m, the longitudinal emittance is $\varepsilon_\parallel = 0.1$ MeV ps.

According to expression (25.111), the radial electric-field component in the near-axial region close to the target is given by

$$E_\rho = \frac{8\pi n_{0a} e Z_a l^2}{3R_\perp^2}\rho, \tag{25.122}$$

i.e., it is a linear function of the radius. From this it follows that the trajectories $\rho(x)$ of the fast particles in the (x, ρ) plane are described by the formula

$$\rho(x) = \rho_0 \exp\left(\sqrt{\frac{2lx}{R_\perp^2}}\right), \tag{25.123}$$

where ρ_0 is the initial radial coordinate of a particle. As a result, the transverse emittance of the fast proton beam, associated with the transverse non-uniformity of the electric field, at the boundary of the acceleration region, at $x = R_\perp$, is $\varepsilon_\perp = \pi\theta d_0$. It can be written out as

$$\varepsilon_\perp = \frac{\pi l d_0 \sqrt{2}}{R_\perp}. \tag{25.124}$$

Let us estimate the energy of the protons, which they can acquire in a double-layer target irradiated by a laser pulse of a given power. The realization of the above-discussed acceleration mode requires a laser with high enough intensity I. The electric field of the wave, expressed via its intensity, $E_{las} = \sqrt{4\pi I/c}$, should exceed the field required to remove the electrons from the focus area, i.e., it must be stronger than the characteristic field of the electric charge separation, which, according to Eq. (25.114), equals $8\pi n_{0\alpha} e Z_\alpha l/3$. Utilizing these relationships and Eq. (25.118) for the electrostatic potential and assuming that the proton initial coordinate is $x_0 = l$, we obtain the relationship between the proton energy and the laser power $\mathcal{P}_{las} = \pi R_\perp^2 I$. It has the form

$$\mathcal{E}_p = \sqrt{\frac{\pi^2 e^2 \mathcal{P}_{las}}{4}}. \tag{25.125}$$

It is convenient to transform this formula into
$\mathcal{E}_p = \pi\sqrt{\frac{\mathcal{P}_{las}}{\overline{\mathcal{P}}}}$, where $\overline{\mathcal{P}} = \frac{2m_e^2c^5}{e^2} = 17.4\,\text{GW}$.

Hence, in order to generate protons with an energy of 200 MeV, laser power on the order of 3 PW and an energy of not less than 100 J are required.

We note that high quality laser accelerated ion beam generation with using the double layer target has been demonstrated with computer simulations (Bulanov et al., 2002; Esirkepov et al., 2002; Morita et al. 2008, 2009) and in the proof-of-principle experiment (Schwoerer et al., 2006).

Using tailored targets for generation of collimated high-energy ion beams has been considered by Bulanov et al. (2000), Sentoku et al., (2000), Ruhl et al. (2001).

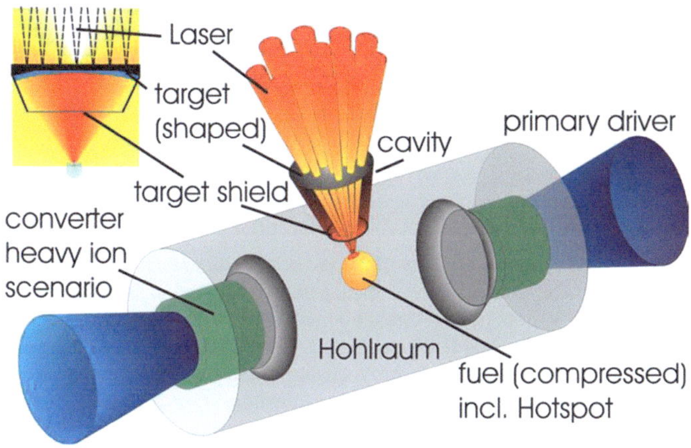

Fig. 25.13 Original idea of indirectly driven fast ignition using a laser accelerated proton beam Roth et al., (2001), Roth et al. (2009). Reprinted figure with APS permission from Roth et al., (2001) as follows: Roth et al. (2001). https://doi.org/10.1103/PhysRevLett.86.436. Copyright (2001) by the American Physical Society

25.11 Several Applications of the Laser Accelerated Ions

25.11.1 Fast Ignition by Intense Laser-Accelerated Proton Beams

The concept of fast ignition in controlled fusion targets is a technique to significantly reduce the constraints to the driving laser or ion beam systems and to simultaneously enhance the gain of the fusion reaction. Fast ignition by intense beams of electrons has been proposed by Tabak et al. (1994). Ability of high-power lasers to generate intense laser-driven proton and ion beams resulted in the proposal by Roth et al., (2001), Bychenkov et al., (2001), Gus'kov (2001) the idea of using the laser accelerated proton beams to ignite a pre-compressed fusion target, as it is illustrated in Fig. 25.13 (for details e.g. see review by Roth (2009) and literature cited therein). The protons were accelerated off the rear surface of the foils and, because of the parabolic geometry, are focused into the compressed fuel. On the focusing of the laser accelerated ions by using the shaped paraboloidal targets see Bulanov et al. (2000) and Ruhl et al. (2001).

25.11.2 Laser Ion Acceleration for Hadron Therapy

The idea of using laser plasma as a source of high-energy ions for the purposes of proton therapy is discussed has been proposed by Bulanov and Khoroshkov (2002)

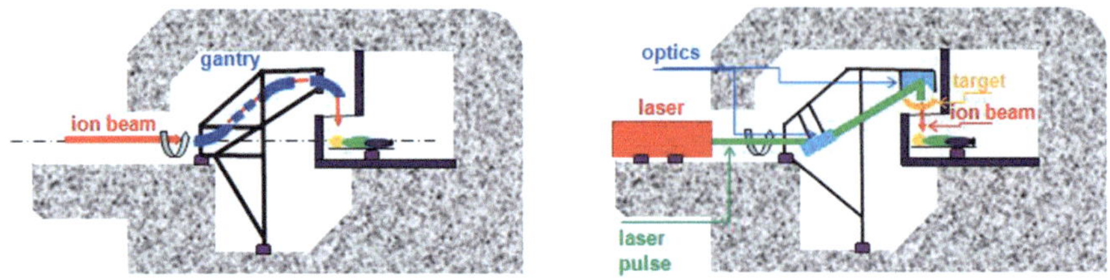

Fig. 25.14 The classical setup of a gantry (left) and (right) its optical counterpart Bulanov and Khoroshkov (2002). **a** The classical setup of a gantry: (1) bending magnets, (2) quadrupole lenses, (3) positioner, (4) system of dose formation and dose monitoring, (5) treatment room, and (6) concrete protection. **b** Individual laser accelerator and the optical setup of the gantry: (1) laser beam, (2) optical system of transportation and focusing of the laser pulse, (3) positioner, (4) dose monitoring system, (5) treatment room, (6) concrete protection, (7) target, and (8) ion beam. In both cases, the casing with the elements mounted on it can be rotated on 180°

(see also review article Bulanov et al., (2014) and references therein). The proposal is based on the efficient ion acceleration in the interaction of high-power laser radiation with gaseous and solid targets (Borghesi et al., 2006; Daido et al., 2012; Macchi et al., 2013; Ziegler et al., 2024). The specific dependence of proton energy losses in biological tissues (the Bragg peak) promotes the solution of one of the main problems of radiation therapy, namely, the irradiation of a malignant tumour with a sufficiently strong and homogeneous dose, ensuring that the irradiation of the surrounding healthy tissues and organs is minimal. In this scheme, a beam of fast ions accelerated by a laser pulse can be integrated in the installations intended for proton therapy as it is illustrated in Fig. 25.14.

25.11.3 Imaging with Laser Accelerated Proton Beams

The use of high energy ion beams, accelerated with conventional accelerators of charged particles, for radiographic applications was proposed a long ago. As far it is concerning the relativistic laser plasma diagnostics, it uses a proton beam to interact with the magnetic field and/or the electric field in the plasma and the intensity profile of the beam is measured on a screen after the interaction. The magnetic and electric fields in the plasma deflect the beam's trajectory and the deflection causes modulation in the intensity profile. From the intensity profile, one can measure the integrated magnetic field and/or electric field providing information with time picosecond and several micron space resolutions (see review article Borghesi et al., (2006) and literature cited therein) (Fig. 25.15).

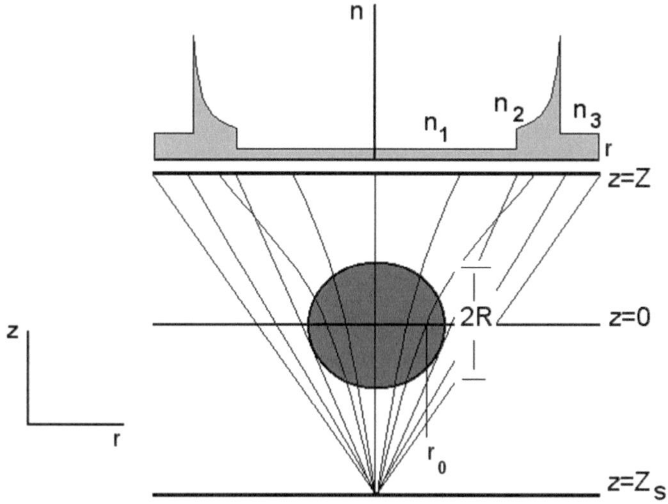

Fig. 25.15 Schematic of the principle of the proton imaging. The proton trajectories are deflected by the localized electric field which results in the proton density modulations on the detector. Reprinted figure with APS permission from Borghesi et al. (2002) as follows: Borghesi et al. (2002). https://doi.org/10.1103/PhysRevLett.88.135002, Copyright (2002) by the American Physical Society

References

Atzeni, S., & Meyer-ter-Vehn, J. (2004). *The physics of inertial fusion: Beam plasma interaction, hydrodynamics, hot dense matter*. OUP Oxford.

Bauer, D., & Mulser, P. (2007). Vacuum heating versus skin layer absorption of intense femtosecond laser pulses. *Physics of Plasmas, 14*, Article 023301.

Borghesi, M., Campbell, D. H., Schiavi, A., et al. (2002). Electric field detection in laser-plasma interaction experiments via the proton imaging technique. *Physics of Plasmas, 9*, 2214.

Borghesi, M., Fuchs, J., Bulanov, S. V., et al. (2006). Fast ions generation by high intensity laser-irradiation of foils and applications. *Fusion Science and Technology, 49*, 412.

Bourdier, A. (1983). Oblique incidence of a strong electromagnetic wave on a cold inhomogeneous electron plasma. Relativistic effects. *Phys. Fluids, 26*, 1804.

Brunel, F. (1987). Not-so-resonant, resonant absorption. *Physical Review Letters, 59*, 52.

Bulanov, S. V., & Sakharov, A, S. (1986). Acceleration of particles captured by a strong potential wave with a curved wave front in a magnetic field. *JETP Letter, 44*, 543.

Bulanov, S. V., Naumova, N. M., Esirkepov, T. Zh., Califano, F., Kato, Y., Liseikina, T. V., Mima, K., Nishihara, K., Sentoku, Y., Pegoraro, F., Ruhl, H., & Ueshima, Y. (2000). On the generation of collimated bunches of relativistic ions during interaction of the laser radiation with plasmas. *JETP Letters, 71*, 407.

Bulanov, S. V., Califano, F., Dudnikova, G. I., Esirkepov, T. Z., Inovenkov, I. N., Kamenets, F. F., Liseikina, T. V., Lontano, M., Mima, K., Naumova, N. M., & Nishihara, K. (2001). Relativistic interaction of laser pulses with plasmas. In V. D. Shafranov (Ed.),*Reviews of plasma physics* (Vol. 22, p. 227). Kluwer Academic.

Bulanov, S. V., & Khoroshkov, V. S. (2002). Feasibility of using laser ion accelerators in proton therapy. *Plasma Physics Reports, 28*, 453.

Bulanov, S. V., Esirkepov, T. Z., Khoroshkov, V. S., Kuznetsov, A. V., & Pegoraro, F. (2002). Oncological hadrontherapy with laser ion accelerators. *Physics Letters A, 299*, 240.

Bulanov, S. V., & Esirkepov, T. Zh. (2007). Comment on "Collimated multi-MeV ion beams from high-intensity laser interactions with underdense plasma". *Physical Review Letters, 98*, 049503.
Bulanov, S. S., Brantov, A., Bychenkov, V. Y., Chvykov, V., Kalinchenko, G., Matsuoka, T., Rousseau, P., Reed, S., Yanovsky, V., Litzenberg, D. W., & Krushelnick, K. (2008). Accelerating monoenergetic protons from ultrathin foils by flat-top laser pulses in the directed-Coulomb-explosion regime. *Physical Review E, 78*, 02641.
Bulanov, S.S., Bychenkov, V.Y., Chvykov, V., Kalinchenko, G., Litzenberg, D.W., Matsuoka, T., Thomas, A.G., Willingale, L., Yanovsky, V., Krushelnick, K., & Maksimchuk, A. (2010). Generation of GeV protons from 1 PW laser interaction with near critical density targets. *Physics of Plasmas, 17*, 043105.
Bulanov, S. V., Esirkepov, T. Zh., Kando, M., Pegoraro, F., Bulanov, S. S., Geddes, C. G. R., Schroeder, C. B., Esarey, E., & Leemans, W. P. (2012a). Ion acceleration from thin foil and extended plasma targets by slow electromagnetic wave and related ion-ion beam instability. *Physics of Plasmas, 19*, 103105.
Bulanov, S. S., Schroeder, C. B., Esarey, E., & Leemans, W. P. (2013). Electromagnetic cascade in high-energy electron, positron, and photon interactions with intense laser pulses. *Physical Review A, 87*, 062110.
Bulanov, S. V., Wilkens, J. J., Molls, M., Esirkepov, T. Zh., Korn, G., Kraft, G., Kraft, S. D., & Khoroshkov, V. S. (2014). Laser ion acceleration for hadron therapy. *Physics Uspekhi, 57*, 1149.
Bulanov, S. V., Yogo, A., Esirkepov, T. Zh., Koga, J. K., Bulanov, S. S., Kondo, K., & Kando, M. (2015). Stochastic regimes in the driven oscillator with a step-like nonlinearity. *Physics of Plasmas, 22*, 063108.
Bulanov, S. S., Esarey, E., Schroeder, C. B., et al. (2016). Radiation pressure acceleration: The factors limiting maximum attainable ion energy. *Physics of Plasmas, 23*, 056703.
Bulanov, S. V., Naumova, N. M., & Pegoraro, F. (1994). Interaction of an ultrashort, relativistically strong laser pulse with an overdense plasma. *Physics of Plasmas, 1*, 745.
Bychenkov, V. Yu., Rozmus, W., Maksimchuk, A., Umstadter, D., & Capjack, C. E. (2001). Fast ignitor concept with light ions. *Plasma Physics Reports, 27*, 1017.
Chandrasekhar, S. (1961). *Hydrodynamic and hydromagnetic stability*. Clarendon Press.
Chernikov, A. A., Sagdeev, R. Z., Usikov, D. A., Zakharov, M. Yu., Zaslavsky, G. M. (1987). Minimal chaos and stochastic web. *Nature, 326*, 559.
Daido, H., Nishiuchi, M., & Pirozhkov, A. S. (2012). Review of laser-driven ion sources and their applications. *Reports on Progress in Physics, 75*, Article 056401.
Debayle, A., Sanz, J., Gremillet, L., & Mima, K. (2013). Toward a self-consistent model of the interaction between an ultra-intense, normally incident laser pulse with an overdense plasma. *Physics of Plasmas, 20*, Article 053107.
Denavit, J. (1992). Absorption of high intensity subpicosecond lasers on solid density targets. *Physical Review Letters, 69*, 3052.
Esirkepov, T. Zh., Bulanov, S. V., Daido, H., Kato, Y., Khoroshkov, V. S., Kitagawa, Y., Mima, K., Nagai, K., Nishihara, K., Sakabe, S., Pegoraro, F., & Tajima, T. (2002). Proposed double-layer target for the generation of high-quality laser-accelerated ion beams. *Physical Review Letters, 89*, 175003.
Esirkepov, T. Zh., Borghesi M., Bulanov S. V., Mourou, G., & Tajima, T. (2004). Highly efficient relativistic-ion generation in the laser-piston regime. *Physical Review Letters, 92*, 175003.
Esirkepov, T. Z., Koga, J.K., Sunahara, A., Morita, T., Nishikino, M., Kageyama, K., Nagatomo, H., Nishihara, K., Sagisaka, A., Kotaki, H., & Nakamura, T. (2014). Prepulse and amplified spontaneous emission effects on the interaction of a petawatt class laser with thin solid targets. *Nuclear Instruments and Methods in Physics Research Section A: Accelerators, Spectrometers, Detectors and Associated Equipment, 745*, 150.
Fukuda, Y., Faenov, A. Y., Tampo, M., et al. (2009). Energy increase in multi-MeV ion acceleration in the interaction of a short pulse laser with a cluster-gas target. *Physical Review Letters, 103*, Article 165002.

Gurevich, A. V., Pariiskaya, L. V., & Pitaevskii, L. P. (1966). Self-similar motion of rarefied plasma. *Soviet Physics ZETP, 22*, 449.

Gus'kov, S. Yu. (2001). Direct ignition of inertial fusion targets by a laser-plasma ion stream. *Quantum Electronics, 31*, 885.

Hakimi, S., Obst-Huebl, L., Huebl, A., et al. (2022). Laser–solid interaction studies enabled by the new capabilities of the iP2 BELLA PW beamline. *Physics of Plasmas, 29*, Article 083102.

Kaplan, A. E., Dubetsky, B. Y., & Shkolnikov, P. L. (2003). Shock shells in coulomb explosions of nanoclusters. *Physical Review Letters, 91*, Article 143401.

Kar, S., Borghesi, M., Bulanov, S. V., Macchi, A., Key, M. H., Liseykina, T. V., Mackinnon, A. J., Patel, P. K., Romagnani, L., Schiavi, A., & Willi, O. (2008). Plasma jets driven by ultra-intense laser interaction with thin foils. *Physical Review Letters, 100*, Article 225004.

Kar, S., Kakolee, K. F., Qiao, B., et al. (2012). Ion acceleration in multispecies targets driven by intense laser radiation pressure. *Physical Review Letters, 109*, Article 185006.

Katsouleas, T., & Dawson, J. M. (1983). Unlimited electron acceleration in laser-driven plasma waves. *Physical Review Letters, 51*, 392.

Kemp, A. J., & Divol, L. (2012a). Interaction physics of multipicosecond petawatt laser pulses with overdense plasma. *Physical Review Letters, 109*, 1.

Kemp, A. J., & Divol, L. (2012b). Interaction physics of multipicosecond petawatt laser pulses with overdense plasma. *Physical Review Letters, 109*, Article 195005.

Kim, I. J., Pae, K. H., Choi, I. W., et al. (2016). Radiation pressure acceleration of protons to 93 MeV with circularly polarized petawatt laser pulses. *Physics of Plasmas, 23*, Article 070701.

Klimo, O., Psikal, J., Limpouch, J., & Tikhonchuk, V. T. (2008). Monoenergetic ion beams from ultrathin foils irradiated by ultrahigh-contrast circularly polarized laser pulses. *Review Special Topics—Accelerators and Beams, 11*, Article 031301.

Kovalev, V. F., & Bychenkov, V. Yu. (2003). Analytic solutions to the Vlasov equations for expanding plasmas. *Physical Review Letters, 90*, 185004.

Krainov, V. P., & Smirnov, M. B. (2002). Cluster beams in the super-intense femtosecond laser pulse. *Physics Reports, 370*, 237.

Krasheninnikov, S. I. (2014). On stochastic heating of electrons by intense laser radiation in the presence of electrostatic potential well. *Physics of Plasmas, 21*, Article 104510.

Krushelnick, K., Clark, E. L., Beg, F. N., Dangor, A. E., Najmudin, Z., Norreys, P. A., Wei, M., & Zepf, M. (2005). High intensity laser-plasma sources of ions—physics and future applications. *Plasma Physics and Controlled Fusion, 47*, B451.

Kuznetsov, A. V., Esirkepov, T. Zh., Kamenets, F. F., & Bulanov, S. V. (2001). On the efficiency of fast ion acceleration by relativistically strong laser pulses in an underdense plasma. *Plasma Physics Reports, 27*, 211.

Landau, L. D., & Lifshitz, E. M. (1987). *Course of theoretical physics, Vol. 6: Fluid mechanics.* Pergamon Press.

Landau, L. D., & Lifshitz, E. M. (1984). *Course of theoretical physics, Vol. 8: Electrodynamics of continuous media.* Pergamon Press

Last, I., Schek, I., & Jortner, J. J. (1997). Energetics and dynamics of Coulomb explosion of highly charged clusters. *Chemical Physics, 107*, 6685.

Liseykina, T., Mulser, P., & Murakami, M. (2015). Collisionless absorption, hot electron generation, and energy scaling in intense laser-target interaction. *Physics of Plasmas, 22*, Article 033302.

Macchi, A., Borghesi, M., & Passoni, M. (2013). Ion acceleration by superintense laser-plasma interaction. *Reviews of Modern Physics, 85*, 751.

Mackenroth, F., & Bulanov, S. S. (2019). Tailored laser pulse chirp to maintain optimum radiation pressure acceleration of ions. *Physics of Plasmas, 26*, Article 023103.

Manheimer, W., Colombant, D., & Ott, E. (1984). Three-dimensional, nonlinear evolution of the Rayleigh–Taylor instability of a thin layer. *Phys. Fluids, 27*, 2164.

Matsukado, K., Esirkepov, T., Daido, H., Utsumi, T., Li, Z., Fukumi, A., Hayashi, Y., Orimo, S., Nishiuchi, M., Bulanov, S. V., Tajima, T., Noda, A., Iwashita, Y., Shirai, T., Takeuchi, T., Nakamura, S., Yamazaki, A., Ikegami, M., Mihara, T., … Kubota, T. (2003). Energetic protons from a few-micron metallic foil evaporated by intense laser pulse. *Physical Review Letters, 91*, Article 215001.

Milant'ev, V. P. (1997). Cyclotron autoresonance and its applications. *Physics of Uspekhi, 40*, 1.

Morita, T., Esirkepov, T. Zh., Bulanov, S. V., Koga, J., & Yamagiwa, M. (2008). Tunable high-energy ion source via oblique laser pulse incident on a double-layer target. *Physical Review Letters, 100*, 145001.

Morita, T., Esirkepov, T. Zh., Koga, J., Yamagiwa, M., & Bulanov, S. V. (2009). The effect of laser pulse incidence angle on the proton acceleration from a double-layer target. *Plasma Physics and Controlled Fusion, 51*, 024002.

Mourou, G., Tajima, T., & Bulanov, S. V. (2006). Optics in the relativistic regime. *Reviews of Modern Physics, 78*, 309.

Nakamura, T., Bulanov, S. V., Esirkepov, T. Z., & Kando, M. (2010). High-energy ions from near-critical density plasmas via magnetic vortex acceleration. *Physical Review Letters, 105*, Article 135002.

Naumova, N. M., Schlegel, T., Tikhonchuk, V. T., Labaune, C., Sokolov, I. V., & Mourou, G. (2009). Hole boring in a DT pellet and fast-ion ignition with ultraintense laser pulses. *Physical Review Letters, 102*, Article 025002.

Nishihara, K., Amitani, H., Murakami, M., Bulanov, S. V., & Esirkepov, T. Zh. (2001). High energy ions generated by laser driven Coulomb explosion of cluster. *Nuclear Instruments and Methods in Physics Research Section A, 464*, 98.

Olver, F. W. J., Lozier, D. W., Boisvert, R. F., & Clark, C. W. (2010). *NIST handbook of mathematical functions*. Cambridge University Press.

Ott, E. (1972). Nonlinear evolution of the Rayleigh-Taylor instability of a thin layer. *Physical Review Letters, 29*, 1429.

Palmer, C. A. J., Schreiber, J., Nagel, S. R., et al. (2012). Rayleigh-Taylor instability of an ultrathin foil accelerated by the radiation pressure of an intense laser. *Physical Review Letters, 108*, 22500.

Paradkar, B. S., Krasheninnikov, S. I., & Beg, F. N. (2012). Mechanism of heating of pre-formed plasma electrons in relativistic laser-matter interaction. *Physics of Plasmas, 19*, 060703

Park, J., Bulanov, S. S., Bin, J., Ji, Q., Steinke, S., Vay, J. L., Geddes, C. G. R., Schroeder, C. B., Leemans, W. P., Schenkel, T., & Esarey, E. (2019). Ion acceleration in laser generated megatesla magnetic vortex. *Physics of Plasmas, 26*, Article 103108.

Pegoraro, F., & Bulanov, S. V. (2007). Photon bubbles and ion acceleration in a plasma dominated by the radiation pressure of an electromagnetic pulse. *Physical Review Letters, 99*, Article 065002.

Robinson, A. P. L., Trines, R., Dover, N. P., et al. (2012). Hole-boring radiation pressure acceleration as a basis for producing high-energy proton bunches. *Plasma Physics and Controlled Fusion, 54*, Article 115001.

Roth, M., Cowan, T. E., Key, M. H., Hatchett, S. P., Brown, C., Fountain, W., Johnson, J., Pennington, D. M., Snavely, R. A., Wilks, S. C., Yatsuike, K., Ruhl, H., Pegoraro, F., Bulanov, S. V., Campbell, E. M., Perry, M. D., & Powell, H. (2001). Fast ignition by intense laser accelerated proton beams. *Physical Review Letters, 86*, 436.

Roth, M., et al. (2009). Review on the current status and prospects of fast ignition in fusion targets driven by intense laser generated proton beams. *iPlasma Physics and Controlled Fusion, 51*, Article 014004.

Ruhl, H., Bulanov, S. V., Cowan, T. E., Liseĭkina, T. V., Nickles, P., Pegoraro, F., Roth, M., & Sandner, W. (2001). Computer simulation of the three-dimensional regime of proton acceleration in the interaction of laser radiation with a thin spherical target. *Plasma Physics Reports, 27*, 363.

Sakabe, S., Shimizu, S., Hashida, M., Sato, F., Tsuyukushi, T., Nishihara, K., Okihara, S., Kagawa, T., Izawa, Y., Imasaki, K., & Iida, T. (2004). Generation of high-energy protons from the Coulomb explosion of hydrogen clusters by intense femtosecond laser pulses. *Physical Review A, 69*, Article 023203.

Schwoerer, H., Pfotenhauer, S., Jäckel, O., Amthor, K.-U., Liesfeld, B., Ziegler, W., Sauerbrey, R., Ledingham, K. W. D., Esirkepov, T. (2006). Laser-plasma acceleration of quasi-monoenergetic protons from microstructured targets. *Nature, 439*, 445.

Sentoku, Y., Liseikina, T.V., Esirkepov, T.Z., Califano, F., Naumova, N.M., Ueshima, Y., Vshivkov, V.A., Kato, Y., Mima, K., Nishihara, K., & Pegoraro, F. (2000). High density collimated beams of relativistic ions produced by petawatt laser pulses in plasmas. *Physical Review E, 62*, 7271.

Sheng, Z.-M., Mima, K., Sentoku, Y., Jovanović, M. S., Taguchi, T., Zhang, J., & Meyer-ter-Vehn, J. (2002). Stochastic heating and acceleration of electrons in colliding laser fields in Plasma. *Physical Review Letters, 88*, 055004.

Tabak, M., et al. (1994). Ignition and high gain with ultrapowerful lasers. *Physics of Plasmas, 1*, 1626.

Taguchi, T., Antonsen, T., & Milchberg, H. (2004). Resonant heating of a cluster plasma by intense laser light. *Physical Review Letters, 92*, 205003.

Vshivkov, V. A., Naumova, N. M., Pegoraro, F., & Bulanov, S. V. (1998). Nonlinear electrodynamics of the interaction of ultra-intense laser pulses with a thin foil. *Physics of Plasmas, 5*, 2727.

Wan, F., Wei-Quan Wang, W. Q., Zhao, Q., Zhang, H., Yu, T.-P., Wang, W.-M., Yan, W. C., Zhao, Y. T., Hatsagortsyan, K. Z., Keitel, C. H., Bulanov, S. V., & Li, J. X. (2022). Quasimonoenergetic proton acceleration via quantum radiative compression. *Physical Review Applied*, *17*, Article 024049.

Wilks, S., Kruer, W., Tabak, M., & Langdon, A. B. (1992). Absorption of ultra-intense laser pulses. *Physical Review Letters, 69*, 1383.

Wilks, S. C., Langdon, A. B., Cowan, T., Roth, M., et al. (2001). Energetic proton generation in ultra-intense laser–solid interactions. *Physics of Plasmas, 8*, 542.

Yogo, A., Mima, K., Iwata, N., Tosaki, S., Morace, A., Arikawa, Y., Fujioka, S., Johzaki, T., Sentoku, Y., Nishimura, H., & Sagisaka, A. (2017). Boosting laser-ion acceleration with multi-picosecond pulses. *Scientific Reports, 7*, 42451.

Yogo, A., Bulanov, S. V., Mori, M., et al. (2016). Ion acceleration via 'nonlinear vacuum heating' by the laser pulse obliquely incident on a thin foil target. *Plasma Physics and Controlled Fusion, 58*, 025003.

Ziegler, T., Göthel, I., Assenbaum, S., et al. (2024). Laser-driven high-energy proton beams from cascaded acceleration regimes. *Nature Physics, 20*, 1211.

Chapter 26
Relativistic Catoptrics

Abstract Catoptrics (from Greek: κατοπτρικός katoptrikós, from Greek: κάτοπτρον katoptron "mirror") deals with the phenomena of reflected light using mirrors. Super-strong laser field interacting with matter can create mirrors moving with relativistic velocity. In plasmas, the relativistic flying mirrors are thin and dense electron or electron–ion layers accelerated by the high intensity electromagnetic waves up to velocity close to the speed of light in vacuum. The reflection of the electromagnetic wave at the relativistic mirror results in its energy and frequency change. In the counter-propagation configuration, the up-shifted frequency of the reflected wave is multiplied by the factor proportional to the gamma-factor squared. This scientific area promises the development of the sources of ultrashort x-ray pulses in attosecond range.

Keywords EM wave reflection at relativistic mirror · Wave frequency upshifting · EM field intensification

Sources of electromagnetic radiation with the wavelength in a range from microwave to X- ray and gamma-ray range are required for a wide variety of the problems in fundamental and applied physics. This regime is of substantial interest for developing compact sources of coherent radiation in the high photon energy range.

One of the approaches to realize compact sources of hard electromagnetic radiation with controllable parameters is based on the relativistic mirror concept. Within the framework of this concept the electromagnetic wave reflected from the counter-propagating relativistic mirror is compressed in the longitudinal direction and its carrier frequency is upshifted as has been shown by Einstein (1905). See also a paper by Abraham (1904) who also considered the electromagnetic wave reflection from relativistic mirror.

The question on realization of relativistic mirrors under the laboratory conditions has been raised repeatedly over a number of years. It is also attracting great attention now. First of all, the relativistic electron beams (Granatstein et al., 1976; Pasour et al., 1977) and propagating with the relativistic speed the ionization fronts (Hashimshony

Y. Kato et al., *High Power Laser and Plasma Science*, Springer Series on Atomic, Optical, and Plasma Physics 130, https://doi.org/10.1007/978-3-032-08930-4_26

et al., 2001; Mori et al., 1995; Semenova, 1967; Zhidkov et al., 2009) have been considered as the candidates.

Such typical parameters of relativistic electron beams and of the ionization waves as the front width and electron density resulted in the choosing the experiment conditions when the electromagnetic wave frequency corresponded to the microwave domain. The optical range of the wave frequency is of special interest because nowadays the lasers are the most powerful sources of optical and infrared coherent radiation. For light reflection it is required the electron density significantly higher than the density with can be achieved in the electron beams.

In this regard, there are questions on whether it is possible to prepare the relativistic mirror of high enough quality for efficient reflection of the light, which can move with substantially large velocity for providing significantly high frequency increasing up to the level corresponding to the photon energy in the x-ray range. The answer on these questions can be found by using the knowledge in the physics of nonlinear processes in relativistic laser plasmas.

Under the conditions typical for interaction of high intensity electromagnetic wave with plasmas nonlinear character of this interaction results in the formation of electron density modulations in the form of high-density bunches and thin layers moving with the relativistic velocity. Interaction of the electromagnetic radiation with these relativistic objects can be considered as the wave reflection from relativistic mirrors (Bulanov et al., 2001; Bulanov et al., 2003; Bulanov et al., 1994; Mourou et al., 2006; Bulanov et al., 2006a, 2006b; Kulagin et al., 2007; Bulanov et al., 2013a, 2013b; Pirozhkov et al., 2007; Naumova et al., 2004; Meyer-ter-Vehn & Wu, 2009; Wu et al., 2010; Bulanov et al., 2010a, 2010b, 2010c, 2010d, 2010e; Bulanov et al., 2012; Esirkepov et al., 2009, 2020; Jeong et al., 2024), which is accompanied by the relativistic frequency shift and change of the reflected wave amplitude. This is an underlining mechanism of high order harmonics generation in relativistic collisionless plasma which is based on the oscillating relativistic mirror concept by Bulanov et al. (1994) (see Lichters et al., 1996, and review articles Bulanov et al., 2013a, 2013b; Teubner & Gibbon, 2009; Bulanov et al., 2015a and literature cited therein), of generation of attosecond pulses of electromagnetic radiation, and frequency upshifting and wave intensification.

We notice a principal difference between the electromagnetic wave scattering by a single relativistic electron corresponding to the Thomson scattering in the limit of classical electrodynamics or to the Compton scattering in the quantum limit, and the wave reflection by a dense enough electron layer moving with relativistic velocity. In both cases, the maximal frequency upshifting is the same been proportional to the square of the electron Lorentz factor. The difference is in the number of reflected photons. If in the first case the intensity of incoherently scattered light is proportional to the number of electrons, in the second case of a dense electron layer it is proportional to the square of the electron number, which is an advantageous property of the mirror-based sources of coherent x-ray radiation under the development for various applications.

26.1 Reflection of Electromagnetic Wave from Relativistic Mirror

The special theory of relativity formulated by A. Einstein a century ago is now a standard working tool for investigating the interaction of ultra-intense electromagnetic fields with matter. In his seminal paper Einstein (1905) in which he introduced special relativity, A. Einstein discussed the problem of the reflection of light from a mirror moving with constant velocity close to the speed of light as an example of the use of relativistic transformations.

26.1.1 Uniformly Moving Mirror

We consider the light reflection at the mirror moving with the constant velocity $v_M = \beta_M c$, where c—is the speed of light in vacuum, as it is illustrated in Fig. 26.1, where we plot the world lines of the mirrors M_1 and M_2, as well as the world lines of the incident and reflected electromagnetic waves. The mirror M_1 moves with the subluminal velocity $\beta_{M1} < 1$. The velocity of the second mirror M_2 is above speed of light in vacuum with $\beta_{M2} > 1$. Superluminal mirrors can be formed by the modulations of the refraction index whose phase velocity could be arbitrarily large, e.g. been formed by the ionization wave.

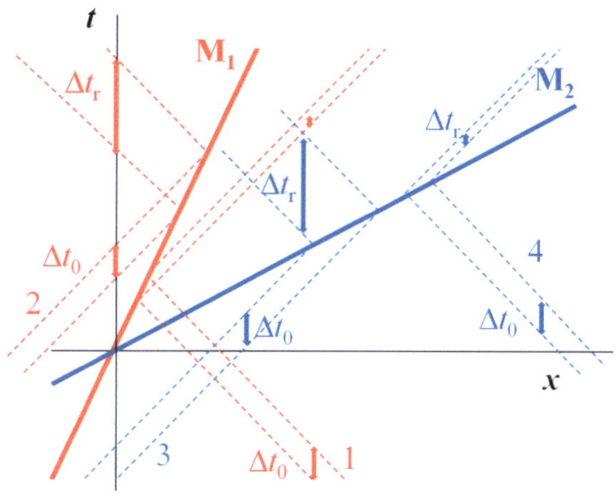

Fig. 26.1 Light reflection at the relativistic mirror. The mirror M_1 moves with the subluminal velocity $\beta_{M1} < 1$. The velocity of the second mirror M_2 is above speed of light in vacuum with $\beta_{M2} > 1$. In cases 1 and 4 the head on interaction of the light with mirrors results in the frequency up-shifting. The light-mirror interaction with the receding mirrors lead to the downshift of the frequency. The wave period changes from $\Delta t_0 = 2\pi/\omega_0$ to $\Delta t_r = 2\pi/\omega_r$

Since the world lines of the points on the subluminal mirror M_1 are connected by the space-like intervals, in the frame of reference moving with the mirror velocity $v_{\mathrm{M1}} = \beta_{\mathrm{M1}} c$ the mirror is at the rest. The light reflection at this mirror corresponds to the change of sign of the normal to mirror component of the wave vector.

In the case of a super-luminal mirror, M_2 it comprises the points connected by the time-like intervals. There is a frame of reference where the mirror M_2 is on the surface $t' = 0, -\infty < x' < +\infty$, i.e. the medium parameters change simultaneously. As we shall see below, the wave "reflection" at the super-luminal mirror results in the change of the sign of the wave frequency and the wave vector remains constant.

26.1.2 Sub-luminal Mirror

Change of the frequency and amplitude of an electromagnetic wave reflected by moving mirror occurs due to the double Doppler effect. Several remarkable features characterize the process of the electromagnetic wave reflection from a relativistic mirror. The frequency of reflected wave depends on the incidence angle and the mirror velocity. In order to find the reflected wave parameters we consider reflection and refraction of the plane monochromatic wave at the plane interface between vacuum region, denoted as a region 1, and the plasma slab (it is the mirror in this case, denoted as a region 2) moving with the velocity $v_{M1} = \beta_{M1} c$ along the x—axis as illustrated in Fig. 26.2a). In the boosted frame of reference, $\mathcal{M}$, where the mirror is at the rest (Fig. 26.2b), the wave reflection and refraction is are described by Snell's law, i.e. the incidence and reflection angles are equal to each other, $\theta_i' = \theta_r'$, and the refraction angle is given by equation $\sin\theta_t' = \sin\theta_i'\, n_1'/n_2'$, where n_1' and n_2' are the refraction indices in the media 1 and 2.

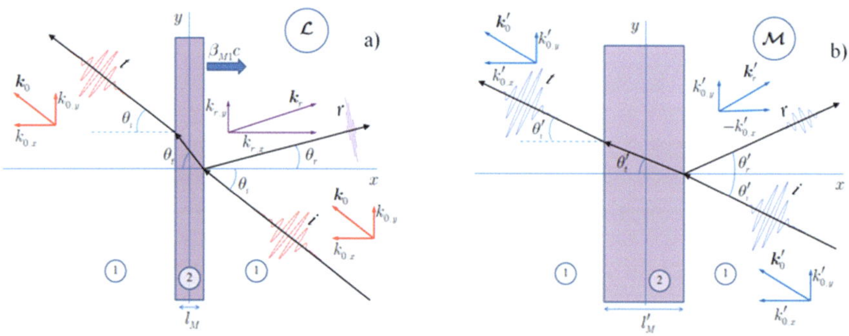

Fig. 26.2 Reflection of the electromagnetic wave at the plasma slab moving with the velocity $v_{M1} = \beta_{M1} c$ along the x–axis. **a** Incident, reflected, and transmitted waves in the laboratory frame of reference $\mathcal{L}$. **b** Incident, reflected, and transmitted waves in the mirror rest frame of reference, $\mathcal{M}$

Assuming that the mediums 1 and 2 are plasma regions with the electron densities $n'_{e,1}$ and $n'_{e,2}$, we obtain

$$\sin\theta'_t = \sin\theta'_i\sqrt{\frac{\omega'^2-\omega_{pe1}^2}{\omega'^2-\omega_{pe2}^2}}. \tag{26.1}$$

Here $\omega_{pe1} = \sqrt{4\pi n'_{e1}e^2/m_e}$, $\omega_{pe2} = \sqrt{4\pi n'_{e2}e^2/m_e}$, and ω' is the wave frequency in the boosted frame of reference. In this frame of reference the wave frequency and the wave number are given by

$$\omega' = \gamma_{M1}(\omega_0 + k_{0,x}\beta_{M1}c),\ k'_x = \gamma_{M1}(k_{0,x}+\omega_0\beta_{M1}/c),\ k'_y = k_{0,y}, \tag{26.2}$$

with $\gamma_{M1} = 1/\sqrt{1-\beta_{M1}^2}$ being the mirror relativistic gamma-factor. As a result of the wave reflection the x component of the wave vector changes sign (Fig. 26.2b). By performing the Lorentz transformation to the laboratory frame of reference, we can find the frequency and wave vector of reflected wave.

Taking into account that $\sin\theta_i = k_{0,y}/k_0$, $\sin\theta_r = k_{0,y}/k_r$ and using relationships (26.2), we obtain

$$\omega_r = \frac{\omega_0(1+\beta_{M1}^2)+2\beta_{M1}\sqrt{\omega_0^2-\omega_{pe}^2}\cos\theta_i}{1-\beta_{M1}^2}. \tag{26.3}$$

Here ω_0 and ω_r are the frequency of the incident on the mirror light and the frequency of reflected waves. The reflection angle, θ_r, is related to the incidence angle as

$$\sin\theta_r = \frac{\omega_0}{\omega_r}\sin\theta_i = \frac{\omega_0\left(1-\beta_{M1}^2\right)\sin\theta_i}{\omega_0(1+\beta_{M1}^2)+2\beta_{M1}\sqrt{\omega_0^2-\omega_{pe}^2}\cos\theta_i}. \tag{26.4}$$

If the reflection coefficient is equal to unity, ideally reflecting mirror, then during the wave – mirror interaction the ratio of the amplitude of the electric field to its frequency is constant:

$$\frac{E_r}{\omega_r} = \frac{E_0}{\omega_0} \tag{26.5}$$

Here and below indices "0" and "r" denote the parameters of the incident on the mirror and reflected radiation in the laboratory frame of reference.

Depending on whether the wave and mirror are co-propagating ($\beta_{M1} < 0$) or counter-propagating ($\beta_{M1} > 0$) in the laboratory frame of reference we have either the frequency downshift or the upshift, the stretching or the compression of the electromagnetic pulse, and decreasing or increasing of the wave energy. In the most simple configuration of the wave propagating in vacuum normally incident on the

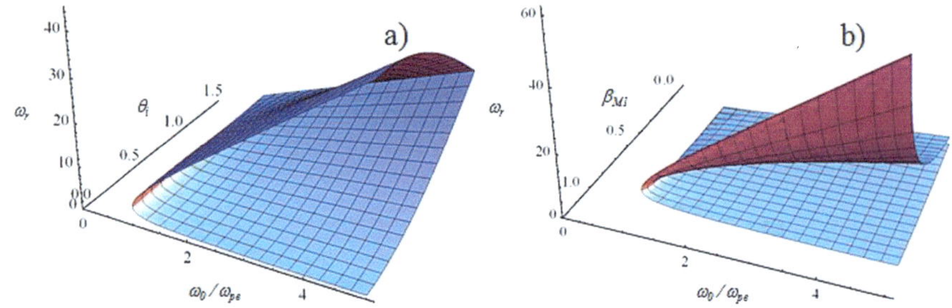

Fig. 26.3 Reflected wave frequency dependence on the frequency ω_0 of the incident wave (the frequency is normalized on ω_{pe}), and **a** on the angle of incidence θ_i for $\beta_M = 0.8$, and **b** on the normalized mirror velocity (β_{M1}) for $\theta_i = 0$. Upper branch corresponds to the frequency upshifting in the counter propagating wave interaction with relativistic mirror. Lower branch corresponds to the frequency down shifting in the co-propagating wave interaction with receding mirror

mirror ($\theta_i = 0$) expression (26.3) yields

$$\omega_r = \omega_0 \frac{1 + \beta_{M1}}{1 - \beta_{M1}} = \omega_0 (1 + \beta_{M1})^2 \gamma_{M1}^2. \tag{26.6}$$

In the ultrarelativistic limit, when $\gamma_{M1} \gg 1$, the reflected wave frequency is by a factor $\approx 4\gamma_{M1}^2$ higher (lower). The reflected wave energy changes accordingly.

Figure 26.3 illustrates the dependence of the reflected wave frequency dependence on the frequency of the incident wave, on the angle of incidence, and on the on the mirror velocity. Upper branch corresponds to the frequency upshifting in the counter propagating wave interaction with relativistic mirror. Lower branch corresponds to the frequency down shifting in the co-propagating wave interaction with the receding relativistic mirror.

In the case of the wave interacting with the receding mirror, i.e. in the co-propagating configuration, we notice that the wave group velocity should not be lower than the mirror velocity (Rosanov et al., 2011),

$$v_g = c\sqrt{1 - \left(\frac{\omega_{pe}}{\omega_0}\right)^2} \geq \beta_{M1} c. \tag{26.7}$$

This is equivalent to the constraint $\omega_0 \geq \omega_{pe}\gamma_{M1}$. If the medium moves, one should also take the Fizeau effect—the light dragging along by the medium.

26.2 Moving Plasma Slab as Relativistic Mirror

As noted above, the paper published by Einstein in (1905) [1] shows that the electromagnetic wave interaction with a counter-propagating relativistic mirror (in the head-on wave-mirror collision configuration) results in the reflected electromagnetic pulse compression and in upshifting of its carrier frequency. Here we analyze the case of the moving with relativistic velocity the plasma-vacuum interface as relativistic mirror. The reflected pulse frequency and amplitude increasing are given by relationships

$$\omega_r = \omega_0 g_M(\beta_M) \quad \text{and} \quad E_r = \rho E_0 g_M(\beta_M), \tag{26.8}$$

where ω_r and ω_0 and E_r and E_0 are the incident and reflected pulse frequency and amplitude, respectively, and $\beta_M = v_M/c$ is the normalized mirror velocity, and ρ is the mirror reflectivity. The frequency upshifting (the pulse intensification) factor $g_M(\beta_M)$ depends on the mirror velocity as

$$g_M(\beta_M) = \left(\frac{1+\beta_M}{1-\beta_M}\right). \tag{26.9}$$

In the ultra-relativistic limit when $\beta_M \to 1$, the factor $g_M(\beta_M)$ is approximately equal to $4\gamma_M^2$ with the mirror gamma-factor given by $\gamma_M = 1/\sqrt{1-\beta_M^2}$.

In the medium, the modulations of the refraction index moving with the relativistic velocity play a role of the mirror. Since the velocity of the modulations is in general a phase velocity it can be subluminal $\beta_M < 1$, luminal $\beta_M = 1$, or superluminal with $\beta_M > 1$. As Bolotovskii and Ginzburg (1972) explain the motion with superluminal velocity of the object emitting or reflecting electromagnetic radiation is not forbidden by the special theory of relativity provided this is not the particle velocity or the group velocity of the wave (see also Ginzburg, 1979; Sutherland & Shepanski, 1986; Jeong et al., 2021; Asenjo, 2024; Esirkepov & Bulanov, 2024a, 2024b). If the normalized velocity of the mirror is larger than unity, $\beta_M > 1$, the frequency upshifting factor equals

$$g_M = \left(\frac{\beta_M+1}{\beta_M-1}\right). \tag{26.10}$$

Figure 26.4 illustrates the electromagnetic pulse reflection (transmission) at subluminal and superluminal mirrors. Depending on the mirror's reflectivity a portion of the pulse is transmitted through the mirror and a portion is reflected appearing in the front of subluminal mirror or propagates through the mirror appearing behind the mirror. In both these cases, the electromagnetic pulse frequency is upshifted according to Eq. (26.8) with the factor $g_M(\beta_M)$ given either by Eq. (26.9) or (26.10).

The light interaction with superluminal relativistic mirrors have been analyzed by Esirkepov et al., 2009, Zhidkov et al., 2009, Bu et al., 2016. Esirkepov et al.,

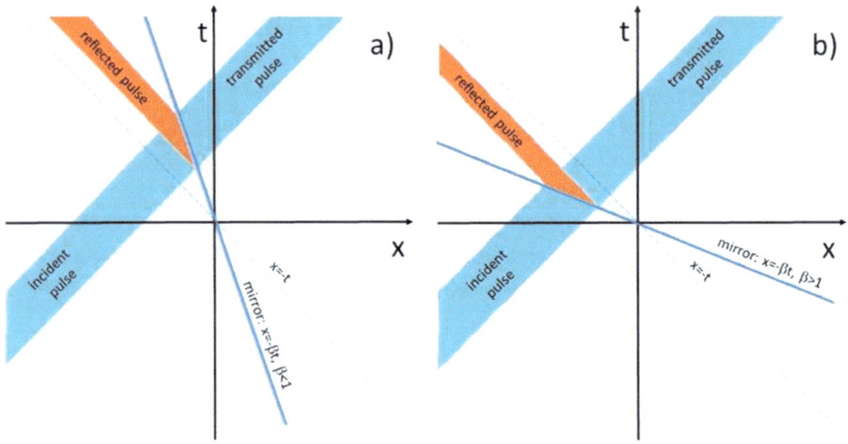

Fig. 26.4 Electromagnetic pulse reflection (transmission) at **a** subluminal and **b** superluminal mirror

2009 pay attention to vanishing of the reflection coefficient at the luminal mirror in the limit $\beta_M \to 1$. This fact can be elucidated by considering the light reflection at the plasma slab moving with relativistic velocity $x = tc\beta_M$. Considering the light reflection at the thin slab plasma layer, for which the density distribution is given by

$$n_e = n_{0e} l \delta(x + tc\beta_M)$$

where $\delta(\cdot)$ is the Dirac delta function. In this case the reflection coefficient $R = |\rho|^2$ equals (see also detailed derivation of the reflection coefficient in Sect. 26.10)

$$R = \frac{(2\pi n_{0e} e^2 l)^2}{(2\pi n_{0e} e^2 l)^2 + (m_e \omega_0 c)^2}. \tag{26.11}$$

In the boosted frame of reference $\mathcal{M}$ at $\beta_M \to 1$ the reflection coefficient tends to zero as

$$R \approx \frac{\omega_{pe}^4 l^2}{16\omega_0^2 c^2 \gamma_M^4} \tag{26.12}$$

In that frame of reference moving with velocity $c\beta_M$, where the plasma slab is at rest, the electromagnetic wave frequency tends to infinity at $\beta_M \to 1$, i.e., at the infinite mirror gamma-factor $\gamma_M \to \infty$, as $\omega = \omega_0 \sqrt{g_M} \approx \omega_0 2\gamma_M$. Here the frequency of the electromagnetic wave incident on the plasma in the laboratory frame of reference equals ω_0. Taking to account that the Langmuir frequency is Lorentz invariant we find that the reflection coefficient tends to zero for $\gamma_M \to \infty$, i.e. we have "zero times infinity" indeterminate situation. In a naïve way, according to Eq. (26.8) we can find that the amplitude of the reflected wave tends to finite value:

$$E_r \approx E_0 \frac{\omega_{pe}^2 l}{\omega_0 c}.$$

We see that the reflection coefficient tends to zero for $\gamma_M \to \infty$, while the amplitude of the incident wave tends to infinity in the mirror frame of reference $\mathcal{M}$ i.e. we have "zero times infinity" indeterminate situation. In a naïve way, we can say that the amplitude of the reflected wave tends to finite value nonetheless the reflection vanishes, *since the wavelength and duration of the reflected pulse shorten proportionally* $\approx 1/4\gamma_M^4$ *the* ***energy*** *of the reflected electromagnetic wave tends to zero as the mirror velocity approaches speed of light in vacuum.* In addition, the thin slab approximation is valid provided that the slab thickness measured in the mirror rest frame reference, l, is smaller than the electromagnetic wave wavelength $\lambda_M = \lambda_0/2\gamma_M$. In the limit $\gamma_M \to \infty$, when l becomes wider than λ_M we should use a semi-infinite plasma mirror approximation considered below in Sect. 26.6.

The luminal mirror case when the mirror velocity equals speed of light in vacuum $\beta_M = 1$ requires special consideration in particular taking into account the finite thickness of the plasma layer. Below we present a solution of the problem of the electromagnetic wave interaction with luminal plasma mirror showing that although the wave reflection at the luminal mirror vanishes, i.e. the electromagnetic pulse cannot overtake the mirror, its frequency spectrum changes.

26.2.1 Electromagnetic Wave Reflection at the Subluminal and Superluminal Relativistic Plasmas

To describe the electromagnetic wave propagation in inhomogeneous non-stationary plasma we use the wave equation

$$\partial_{tt} a - c^2 \partial_{xx} a + \omega_{pe}^2 a = 0 \tag{26.13}$$

with the Langmuir frequency ω_{pe} depending on time and coordinate. Assuming that the plasma slab moves along the x direction with constant velocity $c\beta_M$ towards negative x we find convenient to introduce the independent variables

$$\xi = \frac{ct - x}{\sqrt{2}} \quad \text{and} \quad \eta = \frac{c\beta_M t + x}{\sqrt{2}}. \tag{26.14}$$

The (x, t) variables are expressed via the (η, ξ) variables as

$$t = \frac{\sqrt{2}(\eta + \xi)}{c(1 + \beta_m)} \quad \text{and} \quad x = \frac{\sqrt{2}(\eta - \beta_m \xi)}{(1 + \beta_m)}. \tag{26.15}$$

The Langmuir frequency in the case under consideration depends on the variable η only.

In the (η, ξ) variables, the wave Eq. (26.13) takes the form

$$-(1-\beta_M^2)\partial_{\eta\eta}a + 2(1+\beta_M)\partial_{\eta\xi}a + \frac{2\omega_{pe}^2}{c^2}a = 0 \tag{26.16}$$

In the limit $\beta_M \to 1$ this is a singular differential equation with the small parameter $\epsilon = 1 - \beta_M$ at the front of the high order derivative with respect to the variable η. From the Tikhonov theorem (Tikhonov, 1952) follows that in the limit $\epsilon \to 0$ the solution of Eq. (26.16) tends to the solution of the degenerate equation

$$(1+\beta_M)\partial_{\eta\xi}a + \frac{\omega_{pe}^2(\eta)}{c^2}a = 0. \tag{26.17}$$

First, we analyse the solution of Eq. (26.16) with the aim to find the reflection coefficient of the wave at the plasma-vacuum interface assuming that the Langmuir frequency dependence on the coordinate η is given by

$$\omega_{pe}(\eta) = \Theta(\eta)\omega_{pe}. \tag{26.18}$$

Here $\Theta(\eta)$ is the Heaviside step function equal to zero for negative η and to unity for $\eta > 0$.

In the region $\eta < 0$ where $\omega_{pe}(\eta) = 0$, the solution of the equation

$$-(1-\beta_M^2)\partial_{\eta\eta}a + 2(1+\beta_M)\partial_{\eta\xi}a = 0 \tag{26.19}$$

can be written as a sum of two arbitrary functions

$$f(\xi) \quad \text{and} \quad h((1-\beta_M^2)\xi + 2(1+\beta_M)\eta). \tag{26.20}$$

To find the reflection coefficient it is convenient to write the solution of Eq. (26.19) in the form

$$a(\eta, \xi) = a_0 \exp(-i\Omega\xi + iK\eta). \tag{26.21}$$

Electromagnetic wave reflection at the subluminal relativistic mirror. First, we consider the case of the electromagnetic wave reflection at the relativistic mirror moving with subluminal velocity ($\beta_M < 1$). From Eqs. (26.19) and (26.20) we have for the wavenumber K the dispersion equation

$$(1-\beta_M^2)K^2 + 2(1+\beta_M)\Omega K = 0 \tag{26.22}$$

whose solution is

$$K_+ = 0 \quad \text{and} \quad K_- = -\frac{2\Omega}{1-\beta_M}. \tag{26.23}$$

Using these expressions, we represent the electromagnetic wave in the $\eta < 0$ region in the form of the incident wave of the frequency and amplitude equal to 1 and Ωc, respectively, and of the reflected wave which amplitude equals ρ,

$$a(\eta, \xi) = \exp(-i\Omega\xi) + \rho \exp\left(-i\Omega\left(\xi + \frac{2}{1-\beta_M}\eta\right)\right). \tag{26.24}$$

The frequency of the incident electromagnetic wave Ωc is one of the parameters of the problem under consideration. The phase of the reflected electromagnetic wave is

$$-\Omega\left(\xi + \frac{2}{1-\beta_M}\eta\right) \equiv -\Omega\left(\frac{1+\beta_M}{1-\beta_M}\right)\frac{(ct+x)}{\sqrt{2}}. \tag{26.25}$$

showing the frequency (and wavenumber) upshifting according to Eqs. (26.8) and (26.9).

In the region $\eta > 0$ the wave number of the transmitted electromagnetic wave is given by the dispersion equation

$$(1-\beta_M^2)K^2 + 2(1+\beta_M)\Omega K + \frac{2\omega_{pe}^2}{c^2} = 0. \tag{26.26}$$

Its solution yields

$$K_{\mp} = -\frac{\Omega \pm \sqrt{\Omega^2 - 2\omega_{pe}^2(1-\beta_M)/c^2(1+\beta_M)}}{1-\beta_M}. \tag{26.27}$$

As a result, the transmitted electromagnetic wave depends on the variables (η, ξ) as

$$a(\eta, \xi) = \tau \exp\left(-i\Omega\xi - i\frac{\Omega - \sqrt{\Omega^2 - 2\omega_{pe}^2(1-\beta_M)/c^2(1+\beta_M)}}{1-\beta_M}\eta\right). \tag{26.28}$$

Here we assume that $\Omega^2 > 2\omega_{pe}^2(1-\beta_M^2)/c^2(1+\beta_M)^2$, i.e. the plasma has an "under-critical" density.

At the plasma-vacuum interface, the boundary conditions require the continuity of the function a and its derivative with respect to the variable η. Using these conditions, we find the relationships between the amplitudes of reflected and transmitted waves (they are ρ and τ, respectively).

$$1 + \rho = \tau, \tag{26.29}$$

$$-\frac{2\Omega}{1-\beta_M}\rho = \frac{\Omega - \sqrt{\Omega^2 - 2\omega_{pe}^2(1-\beta_M)/c^2(1+\beta_M)}}{1-\beta_M}\tau. \tag{26.30}$$

This yields for the reflection coefficient $R = |\rho|^2$

$$R = \left|\frac{\Omega - \sqrt{\Omega^2 - 2\omega_{pe}^2(1-\beta_M)/c^2(1+\beta_M)}}{\Omega + \sqrt{\Omega^2 - 2\omega_{pe}^2(1-\beta_M)/c^2(1+\beta_M)}}\right|^2. \tag{26.31}$$

If $\Omega \to (\omega_{pe}/c)\sqrt{2(1-\beta_M)/(1+\beta_M)}$ then the reflection coefficient tends to unity. For $\beta_M \to 1$ the reflection coefficient is approximately equal to

$$R \approx \left|\frac{\omega_{pe}^4(1-\beta_M)^2}{\Omega^4 c^4(1+\beta_M)^2}\right| \tag{26.32}$$

in accordance with Eq. (26.12). It vanishes when the plasma vacuum interface moves with the speed of light.

Electromagnetic wave reflection at the superluminal relativistic mirror. In the case of the electromagnetic wave reflection at the superluminal relativistic mirror, for which $\beta_M > 1$, from Eq. (26.19) follows the dispersion equation for the wavenumber K

$$(\beta_M^2 - 1)K^2 - 2(1+\beta_M)\Omega K = 0 \tag{26.33}$$

whose solution is

$$K = 0 \quad \text{and} \quad K = \frac{2\Omega}{\beta_M - 1}. \tag{26.34}$$

The electromagnetic wave in the $\eta < 0$ region propagates along the x direction with no reflected wave,

$$a(\eta, \xi) = \exp(-i\Omega\xi). \tag{26.35}$$

In the region $\eta > 0$ the wave number of the transmitted electromagnetic wave can be found from the dispersion equation

$$(\beta_M^2 - 1)K^2 - 2(1+\beta_M)\Omega K - \frac{2\omega_{pe}^2}{c^2} = 0. \tag{26.36}$$

Its solution yields

$$K_{\pm} = \frac{\Omega \pm \sqrt{\Omega^2 + 2\omega_{pe}^2(\beta_M - 1)/c^2(1+\beta_M)}}{\beta_M - 1}. \tag{26.37}$$

As a result the transmitted electromagnetic wave depends on the variables (η, ξ) as

$$a(\eta, \xi) = \tau_+ \exp(-i\Omega\xi + iK_+\eta) + \tau_- \exp(-i\Omega\xi + iK_-\eta). \tag{26.38}$$

As in the case of subluminal mirror considered above, using the boundary conditions At the plasma-vacuum interface, which require the continuity of the function a and its derivative with respect to the variable η, we find the relationships between the amplitudes of transmitted waves (they are τ_+ and τ_-, respectively).

$$1 = \tau_+ + \tau_-, \tag{26.39}$$

$$0 = K_+\tau_+ + K_-\tau_-. \tag{26.40}$$

This yields for the coefficients τ_+ and τ_-

$$\tau_+ = -\frac{\Omega - \sqrt{\Omega^2 + 2\omega_{pe}^2(\beta_M - 1)/c^2(1+\beta_M)}}{2\Omega}, \tag{26.41}$$

$$\tau_- = \frac{\Omega + \sqrt{\Omega^2 + 2\omega_{pe}^2(\beta_M - 1)/c^2(1+\beta_M)}}{2\Omega}. \tag{26.42}$$

In the limit $\beta_M \to 1$ the wave amplitude τ_+ vanishes and τ_- becomes equal to unity.

26.2.2 *Electromagnetic Wave "Propagation Through and Reflection at" the Luminal Relativistic Mirror*

In the case $\beta_M = 1$, the electromagnetic wave interaction with the luminal mirror is described by Eq. (26.10) which takes the form

$$\partial_{\eta\xi} a + \frac{\omega_{pe}^2(\eta)}{2c^2} a = 0. \tag{26.43}$$

Taking the Fourier transform of the function $a(\eta, \xi)$ with respect to the variable, η we obtain from Eq. (26.43) for

$$a_{\overline{\Omega}}(\eta) = \frac{1}{\sqrt{2\pi}} \int_{-\infty}^{+\infty} a(\eta, \xi) \exp(i\overline{\Omega}\xi) d\xi \tag{26.44}$$

the equation

$$\partial_\eta a_{\overline{\Omega}} + i \frac{\omega_{pe}^2(\eta)}{2c^2\overline{\Omega}} a_{\overline{\Omega}} = 0, \tag{26.45}$$

whose solution yields

$$a_{\overline{\Omega}}(\eta) = a_0(\overline{\Omega}) \exp\left(-\frac{i}{2c^2\overline{\Omega}} \int_0^\eta \omega_{pe}^2(s) ds \right). \tag{26.46}$$

Here the function $a_0(\overline{\Omega})$ is determined by the initial conditions. If $a_0(\overline{\Omega}) = 2\pi a_0\, \delta(\overline{\Omega} - \Omega)$, where $\delta(\Omega)$ is the Dirac delta-function, the inverse Fourier transform gives the solution of this equation in the form of the constant amplitude wave

$$a(\eta, \xi) = a_0 \exp(-i\Psi(\eta, \xi)). \tag{26.47}$$

whose phase (it is also called the eikonal) reads

$$\Psi(\eta, \xi) = \Omega\xi + \frac{1}{2\Omega c^2} \int_0^\eta \omega_{pe}^2(s) ds. \tag{26.48}$$

If the function ω_{pe}^2 does not depend on the variable η, calculation of the wave frequency $\omega = -\partial_t \Psi(\eta, \xi)$ and wave number $k = \partial_x \Psi(\eta, \xi)$ results in expressions for the phase and group velocity of the wave equal to $v_{ph} = \omega/k$ and $v_g = \partial\omega/\partial k$, respectively. They are

$$v_{ph} = c \frac{2\Omega^2 c^2 + \omega_{pe}^2}{2\Omega^2 c^2 - \omega_{pe}^2} \quad \text{and} \quad v_g = c \frac{2\Omega^2 c^2 - \omega_{pe}^2}{2\Omega^2 c^2 + \omega_{pe}^2}. \tag{26.49}$$

Here we used the relationships between the variables (η, ξ) and (t, x) given by Eq. (26.14). The phase velocity is superluminal, and the group velocity is subluminal. If $2\Omega^2 c^2 > \omega_{pe}^2$, the wave propagates towards positive values of the coordinate x. In the case $2\Omega^2 c^2 < \omega_{pe}^2$, the wave propagates in opposite direction. We may say about the wave reflection. However, the electromagnetic pulse does not overtake the plasma-vacuum interface because its group velocity is subluminal.

A generic expression that describes the electromagnetic wave dependence on the variables η and ξ, i.e. on time and coordinate can be written as

$$a(\eta, \xi) = \frac{1}{\sqrt{2\pi}} \int_{-\infty}^{+\infty} a_0(\overline{\Omega}) \exp\left(-i\overline{\Omega}\xi - \frac{i}{2\overline{\Omega}c^2} \int_0^{\eta} \omega_{pe}^2(s)ds \right) d\overline{\Omega}. \tag{26.50}$$

The integrand can be written in the form $a(\overline{\Omega}) \exp(i\Psi(\overline{\Omega}, \eta, \xi))$ with the phase

$$\Psi(\overline{\Omega}, \eta, \xi) = -\overline{\Omega}\xi - \frac{1}{2\overline{\Omega}c^2} \int_0^{\eta} \omega_{pe}^2(s)ds). \tag{26.51}$$

The long-time asymptotic behavior of the wave can be found by applying the stationary phase method to calculate the Fourier integral (26.50). The extremum of the phase $\Psi(\overline{\Omega}, \eta, \xi)$ given by Eq. (26.51) in the integrand corresponds to

$$\frac{d}{d\overline{\Omega}} \left(\overline{\Omega}\xi + \frac{1}{2\overline{\Omega}c^2} \int_0^{\eta} \omega_{pe}^2(s)ds) \right) \Bigg|_{\overline{\Omega}=\overline{\Omega}*} = 0, \tag{26.52}$$

which gives

$$\overline{\Omega}^* = \sqrt{\frac{1}{2\xi c^2} \int_0^{\eta} \omega_{pe}^2(s)ds}. \tag{26.53}$$

and

$$a(\eta, \xi) \approx \exp\left[-i \sqrt{\frac{\xi}{2c^2} \int_0^{\eta} \omega_{pe}^2(s)ds} \right]. \tag{26.54}$$

Expression (26.50) can be considered as the Fourier components of the electromagnetic pulse, which before interaction with the luminal mirror had the delta-function form.

The expressions obtained above resolve the issue concerning the indefinite value "zero times infinity": inside the luminal mirror the field depends on the the variables η and ξ, which shows that the wave has "reflected" and "transmitted" components, whereas outside the mirror we have only incident and transmitted waves, there is no reflected wave.

26.3 Electromagnetic Wave in Media with the Parameters Changing Abruptly in Time (Time Reflection)

The super-luminal mirror may be an interface between two media (in particular, it is an interface between the plasma and not ionized gas as in the case of the ionization wave). In the frame of reference $\mathcal{M}$ the mirror M_2 is located on the surface $t' = 0$. Then the wave enters the medium behind the mirror. Writing the Lorentz transformation formulae in the form are

$$t' = \gamma_V (t - xV/c^2), \quad \text{and} \quad x' = \gamma_V (x - Vt), \tag{26.55}$$

where $\gamma_V = 1/\sqrt{1 - V^2/c^2}$, and assuming that $x = \beta_{M2} t$, we find for the boost velocity

$$V = c/\beta_{M2}. \tag{26.56}$$

In the frame of reference $\mathcal{M}$ the wave "reflection" becomes the Cauchy problem with initial conditions imposed by the incident wave. In other words, this is a problem of the electromagnetic wave propagation in nonstationary medium.

Electromagnetic wave reflection from the sub-luminal mirror corresponds to a nonadiabatic regime of the wave propagation in inhomogeneous media. Adiabatic regimes of the wave propagation are described by geometric optics. The geometric optics approximation is equivalent to the conservation of adiabatic invariants of the wave field.

In the case of the medium with nonstationary parameters, i.e. with the refraction index depending on time, one may also say that the photon number is conserved during the wave propagation. From the analysis of accuracy of the adiabatic invariant conservation we know that in nonstationary medium it implies a slowness of the wave frequency variations, $\partial_t \omega/\omega^2 \ll 1$. The accuracy of the adiabatic invariant conservation is exponential.

The condition of the adiabatic change of the wave parameters in a spatially inhomogeneous medium imposes a constraint on the wave number, which has a form $\partial_x k/k^2 \ll 1$. In the vicinity of sharp spatial inhomogeneity, where $\partial_x k/k^2 \gg 1$, matching of the solutions of the wave equation describing the wave in the homogeneous regions at the discontinuity yields well known coefficients of the wave reflection and transmission (Landau & Lifshitz, 1984).

During propagation in an inhomogeneous medium the wave frequency remains unchanged meanwhile the wave-vector changes: in the case of reflection and refraction of the wave at a plane interface between two media with different refraction indices, the wave vectors of incident, reflected and transmitted waves are all different.

26.3.1 "Time Reflection"

Now we discuss what happens with plane electromagnetic wave propagating in the medium in which the refraction index changes instantaneously Wilks et al. (1988) (see also Qu et al., 2018).

In this case, for the wave frequency and its time derivative we have the condition $\partial_t \omega/\omega^2 \gg 1$. We consider the small amplitude electromagnetic wave in a collisionless plasma whose frequency and wave number in the boosted frame $\mathcal{M}$ are related to each other according to the dispersion equation $\omega_0'^2 = k_0'^2 c^2 + \omega_{pe}^2$, were

$$\omega_0' = \gamma_V(\omega_0 - k_0 V/c^2), \quad \text{and} \quad k_0' = \gamma_V(\omega_0 - k_0 V) \tag{26.57}$$

with V given by Eq. (26.56).

We assume that the plasma density depends on time. For $t' < 0$ the Langmuir frequency vanishes $\omega_{pe} = 0$, i.e. the wave dispersion equation is $\omega_0'^2 = k_0'^2 c^2$, meanwhile for $t' > 0$ we have $\omega_{pe} \neq 0$ and the dispersion equation is given by $\omega_f'^2 = k_f'^2 c^2 + \omega_{pe}^2$. Here ω_0', k_0' and ω_f', k_f' are the initial and final values of the wave frequency and number (see Fig. 26.5) in the boosted frame of reference $\mathcal{M}$.

At $t' < 0$ the electric and magnetic field components correspond to the linearly polarized plane wave, propagating in the positive direction along the x—axis:

$$\boldsymbol{E}' = E_0' \cos(k_0' x' - \omega_0' t')\boldsymbol{e}_2 = E_0' \cos[k_0'(x' - ct')]\boldsymbol{e}_2, \tag{26.58}$$

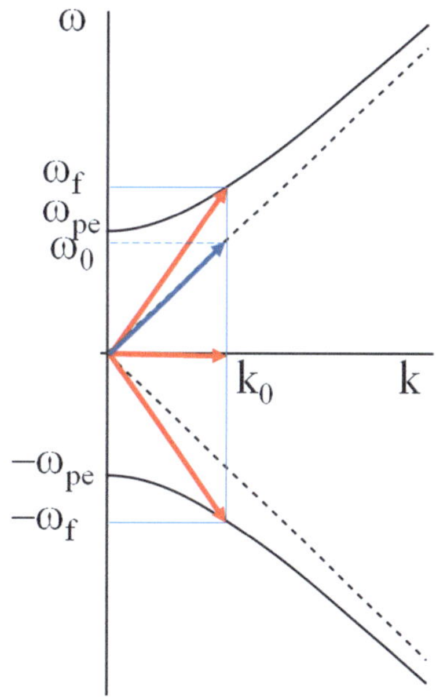

Fig. 26.5 Dispersion relations between frequency and number of the electromagnetic wave propagation in a plasma with the density changing instantaneously. As a result of non-adiabatic behaviour of the refraction index the frequencies of the modes propagating towards left and right change and the zero-frequency mode with a non-zero wave number occurs

$$\boldsymbol{B}' = B_0' \cos(k_0' x' - \omega_0' t')\boldsymbol{e}_3 = B_0' \cos[k_0'(x' - ct')]\boldsymbol{e}_3 \tag{26.59}$$

with $E_0' = B_0'$. Here $\boldsymbol{e}_2$ and $\boldsymbol{e}_3$ are unit vectors along the y– and z– axis, respectively. At $t' = 0$ we have $\boldsymbol{E}' = E_0' \cos(k_0' x')\boldsymbol{e}_2$ and $\boldsymbol{B}' = B_0' \cos(k_0' x')\boldsymbol{e}_3$, that imply the initial condition for the Cauchy problem for the Maxwell equations. Additionally, we have an initial condition for the electron hydrodynamics equations: at $t' = 0$ the electron velocity vanishes, $\boldsymbol{v'}_e(0) = 0$, and the electron density is equal to $n_e' = \omega_{pe}^2 m_e / 4\pi e^2$.

As the result of the Langmuir frequency change from zero at $t < 0$ to ω_{pe} for $t > 0$ we obtain

$$\boldsymbol{E}' = [E_-' \cos(k_f' x' - \omega_f' t') + E_+' \cos(k_f' x' + \omega_f' t')]\boldsymbol{e}_2 \tag{26.60}$$

$$\begin{aligned} \boldsymbol{B}' = &- \frac{k_f'^2 c^2}{k_f'^2 c^2 + \omega_{pe}^2} \left[E\prime_- \cos(k_f' x' - \omega_f' t') - E_+' \cos(k_f' x' + \omega_f' t') \right] \boldsymbol{e}_3 \\ &+ B_{st}' \cos(k_f' x')\boldsymbol{e}_3. \end{aligned} \tag{26.61}$$

Here the wave number $k_f' = k_0'$ remains unchanged by virtue of the medium homogeneity while the wave frequency becomes equal to $\omega_f' = \pm\sqrt{k_0'^2 c^2 + \omega_{pe}^2}$. The later expression can be written in the form

$$\omega_f' = \pm\sqrt{\omega_0'^2 + \omega_{pe}^2}. \tag{26.62}$$

We see that the wave frequency value at $t' > 0$ is larger than ω_0'.

For the electron velocity in the plasma, we obtain

$$\begin{aligned} \boldsymbol{v'}_e = &- \frac{ie}{m_e \sqrt{\omega_0'^2 + \omega_{pe}^2}} \left[E_-' \cos(k_f' x' - \omega_f' t') - E_+' \cos(k_f' x' + \omega_f' t') \right] \boldsymbol{e}_2 \\ &+ v_{st}' \cos(k_f' x')\boldsymbol{e}_2. \end{aligned} \tag{26.63}$$

Equations (26.60)–(26.63) show that the electromagnetic wave is transformed to two counter-propagating waves, whose amplitudes are given by

$$E_\pm' = \frac{E_0'}{2} \left[1 \pm \frac{\omega_0'}{\sqrt{\omega_0'^2 + \omega_{pe}^2}} \right]. \tag{26.64}$$

In addition, the quasi-static magnetic field is generated in plasma. It is equal to

$$\boldsymbol{B}_{st}' = \frac{4\pi n_0 e v_{st}'}{k_f' c} \cos(k_f' x')\boldsymbol{e}_2 = \frac{\omega_{pe}^2}{\omega_0'^2 + \omega_{pe}^2} E_0' \cos(k_f' x')\boldsymbol{e}_2. \tag{26.65}$$

This example shows that non-adiabatic change of the medium refraction index leads to transformation of the initial wave with the frequency and wave number ω_0' and k_0' to two counter-propagating waves of the frequencies $+\sqrt{\omega_0'^2 + \omega_{pe}^2}$ and $-\sqrt{\omega_0'^2 + \omega_{pe}^2}$ with the same wave number $k_f' = k_0'$.

Performing the Lorentz transform back to the laboratory frame of reference $\mathcal{L}$ we obtain for the wave frequency

$$\omega_f = \frac{\omega_0(\beta_{M2} + 1) \pm \beta_{M2}\sqrt{\omega_0^2(\beta_{M2} + 1)^2 + \omega_{pe}^2(\beta_{M2}^2 - 1)}}{\beta_{M2}^2 - 1}, \tag{26.66}$$

and for the wave number

$$k_f = \frac{k_0 c\beta_{M2}(\beta_{M2} + 1) \pm \sqrt{k_0^2 c^2(\beta_{M2} + 1)^2 + \omega_{pe}^2(\beta_{M2}^2 - 1)}}{\beta_{M2}^2 - 1}. \tag{26.67}$$

In the limit $\omega_0 = \omega_{pe}\sqrt{(\beta_{M2} - 1)/(\beta_{M2} + 1)}$ these expressions yield for the value of the wave frequency upshifted at the super-luminal front

$$\omega_f = \omega_0 \frac{\beta_{M2} + 1}{\beta_{M2} - 1}. \tag{26.68}$$

When we have the medium with slowly changing parameters, i.e. when the wave interaction is in an adiabatic regime, $\partial_t\omega/\omega^2 \ll 1$, the accuracy of the adiabatic approximation is exponential. The amplitude of the counter-propagating wave is exponentially small.

26.4 Light Reflection at the Receding Mirror

The light interaction with a co-propagating plasma slab corresponds to the reflection off a receding mirror. It results in the frequency down-shift and the energy decrease of the reflected radiation. The frequency decreases by a factor $(1 - \beta_M)/(1 + \beta_M)$, which in the limit $\beta_M \to 1$ approximately equals $4\gamma_M^2$. Assuming a fully reflecting mirror, we obtain that the energy of the reflected electromagnetic pulse becomes equal to

$$\mathcal{E}_r = \mathcal{E}_{las} \frac{1 - \beta_M}{1 + \beta_M}, \tag{26.69}$$

where $\mathcal{E}_{las}$ is the energy of the laser pulse incident on the mirror. As a result, the energy transferred to the target equals

$$\Delta\mathcal{E} = \mathcal{E}_{las}\frac{2\beta_M}{1+\beta_M}. \tag{26.70}$$

This process corresponds to the ion acceleration by the radiation pressure of the laser light discussed above in Sect. 4.7. As we see, in the limit $\gamma_M >> 1$ almost all the energy of the laser pulse is transferred to the accelerated ions.

In the case of the incident/reflected light pulse propagating in a continuous medium, the light reflection at the moving mirror acquires other properties including the parametric Doppler effect and the effects of finite group velocity of the electromagnetic pulse (e.g. see review article Bulanov et al., 2013a, 2013b and the literature cited therein). This, in particular, becomes important in the case of the ion acceleration by the radiation pressure of 'slow' light, which may be of special interest in the so called double-sided relativistic mirror configuration (Esirkepov et al., 2009).

26.5 Accelerating Mirror

We consider the wave reflection from the accelerating mirror moving with the time-dependent velocity for the case of normal incidence of the electromagnetic wave Birrell and Davies (1982), Grib et al. (1980). The accelerated mirrors play important role in quantum field theory (Grib et al., 1980) and other areas of fundamental physics studies (Chen & Mourou, 2017). The incident and reflected electromagnetic waves in vacuum obey Maxwell's equations, from which follows the wave equation for the transverse component of the vector potential,

$$\partial_{tt}A - c^2\partial_{xx}A = 0. \tag{26.71}$$

The mirror coordinate at time t is determined by equation $\mathcal{M}(x,t) = \text{constant}$, i.e. the mirror is located in the point $x = x_M(t)$ (see Fig. 26.6a).

We assume that the boundary condition for the field at the mirror is reduced to the Dirichlet condition

$$A(x,t)|_{x=x_M(t)} = 0. \tag{26.72}$$

i.e. the mirror is fully reflecting. This boundary condition is given at the non-stationary surface. We change the independent variables (t, x) to new variables (τ, ξ), for which the mirror is at the rest and the Eq. (26.71) remains the wave equation for $A(x,t)$. At first, it is convenient to introduce the advanced and retarded coordinates u and v equal to

$$u = t - x/c \quad \text{and} \quad v = t + x/c. \tag{26.73}$$

In these coordinates, Eq. (26.71) takes the form

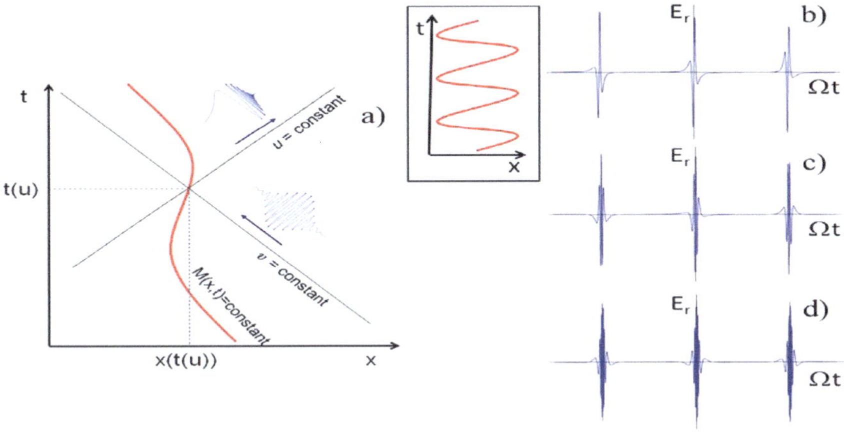

Fig. 26.6 **a** Mirror coordinates (t, x) as a function of u. The incident, v = constant, and reflected, u = constant, light rays. **b–d** Electric field of the wave reflected from the oscillating mirror for $\Omega/\omega_0 = 0.5,\ 1,\ 2$ and $a = 15$, respectively. The insert shows the oscillating mirror position in the plane (t, x). Reproduced from Bulanov et al., 2013a, 2013b @ Uspekhi Fizicheskikh Nauk 2013

$$\partial_{uv} A = 0. \tag{26.74}$$

In coordinates $\overline{u}$ and $\overline{v}$ determined by relationships

$$u = f(\overline{u}) \quad \text{and} \quad v = f(\overline{v}), \tag{26.75}$$

Equation (26.74) preserves its form except for a factor $f'g'$, i.e.

$$\partial_{\overline{uv}} A = 0. \tag{26.76}$$

In new spatial and time coordinates, (ξ, τ), defined as

$$\overline{u} = \tau - \xi/c \quad \text{and} \quad \overline{v} = \tau + \xi/c, \tag{26.77}$$

Equation (26.71) becomes

$$\partial_{\tau\tau} A - c^2 \partial_{\xi\xi} A = 0. \tag{26.78}$$

We choose the functions $f = \tau - \xi/c$ and $g = \tau + \xi/c$ to be such that the mirror is at rest at $\xi = 0$, which implies the equation for the functions f and g,

$$g(\tau) - f(\tau) = 2x_M(g(\tau) + f(\tau)/2), \tag{26.79}$$

transforming the boundary condition (26.72) to

$$A(\xi, \tau)|_{\xi=0} = 0. \tag{26.80}$$

We may choose the function $g(\overline{v})$ to be equal to $g(\overline{v}) = \overline{v}$. This leads to the functional equation for f:

$$\tau - f(\tau) = 2x_M(\tau + f(\tau)/2). \tag{26.81}$$

The solution to the boundary problem for Eq. (26.78) with the boundary condition (26.80) is

$$A(\xi, \tau) = A_0 \exp(\pm i\,\omega_0\tau)\sin(\omega_0\xi/c). \tag{26.82}$$

Using the definition of variables (ξ, τ) given by Eq. (26.77) we find that in the coordinates (t, x) the boundary condition can be written as

$$A(x, t) = A_0[\exp(-i\,\omega_0 v) - \exp(i\psi_r(u))]. \tag{26.83}$$

Here variables u and v are defined by Eq. (26.73) and the function $\psi_r(u)$ is the phase of the reflected wave (the ratio $\psi_r(u)/\omega_0$ is the inverse function of $f(u)$). In other words, the incident and reflected waves are described by transverse component of the vector potential,

$$A(x, t) = A_0(t - x/c) + A_r(t + x/c), \tag{26.84}$$

which is the solution of the wave Eq. (26.78) for the boundary conditions (26.80) at $x = x_M(t)$. They can be written in the form

$$A(x, t) = A_0\left[\exp(-i\omega_0 v) - \exp(-i\int^{u}\omega_r(u)du)\right], \tag{26.85}$$

Using Eq. (26.79) we obtain that the reflected wave phase, $\psi_r(u)$, as a function of variable u is given by

$$\psi_r(u) = \int^{u}\omega_r(u)du = \omega_0[2t(u) - u], \tag{26.86}$$

where we have introduced the frequency of the reflected wave $\omega_r(u)$ related to the phase $\psi_r(u)$. Here $t(u)$ should be found from the equation

$$t(u) = u + x_M(t(u))/c. \tag{26.87}$$

In other words, it can be obtained by finding the position of the light ray intersection with the world line of the mirror in the (t, x) plane. Here $x_M(t(u))$ is the intersection point coordinate, see figure Fig. 26.6a. From these relationships it also follows that $\omega_0 v = \psi_r(u)$.

Differentiating this expression with respect to time, we find

$$\psi_r'(u) = \omega_0 \frac{1 + \beta_M(u)}{1 - \beta_M(u)}, \tag{26.88}$$

where $\beta_M(u) = dx_M(t(u))/cdt$ is the normalized on c mirror velocity. The derivative of the phase with respect to variable u determined by Eq. (26.73) is nothing more than the frequency of the reflected wave, ω_r.

By differentiating expression (26.85) with respect to time, we can find a relation between the electric field $E = -\partial_t A/c$ in the incident and reflected waves,

$$E_r = -E_0 \frac{1 + \beta_M(t)}{1 - \beta_M(t)}, \tag{26.89}$$

where $\beta_M(t) \equiv \beta_M(u(t))$.

When the mirror moves with uniform acceleration, w_M, a dependence of its coordinate on time is given by the relationship Landau and Lifshitz (1980)

$$x_M(t) = \frac{c}{w_M}\left(\sqrt{c^2 + w_M^2 t^2} - 1\right). \tag{26.90}$$

In this case the frequency of the light reflected from the uniformly accelerated mirror, in the limit $t \to \infty$ grows proportionally to the square of the time for positive acceleration, $w_M > 0$, and it decreases inversely proportional to the square of the time for negative acceleration $w_M < 0$:

$$\omega_r = \omega_0 \frac{\sqrt{c^2 + w_M^2 t^2} + w_M t}{\sqrt{c^2 + w_M^2 t^2} - w_M t} \underset{t\to\infty}{\approx} \begin{cases} 4\omega_0 w_M^2 t^2/c^2, & w_M > 0 \\ \omega_0 m^2 c^2/4t^2, & w_M < 0 \end{cases}. \tag{26.91}$$

The amplitude of the reflected wave increases (decays) in a similar way. By Van Meter et al. (2001) these dependences were obtained and with the Rindler transformation to the accelerated frame of reference, Rindler (1966).

Equation (26.91) shows that for negative acceleration, $w_M < 0$, in the limit $w_M t/mc \to \infty$ the frequency of reflected wave tends to zero, i.e. the wave period tends to infinity. In this limit the term "frequency" becomes meaningless. In order to demonstrate this, we consider a plane wave of the frequency ω_0, propagating at a speed of light in vacuum. The world line of uniformly accelerated mirror is given by

$$t = \frac{c}{w_M}\sinh\left(\frac{w_M \tau}{c}\right), \tag{26.92}$$

$$x_M = \frac{c^2}{w_M}\left[\cosh\left(\frac{w_M \tau}{c}\right) - 1\right], \tag{26.93}$$

where $\tau = \int dt/\gamma_M$ is the proper time. As well known, these expressions in special relativity are the relationships between the Cartesian (Minkowski) coordinates (t, x) and the Rindler coordinates (τ, w) (Rindler, 1966). The wave crests are separated by a time interval $t_0 = 2\pi/\omega_0$. We assume that the first wave crest reaches the mirror at time $t = 0$, i.e. in the point $x_M = 0$. The second crest hits the mirror at

$$\tau_0 = -\frac{c}{w_M}\ln\left(1 - \frac{w_M t_0}{c}\right). \tag{26.94}$$

This is the wave period expressed via the proper time. As we see for $w_M > 0$ at $w_M t_0/c \to 1$ it tends to infinity.

In order to find the reflected wave period measured in the laboratory frame of reference we calculate the proper time at which the wave crest cuts the $x = 0$ axis, and express it in terms of the time t. This yields for the reflected wave frequency

$$\omega_r = \omega_0\left(1 - \frac{2\pi w_M}{c\omega_0}\right). \tag{26.95}$$

For $w_M > 0$, which corresponds to the wave interaction with receding mirror, at $2\pi w_M/\omega_0 c \to 1$ the frequency ω_r tends to zero. In the case of the counter-propagation, $w_M < 0$, in the limit $2\pi w_M/\omega_0 c \gg 1$ we have $\omega_r \approx 2\pi w_M/c$. We note that the results obtained are related to the case of the wave reflection at the turnoff point of the uniformly accelerated mirror.

If the mirror is a thin electron layer driven by a normally incident linearly polarized electromagnetic wave with the normalized amplitude $a_0 = eE_0/m_e\omega_0 c$ and frequency ω_0, then the longitudinal acceleration is given by

$$w_M(\psi) = \frac{c\omega_0 a_0^2 \sin(2\psi)}{2 + a_0^2 \sin^2(\psi)}. \tag{26.96}$$

Here $\psi = \omega_0(t - x/c)$ is the phase of the wave. The acceleration (26.96) reaches the maximum values periodically at the points

$$\psi_n = \mathrm{ArcTan}\left(\sqrt{\frac{2}{2 + a_0^2}}\right) + 2\pi n, \quad (n = 1, 2, 3, \ldots), \tag{26.97}$$

where it takes the values

$$\max\{w_M\} = \pm\frac{c\omega_0 a_0^2}{\sqrt{2(2 + a_0^2)}}. \tag{26.98}$$

As we see in the limit of relativistically intense electromagnetic wave with $a_0 \gg 1$ the above discussed condition $2\pi w_M/\omega_0 c \gg 1$ is fulfilled. In other words, when the relativistically intense electromagnetic wave interacts with a thin electron layer the frequency spectrum of the scattered electromagnetic radiation is strongly modified.

Note that Chen and Mourou (2017) considered the accelerated relativistic mirror formed in laser plasma to investigate the black hole information loss paradox. The detailed theoretical analysis of the accelerated relativistic mirror produced by the high intensity short laser pulse in an inhomogeneous plasma can be found in the paper by Lobet et al. (2013).

26.6 Electrodynamics of a Thin Foil

Let us consider in a one-dimensional approximation the interaction of a laser pulse with relativistic mirrors modelled by thin foil targets Bulanov et al. (2013b). Each foil comprises two layers: an ion layer with positive electric charge $en_0 l_0$ and a negatively charged, $-en_0 l_0$, electron layer, l_0 is the thickness of the foil which has equal ion and electron density (see Fig. 26.7). Here and below for the sake of brevity we assume that ions and electrons have equal electric charge and that the layer thickness and density are the same for all layers.

It is convenient to describe the thin foil distribution as a delta-function in both momentum and coordinate. Below we use dimensionless variables with time and space normalized on ω_0^{-1} and c/ω_0 respectively, the density unit is he critical density for a wave with frequency ω_0, and the electromagnetic field is normalized on

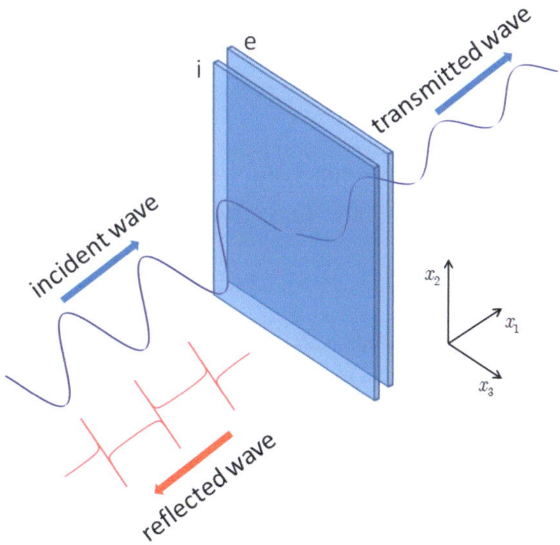

Fig. 26.7 Target comprising thin ion and electron layers. Reproduced from Bulanov et al. (2013b), with the permission of AIP Publishing

$m_e\omega_0 c/e$. The particle velocity and momentum are normalized on c and $m_\alpha c$ where α denotes the species in the $\alpha_{\rm th}$ layer.

The parameter describing the electrodynamic properties of the $\alpha_{\rm th}$ layer is the normalized areal charge density ϵ_α, which being expressed in terms of the dimensional layer density and thickness is given by (see Vshivkov et al., 1998)

$$\epsilon_\alpha = \frac{2\pi n_0 e^2 l_0}{m_\alpha \omega_0 c}. \tag{26.99}$$

The electromagnetic field obeys the Maxwell equations, with the four-vector of the electric current density equal to

$$j^\nu = \sum_\alpha j^\nu_\alpha = \sum_\alpha Z_\alpha(c, \boldsymbol{v}_\alpha) e n_0 l_0 \delta(x - x_\alpha(t)) \tag{26.100}$$

with $\nu = 0, 1, 2, 3$. Here $\delta(x)$ is the Dirac function and $Z_\alpha = \pm 1$. The $\alpha_{\rm th}$ layer velocity is $\boldsymbol{v}_\alpha = v_{1,\alpha}\boldsymbol{e}_1 + v_{2,\alpha}\boldsymbol{e}_2 + v_{3,\alpha}\boldsymbol{e}_3$, and $\boldsymbol{e}_1, \boldsymbol{e}_2, \boldsymbol{e}_3$ are unit vectors in the x, y and z directions, $x_\alpha(t)$ is the $\alpha_{\rm th}$ layer coordinate.

The solution to the Maxwell equations with the four-vector electric current density (26.100) yields Bratman and Samsonov (1995)

$$\boldsymbol{E}_\alpha = 2\pi n_0 l_0 Z_\alpha e \left[s_\alpha(x, \bar{t}_\alpha)\boldsymbol{e}_1 + \frac{v_{2,\alpha}\boldsymbol{e}_2 + v_{3,\alpha}\boldsymbol{e}_3}{c - s_\alpha(x, \bar{t}_\alpha) v_{1,\alpha}(\bar{t}_\alpha)} \right], \tag{26.101}$$

$$\boldsymbol{B}_\alpha = -2\pi n_0 l_0 Z_\alpha e\, s_\alpha(x, \bar{t}_\alpha) \frac{v_{3,\alpha}\boldsymbol{e}_2 - v_{2,\alpha}\boldsymbol{e}_3}{c - s_\alpha(x, \bar{t}_\alpha) v_{1,\alpha}(\bar{t}_\alpha)}, \tag{26.102}$$

for the electric and magnetic field formed by a single $\alpha_{\rm th}$ layer, where $s_\alpha(x, \bar{t}_\alpha) = \mathrm{sgn}(x - x_\alpha(\bar{t}_\alpha))$ with the signum function $\mathrm{sgn}(x) = +1$ for $x > 0$ and $\mathrm{sgn}(x) = -1$ if $x < 0$ and $\mu_\alpha = m_e/m_\alpha$.

For given longitudinal $v_{1,\alpha}$ and transverse $v_{2,\alpha}\boldsymbol{e}_2 + v_{3,\alpha}\boldsymbol{e}_3$ components of the particle velocity, the expressions (26.101) and (26.102) describe the electromagnetic wave emitted by the thin layer, which acts as a one-dimensional electric charge.

The retarded time is determined by the equation.

$$\bar{t}_\alpha = t - \left| x - x_\alpha(\bar{t}_\alpha) \right| / c. \tag{26.103}$$

The relationships (26.101)–(26.103) can also be derived with the Lienard-Wiechert potentials for the four-vector of the electric current density (26.100).

Taking into account that the transverse components of the fields $\boldsymbol{E}_{\alpha,l}$ and $\boldsymbol{B}_{\alpha,l}$ at the $\alpha_{\rm th}$ layer, $x = x_\alpha(t)$ and $\bar{t}_\alpha = t$, are equal to the average of their values at both sides,

$$\boldsymbol{E}_{\alpha,l} = 2\pi n_0 l_0 c Z_\alpha e \frac{v_{2,\alpha}\boldsymbol{e}_2 + v_{3,\alpha}\boldsymbol{e}_3}{c^2 - v^2_{1,\alpha}}, \tag{26.104}$$

$$\boldsymbol{B}_{\alpha,l} = -2\pi n_0 l_0 Z_\alpha e\, v_{1,\alpha} \frac{v_{3,\alpha}\boldsymbol{e}_2 - v_{2,\alpha}\boldsymbol{e}_3}{c^2 - v_{1,\alpha}^2}, \tag{26.105}$$

we can write the expression for the electromagnetic acting on the α_{th} layer as the superposition of the external and self-action fields: $\boldsymbol{E} + \boldsymbol{E}_{\alpha,l}$ and $\boldsymbol{B} + \boldsymbol{B}_{\alpha,l}$, where

$$\boldsymbol{E} = -\frac{1}{c}\partial_t \boldsymbol{A}_{0,\perp}(x_\alpha, t) + \sum_{\alpha\neq\alpha'} \boldsymbol{E}_{\alpha'}(x_\alpha, \bar{t}_{\alpha,\alpha'}(t)), \tag{26.106}$$

$$\boldsymbol{B} = \boldsymbol{e}_1 \times \partial_x \boldsymbol{A}_{0,\perp}(x_\alpha, t) + \sum_{\alpha\neq\alpha'} \boldsymbol{B}_{\alpha'}(x_\alpha, \bar{t}_{\alpha,\alpha'}(t)). \tag{26.107}$$

Here the retarded time $\bar{t}_{\alpha,\alpha'}(t)$ should be found from equation

$$\bar{t}_{\alpha,\alpha'} = t - \left|x_\alpha(t) - x_{\alpha'}(\bar{t}_{\alpha,\alpha'})\right|/c. \tag{26.108}$$

The vector potential $\boldsymbol{A}_{0,\perp}$ corresponds to the external electromagnetic field. In particular it describes the electromagnetic pulse incident on the target. In these expressions ∂_x and ∂_t denote partial derivatives with respect to the coordinate x and the time t.

Using the above-obtained relationships, we can write the equations of the α_{th} layer motion in components as

$$\dot{p}_{1,\alpha} = Z_\alpha \mu_\alpha \left(E_1 + \frac{p_{2,\alpha}B_3 - p_{3,\alpha}B_2}{\gamma_\alpha}\right) - \epsilon_\alpha \frac{p_{1,\alpha}(p_{2,\alpha}^2 + p_{3,\alpha}^2)}{\gamma_\alpha(\gamma_\alpha^2 - p_{1,\alpha}^2)}, \tag{26.109}$$

$$\dot{p}_{2,\alpha} = Z_\alpha \mu_\alpha \left(E_2 - \frac{p_{1,\alpha}B_3}{\gamma_\alpha}\right) - \epsilon_\alpha \frac{p_{2,\alpha}}{\gamma_\alpha}, \tag{26.110}$$

$$\dot{p}_{3,\alpha} = Z_\alpha \mu_\alpha \left(E_3 + \frac{p_{1,\alpha}B_2}{\gamma_\alpha}\right) - \epsilon_\alpha \frac{p_{3,\alpha}}{\gamma_\alpha}. \tag{26.111}$$

Here, a dot, "·", denotes time derivative, $\mu_\alpha = m_e/m_\alpha$, $p_{1,\alpha}$ and $p_{2,\alpha}\boldsymbol{e}_2 + p_{3,\alpha}\boldsymbol{e}_3$ are the longitudinal and perpendicular momenta of the of the particles in the α_{th} layer normalized on $m_\alpha c$. The time and the coordinate are normalized on ω^{-1} and $\lambda\!\!\!^{-} = c/\omega$, respectively. The electric and magnetic field normalization unit is $m_e \omega c/e$.

The layer coordinate $x_\alpha(t)$ depends on time according to equation

$$\dot{x}_\alpha = \frac{p_{1,\alpha}}{\gamma_\alpha}, \tag{26.112}$$

where $\gamma_\alpha = \sqrt{1 + p_{1,\alpha}^2 + p_{2,\alpha}^2 + p_{3,\alpha}^2}$ is the relativistic Lorentz factor. The longitudinal and perpendicular components of the electric field are equal to $(\boldsymbol{e}_1 \cdot \boldsymbol{E})$ and $\boldsymbol{E} - \boldsymbol{e}_1(\boldsymbol{e}_1 \cdot \boldsymbol{E})$, respectively. The last terms on the right-hand sides of Eqs. (26.62),

(26.63), and (26.64) are the longitudinal and perpendicular components of the 1D electrodynamics radiation friction force, respectively.

Multiplying Eqs. (26.109)–(26.111) by $\boldsymbol{v}_\alpha$ and adding them, we obtain the equation

$$\dot{\mathcal{E}}_\alpha = Z_\alpha(\mathbf{E}\cdot\boldsymbol{v}_\alpha) - \epsilon_\alpha \frac{p_{2,\alpha}^2 + p_{3,\alpha}^2}{1 + p_{2,\alpha}^2 + p_{3,\alpha}^2}, \tag{26.113}$$

where $\mathcal{E}_\alpha$ is a kinetic energy of the αth layer. As we see, the rate of radiative energy losses depends only on the momentum component along the layer. The rate of energy loss vanishes at $p_{2,\alpha}^2 + p_{3,\alpha}^2 = 0$ and it is limited by the value of ϵ_α, given by Eq. (26.99), because the layer electric field cannot exceed $2\pi e n_0 l_0$ (in dimensional units).

It is convenient to take the vector potential $\boldsymbol{a}_{0,\perp}$ corresponding to the electromagnetic wave propagating in the positive x direction, i.e., to depend on $t - x$ and to introduce the function

$$h_\alpha = \gamma_\alpha - p_\alpha \tag{26.114}$$

and the variable

$$\tau_\alpha = t - x_{1,\alpha}. \tag{26.115}$$

In the limit of weak radiation friction $\epsilon_\alpha \to 0$ and vanishing longitudinal electric field E_1, the function h_α is an integral of motion. Using these variables, we can rewrite Eqs. (26.109)–(26.111) in the form

$$\frac{dh_\alpha}{d\tau_\alpha} = -Z_\alpha \mu_\alpha E_1 - \epsilon_\alpha \frac{p_{2,\alpha}^2 + p_{3,\alpha}^2}{1 + p_{2,\alpha}^2 + p_{3,\alpha}^2}, \tag{26.116}$$

$$\frac{dx_{1,\alpha}}{d\tau_\alpha} = \frac{1 + p_{2,\alpha}^2 + p_{3,\alpha}^2 - h_\alpha^2}{2h_\alpha^2}, \tag{26.117}$$

$$\frac{dx_{2,\alpha}}{d\tau_\alpha} = \frac{p_{2,\alpha}}{h_\alpha}, \tag{26.118}$$

$$\frac{dx_{3,\alpha}}{d\tau_\alpha} = \frac{p_{3,\alpha}}{h_\alpha} \tag{26.119}$$

with

$$p_{1,\alpha} = \frac{1 + p_{2,\alpha}^2 + p_{3,\alpha}^2 - h_\alpha^2}{2h_\alpha}, \tag{26.120}$$

$$p_{2,\alpha} = a_{2,\alpha} - \epsilon_\alpha \left(x_{2,\alpha} - x_{2,\alpha}\Big|_{\tau_\alpha = -x_{1,\alpha}} \right), \tag{26.121}$$

$$p_{3,\alpha} = a_{3,\alpha} - \epsilon_\alpha \left(x_{3,\alpha} - x_{3,\alpha}\Big|_{\tau_\alpha = -x_{1,\alpha}} \right) \tag{26.122}$$

and

$$\gamma_\alpha = \frac{1 + p_{2,\alpha}^2 + p_{3,\alpha}^2 + h_\alpha^2}{2h_\alpha}. \tag{26.123}$$

26.7 Acceleration of a Single Electron Layer by the Laser Light Pressure

In order to elucidate the basic properties of the 1D electrodynamics formulated above we consider the motion of a single electron layer in the plane electromagnetic wave $\boldsymbol{a}_{0,\perp}(x - t)$. In this case the longitudinal component of the electric field, E_1, on the right hand side of Eq. (26.109) vanishes, and the electric and magnetic fields are equal to $\boldsymbol{E} = -\partial_t \boldsymbol{a}_{0,\perp}(x - t)$ and $\boldsymbol{B} = \boldsymbol{e}_1 \times \partial_x \boldsymbol{a}_{0,\perp}(x - t)$ respectively, with given $\boldsymbol{a}_{0,\perp}$. The electric and magnetic fields are taken at $x = x_\alpha(t)$.

In the case without radiation losses, when $\epsilon_\alpha = 0$, Eqs. (26.109)–(26.111) yield the results which are well known from the book by Landau and Lifshitz (1980). In particular, in this case the function $h_\alpha = \gamma_\alpha - p_\alpha$ determined by Eq. (26.113) is the integral of motion. If the layer before interacting with the electromagnetic pulse is at rest it is equal to unity, $h_\alpha = 1$. Then, for $p_{1,\alpha}$, γ_α, $x_{1,\alpha}$, and τ_α, we have

$$p_{1,\alpha} = \frac{1}{2}\left(a_{0,2}^2(\tau_\alpha) + a_{0,3}^2(\tau_\alpha)\right), \tag{26.124}$$

$$\gamma_\alpha = 1 + p_{1,\alpha} = 1 + \frac{1}{2}\left(a_{0,2}^2(\tau_\alpha) + a_{0,3}^2(\tau_\alpha)\right), \tag{26.125}$$

$$x_{1,\alpha}(\tau_\alpha) = \frac{1}{2}\int_{-\infty}^{\tau_\alpha} d\tau'\left(a_{0,2}^2(\tau'_\alpha) + a_{0,3}^2(\tau'_\alpha)\right), \tag{26.126}$$

and

$$t = \tau_\alpha + x_{1,\alpha}(\tau_\alpha). \tag{26.127}$$

As a result of the interaction of the electron layer with a finite duration electromagnetic pulse, its kinetic energy, $\mathcal{E}_{\text{kin},\alpha} = \gamma_\alpha - 1$, increases from zero to a maximum value equal to $a_m^2/2$ and then decreases to almost zero (an exponentially small value for a pulse longer than its wavelength) after the electromagnetic pulse has overtaken the layer. Here, a_m is the maximum amplitude of the pulse. This fact is referred to as the Lawson–Woodward theorem discussed above. The layer displacement from the initial position is equal to

$$\xi_{1,\alpha} = \frac{1}{2}\int_{-\infty}^{+\infty} d\tau'\big(a_{0,2}^2(\tau'_\alpha) + a_{0,3}^2(\tau'_\alpha)\big). \tag{26.128}$$

In the limit of small but finite radiation losses, we can find the radiation scattered by the layer. Considering ϵ_α a as the small parameter of a perturbation expansion, we calculate the reflected and transmitted waves by using Eqs. (26.104)–(26.105) in which the layer velocity components and retarded time $\bar{t}_\alpha$ are obtained from Eqs. (26.109)–(26.111) for a plane wave linearly polarized along the 2-direction. This yields for the electric field of the wave scattered in forward direction

$$E_{2,\alpha}(x,t) = -\epsilon_\alpha a_2 \sin(\tau_\alpha)|_{\tau_\alpha = t-x}, \tag{26.129}$$

i.e., the transmitted wave is $(1-\epsilon_\alpha)a_2\sin(\tau_\alpha)$. The backward scattered wave, which is the wave reflected from the receding layer, is given by

$$E_{2,\alpha}(x,t) = \left.\frac{\epsilon_\alpha a_2 \sin(\tau_\alpha)}{1 + a_2^2 \sin^2(\tau_\alpha)}\right|_{\tau_\alpha + \frac{a_2^2}{2}\left(\tau_\alpha - \frac{\sin(2\tau_\alpha)}{2}\right) = t+x}. \tag{26.130}$$

Here we consider the interaction of the layer with a sinusoidal electromagnetic wave given by $\boldsymbol{a}_{0,\perp} = a_2\sin(x-t)\boldsymbol{e}_2$ for $t > 0$ and zero before. Figure 26.8 shows the waves emitted in forward and backward directions, respectively.

Due to the double Doppler effect the wavelength of the wave reflected back by the receding layer (Fig. 26.8b) is larger than the incident wavelength. In addition, the reflected wave is not sinusoidal. The minimal electric field where $\partial_x E_2 = 0$ equal to ϵ_α/a_2 in the limit $a_2 \gg 1$. In this limit each half-period, the wave profile becomes singular at the point where $\partial_{t_\alpha} x = 0$. In the vicinity of the singular point, $t+x$ and E_2 depend on the variable τ_α as

$$t + x \approx \tau_\alpha - \frac{a_2^2}{3}\tau_\alpha^3 \quad \text{and} \quad E_2 \approx a_2\tau_\alpha - a_2^2\tau_\alpha^3, \tag{26.131}$$

which gives

$$E_2 \approx a_2(t+x) - \frac{2}{3}a_2^2(t+x). \tag{26.132}$$

The electric field reaches the maximum $E_{2,m} = \sqrt{2}/3$ at $(t+x)_m = a_2/\sqrt{2}$ with the maximum width equal to $(t+x)_m = 1/\sqrt{2}a_2$. As it is seen in Fig. 26.8c, where we show the local structure of the electric field in the wave and the corresponding time dependence of the longitudinal velocity of the layer emitting the wave, spikes

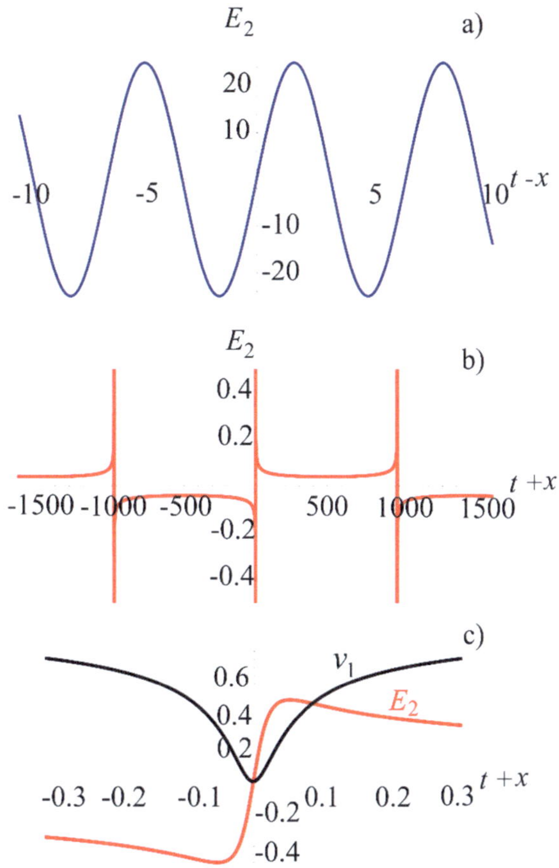

Fig. 26.8 Emitted in the forward direction (**a**) and backward reflected (**b**) waves for $a_2 = 25$. The electric field amplitude is divided by ϵ_α. **c** Local structure of the electric field E_2 and of the longitudinal velocity of the electron layer v_1. Reproduced from Bulanov et al. (2013b), with the permission of AIP Publishing

of the electric field are formed in the reflected wave when the layer stops, i.e., at $v_1 = 0$.

26.8 Oscillating Mirror

When the wave reflects from the relativistic mirror its frequency spectrum extends to the high frequency range and the wave breaks up to short wave packets. According to Eq. (26.3) the wave frequency increases by a factor approximately equal to $4\gamma_M^2$. By the same factor the maximal electric field increases according to Eq. (26.5).

The Relativistic Oscillation Mirror concept has been proposed by Bulanov et al. (1994) as a mechanism of high order harmonic generation when an overdense plasma is irradiated by a relativistically intense laser radiation. The generation of high frequency radiation in thus interaction regime was experimentally demonstrated (see review article Teubner and Gibbon (2009) and literature cited therein).

As an illustration of this process, we consider a thin electron layer oscillating under the action of a linearly polarized electromagnetic wave of the frequency Ω. The wave the electric field is parallel to the y axis:

$$E_y = E_0 \cos(\Omega(t + x/c)). \tag{26.133}$$

The second electromagnetic wave with the amplitude $a_i = eE_0/m_e\omega_0 c$ is reflected from the electron layer as from the mirror. In order to describe the electron layer motion, we use the results of an exact solution of the problem on the electric charge dynamics in the field of an electromagnetic wave, which has been discussed in Chap. 2. This solution yields parametric dependences of the charge coordinates and momentum components on time:

$$\begin{aligned} x_1 &= \frac{a_0^2}{4(2+a_0^2)} \sin 2\eta, \quad x_2 = \frac{a_0}{\sqrt{1+a_0^2/2}} \cos\eta, \quad x_3 = 0, \\ t &= \eta + \frac{a_0^2}{4(2+a_0^2)} \sin 2\eta, \\ p_1 &= \frac{a_0^2}{4\sqrt{1+a_0^2/2}} \cos 2\eta, \quad p_2 = a_0 \sin\eta, \quad p_3 = 0. \end{aligned} \tag{26.134}$$

Here the coordinates and time are normalized on c/Ω and Ω^{-1}, $\eta = \Omega v$, the momentum is measured in the $m_e c$ units, the normalized amplitude of the wave equals $a_0 = eE_0/m_e\Omega c$.

By using solution given by Eq. (26.134) and expressions (26.104)–(26.107) we can calculate the phase and frequency of the reflected wave and find the electric field

$$E_r(t) = E_i \omega_0^{-1} \psi_r'(t) \cos(\psi_r(t)), \tag{26.135}$$

which is shown in Fig. 26.8b for the case of the constant amplitude incident wave. The reflected radiation has a form of a sequence of wave packets with the frequency in the limit $a_0 \gg 1$ approximately equal to $\omega_0 a_0^2/2$ and with the wave packet duration $\approx 2\pi/\Omega a_0$. Here we assume that the second wave propagates in the direction opposite to the direction of the wave which drives the mirror oscillations.

If the second wave and the wave driving the mirror oscillations propagate in the same direction then the electric field in the reflected wave has a form of a sequence of short pulses with the width $\approx \pi/a_0\Omega$ with no high frequency filling. We note that the generation of ultrashort electromagnetic pulses under similar conditions has been observed in the computer simulations of the relativistically strong electromagnetic wave interaction with overdense plasma target Naumova et al. (2004).

The single electron layer moving in the plane electromagnetic wave emits the radiation whose form corresponds to the radiation scattered by the oscillating relativistic mirror. The reflected wave presented in Fig. 26.8 is calculated in the frame of reference where the electron layer moves forward been pushed by the electromagnetic wave. In the frame of reference where the layer is on average at rest, the electric field spikes of the back reflected wave shown in figure. Figure 26.8b are formed at the moment when the mirror reaches its maximum velocity in the backward direction. The velocity of this frame of reference is equal to

$$v_f = \frac{1}{2\pi} \int\limits_{-\pi}^{+\pi} \frac{a_{0,2}^2(\tau_\alpha') + a_{0,3}^2(\tau_\alpha')}{2 + a_{0,2}^2(\tau_\alpha') + a_{0,3}^2(\tau_\alpha')} d\tau'. \tag{26.136}$$

In the case of the linearly polarized wave with $a_{0,2}(\tau) = a_2 \sin \tau$, the layer moves on average with the velocity $v_f = 1 - 1/\sqrt{1 + a_{0,2}^2}$. The spike width and amplitude in this frame of reference changes according to the Lorentz transformation rules.

Now we consider as a good theoretical model elucidating the basic features of the Relativistic Oscillation Mirror concept the configuration when a thin foil made of two layers of electrons and ions irradiated by a high intensity electromagnetic wave. We assume that the ion layer is at the rest at $x = 0$. When the electron layers move with respect to the ion layer an electric field due to charge separation is generated equal to $\boldsymbol{E}(x) = -\epsilon_e \mathrm{sgn}(x)\boldsymbol{e}_1$. We consider an electromagnetic pulse whose form is given by

$$\boldsymbol{a}_{0,\perp}(x, t) = a_0 \exp\left[-\frac{(t - x/c)^2}{t_{EM}^2}\right] \cos\left[\omega_0\left(t - \frac{x}{c}\right)\right]\boldsymbol{e}_2 \tag{26.137}$$

normally incident on the foil. The amplitude and the duration of this linearly polarized short pulse are $a_0 = 25$ and $t_{EM} = 5\pi$, respectively. Before the laser pulse hits the target the electrons are located at $x = 0$ with $\boldsymbol{p}_0 = 0$.

We take the dimensionless parameter ϵ_e, that characterizes both the radiation losses and the electric charge separation electric field, equal to $\epsilon_e = 5$. This choice corresponds to the limit when $\epsilon_e \ll a_0$, and thus in this case the foil is almost transparent for the incident electromagnetic radiation. In this regime of the electromagnetic wave interaction with the target comprising ion and electron layers the radiation pressure pushes the electron layer forwards. The interaction is nonadiabatic with respect to the longitudinal "sawtooth" oscillation excitation. The "sawtooth" oscillations are seen in the longitudinal electron momentum and coordinate dependence on time presented in Fig. 26.9.

As it is seen, the net layer displacement at the end of the interaction of the pulse with the layer provides the initial condition for the "sawtooth" oscillations that are a periodic sequence of hyperbolic motions of the electric charge in the homogeneous electric field due to the ion layer. Within an oscillation half cycle the electron layer momentum depends on time as

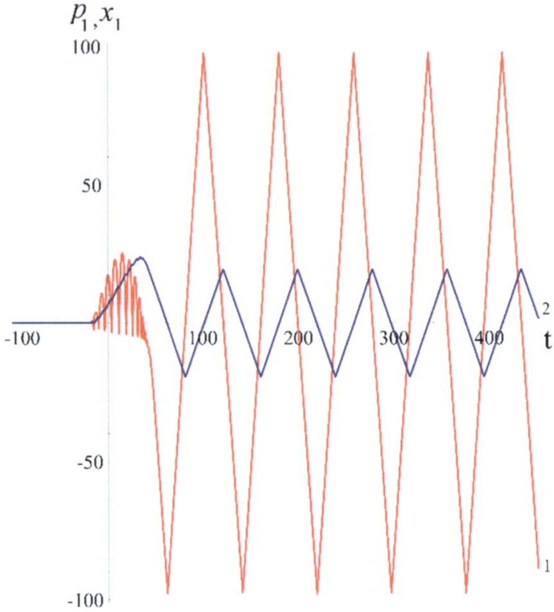

Fig. 26.9 Time dependence of the longitudinal electron momentum, $p_1(t)$, (red curve, 1) and of the layer coordinate, $x_1(t)$, (blue curve, 2) for $a_0 = 25$ and $t_{EM} = 5\pi$, and $\epsilon_e = 5$. Reproduced from Bulanov et al. (2013b), with the permission of AIP Publishing

$$p_1(t) = p_m - \epsilon_e t \tag{26.138}$$

in the time interval $0 < t < t_m$, where t_m is the half-cycle duration equal to $2p_m/\epsilon_e$. The time dependence of the layer coordinate is given by

$$x_1(t) = \frac{1}{\epsilon_e}\left[\sqrt{1 + p_m^2} - \sqrt{1 + (p_m - \epsilon_e t)^2}\right]. \tag{26.139}$$

The transmitted and reflected waves shown in Fig. 26.10 have approximately of the same amplitude level because the receding relativistic mirror becomes less transparent while it is accelerated in the forward direction. This results in a relative enhancement of the reflected wave amplitude. The spectra of the reflected and transmitted radiation contain high order harmonics. The reflected wave has the form of ultrashort spikes. The distance between them corresponds to the stretched wavelength of the incident light due to the double Doppler effect because part of the wave interaction with the oscillating electron layer occurs under the conditions of reflection from a receding mirror. We note that the strongest spike at the rear of the reflected pulse is formed due to interaction with the sawtooth oscillations.

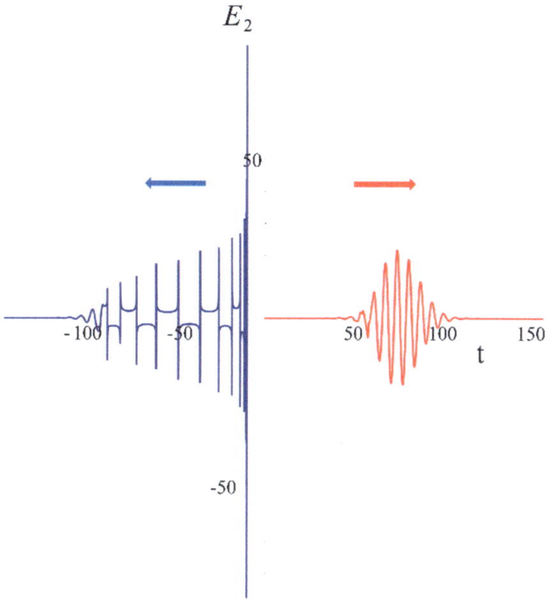

Fig. 26.10 Reflected, $E_2(t + x/c)$, (blue curve) and transmitted, $E_2(t - x/c)$, (red curve) electromagnetic waves for $a_0 = 25$ and $t_{EM} = 5\pi$, and $\epsilon_e = 5$. Reproduced from Bulanov et al. (2013b), with the permission of AIP Publishing

26.9 Relativistic Flying Mirror

A method to generate high frequency radiation based on the concept of the Relativistic Flying Mirror considers a thin plasma shell travelling close to the speed of light as a relativistic mirror. The reflected light undergoes frequency upshift, compression and intensification due to a relativistic double Doppler effect.

Using a double layer thin foil target as a Relativistic Flying Mirror model, we consider the configuration of two counter propagating pulses. The first electromagnetic pulse driver pushes the electron layer forwards with relativistic velocity. The second pulse is relatively week and propagates in the opposite direction. As a result of its head-on collision with the Relativistic Flying Mirror a portion of the photons from this pulse is back reflected. This process is accompanied by the frequency upshifting of the reflected photons and by the modulation of the reflected pulse. When the ponderomotive force of the driver electromagnetic pulse is substantially larger than the force from the electric field due to the electric charge separation, i.e. when $\epsilon_e \ll a_0$, the motion of the relativistic electron layer can be described by Eqs. (26.134). In the case when the electron layer is accelerated by a linearly polarized electromagnetic pulse the phase of the reflected part of the weaker electromagnetic wave is given by

$$\psi_r(u) = \omega_s\left(u + \frac{a_0^2}{2}u - \frac{a_0^2}{4\omega}\sin 2\omega u\right) \tag{26.140}$$

with $u = t - x/c$ and ω_s the frequency of the electromagnetic source pulse. The reflected pulse frequency given by a derivative of the phase ψ_r with respect to time is

$$\omega_r(u) = \omega_s\left(1 + a_0^2 \sin^2 \omega\, u\right). \tag{26.141}$$

The frequency upshifting factor $g = \omega_r/\omega_s$ depends on the longitudinal velocity of the mirror, $v_M = p_1/\gamma$ as

$$g = \frac{\gamma + p_1}{\gamma - p_1}. \tag{26.142}$$

If the source pulse frequency is equal to the driver pulse frequency, $\omega_s = \omega = \omega_0$, the reflected pulse frequency (26.140) changes from ω_0 to $\omega_r(u) = \omega_0\left(1 + a_0^2\right)$. The wave amplitude is modulated accordingly. The reflected radiation consists of a sequence of short high frequency pulses.

Figure 26.11 shows the results of the numerical integration of Eqs. (26.109)–(26.111). The Gaussian electromagnetic pulse driver is linearly polarized with $a_2 = 15$ and $t_{EM} = 5\pi$. The source electromagnetic pulse is linearly polarized in the perpendicular plane, with $a_3 = 1$ and $t_{EM} = 150\pi$.

In Fig. 26.11a we plot the time dependence of the longitudinal momentum, $p_1(t)$, of the electron layer (red curve), of the layer coordinate, $x_1(t)$, (blue curve) and of the frequency upshifting factor $g(t)$ (black curve) for the driver electromagnetic pulse. The foil transparency parameter is equal to $\epsilon_e = 0.1$.

The electron layer, while oscillating back and forth, moves on average forwards with a relativistic velocity. The frequency upshifting factor $g(t)$ oscillates synchronously with the layer momentum $p_1(t)$. According to expressions (26.134) and (26.142) the factor g and the layer longitudinal momentum are related in the limit $a_0 \gg \epsilon_e$ as $g = 1 + 2p_1$, i.e. the factor g scales with the layer energy as $g \propto \gamma_e$. For the chosen electromagnetic pulse driver amplitude equal to 15 the maximum value of the factor g is 226.

In Fig. 26.12 we present the frequency spectrum of the driver and source pulses. Figure 26.12a shows the dependence of the absolute value of the Fourier transform of the E_2 component of the electric field, corresponding to the incident and transmitted electromagnetic radiation of the driver pulse. In Fig. 26.12b we plot the dependence of the absolute value of the Fourier transform of the E_3 component of the electric field, which corresponds to the incident and reflected electromagnetic of the source pulse. The spectrum of the reflected radiation is enriched by high order harmonics. It has a form of the plateau, which extends to the value of the order of $\omega_{\max} \approx g\omega_0$.

The reflected electromagnetic pulse as seen in Fig. 26.11b is approximately shorter by a factor $g = 226$ than the source pulse incident on the foil. Figure 26.11c shows that the reflected wave breaks up into a train of high frequency pulses, which are frequency modulated as seen in Fig. 26.11d, i.e. in general the frequency upshifting and shortening of the reflected pulse is accompanied by the generation of high order harmonics.

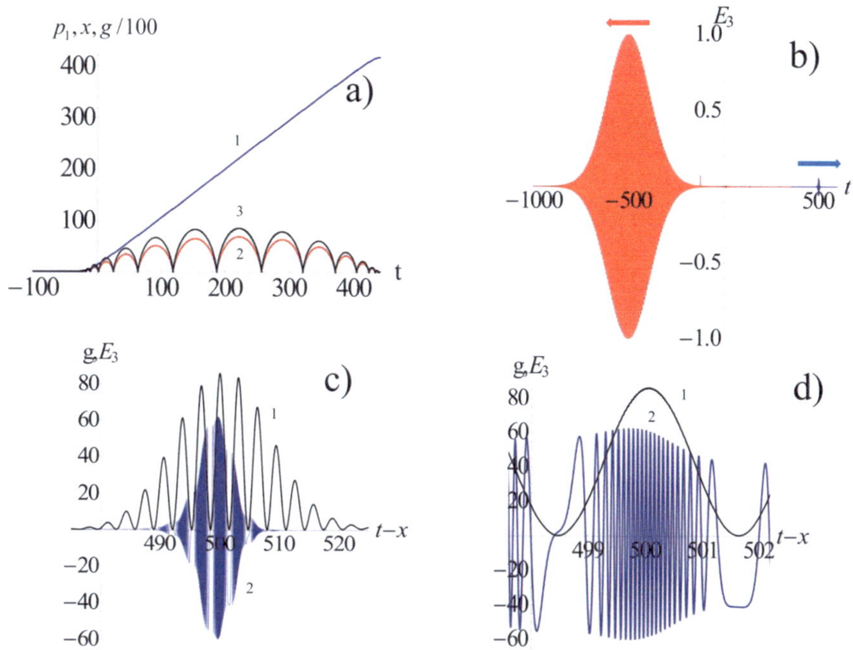

Fig. 26.11 **a** Time dependence of the longitudinal electron layer momentum, $p_1(t)$, (red curve), of the layer coordinate, $x_1(t)$, (blue curve) and of the factor $g(t)$ (black curve) for the driver electromagnetic pulse with $a_2 = 15$ and $t_{EM} = 5\pi$, for $\epsilon_e = 0.1$. Counter-propagating source pulse: **b** Reflected, $E_3(t - x/c)$, (blue curve) and transmitted, $E_3(t + x/c)$, (red curve) waves for $a_3 = 1$ and $t_{EM} = 150\pi$. **c** Reflected pulse (blue curve) and frequency upshifting factor $g(t)$ (black curve). **d** Close up of the reflected pulse (blue curve) and frequency upshifting factor $g(t)$ (black curve). Reproduced from Bulanov et al. (2013b), with the permission of AIP Publishing

The amplitude of the reflected electromagnetic wave is proportional to the amplitude of the incident radiation times the factor g and times the reflection coefficient $\rho(\omega)$. The reflection coefficient is calculated below.

26.10 Light Reflection from a Thin Electron Layer

Here we discuss the reflection of electromagnetic waves from a thin electron layer. In order to calculate the reflection R and transmission T coefficients, we consider the interaction of electromagnetic waves with a strongly nonuniform distribution of electron density in the layer. Let the electromagnetic wave is given by the z component of the vector potential $A_z(x, y, t)$, i.e. it has the so-called s-polarization. The function $A_z(x, y, t)$ obeys the wave equation

$$\partial_{tt} A_z - c^2\left(\partial_{xx} A_z + \partial_{yy} A_z\right) + \omega_{pe}^2(x - \beta_M ct) A_z = 0, \tag{26.143}$$

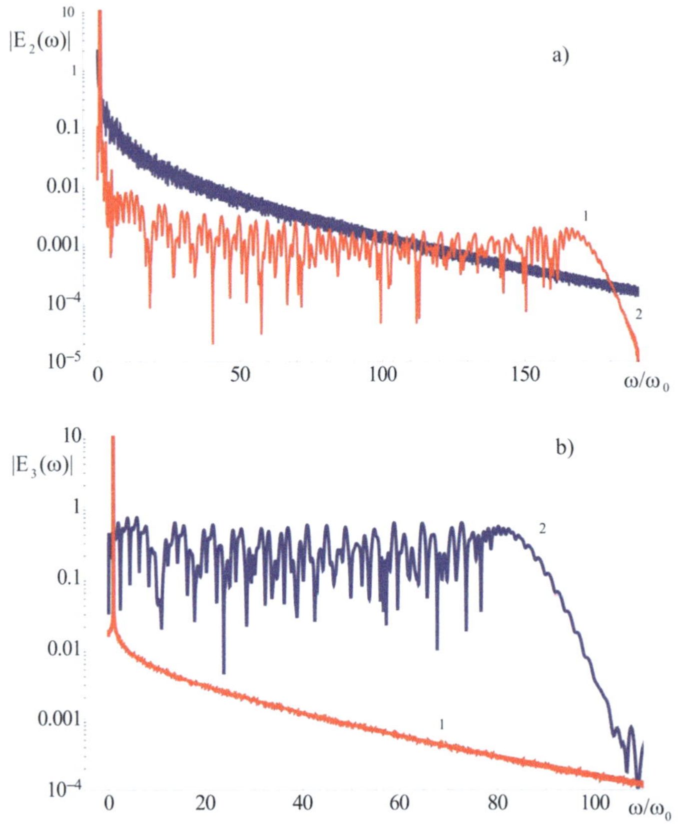

Fig. 26.12 The frequency spectrum of the driver and source pulses. **a** The dependence of the absolute value of the Fourier transform of the E_2 component of the electric field, corresponding to the incident and transmitted electromagnetic of the driver pulse. **b** The dependence of the absolute value of the Fourier transform of the E_3 component of the electric field, which corresponds to the incident and reflected electromagnetic of the source pulse. Reproduced from Bulanov et al. (2013b), with the permission of AIP Publishing

where $\omega_{pe}^2(x - \beta_M ct) = 4\pi e^2 n(x - \beta_M ct)/m_e\gamma_M$ is the plasma frequency with the relativistic gamma-factor equal to γ_M.

We perform the Lorentz transformation to a boosted frame reference moving with the phase velocity of the plasma wave. In this frame of reference, Eq. (26.143) takes the form

$$\frac{d^2a(\zeta)}{d\zeta^2} + \big(s^2 - v(\zeta)\big)a(\zeta) = 0, \tag{26.144}$$

where $s^2 = \left(\omega'/c\right)^2 - k_y^2$, $\zeta = \gamma_M(x - \beta_M ct)$, t', k' and ω' are the square of the wave number, coordinate, time and frequency in the boosted frame of reference, and

$\nu(\zeta) = \omega_{pe}^2(\zeta)/c^2$. The vector potential is normalized on $m_e c^2/e$, and its dependence on time t' and coordinates ζ and y is

$$a(\zeta) = \frac{eA_z(\zeta)}{m_e c^2} \exp\left[-i\left(\omega' t' - k_y y\right)\right] \tag{26.145}$$

The reflection coefficient of electromagnetic radiation from a thin electron layer we find in a similar way as is done in the classical problem of scattering. Assuming that the layer is in the rest frame of reference, we take the distribution of electron density in the form proportional to the Dirac delta function:

$$n(\zeta) = n_0 l_0 \delta(\zeta), \tag{26.146}$$

where l_0 is the electron layer thickness.

In the frame of reference moving with the velocity $v_M = \beta_M c$ Eq. (26.144) can be written as

$$\frac{d^2 a}{d\zeta^2} + \left[s^2 - g_\delta \delta(\zeta)\right] a = 0, \tag{26.147}$$

where $g_\delta = k_p^2 l_0$ and $k_p = c/\omega_{pe}$. Integrating this equation over ζ in the interval $-\epsilon < \zeta < \epsilon$ and letting ϵ to zero, we obtain the condition for the jump of the derivative $da/d\zeta$ on the boundary $\zeta = 0$:

$$\left.\frac{da}{d\zeta}\right|_{+0} - \left.\frac{da}{d\zeta}\right|_{-0} = g_\delta a(0). \tag{26.148}$$

the function $a(\zeta)$ is continuous in the point $\zeta = 0$. Solution to Eq. (26.147), describing the wave reflection from a thin layer can be represented as

$$a(\zeta) = \text{const} \begin{cases} \exp(is\zeta) + \rho \exp(-is\zeta), & \zeta \geq 0, \\ \tau \exp(is\zeta), & \zeta < 0, \end{cases} \tag{26.149}$$

where ρ and τ are related to each other by equations following from the boundary condition (26.148):

$$1 + \rho(s) = \tau(s), \tag{26.150}$$

$$is[1 - \rho(s) - \tau(s)] = g_\delta \tau(s). \tag{26.151}$$

As a result the reflected wave amplitude is equal to

$$\rho(s) = -\frac{g_\delta}{2is + g_\delta} = \tau(s) - 1. \tag{26.152}$$

From this expression we obtain the reflection coefficient for the number of photons

$$\mathrm{R}_\delta = \frac{g_\delta^2}{4s^2 + g_\delta^2}. \quad (26.153)$$

For large values of the electron surface density, $n_0 l_0$, i.e. for $g_\delta^2 \gg 4s^2$, the reflection coefficient close to unity. In the opposite limit, we can neglect g_δ^2 as compared with $4s^2$ in the denominator of the right-hand side of the expression (26.153) and obtain

$$\mathrm{R}_\delta \approx \frac{g_\delta^2}{4s^2} \approx \frac{(n_0 l_0 \lambda_0 r_e)^2}{\gamma_M^2}. \quad (26.154)$$

It is seen that the reflection coefficient is proportional to the square of the electron density in the surface layer, $\sim (n_0 l_0)^2$, i.e. the reflection is coherent.

As shown by Bulanov et al. (2010a, 2010b, 2010c, 2010d, 2010e), sufficiently homogeneous laser pulses are required to form continuous in the transverse direction thin electron layers. When this condition is not fulfilled in the interaction of laser radiation with a thin foil a cloud of ultrashort electron bunched (a swarm of electron bunches) is formed.

26.11 Relativistic Transparency

Let us consider the case of normal incidence of a plane circularly polarized electromagnetic wave on an infinitely thin electron layer in the frame of reference where the layer is on average at the rest. In this frame of reference the wave electric (and magnetic) field of the normalized amplitude a rotates with frequency ω. It is convenient to write the longitudinal and perpendicular components of electron momentum $p_{1,e}$ and $p_{2,e}\mathbf{e}_2 + p_{3,e}\mathbf{e}_3$ as a combination of vectors, which are nonrotating and rotating with angular frequency ω,

$$\begin{pmatrix} p_{1,e} \\ p_{\parallel,e} \\ p_{\perp,e} \end{pmatrix} = \begin{pmatrix} 1 & 0 & 0 \\ 0 & \sin\omega t & -\cos\omega t \\ 0 & \cos\omega t & \sin\omega t \end{pmatrix} \begin{pmatrix} p_{1,e} \\ p_{2,e} \\ p_{3,e} \end{pmatrix}, \quad (26.155)$$

where $p_{\parallel,e}$ and $p_{\perp,e}$ are the components parallel and perpendicular to the electric field. Equations (26.121) and (26.122) yield

$$p_{\parallel,e} = a_0 - \epsilon_e \frac{p_{\perp,e}}{\gamma_e}, \quad (26.156)$$

$$p_{\perp,e} = \epsilon_e \frac{p_{\parallel,e}}{\gamma_e} \quad (26.157)$$

with $\gamma_e = \sqrt{1 + p_{\parallel,e}^2 + p_{\perp,e}^2}$.

Solving this system of algebraic equations we obtain

$$p_{\parallel,e} = \frac{\sqrt{(1-a_0^2+\epsilon_e^2)^2+4a_0^2}-(1-a_0^2+\epsilon_e^2)}{2a_0}, \tag{26.158}$$

$$p_{\perp,e} = \frac{\epsilon_e}{\sqrt{2}a_0}\frac{\sqrt{(1-a_0^2+\epsilon_e^2)^2+4a_0^2}-(1-a_0^2+\epsilon_e^2)}{\sqrt{\sqrt{(1-a_0^2+\epsilon_e^2)^2+4a_0^2}+(1-a_0^2+\epsilon_e^2)}}. \tag{26.159}$$

In the limit of a relatively weak electromagnetic field when $1 \ll a_0 \ll \epsilon_e$ expressions (26.158) and (26.159) have the asymptotic expressions

$$p_{\parallel,e} = \frac{a}{1+\epsilon_e^2} - \frac{\epsilon_e^2}{(1+\epsilon_e^2)^3}a_0^3 + O(a_0^5), \tag{26.160}$$

$$p_{\perp,e} = \frac{\epsilon_e}{1+\epsilon_e^2}a_0 - \frac{\epsilon_e(1-\epsilon_e^2)}{(1+\epsilon_e^2)^3}a_0^3 + O(a_0^5). \tag{26.161}$$

In the opposite limit for $1 \ll \epsilon_e \ll a$ the asymptotic relations are

$$p_{\parallel,e} = a_0 - \frac{\epsilon_e^2}{a_0} + O(a_0^{-3}), \tag{26.162}$$

$$p_{\perp,e} = \epsilon_e - \frac{\epsilon_e(1+\epsilon_e^2)}{2a_0} + O(a_0^{-3}). \tag{26.163}$$

As we can see, in the dissipative range of parameters, which corresponds to a relatively low electromagnetic field amplitude $a_0 \ll \epsilon_e$, the component of the electron momentum $p_{\parallel,e}$ parallel to the electric field is much larger than the perpendicular component $p_{\perp,e}$. In the limit of a strong electric field $a_0 \gg \epsilon_e$ we have $p_{\parallel,e} \gg p_{\perp,e}$, that is, the electron momentum is almost perpendicular to the instantaneous direction of the electric field.

Using obtained expressions (26.152) and (26.159) and Eqs. (26.101) and (26.102) we can find the amplitudes of the reflected and transmitted waves. If the incident wave has moderate intensity, $a_0 \ll \epsilon_e$, amplitudes of reflected and transmitted waves are

$$a_r = \frac{a_0\epsilon_e}{\sqrt{1+\epsilon_e^2}} \quad \text{and} \quad a_t = \frac{a_0}{\sqrt{1+\epsilon_e^2}}. \tag{26.164}$$

We see that for $\epsilon_e \gg 1$ the wave is reflected almost totally, while in the limit $\epsilon_e \ll 1$ the electron layer is transparent.

For ultrahigh intensity, $a_0 \gg 1$, if $a_0 \gg \epsilon_e \gg 1$, the amplitude of the reflected wave equals ϵ_e, while the change in the transmitted wave amplitude, $\sqrt{a_0^2 - \epsilon_e^2}$, is small. This regime can be explained by taking into account the fact that the electric current density in a plasma cannot exceed $j_{\text{lim}} = enc$ since the electron velocity is limited by the speed of light in vacuum.

We see that the condition for the foil to be transparent to the electromagnetic radiation in the limit of moderate intensity is $a_0 \ll 1$. It can be rewritten as $\omega \gg \omega_{pe}(l_0/2\lambda\!\!\!^{-}_p)$, where $\lambda\!\!\!^{-}_p = c/\omega_{pe}$, which differs from the transparency condition for uniform plasmas by the factor $(l_0/2\lambda_p)$.

For relativistically strong waves with $a_0 \gg 1$, a foil with $\epsilon_e \gg 1$ is transparent if $a_0 \gg \epsilon_e$. This condition can be written as

$$\omega \gg \frac{\omega_{pe}^2 l_0}{2ca_0} \tag{26.165}$$

while a uniform plasma is transparent to relativistically strong radiation if written as

$$\omega \gg \frac{\omega_{pe}}{(1 + a_0^2)^{1/4}}. \tag{26.166}$$

26.11.1 Plasma Shutter

The relativistic transparency of a thin foil can be used for shaping the laser pulse at pico-second and sub-picosecond time scale (Vshivkov et al., 1998). As we can see from the above discussed relationships thin plasma slab with the density overcritical for small amplitude electromagnetic wave becomes transparent for relativistically strong radiation. As a result, relatively low amplitude part of a finite size laser pulse is reflected back while the part with high enough amplitude electromagnetic radiation propagates through the foil. This is known as a plasma shutter which has been studied theoretically by Vshivkov et al. (1998), and by Matys et al. (2022). It has been observed in the experiments by Reed et al. (2009), Palaniyappan et al. (2012) and Wei et al. (2017).

Let us now consider a laser pulse with amplitude varying along x. The amplitude is zero at the beginning of the pulse, increases up to its maximum value $a_0 \gg 1$ and then decreases to zero. If $a_0 \gg \epsilon_e$ the portion of the pulse where $a_0 < \epsilon_e$ is reflected by the foil, while the portion with $a_0 > \epsilon_e$ propagates through the foil. This effect is seen in Fig. 26.13 where we plot the amplitude of the incident, transmitted and reflected pulses calculated for the case: $a_0 = 8$ and $\epsilon_e = 4$. We see steepening of the transmitted pulse and flattening of the reflected pulse.

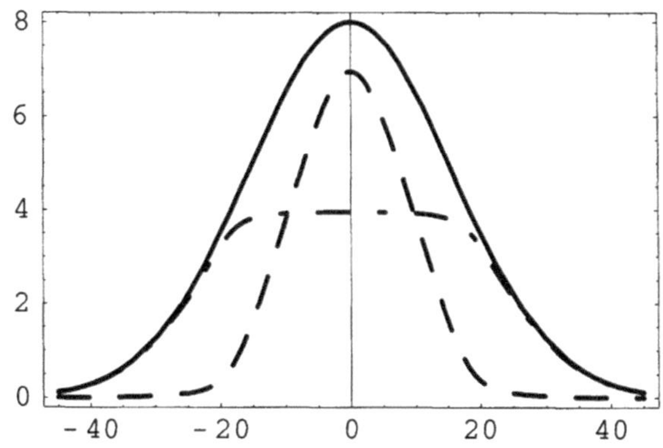

Fig. 26.13 Amplitude of the Gaussian, circularly polarized, incident pulse (solid line), of the transmitted (dashed line) and reflected (dotted-dashed line) pulses in the case of normal incidence $a_0 = 8$ and $\epsilon_e = 4$ showing the steepening of the transmitted pulse and the flattening of the reflected pulse Vshivkov et al. (1998). Reproduced from Vshivkov et al. (1998), with the permission of AIP Publishing

Results of numerical simulations of the laser-foil interaction in the case of normal incidence are shown in Fig. 26.14 (Vshivkov et al. 1998), where the case 1-D simulation is present and in Fig. 26.15 for 2-D PIC simulations. The energy absorbed by the particle in the foil is only a few percent of the total pulse energy. The pulse curves the foil and makes it concave. The modification of the foil shape (row c) acts as a concave mirror and focuses the reflected radiation into a narrow beam with a width much smaller than that of the incident pulse.

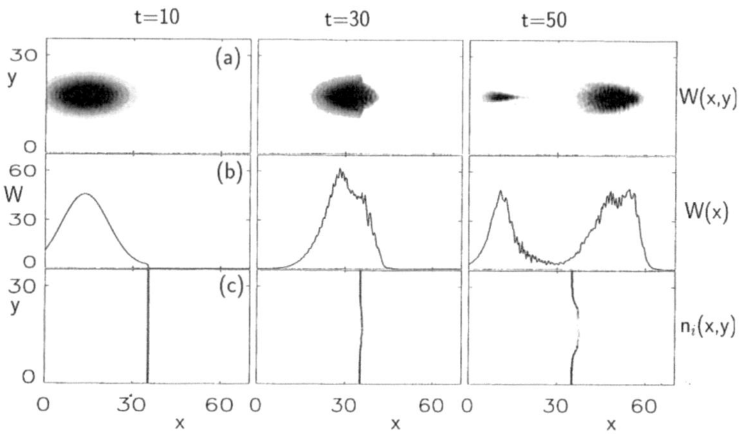

Fig. 26.14 Results of the 1D3V PIC simulations of the laser-foil interaction. The foil is located at the coordinate $x = 0$. The electric field in the pulse is shown as a function of coordinate before (frame a) and after (frame b) the pulse interaction with the foil. Reproduced from Vshivkov et al. (1998), with the permission of AIP Publishing

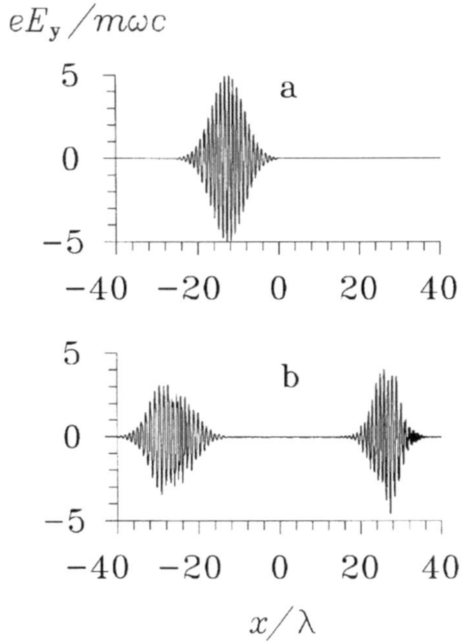

Fig. 26.15 Results of the 2D3V PIC simulations of the laser-foil interaction. The foil is located at the coordinate $x = 30$. In row **a** the distribution of the electromagnetic energy density is shown as a function of the x and y coordinates. In row **b** the distribution of the electromagnetic energy density along the x axis at $y = 0$ is shown. Row **c** shows the thin foil target deformation by the laser pulse pressure. Lengths are normalized on λ and times on $2\pi/\omega$. Reproduced from Vshivkov et al. (1998), with the permission of AIP Publishing

In the case of the laser interaction with thin-foil target in the regime of relativistic transparency the foil "peels" the laser pulse not only in the longitudinal direction but also in transverse direction increasing the contrast of the pulse, which can play an important role for high efficiency ion acceleration and high-order harmonics generation.

26.12 Interaction of Electromagnetic Waves with Nonlinear Wake Waves

A few publications have been devoted to studying of interaction of charged particles and photons with plasma waves. The electron density modulations formed by nonlinear wake plasma waves can be used as the mirrors moving with relativistic velocity. This process refers to the range of problems associated with the interaction of electromagnetic waves with plasma waves, which from the more general point of view includes the study of the transformation of waves of different types and the parametric instabilities and the Raman scattering. Below we shall consider only those aspects of the photon interaction with the wake waves, which have a direct relation to the relativistic plasma mirrors.

26.12.1 The Photon Accelerator

In order to describe the propagation of short enough packet of electromagnetic radiation through the plasma we can use the geometrical optics approximation for the coordinate (the center of the wave packet) and momentum (wave vector). The equations can be written in the Hamiltonian form with the Hamiltonian

$$\omega(x, \boldsymbol{k}; t) = \sqrt{k^2 c^2 + \omega_{pe}^2 (x - v_{ph} t)}. \tag{26.167}$$

This expression is nothing more than a dispersion equation for the dependence of the wave frequency on the wave vector $k^2 = k_\parallel^2 + k_\perp^2$. In the same way as was used above for description of the electron motion, the transition to a variable $X = x - v_{ph} t$ allows to obtain the Hamiltonian for photons

$$\mathcal{W}(x, k_x) = \sqrt{k_x^2 c^2 + \omega_{pe}^2 (X)} - \beta_{ph} k_x c. \tag{26.168}$$

Here we use the conservation of the transverse momentum $k_\perp$ and it is assumed for the sake of simplicity, that it is equal to zero.

Lines of constant level of the Hamiltonian (26.168), which are the trajectories in the phase plane, are shown in Fig. 6.11d (Esirkepov et al., 2006). Due to the obvious similarity of between the photon phase plane and the phase plane of electrons in Fig. 6.12a the process of increasing the frequency of electromagnetic radiation of a short wave packet during its interaction with the wake plasma waves are called the "photon acceleration" Wilks et al., 1988 (see also Mendonca, 2001; Sandberg & Thomas, 2023, Mendonca, 2024). In Sect. 26.3, we have already discussed the photon accelerator like case leading to the formation of the electromagnetic/electron rings in the short pulse laser propagating in an underdense plasma (see Sect. 3.7.1).

Note that in the case of excitation of the wake plasma wave by a short electron beam the phase portrait of the photon is topologically equivalent to the phase portrait of positrons.

On the trajectory corresponding to the value of the Hamiltonian (26.168) equal to

$$\mathcal{W}(X, k_x) = w_0 = \omega_0 - \beta_{ph} k_{x,0} c, \tag{26.169}$$

the dependence of the photon on the coordinate X is given by

$$\omega = \gamma_{ph}^2 w_0 \left(1 \pm \beta_{ph} \sqrt{1 - \omega_{pe}^2 (X) / w_0^2 \gamma_{ph}^2} \right). \tag{26.170}$$

A photon with initial values of frequency and wave number equal to ω_0 and $k_{x,0}$ in the process of the reflection on the first or second maximum of the electron density in the wake wave is experiencing the frequency upshift

$$\tilde{\omega} = \left(2\gamma_{ph}^2 - 1 \right) \omega_0 - 2\beta_{ph} \gamma_{ph}^2 k_{x,0} c. \tag{26.171}$$

The maximum value of the photon frequency $\tilde{\omega}_{\max}$ of the reflected by the wake wave corresponds to the trajectory, which lies on the separatrix separating the transient photons from the reflected. For that separatrix we have $w_0 = \hat{\omega}_{pe}/\gamma_{ph}$, where $\hat{\omega}_{pe} = \omega_{pe}(X_s)$ is a maximum of the plasma frequency максимум плазменной частоты, reached at the points of maximum electron density. From Eq. (26.170) we find that

$$\tilde{\omega}_{\max} = \gamma_{ph}\left(\hat{\omega}_{pe} + \beta_{ph}\sqrt{\hat{\omega}_{pe}^2 - \omega_{pe,0}^2}\right). \tag{26.172}$$

Here $\omega_{pe,0} = \omega_{pe}(+\infty)$ is the plasma frequency in the region far ahead the laser pulse. In the limit $\hat{\omega}_{pe} \gg \omega_{pe,0}$ and $\beta_{ph} \approx 1$ the maximum of the photon frequency equals $\tilde{\omega}_{\max} \approx 2\gamma_{ph}\hat{\omega}_{pe}$.

Similarly to the case of the electron interaction with plasma wave discussed in Chap. 2 (see Fig. 2.2) here we illustrate the various regimes of interaction of a short electromagnetic pulse with an electrostatic wave. In Fig. we present here patterns of phase planes that correspond to the contours of constant levels of the Hamiltonian, given by Eq. (26.168) for different values of the phase velocity of the plasma wave β_w. For simplicity we also assume that the Langmuir frequency $\omega_{pe}^2(X)$ has the form $\omega_{pe}^2(X) = (1 + \delta n \cos X)^2$. Figure 26.16 shows isocontours for the Hamiltonian $\mathcal{W}(x, k_x)$ for $\delta n = 0.1$: (a) subluminal phase velocity $\beta_w = 0.95$; (b) luminal phase velocity $\beta_w = 1$; (c) superluminal phase velocity $\beta_w = 1.1$.

As seen in Fig. 26.16a, in the case of the electrostatic wave propagating with a subluminal phase velocity ($\beta_{ph} = 0.99$) on the phase plane there is a separatrix that encloses the trajectories of trapped "photons" (i.e. the trajectory of trapped short electromagnetic pulses). Below and above the separatrix the trajectories correspond to untrapped "photons". In Fig. 26.16b, we have a phase plot of the luminal plasma wave ($\beta_{ph} = 1$). While the phase velocity of the wave approaches the luminal case the separatrix blows up to infinity. The phase plane that corresponds to the interaction of the "photons" with the electrostatic wave with superluminal phase velocity

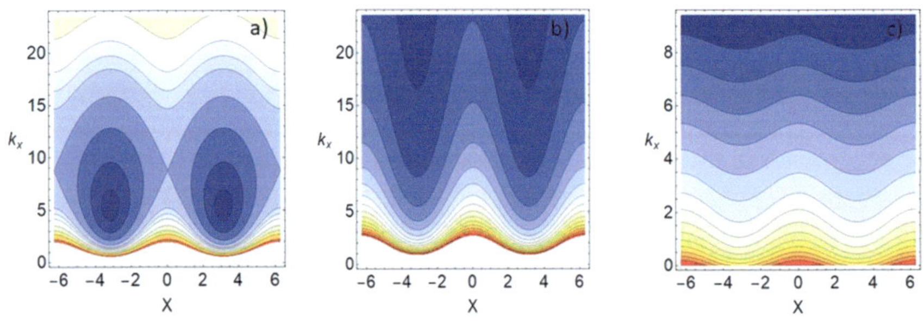

Fig. 26.16 Contours of constant level in the (X, k_x) plane of the Hamiltonian given by Eq. (26.167) for $\delta n = 0.25$: **a** subluminal phase velocity $\beta_{ph} = 0.99$; **b** luminal phase velocity $\beta_{ph} = 1$; **c** superluminal phase velocity $\beta_{ph} = 1.1$

($\beta_{ph} = 1.1$) does not have any separatrix as seen in Fig. 26.16c. All trajectories are the trajectories of the transient "photons", whose energy periodically changes.

Apparently, there is a similarity between the electron acceleration by the wakefield as illustrated in Fig. 6.11 from Esirkepov et al. (2006), and Fig. 2.2.

We note that the frequency upshifted high intense electromagnetic pulse from the "photon accelerator" and XFEL can be considered for reaching the parameters of interest for studying the strong field quantum electrodynamics physics in the collision with high-energy electrons (Russel et al., 2024).

26.12.2 The Electromagnetic Wave Reflection from the Caustics in the Electron Density

As has been noted in this paper, the concept of relativistic flying mirrors is based on an idea that an electromagnetic wave can be reflected by nonlinear plasma waves near the threshold of wave-breaking with small but non-exponentially small reflectivity. Below we discuss the reflection of electromagnetic waves from the electron density maxima with a singularity in the density distribution. Following the widely used terminology, we call these singularities (singularities of Lagrangian mapping) the caustics. In order to calculate the reflection R and transmission T coefficients, consider the interaction of electromagnetic waves with a strongly uninform distribution of electron density in the plasma wave. Let the electromagnetic wave is given by the z component of the vector potential $A_z(x, y, t)$, that obeys the wave equation

$$\partial_{tt}A_z - c^2\left(\partial_{xx}A_z + \partial_{yy}A_z\right) + \omega_{pe}^2(X)A_z = 0, \tag{26.173}$$

where $\omega_{pe}^2(X) = 4\pi e^2 n(X)/m_e\gamma_{ph}$ is the plasma frequency with the relativistic gamma-factor equal to γ_{ph}.

We perform the Lorentz transformation to a boosted frame reference moving with the phase velocity of the plasma wave. In this frame of reference, Eq. (26.171) takes the form

$$\frac{d^2a(\zeta)}{d\zeta^2} + \left(s^2 - \nu(\zeta)\right)a(\zeta) = 0, \tag{26.174}$$

where $s^2 = \left(\omega'/c\right)^2 - k_y^2$, $\zeta = \gamma_{ph}\left(x - v_{ph}t\right)$, t', k' and ω' are the square of the wave number, coordinate, time and frequency in the boosted frame of reference, and $\nu(\zeta) = \omega_{pe}^2(\zeta)/c^2$. The vector potential is normalized on m_ec^2/e, and its dependence on time t' and coordinates ζ and y is

$$a(\zeta) = \frac{eA_z(\zeta)}{m_ec^2}\exp\left[-i\left(\omega't' - k_yy\right)\right] \tag{26.175}$$

We represent the solution of the Eq. (26.173) in the form

$$a(\zeta) = b_+(\zeta)\exp(is\zeta) + b_-(\zeta)\exp(-is\zeta), \tag{26.176}$$

where the unknown functions $b_+(\zeta)$ и $b_-(\zeta)$ are related to the amplitudes of the reflected and transmitted waves. In the limit $\zeta \to -\infty$ the function $b_+(\zeta)$ equal to the amplitude of the incident wave, which is assumed equal to unity, and $b_-(-\infty) = \rho$ corresponds to the amplitude of the reflected wave. For $\zeta \to +\infty$ the function $b_+(\zeta)$ equal to the amplitude of the transmitted wave τ, and $b_-(\zeta)$ vanishes. Therefore, $|b_+(-\infty)|^2 = 1, \quad |b_-(-\infty)|^2 = \mathrm{R}$ and $|b_+(+\infty)|^2 = \mathrm{T}, \quad b_-(+\infty) = 0$.

Since we introduced two unknown functions $b_+(\zeta)$ and $b_-(\zeta)$ instead of one, $a(\zeta)$, it is necessary to impose on them an additional condition. We choose it as a requirement that the derivative $da/d\zeta$ equals

$$\frac{da}{d\varsigma} = is\big[b_+(\zeta)\exp(is\zeta) - b_-(\zeta)\exp(-is\zeta)\big], \tag{26.177}$$

i.e.

$$\frac{db_+}{d\varsigma}\exp(is\zeta) = -\frac{db_-}{d\varsigma}\exp(-is\zeta). \tag{26.178}$$

Substituting expression (26.176) to Eq. (26.174) and taking into account Eqs. (26.177) and (26.178), we obtain a system of equations for the $b_+(\zeta)$ and $b_-(\zeta)$, which is convenient to write in the form

$$\frac{d}{d\zeta}\begin{pmatrix} b_+ \\ b_- \end{pmatrix} = \frac{i\,\nu(\zeta)}{2s}\begin{pmatrix} -1 & -\exp(-2is\zeta) \\ \exp(2is\zeta) & 1 \end{pmatrix}\begin{pmatrix} b_+ \\ b_- \end{pmatrix}. \tag{26.179}$$

We assume that the amplitude of reflected wave is small compared with the amplitude of the incident wave, $\mathrm{R} \ll 1$, which, in particular, is satisfied if $s^2 \gg \nu(\zeta)$. We seek a solution of the system (26.178) within an approach corresponding to the approximation, in which the refraction index variations are considered as a perturbation. As a result, we find

$$\rho = \frac{i}{2s}\int_{-\infty}^{+\infty} \nu(\zeta)\exp(-2is\varsigma)\,d\zeta. \tag{26.180}$$

We shall consider the distribution of electron density typical for a plasma wave near the wave-breaking threshold (as illustrated on Fig. 26.17).

In accordance with the above-presented analysis of the properties of the breaking plasma waves it can be approximated by the function

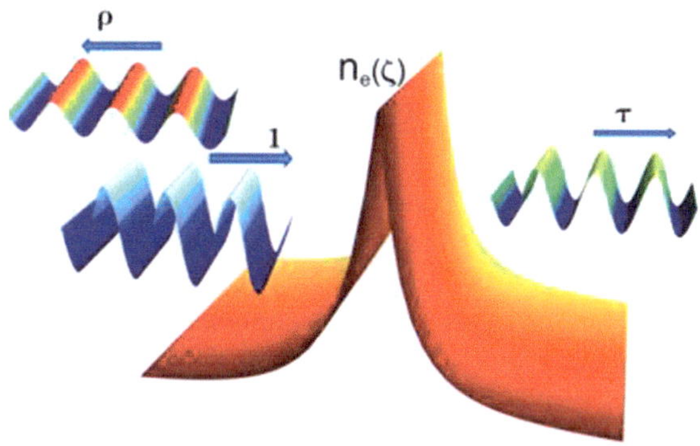

Fig. 26.17 Electromagnetic wave reflection/transmission at/through the caustic in the electron density distribution

$$\nu(\zeta) = \frac{(2/9)^{1/3}(1+a_m^2)k_p^{4/3}\gamma_M^{2/3}}{\zeta^{2/3}}. \tag{26.181}$$

Calculating the integral (26.179) for this integrand, we find

$$\rho(s) = \frac{i3^{1/2}(2/9)^{1/3}\Gamma(1/3)(1+a_m^2)^{1/6}k_p^{4/3}\gamma_M^{2/3}}{(2s)^{4/3}}, \tag{26.182}$$

where $\Gamma(z)$ is the Gamma function. In the limit of a relatively large l, when $sl \gg 1$, over-barrier reflection is exponentially small

$$\rho_{2/3}(s,l) \approx \frac{i\pi}{s^{5/3}l^{2/3}\Gamma(1/3)}\exp(-2sl). \tag{26.183}$$

In the opposite limit, when $sl \to 0$, we have weak but non-exponentially weak reflection

$$\rho_{2/3}(s) = \frac{i3^{1/2}\Gamma(1/3)g_{2/3}}{(2s)^2}. \tag{26.184}$$

Accordingly, the reflection coefficient is equal to

$$\mathrm{R}_{2/3}(s) = \frac{3\Gamma^2(1/3)g_{2/3}^2}{(2s)^4}. \tag{26.185}$$

We emphasize here that these expressions are obtained in the co-moving frame of reference. In the laboratory frame of reference the coefficient R has a meaning of the reflection coefficient giving the relative number of photons.

26.12.3 Thin Electron Layer as a Relativistic Mirror

The interaction of a sufficiently wide electromagnetic pulse with a thin foil provided that the ponderomotive force of the pulse significantly exceeds the force caused by the electric field which occurs due to electric charge separation, can result in formation of a dense electron layer moving in the direction of electromagnetic wave propagation. The second counter-propagating electromagnetic pulse may be partly reflected from a thin electron layer, that in accordance with the above discussed double Doppler effect should lead to compression of the reflected pulse and to its frequency upshift.

To describe the acceleration of a thin electron layer by laser pulse we use the known exact solution of the equations of motion of a charged particle in the field of a plane electromagnetic wave. For a plane wave propagating along the axis x, and given by the vector potential $\boldsymbol{A}_0(t-x/c)$, the transverse component of the generalized momentum of a particle conserves,

$$\boldsymbol{p}_\perp - \frac{e}{c}\boldsymbol{A}_\perp = \text{const}. \tag{26.186}$$

The integral of motion which is a combination of the parallel component of the electron momentum $p_{||}$ and its energy $m_e c^2 \gamma_e$,

$$m_e c^2 \gamma_e - p_{||} c = \text{constant}, \tag{26.187}$$

also conserves. Here m_e and c are the electron mass and speed of light in vacuum. Note that the integral (26.186) is the Hamiltonian of an one-dimensional system describing a particle with momentum $p_{\|}$ and coordinate $u = t - x/c$. In the frame of reference in which the electron layer is at rest before the wave arrival, the constant in the right-hand side of Eq. (26.139) should be set equal to $m_e c^2$. Solution to Eqs. (26.186) and (26.187) gives for the electron kinetic energy $\mathcal{E}_{kin} = m_e c^2(\gamma_e - 1)$ and momentum

$$\mathcal{E}_{kin} = \frac{1}{2} m_e c^2 |\boldsymbol{a}_\perp(u)|^2, \tag{26.188}$$

$$\boldsymbol{p}_\perp = m_e c \boldsymbol{a}_\perp(u), \quad p_{||} = \frac{1}{2} m_e c |\boldsymbol{a}_\perp(u)|^2. \tag{26.189}$$

In these expressions $\boldsymbol{a}_\perp(u) = e\boldsymbol{A}_\perp(u)/m_e c^2$ is the normalized vector potential of the wave.

The electron velocity along the x axis is equal to

$$v_{||} = \frac{p_{||}}{m_e \gamma} = c \frac{|\boldsymbol{a}_\perp(u)|^2}{2 + |\boldsymbol{a}_\perp(u)|^2}. \tag{26.190}$$

Coordinate x of the particle in general case is defined by implicit equations. For a circularly polarized wave, for example, with

$$\boldsymbol{a}_\perp(u) = a(u)\big[\boldsymbol{e}_y \cos(\omega u) + \boldsymbol{e}_z \sin(\omega u)\big], \tag{26.191}$$

the coordinate x should be found from the equation

$$\int^{t-x/c} \big[2 + |a(u)|^2\big] du = 2t. \tag{26.192}$$

If the amplitude of the laser pulse is constant, then from Eq. (26.192) follows the dependence of the particle position on time

$$x(t) = x_0 + c\frac{|a|^2}{2 + |a|^2}\left(t - \frac{x_0}{c}\right), \tag{26.193}$$

where x_0 is the value of coordinate at the moment of arrival of the laser pulse, $t_0 = x_0/c$. The thickness of the electron layer is decreased by a factor $\left(2 + |a|^2\right)/2$.

In order to find the distribution of electrons within the layer, it is necessary to solve the continuity equation for the electron density with the velocity given by Eq. (26.190). It is easy to find that

$$n(u) = \frac{n_0}{1 - \beta_M(u)} = n_0\left(1 + \frac{|\boldsymbol{a}_\perp(u)|^2}{2}\right). \tag{26.194}$$

Here $\beta_M(u) = v_\parallel(u)/c$. As a result of the interaction of a constant amplitude laser pulse with the electron layer the density increases by a factor $\left(2 + |a|^2\right)/2$ in accordance with the found above conclusion that the layer is longitudinally compressed.

Using the expression for the electron velocity given by Eq. (26.190), we find the corresponding gamma factor of the relativistic mirror

$$\gamma_M(u) = \frac{1}{\sqrt{1 - v_\parallel^2(u)/c^2}} = \frac{2 + |\boldsymbol{a}_\perp(u)|^2}{2\sqrt{1 + |\boldsymbol{a}_\perp(u)|^2}}. \tag{26.195}$$

It can be seen that in the limit of large amplitude electromagnetic wave when $|\boldsymbol{a}_\perp(u)| \to \infty$, the gamma-factor is proportional to the first power of $|\boldsymbol{a}_\perp|$.

In the case where the electronic layer is accelerated by the linearly polarized electromagnetic wave, we can find the parameters of the reflected from the layer second wave using the approach formulated in Sect. 26.8. According to relationship (26.146) the phase of reflected wave is given by

$$\psi_r(u) = \omega_0\left(u + \frac{a_0^2}{2}u - \frac{a_0^2}{4\Omega}\sin 2\Omega u\right) \tag{26.196}$$

in the configuration of a head-on collision of the wave with a mirror. The wave frequency and amplitude vary from ω_0 and E_0 to $\omega_0(1 + a_0^2)$ and $E_0(1 + a_0^2)$. In the frame of reference where the mirror on average is at rest, the reflected radiation has a form of the train of short high frequency pulses shown in Fig. 26.8.

If the second wave propagates in the direction of motion of the electron layer, which is the interaction with the receding mirror, the phase of reflected wave is given by

$$\psi_r(v) = \frac{\omega_0}{\Omega}\frac{\text{arctg}\left(\sqrt{1 + a_0^2}\text{tg }\Omega v\right)}{\sqrt{1 + a_0^2}}. \tag{26.197}$$

The wave frequency and amplitude are proportional to $(1 + a_0^2 \sin 2\Omega v)^{-1}$, i.e. they vary from ω_0 and E_0 to $\omega_0(1 + a_0^2)^{-1}$ and $E_0(1 + a_0^2)^{-1}$. The electric field has the form of a sequence of short pulses of width $\approx \pi/a_0\Omega$.

As shown by Bulanov et al. (2010a, 2010b, 2010c, 2010d, 2010e), sufficiently homogeneous laser pulses are required to form continuous in the transverse direction thin electron layers. When this condition is not fulfilled in the interaction of laser radiation with a thin foil a cloud of ultrashort electron bunched (a swarm of electron bunches) is formed. Kiefer et al. (2013) presented the results of experimental realization of the thin layer relativistic mirror.

26.13 Attosecond Electromagnetic Pulses Generated at Nonlinear Plasma Oscillation Excited in the Plasma Resonance Region

As above discussed, an approach to generate high frequency radiation based on the concept of the relativistic flying mirror utilizes a plasma shell traveling close to the speed of light and acting as a relativistic mirror. Reflected light undergoes double Doppler frequency up-shift, compression, intensification due to relativistic effects and focusing (if the plasma shell is curved).

Relativistic plasma oscillation waves can be generated in the plasma in various ways. When the laser radiation interacts with an inhomogeneous plasma, e.g. with the plasma corona formed at the front surface of a solid target, Langmuir oscillations are excited in the vicinity of the plasma critical density surface, which is also called the plasma resonance region (see Chap. 4). These Langmuir oscillations take the form of a localized wave packet whose group velocity is zero while its phase velocity is directed against the gradient of the plasma density. Sufficiently high amplitude oscillations eventually break producing thin dense electron layers, which can play a role of relativistic mirrors for the incident electromagnetic radiation.

Let us consider the p-polarized electromagnetic wave obliquely incident on a slab of an inhomogeneous plasma. As it has been shown in Chap. 4, in the plasma resonance region, where the local plasma frequency ω_{pe} is equal to the frequency of the incident electromagnetic wave ω_0, it forms a singularity where the electric field component parallel to the gradient of the plasma density is given by

$$E_x = \frac{E_d \cos\theta_i}{\epsilon(x)}. \tag{26.198}$$

with E_d is the driver field related to the incident wave and

$$\epsilon(x) = 1 - \frac{\omega_{pe}^2(x')}{\omega'(\omega' + i\nu'(x'))} \approx \frac{x}{L} + is. \tag{26.199}$$

See Eqs. (4.43), (4.44), and (4.46).

Here the parameter $s = \nu_{eff}/\omega_0$ with the effective decay rate ν_{eff} describing the effects of dissipation, weak nonlinearity, dispersion, etc. (see Eq. (4.30)), and L being equal to the plasma inhomogeneity scale-length. The maximum of the x-component of the electric field in the plasma resonance region is about $E_{\max} = E_d/s$. The width of the plasma resonance region is equal to $\Delta x = sL$. Here for the sake of simplicity we consider here small amplitude plasma oscillations. For the analysis of nonlinear plasma oscillations in the resonance region see Chap. 4. The motion of the electron fluid element in the x direction is described by the equation

$$\ddot{\xi}_x = -eE_x/m_e. \tag{26.200}$$

where ξ_x is the x-component of the electron displacement. From this equation, we obtain for ξ_x

$$\xi_x = r_E \frac{s\sin(\omega_0 t) - (x/L)\cos(\omega_0 t)}{(x/L)^2 + s^2} \tag{26.201}$$

with $r_E = eE_d \sin\theta_i/m_e\omega_0^2$ being the amplitude of the electron quiver motion. As we see this equation describes the localized Langmuir wave packet of width $\Delta x = sL$ with the plasma oscillations inside having the form of a wave propagating against the density gradient with the phase velocity equal to

$$v_{ph} = \omega_0 \Delta x = \omega_0 sL \tag{26.202}$$

and the group velocity equal to zero. This causes the modulations of the electron density. Using the known dependence of the function ξ_x on time and coordinate we can calculate the electron density. It is approximately given by

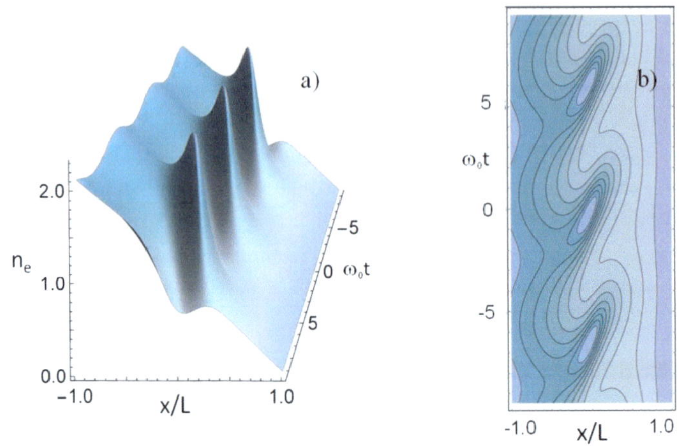

Fig. 26.18 The electron density modulations in the vicinity of the plasma resonance region are given by Eqs. (4.16) and (4.25) for $r_E = 0.05$, $s = 0.3$ and $L = 1$. **a** Electron density n_e versus x and t. **b** Constant density contours in the (x, t) plane. Bulanov et al. (2016)

$$n_e \approx n_0(x)(1 - \partial_{x_0}\xi_x) \tag{26.203}$$

According to Eqs. (26.201) and (26.203) the density modulations have the form of spikes appearing and disappearing in the plasma resonance region as shown in Fig. 26.18. In other words, we have plasma mirrors periodically appearing and disappearing at the plasma critical surface. The electromagnetic wave reflection at such mirrors results in the generation of a train of ultra-short pulses as illustrated in Fig. 26.19. Each pulse duration is about

$$\Delta T = \frac{2\pi}{\omega_0 g_M} \tag{26.204}$$

The frequency upshifting factor g_M according to Eq. (26.10) is $g_M = \gamma_{ph}^2(1+\beta_{ph}^2)$. Here $\beta_{ph} = v_{ph}/c$ is the normalized phase velocity of Langmuir oscillations and $\gamma_{ph} = 1/\sqrt{1-\beta_{ph}^2}$ is the corresponding Lorentz factor.

Above we have seen that the electromagnetic wave reflection is maximal for breaking plasma oscillations. At the wave breaking threshold, the electron velocity equals the phase velocity of the wave, i.e. $\gamma_e = \gamma_{ph}$. In the case of plasma oscillations driven by the relativistically intense electromagnetic wave with the normalized amplitude a_0 the Lorentz factor calculated for the electron quiver energy approximately equals $\gamma_e = a_0$. As a result, we have for the frequency upshifting factor $g_M \approx 4a_0^2$. We note that the train of almost identical short pulses, produced via reflection off the periodically appearing and disappearing relativistic mirrors, has a spectrum consisting of high-order harmonics. The width of an individual harmonic is inversely proportional to the number of short pulses, which is determined by the

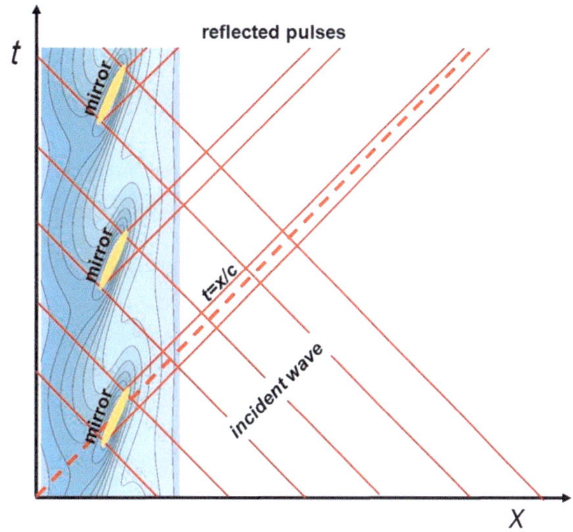

Fig. 26.19 Generation of ultra-short pulses in the wave reflected at periodically appearing and disappearing mirrors. Bulanov et al. (2016)

duration of the incident electromagnetic wave and by the lifetime of the plasma resonance region. In the coordinate space the reflected wave packet is comprised of extremely short (attosecond) pulses (Naumova et al., 2004). The electromagnetic pulse width for optimal conditions scales as $\Delta t = 2\pi/\omega_0 a_0^2$, e.g. $500/a_0^2$ attoseconds Fig. 26.18.

This mechanism of ultra-short electromagnetic pulse generation can be considered as one of the versions of the Relativistic Oscillation Mirror occurrence leading to generation of high order harmonics in the interaction of high intense laser pulse interaction with solid targets (see Bulanov et al., 1994; Naumova et al., 2004; Teubner & Gibbon, 2009).

We note that ultra-short x-ray and gamma-ray pulses can be generated by means of other various methods e.g. by using specially polarized laser pulses (Wang et al., 2020) and thin wire targets (Hadjisolomou et al., 2025).

26.14 Sliding Mirror Model of Generation of High-Energy Attosecond Pulses by the Relativistic-Irradiance Short Laser Pulse Interacting with a Thin Foil

When a thin foil target is irradiated by two laser pulses, or it is irradiated by one pulse under the condition when the charge separation electric field exceeds the laser field, the electron motion in the direction perpendicular to the target surface is suppressed. Electrons can only *slide* along the foil. Using an analytical model, Pirozhkov et al. (2006a), (2006b) describe the so-called Sliding Mirror regime of attosecond pulse generation by the relativistic-irradiance short laser pulse interacting with the sliding mirror. Pulses with the duration of a few hundred attoseconds can be generated with the conversion efficiency up to few percents. The sliding mirror model is illustrated on Figs. 26.20, 26.21 and 26.22.

The frequency spectrum scales with the harmonic number, n, as $S_n = 0.4(4\epsilon_p/n)^2$, where the dimensionless parameter characterizing the foil target is $\epsilon_p = 2\pi n_0 e^2 l_0/m_e\omega_0 c = \pi n_0 l_0/n_{cr}\lambda_0$ (Vshivkov et al., 1998). Here n_0, l_0, n_{cr} are the electron density in the foil, the foil thickness and critical density. It is assumed that the EM pulse is strong enough for the condition $a_0 \geq \epsilon_p$ to be fulfilled. The cut-off frequency of high order harmonic spectrum produced by the sliding mirror is equal to $\omega_0 a_0$.

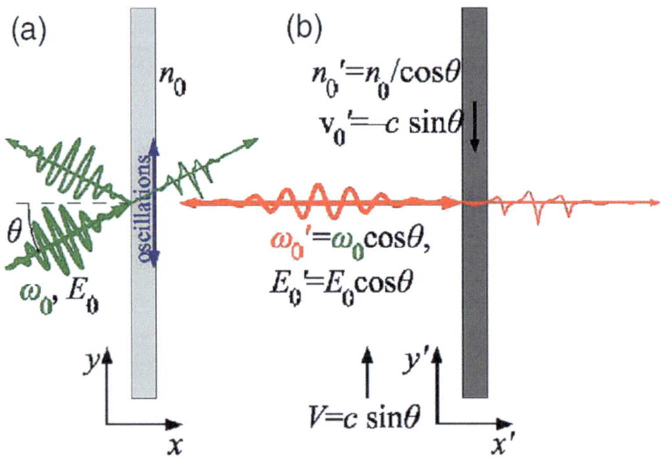

Fig. 26.20 High order harmonics generated in the laser-thin foil interaction within the sliding mirror model. **a** laboratory frame of reference; **b** the laser-foil interaction in the boosted frame reference

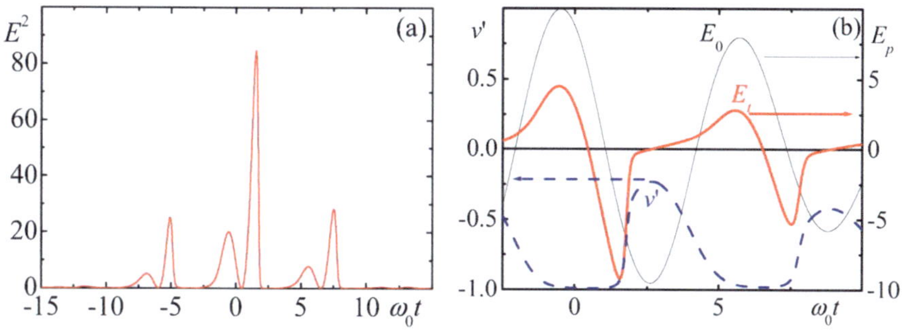

Fig. 26.21 Trains of attosecond pulses generated by *p*-polarized incident pulses with $a_0 = 10$, $\epsilon_p = 2\pi n_0 e^2 l / m_e \omega_0 c = 10$, $\theta_i = 45^o$. **a** Transmitted electric field squared. **b** Dashed line, electron speed in the boosted reference frame; thin and thick lines, the incident and the transmitted electric fields, respectively

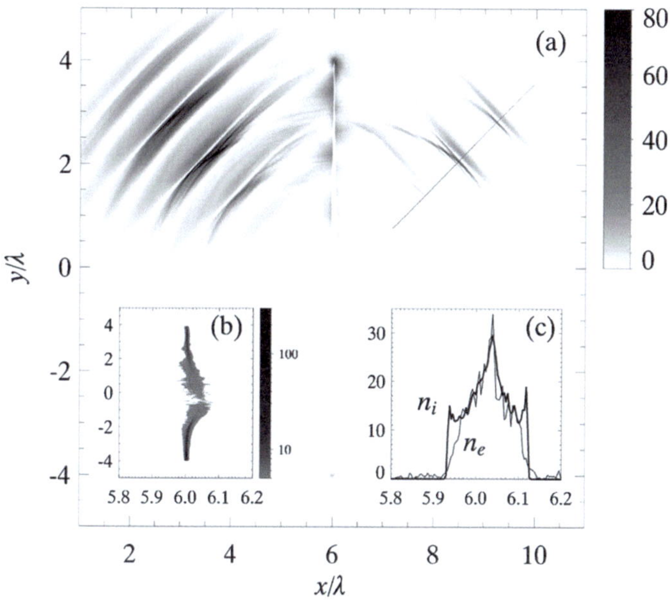

Fig. 26.22 2D PIC simulation results, $t = 26\pi$ (at time $t = 0$, the laser pulse center is situated at $10.8\lambda_0$ along the diagonal before the foil) **a** Distribution of the electric-field squared after the interaction with the laser pulse. Foil position can be traced by the region with small energy density in the central part. **b** Electron-density distribution in the (x, y) plane; note the scale difference between the horizontal and vertical axes. **c** Electron density (thin line) and ion density (thick line) along the line $y = 0$ (the focal spot center)

26.15 Burst Intensification by Singularity Emitting Radiation in Multi-Stream Flow

Burst Intensification by Singularity Emitting Radiation (BISER) (Pirozhkov et al., 2006a, 2006b) is a universal phenomenon, which occurs in media emitting travelling waves. The mechanism is explained using particle-in-cell simulations and catastrophe theory. If a singularity appears in such medium, e.g., due to multi-stream flow, it produces strongly intensified emission due to constructive interference of waves from elementary emitters. BISER results in bright coherent x-ray generation by relativistic plasma singularities driven by multi-terawatt femtosecond lasers focused onto gas jet targets (Pirozhkov et al., 2012, 2014). In plasma, density singularities are produced by the laser pulse which pushes electrons aside, creating a multi-stream flow, typically seen as a bow wave (Esirkepov et al., 2008). These singularities manifest themselves as point-like coherent x-ray emitters. In previous experiments, BISER resulted from relativistic self-focusing, which is laser beam instability. This spontaneous regime exhibited relatively large shot-to-shot fluctuations of the x-ray source position and low overall stability. However, stability is desirable for prospective applications of the new coherent x-ray source.

A regime of relativistic high-order harmonic generation has been discovered by Pirozhkov et al., 2012 in the experiments with multi-terawatt relativistic-irradiance ($> 10^{18}$ W cm^{-2}) femtosecond ($\sim$ 30–50 fs) lasers focused to underdense (few $\times$ 10^{19} cm^{-3}) plasma formed in gas jet targets produce comb-like spectra with hundreds of even and odd harmonic orders reaching the photon energy of 360 eV, including the 'water window' spectral range. Harmonics are generated either by linearly or circularly polarized pulses from the J-KAREN (KPSI, JAEA) and Astra Gemini (CLF, RAL, UK) lasers. The photon number scalability has been demonstrated with a 120 TW laser, producing 40 μJ sr^{-1} per harmonic at 120 eV. The experimental results are explained using particle-in-cell simulations and catastrophe theory. The mechanism of harmonic generation occurs in the laser light interaction with sharp, structurally stable, oscillating electron spikes at the joint of the boundaries of the wake and bow waves (they are described in Chap. 6) excited by a laser pulse in underdense plasma. Detailed descriptions of the experiments, simulations and model are provided and the features are shown, including data obtained with a two-channel spectrograph, harmonic generation by circularly polarized laser pulses and angular distribution.

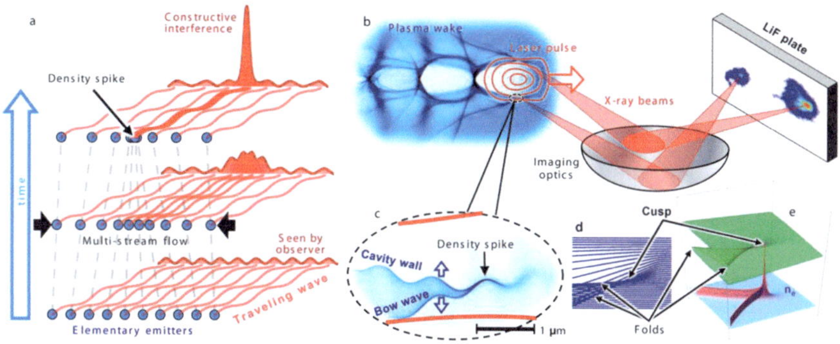

Fig. 26.23 Formation of a singularity (density spike) and corresponding burst intensification of emitted radiation. **a** Multi-stream converging flow consisting of elementary emitters eventually forms a density spike capable of generating a burst of coherent emission. **b** In the experiment, the near-infrared laser pulse propagates in the plasma creating density spikes, which produce x-rays that are collected by imaging optics and recorded by a LiF plate. **c** A density spike appears in a multi-stream flow of electrons pushed aside by the laser field (PIC simulation). **d**, **e** A multi-stream flow model, where the electron transverse displacement forms regions of high concentration, corresponding to density catastrophes (Poston & Stewar, 1978): point-like cusp and folds. Pirozhkov et al. (2017)

This process is called Burst Intensification by Singularity Emitting Radiation (BISER). Singularities in multi-stream flows of emitting media cause constructive interference of emitted travelling waves, forming extremely localized sources of bright coherent emission. The paper by Pirozhkov et al. (2017) demonstrates the extreme localization of BISER by direct observation of nano-scale coherent x-ray sources in a laser plasma. The energy emitted into the spectral range from 60 to 100 eV is up to ~ 100 nJ, corresponding to ~ 1010 photons. Computer simulations with PIC code (Esirkepov, 2001) reveal that these sources emit trains of attosecond *x*-ray pulses.

Figure 26.23 illustrates the underlying physical mechanism of the BISER concept. Results of experiment and PIC simulations are presented in Figs. 26.23 and 26.24.

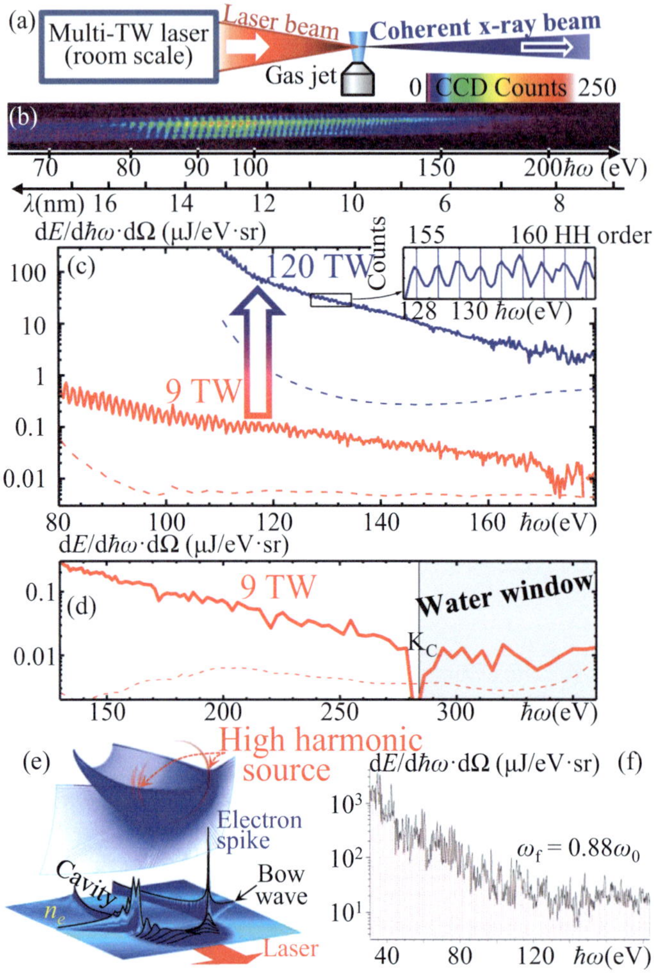

Fig. 26.24 Results of experiment and PIC simulations: **a** Comb-like spectra with hundreds of comparably intense even an odd harmonic orders, **b**, **c** reaching the photon energy of 500 eV, including the 'water window' spectral range, **d**, were generated by either linearly or circularly polarized pulses from the J-KAREN (KPSI, JAEA) and Astra Gemini (CLF, RAL, UK) lasers; (**e**, **f**), at the joint of the wake wave and bow wave boundaries. Importantly, the photon energy and number of photons are scalable with the laser power, frame (**c**). Pirozhkov et al. (2012)

References

Abraham, M. (1904). Zur theorie der strahlung und des strahlungdruckes. *Annalen der Physik (Leipzig)*, *14*, 236.

Asenjo, F. A. (2024). *Different kinds of accelerated propagation of relativistic electromagnetic plasma wavepackets*. arXiv preprint arXiv:2405.12351

Birrell, N. D. P., & Davies, C. W. (1982). *Quantum fields in curved space*. Cambridge University Press.

Bolotovskii, B. M., & Ginzburg, V. L. (1972). The Vavilov-Cerenkov effect and the Doppler effect in the motion of sources with superluminal velocity in vacuum. *Physics Uspekhi, 106*, 577.

Bratman, V. I., & Samsonov, S. V. (1995). Radiation and radiative damping of a charged plane, oscillating with a relativistic velocity. *Physics Letters A, 206*, 377.

Bu, Z., Shen, B., Huang, S., Li, S., & Zhang, H. (2016). Light reversing and folding based on a superluminal flying mirror in a plasma with increasing density. *Plasma Physics and Controlled Fusion, 58*, Article 075008.

Bulanov, S. V., Califano, F., Dudnikova, G. I., Esirkepov, T. Zh., Inovenkov, I. N., Kamenets, F. F., Liseikina, T. V., Lontano, M., Mima, K., Naumova, N. M., & Nishihara, K. (2001). Relativistic interaction of laser pulses with plasmas. In: V. D. Shafranov (Ed.), *Reviews of plasma physics* (Vol. 22, p. 227). Kluwer Academic/Plenum Publishers.

Bulanov, S. S., Bychenkov, V. Y., Chvykov, V., Kalinchenko, G., Litzenberg, D. W., Matsuoka, T., Thomas, A. G., Willingale, L., Yanovsky, V., Krushelnick, K., & Maksimchuk, A. (2010). Generation of GeV protons from 1 PW laser interaction with near critical density targets. *Physics of Plasmas, 17*, 043105.

Bulanov, S. V., Echkina, E. Y., Esirkepov, T. Zh., Inovenkov, I. N., Kando, M., Pegoraro, F., & Korn, G. (2010a). Unlimited ion acceleration by radiation pressure. *Physical Review Letters, 104*, 135003.

Bulanov, S. S., Esirkepov, T. Zh., Thomas, A. G., Koga, J. K., & Bulanov, S. V. (2010b). Schwinger limit attainability with extreme power lasers. *Physical Review Letters, 105*, 220407.

Bulanov, S. V., Esirkepov, T. Zh., Kando, M., Bulanov, S. S., Rykovanov, S. G., & Pegoraro, F. (2013b). Strong field electrodynamics of a thin foil. *Physics of Plasmas, 20*, 123114.

Bulanov, S. V., Esirkepov, T. Zh., Kando, M., Koga, J., Kondo, K., & Korn, G. (2015a). On the problems of relativistic laboratory astrophysics and fundamental physics with super powerful lasers. *Plasma Physics Reports, 41*, 1.

Bulanov, S. V., Esirkepov, T. Zh., Kando, M., & Koga, J. (2016). Relativistic mirrors in laser plasmas (analytical methods). *Plasma Sources Science and Technology, 25*, 053001.

Bulanov, S. S., Esirkepov, T. Zh., Kamenets, F. F., & Pegoraro, F. (2006a). Single-cycle high-intensity electromagnetic pulse generation in the interaction of a plasma wakefield with regular nonlinear structures. *Physical Review E, 73*, 036408.

Bulanov, S. V., Esirkepov, T. Zh., & Tajima, T. (2003). Light intensification towards the Schwinger limit. *Physical Review Letters, 91*, Article 085001.

Bulanov, S. S., Maksimchuk, A., Krushelnick, K., et al. (2010a). Ensemble of ultra-high intensity attosecond pulses from laser–plasma interaction. *Physics Letters A, 374*, 476.

Bulanov, S. S., Maksimchuk, A., Schroeder, C. B., et al. (2012). Relativistic spherical plasma waves. *Physics of Plasmas, 19*, Article 020702.

Bulanov, S. S., Mur, V. D., Narozhny, N. B., Nees, J., & Popov, V. S. (2010d). Multiple colliding electromagnetic pulses: A way to lower the threshold of pair production from vacuum. *Physical Review Letters, 104*, Article 220404.

Bulanov, S. S., Narozhny, N. B., Mur, V. D., & Popov, V. S. (2006b). Electron-positron pair production by electromagnetic pulses. *JETP, 102*, 9.

Bulanov, S. V., Naumova, N. M., & Pegoraro, F. (1994). Interaction of an ultrashort, relativistically strong laser pulse with an overdense plasma. *Physics of Plasmas, 1*, 745.

Bulanov, S. S., Schroeder, C. B., Esarey, E., & Leemans, W. P. (2013a). Electromagnetic cascade in high-energy electron, positron, and photon interactions with intense laser pulses. *Physical Review A, 87*, Article 062110.

Chen, P., & Mourou, G. (2017). Accelerating plasma mirrors to investigate the black hole information loss paradox. *Physical Review Letters, 118*, Article 045001.

Einstein, A. (1905). Zur Elektrodynamik bewegter Körper. *Ann. Phys. (Leipzig), 17*, 891.

Esirkepov, T. Zh. (2001). Exact charge conservation scheme for particle-in-cell simulation with an arbitrary form-factor. *Computer Physics Communications, 135*(2), 144–153.

Esirkepov, T. Zh., Bulanov, S. V., Yamagiwa, M., & Tajima, T. (2006). Electron, positron, and photon Wakefield acceleration: Trapping, wake overtaking, and ponderomotive acceleration. *Physical Review Letters, 96*, 014803.

Esirkepov, T. Zh., Kato, Y., & Bulanov, S. V. (2008). Bow wave from ultraintense electromagnetic pulses in plasmas. *Physical Review Letters, 101*(26), 265001.

Esirkepov, T. Zh., Bulanov, S. V., Kando, M., Pirozhkov, A. S., & Zhidkov, A. G. (2009). Boosted high-harmonics pulse from a double-sided relativistic mirror. *Physical Review Letters, 103*, 025002.

Esirkepov, T. Zh., Mu, J., Gu, Y., Jeong, T. M., Valenta, P., Klimo, O., Koga, J. K., Kando, M., Neely, D., Korn, G., Bulanov, S. V., & Pirozhkov, A. S. (2020). Optical probing of relativistic plasma singularities. *Physics of Plasmas, 27,* 052103.

Esirkepov, T. Z., & Bulanov, S. V. (2024a). Luminal mirror. *Physical Review E, 109*, L023202.

Esirkepov, T. Z., & Bulanov, S. V. (2024b). On the electromagnetic wave interaction with subluminal, luminal, and superluminal mirrors. *Physics Letters A, 526*, 129953.

Ginzburg, V. L. (1979). *Theoretical physics and astrophysics.* Pergamon Press.

Granatstein, V. L., Sprangle, P., Parker, R. K., Pasour, J., Herndon, M., Schlesinger, S. P., & Seftor, J. L. (1976). Realization of a relativistic mirror. *Physical Review A, 14*, 1194.

Grib, A. A., Mamaev, S. G., & Mostepanenko, V. M. (1980). *Quantum effects in intense external fields.* Atomizdat.

Hadjisolomou, P., Jeong, T. M., Valenta, P., Macleod, A. J., & Shaisultanov, R. (2025). Attosecond gamma-ray flashes and electron-positron pairs in dyadic laser interaction with micro-wire. *Physical Review E, 111*, Article 025201.

Hashimshony, D., Zigler, A., & Popadopoulos, K. (2001). Conversion of electrostatic to electromagnetic waves by superluminous ionization fronts. *Physical Review Letters, 86*, 2806.

Jeong, T. M., Bulanov, S. V., Hadjisolomou, P., & Esirkepov, T. M. (2021). Superluminal-subluminal orbital angular momentum femtosecond laser focus. *Optics Express, 29*, 31665.

Jeong, T. M., Bulanov, S. V., Valenta, P., & Hadjisolomou, P. (2024). On the synergic approach toward the experimental realization of interesting fundamental science within the framework of relativistic flying mirror concept. *Reviews of Modern Plasma Physics, 8*, 9.

Kiefer, D., Yeung, M., Dzelzainis, T., Foster, P. S., Rykovanov, S. G., Lewis, C. L. S., Marjoribanks, R. S., Ruhl, H., Habs, D., Schreiber, J., Zepf, M., & Dromey, B. (2013). Relativistic electron mirrors from nanoscale foils for coherent frequency upshift to the extreme ultraviolet. *Nature Communications, 4*, 1763.

Kulagin, V. V., Cherepenin, V. A., Hur, M. R., & Suk, H. (2007). Flying mirror model for interaction of a super-intense nonadiabatic laser pulse with a thin plasma layer: Dynamics of electrons in a linearly polarized external field. *Physics of Plasmas, 14*, Article 113101.

Landau, L. D., & Lifshitz, E. M. (1980). *Course of theoretical physics, Vol. 2: The classical theory of fields.* Pergamon.

Landau, L. D., & Lifshitz, E. M. (1984). *Course of theoretical physics, vol. 8: Electrodynamics of continuous media.* Pergamon Press.

Lichters, R., Meyer-ter-Vehn, & J., Pukhov A. (1996). Short-pulse laser harmonics from oscillating plasma surfaces driven at relativistic intensity. *Physics of Plasmas, 3*, 3425 (1996).

Lobet, M., Kando, M., Koga, J. K., Esirkepov, T. Zh., Nakamura, T., Pirozhkov, A. S., & Bulanov, S. V. (2013). Controlling the generation of high frequency electromagnetic pulses with relativistic flying mirrors using an inhomogeneous plasma. *Physics Letters A, 377*, 1114

Matys, M., Bulanov, S. V., Kucharik, M., Jirka, M., Nikl, J., Kecova, M., Proska, J., Psikal, J., Korn, G., & Klimo, O. (2022). Design of plasma shutters for improved heavy ion acceleration by ultra-intense laser pulses. *New Journal of Physics, 24*, 113046.

Mendonça, J. T. (2001). *Theory of photon acceleration.* Institute of Physics Publishing Bristol and Philadelphia.

Mendonça, J. T. (2024). Photon acceleration by superluminal ionization fronts. *Symmetry, 16*, 112.

Meyer-ter-Vehn, J., & Wu, H.-C. (2009). Coherent Thomson backscattering from laser-driven relativistic ultra-thin electron layers. *The European Physical Journal D, 55*, 433

Mori, W. B., Katsouleas, T., Dawson, J. M., & Lai, C. H. (1995). Conversion of dc fields in a capacitor array to radiation by a relativistic ionization front. *Physical Review Letters, 74*, 542.

Mourou, G., Tajima, T., & Bulanov, S. V. (2006). Optics in the relativistic regime. *Reviews of Modern Physics, 78*, 309.

Naumova, N. M., Nees, J. A., Sokolov, I. V., Hou, B., & Mourou, G. A. (2004). Relativistic generation of isolated attosecond pulses in a λ3 focal volume. *Physical Review Letters, 92*, 063902.

Palaniyappan, S., Hegelich, B., Wu, H., et al. (2012). Dynamics of relativistic transparency and optical shuttering in expanding overdense plasma. *Nature Physics, 8*, 763.

Pasour, A., Granatstein, V. L., & Parker, R. K. (1977). Relativistic mirror" experiment with frequency tuning and energy gain. *Physical Review A, 16*, 2441.

Pirozhkov, A. S., Bulanov, S. V., Esirkepov, T. Zh., Mori, M., Sagisaka, A., and Daido, H. (2006b). Attosecond pulse generation in the relativistic regime of the laser-foil interaction: The sliding mirror model. *Physics of Plasmas, 13*, 013107

Pirozhkov, A. S., Bulanov, S. V., Esirkepov, T. Zh., Mori, M., Sagisaka, A., & Daido, H. (2006). Generation of high-energy attosecond pulses by the relativistic-irradiance short laser pulse interacting with a thin foil. *Physics Letters, A, 349*, 256

Pirozhkov, A. S., Ma, J., Kando, M., Esirkepov, T. Zh., Fukuda, Y., Chen, L.-M., Daito, I., Ogura, K., Homma, T., Hayashi, Y., Kotaki, H., Sagisaka, A., Mori, M., Koga, J. K., Kawachi, T., Daido, H., Bulanov, S. V., Kimura, T., Kato, Y., & Tajima, T. (2007). Frequency multiplication of light back-reflected from a relativistic wake wave. *Physics of Plasmas, 14*, 123106

Pirozhkov, A. S., Kando, M., Esirkepov, T. Zh., Gallegos, P., Ahmed, H., Ragozin, E. N., Faenov, A. Ya., Pikuz, T. A., Kawachi, T., Sagisaka, A., Koga, J. K., Coury, M., Green, J., Foster, P., Brenner, C., Dromey, B., Symes, D. R., Mori, M., Kawase, K., et al. (2012). Soft-X-ray harmonic comb from relativistic electron spikes. *Physical Review Letters, 108*, 135004

Pirozhkov, A.S., Kando, M., Esirkepov, T. Z., Gallegos, P., Ahmed, H., Ragozin, E. N., Faenov, A. Y., Pikuz, T. A., Kawachi, T., Sagisaka, A., & Koga, J. K. (2014). High order harmonics from relativistic electron spikes. *New Journal of Physics, 16*, 093003

Pirozhkov, A. S., Esirkepov, T. Z., Pikuz, T. A., Faenov, A. Y., Ogura, K., Hayashi, Y., Kotaki, H., Ragozin, E. N., Neely, D., Kiriyama, H., & Koga, J. K. (2017), Burst intensification by singularity emitting radiation in multi-stream flows. *Scientific Reports, 7*, 17968

Poston, T., & Stewart, I. (1978). *Catastrophe theory and its applications*. Pitman.

Qu, K., Jia, Q., Matthew, Edwards, M. R., & Fisch, N. J. (2018). Theory of electromagnetic wave frequency up-conversion in dynamic media. *Physical Review E, 98*, 023202

Reed, S. A., Matsuoka, T., Bulanov, S., et al. (2009). Relativistic plasma shutter for ultraintense laser pulses. *Applied Physics Letters, 94*, Article 201117.

Rindler, W. (1966). Kruskal space and the uniformly accelerated frame. *American Journal of Physics, 34*, 1174.

Rosanov, N. N., Vysotina, N. V., & Shatsev, A. N. (2011). Forward light reflection from a moving inhomogeneity. *JETP Letters, 93*, 308.

Russell, B. K., Bulanov, S. S., Qian, Q., Arran, C., Blackburn, T. G., Bulanov, S. V., Grittani, G. M., Mangles, S. P. D., Ridgers, C. P., Seipt, D., & Thomas, A. G. R. (2024). Measuring signatures in photon angular spectra to distinguish nonlinear Compton scattering models. *Physical Review A, 110*, Article 062822.

Sandberg, R. T., & Thomas, A. G. R. (2023). Photon acceleration from optical to XUV. *Physical Review Letters, 130*, 085001.

Semenova, V. I. (1967). Reflection of electromagnetic waves from an ionization front. *Radiophysics and Quantum Electronics, 10*, 599.

Sutherland, R. I., & Shepanski, J. R. (1986). Superluminal reference frames and generalized Lorentz transformations. *Physical Review D, 33*, 2896.

Teubner, U., & Gibbon, P. (2009). High-order harmonics from laser-irradiated plasma surfaces. *Reviews of Modern Physics, 81*, 445.

Tikhonov, A. N. (1952). Systems of differential equations containing small parameters in the derivatives. *Matematicheskii Sbornik (N.S.), 73*, 575

Van Meter, J. R., Carlip, S., & Hartemann, F. V. (2001). Reflection of plane waves from a uniformly accelerating mirror. *American Journal of Physics, 69*, 783.

Vshivkov, V. A., Naumova, N. M., Pegoraro, F., & Bulanov, S. V. (1998). Nonlinear electrodynamics of the interaction of ultra-intense laser pulses with a thin foil. *Physics of Plasmas, 5*, 2727.

Wang, J., Bulanov, S. V., Chen, M., Lei, B., Zhang, Y., Zagidullin, R., Zorina, V., Yu, W., Leng, Y., Li, R., Zepf, M., & Rykovanov, S. G. (2020). Relativistic slingshot: A source for single circularly polarized attosecond x-ray pulses. *Physical Review E, 102*, 061201(R).

Wei, W. Q., et al. (2017). Plasma optical shutter in ultraintense laser-foil interaction. *Physics of Plasmas, 24*, Article 113111.

Wilks, S. C., Dawson, J. M., & Mori, W. B. (1988). Frequency up-conversion of electromagnetic radiation with use of an overdense plasma. *Physical Review Letters, 61*, 337.

Wu, H.-C., Meyer-ter-Vehn, J., Fernández, J., & Hegelich, B. M. (2010). Uniform laser-driven relativistic electron layer for coherent Thomson scattering. *Physical Review Letters, 104*, Article 234801.

Zhidkov, A. G., Esirkepov, T., Fujii, T., et al. (2009). Characteristics of light reflected from a dense ionization wave with a tunable velocity. *Physical Review Letters, 103*, Article 215003.

Chapter 27
Radiation from Relativistic Plasmas

Abstract Here we discuss high efficiency mechanism of the generation of the ultra-short hard photon pulses in laser interaction with thin-foil targets and the mechanism of high brightness X-rays, using on the concept of relativistic flying mirror.

Keywords Gamma ray flash · Relativistic flying mirror · Attosecond pulse generation

27.1 High Power γ-ray Flash Generation in the Laser-Plasma Interaction

Extragalactic gamma-ray bursts are the most energetic events observed in astrophysics since the Big Bang (see Piran, 2005). Within a few seconds, the energy of the emitted gamma radiation (in the high-energy region gamma photons of PeV (10^{15}eV) of Galactic origin are registered by Cao et al. (2021) reaches the erg level. This raises one of the most intriguing questions in astrophysics: how to explain the mechanism of generation of gamma-ray bursts in space. According to current concepts (Mészáros & Rees, 1992; Ruffini et al., 2001), powerful gamma-ray bursts occur when stars collide in binary star systems, followed by the collapse of one of the stars and the formation of black holes. As a result of this process, relativistic jets of lepton-hadron plasma are formed, described within the framework of the fireball paradigm by Mészáros and Rees (1992), the generation of electron–positron pairs, the development of plasma instabilities, magnetic reconnection and collisionless shock waves accelerating charged particles to high energies. Gamma-ray bursts are characterized by an extremely high efficiency of converting the energy of charged particles into gamma-ray energy (Fan & Piran, 2006). Although many hypotheses have been proposed to explain the origin of gamma-ray bursts, there is no doubt that an extremely strong electromagnetic field accelerates high-energy electrons from plasma, which then interact with the strong electromagnetic field, emitting high-energy photons. One of the versions of that scenario is known as the synchrotron self-Compton radiation of high-energy photons (Gould, 1979).

Y. Kato et al., *High Power Laser and Plasma Science*, Springer Series on Atomic, Optical, and Plasma Physics 130, https://doi.org/10.1007/978-3-032-08930-4_27

Extremely high efficiency high-power gamma-ray flash can be generated in the ultra-intense laser-plasma interaction (Gong et al., 2017; Gu et al., 2018; Hadjisolomou, Jeong, Kolenaty et al., 2023a, 2023b; Lezhnin et al., 2018; Nakamura et al., 2012; Ridgers et al., 2012; Vranic et al., 2016). A gamma-ray burst in a laser plasma is also the result of a sequence of processes in which high-energy electrons in the plasma are accelerated by a strong electromagnetic field and then interact with the strong electromagnetic field, emitting high-energy photons. It can be considered as an analogue of the synchrotron self-Compton radiation in an ultra-high-intensity electromagnetic field. These are the processes of strong interests for astro-particle physics.

The gamma-rays, generated in laser plasmas due to the nonlinear Thomson scattering, were observed experimentally (see Chen et al., 1998; Sarri et al. 2014; Yan et al., 2017). The bremsstrahlung mechanism can also generate the gamma-rays in this situation (Gizzi et al., 1996; Courtois et al., 2009; Henderson et al., 2014; Gales et al., 2018). However, the nonlinear Thomson and Compton effects are considerably more effective under conditions, corresponding to the multi-petawatt laser power (Vyskocil et al., 2018; Wu et al., 2018).

As has been discussed in Chap. 19, an electron quivering in the laser field emits photons whose energy is proportional to the cube of the electron Lorentz factor: $\hbar\omega_\gamma \approx 0.3\hbar\omega\gamma_e^3$. For moderate laser amplitude below the radiation friction domination limit, when $a_0 < \varepsilon_{rad}^{-1/3}$, a typical value of the photon frequency is proportional to cube of the laser pulse amplitude a_0^3. In the limit of high laser intensity $a_0 > \varepsilon_{rad}^{-1/3}$, i.e. in the radiation friction dominated regime of the laser-plasma interaction, the emitted photon frequency scales as $\omega_\gamma \approx \omega\,(a_0/\varepsilon_{rad})^{3/4}$. For ten petawatt laser radiation with $a_0 \approx 10^3$ the electron gamma-factor approximately equals $(a_0/\varepsilon_{rad})^{1/4} \approx 500$. As a result, the emitted photon energy is in the range of gamma-ray energy corresponding to 125 MeV.

The analysis of the damping of the electromagnetic wave propagating in the plasma due to the radiation friction conducted in Chap. 23 showed fast decay of the wave with the amplitude above $\varepsilon_{rad}^{-1/3}$ (see Eqs. (23.38), (23.39) and Figs. 23.1 and 23.2). The energy lost by the electromagnetic wave is converted into the energy of the gamma-photons. In the high-density plasma this process leads to the gamma-ray flash generation. Figure 27.1 illustrates the concept of the high-power gamma-ray flash generation in the laser–matter interaction.

27.1.1 Gamma-Ray Generation via the Multi-photon Compton Scattering

When the energy of the photon $\hbar\omega_\gamma$ emitted becomes equal to the electron energy, the recoil effect cannot be neglected. Taking the electron energy to be equal to $m_ec^2a_0$ we find that quantum regime starts at the normalized laser amplitude above $\sqrt{m_ec^2/\hbar\omega_0}$, i.e. at the intensity larger than $5 \times 10^{23}\,\text{W/cm}^2$. The electron, colliding

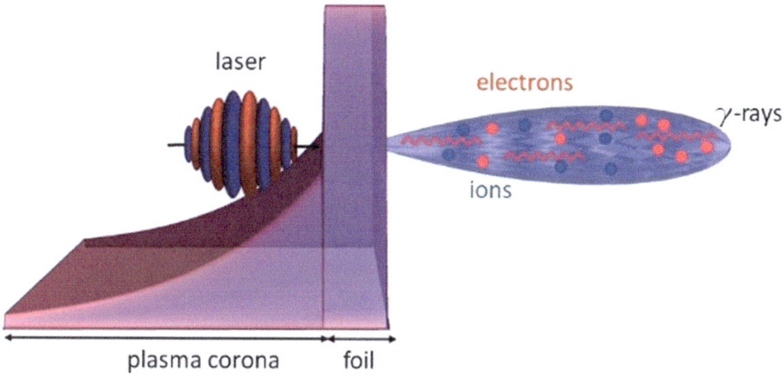

Fig. 27.1 The concept of high-power gamma-ray flash generation in the laser–matter interaction

with the electromagnetic wave, in this case, emits the gamma photons in the nonlinear or multi-photon Compton scattering regime. The required intensity can be reached in the dense corona region, where the laser pulse undergoes the relativistic self-focusing, i.e. at later time of the laser-corona interaction.

At an order of magnitude higher intensity, i.e. at 5×10^{24}W/cm^2, when the dimensionless parameter $\chi_e \approx a_0\gamma_e/a_S$ becomes larger than unity such the QED effects as a recoil can play a significant role for a single electron interacting with the laser field. Here γ_e is the electron gamma-factor and normalized Schwinger field, $a_S = eE_S/m_e\omega_0 c$ with $E_S = m_e^2c^3/e\hbar$, is $a_S = m_ec^2/\hbar\omega_0$. The gamma photon radiation mechanism in this limit is the nonlinear or multi-photon Compton scattering.

For multi petawatt lasers with larger laser amplitudes of the focused light, $a > 3m_ec^2/2\,\hbar\omega_0\gamma_0 \approx 300$, the gamma rays are generated in the multi-photon Compton scattering of ultrarelativistic electrons,

$$s\omega_0 + e^- \rightarrow e^- + \omega_\gamma. \tag{27.1}$$

The energy of photons emitted by ultrarelativistic electrons interacting with the laser light of the frequency ω_0 and amplitude a_0 via the nonlinear Thomson scattering mechanism $\hbar\omega_\gamma$ is proportional to the cube of the electron energy

$$\hbar\omega_\gamma \approx 0.3\,\hbar\omega_0 a_0^3. \tag{27.2}$$

For $a_0 = 300$, i.e. for petawatt class laser radiation, it is in the gamma-ray range.

When the energy of the photon emitted becomes equal to the electron energy, the recoil effect cannot be neglected. Taking the electron energy to be equal to $m_ec^2a_0$, we find from Eq. (27.2) that the quantum regime starts at the laser amplitude above $\sqrt{m_ec^2/\hbar\omega_0}$, i.e., at intensities larger than 5×10^{23}W/cm^2. The electron, colliding with the electromagnetic wave, in this case, emits the gamma photons in the nonlinear or multi-photon Compton scattering regime. The required intensity can be reached in the dense corona region, where the laser pulse undergoes relativistic self-focusing, i.e., at a later time of the laser-corona interaction.

We note that In the one-photon approximation the Compton scattering cross section is given by the Klein-Nishina formula; see Berestetskii et al., (1982). In the ultra-relativistic limit, when the parameter

$$\kappa_C = \left(\frac{\hbar\omega_0}{m_e c^2} + \gamma_{e,0}\right)^2 - \left(\frac{\hbar \boldsymbol{k}_0}{m_e c} + \frac{\hbar \boldsymbol{p}_0}{m_e c}\right)^2 - 1 \tag{27.3}$$

is substantially larger than unity, the total cross-section is given by

$$\sigma_{KN} = \sigma_T \frac{3}{4\kappa_C}\left(\ln \kappa_C + \frac{1}{2}\right) \tag{27.4}$$

where $\gamma_{e,0} = \sqrt{1 + (p_{||,0}/m_e c)^2 + (p_{\perp,0}/m_e c)^2}$ and $p_{||,0}, p_{\perp,0}$ are the electron gamma factor, longitudinal and perpendicular, along and perpendicular to the electromagnetic wave propagation direction, components of the electron momentum, before scattering, respectively. The photon frequency and the wave-vector before scattering equal ω_0 and $\boldsymbol{k}_0$. In the nonrelativistic case, when $\kappa_C \ll 1$, i.e., the electron energy is less than approximately 30 GeV and the Compton scattering cross-section equals the Thomson scattering cross-section $\sigma_T = 8\pi r_e^2/3$.

At an order of magnitude higher intensity, i.e., at 5×10^{24}W/cm^2, when the dimensionless parameter $\chi_e \approx \gamma_e a_0/a_S$ becomes larger than unity, such QED effects as a recoil can play a significant role for a single electron interacting with the laser field. The mechanism of the gamma photon radiation in this limit is the nonlinear or multiphoton Compton scattering.

In the case of Compton scattering on the electron in the field of the electromagnetic wave in a vacuum, the dispersion equation for the wave frequency and the wave vector takes the form $\omega_0^2 = \boldsymbol{k}_0^2 c^2$. In the ultra-relativistic limit, for $p_0 \gg m_e c$, the parameter κ_C given by Eq. (27.3) is approximately equal to $4\hbar\omega_0 p_0/m_e^2 c^3$.

According to Eq. (27.2), an electron interacting with the electromagnetic wave emits high order harmonics. The maximum harmonic number could be equal to a_0^3. In quantum physics, this corresponds to the electron interaction with s photons. Since an electron cannot emit the photon with the energy larger than the electron energy, $\hbar\omega_\gamma < \gamma_e m_e c^2$, we obtain that the photon number is approximately equal to $s \approx \left(m_e c^2/\hbar\omega_0\right) a_0$. In this case, for $p_0 \gg m_e c$, the parameter κ_C becomes equal to $4s\hbar\omega_0 p_0/m_e^2 c^3 = a_0 p_0/m_e c.4$.

Here, we consider the kinematics of the inverse multiphoton Compton scattering process by using the energy and momentum conservation in the electron-photon system. We can find the scattering photon frequency dependence on the electron energy. The energy and momentum conservation equates the sum of electron and s photon energies and momenta before and after the scattering

$$m_e c^2 \gamma_{e,0} + s\hbar\omega_0 = m_e c^2 \gamma_e + \hbar\omega_\gamma, \tag{27.5}$$

$$\boldsymbol{p}_0 + s\hbar\boldsymbol{k}_0 = \boldsymbol{p} + \hbar\boldsymbol{k}. \tag{27.6}$$

Here $\gamma_{e,0} = \sqrt{1 + (\boldsymbol{p}_0/m_e c)^2}$ and $\gamma_e = \sqrt{1 + (\boldsymbol{p}/m_e c)^2}$, $\boldsymbol{p}_0$ and $\boldsymbol{p}$, ω_0 and ω_γ are the electron gamma-factor and the electron momentum before and after interaction, and the laser and gamma ray frequency, respectively, with $\omega_0^2 = \boldsymbol{k}_0^2 c^2$.

Considering the case when the electron and the laser photon momenta before scattering are directed along the x axis, i.e. $\boldsymbol{p}_0 = p_{x,0}\boldsymbol{e}_x$ and $\boldsymbol{k}_0 = k_{x,0}\boldsymbol{e}_x$, we find the solution of Eqs. (27.5) and (27.6) giving for the energy of the gamma-photon,

$$\hbar\omega_\gamma = s\hbar\omega_0 \frac{\gamma_{e,0} m_e c^2 - p_{x,0} c}{s\hbar\omega_0 + \gamma_{e,0} m_e c^2 - (p_{x,0} c - s\hbar k_0 c)\cos\theta}, \tag{27.7}$$

where θ is the angle between the laser pulse and the photon propagation directions. As it is seen, for the receding interaction of the co-propagating ultra-relativistic electron and photon the energy of scattering photon is inverse proportional to the square of the electron energy $\hbar\omega_\gamma \approx 4s\hbar\omega_0\gamma_{e,0}^2$. In the case of the head-on collision of the counter-propagating ultra-relativistic electron and photon the energy of scattering photon is approximately equal to the electron energy, $\hbar\omega_\gamma \approx m_e c^2 \gamma_{e,0}^2$.

In the rest frame of the electron the intensity of radiation is given by

$$\frac{dI}{du} = -\frac{e^2 m_e^2 c^3}{4\pi^2\hbar^2}\frac{u}{(1+u)^3}\left\{\int_z^\infty \mathrm{Ai}(y)dy + \frac{2}{z}\left[1 + \frac{u^2}{2(1+u)}\right]\mathrm{Ai}'(z)\right\}. \tag{27.8}$$

Here

$$u = \chi_\gamma/\chi_e \quad \text{and} \quad z = (u/\chi_e)^{2/3} \tag{27.9}$$

and $\mathrm{Ai}(z)$ is the Airy function (Olver et al., 2010). At small u the intensity is proportional to $u^{1/3}$, i.e. $I \sim \omega_\gamma^{1/3}$, and at large u it decreases exponentially. The maximum of the intensity distribution is at $\chi_\gamma \approx \chi_e^2$ for $\chi_e \ll 1$ and at $\chi_\gamma \approx \chi_e$ if $\chi_e \gg 1$, i.e. at $\omega_\gamma \approx m_e c^2 a/\hbar$. In the laboratory frame of reference, the scattering cross section is maximal in the narrow angle in the backward direction with respect to the laser propagation.

Using expression (27.7) for ω_γ we obtain that the parameter $\chi_\gamma = a\hbar^2\omega_0\omega_\gamma/m_e^2c^4$ equals $\chi_\gamma = 2\gamma_{e,0}a_0\hbar\omega_0/m_e c^2$. At $a = 10^2$ it becomes of the order of unity for $\gamma_{e,0} \approx 2.5 \times 10^3$. We see that the quantum electrodynamics consideration gives the same value $\approx$ 1.25 GeV for the required relativistic electron energy as in the case of the classical approximation corresponding to nonlinear Thomson scattering.

The situation changes when the electromagnetic wave propagates in the plasma, which corresponds to the dispersion equation

$$\omega_0^2 = k_0^2 c^2 + \frac{\omega_{pe}^2}{\sqrt{1 + a_0^2}}$$

(see Eq. (21.20) above), i.e. $\omega_0 \neq k_0 c$. From Eqs. (27.5) and (27.6) follows that the gamma photon energy equals

$$\hbar\omega_\gamma = \frac{s\hbar\omega_0(s\hbar\omega_0 + 2\gamma_{e,0} m_e c^2) + 2 s p_0 \hbar k_0 c^2 - s^2 \hbar^2 k_0^2 c^2}{2[s\hbar\omega_0 + \gamma_{e,0} m_e c^2 - (p_{x,0} c - s\hbar k_0 c)\cos\theta]}. \quad (27.10)$$

For the sake of simplicity we consider the case of the plasma with the near-critical density, for which the electromagnetic wave frequency is equal to the plasma frequency, $\omega_0 = \omega_{pe}/(1 + a_0^2)^{1/4}$ and $k_0 \approx 0$. This yields (Lezhnin et al., 2018)

$$\hbar\omega_\gamma = \frac{s\hbar\omega_0(s\hbar\omega_0 + 2\gamma_{e,0} m_e c^2)}{2(s\hbar\omega_0 + \gamma_{e,0} m_e c^2 - p_{x,0} c \cos\theta)}. \quad (27.11)$$

In the limit $s\hbar\omega_0 \ll m_e c^2/2\gamma_{e,0}$, the energy of the gamma-photon $\hbar\omega_\gamma \approx 2 s\hbar\omega_0 \gamma_{e,0}^2$, is proportional to the electron gamma-factor in the second power. Otherwise, for $s\hbar\omega_0 \ll 2\gamma_{e,0} m_e c^2$ the gamma photon energy is approximately equal to $\hbar\omega_\gamma \approx m_e c^2 \gamma_{e.0}$.

As discussed above, irradiation of plasma targets by high power laser radiation can result in efficient conversion of laser energy to the energy of gamma-ray flash. Publications by Ridgers et al., (2012), Nakamura et al., (2012), Lezhnin et al., (2018), Hadjisolomou et al., (2021) present the results multi-parametric studies of laser-target interaction for generation of a bright gamma-ray-flash. Computer simulations with the kinetic relativistic EPOCH code (Arber et al., 2015) reveal a regime where a significant fraction of laser pulse energy may be converted to gamma rays by optimizing both the pre-plasma and the laser pulse.

The gamma ray collimation is determined by the laser pulse polarization. In the case of linearly polarized laser pulse, the emitted gamma rays have a characteristic two-lobe pattern (Nakamura et al., 2012; Stark et al., 2016; Hadjisolomou et al., 2022). The radially polarized laser radiation irradiating thin-foil target shows substantially collimated gamma radiation. In Figs. 27.2, 27.3 and 27.4 are presented the results of the 3D computer simulations (Hadjisolomou et al., 2022) of the gamma flash generation in the case of the multi-petawatt power laser pulse focused in the lambda-cube regime (Mourou et al., 2002).

The gamma-ray pulse energy, duration and divergence are determined by the laser pulse amplitude and by the plasma target density scale length. Refs. (Hadjisolomou et al., 2022, 2023a, 2023b; Lezhnin et al., 2018; Nakamura et al., 2012; Ridgers et al., 2012) have analyzed the density scale-length effect on the parameters of the emitted gamma-flash.

During interaction of super-high-power laser light with matter the laser pulse is subject to various instabilities. Among them the most important is the relativistic self-focusing resulting in the laser pulse modulation and channeling. This leads to

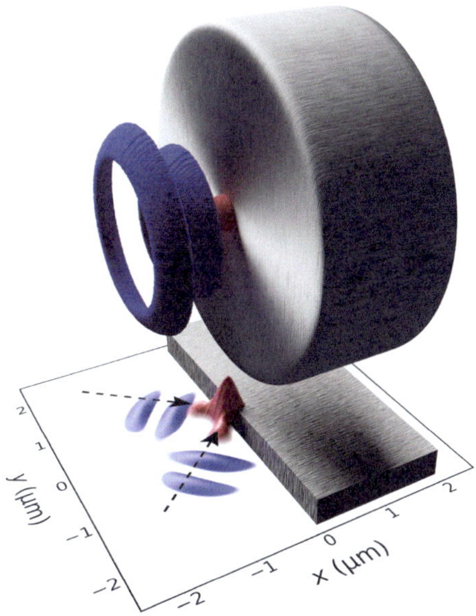

Fig. 27.2 The simulation setup. The gray cylinder shows the target. The blue iso-surface shows the laser pulse at 2 × 1024 W/cm^2, as imported in EPOCH, at ∼ − 4.3 fs. The red iso-surface, at 1025 W/cm^2, is the time-averaged focused laser pulse. The contour plot shows the (x, y) plane of the 3D configuration at z = 0. Reprinted (Fig. 1) with permission from Hadjisolomou et al. (2021) as follows: Hadjisolomou et al. (2021). https://doi.org/10.1103/PhysRevE.104.015203. Copyright (2021) by the American Physical Society

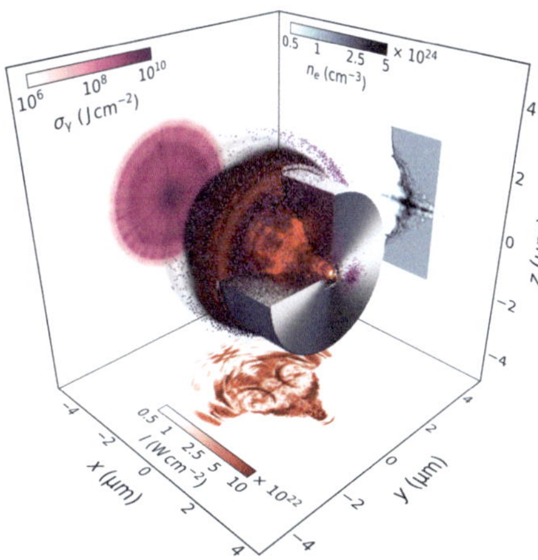

Fig. 27.3 The simulation results at ∼6.7 fs. The intensity and electron density are shown by red and gray colors, respectively; the purple dots are the γ -photons with energy 0.75 GeV. The (x, y) plane and (x, z) plane at z = 0 show the cross sections of the intensity and electron density. The γ –photon surface energy density is shown on the (y, z) plane.. Reprinted (Fig. 1) with permission from Hadjisolomou et al., (2021) as follows: Hadjisolomou et al. (2021). https://doi.org/10.1103/PhysRevE.104.015203. Copyright (2021) by the American Physical Society

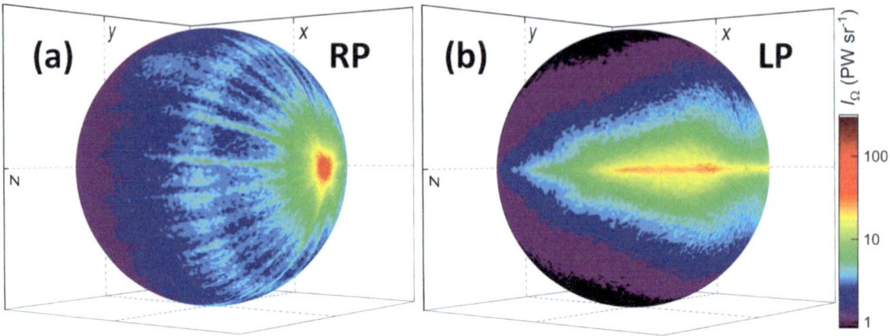

Fig. 27.4 Photon radiant intensity for **a** a radially polarized (RP) laser and **b** a linearly polarized (LP) laser (2021) Reproduced with permission from Hadjisolomou et al. Copyright 2022 Cambridge University Press

an increase of the laser pulse amplitude and to the decrease of the electron density in the interaction region, which changes the laser energy depletion length and the parameters of the gamma rays emitted.

The results presented by Ridgers et al., (2012), Nakamura et al., (2012), Lezhnin et al., (2018) correspond to the studies performed of the laser pulse interaction with high-density targets using Particle in Cell codes where the radiation friction force has been incorporated in the Landau–Lifshitz form. In addition, the quantum electrodynamics radiation friction weakening is taken into account by multiplying the radiation force by the discussed above the Gaunt form-factor $G(\chi_e)$. The plasma target inhomogeneity has been used for optimization of the electromagnetic wave energy conversion to the gamma-ray energy flash.

In the simulations, the laser pulse has the power $\mathcal{P}_{las}$ equal to 100 PW with the pulse duration of $\tau_{las} = 150$ fs.

In the tailored plasma target the density changes exponentially, $n \propto \exp(x/L)$, with the plasma inhomogeneity scale length, L, from $0.1n_{cr}$ to $350n_{cr}$ in the interval of $\approx 20\,\mu$m, and then becomes constant having a thickness of $10\,\mu$m.

Simulation results for the parameters of interest are shown in Fig. 27.5. The 100 PW laser pulse with the normalized amplitude $a_0 = 474$ interacts with the plasma target, whose density inhomogeneity is characterized by a scale length equal to $L = 2.5\,\mu$m.

Hadjisolomou et al., (2023a, 2023b) propose the addition of a constant magnetic field to the laser-solid interaction to obtain the desired γ -photon properties. The γ-ray flash spatial and spectral characteristics are obtained via quantum electrodynamic particle-in-cell simulations. When the constant magnetic field aligns with the laser magnetic field, then the γ -ray emission is significantly enhanced. Moreover, the γ -photon spatial distribution becomes collimated, approximately in the form of a disk.

Kolenatý et al., (2022) studied the interaction of laser-driven γ photons and high-energy charged particles with high-Z targets through Monte Carlo simulations. The interacting particles are taken from particle-in-cell simulations of the interaction of a

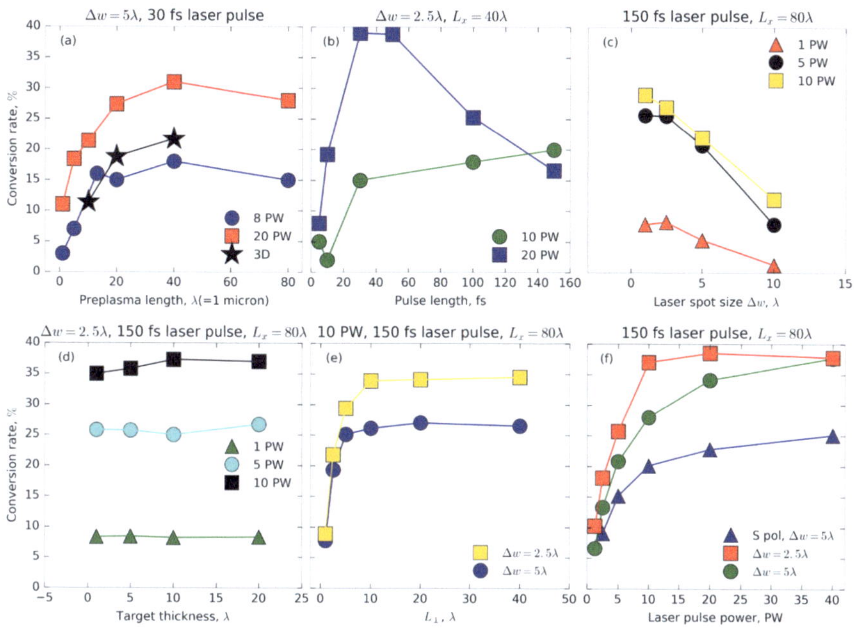

Fig. 27.5 Optimization of laser pulse and hydrogen target parameters for efficient energy conversion to gamma-rays. Each circle, triangle, star, and square correspond to a specific choice of the PIC simulation parameters. Fixing all the parameters but one and varying it, we plot the conversion efficiency dependence on this parameter. The values of the parameters fixed in the simulation are written on each of the frames. Dependence of the conversion efficiency on **a** the preplasma length, in λ; **b** the laser pulse length for two peak laser pulse power values, **c** the laser spot size for two peak laser pulse power values, **d** the high-density slab thickness for two peak laser pulse power values, **e** $L_{\perp}$, which is an effective transverse width of the pre-plasma corona, for two laser spot size values, and **f** the laser pulse power for two laser spot size values. Lezhnin et al., (2018). Lezhnin et al. (2018) Reproduced from, with the permission of AIP Publishing

tightly focused ultra-intense laser pulse with a titanium target. Figure 27.6 illustrates the setup of the gamma ray—high Z target irradiation.

Lead is chosen as the secondary high-Z target because of its high cross-section of the giant dipole resonance and electron–positron pair production. The results reveal an ultra-short, ultra-relativistic collimated positron population and their energy spectra, angular distribution, and temporal profile are found. The authors investigated the target thickness dependence of the resulting total numbers and total kinetic energies of various particle species emitted from the lead target irradiated with laser-generated γ photons and charged particles separately. We plot the charts of residual high-Z nuclides generated by irradiation of the lead target. Because of the short pulse duration, the γ photon (Fig. 27.7).

As discussed in Chap. 22.10, in the Strong Field QED limit key effect to be explored is the nonlinear Breit-Wheeler process: the conversion of high-energy

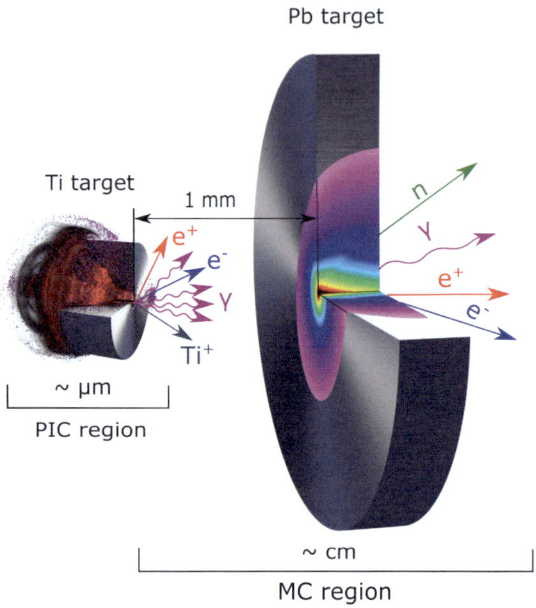

Fig. 27.6 The drawing illustrates the geometrical setup and the scales of particle-in-cell and Monte Carlo simulation regions. The laser-matter interaction resulting in the generation of high-energy particles is computed with the particle-in-cell code using a titanium target. The simulation employs the λ^3 regime, where a radially polarized near-single-cycle pulse is focused to a spherical-like intensity of ~0.5 λ FWHM spatial profile and ~3.4-fs pulse duration. An ~80-PW power laser under the tight focusing scheme used reached an intensity of 10^{25} $\mathrm{Wcm^{-2}}$. The interactions of laser-driven high-energy charged particles and γ photons with high-Z matter and the resulting products were computed with a Monte Carlo code using a lead target. The deposited energy to the lead target is represented by color mapping Kolenatý et al., (2022)

photons into electron–positron pairs through the interaction with a strong electromagnetic field. A major challenge to observing nonlinear Breit-Wheeler pair production experimentally is having a suitable source of high-energy photons. For example, in well-known experiment E144 (Bamber et al., 1999) on the electron–positron pair generation in the collision of ultra-relativistic electrons with high intensity laser pulse the photons were generated via the nonlinear Compton scattering (see left frame on Fig. 27.8). In the paper by Macleod et al., (2023), the authors outline a simple all-optical setup which efficiently generates photons through the γ -flash mechanism by irradiating a solid target with a high-power laser to provide an experimental access to fundamental strong-field quantum electrodynamics processes. Their consider the collision of these photons with a secondary laser and systematically discuss the prospects for exploring the nonlinear Breit-Wheeler process at current and next-generation high-power laser facilities. The right side of Fig. 27.8 illustrates the proposed setup.

As noted above that the gamma-flash mechanism is based on multiphoton Compton scattering in a laser-solid interaction at irradiance typically above

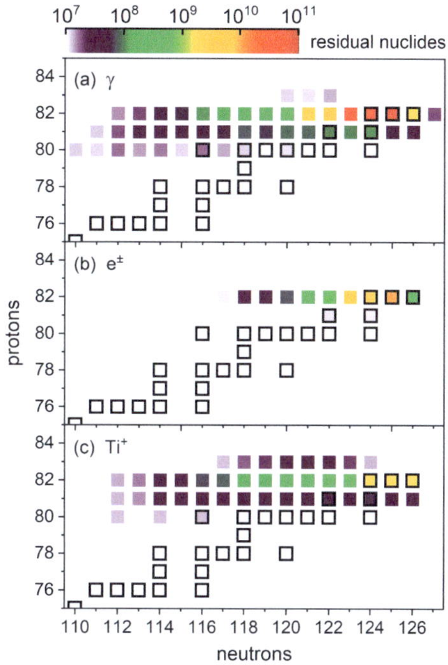

Fig. 27.7 Chart of residual high-Z nuclides generated in the lead target with a thickness of 10 cm irradiated by primary **a** γ photons, **b** electrons and positrons, and **c** titanium ions from the single laser pulse. Stable nuclides are indicated with squares with black edges Kolenatý et al., (2022)

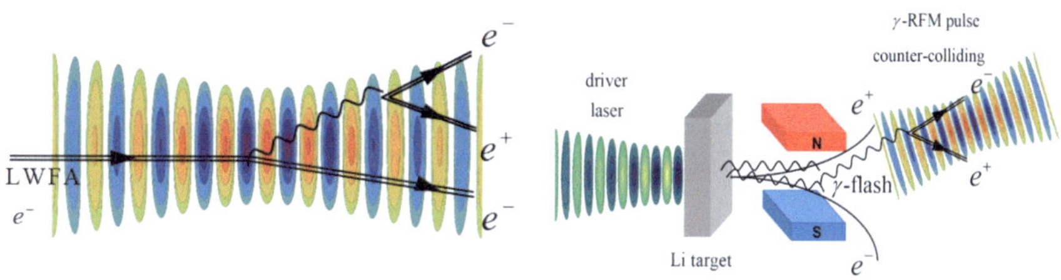

Fig. 27.8 Left hand side: Electron beam collision with counter propagating laser realized in the E144 experiment. Right hand side: Proposed experimental configuration. The total available laser energy Etotal is split into two beams: $\mathcal{E}_{total} = \mathcal{E}_{flash} + \mathcal{E}_{pairs}$. The beam with energy $\mathcal{E}_{flash}$ is used to irradiate a solid lithium target, producing high-energy photons via the so-called γ-flash mechanism. Charged secondary particles from the target are deflected to minimize background. The γ-flash photons propagate from the rear surface of the lithium target and collide with a counterpropagating secondary pulse, energy $\mathcal{E}_{pairs}$, to produce electron–positron pairs via the nonlinear Breit-Wheeler mechanism. Positrons are emitted at an angle ϑ relative to the axis of the colliding counterpropagating laser

10^{23}W/cm^2. As far it is concerning its experimental realization, at lower laser intensity gamma flash is overshadowed by simultaneously generated low-brightness Bremsstrahlung radiation. The work done by Pirozhkov et al., (2024) experimentally differentiates these two mechanisms and demonstrate a gamma flash dominant regime producing several times the number of Bremsstrahlung photons. The authors found steep growth of the gamma flash yield with the laser power and irradiance. The small source size and its high brightness may open a new way towards spatially coherent gamma rays. At high photon energy, the regime is found to be comparable in brightness to astrophysical Gamma Ray Bursts (Gehrels & Mészáros, 2012).

27.2 The Spectrum of the Electromagnetic Radiation Emitted by Relativistic Electrons

In the case of tight focusing of multi-PW laser pulses, the radiation intensity can reach values corresponding to the normalized field amplitude $a_0 = 10^3$. This field amplitude is enough high for the radiation reaction friction force to become dominant. The energy of photons emitted by ultrarelativistic electrons via the nonlinear Thomson scattering mechanism is proportional to the cube of the electron energy. The frequency spectrum of the electromagnetic radiation emitted by a relativistic electron rotating with the frequency ω along the trajectory with the radius of curvature, R_C, is given by the expression:

$$\frac{dI(\omega_\gamma, \mathcal{E})}{d\omega_\gamma} = \frac{\sqrt{3}}{2\pi}\frac{e^2}{c}\left(\frac{\mathcal{E}}{m_e c^2}\right) u(\omega_\gamma, \mathcal{E}) \int\limits_{u(\omega_\gamma, \mathcal{E})}^{\infty} K_{5/3}(x)dx, \tag{27.12}$$

where

$$u(\omega_\gamma, \mathcal{E}) = \frac{2\omega_\gamma}{3\omega}\left(\frac{m_e c^2}{\mathcal{E}}\right)^3, \tag{27.13}$$

as is well known in classical electrodynamics (Landau & Lifshitz, 1980). Here, $K_{5/3}(x)$ is the modified Bessel function. The electron energy is assumed to be $\mathcal{E} \gg \mathrm{m_e c^2}$. The function of the electron energy, $u(\omega_\gamma, \mathcal{E})$, is the ratio of the emitted photon frequency ω_γ to the critical frequency ω_c equal to

$$\omega_c = \frac{3c}{2R_C}\left(\frac{\mathcal{E}}{m_e c^2}\right)^3. \tag{27.14}$$

In the limit $u(\omega_\gamma, \mathcal{E}) \ll 1$, expression (27.12) yields

$$\frac{dI(\omega_\gamma, \mathcal{E})}{d\omega_\gamma} \approx \frac{e^2}{c}\left(\frac{\omega_\gamma R_C}{c}\right)^{1/3}, \tag{27.15}$$

while for $u(\omega_\gamma, \mathcal{E}) \gg 1$ we have

$$\frac{dI(\omega_\gamma, \mathcal{E})}{d\omega_\gamma} \approx \sqrt{\frac{3\pi}{2}\frac{e^2}{c}}\left(\frac{\mathcal{E}}{m_e c^2}\right)^2 u(\omega_\gamma, \mathcal{E})^2 \exp(-u(\omega_\gamma, \mathcal{E})). \quad (27.16)$$

The characteristic energy of a photon emitted via nonlinear Thomson scattering by the electron rotating in the electric field scales with the electron quiver energy as

$$\hbar\omega_c = \frac{3}{2}\hbar\omega_0\left(\frac{\mathcal{E}}{m_e c^2}\right)^3, \quad (27.17)$$

where ω_0 is the laser frequency. The energy of the electron quivering in plasma under the action of an electromagnetic wave with an amplitude of $a_0 = eE/m_e\omega_0 c \gg 1$ is approximately equal to $\mathcal{E} = a_0 m_e c^2$. For a laser frequency of the order of $10^{15}\mathrm{s}^{-1}$ the emitted photon energy is in the gamma ray range if $a_0 > 10^2$, which corresponds to a laser intensity higher than 10^{22} W/cm^2. Computer simulations of multi-petawatt power short laser pulse interaction with a near-critical plasma (Lezhnin et al., 2018; Nakamura et al., 2012; Ridgers et al., 2012) show that the electron energy spectra in the interval $\mathcal{E} < \mathcal{E}_m$ can be approximated by a power-law dependence. For the energy $\mathcal{E} > \mathcal{E}_m$ the spectrum has exponential form. Under the conditions of the simulations presented in Refs. (Lezhnin et al., 2018; Nakamura et al., 2012) the energy, $\mathcal{E}\Theta_m$, is of the order of 50 MeV. The energy distribution of fast electrons can be described by the dependence (see Fig. 27.9).

$$\frac{dN}{d\mathcal{E}} = K\mathcal{E}^{-\kappa}\exp\left(-\frac{\mathcal{E}}{\mathcal{E}_\mathrm{m}}\right) \quad (27.18)$$

The average spectrum of the emitted gamma-photons is given by the integral

$$J(\omega_\gamma) = \int_0^\infty \frac{dI(\omega_\gamma, \mathcal{E})}{d\omega_\gamma}\frac{dN}{d\mathcal{E}}d\mathcal{E}. \quad (27.19)$$

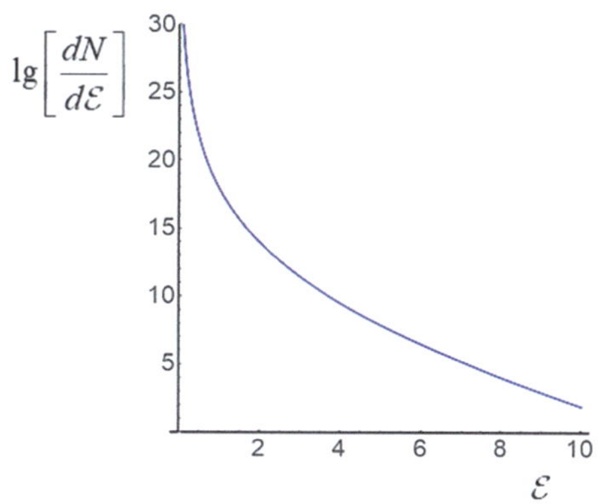

Fig. 27.9 Energy distribution of fast electrons (27.7) for $K = 1$, $\kappa = 2$, and $\mathcal{E}_\mathrm{m} = 0.2$

We analyze the spectrum of the radiation emitted in the process of nonlinear Thomson scattering of a circularly polarized electromagnetic wave, when the function $u(\omega_\gamma, \mathcal{E})$ is

$$u(\omega_\gamma, \mathcal{E}) = \frac{2\omega_\gamma}{3\omega_0}\left(\frac{m_e c^2}{\mathcal{E}}\right)^3, \tag{27.20}$$

Substituting the functions (27.12) and (27.18) into (27.19) and performing integration over $\mathcal{E}$, we obtain the average spectrum of the emitted gamma-photons

$$J(\omega_\gamma) = \frac{\sqrt{3}}{2\pi} K \frac{e^2}{c}\left(\frac{2}{3\omega_0}\right)^{(2-\kappa)/3} (m_e c^2)^{1-\kappa} F_C(\varepsilon_m, \kappa), \tag{27.21}$$

where

$$\varepsilon_m = \frac{\mathcal{E}_m}{m_e c^2}\left(\frac{3\omega_0}{2\omega_\gamma}\right)^{1/3}, \tag{27.22}$$

and

$$F_C(\varepsilon_m, \kappa) = \varepsilon_m^{-(1+\kappa)} \int_0^\infty K_{5/3}(x)\Gamma(-1-\kappa, x^{-1/3}\varepsilon_m^{-1})\, dx. \tag{27.23}$$

Here, $\Gamma(a, z)$ is the incomplete gamma function (Olver et al., 2010).

$$\Gamma(a, x) = \int_x^\infty t^{s-1} e^{-t} dt.$$

The function $F_C(\varepsilon_m, \kappa)$ can be expressed in terms of hypergeometric functions.

In the low frequency range, $\omega_\gamma \gg \omega_0(3/2)(\mathcal{E}_m/m_e c^2)^3$, using the asymptotic representation of the incomplete gamma function at $z \to 0$, i.e. at $x^{-1/3}\varepsilon_m^{-1} \to 0$,

$$\Gamma(-1-\kappa, x^{-1/3}\varepsilon_m^{-1}) = \Gamma(-1-\kappa) + \cdots \tag{27.24}$$

we obtain

$$F_C(\infty, \kappa) = \frac{2^{(\kappa-2)/3}}{1+\kappa}\Gamma\left(\frac{\kappa-1}{6}\right)\Gamma\left(\frac{\kappa+9}{6}\right). \tag{27.25}$$

This expression is valid provided $\kappa > 1$. The spectrum has the power form,

$$J(\omega_\gamma) \propto \omega_\gamma^{(2-\kappa)/3} \tag{27.26}$$

If the power index is greater than 2, $\kappa > 2$, the dependence $J(\omega_\gamma)$ is monotonically decreasing as $\omega_\gamma/\omega_0 \to \infty$. For $1 < \kappa < 2$ the function $J(\omega_\gamma)$ grows with ω_γ. If the power index is less than unity, $\kappa < 1$, we should take into account that the electron energy spectrum is truncated at the high energy end (see Eq. (27.18)).

Analyzing the asymptotic behavior of the function $F_C(\varepsilon_m, \kappa)$ given by Eq. (27.23) as $\varepsilon_m \to \infty$ for $\kappa < 1$ we find the low frequency scaling ($\omega_\gamma \ll (3\omega_0/2)(\mathcal{E}_m/m_e c^2)^3$, $\kappa < 1$)

$$J(\omega_\gamma) \approx \frac{\sqrt{3}}{2\pi} K \frac{e^2}{c} \left(\frac{\mathcal{E}_{\mathrm{m}}}{m_e c^2} \right)^{1-\kappa} 2^{-1/3} \Gamma\left(\frac{2}{3}\right) \Gamma(1-\kappa) \left(\frac{2\omega_\gamma}{3\omega_0} \right)^{1/3}. \tag{27.27}$$

Considering the limit $\varepsilon_m \to 0$, which corresponds to the high frequency range,

$$\omega_\gamma \gg \omega_0 (3/2)(\mathcal{E}_m/m_e c^2)^3,$$

we obtain

$$F_C(\varepsilon_m, \kappa) = \sqrt{\frac{\pi}{2}} \varepsilon_m^{-\kappa} \int\limits_0^\infty x^{(1+2\kappa)/6} \exp\left(-x - \frac{1}{x^{1/3} \varepsilon_m} \right) dx \propto \exp\left(- \frac{4}{3^{3/4} \varepsilon_m^{3/4}} \right), \tag{27.28}$$

i.e., the spectrum decreases exponentially with photon frequency:

$$J(\omega_\gamma) \propto \exp\left(-4 \left(\frac{2\omega_\gamma}{81\omega_0} \left(\frac{m_e c^2}{\mathcal{E}_m} \right)^3 \right)^{1/4} \right). \tag{27.29}$$

As a result, we see that the frequency spectrum has a maximum at

$$\omega_\gamma \approx 0.158 \omega_0 \left(\frac{\mathcal{E}_m}{m_e c^2} \right)^3 \tag{27.30}$$

On Fig. 27.10 we show typical frequency spectra calculated with using the expression (27.10)–(27.12) for

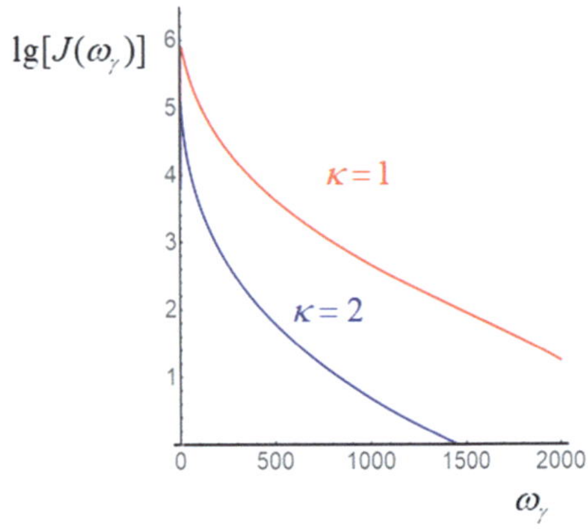

Fig. 27.10 Gamma ray frequency spectra calculated with Eqs. (27.10)–(27.12) for $K = 10^3$, $\kappa = 1$ and 2, and $\mathcal{E}_{\rm m} = 1$

27.3 Source of High Brightness X-rays Within the Relativistic Flying Mirror Concept

27.3.1 The Relativistic Flying Mirror in the Nonlinear Wake Waves

The relativistic flying mirrors produced in the interaction of short laser pulses with subcritical concentration plasmas, as discussed in Chapter 25 (see Figs. 27.11 and 27.12), can be used for developing the compact source of high brightness X-rays. Within the framework of this concept the relativistic mirrors are related to the thin layers of relativistic electrons generated in nonlinear plasma waves which are excited in the plasma by a short strong laser pulse. Surfaces of constant density in the wake of nonlinear waves have the form of paraboloids of revolution. Second, the counter-propagating laser pulse is partially reflected from relativistic mirrors. This results in a frequency upshifting of the reflected pulse, its shortening in the longitudinal direction and focusing by paraboloidal mirrors in the transverse direction, providing the intensification of reflected radiation, even though the reflection coefficient is relatively small.

The relativistic flying mirror with large enough reflection coefficient of the counter-propagating radiation, is formed in the process of the wake wave breaking. Phase velocity of the wake wave v_{ph} equals the group velocity of the driver laser pulse $v_g = \partial\omega/\partial k$, which, in accordance with the dispersion equation for the transverse (electromagnetic) waves, $\omega = \sqrt{k^2c^2 + \omega_{pe}^2}$, is given by formula

$$v_g = c\sqrt{1 - (\omega_{pe}/\omega_0)^2}.$$

In a low-density plasma, where the laser frequency is much higher than the Langmuir frequency $\omega_0 \gg \omega_{pe}$, the group velocity of electromagnetic waves is close

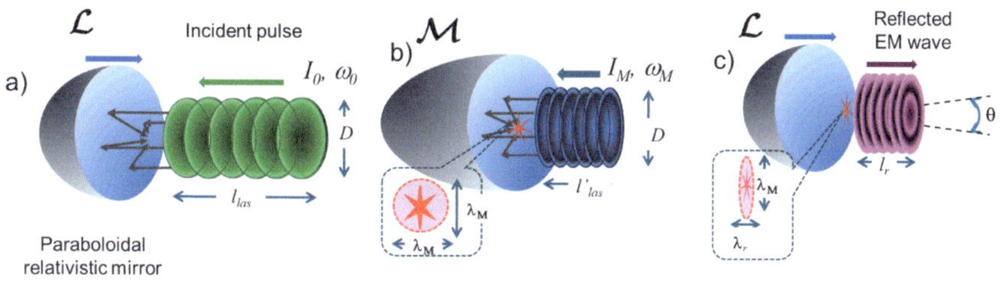

Fig. 27.11 The concept of flying relativistic mirror. **a** Laboratory frame of reference. Before reflecting from the mirror the laser pulse propagates from right to left. **b** Frame of reference where the mirror is at rest. The incident laser pulse has a length $L' \approx L_s/2\gamma_M$ and wavelength $\lambda' \approx \lambda_s/2\gamma_M$. After reflection from a parabolic mirror the radiation is focused to the region of size $\approx \lambda'$. **c** Laboratory frame of reference. The reflected pulse is compressed by a factor $4\gamma_M^2$ its wavelength becomes equal $\lambda'' \approx \lambda_s/4\gamma_M^2$. Moving the focal region has the form of an ellipsoid with the longitudinal λ'' and transverse λ' dimension, respectively. Radiation is collimated within the angle $\theta \approx 1/\gamma_M$

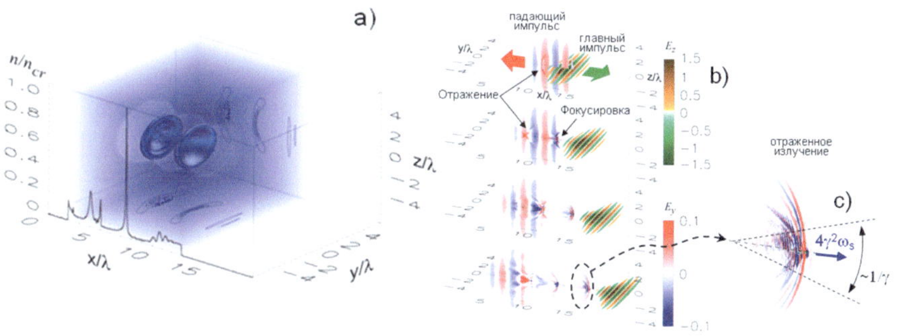

Fig. 27.12 **a** Paraboloidal thin shells of relativistic electrons in the wake wave left behind a strong short laser pulse. **b** Colliding laser pulses having mutually perpendicular polarizations. **c** Insert: the reflected pulse with shortened wavelength is compressed in the longitudinal direction and focused in transverse direction. In **a**–**c** the results of a three-dimensional particle-in-cell computer simulation are shown. Reproduced from Bulanov et al. (2003), with the permission of AIP Publishing

to the speed of light in vacuum. At the threshold of wave breaking the velocity of the electrons in the wake wave becomes comparable in magnitude with the wake wave phase velocity v_{ph}, which is equivalent to the condition for the electron energy $\mathcal{E}_e = m_e c^2 \gamma_e$ and the relativistic gamma-factor calculated for v_{ph}. The wave breaking occurs if

$$\gamma_e \geq \gamma_{ph} \equiv (1 - v_{ph}^2/c^2)^{-1/2} = \omega_0/\omega_{pe}. \quad (27.31)$$

As a result of the plasma wave breaking a singularity in the distribution of electron density occurs, which breaks the geometrical optics approximation and increases the reflection coefficient. If there is no wave-breaking, the reflectivity is exponentially weak.

In the simplest model, when the singularity is approximated by a delta function, the reflection coefficient is given by the formula (25.105). In terms of the number of photons it depends on the density of the plasma as

$$R_\delta \approx 1/2\gamma_{ph}^3 = (\omega_{pe}/\omega_0)^3/2 = (n/n_{cr})^{3/2}/2$$

in the limit $\gamma_{ph} \gg 1$. Taking into account the pulse compression in the transverse direction as a result of focusing by the parabolic mirror with the diameter D_0, we obtain for reflected wave intensity

$$I_r \approx 32 I_0 \left(\frac{D_0}{\lambda_0}\right)^2 \gamma_{ph}^3. \tag{27.32}$$

The reflected wave and power equal to $\mathcal{E}_r \approx 2\mathcal{E}_0/\gamma_{ph}$ and $\mathcal{P}_r \approx 8\mathcal{P}_0\gamma_{ph}$, respectively.

As an example, let us consider the parameters required to achieve the Schwinger limit for the electromagnetic wave reflected by a flying mirror. We assume that the laser pulse of the amplitude $a_0 = 15$ interacts with the plasma with the density of 10^{18} cm^{-3}. Associated with the wake wave relativistic gamma factor γ_{ph}, equals 45, which corresponds the frequency upshifting by a factor equal to 810. The intensity and the transverse size of the exciting wake wave laser pulse is assumed to be 4×10^{20} W/cm^2 and 40 μm. For the counter-propagating laser pulse with the intensity of 2×10^{19} W/cm^2 the beam diameter $D_0 = 40$ μm the intensity of the partially reflected from the mirror radiation appears to be of the order of 5×10^{28} W/cm^2, which corresponds to the Schwinger limit. These parameters correspond to the laser pulse energy of 10 kJ and 50 J, respectively.

If such intensity values are reached this will make possible the production of electron–positron pairs in the vacuum. We note here that the effects of the transverse inhomogeneity of the electromagnetic wave in the focal region and the effects of non-monochromaticity can significantly increase the probability of the pair electron–positron pair creation in vacuum.

The use of spherical Langmuir waves as a relativistic mirror can allow for a much higher intensification of the reflected radiation. Owing to the spherical geometry, the reflected wave with a wavelength $\approx \lambda_0/4\gamma_{ph}^2$ is focused in the volume $\lambda_0^3/48\gamma_{ph}^6$, that gives for the radiation intensity

$$I_r \approx I_0\gamma_{ph}^4 \text{ for the reflection coefficient } R_{2/3} \approx 1/\gamma_{ph}^4$$

and

$$I_r \approx I_0\gamma_{ph}^5 \text{ in the case } R_\delta \approx 1/\gamma_{ph}^3$$

27.3.2 *Experimental Demonstration of a Relativistic Flying Mirror*

The proof of principal experiments of the flying relativistic mirror concept were discussed above in Chap. 23. As the result the generation of soft X-rays with a narrow energy spectrum has been reported. Factor of frequency upshifting reached a value of 100.

In the first experiment, which demonstrated the concept of "relativistic flying mirror" in the laser plasma, two short laser pulses collided at an angle 135° in a supersonic gas jet. The configuration of the experiment is shown on Fig. 27.13. The laser used in this experiment generates pulses of radiation having a wavelength of 820 nm, with energy of 210 mJ and duration of 76 fs, i.e. it has 2.75 TW power.

The horizontally polarized driver laser pulse is focused by off-axis parabolic mirror with a focal length 645 mm in the supersonic gas jet, ejected out of the nozzle of rectangular cross section with the size of the long side equal to 10 mm, and with the short size side equal to 1.26 mm. The laser pulse propagates along the direction of the short side of the nozzle. The second pulse (the source pulse)

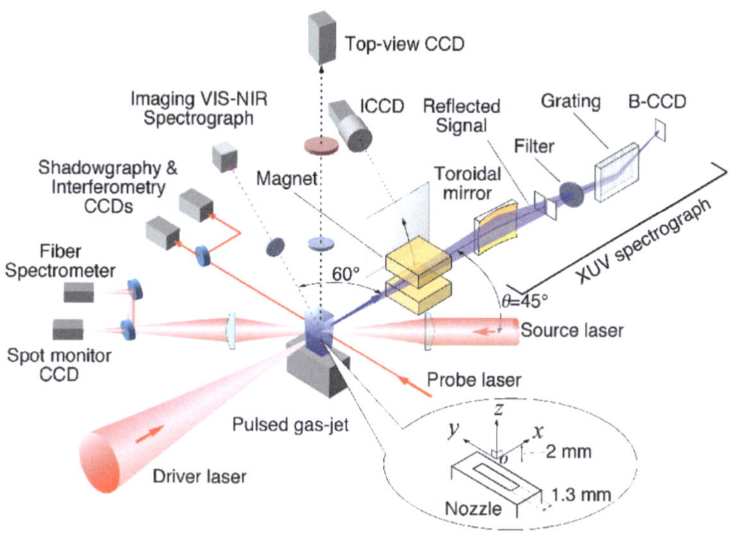

Fig. 27.13 Configuration of the proof of principal experiments for the flying relativistic mirror concept. Strong driver pulse excites a plasma wake wave. As a result of wake wave breaking relativistic mirrors are generated. The second, weaker source laser pulse that propagates at an angle $\theta = 45°$, is focused in the place where the wake wave breaks. The third probe laser pulse is used for high precision alignment. A permanent magnet is used to deflect relativistic electrons accelerated by the wake wave. The electron beam forms an image on the phosphor screen, showing the electron energy spectrum, which is measured by using a CCD (charge-coupled device). The reflected by nonlinear wake wave signal is measured by the grazing incidence spectrometer sensitive to radiation in the soft x-ray range having a wavelength from 5 to 15 nm. Reproduced from Bulanov et al., (2013) @ Uspekhi Fizicheskikh Nauk 2013

is focused by a plano-convex lens with a focal length of 200 mm, which is placed on the five-axle movable lens holder, which provides a spatial overlay with the first pulse-driver. The source pulse propagates at an angle $\theta = 45°$ to avoid laser damage by the radiation propagating in the opposite direction relative to the pulse driver. The driver pulse has the $1/e^2$ focus diameter equal to 27 microns and the intensity of 5×10^{17} W/cm^2. The source pulse energy is 6.3% of the driver's energy, which corresponds to the intensity of $\approx 10^{17}$ W/cm^2.

During the interaction of the driver laser pulse with the gas jet of gas is ionized. The driver excites the wake plasma wave which accelerates the electrons to ultra-relativistic energies. A permanent magnet is used to bend the relativistic electron beam, which forms an image on the phosphor screen, showing the electron energy spectrum measured by using a CCD (charge-coupled device).

In the configuration of the counter-crossing of two laser pulses at the angle $\theta = 45°$ the frequency of the reflected from the relativistic mirror light should be increased by a factor $\omega_r/\omega_s \approx 3.4\gamma_{ph}^2$. The plasma density in the target is approximately equal to 5×10^{19} cm^{-3}. For this plasma density the wake wave gamma factor is $\gamma_{ph} \approx 6.5$ and the theoretically predicted frequency upshifting factor is 140.

For the above written parameters of the laser radiation and the target plasma the laser pulse driver excites the wake wave with large enough amplitude to exceed the wave-breaking threshold. However, the excess over the wave-breaking threshold is not too great, and the regular wake wave structure is not destroyed. Evidence of this regime is the observation of quasi-mono-energetic spectra of accelerated by the wake wave ultrarelativistic electrons with the energy of 20 MeV. In accordance with the theory of wake-field acceleration, the results of which are presented above, the injection of electrons into the accelerating phase of the wake field is the result of the nonlinear plasma wave breaking leading to the multi-stream electron flow. At the next stage, the electrons are accelerated by the electric field of the wake wave, forming a narrow band in the vicinity of the separatrix in phase plane (longitudinal coordinate—the longitudinal momentum), which results in the characteristic shape of the electron energy spectrum. Another indication on the nonlinear character of the wake is the detection of upshifted and downshifted maxima in the frequency spectrum of the scattered at an angle 60° radiation corresponding to the stimulated backward Raman scattering (SRS). In addition, an analysis of interferogram reveals the channel formation in the plasma density, which is also in accordance with wake excitation by the electromagnetic wave instable against the relativistic self-focusing leading to redistribution of the electron density (Fig. 27.14).

Further development of the theory of relativistic flying mirror and experiments with laser light of a higher power allowed to demonstrate high-efficiency regimes, which provides opportunities for developing highly effective sources in the energy range corresponding to the hard X-rays. In this case, the laser is used with an energy of 0.5 J and a power of 15 TW. In contrast to the experiments discussed above, the realized configuration two pulses propagating in opposite directions collide, which provides several orders of magnitude higher the number of reflected photons in the spectral range corresponding to the soft X-rays. To detect the reflected radiation

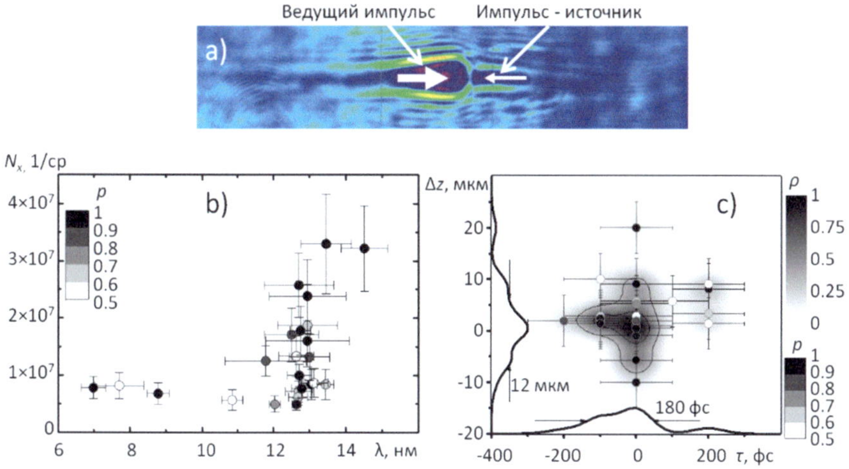

Fig. 27.14 **a** Shadowgram image of two colliding laser pulses. **b** The wavelength and the number of reflected photons measured in 24 different shots. Length of vertical bars characterizes the measurement error, and the horizontal bar shows the spectrum width. **c** The dependence of the number of reflected photons on the laser pulse delay time τ_s and on the deviation of the laser pulse in the vertical direction δz. Grayscale shows the normalized density of the reflected photons n_{ph}, in unites (sterad μm ps)$^{-1}$; the distance between the contours equals 0.33 Kando et al., (2007), Pirozhkov et al., (2007). Reproduced from Bulanov et al., (2013) @ Uspekhi Fizicheskikh Nauk 2013

the used spectrometer has a relatively large aperture for collecting the large enough number of photons from a representative solid angle. Various methods are used to perform collision pulses at a predetermined with high accuracy place in the desired time.

The observed spectrum in the extreme ultraviolet region is presented in Fig. 27.15b.

The number of photons of hard electromagnetic radiation, registered in the experiment allows us to conclude that the agreement with theoretical predictions for the reflection coefficient, defined by the formula (26.137), is achieved.

In addition to the issue on relativistic mirror reflectivity it is of particular interest to determine whether the mirror has a high-quality reflective surface, that is enough smooth. In the experiment the observed radiation in the wavelength range from 12 to 20 nm within the observation angles in the range from 9 to 17° has a sufficiently smooth spectral distribution. Since the factor of the frequency upshifting depends on the angle at which the reflected light propagates and on the local angle of reflection from the mirror surface, this shows that the radiation is reflected from a smooth surface having a curvature. This is also consistent with the predictions of the theory, which is important for further research aimed at demonstrating the sharp focusing of X-rays to reach the high limits of their intensity.

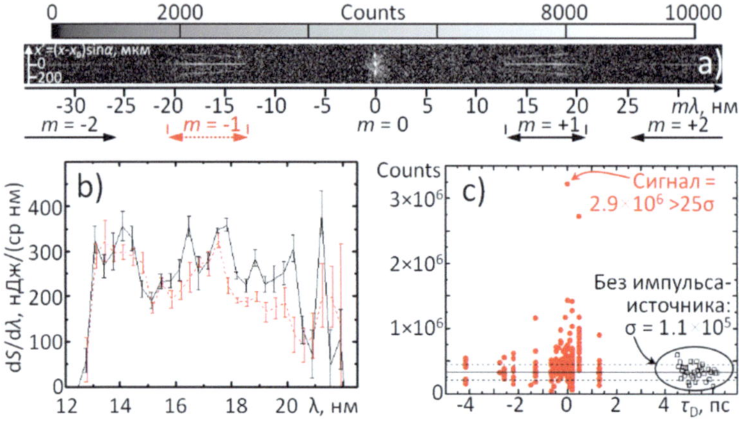

Fig. 27.15 The radiation reflected from a relativistic flying mirrors gamma factor equal to $\gamma_{ph} = 6 \pm 1$. **a** The original CCD image obtained by using normal incidence spectrograph. **b** The spectrum of the reflected radiation in the diffraction order m = 1 (black color curve) and in the order m = − 1 (red curve). **c** The dependence of the CCD in the diffraction order m = 1 on the time delay between laser pulses (red dots) and in the absence of the second pulse source (black dots) Kando et al., (2009). Reproduced from Bulanov et al., (2013) @ Uspekhi Fizicheskikh Nauk 2013

27.4 Other Schemes of High Intensity X-ray Source Based on RFM

There are known several other schemes to develop compact X-ray sources of high intensity based on the use of relativistic mirrors formed in the process of nonlinear interaction of laser radiation with a plasma, whose implementation will create exciting opportunities for experiments on nonlinear electrodynamics of continuous media in the relativistic regime. Nonlinear interaction of Langmuir waves with such regular nonlinear structures in a collisionless plasma, as the relativistic electromagnetic solitons, the electron vortices and the plasma waves are considered as a way towards generating ultra-short electromagnetic pulses of high amplitude. These X-ray pulses are generated due to a partial reflection of low-frequency electromagnetic field from a relativistic mirror. This process is accompanied by frequency upshifting due to the double Doppler effect. This effect leads also to a strong field compression and to the electromagnetic pulse formation with scale-length substantially smaller than of the original non-linear structure.

Dense electron layers accelerated to relativistic velocity by laser radiation interacting with a thin plasma foil are also considered as relativistic mirrors capable of generating beams of coherent X-ray and gamma radiation of high brightness (Kulagin et al., 2007).

References

Arber, T. D., Bennett, K., Brady, C. S., et al. (2015). Contemporary particle-in-cell approach to laser-plasma modelling. *Plasma Physics and Controlled Fusion, 57*, Article 113001.

Bamber, C., Boege, S. J., Koffas, T., Kotseroglou, T., Melissinos, A. C., Meyerhofer, D. D., Reis, D. A., Ragg, W., Bula, C., McDonald, K. T., & Prebys, E. J. (1999). Studies of nonlinear QED in collisions of 46.6 GeV electrons with intense laser pulses. *Physics Review D, 60*, 092004.

Berestetskii, V. B., Lifshitz, E. M., and Pitaevskii, L. P., (1982), *Course of Theoretical Physics, Vol. 4: Quantum Electrodynamics* (Pergamon, Oxford)

Bulanov, S. V., Esirkepov, T., & Tajima, T. (2003). Light intensification towards the Schwinger limit. *Physical Review Letters, 91*, Article 085001.

Bulanov, S. S., Schroeder, C. B., Esarey, E., & Leemans, W. P. (2013). Electromagnetic cascade in high-energy electron, positron, and photon interactions with intense laser pulses. *Physical Review A, 87*, Article 062110.

Cao, Z., Aharonian, F. A., An, Q., et al. (2021). Ultrahigh-energy photons up to 1.4 petaelectronvolts from 12 γ-ray Galactic sources. *Nature, 594*(7861), 33.

Chen, S.-Y., Maksimchuk, A., & Umstadter, D. (1998). Experimental observation of relativistic nonlinear Thomson scattering. *Nature, 396*, 653.

Courtois, C., Compant La Fontaine, A., Landoas, O., Lidove, G., Meot, V., Morel, P., Nuter, R., Lefebvre, E., Boscheron, A., Grenier, J., Aleonard, M. M., Gerbaux, M., Gobet, F., Hannachi, F., Malka, G., Scheurer, J. N., & Tarisien, M. (2009). Effect of plasma density scale length on the properties of bremsstrahlung X-ray sources created by picosecond laser pulses. *Physics of Plasmas, 16*, Article 013105.

Fan, Y., & Piran, T. (2006). Gamma-ray burst efficiency and possible physical processes shaping the early afterglow. *Monthly Notices of Royal Astronomical Society, 369*, 197.

Gales, S., Tanaka, K., Balabanski, D.L., et al. (2018). The extreme light infrastructure-nuclear physics (ELI-NP) facility: new horizons in physics with 10 PW ultra-intense lasers and 20 MeV brilliant gamma beams. *Reports on Progress in Physics, 81*, 094301.

Gehrels, N., & Mészáros, P. (2012). Gamma-ray bursts. *Science, 337*, 932.

Gizzi, L. A., Giulietti, D., Giulietti, A., Audebert, P., Bastiani, S., Geindre, J. P., & Mysyrowicz, A. (1996). Simultaneous measurements of hard X rays and second-harmonic emission in fs laser-target interactions. *Physical Review Letters, 76*, 2278.

Gong, Z., Hu, R. H., Shou, Y. R., Qiao, B., Chen, C. E., He, X. T., Bulanov, S. S., Esirkepov, T. Zh., Bulanov, S. V., & Yan, X. Q. (2017). High-efficiency gamma-ray flash generation via multiple-laser scattering in ponderomotive potential well. *Physical Review E, 95*, 013210.

Gould, R. J. (1979). Compton and synchrotron processes in spherically-symmetric non-thermal sources. *Astronomy and Astrophysics, 76*, 306.

Gu, Y.-J., Klimo, O., Bulanov, S. V., & Weber, S. (2018). Brilliant gamma-ray beam and electron–positron pair production by enhanced attosecond pulses. *Communications Physics, 1*, 93.

Hadjisolomou, P., Jeong, T. M., Valenta, P., Korn, G., and S. V. Bulanov, S. V. (2021), *Gamma-ray flash generation in irradiating a thin foil target by a single-cycle tightly focused extreme power laser pulse*, Physical Review E 104, 015203

Hadjisolomou, P., Jeong, T. M., Kolenaty, D., Macleod, A. J., Olšovcová, V., Versaci, R., Ridgers, C. P., & Bulanov, S. V. (2023a). Gamma-flash generation in multi-petawatt laser–matter interactions. *Physics of Plasmas, 30*, Article 093103.

Hadjisolomou, P., Jeong, T. M., Valenta, P., Kolenatý, D., Versaci, R., Olšovcová, V., & Bulanov, S. V. (2022). Gamma-ray flash in the interaction of a tightly focused single-cycle ultra-intense laser pulse with a solid target. *Journal of Plasma Physics, 88*, 905880104.

Hadjisolomou, P., Shaisultanov, R., Jeong, T. M., et al. (2023b). The effect of ultrastrong magnetic fields on laser-produced Gamma-ray flashes. *Physical Review Research, 5*, Article 043153.

Henderson, A., Liang, E., Riley, N., Yepes, P., Dyer, G., Serratto, K., & Shagin, P. (2014). Ultra-intense gamma-rays created using the Texas Petawatt Laser. *High Energy Density Physics, 12*, 46.

Kando, M., Fukuda, Y., Pirozhkov, A. S., Ma, J., Daito, I., Chen, L.-M., Esirkepov, T. Z., Ogura, K., Homma, T., Hayashi, Y., Kotaki, H., Sagisaka, A., Mori, M., Koga, J. K., Daido, H., Bulanov, S. V., Kimura, T., Kato, Y., & Tajima, T. (2007). Demonstration of laser-frequency upshift by electron-density modulations in a plasma Wakefield. *Physical Review Letters, 99*, Article 135001.

Kando, M., Pirozhkov, A. S., Kawase, K., Esirkepov, T. Z., Fukuda, Y., Kiriyama, H., Okada, H., Daito, I., Kameshima, T., Hayashi, Y., Kotaki, H., Mori, M., Koga, J. K., Daido, H., Faenov, A. Y., Pikuz, T., Ma, J., Chen, L.-M., Ragozin, E. N., … Bulanov, S. V. (2009). Enhancement of photon number reflected by the relativistic flying mirror. *Physical Review Letters, 103*, Article 235003.

Kolenatý, D., Hadjisolomou, P., Versaci, R., Jeong, T. M., Valenta, P., Olšovcová, V., & Bulanov, S. V. (2022). Electron-positron pairs and radioactive nuclei production by irradiation of high-Z target with gamma-photon flash generated by an ultra-intense laser in the lambda-cube regime. *Physical Review Research, 4*, Article 023124. https://doi.org/10.1103/PhysRevResearch.4.023124

Kulagin, V. V., Cherepenin, V. A., Hur, M. R., & Suk, H. (2007). Flying mirror model for interaction of a super-intense nonadiabatic laser pulse with a thin plasma layer: Dynamics of electrons in a linearly polarized external field. *Physics of Plasmas, 14*, Article 113101.

Landau, L. D., & Lifshitz, E. M. (1980). *Course of theoretical physics, Vol. 2: The classical theory of fields*. Pergamon.

Lezhnin, K. V., Sasorov, P. V., Korn, G., & Bulanov, S. V. (2018). High power gamma flare generation in multi-petawatt laser interaction with tailored targets. *Physics of Plasmas, 25*, Article 123105.

MacLeod, A. J., Hadjisolomou, P., Jeong, T. M., & Bulanov, S. V. (2023). All-optical nonlinear Breit-Wheeler pair production with γ-flash photons. *Physics Review A, 107*, 012215.

Mészáros, P., & Rees, M. J. (1992). High-entropy fireballs and jets in gamma-ray burst sources. *Monthly Notices of Royal Astronomical Society, 257*, 29P.

Mourou, G., Chang, Z., Maksimchuk, A., Nees, J., Bulanov, S. V., Bychenkov, V. Y., Esirkepov, T. Z., Naumova, N. M., Pegoraro, F., & Ruhl, H. (2002). On the design of experiments for the study of relativistic nonlinear optics in the limit of single-cycle pulse duration and single-wavelength spot size. *Plasma Physics Reports, 28*, 12.

Nakamura, T., Koga, J. K., Esirkepov, T. Zh., Kando, M., Korn, G., & Bulanov, S. V. (2012). High power Gamma-ray flash generation in the ultra intense laser-plasma interaction. *Physical Review Letters, 108*, 195001.

Olver, F. W. J., Lozier, D. W., Boisvert, R. F., & Clark, C. W. (2010). *NIST Handbook of Mathematical Functions*. Cambridge University Press.

Piran, T. (2005). The physics of gamma-ray bursts. *Review Modern Physics, 76*, 1143.

Pirozhkov, A. S., Ma, J., Kando, M., Esirkepov, T. Zh., Fukuda, Y., Chen, L.-M., Daito, I., Ogura, K., Homma, T., Hayashi, Y., Kotaki, H., Sagisaka, A., Mori, M., Koga, J. K., Kawachi, T., Daido, H., Bulanov, S. V., Kimura, T., Kato, Y., & Tajima, T. (2007). Frequency multiplication of light back-reflected from a relativistic wake wave. *Physics of Plasmas, 14*, 123106.

Pirozhkov, A. S., Sagisaka, A., Ogura, K., Vishnyakov, E. A., Shatokhin, A. N., Armstrong, C. D., Esirkepov, T. Z., Izquierdo, B. G., Pikuz, T. A., Hadjisolomou, P., & Alkhimova, M. A. (2024). Demonstration of the brightest nano-size gamma source. arXiv preprint arXiv:2410.06537

Ridgers, C. P., Brady, C. S., Duclous, R., Kirk, J. G., Bennett, K., Arber, T. D., Robinson, A. P. L., & Bell, A. R. (2012). Dense electron-positron plasmas and ultraintense γ rays from laser-irradiated solids. *Physical Review Letters, 108*, Article 165006.

Ruffini, R., Bianco, C. L., Fraschetti, F., Xue, S. S., & Chardonnet, P. (2001). On the Interpretation of the Burst Structure of Gamma-Ray Bursts. *Astrophysical Journal, 555*, L113.

Sarri, G., Corvan, D. J., Schumaker, W., Cole, J. M., Di Piazza, A., Ahmed, H., Harvey, C., Keitel, C. H., Krushelnick, K., Mangles, S. P. D., Najmudin, Z., Symes, D., Thomas, A. G. R., Yeung, M., Zhao, Z., & Zepf, M. (2014). Ultrahigh brilliance multi-MeV gamma-ray beams from nonlinear relativistic Thomson scattering. *Physical Review Letters, 113*, Article 224801.

Stark, D. J., Toncian, T., & Arefiev, A. V. (2016). Enhanced multi-mev photon emission by a laser-driven electron beam in a self-generated magnetic field. *Physical Review Letters, 116*, 185003.

Vranic, M., Grismayer, T., Fonseca, R. A., & Silva, L. O. (2016). Electron–positron cascades in multiple-laser optical traps. *Plasma Physics for Controlled Fusion, 59*, Article 014040.

Vyskocil, J., Klimo, O., & Weber, S. (2018). Simulations of bremsstrahlung emission in ultra-intense laser interactions with foil targets. *Plasma Physics for Controlled Fusion, 60*, Article 054013.

Wu, D., Yu, W., Zhao, Y. T., Fritzsche, S., & Heet, X. T. (2018). Characteristics of X/γ-ray radiations by intense laser interactions with high-Z solids: The role of bremsstrahlung and radiation reactions. *Matter and Radiation at Extremes, 3*, 293.

Yan, W. C., Fruhling, G., Golovin, D., et al. (2017). High-order multiphoton Thomson scattering. *Nature Photonics, 11*, Article 514520.

Chapter 28
Super Strong Field Limits

Abstract Superstrong electro-magnetic field science is an area attractive for fundamental theoretical physics studies promising novel discovering novel physics and wide range of application.

Keywords Photon-photon scattering · Nonlinear EM waves in QED vacuum · Electron–positron pair creation · Cherenkov radiation in vacuum

According to a widely accepted point of view, classical and quantum electrodynamics represent well-understood and fully complete areas of science whereas one of the forefronts of fundamental physics is in string theory, the predictions of which will bring experimental physics to novel, higher than ever, levels. In the second half of the twentieth century and in the beginning of the twenty-first century, humankind witnessed enormous progress in elementary particle physics with the formulation and experimental proof of the Standard Model, in cosmology with the observational evidence of dark matter and dark energy in the Universe, and observation of gravitation waves. At the same time, classical physics continued its vigorous development. It resulted in the understanding of the nature of chaos in simple mechanical systems with its relation to the turbulence problem and in the achievements in nonlinear wave theory describing the solitons, which led to the formation of a novel area in mathematical physics, thus demonstrating that there cannot be a fully complete area of science. This fact has been known by prominent scientists (Ginzburg, 1963; Zee, 2010). In the beginning of the twenty-first century, there appeared a demand to understand the cooperative behaviour of relativistic quantum systems and to probe the quantum vacuum. It was realized that the vacuum probing becomes possible by using high power lasers (Mourou et al., 2006; Marklund & Shukla, 2006; Di Piazza et al., 2012; King & Heinzl, 2016; Baumann et al., 2019; Zhang et al., 2021; Gonoskov et al., 2022; Yu et al., 2024; Sarri et al., 2025). With increasing of the laser intensity, we shall encounter novel physical processes such as the radiation reaction dominated regimes and then the regime of the quantum electrodynamics (QED) processes. Near the intensity, corresponding to the QED critical electric field, light can generate electron–positron pairs from vacuum and the vacuum begins to act nonlinearly.

Y. Kato et al., *High Power Laser and Plasma Science*, Springer Series on Atomic, Optical, and Plasma Physics 130, https://doi.org/10.1007/978-3-032-08930-4_28

There are several ways to achieve the higher intensity required for revealing these processes.

One of the methods is based on the simultaneous laser frequency upshifting and pulse compression. These two phenomena were demonstrated within the considered above the Relativistic Flying Mirror (RFM) concept, which uses the laser pulse compression, frequency up-shift, and focusing by counter-propagating breaking plasma waves—relativistic parabolic mirrors (see Sect. 3.7 and Bulanov et al., 2003, 2013a, 2013b). In the proof of principle experiments for this concept a narrow band XUV generation was demonstrated (Kando et al., 2007; Pirozhkov et al., 2007) with the high photon number (Kando et al., 2009). According to Bulanov et al., (2003), Koga et al., (2012), Valenta et al., (2020), Jeong et al. (2021) with the future high power laser facilities the QED limits in the electromagnetic wave interaction with matter will be approached.

Another way was demonstrated in large scale experiments (Bamber et al., 1999), where an approximately 50 GeV bunch of electrons from SLAC accelerator of charged particles interacted with a counter-propagating laser pulse of the intensity of approximately of 5×10^{17} W/cm^2. Gamma-rays with the energy of 30 GeV produced in the multi-photon Compton scattering then subsequently interacted with the laser light creating the electron–positron pairs via the nonlinear Breit-Wheeler process.

Probing phenomena predicted by the nonlinear QED theory, e.g. via the Breit-Wheeler process, is one of the central goals of the high-power lasers which will come into operation in several years. In contrast with the aforementioned approach, which has used the laser light interaction with the ultrarelativistic electrons accelerated by the standard accelerator of charged particles, in relativistic laser plasmas the high energy electron beams generated by nonlinear plasma waves, e.g. via the LWFA mechanism, are expected to interact with electromagnetic radiation leading to a copious production of gamma photons, which then produce the electron–positron avalanches. The prospects of implementation of the nonlinear QED regimes in studying the properties of relativistic continuous media and high energy particles has resulted in fast growing scientific directions related to the QED laser plasmas.

28.1 Photon-Photon Scattering

In contrast to classical physics, in quantum electrodynamics the vacuum is not a void substance. It in the presence of an external electromagnetic field shows nonlinear optical properties possessing the properties of a dispersive, absorptive and dichroic medium. These properties are based on the photon-photon interaction in vacuum. In classical electrodynamics, photons in vacuum do not interact due to the linear nature of Maxwell's equations. In quantum electrodynamics, photon-photon interaction occurs via the generation of virtual electron–positron pairs as shown in Fig. 28.1.

In the relatively low energy limit, for photon energies below the electron rest-mass-energy, $\mathcal{E}_\gamma = \hbar\omega_\gamma < m_e c^2$, the total scattering cross-section for the non-polarized photons equals (Karplus & Neuman, 1951; Berestetskii et al., 1982)

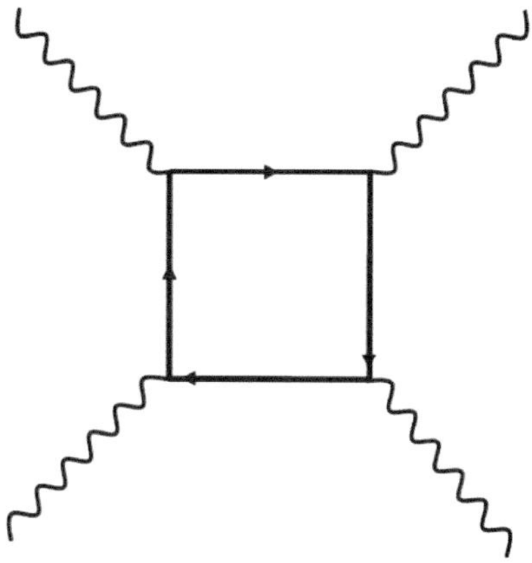

Fig. 28.1 Feynman diagram corresponding to photon-photon scattering in via creation of virtual electron–positron pairs in vacuum

$$\sigma_{\gamma\gamma} = \frac{973}{10125\pi}\alpha^2 r_e^2 \left(\frac{\hbar\omega_\gamma}{m_e c^2}\right)^6 = 1.29 \times 10^{-31}\left(\frac{\hbar\omega_\gamma}{m_e c^2}\right)^6 \mathrm{cm}^2. \tag{28.1}$$

It reaches the maximum at $\hbar\omega_\gamma \approx 1.5 m_e c^2$, where it is approximately equal to $10^{-30}\mathrm{cm}^2$. For the ultrarelativistic energy photons, the scattering cross-section (Akhieser et al., 1936)

$$\sigma_{\gamma\gamma} = 4.7\alpha^2 r_e^2 \left(\frac{m_e c^2}{\hbar\omega_\gamma}\right)^2 = 1.96 \times 10^{-29}\left(\frac{m_e c^2}{\hbar\omega_\gamma}\right)^2 \mathrm{cm}^2. \tag{28.2}$$

Cross section $\sigma_{\gamma\gamma}$ dependence on the photon energy $\hbar\omega/m_e c^2$ is shown in Fig. 28.2.

To find the number of the photon-photon scattering events we estimate the photon number density n_γ in the electromagnetic pulse of the amplitude E and frequency ω_γ. It is equal to the ratio of the electromagnetic field energy density to the photon energy $\hbar\omega_\gamma$. Neglecting numerical coefficients, which depend on the specific geometry of the electromagnetic field, we obtain

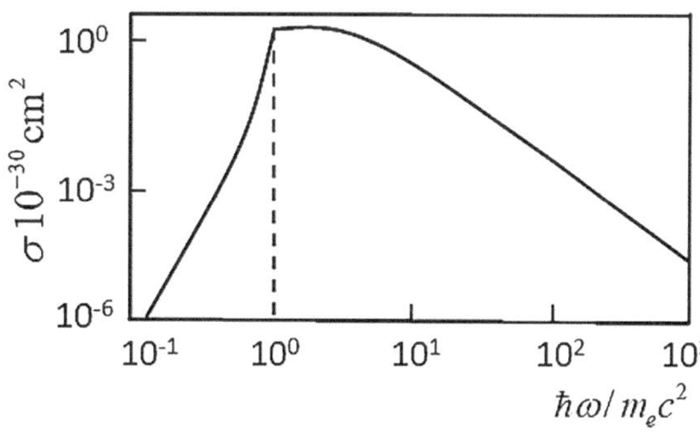

Fig. 28.2 Cross section $\sigma_{\gamma\gamma}$ dependence on the photon energy $\hbar\omega/m_e c^2$

$$n_\gamma = \frac{E^2}{4\pi \hbar\omega_\gamma} = \frac{4\pi a_\gamma^2}{\lambda_\gamma \lambda_C r_e}, \tag{28.3}$$

where

$$a_\gamma = eE/m_e\omega_\gamma c \tag{28.4}$$

and $\lambda_\gamma = 2\pi c/\omega_\gamma$. The photon scattering rate is $n_\gamma c\sigma_{\gamma\gamma}$ and the number of the scattering events of single photon over the distance $\lambda_\gamma/2\pi = \lambda_\gamma$ is given by $n_\gamma \lambda_\gamma \sigma_{\gamma\gamma}$. The total number of photons in the volume $(\lambda_\gamma/2\pi)^3$ is approximately equal to $a_\gamma^2(\lambda_\gamma^2/\lambda_C r_e)$. Multiplying the number of the scattering events per single photon by the total number of photons we obtain the total number of scatterings per $\lambda_\gamma^3\omega_\gamma^{-1}$ 4-volume

$$N_{\gamma\gamma} = a_\gamma^4\left(\frac{\lambda_\gamma}{\lambda_C r_e}\right)^2 \sigma_{\gamma\gamma} = \frac{973}{10125\pi} a_\gamma^4 \alpha^2 \left(\frac{\hbar\omega_\gamma}{m_e c^2}\right)^4. \tag{28.5}$$

Taking into account the relationship between the normalized amplitude a_γ and the photon frequency ω_γ given by Eq. (28.4) we can rewrite Eq. (28.5) as follows

$$N_{\gamma\gamma} = \frac{973}{10125\pi}\alpha^2\left(\frac{E}{E_S}\right)^4 \approx 0.0306\,\alpha^2\left(\frac{I}{I_S}\right)^2, \tag{28.6}$$

where

$$E_S = \frac{m_e^2 c^3}{e\hbar}$$

is the critical electric field of quantum electrodynamics, which is also called the "Schwinger field" and/or "Sauter-Schwinger field". The corresponding to this field intensity of electromagnetic radiation equals

$$I_S = \frac{m_e^4 c^7}{4\pi e^2\hbar^2} = 2.36 \times 10^{29}\,\mathrm{W/cm^2};$$

see Eqs. (19.23), (19.24), and (19.27). As we see, the photon scattering probability per $\lambda_\gamma^3\omega_\gamma^{-1}$ 4-volume does not depend on the frequency (and on the wavelength) of the photons. According to the relationship (28.6) it is substantially smaller than unity for the electromagnetic wave intensity below I_S. Apparently, in the case of substantially long electromagnetic pulse focused into the spot of the size r_f larger than the wavelength, the scattering probability is higher than that given by (28.6) by the factor equal to the square of the number of wave periods per pulse, N_p, which is given by Eq. (28.10), times the square of the pulse waist, i.e. by $N_p^2(r_f/\lambda_\gamma)^2$. Here it is assumed that the electromagnetic pulse length is smaller than the Rayleigh length, $2\pi r_f^2/\lambda_\gamma$, i.e. $N_p^2(r_f/\lambda_\gamma)^2 \leq 4\pi^2(r_f/\lambda_\gamma)^6$.

The photon-photon scattering has been observed in the experiments conducted at the Large Hadron Collider in ultraperipheral collisions of heavy ions (see ATLAS Collaboration (2017), d'Enteria and da Silveira (2013) as well as Zha et al., 2023; Zha & Tang, 2021; Baur et al., 2002).

At the energy of colliding photons exceeding $m_e c^2$, i.e.

$$\hbar\omega_1\ \hbar\omega_2 > m_e^2 c^4, \tag{28.7}$$

where $\hbar\omega_1$ and $\hbar\omega_2$ are the photon energies, electron–positron pair creation in the Breit and Wheeler (1934) process

$$\gamma + \gamma \to e^+ + e^- \tag{28.8}$$

becomes possible. The cross section of the electron–positron pair production is given by Akhiezer and Berestetskii (1965), Gould and Schreder (1967), Nikishov (1962)

$$\sigma_{\gamma\gamma\to e^+e^-} = \frac{1}{2}\pi r_e^2(1-\beta_e^2)\left\{(3-\beta_e^4)\log\left(\frac{1+\beta_e}{1-\beta_e}\right) - 2\beta_e(2-\beta_e^2)\right\} \tag{28.9}$$

where $r_e = e^2/m_e c^2 = 2.8\times 10^{-13}$ cm is the classical electron radius and

$$\beta_e = \sqrt{1-\frac{1}{s}} \tag{28.10}$$

with

$$s = \frac{\hbar^2\omega_1\omega_2}{2m_e^2c^4}(1-\cos\vartheta) \tag{28.11}$$

which is the square of the gamma-quanta energy in the center of mass system and ϑ is the angle between the photon momenta. For the pair creation to occur, the parameter s must be above unity, $s > 1$. The cross section $\sigma_{\gamma\gamma\to e^+e^-}$ dependence on the photon energy $\hbar\omega/m_e c^2$ assuming $\omega_1 = \omega_2 = \omega$ and $\vartheta = \pi$ is shown in Fig. 28.3. Near the threshold $s \to 1$, the electron–positron creation cross section has a form

$$\sigma_{\gamma\gamma\to e^+e^-} \approx \pi r_e^2\sqrt{s-1} \tag{28.12}$$

The cross section tends to zero at $s \to 1$. It reaches the maximum $\sigma_{\gamma\gamma\to e^+e^-,m} = 5.997$ at $\hbar\omega/m_e c^2 \approx 1.1945$.

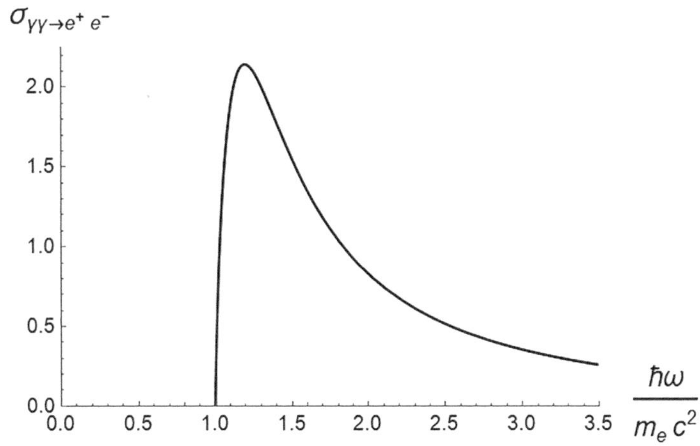

Fig. 28.3 Cross section $\sigma_{\gamma\gamma\to e^+e^-}$ dependence on the photon energy $\hbar\omega/m_e c^2$ assuming $\omega_1 = \omega_2 = \omega$ and $\vartheta = \pi$

28.2 Invariant 'Photon Mass' in a Strong Electromagnetic Field

The photon-photon scattering results in the vacuum polarization when the vacuum acquires the properties of the medium with the refraction index depending on the electromagnetic wave intensity. In the case of classical electrodynamics described by the Maxwell equations, the electromagnetic wave frequency and wave vector are related to each other as $\omega^2 = \boldsymbol{k}^2 c^2$, i.e. the velocity of the electromagnetic wave propagation is equal to the speed of light in vacuum. This fact corresponds to zero photon mass. The vacuum polarization leads to the electromagnetic wave propagation velocity below the speed of light in vacuum, which modifies the dispersion equation for the wave frequency and wave number. In the case of the electromagnetic wave colliding with the crossed electromagnetic field in a quantum vacuum, the dispersion equation is (Narozhny, 1969; Ritus, 1970)

$$\omega^2 = \boldsymbol{k}^2 c^2 - \mu_{||,\perp}^2 c^2/\hbar^2 \tag{28.13}$$

Here, $\mu_{||,\perp}$ is the 'invariant photon mass'. The signs $||, \perp$ in a subscript of $\mu_{||,\perp}$ denote the parallel and perpendicular polarizations of the colliding electromagnetic waves. The invariant mass depends on the photon frequency (it is the photon energy expressed in terms of the quantum parameter χ_γ). According to (Ritus, 1970) the square of the photon mass is given by

$$\mu_{||,\perp}^2 = -\frac{\alpha m_e^2}{6\sqrt{\pi}} \int_1^\infty du \frac{8u + 1 \mp 3}{\zeta u \sqrt{u(u-1)}} \frac{df}{d\zeta} \tag{28.14}$$

with

$$\zeta = \left(\frac{4u}{\chi_\gamma}\right)^{2/3} \tag{28.15}$$

and

$$f(\zeta) = i\int_0^\infty dt \exp\left[-i\left(\zeta t + \frac{t^3}{3}\right)\right] = \pi[iAi(\zeta) + Gi(\zeta)]. \tag{28.16}$$

Here Ai(ζ) and Gi(ζ) are the Airy function and the inhomogeneous Airy function (it is also known as the Scorer function) (Olver et al., 2010), respectively. The quantum parameter χ_γ for the photon counter-colliding with the crossed field of the amplitude E is equal to

$$\chi_\gamma = \frac{E}{E_S}\frac{\hbar(\omega + k_x c)}{m_e c^2} \tag{28.17}$$

where k_x is the x-component of the wave vector. Here we assume here that the Poynting vector of the strong low frequency wave $c\boldsymbol{E} \times \boldsymbol{B}/4\pi$ is directed in the negative direction along the x-axis. For definiteness we set $\boldsymbol{E} = E_0\boldsymbol{e}_z$ and $\boldsymbol{B} = E_0\boldsymbol{e}_y$. The high frequency electromagnetic wave propagates in the positive direction along the x-axis.

Using analytical properties of the functions Ai(ζ) and Gi(ζ) we can obtain the integral in the r.h.s. of Eq. (28.14) in the limit $\chi_\gamma \to 0$. These yields

$$\mu^2_{||,\perp} = -\alpha m_e^2\left\{\frac{11 \pm 3}{90\pi}\left[\chi_\gamma^2 + \frac{1}{3}\chi_\gamma^4\right] + i\sqrt{\frac{3}{2}}\frac{3 \pm 1}{16}\exp\left(-\frac{8}{3\chi_\gamma^2}\right)\right\}. \tag{28.18}$$

For $\chi_\gamma >> 1$ the square of the photon mass is

$$\mu^2_{||,\perp} = \alpha m_e^2\left[\frac{5 \pm 1}{28\pi^2}\sqrt{3}\Gamma^4\left(\frac{2}{3}\right)(1 - i\sqrt{3})(3\chi_\gamma)^{2/3}\right]. \tag{28.19}$$

Here, $\Gamma(x)$ is the Euler gamma function (Olver et al., 2010). From these expressions, we see that the nonlinear QED vacuum is dissipative and dispersive optical medium. The imaginary part of the invariant mass square, $\mu^2_{||,\perp}$, gives a probability of the electron–positron pair creation in the multiphoton Breit-Wheeler process. The dispersion effects in the limit $\chi_\gamma \ll 1$ allow for the electromagnetic solitons to occur (see Bulanov et al. 2020; Pegoraro & Bulanov, 2021). As noted by Narozhny (1969) and Ritus (1970), for large χ_γ, at $\alpha\chi_\gamma^{2/3} \approx 1$, the photon mass becomes of the order of the electron mass. In this limit, a new physics may come into play (Yakimenko et al., 2019; Fedotov et al., 2023).

Figure 28.4 shows the real and imaginary parts of the square of the photon invariant mass

$$\mu^2_{||,\perp} = \mathcal{R}e\{\mu^2_{||,\perp}(\chi_\gamma)\} + i\mathcal{I}m\{\mu^2_{||,\perp}(\chi_\gamma)\},$$

given by Eq. (28.14) versus the parameter χ_γ within the interval $0 < \chi_\gamma < 0$. The real and imaginary parts of $\mu_{||,\perp}$ tend to zero at $\chi_\gamma \to 0$ according to Eq. (28.18). The real parts vanish at $\chi_\gamma \to 16.829$ and $\chi_\gamma \to 14.395$ for $\mu_{||}$ and $\mu_\perp$, where $\mathcal{I}m\{\mu^2_{||}(16.829)\} = 12.93$ and $\mathcal{I}m\{\mu^2_\perp(14.395)\} = 14.26$, respectively.

In the limit of small χ_γ, substituting the real part of the photon mass given by the first two terms in Eq. (28.18) into the dispersion Eq. (28.13) we obtain for the relationship between the electromagnetic wave frequency and the wave number

$$\omega^2 - k^2 + \kappa_{||,\perp}\left[(\omega + k_x)^2 + \frac{1}{3}W_0^2(\omega + k_x)^4\right] + i\rho_{||,\perp}(\omega + k_x)\exp\left(-\frac{8}{3W_0(\omega + k_x)}\right) = 0, \tag{28.20}$$

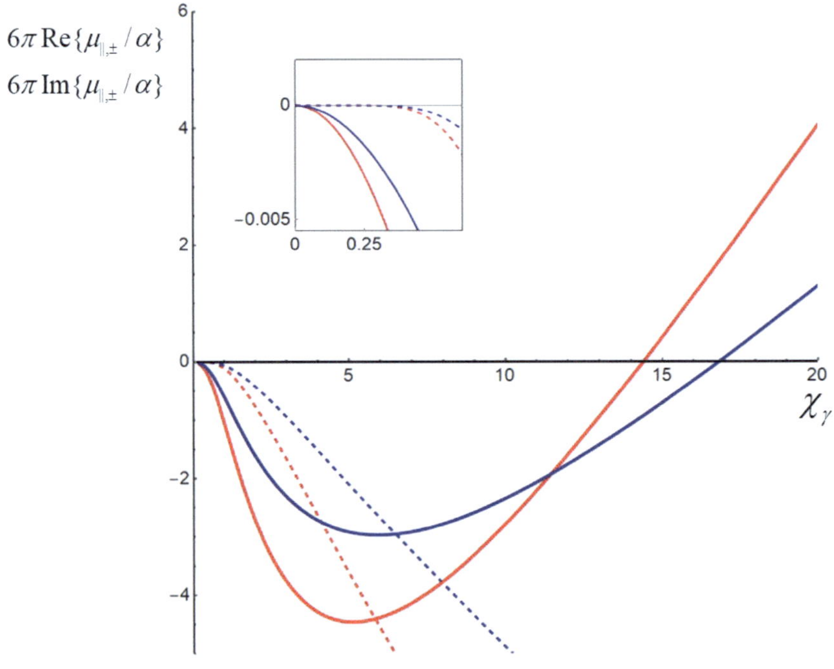

Fig. 28.4 Real and imaginary parts of the square the electron invariant mass $\mu^2_{||,\perp}$ versus χ_γ. Real parts, $\mathcal{R}e\{\mu^2_{||,\perp}(\chi_\gamma)\}$, are shown by solid and imaginary parts, $\mathcal{I}m\{\mu^2_{||,\perp}(\chi_\gamma)\}$, are shown by dashed curves. Blue corresponds to parallel polarization, and red corresponds to perpendicular polarization. The square the electron invariant mass is normalized on $\alpha/6\pi$. The inset shows a close-up at low values of χ_γ. See for details the paper by Ritus (1970)

where

$$\kappa_{||,\perp} = \frac{\alpha(11 \pm 3)}{90\pi} W_0^2, \quad \rho_{||,\perp} = \alpha W_0 \sqrt{\frac{3}{2}} \frac{(3 \pm 1)}{16},$$

and $W_0 = E_0/E_S$ is the electric field of the cross field electromagnetic wave normalized on E_S, $k^2 = k_x^2 + k_y^2$, the wave number and frequency are normalized on $\lambda_C^{-1} = m_e c/\hbar$ and $c/\lambda_C = m_e c^2/\hbar$, i.e. the wave propagation velocity is normalized on c.

The term k_y^2 in the left hand side of Eq. (28.20) describes the diffraction effects. As we may see from Eq. (28.20) the co-propagating with the crossed field wave electromagnetic wave has the velocity equal to the speed of light in vacuum. Assuming $k_y = 0$ we find that the dispersion Eq. (28.20) has two solutions. The first, $\omega = -k_x$, corresponds to the electromagnetic wave co-propagating with the crossed field. The second is a solution of the equation

$$\omega(1 + \kappa_{||,\perp}) - k_x(1 - \kappa_{||,\perp}) + = i\rho_{||,\perp} \exp\left(-\frac{8}{3W_0(\omega + k_x)}\right). \tag{28.21}$$

Representing the wave frequency as

$$\omega = \omega' + i\omega'' \tag{28.22}$$

and assuming that the imaginary part of the frequency is much less that the real part, $\omega'' \ll \omega'$, we find that

$$\omega' = k_x \frac{1 - \kappa_{||,\perp}}{1 + \kappa_{||,\perp}} \tag{28.23}$$

and

$$\omega'' = \rho_{||,\perp} \exp\left(-\frac{4}{3W_0\omega'}\right). \tag{28.24}$$

The rate of the electron–positron pair production via the nonlinear Breit-Wheeler process is proportional to the imaginary part of the frequency given by (28.24) (see Nikishov & Ritus, 1970).

The counter-propagating wave phase and group velocities are equal to each other. They are smaller than the speed of light in vacuum being approximately equal to

$$\beta_{||,\perp} = 1 - \alpha \frac{11 \pm 3}{45\pi} W_0^2 + \cdots, \tag{28.25}$$

where $\beta_{||,\perp}$ is the normalized on c wave velocity. We note that in the limit of relatively low wave amplitude (when $\chi_\gamma \ll 1$) and frequency the QED vacuum is dispersionless medium, i.e. the phase and group velocity are equal. This fact has two consequences:

(1) as a result of counter-propagating waves the wave phase changes proportionally to the square of the crossed field amplitude, $(E_0/E_S)^2$ (see Tommasini et al., 2008; King & Heinzl, 2016; Gies et al., 2018; Rosso et al., 2021; MacLeod et al., 2023a, 2023b); (2) the propagation velocity of the waves with different polarization are different leading to the wave polarization rotation (Heinzl et al., 2006; Karbstein et al., 2021; King & Heinzl, 2016; Shen et al., 2018). In other words, the nonlinear QED vacuum is a birefringent optical medium.

In the limit $\chi_\gamma \gg 1$, Eqs. (28.13) and (28.19) give

$$\omega^2 - k^2 = \upsilon_{||,\perp}(1 - i\sqrt{3})(\omega + k_x)^{2/3} \tag{28.26}$$

with

$$\upsilon_{||,\perp} = \alpha\left(\frac{5 \pm 1}{28\pi^2}\right)3^{5/6}\Gamma^4\left(\frac{2}{3}\right)W_0^{2/3}. \tag{28.27}$$

As in the case of small χ_γ Eq. (28.26) has a solution $\omega = -k_x$ for the wave co-propagating with the strong wave, i.e. this wave has the phase and group velocity equal to speed of light in vacuum. For the counter-propagating electromagnetic wave the phase and group velocities are larger and smaller than speed of light in vacuum, respectively, being approximately equal to

$$\beta_{ph\ ||,\perp} \approx 1 + \upsilon_{||,\perp}\frac{1}{(2k_x)^{4/3}} \cdots \tag{28.28}$$

and

$$\beta_{g\ ||,\perp} \approx 1 - \upsilon_{||,\perp}\frac{1}{3(2k_x)^{4/3}} \cdots . \tag{28.29}$$

The wave decays with the decrement equal to

$$\omega'' = \sqrt{3}\upsilon_{||,\perp}(2k_x)^{-1/3}. \tag{28.30}$$

28.2.1 Frederick's Diagrams

In the long-wavelength limit, when $k_x \to 0$ one can neglect the last two terms in the l.h.s. of Eq. (28.20), i.e., neglect the dispersion and dissipation but retaining the diffraction and vacuum polarization effects taking $\omega'' = 0$ and $\omega' = \omega$. Then the dispersion equation takes the form

$$\omega^2 - k^2 + \kappa_{||,\perp}(\omega + k\ \cos\theta)^2 = 0, \tag{28.31}$$

where θ is the angle between the wave vector direction and the x-axis in the polar coordinate system.

Figure 28.5 (left) shows the electromagnetic waves in the (x, t) plane corresponding to the wave equation solution for which the dispersion equation is given by (28.31) for $\theta = \pi/2$. The solution is determined by the initial conditions

$$\dot{a}_0 = 0 \quad \text{and} \quad a_0(x) = \exp\left(-x^2/l^2\right)$$

with $l = 0.125$ and $\kappa_{||} = 1/3$, i.e., $v = 0.5c$. There are two waves. One of them propagates from the right to the left with the speed of light in vacuum. The other wave propagates from left to the right with speed equal to v. The ratio of their amplitudes is equal to v/c.

The solution of the dispersion Eq. (28.26) gives the wave frequency

$$\omega = -k\frac{\kappa_{||,\perp}\cos\theta + \sqrt{1+\kappa_{||,\perp}\sin^2\theta}}{1+\kappa_{||,\perp}}, \tag{28.32}$$

This expression yields the phase diagram representing the dependence of normalized phase velocity $\beta_{ph} = \omega/kc$ on the angle θ,

$$\beta_{ph} = -\frac{\kappa_{||,\perp}\cos\theta + \sqrt{1+\kappa_{||,\perp}\sin^2\theta}}{1+\kappa_{||,\perp}}. \tag{28.33}$$

It is shown in Fig. 28.5 (right).

Frederick's diagram (it is the polar diagram for group velocity of the wave, for details, e.g. see Kadomtsev, 2001) can be obtained by calculating the group velocity $\boldsymbol{v}_g = \partial_{\boldsymbol{k}}\omega$.

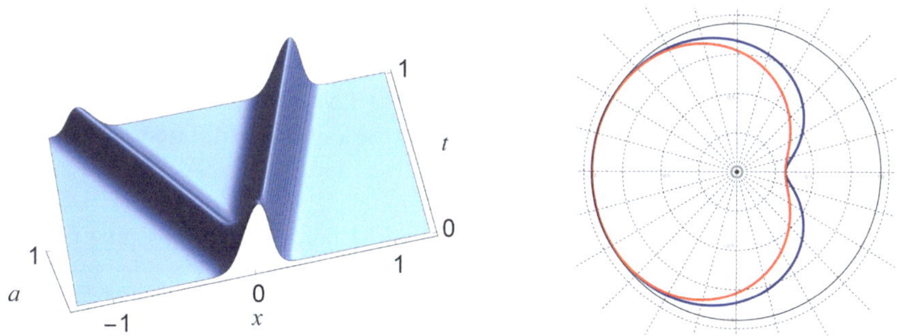

Fig. 28.5 (left) Electromagnetic waves in the (x, t) plane. The solution is determined by the initial conditions $\dot{a}_0 = 0$ and $a_0(x) = \exp(-x^2/l^2)$ with $l = 0.125$ and $\kappa_{||} = 1/3$, i.e., $v = 0.5c$. (right) Polar phase diagrams for group velocity β_g (blue), phase velocity β_{ph} (red) and speed of light in vacuum $\beta = 1$ (black) for $\kappa_{||} = 0.5$

Calculating $\partial_{\boldsymbol{k}}\omega$ we obtain for the normalized value of the group velocity $\beta_g = |\partial_{\boldsymbol{k}}\omega|$ as a function of the angle θ

$$\beta_g = -\frac{\sqrt{\left(1+\kappa_{||,\perp}\right)^2 \sin^2\theta + \left(\cos\theta + \kappa_{||,\perp}\sqrt{1+\kappa_{||,\perp}\sin^2\theta}\right)^2}}{(1+\kappa_{||,\perp})\sqrt{1+\kappa_{||,\perp}\sin^2\theta}}, \tag{28.34}$$

Figure 28.5 (right) presents the polar phase diagrams for the phase β_{ph} velocity and the group velocity β_g and speed of light in vacuum for $\kappa_{||} = 0.5$. As it is seen the phase and group velocities are equal to each other for waves propagating along the x-axis being equal to speed of light in vacuum for co-propagating waves, i.e., for $\theta = 0$ when $\beta_{ph} = \beta_g = (1-\kappa_{||})/(1+\kappa_{||}) < 1$ and $\beta_{ph} = \beta_g = -1$ for $\theta = \pi$. In the cases $\theta \neq 0$ and $\theta \neq \pi$, the phase velocity of the electromagnetic wave is smaller than the group velocity.

28.3 Synergic Cherenkov-Compton Radiation

Since strong electromagnetic wave induces the vacuum polarization, in the long wavelength limit, the QED vacuum behaves as a medium with a refraction index larger than unity, i.e. the speed of propagation of the interacting electromagnetic waves is below the speed of light in vacuum as follows from Eq. (28.25). One of the consequences of this fact is a possibility of the Cherenkov radiation of the high-energy electrons traversing the electromagnetic field (Narozhny, 1969; Ritus, 1970; Schwinger et al., 1976; Ginzburg & Tsitovich, 1979; Dremin, 2002). In Refs. Dremin (2002), Macleod et al. (2019), Bulanov et al., (2019) the electromagnetic field was considered to be generated by extremely high-power lasers.

When ultra-relativistic electron collides with the EM wave it undergoes radiation losses which can prevent the electron from reaching the high intensity EM field region as it has been discussed above (see also Jirka et al., 2021, 2022). To describe the one-dimensional relativistic electron dynamics in the EM field, following to approach formulated by Bulanov et al. (2013a, 2013b), we use a system of differential equations for the distribution functions of electrons, positrons and photons. We emphasize here that electron-EM field interaction depends on initial electron/positron/photon energy and the instantaneous (at time t) value of the EM field. Here we assume that the electron, positron, and photon dynamics is dominated by the longitudinal motion.

For a multi-petawatt laser focused to a one-lambda (it is of the one-micron size) spot the normalized amplitude approximately equals to $a_0 \approx 10^3$. Here we assume that the characteristic size of the high intensity region is equal to one lambda. As the energy of the laser wake field accelerated (LWFA) electrons according to the LWFA scaling (3.5.103), Lu et al., (2007), Bulanov et al., (2016), is $\mathcal{E}_e = 10(\mathcal{P}/1\ \text{PW})$ GeV, then for a 10 PW laser it can reach 100 GeV, i.e. $\gamma_{in} = 2 \times 10^5$. Solution by Jirka et al., (2021) of a system of equation for the electron, positron and photon distribution functions for these initial parameters and assuming the electron beam is mono-energetic before the interaction; we obtain a broad spectrum of electrons after the interaction. The electron beam lost approximately half of its energy; however, a substantial part of electrons has energy above 40 GeV. The situation at the maximum of the EM field is slightly different. Approximately 7% of the initial electrons reach the maximum of the EM field without emitting a photon. Moreover, around 60% of electrons have energy higher than 40 GeV. The solution shows that a significant number of electrons with very high energies can reach the maximum of the electromagnetic field of the tightly focused laser pulse. As shown by Jirka and Bulanov (2024) the optimal choice of laser pulse polarization (it this case it has the radial polarization) can enhance substantially the number of created electron–positron pairs.

Here, we consider the kinematics of an inverse multiphoton scattering process when an ultra-relativistic electron collides with an electromagnetic wave. Before and after the collision the electron has the momentum $\boldsymbol{p}_0$ and energy $m_e c^2 \gamma_0 = \sqrt{m_e^2 c^4 + p_0^2 c^2}$ and the momentum $\boldsymbol{p}$ and energy $m_e c^2 \gamma_e = \sqrt{m_e^2 c^4 + p^2 c^2}$, respectively. The scattered photons have the frequency ω_0 before collision and ω_γ after it. By using the energy and momentum conservation in the electron-photon system, we can find the scattering photon frequency dependence on the electron energy, the wave amplitude, and the scattering angle. As the electron interacts with s photons from the wave and emits high-energy photon, the energy and momenta of all particles are connected by the conservation laws, which yield:

$$m_e c^2 \gamma_0 + s\hbar\omega_0 = m_e c^2 \gamma_e + \hbar\omega_\gamma, \tag{28.35}$$

$$\boldsymbol{p}_0 + s\hbar\boldsymbol{k}_0 = \boldsymbol{p} + \hbar\boldsymbol{k}_\gamma, \tag{28.36}$$

where $\hbar\boldsymbol{k}_0$ is the electromagnetic wave photon momentum and $\hbar\boldsymbol{k}_\gamma$ is the emitted photon momentum, respectively, and s is the number of scattered photons. Figure 28.6 illustrates the electron - s photon scattering, which results in generating the photon with the momentum $\hbar k_\gamma$.

We consider the case when photon-electron interaction occurs in the (x, y) plane, i.e., $\boldsymbol{p}_0 = p_{x,0}\boldsymbol{e}_x + p_{y,0}\boldsymbol{e}_y$ and $\boldsymbol{p} = p_x\boldsymbol{e}_x + p_y\boldsymbol{e}_y$. For the perpendicular component of the momentum $p_{\perp,0}$ we may assume that it is equal to the quiver electron momentum in the electromagnetic field of the amplitude a_0, i.e. the electromagnetic wave is linearly polarized with the electric field along the y axis.

Above discussed processes yield that, in the limit $\chi_\gamma \ll 1$, a difference between the refraction index value and unity, $\Delta n_{||,\perp} = n_{||,\perp} - 1$ is equal to

$$\Delta n_{||,\perp} = \alpha \frac{11 \pm 3}{45\pi} W_0^2 + \cdots, \tag{28.37}$$

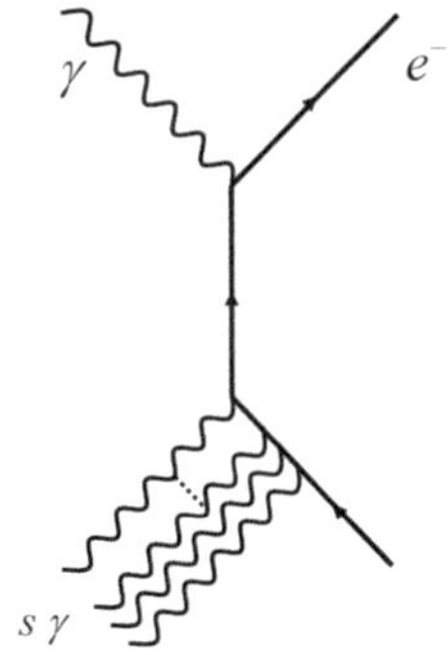

Fig. 28.6 Scattering of the electron on s photons

i.e. $\Delta n_{||,\perp} = 2\kappa_{||,\perp}$. From this expression it follows that in the low photon energy limit when $\chi_\gamma \ll 1$, the normalized phase velocity (and equal to it group velocity), $\beta_{||,\perp} = v_{||,\perp}/c$, of the linearly polarized counter-propagating electromagnetic wave with the electric field parallel or perpendicular to the y axis equals $\beta_{||,\perp} = 1-\varepsilon_{||,\perp} W_0^2$, where the coefficients $\varepsilon_{||,\perp} = \alpha(11 \pm 3)/45\pi$ is approximately equal to 10^{-3}.

Now for the sake of brevity we assume that the transverse component of electron momentum before interaction vanishes, i.e. $\boldsymbol{p}_0 = p_{x,0}\boldsymbol{e}_x$. From Eqs. (28.18, 28.19, 28.37) follows that, as a consequence of the vacuum polarization, the relationship between wave number and frequency of the electromagnetic wave corresponding to the wave propagating in a medium with the refraction index not equal to unity can be written as

$$\boldsymbol{k} = \frac{\boldsymbol{k}}{|\boldsymbol{k}|} n_{||,\perp}\omega. \tag{28.38}$$

with $\boldsymbol{k} = |\boldsymbol{k}|(\boldsymbol{e}_x \cos\theta + \boldsymbol{e}_y \sin\theta)$, where θ is the angle between the scattered photon wave number and the x-axis. Using this relationship to solve Eqs. (28.35) and (28.36), we obtain for the energy of scattered photon

$$\hbar\omega_\gamma = g \pm \sqrt{g^2 + 2s\hbar\omega_0 \left(\frac{m_e c^2 \gamma_0 + p_{x,0}c}{n_{||,\perp}^2 - 1}\right)}, \tag{28.39}$$

where

$$g = \frac{(p_{x,0}c - s\hbar\omega_0)n_{||,\perp}\cos\theta - m_e c^2\gamma_0 - s\hbar\omega_0}{n_{||,\perp}^2 - 1}. \tag{28.40}$$

Only positive solution for the photon frequency ω_γ is relevant to our problem.

Equation (28.39) describes kinematics of the Synergic Cherenkov-Compton Radiation process in the QED vacuum, whose optical properties are modified by the strong electromagnetic wave.

In the case, when the modification of the vacuum refraction index is weak enough, i.e. when $0 < n_{||,\perp}^2 - 1 \ll 1$, the dependence on the parameters of the emitted photon frequency ω_γ given by Eq. (28.34) can be presented as a combination of two modes with a continuous transition between them. When the function g given by Eq. (28.39) is positive and $s\hbar\omega_0 \ll m_e c^2 \gamma_e$ the photon energy can be found to be

$$\hbar\omega_{Ch} \approx 2g + \left[\frac{m_e c^2 \gamma_0 + p_{x,0} c}{g(n_{||,\perp}^2 - 1)}\right], \tag{28.41}$$

This expression corresponds to the Cherenkov radiation with a notation ω_{Ch} used for the frequency of the photon emitted in this regime.

In the opposite limit, when the function g given by Eq. (28.40) is negative, in the limit $s\hbar\omega_0 << m_e c^2 \gamma_e$, Eq. (3.10.39) can be rewritten as

$$\hbar\omega_C \approx -\frac{s\hbar\omega_0(m_e c^2 \gamma_0 + p_{x,0} c)}{m_e c^2 \gamma_0 + s\hbar\omega_0 - (p_{x,0} c - s\hbar\omega_0) n_{||,\perp} \cos\theta}, \tag{28.42}$$

corresponding to the Compton scattering mode with the photon frequency equal to ω_C. In the limit $s \ll s_m$, were

$$s_m = -\frac{m_e c^2}{4\hbar\omega_0 \gamma_0}, \tag{28.43}$$

with $\gamma_0 = \sqrt{1 + p_{x,0}^2/m_e^2 c^2}$. The photon energy equals

$$\hbar\omega_C \approx 4s\hbar\omega_0 \gamma_0^2.$$

At $s \approx s_m$ we have $\hbar\omega_C \approx m_e c^2 \gamma_0$. The condition $s \approx s_m$ shows when the recoil effects due to the photon emission become dominant. Equation (28.43) gives the photon number absorbed by the electron with the energy $m_e c^2 \gamma_0$ from the electromagnetic wave for radiating one high-energy photon.

Assumption $0 < n_{||,\perp}^2 - 1 \ll 1$ in Eq. (28.41) yields a condition of Cherenkov radiation,

$$n_{||,\perp} > \frac{\sqrt{m_e^2 c^2 + p_{x,0}^2}}{p_{x,0}}, \tag{28.44}$$

imposing the requirement on the electron energy (Ginzburg, 1940; Sokolov, 1940; Ginzburg & Tsitovich, 1979; Dremin, 2002). We note that this condition is written assuming that

$$s\hbar\omega_0 \ll \frac{\sqrt{m_e^2c^4 + p_{x,0}^2c^2} - p_{x,0}cn_{||,\perp}\cos\theta}{1 + n_{||,\perp}\cos\theta}, \tag{28.45}$$

The electron energy should be large enough to be above the Cherenkov radiation threshold, $m_e c^2 \gamma_{Ch}$:

$$\gamma_0 > \gamma_{Ch} = \frac{1}{\sqrt{2(n_{||,\perp} - 1)}} = \sqrt{\frac{45\pi E_S^2}{\alpha(11 \pm 3)E_0^2}} \approx 42\sqrt{\frac{I_S}{I_0}}, \tag{28.46}$$

Here the laser intensity $I_0 = cE_0^2/4\pi$, which in the focus region of 10 PW laser approximately is equal to $10^{24}\,\mathrm{W/cm^2}$, and $I_S = cE_S^2/4\pi \approx 10^{29}\mathrm{W/cm^2}$, i.e. the Cherenkov radiation threshold is exceeded for the electron energy above 10 GeV.

We can see it from Eqs. (28.41) and (28.42) that photons emitted via Cherenkov and Compton mechanisms have different angular distribution. The photons emitted via the Cherenkov radiation process are confined within the Cherenkov cone with the angle $\theta_{Ch} = 2\sqrt{\varepsilon_{||,\perp}I_0/I_S}$. In the focus of ten-petawatt laser it is approximately equal to 2×10^{-5}. The Compton scattered photons are within the cone with the angle $\theta_C \approx 1/\gamma_0$. Although characteristic angle values for these processes are of the same order, the dependence of the photon energy on the angle is different as it is seen from Fig. 28.7, where we show the angle distribution of the photon energy logarithm for Cherenkov radiation and for Compton scattering. Photons of substantially low energy emitted via the Compton scattering are localized within the relatively wide cone (blue curve). The narrower beam of the Cherenkov radiation photons with higher energy is shown with the red color. The magenta lines $\theta = \pm\theta_{Ch}$ show the Cherenkov cone.

According to the Cherenkov radiation theory (Frank & Tamm, 1937) the rate of the energy loss due to the Cherenkov radiation friction force along the electron trajectory is

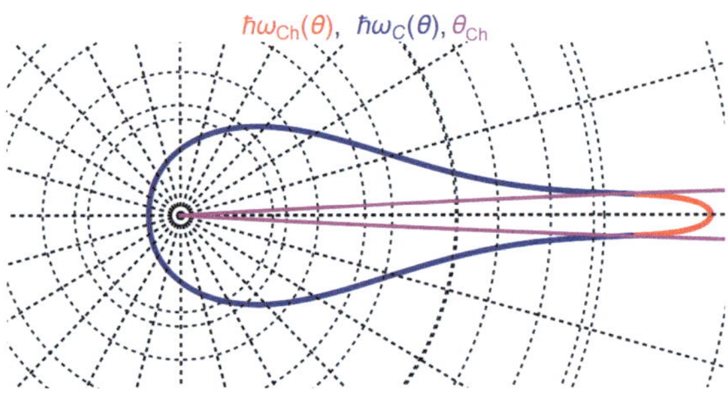

Fig. 28.7 Angle distribution of the energy logarithm for photon radiated by the Synergic Cherenkov-Compton Radiation mechanism. Blue color used for Compton scattering, $\hbar\omega_C(\theta)$, and red color for the Cherenkov radiation, $\hbar\omega_{Ch}(\theta)$. Magenta lines $\theta = \pm\theta_{Ch}$ show the Cherenkov cone. Reprinted figure with permission from Bulanov et al. (2019). Copyright (2019) by the American Physical Society. https://doi.org/10.1103/PhysRevD.100.016012

$$\frac{d\mathcal{E}_e}{dx} = -\left(\frac{e}{c}\right)^2 \int\limits_{\beta_e n > 1} \left(1 - \frac{1}{\beta_e^2 n_{||,\perp}^2}\right)\omega d\omega \approx -\frac{e^2}{\lambda_C^2}\varepsilon_{||,\perp}\left(\frac{E_0}{E_S}\right)^2. \tag{28.47}$$

Integration is done over the region where $\beta_e n_{||,\perp} > 1$ where $\beta_e = p_{x,0}/m_e c\gamma_0$ is the normalized electron velocity. From this equation follows that the number of photons emitted by the electron propagating over the distance λ_1 in the spectral interval between the wavelengths λ_1 and λ_2 equals

$$N_{ph} = 2\pi\alpha l\left(\frac{1}{\lambda_2} - \frac{1}{\lambda_1}\right)\left(1 - \frac{1}{\beta_e^2 n_{||,\perp}^2}\right), \tag{28.48}$$

where $n_{||,\perp}$ is the average refraction index value.

The formation length (Baier & Katkov, 2005) in the case of Cherenkov radiation is given by

$$l_{Ch} \approx \gamma_0 \gamma_{Ch} \lambda_{Ch} \approx \lambda_C \gamma_0. \tag{28.49}$$

Here

$$\lambda_C = \hbar/m_e c = 3.8 \times 10^{-11}\ \text{cm}$$

is the Compton scattering wavelength, $\lambda_{Ch} = \lambda_C/\gamma_{Ch}$ is the wavelength of radiation emitted via the Cherenkov radiation mechanism, and γ_{Ch} is the normalized threshold electron energy determined by expression (28.43). For 10 PW laser radiation the electron threshold energy is approximately equal to 10 GeV. According to Eq. (28.45) the radiation formation length is $l_{Ch} \approx 5 \times 10^{-5}$ cm, i.e. it is comparable in magnitude with the laser wavelength. Traversing the laser focus region the electron emits 0.2 photons. Assuming the electric charge of the LWFA electron bunch of 100 pC we obtain 10^4 photons (the efficiency of the electron energy conversion to the Cherenkov radiation is 10^{-4}). Photons produced in Compton scattering have approximately the same frequency. Interaction of the Compton scattering photons with the laser field will result in the nonlinear Breit-Wheeler electron–positron pair plasma generation.

In the limit $\chi_\gamma \gg 1$, since according to Eq. (28.28) the phase velocity of electromagnetic wave is above the speed of light in vacuum the electron cannot emit the Cherenkov radiation.

28.4 Nonlinear Electromagnetic Waves in Quantum Vacuum

28.4.1 Heisenberg-Euler Electrodynamics

In the low photon energy range, when $\mathcal{E}_\gamma = \hbar\omega_\gamma < m_e c^2$, which is the applicability range of Eq. (28.1), the nonlinear vacuum electrodynamics can be described by the Heisenberg-Euler Lagrangian (Heisenberg & Euler, 1936; Sakharov, 1991; Weisskopf, 1936),

$$\mathcal{L} = \mathcal{L}_0 + \mathcal{L}'. \tag{28.50}$$

Here

$$\mathcal{L}_0 = -\frac{m_e^4}{16\pi^2\alpha} F_{\mu\nu}F^{\mu\nu}. \tag{28.51}$$

is the Lagrangian of classical electrodynamics with the 4-tensor of the electromagnetic field $F^{\mu\nu}$ given by Eq. (19.21). Its variation yields the Maxwell equations in vacuum. In the Euler–Heisenberg theory, the quantum electrodynamics radiation corrections are described by the Lagrangian $\mathcal{L}'$ on the right-hand side of Eq. (28.50). It has real $\mathcal{R}e\{\mathcal{L}'\}$ and imaginary $\mathcal{I}m\{\mathcal{L}'\}$ parts (see Berestetskii et al., 1982):

$$\begin{aligned}\mathcal{R}e\{\mathcal{L}'\} &= -\frac{m_e^4}{8\pi^2}\mathcal{M}(\mathfrak{e}, \mathfrak{b}) \\ &= -\frac{m_e^4}{8\pi^2}\int_0^\infty \frac{\exp(-\eta)}{\eta^3}\left\{1 - \frac{\eta^2}{3}\left(\mathfrak{e}^2 - \mathfrak{b}^2\right) - [\eta\mathfrak{e}\cot(\eta\mathfrak{e})][\eta\mathfrak{b}\coth(\eta\mathfrak{b})]\right\}d\eta\end{aligned} \tag{28.52}$$

$$\mathcal{I}m\{\mathcal{L}\} = \frac{m_e^4}{4\pi^3}\mathfrak{e}^2\exp\left(-\frac{\pi}{\mathfrak{e}}\right). \tag{28.53}$$

Here we use the units $e = c = \hbar = 1$ with the electromagnetic field normalized on the critical field of quantum electrodynamics E_S given by Eq. (19.16). Written in these units the critical field is $E_S = m_e^2/e$. The invariant electric and magnetic fields $\mathfrak{e}$ and $\mathfrak{b}$ can be expressed via the normalized Poincare invariants

$$\mathfrak{F} = \frac{\boldsymbol{B}^2 - \boldsymbol{E}^2}{2E_S^2} \quad \text{and} \quad \mathfrak{G} = \frac{\boldsymbol{E}\cdot\boldsymbol{B}}{E_S^2}. \tag{28.54}$$

as

$$\mathfrak{e} = \sqrt{\sqrt{\mathfrak{F}^2 + \mathfrak{G}^2} - \mathfrak{F}} \quad \text{and} \quad \mathfrak{b} = \sqrt{\sqrt{\mathfrak{F}^2 + \mathfrak{G}^2} + \mathfrak{F}}. \tag{28.55}$$

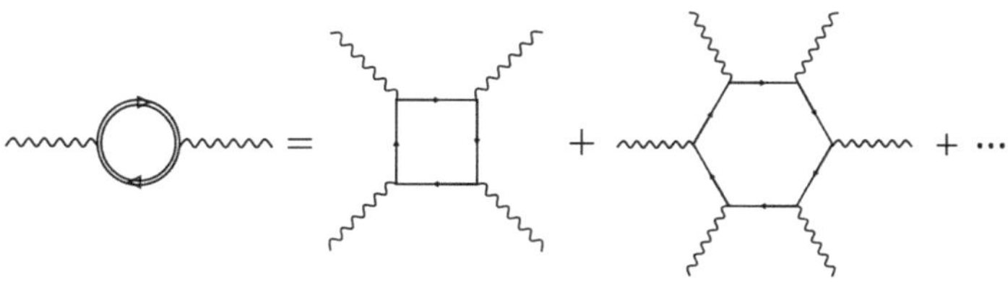

Fig. 28.8 Vacuum polarization as the four-, six- etc. photon scattering

The Heisenberg–Euler Lagrangian in the form given by Eq. (28.47) should be used for obtaining an asymptotic series over the invariant electric and magnetic fields $\mathfrak{e}$ and $\mathfrak{b}$ assuming that they are small. In this limit, we can expand the function $\mathcal{M}(\mathfrak{e}, \mathfrak{b})$ in Eq. (28.52) for small arguments as (e.g. see Sasorov et al., 2021)

$$\mathcal{M}(\mathfrak{e}, \mathfrak{b}) = \frac{\Gamma(2)}{45}\left[\left(\mathfrak{e}^4 + \mathfrak{b}^4\right) + 5\mathfrak{e}^2\mathfrak{b}^2\right] - \frac{\Gamma(4)}{945}\left[2\left(\mathfrak{b}^6 - \mathfrak{e}^6\right) + 7\mathfrak{e}^2\mathfrak{b}^2\left(\mathfrak{b}^2 - \mathfrak{e}^2\right)\right] + \frac{\Gamma(6)}{7 \times 45^2}\left[3\left(\mathfrak{e}^8 + \mathfrak{b}^8\right) + 10\mathfrak{e}^2\mathfrak{b}^2\left(\mathfrak{b}^4 + \mathfrak{e}^4\right) - 7\mathfrak{e}^4\mathfrak{b}^4\right] + \cdots \qquad (28.56)$$

Here $\Gamma(x)$ is the Euler gamma function (Olver et al., 2010). The properties of the expansion of $\mathcal{M}(\mathfrak{e}, \mathfrak{b})$ in series of powers of the fields fields $\mathfrak{e}$ and $\mathfrak{b}$ are discussed by Dunne (2005). In the weak field approximation, the expression (28.56) yields for the Lagrangian $\mathcal{L}'$ (see Heyl & Hernquist, 1997)

$$\mathcal{L}' = \frac{\kappa}{45}\left(4\mathfrak{F}^2 + 7\mathfrak{G}^2\right) - \frac{\kappa}{315}\mathfrak{F}\left(8\mathfrak{F}^2 + 13\mathfrak{G}^2\right) + \frac{8\kappa}{945}\left(48\mathfrak{F}^4 + 88\mathfrak{F}^2\mathfrak{G}^2 + 19\mathfrak{G}^4\right) + \cdots, \qquad (28.57)$$

where $\kappa = m_e^4/8\pi^2$. In the Lagrangian $\mathcal{L}'$ the first, second and third, etc. terms on the right-hand side correspond to four-, six- and eight- etc.- photon interactions, respectively. This fact is illustrated in Fig. 28.8.

28.4.2 Vacuum Polarization

The Heisenberg-Euler electrodynamics Eqs. (28.50–28.53) is used to describe the vacuum polarization as a consequence of the virtual electron–positron pair response on the action of strong electromagnetic field. By virtue of the fact that the relationship between the electric and magnetic field, $\boldsymbol{E}$ and $\boldsymbol{B}$, with the potentials $\boldsymbol{A}$ and φ remains unchanged, the first pair of Maxwell's equations is not modified. The second pair can be written in the form

$$\nabla \times (\boldsymbol{B} - 4\pi \boldsymbol{M}) = \frac{1}{c}\partial_t(\boldsymbol{E} - 4\pi \boldsymbol{P}), \tag{28.58}$$

$$\nabla \cdot (\boldsymbol{E} + 4\pi \boldsymbol{P}) = 0, \tag{28.59}$$

with the vectors of electric and magnetic polarization of vacuum

$$\boldsymbol{P} = \frac{\partial \mathcal{L}'}{\partial \boldsymbol{E}} \quad \text{and} \quad \boldsymbol{M} = \frac{\partial \mathcal{L}'}{\partial \boldsymbol{B}}, \tag{28.60}$$

respectively.

The vectors $\boldsymbol{P}$ and $\boldsymbol{M}$, the case of plane electromagnetic wave, vanish, because in this case both the Poincare invariants $\mathfrak{F} = (\boldsymbol{E}^2 - \boldsymbol{B}^2)/2$ and $\mathfrak{G} = \boldsymbol{E} \cdot \boldsymbol{B}$ are equal to zero. In other words, the plane electromagnetic wave does not polarize the vacuum because the co-propagating photons do not interact with each other.

The week field, $E \ll E_S$, Lagranian (28.50) can be written as

$$\mathcal{L} = -\frac{1}{4}F^{\mu\nu}F_{\mu\nu} + \frac{2}{45}\frac{\alpha^2\hbar^3}{m_e^4 c^5}\left[4\left(F^{\mu\nu}F_{\mu\nu}\right)^2 + 7\left(\varepsilon^{\kappa\lambda\mu\nu}F_{\kappa\lambda}F_{\mu\nu}\right)^2\right]. \tag{28.61}$$

It leads to the field equations

$$\partial_\mu\left\{F^{\mu\nu} - \frac{2}{45}\frac{\alpha^2}{m_e^4}\left[4\left(F_{\kappa\lambda}F^{\kappa\lambda}\right)F^{\mu\nu} + 7\left(F_{\kappa\lambda}{}^*F^{\kappa\lambda}\right){}^*F^{\mu\nu}\right]\right\} = 0, \tag{28.62}$$

where the tensor ${}^*F^{\alpha\beta}$ is dual to $F^{\alpha\beta}$. Here and below in this section for the sake of brevity we use the QED notations: $\hbar = c = 1$.

Following Aleksandrov et al. (1985), Rosanov (1993), (1998) the field Eq. (28.62) can be solved by successive approximations assuming

$$F_{\mu\nu} = F_{\mu\nu}^{(0)} + F_{\mu\nu}^{(1)} + \cdots, \; \partial_\mu F^{(0)\mu\nu} = 0, \; \partial_\mu F^{(1)\mu\nu} = 4\pi j^\nu \tag{28.63}$$

with the effective 4-current density resulting from the photon-photon interaction

$$j^\nu = -\frac{2}{180\pi}\frac{\alpha^2}{m_e^4}\partial_\mu\left[4\left(F_{\kappa\lambda}^{(0)}F^{(0)\kappa\lambda}\right)F^{(0)\mu\nu} + 7\left(F_{\kappa\lambda}^{(0)*}F^{(0)\kappa\lambda}\right){}^*F^{(0)\mu\nu}\right]. \tag{28.64}$$

In 3D notations, using expressions (28.60) for the vectors of electric and magnetic polarization of vacuum, the electric current density can be written as

$$\boldsymbol{j}_{\text{vac}} = \partial_{tt}\boldsymbol{P} - \nabla(\nabla \cdot \boldsymbol{P}) + \nabla \times \partial_t \boldsymbol{M}. \tag{28.65}$$

The first order contribution to the vector potential, according to Eqs. (28.60) and (28.65), is given by the retarded potentials

$$A^{(1)}(\boldsymbol{R}, t) = \frac{1}{4\pi} \int_V d^3\boldsymbol{r} \frac{\boldsymbol{j}_{\rm vac}(\boldsymbol{r}, t - |\boldsymbol{R} - \boldsymbol{r}|)}{|\boldsymbol{R} - \boldsymbol{r}|}, \tag{28.66}$$

where the integration performed over the volume of the electric current density location.

28.4.3 Electromagnetic Field Configuration in the Dirac Light-Cone Coordinates

Here we consider two counter propagating waves along the x-axis. The vector potential corresponding to the electromagnetic field can be presented in the form

$$\boldsymbol{A}(x, t) = A_y(x, t)\boldsymbol{e}_y + A_z(x, t)\boldsymbol{e}_z, \tag{28.67}$$

We write the electromagnetic field equations in Dirac's light-cone coordinates x^+ and x^- defined as (see e.g. Kim & Noz, 1982)

$$x^+ = \frac{t + x}{\sqrt{2}} \quad \text{and} \quad x^- = \frac{t - x}{\sqrt{2}}. \tag{28.68}$$

The Lorentz transform of the light-cone variables under a boost along x with velocity β is given by

$$x'^+ = \frac{t' + x'}{\sqrt{2}} = e^{-\eta} \frac{t + x}{\sqrt{2}} = e^{-\eta} x^+, \tag{28.69}$$

$$x'^- = \frac{t' - x'}{\sqrt{2}} = e^{+\eta} \frac{t - x}{\sqrt{2}} = e^{+\eta} x^- \tag{28.70}$$

with a prime denoting variable in the boosted frame of reference and η equal to

$$\eta = \ln \sqrt{\frac{1 + \beta}{1 - \beta}}. \tag{28.71}$$

The following shorthand stands for the partial derivatives of a function f with respect to variables x^+ and x^-:

$$\partial_+ f = \left(\frac{\partial f}{\partial x^+}\right)_{x^-} \quad \text{and} \quad \partial_- f = \left(\frac{\partial f}{\partial x^-}\right)_{x^+}. \tag{28.72}$$

Then the derivatives are transformed as

$$(\partial_+) = e^{+\eta}\partial_+ \quad \text{and} \quad (\partial_-) = e^{-\eta}\partial_-. \tag{28.73}$$

Considering for simplicity interacting electromagnetic waves of the same polarization we find that the Poincare invariant $\mathfrak{G}$ defined by Eq. (28.54) vanishes identically. The coordinate system can be chosen so that in Eq. (28.58) the y-component of the vector potential vanishes, $A_y = 0$. We will use a notation $a(x,t)$ for $A_z(x,t)$. Calculating the derivatives of $a(x,t)$ with respect to the coordinates x^+ and x^-,

$$u = \partial_+ a \quad \text{and} \quad w = \partial_- a \tag{28.74}$$

we find the relationships between u and w, and the electric, $e_z = -\partial_t a$, and magnetic, $b_y = -\partial_x a$, field components. They read

$$u = -\frac{e_z - b_y}{\sqrt{2}} \quad \text{and} \quad w = -\frac{e_z + b_y}{\sqrt{2}}. \tag{28.75}$$

Due to commutation of partial derivatives, $\partial_{+-}a = \partial_{-+}a$, the field variables u and w are related to each other as

$$\partial_+ u = \partial_- w. \tag{28.76}$$

The Lorentz transformation of the fields u and w

$$u' = -\frac{e'_z - b'_y}{\sqrt{2}} = -e^{-\eta}\frac{e_z - b_y}{\sqrt{2}} = e^{-\eta}u \tag{28.77}$$

$$w' = -\frac{e'_z + b'_y}{\sqrt{2}} = -e^{+\eta}\frac{e_z + b_y}{\sqrt{2}} = e^{+\eta}w \tag{28.78}$$

shows that the w and u are transformed as the $-$ and $+$ components of a contravariant four-vector, respectively. The field product uw is Lorentz invariant

$$u'w' = uw. \tag{28.79}$$

and is equal to the first Poincare invariant

$$\mathfrak{F} = (b_y^2 - e_z^2)/2.$$

28.4.4 Nonlinear Wave Equation

For simplicity, we consider the case when interacting electromagnetic pulses have the same polarization. Then the Lagrangian given by Eqs. (28.50) and (28.57) can be written in terms of the field variables u and w as

$$\mathcal{L} = -\frac{m_e^4}{8\pi^2\alpha}[uw + \alpha Q(uw)]. \tag{28.80}$$

In the small amplitude field limit $uw \ll 1$, the function $Q(uw)$ equals

$$Q(uw) = -\frac{1}{2\pi}\mathcal{M}(0, \sqrt{2uw}). \tag{28.81}$$

Expansion of the function $Q(\varsigma)$ in asymptotic series yields (Dunne, 2005; Sasorov et al., 2021)

$$Q(\varsigma) = \sum_{m=2}^{\infty} b_m \varsigma^m \tag{28.82}$$

where the coefficients

$$b_m = \frac{2^{3(m-1)}B_{2m}}{\pi m(2m-1)(m-1)}. \tag{28.83}$$

are proportional to the Bernoulli numbers B_n (Olver et al., 2010).

The Euler–Lagrange formalism applied to the Lagrangian (28.80)

$$\partial_-(\partial_u\mathcal{L}) + \partial_+(\partial_w\mathcal{L}) = 0 \tag{28.84}$$

with Eq. (28.76) constitute the electromagnetic field equations. They read

$$\begin{aligned}\left[1 + 4\varepsilon_2 uw + 9\varepsilon_3 u^2w^2 + 16\varepsilon_4 u^3w^3 + \cdots\right]\partial_+ u \\ = \left[\varepsilon_2 u^2 + 3\varepsilon_3 u^3 w + 6\varepsilon_4 u^4 w^2 + \cdots\right]\partial_+ w \\ + \left[\varepsilon_2 w^2 + 3\varepsilon_3 uw^3 + 6\varepsilon_4 u^2 w^4 + \cdots\right]\partial_- u\end{aligned} \tag{28.85}$$

with $\varepsilon_2 = -(2/45\pi)\alpha$, $\varepsilon_3 = (16/315\pi)\alpha$, and $\varepsilon_4 = -(64/315\pi)\alpha$.

28.4.5 Breaking of Strong Electromagnetic Wave in Quantum Vacuum

In the media where the refraction index dependence on the field amplitude leads to the nonlinear response, the electromagnetic wave can evolve into a configuration with singularities (Landau & Lifshitz, 1987). The appearance of singularities in the Heisenberg-Euler electrodynamics is noticed by Lutzky and Toll (1959), Zheleznyakov and Fabrikant (1982), Kadlecova et al. (2019), Pegoraro and Bulanov (2019) who obtained singular particular solutions of equations derived from the Heisenberg–Euler Lagrangian. Böhl et al., (2015) demonstrated the electromagnetic wave steepening by numerical integration of nonlinear quantum vacuum electrodynamics equations.

Retaining for simplicity in Eq. (28.85) the terms corresponding to 4-photon interaction only, we write it as

$$(1 - 4\varepsilon_2 uw)\partial_+ u = \varepsilon_2 w^2 \partial_- u + \varepsilon_2 u^2 \partial_+ w. \tag{28.86}$$

Using Eq. (28.76) we obtain the second equation,

$$\partial_+ u = \partial_- w. \tag{28.87}$$

These quasilinear equations (they are the equations linear with respect to partial derivatives) admit the self-similar solutions. Here we seek a self-similar solution to Eqs. (28.86) and (28.87) of a simple wave (e.g., see Witham, 1974; Landau & Lifshitz, 1987; Kadomtsev, 2001), in which w is considered as a function of u: $w(u)$ (or equivalently, u is a function of w: $u(w)$). The simple wave (or Riemann wave) represents an exact solution of self–similar type of the nonlinear wave equations describing the finite amplitude wave propagating in a continuous media. With this assumption, we obtain from Eqs. (28.86) and (28.87) the system of equations

$$[(1 - 4\varepsilon_2 uw)J - \varepsilon_2 u^2]\partial_+ w - \varepsilon_2 w^2 J \partial_- w = 0, \tag{28.88}$$

$$J\partial_+ w - \partial_- w = 0, \tag{28.89}$$

where

$$J = du/dw \tag{28.90}$$

is Jacobian. The system of Eqs. (28.79) and (28.80) has nontrivial solution provided that the determinant

$$\begin{vmatrix} (1 - 4\varepsilon_2 uw)J - \varepsilon_2 u^2 & -\varepsilon_2 w^2 J \\ J & -1 \end{vmatrix}, \tag{28.91}$$

is equal to zero. This condition yields for the Jacobian

$$J_{1,2} = \frac{1 - 4\varepsilon_2 uw \pm \sqrt{(1 - 4\varepsilon_2 uw)^2 - 4\varepsilon_2^2 u^2 w^2}}{2\varepsilon_2 w^2}. \tag{28.92}$$

Assuming smallness of the field amplitude, $\varepsilon_2 uw \ll 1$, we find that

$$J_1 \approx \frac{1}{\varepsilon_2 w^2} \quad \text{and} \quad J_2 \approx \varepsilon_2 u^2, \tag{28.93}$$

which gives for the relationships between the functions u and w

$$u = -\frac{1}{\varepsilon_2 w} + c_1 \quad \text{and} \quad w = -\frac{1}{\varepsilon_2 u} + c_2, \tag{28.94}$$

where c_1 and c_2 are constants.

Substituting the Jacobian (28.93) to Eq. (28.89), we obtain the nonlinear wave equation

$$\partial_+ w - \varepsilon_2 w^2 \partial_- w = 0. \tag{28.95}$$

A solution to this equation can be obtained in a standard manner. According to this solution, the function $w(x^+, x^-)$ transfers along the characteristic x_0^- without distortion

$$w = w_0(x_0^-). \tag{28.96}$$

Here the function $w_0(x_0^-)$ is given at $x^+ = 0$.

The characteristic equation for Eq. (28.95) is

$$\frac{dx^-}{dx^+} = -\varepsilon_2 w^2. \tag{28.97}$$

It has a solution

$$x^- = x_0^- - \varepsilon_2 w_0^2(x_0^-) x^+. \tag{28.98}$$

As seen, each of the points on the wave profile $w_0(x_0^-)$ propagates with the velocity depending on the wave amplitude: the higher the amplitude, the lower velocity (see Eq. (28.25)). This leads to the wave profile deformation. In the region, where the gradient dw_0^2/dx_0^- is positive, occurs the wave steepening leading to the wave breaking, when the gradient tends to infinity. The coordinate x_{br}^+, at which the wave breaks, equals

$$x_{br}^{+} = \frac{1}{2\varepsilon_2 w_0(x_{0,m}^{-}) w_0'(x_{0,m}^{-})}. \tag{28.99}$$

The coordinate $x_{0,m}^{-}$ corresponds to the point where the gradient is maximal. A prime here denotes a differentiation with respect to the Lagrange coordinate x_0^{-}.

We note that the singularity formed at the electromagnetic wave breaking in the QED vacuum corresponds to the rarefaction shock wave formation (the wave steepens and breaks in the backwards direction, see Kadlecova et al. (2019), because the wave crest propagates with a speed lower than the propagation speed of the part of the pulse with lower amplitude.

28.4.6 Electromagnetic Solitons in the QED Vacuum

As we have discussed above, the refraction index of nonlinear QED vacuum depends nonlinearly on the colliding electromagnetic wave amplitude, and the resulting wave self-action can lead to wave steepening and wave breaking. In the long wavelength limit the QED vacuum is a dispersionless medium, i.e., the phase and group velocity of the electromagnetic wave are equal. The vacuum dispersion effects are formally described by the term in the r.h.s. of Eq. (28.18) in square brackets proportional to χ_γ^4. In general, the nonlinearity and dispersion balance provides the condition for the formation of solitary waves (Witham, 1974; Zakharov et al., 1984), which can propagate over large distance without changing their shape. There is a wide class of the solitons described by the Korteweg-de-Vries (KdV) equation (Korteweg & De Vries, 1885) (a generalization of the KdV equation to the multidimensional case is known as the Kadomtsev–Petviashvili (KP) equation, Kadomtsev & Petviashvili, 1970). Bulanov et al. (2020) have shown that the KdV and KP solitons can exist in the QED vacuum. To demonstrate this, using Eqs. (28.20) and (28.86) we obtain the equation for the vector potential

$$A(x^+, x^-, y) = W_0 x^+ + a(x^+, x^-, y), \tag{28.100}$$

$$\partial_-[\partial_+ w - \varepsilon_2(W_0^2 + 2W_0 w)\partial_- w + \mu\, \partial_{---} w] = \partial_{yy} w, \tag{28.101}$$

where $\mu = \varepsilon_2 W_0^4/3$, $w = \partial_+ a$ and assumed that $w \ll W_0$. It has a form of the Kadomtsev–Petviashvili (KP) equation. For discussions of the properties and solutions of the KP equations see Biondini and Pelinovsky (2008) and references therein).

If we neglect the diffraction effects, i.e. assuming that $\partial_{yy} w = 0$, Eq. (28.101) can be cast to the form

$$\partial_+ w - \varepsilon_2(W_0^2 + 2W_0 w)\partial_- w + \mu\, \partial_{---} w = 0. \tag{28.102}$$

This is the KdV equation, which is an equation in partial derivatives. We seek its solution depending on the variable

$$\psi = R x^{+} + S x^{-} \tag{28.103}$$

with constant R and S. For the independed variable ψ Eq. (28.102) becomes an ordinary differential equation

$$R w' - \varepsilon_2 (W_0^2 + 2 W_0 w) S w' + \mu S^3 w''' = 0, \tag{28.104}$$

where a prime stands for differentiation with respect to ψ. Its integration yields

$$(R - \varepsilon_2 S W_0^2) w - \varepsilon_2 W_0 S w^2 + \mu S^3 w'' = 0. \tag{28.105}$$

This equation has an integral, which can be written as

$$(R - \varepsilon_2 S W_0^2) w^2 - \frac{2}{3} \varepsilon_2 W_0 S w^3 + \mu S^3 (w')^2 = 0. \tag{28.106}$$

Here the integration constants are chosen to have the localized solution. It is the soliton

$$w = -w_s \mathrm{Sech}^2 \left[\frac{(R - \varepsilon_2 S W_0^2)}{2\sqrt{S^3 \mu}} \psi \right] \tag{28.107}$$

whose amplitude and width are equal to

$$w_s = -\frac{3}{2} \frac{R - \varepsilon_2 S W_0^2}{\varepsilon_2 S W_0} \quad \text{and} \quad l_s = \frac{2\sqrt{S^3 \mu}}{R - \varepsilon_2 S W_0^2}, \tag{28.108}$$

respectively. Equation (28.103) gives a relationship between the coordinate x and time t

$$x = \frac{S - R}{S + R} t + \frac{\sqrt{2}}{S + R} \psi, \tag{28.109}$$

which shows that the velocity of the soliton propagation $v_s \approx (1 - 2R/S)$ depends on the soliton amplitude as

$$v_s \approx 1 - 2\varepsilon_2 W_0^2 - \frac{4}{3} \varepsilon_2 W_0 w_m. \tag{28.110}$$

The soliton width is approximately equal to $l_s \approx W_0/(w_m \sqrt{\varepsilon_2})$. In dimensional form it is $l_s \approx \lambda_C W_0/(w_m \sqrt{\varepsilon_2})$, i.e. for $W_0/w_m = 10^{-2}$ it is approximately equal to 10 nm.

The solitons obtained above correspond to the approximation using the Langangian with high derivatives (Pegoraro & Bulanov, 2021). In nonlinear quantum electrodynamics the high order Lagrangians have been considered by Lee and Wicks (1970) and Mamaev et al. (1981). The physical interpretation of high order derivatives Lagrangians presents some difficulties as these Lagrangians lead to "ghost" degrees of freedom and to instabilities. In 1850 Ostrogradsky proved in the context of classical mechanics that a Lagrangian of the form $\mathcal{L}(q, \dot{q}, \ddot{q})$, which requires four initial conditions and thus involves four canonical variables, leads to a Hamiltonian that is not bounded from below with respect to a "ghost" degree of freedom. Pegoraro and Bulanov (2021) have found that in the context of the solitonic solutions of nonlinear QED wave equations these requires the photon energy beyond the limits of applicability of the Heisenberg-Euler electrodynamics.

28.4.7 Generation of High Order Harmonics in Quantum Vacuum

High order harmonics generation in vacuum in the course of the collision of the extreme intensity laser pulses has attracted substantial attention (Rosanov, 1993; Kaplan & Ding, 2000; Di Piazza et al., 2005; Lundström et al., 2006; Narozhny & Fedotov, 2007, Fillion-Gourdeau et al., 2015, Sasorov et al., 2021) because it sheds light on the dynamical properties of quantum electrodynamics vacuum in strong electromagnetic fields. When a wave approaches the wave breaking point, wave steepening is equivalent to harmonics generation with higher numbers. In the configuration of two head on colliding electromagnetic pulses, the generation of high order harmonics are described by Eqs. (28.76) and (28.84). Their solutions can be found within the framework of the perturbation theory approximation seeking the solution in the form

$$u = u_0 + \varepsilon_2 u_1 + \varepsilon_3 u_2 + \varepsilon_4 u_3 \ldots, \tag{28.111}$$

$$w = w_0 + \varepsilon_2 w_1 + \varepsilon_3 w_2 + \varepsilon_4 w_4 + \cdots . \tag{28.112}$$

In zeroth order on small parameters ε_i the functions u_0 and w_0 obey equations

$$\partial_+ u_0 = 0 \quad \text{and} \quad \partial_- w_0 = 0 \tag{28.113}$$

from which follows that u_0 and w_0 are arbitrary functions of variables x^- and x^+, respectively, i.e. $u_0(x^-)$ and $w_0(x^+)$. For localized $u_0(x^-)$ and $w_0(x^+)$ they describe the counter-propagating electromagnetic pulses before collision. For the fields $u(x^+, x^-)$ and $w(x^+, x^-)$ the perturbation theory yields

$$u(x^-,x^+) = u_0(x^-) + \varepsilon_2\left(u_0^2(x^-)w_0(x^+) + \partial_- u_0(x^-)\int^{x^+} w_0^2(s)ds\right)$$
$$-\frac{3}{2}\varepsilon_3\left(u_0^3(x^-)w_0^2(x^+) + 2u_0(x^-)\partial_- u_0(x^-)\int^{x^+} w_0^3(s)ds\right)$$
$$+\varepsilon_4\left(2u_0^4(x^-)w_0^3(x^+) + 6u_0^2(x^-)\partial_- u_0(x^-)\int^{x^+} w_0^4(s)ds\right) + \cdots, \tag{28.114}$$

$$w(x^-,x^+) = w_0(x^+) + \varepsilon_2\left(u_0(x^-)w_0^2(x^+) + \partial_+ w_0(x^+)\int^{x^-} u_0^2(s)ds\right)$$
$$-\frac{3}{2}\varepsilon_3\left(u_0^2(x^-)w_0^3(x^+) + 2w_0(x^+)\partial_+ w_0(x^+)\int^{x^-} u_0^3(s)ds\right)$$
$$+\varepsilon_4\left(2u_0^3(x^-)w_0^4(x^+) + 6w_0^2(x^-)\partial_- w_0(x^-)\int^{x^-} u_0^4(s)ds\right) + \cdots. \tag{28.115}$$

We consider the case when the functions $u_0(x^-)$ and $w_0(x^+)$ have a form of Gaussian pulses

$$u_0(x^-) = U_0 \exp\left[-\frac{1}{2}\left(\frac{x^-}{l^-}\right)^2\right]\cos\left(k^- x^- + \varphi^-\right) \tag{28.116}$$

and

$$w_0(x^+) = W_0 \exp\left[-\frac{1}{2}\left(\frac{\mathrm{x}^+}{\mathrm{l}^+}\right)^2\right]\cos\left(\mathrm{k}^+\mathrm{x}^+ + \varphi^+\right) \tag{28.117}$$

with the amplitudes U_0 and W_0, pulse widths l^- and l^+, carrying wave numbers k^- and k^+, and phases φ^- and φ^+.

As we may see, in the right-hand side of Eqs. (28.114) and (28.115) the second terms of the order of ε_2 in the brackets describe the phase shift of the wave interacting with the counter-propagating wave as has been discussed above. The second term in the right hand side of Eqs. (28.114) and (28.115) proportional to ε_3 describe the second harmonic generation. For the interacting electromagnetic pulses of finite length, which are given by expressions (28.116) and (28.117), the second (and even high number harmonics) harmonic amplitude is not exponentially small provided the pulses are short enough. In the case of the scattered pulse given by Eq. (28.114), it should be $k^+l^+ \ll 1$. For the second harmonic given by Eq. (28.115) to be not exponentially small we should have $k^-l^- \ll 1$. In its turn, the second term in the r. h. s. of Eqs. (28.114) and (28.115) proportional to ε_4 describe the third harmonic generation. More detailed consideration of the high order harmonic generation in the QED vacuum can be found in publication by Sasorov et al. (2021).

It is not surprising that the dynamics of the electron interacting with three-, four-, etc. colliding pulses are even more complicated and rich with novel patterns (Vranic et al., 2016; Gong et al., 2017, Bulanov et al., 2017).

28.5 Laser Driven Lepton-Gamma (e^- e^+ γ) Plasma

High-energy electrons and gamma rays interacting with matter create positrons via the trident and Bether-Heitler processes in collisions of electrons (positrons) with electrons and nucleus (Bethe & Heitler, 1934; Bhabha, 1935; Gryaznykh et al., 1998; Heitler, 1954), the Breit-Wheeler process in collision of two gamma-photons (see Eq. (28.9) and Akhiezer & Berestetskii, 1965; Nikishov, 1962; Gould & Schreder, 1967; Bisnovatyi-Kogan et al., 1971; Ruffini et al., 2016; Yu et al., 2024), the multi-photon Breit-Wheeler process (Nikishov & Ritus, 1970), and by the Schwinger mechanism (see Eq. (28.53) and Berestetskii et al., 1982; Popov, 2004).

As well known, positron creation is a threshold process. In the laboratory frame of reference, the threshold for creating the electron–positron pair is $3m_ec^2$ in the electron collision with heavy particle with the electric charge equal to Ze, and is $7m_ec^2$ for the electron collision with the electron at the rest. The cross section for positron creation is (Berestetskii et al., 1982)

$$\varepsilon_{eZ\to e^-e^+} = \frac{28}{27\pi}(\alpha Z)^2 r_e^2 \ln^3(\beta\gamma), \tag{28.118}$$

which is approximately equal to $Z^2 10^{-30}\text{cm}^2$. The positrons generated in the collision of relativistic laser accelerated electrons with high Z target have been observed in the experiments by Gahn et al., (2002), Chen et al., (2010), Sarri et al., (2015).

When the gamma photon (it can be generated via the bremsstrahlung radiation mechanism or due to the gamma-flash process discussed above) collides with the heavy nuclei of the charge Ze the total cross section of the electron–positron pair creation is given by (see Berestetskii et al., 1982)

$$\varepsilon_{\gamma Z\to e^-e^+} = \frac{28}{9}\alpha Z^2 r_e^2 \ln\left(\frac{\hbar\omega_\gamma}{m_ec^2}\right), \tag{28.119}$$

where assumed that $\hbar\omega_\gamma \gg m_ec^2$.

Similar processes result in the di-muon (μ^+, μ^-) generation. Since the muon mass equals $m_\mu = 105.66\ \text{MeV}/\text{c}^2$, i.e. $m_\mu \approx 206.768\ m_e$ the classical muon radius

$r_\mu = e^2/m_\mu c^2 \approx (m_e/m_\nu)r_e$ is approximately 207 times smaller than the classical electron radius. As a result, the probability of the di-muon (μ^+, μ^-) generation is approximately 50,000 times less than the probability of the electron–positron pair generation (see Ruffini et al., 2009; Titov et al., 2009; Yu et al., 2024).

In the case of the Breit-Wheeler process of the positron creation in collision of two gamma-photons the threshold for the energy of colliding photons with the energy $\hbar\omega_{\gamma,1}$ and $\hbar\omega_{\gamma,2}$ is

$$\sqrt{\omega_{\gamma,1}\omega_{\gamma,2}} > \frac{m_e c^2}{\hbar}. \tag{28.120}$$

The product $\omega_{\gamma,1}\omega_{\gamma,2}$ is Lorentz invariant.

Here we discuss the requirements for realization of abundant electron–positron pair creation in the petawatt-class laser interaction with relativistic electron beam, as discussed e. g. by Bamber et al., (1999), Bulanov et al. (2011b), Thomas et al., (2012), Di Piazza et al., (2012), Abramowicz et al., (2021), Macleod (2022).

The positrons were observed in the E-144 experiments (Bamber et al., 1999) on the 527 nm terawatt laser interaction with 46.6 GeV electrons from the SLAC accelerator beam. The positrons were generated in a two-step process in which laser photons were scattered by the electrons and then the generated high-energy photons collided with several laser photons to produce an electron–positron pair. This corresponds to the multi-photon Breit-Wheeler process,

$$s\,\omega_0 + \omega_\gamma \rightarrow e^+ + e^-, \tag{28.121}$$

where s is the number of laser photons colliding with the gamma-photon to produce the pair.

The parameter χ_γ defined by Eq. (28.17), which determines the probability of the electron–positron pair creation by photons, for the photon colliding with a plane electromagnetic wave in the counter-propagating configuration can be written as

$$\chi_\gamma \approx 2\frac{E}{E_s}\frac{\hbar\omega_\gamma}{m_e c^2} = 2a_0\frac{\hbar^2\omega_0\omega_\gamma}{m_e^2 c^4} = 2a_0\frac{\hbar^2\overline{\omega}_0\overline{\omega}_\gamma}{m_e^2 c^4}. \tag{28.122}$$

The right hand side term of this expression, in which the bars indicate frequencies in another frame of reference, explicitly shows the relativistic invariance of χ_γ. It becomes of the order of unity for the energy of the photon, which creates an electron–positron pair, equal to

$$\hbar\omega_\gamma = m_e c^2 \frac{m_e c^2}{2\,\hbar\omega_0 a}. \tag{28.123}$$

Using this expression, we can find the number of laser photons s required to create the electron–positron pair. In the reference frame where the electron–positron pair is at the rest we have the relationship $s\tilde{\omega}_0 = \tilde{\omega}_\gamma$, which gives

$s = \tilde{\omega}_\gamma/\tilde{\omega}_0 \approx \omega_\gamma/4\omega_0\tilde{\gamma}^2$. Here the tilde is used for quantities in the pair rest frame and $\tilde{\gamma}$ corresponds to the Lorentz transformation to this frame. For the photon energy, we have $s\hbar\tilde{\omega}_0 = m_e c^2$. Since $\hbar\tilde{\omega}_\gamma = m_e c^2$ and $\tilde{\omega}_0\tilde{\omega}_\gamma = \omega_0\omega_\gamma = m_e^2 c^4 \chi_\gamma/\hbar^2 a_0$ we find $\tilde{\gamma} = \chi_\gamma m_e c^2/2\hbar\omega_0 a_0$, which yields

$$s = \frac{a_0}{\chi_\gamma}. \tag{28.124}$$

Prolific electron–positron pair creation leading to an avalanche type discharge with lepton- gamma plasma generation via the multi-photon Breit-Wheeler process (Breit & Wheeler, 1934) discussed in Refs. (Bell & Kirk, 2008; Fedotov et al., 2010; Heinzl et al., 2010; Nikishov & Ritus, 1970; Reiss, 1962; Ritus, 1979) requires high enough values of the parameters χ_e and χ_γ defined by Eqs. (19.20) and (19.23).

In the limit of small parameter $\chi_\gamma \ll 1$ the rate of the electron–positron pair creation is exponentially small,

$$W_{||}(\chi_\gamma) = \frac{3}{32}\frac{m_e^4 c^5}{\hbar^4\omega_\gamma}\left(\frac{\chi_\gamma}{2\pi}\right)^{3/2}\exp\left(-\frac{8}{3\chi_\gamma}\right). \tag{28.125}$$

In the limit $\chi_\gamma \gg 1$ the pair creation rate is given by

$$W_{||}(\chi_\gamma) = \frac{27\,\Gamma^7(2/3)}{56\pi^5}\frac{e^2 m_e^2 c^3}{\hbar^3\omega_\gamma}\left(\frac{3\chi_\gamma}{2}\right)^{2/3}. \tag{28.126}$$

Here $\hbar\omega_\gamma$ is the energy of the photon with which the Breit-Wheeler process creates an electron–positron pair.

The probability per unit time of the pair creation by photons in the region of $\chi_\gamma \approx 1$ is of the order of

$$W \approx 0.1\frac{e^2 m_e^2 c^3}{\hbar^3\omega_\gamma} = 0.1\alpha\left(\frac{m_e c^2}{\hbar}\right)\left(\frac{\hbar\omega_0}{m_e c^2}\right)a_0 = 0.1\alpha\omega_0 a_0.$$

The mean free path of the photon of the energy $\hbar\omega_\gamma$ before pair creation is equal to $l_{m.f.p} = c/W$. It is equal to

$$l_{m.f.p} = \frac{\lambda_0}{0.2\pi\alpha\, a_0} \approx 220\frac{\lambda_0}{a_0}. \tag{28.127}$$

This is approximately equal to $2\,\mu m$ for the laser amplitude $a_0 = 10^2$.

Now we estimate the value of the parameter χ_e, which characterizes the nonlinear Compton scattering. We consider the electron interacting with the plane electromagnetic wave. As a result of the interaction with the laser pulse, the electron, being inside the pulse, acquires longitudinal and transverse momentum components. The component of the electron momentum along the direction of the laser pulse propagation can be found from Eqs. (20.28)–(20.31) which yield the equation

$$\sqrt{m_e^2c^2 + m_e^2c^2a_0^2 + p_x^2} - p_x = \sqrt{m_e^2c^2 + p_0^2} - p_0 \tag{28.128}$$

with initial electron momentum p_0.

We use the conservation of generalized momentum which yields for the component of the electron momentum parallel to the laser electric field: $p_y = a\, m_e c$, provided the radiation friction effects are negligibly small. The x-component of the electron momentum before collision with the laser pulse is negative and equal to $-|p_0|$. For a plane electromagnetic wave propagating along the x-axis with $\boldsymbol{E} = E(x-ct)\boldsymbol{e}_y$ and $\boldsymbol{B} = E(x-ct)\boldsymbol{e}_z$, where $E = c^{-1}A'(x-ct)$ and $A(x-ct)$ is the y-component of the 4-vector potential. Prime denotes differentiation with respect to the variable $x - ct$. These yield that the invariant χ_e takes the form

$$\chi_e = \frac{E}{E_S}\left(\gamma_e - \frac{p_x}{m_e c}\right). \tag{28.129}$$

Using relationships (28.118) and (28.119) we obtain

$$\chi_e = \frac{E}{E_S}\left[\frac{(m_e^2c^2 + p_0^2)^{1/2} - p_0}{m_e c}\right] = a\frac{\hbar\omega_0}{m_e c^2}\left[\frac{(m_e^2c^2 + p_0^2)^{1/2} - p_0}{m_e c}\right]. \tag{28.130}$$

For 1 μm petawatt laser pulse focused to a few micron focus spot (this corresponds to the electromagnetic wave amplitude equal to $a_0 = 10^2$) the parameter χ_e becomes equal to unity for $\gamma_0 = 2.5 \times 10^3$, i.e. for the electron energy approximately equal to 1.3 GeV.

Since for the large gamma factors γ_e, the photon is emitted by the electron (positron) in a narrow angle almost parallel to the electron momentum with the energy approximately equal to the electron energy, the parameters χ_e and χ_γ are approximately equal to each other, although this is not necessarily so in the limit when the electron emits photons according to classical electrodynamics, i.e. when

$$\gamma_e < \sqrt{\frac{m_e c^2}{\hbar\omega_0}} = a_S^{1/2}.$$

In order to find the threshold for the avalanche development we need to estimate the quantum electrodynamics parameter χ_e. The condition for avalanche development corresponding to this parameter is that it should become of the order of unity within one tenth of the electromagnetic field period [e.g. see Refs. (Bell & Kirk, 2008; Fedotov et al., 2010)].

In the limit of weak radiation damping, $a_0 \ll \varepsilon_{rad}^{-1/3}$, the absolute value of the electron momentum is proportional to the electric field magnitude,

$$p_\perp = m_e c a_0 \quad \text{with} \quad p_\parallel = m_e c \varepsilon_{rad} a_0^4, \tag{28.131}$$

i.e. $p_\perp \gg p_{||}$, while in the regime of dominant radiation damping effects, i.e. at $a_0 \gg s\varepsilon_{rad}^{-1/3}$, it is given by

$$p_{||} = m_e c (a_0/\varepsilon_{rad})^{1/4} \quad \text{and} \quad p_\perp = m_e c (a_0 \varepsilon_{rad})^{-1/2}, \tag{28.132}$$

i.e. $p_\perp \ll p_{||}$. Here $p_\perp$ and $p_{||}$ are the momentum components perpendicular and parallel to the electric field. For the momentum dependence given by this expression the power radiated by the electron is $m_e c^2 \omega_0 a_0$, i.e. the energy obtained from the driving electromagnetic wave is almost completely re-emitted in the form of high energy gamma rays. At $a_0 \approx \varepsilon_{rad}^{-1/3}$ we have for the gamma photon energy

$$\hbar\omega_\gamma \approx 0.45\,\hbar \frac{m_e c^3}{e^2}. \tag{28.133}$$

For example, if $a_0 \approx \varepsilon_{rad}^{-1/3} \approx 450$ the circularly polarized laser pulse with the wavelength of 0.8 μm and intensity of $I \approx 2.2 \times 10^{23}$ W/cm^2 generates a burst of gamma photons with the energy approximately equal to 20 MeV with the duration determined either by the laser pulse duration or by the decay time of the laser pulse in a plasma.

In the regime when the radiation friction effects are dominant, i.e. when $a_0 \gg \varepsilon_{rad}^{-1/3}$, the angle between the electron momentum and the electric field is small being approximately equal to $(\varepsilon_{rad} a_0^3)^{-1/4}$, i.e. the electron moves almost opposite to the electric field direction. The electron momentum is given by expression (28.17). This yields an estimation

$$\chi_e \approx \frac{a_0 p_\perp}{a_S m_e c} = \left(\frac{a_0}{a_S^2 \varepsilon_{rad}}\right)^{1/2}. \tag{28.134}$$

This becomes greater than unity for $a_0 > a_S^2 \varepsilon_{rad} \approx 1.6 \times 10^3$, which corresponds to the laser intensity equal to 5.5×10^{24} W/cm^2. Without radiation friction taken into account the intensity requirements are much softer. This is the condition of the electron–positron avalanche onset. We see that the radiation friction effects do not prevent the electron–positron gamma-ray plasma cascade occurring in the case of circularly polarized colliding laser pulses. Extending our discussion on the XFEL generated photon beams, for which $\varepsilon_{rad} \approx 8 \times 10^{-5}$, we find that the condition $\chi_e > 1$ requires $a_0 > 0.3$. As we see the condition corresponds to the electric field for which the intensity is equal to 5.5×10^{24} W/cm^2. The charcteristic field for the avalanche onset is $E_\chi \approx \alpha E_S$, i.e. the quantum radiation effects become dominant at the field strength a factor 1/137 smaller than the QED critical field E_S.

28.5.1 Electron–positron Pair Creation from Vacuum

The imaginary part of the Lagrangian (28.3) is proportional to the probability of the electron positron pair creation, $w = 2\mathrm{Im}\{\mathcal{L}'\}$ (see Berestetskii et al., 1982). It is convenient to express the probability of the pair creation via the normalized Poincare invariants

$$\mathfrak{f} = \frac{\mathfrak{F}}{E_S^2} = \frac{\boldsymbol{B}^2 - \boldsymbol{E}^2}{2E_S^2} \quad \text{and} \quad \mathfrak{g} = \frac{\mathfrak{G}}{E_S^2} = \frac{\boldsymbol{E}\cdot\boldsymbol{B}}{2E_S^2}. \tag{28.135}$$

Electron–positron pair creation from vacuum by the electromagnetic field (Berestetskii et al., 1982; Popov, 2004) is a tunnelling process, which requires the parameter

$$\gamma_K = 1/a$$

to be much less than unity, i.e. $a \ll 1$, and

$$K_0 = 2\frac{m_e c^2}{\hbar\omega} \equiv 2a_S$$

to be much larger than unity (Popov, 2004), with the number of pairs created per unit volume in unit time given by

$$W_{e^+e^-} = \frac{e^2 E_S^2}{4\pi^2\hbar^2 c}\mathfrak{e}\,\mathfrak{b}\,\coth\left(\frac{\pi\mathfrak{b}}{\mathfrak{e}}\right)\exp\left(-\frac{\pi}{\mathfrak{e}}\right). \tag{28.136}$$

Here $\mathfrak{e}$ and $\mathfrak{b}$ are the normalized invariant electric and magnetic fields in the frame of reference, where they are parallel:

$$\mathfrak{e} = \sqrt{\sqrt{\mathfrak{f}^2 + \mathfrak{g}^2} - \mathfrak{f}} \quad \text{and} \quad \mathfrak{b} = \sqrt{\sqrt{\mathfrak{f}^2 + \mathfrak{g}^2} + \mathfrak{f}}. \tag{28.137}$$

When the invariant electric field vanishes, $\mathfrak{e} \to 0$, the pair creation rate $W_{e^+e^-}$ tends to zero. For invariant magnetic field equal to zero, $\mathfrak{b} \to 0$, and $\mathfrak{e} > 0$ we have

$$W_{e^+e^-} = \frac{e^2 E_S^2}{4\pi^3\hbar^2 c}\mathfrak{e}^2\exp\left(-\frac{\pi}{\mathfrak{e}}\right) = \frac{c}{4\pi^3 \lambda_C^4}\mathfrak{e}^2\exp\left(-\frac{\pi}{\mathfrak{e}}\right). \tag{28.138}$$

If the electromagnetic field varies sufficiently slowly in space and time, i.e. the characteristic scale of this variation is much larger than the characteristic scale of pair production process, which is equal to the Compton length, then the field can be considered as constant in each point of 4-volume and the number of pairs is equal to

$$N_{e^+e^-} = \int W_{e^+e^-} d^4x, \tag{28.139}$$

where the integration is performed over the 4-volume.

Since the pairs are produced near the maximum of the electric field $\mathfrak{e} = \mathfrak{e}_m$, we express the dependence of the invariant electric field on time and space coordinates locally as

$$\mathfrak{e} \approx \mathfrak{e}_m \left(1 - \frac{x^2}{2\delta x^2} - \frac{y^2}{2\delta y^2} - \frac{z^2}{2\delta z^2} - \frac{t^2}{2\delta t^2}\right), \tag{28.140}$$

and substitute it into Eq. (28.138) for the rate $W_{e^+e^-}$; here δx, δy, δz, and δt are the characteristic sizes and duration of high-intensity part of electromagnetic field. Integrating it over the 4-volume we obtain for the number of pairs (Narozhny et al., 2004; Bulanov et al., 2006; Esposti & Torgrimsson, 2024)

$$N_{e^+e^-} = \frac{c\delta t \delta x \delta y \delta z}{64\pi^4 \lambda_C^4} \mathfrak{e}_m^2 \exp\left(-\frac{\pi}{\mathfrak{e}_m}\right). \tag{28.141}$$

We note here that in the case when the parameter $\gamma_K = 1/a$ characterizing the adiabaticity of the process is large, $\gamma_K = 1/a \gg 1$, i.e. $a \ll 1$, which corresponds to the high frequency electromagnetic radiation, and $K = 2m_e c^2/\hbar\omega$ is large as well, the pair creation occurs in the multiphoton regime with the number of pairs created per unit volume in unit time given by (Popov, 2004)

$$W_{e^+e^-} \approx \frac{2c}{\pi^3 \lambda_C^4 K^{5/2}} \left(\frac{4\gamma_K}{e_N}\right)^{-2K}. \tag{28.142}$$

Here $e_N = 2.718281828\ldots$ is the Napier- or the Euler number. Assuming the field distribution is given by formula (28.140) with characteristic size of the order of λ and integrating Eq. (28.138) over the 4-volume, we obtain for the number of pairs generated per period in the multiphoton regime

$$N_{e^+e^-} = \frac{\pi}{2^7 K^{9/2}} \left(\frac{e_N\, a_m}{4}\right)^{2K}. \tag{28.143}$$

As we see it is always small, provided $a_m \ll 1$, and $K \gg 1$.

28.5.2 Configuration of 4π-Dipole Electromagnetic Field

Pair creation requires the first invariant $\mathfrak{f}$ given by Eq. (28.137) be negative. This condition can be fulfilled in the vicinity of the antinodes of colliding plane electromagnetic waves, or/and in the configuration formed by several focused electromagnetic pulses (Bulanov et al., 2006, 2010a, 2010b, 2010c). Near the focus region this configuration can be modelled by the axially symmetric three-dimensional (3D) electromagnetic field comprised of time-dependent electric and magnetic fields as proposed by Bulanov et al., (2010a, 2010b, 2010c). As is known, in the 3D case the analogy of a linearly polarized electromagnetic wave is the TM mode (it is a so-called transverse magnetic mode), with poloidal electric and toroidal magnetic fields as described by

$$\boldsymbol{B}(R, \theta, t) = \boldsymbol{e}_\phi \frac{a_m \exp(i\omega_0 t)}{(8\pi R)^{1/2}} J_{n+1/2}(k_0 R)\partial_\theta L_n^l(\cos(\theta)). \tag{28.144}$$

Here R, θ, ϕ are spherical coordinates, $k_0 = \omega_0/c$, $J_\nu(x)$ and $L_n^l(x)$ are the Bessel function and associated Legendre polynomials, respectively. The electric field is equal to

$$\boldsymbol{E} = ik_0^{-1}(\nabla \times \boldsymbol{B}). \tag{28.145}$$

This configuration approximating the electromagnetic fields found in Bulanov et al., (2010a, 2010b, 2010c) near the field maximum are described by the solution to Maxwell's equations in vacuum. The electromagnetic fields for the TM mode and the first Poincaré invariant are shown in Fig. 28.9. There we use the cylindrical coordinates r, ϕ, z, and $\boldsymbol{e}_r, \boldsymbol{e}_\phi, \boldsymbol{e}_z$ are the unit vectors in the r, ϕ, z directions. See also Eqs. (23.93)–(23.107) and Fig. 5.6.

The second invariant, $\mathfrak{g}$, is equal to zero, i.e., for this electromagnetic configuration the invariant magnetic field vanishes, $\mathfrak{b} = 0$.

For the TM mode, in the vicinity of the electric field maximum, the component of the electric field oscillates in the vertical direction, the radial component of the electric field is relatively small,

$$\boldsymbol{E} \approx \boldsymbol{e}_z E_m \cos(\omega_0 t) + \boldsymbol{e}_r \frac{E_m}{8} k_0^2 rz \cos(\omega_0 t) \tag{28.146}$$

and the ϕ component of the magnetic field vanishes on the axis being linearly proportional to the radius,

$$\boldsymbol{B} \approx \boldsymbol{e}_\phi \frac{E_m}{8} k_0 r \sin(\omega_0 t) \tag{28.147}$$

Using expression for the probability of electron–positron pair creation given by Eq. (28.141) and expanding the first Poincaré invariant, $\mathfrak{F}_{TM}$, in the vicinity of its

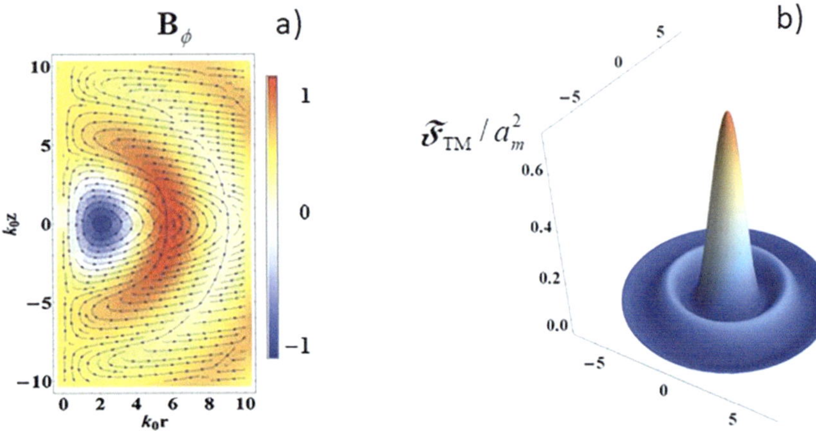

Fig. 28.9 **a** The vector field shows r and z components of the poloidal electric field in the (r, z) plane for the TM mode. The color density shows the toroidal magnetic fields distribution, $B_\phi(r, z)$. **b** The first Poincare invariant $-\mathfrak{F}_{TM}(r, z = 0, t = 0)$ normalized on the square of the dimensionless laser amplitude a_m^2 and taken with minus sign. Reprinted figure with APS permission from Bulanov et al. (2010a, 2010b, 2010c) as follows: Bulanov et al. (2010a, 2010b, 2010c). https://doi.org/10.1103/PhysRevLett.105.220407, Copyright (2010) by the American Physical Society

maximum, we find that the pairs are created in a small 4-volume near the electric field maximum with characteristic size

$$\pi \Delta r^2 \Delta z \Delta t \approx \left(\frac{5^{3/2}\lambda_0^4}{16\pi^5 c}\right)\left(\frac{a_m}{a_S}\right)^2. \tag{28.148}$$

This yields the number of pairs produced per wave period,

$$N_{e^+e^-} = \left(\frac{5^{3/2}}{16\pi^3}\right) a_m^2 \exp\left(-\pi \frac{a_S}{a_m}\right), \tag{28.149}$$

i.e., the first pairs can be observed for a $\lambda_0 = 1$ μm wavelength laser at an intensity of the order of 2×10^{27} W/cm^2 (see Bulanov et al. (2006), Bulanov et al., (2010a, 2010b, 2010c)), which corresponds to $a_m = 0.075 a_S$, i.e., a characteristic size, Δr, approximately equal to $0.04\lambda_0$.

28.6 Approaching the Schwinger Field Limit with Relativistic Mirror

The concept of the flying mirror has been formulated by Bulanov et al., (2003) as a way to approach the critical QED electric field, the Schwinger field limit. This concept uses the fact that an electromagnetic wave reflected off a moving mirror due to the double Doppler effect undergoes frequency multiplication and compactification. This makes this effect an attractive basis for a source of powerful high-frequency radiation. There are several schemes for developing compact, intense, brilliant, tunable X-ray sources by using the relativistic mirrors formed in nonlinear interactions in laser plasmas, whose realization will open new ways in nonlinear electrodynamics of continuous media in the relativistic regime. In the literature, we can find numerous publications presenting the results of theoretical, computer simulation, and experimental studies of this problem, e.g. see the review article Bulanov et al. (2003, 2013a, 2013b) and references therein, and Naumova et al., (2004), Gordienko et al., (2005), Kiefer et al., (2013), Kulagin et al., (2007), Bulanov et al., (2012), Koga et al. (2012), Kahaly et al. (2013), Vincenti (2019), Valenta et al., (2020), Lamač et al. (2024), Jeong et al. (2025).

28.6.1 Overdense Plasma Slab as Relativistic Mirror

To illustrate basic approach towards the electromagnetic wave intensification we consider the *relativistic mirror* based on the utilization of the overdense plasma slab moving with relativistic velocity (see Esirkepov et al. (2009) and Sects. 7.4 and 8.2 above). As shown by Esirkepov et al. (2004), thin overcritical density foil target can be accelerated up to relativistic velocity by the radiation pressure of high-power laser pulse. According to the energy scaling of the RPDA acceleration mechanism given by Eq. (25.64) ultra-relativistic slab velocity can be reached by irradiating the thin foil target by kJ energy short pulse lasers.

According to relationships presented in Sect. 8.2, the reflected pulse frequency and amplitude increasing are given by relationships $\omega_r = \omega_0 g_M(\beta_M)$ and $E_r = \rho E_0 g_M(\beta_M)$, where ω_r and ω_0 and E_r and E_0 are the incident and reflected pulse frequency and amplitude, respectively, and $\beta_M = v_M/c$ is the normalized mirror velocity, and ρ is the mirror reflectivity. The frequency upshifting (the pulse intensification) factor $g_M(\beta_M)$ depends on the mirror velocity as $g_M(\beta_M) = (1+\beta_M)/(1-\beta_M)$. In the ultra-relativistic limit when $\beta_M \to 1$, the factor $g_M(\beta_M)$ is approximately equal to $4\gamma_M^2$ with the mirror gamma-factor given by $\gamma_M = 1/\sqrt{1-\beta_M^2}$.

To calculate the amplitude of the reflected radiation one should know the reflection coefficient, which has been calculated in Sect. 8. As above noted, Esirkepov et al. (2009) and Esirkepov and Bulanov (2024) pay attention to vanishing of the reflection coefficient when the mirror velocity approaches the value of speed of light in vacuum, $\beta_M \to 1$. For the light reflection at the interface between vacuum region at $x < -tc\beta_M$ and plasma region at $x > -tc\beta_M$, we use the Fresnel formula (26.11) for the reflection coefficient $R = |\rho|^2$ equal to

$$R = \left(\frac{1-n}{1+n}\right)^2 \tag{28.150}$$

with the plasma refraction index $n = \sqrt{1 - \Omega_{pe}^2/\omega^2}$. Here we consider the case

$$\Omega_{pe} = \sqrt{\frac{4\pi n_e e^2}{m_e\sqrt{1+a_0^2}}} = \frac{\omega_{pe}}{(1+a_0^2)^{1/4}} \tag{28.151}$$

is the Langmuir frequency modified in the case of relativistically strong electromagnetic wave interacting with plasma [see Eq. (3.20) and Akhiezer & Polovin, 1956]. In the case of relatively low-density plasma and/or high amplitude electromagnetic field, where $\Omega_{pe}^2/\omega^2 \ll 1$, the reflection coefficient is approximately equal to

$$R_M \approx \frac{\omega_{pe}^4}{16\omega_M^4(1+a_0^2)} \tag{28.152}$$

A subscription M denotes that the reflection coefficient is written for the frame of reference where the relativistic mirror is at rest; the Langmuir frequency ω_{pe} and normalized amplitude of the laser pulse a_0 are Lorentz invariant. If in the laboratory frame of reference $\mathcal{L}$ the frequency and the amplitude of the electromagnetic wave incident on the mirror equal, ω_0, and E_0, respectively, then in the mirror frame of reference $\mathcal{M}$ they are equal to $\omega_M \approx 2\gamma_M \omega_0$ and $E_M \approx 2\gamma_M E_0$.

The amplitude of the reflected electromagnetic wave equals

$$E_r \approx 2\gamma_M \rho_M E_M, \tag{28.153}$$

where $\rho_M = \sqrt{R_M}$. We see that the reflected electromagnetic wave amplitude does not depend on the Lorentz factor γ_M of the mirror being equal to

$$E_r \approx \frac{E_0}{\sqrt{(m_e\omega_0 c/e)^2 + E_0^2}}\left(\frac{\pi n_e e c}{\omega_0}\right), \tag{28.154}$$

i.e. in the limit $a_0 \gg 1$ we have $E_r \approx \pi e n_e c/\omega_0$, which *does not depend on the specific value of the amplitude of the laser pulse nor on the mirror gamma-factor*.

The reflected electromagnetic wave has an amplitude larger than the incident wave amplitude provided

$$\frac{\omega_{pe}^2}{4a_0\omega_0^2} = \frac{n_e}{2\overline{n}_{cr}} > 1, \tag{28.155}$$

i.e. if the plasma density is above the twice critical plasma density, $\overline{n}_{cr} = m_e\omega_0^2\sqrt{a_0}/4\pi e^2$, for which the relativistic effects are considered. i.e. if the plasma density (here it is the density in the rest frame of the mirror) is above the twice critical plasma density,

$$\bar{n}_{cr} = \frac{m_e \omega_0^2 \sqrt{1 + a_0^2}}{4\pi e^2},$$

for which the relativistic effects are considered.

Note that for opaque target when $\Omega_{pe}^2/\omega_M^2 \geq 1$ the reflection coefficient equals unity, i.e. $\rho = 1$ in Eq. (2), the electric field in the reflected electromagnetic wave equals $E_r = E_0 g_M(\beta_M) \approx 4\gamma_M^2 E_0$ as has been found by Einstein (1905). This regime assumes that the mirror gamma-factor is relatively small $\gamma_M < \Omega_{pe}/2\omega_0$.

28.6.2 Concave (Paraboloidal) Relativistic Mirror

Further intensification of the electromagnetic field can be achieved with the tailored targets of paraboloidal form. Such the target can focus the reflected wave into the spot of the size of the electromagnetic wave wavelength in the mirror frame of reference $\mathcal{M}$, where it is equal to $\lambda_M = \lambda_0/2\gamma_M$. This configuration is depicted in Fig. 9.11c, where we see the wave focused to the focus spot moving with relativistic velocity. As a result, the amplitude of intensified electromagnetic waves can reach

$$E_r^f \approx \frac{2\pi n_e e c \gamma_M}{\omega_0} \left(\frac{D}{\lambda_0} \right). \tag{28.156}$$

Here $D \gg \lambda_0$ is the waist of the incident laser beam. The reflected electromagnetic wave has an amplitude larger than the incident wave amplitude provided $D \gg \lambda_0$.

Calculating the ratio of this field to the Sauter-Schwinger field $E_S = m_e^2 c^3/e\hbar$, we obtain

$$\frac{E_r^f}{E_S} \approx \frac{2\pi n_e e^2 \gamma_M \hbar}{m_e^2 c^3 \omega_0} \left(\frac{D}{\lambda_0} \right) = n_e r_e \lambda\!\!\!^{-}_C D \gamma_M \approx 10^{-23}_{[cm^2]} n_e D \gamma_M . \tag{28.157}$$

In the case under consideration, the invariant electric field is a factor $1/2\gamma_M$ smaller than the field given by expression (28.137).

28.6.3 Electron–Positron Pair Creation by Colliding Electromagnetic Pulses

Even if the invariant electric field determined by Eq. (28.137) is a factor $1/2\gamma_M$ smaller than the field given by expression (28.10) the electric field of the electromagnetic wave reflected at the relativistic mirror can be *larger than the Sauter-Schwinger field* can create the electron–positron pairs efficiently in the electromagnetic field configuration of two counter-propagating head-on-colliding pulses, as has been considered by Jeong et al. (2021), Kohlfürst et al. (2022), Jeong et al. (2024), (2025), Mercuri-Baron et al. (2025).

Representing two head-on-colliding electromagnetic pulses by the vector-potential

$$\boldsymbol{A}(x,t) = A(x,t)\boldsymbol{e}_y$$

with

$$A(x,t) = A_0\left[\exp\left(-\frac{(x-ct)^2}{2L^2}\right)\sin(k(x-ct)) - \exp\left(-\frac{(x+ct)^2}{2L^2}\right)\sin(k(x+ct))\right], \tag{28.158}$$

where A_0 is the electromagnetic wave amplitude, L is the width of the electromagnetic pulse, k is the wavenumber, and $\boldsymbol{e}_y$ is the unit vector along the y axis, we can calculate the first Poincare invariant

$$\mathfrak{F}(x,t) = \frac{1}{2}\left[\boldsymbol{B}^2(x,t) - \boldsymbol{E}^2(x,t)\right]. \tag{28.159}$$

It is equal to

$$\begin{aligned}\mathfrak{F}(x,t) = &-\frac{2A_0^2}{L^4}\exp\left(-\frac{x^2+c^2t^2}{L^2}\right)\\ &\left[kL^2\cos(k(x-ct)) - (x-ct)\sin(k(x-ct))\right]\\ &\left[kL^2\cos(k(x+ct)) - (x+ct)\sin(k(x+ct))\right].\end{aligned} \tag{28.160}$$

In Fig. 28.10a we show the electromagnetic field of two head-on-colliding electromagnetic waves given by $A(x,t)$ described by Eq. (28.158), and in Fig. 28.10b we present the electromagnetic the first Poincare invariant $\mathfrak{F}(x,t)$ given by Eq. (28.160). We see that the Poincare invariant is not exponentially small in the region $x^2+c^2t^2 \leq L^2$, where it can be approximated as

$$\mathfrak{F}(x,t) \approx -2A_0^2k^2\left(1-(1+k^2L^2)\frac{x^2+c^2t^2}{L^2}\right). \tag{28.161}$$

Using Eqs. (28.159), (28.160) and (28.137) we can find the invariant electric field $\mathfrak{e}$ and the number of electron–positron pairs created by head-on-colliding electromagnetic pulses. According to expression (28.141) it is approximately equal to

$$N_{e^+e^-} = \frac{L^4}{4\pi^2\lambda_C^4(1+k^2L^2)^2}\mathfrak{E}_0^2\exp\left(-\frac{\pi}{\mathfrak{E}_0}\right), \tag{28.162}$$

where $\mathfrak{E}_0 = kA_0/E_S$.

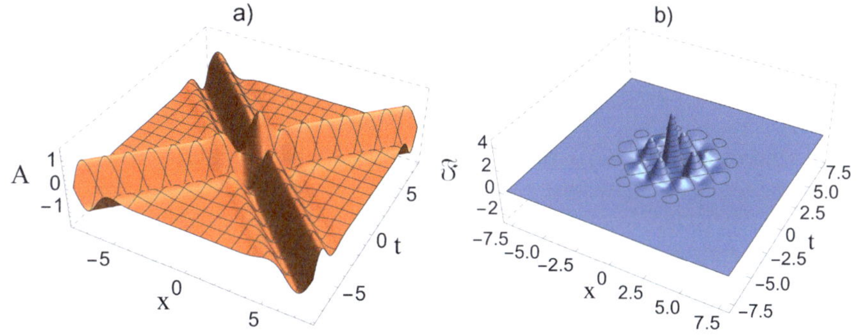

Fig. 28.10 **a** Two head-on-colliding electromagnetic waves given by $A(x, t)$ for $A_0 = 1$, $L = 1.5$, $k = 1.5$, $c = 1$. **b** The first Poincare invariant -$\mathfrak{F}(x, t)$

In Fig. 28.11 we show the number of electron–positron pairs versus time, $N_{e^+e^-}(t)$, in the plane $x = 0$ for (a) for $A_0 = 1$, $k = 1$, $L = 1, 2, 3, 4, 5, 6$, and $c = 1$; (b) for $A_0 = 0.5$, $k = 1, 2, 3, 4, 5, 6$, $L = 1$, and $c = 1$. The number of electron–positron pairs is normalized on $D^2L^2/4\pi^2\lambda_C{}^4$. The electric field according to Eqs. (28.135), (28.137) and (28.160) is

$$\mathfrak{E}(x = 0, t) = \frac{\sqrt{2}A_0}{L^2 E_S} \exp\left(-\frac{c^2t^2}{2L^2}\right)\left[kL^2 \cos(kct) - t \sin(kct)\right] \tag{28.163}$$

being normalized on E_S.

As mentioned above, within the frameworks of the Relativistic Mirror paradigm, the intensified electromagnetic wave can have, in the laboratory frame of reference, the amplitude *larger than the Sauter-Schwinger field.* This requires developing theoretical description of the electromagnetic field screening and dissipation due to the electron–positron plasma creation (Bulanov et al., 2005, 2010a, 2010b, 2010c; Nerush et al., 2011; Benedetti et al., 2011).

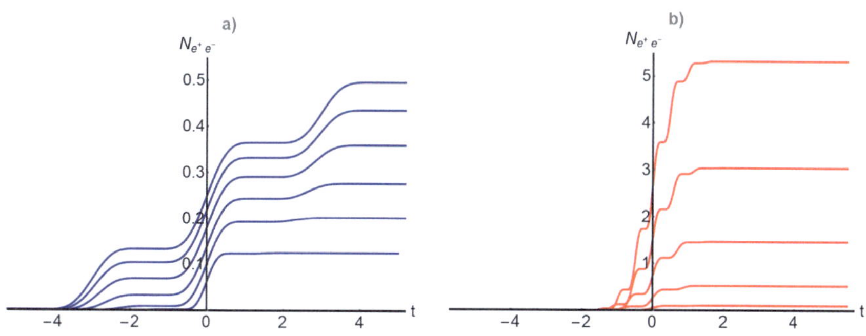

Fig. 28.11 The number of electron–positron pairs versus time in the plane $x = 0$ for **a** for $A_0 = 1$, $k = 1$, $L = 1, 2, 3, 4, 5, 6$, and $c = 1$; **b** for $A_0 = 0.5$, $k = 1, 2, 3, 4, 5, 6$, $L = 1$, and $c = 1$

28.6.4 Discussions and Conclusion

In the relativistic transparency regime, a scaling that gives the relationship between the parameters of the electromagnetic pulse intensified during reflection from a plasma mirror moving with a relativistic velocity. This scaling shows that the intensity of the reflected electromagnetic pulse can reach and even exceed the intensity corresponding to the Sauter-Schwinger field, and this can be done by using the next generation of ultra-powerful laser devices. Although in this scheme the intensity of electromagnetic radiation in the focal region can be ultra-high, on the plasma mirror the electromagnetic wave amplitude can be acceptably low so as not to significantly damage the mirror (Valenta et al., 2020). The duration of the reflected electromagnetic pulse tends to zero when the mirror velocity approaches speed of light. We have considered the generation of an electron–positron pair in a vacuum by two counter-propagating electromagnetic pulses intensified by reflection from relativistic plasma mirror of paraboloidal form.

The intensified electromagnetic field can be considered for another approach towards studying strong field QED effects in the collision of the electromagnetic pulse intensified by the relativistic mirrors with ultra-relativistic electrons (e.g. see have been considered by Zaïm et al., 2024). In this regime the nonlinear Compton scattering generates high-energy gamma-photons subsequently colliding with high-intensity electromagnetic wave, which leads to electron–positron pairs are created via the multiphoton Breit-Wheeler process as in the case of E144 experiment (Bamber et al., 1999). Consideration of the Sauter-Schwinger field limit attainability when the Breit-Wheeler pair creation process can develop is presented in the paper by Bulanov et al. (2010a, 2010b, 2010c).

Relativistic mirrors intensifying laser light can provide the conditions for studying the electron–positron pair created via the multiphoton Breit-Wheeler process in the all-optical setup (MacLeod et al., 2023a, 2023b), which uses the collision of the gamma-photons generated in the gamma-flash regime (see Hadjisolomou et al., 2023 and references therein) with strong electromagnetic field.

Electromagnetic pulses generated by multi-petawatt lasers in the pulse colliding configuration, forming relativistic plasma mirrors and accelerating high-energy electrons, have been proposed for studying Cherenkov radiation in quantum vacuum (Dremin, 2002; Macleod et al., 2019; Bulanov et al., 2019; Jirka et al., 2023).

References

Abramowicz, H., et al. (2021). Conceptual design report for the LUXE experiment. *The European Physical Journal Special Topics, 230*, 2445.

Akhieser, A., Landau, L., & Pomeranchook, I. (1936). Scattering of light by light. *Nature, 138*, 206. [citation added above in Chapter 28]

Akhiezer, A. I., & Berestetskii, V. B. (1965). *Quantum electrodynamics*. Interscience Publishers.

Akhiezer, A. I., & Polovin, R. V. (1956). Theory of wave motion of an electron plasma. *Soviet Physics JETP, 3*, 696.

Aleksandrov, E. B., Ansel'm, A. A., & Moskalev, A. N. (1985). Vacuum birefringence in an intense laser radiation field. *Sov. Phys. JETP, 62*, 680.

ATLAS Collaboration. (2017). *Nature Physics 13*, 852.

Baier, V. N., & Katkov, V. M. (2005). Concept of formation length in radiation theory. *Physics Reports, 409*, 261.

Bamber, C., Boege, S. J., Koffas, T., Kotseroglou, T., Melissinos, A. C., Meyerhofer, D. D., Reis, D. A., Ragg, W., Bula, C., McDonald, K. T., & Prebys, E. J. (1999). Studies of nonlinear QED in collisions of 46.6 GeV electrons with intense laser pulses. *Physical Review D, 60*, 092004.

Baumann, C., Nerush, E. N., Pukhov, A., & Kostyukov, I. Yu. (2019). Probing non-perturbative QED with electron-laser collisions. *Scientific Reports, 9*, 9407.

Baur, G., Hencken, K., Trautmann, D., Sadovsky, S., & Kharlov, Yu. (2002). Coherent $\gamma\gamma$ and γA interactions in very peripheral collisions at relativistic ion colliders. *Physics Reports, 364*, 359.

Bell, A. R., & Kirk, J. G. (2008). Possibility of prolific pair production with high-power lasers. *Physical Review Letters, 101*, Article 200403.

Benedetti, A., Han, W.-B., Ruffini, R., & Vereshchagin, G. V. (2011). On the frequency of oscillations in the pair plasma generated by a strong electric field. *Physics Letters B, 698*, 75.

Berestetskii, V. B., Lifshitz, E. M., & Pitaevskii, L. P. (1982). *Course of theoretical physics, Vol. 4: Quantum electrodynamics*. Pergamon.

Bethe, H., & Heitler, W. (1934). On the stopping of fast particles and on the creation of positive electrons. *Proceedings of the Royal Society of London a: Mathematical, Physical and Engineering Sciences, 146*, 83.

Bhabha, H. J. (1935). The creation of electron pairs by fast charged particles. *Proceedings of the Royal Society of London. Series A, 152*, 559.

Biondini, G., & Pelinovsky, D. E. (2008). Kadomtsev-Petviashvili equation. *Scholarpedia, 3*, 6539.

Bisnovatyi-Kogan, G. S., Zel'dovich, Ya. B., & Syunyaev, R. A. (1971). Physical processes in a low-density relativistic plasma. *Soviet Astronomy-AJ, 15*, 17.

Böhl, P., King, B., & Ruhl, H. (2015). Vacuum high-harmonic generation in the shock regime. *Physical Review A, 92*, Article 032115.

Breit, G., & Wheeler, J. A. (1934). Collision of two light quanta. *Physical Review, 46*, 1087.

Bulanov, S. V., Echkina, E. Y., Esirkepov, T. Z., Inovenkov, I. N., Kando, M., Pegoraro, F., & Korn, G. (2010). Unlimited ion acceleration by radiation pressure. *Physical Review Letters, 104*, 135003.

Bulanov, S. S., Esirkepov, T. Z., Thomas, A. G., Koga, J. K., & Bulanov, S. V. (2010). Schwinger limit attainability with extreme power lasers. *Physical Review Letters, 105*, 220407.

Bulanov, S. V., Esirkepov, T. Z., Hayashi, Y., Kando, M., Kiriyama, H., Koga, J. K., Kondo, K., Kotaki, H., Pirozhkov, A. S., Bulanov, S. S. & Zhidkov, A. G. (2011b). On the design of experiments for the study of extreme field limits in the interaction of laser with ulrarelativistic electron beam. *Nuclear Instruments and Methods in Physics Research Section A, 660*, 31.

Bulanov, S. V., Esirkepov, T. Zh., Kando, M., Bulanov, S. S., Rykovanov, S. G., & Pegoraro, F. (2013b). Strong field electrodynamics of a thin foil. *Physics of Plasmas, 20*, 123114.

Bulanov, S. V., Esirkepov, T. Zh., Kando, M., Pirozhkov, A. S., & Rosanov, N. N. (2013a). Relativistic mirrors in plasmas–novel results and perspectives. *Physics Uspekhi, 56*, 429.

Bulanov, S. V., Esirkepov, T. Zh., Koga, J., Bulanov, S. S., Gong, Z., Yan, X. Q., & Kando, M. (2017). Charged particle dynamics in multiple colliding electromagnetic waves. Survey of random walk, Lévy flights, limit circles, attractors and structurally determinate patterns. *Journal of Plasma Physics, 83*, 905830202.

Bulanov, S. V., Sasorov, P. V., F. Pegoraro, F., H. Kadlecová, H., Bulanov, S. S., Esirkepov, T. Zh., Rosanov, N. N., & Korn, G. (2020). Electromagnetic solitons in quantum vacuum. *Physical Review D, 101*, 016016.

Bulanov, S. V. (2017). Magnetic Reconnection: From MHD to QED. *Plasma Physics and Controlled Fusion, 59*, Article 014029.

Bulanov, S. V. (2021). Electron dynamics in the field of strong plasma and electromagnetic waves: A review. *Physics of Wave Phenomena, 29*, 1.

Bulanov, S. S., Esarey, E., Schroeder, C. B., et al. (2016). Radiation pressure acceleration: The factors limiting maximum attainable ion energy. *Physics of Plasmas, 23*, Article 056703.

Bulanov, S. V., Esirkepov, T., & Tajima, T. (2003). Light intensification towards the Schwinger limit. *Physical Review Letters, 91*, Article 085001.

Bulanov, S. S., Fedotov, A. M., & Pegoraro, F. (2005). Damping of electromagnetic waves due to electron-positron pair production. *Physical Review E, 71*, Article 016404.

Bulanov, S. S., Maksimchuk, A., Schroeder, C. B., et al. (2012). Relativistic spherical plasma waves. *Physics of Plasmas, 19*, Article 020702.

Bulanov, S. S., Mur, V. D., Narozhny, N. B., Nees, J., & Popov, V. S. (2010b). Multiple colliding electromagnetic pulses: A way to lower the threshold of pair production from vacuum. *Physical Review Letters, 104*, Article 220404.

Bulanov, S. S., Narozhny, N. B., Mur, V. D., & Popov, V. S. (2006). Electron-positron pair production by electromagnetic pulses. *JETP, 102*, 9.

Bulanov, S. V., Sasorov, P. V., Bulanov, S. S., & Korn, G. (2019). Synergic Cherenkov-Compton radiation. *Physical Review D, 100*, Article 016012.

Chen, H., Wilks, S. C., Meyerhofer, D. D., Bonlie, J., Chen, C. D., Chen, S. N., Courtois, C., Elberson, L., Gregori, G., Kruer, W., Landoas, O., Mithen, J., Myatt, J., Murphy, C. D., Nilson, P., Price, D., Schneider, M., Shepherd, R., Stoeckl, C., … Beiersdorfer, P. (2010). Relativistic Quasi-monoenergetic Positron Jets from intense laser-solid interactions. *Physical Review Letters, 105*, Article 015003.

D'Enterria, D., & Da Silveira, G. G. (2013). Observing light-by-light scattering at the large hadron collider. *Physical Review Letters, 111*, 080405.

Di Piazza, A., Hatsagortsyan, K. Z., & Keitel, C. H. (2005). Harmonic generation from laser-driven vacuum. *Physical Review D, 72*, Article 085005.

Di Piazza, A., Muller, C., Hatsagortsyan, K. Z., & Keitel, C. H. (2012). Extremely high-intensity laser interactions with fundamental quantum systems. *Reviews of Modern Physics, 84*, 1177.

Dremin, I. M. (2002). Cherenkov radiation and pair production by particles traversing laser beams. *JETP Letters, 76*, 151.

Dunne, G. V. (2005). Heisenberg–Euler effective Lagrangians: Basics and extensions from fields to strings. In M. Shifman, A. Vainshtein & J. Wheater (Eds.), *Circumnavigating theoretical physics* (Ian Kogan Memorial Collection) (p. 445). World Scientific.

Einstein, A. (1905). Zur Elektrodynamik bewegter Körper. *Annals of Physics (Leipzig), 17*, 891.

Esirkepov, T. Zh., Borghesi M., Bulanov S. V., Mourou, G., & Tajima, T. (2004). Highly efficient relativistic-ion generation in the laser-piston regime. *Physical Review Letters, 92*, 175003.

Esirkepov, T. Zh., Bulanov, S. V., Kando, M., Pirozhkov, A. S., & Zhidkov, A. G. (2009). Boosted high-harmonics pulse from a double-sided relativistic mirror. *Physical Review Letters, 103*, 025002.

Esirkepov, T. Z., & Bulanov, S. V. (2024). Luminal mirror. *Physical Review E, 109*, L023202.

Esposti, G. D., & Torgrimsson, G. (2024). Momentum spectrum of Schwinger pair production in four-dimensional e-dipole fields. *Physical Review D, 109*, 016013.

Fedotov, A., Ilderton, A., Karbstein, F., et al. (2023). Advances in QED with intense background fields. *Physics Reports, 1010*, 1.

Fedotov, A. M., Narozhny, N. B., Mourou, G., & Korn, G. (2010). Limitations on the attainable intensity of high-power lasers. *Physical Review Letters, 105*, Article 080402.

Fillion-Gourdeau, F., Lefebvre, C., & MacLean, S. (2015). Scheme for the detection of mixing processes in vacuum. *Physical Review A, 91*, Article 031801.
Frank, I. M., & Tamm, I. E. (1937). Coherent visible radiation of fast electrons passing through matter. *Comptes Rendus De L'accademie des Sciences De l'USSR, 14*, 109.
Gahn, C., Tsakiris, G. D., Pretzler, G., & Witte, K. J. (2002). Generation of MeV electrons and positrons with femtosecond pulses from a table-top laser system. *Physics of Plasmas, 9*, 987.
Gies, H., Karbstein, F., Kohlfürst, C., & Seegert, N. (2018). Photon-photon scattering at the high-intensity frontier. *Physical Review D, 97*, Article 076002.
Ginzburg, V. L. (1940). The quantum theory of radiation of an electron uniformly moving in a medium. *Soviet Journal of Physics, 2*, 441.
Ginzburg, V. L. (1963). In memory of Niels Bohr. *Soviet Physics Uspekhi, 80*, 207.
Ginzburg, V. L., & Tsytovich, V. N. (1979). Several problems of the theory of transition radiation and transition scattering. *Physics Reports, 49*, 1.
Gong, Z., Hu, R. H., Shou, Y. R., Qiao, B., Chen, C. E., He, X. T., Bulanov, S. S., Esirkepov, T. Zh., Bulanov, S. V., Yan, X. Q., (2017). High-efficiency gamma-ray flash generation via multiple-laser scattering in ponderomotive potential well. *Physical Review E, 95*, 013210.
Gonoskov, A., Blackburn, T. G., Marklund, M., & Bulanov, S. S. (2022). Charged particle motion and radiation in strong electromagnetic fields. *Reviews of Modern Physics, 94*, Article 045001.
Gordienko, S., Pukhov, A., Shorokhov, O., & Baeva, T. (2005). Coherent focusing of high harmonics: A new way towards the extreme intensities. *Physical Review Letters, 94*, Article 103903.
Gould, R. J., & Schreder, G. P. (1967). Pair production in photon-photon collisions. *Physical Review, 155*, 1404.
Gryaznykh, D. A., Kandiev, Ya. Z., Lykov, V. A. (1998). Estimates of electron-positron pair production in the interaction of high-power laser radiation with high-Z targets. *Journal of Experimental and Theoretical Physics Letters, 67*, 257.
Hadjisolomou, P., Jeong, T. M., Kolenaty, D., Macleod, A. J., Olšovcová, V., Versaci, R., Ridgers, C. P., & Bulanov, S. V. (2023). Gamma-flash generation in multi-petawatt laser–matter interactions. *Physics of Plasmas, 30*, Article 093103.
Heinzl, T., Ilderton, A., & Marklund, M. (2010). Finite size effects in stimulated laser pair production. *Physics Letters B, 692*, 250.
Heinzl, T., Liesfeld, B., Amthor, K., Schwoerer, H., Sauerbrey, R., & Wipf, A. (2006). On the observation of vacuum birefringence. *Optics Communications, 267*, 318.
Heisenberg, W., & Euler, H. (1936). Folgerungen aus der diracschen theorie des positrons. *Zeitschrift Für Physik, 98*, 714.
Heitler, W. (1954). *The quantum theory of radiation.* Oxford University Press.
Heyl, J. S., & Hernquist, L. (1997). Analytic form for the effective Lagrangian of QED and its application to pair production and photon splitting. *Physical Review D, 55*, 2449.
Jeong, T. M., Bulanov, S. V., Valenta, P., Korn, Korn, G., Esirkepov, T. Zh., Koga, J. K., Pirozhkov, A. S., Kando, M., & Bulanov, S. S. (2021). Relativistic flying laser focus by a laser-produced parabolic plasma mirror. *Physical Review A, 104*, 053533.
Jeong, T. M., Bulanov, S. V., Shaisultanov, R., Hadjisolomou, P., & Esirkepov, T. M. (2025). Toward super-strong field with the relativistic curved plasma mirror. *Physical Review* A ###.
Jeong, T. M., Bulanov, S. V., Valenta, P., & Hadjisolomou, P. (2024). On the synergic approach toward the experimental realization of interesting fundamental science within the framework of relativistic flying mirror concept. *Reviews of Modern Plasma Physics, 8*, 9.
Jirka, M., & Bulanov, S. V. (2024). Effects of colliding laser pulses polarization on $e^- e^+$ cascade development in extreme focusing. *Physical Review Letters, 133*, Article 125001.
Jirka, M., Sasorov, P. V., & Bulanov, S. V. (2022). On the new and old physics in the interaction of a radiating electron with the extreme electromagnetic field. *Physical Review D, 105*, Article 113004.
Jirka, M., Sasorov, P., & Bulanov, S. V. (2023). Radiation from a polarized vacuum in a laser-particle collision. *Physical Review A, 107*, Article 052805.

Jirka, M., Sasorov, P. V., Bulanov, S. S., Korn, G., Rus, B., & Bulanov, S. V. (2021). Reaching high laser intensity by a radiating electron. *Physical Review A, 103*, Article 053114.

Kadlecová, H., Korn, G., & Bulanov, S. V. (2019). Electromagnetic shocks in the quantum vacuum. *Physical Review D, 99*, Article 036002.

Kadomstev, B. B. (2001). Cooperative effects in plasmas. In V. D. Shafranov (Ed.), *Reviews of plasma physics* (Vol. 22, p. 1). Kluwer Academic/Plenum Publishers.

Kadomtsev, B. B., & Petviashvili, V. I. (1970). On the stability of solitary waves in weakly dispersive media. *Soviet Physics—Doklady, 15*, 539.

Kahaly, S., Monchocé, S., Vincenti, H., Dzelzainis, T., Dromey, B., Zepf, M., Martin, P., & Quéré, F. (2013). Direct observation of density-gradient effects in harmonic generation from plasma mirrors. *Physical Review Letters, 110*, 175001.

Kando, M., Fukuda, Y., Pirozhkov, A. S., Ma, J., Daito, I., Chen, L.-M., Esirkepov, T. Z., Ogura, K., Homma, T., Hayashi, Y., Kotaki, H., Sagisaka, A., Mori, M., Koga, J. K., Daido, H., Bulanov, S. V., Kimura, T., Kato, Y., & Tajima, T. (2007). Demonstration of laser-frequency upshift by electron-density modulations in a plasma Wakefield. *Physical Review Letters, 99*, Article 135001.

Kando, M., Pirozhkov, A. S., Kawase, K., Esirkepov, T. Z., Fukuda, Y., Kiriyama, H., Okada, H., Daito, I., Kameshima, T., Hayashi, Y., Kotaki, H., Mori, M., Koga, J. K., Daido, H., Faenov, A. Y., Pikuz, T., Ma, J., Chen, L.-M., Ragozin, E. N., … Bulanov, S. V. (2009). Enhancement of photon number reflected by the relativistic flying mirror. *Physical Review Letters, 103*, Article 235003.

Kaplan, A. E., & Ding, Y. J. (2000). Field-gradient-induced second-harmonic generation in magnetized vacuum. *Physical Review A, 62*, Article 043805.

Karbstein, F., Sundqvist, C., Schulze, K. S., Uschmann, I., Gies, H., & Paulus, G. G. (2021). Vacuum birefringence at X-ray free-electron lasers. *New Journal of Physics, 23*, Article 095001.

Karplus, R., & Neuman, M. (1951). The scattering of light by light. *Physical Review, 83*, 776.

Kiefer, D., Yeung, M., Dzelzainis, T., Foster, P. S., Rykovanov, S. G., Lewis, C. L. S., Marjoribanks, R. S., Ruhl, H., Habs, D., Schreiber, J., Zepf, M., & Dromey, B. (2013). Relativistic electron mirrors from nanoscale foils for coherent frequency upshift to the extreme ultraviolet. *Nature Communications, 4*, 1763.

Kim, Y. S., & Noz, M. E. (1982). Dirac's light-cone coordinate system. *American Journal of Physics, 50*, 721.

King, B., & Heinzl, T. (2016). Measuring vacuum polarization with high-power lasers. *High Power Laser Science and Engineering, 4*, 1.

Koga, J., Bulanov, S. V., Esirkepov, T. Zh., Pirozhkov, A. S., Kando, M., & Rosanov, N. N. (2012). Possibility of measuring photon-photon scattering via relativistic mirrors. *Physical Review A, 86*, 053823.

Kohlfürst, C., Ahmadiniaz, N., Oertel, J., & Schützhold, R. (2022). Sauter-Schwinger effect for colliding laser pulses. *Physical Review Letters, 129*, Article 241801.

Korteweg, D. J., & de Vries, G. (1885). On the change of form of long waves advancing in a rectangular canal, and on a new type of long stationary waves. *Philosophical Magazine, 39*, 422.

Kulagin, V. V., Cherepenin, V. A., Hur, M. R., & Suk, H. (2007). Flying mirror model for interaction of a super-intense nonadiabatic laser pulse with a thin plasma layer: Dynamics of electrons in a linearly polarized external field. *Physics of Plasmas, 14*, 113101.

Lamač, M., Valenta, P., Nejdl, J., Chaulagain, U., Jeong, T. M., & Bulanov, S. V. (2024). Bright coherent attosecond X-ray pulses from beam-driven relativistic mirrors. arXiv preprint arXiv: 2403.03277

Landau, L. D., & Lifshitz, E. M. (1987). *Course of theoretical physics, Vol. 6: Fluid mechanics*. Pergamon Press.

Lee, T. D., & Wick, G. C. (1970). Finite theory of quantum electrodynamics. *Physical Review D, 2*, 1033.

Lu, W., Tzoufras, M., Joshi, C., Tsung, F. S., Mori, W. B., Vieira, J., Fonseca, R. A., & Silva, L. O. (2007). Generating multi-gev electron bunches using single stage laser Wakefield acceleration in a 3d nonlinear regime. *Physical Review Special Topics-Accelerators and Beams, 10*, Article 061301.

Lundström, E., Brodin, G., Lundin, J., Marklund, M., Bingham, R., Collier, J., Mendonca, J. T., & Norreys, P. (2006). Using high-power lasers for detection of elastic photon–photon scattering. *Physical Review Letters, 96*, Article 083602.

Lutzky, M., & Toll, J. S. (1959). Formation of discontinuities in classical nonlinear electrodynamics. *Physical Review, 113*, 1649.

Macleod, A. J. (2022). From theory to precision modelling of strong-field QED in the transition regime. *Journal of Physics: Conference Series, 2249*, 012022.

Macleod, A. J., Edwards, J. P., Heinzl, T., King, B., & Bulanov, S. V. (2023a). Strong-field vacuum polarisation with high energy lasers. *New Journal of Physics, 25*, Article 093002.

MacLeod, A. J., Hadjisolomou, P., Jeong, T. M., & Bulanov, S. V. (2023b). All-optical nonlinear Breit-Wheeler pair production with γ-flash photons. *Physical Review A, 107*, Article 012215.

Macleod, A. J., Noble, A., & Jaroszynski, D. A. (2019). Cherenkov radiation from the quantum vacuum. *Physical Review Letters, 122*, Article 161601.

Mamaev, S. G., Mostepanenko, V. M., & Eides, M. I. (1981). Effective action for a non-stationary electromagnetic field. *Soviet Journal of Nuclear Physics, 33*, 569.

Marklund, M., & Shukla, P. (2006). Nonlinear collective effects in photon-photon and photon-plasma interactions. *Reviews of Modern Physics, 78*, 591.

Mercuri-Baron, A., Mironov, A. A., Riconda, C., Grassi, A., Grech, M. (2025). Growth rate of self-sustained QED cascades induced by intense lasers. *Physical Review X, 15*, 011062.

Mourou, G., Tajima, T., & Bulanov, S. V. (2006). Optics in the relativistic regime. *Reviews of Modern Physics, 78*, 309.

Narozhny, N. B. (1969). Propagation of plane electromagnetic waves in a constant field. *Soviet Physics JETP, 28*, 371.

Narozhny, N. B., Bulanov, S. S., Mur, V. D., & Popov, V. S. (2004). E+ e−-pair production by a focused laser pulse in vacuum. *Physics Letters A, 330*, 1.

Narozhny, N. B., & Fedotov, A. M. (2007). Third-harmonic generation in a vacuum at the focus of a high-intensity laser beam. *Laser Physics, 17*, 350.

Naumova, N. M., Nees, J. A., Sokolov, I. V., Hou, B., & Mourou, G. A. (2004). Relativistic generation of isolated attosecond pulses in a λ3 focal volume. *Physical Review Letters, 92*, Article 063902.

Nerush, E. N., Kostyukov, I. Y., Fedotov, A. M., Narozhny, N. B., Elkina, N. V., & Ruhl, H. (2011). Laser field absorption in self-generated electron-positron pair plasma. *Physical Review Letters, 106*, Article 035001.

Nikishov, A. I. (1962). Absorption of high-energy photons in the universe. *Soviet Physics JETP, 14*, 393.

Nikishov, A. I., & Ritus, V. I. (1970). Interaction of electrons and photons with a very strong electromagnetic field. *Soviet Physics Uspekhi, 13*, 303.

Olver, F. W. J., Lozier, D. W., Boisvert, R. F., & Clark, C. W. (2010). *NIST handbook of mathematical functions*. Cambridge University Press.

Pegoraro, F., & Bulanov, S. V. (2019). Hodograph solutions of the wave equation of nonlinear electrodynamics in the quantum vacuum. *Physical Review D, 100*, Article 036004.

Pegoraro, F., & Bulanov, S. V. (2021). Nonlinear waves in a dispersive vacuum described with a high order derivative electromagnetic Lagrangian. *Physical Review D, 103*, Article 096012. Erratum: (2025). *Physical Review D, 112*, Article 079903.

Pirozhkov, A. S., Ma, J., Kando, M., Esirkepov, T. Zh., Fukuda, Y., Chen, L.-M., Daito, I., Ogura, K., Homma, T., Hayashi, Y., Kotaki, H., Sagisaka, A., Mori, M., Koga, J. K., Kawachi, T., Daido, H., Bulanov, S. V., Kimura, T., Kato, Y., & Tajima, T. (2007). Frequency multiplication of light back-reflected from a relativistic wake wave. *Physics of Plasmas, 14*, 123106.

Popov, V. S. (2004). Tunnel and multiphoton ionization of atoms and ions in a strong laser field (Keldysh theory). *Physics Uspekhi, 47*, 855.

Reiss, H. R. (1962). Absorption of light by light. *Journal of Mathematics and Physics, 59*, 3.
Ritus, V. I. (1970). Radiative effects and their enhancement in an intense electromagnetic field. *Soviet Physics JETP, 30*, 1181.
Ritus, V. I. (1979). Quantum effects of the interaction of elementary particles with an intense electromagnetic field. *Trudy Fiziko-Matematicheskogo Instituta, Akademiya Nauk SSSR, 111*, 5.
Rosanov, N. N. (1998). Self-action of intense electromagnetic radiation in an electron-positron vacuum. *JETP, 86*, 284.
Rosanov, N. N. (1993). Four-wave interactions of intense radiation in vacuum. *JETP, 76*, 991.
Rosso, L., Lera, R., Ravichandran, S., Longman, A., He, C. Z., Perez-Hernández, J. A., Apiñaniz, J., Smith, L. D., Fedosejevs, R., & Hill, W. T., III. (2021). Towards a direct measurement of the quantum-vacuum Lagrangian coupling coefficients using two counter-propagating super-intense laser pulses. *New Journal of Physics, 24*, Article 025010.
Ruffini, R., Vereshchagin, G., & Xue, S. S. (2009). Electron-positron pairs in physics and astrophysics: From heavy nuclei to black holes. *Physics Reports, 487*, 1.
Ruffini, R., Vereshchagin, G., & Xue, S. S. (2016). Cosmic absorption of ultra high energy particles. *Astrophysics and Space Science, 361*, 82.
Sakharov, A. D. (1991). Spectral density of eigenvalues of the wave equation and vacuum polarization. *Soviet Physics Uspekhi, 34*, 395.
Sarri, G., Poder, K., Cole, J. M., Schumaker, W., Di Piazza, A., Reville, B., Dzelzainis, T., Doria, D., Gizzi, L. A., Grittani, G., & Kar, S. (2015). Generation of neutral and high-density electron–positron pair plasmas in the laboratory. *Nature Communications, 6*, 1.
Sarri, G., King, B., Blackburn, T., Ilderton, A., Boogert, S., Bulanov, S. S., et al. (2025), Input to the European strategy for particle physics: strong-field quantum electrodynamics. *The European Physical Journal Plus, 140*, 1151.
Sasorov, P., Pegoraro, Esirkepov, T., & Bulanov, S. V. (2021). Generation of high order harmonics in heisenberg-euler electrodynamics. *New Journal of Physics, 23*, 105003.
Schwinger, J., Tsai, W. Y., & Erber, T. (1976). Classical and Quantum Theory of Synergic Synchrotron-Čerenkov Radiation. *Annalen der Physik, 96*, 303.
Shen, B., Bu, Z., Xu, J., Xu, T., et al. (2018). Exploring vacuum birefringence based on a 100 PW laser and an X-ray free electron laser beam. *Plasma Physics and Controlled Fusion, 60*, Article 044002.
Sokolov, A. A. (1940). Quantum theory of Cerenkov adiation. *DAN SSSR, 28*, 415.
Thomas, A. G. R., Ridgers, C. P., Bulanov, S. S., Griffin, B. J., & Mangles, S. P. D. (2012). Strong radiation-damping effects in a gamma-ray source generated by the interaction of a high-intensity laser with a Wakefield-accelerated electron beam. *Physical Review X, 2*, Article 041004.
Titov, A. I., Kämpfer, B., & Takabe, H. (2009). Dimuon production by laser-Wakefield accelerated electrons. *Physical Review Accelerators and Beams, 12*, Article 111301.
Tommasini, D., Ferrando, A., Michinel, H., & Seco, M. (2008). Detecting photon-photon scattering in vacuum at exawatt lasers. *Physical Review A, 77*, Article 042101.
Valenta, P., Esirkepov, T. Zh., Koga, J. K., Pirozhkov, A. S., Kando, M., Kawachi, T., Liu, Y.-K., Fang, P., Chen, P., Mu, J., Korn, G., Klimo, O., & Bulanov, S. V. (2020). Recoil effects on reflection from relativistic mirrors in laser plasmas. *Physics of Plasmas, 27*, 032109.
Vincenti, H. (2019). Achieving extreme light intensities using optically curved relativistic plasma mirrors. *Physical Review Letters, 123*, Article 105001.
Vranic, M., Grismayer, T., Fonseca, R. A., & Silva, L. O. (2016). Electron–positron cascades in multiple-laser optical traps. *Plasma Physics and Controlled Fusion Plasma, 59*, Article 014040.
Weisskopf, V. (1936). Uber die Elektrodynamik des Vakuums auf Grund der Quantentheorie des Elektrons. *Kongelige Danske Videnskabernes Selskab Matematisk Fysiske Meddelelser, 24*, 3.
Whitham, G. B. (1974). *Linear and nonlinear waves*. Wiley.
Yakimenko, V., Meuren, S., Del Gaudio, F., et al. (2019). Prospect of studying nonperturbative QED with beam-beam collisions. *Physical Review Letters, 122*, Article 190404.

Yu, T.-P., Liu, K., Zhao, J., Zhu, X. L., Lu, Y., Cao, Y., Zhang, H., Shao, F.-Q., & Sheng, Z.-M. (2024). Bright X/gamma-ray emission and lepton pair production by strong laser fields: A review. *Reviews of Modern Plasma Physics, 8*, 24.

Zaïm, N., Sainte-Marie, A., Fedeli, L., Bartoli, P., Huebl, A., Leblanc, A., Vay, J.-L., & Vincenti, H. (2024). Light-matter interaction near the schwinger limit using tightly focused doppler-boosted lasers. *Physical Review Letters, 132*, Article 175002.

Zakharov, V. E., Manakov, S. V., Novikov, S. P., & Pitaevskii, L. P. (1984). *Theory of Solitons.* Consultants Bureau.

Zee, A. (2010). *Quantum field theory in a Nutshell.* Princeton University.

Zha, W., Brandenburg, J. D., & Xu, Z. (2023). The Breit-Wheeler process in relativistic heavy-ion collisions: Creating matter from pure energy. *Nuclear Physics News, 33*(3), 27.

Zha, W., & Tang, Z. (2021). Discovery of higher-order quantum electrodynamics effect for the vacuum pair production. *Journal of High Energy Physics, 8*, 1.

Zhang, H., Zhao, J., Hu, Y., et al. (2021). Efficient bright g-ray vortex emission from a laser-illuminated light-fan-in-channel target. *High Power Laser Science and Engineering, 9*, Article e43.

Zheleznyakov, V. V., & Fabrikant, A. L. (1982). Electromagnetic shock waves in a magnetized vacuum. *JETP, 55*, 794.

Index

Y. Kato et al., *High Power Laser and Plasma Science*, Springer Series on Atomic, Optical, and Plasma Physics 130, https://doi.org/10.1007/978-3-032-08930-4

E

F

G

Q

R

S

MIX
Papier aus verantwortungsvollen Quellen
Paper from responsible sources
FSC® C105338
FSC
www.fsc.org

If you have any concerns about our products,
you can contact us on
ProductSafety@springernature.com

In case Publisher is established outside the EU,
the EU authorized representative is:
Springer Nature Customer Service Center GmbH
Europaplatz 3, 69115 Heidelberg, Germany

Printed by Libri Plureos GmbH
in Hamburg, Germany